essential cell biology

second edition

essential cell biology

second edition

Bruce Alberts • Dennis Bray

Karen Hopkin • Alexander Johnson

Julian Lewis • Martin Raff

Keith Roberts • Peter Walter

GS Garland Science
Taylor & Francis Group

NEW YORK AND LONDON

Garland Science
Vice President: Denise Schanck
Managing Production Editor: Emma Hunt
Proofreader and Layout: Emma Hunt
Senior Editorial Assistants: Kirsten Jenner, Frances Morgan
Editorial Assistant: Lilith Wood
Text Editors: Elizabeth Zayatz, Eleanor Lawrence
Copy Editors: Richard Mickey, Christine Gregory
Book and Cover Design: Blink Studio, London
Illustrator: Nigel Orme
Cover Illustration: Jose Ortega
Indexer: Merrall-Ross International Ltd.
Manufacturing: Nigel Eyre and Marion Morrow

Essential Cell Biology 2 Interactive
Artistic and Scientific Direction: Peter Walter
Narrated by: Julie Theriot
Production Design and Development: Michael Morales

Bruce Alberts received his Ph.D. from Harvard University and is President of the National Academy of Sciences and Professor of Biochemistry and Biophysics at the University of California, San Francisco. **Dennis Bray** received his Ph.D. from Massachusetts Institute of Technology and is currently a Medical Research Council Fellow in the Department of Zoology, University of Cambridge. **Karen Hopkin** received her Ph.D. in biochemistry from the Albert Einstein College of Medicine in the Bronx and is a science writer in Cambridge, Massachusetts. **Alexander Johnson** received his Ph.D. from Harvard University and is a Professor of Microbiology and Immunology and Co-Director of the Biochemistry and Molecular Biology Program at the University of California, San Francisco. **Julian Lewis** received his D.Phil. from the University of Oxford and is a Principal Scientist at the London Research Institute of Cancer Research UK. **Martin Raff** received his M.D. from McGill University and is at the Medical Research Council Laboratory for Molecular Cell Biology and Cell Biology Unit and in the Biology Department at University College London. **Keith Roberts** received his Ph.D. from the University of Cambridge and is Associate Research Director at the John Innes Centre, Norwich. **Peter Walter** received his Ph.D. from The Rockefeller University in New York and is Professor and Chairman of the Department of Biochemistry and Biophysics at the University of California, San Francisco, and an Investigator of the Howard Hughes Medical Institute.

Published by Garland Science, a member of the Taylor & Francis Group, 270 Madison Avenue, New York, NY 10016, USA and 2 Park Square, Milton Park, Abingdon, OX14 4RN, UK.

Printed in United States of America.

15 14 13 12 11 10 9 8 7 6 5 4 3

Library of Congress Cataloging-in-Publication Data
Essential cell biology / Bruce Alberts ... [et al.].-- 2nd ed.
  p. cm.
ISBN 0-8153-3480-X.—ISBN 0-8153-3481-8 (pbk.)
1. Cytology. 2. Molecular biology. 3. Biochemistry. I. Alberts, Bruce.
QH581.2.E78 2003
571.6—dc21
                                                    2003011505

Preface

"The key to every biological problem must finally be sought in the cell, for every living organism is, or at sometime has been, a cell."

E. B. Wilson, cell biologist (1925)

The eight decades since Wilson wrote these words have seen a transformation in our understanding of how cells work. The key to the central mystery—how cells store, use, and transmit hereditary information—has been found, and it has opened up a new world. This revolution in biology is among the great adventures of human discovery. Through the perspective of cell biology, we can now explain the fundamental machinery of life and begin to map out a unified picture of the astonishing diversity of organisms and phenomena that it gives rise to, from the chemistry of a bacterium or the shaping of a leaf to the processes that allow us to move, think, talk, and experience the world around us. We can trace the ancestry of our own chemical components through the genetic instructions that specify them—instructions that we share with other organisms to an extent that Wilson could never have imagined. In this way, we have learned to see ourselves in a new light, as close cousins to all other living things.

The new knowledge has brought many practical benefits, including improvements in human health and prosperity. At the same time, it has led to ethical debates and controversy over issues such as genetic testing for inherited diseases, the balancing of environmental risks with benefits, genetic modification of crops and animals, the use of DNA fingerprinting in court cases, and the possibility of human reproductive cloning. These are only a few of the biology-based issues that we have to grapple with today. The successful application of the new knowledge will require many difficult decisions for us as citizens. A basic understanding of cell biology is needed if these decisions are to be intelligent ones.

Our original purpose in writing this book was to provide a straightforward explanation of the workings of a living cell. By "workings," we mean principally the way in which the molecules of the cell—especially the protein, DNA, and RNA molecules—cooperate to create a system that feeds, moves, grows, divides, and responds to stimuli—one, in short, that is alive. By "straightforward," we mean an account that can be easily understood by a reader approaching modern biology for the first time. The need for a short, clear account of the essentials of cell biology became apparent to us while we were writing *Molecular Biology of the Cell (MBoC)*, which is now in its fourth edition. *MBoC* is a large book aimed at advanced undergraduates and graduate students specializing in the life sciences or medicine. Many students and educated lay people who require an introductory account of cell biology would find this text too detailed for their needs. *Essential Cell Biology (ECB)*, in contrast, is designed to provide the fundamentals of cell biology that are required by anyone to understand the biomedical, as well as the broader biological issues that affect our lives.

ECB is as short and simple as we can make it, and we have reduced technical vocabulary to a minimum. In this second edition, we have brought the book completely up to date, with a new emphasis on genomes, including an overview of the human genome sequence and a new chapter on How Genes and Genomes Evolve. In response to requests from many users of the first edition, we have added a chapter on Genetics, Meiosis, and the Molecular Basis of Heredity. There are also new sections on many topics that are frequently in the news, including stem cells, cloning, DNA microarrays, programmed cell death, and cancer. The second edition also features a new series of "How We Know" sections, one for each chapter. Using both classical and current experiments, these sections illustrate how biologists tackle important questions and how their experimental results shape future ideas. As before, the diagrams in *ECB* emphasize central concepts and are stripped of unnecessary details. The key terms introduced in each chapter are highlighted when they first appear and are collected together at the end of the book in a large, illustrated glossary. We have not listed references for further reading: those wishing to explore a subject in greater depth are encouraged to consult the extensive reading lists in *MBoC4* and on the *MBoC4* Web site.

A central feature of the book is the many questions that are presented in the text margins and at the end of each chapter. These are designed to provoke students to think about what they have read and to encourage them to pause and test their understanding. Many questions challenge the student to place the newly acquired information in a broader biological context, and some have more than one valid answer. Others invite speculation. Answers to all the questions are given at the end of the book; in many cases these provide a commentary or an alternative perspective on material presented in the main text.

For those who want to develop their active grasp of cell biology further and to get a deeper understanding of how cell biologists extract conclusions from experiments, we recommend *Molecular Biology of the Cell, Fourth Edition: A Problems Approach*, by John Wilson and Tim Hunt. Though written as a companion to *MBoC*, this contains questions at all levels of difficulty and is a goldmine of thought-provoking problems for teachers and students. We have drawn upon it for some of the questions in *ECB*, and we are very grateful to its authors.

As never before, new imaging and computer technologies have increased our access to the inner workings of living cells. We have tried to capture some of the excitement of these advances in *Essential Cell Biology 2 (ECB2) Interactive*, a CD-ROM disk that is included with each book. It contains over one hundred video clips, animations, molecular structures, and high-resolution micrographs—all designed to complement the material in individual book chapters. One cannot watch cells crawling, dividing, segregating their chromosomes, or rearranging their surface without a sense of wonder at the molecular mechanisms that underlie these processes. We hope that *ECB2 Interactive* will motivate and intrigue students while reinforcing basic concepts covered in the text, and thereby will make the learning of cell biology both easier and more rewarding. Each chapter of the book concludes with a list of multimedia highlights from *ECB2 Interactive*.

The authors of the first edition are pleased to welcome Karen Hopkin to the team. Trained as a biochemist, Karen now writes about science, and her main responsibility has been to make the book clear, accessible, and fun to read. As with *MBoC*, each chapter of *ECB* is the product of communal effort, with individual drafts circulating from one author to another. In addition, many people have helped us, and these are credited in the Acknowledgements that follow. We would be remiss, however, if we did not offer special thanks to Bill Sullivan: as an experienced teacher of genetics, he was instrumental in shaping the new chapter on Genetics, Meiosis, and the Molecular Basis of Heredity.

Despite our best efforts, it is inevitable that there will be errors in the book. We encourage readers who find them to let us know at science@garland.com, so that we can correct these errors in the next printing.

Acknowledgments

The authors acknowledge the many contributions of professors from around the world in the creation of this Second Edition. In particular, more than 200 faculty members who taught with the first edition guided this revision by suggesting ways in which we might improve it. Their thoughtful comments are most appreciated.

We would especially like to thank Dr. Linda Matsuuchi, whose review of the entire first edition provided especially important insights and helped to create the revision plan.

The "How We Know" sections were reviewed by the following professors:

John Andersland, Western Kentucky University
James D. Berger, University of British Columbia
David Bourgaize, Whittier College
Anand Chandrasekhar, University of Missouri
David L. Gard, University of Utah
Paul Overvoorde, Macalester College
Neil Simister, Brandeis University
Efthimios M.C. Skoulakis, Texas A&M University
Lawrence Wiseman, College of William and Mary

We are also grateful to the following expert reviewers, whose advice in specific areas of research improved this edition:

Scott Hawley, Stowers Institute, Missouri
Maynard Olsen, University of Washington

We received invaluable help from many Garland staff. Nigel Orme took original drawings created by author Keith Roberts and redrew them on a computer, or occasionally by hand, with skill and flair. Emma Hunt was responsible for the layout of the entire book and oversaw its production. Michael Morales produced and edited *Essential Cell Biology 2 Interactive*. Elizabeth Zayatz did the developmental editing of each chapter, and Eleanor Lawrence read the entire book for clarity and consistency. Adam Sendroff and Jane Mackarell provided user feedback and collected opinions and suggestions from many teachers. Denise Schanck, the Vice President of Garland Science, orchestrated all of this with great taste and diplomacy, with the aid of Lilith Wood, Garland editorial assistant. Last but not least, we are grateful to our families and our colleagues for their support and tolerance.

Instructor's Resources

Art of Essential Cell Biology, Second Edition

This CD-ROM contains all of the images from the book, available in two convenient formats: PowerPoint and JPEG. The images have been pre-loaded into PowerPoint presentations, one presentation for each chapter of the book. The images are also available as individual JPEG files, which are contained in separate folders from the PowerPoint presentations. The individual JPEG files have been optimized for printing and Web display.

The CD also contains "concept building" PowerPoint slides, which are new to this edition. These slides break down individual pieces of art into their component parts, displaying each component, one at a time, as a separate PowerPoint slide, until the figure is completed.

Cell Biology for Life

Created by Katayoun Chamany, founder of the "Science, Technology and Society Program" at Eugene Lang College of the New School University, *Cell Biology for Life (CBL)* is a Web-accessible curricular supplement that encourages undergraduates to view cell biology as a dynamic field of study whose applications can be seen in everyday life. Inquiry-based assignments and classroom activities highlight the fundamental principles and methods of cell biology through an exploration of three key topics: botulinum toxin use and abuse, the stem cell research debates, and the relationships between the human papilloma virus and cancer. The materials developed for each topic include teaching suggestions and assessment tools, primary and secondary literature, visual media, and detailed references. For additional information and access to the Web site, please visit www.garlandscience.com.

Testbank

The new testbank has been written by Kirsten R. Benjamin, Research Fellow, Center for Genomic Experimentation and Computation, Molecular Sciences Institute, Berkeley and Linda Huang, Assistant Professor of Biology, University of Massachusetts, Boston.

The Second Edition testbank includes a variety of question formats, from multiple choice and fill-in-the-blank to discussion questions, all keyed to concepts in the textbook. It also includes more challenging "thought" questions. Drawing upon the new "How We Know" sections of the book, a selection of questions are written to test students' understanding of modern approaches in experimental biology. The testbank is based on the philosophy that a good exam should do much more than simply test students' ability to memorize information: it should require them to reflect upon and integrate information as a part of a sound understanding. This testbank provides a comprehensive sampling of questions that can be used either directly or as inspiration for instructors to write their own test questions. For more information about accessing the testbank, contact science@garland.com.

Classwire™

The Classwire course management system, available at www.classwire.com/garlandscience, allows instructors to build Web sites for their courses easily. It also serves as an online archive for instructor's resources. After registering for Classwire, you will be able to download all the figures from the book, as well as the animations, videos and molecular structures from the CD. Additional instructor's resources for Garland Science textbooks are also available on Classwire. Please contact science@garland.com for additional information on accessing the Classwire system. (Classwire™ is a trademark of Chalkfree, Inc.)

Instructor's DVD

This DVD contains 90 videos, animations, and molecular models from the *Essential Cell Biology 2 Interactive* CD-ROM. The selected movies from the CD have been rendered in a full-screen, DVD format for optimal display in the classroom or a large teaching theater. The DVD may be played on either a computer with a DVD drive and software, or on a standard DVD player.

Overhead Transparencies

250 full-color overhead transparencies for classroom projection are available to qualified instructors.

Contents and Special Features

Chapter 1	Introduction to Cells	1
Panel 1–1	Light and electron microscopy	8
Panel 1–2	Cells: the principal features of animal, plant, and bacterial cells	25
How We Know	Life's common mechanisms	30
Chapter 2	Chemical Components of Cells	39
How We Know	What are macromolecules?	60
Panel 2–1	Chemical bonds and groups	66
Panel 2–2	The chemical properties of water	68
Panel 2–3	An outline of some of the types of sugars	70
Panel 2–4	Fatty acids and other lipids	72
Panel 2–5	The 20 amino acids found in proteins	74
Panel 2–6	A survey of the nucleotides	76
Panel 2–7	The principal types of weak noncovalent bonds	78
Chapter 3	Energy, Catalysis, and Biosynthesis	83
Panel 3–1	Free energy and biological reactions	96
How We Know	Using kinetics to model and manipulate metabolic pathways	103
Chapter 4	Protein Structure and Function	119
Panel 4–1	A few examples of some general protein functions	120
How We Know	Probing protein structure	129
Panel 4–2	Four different ways of depicting a small protein	132
Panel 4–3	Cell breakage and initial fractionation of cell extracts	160
Panel 4–4	Protein separation by chromatography	162
Panel 4–5	Protein separation by electrophoresis	163
Panel 4–6	Making and using antibodies	164
Chapter 5	DNA and Chromosomes	169
How We Know	Genes are made of DNA	172
Chapter 6	DNA Replication, Repair, and Recombination	195
How We Know	Finding replication origins	198
Chapter 7	From DNA to Protein: How Cells Read the Genome	229
How We Know	Cracking the genetic code	246
Chapter 8	Control of Gene Expression	267
How We Know	Gene regulation—the story of *eve*	282
Chapter 9	How Genes and Genomes Evolve	293
How We Know	Counting genes	314

| Chapter 10 | Manipulating Genes and Cells | 323 |
| How We Know | Sequencing the human genome | 334 |

| Chapter 11 | Membrane Structure | 365 |
| How We Know | Measuring membrane flow | 384 |

| Chapter 12 | Membrane Transport | 389 |
| How We Know | Squid reveal secrets of membrane excitability | 414 |

Chapter 13	How Cells Obtain Energy from Food	427
Panel 13–1	Details of the 10 steps of glycolysis	432
How We Know	Unraveling the citric acid cycle	442
Panel 13–2	The complete citric acid cycle	450

Chapter 14	Energy Generation in Mitochondria and Chloroplasts	453
How We Know	How chemiosmotic coupling drives ATP synthesis	460
Panel 14–1	Redox potentials	471

| Chapter 15 | Intracellular Compartments and Transport | 497 |
| How We Know | Tracking protein and vesicle transport | 520 |

| Chapter 16 | Cell Communication | 533 |
| How We Know | Untangling cell signaling pathways | 561 |

| Chapter 17 | Cytoskeleton | 573 |
| How We Know | Pursuing motor proteins | 586 |

| Chapter 18 | Cell-Cycle Control and Cell Death | 611 |
| How We Know | Discovery of cyclins and Cdks | 618 |

Chapter 19	Cell Division	637
Panel 19–1	The principal stages of M phase in an animal cell	642
How We Know	Building the mitotic spindle	646

Chapter 20	Genetics, Meiosis, and the Molecular Basis of Heredity	659
How We Know	Reading genetic linkage maps	682
Panel 20–1	Some essentials of classical genetics	685

Chapter 21	Tissues and Cancer	697
Panel 21–1	The cell types and tissues from which higher plants are constructed	700
How We Know	Making sense of the genes that are critical for cancer	734

| Answers to Questions | | A:1 |

| Glossary | | G:1 |

| Index | | I:1 |

Detailed Contents

Chapter 1   Introduction to Cells   1

Unity and Diversity of Cells   1

Cells Vary Enormously in Appearance and
Function   2

Living Cells All Have a Similar Basic Chemistry   3

All Present-Day Cells Have Apparently Evolved
from the Same Ancestor   4

Genes Provide the Instructions for Cellular Form,
Function, and Complex Behavior   5

Cells Under the Microscope   5

The Invention of the Light Microscope Led to
the Discovery of Cells   6

Cells, Organelles, and Even Molecules Can Be
Seen Under the Microscope   7

The Procaryotic Cell   11

Procaryotes Are the Most Diverse of Cells   14

The World of Procaryotes Is Divided into Two
Domains: Eubacteria and Archaea   15

The Eucaryotic Cell   16

The Nucleus Is the Information Store of the Cell   16

Mitochondria Generate Energy from Food
to Power the Cell   17

Chloroplasts Capture Energy from Sunlight   18

Internal Membranes Create Intracellular
Compartments with Different Functions   19

The Cytosol Is a Concentrated Aqueous Gel
of Large and Small Molecules   22

The Cytoskeleton Is Responsible for Directed
Cell Movements   22

The Cytoplasm Is Far from Static   23

Eucaryotic Cells May Have Originated as
Predators   24

Model Organisms   27

Molecular Biologists Have Focused on *E. coli*   28

Brewer's Yeast Is a Simple Eucaryotic Cell   28

Arabidopsis Has Been Chosen Out of 300,000
Species as a Model Plant   28

The World of Animals Is Represented by a Fly,
a Worm, a Mouse, and *Homo sapiens*   29

Comparing Genome Sequences Reveals
Life's Common Heritage   33

Chapter 2   Chemical Components of Cells   39

Chemical Bonds   39

Cells Are Made of Relatively Few Types of Atoms   40

The Outermost Electrons Determine How Atoms
Interact   41

Ionic Bonds Form by the Gain and Loss
of Electrons   43

Covalent Bonds Form by the Sharing of Electrons   45

Covalent Bonds Vary in Strength   46

There Are Different Types of Covalent Bonds   47

Water Is Held Together by Hydrogen Bonds   48

Some Polar Molecules Form Acids and Bases
in Water   49

Molecules in Cells   50

A Cell Is Formed from Carbon Compounds   50

Cells Contain Four Major Families of Small
Organic Molecules   51

Sugars Are Energy Sources for Cells and Subunits
of Polysaccharides   52

Fatty Acids Are Components of Cell Membranes   53

Amino Acids Are the Subunits of Proteins   55

Nucleotides Are the Subunits of DNA and RNA   56

Macromolecules in Cells   58

Macromolecules Contain a Specific Sequence
of Subunits   59

Noncovalent Bonds Specify the Precise Shape
of a Macromolecule   62

Noncovalent Bonds Allow a Macromolecule
to Bind Other Selected Molecules   63

Chapter 3   Energy, Catalysis, and Biosynthesis   83

Catalysis and the Use of Energy by Cells   84

Biological Order Is Made Possible by the Release of Heat Energy from Cells   85

Photosynthetic Organisms Use Sunlight to Synthesize Organic Molecules   88

Cells Obtain Energy by the Oxidation of Organic Molecules   89

Oxidation and Reduction Involve Electron Transfers   90

Enzymes Lower the Barriers That Block Chemical Reactions   91

The Free-Energy Change for a Reaction Determines Whether It Can Occur   93

The Concentration of Reactants Influences the Free-Energy Change and a Reaction's Direction   94

The Equilibrium Constant Indicates the Strength of Molecular Interactions   95

For Sequential Reactions, the Changes in Free Energy Are Additive   98

Rapid Diffusion Allows Enzymes to Find Their Substrates   100

V_{max} and K_M Measure Enzyme Performance   101

Activated Carrier Molecules and Biosynthesis   106

The Formation of an Activated Carrier Is Coupled to an Energetically Favorable Reaction   106

ATP Is the Most Widely Used Activated Carrier Molecule   107

Energy Stored in ATP Is Often Harnessed to Join Two Molecules Together   108

NADH and NADPH Are Important Electron Carriers   109

There Are Many Other Activated Carrier Molecules in Cells   111

The Synthesis of Biological Polymers Requires an Energy Input   112

Chapter 4   Protein Structure and Function   119

The Shape and Structure of Proteins   119

The Shape of a Protein Is Specified by Its Amino Acid Sequence   121

Proteins Fold into a Conformation of Lowest Energy   124

Proteins Come in a Wide Variety of Complicated Shapes   125

The α Helix and the β Sheet Are Common Folding Patterns   126

Helices Form Readily in Biological Structures   134

β Sheets Form Rigid Structures at the Core of Many Proteins   135

Proteins Have Several Levels of Organization   136

Few of the Many Possible Polypeptide Chains Will Be Useful   137

Proteins Can Be Classified into Families   138

Large Protein Molecules Often Contain More Than One Polypeptide Chain   139

Proteins Can Assemble into Filaments, Sheets, or Spheres   140

Some Types of Proteins Have Elongated Fibrous Shapes   141

Extracellular Proteins Are Often Stabilized by Covalent Cross-Linkages   142

How Proteins Work   143

All Proteins Bind to Other Molecules   143

The Binding Sites of Antibodies Are Especially Versatile   144

Enzymes Are Powerful and Highly Specific Catalysts   145

Lysozyme Illustrates How an Enzyme Works   146

Tightly Bound Small Molecules Add Extra Functions to Proteins   149

How Proteins Are Controlled   150

The Catalytic Activities of Enzymes Are Often Regulated by Other Molecules   151

Allosteric Enzymes Have Two Binding Sites That Influence One Another   151

Phosphorylation Can Control Protein Activity by Triggering a Conformational Change   153

GTP-Binding Proteins Are Also Regulated by the Cyclic Gain and Loss of a Phosphate Group   154

Nucleotide Hydrolysis Allows Motor Proteins to Produce Large Movements in Cells   155

Proteins Often Form Large Complexes That Function as Protein Machines   156

Large-Scale Studies of Protein Structure and Function Are Increasing the Pace of Discovery   157

Chapter 5   DNA and Chromosomes   169

The Structure and Function of DNA   170

A DNA Molecule Consists of Two Complementary Chains of Nucleotides   171

The Structure of DNA Provides a Mechanism for Heredity   176

The Structure of Eucaryotic Chromosomes   177

Eucaryotic DNA Is Packaged into Chromosomes   178

Chromosomes Contain Long Strings of Genes   179

Chromosomes Exist in Different States Throughout the Life of a Cell   181

Interphase Chromosomes Are Organized Within the Nucleus   183

The DNA in Chromosomes Is Highly Condensed   183

Nucleosomes Are the Basic Units of Chromatin Structure   184

Chromosomes Have Several Levels of DNA Packing   186

Interphase Chromosomes Contain Both Condensed and More Extended Forms of Chromatin   187

Changes in Nucleosome Structure Allow Access to DNA   189

Chapter 6   DNA Replication, Repair, and Recombination   195

DNA Replication   196

Base-Pairing Enables DNA Replication   196

DNA Synthesis Begins at Replication Origins   197

New DNA Synthesis Occurs at Replication Forks   201

The Replication Fork Is Asymmetrical   202

DNA Polymerase Is Self-correcting   203

Short Lengths of RNA Act as Primers for DNA Synthesis   204

Proteins at a Replication Fork Cooperate to Form a Replication Machine   206

Telomerase Replicates the Ends of Eucaryotic Chromosomes   207

DNA Replication Is Relatively Well Understood   208

DNA Repair   209

Mutations Can Have Severe Consequences for an Organism   209

A DNA Mismatch Repair System Removes Replication Errors That Escape the Replication Machine   210

DNA Is Continually Suffering Damage in Cells   212

The Stability of Genes Depends on DNA Repair   213

The High Fidelity of DNA Maintenance Allows Closely Related Species to Have Proteins with Very Similar Sequences   214

DNA Recombination   215

Homologous Recombination Results in an Exact Exchange of Genetic Information   215

Recombination Can Also Occur Between Nonhomologous DNA Sequences   216

Mobile Genetic Elements Encode the Components They Need for Movement   217

A Large Fraction of the Human Genome Is Composed of Two Families of Transposable Sequences   218

Viruses Are Fully Mobile Genetic Elements That Can Escape from Cells   219

Retroviruses Reverse the Normal Flow of Genetic Information   221

Chapter 7   From DNA to Protein: How Cells Read the Genome   229

From DNA to RNA   230

Portions of DNA Sequence Are Transcribed into RNA   230

Transcription Produces RNA Complementary to One Strand of DNA   231

Several Types of RNA Are Produced in Cells   233

Signals in DNA Tell RNA Polymerase Where to Start and Finish   234

Eucaryotic RNAs Are Transcribed and Processed Simultaneously in the Nucleus   236

Eucaryotic Genes Are Interrupted by Noncoding Sequences   237

Introns Are Removed by RNA Splicing   238

Mature Eucaryotic mRNAs Are Selectively Exported from the Nucleus   241

mRNA Molecules Are Eventually Degraded by the Cell   242

The Earliest Cells May Have Had Introns in Their Genes   242

From RNA to Protein   243

An mRNA Sequence Is Decoded in Sets of Three Nucleotides   244

tRNA Molecules Match Amino Acids to Codons in mRNA   245

Specific Enzymes Couple tRNAs to the Correct Amino Acid   248

The RNA Message Is Decoded on Ribosomes   248

The Ribosome Is a Ribozyme   251

Codons in mRNA Signal Where to Start and to Stop Protein Synthesis   253

Proteins Are Made on Polyribosomes   254

Inhibitors of Procaryotic Protein Synthesis Are Used as Antibiotics   255

Carefully Controlled Protein Breakdown Helps Regulate the Amount of Each Protein in a Cell   256

There Are Many Steps Between DNA and Protein   257

RNA and the Origins of Life   258

Life Requires Autocatalysis   259

RNA Can Both Store Information and Catalyze Chemical Reactions   259

RNA Is Thought to Predate DNA in Evolution   261

Chapter 8   Control of Gene Expression   267

An Overview of Gene Expression   268

The Different Cell Types of a Multicellular Organism Contain the Same DNA   268

Different Cell Types Produce Different Sets of Proteins   268

A Cell Can Change the Expression of Its Genes in Response to External Signals   270

Gene Expression Can Be Regulated at Many of the Steps in the Pathway from DNA to RNA to Protein   270

How Transcriptional Switches Work   271

Transcription Is Controlled by Proteins Binding to Regulatory DNA Sequences   271

Repressors Turn Genes Off, Activators Turn Them On   273

An Activator and a Repressor Control the *lac* Operon   275

Initiation of Eucaryotic Gene Transcription Is a Complex Process   275

Eucaryotic RNA Polymerase Requires General Transcription Factors   276

Eucaryotic Gene Regulatory Proteins Control Gene Expression from a Distance   278

Packing of Promoter DNA into Nucleosomes Can Affect Initiation of Transcription   279

The Molecular Mechanisms that Create Specialized Cell Types   280

Eucaryotic Genes Are Regulated by Combinations of Proteins   281

The Expression of Different Genes Can Be Coordinated by a Single Protein   281

Combinatorial Control Can Create Different Cell Types   285

Stable Patterns of Gene Expression Can Be Transmitted to Daughter Cells   286

The Formation of an Entire Organ Can Be Triggered by a Single Gene Regulatory Protein   288

Chapter 9:   How Genes and Genomes Evolve   293

Generating Genetic Variation   293

Five Main Types of Genetic Change Contribute
to Evolution   295

Genome Alterations Are Caused by Failures
of the Normal Mechanisms for Copying and
Maintaining DNA   296

DNA Duplications Give Rise to Families of
Related Genes Within a Single Cell   297

The Evolution of the Globin Gene Family
Shows How DNA Duplications Contribute
to the Evolution of Organisms   298

Gene Duplication and Divergence Provide
a Critical Source of Genetic Novelty for
Evolving Organisms   299

New Genes Can Be Generated by Repeating
the Same Exon   300

Novel Genes Can Also Be Created by
Exon Shuffling   300

The Evolution of Genomes Has Been
Accelerated by the Movement of
Transposable Elements   301

Genes Can Be Exchanged Between Organisms
by Horizontal Gene Transfer   302

Reconstructing Life's Family Tree   304

Genetic Changes That Offer an Organism
a Selective Advantage Are the Most Likely
to Be Preserved   304

The Genome Sequences of Two Species Differ
in Proportion to the Length of Time That
They Have Evolved Separately   305

Humans and Chimpanzee Genomes Are Similar in
Organization as Well as Detailed Sequence   306

Functionally Important Sequences Show Up as
Islands of DNA Sequence Conservation   307

Genome Comparisons Suggest That "Junk DNA"
is Dispensable   308

Sequence Conservation Allows Us to Trace Even
the Most Distant Evolutionary Relationships   309

Examining the Human Genome   311

The Nucleotide Sequence of the Human Genome
Shows How Our Genes Are Arranged   311

Genetic Variation Within the Human Genome
Contributes to Our Individuality   313

Comparing Our DNA with That of Related
Organisms Helps Us to Interpret the Human
Genome   316

The Human Genome Contains Copious
Information Yet to Be Deciphered   317

Chapter 10   Manipulating Genes and Cells   323

Isolating Cells and Growing Them
in Culture   324

A Uniform Population of Cells Can Be Obtained
from a Tissue   325

Cells Can Be Grown in a Culture Dish   325

Maintaining Eucaryotic Cells in Culture Poses
Special Challenges   326

How DNA Molecules Are Analyzed   327

Restriction Nucleases Cut DNA Molecules
at Specific Sites   328

Gel Electrophoresis Separates DNA Fragments
of Different Sizes   329

The Nucleotide Sequence of DNA Fragments
Can Be Determined   331

Genome Sequences Are Searched to
Identify Genes   333

Nucleic Acid Hybridization   336

DNA Hybridization Facilitates the Diagnosis
of Genetic Diseases   336

Hybridization on DNA Microarrays Monitors the
Expression of Thousands of Genes at Once   338

In Situ Hybridization Locates Nucleic Acid
Sequences in Cells or on Chromosomes   340

DNA Cloning   341

DNA Ligase Joins DNA Fragments Together to
Produce a Recombinant DNA Molecule   341

Recombinant DNA Can Be Copied Inside
Bacterial Cells   341

Specialized Plasmid Vectors Are Used to
Clone DNA   342

Human Genes Are Isolated by DNA Cloning   343

cDNA Libraries Represent the mRNA Produced
by a Particular Tissue   346

The Polymerase Chain Reaction Amplifies
Selected DNA Sequences   347

DNA Engineering   352

Completely Novel DNA Molecules Can Be
Constructed   352

Rare Cellular Proteins Can Be Made in Large
Amounts Using Cloned DNA   352

Engineered Genes Can Reveal When and
Where a Gene Is Expressed   353

Mutant Organisms Best Reveal the Function
of a Gene   355

Animals Can Be Genetically Altered   356

Transgenic Plants Are Important for Both
Cell Biology and Agriculture   359

Chapter 11   Membrane Structure   365

The Lipid Bilayer   366
Membrane Lipids Form Bilayers in Water   367
The Lipid Bilayer Is a Two-dimensional Fluid   370
The Fluidity of a Lipid Bilayer Depends on
  Its Composition   371
The Lipid Bilayer Is Asymmetrical   373
Lipid Asymmetry Is Generated Inside the Cell   373

Membrane Proteins   374
Membrane Proteins Associate with the Lipid
  Bilayer in Various Ways   375

A Polypeptide Chain Usually Crosses the
  Bilayer as an α Helix   376
Membrane Proteins Can Be Solubilized
  in Detergents and Purified   377
The Complete Structure Is Known for a Few
  Membrane Proteins   378
The Plasma Membrane Is Reinforced
  by the Cell Cortex   380
The Cell Surface Is Coated with Carbohydrate   381
Cells Can Restrict the Movement of
  Membrane Proteins   383

Chapter 12   Membrane Transport   389

Principles of Membrane Transport   389
The Ion Concentrations Inside a Cell Are Very
  Different from Those Outside   390
Lipid Bilayers Are Impermeable to Solutes
  and Ions   391
Membrane Transport Proteins Fall into Two
  Classes: Carriers and Channels   391
Solutes Cross Membranes by Passive or Active
  Transport   392

Carrier Proteins and Their Functions   393
Concentration Gradients and Electrical
  Forces Drive Passive Transport   393
Active Transport Moves Solutes Against Their
  Electrochemical Gradients   395
Animal Cells Use the Energy of ATP Hydrolysis
  to Pump Out Na^+ 396
The Na^+-K^+ Pump Is Driven by the Transient
  Addition of a Phosphate Group   397
Animal Cells Use the Na^+ Gradient to Take Up
  Nutrients Actively   397
The Na^+-K^+ Pump Helps Maintain the Osmotic
  Balance of Animal Cells   399
Intracellular Ca^{2+} Concentrations Are Kept
  Low by Ca^{2+} Pumps   401
H^+ Gradients Are Used to Drive Membrane
  Transport in Plants, Fungi, and Bacteria   402

Ion Channels and the Membrane
  Potential   403
Ion Channels Are Ion-Selective and Gated   403
Ion Channels Randomly Snap Between Open
  and Closed States   405
Different Types of Stimuli Influence the Opening
  and Closing of Ion Channels   407
Voltage-gated Ion Channels Respond to the
  Membrane Potential   407
Membrane Potential Is Governed by Membrane
  Permeability to Specific Ions   408

Ion Channels and Signaling in
  Nerve Cells   411
Action Potentials Provide for Rapid Long-Distance
  Communication   411
Action Potentials Are Usually Mediated by
  Voltage-gated Na^+ Channels   412
Voltage-gated Ca^{2+} Channels Convert
  Electrical Signals into Chemical Signals at
  Nerve Terminals   417
Transmitter-gated Channels in Target Cells Convert
  Chemical Signals Back into Electrical Signals   417
Neurons Receive Both Excitatory and Inhibitory
  Inputs   419
Transmitter-gated Ion Channels Are Major
  Targets for Psychoactive Drugs   419
Synaptic Connections Enable You to Think,
  Act, and Remember   420

Chapter 13   How Cells Obtain Energy from Food   427

The Breakdown of Sugars and Fats   428

Food Molecules Are Broken Down in Three
  Stages   428
Glycolysis Is a Central ATP-producing Pathway   430
Fermentations Allow ATP to Be Produced in the
  Absence of Oxygen   431
Glycolysis Illustrates How Enzymes Couple
  Oxidation to Energy Storage   434
Sugars and Fats Are Both Degraded to
  Acetyl CoA in Mitochondria   435
The Citric Acid Cycle Generates NADH by
  Oxidizing Acetyl Groups to CO_2 439

Electron Transport Drives the Synthesis of
  the Majority of the ATP in Most Cells   441

Storing and Utilizing Food   444

Organisms Store Food Molecules in Special
  Reservoirs   444
Chloroplasts and Mitochondria Collaborate
  in Plant Cells   446
Many Biosynthetic Pathways Begin with
  Glycolysis or the Citric Acid Cycle   447
Metabolism Is Organized and Regulated   448

Chapter 14   Energy Generation in Mitochondria and Chloroplasts   453

Cells Obtain Most of Their Energy by a
  Membrane-based Mechanism   453

Mitochondria and Oxidative
Phosphorylation   455

A Mitochondrion Contains an Outer
  Membrane, an Inner Membrane, and
  Two Internal Compartments   455
High-Energy Electrons Are Generated via
  the Citric Acid Cycle   457
A Chemiosmotic Process Converts
  Oxidation Energy into ATP   458
Electrons Are Transferred Along a Chain of
  Proteins in the Inner Mitochondrial Membrane   459
Electron Transport Generates a Proton
  Gradient Across the Membrane   462
The Proton Gradient Drives ATP Synthesis   464
Coupled Transport Across the Inner
  Mitochondrial Membrane Is Driven by the
  Electrochemical Proton Gradient   466
Proton Gradients Produce Most of the
  Cell's ATP   466
The Rapid Conversion of ADP to ATP in
  Mitochondria Maintains a High ATP/ADP Ratio
  in Cells   468

Electron-Transport Chains and Proton
Pumping   468

Protons Are Readily Moved by the Transfer of
  Electrons   468
The Redox Potential Is a Measure of Electron
  Affinities   469

Electron Transfers Release Large Amounts
  of Energy   470
Metals Tightly Bound to Proteins Form Versatile
  Electron Carriers   472
Cytochrome Oxidase Catalyzes Oxygen
  Reduction   474
The Mechanism of H^+ Pumping Will Soon Be
  Understood in Atomic Detail   475
Respiration Is Amazingly Efficient   476

Chloroplasts and Photosynthesis   478

Chloroplasts Resemble Mitochondria but
  Have an Extra Compartment   478
Chloroplasts Capture Energy from Sunlight
  and Use It to Fix Carbon   480
Excited Chlorophyll Molecules Funnel Energy
  into a Reaction Center   481
Light Energy Drives the Synthesis of ATP
  and NADPH   482
Carbon Fixation Is Catalyzed by Ribulose
  Bisphosphate Carboxylase   485
Carbon Fixation in Chloroplasts Generates
  Sucrose and Starch   486

The Origins of Chloroplasts and
Mitochondria   487

Oxidative Phosphorylation Gave Ancient
  Bacteria an Evolutionary Advantage   488
Photosynthetic Bacteria Made Even Fewer
  Demands on Their Environment   489
The Lifestyle of *Methanococcus* Suggests That
  Chemiosmotic Coupling Is an Ancient Process   490

Chapter 15   Intracellular Compartments and Transport   497

Membrane-enclosed Organelles	498
Eucaryotic Cells Contain a Basic Set of Membrane-enclosed Organelles	498
Membrane-enclosed Organelles Evolved in Different Ways	500
Protein Sorting	502
Proteins Are Imported into Organelles by Three Mechanisms	502
Signal Sequences Direct Proteins to the Correct Compartment	503
Proteins Enter the Nucleus Through Nuclear Pores	504
Proteins Unfold to Enter Mitochondria and Chloroplasts	506
Proteins Enter the Endoplasmic Reticulum While Being Synthesized	507
Soluble Proteins Are Released into the ER Lumen	509
Start and Stop Signals Determine the Arrangement of a Transmembrane Protein in the Lipid Bilayer	510
Vesicular Transport	512
Transport Vesicles Carry Soluble Proteins and Membrane Between Compartments	512

Vesicle Budding Is Driven by the Assembly of a Protein Coat	513
The Specificity of Vesicle Docking Depends on SNAREs	515
Secretory Pathways	516
Most Proteins Are Covalently Modified in the ER	516
Exit from the ER Is Controlled to Ensure Protein Quality	517
Proteins Are Further Modified and Sorted in the Golgi Apparatus	518
Secretory Proteins Are Released from the Cell by Exocytosis	519
Endocytic Pathways	523
Specialized Phagocytic Cells Ingest Large Particles	523
Fluid and Macromolecules Are Taken Up by Pinocytosis	525
Receptor-mediated Endocytosis Provides a Specific Route into Animal Cells	525
Endocytosed Macromolecules Are Sorted in Endosomes	526
Lysosomes Are the Principal Sites of Intracellular Digestion	527

Chapter 16   Cell Communication   533

General Principles of Cell Signaling	533
Signals Can Act over Long or Short Range	534
Each Cell Responds to a Limited Set of Signals	536
Receptors Relay Signals via Intracellular Signaling Pathways	538
Nitric Oxide Crosses the Plasma Membrane and Activates Intracellular Enzymes Directly	540
Some Hormones Cross the Plasma Membrane and Bind to Intracellular Receptors	541
Cell-Surface Receptors Fall into Three Main Classes	542
Ion-channel–linked Receptors Convert Chemical Signals into Electrical Ones	544
Many Intracellular Signaling Proteins Act as Molecular Switches	545
G-protein–linked Receptors	546
Stimulation of G-protein–linked Receptors Activates G-Protein Subunits	546
Some G Proteins Regulate Ion Channels	548
Some G Proteins Activate Membrane-bound Enzymes	549

The Cyclic AMP Pathway Can Activate Enzymes and Turn On Genes	550
The Inositol Phospholipid Pathway Triggers a Rise in Intracellular Ca^{2+}	552
A Ca^{2+} Signal Triggers Many Biological Processes	554
Intracellular Signaling Cascades Can Achieve Astonishing Speed, Sensitivity, and Adaptability: A Look at Photoreceptors in the Eye	555
Enzyme-linked Receptors	557
Activated Receptor Tyrosine Kinases Assemble a Complex of Intracellular Signaling Proteins	557
Receptor Tyrosine Kinases Activate the GTP-binding Protein Ras	559
Some Enzyme-linked Receptors Activate a Fast Track to the Nucleus	560
Protein Kinase Networks Integrate Information to Control Complex Cell Behaviors	565
Multicellularity and Cell Communication Evolved Independently in Plants and Animals	566

Intermediate Filaments   574

Intermediate Filaments Are Strong and Ropelike   575

Intermediate Filaments Strengthen Cells Against Mechanical Stress   576

The Nuclear Envelope Is Supported by a Meshwork of Intermediate Filaments   578

Microtubules   579

Microtubules Are Hollow Tubes with Structurally Distinct Ends   579

The Centrosome Is the Major Microtubule-organizing Center in Animal Cells   580

Growing Microtubules Show Dynamic Instability   581

Microtubules Are Maintained by a Balance of Assembly and Disassembly   582

Microtubules Organize the Interior of the Cell   583

Motor Proteins Drive Intracellular Transport   584

Organelles Move Along Microtubules   585

Cilia and Flagella Contain Stable Microtubules Moved by Dynein   590

Actin Filaments   592

Actin Filaments Are Thin and Flexible   593

Actin and Tubulin Polymerize by Similar Mechanisms   593

Many Proteins Bind to Actin and Modify Its Properties   594

An Actin-rich Cortex Underlies the Plasma Membrane of Most Eucaryotic Cells   594

Cell Crawling Depends on Actin   595

Actin Associates with Myosin to Form Contractile Structures   598

Extracellular Signals Control the Arrangement of Actin Filaments   599

Muscle Contraction   600

Muscle Contraction Depends on Bundles of Actin and Myosin   600

During Muscle Contraction Actin Filaments Slide Against Myosin Filaments   601

Muscle Contraction Is Triggered by a Sudden Rise in Ca^{2+} 603

Muscle Cells Perform Highly Specialized Functions in the Body   605

Overview of the Cell Cycle   612

The Eucaryotic Cell Cycle Is Divided into Four Phases   613

A Central Control System Triggers the Major Processes of the Cell Cycle   614

The Cell-Cycle Control System   615

The Cell-Cycle Control System Depends on Cyclically Activated Protein Kinases   616

Cyclin-dependent Protein Kinases Are Regulated by the Accumulation and Destruction of Cyclins   617

The Activity of Cdks Is Also Regulated by Phosphorylation and Dephosphorylation   617

Different Cyclin–Cdk Complexes Trigger Different Steps in the Cell Cycle   620

S-Cdk Initiates DNA Replication and Helps Block Rereplication   621

Cdks Are Inactive Through Most of G_1 622

The Cell-Cycle Control System Can Arrest the Cycle at Specific Checkpoints   622

Cells Can Dismantle Their Control System and Withdraw from the Cell Cycle   624

Programmed Cell Death (Apoptosis)   625

Apoptosis Is Mediated by an Intracellular Proteolytic Cascade   626

The Death Program Is Regulated by the Bcl-2 Family of Intracellular Proteins   627

Extracellular Control of Cell Numbers and Cell Size   628

Animal Cells Require Extracellular Signals to Divide, Grow, and Survive   629

Mitogens Stimulate Cell Division   629

Extracellular Growth Factors Stimulate Cells to Grow   631

Animal Cells Require Survival Factors to Avoid Apoptosis   631

Some Extracellular Signal Proteins Inhibit Cell Growth, Division, or Survival   632

Chapter 19   Cell Division   637

An Overview of M Phase   638

In Preparation for M Phase, DNA-binding Proteins Configure Replicated Chromosomes for Segregation   638

The Cytoskeleton Carries Out Both Mitosis and Cytokinesis   639

Centrosomes Duplicate To Help Form the Two Poles of the Mitotic Spindle   640

M Phase Is Conventionally Divided into Six Stages   640

Mitosis   641

Microtubule Instability Facilitates the Formation of the Mitotic Spindle   641

The Mitotic Spindle Starts to Assemble in Prophase   644

Chromosomes Attach to the Mitotic Spindle at Prometaphase   645

Chromosomes Line Up at the Spindle Equator at Metaphase   648

Daughter Chromosomes Segregate at Anaphase   649

The Nuclear Envelope Re-forms at Telophase   651

Some Organelles Fragment at Mitosis   651

Cytokinesis   652

The Mitotic Spindle Determines the Plane of Cytoplasmic Cleavage   652

The Contractile Ring of Animal Cells Is Made of Actin and Myosin   653

Cytokinesis in Plant Cells Involves New Cell-Wall Formation   654

Gametes Are Formed by a Specialized Kind of Cell Division   655

Chapter 20   Genetics, Meiosis, and the Molecular Basis of Heredity   659

The Benefits of Sex   660

Sexual Reproduction Involves Both Diploid and Haploid Cells   661

Sexual Reproduction Gives Organisms a Competitive Advantage   662

Meiosis   663

Haploid Cells Are Produced From Diploid Cells Through Meiosis   664

Meiosis Involves a Special Process of Chromosome Pairing   664

Extensive Recombination Occurs Between Maternal and Paternal Chromosomes   665

Chromosome Pairing and Recombination Ensure the Proper Segregation of Homologs   667

The Second Meiotic Division Produces Haploid Daughter Cells   667

The Haploid Cells Contain Extensively Reassorted Genetic Information   668

Meiosis Is Not Flawless   670

Fertilization Reconstitutes a Complete Genome   671

Mendel and the Laws of Inheritance   672

Mendel Chose to Study Traits That Are Inherited in a Discrete Fashion   673

Mendel Could Disprove the Alternative Theories of Inheritance   674

Mendel's Experiments Were the First to Reveal the Discrete Nature of Heredity   674

Each Gamete Carries a Single Allele for Each Character   675

Mendel's Law of Segregation Applies to All Sexually Reproducing Organisms   676

Alleles for Different Traits Segregate Independently   677

The Behavior of Chromosomes During Meiosis Underlies Mendel's Laws of Inheritance   678

The Frequency of Recombination Can Be Used to Order Genes on Chromosomes   680

The Phenotype of the Heterozygote Reveals Whether an Allele is Dominant or Recessive   681

Mutant Alleles Sometimes Confer a Selective Advantage   684

Genetics as an Experimental Tool   686

The Classical Approach Begins with Random Mutagenesis   686

Genetic Screens Identify Mutants Deficient in Cellular Processes   687

A Complementation Test Reveals Whether Two Mutations Are in the Same Gene   688

Human Genes Are Inherited in Haplotype Blocks, Which Can Aid in the Search for Mutations That Cause Disease   689

Complex Traits Are Influenced by Multiple Genes   691

Is Our Fate Encoded in Our DNA?   692

Extracellular Matrix and Connective Tissues   698

Plant Cells Have Tough External Walls   698

Cellulose Fibers Give the Plant Cell Wall Its Tensile Strength   702

Animal Connective Tissues Consist Largely of Extracellular Matrix   703

Collagen Provides Tensile Strength in Animal Connective Tissues   704

Cells Organize the Collagen That They Secrete   705

Integrins Couple the Matrix Outside a Cell to the Cytoskeleton Inside It   706

Gels of Polysaccharide and Protein Fill Spaces and Resist Compression   706

Epithelial Sheets and Cell-Cell Junctions   709

Epithelial Sheets Are Polarized and Rest on a Basal Lamina   709

Tight Junctions Make an Epithelium Leak-proof and Separate Its Apical and Basal Surfaces   711

Cytoskeleton-linked Junctions Bind Epithelial Cells Robustly to One Another and to the Basal Lamina   712

Gap Junctions Allow Ions and Small Molecules to Pass from Cell to Cell   715

Tissue Maintenance and Renewal   717

Tissues Are Organized Mixtures of Many Cell Types   718

Different Tissues Are Renewed at Different Rates   720

Stem Cells Generate a Continuous Supply of Terminally Differentiated Cells   721

Stem Cells Can Be Used to Repair Damaged Tissues   722

Nuclear Transplantation Provides a Way to Generate Personalized ES Cells: the Strategy of Therapeutic Cloning   725

Cancer   726

Cancer Cells Proliferate, Invade, and Metastasize   726

Epidemiology Identifies Preventable Causes of Cancer   727

Cancers Develop by an Accumulation of Mutations   728

Cancers Evolve Properties That Give Them a Competitive Advantage   729

Many Diverse Types of Genes Are Critical for Cancer   731

Colorectal Cancer Illustrates How Loss of a Gene Can Lead to Growth of a Tumor   732

An Understanding of Cancer Cell Biology Opens the Way to New Treatments   736

Introduction to Cells

1

What does it mean to be living? People, petunias, and pond scum are all alive; stones, sand, and summer breezes are not. But what are the fundamental properties that characterize living things and distinguish them from nonliving matter?

The answer begins with a basic fact that is taken for granted by biologists now, but marked a revolution in thinking when first established 170 years ago. All living things are made of **cells**: small, membrane-enclosed units filled with a concentrated aqueous solution of chemicals and endowed with the extraordinary ability to create copies of themselves by growing and dividing in two. The simplest forms of life are solitary cells. Higher organisms, including ourselves, are communities of cells derived by growth and division from a single founder cell: each animal, plant, or fungus is a vast colony of individual cells that perform specialized functions coordinated by intricate systems of communication.

Cells, therefore, are the fundamental units of life, and it is to *cell biology* that we must look for an answer to the question of what life is and how it works. With a deeper understanding of the structure, function, behavior, and evolution of cells, we can begin to tackle the grand historical problems of life on Earth: its mysterious origins, its stunning diversity, its invasion of every conceivable habitat. At the same time, cell biology can provide us with answers to the questions we have about ourselves: Where did we come from? How do we develop from a single fertilized egg cell? How is each of us different from every other person on Earth? Why do we get sick, grow old, and die?

In this chapter we begin by looking at the great variety of forms that cells can show, and we also take a preliminary glimpse at the chemical machinery that all cells have in common. We then consider how cells are made visible under the microscope and what we see when we peer inside them. Finally, we will discuss how we can exploit the similarities of living things to achieve a coherent understanding of all the forms of life on Earth—from the tiniest bacterium to the mightiest oak.

Unity and Diversity of Cells

Cell biologists often speak of "the cell" without specifying any particular cell. But cells are not all alike; in fact, they can be wildly different. It is estimated that there are at least 10 million—perhaps 100 million—distinct species of living things in the world. Before delving deeper into cell biology, we must take stock: what do the cells of these species have in common—the bacterium and the butterfly, the rose and the dolphin? And in what ways do they differ?

Unity and Diversity of Cells
Cells Vary Enormously in Appearance and Function
Living Cells All Have a Similar Basic Chemistry
All Present-Day Cells Have Apparently Evolved from the Same Ancestor
Genes Provide the Instructions for Cellular Form, Function, and Complex Behavior

Cells Under the Microscope
The Invention of the Light Microscope Led to the Discovery of Cells
Cells, Organelles, and Even Molecules Can Be Seen Under the Microscope

The Procaryotic Cell
Procaryotes Are the Most Diverse of Cells
The World of Procaryotes Is Divided into Two Domains: Eubacteria and Archaea

The Eucaryotic Cell
The Nucleus Is the Information Store of the Cell
Mitochondria Generate Energy from Food to Power the Cell
Chloroplasts Capture Energy from Sunlight
Internal Membranes Create Intracellular Compartments with Different Functions
The Cytosol Is a Concentrated Aqueous Gel of Large and Small Molecules
The Cytoskeleton Is Responsible for Directed Cell Movements
The Cytoplasm Is Far from Static
Eucaryotic Cells May Have Originated as Predators

Model Organisms
Molecular Biologists Have Focused on *E. coli*
Brewer's Yeast Is a Simple Eucaryotic Cell
Arabidopsis Has Been Chosen Out of 300,000 Species as a Model Plant
The World of Animals Is Represented by a Fly, a Worm, a Mouse, and *Homo sapiens*
Comparing Genome Sequences Reveals Life's Common Heritage

Cells Vary Enormously in Appearance and Function

Let us begin with size. A bacterial cell—say a *Lactobacillus* in a piece of cheese—is a few **micrometers**, or µm, in length. A frog's egg—which is also a single cell—has a diameter of about 1 millimeter. If we scaled them up so that the *Lactobacillus* were the size of a person, the frog's egg would be half a mile high.

Cells vary no less widely in their shapes and functions. Consider the gallery of cells displayed in Figure 1–1. A typical nerve cell in your brain is enormously extended; it sends out its electrical signals along a fine protrusion that is 10,000 times longer than it is thick, and it receives signals from other cells through a mass of shorter processes that sprout from its body like the branches of a tree. A *Paramecium* in a drop of pond water is shaped like a submarine and is covered with tens of thousands of *cilia*—hairlike extensions whose sinuous beating sweeps the

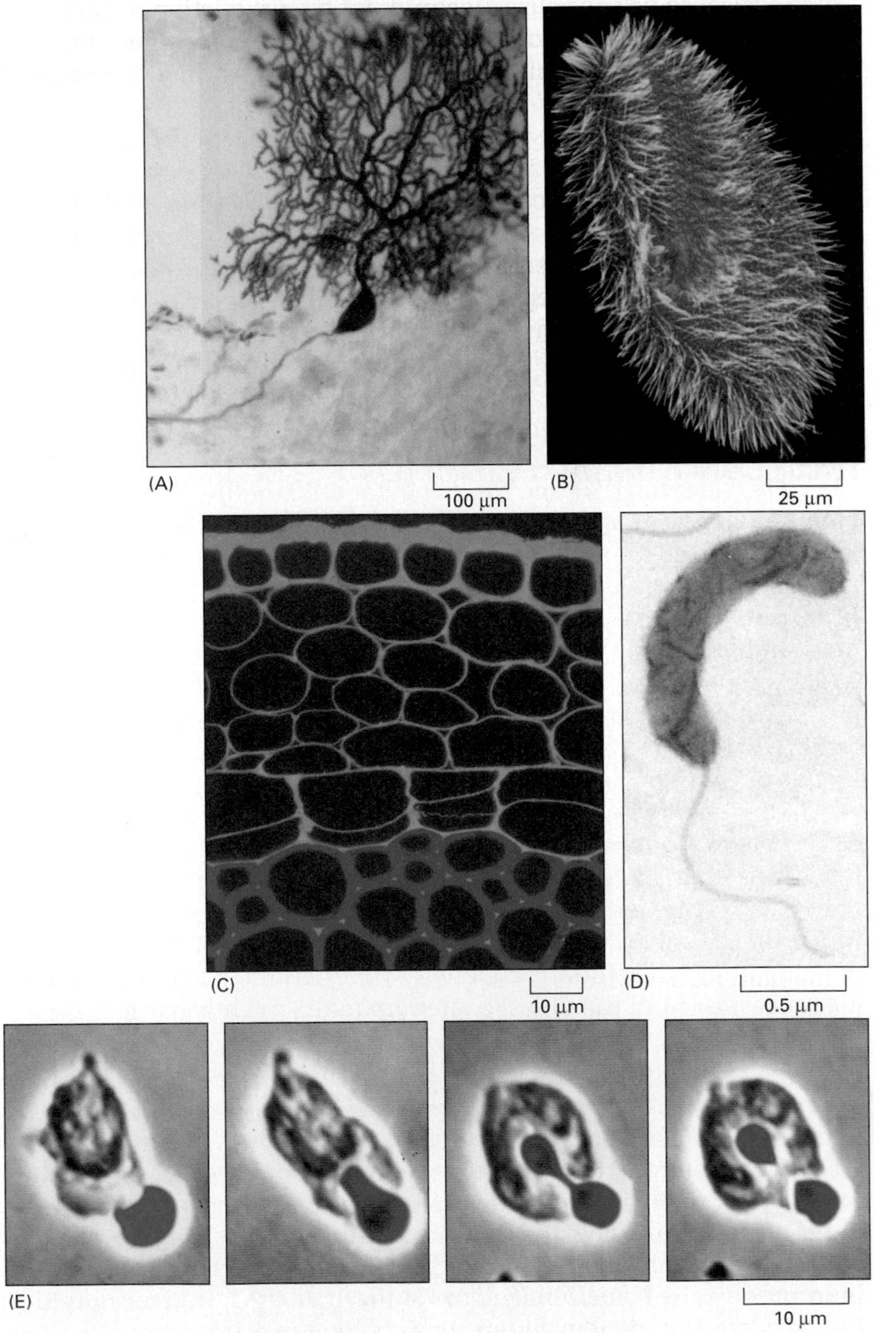

Figure 1–1 **Cells come in a variety of shapes and sizes.** (A) A nerve cell from the cerebellum (a part of the brain that controls movement). This cell has a huge branching tree of processes, through which it receives signals from as many as 100,000 other nerve cells.
(B) *Paramecium.* This protozoan—a single giant cell—swims by means of the beating cilia that cover its surface. (C) A section of a young plant stem in which cellulose is stained *red* and another cell wall component, pectin, is stained *orange.* The outermost layer of cells is at the top of the photo. (D) A tiny bacterium, *Bdellovibrio bacteriovorus,* that uses a single terminal flagellum to propel itself. This bacterium attacks, kills, and feeds on other, larger bacteria. (E) A human white blood cell (a neutrophil) approaching and engulfing a red blood cell. (A, courtesy of Constantino Sotelo; B, courtesy of Anne Fleury, Michel Laurent, and André Adoutte; D, courtesy of Murry Stein; E, courtesy of Stephen E. Malawista and Anne de Boisfleury Chevance.)

cell forward, rotating as it goes. A cell in the surface layer of a plant is a squat, immobile prism that surrounds itself in a rigid box of cellulose, with an outer waterproof coating of wax. A *Bdellovibrio* bacterium is a sausage-shaped torpedo driven forward by a rotating corkscrew-like *flagellum* that is attached to its stern, where it acts as a propeller. A neutrophil or a macrophage in the body of an animal crawls through tissues, constantly pouring itself into new shapes and engulfing debris, foreign microorganisms, and dead or dying cells.

Some cells are clad only in a flimsy *plasma membrane*; others augment this membranous cover by cloaking themselves in an outer layer of slime, building themselves rigid *cell walls*, or surrounding themselves with a hard, mineralized material, such as that found in bone.

Cells are also enormously diverse in their chemical requirements and activities. Some require oxygen to live; for others it is deadly. Some consume little more than air, sunlight, and water as their raw materials; others need a complex mixture of molecules produced by other cells. Some appear to be specialized factories for the production of particular substances, such as hormones, starch, fat, latex, or pigments. Some are engines, like muscle, burning fuel to do mechanical work; or electricity generators, like the modified muscle cells in the electric eel.

Some modifications specialize a cell so much that they spoil its chances of leaving any descendants. Such specialization would be senseless for a species of cell that lived a solitary life. In a multicellular organism, however, there is a division of labor among cells, allowing some cells to become specialized to an extreme degree for particular tasks and leaving them dependent on their fellow cells for many basic requirements. Even the most basic need of all, that of passing on the genetic instructions to the next generation, is delegated to specialists—the egg and the sperm.

Living Cells All Have a Similar Basic Chemistry

Despite the extraordinary diversity of plants and animals, people have recognized from time immemorial that these organisms have something in common, something that entitles them all to be called living things. With the invention of the microscope, it became clear that plants and animals are assemblies of cells, that cells can also exist as independent organisms, and that cells individually are living in the sense that they can grow, reproduce, convert energy from one form into another, respond to their environment, and so on. But while it seemed easy enough to recognize life, it was remarkably difficult to say in what sense all living things were alike. Textbooks had to settle for defining life in abstract general terms related to growth and reproduction.

The discoveries of biochemistry and molecular biology have made this problem disappear in a most spectacular way. Although they are infinitely varied when viewed from the outside, all living things are fundamentally similar inside. We now know that cells resemble one another to an astonishing degree in the details of their chemistry, sharing the same machinery for the most basic functions. All cells are composed of the same sorts of molecules that participate in the same types of chemical reactions (discussed in Chapter 2). In all living things, genetic instructions—*genes*—are stored in *DNA* molecules, written in the same chemical code, constructed out of the same chemical building blocks, interpreted by essentially the same chemical machinery, and duplicated in the same way to allow the organism to reproduce. Thus, in every cell, the long DNA polymer chains are made from the same set of four monomers, called *nucleotides,* strung together in different sequences like the letters of an alphabet to convey different information. In every

Question 1–1

"Life" is easy to recognize but difficult to define. The dictionary defines life as "The state or quality that distinguishes living beings or organisms from dead ones and from inorganic matter, characterized chiefly by metabolism, growth, and the ability to reproduce and respond to stimuli." Biology textbooks usually elaborate slightly; for example, according to a popular text, living things:

1. Are highly organized compared to natural inanimate objects.
2. Display homeostasis, maintaining a relatively constant internal environment.
3. Reproduce themselves.
4. Grow and develop from simple beginnings.
5. Take energy and matter from the environment and transform it.
6. Respond to stimuli.
7. Show adaptation to their environment.

Score yourself, a vacuum cleaner, and a potato with respect to these characteristics.

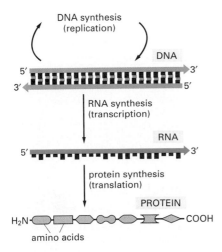

Figure 1–2 **In all living cells, genetic information flows from DNA to RNA (transcription) and from RNA to protein (translation).** Together these processes are known as gene expression.

cell, the instructions in the DNA are read out, or *transcribed*, into a chemically related set of molecules, made of *RNA* (Figure 1–2). The messages carried by the RNA molecules are in turn *translated* into yet another chemical form: they are used to direct the synthesis of a huge variety of large *protein* molecules that dominate the behavior of the cell, serving as structural supports, chemical catalysts, molecular motors, and so on. In every living thing, the same set of 20 *amino acids* is used to make proteins. But the amino acids are linked in different sequences, conferring different chemical properties on the protein molecules, just as different sequences of letters spell different words. In this way the same basic biochemical machinery has served to generate the whole gamut of living things (Figure 1–3). A more detailed discussion of the structure and function of proteins, RNA, and DNA is presented in Chapters 4 through 8.

If cells are the fundamental unit of living matter, then nothing less than a cell can truly be called living. Viruses, for example, contain some of the same types of molecules as cells but have no ability to reproduce themselves by their own efforts; they get themselves copied only by parasitizing the reproductive machinery of cells that they invade. Thus, viruses are chemical zombies, inert and inactive outside of their host cells, but exerting a malign control once they gain entry.

All Present-Day Cells Have Apparently Evolved from the Same Ancestor

A cell reproduces by duplicating its DNA and then dividing in two, passing a copy of the genetic instructions encoded in the DNA to each of its daughter cells. That is why the daughter cells resemble the parent cell. The copying is not always perfect, and the instructions are occasionally

Figure 1–3 **All living organisms are constructed from cells.** A bacterium, a butterfly, a rose, and a dolphin are all made of cells that have a fundamentally similar chemistry and operate according to the same basic principles. (A, courtesy of Tony Brain and Science Photo Library; B, courtesy of J.S. and E.J. Woolmer, © Oxford Scientific Films; C, courtesy of the John Innes Foundation; D, courtesy of Jonathan Gordon, IFAW.)

(A) (B) (C) (D)

corrupted. That is why the daughters do not always match the parent exactly. *Mutations*—changes in the DNA—can create offspring that are changed for the worse (in that they are less able to survive and reproduce); changed for the better (in that they are better able to survive and reproduce); or changed neutrally (in that they are genetically different, but equally viable). The struggle for survival eliminates the first, favors the second, and tolerates the third. The genes of the next generation will be the genes of the survivors. Intermittently, the pattern of descent may be complicated by sexual reproduction, in which two cells of the same species fuse, pooling their DNA; the genetic cards are then shuffled, re-dealt, and distributed in new combinations to the next generation, to be tested again for their survival value.

These simple principles of change and selection, applied repeatedly over billions of cell generations, are the basis of **evolution**—the process by which living species become gradually modified and adapted to their environment in more and more sophisticated ways. Evolution offers a startling but compelling explanation of why present-day cells are so similar in their fundamentals: they have all inherited their genetic instructions from the same common ancestor. It is estimated that this ancestral cell existed between 3.5 billion and 3.8 billion years ago, and we must suppose that it contained a prototype of the universal machinery of all life on Earth today. Through mutation, its descendants have gradually diverged to fill every habitat on Earth with living things, exploiting the potential of the machinery in an endless variety of ways.

Question 1–2

Mutations are mistakes in the DNA that change the genetic plan from the previous generation. Envision a shoe factory. Would you expect mistakes (i.e., unintentional changes) in copying the shoe design to lead to improvements in the shoes produced? Explain your answer.

Genes Provide the Instructions for Cellular Form, Function, and Complex Behavior

A cell's **genome**—that is, the entire library of genetic information in its DNA—provides a genetic program that instructs the cell how to function, and, for plant and animal cells, how to grow into an organism with hundreds of different cell types. Within an individual plant or animal, these cells can be extraordinarily varied, as we shall discuss in Chapter 21. Fat cells, skin cells, bone cells, and nerve cells seem as dissimilar as any cells could be. Yet all these *differentiated cell types* are generated during embryonic development from a single fertilized egg cell, and all contain identical copies of the DNA of the species. Their varied characters stem from the way that individual cells use their genetic instructions. Different cells *express* different genes—that is, they turn on production of some proteins and not others, depending on the cues that they and their ancestor cells have received from their surroundings.

The DNA, therefore, is not just a shopping list specifying the molecules that every cell must have, and a cell is not just an assembly of all the items on the list. Each cell is capable of carrying out a variety of biological tasks, depending on its environment and its history, using the information encoded in its DNA to guide its activities. Later in this book, we shall see in detail how DNA defines both the parts list of the cell and the rules that decide when and where these parts are to be made.

Cells Under the Microscope

Today we have the technology to decipher the underlying principles that govern the structure and activity of the cell. But cell biology began without these tools. To appreciate the predicament facing those who first glimpsed cells, imagine the perplexity of a scientist of a bygone era—a Leonardo da Vinci, let us say—trying to grasp the workings of a modern-day laptop computer. He would have no way of knowing that the key to understanding how this machine works lies in identifying and decoding its resident programs. After examining the laptop's external case, lifting the screen, and poking the keys, this learned and curious individual might pry the thing open to see what lay inside: no gears or levers, no tiny imp writing messages on the screen. Instead, he would confront mysterious boards covered with metallic tracks and studded with rectangular black wafers; a heavy, bricklike object that gives off small sparks when poked with a pair of metal tweezers; and various other deeply puzzling bits and pieces. The earliest cell biologists engaged in a similar kind of exploration. They began by simply looking at tissues and cells, then breaking them open or slicing them up and attempting to peer inside. What they saw was to them, as to the Renaissance scholar confronted with the computer, profoundly baffling. Nevertheless, this type of visual investigation was the first step toward understanding, and it remains essential in the study of cell biology.

Cells, in general, are very small—too small to be seen with the naked eye. They were not made visible until the seventeenth century, when the **microscope** was invented. For hundreds of years afterward, all that was known about cells was discovered using this instrument. *Light microscopes*, which use visible light to illuminate specimens, are still vital pieces of equipment in the cell biology laboratory.

Although these instruments now incorporate many sophisticated improvements, the properties of light itself set a limit to the fineness of detail they can reveal. *Electron microscopes*, invented in the 1930s, go beyond this limit by using beams of electrons instead of beams of light as the source of illumination, greatly extending our ability to see the fine details of cells and even making some of the larger molecules visible individually. A survey of the principal types of microscopy used to examine cells is given in Panel 1–1 (pp. 8–9).

The Invention of the Light Microscope Led to the Discovery of Cells

The development of the light microscope depended on advances in the production of glass lenses. By the seventeenth century, lenses were refined to the point that they could serve to make simple microscopes. Using such an instrument, Robert Hooke examined a piece of cork and in 1665 reported to the Royal Society of London that the cork was composed of a mass of minute chambers, which he called "cells." The name "cell" stuck, even though the structures Hooke described were only the cell walls that remained after the living plant cells inside them had died. Later, Hooke and some of his contemporaries were able to see living cells.

For almost 200 years, the light microscope remained an exotic instrument, available only to a few wealthy individuals. It was not until the nineteenth century that it began to be widely used to look at cells. The emergence of cell biology as a distinct science was a gradual process to which many individuals contributed, but its official birth is generally said to be signaled by two publications: one by the botanist Matthias Schleiden in 1838 and the other by the zoologist Theodor Schwann in 1839. In these papers, Schleiden and Schwann documented the results of a systematic investigation of plant and animal tissues with the light microscope, showing that cells were the universal building blocks of all living tissues. Their work, and that of other nineteenth-century microscopists, slowly led to the realization that all living cells are formed by the division of existing cells—a principle sometimes referred

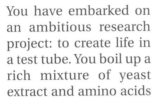

Question 1–3

You have embarked on an ambitious research project: to create life in a test tube. You boil up a rich mixture of yeast extract and amino acids in a flask along with a sprinkling of the inorganic salts known to be essential for life. You seal the flask and allow it to cool. After several months, the liquid is as clear as ever, and there are no signs of life. A friend suggests that excluding the air was a mistake, since most life as we know it requires oxygen. You repeat the experiment, but this time you leave the flask open to the atmosphere. To your great delight, the liquid becomes cloudy after a few days and under the microscope you see beautiful small cells that are clearly growing and dividing. Does this experiment prove that you managed to generate a novel life form? How might you redesign your experiment to allow air into the flask, yet eliminate the possibility that contamination is the explanation for the results? (For a ready-made answer, look up the experiments of Louis Pasteur.)

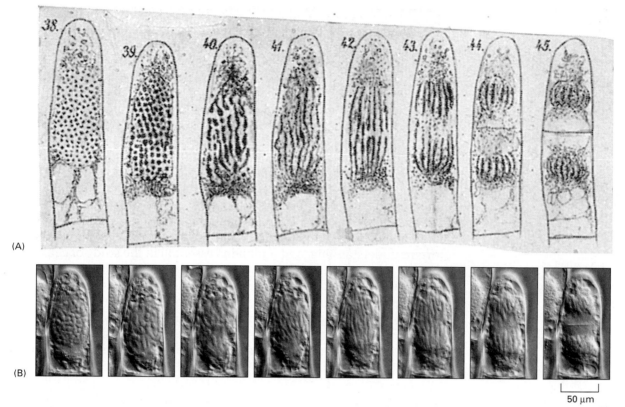

(A)

(B)

50 µm

to as the *cell theory* (Figure 1–4). The implication that living organisms do not arise spontaneously but can be generated only from existing organisms was hotly contested, but it was finally confirmed by experiments performed in the 1860s by Louis Pasteur.

The principle that cells are generated only from preexisting cells and inherit their characteristics from them underlies all of biology and gives the subject a unique flavor: in biology, questions about the present are inescapably linked to questions about the past. To understand why present-day cells and organisms behave as they do, we need to understand their history, all the way back to the misty origins of the first cells on Earth. Darwin's theory of evolution, published in 1859, provided the key insight that makes this history comprehensible, by showing how random variation and natural selection can drive the production of organisms with novel features, adapted to new ways of life. The theory of evolution explains how diversity has arisen among organisms that share a common ancestry. When combined with the cell theory, it leads us to a view of all life, from its beginnings to the present day, as one vast family tree of individual cells. Although this book is primarily about how cells work today, we shall encounter the theme of evolution again and again.

Figure 1–4 Early microscopes revealed new cells forming by division of existing cells. (A) In 1880, Eduard Strasburger drew a living plant cell (a hair cell from a *Tradescantia* flower), which he observed dividing into two daughter cells over a period of 2.5 hours. (B) A comparable living cell photographed recently through a modern light microscope. (B, courtesy of Peter Hepler.)

Cells, Organelles, and Even Molecules Can Be Seen Under the Microscope

If you cut a very thin slice of a suitable plant or animal tissue and place it under a light microscope, you will see that the tissue is divided into thousands of small cells. These may be either closely packed or separated from one another by an *extracellular matrix*, a dense material often made of protein fibers embedded in a polysaccharide gel (Figure 1–5). Each cell is typically about 5–20 µm in diameter (Figure 1–6). If you have taken care to keep your specimen under the right conditions, you will see that the cells show signs of life: particles move around inside them, and if you watch patiently, you may see a cell slowly change shape

Panel 1–1 Light and electron microscopy

THE LIGHT MICROSCOPE

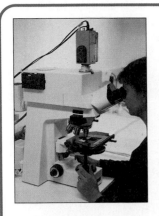

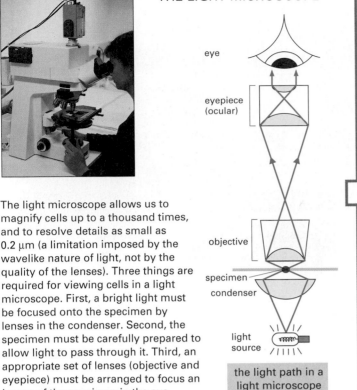

eye

eyepiece (ocular)

objective

specimen

condenser

light source

the light path in a light microscope

The light microscope allows us to magnify cells up to a thousand times, and to resolve details as small as 0.2 μm (a limitation imposed by the wavelike nature of light, not by the quality of the lenses). Three things are required for viewing cells in a light microscope. First, a bright light must be focused onto the specimen by lenses in the condenser. Second, the specimen must be carefully prepared to allow light to pass through it. Third, an appropriate set of lenses (objective and eyepiece) must be arranged to focus an image of the specimen in the eye.

FLUORESCENCE MICROSCOPY

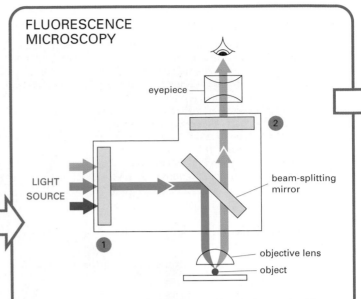

eyepiece

2

beam-splitting mirror

LIGHT SOURCE

objective lens

object

1

Fluorescent dyes used for staining cells are detected with the aid of a *fluorescence microscope*. This is similar to an ordinary light microscope except that the illuminating light is passed through two sets of filters. The first (1) filters the light before it reaches the specimen, passing only those wavelengths that excite the particular fluorescent dye. The second (2) blocks out this light and passes only those wavelengths emitted when the dye fluoresces. Dyed objects show up in bright color on a dark background.

LOOKING AT LIVING CELLS

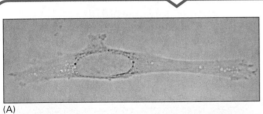

(A)

(B)

(C)

50 μm

The same unstained, living animal cell (fibroblast) in culture viewed with (A) straightforward (bright-field) optics; (B) phase-contrast optics; (C) interference-contrast optics. These latter systems exploit differences in the way light travels through regions of the cell with differing refractive indexes. All three images can be obtained on the same microscope simply by interchanging optical components.

FIXED SAMPLES

Most tissues are neither small enough nor transparent enough to examine directly in the microscope. Typically, therefore, they are chemically fixed and cut into very thin slices, or *sections*, that can be mounted on a glass microscope slide and subsequently stained to reveal different components of the cells. A stained section of a plant root tip is shown here (D). (Courtesy of Catherine Kidner)

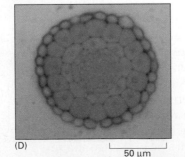

(D)

50 μm

FLUORESCENT PROBES

Dividing cells seen with a fluorescence microscope after staining with specific fluorescent dyes.

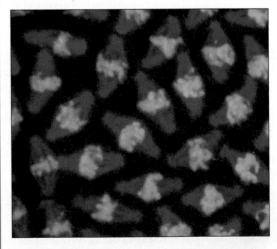

Fluorescent dyes absorb light at one wavelength and emit it at another, longer wavelength. Some such dyes bind specifically to particular molecules in cells and can reveal their location when examined with a fluorescence microscope. An example is the stain for DNA shown here (*green*). Other dyes can be coupled to antibody molecules, which then serve as highly specific and versatile staining reagents that bind selectively to particular macromolecules, allowing us to see their distribution in the cell. In the example shown, a microtubule protein in the mitotic spindle is stained *red* with a fluorescent antibody. (Courtesy of William Sullivan.)

CONFOCAL MICROSCOPY

A confocal microscope is a fluorescence microscope with a laser as its source of illumination. This is focused onto a single point at a specific depth in the specimen, and a pinhole aperture in the detector allows only fluorescence emitted from the exact point of focus to be included in the image. Scanning the laser beam across the specimen generates a sharp two-dimensional image of the plane of focus. A series of optical sections at different depths allows a three-dimensional image to be constructed. An intact insect embryo is shown here stained with a fluorescent probe for actin (a type of protein). (A) Conventional fluorescence microscopy generates a blurry image due to the presence of fluorescent structures above and below the plane of focus. (B) Confocal microscopy provides a crisp optical section of the cells in the embryo. (A, courtesy of Richard Warn; B, courtesy of Peter Shaw.)

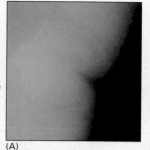 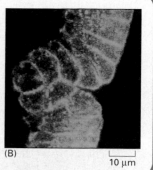

(A)   (B)   10 μm

TRANSMISSION ELECTRON MICROSCOPY

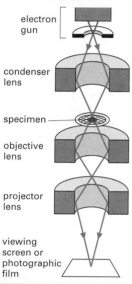

electron gun

condenser lens

specimen

objective lens

projector lens

viewing screen or photographic film

Courtesy of Philips Electron Optics

The electron micrograph below shows a small region of a cell in a piece of testis. The tissue has been chemically fixed, embedded in plastic, and cut into very thin sections that have then been stained with salts of uranium and lead. (Courtesy of Daniel S. Friend.)

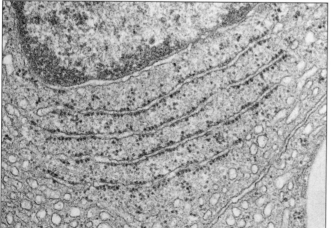

0.5 μm

The transmission electron microscope (TEM) is in principle similar to an inverted light microscope, but it uses a beam of electrons instead of a beam of light, and magnetic coils to focus the beam instead of glass lenses. The specimen, which is placed in a vacuum, must be very thin. Contrast is usually introduced by electron-dense heavy-metal stains that locally absorb or scatter electrons, removing them from the beam as it passes through the specimen. The TEM has a useful magnification of up to a million-fold and with biological specimens can resolve details as small as about 2 nm.

SCANNING ELECTRON MICROSCOPY

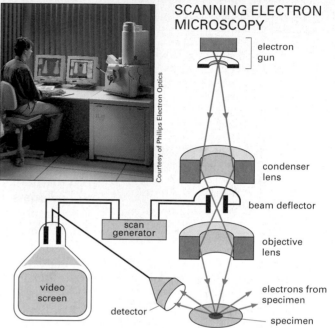

Courtesy of Philips Electron Optics

electron gun

condenser lens

beam deflector

scan generator

objective lens

video screen

detector

electrons from specimen

specimen

In the scanning electron microscope (SEM) the specimen, which has been coated with a very thin film of a heavy metal, is scanned by a beam of electrons brought to a focus on the specimen by the electromagnetic coils that, in electron microscopes, act as lenses. The quantity of electrons scattered or emitted as the beam bombards each successive point on the surface of the specimen is measured by the detector, and is used to control the intensity of successive points in an image built up on a video screen. The microscope creates striking images of three-dimensional objects with great depth of focus and can resolve details down to somewhere between 3 nm and 20 nm, depending on the instrument.

5 μm

1 μm

Scanning electron micrograph of the stereocilia projecting from a hair cell in the inner ear (*left*). For comparison, the same structure is shown by light microscopy, at the limit of its resolution (*above*). (Courtesy of Richard Jacobs and James Hudspeth.)

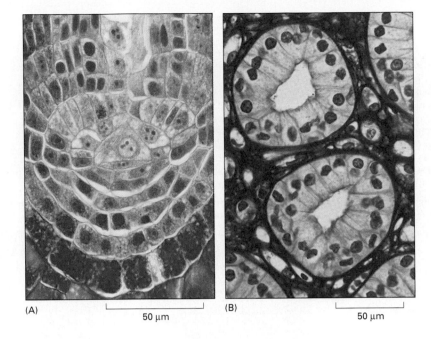

(A) ⊢————⊣ 50 μm   (B) ⊢————⊣ 50 μm

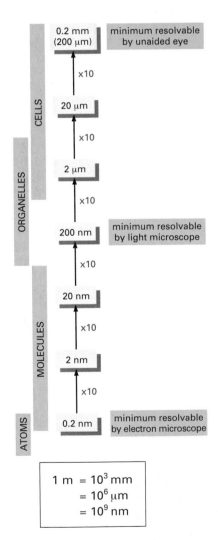

Figure 1–6 **What can we see?** This schematic shows the sizes of cells and of their component parts, and the units in which they are measured.

and divide into two (see Figure 1–4). (Some speeded-up movies of cell division are included on the CD-ROM that accompanies this book.)

To see the internal structure of a cell is difficult, not only because the parts are small but also because they are transparent and mostly colorless. One approach is to stain cells with dyes that color particular components differently (see Figure 1–5). Alternatively, one can exploit the fact that cell components differ slightly from one another in refractive index, just as glass differs in refractive index from water, causing light rays to be deflected as they pass from the one medium into the other. The small differences in refractive index can be made visible by sophisticated optical techniques, and the resulting images can be enhanced further by electronic processing (see Panel 1–1, pp. 8–9).

The cell thus revealed has a distinct anatomy (Figure 1–7). It has a sharply defined boundary, indicating the presence of an enclosing membrane. In the middle, a large, round body, the *nucleus*, is prominent. Around the nucleus and filling the cell's interior lies the **cytoplasm**, a transparent substance crammed with what seems at first to be a jumble of miscellaneous tiny objects. With a good light microscope, one can begin to distinguish and classify specific components in the cytoplasm (Figure 1–7B). However, structures smaller than about 0.2 μm—about half the wavelength of visible light—cannot be resolved (points closer than this are not distinguishable but appear as a single blur).

For higher magnification and better resolution one must turn to an electron microscope, which can reveal details down to a few **nanometers**, or nm (see Figure 1–6). Cell samples for the electron microscope require painstaking preparation. Even for light microscopy, a tissue usually has to be *fixed* (that is, preserved by pickling in a reactive chemical solution), supported by *embedding* in a solid wax or resin, *sectioned* into thin slices, and *stained* before it is viewed. For electron microscopy, similar procedures are required, but the sections have to be much thinner and there is no possibility of looking at living, wet cells.

When the sections are cut, stained, and placed in the electron microscope, much of the jumble of cell components becomes sharply resolved into distinct **organelles**—separate, recognizable substructures that are only hazily defined under the light microscope. A delicate membrane, about 5 nm thick, is visible enclosing the cell, and similar

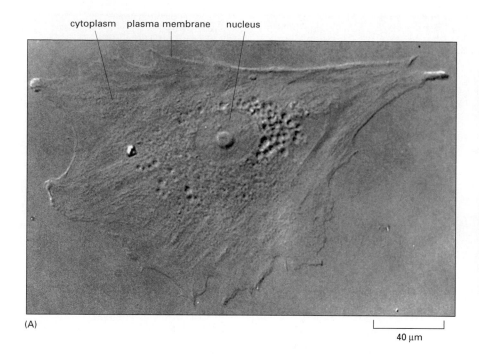

cytoplasm plasma membrane nucleus

(A)

|_____|
40 µm

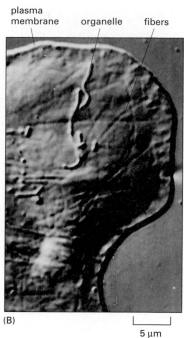

plasma
membrane organelle fibers

(B)

|_____|
5 µm

membranes form the boundary of many of the organelles inside (Figure 1–8A, B). The external membrane is called the *plasma membrane*, while the membranes surrounding organelles are called *internal membranes*. With an electron microscope, even some of the individual large molecules in a cell can be seen (Figure 1–8C).

The type of electron microscope used to look at thin sections of tissue is known as a *transmission electron microscope*. This is in principle similar to a light microscope, only it transmits a beam of electrons rather than a beam of light through the sample. Another type of electron microscope—the *scanning electron microscope*—scatters electrons off the sample and so is used to look at the surface detail of cells and other structures (see Panel 1–1, pp. 8–9). Electron microscopy enables biologists to see the structure of biological membranes, which are only two (large) molecules thick (described in detail in Chapters 11 and 12). Even with the most powerful electron microscopes, however, one cannot see the individual atoms that make up molecules (Figure 1–9).

The microscope is not the only tool that modern biologists use to study the details of cell components. Techniques such as X-ray crystallography, for example, can be used to determine the three-dimensional structure of protein molecules (discussed in Chapter 4). We shall describe other methods for probing the inner workings of cells as they arise throughout the book.

The Procaryotic Cell

Of all the types of cells revealed by the microscope, **bacteria** have the simplest structure and come closest to showing us life stripped down to its essentials. Indeed, bacteria contain essentially no organelles—not even a nucleus to hold their DNA. This property—the presence or absence of a nucleus—is used as the basis for a simple but fundamental classification of all living things. Organisms whose cells have a nucleus are called **eucaryotes** (from the Greek words *eu*, meaning "well" or "truly," and *karyon*, a "kernel" or "nucleus"). Organisms whose cells do not have a nucleus are called **procaryotes** (from *pro*, meaning "before"). The terms "bacterium" and "procaryote" are often used interchangeably, although we shall see that the category of procaryotes also

Figure 1–7 The internal structures of a living cell can be seen under a light microscope. (A) A cell taken from human skin and growing in tissue culture was photographed through a light microscope. Fibers and organelles, particularly the nucleus, can be distinguished. (B) Detail of part of a newt cell growing in culture. The video image, at high magnification, has been computer-enhanced, and numerous organelles and fibers can be seen. (A, courtesy of Casey Cunningham; B, courtesy of Lynne Cassimeris.)

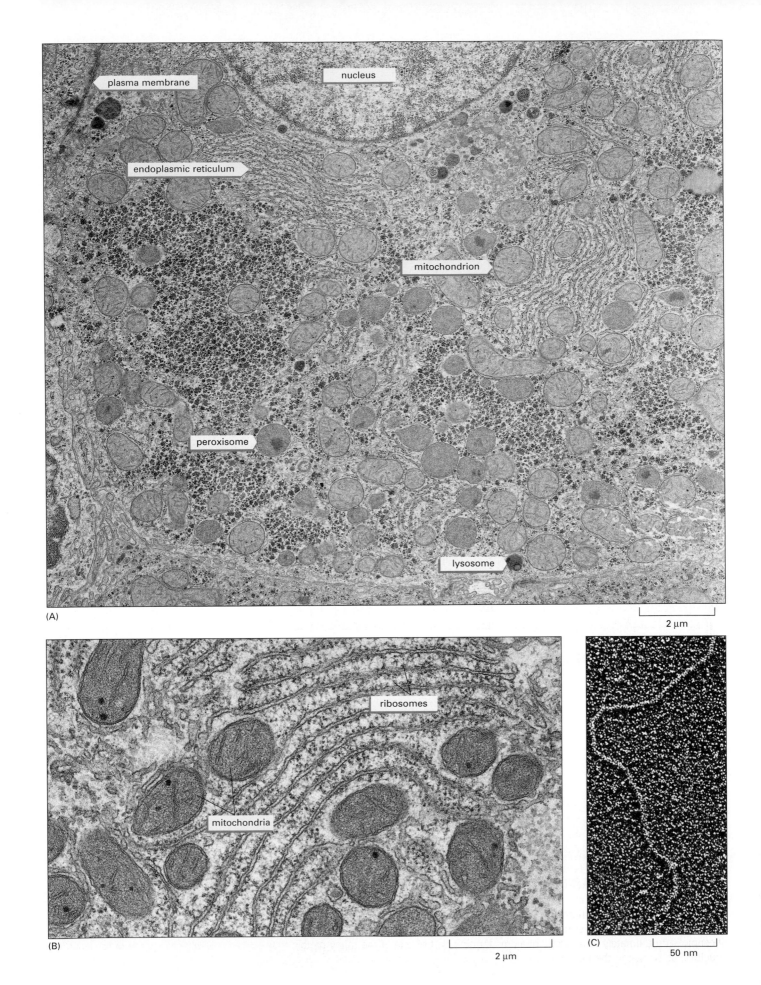

(A)

plasma membrane

nucleus

endoplasmic reticulum

mitochondrion

peroxisome

lysosome

2 µm

(B)

ribosomes

mitochondria

2 µm

(C)

50 nm

Figure 1–8 (opposite page) The fine structure of a cell can be seen in a transmission electron microscope. (A) Thin section of a liver cell showing the enormous amount of detail that is visible. Some of the components to be discussed later in the chapter are labeled; they are identifiable by their size and shape. (B) A small region of the cytoplasm at somewhat higher magnification. The smallest structures that are clearly visible are the ribosomes, each of which is made of 80–90 or so individual large molecules. (C) Portion of a long, threadlike DNA molecule isolated from a cell and seen by electron microscopy. (A and B, courtesy of Daniel S. Friend; C, courtesy of Mei Lie Wong.)

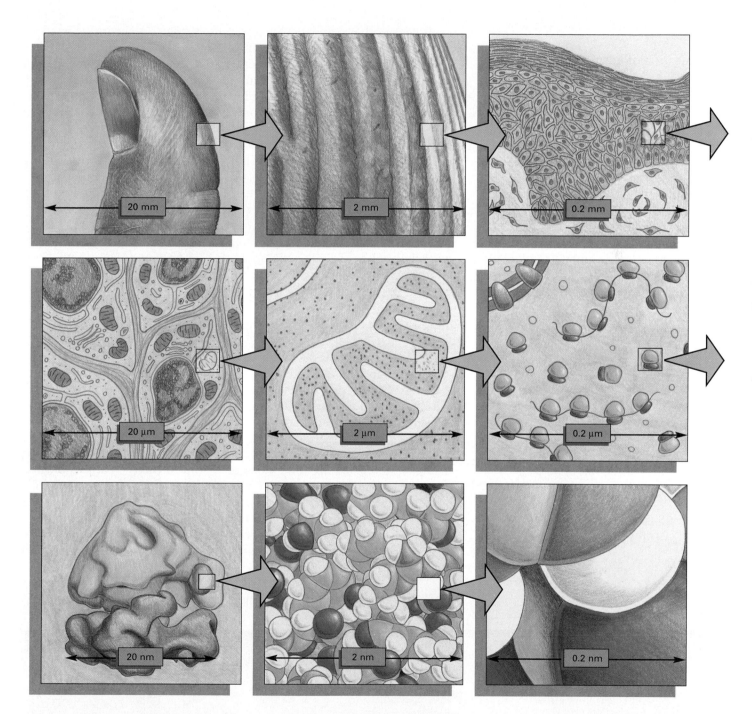

Figure 1–9 How big is a cell and how big are its parts? This diagram conveys a sense of scale between living cells and atoms. Each panel shows an image that is then magnified by a factor of 10 in an imaginary progression from a thumb, through skin cells, to a ribosome, and ultimately to a cluster of atoms forming part of one of the many protein molecules in our bodies. Details of molecular structure, as shown in the last two panels, are beyond the power of the electron microscope.

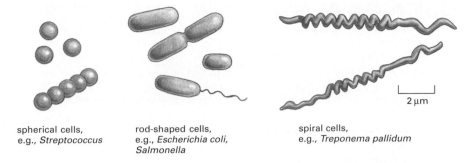

Figure 1–10 Bacteria come in different shapes and sizes. Typical spherical, rodlike, and spiral-shaped bacteria are drawn to scale. The spiral cells shown are the organisms that cause syphilis.

spherical cells,
e.g., *Streptococcus*

rod-shaped cells,
e.g., *Escherichia coli,
Salmonella*

spiral cells,
e.g., *Treponema pallidum*

2 μm

Note on biological names

Species of living organisms are officially identified by a pair of Latin words, usually printed in italics, analogous to a person's given name and surname. The genus (*Escherichia*, corresponding to a surname) is stated first; the second term (*coli*) qualifies this, identifying a particular species belonging to that genus. For short, the genus name may be abbreviated (*E. coli*), or the species label may be dropped (so that we often speak of the fly *Drosophila*, meaning *Drosophila melanogaster*).

includes another class of cells, so remotely related to ordinary bacteria that they are given a separate name.

Bacteria are typically spherical, rodlike, or corkscrew-shaped, and small—just a few micrometers long (Figure 1–10). They often have a tough protective coat, called a cell wall, surrounding the plasma membrane, which encloses a single compartment containing the cytoplasm and the DNA. In the electron microscope this cell interior typically appears as a matrix of varying texture without any obvious organized internal structure (Figure 1–11). The cells reproduce quickly by dividing in two. Under optimum conditions, when food is plentiful, a procaryotic cell can duplicate itself in as little as 20 minutes. In less than 11 hours, by repeated divisions, a single procaryote can give rise to 5 billion progeny (which is approximately equal to the total number of humans presently on earth). Thanks to their large numbers, rapid growth rates, and ability to exchange bits of genetic material by a process akin to sex, populations of procaryotic cells can evolve fast, rapidly acquiring the ability to use a new food source or to resist being killed by a new antibiotic.

Procaryotes Are the Most Diverse of Cells

Most procaryotes live as single-celled organisms, although some join together to form chains, clusters, or other organized multicellular structures. In shape and structure procaryotes may seem simple and limited, but in terms of chemistry they are the most diverse and inventive class of cells. These creatures exploit an enormous range of habitats, from hot puddles of volcanic mud to the interiors of other living cells, and they vastly outnumber other living organisms on Earth. Some are aerobic, using oxygen to oxidize food molecules; some are strictly anaerobic and are killed by the slightest exposure to oxygen. As we will see later in this chapter, *mitochondria*—the organelles that generate energy for the eucaryotic cell—are thought to have evolved from aerobic bacteria that took to living inside the anaerobic ancestors of today's eucaryotic cells. Thus our own oxygen-based metabolism can be regarded as a product of the activities of bacterial cells.

Virtually any organic material, from wood to petroleum, can be used as food by one sort of bacterium or another. Still more remarkable,

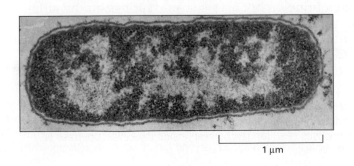

Figure 1–11 The bacterium *Escherichia coli (E. coli)* is understood more thoroughly than any other living organism. An electron micrograph of a longitudinal section is shown here; the cell's DNA is concentrated in the lightly stained region. (Courtesy of E. Kellenberger.)

1 μm

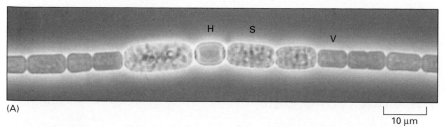

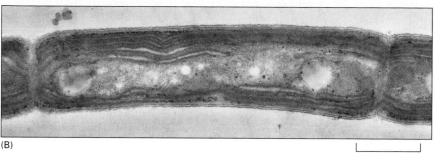

(A)

10 μm

(B)

1 μm

Figure 1–12 **Some bacteria are photosynthetic.** (A) *Anabaena cylindrica* forms long, multicellular filaments. This light micrograph shows specialized cells that either fix nitrogen (that is, capture N_2 from the atmosphere and incorporate it into organic compounds; labeled *H)*, fix CO_2 (through photosynthesis; *V),* or become resistant spores (*S*). (B) An electron micrograph of *Phormidium laminosum* shows the intracellular membranes where photosynthesis occurs. Note that even some procaryotes can form simple multicellular organisms. (A, courtesy of David Adams; B, courtesy of D.P. Hill and C.J. Howe.)

some procaryotes can live entirely on inorganic substances: they get their carbon from CO_2 in the atmosphere, their nitrogen from atmospheric N_2, and their oxygen, hydrogen, sulfur, and phosphorus from air, water, and inorganic minerals. Some of these procaryotic cells, like plant cells, perform *photosynthesis*, getting the energy they need for biosynthesis from sunlight (Figure 1–12); others derive energy from the chemical reactivity of inorganic substances in the environment (Figure 1–13). In either case, such procaryotes play a unique and fundamental part in the economy of life on Earth: other living things depend on the organic compounds that these cells generate from inorganic materials.

Plants, too, can capture energy from sunlight and carbon from atmospheric CO_2. But plants unaided by bacteria cannot capture N_2 from the atmosphere, and in a sense even plants depend on bacteria for photosynthesis. It is almost certain that the organelles in the plant cell that perform photosynthesis—the *chloroplasts*—have evolved from photosynthetic bacteria that found a home inside the plant cell's cytoplasm.

The World of Procaryotes Is Divided into Two Domains: Eubacteria and Archaea

Traditionally, all procaryotes have been classified together in one large group. But molecular studies reveal that there is a gulf within the class of procaryotes, dividing it into two distinct *domains*, called the *eubacteria* (or simply *bacteria)* and the *archaea*. Remarkably, at a molecular level, the members of these two domains differ as much from one another as either does from the eucaryotes. Most of the procaryotes familiar from everyday life—the species that live in the soil or make us ill—are eubacteria. Archaea are not only found in these habitats, but also in environments hostile to most other cells: there are species that live in concentrated brine, in hot acid volcanic springs, in the airless depths of marine sediments, in the sludge of sewage treatment plants, in pools beneath the frozen surface of Antarctica, and in the acidic, oxygen-free environment of a cow's stomach, where they break down

Question 1–4

A bacterium weighs about 10^{-12} g and can divide every 20 minutes. If a single bacterial cell carried on dividing at this rate, how long would it take before the mass of bacteria would equal that of the Earth (6 $\times 10^{24}$ kg)? Contrast your result with the fact that bacteria originated at least 3.5 billion years ago and have been dividing ever since. Explain the apparent paradox. (The number of cells N in a culture at time t is described by the equation $N = N_0 \times 2^{t/G}$, where N_0 is the number of cells at zero time and G is the population doubling time.)

Figure 1–13 **A sulfur bacterium gets its energy from H$_2$S.** *Beggiatoa*, a procaryote that lives in sulfurous environments, oxidizes H_2S and can fix carbon even in the dark. In this light micrograph, yellow deposits of sulfur can be seen inside the cells. (Courtesy of Ralph W. Wolfe.)

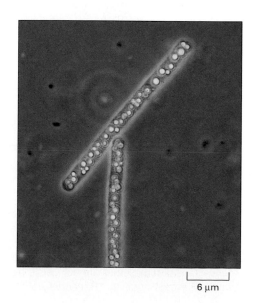

6 μm

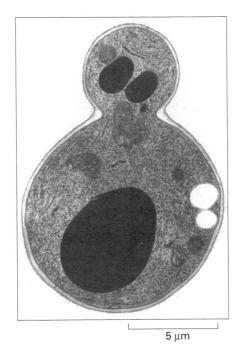

Figure 1–14 **Yeasts are simple free-living eucaryotes.** The cell shown in this light micrograph belongs to the same species that makes dough rise and turns malted barley juice into beer. It reproduces by forming a bud and then dividing asymmetrically into a large and a small daughter cell. (Courtesy of Soren Mogelsvang and Natalia Gomez-Ospina.)

5 μm

cellulose and generate methane gas. Many of these environments resemble the harsh conditions that must have existed on the primitive Earth, where living things first evolved, before the atmosphere became rich in oxygen.

The Eucaryotic Cell

Eucaryotic cells, in general, are bigger and more elaborate than bacteria and archaea. Some live independent lives as single-celled organisms, such as amoebae and yeasts (Figure 1–14); others live in multicellular assemblies. All of the more complex multicellular organisms—including plants, animals, and fungi—are formed from eucaryotic cells.

By definition, all eucaryotic cells have a nucleus. But possession of a nucleus goes hand-in-hand with possession of a variety of other organelles, most of which are likewise common to all these eucaryotic organisms. We will now take a look at the main organelles found in eucaryotic cells from the point of view of their functions.

The Nucleus Is the Information Store of the Cell

The **nucleus** is usually the most prominent organelle in a eucaryotic cell (Figure 1–15). It is enclosed within two concentric membranes that

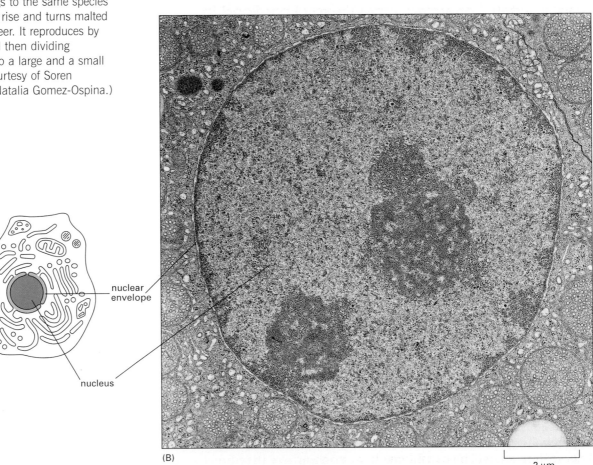

nuclear envelope

(A)

nucleus

(B)

2 μm

Figure 1–15 **The nucleus contains most of the DNA in a eucaryotic cell.** (A) In this schematic diagram of a typical animal cell—complete with its extensive system of membrane-enclosed organelles—the nucleus is colored *brown*, the nuclear envelope is *green*, and the cytoplasm (the interior of the cell outside the nucleus) is *white*. (B) The nucleus is the most prominent organelle in this thin section of a mammalian cell examined in the electron microscope. Individual chromosomes are not visible because the DNA is dispersed as fine threads throughout the nucleus at this stage of the cell's growth. (B, courtesy of Daniel S. Friend.)

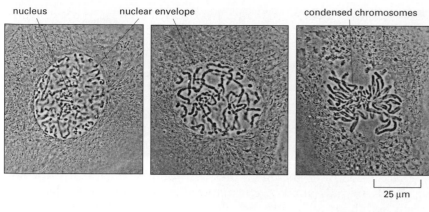

nucleus   nuclear envelope   condensed chromosomes

25 µm

Figure 1–16 **Chromosomes become visible when a cell is about to divide.** As a cell prepares to divide, its DNA condenses into threadlike chromosomes that can be distinguished in the light microscope. The photographs show three successive steps in this process in a cultured cell from a newt's lung. (Courtesy of Conly L. Rieder.)

form the *nuclear envelope*, and it contains molecules of DNA—extremely long polymers that encode the genetic information of the organism. In the light microscope, these giant DNA molecules become visible as individual **chromosomes** when they become more compact as a cell prepares to divide into two daughter cells (Figure 1–16). DNA also stores the genetic information in procaryotic cells; these cells lack a distinct nucleus not because they lack DNA, but because they do not keep it inside a nuclear envelope, segregated from the rest of the cell contents.

Mitochondria Generate Usable Energy from Food to Power the Cell

Among the most conspicuous organelles in the cytoplasm, **mitochondria** are present in essentially all eucaryotic cells (Figure 1–17). These organelles have a very distinctive structure when seen under the electron microscope: each mitochondrion appears sausage- or worm-shaped, from one to many micrometers long; and each is enclosed in two separate membranes. The inner membrane is formed into folds that project into the interior of the mitochondrion (Figure 1–18). Mitochondria contain their own DNA and reproduce by dividing in two. Because mitochondria resemble bacteria in so many ways, they are thought to derive from bacteria that were engulfed by some ancestor of present-day eucaryotic cells (Figure 1–19). This evidently created a *symbiotic* relationship—one in which the host eucaryote and the engulfed bacterium helped one another to survive and reproduce.

Observation under the microscope by itself gives little indication of what mitochondria do. Their function was discovered by breaking open cells and then spinning the soup of cell fragments in a centrifuge; this separates the organelles according to their size, shape, and density. Purified mitochondria were then tested to see what chemical processes they could perform. This revealed that mitochondria are generators of chemical energy for the cell. They harness the energy from the oxidation of food molecules, such as sugars, to produce *adenosine triphosphate*, or ATP—the basic chemical fuel that powers most of the cell's activities. Because the mitochondrion consumes oxygen and releases carbon dioxide in the course of this activity, the entire process is called *cellular respiration*—essentially, breathing on a cellular level. The process of cellular respiration will be considered in more detail in Chapter 14.

Without mitochondria, animals, fungi, and plants would be unable to use oxygen to extract the maximum amount of energy from the food molecules that nourish them. Oxygen would be a poison for them, rather than an essential requirement. There are, in fact, a few anaerobic eucaryotes that lack mitochondria and live only in environments that are oxygen-free.

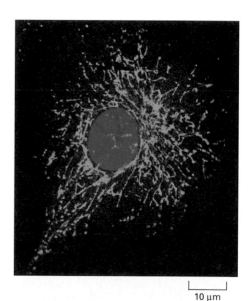

10 µm

Figure 1–17 **Mitochondria serve as cellular powerhouses.** These organelles, seen with a light microscope, are power generators that oxidize food molecules to produce useful chemical energy in almost all eucaryotic cells. Mitochondria are quite variable in shape; in this cultured mammalian cell they are stained *green* with a fluorescent dye and appear wormlike. The nucleus is stained *blue*. (Courtesy of Lan Bo Chen.)

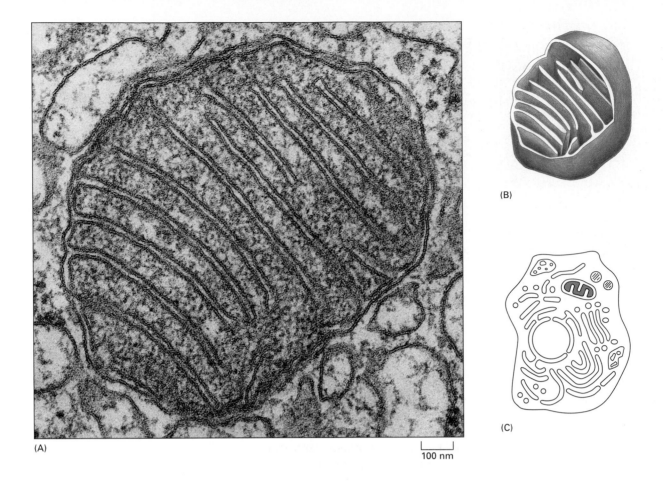

100 nm

(A)

(B)

(C)

Figure 1–18 The electron microscope reveals the folds in the mitochondrial membrane. (A) A cross section of a mitochondrion. (B) This three-dimensional representation of the arrangement of the mitochondrial membranes shows the smooth outer membrane and the highly convoluted inner membrane. The inner membrane contains most of the proteins responsible for cellular respiration, and it is highly folded to provide a large surface area for this activity. (C) In this schematic cell, the interior space of the mitochondrion is colored. (A, courtesy of Daniel S. Friend.)

Chloroplasts Capture Energy from Sunlight

Chloroplasts are large green organelles that are found only in the cells of plants and algae, not in the cells of animals or fungi. These organelles have an even more complex structure than mitochondria: in addition to their two surrounding membranes, chloroplasts possess internal stacks of membranes containing the green pigment *chlorophyll* (Figure 1–20). When a plant is kept in the dark, its greenness fades; when put back in the light, its greenness returns. This suggests that the chlorophyll—and the chloroplasts that contain it—are crucial to the special relationship that plants and algae have with light. But what is that relationship?

Figure 1–19 Mitochondria most likely evolved from engulfed bacteria. It is virtually certain that mitochondria originate from bacteria that were engulfed by an ancestral eucaryotic cell and survived inside it, living in symbiosis with their host.

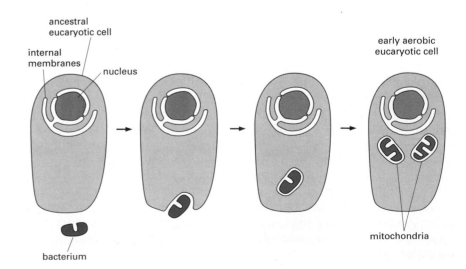

ancestral
eucaryotic cell

internal
membranes

nucleus

early aerobic
eucaryotic cell

bacterium

mitochondria

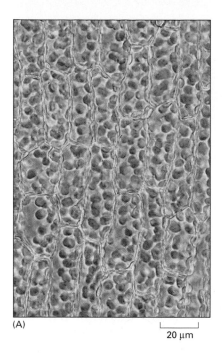
(A)

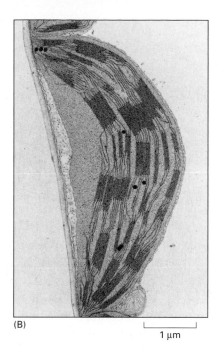
(B)

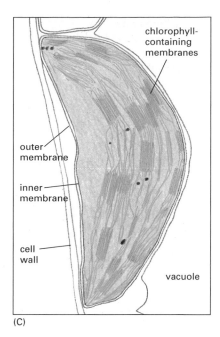

chlorophyll-containing membranes

outer membrane

inner membrane

cell wall

vacuole

(C)

20 μm

1 μm

Animals and plants all need energy to live, grow, and reproduce. Animals can use only the chemical energy they obtain by feeding on the products of other living things. But plants can get their energy directly from sunlight, and chloroplasts are the organelles that enable them to do so. From the standpoint of life on Earth, chloroplasts carry out an even more essential task than mitochondria: they perform photosynthesis—that is, they trap the energy of sunlight in chlorophyll molecules and use this energy to drive the manufacture of energy-rich sugar molecules. In the process they release oxygen as a molecular by-product. Plant cells can then extract this stored chemical energy when they need it, by oxidizing these sugars in their mitochondria, just as animal cells can. Chloroplasts thus generate both the food molecules and the oxygen that all mitochondria use. How they do so will be explained in Chapter 14.

Like mitochondria, chloroplasts contain their own DNA, reproduce by dividing in two, and are thought to have evolved from bacteria—in this case from photosynthetic bacteria that were somehow engulfed by an early eucaryotic cell (Figure 1–21).

Internal Membranes Create Intracellular Compartments with Different Functions

Nuclei, mitochondria, and chloroplasts are not the only membrane-enclosed organelles inside eucaryotic cells. The cytoplasm contains a profusion of other organelles—most of them enclosed by single membranes—that perform many distinct functions. Most of these structures are involved with the cell's ability to import raw materials and to export manufactured substances and waste products. Some of these membrane-enclosed organelles are enormously enlarged in cells that are specialized for secretion of proteins; others are particularly plentiful in cells specialized for digestion of foreign bodies.

The *endoplasmic reticulum (ER)*—an irregular maze of interconnected spaces enclosed by a folded membrane (Figure 1–22)—is the site at which most cell membrane components, as well as materials destined for export from the cell, are made. Stacks of flattened membrane-enclosed sacs constitute the *Golgi apparatus* (Figure 1–23), which

Figure 1–20 Chloroplasts capture the energy of sunlight in plant cells. (A) Leaf cells in a moss, viewed in a light microscope, each contain many green chloroplasts. (B) Electron micrograph of a chloroplast in a grass leaf shows the organelle's extensive system of internal membranes. The flattened sacs of membrane contain chlorophyll and are arranged in stacks. (C) The sketch highlights the features seen in (B). (B, courtesy of Eldon Newcomb.)

Question 1–5

According to Figure 1–19, why does the mitochondrion have both an outer and an inner membrane? Which of the two mitochondrial membranes should be—in evolutionary terms—derived from the cell membrane of the ancestral eucaryotic cell? In the electron micrograph of a mitochondrion in Figure 1–18A, identify the space that contains the mitochondrial DNA, i.e., the space that corresponds to the cytosol of the bacterium that was internalized by the ancestral eucaryotic cell shown in Figure 1–19.

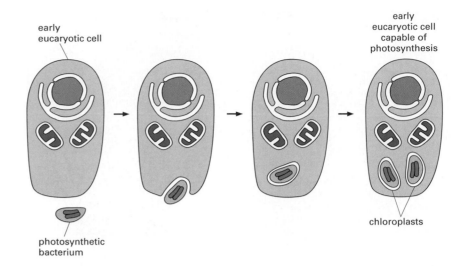

Figure 1–21 Chloroplasts, like mitochondria, have evolved from engulfed bacteria. Chloroplasts are thought to have originated from symbiotic photosynthetic bacteria, which were taken up by early eucaryotic cells that already contained mitochondria.

early eucaryotic cell

early eucaryotic cell capable of photosynthesis

chloroplasts

photosynthetic bacterium

receives and often modifies chemically the molecules made in the endoplasmic reticulum, and then directs them to the exterior of the cell or to various other locations. *Lysosomes* are small, irregularly shaped organelles in which intracellular digestion occurs, releasing nutrients from food particles and breaking down unwanted molecules for recycling or excretion. And *peroxisomes* are small, membrane-enclosed vesicles that provide a contained environment for reactions in which hydrogen peroxide, a dangerously reactive chemical, is generated and degraded. Membranes also form many different types of small *vesicles* involved in the transport of materials between one membrane-enclosed organelle and another. This whole system of related organelles is sketched in Figure 1–24A.

A continual exchange of materials takes place between the endoplasmic reticulum, the Golgi apparatus, the lysosomes, and the outside of the cell. The exchange is mediated by small membrane-enclosed vesicles that pinch off from the membrane of one organelle and fuse with another, like tiny soap bubbles budding from and rejoining larger bubbles. At the surface of the cell, for example, portions of the plasma membrane tuck inward and pinch off to form vesicles that carry into the cell material captured from the external medium (Figure 1–25). These

Figure 1–22 Many cellular components are produced in the endoplasmic reticulum (ER). (A) Schematic diagram of an animal cell shows the endoplasmic reticulum in *green*. (B) Electron micrograph of a thin section of a mammalian pancreatic cell shows a small part of the endoplasmic reticulum, of which there are vast tracts in this cell type, which is specialized for protein secretion. Note that the ER is continuous with the membrane of the nuclear envelope. The black particles studding the particular region of the ER shown here are ribosomes—the molecular assemblies that perform protein synthesis. Because of its appearance, ribosome-coated ER is often called "rough ER." (B, courtesy of Lelio Orci.)

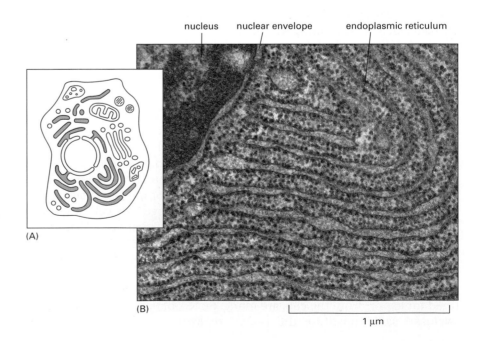

nucleus   nuclear envelope   endoplasmic reticulum

(A)

(B)

1 μm

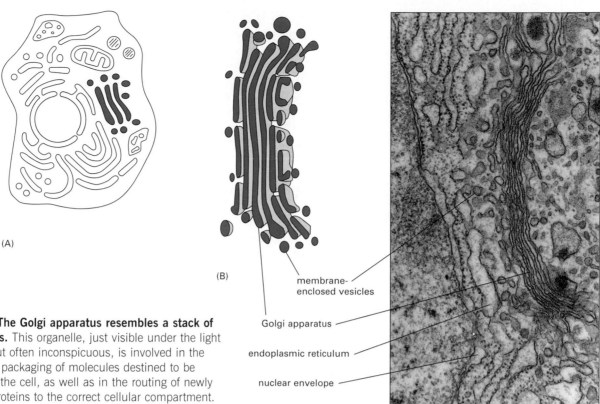

(A)

(B)

(C)

membrane-
enclosed vesicles

Golgi apparatus

endoplasmic reticulum

nuclear envelope

1 μm

Figure 1–23 The Golgi apparatus resembles a stack of flattened discs. This organelle, just visible under the light microscope but often inconspicuous, is involved in the synthesis and packaging of molecules destined to be secreted from the cell, as well as in the routing of newly synthesized proteins to the correct cellular compartment. (A) Schematic diagram of an animal cell with the Golgi apparatus colored *red.* (B) Drawing of the Golgi apparatus reconstructed from electron microscope images. The organelle is composed of flattened sacs of membrane stacked in layers, from which small vesicles pinch off and fuse. Only one stack is shown here, but several can be present in each cell. (C) Electron micrograph of the Golgi apparatus from a typical animal cell. (C, courtesy of Brij J. Gupta.)

generally fuse with lysosomes, where the imported material is digested. Animal cells can engulf very large particles, or even entire foreign cells, by this process of *endocytosis.* The reverse process, *exocytosis,* whereby vesicles from inside the cell fuse with the plasma membrane and release their contents into the external medium, is also a common cellular activity (see Figure 1–25). Hormones, neurotransmitters, and other signaling molecules are secreted from cells by exocytosis. How the membrane-enclosed organelles transport proteins and other molecules from place to place inside the cell will be discussed in more detail in Chapter 15.

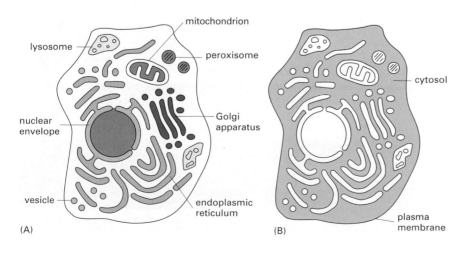

mitochondrion

lysosome

peroxisome

cytosol

nuclear
envelope

Golgi
apparatus

vesicle

endoplasmic
reticulum

plasma
membrane

(A)

(B)

Figure 1–24 Membrane-enclosed organelles are distributed throughout the cytoplasm. (A) A variety of membrane-enclosed compartments exist within eucaryotic cells, each specialized to perform a different function. (B) The rest of the cell, excluding all these organelles, is called the cytosol (colored *blue).* This region is the site of many vital cellular activities.

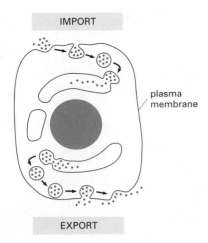

plasma
membrane

Figure 1–25 Cells engage in endocytosis and exocytosis. Cells can import materials from the external medium by capturing them in vesicles that pinch off from the plasma membrane. The vesicles ultimately fuse with lysosomes, where intracellular digestion occurs. By a converse process, cells export materials they have synthesized in intracellular compartments: the materials are stored in the intracellular vesicles and released to the exterior when these vesicles fuse with the plasma membrane.

Question 1–6

Suggest a reason why it would be advantageous for eucaryotic cells to evolve elaborate internal membrane systems that allow them to import substances from the outside, as shown in Figure 1–25.

The Cytosol Is a Concentrated Aqueous Gel of Large and Small Molecules

If we were to strip the plasma membrane from a eucaryotic cell and then remove all of its membrane-enclosed organelles, including nucleus, ER, Golgi apparatus, mitochondria, and chloroplasts, we would be left with the **cytosol** (Figure 1–24B). In most cells the cytosol fills the largest single compartment, which in bacteria is generally the only intracellular compartment. The cytosol contains a host of large and small molecules, crowded together so closely that it behaves more like a water-based gel than a liquid solution (Figure 1–26). It is the site of many chemical reactions that are fundamental to the cell's existence. The early steps in the breakdown of nutrient molecules take place in the cytosol, for example, and it is here too that the cell performs one of its key synthetic processes—the manufacture of proteins. **Ribosomes**, the tiny molecular machines that make the protein molecules, are visible with the electron microscope as small particles in the cytosol, often attached to the cytosolic face of the ER (see Figures 1–8B and 1–22B).

The Cytoskeleton Is Responsible for Directed Cell Movements

The cytoplasm is not just a structureless soup of chemicals and organelles. Under the electron microscope one can see that in eucaryotic cells (but not in bacteria), the cytosol is crisscrossed by long, fine filaments of protein. Frequently the filaments can be seen to be anchored at one end to the plasma membrane or to radiate out from a central site adjacent to the nucleus. This system of filaments is called the **cytoskeleton** (Figure 1–27). The thinnest of the filaments are *actin filaments,* which are present in all eucaryotic cells but occur in especially large numbers inside muscle cells, where they serve as part of the machinery that generates contractile forces. The thickest filaments are called *microtubules,* because they have the form of minute hollow tubes. They become reorganized into spectacular arrays in dividing cells, where they help pull the duplicated chromosomes in opposite directions and distribute them equally to the two daughter cells (Figure 1–28). Intermediate in thickness between actin filaments and microtubules are the *intermediate filaments,* which serve to strengthen the cell mechanically. These three types of filaments, together with other proteins that attach to them, form a system of girders, ropes, and motors that gives the cell its mechanical strength, controls its shape, and drives and guides its movements.

Figure 1–26 The cytoplasm is stuffed with organelles and a host of large and small molecules. This schematic drawing, based on the known sizes and concentrations of molecules in the cytosol, shows how crowded the cytoplasm is. The panorama begins on the far left at the cell surface; moves through the endoplasmic reticulum, Golgi apparatus, and a mitochondrion; and ends on the far right in the nucleus. Note that some ribosomes (*large pink objects*) are free in the cytosol, while others are attached to the ER. (Courtesy of D. Goodsell.)

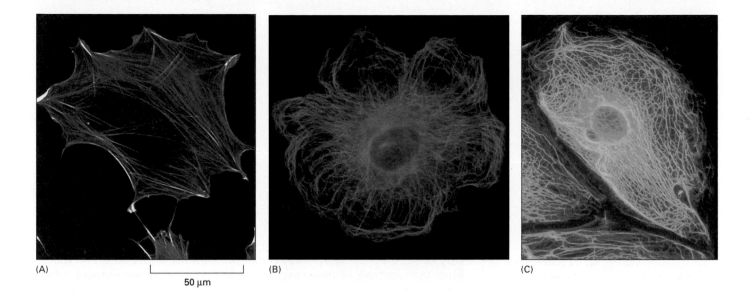

(A)

50 μm

(B)

(C)

Because the cytoskeleton governs the internal organization of the cell as well as its external features, it is as necessary to a plant cell—boxed in by a tough wall of extracellular matrix—as it is to an animal cell that freely bends, stretches, swims, or crawls. In a plant cell, for example, organelles such as mitochondria are driven in a constant stream around the cell interior along cytoskeletal tracks. And animal cells and plant cells alike depend on the cytoskeleton to separate their internal components into two daughter sets during cell division. We will examine the cytoskeleton in detail in Chapter 17. We will examine its role in cell division in Chapters 18 and 19, and in Chapter 16 see how signals from the environment alter its structure.

The Cytoplasm Is Far from Static

It is helpful to have a sense of the pace of movements inside a cell. The cytoskeleton itself is constantly changing, a dynamic jungle of ropes and rods that are continually being strung together and taken apart; filaments can assemble and then disappear in a matter of minutes. Along these tracks and cables, organelles and vesicles hurry to and fro, racing across the width of the cell in a fraction of a second. The ER and the molecules that fill every free space are in frantic thermal commotion—with unattached proteins buzzing around so fast that, even though they move at random, they visit every corner of the cell within a few seconds, constantly colliding with an even more tumultuous dust storm of smaller organic molecules.

Of course, neither the bustling nature of the cell's interior nor the details of cell structure were appreciated when scientists first peered into a microscope; our knowledge of cell structure accumulated slowly. A few of the key discoveries are listed in Table 1–1. Panel 1–2 summarizes the differences between animal, plant, and bacterial cells.

Figure 1–27 The cytoskeleton is a network of filaments that helps define a cell's shape. Filaments made of protein provide all eucaryotic cells with an internal framework that helps organize the internal activities of the cell and underlies its movements and changes of shape. Different types of filaments can be detected using different fluorescent stains. Shown here are (A) actin filaments, (B) microtubules, and (C) intermediate filaments. (A, courtesy of Simon Barry and Chris D'Lacey; B, courtesy of Nancy Kedersha; C, courtesy of Clive Lloyd.)

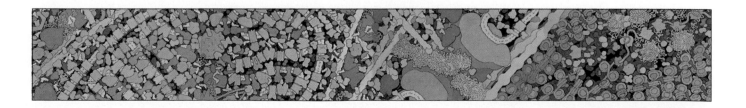

Figure 1–28 Microtubules help distribute the chromosomes in a dividing cell. When a cell divides, its nuclear envelope breaks down and its DNA condenses into pairs of visible chromosomes, which are pulled apart into separate cells by microtubules. The microtubules radiate from foci at opposite ends of the dividing cell. (Photograph courtesy of Conly L. Rieder.)

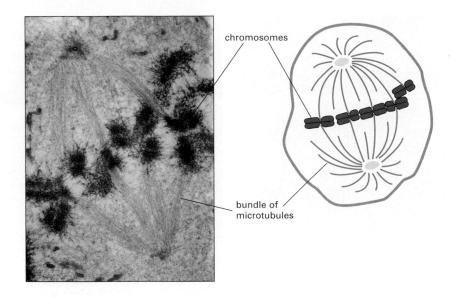

chromosomes

bundle of microtubules

Question 1–7

Discuss the relative advantages and disadvantages of light and electron microscopy. How could you best visualize (a) a living skin cell, (b) a yeast mitochondrion, (c) a bacterium, and (d) a microtubule?

Eucaryotic Cells May Have Originated as Predators

Eucaryotic cells are typically 10 times the length and 1000 times the volume of procaryotic cells (although there is huge size variation within each category). As we have seen, eucaryotes possess in addition a whole collection of other features—a cytoskeleton, mitochondria, and other organelles—that set them apart from bacteria and archaea.

When and how eucaryotes evolved these systems remains something of a mystery. Although eucaryotes, bacteria, and archaea must have diverged from one another very early in the history of life on Earth (discussed in Chapter 14), the eucaryotes did not acquire all of their distinctive features at the same time (Figure 1–29). According to one theory, the ancestral eucaryotic cell was a predator that fed by capturing

---

Table 1–1 Historical Landmarks in Determining Cell Structure

1665	Hooke uses a primitive microscope to describe small pores in sections of cork that he calls "**cells**."
1674	Leeuwenhoek reports his discovery of **protozoa**. Nine years later, he sees **bacteria** for the first time.
1833	Brown publishes his microscopic observations of orchids, clearly describing the cell **nucleus**.
1838	Schleiden and Schwann propose the **cell theory**, stating that the nucleated cell is the universal building block of plant and animal tissues.
1857	Kölliker describes **mitochondria** in muscle cells.
1879	Flemming describes with great clarity **chromosome behavior during mitosis** in animal cells.
1881	Cajal and other histologists develop staining methods that reveal the structure of **nerve cells** and the organization of neural tissue.
1898	Golgi first sees, and describes, the **Golgi apparatus** by staining cells with silver nitrate.
1902	Boveri links **chromosomes and heredity** by observing chromosome behavior during sexual reproduction.
1952	Palade, Porter, and Sjöstrand develop methods of **electron microscopy** that enable many intracellular structures to be seen for the first time. In one of the first applications of these techniques, Huxley shows that muscle contains arrays of protein filaments—the first evidence of a **cytoskeleton**.
1957	Robertson describes the bilayer structure of the **cell membrane**, seen for the first time in the electron microscope.
1960	Kendrew describes the first detailed **protein structure** (sperm whale **myoglobin**) to a resolution of 0.2 nm using **X-ray crystallography**. Perutz proposes a lower-resolution structure for **hemoglobin**.
1968	Petran and collaborators make the first **confocal microscope**.
1974	Lazarides and Weber develop the use of **fluorescent antibodies** to stain the cytoskeleton.
1994	Chalfie and collaborators introduce **green fluorescent protein** (**GFP**) as a marker in microscopy.

Panel 1–2 Cells: the principal features of animal, plant, and bacterial cells

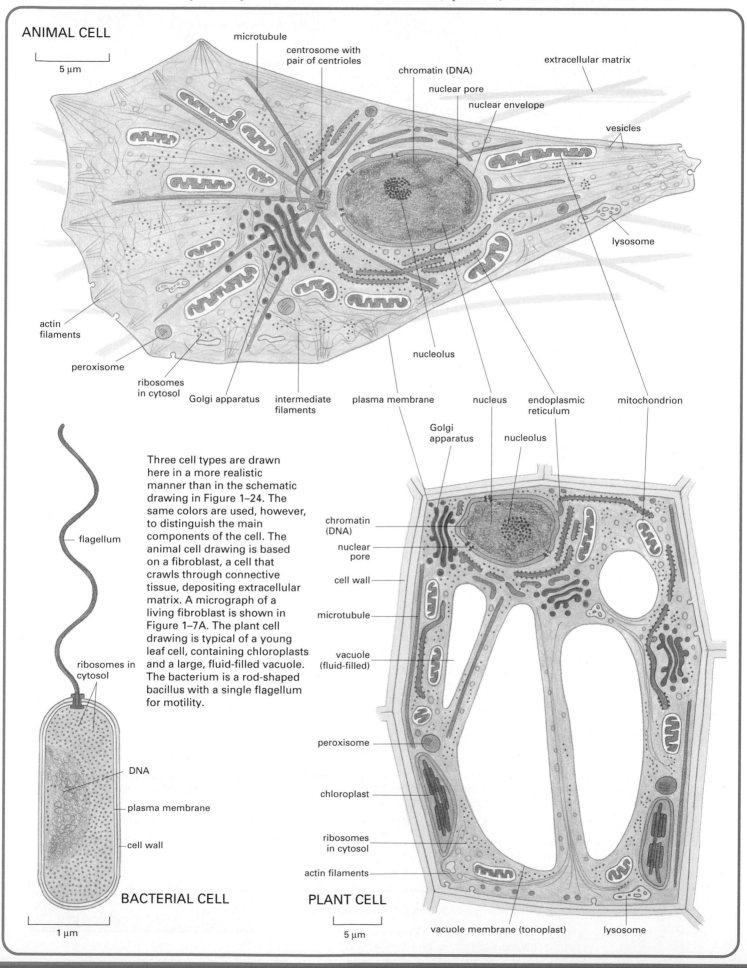

ANIMAL CELL

5 μm

microtubule

centrosome with
pair of centrioles

chromatin (DNA)

nuclear pore

nuclear envelope

extracellular matrix

vesicles

lysosome

actin
filaments

peroxisome

ribosomes
in cytosol

Golgi apparatus

intermediate
filaments

plasma membrane

nucleolus

nucleus

endoplasmic
reticulum

mitochondrion

Golgi
apparatus

nucleolus

Three cell types are drawn
here in a more realistic
manner than in the schematic
drawing in Figure 1–24. The
same colors are used, however,
to distinguish the main
components of the cell. The
animal cell drawing is based
on a fibroblast, a cell that
crawls through connective
tissue, depositing extracellular
matrix. A micrograph of a
living fibroblast is shown in
Figure 1–7A. The plant cell
drawing is typical of a young
leaf cell, containing chloroplasts
and a large, fluid-filled vacuole.
The bacterium is a rod-shaped
bacillus with a single flagellum
for motility.

flagellum

ribosomes in
cytosol

DNA

plasma membrane

cell wall

BACTERIAL CELL

1 μm

chromatin
(DNA)

nuclear
pore

cell wall

microtubule

vacuole
(fluid-filled)

peroxisome

chloroplast

ribosomes
in cytosol

actin filaments

PLANT CELL

5 μm

vacuole membrane (tonoplast)

lysosome

Figure 1–29 **Where did eucaryotes come from?** The eucaryotic, eubacterial, and archaean lineages diverged from one another very early in the evolution of life on Earth. Some time later, the eucaryotes acquired mitochondria; later still, a subset of eucaryotes acquired chloroplasts. Mitochondria are essentially the same in plants, animals, and fungi, and therefore are thought to have been acquired before these lines diverged.

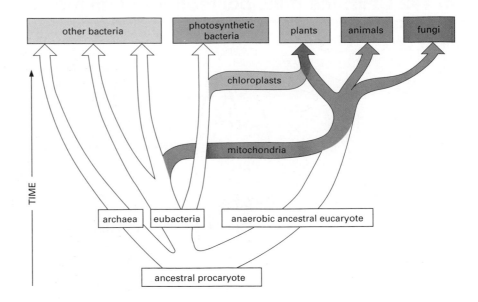

other cells. Such a way of life requires a large size, a flexible membrane, and a cytoskeleton to help the cell move and eat. The nuclear compartment may have evolved to protect the fragile DNA from being damaged by the movement of the cytoskeleton.

Such a primitive eucaryote, with a nucleus and cytoskeleton, was most likely the sort that engulfed the free-living, oxygen-metabolizing eubacteria that were the ancestors of the mitochondria. This partnership is thought to have been established 1.5 billion years ago, when the Earth's atmosphere first became rich in oxygen. A subset of these cells later acquired chloroplasts by engulfing photosynthetic bacteria (see Figure 1–29).

That single-celled eucaryotes can prey upon and swallow other cells is borne out by the behavior of many of the free-living actively motile microorganisms called **protozoans**. *Didinium*, for example, is a large, carnivorous protozoan with a diameter of about 150 μm—perhaps 10 times that of an average human cell. It has a globular body encircled by two fringes of cilia, and its front end is flattened except for a single protrusion rather like a snout (Figure 1–30). *Didinium* swims at high speed by means of its beating cilia. When it encounters a suitable prey, usually another type of protozoan, it releases numerous small, paralyzing darts from its snout region. Then *Didinium* attaches to and

Figure 1–30 **One protozoan eats another.** (A) The micrograph shows *Didinium* on its own, with its circumferential rings of beating cilia and its "snout" at the top. (B) *Didinium* is seen ingesting another ciliated protozoan, *Paramecium*. (Courtesy of D. Barlow.)

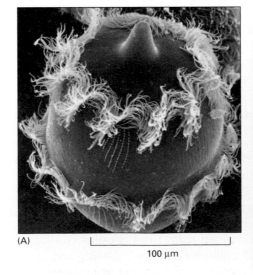

(A)

100 μm

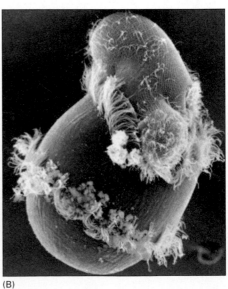

(B)

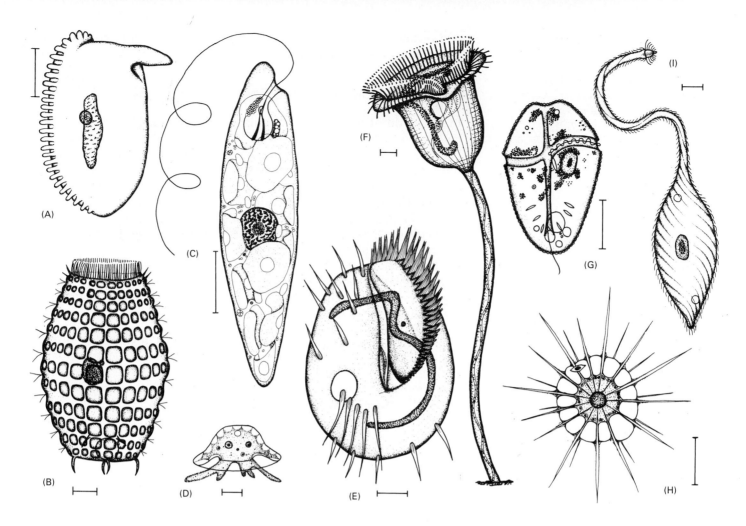

devours the other cell, inverting like a hollow ball to engulf its victim, which is almost as large as itself.

Protozoans include some of the most complex cells known. Figure 1–31 conveys something of the variety of forms of protozoans, and their behavior is just as diverse: they can be photosynthetic or carnivorous, motile or sedentary. Their cellular anatomy is often elaborate and includes such structures as sensory bristles, photoreceptors, beating cilia, stalklike appendages, mouthparts, stinging darts, and musclelike contractile bundles. Although they are single cells, protozoans can be as intricate and versatile as many multicellular organisms.

Figure 1–31 An assortment of protozoans illustrates the enormous variety within this class of single-celled microorganisms. These drawings are done to different scales, but in each case the scale bar represents 10 μm. The organisms in (A), (B), (E), (F), and (I) are ciliates; (C) is a euglenoid; (D) is an amoeba; (G) is a dinoflagellate; and (H) is a heliozoan. (From M.A. Sleigh, The Biology of Protozoa. London: Edward Arnold, 1973.)

Model Organisms

Because all cells are descended from a common ancestor and their fundamental properties have been conserved through evolution, knowledge gained from the study of one organism contributes to our understanding of others, including ourselves. But certain organisms are easier than others to study in the laboratory. Some reproduce rapidly and yield readily to genetic manipulations; others are, for example, multicellular but transparent, so that one can directly match the development of all their internal tissues and organs. For these reasons, large communities of biologists have become dedicated to studying different aspects of the biology of a few chosen species, pooling their knowledge so as to gain a deeper understanding than could be achieved if their efforts were spread over many different species. Information obtained from these studies contributes to our understanding of how all cells work. In the following sections, we will examine some of these representative **model organisms** and review the benefits each offers to the study of cell biology and, in many cases, to the promotion of human health.

Molecular Biologists Have Focused on *E. coli*

In the world of bacteria, the spotlight of molecular biology has fallen chiefly on just one species: *Escherichia coli*, or *E. coli* for short (see Figure 1–11). This small, rod-shaped eubacterial cell normally lives in the gut of humans and other vertebrates, but it can be grown easily in a simple nutrient broth in a culture bottle. *E. coli* copes well with variable chemical conditions in its environment, and it reproduces rapidly. Its genetic instructions are contained in a single, circular, double-stranded molecule of DNA, approximately 4.6 million nucleotide pairs long, and it makes 4300 different kinds of proteins.

In molecular terms, we understand the workings of *E. coli* more thoroughly than those of any other living organism. Most of our knowledge of the fundamental mechanisms of life—including how cells replicate their DNA and how they decode these genetic instructions to make proteins—has come from studies of *E. coli*. Subsequent research has confirmed that these basic processes occur in essentially the same way in our own cells as they do in *E. coli*.

Brewer's Yeast Is a Simple Eucaryotic Cell

We tend to be preoccupied with eucaryotes because we are eucaryotes ourselves. But human cells are complicated and difficult to work with, and if we want to understand the fundamentals of eucaryotic cell biology, it is often more effective to concentrate on a species that, like *E. coli* among the bacteria, is simple and robust and reproduces rapidly. The popular choice for this role of minimal model eucaryote has been the yeast *Saccharomyces cerevisiae* (Figure 1–32)—the same microorganism that is used for brewing beer and baking bread.

S. cerevisiae is a small, single-celled fungus and thus, according to modern views, is at least as closely related to animals as it is to plants. Like other fungi, it has a rigid cell wall, is relatively immobile, and possesses mitochondria but not chloroplasts. When nutrients are plentiful, it reproduces almost as rapidly as a bacterium. As its nucleus contains only about 2.5 times as much DNA as *E. coli*, the yeast is also a good subject for genetic analysis. Even though its genome is small (by eucaryotic standards), the yeast carries out all the basic tasks every eucaryotic cell must perform. Genetic and biochemical studies in yeast have been crucial to understanding many basic mechanisms in eucaryotic cells, including the cell-division cycle—the chain of events by which the nucleus and all the other components of a cell are duplicated and parceled out to create two daughter cells. In fact, the machinery that governs cell division has been so well conserved over the course of evolution that many of its components can function interchangeably in yeast and human cells. If a mutant yeast lacks a gene essential for cell division, providing it with a copy of the corresponding gene from a human will cure the yeast's defect and enable it to divide normally (see How We Know, pp. 30–31).

Arabidopsis Has Been Chosen Out of 300,000 Species as a Model Plant

The large multicellular organisms that we see around us—the flowers and trees and animals—seem fantastically varied, but they are much

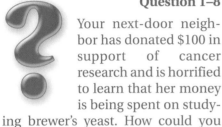

Question 1–8

Your next-door neighbor has donated $100 in support of cancer research and is horrified to learn that her money is being spent on studying brewer's yeast. How could you put her mind at ease?

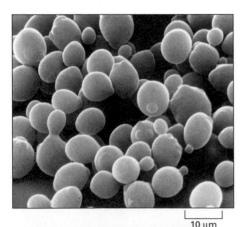

10 μm

Figure 1–32 The yeast *Saccharomyces cerevisiae* is a model eucaryote. In this scanning electron micrograph a few yeast cells are seen in the process of dividing. Another micrograph of the same species of cells is shown in Figure 1–14. (Courtesy of Ira Herskowitz and Eric Schabatach.)

closer to one another in their evolutionary origins, and more similar in their basic cell biology, than the great host of microscopic single-cell organisms. Whereas bacteria and eucaryotes separated from each other more than 3 billion years ago, plants, animals, and fungi are separated by only about 1.5 billion years, fish and mammals by only about 450 million years, and the different species of flowering plants by less than 200 million years.

The close evolutionary relationship among all flowering plants means that we can get insight into the cell and molecular biology of flowering plants by focusing on just a few convenient species for detailed analysis. Out of the several hundred thousand species of flowering plants on Earth today, molecular biologists have recently focused their efforts on a small weed, the common wall cress *Arabidopsis thaliana* (Figure 1–33), which can be grown indoors in large numbers and produces thousands of offspring per plant within 8 to 10 weeks. *Arabidopsis* has a genome of approximately 110 million nucleotide pairs, about 8 times as many as yeast, and its complete sequence is known. By examining the genetic instructions that *Arabidopsis* carries, we are beginning to learn more about the genetics, molecular biology, and evolution of flowering plants, which dominate nearly every ecosystem on land. Because genes found in *Arabidopsis* have counterparts in agricultural species, studying this simple weed provides insights into the development and physiology of the crop plants upon which our lives depend, as well as all the other plant species that are our companions on Earth.

The World of Animals Is Represented by a Fly, a Worm, a Mouse, and *Homo sapiens*

Multicellular animals account for the majority of all named species of living organisms, and the majority of animal species are insects. It is fitting, therefore, that an insect, the small fruit fly *Drosophila melanogaster* (Figure 1–34), should occupy a central place in biological research. In fact, the foundations of classical genetics were built to a large extent on studies of this insect. More than 80 years ago, for example, studies of the fruit fly provided definitive proof that genes—the units of heredity—are carried on chromosomes. In more recent times, a concentrated systematic effort has been made to elucidate the genetics

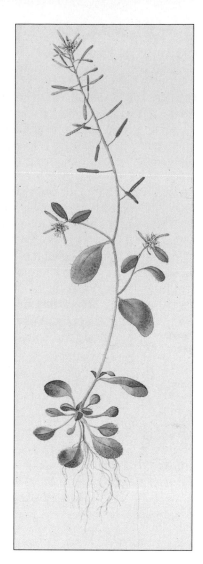

Figure 1–33 *Arabidopsis thaliana*, the common wall cress, is a model plant. This small weed has become the favorite organism of plant molecular and developmental biologists. (Courtesy of Toni Hayden and the John Innes Centre.)

Figure 1–34 *Drosophila melanogaster* is a favorite among developmental biologists and geneticists. Molecular genetic studies on this small fly have provided a key to the understanding of how all animals develop. (Courtesy of E.B. Lewis.)

How We Know: Life's Common Mechanisms

All living things are made of cells, and cells—as we have discussed in this chapter—are all fundamentally similar inside: they store their genetic instructions in DNA molecules, which direct the production of proteins, and proteins in turn carry out the cell's chemical reactions, give it its shape, and control its behavior. But how deep do these similarities really run? Are parts from one type of cell interchangeable with parts from another? Would an enzyme that digests glucose in a bacterium be able to break down the same sugar if it were asked to function inside a yeast, a lobster, or a human? What about the molecular machines that copy and interpret genetic information? Are they functionally equivalent from one organism to another? Are their component molecules interchangeable? Answers have come from many sources, but most strikingly from experiments on one of the most fundamental processes of life: cell division.

Divide or die

All cells come from other cells, and the only way to make a new cell is through division of a preexisting cell. To reproduce, a parent cell must execute an orderly sequence of reactions through which it duplicates its contents and divides in two. This critical process of duplication and division, known as the cell cycle, is complex and carefully controlled. Defects in any of the proteins involved in the cell cycle can be fatal.

Unfortunately, the lethal effects of cell-cycle mutations present a problem if one wants to discover the components of the cell-cycle control machinery and find out how they work. Scientists depend on mutants to identify genes and proteins on the basis of their functions: if a gene is essential for a given process, a mutation that disrupts the gene will show up as a disturbance of that process. By analyzing the misbehavior of the mutant organism, one can pinpoint the function for which the gene is needed, and by analyzing the DNA of the mutant one can track down the gene itself.

For such an analysis, however, a single mutant cell is not enough: one needs a large colony of cells carrying the mutation. And this is the problem. If the mutation disrupts a process critical to life, such as cell division, how can one ever obtain such a colony? Geneticists have found an ingenious solution. Mutants defective in cell-cycle genes can be maintained and studied if their defect is conditional—that is, if the gene product fails to function only under certain specific conditions. In particular, one can often find mutations that are temperature-sensitive: the mutant protein functions correctly when the organism is kept cool, allowing the cells to reproduce, but fails when the temperature is warmer, allowing the cells to display their interesting defect (Figure 1–35). The study of such conditional mutants in yeast has allowed the discovery of the genes that control the cell-division cycle—the *cdc* genes—and has led to an understanding of how they work.

The same temperature-sensitive mutants, it turns out, offer an opportunity to see whether proteins from one organism can function interchangeably in another. Can a protein from a different organism cure a cell-cycle defect in a mutant yeast and enable it to reproduce normally? The first experiment was performed using two different species of yeast.

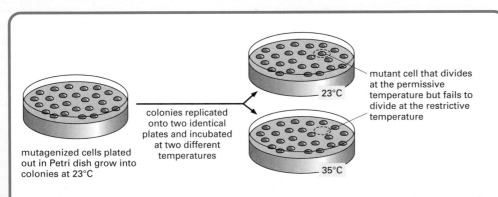

Figure 1–35 Yeast cells that contain a temperature-sensitive mutation can be generated in the laboratory. Yeast are incubated with a chemical that generates mutations in their DNA. These cells are spread onto a plate and allowed to grow at a permissive temperature, that is, one at which the cells divide normally. The colonies are transferred to two identical Petri dishes using a technique called replica plating. One of these plates is incubated at the cooler, "permissive" temperature, the other at a higher temperature. Cells containing a temperature-sensitive mutation in a gene essential for proliferation can divide at the permissive temperature but fail to divide at the warmer, nonpermissive temperature.

Labels in figure:
mutagenized cells plated out in Petri dish grow into colonies at 23°C

colonies replicated onto two identical plates and incubated at two different temperatures

mutant cell that divides at the permissive temperature but fails to divide at the restrictive temperature

23°C

35°C

Next of kin

Yeasts—unicellular fungi—are popular organisms for studies of cell division because they are eucaryotes, like us, yet they are small, simple, rapidly reproducing, and easy to manipulate experimentally. *Saccharomyces cerevisiae,* the most widely studied yeast, divides by forming a small bud that grows steadily until it separates from the mother cell (see Figures 1–14 and 1–32). A second species of yeast, *Schizosaccharomyces pombe*, is also popular for studies of cell growth and division. Named after the African beer from which it was first isolated, *S. pombe* is a rod-shaped yeast that grows by elongation at its ends and divides by fission of this rod into two, through the formation of a partition in the center of the rod.

Although they differ in their style of cell division, both yeasts must copy their DNA and parcel this material to their progeny. To establish whether the proteins controlling the whole process in *S. cerevisiae* and *S. pombe* are functionally equivalent, Paul Nurse and his colleagues set out to determine whether *S. pombe* cell-cycle mutants could be rescued by a gene from *S. cerevisiae*. The starting point was a colony of temperature-sensitive *S. pombe* mutants that were incapable of proceeding through the cell cycle when grown at a warm 35°C. These mutant cells had a defect in a gene called *cdc2*, which is required to trigger several key events in the cell division cycle. The researchers then introduced into these defective cells a collection of DNA fragments prepared from *S. cerevisiae* (Figure 1–36).

When these cultures were incubated at 35°C, the researchers found that some of the cells had regained the ability to reproduce: spread onto a plate of medium, these cells could divide again and again, forming small colonies containing millions of yeast cells (see Figure 1–35). These

human	FGLARAFGIPIRVYTHEVVTLWYRSPEVLLGS
S. pombe	FGLARSFGVPLRNYTHEIVTLWYRAPEVLLGS
S. cerevisiae	FGLARAFGVPLRAYTHEIVTLWYRAPEVLLGG
human	ARYSTPVDIWSIGTIFAELATKLPLFHGDSEI
S. pombe	RHYSTGVDIWSVGCIFAENIRRSPLFPGDSEI
S. cerevisiae	KQYSTGVDTWSIGCIFAEHCNRLPIFSGDSEI

Figure 1–37 The cell-division-cycle proteins from yeasts and human are very similar in their amino acid sequences. Identities between the amino acid sequences of a region of the human CDC2 protein, the cdc2 protein of *S. pombe,* and cdc28 of *S. cerevisiae* are boxed. Each amino acid is represented by a single letter.

"cured" yeast cells, the researchers discovered, had received a fragment of DNA containing *cdc28*, a gene from *S. cerevisiae* that was already familiar from pioneering cell-division-cycle studies (by Lee Hartwell and colleagues) in the budding yeast. The *cdc28* gene encodes a protein that performs the same function in budding yeast as *cdc2* does in the fission yeast.

Perhaps the result is not all that surprising. How different can one yeast be from another? What about more distant relatives? To find out, the researchers performed the same experiment, this time using human DNA to rescue the yeast cell-cycle mutants. The results were the same. A human gene, which the investigators dubbed *CDC2,* could substitute for its equivalent in yeast, enabling the cells to divide normally.

Reading genes
Not only are the human and yeast proteins functionally equivalent, they are almost the exact same size and closely similar in the order of the amino acids of which they are made. When the Nurse team examined the amino acid sequences of the proteins, it found that human CDC2 is identical to the *S. pombe* cdc2 protein in 63% of its amino acids and 58% identical to CDC28 from *S. cerevisiae* (Figure 1–37).

These experiments show that proteins from different organisms can be functionally interchangeable. In fact, the molecules that orchestrate cell division in eucaryotes are so fundamentally important that they have been conserved almost unchanged over more than a billion years of eucaryotic evolution.

The same experiment highlights another, even more basic point. The mutant yeast was rescued, not by direct injection of the human protein, but by introduction of a piece of human DNA. The yeast could read and use this information correctly, because the molecular machinery for these fundamental processes is also similar from cell to cell and from organism to organism. A yeast cell has all the equipment it needs to interpret the instructions encoded in a human gene and to use that information to direct the production of a fully functional human protein.

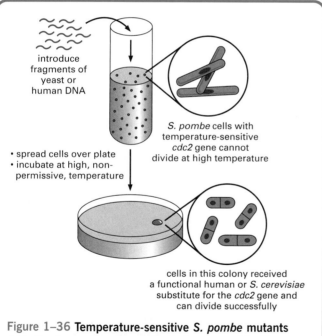

introduce fragments of yeast or human DNA

- spread cells over plate
- incubate at high, non-permissive, temperature

S. pombe cells with temperature-sensitive *cdc2* gene cannot divide at high temperature

cells in this colony received a functional human or *S. cerevisiae* substitute for the *cdc2* gene and can divide successfully

Figure 1–36 Temperature-sensitive *S. pombe* mutants defective in a cell-cycle gene can be rescued by the equivalent gene from *S. cerevisiae*—or from humans. DNA is collected from *S. cerevisiae* (or from humans) and broken into large fragments, which are introduced into a culture of temperature-sensitive *S. pombe* mutants. We will learn how DNA can be manipulated and moved into different cell types in Chapter 10. The yeast cells that received foreign DNA are then spread on to a plate and incubated at the non-permissive temperature. The rare cells that survive and grow on these plates contain a gene that allows the cells to divide normally.

of *Drosophila*, and especially the genetic mechanisms underlying its embryonic and larval development. Through this work on *Drosophila*, we are at last beginning to understand in detail how living cells achieve their most spectacular feat: how a single fertilized egg cell (or zygote) develops into a multicellular organism comprising vast numbers of cells of differing types, organized in an exactly predictable way. *Drosophila* mutants with body parts strangely misplaced or oddly patterned have provided the key to identifying and characterizing the genes that are needed to make a properly structured adult body, with gut, wings, legs, eyes, and all of the other bits in their correct places. These genes—which are copied and passed on to every cell in the body—define how each cell will behave in its social interactions with its sisters and cousins, and in that way they control the structures that the cells create. *Drosophila*, more than any other organism, has shown us how to trace the chain of cause and effect from the genetic instructions encoded in the DNA to the structure of the adult multicellular organism. Moreover, the genes of *Drosophila* have turned out to be amazingly similar to those of humans—far more similar than one would suspect from outward appearances. Thus the fly serves as a model for studying human development and disease. The fly genome—185 million nucleotide pairs encoding just over 13,000 genes—contains counterparts for most of the genes known to be critical in human diseases.

Another widely studied organism, smaller and simpler than *Drosophila*, is the nematode worm *Caenorhabditis elegans* (Figure 1–38), a harmless relative of the eelworms that attack the roots of crops. This creature develops with clockwork precision from a fertilized egg cell into an adult with exactly 959 body cells (plus a variable number of egg and sperm cells)—an unusual degree of regularity for an animal. We now have a minutely detailed description of the sequence of events by which this occurs—as the cells divide, move, and become specialized, according to strict and predictable rules. Its genome—some 97 million nucleotide pairs containing about 19,000 genes—has also been sequenced, and a wealth of mutants is available for testing how these genes function. It appears that 70% of human proteins have some counterpart in the worm, and *C. elegans* like *Drosophila* has proven to be a valuable model for many of the processes that occur in our own bodies. Studies of nematode development, for example, have led to an understanding of *programmed cell death*, a process by which surplus cells are disposed of in the body—a topic of importance for cancer research (discussed in Chapters 18 and 21).

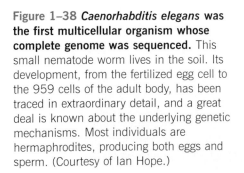

Figure 1–38 *Caenorhabditis elegans* was the first multicellular organism whose complete genome was sequenced. This small nematode worm lives in the soil. Its development, from the fertilized egg cell to the 959 cells of the adult body, has been traced in extraordinary detail, and a great deal is known about the underlying genetic mechanisms. Most individuals are hermaphrodites, producing both eggs and sperm. (Courtesy of Ian Hope.)

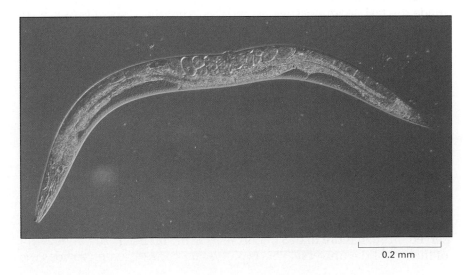

0.2 mm

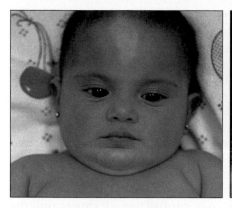

Figure 1–39 **Different living species share similar genes.** The human baby and the mouse shown here have similar white patches on their foreheads because they both have defects in the same gene (called *Kit*), required for the development and maintenance of pigment cells. (Courtesy of R.A. Fleischman, from *Proc. Natl. Acad. Sci. U.S.A.* 88:10885–10889, 1991. © National Academy of Sciences.)

At the other extreme, mammals are among the most complex of animals, with 2 to 3 times as many genes as *Drosophila*, 25 times as much DNA per cell, and millions of times more cells in their adult bodies. The mouse has long been used as the model organism in which to study mammalian genetics, development, immunology, and cell biology. New techniques have given it even greater importance. It is now possible to breed mice with deliberately engineered mutations in any specific gene, or with artificially constructed genes introduced into them. In this way, one can test what a given gene is required for and how it functions. And almost every human gene has a counterpart in the mouse, with similar DNA sequence and function.

But humans are not mice—or worms or flies or yeast—and so we also study human beings themselves. Research in many areas of cell biology has been largely driven by medical interests, and a great deal of what we know has come from studies of human cells. The medical database on human cells is enormous, and although naturally occurring mutations in any given gene are rare, the consequences of mutations in thousands of different genes are known without resort to genetic engineering. This is because humans demonstrate the unique behavior of reporting on and recording their own genetic defects; in no other species are billions of individuals so intensively examined, described, and investigated.

Nevertheless, the extent of our ignorance is still daunting. The mammalian body is enormously complex, and one might despair of ever understanding how the DNA in a fertilized mouse egg cell makes it generate a mouse, or how the DNA in a human egg cell directs the development of a human. Yet, the revelations of molecular biology have made the task seem possible. As much as anything, this new optimism has come from the realization that the genes of one type of animal have close counterparts in most other types of animals, apparently serving similar functions (Figure 1–39). We all have a common evolutionary origin, and under the surface it seems that we share the same molecular mechanisms. Flies, worms, mice, and humans thus provide a key to understanding how animals in general are made and how their cells operate.

Comparing Genome Sequences Reveals Life's Common Heritage

At a molecular level, evolutionary change has been remarkably slow. We can see in present-day organisms many features that have been preserved through more than 3 billion years of life on Earth, or about a fifth of the age of the universe. This evolutionary conservatism provides the foundation on which the study of molecular biology is built. To set the scene for the chapters that follow, therefore, we end our introduction by considering a little more closely the family relationships and basic similarities among all living things. This topic has been dramatically clarified in the past few years by analysis of genome sequences—the sequences in which the four universal nucleotides are strung together to form the DNA of a given species (as discussed in more detail in Chapter 9). DNA sequencing has made it easy to detect family resemblances between genes: if two genes from different organisms have closely similar DNA sequences, it is highly probable that both genes are descended from a common ancestral gene. Genes (and gene products) related in this way are said to be **homologous**. Given the complete genome sequences of representative organisms from all three domains of life—archaea, eubacteria, and eucaryotes—one can search systematically for homologies that span this enormous evolutionary divide. In this way, we can begin to take stock of the common inheritance of all living things and to trace life's origins back to the earliest ancestral cells. There are difficulties in this enterprise: some ancestral genes are lost, and some have changed so much that they are not readily recognizable as relatives. Despite these uncertainties, comparing genome sequences from the most widely separated branches of the tree of life can give us a sense of which genes are fundamental necessities for living cells.

A comparison of the complete genomes of five eubacteria, one archaean, and one eucaryote (a yeast) revealed a core set of 239 families of protein-coding genes that have representatives in all three domains. Most of these genes can be assigned a function, with the largest number of shared gene families being involved in amino acid metabolism and transport, and in the production and function of ribosomes. Thus the minimum number of genes needed for a cell to be viable in today's environments is probably not much less than 200–300.

Most organisms possess significantly more than this. Even procaryotes—frugal cells that carry very little superfluous genetic baggage—typically have genomes that contain at least 1 million nucleotide pairs and encode 1000 to 8000 genes (468 genes, in the bacterium *Mycoplasma genitalium*, is the minimum so far recorded for any species). With these few thousand genes, bacteria are able to thrive in even the most hostile environments on Earth.

The compact genomes of typical bacteria are dwarfed by the genomes of typical eucaryotes. The human genome, for example, contains about 700 times more DNA than the *E. coli* genome, and the genome of a fern contains about 100 times more than that of a human (Figure 1–40). In terms of gene numbers, however, the differences are not so great. We have only about seven times as many genes as *E. coli*, if we count a gene as the stretch of DNA that contains the specifications for a protein molecule. Moreover, many of our 30,000 genes and corresponding proteins themselves fall into closely related family groups, such as the family of hemoglobins, which has nine closely related members in humans. The number of fundamentally different proteins in a human is thus not very many times more than in a bacterium, and the number of human genes that have identifiable counterparts in the bacterium is a significant fraction of the total.

The rest of our human DNA—the vast bulk that does not code for protein—is a mixture of sequences that help to regulate the expression of the genes, and sequences that seem to be dispensable junk, retained like a mass of old papers because, if there is no pressure to keep an archive small, it is easier to save everything than to sort out the valuable information and discard the rest. The large quantity of regulatory DNA allows for enormous complexity and sophistication in the way different

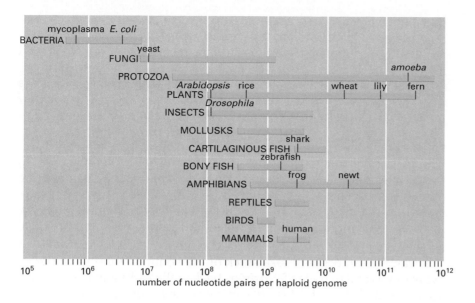

Figure 1–40 Organisms vary enormously in the sizes of their genomes. Genome size is measured in nucleotide pairs of DNA per haploid genome, that is, per single copy of the genome. (The cells of sexually reproducing organisms such as ourselves are generally diploid: they contain two copies of the genome, one inherited from the mother, the other from the father.) Closely related organisms can vary widely in the quantity of DNA in their genomes, even though they contain similar numbers of functionally distinct genes. (Data from W.-H. Li, Molecular Evolution, pp. 380–383. Sunderland, MA: Sinauer, 1997.)

genes in a eucaryotic multicellular organism are brought into action at different times and places. But the basic list of parts—the set of proteins that our cells can make, as specified by the DNA—is not much longer than the parts list of an automobile, and many of those parts are common not only to all animals, but to the entire living world.

That a length of DNA can program the growth, development, and reproduction of living cells and complex organisms is truly an amazing phenomenon. In the rest of this book, we will try to explain how cells work—in part by examining their component parts, in part by investigating how their genomes direct the manufacture of these components so as to reproduce and run each living thing.

Essential Concepts

- Cells are the fundamental units of life. All present-day cells are believed to have evolved from an ancestral cell that existed more than 3 billion years ago.
- All cells, and hence all living things, grow, convert energy from one form to another, sense and respond to their environment, and reproduce themselves.
- All cells are enclosed by a plasma membrane that separates the inside of the cell from the environment.
- All cells contain DNA as a store of genetic information and use it to guide the synthesis of proteins.
- Cells in a multicellular organism, though they all contain the same DNA, can be very different. They use their genetic information to direct their biochemical activities according to cues they receive from their environment.
- Cells of animal and plant tissues are typically 5–20 μm in diameter and can be seen with a light microscope, which also reveals some of their internal components, or organelles. The electron microscope permits the smaller organelles and even individual molecules to be seen, but specimens require elaborate preparation and cannot be viewed alive.
- Bacteria, the simplest of present-day living cells, are procaryotes: although they contain DNA, they lack a nucleus and other organelles and probably resemble most closely the ancestral cell.

- Different species of procaryotes are diverse in their chemical capabilities and inhabit an amazingly wide range of habitats. Two fundamental evolutionary subdivisions are recognized: eubacteria and archaea.
- Eucaryotic cells possess a nucleus. They probably evolved in a series of stages from cells more similar to bacteria. An important step appears to have been the acquisition of mitochondria, originating as engulfed bacteria living in symbiosis with larger anaerobic cells.
- The nucleus is the most prominent organelle in most plant and animal cells. It contains the genetic information of the organism, stored in DNA molecules. The rest of the cell's contents, apart from the nucleus, constitute the cytoplasm.
- Within the cytoplasm, plant and animal cells contain a variety of internal membrane-enclosed organelles with specialized chemical functions. Mitochondria carry out the oxidation of food molecules. In plant cells, chloroplasts perform photosynthesis. The endoplasmic reticulum, the Golgi apparatus, and lysosomes permit cells to synthesize complex molecules for export from the cell and for insertion in cell membranes, and to import and digest large molecules.
- The remaining intracellular component, excluding the membrane-enclosed organelles, is the cytosol. This contains a concentrated mixture of large and small molecules that carry out many essential biochemical processes.
- A system of protein filaments called the cytoskeleton extends throughout the cytosol. This governs cell shape and movement and enables organelles and molecules to be transported from one location to another in the cytoplasm.
- Free-living single-celled eucaryotic microorganisms include some of the most complex eucaryotic cells known, and they are able to swim, mate, hunt, and devour food. Other types of eucaryotic cells, derived from a fertilized egg, cooperate to form large, complex multicellular organisms composed of thousands of billions of cells.
- Biologists have chosen a small number of organisms as a focus for intense investigation. These include the bacterium *E. coli*, brewer's yeast, a nematode worm, a fly, a small plant, a mouse, and the human species itself.
- Although the minimum number of genes needed for a viable cell is probably less than 400, most cells contain significantly more. Yet even such a complex organism as a human has only about 30,000 genes—twice as many as a fly, seven times as many as *E. coli*.

Key Terms

bacterium	eucaryote	model organism
cell	evolution	nanometer
chloroplast	genome	nucleus
chromosome	homologous	organelle
cytoplasm	micrometer	procaryote
cytoskeleton	microscope	protozoan
cytosol	mitochondrion	ribosome

Questions

Question 1–9

By now you should be familiar with the following cellular components. Briefly define what they are and what function they provide for cells.

  A. cytosol
  B. cytoplasm
  C. mitochondria
  D. nucleus
  E. chloroplasts
  F. lysosomes
  G. chromosomes
  H. Golgi apparatus
  I. peroxisomes
  J. plasma membrane
  K. endoplasmic reticulum
  L. cytoskeleton

Question 1–10

Which of the following statements are correct? Explain your answers.

  A. The hereditary information of a cell is passed on by its proteins.
  B. Bacterial DNA is found in the cytosol.
  C. Plants are composed of procaryotic cells.
  D. All cells of the same organism have the same number of chromosomes (with the exception of egg and sperm cells).
  E. The cytosol contains membrane-enclosed organelles, such as lysosomes.
  F. Nuclei and mitochondria are surrounded by a double membrane.
  G. Protozoans are complex organisms with a set of specialized cells that form tissues, such as flagella, mouthparts, stinging darts, and leglike appendages.
  H. Lysosomes and peroxisomes are the site of degradation of unwanted materials.

Question 1–11

To get a feeling for the size of cells (and to practice the use of the metric system) consider the following: the human brain weighs about 1 kg and contains about 10^{12} cells. Calculate the average size of a brain cell (although we know that their sizes vary widely), assuming that each cell is entirely filled with water (1 cm^3 of water weighs 1 g). What would be the length of one side of this average-sized brain cell if it were a simple cube? If the cells were spread out as a thin layer that is only a single cell thick, how many pages of this book would this layer cover?

Question 1–12

Identify the different organelles indicated with letters in the electron micrograph shown in Figure Q1–12. Estimate the length of the scale bar in the figure.

Question 1–13

There are three major classes of filaments that make up the cytoskeleton. What are they and what are the differences in their functions? Which cytoskeletal filaments would be most plentiful in a muscle cell or in an epidermal cell making up the outer layer of the skin? Explain your answers.

Question 1–14

Natural selection is such a powerful force in evolution because cells with even a small growth advantage quickly outgrow their competitors. To illustrate this process, consider a cell culture that contains 1 million bacterial cells that double every 20 minutes. A single cell in this culture acquires a mutation that allows it to divide faster, with a generation time of only 15 minutes. Assuming that there is an unlimited food supply and no cell death, how long would it take before the progeny of the mutated cell became predominant in the culture? (Before you go through the calculation, make a guess: do you think it would take about a day, a week, a month, or a year?) How many cells of either type are present in the culture at this time? (The number of cells N in the culture at time t is described by the equation $N = N_0 \times 2^{t/G}$, where N_0 is the number of cells at zero time and G is the generation time.)

Question 1–15

When bacteria are grown under adverse conditions, i.e., in the presence of a poison such as an antibiotic, most cells grow slowly. But it is not uncommon that the growth rate of a bacterial culture kept in the presence of the poison is restored after a few days to that observed in its absence. Suggest why this may be the case.

Question 1–16

Apply the principle of exponential growth as described in Question 1–14 to the cells in a multicellular organism, such as yourself. There are about 10^{13} cells in your body. Assume that one cell acquires a mutation that allows it to divide in an uncontrolled manner (that is,

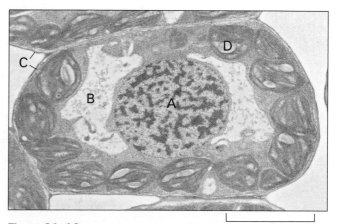

Figure Q1–12
? µm

it becomes a cancer cell). Some cancer cells can grow with a generation time of about 24 hours. If none of the cancer cells died, how long would it take before 10^{13} cells in your body would be cancer cells? (Use the equation $N = N_0 \times 2^{t/G}$, with t, the time, and G, the length of each generation. Hint: $10^{13} \cong 2^{43}$.)

Question 1–17

Discuss the following statement: "The structure and function of a living cell are dictated by the laws of physics and chemistry."

Question 1–18

What, if any, are the advantages in being multicellular?

Question 1–19

Draw to scale the outline of two spherical cells, one a bacterium with a diameter of 1 μm, the other an animal cell with a diameter of 15 μm. Calculate the volume, surface area, and surface-to-volume ratio for each cell. How would this latter value change if you included the internal membranes of the cell in the calculation of surface area (assume internal membranes have 15 times the area of the plasma membrane)? (The volume of a sphere is given by $4\pi R^3/3$ and its surface by $4\pi R^2$, where R is its radius.) Discuss the following hypothesis: "Internal membranes allowed bigger cells to evolve."

Question 1–20

What are the arguments that all living cells evolved from a common ancestor cell? Imagine the very early days of evolution of life on earth. Would you assume that the primordial ancestor cell was the first and only cell to form?

Question 1–21

In Figure 1–26, proteins are blue, nucleic acids are orange or red, lipids are yellow, and polysaccharides are green. Identify major organelles and other important cellular structures shown in this slice through a eucaryotic cell.

Highlights from *Essential Cell Biology 2 Interactive* CD-ROM

1.1 Keratocyte Dance
1.2 Crawling Amoeba

Chemical Components of Cells

It is at first sight difficult to accept that living creatures are merely chemical systems. Their incredible diversity of form, their seemingly purposeful behavior, and their ability to grow and reproduce all seem to set them apart from the world of solids, liquids, and gases that chemistry normally describes. Indeed, until the nineteenth century it was widely accepted that animals contained a vital force—an "animus"—that was responsible for their distinctive properties.

We now know that there is nothing in living organisms that disobeys chemical or physical laws. However, the chemistry of life is indeed a special kind. First, it is based overwhelmingly on carbon compounds, the study of which is known as *organic chemistry*. Second, it depends almost exclusively on chemical reactions that take place in a watery, or *aqueous*, solution and in the relatively narrow range of temperatures experienced on Earth. Third, it is enormously complex: even the simplest cell is vastly more complicated in its chemistry than any other chemical system known. Fourth, it is dominated and coordinated by enormous *polymeric molecules*—chains of chemical subunits linked end-to-end—whose unique properties enable cells and organisms to grow and reproduce and to do all the other things that are characteristic of life. Finally, it is tightly regulated: cells deploy a variety of mechanisms to make sure that all their chemical reactions occur at the proper place and time.

Chemistry, in a sense, dictates all of biology. In this chapter, therefore, we briefly survey the chemistry of the living cell. We will meet the molecules from which cells are made and examine their structures, their shapes, and their chemical properties. These molecules determine the size, structure, and function of living cells. By understanding how these molecules interact, we can begin to see how cells exploit the laws of chemistry and physics to stay alive.

Chemical Bonds

Matter is made of combinations of *elements*—substances such as hydrogen or carbon that cannot be broken down or converted into other substances by chemical means. The smallest particle of an element that still retains its distinctive chemical properties is an *atom*. The characteristics of substances other than pure elements—including the materials from which living cells are made—depend on which atoms they contain, and the way these atoms are linked together in groups to form *molecules*. In order to understand how living organisms are built from inanimate matter, therefore, it is crucial to know how the chemical bonds that hold atoms together in molecules are formed.

Chemical Bonds

Cells Are Made of Relatively Few Types of Atoms

The Outermost Electrons Determine How Atoms Interact

Ionic Bonds Form by the Gain and Loss of Electrons

Covalent Bonds Form by the Sharing of Electrons

Covalent Bonds Vary in Strength

There Are Different Types of Covalent Bonds

Water Is Held Together by Hydrogen Bonds

Some Polar Molecules Form Acids and Bases in Water

Molecules in Cells

A Cell Is Formed from Carbon Compounds

Cells Contain Four Major Families of Small Organic Molecules

Sugars Are Energy Sources for Cells and Subunits of Polysaccharides

Fatty Acids Are Components of Cell Membranes

Amino Acids Are the Subunits of Proteins

Nucleotides Are the Subunits of DNA and RNA

Macromolecules in Cells

Macromolecules Contain a Specific Sequence of Subunits

Noncovalent Bonds Specify the Precise Shape of a Macromolecule

Noncovalent Bonds Allow a Macromolecule to Bind Other Selected Molecules

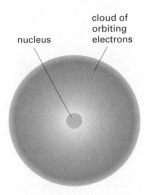

Figure 2–1 An atom consists of a nucleus surrounded by an electron cloud. The dense, positively charged nucleus contains most of the atom's mass. The much lighter and negatively charged electrons occupy space around the nucleus, as governed by the laws of quantum mechanics. The electrons are depicted as a continuous cloud, as there is no way of predicting exactly where an electron is at any given instant of time. The density of shading of the cloud is an indication of the probability that electrons will be found there. The diameter of the electron cloud ranges from about 0.1 nm (for hydrogen) to about 0.4 nm (for atoms of high atomic number). The nucleus is very much smaller: about 2×10^{-5} nm for carbon, for example.

Figure 2–2 The number of protons in an atom determines its atomic number. Schematic representations of an atom of carbon and an atom of hydrogen. The nucleus of every atom except hydrogen consists of both positively charged protons and electrically neutral neutrons. The number of electrons in an atom is equal to the number of protons, so that the atom has no net charge. In contrast to Figure 2–1, the electrons are shown here as individual particles. The concentric *black circles* represent in a highly schematic form the "orbits"(that is, the different distributions) of the electrons. The neutrons, protons, and electrons are in reality minute in relation to the atom as a whole; their size is greatly exaggerated here.

Cells Are Made of Relatively Few Types of Atoms

Each **atom** has at its center a dense, positively charged nucleus, which is surrounded at some distance by a cloud of negatively charged **electrons,** held in orbit by electrostatic attraction to the nucleus (Figure 2–1). The nucleus consists of two kinds of subatomic particles: **protons**, which are positively charged, and **neutrons**, which are electrically neutral. The number of protons present in an atomic nucleus determines its **atomic number**. An atom of hydrogen has a nucleus composed of a single proton; so hydrogen, with an atomic number of 1, is the lightest element. An atom of carbon has six protons in its nucleus and an atomic number of 6 (Figure 2–2). The electric charge carried by each proton is exactly equal and opposite to the charge carried by a single electron. Because the whole atom is electrically neutral, the number of negatively charged electrons surrounding the nucleus is equal to the number of positively charged protons that the nucleus contains; thus the number of electrons in an atom also equals the atomic number. All atoms of a given element have the same atomic number, and we shall shortly see that this number dictates the chemical behavior of the element.

Neutrons are uncharged subatomic particles with essentially the same mass as protons. They contribute to the structural stability of the nucleus—if there are too many or too few, the nucleus may disintegrate by radioactive decay—but they do not alter the chemical properties of the atom. Thus an element can exist in several physically distinguishable but chemically identical forms, called *isotopes*, each isotope having a different number of neutrons but the same number of protons. Multiple isotopes of almost all the elements occur naturally, including some that are unstable. For example, while most carbon on Earth exists as the stable isotope carbon 12, with six protons and six neutrons, there are also small amounts of an unstable isotope, the radioactive carbon 14, whose atoms have six protons and eight neutrons. Carbon 14 undergoes radioactive decay at a slow but steady rate, which is the basis for the technique of carbon 14 dating of organic material in archaeology.

The **atomic weight** of an atom, or the **molecular weight** of a molecule, is its mass relative to that of a hydrogen atom. This is essentially equal to the number of protons plus neutrons that the atom or molecule contains, because the electrons are so light that they contribute almost nothing to the total mass. Thus the major isotope of carbon has an atomic weight of 12 and is symbolized as ^{12}C. The unstable carbon isotope just mentioned has an atomic weight of 14 and is written as ^{14}C. The mass of an atom or a molecule is often specified in *daltons*, one dalton being an atomic mass unit approximately equal to the mass of a hydrogen atom.

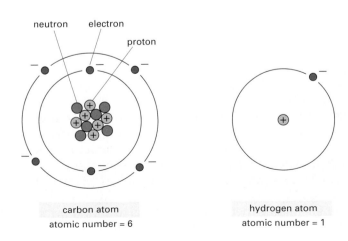

Atoms are so small that it is hard to imagine their size. An individual carbon atom is roughly 0.2 nm in diameter, so that it would take about 5 million of them, laid out in a straight line, to span a millimeter. One proton or neutron weighs approximately $1/(6 \times 10^{23})$ gram. Hydrogen has only one proton, with an atomic weight of one, so one gram of hydrogen contains 6×10^{23} atoms. For carbon, with an atomic weight of twelve, 12 grams of carbon contain 6×10^{23} atoms. This huge number (6×10^{23}, called **Avogadro's number**) is the key scale factor describing the relationship between everyday quantities and numbers of individual atoms or molecules. If a substance has a molecular weight of M, a mass of M grams of the substance will contain 6×10^{23} molecules. This quantity is called one *mole* of the substance (Figure 2–3). The concept of mole is used widely in chemistry as a way to represent the number of molecules that are available to participate in chemical reactions.

There are 92 naturally occurring elements, each differing from the others in the number of protons and electrons in its atoms. Living organisms, however, are made of only a small selection of these elements, four of which—carbon (C), hydrogen (H), nitrogen (N), and oxygen (O)—make up 96.5% of an organism's weight. This composition differs markedly from that of the nonliving inorganic environment (Figure 2–4) and is evidence of a distinctive type of chemistry.

The Outermost Electrons Determine How Atoms Interact

To understand how atoms bond together to form the molecules that make up living organisms, we have to pay special attention to their electrons. Protons and neutrons are welded tightly to one another in the

A mole is X grams of a substance, where X is its relative molecular mass (molecular weight). It will contain 6×10^{23} molecules of the substance.

1 mole of carbon weighs 12 g
1 mole of glucose weighs 180 g
1 mole of sodium chloride weighs 58 g

Molar solutions have a concentration of 1 mole of the substance in 1 liter of solution. A molar solution (1 M) of glucose, for example, has 180 g/l, while a millimolar solution (1 mM) has 180 mg/l.

The standard abbreviation for gram is g; the abbreviation for liter is l.

Figure 2–3 What's a mole? Some sample calculations of moles and molar solutions.

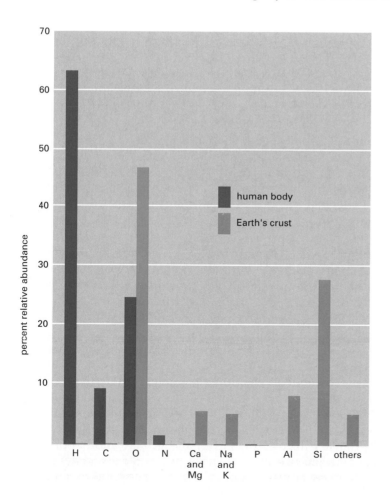

Figure 2–4 The elements abundant in the Earth's crust differ radically from those abundant in the tissues of an animal. The abundance of each element is expressed as a percentage of the total number of atoms present in a sample, including water. Thus, for example, more than 60% of the atoms in a living organism are hydrogen atoms. The relative abundance of elements is similar in all living things.

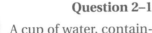
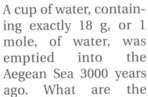

Figure 2–5 An element's chemical reactivity is based on how its outermost electron shell is filled. All of the elements commonly found in living organisms have unfilled outermost shells (*red*) and can thus participate in chemical reactions with other atoms. Inert gases (*yellow*), in contrast, have only filled shells and are chemically unreactive.

atomic number	element	I	II	III	IV
			energy level (electron shell)		
1	Hydrogen	1			
2	Helium	2			
6	Carbon	2	4		
7	Nitrogen	2	5		
8	Oxygen	2	6		
10	Neon	2	8		
11	Sodium	2	8	1	
12	Magnesium	2	8	2	
15	Phosphorus	2	8	5	
16	Sulfur	2	8	6	
17	Chlorine	2	8	7	
18	Argon	2	8	8	
19	Potassium	2	8	8	1
20	Calcium	2	8	8	2

Question 2–1

A cup of water, containing exactly 18 g, or 1 mole, of water, was emptied into the Aegean Sea 3000 years ago. What are the chances that the same quantity of water, scooped today from the Pacific Ocean, would include one of these "Greek" water molecules? Assume perfect mixing and an approximate volume for the world's oceans of 1.5 billion cubic kilometers (1.5×10^9 km³).

nucleus and change partners only under extreme conditions—during radioactive decay, for example, or in the interior of the sun or of a nuclear reactor. In living tissues, only the electrons of an atom undergo rearrangements. They form the accessible part of the atom and specify the rules of chemistry by which atoms combine to form molecules.

Electrons are in continuous motion around the nucleus, but motions on this submicroscopic scale obey different laws from those we are familiar with in everyday life. These laws dictate that electrons in an atom can exist only in certain discrete regions of movement—roughly speaking, discrete orbits—and that there is a strict limit to the number of electrons that can be accommodated in an orbit of a given type, a so-called *electron shell*. The electrons closest on average to the positive nucleus are attracted most strongly to it and occupy the inner, most tightly bound shell. This innermost shell can hold a maximum of two electrons. The second shell is farther away from the nucleus, and its electrons are less tightly bound. This second shell can hold up to eight electrons. The third shell contains electrons that are even less tightly bound; it can also hold up to eight electrons. The fourth and fifth shells can hold 18 electrons each. Atoms with more than four shells are very rare in biological molecules.

The arrangement of electrons in an atom is most stable when all the electrons are in the most tightly bound states that are possible for them—that is, when they occupy the innermost shells, closest to the positively charged nucleus. Therefore, with certain exceptions in the larger atoms, the electrons of an atom fill the shells in order—the first before the second, the second before the third, and so on. An atom whose outermost shell is entirely filled with electrons is especially stable and therefore chemically unreactive. Examples are helium with 2 electrons (and an atomic number of 2), neon with 2 + 8 (atomic number 10), and argon with 2 + 8 + 8 (atomic number 18); these are all inert gases. Hydrogen, by contrast, has only one electron, which leaves its outermost shell half-filled, so it is highly reactive. The atoms found in living tissues all have incomplete outer electron shells and are therefore able to react with one another to form molecules (Figure 2–5).

Because an unfilled electron shell is less stable than a filled one, atoms with incomplete outer shells have a strong tendency to interact with other atoms so as to either gain or lose enough electrons to achieve a completed outermost shell. This electron exchange can be achieved either by transferring electrons from one atom to another or by sharing electrons between two atoms. These two strategies generate the two types of **chemical bonds** that bind atoms to one another: an *ionic bond* is formed when electrons are donated by one atom to another, whereas a *covalent bond* is formed when two atoms share a pair of electrons (Figure 2–6). In the case of the covalent bond, the pair of electrons is often shared unequally, with one atom attracting the shared electrons more than the other; this results in a *polar covalent bond*, as we will discuss later.

An H atom, which needs only one more electron to fill its shell, generally acquires it by sharing—forming one covalent bond with another atom; in many cases this bond is polar. The other most common elements in living cells—C, N, and O, which have an incomplete second shell, and P and S, which have an incomplete third shell (see Figure 2–5)—generally share electrons and achieve a filled outer shell of eight electrons by forming several covalent bonds. The number of electrons an atom must acquire or lose (either by sharing or by transfer) to attain a filled outer shell is known as its *valence*.

Because the state of the outer electron shell determines the chemical properties of an element, when the elements are listed in order of their atomic number we see a periodic recurrence of elements with similar properties: an element with, say, an incomplete second shell containing one electron will behave in much the same way as an element that has filled its second shell and has an incomplete third shell containing one electron. The metals, for example, all have incomplete outer shells with just one or a few electrons, whereas, as we have just seen, the inert gases have full outer shells. This arrangement gives rise to the famous *periodic table* of the elements, which is outlined in Figure 2–7. Elements found in living organisms are highlighted.

Ionic Bonds Form by the Gain and Loss of Electrons

Ionic bonds are most likely to be formed by atoms that have just one or two electrons in their unfilled outer shell or are just one or two electrons short of acquiring a filled outer shell. These atoms can generally attain a completely filled outer electron shell most easily by giving electrons to—or accepting electrons from—another atom, rather than by sharing them. For example, returning to Figure 2–5, we see that a sodium (Na) atom, with atomic number 11, can strip itself down to a filled shell by giving up the single electron external to its second shell. By contrast, a chlorine (Cl) atom, with atomic number 17, can complete its outer shell by gaining just one electron. Consequently, if a Na atom encounters a Cl atom, an electron can jump from the Na to the Cl, leaving both atoms

Question 2–2

A carbon atom contains six protons and six neutrons.

A. What are its atomic number and atomic weight?

B. How many electrons does it have?

C. How many additional electrons must it add to fill its outermost shell? How does this affect carbon's chemical behavior?

D. Carbon with an atomic weight of 14 is radioactive. How does it differ in structure from nonradioactive carbon? How does this difference affect its chemical behavior?

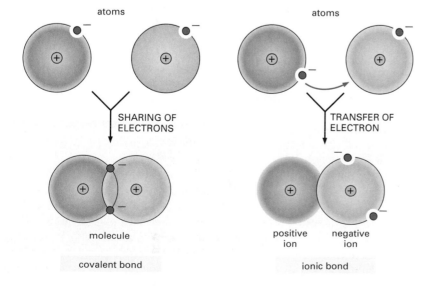

Figure 2–6 Atoms can attain a more stable arrangement of electrons in their outermost shell by interacting with one another. A covalent bond is formed when electrons are shared between atoms. An ionic bond is formed when electrons are transferred from one atom to the other. The two cases shown represent extremes; often, covalent bonds form with a partial transfer (unequal sharing of electrons), resulting in a polar covalent bond (see, for example, Figure 3–12).

Figure 2–7 Elements ordered by their atomic number form the periodic table.

Elements fall into groups that show similar properties based on the number of electrons each element possesses in its outer shell. For example, Mg and Ca tend to give away the two electrons in their outer shells; C, N, and O complete their second shells by sharing electrons. The four elements highlighted in *red* constitute 99% of the total number of atoms present in the human body. An additional seven elements, highlighted in *blue*, together represent about 0.9% of the total. Other elements, shown in *green*, are required in trace amounts by humans. It remains unclear whether those elements shown in *yellow* are essential in humans or not. The chemistry of life, it seems, is therefore predominantly the chemistry of lighter elements.

Atomic weights, given by the sum of the protons and neutrons in the atomic nucleus, can vary with the particular isotope of the element. The atomic weights shown here are those of the most common isotope of each element.

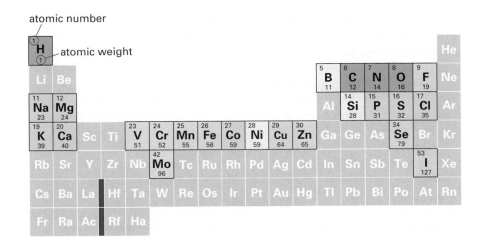

atomic number

atomic weight

with filled outer shells. The offspring of this marriage between sodium, a soft and intensely reactive metal, and chlorine, a toxic green gas, is table salt (NaCl).

When an electron jumps from Na to Cl, both atoms become electrically charged **ions**. The Na atom that lost an electron now has one less electron than it has protons in its nucleus; it therefore has a net single positive charge (Na^+). The Cl atom that gained an electron now has one more electron than it has protons and has a single negative charge (Cl^-). Positive ions are called *cations*, and negative ions, *anions*. Ions can be further classified according to how many electrons are lost or gained. Na and potassium (K) have one electron to lose, so they form cations with a single positive charge (Na^+ and K^+); magnesium and calcium have two electrons to lose and form cations with two positive charges (Mg^{2+} and Ca^{2+}).

Because of their opposite charges, Na^+ and Cl^- are attracted to each other and are thereby held together in an **ionic bond**. A salt crystal contains astronomical numbers of Na^+ and Cl^- packed together in a precise three-dimensional array with their opposite charges exactly balanced—a crystal only 1 mm across contains about 2×10^{19} ions of each type (Figure 2–8). Substances such as NaCl, which are held together solely by ionic bonds, are generally called *salts* rather than molecules.

Figure 2–8 Sodium chloride is held together by ionic bond formation.

(A) An atom of sodium (Na) reacts with an atom of chlorine (Cl). Electrons of each atom are shown in their different energy levels; electrons in the chemically reactive (incompletely filled) shells are shown in *red*. The reaction takes place with transfer of a single electron from sodium to chlorine, forming two electrically charged atoms, or ions, each with complete sets of electrons in their outermost levels. The two ions with opposite charge are held together by electrostatic attraction. (B) The product of the reaction between sodium and chlorine, crystalline sodium chloride, contains sodium and chloride ions packed closely together in a regular array in which the charges are exactly balanced. (C) Color photograph of crystals of sodium chloride.

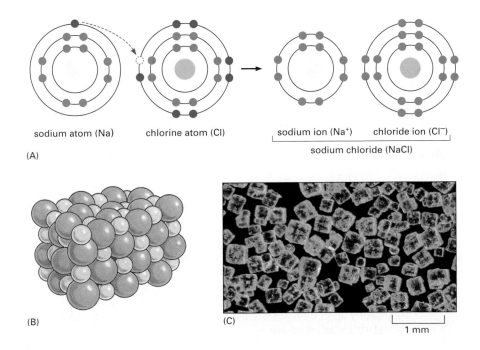

Chapter 2: Chemical Components of Cells

Ionic bonds are just one of several types of *noncovalent bonds* that can exist between atoms. We will meet another example of a noncovalent bond, the *hydrogen bond*, later in this chapter. Because of a favorable interaction between ions and water molecules (which are polar), many salts (including NaCl) are highly soluble in water. They dissociate into individual ions (such as Na^+ and Cl^-), each surrounded by a group of water molecules. For the same reason, the strength of a hydrogen bond between two molecules is significantly reduced in water. In contrast, covalent bond strengths are not affected by an interaction with water.

Covalent Bonds Form by the Sharing of Electrons

All of the characteristics of a cell depend on the molecules it contains. A **molecule** is a cluster of atoms held together by **covalent bonds**, in which electrons are shared rather than transferred between atoms. The shared electrons complete the outer shells of both atoms. In the simplest possible molecule—a molecule of hydrogen (H_2)—two H atoms, each with a single electron, share their two electrons, thus filling their outermost shells. The shared electrons form a cloud of negative charge that is densest between the two positively charged nuclei. This electron density helps to hold the nuclei together by opposing the mutual repulsion between the like charges that would otherwise force them apart. The attractive and repulsive forces are in balance when the nuclei are separated by a characteristic distance, called the *bond length* (Figure 2–9).

Whereas an H atom can form only a single covalent bond, the other common atoms that form covalent bonds in cells—O, N, S, and P, as well as the all-important C—can form more than one. The outermost shells of these atoms, as we have seen, can accommodate up to eight electrons, and they form covalent bonds with as many other atoms as necessary to reach this number. Oxygen, with six electrons in its outer shell, is most stable when it acquires two extra electrons by sharing with other atoms and it therefore forms up to two covalent bonds. Nitrogen, with five outer electrons, forms a maximum of three covalent bonds, while carbon, with four outer electrons, forms up to four covalent bonds—thus sharing four pairs of electrons (see Figure 2–5).

When one atom forms covalent bonds with several others, these multiple bonds have definite orientations in space relative to one another, reflecting the orientations of the orbits of the shared electrons. Covalent bonds between multiple atoms are therefore characterized by specific bond angles as well as bond lengths and bond energies (Figure 2–10). The four covalent bonds that can form around a carbon atom, for example, are arranged as if pointing to the four corners of a regular tetrahedron. The precise orientation of the covalent bonds around carbon is the basis for the three-dimensional geometry of organic molecules.

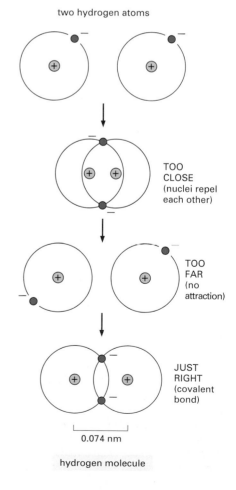

two hydrogen atoms

TOO CLOSE (nuclei repel each other)

TOO FAR (no attraction)

JUST RIGHT (covalent bond)

0.074 nm

hydrogen molecule

Figure 2–9 **The hydrogen molecule is held together by a covalent bond.** Each hydrogen atom in isolation has a single electron, which means that its first (and only) electron shell is incompletely filled. By coming together the two atoms are able to share the two electrons, and each obtains a completely filled first shell, with the shared electrons adopting modified orbits around the two nuclei. The covalent bond between the two atoms has a definite length. If the atoms were closer together, the positive nuclei would repel each other; if they were farther apart than this distance, they would not be able to share electrons as effectively.

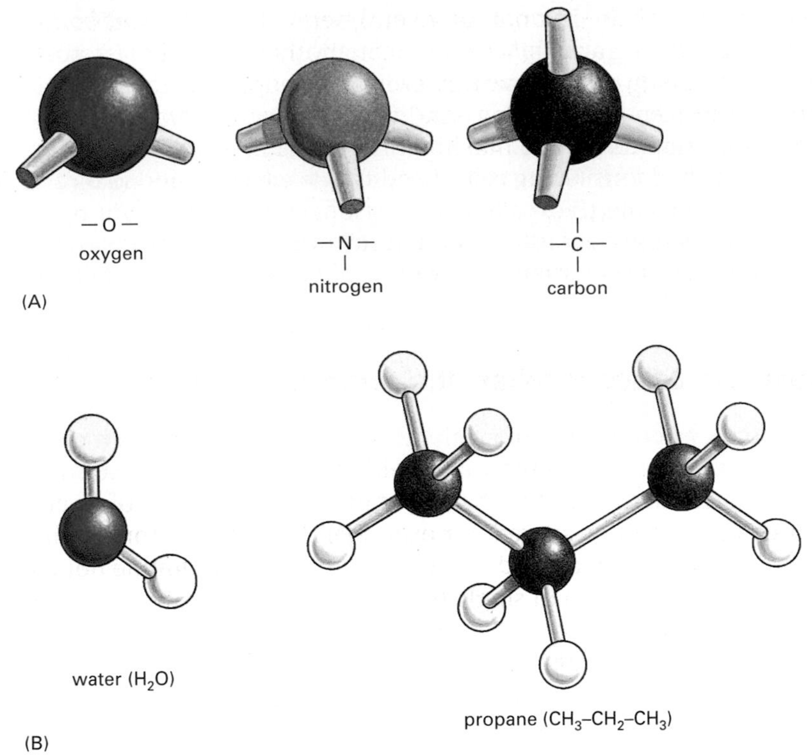

Figure 2–10 Covalent bonds are characterized by particular geometries. (A) The spatial arrangement of the covalent bonds that can be formed by oxygen, nitrogen, and carbon. (B) Molecules formed from these atoms therefore have a precise three-dimensional structure, as shown here for water and propane, defined by the bond angles and bond lengths for each covalent linkage. A water molecule, for example, forms a "V" shape with an angle close to 109°. In these ball-and-stick models, the different *colored balls* are the atoms, and the *sticks* are the covalent bonds. The colors traditionally used to represent the different atoms—*black* for carbon, *white* for hydrogen, *blue* for nitrogen, and *red* for oxygen—were established by the chemist August Wilhelm Hofmann in 1865 when he used a set of colored croquet balls to build molecular models for a public lecture on the "combining power" of atoms.

$-O-$
oxygen

$-N-$
nitrogen

$-C-$
carbon

(A)

water (H_2O)

propane (CH_3–CH_2–CH_3)

(B)

Covalent Bonds Vary in Strength

We have already seen that the covalent bond between two atoms has a characteristic length that depends on the atoms involved. A further crucial property of any bond—covalent or noncovalent—is its strength. *Bond strength* is measured by the amount of energy that must be supplied to break a bond, usually expressed today in units of either kilocalories per mole (kcal/mole) or kilojoules per mole (kJ/mole). A kilocalorie is the amount of energy needed to raise the temperature of one liter of water by one degree centigrade. Thus if 1 kilocalorie of energy must be supplied to break 6×10^{23} bonds of a specific type (that is, 1 mole of these bonds), then the strength of that bond is 1 kcal/mole. The other unit, kJ/mole, derived from the SI units (Système Internationale d'Unités) universally employed by physical scientists, is increasingly accepted by cell biologists. One kilocalorie is equal to about 4.2 kilojoules. Typical strengths and lengths of the main classes of chemical bonds are given in Table 2–1.

To get an idea of what bond strengths mean, it is helpful to compare them with the average energies of the impacts that molecules continually undergo owing to collisions with other molecules in their

Table 2–1 Covalent and noncovalent chemical bonds have different lengths and strengths. Bond strengths are measured by the energy required to break them, in kilocalories or kilojoules per mole (see Glossary for definitions of these units). The length of a hydrogen bond X–H–X is defined as the distance between the two nonhydrogen atoms (X). The bond strengths and lengths listed are approximate, because the exact values will depend on the atoms involved. The different types of noncovalent bonds are described later in the chapter (see Panel 2–7, pp. 78–79).

BOND TYPE		LENGTH (nm)	STRENGTH (kcal/mole)	
			IN VACUUM	IN WATER
Covalent		0.15	90 (377)*	90 (377)
Noncovalent:	ionic	0.25	80 (335)	3 (12.6)
	hydrogen	0.30	4 (16.7)	1 (4.2)
	van der Waals attraction (per atom)	0.35	0.1 (0.4)	0.1 (0.4)

*Values in parentheses are kJ/mole. 1 calorie = 4.184 joules.

environment—their thermal, or heat, energy. Typical covalent bonds are stronger than these thermal energies by a factor of 100, so they are resistant to being pulled apart by thermal motions—heating—and are normally broken only during specific chemical reactions with other atoms and molecules. The making and breaking of covalent bonds are violent events, and in living cells these events are carefully controlled by highly specific catalysts, called *enzymes*. Noncovalent bonds as a rule are much weaker; we shall see later that they are critically important in the cell in the many situations where molecules have to associate and dissociate readily to carry out their functions.

There Are Different Types of Covalent Bonds

Most covalent bonds involve the sharing of two electrons, one donated by each participating atom; these are called *single bonds*. Some covalent bonds, however, involve the sharing of more than one pair of electrons. Four electrons can be shared, for example, two coming from each participating atom; such a bond is called a *double bond*. Double bonds are shorter and stronger than single bonds and have a characteristic effect on the three-dimensional geometry of molecules containing them. A single covalent bond between two atoms generally allows the rotation of one part of a molecule relative to the other around the bond axis. A double bond prevents such rotation, producing a more rigid and less flexible arrangement of atoms (Figure 2–11). This restriction has a major influence on the three-dimensional shape of many macromolecules. Panel 2–1 (pp. 66–67) reviews the chemical bonds commonly encountered in biological molecules.

Some molecules contain atoms that share electrons in a way that produces bonds that are intermediate in character between single and double bonds. The highly stable benzene molecule, for example, is made up of a ring of six carbon atoms in which the bonding electrons are evenly distributed (although the arrangement is sometimes depicted as an alternating sequence of single and double bonds, as shown in Panel 2–1).

When the atoms joined by a single covalent bond belong to different elements, the two atoms usually attract the shared electrons to different degrees. Compared with a C atom, for example, O and N atoms attract electrons relatively strongly, whereas an H atom attracts electrons relatively weakly (because of the relative differences in their positive charges). By definition, a **polar** structure (in the electrical sense) is one in which the positive charge is concentrated toward one end of the molecule (the positive pole) and the negative charge is concentrated toward the other end (the negative pole). Covalent bonds in which the electrons are shared unequally in this way are therefore known as polar covalent bonds. For example, the covalent bond between oxygen and hydrogen, –O–H, or between nitrogen and hydrogen, –N–H, is polar, whereas the bond between carbon and hydrogen, –C–H, has the electrons attracted much more equally by both atoms and is relatively nonpolar (Figure 2–12).

Polar covalent bonds are extremely important in biology because they allow molecules to interact through electrical forces. Any large molecule with many polar groups will have a pattern of partial positive

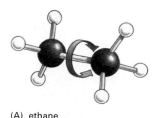

(A) ethane

(B) ethene

Figure 2–11 Carbon–carbon double bonds are shorter and more rigid than C–C single bonds. (A) The ethane molecule, with a single covalent bond between the two carbon atoms, shows the tetrahedral arrangement of single covalent bonds formed by carbon. One of the CH$_3$ groups joined by the covalent bond can rotate relative to the other around the bond axis. (B) The double bond between the two carbon atoms in a molecule of ethene (ethylene) alters the bond geometry of the carbon atoms and brings all the atoms into the same plane; the double bond prevents the rotation of one CH$_2$ group relative to the other.

water

oxygen

Figure 2–12 In polar covalent bonds, the electrons are shared unequally. Comparison of electron distributions in polar molecules such as water (H$_2$O) and nonpolar molecules such as oxygen (O$_2$). δ^+ indicates partial positive charge; δ^- indicates partial negative charge.

Question 2–3

Discuss whether the following statement is correct: "An ionic bond can, in principle, be thought of as a very polar covalent bond. Polar covalent bonds, then, fall somewhere between ionic bonds at one end of the spectrum and nonpolar covalent bonds at the other end."

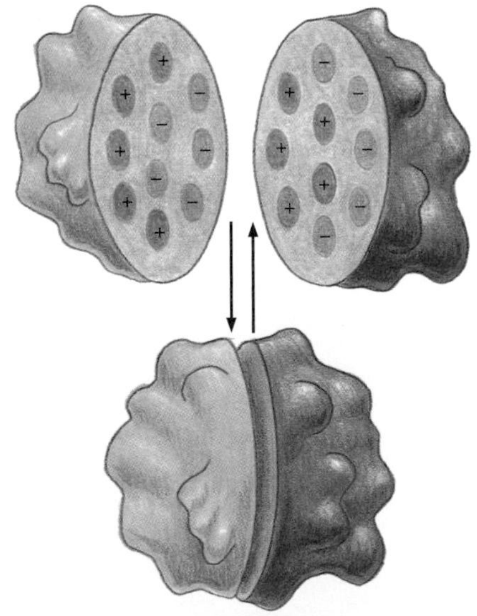

and negative charges on its surface. When such a molecule encounters a second molecule with a complementary set of charges, the two will be attracted to each other by weak noncovalent ionic bonds that resemble (but are weaker than) the ionic bonds that hold together salts such as NaCl. When enough of these weak noncovalent bonds form between two large molecules, their surfaces will stick specifically to each other, as illustrated in Figure 2–13.

Water Is Held Together by Hydrogen Bonds

Water accounts for about 70% of a cell's weight, and most intracellular reactions occur in an aqueous environment. Life on Earth is thought to have begun in the ocean, and the conditions in that primeval environment put a permanent stamp on the chemistry of living things. Life therefore hinges on the properties of water.

In each molecule of water (H_2O) the two H atoms are linked to the O atom by covalent bonds. The two bonds are highly polar because the O is strongly attractive for electrons, whereas the H is only weakly attractive. Consequently, there is an unequal distribution of electrons in a water molecule, with a preponderance of positive charge on the two H atoms and negative charge on the O (see Figure 2–12 and Panel 2–2, pp. 68–69). When a positively charged region of one water molecule (that is, one of its H atoms) comes close to a negatively charged region (that is, the O) of a second water molecule, the electrical attraction between them can establish a weak bond called a **hydrogen bond**. These bonds are much weaker than covalent bonds and are easily broken by the random thermal motions due to the heat energy of the molecules, so each bond lasts only an exceedingly short time. But the combined effect of many weak bonds is far from trivial. Each water molecule can form hydrogen bonds through its two H atoms to two other water molecules, producing a network in which hydrogen bonds are being continually broken and formed. It is because of these interlocking hydrogen bonds that water at room temperature is a liquid—with a high boiling point and high surface tension—and not a gas. Without hydrogen bonds, life as we know it could not exist. The biologically significant properties of water are reviewed in Panel 2–2.

Not all hydrogen atoms form hydrogen bonds. In general, a hydrogen bond can form whenever a positively charged H held in one molecule by a polar covalent linkage comes close to a negatively charged atom—typically an oxygen or a nitrogen—belonging to another molecule. Hydrogen bonds can also occur between different parts of a single large molecule, where they often help create special shapes. But the hydrogen bond is just one member of a family of weak noncovalent bonds that play a crucial role in allowing large molecules to fold up in unique ways and to bind selectively to other molecules, as we will discuss later in this chapter.

Molecules, such as alcohols, that contain polar bonds and that can form hydrogen bonds mix well with water. As mentioned previously, molecules carrying positive or negative charges (ions) likewise dissolve readily in water. Such molecules are termed **hydrophilic**, meaning that they are "water-loving." A large proportion of the molecules in the aqueous environment of a cell necessarily fall into this category, including sugars, DNA, RNA, and a majority of proteins. **Hydrophobic** (water-fearing) molecules, by contrast, are uncharged and form few or no hydrogen bonds, and so do not dissolve in water. Hydrocarbons are an important example of hydrophobic cellular constituents (see Panel 2–1, pp. 66–67). In these molecules the H atoms are covalently linked to C atoms by a largely nonpolar bond. Because the H atoms have almost no

Figure 2–13 Proteins can bind to one another through complementary charges on their surfaces.

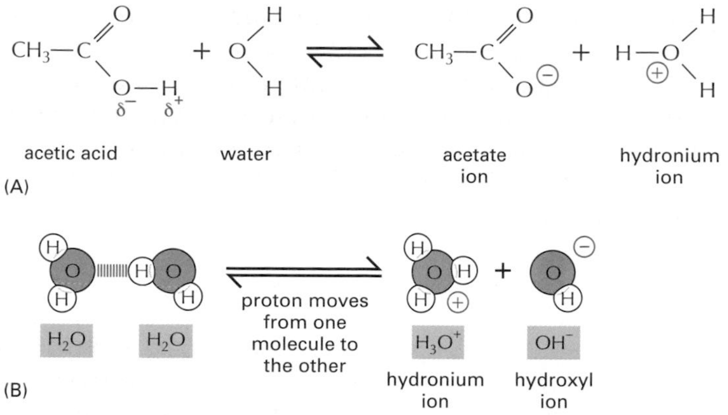

acetic acid   water   acetate ion   hydronium ion

(A)

H_2O H_2O proton moves from one molecule to the other H_3O^+ hydronium ion OH^- hydroxyl ion

(B)

Figure 2–14 Protons are on the move in aqueous solutions. (A) The reaction that takes place when a molecule of acetic acid dissolves in water. (B) Water molecules are continually exchanging protons with each other to form hydronium and hydroxyl ions. These ions in turn rapidly recombine to form water molecules.

net positive charge, they cannot form effective hydrogen bonds to other molecules. This makes the hydrocarbon as a whole hydrophobic—a property that is exploited by cells, whose membranes are constructed from molecules that have long hydrocarbon tails, as we shall see in Chapter 11. Because they do not dissolve in water, the hydrophobic hydrocarbons can form the thin membrane barriers that keep the aqueous interior of the cell separate from the surrounding, also aqueous, environment.

Some Polar Molecules Form Acids and Bases in Water

One of the simplest kinds of chemical reaction, and one that has profound significance in cells, takes place when a molecule possessing a highly polar covalent bond between a hydrogen and another atom dissolves in water. The hydrogen atom in such a molecule has given up its electron almost entirely to the companion atom, and so exists as an almost naked positively charged hydrogen nucleus—in other words, a *proton (H+)*. When the polar molecule becomes surrounded by water molecules, the proton will be attracted to the partial negative charge on the O atom of an adjacent water molecule; this proton can dissociate from its original partner and associate instead with the oxygen atom of the water molecule, generating a **hydronium ion** (H_3O^+) (Figure 2–14A). The reverse reaction also takes place very readily, so one has to imagine an equilibrium state in which billions of protons are constantly flitting to and fro, between one molecule in the aqueous solution and another.

Substances that release protons when they dissolve in water, thus forming H_3O^+, are termed **acids**. The higher the concentration of H_3O^+, the more acidic the solution. H_3O^+ is present even in pure water, at a concentration of 10^{-7} M, as a result of the movement of protons from one water molecule to another (Figure 2–14B). By tradition, the H_3O^+ concentration is usually referred to as the H^+ concentration, even though most protons in an aqueous solution are present as H_3O^+.

Acids are characterized as being strong or weak, depending on how readily they give up their protons to water. Strong acids, such as HCl, lose their protons quickly. Acetic acid, on the other hand, is a weak acid because it tends to hold onto its proton when dissolved in water. Many of the acids important in the cell—such as molecules containing a carboxyl (COOH) group—are weak acids (see Panel 2–2, pp 68–69). Their tendency to dissociate with some reluctance is a useful characteristic in the cellular environment.

Because the proton of a hydronium ion can be passed readily to many types of molecules in cells, altering their character, the concentration of H_3O^+ inside a cell (the acidity) must be closely regulated. Acids—especially weak acids—will give up their protons more readily if

Question 2–4

What, if anything, is wrong with the following statement: "When NaCl is dissolved in water, the water molecules closest to the ions will tend to preferentially orient themselves so that their oxygen atoms face the sodium ions and face away from the chloride ions." Explain your answer.

the concentration of H_3O^+ in solution is low and will tend to receive them back if the concentration in solution is high.

The opposite of an acid is a **base**. Any molecule capable of accepting a proton is called a base. Just as the defining property of an acid is that it raises the concentration of H_3O^+ ions by donating a proton to a water molecule, so the defining property of a base is that it raises the concentration of hydroxyl (OH^-) ions by removing a proton from a water molecule. Thus sodium hydroxide (NaOH) is basic (the term *alkaline* is also used) because it dissociates in aqueous solution to form Na^+ ions and OH^- ions. Because NaOH dissociates readily in water, it is called a strong base. More important in living cells, however, are the weak bases—those that have a weak tendency to reversibly accept a proton from water. Many biologically important weak bases contain an amino (NH_2) group. This group can generate OH^- by taking a proton from water: $-NH_2 + H_2O \rightarrow -NH_3^+ + OH^-$ (see Panel 2–2, pp. 68–69).

Because an OH^- ion combines with a H_3O^+ ion to form two water molecules, an increase in the OH^- concentration forces a decrease in the concentration of H_3O^+, and vice versa. A pure solution of water contains an equally low concentration (10^{-7} M) of both ions; it is neither acidic nor basic and is therefore said to be *neutral*.

To avoid the use of unwieldy numbers, the concentration of H_3O^+ is expressed using a logarithmic scale called the **pH scale**, as illustrated in Panel 2–2. Pure water has a pH of 7.0 and the inside of cells is also kept close to neutrality. Acidic solutions have a pH < 7, and basic solutions a pH > 7.

Question 2–5

A. Are there any H_3O^+ ions present in pure water at neutral pH (i.e., at pH = 7.0)? If so, how are they formed?

B. If they exist, what is the ratio of H_3O^+ ions to H_2O molecules at neutral pH? (Hint: the molecular weight of water is 18, and 1 liter of water weighs 1 kg.)

Molecules in Cells

Having looked at the ways atoms combine into small molecules, and how these molecules behave in an aqueous environment, we now examine the main classes of small molecules found in cells and their biological roles. Amazingly, we will see that a few basic categories of molecules, formed from a handful of different elements, give rise to all the extraordinary richness of form and behavior shown by living things.

A Cell Is Formed from Carbon Compounds

If we disregard water, nearly all of the molecules in a cell are based on carbon. Carbon is outstanding among all the elements in its ability to form large molecules; silicon—an element with the same electron configuration in its outer shell—is a poor second. Because carbon is small and has four electrons and four vacancies in its outermost shell, a carbon atom can form four covalent bonds with other atoms. Most important, one carbon atom can join to other carbon atoms through highly stable covalent C–C bonds to form chains and rings and hence generate large and complex molecules with no obvious upper limit to their size (see Panel 2–1, pp. 66–67). The small and large carbon compounds made by cells are called *organic molecules*. All other molecules, including water, are said to be *inorganic* by contrast.

Certain combinations of atoms, such as the methyl ($-CH_3$), hydroxyl ($-OH$), carboxyl ($-COOH$), carbonyl ($-C=O$), phosphoryl ($-PO_3^{2-}$), and amino ($-NH_2$) groups, occur repeatedly in organic molecules. Each such group has distinct chemical and physical properties that influence the behavior of the molecule in which the group occurs—whether they tend to gain or lose protons and which molecules they interact with, for example. Becoming familiar with these groups and their chemical properties greatly simplifies one's view of the chemistry

Table 2–2 The Approximate Chemical Composition of a Bacterial Cell

	PERCENT OF TOTAL CELL WEIGHT	NUMBER OF TYPES OF EACH MOLECULE
Water	70	1
Inorganic ions	1	20
Sugars and precursors	1	250
Amino acids and precursors	0.4	100
Nucleotides and precursors	0.4	100
Fatty acids and precursors	1	50
Other small molecules	0.2	~300
Macromolecules (proteins, nucleic acids, and polysaccharides)	26	~3000

of life. The most common **chemical groups** and some of their properties are summarized in Panel 2–1 (pp. 66–67).

Cells Contain Four Major Families of Small Organic Molecules

The small organic molecules of the cell are carbon compounds with molecular weights in the range 100 to 1000 that contain up to 30 or so carbon atoms. They are usually found free in solution in the cytoplasm and have many different fates. Some are used as *monomer* subunits to construct the giant polymeric *macromolecules*—the proteins, nucleic acids, and large polysaccharides—of the cell. Others act as energy sources and are broken down and transformed into other small molecules in a maze of intracellular metabolic pathways. Many small molecules have more than one role in the cell—acting, for example, as both a potential subunit for a macromolecule and as an energy source. It is critical to recognize that small organic molecules are much less abundant than the organic macromolecules in living organisms, accounting for only about one-tenth of the total mass of organic matter in a cell (Table 2–2). As a rough guess, there may be a thousand different kinds of these small molecules in a typical cell.

All organic molecules are synthesized from—and are broken down into—the same set of simple compounds. Both their synthesis and their breakdown occur through sequences of simple chemical changes that are limited in variety and follow definite step-by-step rules. As a consequence, the compounds in a cell are chemically related and most can be classified into a small number of distinct families. Broadly speaking, cells contain four major families of small organic molecules: the *sugars*, the *fatty acids*, the *amino acids*, and the *nucleotides* (Figure 2–15). Although many compounds present in cells do not fit into these

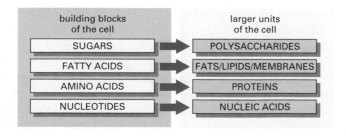

Figure 2–15 **Sugars, fatty acids, amino acids, and nucleotides constitute the four main families of small organic molecules in cells.** They form the monomeric building blocks, or subunits, for most of the macromolecules and other assemblies of the cell. Some, like the sugars and the fatty acids, are also energy sources.

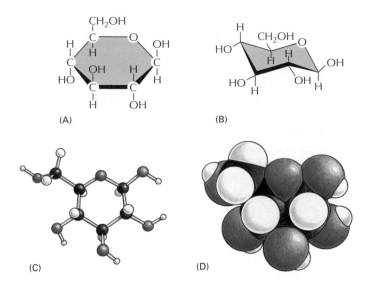

Figure 2–16 The structure of glucose, a simple sugar, can be represented in several ways. In the structural formulas shown in (A), the atoms are shown as chemical symbols linked together by *solid lines,* which represent the covalent bonds. The *thickened lines* are used to indicate the plane of the sugar ring and to show that the –H and –OH groups are not in the same plane as the ring. (B) Another kind of structural formula that shows the three-dimensional structure of glucose in the so-called "chair configuration." (C) A ball-and-stick model in which the three-dimensional arrangement of the atoms in space is indicated. (D) A space-filling model, which, as well as depicting the three-dimensional arrangement of the atoms, also gives some idea of their relative sizes and of the surface contours of the molecule. The atoms in (C) and (D) are colored as follows: C, *black;* H, *white;* O, *red.* This is the conventional color coding for these atoms (see Figure 2–10) and will be used throughout this book.

Question 2–6

Have a close look at the ball-and-stick and the space-filling representations of the glucose molecule shown in Figure 2–16C and D. Note that in both illustrations there are hydrogen atoms of two different sizes. Do we need to apologize because the artist made a mistake? Explain your answer.

categories, these four families of small organic molecules, together with the macromolecules made by linking them into long chains, account for a large fraction of a cell's mass (see Table 2–2).

Sugars Are Energy Sources for Cells and Subunits of Polysaccharides

The simplest **sugars**—the *monosaccharides*—are compounds with the general formula $(CH_2O)_n$, where n is usually 3, 4, 5, or 6. Sugars, and the molecules made from them, are also called *carbohydrates* because of this simple formula. Glucose, for example, has the formula $C_6H_{12}O_6$ (Figure 2–16). The formula, however, does not fully define the molecule: the same set of carbons, hydrogens, and oxygens can be joined together by covalent bonds in a variety of ways, creating structures with different shapes. Glucose, for example, can be converted into a different sugar—mannose or galactose—simply by switching the orientations of specific OH groups relative to the rest of the molecule (Panel 2–3, pp. 70–71). Each of these sugars, moreover, can exist in either of two forms, called the D-form and the L-form, which are mirror images of each other. Sets of molecules with the same chemical formula but different structures are called *isomers,* and mirror-image pairs of molecules are called *optical isomers.* Isomers are widespread among organic molecules in general, and they play a major part in generating the enormous variety of sugars. A more complete outline of sugar structures and chemistry is presented in Panel 2–3.

Monosaccharides can be linked by covalent bonds to form larger carbohydrates. Two monosaccharides linked together make a disaccharide, such as sucrose, which is composed of a glucose and a fructose unit. Larger sugar polymers range from the *oligosaccharides* (trisaccharides, tetrasaccharides, and so on) up to giant *polysaccharides,* which can contain thousands of monosaccharide units. In most cases, the prefix "oligo-" is used to refer to macromolecules made of a small number of monomers, between 3 and 50 or so. Polymers, in contrast, can contain hundreds or thousands of subunits.

The way that sugars are linked together illustrates some common features of biochemical bond formation. A bond is formed between an –OH group on one sugar and an –OH group on another by a **condensation reaction**, in which a molecule of water is expelled as the bond is formed (Figure 2–17). Subunits in other biological polymers, such as nucleic acids and proteins, are also linked by condensation reactions in which water is expelled. The bonds created by all of these condensation reactions can be broken by the reverse process of **hydrolysis**, in which a molecule of water is consumed (see Figure 2–17).

Because each monosaccharide has several free hydroxyl groups that can form a link to another monosaccharide (or to some other compound), sugar polymers can be branched, and the number of possible polysaccharide structures is extremely large. For this reason it is much more difficult to determine the arrangement of sugars in a polysaccharide than to determine the nucleotide sequence of a DNA molecule, where each unit is joined to the next in exactly the same way.

The monosaccharide *glucose* has a central role as an energy source for cells. It is broken down to smaller molecules in a series of reactions, releasing energy that the cell can harness to do useful work, as we will explain in Chapter 13. Cells use simple polysaccharides composed only of glucose units—principally *glycogen* in animals and *starch* in plants—as long-term stores of glucose, held in reserve for energy production.

Sugars do not function exclusively in the production and storage of energy. They are also used, for example, to make mechanical supports. The most abundant organic molecule on Earth—the *cellulose* that forms plant cell walls—is a polysaccharide of glucose. Another extraordinarily abundant organic substance, the *chitin* of insect exoskeletons and fungal cell walls, is also a polysaccharide—in this case a linear polymer of a sugar derivative called *N*-acetylglucosamine (see Panel 2–3, pp. 70–71). Other polysaccharides, with their tendency to be slippery when wet, are the main components of slime, mucus, and gristle.

Smaller oligosaccharides can be covalently linked to proteins to form glycoproteins, or to lipids to form *glycolipids* (Panel 2–4, pp. 72–73), which are both found in cell membranes. The surfaces of most cells are decorated with sugar polymers that belong to glycoproteins and glycolipids in the plasma membrane. These sugar side chains are often recognized selectively by other cells. Differences in the types of cell-surface sugars form the molecular basis for different human blood groups.

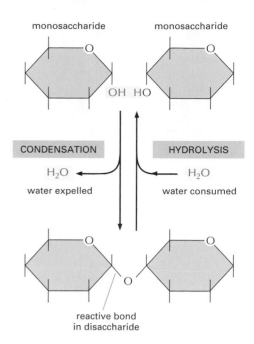

Figure 2–17 Two monosaccharides can be linked to form a disaccharide. This reaction belongs to a general category of reactions termed *condensation reactions,* in which two molecules join together due to the loss of a water molecule. The reverse reaction (in which water is added) is termed *hydrolysis.*

Fatty Acids Are Components of Cell Membranes

A **fatty acid** molecule, such as *palmitic acid* (Figure 2–18), has two chemically distinct regions. One is a long hydrocarbon chain, which is

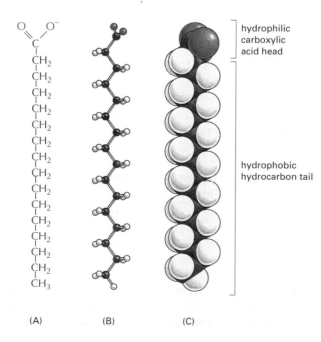

Figure 2–18 Fatty acids have both hydrophobic and hydrophilic components. The hydrophobic hydrocarbon chain is attached to a hydrophilic carboxylic acid group. Palmitic acid is shown here. Different fatty acids have different hydrocarbon tails. (A) Structural formula. The carboxylic acid head group is shown in its ionized form. (B) Ball-and-stick model. (C) Space-filling model.

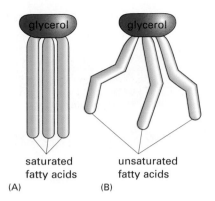

saturated
fatty acids

(A)

unsaturated
fatty acids

(B)

Figure 2–19 The properties of fats depend on the fatty acid side chains they carry. Fatty acids are stored in the cytoplasm of many cells in the form of droplets of *triacylglycerol* compounds made of three fatty acid chains joined to a glycerol molecule. (A) Saturated fats, such as tristearate, are found in meat and dairy products. The lack of double bonds in the fatty acid chains allows these molecules to pack together tightly, which is why butter and lard are solid at room temperature. (B) Plant oils, such as the corn oil, contain unsaturated fatty acids,which may be monounsaturated (contains one double bond) or polyunsaturated (containing multiple double bonds). The double bonds produce kinks in the fatty acid chains that prevents the fats from packing close together; for this reason, plant oils are liquid at room temperature. Although fats are essential in the diet, saturated fats raise the concentrations of cholesterol in the blood and can cause arteries to become clogged with fat, a condition that can lead to heart disease.

hydrophobic and not very reactive chemically. The other is a carboxyl (–COOH) group, which behaves as an acid (carboxylic acid): it is ionized in solution (–COO⁻), extremely hydrophilic, and chemically reactive. Almost all the fatty acid molecules in a cell are covalently linked to other molecules by their carboxylic acid group (see Panel 2–4, pp. 72–73). Molecules such as fatty acids, which possess both hydrophobic and hydrophilic regions, are termed *amphipathic*.

The hydrocarbon tail of palmitic acid is *saturated:* it has no double bonds between its carbon atoms and contains the maximum possible number of hydrogens. Stearic acid, another one of the common fatty acids in animal fat, is also saturated. Some other fatty acids, such as oleic acid, have *unsaturated* tails, with one or more double bonds along their length. The double bonds create kinks in the molecules, interfering with their ability to pack together in a solid mass. How tightly the fatty acids found in cell membranes pack affects the fluidity of the membrane. And it is the absence or presence of these double bonds that accounts for the difference between hard (saturated) and soft (polyunsaturated) margarine. The many different fatty acids found in cells differ only in the length of their hydrocarbon chains and in the number and position of the carbon–carbon double bonds (see Panel 2–4).

Fatty acids serve as a concentrated food reserve in cells: they can be broken down to produce about six times as much usable energy, weight for weight, as glucose. Fatty acids are stored in the cytoplasm of many cells in the form of droplets of *triacylglycerol* molecules—compounds made of three fatty acid chains joined to a glycerol molecule (see Panel 2–4). These molecules are the animal fats found in meat, butter, and cream, and the plant oils such as corn oil and olive oil (Figure 2–19). When a cell needs energy, the fatty acid chains can be released from triacylglycerols and broken down into two-carbon units. These two-carbon units are identical to those derived from the breakdown of glucose, and they enter the same energy-yielding reaction pathways, as will be described in Chapter 13.

Fatty acids and their derivatives, including triacylglycerols, are examples of *lipids*. This class of biological molecules is a loosely defined collection with the common feature that they are insoluble in water and soluble in fat and organic solvents such as benzene. Lipids typically contain long hydrocarbon chains, as in the fatty acids and *isoprenes*—or multiple linked aromatic rings, as in the *steroids* (see Panel 2–4).

The most important function of fatty acids in cells is in the construction of cell membranes. These thin sheets enclose all cells and

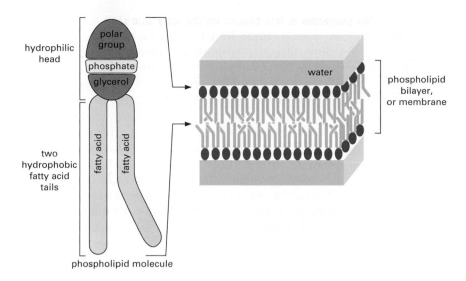

Figure 2–20 **Phospholipids aggregate to form cell membranes.** In an aqueous environment, the hydrophobic tails of phospholipids pack together to exclude water, forming a bilayer with the hydrophilic head of each phospholipid facing the aqueous environment.

surround their internal organelles. They are composed largely of *phospholipids*, which are small molecules that, like triacylglycerols, are constructed mainly from fatty acids and glycerol. In phospholipids the glycerol is joined to two fatty acid chains, rather than to three as in triacylglycerols. The "third" site on the glycerol is linked to a hydrophilic phosphate group, which is in turn attached to a small hydrophilic compound such as choline (see Panel 2–4, pp. 72–73). Phospholipids are strongly amphipathic: each phospholipid molecule has a hydrophobic tail, composed of the two fatty acid chains, and a hydrophilic head, where the phosphate is located. This gives them different physical and chemical properties from triacylglycerols, which are predominantly hydrophobic. Other lipids present in the cell membrane contain one or more sugars instead of a phosphate group. Several of these *glycolipids* play an important role in intracellular cell signaling, as we will see in Chapter 16.

The membrane-forming property of phospholipids results from their amphipathic nature. Phospholipids will spread over the surface of water to form a monolayer of phospholipid molecules, with the hydrophobic tails facing the air and the hydrophilic heads in contact with the water. Two such molecular layers can readily combine tail-to-tail in water to make a phospholipid sandwich, or *lipid bilayer*, which forms the structural basis of all cell membranes (Figure 2–20; discussed further in Chapter 11).

Amino Acids Are the Subunits of Proteins

Amino acids are a varied class of molecules with one defining property: they all possess a carboxylic acid group and an amino group, both linked to the same carbon atom called the α-carbon (Figure 2–21). Their chemical variety comes from the side chain that is also attached to the

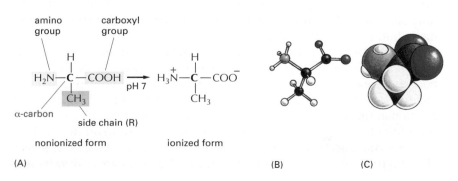

Figure 2–21 **Alanine is one of the simplest amino acids.** (A) In the cell, where the pH is close to 7, the free amino acid exists in its ionized form; but when it is incorporated into a polypeptide chain, the charges on the amino and carboxyl groups disappear. (B) A ball-and-stick model and (C) a space-filling model of alanine (C, *black*; H, *white*; O, *red*; N, *blue*).

α-carbon. Cells use amino acids to build **proteins**, which are polymers of amino acids joined head-to-tail in a long chain that is then folded into a three-dimensional structure unique to each type of protein.

The covalent linkage between two adjacent amino acids in a protein chain is called a *peptide bond*; the chain of amino acids is also known as a polypeptide (Figure 2–22). Peptide bonds are formed by condensation reactions that link one amino acid to the next. Regardless of the specific amino acids from which it is made, the polypeptide always has an amino (NH_2) group at one end (its *N-terminus*) and a carboxyl (COOH) group at its other end (its *C-terminus*). This gives a protein or polypeptide a definite directionality—a structural (as opposed to electrical) polarity.

Twenty types of amino acids are commonly found in proteins, each with a different side chain attached to the α-carbon atom (Panel 2–5, pp. 74–75). The same 20 amino acids occur over and over again in all proteins, whether they hail from bacteria, plants, or animals. How this precise set of 20 amino acids came to be chosen is one of the mysteries surrounding the evolution of life; there is no obvious chemical reason why other amino acids could not have served just as well. But once the selection had been locked into place, it could not be changed; too much chemistry had evolved to exploit it. Switching the types of amino acids used by cells would require every living creature to retool its entire metabolism, and genetic code, to cope with the new building blocks.

Like sugars, all amino acids (except glycine) exist as optical isomers in D- and L-forms (see Panel 2–5). But only L-forms are ever found in proteins (although D-amino acids occur as part of bacterial cell walls and in some antibiotics). The origin of this exclusive use of L-amino acids to make proteins is another evolutionary mystery.

The chemical versatility that the 20 standard amino acids provide is vitally important to the function of proteins. Five of the 20 amino acids have side chains that can form ions in solution and can therefore carry a charge (lysine and glutamic acid, for example, shown in Figure 2–22). The others are uncharged. Some amino acids are polar and hydrophilic, and some are nonpolar and hydrophobic (see Panel 2–5). As we will discuss in Chapter 4, the collective properties of the amino acid side chains underlie all the diverse and sophisticated functions of proteins.

Nucleotides Are the Subunits of DNA and RNA

A **nucleoside** is a molecule made of a nitrogen-containing ring compound linked to a five-carbon sugar, which can be either ribose or deoxyribose (Panel 2–6, pp. 76–77). A nucleoside sporting one or more phosphate groups attached to its sugar is called a **nucleotide**. Nucleotides containing ribose are known as ribonucleotides, and those containing deoxyribose as deoxyribonucleotides.

The nitrogen-containing rings are generally referred to as *bases* for historical reasons: under acidic conditions they can each bind a H^+ (proton) and thereby increase the concentration of OH^- ions in aqueous solution. There is a strong family resemblance between the different nucleotide bases. *Cytosine* (C), *thymine* (T), and *uracil* (U) are called *pyrimidines* because they all derive from a six-membered pyrimidine ring; *guanine* (G) and *adenine* (A) are *purine* compounds, which bear a second, five-membered ring fused to the six-membered ring. Each nucleotide is named after the base it contains (see Panel 2–6).

Nucleotides can act as short-term carriers of chemical energy. Above all others, the ribonucleotide **adenosine triphosphate**, or **ATP** (Figure 2–23), participates in the transfer of energy in hundreds of cellular reactions. ATP is formed through reactions that are driven by the

Question 2–7

Why do you suppose only L-amino acids and not a random mixture of the L- and D-forms of each amino acid are used to make proteins?

N-terminus of
polypeptide chain

Phe

Ser

Glu

Lys

C-terminus of
polypeptide chain

Figure 2–22 Proteins are held together by peptide bonds. The four amino acid residues shown are linked together by three peptide bonds, one of which is highlighted in *yellow*. One of the amino acids is shaded in *gray*. The amino acid side chains are shown in *red*. The two ends of a polypeptide chain are chemically distinct. One end, the N-terminus, is capped by an amino group, and the other, the C-terminus, ends in a carboxyl group. The sequence of amino acid residues in a protein or polypeptide is abbreviated using either a three-letter or a one-letter code, and the sequence is always read from the N-terminus (see Panel 2–5). In the example given, the sequence is Phe-Ser-Glu-Lys (or FSEK).

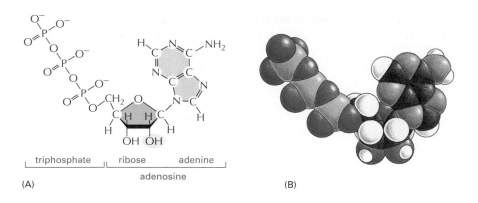

Figure 2–23 Adenosine triphosphate (ATP) is a nucleotide whose reactivity resides in its terminal phosphate groups. (A) Structural formula. (B) Space-filling model. In (B) the colors of the atoms are C, *black;* H, *white;* N, *blue;* O, *red;* and P, *green.* The deoxyribonucleotide version of adenosine triphosphate (dATP) differs only in that a hydrogen atom replaces the hydroxyl group shaded in *yellow* in (A).

energy released by the oxidative breakdown of foodstuffs. Its three phosphates are linked in series by two *phosphoanhydride bonds* (see Panel 2–6, pp. 76–77). Rupture of these bonds releases large amounts of useful energy. The terminal phosphate group in particular is frequently split off by hydrolysis (Figure 2–24). In many situations, transfer of this phosphate to other molecules releases energy that drives energy-requiring biosynthetic reactions. Other nucleotide derivatives serve as carriers for the transfer of other chemical groups, as will be described in Chapter 3.

The most fundamental role of nucleotides in the cell is in the storage and retrieval of biological information. Nucleotides serve as building blocks for the construction of *nucleic acids*—long polymers in which nucleotide subunits are covalently linked by the formation of a *phosphodiester bond* between the phosphate group attached to the sugar of one nucleotide and a hydroxyl group on the sugar of the next nucleotide (Figure 2–25). Nucleic acid chains are synthesized from energy-rich nucleoside triphosphates by a condensation reaction that releases inorganic pyrophosphate during phosphodiester bond formation (see Panel 2–6).

There are two main types of nucleic acids, which differ in the type of sugar they use in their sugar-phosphate backbone. Those based on the sugar ribose are known as **ribonucleic acids**, or **RNA**, and contain the bases A, G, C, and U. Those based on deoxyribose (in which the hydroxyl at the 2′ position of the ribose carbon ring is replaced by a hydrogen; see Panel 2–6) are known as **deoxyribonucleic acids**, or **DNA**,

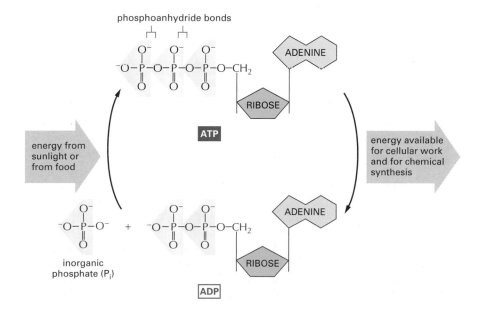

Figure 2–24 ATP serves as an energy carrier in cells. The energy-requiring formation of ATP from ADP and inorganic phosphate is coupled to the energy-yielding oxidation of foodstuffs (in animal cells, fungi, and some bacteria) or to the capture of light (in plant cells and some bacteria). The hydrolysis of this ATP back to ADP and inorganic phosphate in turn provides the energy to drive many cellular reactions. Together these reactions form the ATP cycle.

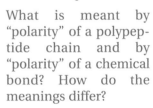

5' end

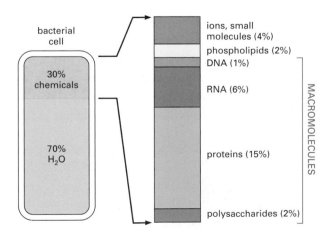

G

A

T

C

3' end

Figure 2–25 A short length of one chain of a deoxyribonucleic acid (DNA) molecule shows the bonds linking four consecutive nucleotide residues. One of the phosphates linking adjacent nucleotides by phosphodiester bonds is highlighted in *yellow,* and one of the nucleotides is enclosed in a *gray box.* Nucleotides are joined by a phosphodiester linkage between specific carbon atoms of the sugar ring, known as the 5' and 3' atoms. For this reason, one end of a polynucleotide chain, the 5' end, will have a free phosphate group and the other, the 3' end, a free hydroxyl group. The linear sequence of nucleotide residues in a polynucleotide chain is commonly abbreviated by a one-letter code, and the sequence is always read from the 5' end. In the example illustrated the sequence is G-A-T-C.

and contain the bases A, G, C, and T (T is chemically similar to the U in RNA) (see Figure 2–25). RNA usually occurs in cells in the form of a single-stranded polynucleotide chain, but DNA is virtually always in the form of a double-stranded molecule, the DNA double helix that is composed of two polynucleotide chains running antiparallel to each other and held together by hydrogen-bonding between the bases of the two chains (Panel 2–7, pp. 78–79).

The linear sequence of nucleotides in a DNA or an RNA encodes genetic information. The two nucleic acids, however, have somewhat different roles in the cell. DNA, with its more stable, hydrogen-bonded helices, acts as a long-term repository for hereditary information, while single-stranded RNA is usually a more transient carrier of molecular instructions. The ability of the bases in different nucleic acid molecules to recognize and pair with each other by hydrogen-bonding (called *base-pairing*)—G with C, and A with either T or U—underlies all of heredity and evolution, as explained in Chapter 5.

Macromolecules in Cells

On the basis of weight, macromolecules are by far the most abundant of the carbon-containing molecules in a living cell (Figure 2–26). They are the principal building blocks from which a cell is constructed and also the components that confer the most distinctive properties of living things. Intermediate in size and complexity between small molecules and cell organelles, **macromolecules** are polymers that are constructed simply by covalently linking small organic molecules (called **monomers**, or *subunits)* into long chains, or **polymers** (Figure 2–27 and How We Know, pp. 60–61). Yet they have many unexpected properties

Question 2–8

What is meant by "polarity" of a polypeptide chain and by "polarity" of a chemical bond? How do the meanings differ?

Figure 2–26 Macromolecules are abundant in cells. The approximate composition of a bacterial cell is shown. The composition of an animal cell is similar.

bacterial cell

30% chemicals

70% H₂O

ions, small molecules (4%)
phospholipids (2%)
DNA (1%)
RNA (6%)
proteins (15%)
polysaccharides (2%)

MACROMOLECULES

that could not have been predicted from their simple constituents. For example, DNA and RNA molecules (the nucleic acids) store and transmit hereditary information.

Proteins are especially versatile and perform thousands of distinct functions in cells. Many proteins serve as enzymes that catalyze the chemical reactions that take place in the cell. All of the reactions whereby cells extract energy from food molecules are catalyzed by proteins serving as enzymes. Enzymes also synthesize important molecules. For example, an enzyme called ribulose bisphosphate carboxylase, found in chloroplasts, converts CO_2 to sugars in plants; this protein thereby creates most of the organic matter used by the rest of the living world. Other proteins are used to build structural components: tubulin self-assembles to make the cell's long, stiff microtubules (see Figure 1–27). Histone proteins pack the cell's DNA in chromosomes. Yet other proteins act as molecular motors to produce force and movement, as in the case of myosin in muscle. Proteins also have a wide variety of other functions, and we will examine the molecular basis of many of them later in this book. Here we consider only some general principles of macromolecular chemistry that make such functions possible.

Macromolecules Contain a Specific Sequence of Subunits

Although the chemical reactions for adding subunits to each polymer are different in detail for proteins, nucleic acids, and polysaccharides, they share important features. Each polymer grows by the addition of a monomer onto the end of a growing polymer chain via a condensation reaction, in which a molecule of water is lost with each subunit added (Figure 2–28; see also Figure 2–17). In all cases the reactions are catalyzed by specific enzymes, which ensure that only monomers of the appropriate type are incorporated.

The stepwise polymerization of monomers into a long chain is a simple way to manufacture a large, complex molecule, because the subunits are added by the same reaction performed over and over again by the same set of enzymes. In a sense, the process resembles the repetitive operation of a machine in a factory—except in one crucial respect. Apart from some of the polysaccharides, most macromolecules are made from a set of monomers that are slightly different from one another, for example, the 20 different amino acids from which proteins are made (see Panel 2–5, pp. 74–75). Most important, the polymer chain is not assembled at random from these subunits; instead the subunits are added in a particular order, or **sequence**.

The mechanisms that specify polymer sequence in the cell are discussed in Chapters 6 and 7. These mechanisms are central to biology because the biological function of proteins, nucleic acids, and many polysaccharides is absolutely dependent on the particular sequence of subunits in the linear chain. The possibility of varying the sequence of subunits creates enormous diversity in the polymeric molecules that can be produced. Thus, for a protein chain 200 amino acids long, there are 20^{200} possible combinations ($20 \times 20 \times 20 \times 20 \ldots$ multiplied 200 times), while for a DNA molecule 10,000 nucleotides long (small by DNA

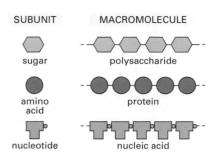

Figure 2–27 Macromolecules are made from monomeric subunits. Each macromolecule is a polymer formed from small molecules (called monomers or subunits) linked together by covalent bonds.

Question 2–9

In principle, there are many different, chemically diverse ways in which small molecules can be linked to form polymers. For example, the small molecule ethene ($CH_2=CH_2$) is used commercially to make the plastic polyethylene ($\ldots–CH_2–CH_2–CH_2–CH_2–CH_2–\ldots$). The individual subunits of the three major classes of biological macromolecules, however, are all linked by similar reaction mechanisms, i.e., by condensation reactions that eliminate water. Can you think of any benefits that this chemistry offers and why it might have been selected in evolution?

Figure 2–28 Macromolecules are formed by adding subunits to one end. In a condensation reaction, a molecule of water is lost with the addition of each monomer to one end of the growing chain. The reverse reaction—the breakdown of the polymer—occurs by the simple addition of water (hydrolysis).

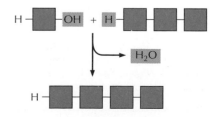

How We Know: What Are Macromolecules?

The idea that proteins, polysaccharides, and nucleic acids are large molecules that are constructed from smaller subunits, linked one after another into long molecular chains, may seem fairly obvious today. But this was not always the case. In the early part of the twentieth century, few scientists believed in the existence of such macromolecules—polymers built from repeating units held together by covalent bonds. The notion that such "frighteningly large" compounds could be assembled from simple building blocks was considered "downright shocking" by chemists of the day. Instead, they thought that proteins and other seemingly large molecules were simply heterogeneous aggregates of small molecules held together by weak "association forces" (Figure 2–29).

The idea that proteins and other polymers were large came from observing their behavior in solution. At the time, scientists were working with a variety of proteins and carbohydrates derived from foodstuffs and natural materials—albumin from egg whites, casein from milk, collagen from gelatin, and cellulose from wood. Their chemical composition seemed simple enough—like other organic molecules they contained carbon, hydrogen, oxygen, and, in the case of proteins, nitrogen. But they behaved oddly in solution, showing, for example, an inability to diffuse through a fine filter.

What was not clear, however, was why these molecules misbehaved in solution. Were they really giant molecules, composed of an unusual number of covalently linked atoms? Or were they more like a colloidal suspension of particles—a big, sticky hodgepodge of simpler molecules that associate only loosely?

One way to distinguish between the two possibilities was to determine the actual size of one of these molecules. If a substance such as serum albumin was made of molecules of uniform size, that would support the existence of true macromolecules. If albumin were instead a miscellaneous conglomeration of peptides, a solution of it should harbor molecules of a variety of sizes.

Unfortunately, the techniques available to scientists in the early 1900s were not ideal for measuring the sizes of such large molecules. Some chemists estimated a protein's size by determining how much it would depress a solution's freezing point; others measured the osmotic pressure of protein solutions. These methods were susceptible to experimental error and gave variable results. Different techniques, for example, suggested that cellulose was anywhere from 6000 to 103,000 daltons in mass. Such variation helped to fuel the hypothesis that proteins and carbohydrates were loose aggregates rather than macromolecules.

Many scientists simply had trouble believing that molecules heavier than about 4000 daltons—the largest compound that had been synthesized by organic chemists—could exist at all. Take hemoglobin, the oxygen-carrying protein in red blood cells. Researchers tried to estimate its size by breaking it down into its chemical components. In addition to carbon, hydrogen, nitrogen, and oxygen, hemoglobin contains a small amount of iron. Working out the percentages, it appeared that hemoglobin had one atom of iron for every 712 atoms of carbon—and a minimum weight of 16,700 daltons. Could a molecule with hundreds of carbon atoms in one long chain remain intact in a cell and perform specific functions? Emil Fischer, the organic chemist who determined that the amino acids in proteins are linked by peptide bonds, thought that a polypeptide chain could grow no longer than about 30 or 40 amino acids. As for hemoglobin with its purported 700 carbon atoms, the existence of molecular chains of such "truly fantastic lengths" was deemed "very improbable" by leading chemists.

Definitive resolution of the debate had to await the development of new techniques. Convincing evidence that proteins are macromolecules came from studies using the ultracentrifuge—a device that uses centrifugal force to separate molecules according to their size (Figure 2–30; see also Panel 4–3, pp. 160–161). Theodor Svedberg, who

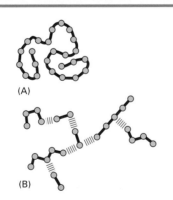

Figure 2–29 What might a macromolecule look like?
Chemists in the early part of the twentieth century debated whether proteins, polysaccharides, and other apparently large molecules were (A) discrete particles made of an unusually large number of covalently linked atoms or (B) a loose aggregation of heterogeneous small molecules held together by weak association forces.

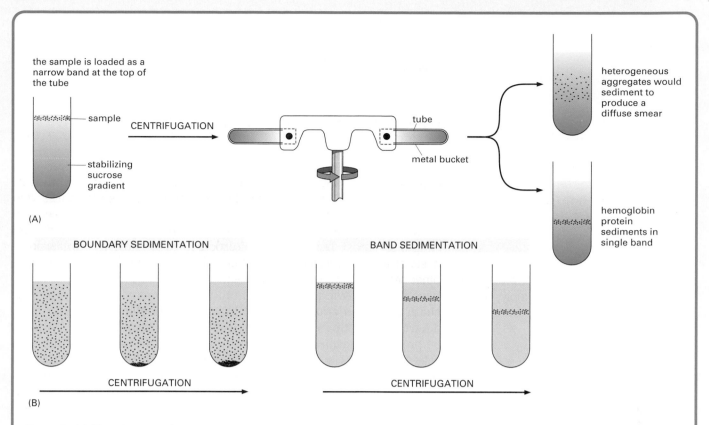

Figure 2–30 The ultracentrifuge helps to settle the macromolecular debate. In the ultracentrifuge, centrifugal forces exceeding 500,000 times the force of gravity can be used to separate proteins or other large molecules. (A) For example, samples can be loaded in a thin layer on top of a gradient of sucrose solution formed in a tube. The tube can be placed in a metal rotor that is rotated at high speed in an ultracentrifuge. Molecules of different sizes sediment at different rates, and these molecules will therefore move as distinct bands in the sample tube. If hemoglobin were a loose aggregate of heterogeneous peptides, it would show a broad smear of sizes after centrifugation *(top)*. Instead it appears as a sharp band with a molecular weight of 68,000 daltons *(bottom)*.

Although the ultracentrifuge is now a standard, almost mundane, fixture in most biochemistry labs, its construction was a huge technological challenge. The centrifuge rotor must be capable of spinning at high speeds for many hours at constant temperature and with high stability; otherwise convection occurs in the sedimenting solution and ruins the experiment. In 1926 Svedberg won the Nobel Prize in Chemistry for his ultracentrifuge design and its application to chemistry.

(B) In his actual experiment, Svedberg filled a special cell in the centrifuge with a homogeneous solution of hemoglobin; by shining light through the cell, he then carefully monitored the moving boundary between the sedimenting protein molecules and the clear aqueous solution left behind (so-called boundary sedimentation). The more recently developed method shown in (A) is a form of "band sedimentation."

designed the machine in 1925, performed the first studies. If a protein were really an aggregate of smaller molecules, he reasoned, it would appear as a smear of molecules of different size when sedimented in an ultracentrifuge. Using hemoglobin as his test protein, Svedberg found that the centrifuged sample revealed a single, sharp band with a molecular weight of 68,000 daltons. His results strongly supported the theory that proteins are true macromolecules.

Additional evidence continued to accumulate throughout the 1930s, as other researchers began to prepare crystals of pure protein that could be studied by X-ray diffraction. Only molecules with a uniform size and shape can form highly ordered crystals and diffract X-rays in such a way that their three-dimensional structure can be determined,

as we shall see in Chapter 4. A heterogeneous suspension could not be studied in this way.

We now take it for granted that large macromolecules carry out many of the most important activities in living cells. But respected chemists once viewed the existence of such polymers with the same sort of skepticism that a zoologist might show on being told that somewhere in Africa live elephants that are 500 feet long and 100 feet high. It took decades for researchers to master the techniques they needed to convince everyone that molecules 10 times larger than anything they had ever encountered were a cornerstone of biology. As we shall see throughout this book, such a labored pathway to discovery is not unusual, and progress in science is often driven by advances in measurement technologies.

standards), with its four different nucleotides there are $4^{10,000}$ different possibilities, an unimaginably large number. Thus the machinery of polymerization must be subject to a sensitive control that allows it to specify exactly which subunit should be added next to the growing polymer.

Noncovalent Bonds Specify the Precise Shape of a Macromolecule

Most of the single covalent bonds in a macromolecule allow rotation of the atoms they join, so that the polymer chain has great flexibility. In principle, this allows a macromolecule to adopt an almost unlimited number of shapes, or **conformations**, as the polymer chain writhes and rotates under the influence of random thermal energy. However, the shapes of most biological macromolecules are highly constrained because of weaker **noncovalent bonds** that form between different parts of the molecule. If these weaker bonds are formed in sufficient numbers, they will prevent the random movements and the polymer chain may then adopt preferentially one particular conformation, as determined by the linear sequence of monomers in its chain. Virtually all protein molecules and many of the small RNA molecules found in cells fold tightly into one highly preferred conformation in this way (Figure 2–31).

The noncovalent bonds important in biological molecules include two types described earlier in this chapter—ionic bonds and hydrogen bonds (Panel 2–7, pp. 78–79). Ionic bonds, although strong on their own, are quite weak in water. This is because charged groups are shielded by their interactions with water molecules or with other salts present in the aqueous solution. Ionic bonds, however, are very important in biological systems. An enzyme that binds a positively charged substrate will often use a negatively charged amino acid side chain to guide its substrate into the proper position. And we have already mentioned the importance of hydrogen bonds in establishing the unique properties of water. Hydrogen bonds also hold two strands of the DNA double helix together. Because individual hydrogen bonds are weak, enzymes can easily unzip the helix—for example, when a cell needs to copy its genetic material.

A third type of weak bond results from *van der Waals attractions*, which are a form of electrical attraction caused by fluctuating electric charges that arise whenever two atoms come within a very short distance

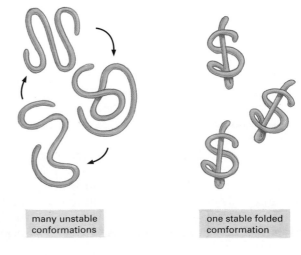

many unstable
conformations

one stable folded
comformation

Figure 2–31 **Most proteins and many RNA molecules fold into only one stable conformation.** If the weak bonds maintaining this stable conformation are disrupted, the molecule becomes a flexible chain that usually has no biological value.

of each other. Although van der Waals interactions are weaker than hydrogen bonds, in large numbers they play an important role in the attraction between large molecules with complementary shapes. All of these noncovalent forces are reviewed in Panel 2–7, pp. 78–79.

Another important noncovalent force is created by the three-dimensional structure of water, which forces hydrophobic groups together in order to minimize their disruptive effect on the hydrogen-bonded network of water molecules (see Panel 2–7, and Panel 2–2, pp. 68–69). This expulsion from the aqueous solution generates what is sometimes thought of as a fourth kind of weak noncovalent bond, called a *hydrophobic interaction*. This interaction forces phospholipid molecules together in cell membranes, and it also gives most protein molecules a compact, globular shape.

Question 2–10

Why could covalent bonds not be used in place of noncovalent bonds to mediate most of the interactions of macromolecules?

Noncovalent Bonds Allow a Macromolecule to Bind Other Selected Molecules

Although noncovalent bonds are individually very weak, they can add up to create a strong attraction between two molecules when these molecules fit together very closely, like a hand in a glove, with many noncovalent bonds between them (see Panel 2–7). This form of molecular interaction provides for great specificity in the binding of macromolecules to other molecules, because the multipoint contacts required for strong binding make it possible for a macromolecule to select—through binding interactions—just one of the many thousands of different molecules present inside a cell. Moreover, because the strength of the binding depends on the number of noncovalent bonds that are formed, interactions of almost any strength are possible.

Binding of this type underlies all biological catalysis, making it possible for proteins to function as enzymes. Noncovalent bonds can also stabilize associations between two different macromolecules if their surfaces match closely (Figure 2–32). These bonds thereby allow macromolecules to be used as building blocks for the formation of much larger structures. For example, proteins often bind together into multiprotein complexes, thereby forming intricate machines with multiple

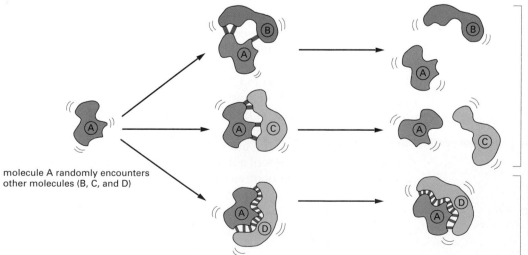

molecule A randomly encounters other molecules (B, C, and D)

the surfaces of molecules A and B, and A and C, are a poor match and are capable of forming only a few weak bonds; thermal motion rapidly breaks them apart

the surfaces of molecules A and D match well and therefore can form enough weak bonds to withstand thermal motion; they therefore stay bound to each other

Figure 2–32 Noncovalent bonds mediate interactions between macromolecules.

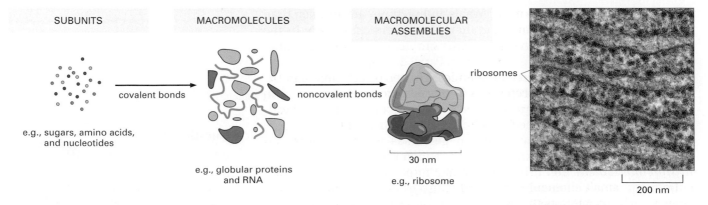

SUBUNITS   MACROMOLECULES   MACROMOLECULAR ASSEMBLIES

covalent bonds   noncovalent bonds   ribosomes

e.g., sugars, amino acids, and nucleotides

e.g., globular proteins and RNA

30 nm

e.g., ribosome

200 nm

Figure 2–33 Small molecules join together to form macromolecules, which can assemble into large macromolecular complexes. Subunits, proteins, and a ribosome are drawn to scale. Ribosomes are part of the machinery the cell uses to make proteins. Each ribosome is composed of about 90 macromolecules (proteins and RNA molecules), and is large enough to see in the electron microscope. The micrograph on the right shows numerous ribosomes attached to membranes in the cell. (Courtesy of Lelio Orci.)

moving parts that perform such complex tasks as DNA replication and protein synthesis (Figure 2–33). Thus noncovalent bonds account for much of the specificity that we associate with living cells.

Essential Concepts

- Living cells obey the same chemical and physical laws as nonliving things. Like all other forms of matter, they are composed of atoms, which are the smallest units of chemical elements that retain distinctive chemical properties.
- Atoms are made up of smaller particles. The nucleus of an atom contains protons, which are positively charged, and uncharged neutrons. The nucleus is surrounded by a cloud of negatively charged electrons.
- The number of electrons in an atom is equal to the number of protons in its nucleus. The nuclei of different isotopes of the same element contain the same number of protons but different numbers of neutrons.
- Living cells are made up of a limited number of elements, four of which—C,H,N,O—make up 96.5% of their mass.
- The chemical properties of an atom are determined by the number and arrangement of its electrons. An atom is most stable when all of its electrons are at their lowest possible energy level and when each electron shell is completely filled.
- Chemical bonds form between atoms as electrons move to reach a more stable arrangement. Clusters of two or more atoms held together by chemical bonds are known as molecules.
- When an electron jumps from one atom to another, two ions of opposite charge are formed; ionic bonds then arise by the mutual attraction of these charged atoms.
- A covalent bond consists of a pair of electrons shared between adjacent atoms. If two pairs of electrons are shared, a double bond is formed.

- Living organisms contain a distinctive and restricted set of small carbon-based molecules that are essentially the same for every living species. The main categories are sugars, fatty acids, amino acids, and nucleotides.
- Sugars are a primary source of chemical energy for cells and can be incorporated into polysaccharides for energy storage.
- Fatty acids are also important for energy storage, but their most essential function is in the formation of cell membranes.
- The vast majority of the dry mass of a cell consists of macromolecules, formed as polymers of sugars, amino acids, or nucleotides.
- Macromolecules are intermediate both in size and complexity between small molecules and cell organelles. They have many remarkable properties that are not easily deduced from the subunits from which they are made.
- Polymers consisting of amino acids constitute the remarkably diverse and versatile class of macromolecules known as proteins.
- Nucleotides play a central part in energy transfer and are the subunits from which the informational macromolecules, RNA and DNA, are made.
- Macromolecules are made as polymers of subunits by repetitive condensation reactions. Their remarkable diversity arises from the fact that each macromolecule has a unique sequence of subunits.
- Weak noncovalent bonds form between different regions of a macromolecule. These can cause the macromolecule to fold into a unique three-dimensional shape with a special chemistry, as seen most conspicuously in proteins.

Key Terms

acid	electron	noncovalent bond
amino acid	fatty acid	nucleotide
atom	hydrogen bond	pH scale
atomic weight	hydrolysis	polar
ATP	hydronium ion	polymer
Avogadro's number	hydrophilic	protein
base	hydrophobic	proton
chemical bond	ion	RNA
chemical group	ionic bond	sequence
condensation	macromolecule	subunit
conformation	molecular weight	sugar
covalent bond	molecule	valence
DNA	monomer	

Panel 2–1 Chemical bonds and groups

CARBON SKELETONS

Carbon has a unique role in the cell because of its ability to form strong covalent bonds with other carbon atoms. Thus carbon atoms can join to form chains

or branched trees

or rings.

also written as

also written as

also written as

COVALENT BONDS

A covalent bond forms when two atoms come very close together and share one or more of their electrons.

Each atom forms a fixed number of covalent bonds in a defined spatial arrangement.

SINGLE BONDS: 2 electrons shared/bond

DOUBLE BONDS: 4 electrons shared/bond

The precise spatial arrangement of covalent bonds influence the three-dimentional structure—and chemistry—of molecules. In this review panel, we see how covalent bonds are used in a variety of biological molecules.

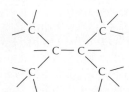

Atoms joined by two or more covalent bonds cannot rotate freely around the bond axis. This restriction is a major influence on the three-dimensional shape of many macromolecules.

C–H COMPOUNDS

Carbon and hydrogen together make stable compounds (or groups) called hydrocarbons. These are nonpolar, do not form hydrogen bonds, and are generally insoluble in water.

H–C–H with H above and below (methane)

H–C– with H above and below (methyl group)

methane

methyl group

part of the hydrocarbon "tail" of a fatty acid molecule

ALTERNATING DOUBLE BONDS

The carbon chain can include double bonds. If these are on alternate carbon atoms, the bonding electrons move within the molecule, stabilizing the structure by a phenomenon called resonance.

the truth is somewhere between these two structures

Alternating double bonds in a ring can generate a very stable structure.

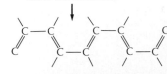

benzene

often written as

C–O COMPOUNDS

Many biological compounds contain a carbon bonded to an oxygen. For example,

alcohol

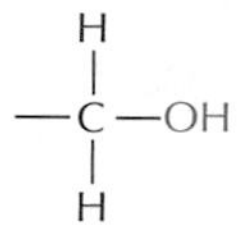

The –OH is called a hydroxyl group.

aldehyde

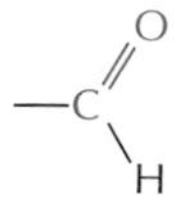

ketone

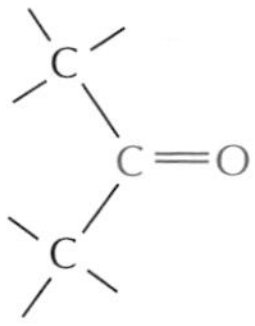

The C=O is called a carbonyl group.

carboxylic acid

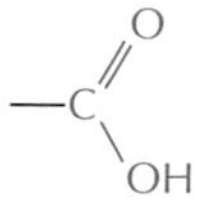

The –COOH is called a carboxyl group. In water this loses an H^+ ion to become –COO⁻.

esters

Esters are formed by combining an acid and an alcohol.

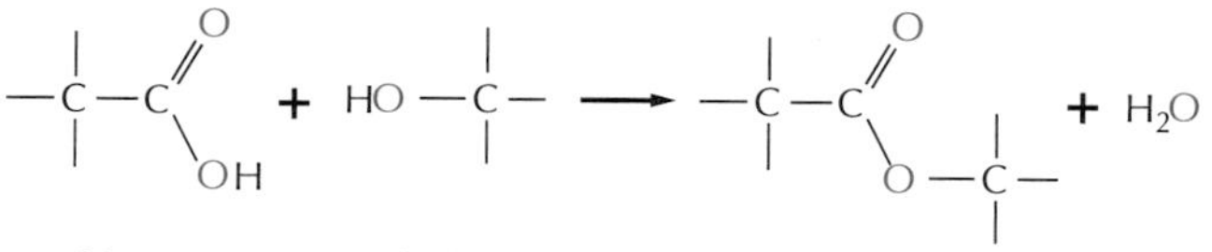

acid   alcohol   ester

C–N COMPOUNDS

Amines and amides are two important examples of compounds containing a carbon linked to a nitrogen.

Amines in water combine with an H^+ ion to become positively charged.

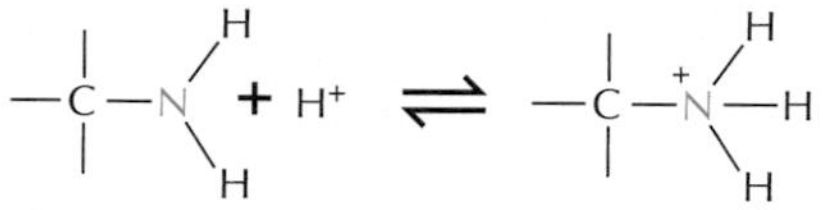

Amides are formed by combining an acid and an amine. Unlike amines, amides are uncharged in water. An example is the peptide bond that joins amino acids in a protein.

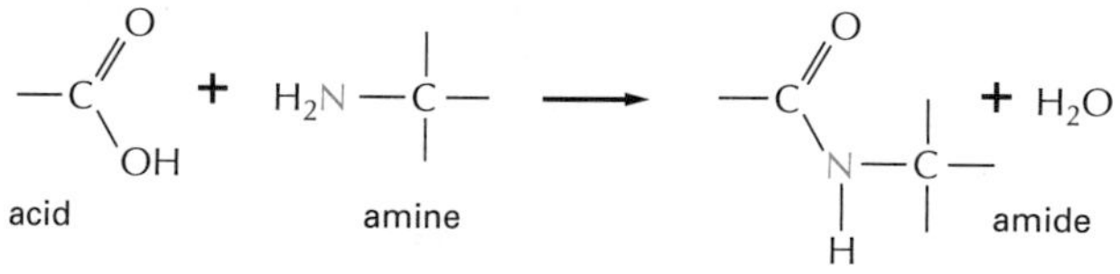

acid   amine   amide

Nitrogen also occurs in several ring compounds, including important constituents of nucleic acids: purines and pyrimidines.

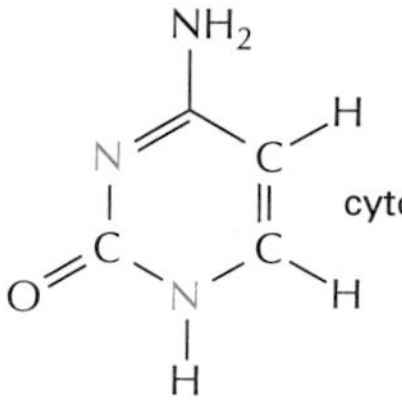

cytosine (a pyrimidine)

PHOSPHATES

Inorganic phosphate is a stable ion formed from phosphoric acid, H_3PO_4. It is often written as P_i.

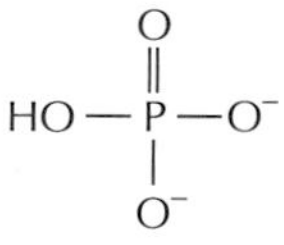

Phosphate esters can form between a phosphate and a free hydroxyl group. Phosphoryl groups are often attached to proteins in this way.

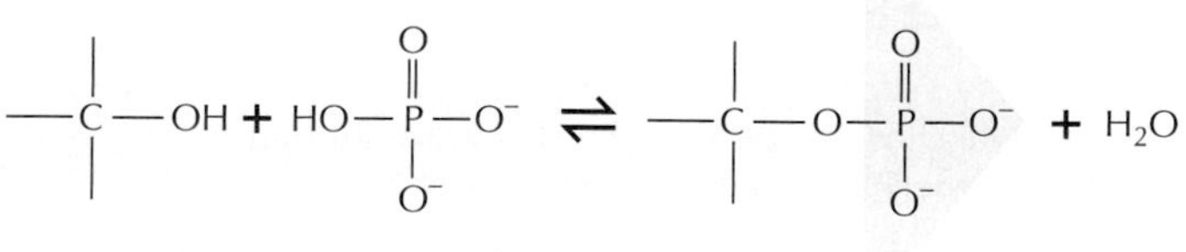

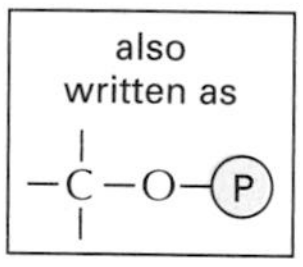

also written as

The combination of a phosphate and a carboxyl group, or two or more phosphate groups, gives an acid anhydride.

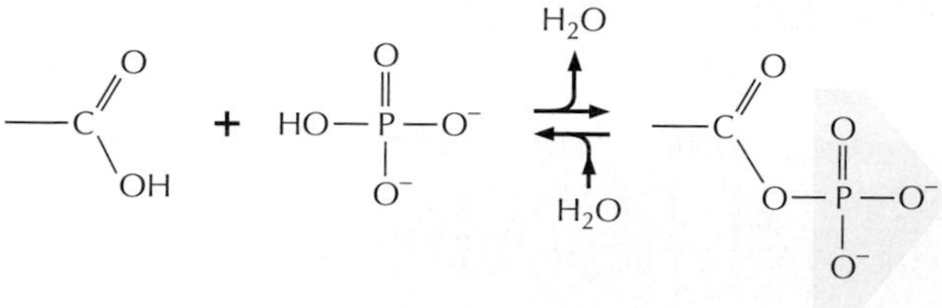

high-energy acyl phosphate bond (carboxylic–phosphoric acid anhydride) found in some metabolites

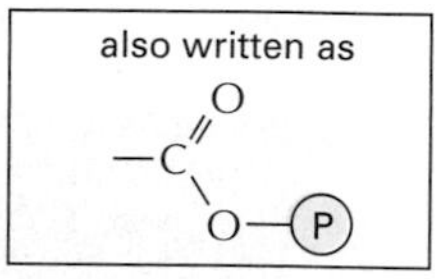

also written as

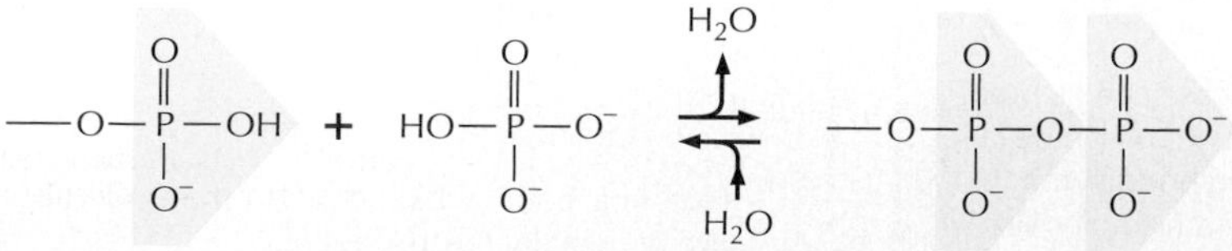

phosphoanhydride—a high-energy bond found in molecules such as ATP

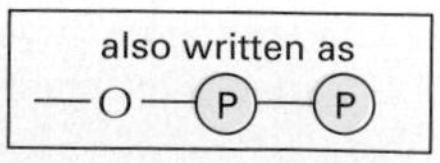

also written as

Panel 2–2 The chemical properties of water

HYDROGEN BONDS

Because they are polarized, two adjacent H_2O molecules can form a linkage known as a hydrogen bond. Hydrogen bonds have only about 1/20 the strength of a covalent bond.

Hydrogen bonds are strongest when the three atoms lie in a straight line.

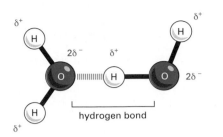

bond lengths

hydrogen bond
0.27 nm

0.10 nm
covalent bond

WATER

Two atoms connected by a covalent bond may exert different attractions for the electrons of the bond. In such cases the bond is polar, with one end slightly negatively charged (δ^-) and the other slightly positively charged (δ^+).

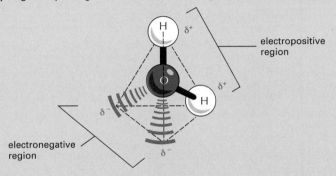

electropositive region

electronegative region

Although a water molecule has an overall neutral charge (having the same number of electrons and protons), the electrons are asymmetrically distributed, making the molecule polar. The oxygen nucleus draws electrons away from the hydrogen nuclei, leaving these nuclei with a small net positive charge. The excess of electron density on the oxygen atom creates weakly negative regions at the other two corners of an imaginary tetrahedron. On these pages we review the chemical properties of water and see how water influences the behavior of biological molecules.

WATER STRUCTURE

Molecules of water join together transiently in a hydrogen-bonded lattice.

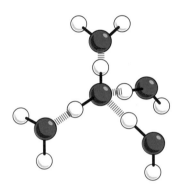

The cohesive nature of water is responsible for many of its unusual properties, such as high surface tension, specific heat, and heat of vaporization.

HYDROPHILIC MOLECULES

Substances that dissolve readily in water are termed hydrophilic. They are composed of ions or polar molecules that attract water molecules through electrical charge effects. Water molecules surround each ion or polar molecule on the surface of a solid substance and carry it into solution.

Ionic substances such as sodium chloride dissolve because water molecules are attracted to the positive (Na^+) or negative (Cl^-) charge of each ion.

Polar substances such as urea dissolve because their molecules form hydrogen bonds with the surrounding water molecules

HYDROPHOBIC MOLECULES

Molecules that contain a preponderance of non-polar bonds are usually insoluble in water and are termed hydrophobic. Water molecules are not attracted to such molecules and so have little tendency to surround them and carry them into solution.

Hydrocarbons, which contain many C–H bonds, are especially hydrophobic.

WATER AS A SOLVENT

Many substances, such as household sugar, dissolve in water. That is, their molecules separate from each other, each becoming surrounded by water molecules.

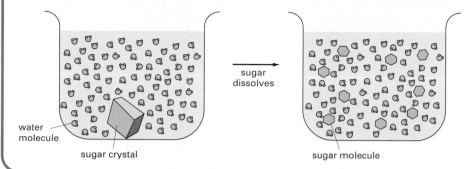

water molecule

sugar crystal

sugar dissolves

sugar molecule

When a substance dissolves in a liquid, the mixture is termed a solution. The dissolved substance (in this case sugar) is the solute, and the liquid that does the dissolving (in this case water) is the solvent. Water is an excellent solvent for many substances because of its polar bonds.

ACIDS

Substances that release hydrogen ions into solution are called acids.

$$HCl \longrightarrow H^+ + Cl^-$$

hydrochloric acid (strong acid)   hydrogen ion   chloride ion

Many of the acids important in the cell are only partially dissociated, and they are therefore weak acids—for example, the carboxyl group (–COOH), which dissociates to give a hydrogen ion in solution.

$$-C{\overset{O}{\underset{OH}{}}} \rightleftharpoons H^+ + -C{\overset{O}{\underset{O^-}{}}}$$

(weak acid)

Note that this is a reversible reaction.

HYDROGEN ION EXCHANGE

Positively charged hydrogen ions (H⁺) can spontaneously move from one water molecule to another, thereby creating two ionic species.

hydronium ion (water acting as a weak base)   hydroxyl ion (water acting as a weak acid)

often written as: $H_2O \rightleftharpoons H^+ + OH^-$

hydrogen ion   hydroxyl ion

Because the process is rapidly reversible, hydrogen ions are continually shuttling between water molecules. Pure water contains a steady-state concentration of hydrogen ions and hydroxyl ions (both 10^{-7} M).

pH

The acidity of a solution is defined by the concentration of H⁺ ions it possesses. For convenience we use the pH scale, where

$$pH = -\log_{10}[H^+]$$

For pure water

$$[H^+] = 10^{-7} \text{ moles/liter}$$

$$pH = 7.0$$

H⁺ conc. moles/liter	pH
10^{-1}	1
10^{-2}	2
10^{-3}	3
10^{-4}	4
10^{-5}	5
10^{-6}	6
10^{-7}	7
10^{-8}	8
10^{-9}	9
10^{-10}	10
10^{-11}	11
10^{-12}	12
10^{-13}	13
10^{-14}	14

ACIDIC

ALKALINE

BASES

Substances that reduce the number of hydrogen ions in solution are called bases. Some bases, such as ammonia, combine directly with hydrogen ions.

$$NH_3 + H^+ \longrightarrow NH_4^+$$

ammonia   hydrogen ion   ammonium ion

Other bases, such as sodium hydroxide, reduce the number of H⁺ ions indirectly, by making OH⁻ ions that then combine directly with H⁺ ions to make H_2O.

$$NaOH \longrightarrow Na^+ + OH^-$$

sodium hydroxide (strong base)   sodium ion   hydroxyl ion

Many bases found in cells are partially dissociated and are termed weak bases. This is true of compounds that contain an amino group (–NH₂), which has a weak tendency to reversibly accept an H⁺ ion from water, increasing the quantity of free OH⁻ ions.

$$-NH_2 + H^+ \rightleftharpoons -NH_3^+$$

Panel 2–3 An outline of some of the types of sugars

MONOSACCHARIDES

Monosaccharides usually have the general formula $(CH_2O)_n$, where n can be 3, 4, 5, or 6, and have two or more hydroxyl groups. They either contain an aldehyde group ($-C\lessgtr^O_H$) and are called aldoses or a ketone group ($\gtr C=O$) and are called ketoses.

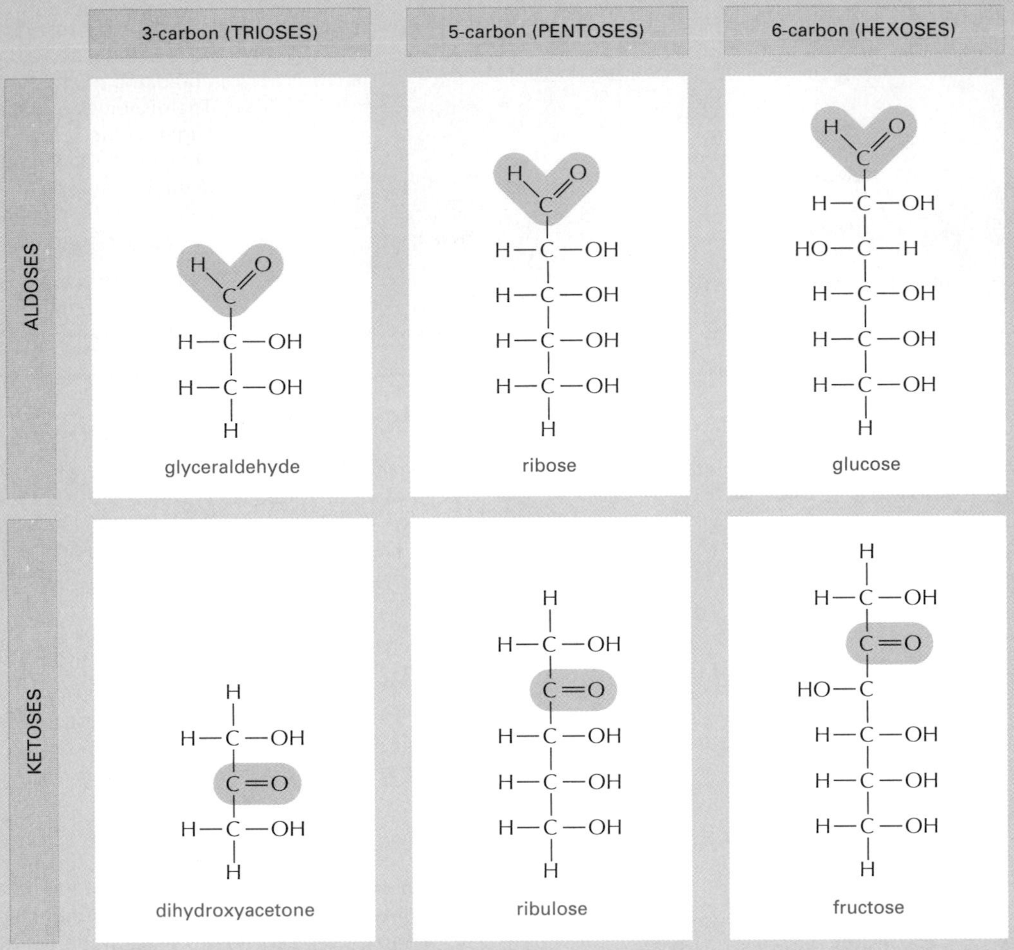

	3-carbon (TRIOSES)	5-carbon (PENTOSES)	6-carbon (HEXOSES)
ALDOSES	glyceraldehyde	ribose	glucose
KETOSES	dihydroxyacetone	ribulose	fructose

RING FORMATION

In aqueous solution, the aldehyde or ketone group of a sugar molecule tends to react with a hydroxyl group of the same molecule, thereby closing the molecule into a ring.

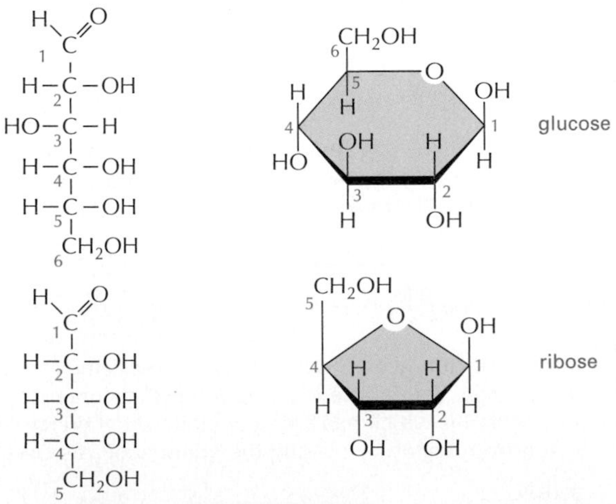

glucose

ribose

Note that each carbon atom has a number.

ISOMERS

Many monosaccharides differ only in the spatial arrangement of atoms—that is, they are isomers. For example, glucose, galactose, and mannose have the same formula ($C_6H_{12}O_6$) but differ in the arrangement of groups around one or two carbon atoms.

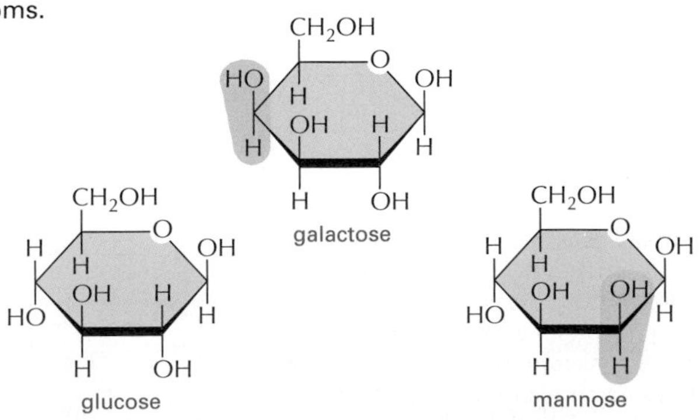

galactose

glucose

mannose

These small differences make only minor changes in the chemical properties of the sugars. But they are recognized by enzymes and other proteins and therefore can have important biological effects.

α AND β LINKS

The hydroxyl group on the carbon that carries the aldehyde or ketone can rapidly change from one position to the other. These two positions are called α and β.

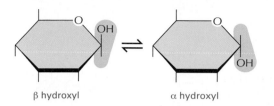

β hydroxyl   α hydroxyl

As soon as one sugar is linked to another, the α or β form is frozen.

SUGAR DERIVATIVES

The hydroxyl groups of a simple monosaccharide can be replaced by other groups. For example,

glucosamine

glucuronic acid

N-acetylglucosamine

DISACCHARIDES

The carbon that carries the aldehyde or the ketone can react with any hydroxyl group on a second sugar molecule to form a disaccharide. Three common disaccharides are

  maltose (glucose + glucose)
  lactose (galactose + glucose)
  sucrose (glucose + fructose)

The reaction forming sucrose is shown here.

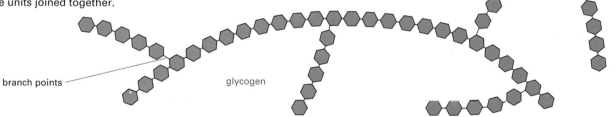

α glucose   β fructose

H_2O

sucrose

OLIGOSACCHARIDES AND POLYSACCHARIDES

Large linear and branched molecules can be made from simple repeating units. Short chains are called oligosaccharides, while long chains are called polysaccharides. Glycogen, for example, is a polysaccharide made entirely of glucose units joined together.

branch points   glycogen

COMPLEX OLIGOSACCHARIDES

In many cases a sugar sequence is nonrepetitive. Many different molecules are possible. Such complex oligosaccharides are usually linked to proteins or to lipids, as is this oligosaccharide, which is part of a cell-surface molecule that defines a particular blood group.

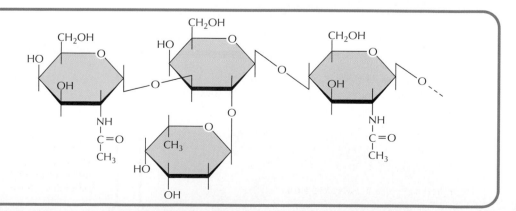

Panel 2–4 Fatty acids and other lipids

FATTY ACIDS

All fatty acids have carboxyl groups with long hydrocarbon tails.

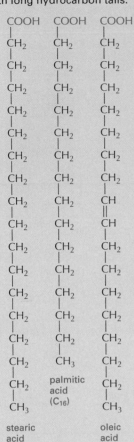

COOH	COOH	COOH
CH₂	CH₂	CH₂
CH₂	CH₂	CH₂
CH₂	CH₂	CH₂
CH₂	CH₂	CH₂
CH₂	CH₂	CH₂
CH₂	CH₂	CH₂
CH₂	CH₂	CH₂
CH₂	CH₂	CH
CH₂	CH₂	CH
CH₂	CH₂	CH₂
CH₂	CH₂	CH₂
CH₂	CH₂	CH₂
CH₂	CH₂	CH₂
CH₂	CH₂	CH₂
CH₂	CH₃	CH₂
CH₂	palmitic acid (C₁₆)	CH₂
CH₃		CH₃
stearic acid (C₁₈)		oleic acid (C₁₈)

Hundreds of different kinds of fatty acids exist. Some have one or more double bonds in their hydrocarbon tail and are said to be unsaturated. Fatty acids with no double bonds are saturated.

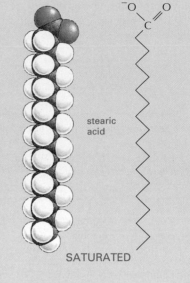

oleic acid

This double bond is rigid and creates a kink in the chain. The rest of the chain is free to rotate about the other C–C bonds.

space-filling model   carbon skeleton

UNSATURATED

stearic acid

SATURATED

TRIACYLGLYCEROLS

Fatty acids are stored as an energy reserve (fats and oils) through an ester linkage to glycerol to form triacylglycerols.

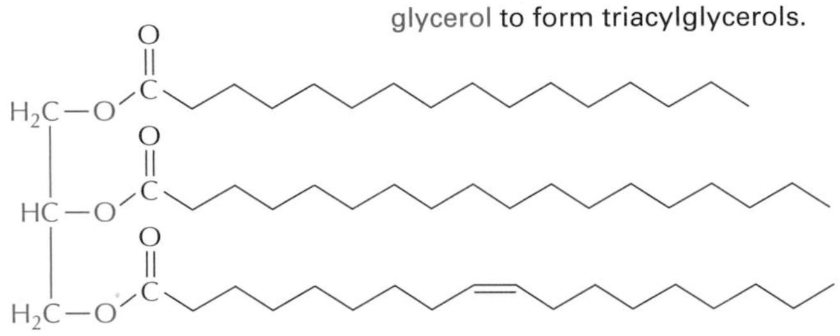

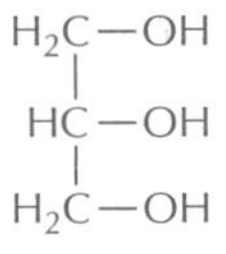

H₂C—OH
HC—OH
H₂C—OH

glycerol

CARBOXYL GROUP

If free, the carboxyl group of a fatty acid will be ionized.

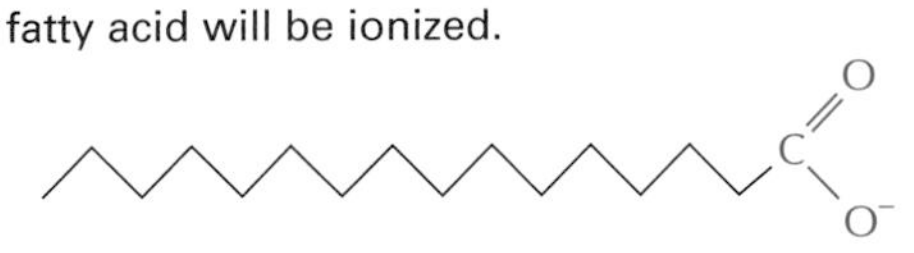

But more often it is linked to other groups to form either esters

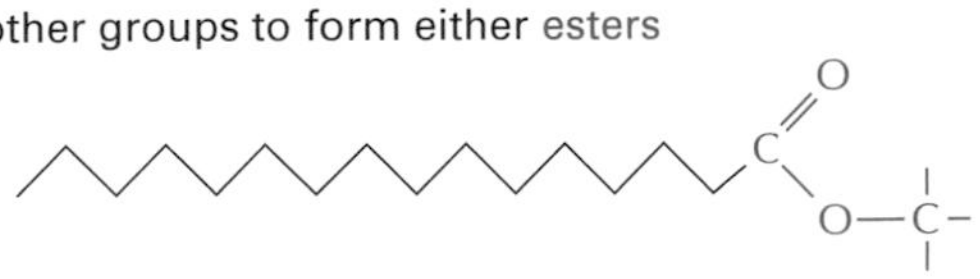

or amides.

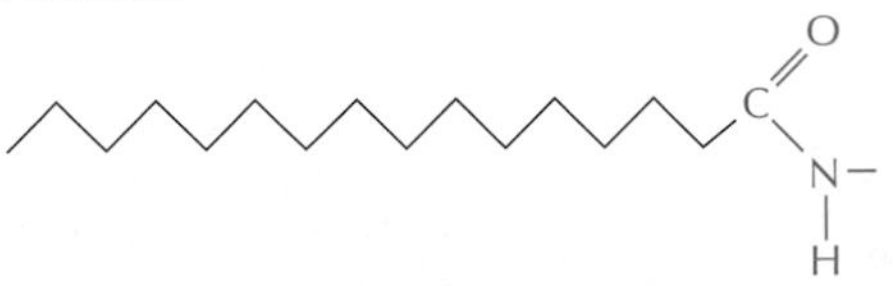

PHOSPHOLIPIDS

Phospholipids are the major constituents of cell membranes.

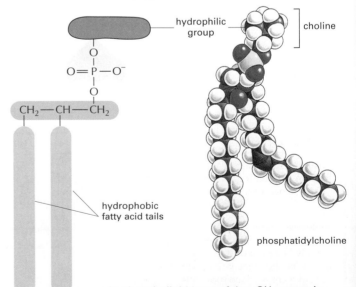

hydrophilic group

choline

O=P—O⁻

CH₂—CH—CH₂

hydrophobic fatty acid tails

general structure of a phospholipid

phosphatidylcholine

In phospholipids two of the –OH groups in glycerol are linked to fatty acids, while the third –OH group is linked to phosphoric acid. The phosphate is further linked to one of a variety of small polar groups (alcohols).

LIPID AGGREGATES

Fatty acids have a hydrophilic head and a hydrophobic tail.

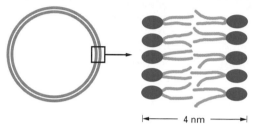

In water they can form a surface film or form small micelles.

micelle

Their derivatives can form larger aggregates held together by hydrophobic forces:

Triglycerides form large spherical fat droplets in the cell cytoplasm.

200 nm or more

Phospholipids and **glycolipids** form self-sealing lipid bilayers that are the basis for all cellular membranes.

4 nm

OTHER LIPIDS

Lipids are defined as the water-insoluble molecules in cells that are soluble in organic solvents. Two other common types of lipids are steroids and polyisoprenoids. Both are made from isoprene units.

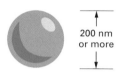

isoprene

STEROIDS

Steroids have a common multiple-ring structure.

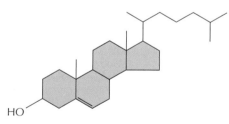

HO

cholesterol—found in many membranes

OH

O

testosterone—male steroid hormone

GLYCOLIPIDS

Like phospholipids, these compounds are composed of a hydrophobic region, containing two long hydrocarbon tails, and a polar region, which, however, contains one or more sugar residues and no phosphate.

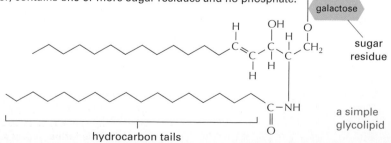

galactose

sugar residue

a simple glycolipid

hydrocarbon tails

POLYISOPRENOIDS

long-chain polymers of isoprene

dolichol phosphate—used to carry activated sugars in the membrane-associated synthesis of glycoproteins and some polysaccharides

Panel 2–5 The 20 amino acids found in proteins

FAMILIES OF AMINO ACIDS

The common amino acids are grouped according to whether their side chains are

    acidic
    basic
    uncharged polar
    nonpolar

These 20 amino acids are given both three-letter and one-letter abbreviations.

Thus: alanine = Ala = A

BASIC SIDE CHAINS

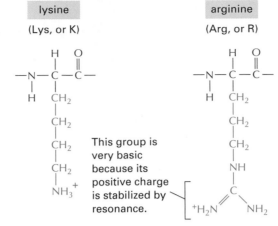

lysine	arginine	histidine
(Lys, or K)	(Arg, or R)	(His, or H)

This group is very basic because its positive charge is stabilized by resonance.

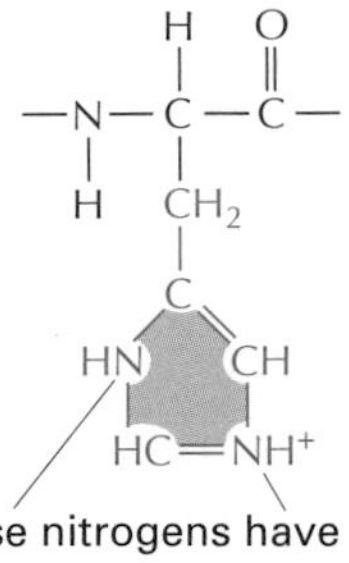

These nitrogens have a relatively weak affinity for an H+ and are only partly positive at neutral pH.

THE AMINO ACID

The general formula of an amino acid is

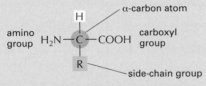

amino group — H_2N
carboxyl group — COOH
α-carbon atom
side-chain group
R

R is commonly one of 20 different side chains. At pH 7 both the amino and carboxyl groups are ionized.

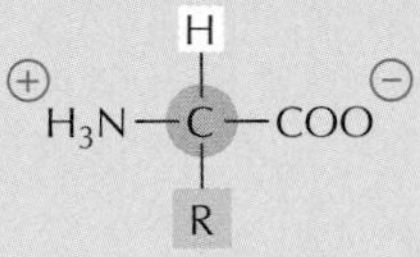

$\oplus$ H_3N — C — COO $\ominus$

These pages present the amino acids found in proteins and show how they are linked.

OPTICAL ISOMERS

The α-carbon atom is asymmetric, allowing for two mirror-image (or stereo-) isomers, L and D.

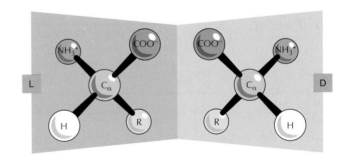

Proteins consist exclusively of L-amino acids.

PEPTIDE BONDS

Amino acids are commonly joined together by an amide linkage, called a peptide bond.

The four atoms in each peptide bond (*gray box*) form a rigid planar unit. There is no rotation around the C–N bond.

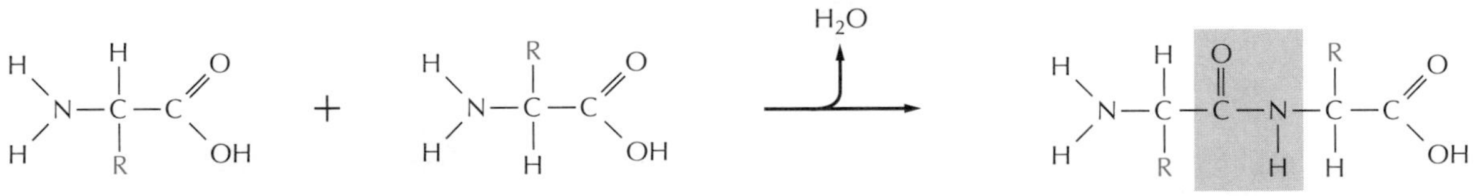

H_2O

Proteins are long polymers of amino acids linked by peptide bonds, and they are always written with the N-terminus toward the left. The sequence of this tripeptide is histidine-cysteine-valine.

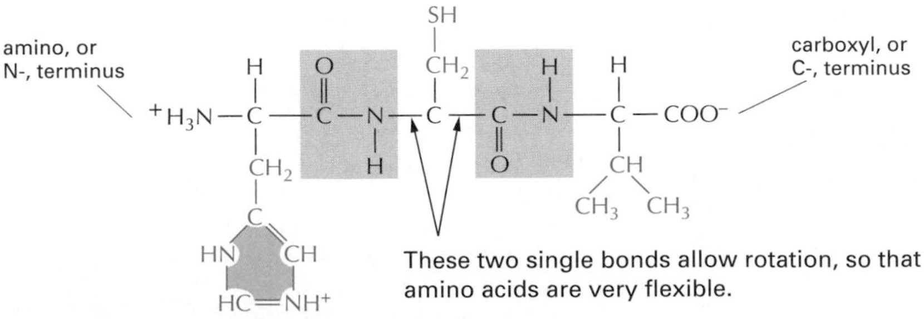

amino, or N-, terminus

carboxyl, or C-, terminus

These two single bonds allow rotation, so that long chains of amino acids are very flexible.

ACIDIC SIDE CHAINS

aspartic acid
(Asp, or D)

glutamic acid
(Glu, or E)

UNCHARGED POLAR SIDE CHAINS

asparagine
(Asn, or N)

glutamine
(Gln, or Q)

Although the amide N is not charged at neutral pH, it is polar.

serine
(Ser, or S)

threonine
(Thr, or T)

tyrosine
(Tyr, or Y)

The –OH group is polar.

NONPOLAR SIDE CHAINS

alanine
(Ala, or A)

valine
(Val, or V)

leucine
(Leu, or L)

isoleucine
(Ile, or I)

proline
(Pro, or P)

(actually an imino acid)

phenylalanine
(Phe, or F)

methionine
(Met, or M)

tryptophan
(Trp, or W)

glycine
(Gly, or G)

cysteine
(Cys, or C)

Disulfide bonds can form between two cysteine side chains in proteins.

$$- -CH_2 - S - S - CH_2 - -$$

Panel 2–6 A survey of the nucleotides

BASES

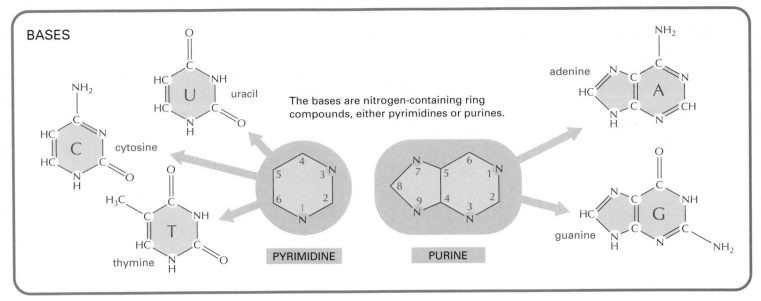

The bases are nitrogen-containing ring compounds, either pyrimidines or purines.

cytosine

uracil

thymine

PYRIMIDINE

PURINE

adenine

guanine

PHOSPHATES

The phosphates are normally joined to the C5 hydroxyl of the ribose or deoxyribose sugar (designated 5'). Mono-, di-, and triphosphates are common.

as in AMP

as in ADP

as in ATP

The phosphate makes a nucleotide negatively charged.

NUCLEOTIDES

A nucleotide consists of a nitrogen-containing base, a five-carbon sugar, and one or more phosphate groups.

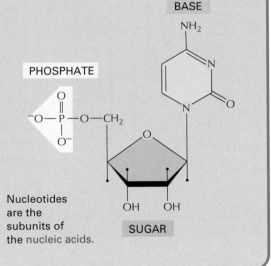

BASE

PHOSPHATE

SUGAR

Nucleotides are the subunits of the nucleic acids.

BASIC SUGAR LINKAGE

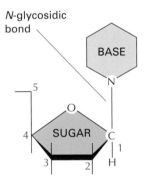

N-glycosidic bond

BASE

SUGAR

The base is linked to the same carbon (C1) used in sugar–sugar bonds.

SUGARS

PENTOSE

a five-carbon sugar

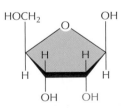

two kinds are used

β-D-ribose used in ribonucleic acid

β-D-2-deoxyribose used in deoxyribonucleic acid

Each numbered carbon on the sugar of a nucleotide is followed by a prime mark; therefore, one speaks of the "5-prime carbon," etc.

NOMENCLATURE

The names can be confusing, but the abbreviations are clear.

BASE	NUCLEOSIDE	ABBR.
adenine	adenosine	A
guanine	guanosine	G
cytosine	cytidine	C
uracil	uridine	U
thymine	thymidine	T

Nucleotides are abbreviated by three capital letters. Some examples follow:

AMP = adenosine monophosphate
dAMP = deoxyadenosine monophosphate
UDP = uridine diphosphate
ATP = adenosine triphosphate

BASE + SUGAR = NUCLEOSIDE

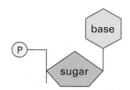

BASE + SUGAR + PHOSPHATE = NUCLEOTIDE

NUCLEIC ACIDS

Nucleotides are joined together by a phosphodiester linkage between 5' and 3' carbon atoms to form nucleic acids. The linear sequence of nucleotides in a nucleic acid chain is commonly abbreviated by a one-letter code, A—G—C—T—T—A—C—A, with the 5' end of the chain at the left.

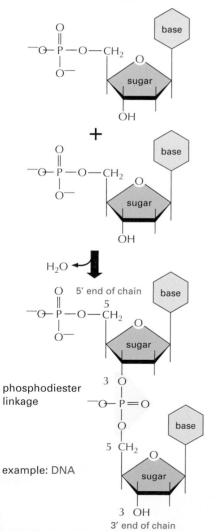

phosphodiester linkage

example: DNA

NUCLEOTIDES HAVE MANY OTHER FUNCTIONS

1 They carry chemical energy in their easily hydrolyzed phosphoanhydride bonds.

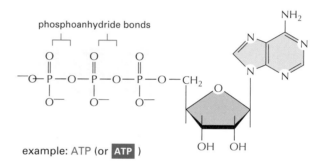

phosphoanhydride bonds

example: ATP (or ATP)

2 They combine with other groups to form coenzymes.

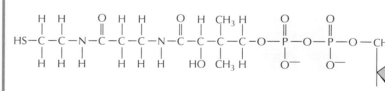

example: coenzyme A (CoA)

3 They are used as signaling molecules in the cell.

example: cyclic AMP

Panel 2–7 The principal types of weak noncovalent bonds

WEAK CHEMICAL BONDS

Organic molecules can interact with other molecules through short-range noncovalent forces.

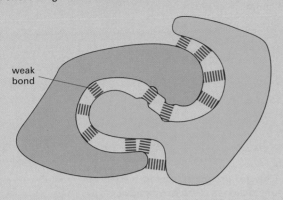

weak bond

Weak chemical bonds have less than 1/20 the strength of a strong covalent bond. They are strong enough to provide tight binding only when many of them are formed simultaneously.

HYDROGEN BONDS

As already described for water (see Panel 2–2, pp. 68–69) hydrogen bonds form when a hydrogen atom is "sandwiched" between two electron-attracting atoms (usually oxygen or nitrogen).

Hydrogen bonds are strongest when the three atoms are in a straight line:

$$\diagdown\!\!O\!-\!H\;||||||||||\;O\diagup \qquad \diagdown\!\!N\!-\!H\;||||||||||\;O\!=$$

Examples in macromolecules:

Amino acids in polypeptide chains hydrogen-bonded together.

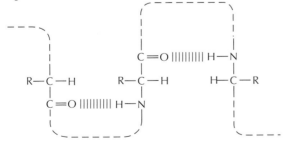

Two bases, G and C, hydrogen-bonded in DNA or RNA.

VAN DER WAALS ATTRACTIONS

If two atoms are too close together they repel each other very strongly. For this reason, an atom can often be treated as a sphere with a fixed radius. The characteristic "size" for each atom is specified by a unique *van der Waals radius*. The contact distance between any two non-covalently bonded atoms is the sum of their van der Waals radii.

| 0.12 nm radius | 0.2 nm radius | 0.15 nm radius | 0.14 nm radius |

At very short distances any two atoms show a weak bonding interaction due to their fluctuating electrical charges. The two atoms will be attracted to each other in this way until the distance between their nuclei is approximately equal to the sum of their van der Waals radii. Although they are individually very weak, van der Waals attractions can become important when two macromolecular surfaces fit very close together, because many atoms are involved.

Note that when two atoms form a covalent bond, the centers of the two atoms (the two atomic nuclei) are much closer together than the sum of the two van der Waals radii. Thus,

| 0.4 nm two non-bonded carbon atoms | 0.15 nm single-bonded carbons | 0.13 nm double-bonded carbons |

HYDROGEN BONDS IN WATER

Any molecules that can form hydrogen bonds to each other can alternatively form hydrogen bonds to water molecules. Because of this competition with water molecules, the hydrogen bonds formed between two molecules dissolved in water are relatively weak.

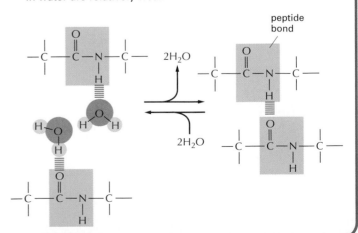

peptide bond

$2H_2O$

$2H_2O$

HYDROPHOBIC FORCES

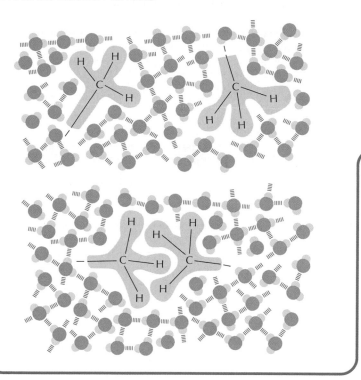

Water forces hydrophobic groups together in order to minimize their disruptive effects on the hydrogen-bonded water network. Hydrophobic groups held together in this way are sometimes said to be held together by "hydrophobic bonds," even though the attraction is actually caused by a repulsion from the water.

IONIC BONDS IN AQUEOUS SOLUTIONS

Charged groups are shielded by their interactions with water molecules. Ionic bonds are therefore quite weak in water.

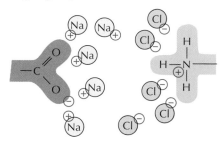

Similarly, other ions in solution can cluster around charged groups and further weaken ionic bonds.

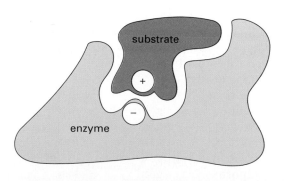

Ionic bonds are very important in biological systems; many enzymes guide substrates into position using ionic interactions.

IONIC BONDS

Ionic interactions occur either between fully charged groups (ionic bond) or between partially charged groups.

The force of attraction between the two charges, δ^+ and δ^-, falls off rapidly as the distance between the charges increases.

In the absence of water, ionic forces are very strong. They are responsible for the strength of such minerals as marble and agate.

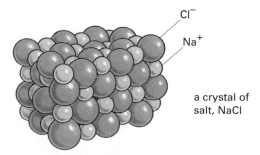

Cl⁻

Na⁺

a crystal of salt, NaCl

substrate

enzyme

Questions

Question 2–11

Which of the following statements are correct? Explain your answers.

A. An atomic nucleus contains protons and neutrons.
B. An atom has more electrons than protons.
C. The nucleus is surrounded by a double membrane.
D. All atoms of the same element have the same number of neutrons.
E. The number of neutrons determines whether the nucleus of an atom is stable or radioactive.
F. Both fatty acids and polysaccharides can be important energy stores in the cell.
G. Hydrogen bonds are weak and can be broken by thermal energy, yet they contribute significantly to the specificity of interactions between macromolecules.

Question 2–12

To gain a better feeling for atomic dimensions, assume that the page on which this question is printed is made entirely of the polysaccharide cellulose, whose molecules are described by the formula $(C_nH_{2n}O_n)$, where n can be a quite large number and is variable from one molecule to another. The atomic weights of carbon, hydrogen, and oxygen are 12, 1, and 16, respectively, and this page weighs 5 g.

A. How many carbon atoms are there in this page?
B. In cellulose, how many carbon atoms would be stacked on top of each other to span the thickness of this page (the size of the page is 21 cm × 27.5 cm, and it is 0.07 mm thick)?
C. Now consider the problem from a different angle. Assume that the page is composed only of carbon atoms. A carbon atom has a diameter of 2×10^{-10} m (0.2 nm); how many carbon atoms of 0.2 nm diameter would it take to span the thickness of the page?
D. Compare your answers from parts B and C and explain any differences.

Question 2–13

A. How many electrons can be accommodated in the first, second, and third electron shells of an atom?
B. How many electrons would atoms of the elements listed below preferentially gain or lose in order to obtain completely filled sets of energy levels?

hydrogen	gain __	lose __
helium	gain __	lose __
oxygen	gain __	lose __
carbon	gain __	lose __
sodium	gain __	lose __
chlorine	gain __	lose __

C. What do the answers tell you about the chemical properties of the elements and the bonds that can form between sodium and chlorine, between oxygen and hydrogen, between carbon and oxygen, and between carbon and hydrogen?

Question 2–14

Oxygen and sulfur have similar chemical properties because both elements have six electrons in their outermost electron shells. Indeed, both elements form molecules with two hydrogen atoms, water (H_2O) and hydrogen sulfide (H_2S). Surprisingly, water is a liquid, yet H_2S is a gas, despite the fact that sulfur is much larger and heavier than oxygen. Explain why this might be the case.

Question 2–15

Write the chemical formula for a condensation reaction of two amino acids to form a peptide bond. Write the formula for its hydrolysis.

Question 2–16

Which of the following statements are correct? Explain your answers.

A. Proteins are so remarkably diverse because each is made from a unique mixture of amino acids that are linked in random order.
B. Lipid bilayers are macromolecules that are made up mostly of phospholipid subunits.
C. Nucleic acids contain sugar groups.
D. Many amino acids have hydrophobic side chains.
E. The hydrophobic tails of phospholipid molecules are repelled from water.
F. DNA contains the four different bases A, G, U, and C.

Question 2–17

A. How many different molecules composed of (a) two, (b) three, and (c) four amino acids, linked together by peptide bonds, can be made from the set of 20 naturally occurring amino acids?
B. Assume you were given a mixture consisting of one molecule each of all possible sequences of a smallish protein of molecular weight 4800. How big a container would you need to hold this sample? Assume that the average molecular weight of an amino acid is 120.
C. What does this calculation tell you about the fraction of possible proteins that are currently in use by living organisms (the average molecular weight of proteins is about 30,000)?

Question 2–18

This is a biology textbook. Explain why the chemical principles that are described in this chapter are important in the context of modern cell biology.

Question 2–19

A. Describe the similarities and differences between van der Waals attractions and hydrogen bonds.

B. Which of the two bonds would form (a) between two hydrogens bound to carbon atoms, (b) between a nitrogen atom and a hydrogen bound to a carbon atom, and (c) between a nitrogen atom and a hydrogen bound to an oxygen atom?

Question 2–20

What are the forces that determine the folding of a macromolecule into a unique shape?

Question 2–21

Fatty acids are said to be "amphipathic." What is meant by this term, and how does an amphipathic molecule behave in water? Draw a diagram to illustrate your answer.

Question 2–22

Are the formulas in Figure Q2–22 correct or incorrect? Explain your answer in each case.

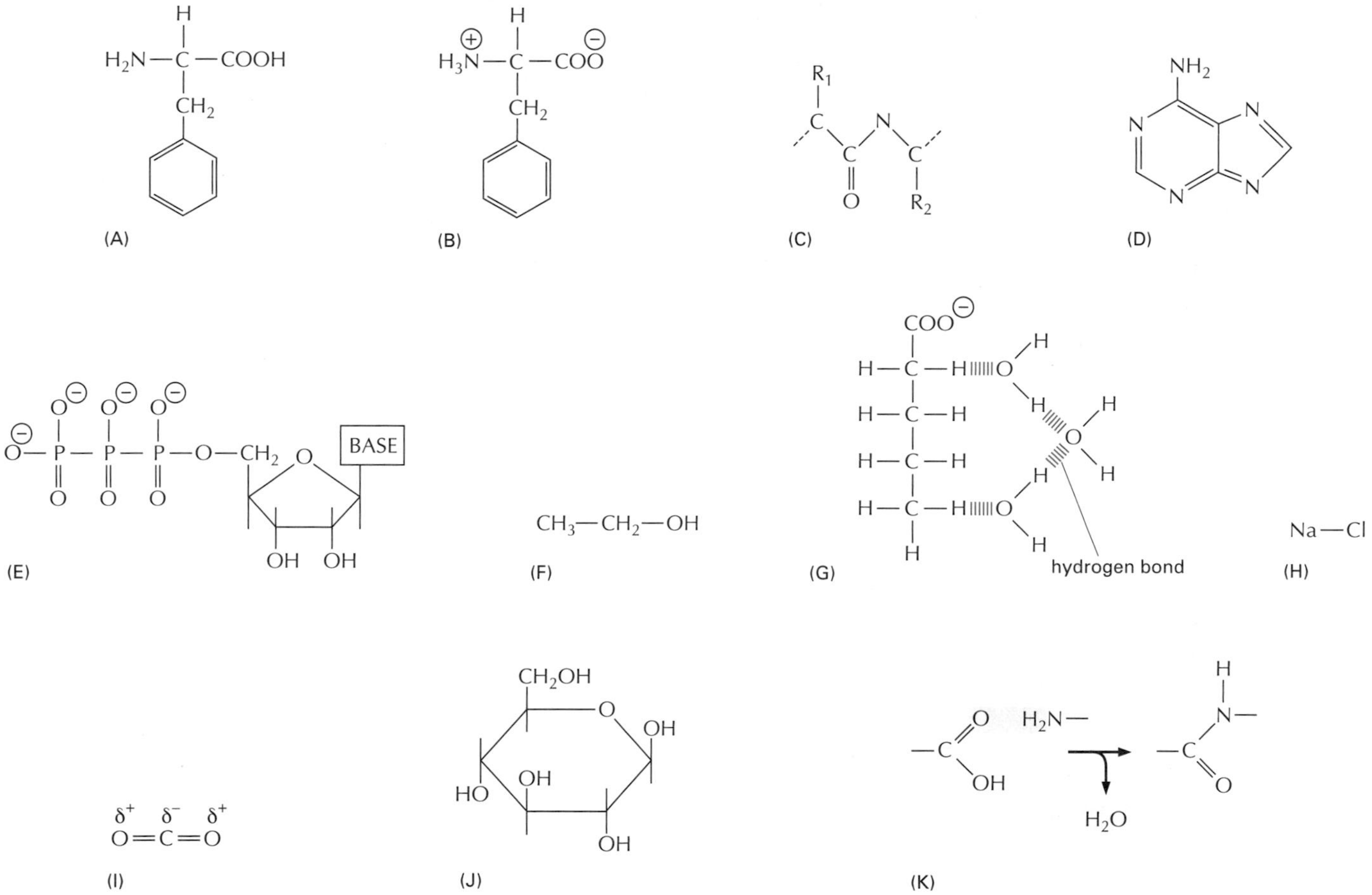

Figure Q2–22

Highlights from *Essential Cell Biology 2 Interactive* CD-ROM

2.1 Glucose Molecule

2.2 Palmitic Acid

2.3 ATP

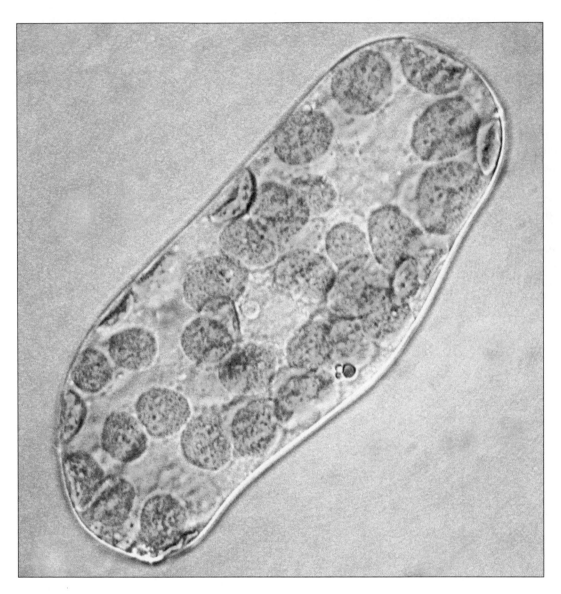

Chloroplasts. Most of the energy that powers life on Earth derives ultimately from sunlight. Electromagnetic energy from the sun is trapped by chlorophyll molecules in the chloroplasts of plants and is converted to chemical bond energy. Fifty or so of these green disc-like chloroplasts can be seen here in a single cell that has been isolated from the interior of a leaf. (Courtesy of Preeti Dahiya.)

Energy, Catalysis, and Biosynthesis

One property above all makes living things seem almost miraculously different from nonliving matter: they create and maintain order, in a universe that is tending always toward greater disorder. To create this order, the cells in a living organism must carry out a never-ending stream of chemical reactions. In some of these reactions, small organic molecules—amino acids, sugars, nucleotides, and lipids—are taken apart or modified to supply the many other small molecules that the cell requires. In other reactions, these small molecules are used to construct an enormously diverse range of proteins, nucleic acids, and other macromolecules that endow living systems with all of their most distinctive properties. Each cell can be viewed as a tiny chemical factory, performing many millions of reactions every second.

To carry out the many chemical reactions needed to sustain it, a living organism requires not only a source of atoms in the form of food molecules, but also a source of energy. Both the atoms and the energy must come, ultimately, from the nonliving environment. In this chapter we discuss why cells require energy, and how they use this energy and the atoms from their environment to create the molecular order that makes life possible.

The chemical reactions that every cell performs would normally occur only at temperatures that are much higher than those that exist inside cells. For this reason, each reaction requires a specific boost in chemical reactivity. But rather than being an inconvenience, this prerequisite is a benefit: not only does it allow reactions to proceed at temperatures that occur inside a cell, but it allows the cell to precisely control its metabolism—a feature central to the chemistry of life. Both the boost in reactivity and the precise chemical control are provided by specialized proteins called *enzymes*, each of which accelerates, or *catalyzes*, just one of the many possible kinds of reactions that a particular molecule might undergo. Enzyme-catalyzed reactions are usually connected in series, so that the product of one reaction becomes the starting material, or *substrate*, for the next (Figure 3–1). These long linear reaction pathways, or metabolic pathways, are in turn linked to one another, forming a complex web of interconnected reactions that enable the cell to survive, grow, and reproduce (Figure 3–2).

Two opposing streams of chemical reactions occur in cells. The *catabolic* pathways break down foodstuffs into smaller molecules, thereby generating both a useful form of energy for the cell and some of the small molecules that the cell needs as building blocks; and the *anabolic*, or *biosynthetic*, pathways use the energy harnessed by catabolism to drive the synthesis of the many molecules that form the cell. Together these two sets of reactions constitute the **metabolism** of the cell (Figure 3–3).

Catalysis and the Use of Energy by Cells

Biological Order Is Made Possible by the Release of Heat Energy from Cells

Photosynthetic Organisms Use Sunlight to Synthesize Organic Molecules

Cells Obtain Energy by the Oxidation of Organic Molecules

Oxidation and Reduction Involve Electron Transfers

Enzymes Lower the Barriers That Block Chemical Reactions

The Free-Energy Change for a Reaction Determines Whether It Can Occur

The Concentration of Reactants Influences the Free-Energy Change and a Reaction's Direction

The Equilibrium Constant Indicates the Strength of Molecular Interactions

For Sequential Reactions, the Changes in Free Energy Are Additive

Rapid Diffusion Allows Enzymes to Find Their Substrates

V_{max} and K_M Measure Enzyme Performance

Activated Carrier Molecules and Biosynthesis

The Formation of an Activated Carrier Is Coupled to an Energetically Favorable Reaction

ATP Is the Most Widely Used Activated Carrier Molecule

Energy Stored in ATP Is Often Harnessed to Join Two Molecules Together

NADH and NADPH Are Important Electron Carriers

There Are Many Other Activated Carrier Molecules in Cells

The Synthesis of Biological Polymers Requires an Energy Input

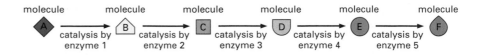

Figure 3–1 A series of enzyme-catalyzed reactions forms a metabolic pathway. Each enzyme catalyzes a particular chemical reaction, which leaves the enzyme unchanged. In this example, a set of enzymes acting in series converts molecule A to molecule F, forming a metabolic pathway.

Cell metabolism is the subject matter of *biochemistry,* and the details need not concern us here. But the general principles by which cells obtain energy from their environment and use it to create order are central to cell biology. These principles are outlined in this chapter, starting with a discussion of why a constant input of energy is needed to sustain living organisms.

Catalysis and the Use of Energy by Cells

Nonliving things left to themselves eventually become disordered: buildings crumble and dead organisms decay. Living cells, by contrast, not only maintain but actually generate order at every level, from the large-scale structure of a butterfly or a flower down to the organization of the atoms in the molecules from which these organisms are made (Figure 3–4). This property of life is made possible by elaborate cellular mechanisms that extract energy from the environment and convert it

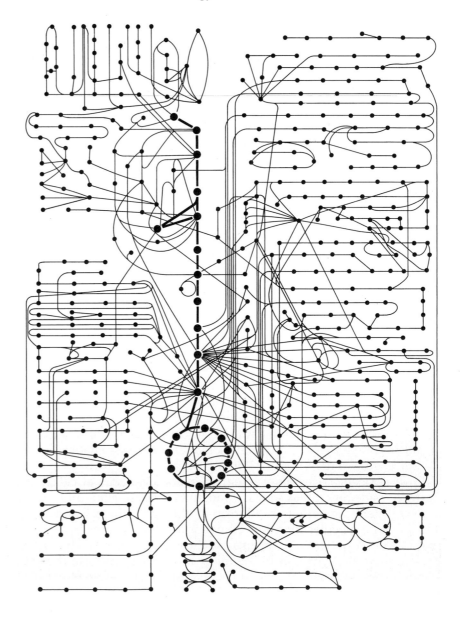

Figure 3–2 In a typical cell, the metabolic pathways are linked to generate a complex interconnected network. About 500 common metabolic reactions are shown diagrammatically, with each molecule in a metabolic pathway represented by a filled circle.

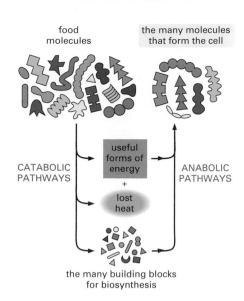

Figure 3–3 **Catabolic and anabolic pathways together constitute the cell's metabolism.** Note that because a major portion of the energy stored in the chemical bonds of food molecules is dissipated as heat, the mass of food required by any organism that derives all of its energy from catabolism is much greater than the mass of the molecules that can be produced by anabolism.

food molecules

the many molecules that form the cell

CATABOLIC PATHWAYS

useful forms of energy + lost heat

ANABOLIC PATHWAYS

the many building blocks for biosynthesis

into the energy stored in chemical bonds, which can be used by the cell to drive the constant generation of biological order. This cellular manipulation of energy guarantees that biological structures maintain their form, even as the materials of which they are made are replaced and recycled: your body today has the same basic structure as it did 10 years ago, even though you now contain atoms that, for the most part, were not in your body then.

Biological Order Is Made Possible by the Release of Heat Energy from Cells

The universal tendency of things to become disordered is expressed in a fundamental law of physics—the *second law of thermodynamics*—which states that in the universe, or in any isolated system (a collection of matter that is completely isolated from the rest of the universe), the degree of disorder (or *entropy*) can only increase. This law has such profound implications for all living things that it is worth restating in several ways.

For example, we can present the second law in terms of probability and state that *systems will change spontaneously toward those arrangements that have the greatest probability*. If we consider, for example, a box of 100 coins all lying heads up, a series of accidents that disturbs the box will tend to move the arrangement toward a mixture of 50 heads and 50 tails. The reason is simple: there are a huge number of possible arrangements of the individual coins in the mixture that can achieve the 50–50 result, but only one possible arrangement that keeps all of the coins oriented heads up. Because the 50–50 mixture accommodates a greater number of possibilities and places fewer constraints on the detailed configuration of each individual coin, we say that it is more "disordered." For the same reason, it is a common experience that one's living space will become increasingly disordered without intentional effort: the movement toward disorder is a spontaneous process, requiring a periodic effort to reverse it (Figure 3–5).

The amount of disorder in a system can be quantified. The measure of a system's disorder is called the **entropy** of the system: the greater the disorder, the greater the entropy. Thus, another way to express the second law of thermodynamics is to say that systems will change spontaneously toward arrangements with greater entropy.

Figure 3–4 **Biological structures are highly ordered.** Well-defined, ornate, and beautiful spatial patterns can be found at every level of organization in living organisms. In order of increasing size: (A) protein molecules in the coat of a virus (a parasite that, although not technically alive, contains the same types of molecules as those found in living cells); (B) the regular array of microtubules seen in a cross section of a sperm tail; (C) surface contours of a pollen grain (a single cell); (D) close-up of the wing of a butterfly showing the pattern created by scales, each scale being the product of a single cell; (E) spiral array of seeds, made of millions of cells, in the head of a sunflower. (A, courtesy of Robert Grant, Stéphane Crainic, and James M. Hogle; B, courtesy of Lewis Tilney; C, courtesy of Colin MacFarlane and Chris Jeffree; D and E, courtesy of Kjell B. Sandved.)

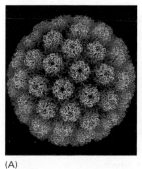

(A)

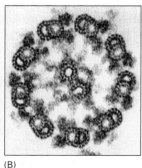

(B)

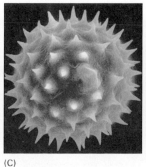

(C)

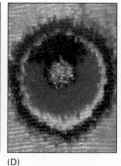

(D)

(E)

Figure 3–5 The spontaneous drive toward disorder is an everyday experience. Reversing this natural tendency toward disorder requires an intentional effort and an input of energy: the restoration of order does not occur spontaneously. In fact, from the second law of thermodynamics, we can be certain that the human intervention required will release enough heat to the environment to more than compensate for the reordering of the items in this room.

ORGANIZED EFFORT REQUIRING ENERGY INPUT

Living cells—by surviving, growing, and forming complex organisms—are generating order and thus might appear to defy the second law of thermodynamics. But this is not the case. A cell is not an isolated system: it takes in energy from its environment—in the form of food, or photons from the sun (or even, for some chemosynthetic bacteria, from inorganic molecules alone)—and it then uses this energy (known as *free energy*) to generate order within itself, forging new chemical bonds or building large macromolecules. In the course of performing the chemical reactions that generate order, chemical bond energy is converted into heat. Heat is energy in its most disordered form—the random jostling of molecules. (Think of the coins in the box.) Because the cell is not an isolated system, the heat energy that its reactions generate is quickly dispersed into the cell's surroundings. There the heat increases the intensity of the thermal motions of the resident molecules, thereby increasing the randomness, or disorder, of the environment (Figure 3–6).

The amount of heat released by the cell must be such that the order generated inside a cell is more than compensated for by a greater decrease in order in the environment. Only in this case is the second law of thermodynamics satisfied, because the total entropy of the system—that of the cell plus its environment—will increase as a result of the chemical reactions inside the cell.

Figure 3–6 Living cells do not defy the second law of thermodynamics. In the diagram on the left, the molecules of both the cell and the rest of the universe (the environment) are depicted in a relatively disordered state. In the diagram on the right, the cell has taken in energy from food molecules and released heat by carrying out a reaction that orders the molecules that the cell contains. Because the heat increases the disorder in the environment around the cell (depicted by the *jagged arrows* and distorted molecules, indicating increased molecular motions), the second law of thermodynamics, which states that the amount of disorder in the universe must always increase, is satisfied even as the cell grows and divides.

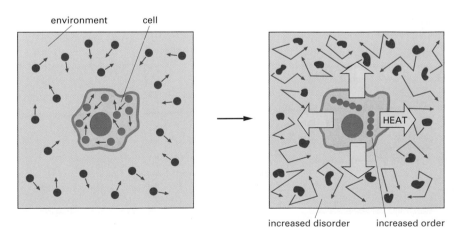

environment   cell

HEAT

increased disorder   increased order

Where does the heat that the cell releases come from? Here we encounter another important law of thermodynamics. The *first law of thermodynamics* states that energy can be converted from one form to another, but that it cannot be created or destroyed. Some forms of energy are illustrated in Figure 3–7. The amount of energy present in different forms will change as a result of the chemical reactions inside the cell, but the first law tells us that the total amount of energy in the universe must always be the same. For example, an animal cell takes in foodstuffs and converts some of the energy present in the chemical bonds between the atoms of these food molecules (chemical bond energy) into the random thermal motion of molecules (heat energy). As we have explained, this conversion of chemical energy into heat energy is essential if the reactions inside the cell are to cause the universe as a whole to become more disordered—as required by the second law.

The cell cannot derive any benefit from the heat energy it produces, however, unless the heat-generating reactions inside the cell are directly linked to the processes that maintain molecular order. It is the tight coupling of heat production to an increase in order that distinguishes the metabolism of a cell from the wasteful burning of fuel in a fire. Later

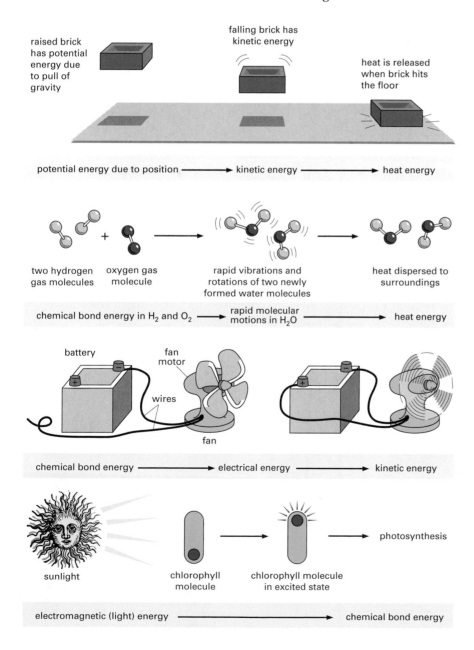

Figure 3–7 Different forms of energy are interconvertible. Energy can be converted from one form to another, but in the processes the total amount of energy must be conserved. Thus, for example, from the height and weight of the brick in the first sketch, we can predict exactly how much heat will be released when it hits the floor. In the second sketch, note that the large amount of chemical bond energy released when water is formed is initially converted to very rapid thermal motions in the two new water molecules; however, collisions with other molecules almost instantaneously spread this kinetic energy evenly throughout the surroundings (heat transfer), making the new molecules indistinguishable from all the rest. Cells can convert chemical bond energy into kinetic energy to drive, for example, molecular motors, although without the intermediate conversion to electrical energy that an appliance such as a fan requires. They also harvest light energy to form chemical bonds.

Figure 3–8 With few exceptions, the radiant energy of sunlight sustains all of the life around us. Trapped by plants and some microorganisms through photosynthesis, light from the sun is the ultimate source of all energy for humans and other animals. (Courtesy of Museum Folkwang, Essen.)

in this chapter, we shall illustrate how this coupling occurs. For the moment, it is sufficient to recognize that a direct linkage of the "burning" of food molecules to the generation of biological order is required if cells are to be able to create and maintain an island of order in a universe tending toward chaos.

Photosynthetic Organisms Use Sunlight to Synthesize Organic Molecules

All animals live on energy stored in the chemical bonds of organic molecules made by other organisms, which they take in as food. The molecules in food also provide the atoms that animals need to construct new living matter. Some animals obtain their food by eating other animals. But at the bottom of the animal food chain are animals that eat plants. These plants, before being consumed, were busy trapping energy from sunlight. As a result, all of the energy used by animal cells is derived ultimately from the sun (Figure 3–8).

Solar energy enters the living world through **photosynthesis** carried out by plants and photosynthetic bacteria. Photosynthesis allows the electromagnetic energy in sunlight to be converted into chemical bond energy in the cell. Plants are able to obtain all of the atoms they need from inorganic sources: carbon from atmospheric carbon dioxide, hydrogen and oxygen from water, nitrogen from ammonia and nitrates in the soil, and other elements needed in smaller amounts from inorganic salts in the soil. They use the energy they derive from sunlight to form chemical bonds between these atoms, linking them into small chemical building blocks such as sugars, amino acids, nucleotides, and fatty acids. These small molecules in turn are converted into the macromolecules—the proteins, nucleic acids, polysaccharides, and lipids—that form the plant. All of these substances serve as food molecules for animals, and for fungi or bacteria, when these plants are later consumed.

The reactions of photosynthesis take place in two stages (Figure 3–9). In the first, light-dependent stage, energy from sunlight is captured and transiently stored as chemical bond energy in specialized small molecules that carry energy in their reactive chemical groups. (We discuss these activated carrier molecules in more detail later.) Molecular oxygen (O_2 gas), derived from the splitting of water by light, is released as a by-product of this first stage.

In the second stage of photosynthesis, the molecules that serve as energy carriers are used to help drive a *carbon-fixation* process in which sugars are manufactured from carbon dioxide gas (CO_2) and water (H_2O). By producing sugars, these light-independent reactions generate a critical source of stored chemical bond energy and materials—both for the plant itself and for any animals that eat it. We describe the elegant mechanisms that underlie these two stages of photosynthesis in Chapter 14.

Figure 3–9 Photosynthesis takes place in two stages. The energy carriers created in the first stage are two molecules that we will discuss shortly—ATP and NADPH.

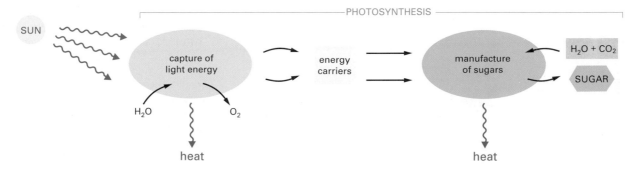

The net result of the entire process of photosynthesis, so far as the green plant is concerned, can be summarized simply in the equation

$$\text{light energy} + CO_2 + H_2O \rightarrow \text{sugars} + O_2 + \text{heat energy}$$

The sugars produced are then used both as a source of chemical bond energy and as a source of materials to make the many other small and large organic molecules that are essential to the plant cell.

Cells Obtain Energy by the Oxidation of Organic Molecules

All animal and plant cells are powered by chemical energy stored in the chemical bonds of organic molecules—either the sugars that a plant has photosynthesized as food for itself or the mixture of large and small molecules that an animal has eaten. To use this energy to live, grow, and reproduce, organisms must extract it in a usable form. In both plants and animals, energy is extracted from food molecules by a process of gradual oxidation, or controlled burning.

The Earth's atmosphere contains a great deal of oxygen, and in the presence of oxygen the most energetically stable form of carbon is CO_2 and that of hydrogen is H_2O. A cell is therefore able to obtain energy from sugars or other organic molecules by allowing their carbon and hydrogen atoms to combine with oxygen to produce CO_2 and H_2O, respectively—a process known as cellular **respiration**.

Photosynthesis and respiration are complementary processes (Figure 3–10). This means that the transactions between plants and animals are not all one way. Plants, animals, and microorganisms have existed together on this planet for so long that many of them have become an essential part of each others' environments. The oxygen released by photosynthesis is consumed in the combustion of organic molecules by nearly all organisms. And some of the CO_2 molecules that are fixed today into organic molecules by photosynthesis in a green leaf were released yesterday into the atmosphere by the respiration of an animal—or by a fungus or bacterium decomposing dead organic matter. We therefore see that carbon utilization forms a huge cycle that involves the *biosphere* (all of the living organisms on Earth) as a whole, crossing boundaries between individual organisms (Figure 3–11). Similarly, atoms of nitrogen, phosphorus, and sulfur move between the living and nonliving worlds in cycles that involve plants, animals, fungi, and bacteria. Procaryotes, in fact, are estimated to contain nearly half the carbon stored in living organisms. And they are the single largest reservoir of nitrogen and phosphorus on Earth, containing 10 times more of these nutrients than plants.

Question 3–1

Consider the equation

$$\text{light energy} + CO_2 + H_2O \rightarrow \text{sugars} + O_2 + \text{heat energy}$$

Would you expect this reaction to be catalyzed by a single enzyme? Why is heat generated in the reaction? Explain your answers.

Figure 3–10 Photosynthesis and cellular respiration are complementary processes in the living world. Photosynthesis uses the energy of sunlight to produce sugars and other organic molecules. These molecules in turn serve as food for other organisms. Many of these organisms carry out cellular respiration, a process that uses O_2 to form CO_2 from the same carbon atoms that had been taken up as CO_2 and converted into sugars by photosynthesis. In the process, the organisms obtain the chemical bond energy that they need to survive. The first cells on the Earth are thought to have been capable of neither photosynthesis nor respiration (discussed in Chapter 14). However, photosynthesis must have preceded respiration on the Earth, because there is strong evidence that billions of years of photosynthesis were required before O_2 had been released in sufficient quantity to create an atmosphere rich in this gas to support respiration. (The Earth's atmosphere presently contains 20% O_2.)

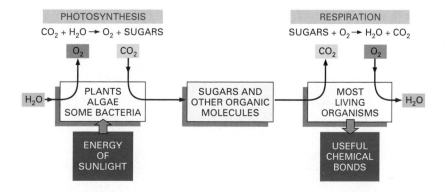

Oxidation and Reduction Involve Electron Transfers

The cell does not oxidize organic molecules in one step, as occurs when organic material is burned in a fire. Through the use of enzyme catalysts, metabolism carries the molecules through a large number of reactions that only rarely involve the direct addition of oxygen. Before we consider some of these reactions and the purpose behind them, we need to discuss what is meant by oxidation.

Oxidation does not mean only the addition of oxygen atoms; the term also applies more generally to any reaction in which electrons are transferred from one atom to another. Oxidation, in this sense, refers to the removal of electrons, and **reduction**—the converse of oxidation—refers to the addition of electrons. Thus, Fe^{2+} is oxidized if it loses an electron to become Fe^{3+}, and a chlorine atom is reduced if it gains an electron to become Cl^-. Because the number of electrons is conserved in a chemical reaction (there is no net loss or gain), oxidation and reduction always occur simultaneously: that is, if one molecule gains an electron in a reaction (reduction), a second molecule loses the electron (oxidation). When a sugar molecule is oxidized to CO_2 and H_2O, for example, the O_2 molecules involved in forming H_2O gain electrons and thus are said to have been reduced.

The terms "oxidation" and "reduction" apply even when there is only a partial shift of electrons between atoms linked by a covalent bond (Figure 3–12A). When a carbon atom becomes covalently bonded to an atom with a strong affinity for electrons—oxygen, chlorine, or sulfur, for example—it gives up more than its equal share of electrons and forms a *polar* covalent bond. The positive charge of the carbon nucleus now slightly exceeds the negative charge of its electrons: the carbon atom therefore acquires a partial positive charge and is said to be oxidized. Conversely, a carbon atom in a C–H linkage has somewhat more than its share of electrons, and so it is said to be reduced (Figure 3–12B).

When a molecule in a cell picks up an electron (e^-), it often picks up a proton (H^+) at the same time (protons being freely available in water). The net effect in this case is to add a hydrogen atom to the molecule

$$A + e^- + H^+ \rightarrow AH$$

Even though a proton plus an electron is involved (instead of just an electron), such *hydrogenation* reactions are reductions, and the reverse, *dehydrogenation* reactions, are oxidations. It is especially easy to tell whether an organic molecule is being oxidized or reduced: reduction is occurring if its number of C–H bonds increases, whereas oxidation is occurring if its number of C–H bonds decreases.

Cells use enzymes to catalyze the oxidation of organic molecules in small steps, through a sequence of reactions that allows useful energy to be harvested.

Figure 3–11 Carbon atoms cycle continuously through the biosphere. Individual carbon atoms are incorporated into organic molecules of the living world by the photosynthetic activity of plants, bacteria, and marine algae. They pass to animals, microorganisms, and organic material in soil and oceans in cyclic pathways. CO_2 is restored to the atmosphere when organic molecules are oxidized by cells or burned by humans as fossil fuels.

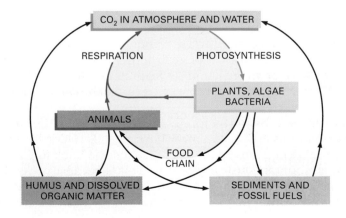

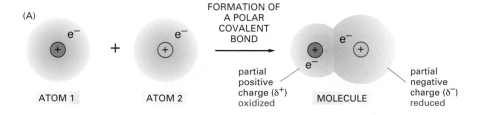

(A)

FORMATION OF
A POLAR
COVALENT
BOND

ATOM 1 + ATOM 2 → MOLECULE

partial
positive
charge (δ⁺)
oxidized

partial
negative
charge (δ⁻)
reduced

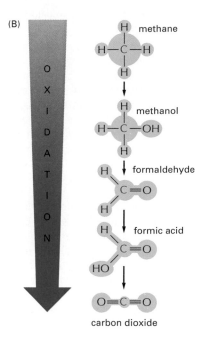

(B)

O X I D A T I O N

methane

methanol

formaldehyde

formic acid

carbon dioxide

Figure 3–12 Oxidation and reduction involve a shift in the balance of electrons. (A) When two atoms form a polar covalent bond (as discussed in Chapter 2), the atom that ends up with a greater share of electrons is said to be reduced, while the other atom, with a lesser share of electrons, is said to be oxidized. The reduced atom has acquired a partial negative charge (δ^-); conversely, the oxidized atom has acquired a partial positive charge (δ^+), as the positive charge on the atomic nucleus now exceeds the total charge of the electrons surrounding it. (B) A simple reduced carbon compound, such as methane, can be oxidized in a stepwise fashion by the successive replacement of its covalently bonded hydrogen atoms with oxygen atoms. With each step, electrons (represented by the *blue* clouds) are shifted away from the carbon, and the carbon atom becomes progressively more oxidized. Each of these steps is catalyzed by a cellular enzyme and is energetically favorable under the conditions present inside a cell.

Enzymes Lower the Barriers That Block Chemical Reactions

Consider the reaction:

$$\text{paper} + O_2 \rightarrow \text{smoke} + \text{ashes} + \text{heat} + CO_2 + H_2O$$

The paper burns readily, releasing into the atmosphere both water and carbon dioxide as gases, and energy as heat. But the reaction is one-way: smoke and ashes never spontaneously gather carbon dioxide and water from the heated atmosphere and reconstitute themselves into paper. When paper burns, its chemical energy is dissipated as heat—not lost from the universe, since energy can never be created or destroyed, but irretrievably dispersed in the chaotic random thermal motions of molecules. At the same time, the atoms and molecules of the paper become dispersed and disordered. In the language of thermodynamics, there has been a release of *free energy*, that is, of energy that can be harnessed to do work or drive chemical reactions. This loss reflects a loss of orderliness in the way the energy and molecules had been stored in the paper. We will discuss free energy in more detail shortly, but the general principle is clear enough intuitively: chemical reactions proceed only in the direction that leads to a loss of free energy; in other words, the spontaneous direction for any reaction is the direction that goes "downhill." A "downhill" reaction in this sense is often said to be *energetically favorable*.

Although the most energetically favorable form of carbon under ordinary conditions is CO_2, and that of hydrogen is H_2O, a living organism will not disappear in a puff of smoke, and the book in your hands will not burst spontaneously into flames. This is because the molecules both in the living organism and in the book are in a relatively stable state, and they cannot be changed to lower-energy states without an

Question 3–2

In which of the following reactions does the red atom undergo an oxidation?

A. $Na \rightarrow Na^+$ (Na atom → Na⁺ ion)
B. $Cl \rightarrow Cl^-$ (Cl atom → Cl⁻ ion)
C. $CH_3CH_2OH \rightarrow CH_3CHO$ (ethanol → acetaldehyde)
D. $CH_3CHO \rightarrow CH_3COOH$ (acetaldehyde → acetic acid)
E. $CH_2{=}CH_2 \rightarrow CH_3CH_3$ (ethene → ethane)

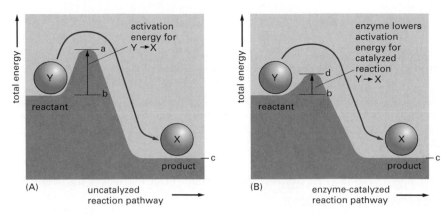

Figure 3–13 Even energetically favorable reactions require activation energy to get them started. (A) Compound Y (a reactant) is in a relatively stable state, and energy is required to convert it to compound X (a product), even though X is at a lower overall energy level than Y. This conversion will not take place, therefore, unless compound Y can acquire enough activation energy (*energy a minus energy b*) from its surroundings to undergo the reaction that converts it into compound X. This energy may be provided by means of an unusually energetic collision with other molecules. For the reverse reaction, X → Y, the activation energy will be much larger (*energy a minus energy c*); this reaction will therefore occur much more rarely. Activation energies are always positive; note, however, that the total energy change for the energetically favorable reaction Y → X is *energy c minus energy b,* a negative number. (B) Energy barriers for specific reactions can be lowered by catalysts, as indicated by the line marked *d.* Enzymes are particularly effective catalysts because they greatly reduce the activation energy for the reactions they perform.

Figure 3–14 Lowering the activation energy greatly increases the probability that a reaction will occur. At any given instant, a population of identical substrate molecules will have a range of energies, distributed as shown on the graph. The varying energies come from collisions with surrounding molecules, which make the substrate molecules jiggle, vibrate, and spin. For a molecule to undergo a chemical reaction, the energy of the molecule must exceed the activation energy barrier for that reaction *(dashed lines);* for most biological reactions, this almost never happens without enzyme catalysis. Even with enzyme catalysis, the substrate molecules must experience a particularly energetic collision to react *(red shaded* area). Raising the temperature can also increase the number of molecules with sufficient energy to overcome the activation energy needed for a reaction; but in contrast to enzyme catalysis, this effect is nonselective, speeding up all reactions.

initial input of energy. In other words, a molecule requires activation energy—a boost over an energy barrier—before it can undergo a chemical reaction that moves it to a lower-energy (more stable) state (Figure 3–13A). In the case of a burning book, the **activation energy** is provided by the heat of a lighted match. For the molecules in the watery solution inside a cell, the boost is delivered by an unusually energetic random collision with surrounding molecules—collisions that become more violent as the temperature is raised.

In a living cell, the push over the energy barrier is greatly aided by a specialized class of proteins—the **enzymes**. Each enzyme binds tightly to one or two molecules, called **substrates**, and holds them in a way that greatly reduces the activation energy needed to facilitate a specific chemical interaction between them (Figure 3–13B). A substance that can lower the activation energy of a reaction is termed a **catalyst**; catalysts increase the rate of chemical reactions because they allow a much larger proportion of the random collisions with surrounding molecules to kick the substrates over the energy barrier, as illustrated in Figure 3–14 and Figure 3–15A. Enzymes are among the most effective catalysts known, often speeding up reactions by a factor of as much as 10^{14} (trillions of times faster than the same reactions would proceed without an enzyme catalyst). Enzymes thereby allow reactions that would not otherwise occur to proceed rapidly at normal temperatures. Without enzymes, life could not exist.

Enzymes are also highly selective. Each usually catalyzes only one particular reaction: in other words, it selectively lowers the activation energy for only one of the several possible chemical reactions that its bound substrate molecules could undergo. In this way enzymes direct each of the many different molecules in a cell along specific reaction pathways (Figure 3–15B and C).

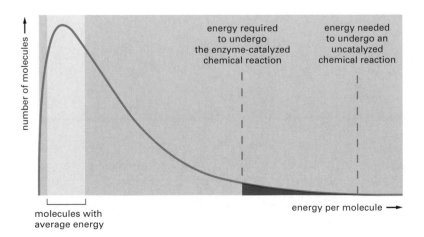

The success of living organisms is attributable to the cell's ability to make enzymes of many types, each with precisely specified properties. Each enzyme has a unique shape containing an *active site*, a pocket or groove in the enzyme into which only particular substrates will fit (Figure 3–16). Like all other catalysts, enzyme molecules themselves remain unchanged after participating in a reaction and therefore can function over and over again. In Chapter 4, we will discuss further how enzymes work, after we have looked in detail at the molecular structure of proteins.

The Free-Energy Change for a Reaction Determines Whether It Can Occur

Although enzymes speed up reactions, they cannot by themselves force energetically unfavorable reactions to occur. To use a water analogy, enzymes by themselves cannot make water run uphill. Cells, however, must do just that in order to grow and divide: they must build highly ordered and energy-rich molecules from small and simple ones. We will

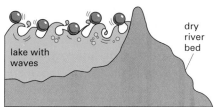

uncatalyzed reaction—waves not large enough to surmount barrier

catalyzed reaction—waves often surmount barrier

(A)

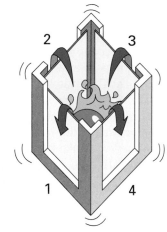

(B)   uncatalyzed

enzyme catalysis of reaction 1

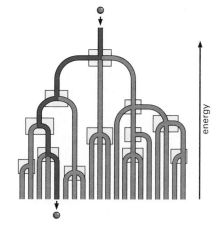

energy

(C)

Figure 3–15 Enzymes catalyze reactions by lowering the activation energy barrier. (A) The dam represents the activation energy, which is lowered by enzyme catalysis. The *green ball* represents a potential substrate that is bouncing up and down in energy level due to constant encounters with waves, an analogy for the thermal bombardment of the substrate with the surrounding water molecules. When the barrier—activation energy—is lowered significantly, it allows the balls (substrates) with sufficient energy to roll downhill, an energetically favorable movement. (B) The four walls of the box represent the activation energy barriers for four different chemical reactions that are all energetically favorable because the products are at lower energy levels than the substrates. In the left-hand box, none of these reactions occurs because even the largest waves are not large enough to surmount any of the energy barriers. In the right-hand box, enzyme catalysis lowers the activation energy for reaction number 1 only; now the jostling of the waves allows passage of the molecule over only this energy barrier, inducing reaction 1. (C) A branching river with a set of barrier dams (*yellow boxes*) serves to illustrate how a series of enzyme-catalyzed reactions determines the exact reaction pathway followed by each molecule inside the cell by controlling specifically which reaction will be allowed at each junction.

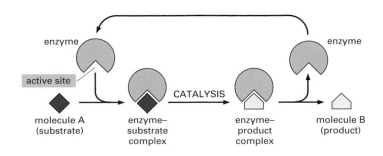

Figure 3–16 Enzymes convert substrates to products while remaining unchanged themselves. Each enzyme has an active site to which one or two substrate molecules bind, forming an enzyme–substrate complex. A reaction occurs at the active site, generating an enzyme–product complex. The product is then released, allowing the enzyme to bind additional substrate molecules.

see that this is done through enzymes that directly *couple* energetically favorable reactions, which release energy and produce heat, to energetically unfavorable reactions, which produce biological order.

Before examining how such coupling is achieved, we must consider more carefully the term "energetically favorable." According to the second law of thermodynamics, a chemical reaction can proceed only if it results in a net (or overall) increase in the disorder of the universe (see Figure 3–6). Disorder increases when useful energy that could be harnessed to do work is dissipated as heat. The criterion for an increase of disorder can be expressed most conveniently in terms of the **free energy**, *G*, of a system. The value of *G* is of interest only when a system undergoes a change, so that the **free-energy change**, denoted ΔG ("delta *G*"), can be specified. Suppose that the system being considered is a collection of molecules. Because of the way free energy is defined, ΔG measures the amount of disorder created in the universe when a reaction takes place that involves these molecules. *Energetically favorable reactions*, by definition, are those that create disorder by decreasing the free energy of the system to which they belong; in other words, they have a *negative* ΔG (Figure 3–17).

A familiar example of an energetically favorable reaction on a macroscopic scale is the "reaction" by which a compressed spring relaxes to an expanded state, releasing its stored elastic energy as heat to its surroundings; an example on a microscopic scale is the dissolving of salt in water. Conversely, *energetically unfavorable reactions*, with a *positive* ΔG—such as those in which two amino acids are joined together to form a peptide bond—by themselves create order in the universe. Because these reactions require energy, they can take place only if they are coupled to a second reaction with a negative ΔG so large that the net ΔG of the entire process is negative (Figure 3–18). These concepts are summarized, with examples, in Panel 3–1 (pp. 96–97).

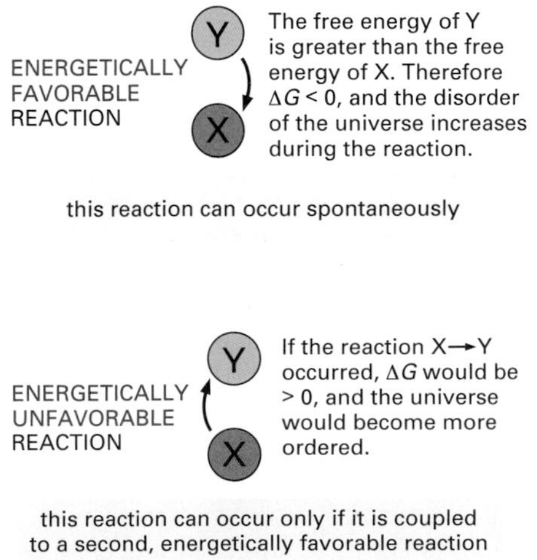

The Concentration of Reactants Influences the Free-Energy Change and a Reaction's Direction

As we have just described, a reaction $Y \rightleftharpoons X$ will go in the direction $Y \rightarrow X$ when the associated free-energy change, ΔG, is negative, just as a tensed spring left to itself will relax and lose its stored energy to its surroundings as heat. For a chemical reaction, however, ΔG depends not only on the energy stored in each individual molecule, but also on the concentrations of the molecules in the reaction mixture. Remember that ΔG reflects the degree to which a reaction creates a more disordered—in other words, a more probable—state of the universe. Recalling our coin analogy, it is very likely that a coin will flip from a head to a tail orientation if a jiggling box contains 90 heads and 10 tails, but this is a less probable event if the box contains 10 heads and 90 tails. For exactly the same reason, for a reversible reaction $Y \rightleftharpoons X$, a large excess of Y over X will tend to drive the reaction in the direction $Y \rightarrow X$; that is, there will

ENERGETICALLY
FAVORABLE
REACTION

Y
X

The free energy of Y is greater than the free energy of X. Therefore $\Delta G < 0$, and the disorder of the universe increases during the reaction.

this reaction can occur spontaneously

ENERGETICALLY
UNFAVORABLE
REACTION

Y
X

If the reaction X→Y occurred, ΔG would be > 0, and the universe would become more ordered.

this reaction can occur only if it is coupled to a second, energetically favorable reaction

Figure 3–17 Energetically favorable reactions have a negative ΔG and energetically unfavorable reactions a positive ΔG.

be a tendency for there to be more molecules making the transition $Y \rightarrow X$ than there are molecules making the transition $X \rightarrow Y$. Therefore, the ΔG becomes more negative for the transition $Y \rightarrow X$ (and more positive for the transition $X \rightarrow Y$) as the ratio of Y to X increases.

How much of a concentration difference is needed to compensate for a given decrease in chemical bond energy (and an accompanying heat release)? The answer is not intuitively obvious, but it can be determined from a thermodynamic analysis that separates the concentration-dependent and the concentration-independent parts of the free-energy change. The ΔG for a given reaction can thereby be written as the sum of two parts: the first, called the *standard free-energy change, $\Delta G°$*, depends on the intrinsic characters of the reacting molecules, based on their behavior under ideal conditions; the second depends on their concentrations. For the simple reaction $Y \rightarrow X$ at 37°C,

$$\Delta G = \Delta G° + 0.616 \ln \frac{[X]}{[Y]}$$

where ΔG is in kilocalories per mole, [Y] and [X] denote the concentrations of Y and X, 0.616 is a constant, and ln is the natural logarithm.

Note that ΔG equals the value of $\Delta G°$ when the molar concentrations of Y and X are equal (ln 1 = 0). As expected, ΔG becomes more negative as the ratio of X to Y decreases (the ln of a number < 1 is negative).

Chemical **equilibrium** is reached when the forward and reverse reaction rates are equal, resulting in a state at which the ratio of reactant to product remains constant. In other words,

$$K = \frac{[X]}{[Y]}$$

where K is the **equilibrium constant**. This expression portrays the point at which the concentration effect just balances the push given to the reaction by $\Delta G°$, so that there is no net change of free energy to drive the reaction in either direction (Figure 3–19). Here $\Delta G = 0$, and so the concentrations of Y and X are such that

$$-0.616 \ln \frac{[X]}{[Y]} = \Delta G° = -1.42 \log \frac{[X]}{[Y]}$$

Because ln is $\log_e$, this means that there is chemical equilibrium at 37°C (our body temperature) when

$$\frac{[X]}{[Y]} = e^{-\Delta G°/0.616} = 10^{-\Delta G°/1.42}$$

Table 3–1 shows how the equilibrium ratio of Y to X (expressed as an equilibrium constant, K) depends on the intrinsic character of the molecules, as expressed in the value of $\Delta G°$.

We have thus far discussed the simplest of reactions, $Y \rightarrow X$, in which a single substrate is converted into a product. But what happens in the more common situation, where two reactants combine to form a single product, $A + B \rightleftharpoons AB$?

The same principles apply, except that now the equilibrium constant, K, depends on the concentrations of both of the reactants in addition to the product:

$$K = [AB]/[A][B]$$

The concentrations of both substrates are multiplied because the formation of product AB depends on the collision of A and B, and these encounters occur at a rate that is proportional to $[A] \times [B]$.

The Equilibrium Constant Indicates the Strength of Molecular Interactions

Because the equilibrium constant (K) of a reaction is related directly to the standard free-energy change ($\Delta G°$), it is commonly employed as a

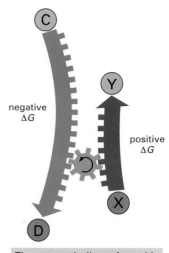

The energetically unfavorable reaction X → Y is driven by the energetically favorable reaction C → D, because the free-energy change for the pair of coupled reactions is less than zero.

Figure 3–18 Reaction coupling can drive an energetically unfavorable reaction.

Question 3–3

Consider again the analogy of the jiggling box containing coins that was described in the text. The reaction, the flipping of coins that either face heads up (H) or tails up (T), is described by the equation H $\rightleftharpoons$ T.

A. What are ΔG and $\Delta G°$ in this analogy?

B. What corresponds to the temperature at which the reaction proceeds? What corresponds to the activation energy of the reaction? Assume you have an "enzyme," called jigglase, that catalyzes this reaction. What would the effect of jigglase be and what, mechanically, might jigglase do in this analogy?

Panel 3–1 Free energy and biological reactions

FREE ENERGY

This panel reviews the concept of free energy and offers examples showing how changes in free energy determine whether—and how—biological reactions occur.

The molecules of a living cell possess energy because of their vibrations, rotations, and movement through space, and because of the energy that is stored in the bonds between individual atoms.

The free energy, G (in kcal/mole or kJ/mole; 1 kilocalorie is equal to 4.184 kilojoules), measures the energy of a molecule that could in principle be used to do useful work at constant temperature (as in a living cell).

REACTIONS CAUSE DISORDER

Think of a chemical reaction occurring in an isolated cell with constant temperature and volume. This reaction can produce disorder in two ways.

1 Changes of bond energy of the reacting molecules can cause heat to be released, which disorders the environment.

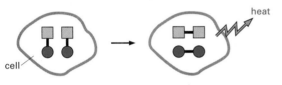

2 The reaction can decrease the amount of order in the reacting molecules—for example, by breaking apart a long chain of molecules, or by disrupting an interaction that prevents bond rotations.

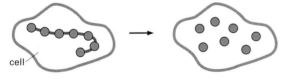

REACTION RATES

A spontaneous reaction is not necessarily an instantaneous reaction: a reaction with a negative free-energy change (ΔG) will not necessarily occur rapidly by itself. For the combustion of glucose in oxygen:

$$\text{glucose} + 6O_2 \longrightarrow 6CO_2 + 6H_2O$$

$$\Delta G° = -686 \text{ kcal/mole}$$

But even this highly favorable reaction may not occur for centuries unless there are enzymes to speed up the process. Conversely, enzymes are able to catalyze reactions and speed up their rate, but they do not change the $\Delta G°$ of the reaction.

ΔG ("DELTA G")

Changes in free energy occurring in a reaction are denoted by ΔG, where "Δ" indicates a difference. Thus for the reaction:

$$A + B \longrightarrow C + D$$

ΔG = free energy (C + D) minus free energy (A + B)

ΔG measures the amount of disorder caused by a reaction: the change in order inside the cell, plus the change in order of the surroundings caused by the heat released.

ΔG is useful because it measures how far away from equilibrium a reaction is. Thus the reaction

has a large negative ΔG because cells keep it a long way from equilibrium by continually making fresh ATP. However, if the cell dies, then most of its ATP becomes hydrolyzed, until equilibrium is reached (forward and backward reactions occur at equal rates) and $\Delta G = 0$.

SPONTANEOUS REACTIONS

From the second law of thermodynamics, we know that the disorder of the universe can only increase. ΔG is *negative* if the disorder of the universe (reaction plus surroundings) *increases*.

In other words, a chemical reaction that occurs spontaneously must have a negative ΔG:

$$G_{products} - G_{reactants} = \Delta G < 0$$

EXAMPLE: The difference in free energy of 100 ml of 10 mM sucrose (common sugar) and 100 ml of 10 mM glucose plus 10 mM fructose is about –5.5 calories. Therefore, the hydrolysis reaction (sucrose ⟶ glucose + fructose) can proceed spontaneously.

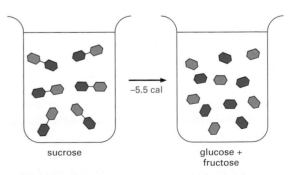

In contrast, the reverse reaction (glucose + fructose ⟶ sucrose), which has a ΔG of +5.5 calories, could not occur without an input of energy from a coupled reaction.

PREDICTING REACTIONS

To predict the outcome of a reaction (will it proceed to the right or to the left? at what point will it stop?), we must measure its standard free-energy change (ΔG°). This quantity represents the gain or loss of free energy as one mole of reactant is converted to one mole of product under "standard conditions" (all molecules present at a concentration of 1 M and pH 7.0).

ΔG° for some reactions

driving force

glucose-1-P → glucose-6-P	–1.7 kcal/mole
sucrose → glucose + fructose	–5.5 kcal/mole
ATP → ADP + P$_i$	–7.3 kcal/mole
glucose + 6O$_2$ → 6CO$_2$ + 6H$_2$O	–686 kcal/mole

CHEMICAL EQUILIBRIA

A fixed relationship exists between the standard free-energy change of a reaction, ΔG°, and its equilibrium constant K. For example, the reversible reaction

$$Y \rightarrow X$$

will proceed until the ratio of concentrations [X]/[Y] is equal to K (note: square brackets [ ] indicate concentration). At this point the free energy of the system will have its lowest value.

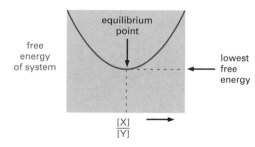

free energy of system

equilibrium point

lowest free energy

$$\frac{[X]}{[Y]} \longrightarrow$$

At 37°C, $\Delta G^\circ = -1.42 \log_{10}K$

$K = 10^{-\Delta G^\circ/1.42}$

For example, the reaction

glucose-1-P   glucose-6-P

has $\Delta G^\circ = -1.74$ kcal/mole. Therefore, its equilibrium constant

$$K = 10^{(1.74/1.42)} = 10^{(1.23)} = 17$$

So the reaction will reach steady state when

$$[\text{glucose-6-P}]/[\text{glucose-1-P}] = 17$$

COUPLED REACTIONS

Reactions can be "coupled" together if they share one or more intermediates. In this case, the overall free-energy change is simply the sum of the individual ΔG° values. A reaction that is unfavorable (has a positive ΔG°) can for this reason be driven by a second, highly favorable reaction.

SINGLE REACTION

glucose   fructose   sucrose

$\Delta G^\circ = +5.5$ kcal/mole

NET RESULT: will not occur!

COUPLED REACTION

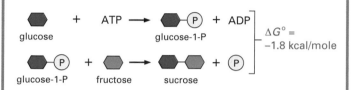

glucose   ATP   glucose-1-P   ADP

glucose-1-P   fructose   sucrose   P

$\Delta G^\circ = -1.8$ kcal/mole

NET RESULT: Sucrose is made in a reaction driven by the hydrolysis of ATP.

HIGH-ENERGY BONDS

One of the most common reactions in the cell is hydrolysis, in which a covalent bond is split by adding water.

The ΔG° for this reaction is sometimes loosely termed the "bond energy." Compounds such as acetyl phosphate and ATP that have a large negative ΔG° of hydrolysis are said to have "high-energy" bonds.

	ΔG° (kcal/mole)
acetyl P → acetate + P$_i$	–10.3
ATP → ADP + P$_i$	–7.3
glucose-6-P → glucose + P$_i$	–3.3

(Note that, for simplicity, water is omitted from the above equations.)

Figure 3–19 Reactions will eventually reach a chemical equilibrium. When a reaction reaches equilibrium, the forward and the backward flux of reacting molecules are equal and opposite.

THE REACTION

The formation of X is energetically favored in this example. But due to thermal bombardments, there will always be some X converting to Y and vice versa.

SUPPOSE WE START WITH AN EQUAL NUMBER OF Y AND X MOLECULES

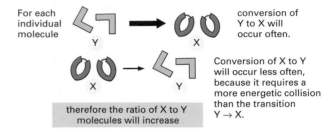

For each individual molecule

conversion of Y to X will occur often.

Conversion of X to Y will occur less often, because it requires a more energetic collision than the transition Y → X.

therefore the ratio of X to Y molecules will increase

EVENTUALLY there will be a large enough excess of X over Y to just compensate for the slow rate of X → Y. Equilibrium will then be attained.

AT EQUILIBRIUM the number of Y molecules being converted to X molecules each second is exactly equal to the number of X molecules being converted to Y molecules each second, so that there is no net change in the ratio of Y to X.

Table 3–1 Relationship Between the Standard Free-Energy Change, $\Delta G°$, and Equilibrium Constant

EQUILIBRIUM CONSTANT $\frac{[X]}{[Y]} = K$	FREE ENERGY OF X MINUS FREE ENERGY OF Y (kcal/mole)
10^5	–7.1 (–29.7)
10^4	–5.7 (–23.8)
10^3	–4.3 (–18.0)
10^2	–2.8 (–11.7)
10	–1.4 (–5.9)
1	0 (0)
10^{-1}	1.4 (5.9)
10^{-2}	2.8 (11.7)
10^{-3}	4.3 (18.0)
10^{-4}	5.7 (23.8)
10^{-5}	7.1 (29.7)

Values of the equilibrium constant were calculated for the simple chemical reaction Y ⇌ X using the equation given in the text.

The $\Delta G°$ given here is in kilocalories per mole at 37°C, with kilojoules per mole in parentheses (1 kilocalorie is equal to 4.184 kilojoules). As explained in the text, $\Delta G°$ represents the free-energy difference under standard conditions (where all components are present at a concentration of 1.0 mole/liter).

From this table, we see that if there is a favorable free-energy change of –4.3 kcal/mole (–18.0 kJ/mole) for the transition Y → X, there will be 1000 times more molecules in state X than in state Y.

measure of the binding strength between molecules. This value is very useful as it indicates the specificity of the interactions between the molecules.

Consider the reaction shown in Figure 3–20, where molecule A interacts with molecule B. The reaction proceeds until it reaches equilibrium, at which point the number of association events precisely equals the number of dissociation events; at this point, the concentrations of reactants and of the complex AB can be used to determine the equilibrium constant.

This equilibrium constant becomes larger as the binding energy between the two molecules increases. And the larger the equilibrium constant, the greater the difference in free energy between the associated and dissociated states (Figure 3–20B). Stronger binding energy therefore favors the interaction of the substrates. Even a change of a few noncovalent bonds can have a striking effect on a binding interaction, as illustrated in Figure 3–21. In this example, eliminating a few hydrogen bonds from a binding interaction causes a dramatic decrease in the amount of complex that exists at equilibrium.

The equilibrium constant governs all of the many associations and dissociations of macromolecules and small molecules inside a living cell—including the binding of enzymes to their substrates.

For Sequential Reactions, the Changes in Free Energy Are Additive

The course of most reactions can be predicted quantitatively. A large body of thermodynamic data has been collected that makes it possible to calculate the standard change in free energy, $\Delta G°$, for most of the important metabolic reactions of the cell. The overall free-energy

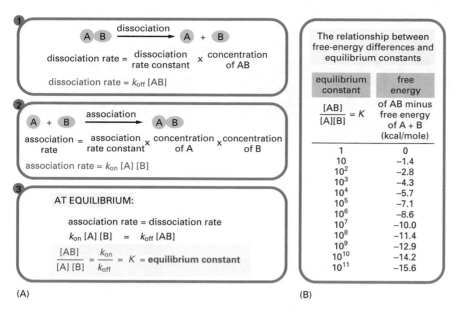

equilibrium constant	free energy
$\frac{[AB]}{[A][B]} = K$	of AB minus free energy of A + B (kcal/mole)
1	0
10	−1.4
10^2	−2.8
10^3	−4.3
10^4	−5.7
10^5	−7.1
10^6	−8.6
10^7	−10.0
10^8	−11.4
10^9	−12.9
10^{10}	−14.2
10^{11}	−15.6

(A)   (B)

Figure 3–20 The energy of binding interactions is reflected in the equilibrium constant. (A) The equilibrium between molecules A and B and the complex AB is maintained by a balance between the two opposing reactions shown in (1) and (2). Molecules A and B must collide in order to react, and the association rate is therefore proportional to the product of their individual concentrations [A] × [B], where the symbol "[ ]" indicates concentration. As shown in (3), the ratio of the rate constants for the association and the dissociation reactions is equal to the equilibrium constant (K) for the reaction. (B) The equilibrium constant in (3) is that for the reaction A + B ⇌ AB, and the larger its value, the stronger is the binding between A and B. For every 1.4 kcal/mole (5.9 kJ/mole) of free-energy difference, the equilibrium constant changes by a factor of 10.

change for a metabolic pathway is then simply the sum of the free-energy changes in each of its component steps. Consider, for example, two sequential reactions

$$X \rightarrow Y \text{ and } Y \rightarrow Z$$

where the $\Delta G°$ values are +5 and –13 kcal/mole, respectively. (Recall that a mole is 6×10^{23} molecules of a substance.) If these two reactions occur sequentially, the $\Delta G°$ for the coupled reaction will be –8 kcal/mole. Thus, the unfavorable reaction $X \rightarrow Y$, which will not occur spontaneously, can be driven by the favorable reaction $Y \rightarrow Z$, provided that the second reaction follows the first.

Cells can therefore cause the energetically unfavorable transition, $X \rightarrow Y$, to occur if an enzyme catalyzing the $X \rightarrow Y$ reaction is supplemented by a second enzyme that catalyzes the energetically favorable reaction, $Y \rightarrow Z$. In effect, the reaction $Y \rightarrow Z$ acts as a "siphon," pulling the conversion of all of molecule X to molecule Y, and thence to molecule Z (Figure 3–22). For example, several of the reactions in the long pathway that converts sugars into CO_2 and H_2O are energetically unfavorable. But the pathway nevertheless proceeds rapidly to completion because the total $\Delta G°$ for the series of sequential reactions has a large negative value.

But forming a sequential pathway is not adequate for many purposes. Often the desired pathway is simply $X \rightarrow Y$, without further conversion of Y to some other product. Fortunately, there are other, more

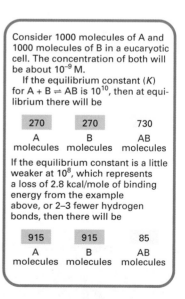

Figure 3–21 Small changes in the number of weak bonds can have drastic effects on a binding interaction. This example illustrates the dramatic effect of the presence or absence of a few weak noncovalent bonds in a biological context.

Figure 3–22 **An energetically unfavorable reaction can be driven by a second reaction, which acts as a chemical siphon.** (A) At equilibrium, there are twice as many X molecules as Y molecules, because X is of lower energy than Y. (B) At equilibrium, there are 25 times more Z molecules than Y molecules, because Z is of much lower energy than Y. (C) If the reactions in (A) and (B) are coupled, nearly all of the X molecules will be converted to Z molecules, as shown.

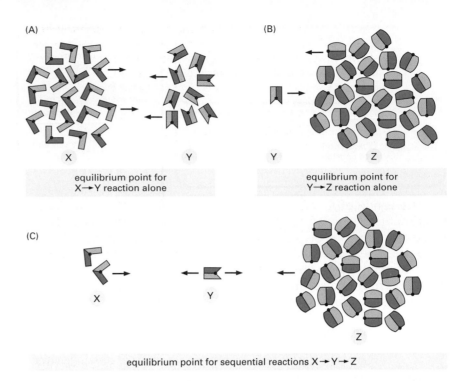

equilibrium point for sequential reactions X → Y → Z

Question 3–4

Look carefully at Figure 3–22. Sketch an energy diagram similar to that in Figure 3–13 for the two reactions alone and for the combined reactions. Indicate the standard free-energy changes for the reactions X → Y, Y → Z, and X → Z in the graph. Indicate how enzymes that catalyze these reactions would change the energy diagram.

general ways of using enzymes to couple reactions together, involving production of activated carrier molecules that can shuttle energy from one reaction site to another. We will see shortly how these systems work.

Rapid Diffusion Allows Enzymes to Find Their Substrates

Enzymes and their substrates are both present in relatively small numbers in a cell. Yet a typical enzyme can capture and process about a thousand substrate molecules every second. This means that an enzyme must be able to release its product and bind a new substrate in a fraction of a millisecond. How can these molecules find each other so quickly inside the cell?

Rapid binding is possible because motions are enormously fast at the molecular level. Because of heat energy, molecules are in constant motion and consequently will explore the space inside the cell very efficiently by wandering randomly through it—a process called **diffusion**. In this way, every molecule in a cell collides with a huge number of other molecules each second. As the molecules in a liquid collide and bounce off one another, an individual molecule moves first one way and then another, its path constituting a *random walk* (Figure 3–23). In such a walk, the average distance that each molecule travels (as the crow flies) from its starting point is proportional to the square root of the time it takes: that is, if it takes a molecule 1 second on average to travel 1 μm, it takes 4 seconds to travel 2 μm, 100 seconds to travel 10 μm, and so on. Thus diffusion only works well for very short distances. To quickly move molecules over larger distances, cells need to rely on more active and directed methods of transport—processes that inevitably require an expenditure of cellular energy.

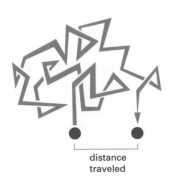

distance traveled

Figure 3–23 **A molecule traverses the cell by taking a random walk.** Molecules in solution move in a random fashion due to the continual buffeting they receive in collisions with other molecules. This movement allows small molecules to diffuse rapidly from one part of the cell to another.

Figure 3–24 The cell cytoplasm is crowded with various molecules. The drawing is approximately to scale. Only the macromolecules are shown: RNAs are shown in *blue*, ribosomes in *green*, and proteins in *red*. Enzymes and other macromolecules diffuse relatively slowly in the cytoplasm, in part because they interact with many other macromolecules; small molecules, by contrast, diffuse nearly as rapidly as they do in water. (Adapted from D.S. Goodsell, *Trends Biochem. Sci.* 16:203–206, 1991.)

100 nm

The inside of a cell is very crowded (Figure 3–24). Nevertheless, experiments in which fluorescent dyes and other labeled molecules are injected into cells show that small organic molecules diffuse through the dense aqueous gel of the cytosol nearly as rapidly as they do through water. A small organic molecule, a substrate, for example, takes only about one-fifth of a second on average to diffuse a distance of 10 μm. Diffusion is therefore an efficient way for small molecules to move limited distances in the cell.

Enzymes and other macromolecules, however, diffuse through the cytoplasm much more slowly than do small molecules. Because enzymes move more slowly than their substrates, we can think of them as sitting still. The rate of encounter of each enzyme molecule with its substrate thus depends on the concentration of the substrate molecule. For example, some abundant substrates are present in the cell at a concentration of 0.5 mM. Because pure water is 55 M, there is only about one such substrate molecule in the cell for every 10^5 water molecules. Nevertheless, the active site on an enzyme molecule that binds this substrate will be bombarded by about 500,000 random collisions with the substrate molecule per second. For a substrate concentration tenfold lower (0.05 mM), the number of collisions drops to 50,000 per second, and so on. A random encounter between the surface of an enzyme and the matching surface of its substrate molecule often leads immediately to the formation of an enzyme–substrate complex that is ready to react. A reaction in which a covalent bond is broken or formed can now occur extremely rapidly. Once one appreciates how quickly molecules move and react, the observed rates of enzymatic catalysis do not seem so amazing.

When an enzyme and substrate have collided and snuggled together properly at the active site, they form multiple weak bonds with each other that persist until random thermal motion causes the molecules to dissociate again. These weak interactions can include hydrogen bonds, van der Waals attractions, and ionic bonds (as discussed in Chapter 2). In general, the stronger the binding of the enzyme and substrate, the slower their rate of dissociation. When two colliding molecules have poorly matching surfaces, few noncovalent bonds are formed and their total energy is negligible compared with that of thermal motion. In this case the two molecules dissociate as rapidly as they come together (see Figure 2–32). This is what prevents incorrect and unwanted associations from forming between mismatched molecules, such as between an enzyme and the wrong substrate.

V_{max} and K_M Measure Enzyme Performance

To catalyze a reaction, an enzyme first binds its substrate. The substrate then undergoes a reaction to form the product, which initially remains bound to the enzyme. Finally, the product is released and diffuses away, leaving the enzyme free to bind to another substrate molecule and

Question 3–5

The enzyme carbonic anhydrase is one of the speediest enzymes known. It catalyzes the hydration of CO_2 to HCO_3^- ($CO_2 + H_2O \rightleftharpoons HCO_3^- + H^+$). The rapid conversion of CO_2 gas into the much more soluble bicarbonate ion (HCO_3^-) is very important for the efficient transport of CO_2 from tissue, where CO_2 is produced by respiration, to the lungs, where it is exhaled. Carbonic anhydrase accelerates the reaction 10^7-fold, hydrating 10^5 CO_2 molecules per second at its maximal speed. What do you suppose limits the speed of the enzyme? Sketch a diagram analogous to the one shown in Figure 3–14 and indicate which portion of your diagram has been designed to display the 10^7-fold acceleration.

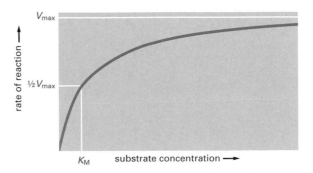

Figure 3–25 **An enzyme's performance depends on how rapidly it can process its substrate.** The rate of an enzyme reaction (*V*) increases as the substrate concentration increases until a maximum value (V_{max}) is reached. At this point all substrate-binding sites on the enzyme molecules are fully occupied, and the rate of reaction is limited by the rate of the catalytic process on the enzyme surface. For most enzymes the concentration of substrate at which the reaction rate is half-maximal (K_M) is a direct measure of how tightly the substrate is bound, with a large value of K_M (a large amount of substrate needed) corresponding to weak binding.

Question 3–6

In cells, an enzyme catalyzes the reaction AB → A + B. It was isolated, however, as an enzyme that carries out the opposite reaction A + B → AB. Explain the paradox.

catalyze another reaction (see Figure 3–16). The rates of these different steps vary widely from one enzyme to another, and they can be measured by mixing purified enzymes and substrates together under carefully defined conditions.

If the concentration of the substrate is increased progressively from a very low value, the concentration of the enzyme–substrate complex—and therefore the rate at which product is formed—initially increases in a linear fashion in direct proportion to substrate concentration. However, as more and more enzyme molecules become occupied by substrate, this rate increase tapers off, until at a very high concentration of substrate it reaches a maximum value, termed V_{max}. At this point, the active sites of all enzyme molecules in the sample are fully occupied with substrate, and the rate of product formation depends only on how rapidly the substrate molecule can be processed. For many enzymes, this **turnover number** is of the order of 1000 substrate molecules per second, although turnover numbers between 1 and 10,000 have been measured.

The concentration of substrate needed to make the enzyme work efficiently is often measured by a different parameter, the Michaelis' constant, K_M, named after one of the biochemists who worked out this relationship. An enzyme's K_M is the concentration of substrate at which the enzyme works at half its maximum speed (0.5 V_{max}; Figure 3–25). In general, a low value of K_M indicates that a substrate binds very tightly to the enzyme, and a large value corresponds to weak binding. For a discussion of how we measure these parameters, and how we can use them to model biochemical pathways—and potentially to design better catalysts—see How We Know, pp. 103–105.

It is important to recognize that when an enzyme (or any catalyst) lowers the activation energy for the reaction Y → X, it also lowers the activation energy for the reaction X → Y by exactly the same amount (see Figure 3–13). The forward and backward reactions will therefore be accelerated by the same factor by an enzyme, and the equilibrium point for the reaction (and $\Delta G°$) will remain unchanged (Figure 3–26).

Figure 3–26 **Enzymes cannot change the equilibrium point for reactions.** Enzymes, like all catalysts, speed up the forward and backward rates of a reaction by the same factor. Therefore, for both the catalyzed and the uncatalyzed reactions shown here, the number of molecules undergoing the transition X → Y is equal to the number of molecules undergoing the transition Y → X when the ratio of Y molecules to X molecules is 3.5 to 1. In other words, both the catalyzed and uncatalyzed reactions will eventually reach the same equilibrium point.

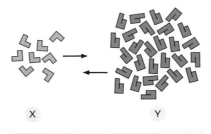

(A) UNCATALYZED REACTION

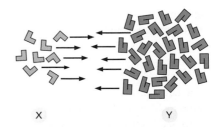

(B) ENZYME-CATALYZED REACTION

How We Know: Using Kinetics to Model and Manipulate Metabolic Pathways

At first glance, it seems that a cell's metabolic pathways are pretty well mapped out. If the complex web of reactions outlined in Figure 3–2 is any indication, each reaction appears to proceed predictably to the next—substrate X is converted to product Y, which is passed along to enzyme Z. So why would anyone need to know exactly how tightly a particular enzyme clutches its substrate? Or whether it can process 100 or 1000 substrate molecules every second?

In reality, these elaborate maps merely suggest which pathways a cell *might* follow as it converts nutrients into small molecules, chemical energy, and the larger building blocks of life. They do not reveal precisely how a cell will behave under a particular set of conditions—which pathways it will take when it is starving, when it is well fed, when oxygen is scarce, when it is stressed, or when it decides to divide. The study of an enzyme's kinetics—how fast it operates, how it handles its substrate, how its activity is controlled—allows one to predict how an individual catalyst will perform, and how it will interact with other enzymes in a network. Such knowledge leads to a deeper understanding of cell biology, and it opens the door to learning how to harness enzymes to perform desired reactions.

Speed

The first step to understanding how an enzyme performs involves determining the maximal velocity, V_{max}, for the reaction it catalyzes. This is accomplished by measuring,

in a test tube, how rapidly the reaction proceeds in the presence of different concentrations of substrate: the rate should increase as the amount of substrate rises until the reaction reaches its V_{max}. The velocity of the reaction is measured by monitoring how quickly the substrate is consumed or how rapidly the product accumulates. In many cases, the appearance of product or the disappearance of substrate can be observed directly with a spectrophotometer. This instrument detects the presence of molecules that absorb light at a selected wavelength; NADH, for example, absorbs light at 340 nm, while its oxidized counterpart NAD^+ does not. So a reaction that generates NADH (by reducing NAD^+) can be monitored by following the formation of NADH spectrophotometrically.

To determine the V_{max} of a reaction, one sets up a series of test tubes, each tube containing a different concentration of substrate. For each tube, one adds the same amount of enzyme and then measures the velocity of the reaction—the number of micromoles of substrate consumed or product generated per minute. Because these numbers will tend to decrease over time, the rate used is the velocity measured early in the reaction. These initial velocity values (v) are then plotted against the substrate concentration, yielding a curve like the one shown in Figure 3–27.

Looking at this plot, however, it is difficult to determine the exact value of V_{max}, as it is not clear where the reaction rate will reach its plateau. To get around this problem, the data

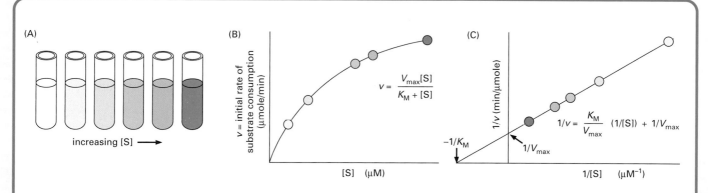

Figure 3–27 Reaction rate data are plotted to determine V_{max} and K_M. (A) A series of substrate concentrations is prepared, enzyme added, and initial velocities determined. (B) The initial velocities are plotted against the substrate concentrations. The equation describing such a hyperbola is $y = ax/(b + x)$. Substituting in the kinetic terms, the equation becomes rate = $V_{max}[S]/(K_M + [S])$, where v is the initial velocity, V_{max} is the asymptote of the curve (the value of y at an infinite value of x), and K_M is equal to the substrate concentration, where v is one-half V_{max}. This is called the *Michaelis–Menten equation,* named for the biochemists who provided evidence for this enzymatic relationship. (C) In the double-reciprocal plot, $1/v$ is plotted against $1/[S]$. The equation describing this straight line is $1/v = (K_M/V_{max}) \times 1/[S] + 1/V_{max}$. When $1/[S] = 0$, the y intercept ($1/v$) is $1/V_{max}$. When $1/v = 0$, the x intercept ($1/[S]$) is $-1/K_M$. By convention, lowercase letters are used for variables (hence v for velocity) and uppercase letters are used for constants (hence V_{max}).

are converted to their reciprocals and graphed in a "double-reciprocal plot," where the inverse of the velocity ($1/v$) appears on the y axis and the inverse of the substrate concentration ($1/[S]$) on the x axis (see Figure 3–27C). This graph yields a straight line whose y intercept (the point where the line crosses the y axis) represents $1/V_{max}$ and whose x intercept corresponds to $-1/K_M$. These values are then easily converted to values for V_{max} and K_M.

Enzymologists use this technique to determine the kinetic parameters of many enzyme-catalyzed reactions (although these days computer programs automatically plot the data and spit out the sought-after values). Some reactions, however, happen too fast to be monitored in this way; the action is essentially complete—the substrate entirely consumed—within thousandths of a second. For these reactions, a special piece of equipment must be used to follow what happens during the first few milliseconds when an enzyme and substrate meet (Figure 3–28).

Control

Substrates are not the only molecules that influence how well or how quickly an enzyme works. Many enzymes can also be controlled by products, alternative substrates, substrate look-alikes, inhibitors, toxins, and other small molecules that either increase or decrease their activity. This regulation allows cells to control when and how rapidly various reactions occur, a process we will consider in more detail in Chapter 4.

Determining how an inhibitor decreases an enzyme's activity can reveal how a metabolic pathway is regulated—and can suggest how those control points can be circumvented by carefully designed mutations in specific genes.

The effect that an inhibitor has on an enzyme's activity is monitored in the same way that we measured the enzyme's kinetics. A curve is generated showing the velocity of the reaction between enzyme and substrate, as described previously. But now additional curves are also produced for reactions in which an inhibitor molecule has been included in the mix.

Comparing these curves, with and without inhibitor, can also reveal how a particular inhibitor impedes enzyme activity. For example, some inhibitors bind to the same site on an enzyme as its substrate. These *competitive inhibitors* block enzyme activity by competing directly with the substrate for the enzyme's attention. They resemble the substrate enough to tie up the enzyme, but they differ enough in structure to avoid getting converted to product. This blockage can be overcome by adding enough substrate so that enzymes are more likely to encounter a substrate molecule than an inhibitor. From the kinetic data, we can see that competitive inhibitors do not change the V_{max} of a reaction; in other words, add enough substrate and the enzyme will encounter mostly substrate molecules and will reach its maximum velocity (Figure 3–29).

Competitive inhibition can be used to treat patients who have been poisoned by ethylene glycol, an ingredient in commercially available antifreeze. Although ethylene glycol is itself not fatally toxic, a by-product of its metabolism, oxalic acid, can be lethal. To prevent oxalic acid from

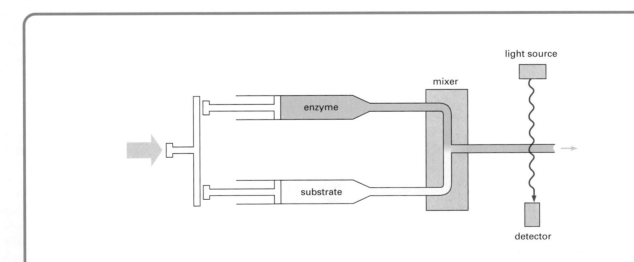

Figure 3–28 A stopped-flow apparatus is used to observe reactions during the first few milliseconds. In this piece of equipment, the enzyme and substrate are rapidly injected into a mixing chamber through two syringes. The enzyme and its substrate meet as they shoot through the mixing tube at flow rates that can easily reach 1000 cm/sec. They then pass through another tube and zoom past a detector that monitors, say, the appearance of product. If the detector is located within a centimeter of where the enzyme and substrate first meet, one can observe reactions when they are only a few milliseconds old.

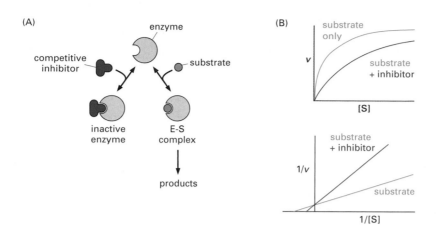

Figure 3–29 A competitive inhibitor directly blocks substrate binding. (A) Diagram showing a competitive inhibitor and a substrate competing to bind to the same site on an enzyme. (B) Competitive inhibitors do not change the V_{max} of a reaction, because the inhibition can be overcome by adding more substrate.

forming, the patient is given a large (though not quite intoxicating) dose of ethanol. Ethanol competes with the ethylene glycol for binding to alcohol dehydrogenase, the first enzyme in the pathway to oxalic acid formation. The nonmetabolized ethylene glycol is then safely eliminated from the body.

Other types of inhibitors may interact with sites on the enzyme distant from where the substrate binds. For example, chelating agents that bind reversibly to ions such as Mg^{2+} will inhibit enzymes that require such metals for their activity. In this case, the substrate can bind, but the enzyme–substrate complex may not form as quickly as it would in the absence of inhibitor. Such inhibition would not be overcome by the addition of more substrate.

Design

With the kinetic data in hand, one can take advantage of modeling programs to predict how an enzyme will perform, and even how a cell will respond when exposed to different conditions—such as the addition of a particular sugar or amino acid to the culture medium, or the addition of a poison or a pollutant. Seeing how a cell manages its resources—which pathways it favors for dealing with particular biochemical challenges—can also suggest strategies we can follow to design better catalysts for reactions of medical or commercial importance: producing drugs or detoxifying industrial waste, for example. Using such tactics, bacteria have even been engineered to produce large amounts of indigo—the dye, originally extracted from plants, that makes your blue jeans blue.

Several computer programs have been developed to facilitate the dissection of complex reaction pathways. One such program provides information about individual reactions—velocities, concentrations of enzymes, substrates, products, inhibitors, and other regulatory molecules—and the program predicts how molecules will flow through the pathway, which products will be generated, and where any bottlenecks might be. The process is not unlike balancing an algebraic equation in which every atom of carbon, nitrogen, oxygen, and so on, must be properly accounted for. Such careful accounting allows one to rationally design ways to manipulate the pathway—rerouting it around a bottleneck, eliminating an important inhibitor, redirecting the reactions to favor the generation of predominantly one product, or extending the pathway to produce a novel molecule. Of course such models must be tested and validated in cells, which may not always behave as predicted.

Implementing most such designs requires using genetic engineering techniques to introduce the gene or genes of choice into a cell that can be manipulated and maintained in the laboratory, usually a bacterium. We discuss these methods at greater length in Chapter 10. Harnessing the power of cell biology for commercial purposes—even to produce something as simple as an amino acid such as tryptophan—is currently a billion-dollar industry. And as more genome data come in, presenting us with more enzymes to exploit, it may not be long before vats of custom-made bacteria are churning out drugs and chemicals that represent the biological equivalent of pure gold.

Activated Carrier Molecules and Biosynthesis

The energy released by the oxidation of food molecules must be stored temporarily before it can be channeled into the construction of either other small organic molecules or the larger and more complex molecules needed by the cell. In most cases, the energy is stored as chemical bond energy in a small set of activated "carrier molecules," which contain one or more energy-rich covalent bonds. These molecules diffuse rapidly throughout the cell and thereby carry their bond energy from the sites of energy generation to the sites where energy is used for biosynthesis and other needed cell activities (Figure 3–30).

Activated carriers store energy in an easily exchangeable form, either as a readily transferable chemical group or as high-energy electrons, and they can serve a dual role as a source of both energy and chemical groups for biosynthetic reactions. The most important of the activated carrier molecules are ATP and two molecules that are closely related to each other, NADH and NADPH—as we discuss in detail shortly. We shall see that cells use activated carrier molecules like money to pay for reactions that otherwise could not take place.

The Formation of an Activated Carrier Is Coupled to an Energetically Favorable Reaction

When a fuel molecule such as glucose is oxidized in a cell, enzyme-catalyzed reactions ensure that a large part of the free energy that is released by oxidation is captured in a chemically useful form, rather than being released wastefully as heat. (Burning sugar in a cell allows you to power metabolic reactions, while burning a chocolate bar in the street will get you nowhere, energetically speaking.) In living systems, this energy capture is achieved by means of a **coupled reaction**, in which an energetically favorable reaction is used to drive an energetically unfavorable one that produces an activated carrier molecule or some other useful molecule. Coupling mechanisms require enzymes, and they are fundamental to all of the energy transactions of the cell.

The nature of a coupled reaction is illustrated by a mechanical analogy in Figure 3–31, in which an energetically favorable chemical reaction is represented by rocks falling from a cliff. The energy of falling rocks would normally be entirely wasted in the form of heat generated by friction when the rocks hit the ground (Figure 3–31A). By careful design, however, part of this energy could be used to drive a paddle wheel that lifts a bucket of water (Figure 3–31B). Because the rocks can now reach the ground only after moving the paddle wheel, we say that the energetically favorable reaction of rocks falling has been directly

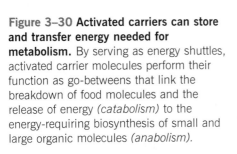

Figure 3–30 **Activated carriers can store and transfer energy needed for metabolism.** By serving as energy shuttles, activated carrier molecules perform their function as go-betweens that link the breakdown of food molecules and the release of energy *(catabolism)* to the energy-requiring biosynthesis of small and large organic molecules *(anabolism)*.

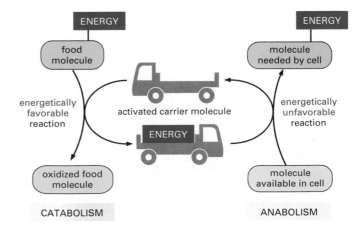

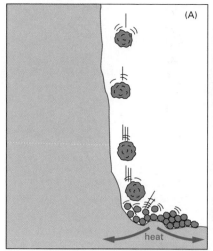
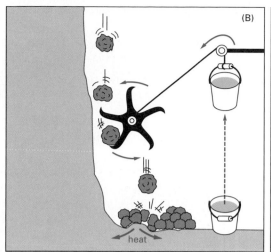
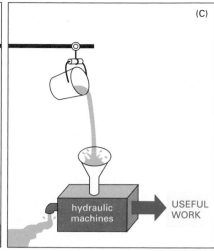

(A)

kinetic energy transformed into heat
energy only

(B)

part of the kinetic energy is used to lift
a bucket of water, and a correspondingly
smaller amount is transformed into heat

(C)

USEFUL
WORK

hydraulic
machines

the potential kinetic energy stored in
the raised bucket of water can be
used to drive hydraulic machines that
carry out a variety of useful tasks

coupled to the energetically unfavorable reaction of lifting the bucket of water. Note that because part of the energy is used to do work in (B), the rocks hit the ground with less velocity than in (A), and correspondingly less energy is wasted as heat.

Analogous processes occur in cells, where enzymes play the role of the paddle wheel in our example. By mechanisms that will be discussed in Chapter 13, enzymes couple an energetically favorable reaction, such as the oxidation of foodstuffs, to an energetically unfavorable reaction, such as the generation of an activated carrier molecule. As a result, the amount of heat released by the oxidation reaction is reduced by exactly the amount of energy that is stored in the energy-rich covalent bonds of the activated carrier molecule. The activated carrier molecule in turn picks up a packet of energy that is large enough to power a chemical reaction elsewhere in the cell.

Figure 3–31 A mechanical model illustrates the principle of coupled chemical reactions. The spontaneous reaction shown in (A) could serve as an analogy for the direct oxidation of glucose to CO_2 and H_2O, which produces heat only. In (B) the same reaction is coupled to a second reaction; this second reaction could serve as an analogy for the synthesis of activated carrier molecules. The energy produced in (B) is in a more useful form than in (A) and can be used to drive a variety of otherwise energetically unfavorable reactions (C).

ATP Is the Most Widely Used Activated Carrier Molecule

The most important and versatile of the activated carriers in cells is **ATP** (adenosine 5′-triphosphate). Just as the energy stored in the raised bucket of water in Figure 3–31B can be used to drive a wide variety of hydraulic machines, ATP serves as a convenient and versatile store, or currency, of energy to drive a variety of chemical reactions in cells. As shown in Figure 3–32, ATP is synthesized in an energetically unfavorable phosphorylation reaction in which a phosphate group is added to **ADP** (adenosine 5′-diphosphate). When required, ATP gives up this energy packet in an energetically favorable hydrolysis to ADP and inorganic phosphate. The regenerated ADP is then available to be used for another round of the phosphorylation reaction that forms ATP, creating an ATP cycle in the cell.

The energetically favorable reaction of ATP hydrolysis is coupled to many otherwise unfavorable reactions through which other molecules are synthesized. We shall encounter several of these reactions later in this chapter and see exactly how this is done. Such hydrolysis reactions often involve the transfer of the terminal phosphate in ATP to another molecule, as illustrated in Figure 3–33. Any reaction that involves the transfer of a phosphate group to a molecule is termed a *phosphorylation* reaction. Phosphorylation reactions are involved in many important cellular functions: they activate substrates, they facilitate the

Question 3–7

Use Figure 3–31B to illustrate the following reaction driven by hydrolysis of ATP:

$$X + ATP \rightarrow Y + ADP + \textcircled{P}$$

A. In this case, which molecule or molecules would be analogous to (i) rocks at top of cliff, (ii) broken debris at bottom of cliff, (iii) bucket at its highest point, (iv) bucket on the ground?

B. What would be analogous to (i) the rocks hitting the ground in the absence of the paddle wheel in Figure 3–31A and (ii) the hydraulic machine in Figure 3–31C?

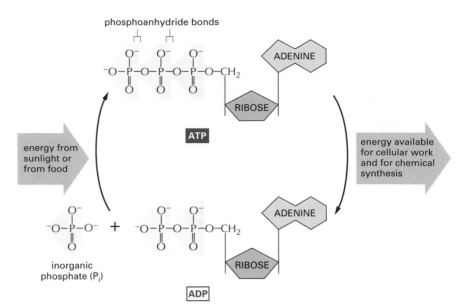

Figure 3–32 The interconversion of ATP and ADP occurs in a cycle. The two outermost phosphates in ATP are held to the rest of the molecule by high-energy phosphoanhydride bonds and are readily transferred. Water can be added to ATP to form ADP and inorganic phosphate (P_i). This hydrolysis of the terminal phosphate of ATP yields between 11 and 13 kcal/mole of usable energy (46 to 54 kJ/mole), depending on the intracellular conditions. The large negative ΔG of this reaction arises from a number of factors. Release of the terminal phosphate group removes an unfavorable repulsion between adjacent negative charges; in addition, the inorganic phosphate ion (P_i) released is stabilized by resonance and by favorable hydrogen-bond formation with water. The formation of ATP from ADP and P_i reverses the hydrolysis reaction. Because this condensation is energetically unfavorable, it must be coupled to an energetically more favorable reaction to occur.

Question 3–8

The phosphoanhydride bond that links two phosphate groups in ATP in a high-energy linkage has a $\Delta G°$ of –7.3 kcal/mole. Hydrolysis of this bond liberates from 11 to 13 kcal/mole of usable energy. How can this be? Why do you think a range of energies is given, rather than a precise number as for $\Delta G°$?

exchange of chemical energy, and they help to control cell signaling processes.

ATP is the most abundant energy carrier in cells. It is used, for example, to supply energy for many of the pumps that transport substances into and out of the cell (discussed in Chapter 12); it also powers the molecular motors that enable muscle cells to contract and nerve cells to transport materials from one end of their long axons to another (discussed in Chapter 17). Why evolution selected this particular nucleotide over the others as the major carrier of energy, however, remains a mystery.

Energy Stored in ATP Is Often Harnessed to Join Two Molecules Together

We have previously discussed one way in which an energetically favorable reaction, Y → Z, can be coupled to an energetically unfavorable reaction, X → Y, so as to enable it to occur. In that scheme a second enzyme catalyzes the energetically favorable reaction Y → Z, pulling all of the X to Y in the process (see Figure 3–22). But when the required product is Y and not Z, this mechanism is not useful.

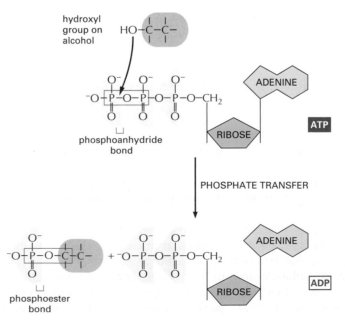

Figure 3–33 The terminal phosphate of ATP can be readily transferred to other molecules. Because an energy-rich phosphoanhydride bond in ATP is converted to a phosphoester bond, this reaction is energetically favorable, having a large negative ΔG. Phosphorylation reactions of this type are involved in the synthesis of phospholipids and in the initial steps of the reactions that catabolize sugars, as well as in many other metabolic events.

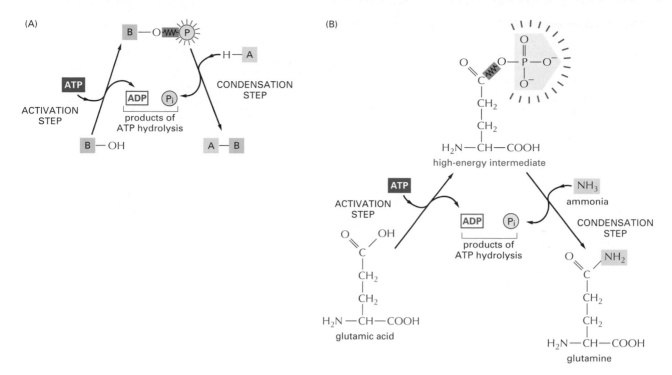

A frequent type of reaction that is needed for biosynthesis is one in which two molecules, A and B, are joined together to produce A–B in the energetically unfavorable condensation reaction

$$A–H + B–OH \rightarrow A–B + H_2O$$

There is an indirect pathway that allows A–H and B–OH to form A–B, in which a coupling to ATP hydrolysis makes the reaction go. Here energy from ATP hydrolysis is first used to convert B–OH to a higher-energy intermediate compound, which then reacts directly with A–H to give A–B. The simplest possible mechanism involves the transfer of a phosphate from ATP to B–OH to make B–O–PO$_3$, in which case the reaction pathway contains only two steps:

1. B–OH + ATP → B–O–PO$_3$ + ADP
2. A–H + B–O–PO$_3$ → A–B + P$_i$

Net result: B–OH + ATP + A–H → A–B + ADP + P$_i$

The condensation reaction, which by itself is energetically unfavorable, has been forced to occur by being directly coupled to ATP hydrolysis in an enzyme-catalyzed reaction pathway (Figure 3–34A).

A biosynthetic reaction of exactly this type is employed to synthesize the amino acid glutamine, as illustrated in Figure 3–34B. We will see shortly that very similar (but more complex) mechanisms are also used to produce nearly all of the large molecules of the cell.

NADH and NADPH Are Important Electron Carriers

Other important activated carrier molecules participate in oxidation–reduction reactions and are commonly part of coupled reactions in cells. These activated carriers are specialized to carry both high-energy electrons and hydrogen atoms. The most important of these electron carriers are **NAD**$^+$ (nicotinamide adenine dinucleotide) and the closely related molecule **NADP**$^+$ (nicotinamide adenine dinucleotide phosphate). Later, we will examine some of the reactions in which they participate. NAD$^+$ and NADP$^+$ each pick up a "packet of energy" in the form of two high-energy electrons plus a proton (H$^+$), becoming **NADH**

Figure 3–34 An energetically unfavorable biosynthetic reaction can be driven by ATP hydrolysis. (A) Schematic illustration of the formation of A–B in the condensation reaction described in the text. (B) The biosynthesis of the amino acid glutamine. Glutamic acid is first converted to a high-energy phosphorylated intermediate (corresponding to the compound B–O–PO$_3$ described in the text), which then reacts with ammonia (corresponding to A–H) to form glutamine. In this example both steps occur on the surface of the same enzyme, glutamine synthase. Note that, for clarity, the amino acids are shown in their uncharged form.

Figure 3–35 NADPH is an important carrier of electrons. (A) NADPH is produced in reactions of the general type shown on the left, in which two hydrogen atoms are removed from a substrate. The oxidized form of the carrier molecule, NADP$^+$, receives one hydrogen atom plus an electron (a hydride ion), and the proton (H$^+$) from the other H atom is released into solution. Because NADPH holds its hydride ion in a high-energy linkage, the added hydride ion can easily be transferred to other molecules, as shown on the right. (B) The structure of NADP$^+$ and NADPH. The part of the NADP$^+$ molecule known as the nicotinamide ring accepts two electrons together with a proton (the equivalent of a hydride ion, H$^-$), forming NADPH. The molecules NAD$^+$ and NADH are identical in structure to NADP$^+$ and NADPH, respectively, except that the indicated phosphate group is absent from both.

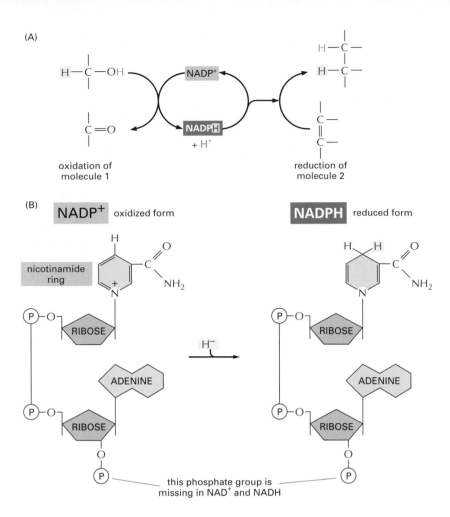

(reduced nicotinamide adenine dinucleotide) and **NADPH** (reduced nicotinamide adenine dinucleotide phosphate), respectively. These molecules can therefore also be regarded as carriers of hydride ions (the H$^+$ plus two electrons, or H$^-$).

Like ATP, NADPH is an activated carrier that participates in many important biosynthetic reactions that would otherwise be energetically unfavorable. NADPH is produced according to the general scheme shown in Figure 3–35. During a special set of energy-yielding catabolic reactions, a hydrogen atom and two electrons are removed from the substrate molecule and added to the nicotinamide ring of NADP$^+$ to form NADPH. This is a typical oxidation–reduction reaction; the substrate is oxidized and NADP$^+$ is reduced.

The hydride ion carried by NADPH is given up readily in a subsequent oxidation–reduction reaction, because the ring can achieve a more stable arrangement of electrons without it. In this subsequent reaction, which regenerates NADP$^+$, the NADPH becomes oxidized and the substrate becomes reduced—thus completing the NADPH cycle. NADPH is efficient at donating its hydride ion to other molecules for the same reason that ATP readily transfers a phosphate: in both cases the transfer is accompanied by a large negative free-energy change. One example of the use of NADPH in biosynthesis is shown in Figure 3–36.

The difference of a single phosphate group has no effect on the electron-transfer properties of NADPH compared with NADH, but it is crucial for their distinctive roles. The extra phosphate group on NADPH is far from the region involved in electron transfer (see Figure 3–35B). But it serves to give a molecule of NADPH a slightly different shape from that of NADH, making it possible for NADPH and NADH to bind as substrates to different sets of enzymes. These two types of carriers are thereby used to deliver electrons (or hydride ions) to different destinations.

Why should there be this division of labor? The answer lies in the need to regulate two sets of electron-transfer reactions independently. NADPH operates chiefly with enzymes that catalyze anabolic reactions, supplying the high-energy electrons needed to synthesize energy-rich biological molecules. NADH, by contrast, has a special role as an intermediate in the catabolic system of reactions that generate ATP through the oxidation of food molecules, as we discuss in Chapter 13. The genesis of NADH from NAD$^+$ and that of NADPH from NADP$^+$ occur by different pathways and are independently regulated, so that the cell can adjust the supply of electrons for these two contrasting purposes. Thus, inside the cell the ratio of NAD$^+$ to NADH is kept high, whereas the ratio of NADP$^+$ to NADPH is kept low. This provides plenty of NAD$^+$ to act as an oxidizing agent and plenty of NADPH to act as a reducing agent—as required for their special roles in catabolism and anabolism, respectively.

There Are Many Other Activated Carrier Molecules in Cells

Other activated carriers also pick up and carry a chemical group in an easily transferred, high-energy linkage (Table 3–2). For example, coenzyme A carries an acetyl group in a readily transferable linkage and in this activated form is known as **acetyl CoA** (acetyl coenzyme A). The structure of acetyl CoA is illustrated in Figure 3–37; it is used to add two carbon units in the biosynthesis of larger molecules.

In acetyl CoA and the other carrier molecules in Table 3–2, the transferable group makes up only a small part of the molecule. The rest consists of a large organic portion that serves as a convenient "handle," facilitating the recognition of the carrier molecule by specific enzymes. As with acetyl CoA, this handle portion very often contains a nucleotide. This curious fact may be a relic from an early stage of cell evolution. It is thought that the main catalysts for early life-forms on the Earth were RNA molecules (or their close relatives) and that proteins were a later evolutionary addition (discussed in Chapter 7). It is therefore tempting to speculate that many of the carrier molecules that we find today originated in an earlier RNA world, where their nucleotide portions would have been useful for binding these carriers to RNA enzymes.

In addition to the transfer reactions catalyzed by the activated carrier molecules ATP (transfer of phosphate) and NADPH (transfer of electrons and hydrogen), other important reactions involve the transfers of methyl, carboxyl, and glucose groups from activated carrier molecules for the purpose of biosynthesis. The activated carriers are usually

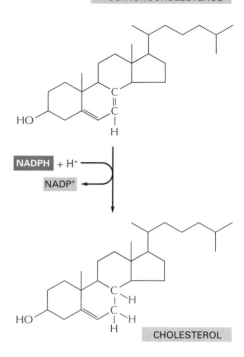

Figure 3–36 NADPH participates in the final stage in one of the biosynthetic routes leading to cholesterol. As in many other biosynthetic reactions, the reduction of the C=C bond is achieved by the transfer of a hydride ion from the carrier molecule NADPH, plus a proton (H$^+$) from the solution.

Table 3–2 Some Activated Carrier Molecules Widely Used in Metabolism

ACTIVATED CARRIER	GROUP CARRIED IN HIGH-ENERGY LINKAGE
ATP	phosphate
NADH, NADPH, FADH$_2$	electrons and hydrogens
Acetyl CoA	acetyl group
Carboxylated biotin	carboxyl group
S-adenosylmethionine	methyl group
Uridine diphosphate glucose	glucose

Figure 3–37 Acetyl coenzyme A (CoA) is another important activated carrier molecule.

A space-filling model is shown above the structure of acetyl CoA. The sulfur atom *(yellow)* forms a thioester bond to acetate. Because the thioester bond is a high-energy linkage, it releases a large amount of free energy when it is hydrolyzed; thus the acetyl group carried by CoA can be readily transferred to other molecules.

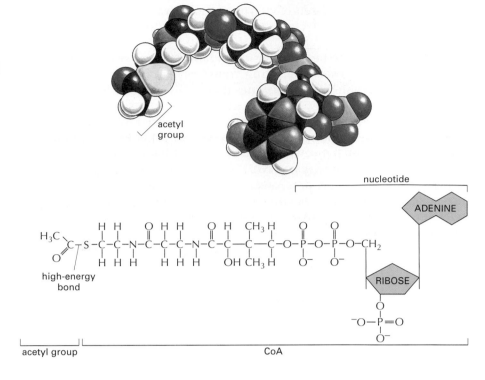

Figure 3–38 An activated carrier molecule transfers a carboxyl group to a substrate molecule.

Carboxylated biotin is used by the enzyme *pyruvate carboxylase* to transfer a carboxyl group for the production of oxaloacetate, a molecule needed in the citric acid cycle. The acceptor molecule for this group transfer reaction is pyruvate. Other enzymes use biotin to transfer carboxyl groups to other acceptor molecules. Note that synthesis of carboxylated biotin requires energy derived from ATP—a general feature of many activated carriers.

generated in reactions coupled to ATP hydrolysis (Figure 3–38). Therefore, the energy that enables their groups to be used for biosynthesis ultimately comes from the catabolic reactions that generate ATP. Similar processes occur in the synthesis of the very large molecules of the cell—the nucleic acids, proteins, and polysaccharides—which we discuss next.

The Synthesis of Biological Polymers Requires an Energy Input

The macromolecules of the cell constitute the vast majority of its dry mass—that is, the mass not due to water. These molecules are made

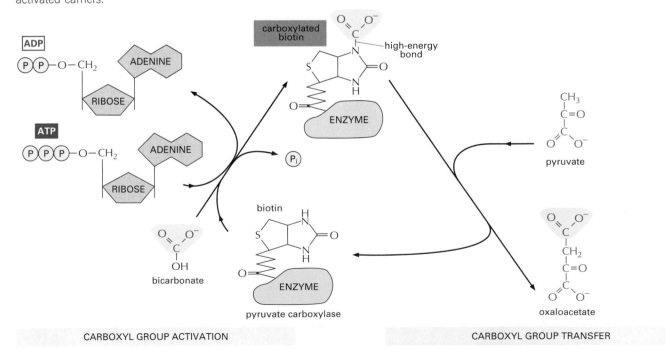

CARBOXYL GROUP ACTIVATION   CARBOXYL GROUP TRANSFER

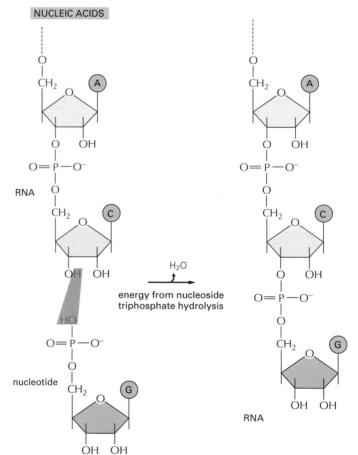

Figure 3–39 **Condensation and hydrolysis are opposite reactions.** The macromolecules of the cell are polymers that are formed from subunits (or monomers) by a condensation reaction and broken down by hydrolysis. Condensation reactions are all energetically unfavorable.

from subunits (or monomers) that are linked together by a *condensation* reaction, in which the constituents of a water molecule (OH plus H) are removed from the two reactants. Consequently, the reverse reaction—the breakdown of polymers—occurs through the enzyme-catalyzed addition of water *(hydrolysis)*. This hydrolysis reaction is energetically favorable, whereas the biosynthetic reactions require an energy input and are more complex (Figure 3–39).

The nucleic acids (DNA and RNA), proteins, and polysaccharides are all polymers that are produced by the repeated addition of a *subunit* onto one end of a growing chain. The mode of synthesis of each of these types of macromolecules is outlined in Figure 3–40. As indicated, the condensation step in each case depends on energy from hydrolysis of a

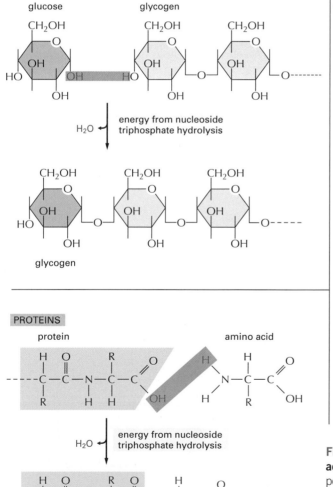

Figure 3–40 **The synthesis of polysaccharides, proteins, and nucleic acids requires an input of energy.** Synthesis of each kind of biological polymer involves the loss of water in a condensation reaction. Not shown is the consumption of high-energy nucleoside triphosphates that is required to activate each subunit prior to its addition. In contrast, the reverse reaction—the breakdown of all three types of polymers—occurs through the simple addition of water (hydrolysis).

nucleoside triphosphate. And yet, except for the nucleic acids, there are no phosphate groups left in the final product molecules. How, then, are the reactions that release the energy of ATP hydrolysis coupled to polymer synthesis?

For each type of macromolecule, an enzyme-catalyzed pathway exists which resembles that discussed previously for the synthesis of the amino acid glutamine (see Figure 3–34). The principle is exactly the same, in that the OH group that will be removed in the condensation reaction is first activated by becoming involved in a high-energy linkage to a second molecule. But the mechanisms used to link ATP hydrolysis to the synthesis of proteins and polysaccharides are more complex than that used for glutamine synthesis. In the biosynthetic pathways leading to these macromolecules, a series of high-energy intermediates generates the final high-energy bond that is broken during the condensation step (as discussed in Chapter 7 for protein synthesis).

There are limits to what each activated carrier can do in driving biosynthesis. For example, the ΔG for the hydrolysis of ATP to ADP and inorganic phosphate (P_i) depends on the concentrations of all of the reactants, and under the usual conditions in a cell it is between –11 and –13 kcal/mole. In principle, this hydrolysis reaction can be used to drive an unfavorable reaction with a ΔG of, perhaps, +10 kcal/mole, provided that a suitable reaction path is available. For some biosynthetic reactions, however, even –13 kcal/mole may not be enough. In these cases the path of ATP hydrolysis can be altered so that it initially produces AMP and pyrophosphate (PP_i), which is itself then hydrolyzed in a subsequent step (Figure 3–41). The whole process makes available a total free-energy change of about –26 kcal/mole. An important biosynthetic reaction that is driven in this way, nucleic acid (polynucleotide) synthesis, is illustrated in Figure 3–42.

Question 3–9

Which of the following reactions will occur only if it is coupled to a second, energetically favorable reaction?

A.   glucose + $O_2 \rightarrow CO_2 + H_2O$

B. $CO_2 + H_2O \rightarrow$ glucose + O_2

C.   nucleoside triphosphates $\rightarrow$ DNA

D.   nucleotide bases $\rightarrow$ nucleoside triphosphates

E.   ADP + ⓟ $\rightarrow$ ATP

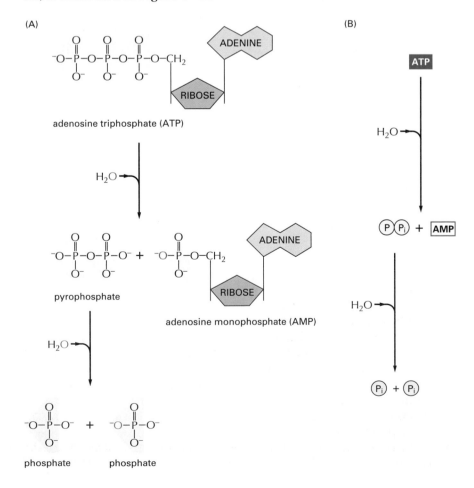

Figure 3–41 In an alternative route for the hydrolysis of ATP, pyrophosphate is first formed and then hydrolyzed. This route releases about twice as much free energy as the reaction shown earlier in Figure 3–32. (A) In the two successive hydrolysis reactions, oxygen atoms from the participating water molecules are retained in the products whereas the hydrogen atoms dissociate to form free hydrogen ions, H⁺. (B) Overall reaction shown in summary form.

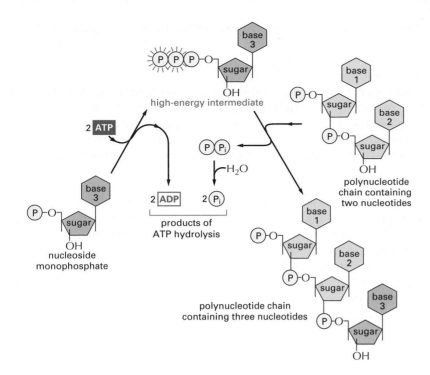

Figure 3–42 Synthesis of a polynucleotide, RNA or DNA, is a multistep process driven by ATP hydrolysis. In the first step a nucleoside monophosphate is activated by the sequential transfer of the terminal phosphate groups from two ATP molecules. The high-energy intermediate formed—a nucleoside triphosphate—exists free in solution until it reacts with the growing end of an RNA or a DNA chain with release of pyrophosphate. Hydrolysis of the latter to inorganic phosphate is highly favorable and helps to drive the overall reaction in the direction of polynucleotide synthesis.

Essential Concepts

- Living organisms are able to exist because of a continual input of energy. Part of this energy is used to carry out essential functions—reactions that support cellular metabolism, growth, and reproduction—and the remainder is lost in the form of heat.

- The primary source of energy for most living organisms is the sun. Plants and photosynthetic bacteria use solar energy to produce organic molecules from carbon dioxide. Animals obtain food by eating plants or by eating animals that feed on plants.

- Each of the many hundreds of chemical reactions that occur in a cell is specifically catalyzed by an enzyme. Large numbers of different enzymes work in sequence to form chains of reactions, called metabolic pathways, each performing a particular set of functions in the cell.

- Catabolic reactions break down food molecules through oxidative pathways and release energy. Anabolic reactions generate the many complex molecules needed by the cell, and they require an energy input. In animal cells, both the building blocks and the energy required for the anabolic reactions are obtained by catabolism.

- Enzymes catalyze reactions by binding to particular substrate molecules in a way that lowers the activation energy required for making and breaking specific covalent bonds.

- The rate at which an enzyme catalyzes a reaction depends on how rapidly it finds its substrates and how quickly the product forms and then diffuses away. These rates vary widely from one enzyme to another, and they can be measured after mixing purified enzymes and substrates together under a set of defined conditions.

- The only chemical reactions possible are those that increase the total amount of disorder in the universe. The free-energy change for a reaction, ΔG, measures this disorder, and it must be less than zero for a reaction to proceed.

- The free-energy change for a chemical reaction, ΔG, depends on the concentrations of the reacting molecules, and it may be calculated from these concentrations if the equilibrium constant (K) of the reaction (or the standard free-energy change $\Delta G°$ for the reactants) is known.

- Equilibrium constants govern all of the associations (and dissociations) that occur between macromolecules and small molecules in the cell. The larger the binding energy between two molecules, the larger the equilibrium constant and the more likely that these molecules will be paired.
- By creating a reaction pathway that couples an energetically favorable reaction to an energetically unfavorable one, enzymes cause otherwise impossible chemical transformations to occur.
- A small set of activated carrier molecules, in particular, ATP, NADH, and NADPH, plays a central part in these coupling events. ATP carries high-energy phosphate groups, whereas NADH and NADPH carry high-energy electrons.
- Food molecules provide the carbon skeletons for the formation of larger molecules. The covalent bonds of these larger molecules are typically produced in reactions that are coupled to energetically favorable bond changes in activated carrier molecules such as ATP and NADPH.

Key Terms

acetyl CoA	enzyme	$NADP^+$, NADPH
activated carrier	equilibrium	oxidation
activation energy	equilibrium constant, K	photosynthesis
ADP, ATP	free energy	reduction
catalyst	ΔG, $\Delta G°$	respiration
condensation	hydrolysis	substrate
coupled reaction	K_m	turnover number
diffusion	metabolism	V_{max}
entropy	NAD^+, NADH	

Questions

Question 3–10

Which of the following statements are correct? Explain your answers.

A. Some enzyme-catalyzed reactions cease completely if their enzyme is absent.

B. High-energy electrons (such as those found in the activated carriers NADH and NADPH) move faster around the atomic nucleus.

C. ATP hydrolysis to form AMP can provide almost twice the energy of ATP hydrolysis to form ADP.

D. A partially oxidized carbon atom has a smaller diameter than a more reduced one.

E. Some activated carrier molecules can transfer energy and chemical groups.

F. The rule that oxidations release energy, whereas reductions require energy input, applies to all chemical reactions, not just those that occur in living cells.

G. Cold-blooded animals have an energetic disadvantage because they give less heat to the environment than warm-blooded animals do. This slows their ability to make ordered macromolecules.

H. Linking the reaction X → Y to a second, energetically favorable reaction Y → Z will shift the equilibrium constant of the first reaction.

Question 3–11

Consider the transition of X → Y in Figure 3–19. Assume that the only difference between X and Y is the presence of three hydrogen bonds in Y that are absent in X. What is the ratio of X to Y when the reaction is in equilibrium? Approximate your answer by using Table 3–1 (p. 98), with 1 kcal/mole as the energy of each hydrogen bond. Assume Y has three additional hydrogen bonds, i.e., a total of six, that distinguish it from X. How would that change the ratio?

Question 3–12

Protein A binds to protein B to form a complex, AB. A cell contains an equilibrium mixture of protein A at a concentration of 1 μM, protein B at a concentration of 1 μM, and protein AB (produced when A binds to B) also at 1 μM.

A. Referring to Figure 3–20, calculate the equilibrium constant for the reaction A + B $\rightleftharpoons$ AB.
B. What would the equilibrium constant be if A, B, and AB were each present in equilibrium at the much lower concentrations of 1 nM each?
C. How many extra hydrogen bonds would be needed to hold A to B at this lower concentration so that a similar portion of the molecules are found in the AB complex? (Remember that each hydrogen bond contributes about 1 kcal/mole.)

Question 3–13

Discuss the statement: "Whether the ΔG for a reaction is larger, smaller, or the same as $\Delta G°$ depends on the concentration of the compounds that participate in the reaction."

Question 3–14

A. How many ATP molecules could maximally be generated from one molecule of glucose, if the complete oxidation of 1 mole of glucose to CO_2 and H_2O yields 686 kcal and the useful chemical energy available in the high-energy phosphate bond of 1 mole of ATP is 12 kcal?
B. Respiration produces 30 moles of ATP from 1 mole of glucose. Compare this number with your answer above. What is the overall efficiency of ATP production from glucose?
C. If the cells of your body oxidize 1 mole of glucose, by how much would the temperature of your body (assume that your body consists of 75 kg of water) increase if the heat were not dissipated into the environment? (Recall that a kilocalorie [kcal] is defined as that amount of energy that heats 1 kg of water by 1°C.)
D. What would the consequences be if the cells of your body could convert the energy in food substances with only 20% efficiency? Would your body—as it is presently constructed—(a) work just fine, (b) overheat, or (c) freeze?
E. A resting human hydrolyzes about 40 kg of ATP every 24 hours. The oxidation of how much glucose would produce this amount of energy? (Hint: look up the structure of ATP in Figure 2–23 to calculate its molecular weight; the atomic weights of H, C, N, O, and P are 1, 12, 14, 16, and 31, respectively.)

Question 3–15

A prominent scientist claims to have isolated mutant cells that can convert 1 molecule of glucose into 57 molecules of ATP. Should this discovery be celebrated, or do you suppose that something might be wrong with it? Explain your answer.

Question 3–16

In a simple reaction A $\rightleftharpoons$ A*, a molecule is interconvertible between two forms that differ in standard free energy by 4.3 kcal/mole, with A* having the higher $G°$. (A) Use Table 3–1 (p. 98) to find how many more molecules will be in state A* compared with state A at equilibrium. (B) If an enzyme lowered the activation energy of the reaction by 2.8 kcal/mole, how would the ratio of A to A* change?

Question 3–17

A reaction in a single-step biosynthetic pathway that converts a metabolite into a particularly vicious poison is energetically highly unfavorable (metabolite $\rightleftharpoons$ poison). The reaction is normally driven by ATP hydrolysis. Assume that a mutation in the enzyme that catalyzes the reaction prevents it from utilizing ATP, but still allows it to catalyze the reaction.

A. Do you suppose it might be safe for you to eat the mutant organism? Base your answer on an estimation of how much less poison the organism would produce, assuming the reaction is in equilibrium and most of the energy stored in ATP is used to drive the unfavorable reaction.
B. Would your answer be different for another mutant enzyme that couples the reaction to ATP hydrolysis but works 100 times slower?

Question 3–18

Consider the effects of two enzymes. Enzyme A catalyzes the reaction
$$ATP + GDP \rightleftharpoons ADP + GTP$$
whereas enzyme B catalyzes the reaction
$$NADH + NADP^+ \rightleftharpoons NAD^+ + NADPH$$
Discuss whether the enzymes would be beneficial or detrimental to cells.

Question 3–19

Discuss the following statement: "Enzymes and heat are alike in that both can speed up reactions that—although thermodynamically feasible—do not occur at an appreciable rate because they require a high activation energy. Diseases that are treated by the careful application of heat, such as by ingestion of hot chicken soup, are therefore likely to be due to the insufficient function of an enzyme."

Question 3–20

The curve shown in Figure 3–25 is described by the Michaelis–Menten equation:
$$\text{rate} = V_{\text{max}} [S]/([S] + K_M)$$
Can you convince yourself that the features qualitatively described in the text are accurately represented

by this equation? In particular, how can the equation be simplified when the substrate concentration is in one of the following ranges: (A) the substrate concentration [S] is much smaller than the K_M, (B) the substrate concentration [S] equals the K_M, and (C) the substrate concentration [S] is much larger than the K_M?

Question 3–21

The rate of a simple enzyme reaction is given by the standard Michaelis–Menten equation

$$\text{rate} = V_{max}\,[S]/([S] + K_M)$$

If the V_{max} of an enzyme is 100 μmole/sec and the K_M is 1 mM, at what substrate concentration is the rate 50 μmole/sec? Plot a graph of rate versus substrate concentration (S) for [S] = 0 to 10 mM. Convert this to a plot of 1/rate versus 1/[S]. Why is the latter plot a straight line?

Question 3–22

Select the correct options in the following and explain your choices. If [S] is much smaller than K_M, the active site of the enzyme is mostly occupied/unoccupied. If [S] is very much greater than K_M, the reaction rate is limited by the enzyme/substrate concentration.

Question 3–23

A. The reaction rates of the reaction S → P catalyzed by enzyme E were determined under conditions such that only very little product was formed. The following data were measured:

Substrate Concentration (μM)	Reaction Rate (μmole/min)
0.08	0.15
0.12	0.21
0.54	0.7
1.23	1.1
1.82	1.3
2.72	1.5
4.94	1.7
10.00	1.8

Plot the above data as a graph. Use this graph to find the K_M and the V_{max} for this enzyme.

B. As described in How We Know (pp. 103–105), to determine these values more precisely, a trick is generally used in which the Michaelis–Menten equation is transformed so that it is possible to plot the data as a straight line. A simple rearrangement yields

$$1/\text{rate} = (K_M/V_{max})\,(1/[S]) + 1/V_{max}$$

that is, an equation of the form $y = ax + b$. Calculate 1/rate and 1/[S] for the data given in part (A) and then plot 1/rate versus 1/[S] as a new graph. Determine K_M and V_{max} from the intercept of the line with the axis, where 1/[S] = 0, combined with the slope of the line. Do your results agree with the estimates made from the first graph of the raw data?

C. Note that part (A) states that only very little product was formed under the reaction conditions. Why is this important?

D. Assume the enzyme is regulated: upon phosphorylation, its K_M increases by a factor of 3 without changing its V_{max}. Is this an activation or inhibition? Plot the data you would expect for the phosphorylated enzyme in both the graph for (A) and the graph for (B).

Highlights from *Essential Cell Biology 2 Interactive* CD-ROM

3.1 Analogy of Enzyme Catalysis

3.2 Random Walk

Protein Structure and Function

When we look at a cell through a microscope or analyze its electrical or biochemical activity, we are, in essence, observing proteins. Proteins are the building blocks from which cells are assembled, and they constitute most of the cell's dry mass. But in addition to providing the cell with shape and structure, proteins also execute nearly all its myriad functions. Enzymes promote intracellular chemical reactions by providing intricate molecular surfaces, contoured with particular bumps and crevices, that can cradle or exclude specific molecules. Proteins embedded in the plasma membrane form the channels and pumps that control the passage of nutrients and other small molecules into and out of the cell. Other proteins carry messages from one cell to another, or act as signal integrators that relay information from the plasma membrane to the nucleus of individual cells. Yet others serve as tiny molecular machines with moving parts: some proteins, such as kinesin, propel organelles through the cytoplasm; others, such as topoisomerase, untangle knotted DNA molecules. Specialized proteins also act as antibodies, toxins, hormones, antifreeze molecules, elastic fibers, or luminescence generators. Before we can hope to understand how genes work, how muscles contract, how nerves conduct electricity, how embryos develop, or how our bodies function, we must understand proteins.

The multiplicity of functions performed by proteins (Panel 4–1, p. 120) arises from the huge number of different shapes they adopt: structure dictates function. So we begin our description of these remarkable macromolecules by discussing their three-dimensional structures and the properties that these structures confer. We next look at how proteins work: how enzymes catalyze chemical reactions, how some proteins act as molecular switches, and how others generate coherent movement. Finally, we examine how protein activity is regulated in cells, and how changes in a protein's structure affect its function. In this chapter, we also present a brief guide to the techniques that biologists use to study proteins, including methods for purifying proteins and determining their structures.

The Shape and Structure of Proteins

From a chemical point of view, proteins are by far the most structurally complex and functionally sophisticated molecules known. This is perhaps not surprising, considering that the structure and activity of each protein has been developed and fine-tuned over billions of years of evolutionary history. We start by considering how the position of each amino acid in the long string of amino acids that forms a protein determines its three-dimensional shape, a structure stabilized by noncovalent interactions between different parts of the molecule. Understanding

The Shape and Structure of Proteins

The Shape of a Protein Is Specified by Its Amino Acid Sequence

Proteins Fold into a Conformation of Lowest Energy

Proteins Come in a Wide Variety of Complicated Shapes

The α Helix and the β Sheet Are Common Folding Patterns

Helices Form Readily in Biological Structures

β Sheets Form Rigid Structures at the Core of Many Proteins

Proteins Have Several Levels of Organization

Few of the Many Possible Polypeptide Chains Will Be Useful

Proteins Can Be Classified into Families

Large Protein Molecules Often Contain More Than One Polypeptide Chain

Proteins Can Assemble into Filaments, Sheets, or Spheres

Some Types of Proteins Have Elongated Fibrous Shapes

Extracellular Proteins Are Often Stabilized by Covalent Cross-Linkages

How Proteins Work

All Proteins Bind to Other Molecules

The Binding Sites of Antibodies Are Especially Versatile

Enzymes Are Powerful and Highly Specific Catalysts

Lysozyme Illustrates How an Enzyme Works

Tightly Bound Small Molecules Add Extra Functions to Proteins

How Proteins Are Controlled

The Catalytic Activities of Enzymes Are Often Regulated by Other Molecules

Allosteric Enzymes Have Two Binding Sites That Influence One Another

Phosphorylation Can Control Protein Activity by Triggering a Conformational Change

GTP-Binding Proteins Are Also Regulated by the Cyclic Gain and Loss of a Phosphate Group

Nucleotide Hydrolysis Allows Motor Proteins to Produce Large Movements in Cells

Proteins Often Form Large Complexes That Function as Protein Machines

Large-Scale Studies of Protein Structure and Function Are Increasing the Pace of Discovery

Panel 4–1 A few examples of some general protein functions

ENZYME

function: Catalyzes covalent bond breakage or formation.

examples: Living cells contain thousands of different enzymes, each of which catalyzes (speeds up) one particular reaction. Examples include: *tryptophan synthetase*—makes the amino acid tryptophan; *pepsin*—degrades dietary proteins in the stomach; *ribulose bisphosphate carboxylase*—helps convert carbon dioxide into sugars in plants; *DNA polymerase*—copies DNA; *protein kinase*—adds a phosphate group to a protein molecule.

STRUCTURAL PROTEIN

function: Provides mechanical support to cells and tissues.

examples: Outside cells, *collagen* and *elastin* are common constituents of extracellular matrix and form fibers in tendons and ligaments. Inside cells, *tubulin* forms long, stiff microtubules and *actin* forms filaments that underlie and support the plasma membrane; α-*keratin* forms fibers that reinforce epithelial cells and is the major protein in hair and horn.

TRANSPORT PROTEIN

function: Carries small molecules or ions.

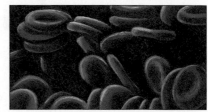

examples: In the bloodstream, *serum albumin* carries lipids, *hemoglobin* carries oxygen, and *transferrin* carries iron. Many proteins embedded in cell membranes transport ions or small molecules across the membrane. For example, the bacterial protein *bacteriorhodopsin* is a light-activated proton pump that transports H^+ ions out of the cell; the *glucose carrier* shuttles glucose into and out of liver cells; and a *Ca^{2+} pump* in muscle cells pumps the calcium ions needed to trigger muscle contraction into the endoplasmic reticulum, where they are stored.

MOTOR PROTEIN

function: Generates movement in cells and tissues.

examples: *Myosin* in skeletal muscle cells provides the motive force for humans to move; *kinesin* interacts with microtubules to move organelles around the cell; *dynein* enables eucaryotic cilia and flagella to beat.

STORAGE PROTEIN

function: Stores small molecules or ions.

examples: Iron is stored in the liver by binding to the small protein *ferritin*; *ovalbumin* in egg white is used as a source of amino acids for the developing bird embryo; *casein* in milk is a source of amino acids for baby mammals.

SIGNAL PROTEIN

function: Carries signals from cell to cell.

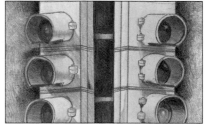

examples: Many of the hormones and growth factors that coordinate physiological function in animals are proteins; *insulin*, for example, is a small protein that controls glucose levels in the blood; *netrin* attracts growing nerve cells in a specific direction in a developing embryo; *nerve growth factor (NGF)* stimulates some types of nerve cells to grow axons; *epidermal growth factor (EGF)* stimulates the growth and division of epithelial cells.

RECEPTOR PROTEIN

function: Detects signals and transmits them to the cell's response machinery.

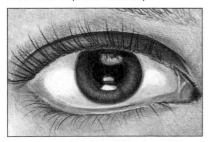

examples: *Rhodopsin* in the retina detects light; the *acetylcholine receptor* in the membrane of a muscle cell receives chemical signals released from a nerve ending; the *insulin receptor* allows a liver cell to respond to the hormone insulin by taking up glucose; the *adrenergic receptor* on heart muscle increases the rate of heartbeat when it binds to adrenaline.

GENE REGULATORY PROTEIN

function: Binds to DNA to switch genes on or off.

examples: The *lactose repressor* in bacteria silences the gene for the enzymes that degrade the sugar lactose; many different *homeodomain proteins* act as genetic switches to control development in multicellular organisms, including humans.

SPECIAL-PURPOSE PROTEIN

function: Highly variable.

examples: Organisms make many proteins with highly specialized properties. These molecules illustrate the amazing range of functions that proteins can perform. The *antifreeze proteins* of Arctic and Antarctic fishes protect their blood against freezing; *green fluorescent protein* from jellyfish emits a green light; *monellin*, a protein found in an African plant, has an intensely sweet taste; mussels and other marine organisms secrete *glue proteins* that attach them firmly to rocks, even when immersed in seawater.

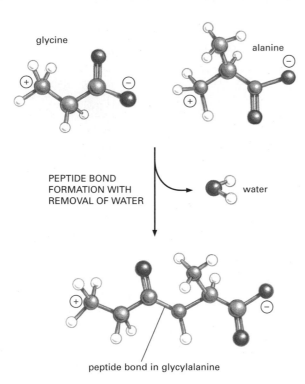

glycine

alanine

PEPTIDE BOND
FORMATION WITH
REMOVAL OF WATER

water

peptide bond in glycylalanine

Figure 4–1 Amino acids are linked together by peptide bonds. A covalent peptide bond forms when the carbon atom from the carboxyl group of one amino acid shares electrons with the nitrogen atom *(blue)* from the amino group of a second amino acid. As indicated, a molecule of water is generated during this condensation reaction.

protein structure at the atomic level will allow us to describe how the precise shape of each protein determines its function in the cell.

The Shape of a Protein Is Specified by Its Amino Acid Sequence

Proteins, as you may recall from Chapter 2, are assembled from a set of 20 different amino acids, each with different chemical properties. A **protein** molecule is made from a long chain of these amino acids, each linked to its neighbor through a covalent peptide bond (Figure 4–1). Proteins, therefore, are also called **polypeptides**. Each type of protein has a unique sequence of amino acids, exactly the same from one molecule to the next. One molecule of insulin, for example, has the same amino acid sequence as every other molecule of insulin. Many thousands of different proteins have been identified; each has its own distinct amino acid sequence.

Each polypeptide chain consists of a backbone that supports the different amino acid side chains. The **polypeptide backbone** is formed from the repeating sequence of atoms along the polypeptide chain. Attached to this repetitive chain are any of the 20 different amino acid **side chains**—the parts of the amino acids that are not involved in forming the peptide bond (Figure 4–2). These side chains give each amino acid its unique properties. Some are nonpolar and hydrophobic ("water-fearing"), some are negatively or positively charged, some are chemically reactive, and so on. The atomic structures of the 20 amino acids in proteins are presented in Panel 2–5 (pp. 74–75), and a brief list of amino acids, with their abbreviations, is provided in Figure 4–3.

Long polypeptide chains are very flexible: many of the covalent bonds that link carbon atoms in an extended chain of amino acids allow free rotation of the atoms they join. Thus proteins can in principle fold in an enormous number of ways. Each folded chain is constrained by many different sets of weak *noncovalent bonds* that form within proteins. These bonds involve atoms in the polypeptide backbone as well as

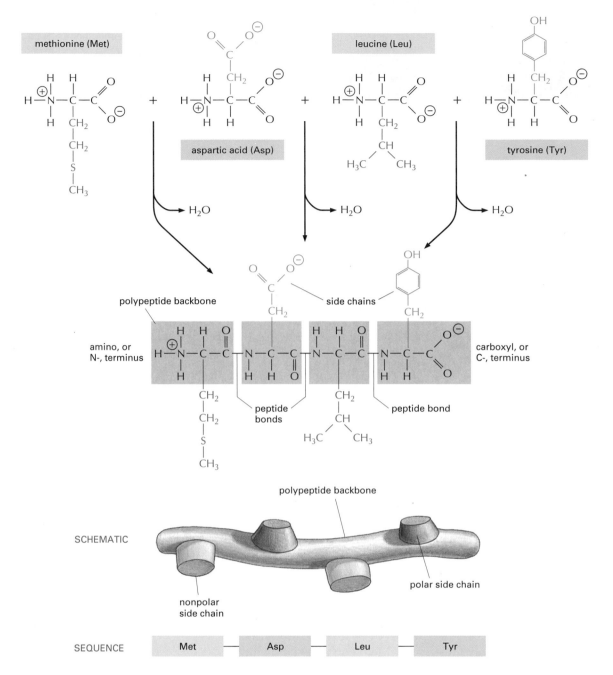

Figure 4–2 Proteins are made of a polypeptide backbone with attached side chains. Each type of protein differs in its amino acid sequence. Thus the sequential position of the chemically distinct side chains gives each protein its individual properties. The two ends of each polypeptide chain are chemically different: the end that carries the free amino group (NH_3^+, also written NH_2) is called the amino, or N-, terminus; and the end carrying the free carboxyl group (COO^-, also written $COOH$) is the carboxyl, or C-, terminus. The amino acid sequence of a protein is always presented in the N to C direction, reading from left to right.

atoms in the amino acid side chains. The noncovalent bonds that help proteins maintain their shape include *hydrogen bonds, ionic bonds,* and *van der Waals attractions,* which are described in Chapter 2 (see Panel 2–7, pp. 78–79). Because individual noncovalent bonds are much weaker than covalent bonds, it takes many noncovalent bonds to hold two regions of a polypeptide chain together tightly. The stability of each folded shape will therefore be affected by the combined strength of large numbers of noncovalent bonds (Figure 4–4).

AMINO ACID			SIDE CHAIN	AMINO ACID			SIDE CHAIN
Aspartic acid	Asp	D	negative	Alanine	Ala	A	nonpolar
Glutamic acid	Glu	E	negative	Glycine	Gly	G	nonpolar
Arginine	Arg	R	positive	Valine	Val	V	nonpolar
Lysine	Lys	K	positive	Leucine	Leu	L	nonpolar
Histidine	His	H	positive	Isoleucine	Ile	I	nonpolar
Asparagine	Asn	N	uncharged polar	Proline	Pro	P	nonpolar
Glutamine	Gln	Q	uncharged polar	Phenylalanine	Phe	F	nonpolar
Serine	Ser	S	uncharged polar	Methionine	Met	M	nonpolar
Threonine	Thr	T	uncharged polar	Tryptophan	Trp	W	nonpolar
Tyrosine	Tyr	Y	uncharged polar	Cysteine	Cys	C	nonpolar

POLAR AMINO ACIDS ───── NONPOLAR AMINO ACIDS ─────

A fourth weak force also plays a central role in determining the shape of a protein. As discussed in Chapter 2, hydrophobic molecules, including the nonpolar side chains of particular amino acids, tend to be forced together in an aqueous, watery environment to minimize their disruptive effect on the hydrogen-bonded network of the surrounding water molecules (see p. 48 and Panel 2–2, pp. 68–69). Therefore, an important factor governing the folding of any protein is the distribution of its polar and nonpolar amino acids. The nonpolar (hydrophobic) side chains—which belong to amino acids such as phenylalanine, leucine, valine, and tryptophan—tend to cluster in the interior of the folded protein (just as hydrophobic oil droplets coalesce to form one large droplet). Tucked away inside the folded protein, hydrophobic side chains can avoid contact with the aqueous cytosol that surrounds them inside a cell. In contrast, polar side chains—such as those belonging to arginine, glutamine, and histidine—tend to arrange themselves near the outside of the folded protein, where they can form hydrogen bonds with water and with other polar molecules (Figure 4–5). When polar amino acids are buried within the protein, they are usually hydrogen-bonded to other polar amino acids or to the polypeptide backbone (Figure 4–6).

Figure 4–3 Twenty different amino acids are found in proteins. Both three-letter and one-letter abbreviations are presented. As shown, there are an equal number of polar and nonpolar side chains. For atomic structures, see Panel 2–5 (pp. 74–75).

Figure 4–4 Three types of noncovalent bonds help proteins fold. Although a single one of these bonds is quite weak, many of them often form together to create a strong bonding arrangement, as in the small peptide shown (center). R is a general designation for a side chain.

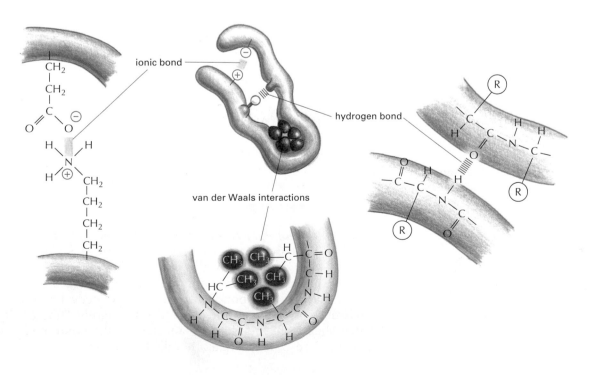

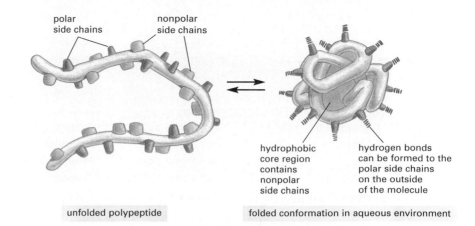

Figure 4–5 Hydrophobic forces help proteins fold into compact conformations. The polar amino acid side chains tend to gather on the outside of the folded protein, where they can interact with water; the nonpolar amino acid side chains are buried on the inside to form a highly packed hydrophobic core of atoms that are hidden from water. In this very schematic drawing, the protein contains only about 30 amino acids.

polar side chains

nonpolar side chains

hydrophobic core region contains nonpolar side chains

hydrogen bonds can be formed to the polar side chains on the outside of the molecule

unfolded polypeptide

folded conformation in aqueous environment

Proteins Fold into a Conformation of Lowest Energy

Each type of protein has a particular three-dimensional structure, which is determined by the order of the amino acids in its chain. The final folded structure, or **conformation**, adopted by any polypeptide chain is determined by energetic considerations: a protein generally folds into the shape in which the free energy (*G*) is minimized. Protein folding has been studied in the laboratory using highly purified proteins. A protein can be unfolded, or *denatured,* by treatment with certain solvents that disrupt the noncovalent interactions holding the folded chain together. This treatment converts the protein into a flexible polypeptide chain that has lost its natural shape. When the denaturing solvent is removed, the protein often refolds spontaneously, or *renatures,* into its original conformation (Figure 4–7). The fact that a renatured protein can, on its own, regain the correct conformation indicates that all the information necessary to specify the three-dimensional shape of a protein is contained in its amino acid sequence.

Each protein normally folds into a single stable conformation. This conformation, however, often changes slightly when the protein interacts with other molecules in the cell. This change in shape is crucial to the function of the protein, as we shall see later in this chapter.

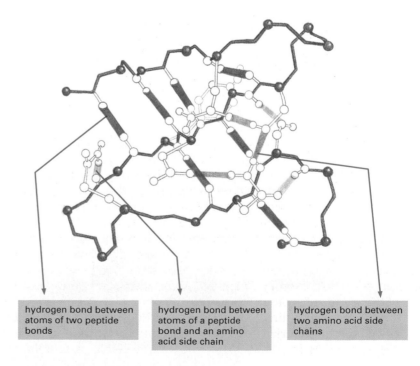

Figure 4–6 Hydrogen bonds can form within a protein molecule. Large numbers of hydrogen bonds form between adjacent regions of the folded polypeptide chain and help stabilize its three-dimensional shape. The protein depicted is a portion of the enzyme lysozyme, and the hydrogen bonds between the three possible pairs of partners have been differently colored, as indicated. (After C.K. Mathews, K.E. van Holde, and K.G. Ahern, Biochemistry, 3rd edn. San Francisco: Benjamin Cummings, 2000.)

hydrogen bond between atoms of two peptide bonds

hydrogen bond between atoms of a peptide bond and an amino acid side chain

hydrogen bond between two amino acid side chains

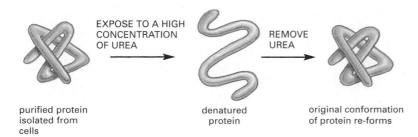

purified protein
isolated from
cells

EXPOSE TO A HIGH
CONCENTRATION
OF UREA

denatured
protein

REMOVE
UREA

original conformation
of protein re-forms

Figure 4–7 Denatured proteins can recover their natural shapes. This type of experiment demonstrates that the conformation of a protein is determined solely by its amino acid sequence. Renaturation works best for small proteins.

When proteins fold improperly, they can form aggregates that can damage cells and even whole tissues. Aggregated proteins underlie a number of neurodegenerative disorders, including Alzheimer's disease and Huntington's disease. Prion diseases—such as scrapie in sheep, bovine spongiform encephalopathy (BSE, or "mad cow" disease) in cattle, and Creutzfeldt–Jacob disease (CJD) in humans—are also caused by protein aggregates. The prion protein, PrP, can adopt a special misfolded form that is considered "infectious," as it can convert properly folded prion proteins into the abnormal conformation (Figure 4–8). This allows the misfolded form of PrP to spread rapidly from cell to cell in the brain, causing the death of the infected animal or human.

Although a protein chain can fold into its correct conformation without outside help, protein folding in a living cell is generally assisted by special proteins called *molecular chaperones*. These proteins bind to partly folded chains and help them to fold along the most energetically favorable pathway. Chaperones are vital in the crowded conditions of the cytoplasm, because they prevent newly synthesized protein chains from associating with the wrong partners. However, the final three-dimensional shape of the protein is still specified by its amino acid sequence: chaperones merely make the folding process more efficient and reliable.

Proteins Come in a Wide Variety of Complicated Shapes

Proteins are the most structurally diverse macromolecules in the cell. Although they range in size from about 30 amino acids to more than 10,000, the vast majority of proteins are between 50 and 2000 amino acids long. Proteins can be globular or fibrous; they can form filaments, sheets, rings, or spheres. Figure 4–9 presents a sampling of proteins whose exact structures are known. We will encounter many of these proteins later in this chapter and throughout the book.

Resolving a protein's structure often begins with determining its **amino acid sequence**. First, cells are broken open and the proteins separated and purified. The precise order of the amino acids in a pure protein can then be established in a number of ways. For many years, protein sequencing was accomplished by analyzing the amino acids in the protein directly, the first protein sequenced being the hormone *insulin*, in 1955. Today we can determine the order of amino acids in a protein more easily by sequencing the gene that encodes it (discussed in Chapter 10). Although indirect, this method is much faster. Once the order of the nucleotides in the DNA that encodes a protein is known, it can then be converted into an amino acid sequence by applying the genetic code (discussed in Chapter 7). The amino acid sequences of tens of thousands of proteins have already been determined in this way.

A combination of direct and indirect methods can also be used to identify and characterize unknown proteins. Now that complete genome sequences for many organisms are available, a protein can

Question 4–1

Urea used in the experiment shown in Figure 4–7 is a molecule that disrupts the hydrogen-bonded network of water molecules. Why might high concentrations of urea unfold proteins? The structure of urea is

$$H_2N \diagdown \underset{\underset{\displaystyle NH_2}{}}{\overset{\overset{\displaystyle O}{\|}}{C}}$$

often be identified by determining the amino acid sequence of a few short peptide fragments and then using those fingerprints to search the full DNA sequence of the organism for its gene.

Although all of the information required for a polypeptide chain to fold is contained in its amino acid sequence, we have not yet learned how to predict reliably a protein's detailed three-dimensional conformation—the spatial arrangement of its atoms. At present, the only way to discover the precise folding pattern of any protein is by experiment, using either X-ray or nuclear magnetic resonance methods (see How We Know, pp. 129–131). So far, the structures of about 10,000 different proteins have been completely analyzed by these techniques. Most have a three-dimensional conformation so intricate and irregular that their structure would require an entire chapter to describe in detail.

We can attempt to illustrate the intricacies of protein conformation by examining the structure of an unusually small protein. Panel 3–2 (pp. 132–133) presents four different depictions of a protein domain called SH2, which has important functions in eucaryotic cells. Constructed from a string of 100 amino acids, the structure is displayed as (A) a polypeptide backbone model, (B) a ribbon model, (C) a wire model that includes the amino acid side chains, and (D) a space-filling model. As indicated in the panel, each model emphasizes different features of the protein. The three horizontal rows show the protein in different orientations, and the images are colored to distinguish the path of the polypeptide chain, from its N-terminus *(purple)* to its C-terminus *(red)*. We will describe the different structural elements in this protein shortly.

Clearly a protein's conformation is amazingly complex, even for a protein as small as HPr. To visualize such complicated structures, scientists have developed various graphical and computer-based tools that allow them to generate a variety of images of selected proteins that can be displayed and rotated on the screen, such as those depicted in Panel 4–2 (see CD-ROM). In addition, describing and presenting such complex protein structures is made easier by recognizing that several common folding patterns underlie these conformations, as we discuss in the next sections.

The α Helix and the β Sheet Are Common Folding Patterns

When the three-dimensional structures of many different protein molecules are compared, it becomes clear that, although the overall conformation of each protein is unique, two regular folding patterns are often found in parts of them. Both were discovered about 50 years ago from studies of hair and silk. The first folding pattern to be discovered, called the **α helix**, was found in the protein *α-keratin*, which is abundant in skin and its derivatives—such as hair, nails, and horns. Within a year of the discovery of the α helix, a second folded structure, called a **β sheet**, was found in the protein *fibroin*, the major constituent of silk. (Biologists often use Greek letters to name their discoveries, with the first example receiving the designation α, the second β, and so on.)

These two folding patterns are particularly common because they result from hydrogen bonds forming between the N–H and C=O groups in the polypeptide backbone. Because the amino acid side chains are not involved in forming these hydrogen bonds, α helices and β sheets can be generated by many different amino acid sequences. In each case, the protein chain adopts a regular, repeating form or motif. Both structural features, and the shorthand cartoon symbols that are used to represent them in ribbon models of proteins, are presented in Figure 4–10.

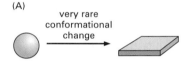

(A)

very rare conformational change

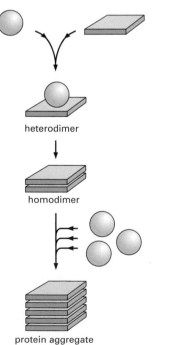

(B) infectious seeding of new protein aggregate

heterodimer

homodimer

protein aggregate

Figure 4–8 Prion diseases are caused by rare proteins whose misfolding is infectious. The protein PrP is the best known of these proteins, but other examples are known.

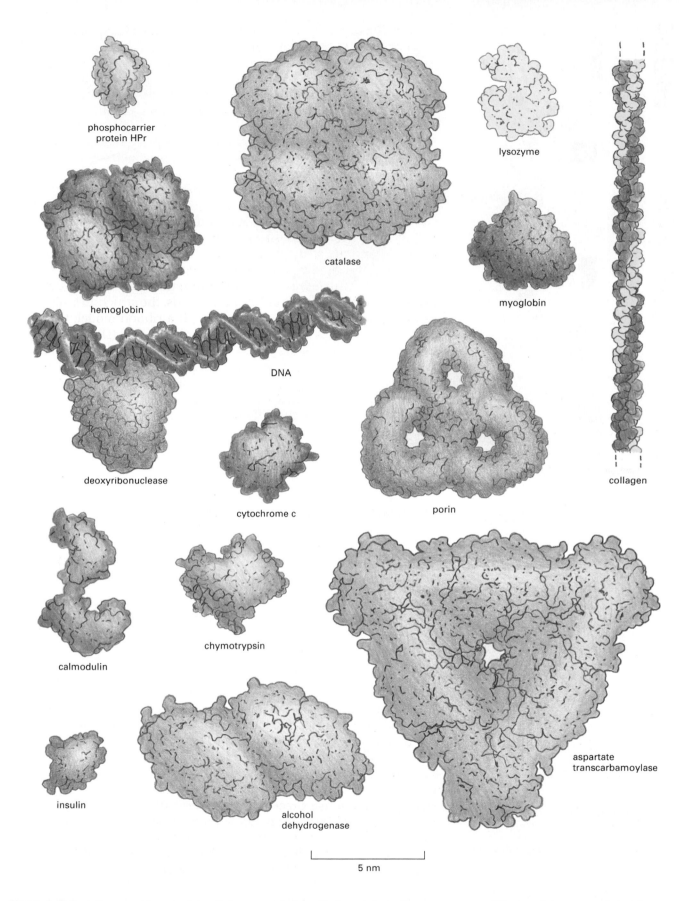

phosphocarrier
protein HPr

catalase

lysozyme

hemoglobin

myoglobin

DNA

deoxyribonuclease

porin

collagen

cytochrome c

calmodulin

chymotrypsin

aspartate
transcarbamoylase

insulin

alcohol
dehydrogenase

5 nm

Figure 4–9 Proteins come in a variety of shapes and sizes. Each protein is shown as a space-filling model, represented at the same scale. In the *top left corner* is the phosphocarrier protein HPr, which is featured in greater detail in Panel 4–2 (pp. 132–133). Part of a molecule of DNA *(gray)* is shown for comparison. (After David S. Goodsell, *Our Molecular Nature.* New York: Springer-Verlag, 1996.)

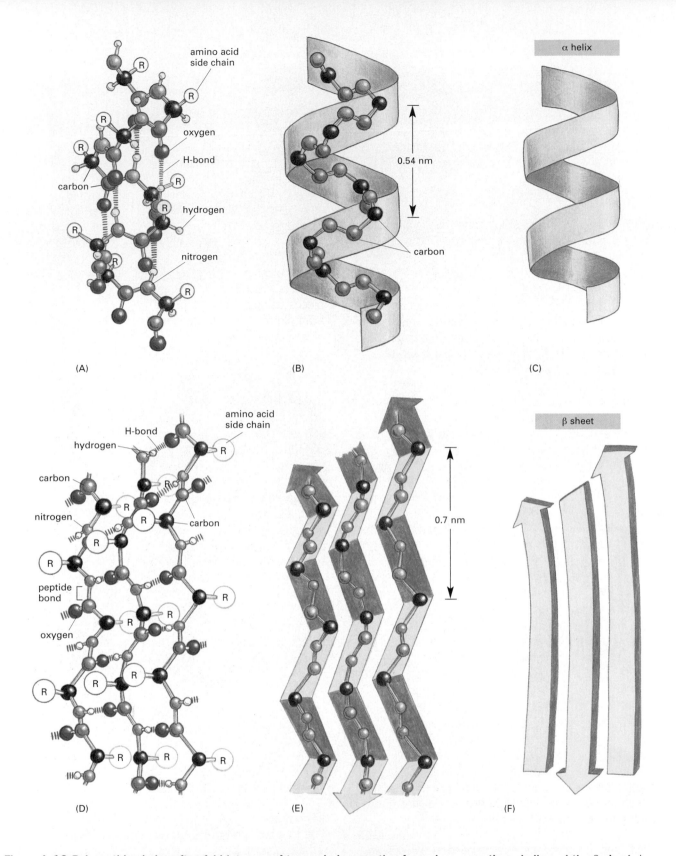

(A)

amino acid side chain

oxygen

H-bond

carbon

hydrogen

nitrogen

(B)

0.54 nm

carbon

(C)

α helix

(D)

H-bond

hydrogen

amino acid side chain

R

carbon

nitrogen

carbon

R

peptide bond

R

oxygen

R

R

R

(E)

0.7 nm

(F)

β sheet

Figure 4–10 Polypeptide chains often fold into one of two orderly repeating forms known as the α helix and the β sheet. In an α helix (A, B, and C), the N–H of every peptide bond is hydrogen-bonded to the C=O of a neighboring peptide bond located four peptide bonds away in the same chain. In the case of the β sheet (D, E, and F), the individual polypeptide chains (strands) in the sheet are held together by hydrogen-bonding between peptide bonds in different strands, and the amino acid side chains in each strand project alternately above and below the plane of the sheet. In the example shown, the adjacent peptide chains run in opposite directions, forming an antiparallel β sheet. (A) and (D) show all of the atoms in the polypeptide backbone, but the amino acid side chains are denoted by R. (B) and (E) show the backbone atoms only, while (C) and (F) display the shorthand cartoon symbols that are used to represent the α helix and the β sheet in ribbon drawings of proteins (see Panel 4–2B, p. 132).

How We Know: Probing Protein Structure

Coiled-coils α helices and β sheets: how do we know that these structures exist? How can we see them? And how can we determine which of these or other structures a particular protein adopts—and whether a protein's shape changes as it does its job?

Most proteins are small—too small to be seen in any detail, even with a powerful electron microscope. To follow the paths that the string of amino acids takes inside a good-sized protein molecule, you need to be able to see the atoms that form the individual amino acids. For that job, you generally need X-rays. Like light, X-rays are a form of electromagnetic radiation. But they have a wavelength of 0.1 nanometer (nm)—the approximate diameter of a hydrogen atom—so they allow you to look at very small and detailed structures.

Before you can examine your protein with X-rays, you have to isolate it in a pure form. You must also determine its amino acid sequence, so that the X-ray data (or other types of structural information) are easier to interpret. With that in mind, we will now review how to go about solving the three-dimensional structure of a protein.

Isolation

Maybe you are interested in studying a set of proteins that is involved in copying DNA, processing RNA, or degrading damaged proteins. Each of these cellular processes is carried out by a molecular machine that contains many different proteins, as we discuss in Chapters 6 and 7.

If you know the identity of any one of the proteins in such a complex, you may be able to identify some of the other proteins in the complex by protein affinity chromatography. Essentially, you break open cells, separate all of the soluble proteins by centrifugation, and then pass those proteins through a column matrix that contains either the pure target protein, or an antibody that binds this protein tightly. In either case, protein complexes will collect on the column and often can be eluted with salt or by changing the pH of the solution. Proteins that are physically associated with the target protein can thereby be identified and isolated. These techniques are described in greater detail in Panels 4–3 through 4–6 (pp. 160–165).

The next step involves visualizing, and then isolating, these proteins by electrophoresis through a polymer gel, which separates polypeptides on the basis of their sizes. If the purification yields a great many proteins, or if the proteins are very similar in size, they can be resolved using two-dimensional gel electrophoresis, which separates proteins by both size and overall electrical charge (see Panel 4–5, p. 163). Both techniques yield a number of bands or spots, each one containing a different protein.

Once a protein has been selected for further study, you are ready to determine its amino acid sequence. The small amount of protein that is present in the gel actually provides enough material for this analysis.

Identification

Before a protein is sequenced (i.e., the order of its amino acids is determined), it is generally broken into smaller pieces using a selective protease. The enzyme trypsin, for example, cleaves polypeptide chains on the carboxyl side of lysine or arginine residues. So if a protein has nine lysines and seven arginines, digestion with trypsin should cut it into 17 peptide fragments.

Mass spectrometry can then be used to determine the exact mass of each peptide fragment—information that will allow you to identify your protein from the list of all proteins produced by the organism as determined from the DNA sequence of its genome. The process works like this. The peptides from the tryptic digest are dried onto a metal plate. The sample is then blasted with a laser, which heats the peptides, causing them to become "ionized"—ejected from the plate in the form of a gas. Accelerated by a powerful electric field, the peptide ions then fly toward a detector, and the time it takes them to get there is related to their mass and their charge. (The larger the peptide is, the slower it moves, while the more highly charged it is, the faster it moves.) Knowledge of the exact mass of each of the protein fragments produced by trypsin cleavage serves as a "fingerprint," which allows the identification of the protein's gene (Figure 4–11) and thus its amino acid sequence.

You now need to produce enough protein to do a structural analysis. Using recombinant DNA technology (which we discuss in detail in Chapter 10), you insert the gene into cells, usually bacteria, and get the cells to produce large amounts of the protein. Once the protein is purified (by the techniques described in Panel 4–3, pp. 160–161), you are ready to attempt to solve its structure.

Interrogation

The toughest part is yet to come. To determine a protein's structure using X-ray crystallography, you first need to coax the protein into forming crystals—large, highly ordered arrays in which every protein has the same conformation and is perfectly aligned with its neighbors. Growing such crystals is still something of an art, as it requires a trial-and-error process of determining the proper conditions for forming the highest-quality crystals—selecting the right ions, the optimal temperature, and so on.

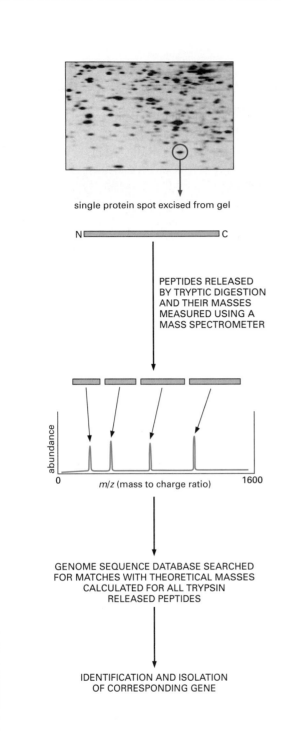

single protein spot excised from gel

N ▭ C

PEPTIDES RELEASED
BY TRYPTIC DIGESTION
AND THEIR MASSES
MEASURED USING A
MASS SPECTROMETER

abundance

0   m/z (mass to charge ratio)   1600

GENOME SEQUENCE DATABASE SEARCHED
FOR MATCHES WITH THEORETICAL MASSES
CALCULATED FOR ALL TRYPSIN
RELEASED PEPTIDES

IDENTIFICATION AND ISOLATION
OF CORRESPONDING GENE

Figure 4–11 **Mass spectrometry can be used to identify proteins by determining the precise masses of peptides derived from them.** In this example, the protein of interest is excised from a two-dimensional polyacrylamide gel and then digested with trypsin. The peptide fragments are loaded onto the mass spectrometer and their exact masses measured. Sequence databases are then searched to find the protein whose calculated tryptic digest profile matches these values. (Micrograph courtesy of Patrick O'Farrell.)

With crystals in hand, you are ready for the X-ray analysis. When a narrow beam of X-rays is directed at a protein crystal, the atoms in the protein molecules scatter the incoming X-rays. These scattered waves either reinforce or cancel one another, producing a complex diffraction pattern that is collected by electronic detectors. The position and intensity of each spot in the diffraction pattern contains information about the position of the atoms in the protein crystal (Figure 4–12).

Because these patterns are so complex—even a small protein can generate 25,000 discrete spots—computers are used to interpret them. By combining information obtained from such maps with the amino acid sequence of the protein, you can generate an atomic model of the protein's structure. To determine if the protein undergoes conformational changes in its structure when it binds a ligand that boosts its activity, you might subsequently try crystallizing it in the presence of the ligand. With crystals of sufficient quality, even small atomic movements can be detected by comparing the structures obtained in the presence and absence of stimulatory or inhibitory ligands.

There is a different way to solve the structure of your protein, one that does not require obtaining protein crystals. If the protein is small—say, 40,000 daltons or less—you can determine its structure by nuclear magnetic resonance (NMR) spectroscopy. This technique takes advantage of the fact that the nuclei of many atoms are intrinsically magnetic and that their behavior is influenced by surrounding atoms. In NMR spectroscopy, a solution of pure protein is placed in a strong magnetic field and then bombarded with radio waves of different frequencies. The hydrogen nuclei in the protein will generate an NMR signal that can be used to determine the distances between the amino acids and between different parts of the protein. Again, combined with the known amino acid sequence, an NMR spectrum can allow you to compute the three-dimensional structure of the protein (Figure 4–13).

If the protein is larger than 40,000 daltons, you can try to break the polypeptide up into its constituent functional domains and then perform this type of NMR analysis on each domain.

Recently, X-ray crystallography was used to determine the structure of the ribosome, a complex cellular machine made of several RNAs and more than 50 proteins. In the future, improvements in X-ray crystallography and NMR spectroscopy should permit rapid analysis of many more proteins and protein machines. And once enough structures have been determined, we may be able to generate algorithms for accurately predicting structure based solely on the amino acid sequence itself. After all, it is the sequence of the amino acids alone that determines how each protein folds up into its three-dimenionsal structure.

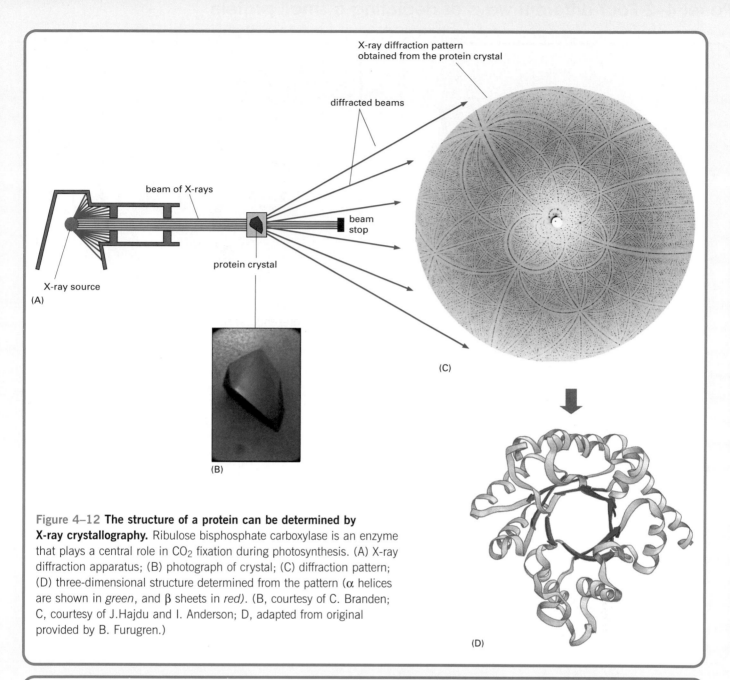

Figure 4–12 The structure of a protein can be determined by X-ray crystallography. Ribulose bisphosphate carboxylase is an enzyme that plays a central role in CO_2 fixation during photosynthesis. (A) X-ray diffraction apparatus; (B) photograph of crystal; (C) diffraction pattern; (D) three-dimensional structure determined from the pattern (α helices are shown in *green*, and β sheets in *red*). (B, courtesy of C. Branden; C, courtesy of J.Hajdu and I. Anderson; D, adapted from original provided by B. Furugren.)

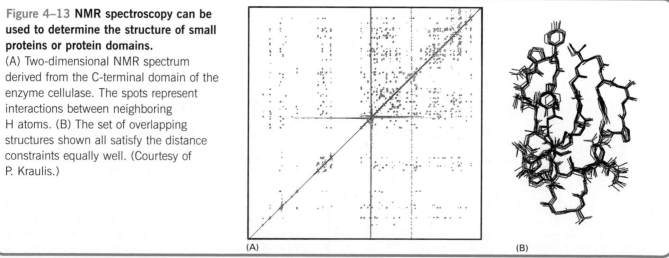

Figure 4–13 NMR spectroscopy can be used to determine the structure of small proteins or protein domains.
(A) Two-dimensional NMR spectrum derived from the C-terminal domain of the enzyme cellulase. The spots represent interactions between neighboring H atoms. (B) The set of overlapping structures shown all satisfy the distance constraints equally well. (Courtesy of P. Kraulis.)

Panel 4–2 Four different ways of depicting a small protein

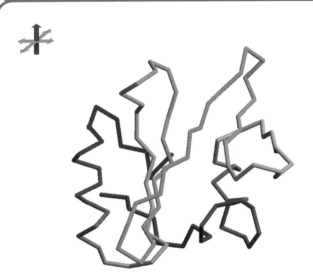

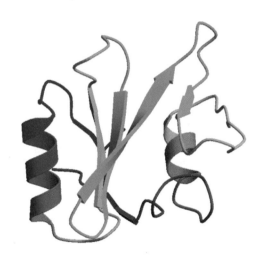

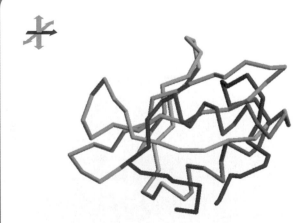

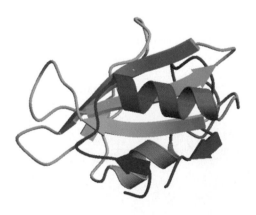

(A) Backbone: Shows the overall organization of the polypeptide chain; a clean way to compare structures of related proteins.

(B) Ribbon: Easy way to visualize secondary structures, such as α helices and β sheets.

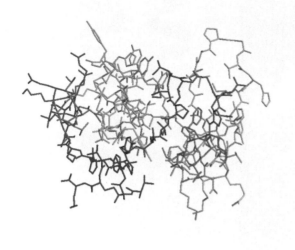

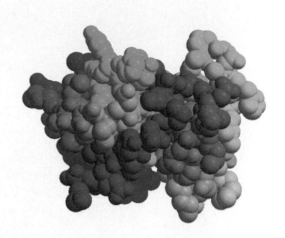

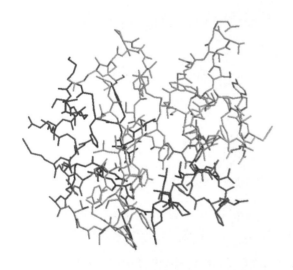

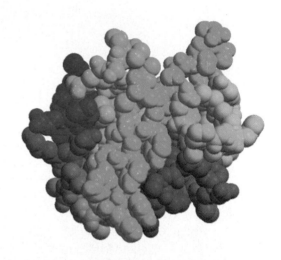

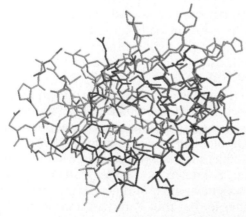

(C) Wire: Highlights side chains and their relative proximities; useful for predicting which amino acids might be involved in a protein's activity, particularly if the protein is an enzyme.

(D) Space-filling: Provides contour map of the protein; gives a feel for the shape of the protein and shows which amino acid side chains are exposed on its surface. Shows how the protein might look to a small molecule, such as water, or to another protein.

(Courtesy of David Lawson.)

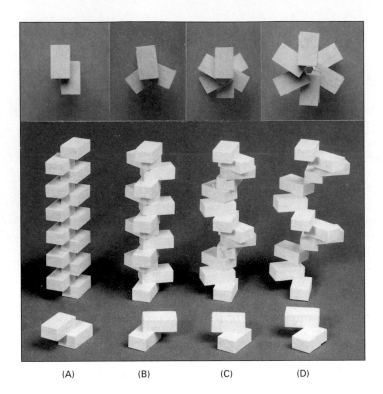

left-
handed

right-
handed

(E)

(A) (B) (C) (D)

Figure 4–14 The helix is a regular biological structure. A helix will form when a series of subunits bind to each other in a regular way (A–D). At the bottom, the interaction between two subunits is shown; behind them are the helices that result. These helices have two (A), three (B), or six (C and D) subunits per helical turn. At the top, the arrangement of subunits has been photographed from directly above the helix. Note that the helix in (D) has a wider path than that in (C), but the same number of subunits per turn. (E) A helix can be either right-handed or left-handed. As a reference, it is useful to remember that standard metal screws, which insert when turned clockwise, are right-handed. Note that a helix preserves the same handedness when it is turned upside down.

Question 4–2

Look at the models of the small protein in Panel 4–2, pp. 132–133. Are the α helices right- or left-handed? Are the three chains that form the largest region of the β sheet parallel or antiparallel? Starting at the N-terminus (the *purple* end), trace your finger along the peptide backbone. Are there any knots? Why, or why not?

Helices Form Readily in Biological Structures

The abundance of helices in proteins is, in a way, not surprising. A **helix** is an unexceptional regular structure that resembles a spiral staircase, as illustrated in Figure 4–14. It is generated simply by placing many similar subunits next to each other, each in the same strictly repeated relationship to the one before. Because it is very rare for subunits to join up in a straight line, this arrangement will generally result in a helix. Depending on the twist of the staircase, a helix is said to be either right-handed or left-handed (Figure 4–14E). Handedness is not affected by turning the helix upside down, but it is reversed if the helix is reflected in the mirror.

An α helix is generated when a single polypeptide chain turns around itself to form a structurally rigid cylinder. A hydrogen bond is made between every fourth peptide bond, linking the C=O of one peptide bond to the N–H of another (see Figure 4–10A). This gives rise to a regular helix with a complete turn every 3.6 amino acids.

Short regions of α helix are especially abundant in the proteins located in cell membranes, such as transport proteins and receptors. We will see in Chapter 11 that those portions of a transmembrane protein that cross the lipid bilayer usually form an α helix that is composed largely of amino acids with nonpolar side chains. The polypeptide backbone, which is hydrophilic, is hydrogen-bonded to itself in the α helix, and it is shielded from the hydrophobic lipid environment of the membrane by its protruding nonpolar side chains (Figure 4–15).

Sometimes a pair of α helices will wrap around one another to form a particularly stable structure, known as a **coiled-coil**. This structure forms when the two α helices have most of their nonpolar (hydrophobic) side chains on one side, so that they can twist around each other with these side chains facing inward—minimizing their contact with the aqueous cytosol (Figure 4–16). Long, rodlike coiled-coils form the structural framework for many elongated proteins. Examples include α-keratin, which forms the intracellular fibers that reinforce the outer layer of the skin, and myosin, the protein responsible for muscle contraction (discussed in Chapter 17).

β Sheets Form Rigid Structures at the Core of Many Proteins

HPr, the small protein we examined in Panel 4–2, contains both α helix and β sheet structures. As shown previously in Figure 4–10D, β sheets are made when hydrogen bonds form between segments of polypeptide chains lying side by side. When the structure consists of neighboring polypeptide chains that run in the same orientation (say, from the N-terminus to the C-terminus), it is considered a *parallel β sheet;* when it forms from a polypeptide chain that folds back and forth upon itself—with each section of the chain running in the direction opposite to that of its immediate neighbors—the structure is an *antiparallel β sheet* (Figure 4–17). Both types of β sheet produce a very rigid, pleated structure, and they form the core of many proteins.

β sheets give silk fibers their remarkable tensile strength. And they can help keep insects from freezing in the cold. In one type of antifreeze protein, isolated from beetles that frequent cold climates, a series of parallel β sheets forms a beautifully flat surface along one side of the protein molecule (Figure 4–18). This array appears to offer a perfect platform for binding to the regularly spaced water molecules that are present in an ice lattice. By adhering to the ice crystals that form when water is cooled below its freezing point, the antifreeze protein prevents the ice crystals from growing—thereby keeping the insects' cells from freezing solid.

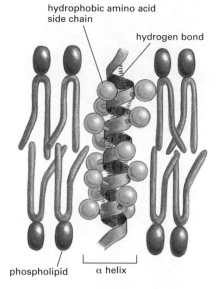

Figure 4–15 A segment of α helix can cross a lipid bilayer. The hydrophobic side chains of the amino acids forming the α helix contact the hydrophobic hydrocarbon tails of the phospholipid molecules, while the hydrophilic parts of the polypeptide backbone form hydrogen bonds with one another in the interior of the helix. About 20 amino acids are required to span a membrane in this way.

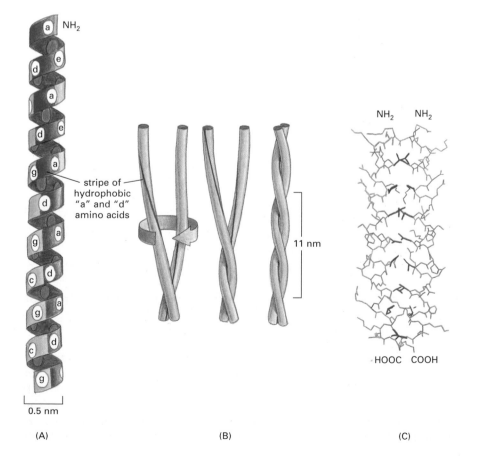

(A)   (B)   (C)

Figure 4–16 Intertwined α helices can form a coiled-coil. In (A) a single α helix is shown, with successive amino acid side chains labeled in a sevenfold sequence "abcdefg." Amino acids "a" and "d" in such a sequence lie close together on the cylinder surface, forming a stripe *(shaded in red)* that winds slowly around the α helix. Proteins that form coiled-coils typically have nonpolar amino acids at positions "a" and "d." Consequently, as shown in (B), the two α helices can wrap around each other with the nonpolar side chains of one α helix interacting with the nonpolar side chains of the other, while the more hydrophilic amino acid side chains are left exposed to the aqueous environment. (C) The atomic structure of a coiled-coil determined by X-ray crystallography. The *red* side chains are nonpolar.

(A)

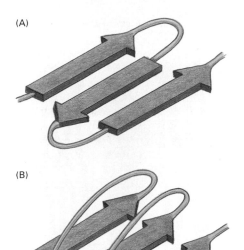

(B)

Figure 4–17 β sheets come in two varieties. (A) Antiparallel β sheet (see also Figure 4–10D). (B) Parallel β sheet. Both of these structures are common in proteins. By convention, the arrows point toward the C-terminus of the polypeptide chain.

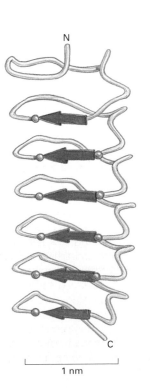

1 nm

Question 4–3

Remembering that the side chains projecting from each polypeptide backbone in a β sheet point alternately above and below the plane of the sheet (see Figure 4–10D), consider the following protein sequence: Leu-Lys-Val-Asp-Ile-Ser-Leu-Arg-Leu-Lys-Ile-Arg-Phe-Glu. Do you find anything remarkable about the arrangement of the amino acids in this sequence when incorporated into a β sheet? Can you make any predictions as to how the β sheet might be arranged in a protein? (Hint: consult the properties of the amino acids listed in Figure 4–3.)

Proteins Have Several Levels of Organization

A protein's structure does not end with α helices and β sheets; there are also higher levels of organization. These levels are not independent, but are built one upon the next until the three-dimensional structure of the entire protein has been fully defined. A protein's structure begins with its amino acid sequence, which is thus considered its *primary structure*. The next level of organization includes the α helices and β sheets that form within certain segments of a polypeptide chain; these folds are elements of the protein's *secondary structure*. The full, three-dimensional conformation formed by an entire polypeptide chain—including the α helices, β sheets, random coils, and any other loops and folds that form between the N- and C-termini—is sometimes referred to as the *tertiary structure* (see the structures shown in Panel 4–2, for example). Finally, if a particular protein molecule is formed as a complex of more than one polypeptide chain, then the complete structure is designated its *quaternary structure.*

Studies of the conformation, function, and evolution of proteins have also revealed the importance of a level of organization distinct from those just described. This is the **protein domain**, which is defined as any segment of a polypeptide chain that can fold independently into a compact, stable structure. A domain usually consists of 100 to 250 amino acids (folded into α helices and β sheets and other elements of secondary structure), and it is the modular unit from which many larger proteins are constructed (Figure 4–19). The different domains of a protein are often associated with different functions. For example, the bacterial *catabolite activator protein (CAP)*, illustrated in Figure 4–19, has two domains: the small domain binds to DNA, while the large domain binds cyclic AMP, an intracellular signaling molecule. When the large domain binds cyclic AMP, it causes a conformational change in the protein that enables the small domain to bind to a specific DNA sequence and promote expression of adjacent genes.

A small protein molecule like HPr contains only a single domain. Larger proteins can contain as many as several dozen domains, which are usually connected by relatively unstructured lengths of polypeptide chain. Ribbon models of three differently organized domains are presented in Figure 4–20.

Figure 4–18 β sheets provide an ideal ice-binding surface in an antifreeze protein. The six parallel β strands, shown here in *red,* form a flat surface with 10 hydroxyl groups *(blue)* arranged at distances that correspond to water molecules in an ice lattice. The protein can therefore bind to ice crystals, preventing their growth. (After Y.C. Liou et al., *Nature* 406:322–324, 2000.)

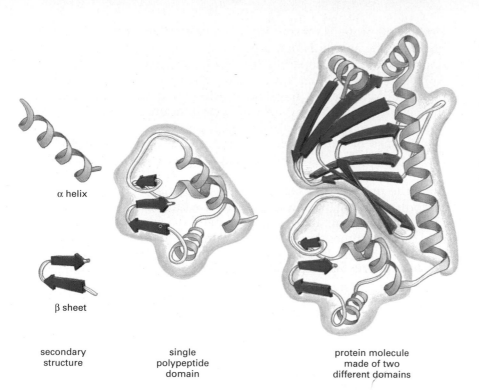

Figure 4–19 Many proteins are composed of separate functional domains. Elements of secondary structure such as α helices and β sheets pack together into stable, independently folding globular elements called domains. A typical protein molecule is built from one or more domains, often linked through relatively unstructured regions of polypeptide chain. The ribbon diagram on the right is of the bacterial gene regulatory protein CAP, with one large (outlined in *blue)* and one small (outlined in *gray*) domain.

α helix

β sheet

secondary structure

single polypeptide domain

protein molecule made of two different domains

Few of the Many Possible Polypeptide Chains Will Be Useful

In theory, a vast number of different polypeptide chains could be made. Because each of the 20 amino acids is chemically distinct and each can, in principle, occur at any position in a protein chain, a polypeptide chain four amino acids long has $20 \times 20 \times 20 \times 20 = 160,000$ different possible sequences. In other words, for a polypeptide that is n amino acids long, 20^n different chains are possible. For a typical protein length of 300 amino acids, more than 20^{300} (that's 10^{390}) structurally different polypeptide chains could theoretically be made.

However, only a very small fraction of this unimaginably large number of polypeptide chains would adopt a single stable three-dimensional conformation. The vast majority of individual protein molecules would have many different conformations of roughly equal stability, each conformation having different chemical properties. So why do

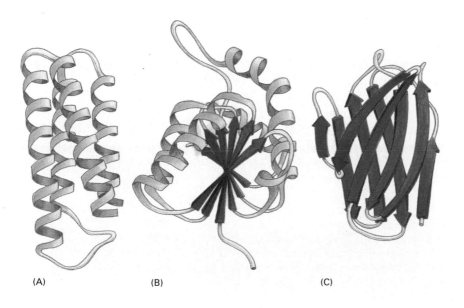

(A)

(B)

(C)

Figure 4–20 Ribbon models show three different protein domains. (A) Cytochrome b_{562}, a single-domain protein involved in electron transfer in *E. coli,* is composed almost entirely of α helices. (B) The NAD-binding domain of the enzyme lactic dehydrogenase is composed of a mixture of α helices and β sheets. (C) The variable domain of an immunoglobulin (antibody) light chain is composed of a sandwich of two β sheets. In these examples, the α helices are shown in *green,* while strands organized as β sheets are denoted by *red arrows.* The protruding loop regions *(yellow)* often form the binding sites for other molecules. (Drawings courtesy of Jane Richardson.)

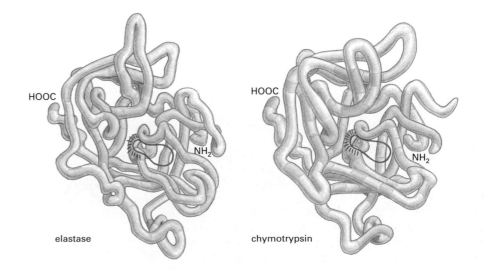

Question 4–4

Random mutations only very rarely result in changes in a protein that improve its usefulness for the cell, yet useful mutations are selected in evolution. Because these changes are so rare, for each useful mutation there are innumerable mutations that lead to either no improvement or inactive proteins. Why, then, do cells not contain millions of different proteins that are of no use?

virtually all proteins present in cells adopt unique and stable conformations? The answer is that a protein with many different conformations and variable properties would not be biologically useful, for it would be like a tool that unexpectedly changes its function. Such proteins would therefore have been eliminated by natural selection through the long trial-and-error process that underlies cellular evolution (discussed in Chapter 9).

Because of natural selection, the amino acid sequence of each present-day protein has evolved to guarantee that the polypeptide will adopt an extremely stable conformation—a structure that bestows upon the protein the exact chemical properties that will enable it to perform a particular catalytic or structural function in the cell. Proteins are so precisely built that the change of even a few atoms in one amino acid can sometimes disrupt the structure of a protein and thereby eliminate its function. In fact, many protein structures are so stable and effective that they have been conserved throughout evolution among many diverse organisms. The three-dimensional structures of the DNA-binding domains from the yeast α2 protein and the *Drosophila* engrailed protein, for example, are almost completely superimposable even though these organisms are separated by more than a billion years of evolution.

Proteins Can Be Classified into Families

Once a protein had evolved a stable conformation with useful properties, its structure could be modified over time to enable it to perform new functions. We know that this occurred quite often during evolution, because many present-day proteins can be grouped into **protein families**, in which each family member has an amino acid sequence and a three-dimensional conformation that closely resembles that of the other family members.

Consider, for example, the *serine proteases,* a family of protein-cleaving (proteolytic) enzymes that includes the digestive enzymes chymotrypsin, trypsin, and elastase, as well as several proteases involved in blood clotting. When any two of these enzymes are compared, portions of their amino acid sequences are found to be nearly the same. The similarity of their three-dimensional conformations is even more striking: most of the detailed twists and turns in their polypeptide chains, which are several hundred amino acids long, are virtually identical (Figure 4–21). The various serine proteases nevertheless have distinct enzymatic activities, each cleaving different proteins or the peptide bonds

Figure 4–21 Serine proteases belong to a family of proteolytic enzymes. The backbone conformations of elastase and chymotrypsin. Although only those amino acids in the polypeptide chain shaded in *green* are the same in the two proteins, the two conformations are very similar nearly everywhere. The active site of each enzyme is circled in *red;* this is where the peptide bonds of the proteins that serve as substrates are bound and cleaved by hydrolysis. The serine proteases derive their name from the amino acid serine, whose side chain is part of the active site of each enzyme and directly participates in the cleavage reaction.

elastase   chymotrypsin

between different types of amino acids. Slight differences in structure allow each of these proteases to prefer different substrates; thus each carries out a distinct function in an organism.

Large Protein Molecules Often Contain More Than One Polypeptide Chain

The same weak noncovalent bonds that enable a polypeptide chain to fold into a specific conformation also allow proteins to bind to each other to produce larger structures in the cell. Any region on a protein's surface that interacts with another molecule through sets of noncovalent bonds is termed a *binding site*. A protein can contain binding sites for a variety of molecules, large and small. If a binding site recognizes the surface of a second protein, the tight binding of two folded polypeptide chains at this site will create a larger protein molecule with a precisely defined geometry. Each polypeptide chain in such a protein is called a **subunit**. Each of these protein subunits may contain more than one domain, a portion of the polypeptide chain that folds up separately.

In the simplest case, two identical folded polypeptide chains will bind to each other in a "head-to-head" arrangement, forming a symmetrical complex of two protein subunits (called a *dimer)* that is held together by interactions between two identical binding sites. The CAP protein discussed previously is in fact a dimeric protein in the bacterial cell (Figure 4–22A); it is formed from two identical copies of the protein subunit shown previously in Figure 4–19. Many other symmetrical

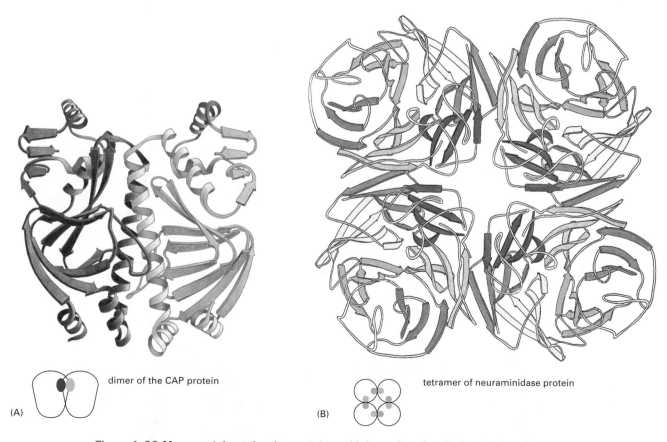

dimer of the CAP protein

(A)

tetramer of neuraminidase protein

(B)

Figure 4–22 Many protein molecules contain multiple copies of a single protein subunit. (A) A symmetrical dimer. The CAP protein exists as a complex of two identical polypeptide chains (see also Figure 4–19). (B) A symmetrical tetramer. The enzyme neuraminidase exists as a ring of four identical polypeptide chains. For both (A) and (B), a small schematic below the structure emphasizes how the repeated use of the same binding interaction forms the structure.

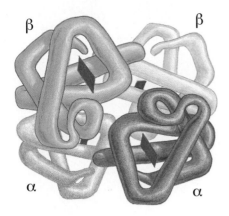

β   β

α   α

protein complexes, formed from multiple copies of a single polypeptide chain, are commonly found in cells. The enzyme *neuraminidase*, for example, consists of a ring of four identical protein subunits (Figure 4–22B).

Other proteins contain two or more different types of polypeptide chains. *Hemoglobin*, the protein that carries oxygen in red blood cells, is a particularly well-studied example (Figure 4–23). The protein contains two identical α-globin subunits and two identical β-globin subunits, symmetrically arranged. Such multisubunit proteins are numerous in cells and can be very large.

Proteins Can Assemble into Filaments, Sheets, or Spheres

Proteins can form even larger assemblies than those discussed so far. Most simply, a chain of identical protein molecules can be formed if the binding site on one protein molecule is complementary to another region on the surface of another protein molecule of the same type. Because each protein molecule is bound to its neighbor in an identical way, the molecules will often be arranged in a helix that can be extended indefinitely (Figure 4–24). This type of arrangement can produce an extended protein filament. An actin filament, for example, is a long helical structure formed from many molecules of the protein actin (Figure 4–25). Actin is extremely abundant in eucaryotic cells, where it forms one of the major filament systems of the cytoskeleton (discussed in Chapter 17). Other sets of proteins associate to form extended sheets or tubes (Figure 4–26), as in the microtubules of the cell cytoskeleton; or cagelike spherical shells, as in the protein coats of virus particles (Figure 4–27).

Many large structures, such as viruses and ribosomes, are built from a mixture of one or more types of protein plus RNA or DNA molecules.

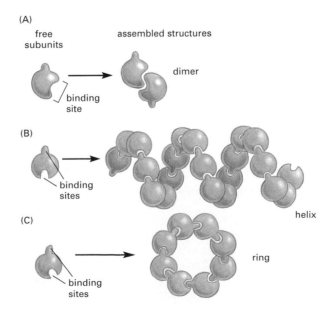

(A)
free subunits   assembled structures
                                    dimer
binding site

(B)
binding sites   helix

(C)
binding sites   ring

Figure 4–24 Proteins can assemble into complex structures. (A) A protein with just one binding site can form a dimer with another identical protein. (B) Identical proteins with two different binding sites will often form a long helical filament. (C) If the two binding sites are disposed appropriately in relation to each other, the protein subunits will form a closed ring instead of a helix (see also Figure 4–22B).

actin helix

These structures can be isolated in pure form and dissociated into their constituent macromolecules. It is often possible to mix the isolated components back together and watch them reassemble spontaneously into the original structure. This demonstrates that all the information needed for assembly of the complicated structure is contained in the macromolecules themselves. Experiments of this type show that much of the structure of a cell is self-organizing: if the required proteins are produced in the right amounts, the appropriate structures will form.

Some Types of Proteins Have Elongated Fibrous Shapes

Most of the proteins we have discussed so far are **globular proteins**, in which the polypeptide chain folds up into a compact shape like a ball with an irregular surface. Enzymes tend to be globular proteins: even though many are large and complicated, with multiple subunits, most have an overall rounded shape (see Figure 4–9). In contrast, other proteins have roles in the cell which require that they span a large distance. These proteins generally have a relatively simple, elongated three-dimensional structure and are commonly referred to as **fibrous proteins**.

One large class of intracellular fibrous proteins resembles α-keratin, which we met earlier. Keratin filaments are extremely stable: long-lived structures such as hair, horn, and nails are composed mainly of this protein. An α-keratin molecule is a dimer of two identical subunits, with the long α helices of each subunit forming a coiled-coil (see Figure 4–16). These coiled-coil regions are capped at either end by globular domains containing binding sites. These sites allow the molecules in this class to assemble into ropelike *intermediate filaments*—a component of the cell cytoskeleton that creates a structural scaffold for the cell's interior (discussed in Chapter 17).

Fibrous proteins are especially abundant outside the cell, where they form the gel-like *extracellular matrix* that helps cells bind together to form tissues. These proteins are secreted by the cells into their surroundings, where they often assemble into sheets or long fibrils. *Collagen* is the most abundant of these fibrous proteins in animal tissues. The collagen molecule consists of three long polypeptide chains, each containing the nonpolar amino acid glycine at every third position. This regular structure allows the chains to wind around one

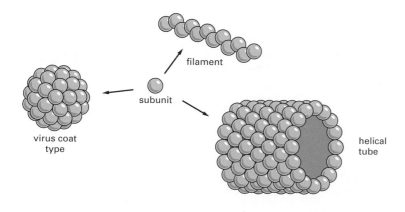

filament

subunit

virus coat type

helical tube

Figure 4–26 Single protein subunits can pack to form a filament, a tube, or a spherical shell.

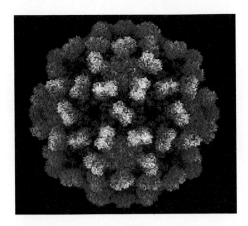

Figure 4–27 **Viral capsids are made of spherical protein assemblies.** The structure of tomato bushy stunt virus, shown here, was determined by X-ray crystallography and is known in atomic detail. (Courtesy of Robert Grant, Stephan Crainic, and James M. Hogle.)

another to generate a long regular triple helix (Figure 4–28A). Many collagen molecules then bind to one another side-by-side and end-to-end to create long overlapping arrays—thereby generating the extremely strong collagen fibrils that hold tissues together, as described in Chapter 21.

In complete contrast to collagen is another protein in the extracellular matrix, *elastin*. Elastin molecules are formed from relatively loose and unstructured polypeptide chains that are covalently cross-linked into a rubberlike elastic meshwork. The resulting elastic fibers enable skin and other tissues, such as arteries and lungs, to stretch and recoil without tearing. As illustrated in Figure 4–28B, the elasticity is due to the ability of the individual protein molecules to uncoil reversibly whenever they are stretched.

Extracellular Proteins Are Often Stabilized by Covalent Cross-Linkages

Many protein molecules are either attached to the outside of a cell's plasma membrane or secreted as part of the extracellular matrix. All such proteins are directly exposed to extracellular conditions. To help maintain their structures, the polypeptide chains in such proteins are often stabilized by covalent cross-linkages. These linkages can tie together two amino acids in the same protein, or connect different polypeptide chains in a multisubunit protein. The most common cross-links in proteins are covalent sulfur–sulfur bonds. These **disulfide bonds** (also called *S–S bonds*) form as proteins are being exported from cells. Their formation is catalyzed in the endoplasmic reticulum by a special enzyme that links together two –SH groups from cysteine side chains that are adjacent in the folded protein (Figure 4–29). Disulfide bonds do not change the conformation of a protein, but instead act as a sort of "atomic staple" to reinforce its most favored conformation. For example, lysozyme—an enzyme in tears that can dissolve bacterial cell

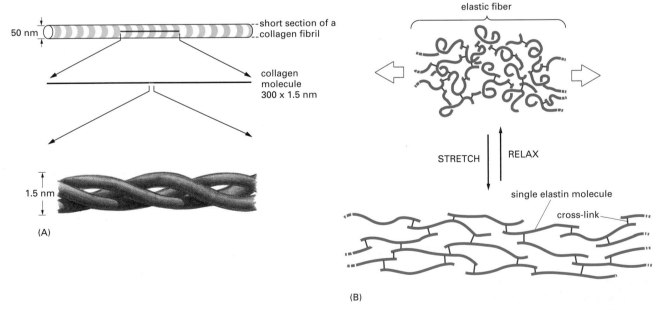

Figure 4–28 **Collagen and elastin are abundant fibrous proteins.** (A) Collagen is a triple helix formed by three extended protein chains that wrap around one another. Many rodlike collagen molecules are cross-linked together in the extracellular space to form collagen fibrils *(top)* that have the tensile strength of steel. The striping on the collagen fibril is caused by the regular repeating arrangement of the collagen molecules within the fibril. (B) Elastin polypeptide chains are cross-linked together to form rubberlike, elastic fibers. Each elastin molecule uncoils into a more extended conformation when the fiber is stretched and will recoil spontaneously as soon as the stretching force is relaxed.

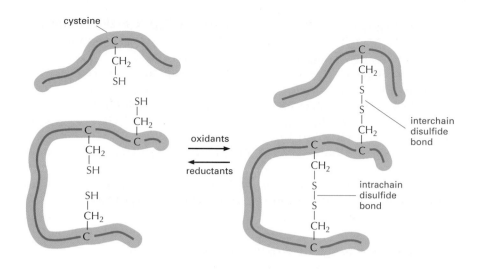

cysteine

oxidants
reductants

interchain disulfide bond

intrachain disulfide bond

Figure 4–29 **Disulfide bonds help stabilize a favored protein conformation.** This diagram illustrates how covalent disulfide bonds form between adjacent cysteine side chains. As indicated, these cross-linkages can join either two parts of the same polypeptide chain or two different polypeptide chains. Because the energy required to break one covalent bond is much larger than the energy required to break even a whole set of noncovalent bonds (see Table 2–1), a disulfide bond can have a major stabilizing effect on a protein.

walls—retains its antibacterial activity for a long time because it is stabilized by such cross-linkages.

Disulfide bonds generally do not form in the cell cytosol, where a high concentration of reducing agents converts such bonds back to cysteine –SH groups. Apparently, proteins do not require this type of structural reinforcement in the relatively mild environment inside the cell.

How Proteins Work

Proteins are not inert lumps of material. Because of their different amino acid sequences, proteins come in an enormous variety of different shapes—each with a unique surface topography of chemical groups. And a protein's conformation endows it with a unique function based on its chemical properties and its precisely engineered parts whose actions are coupled to chemical events. This union of structure, chemistry, and activity gives proteins the extraordinary ability to orchestrate the dynamic processes that occur in living cells.

For proteins, then, form and function are inexorably linked. But the fundamental question remains: how do proteins accomplish their function? In this part of the chapter, we will see that the activity of proteins depends on their ability to bind specifically to other molecules, allowing them to act as catalysts, signal receptors, and tiny motors. The examples we review here by no means exhaust the vast functional repertoire of proteins. However, the specialized functions of many of the proteins you will encounter elsewhere in this book are based on similar principles.

All Proteins Bind to Other Molecules

The biological properties of a protein molecule depend on its physical interaction with other molecules. Antibodies attach to viruses or bacteria as a signal to the body's defenses, the enzyme hexokinase binds glucose and ATP to catalyze a reaction between them, actin molecules bind to each other to assemble into long filaments, and so on. Indeed, all proteins stick, or *bind*, to other molecules. In some cases this binding is very tight; in others it is weak and short-lived. But the binding always shows great *specificity,* in the sense that each protein molecule can bind to just one or a few molecules out of the many thousands of different molecules it encounters. Any substance that is bound by a protein— whether it is an ion, a small molecule, or a macromolecule—is referred to as a **ligand** for that protein (from the Latin *ligare,* "to bind").

Question 4–5

Hair is composed largely of fibers of the protein keratin. Individual keratin fibers are covalently attached to one another (cross-linked) by many disulfide bonds. If curly hair is treated with mild reducing agents that break a few of the cross-links, pulled straight, and then oxidized again, it remains straight. Draw a diagram that illustrates the different stages of this chemical and mechanical process at the molecular level, focusing on the disulfide bonds. What do you think would happen if hair were treated with strong reducing agents that break all disulfide bonds?

The ability of a protein to bind selectively and with high affinity to a ligand is due to the formation of a set of weak, noncovalent bonds—hydrogen bonds, ionic bonds, and van der Waals attractions—plus favorable hydrophobic interactions (see Panel 2–7, pp. 78–79). Each individual bond is weak, so that an effective interaction requires that many weak bonds be formed simultaneously. This is possible only if the surface contours of the ligand molecule fit very closely to the protein, matching it like a hand in a glove (Figure 4–30).

When molecules have poorly matching surfaces, few noncovalent bonds are formed and the two molecules dissociate as rapidly as they come together. This is what prevents incorrect and unwanted associations from forming between mismatched molecules. At the other extreme, when many noncovalent bonds are formed, the association can persist for a very long time. Strong interactions occur in cells whenever a biological function requires that molecules remain tightly associated for a long time—for example, when a group of macromolecules come together to form a subcellular structure such as a ribosome.

The region of a protein that associates with a ligand, known as its **binding site**, usually consists of a cavity in the protein surface formed by a particular arrangement of amino acids. These amino acids belong to widely separated regions of the polypeptide chain that are brought together when the protein folds (Figure 4–31). Other regions on the surface often provide binding sites for different ligands, allowing the protein's activity to be regulated, as we shall see later. Yet other parts of the protein may be required to attract or attach the protein to a particular location in the cell—for example, the hydrophobic α helix of a membrane-spanning protein allows it to be inserted into the lipid bilayer of a cell membrane (discussed in Chapter 11).

Although the atoms buried in the interior of the protein have no direct contact with the ligand, they provide an essential scaffold that gives the surface its contours and chemical properties. Even small changes to the amino acids in the interior of a protein molecule often change its three-dimensional shape and destroy the protein's ability to function.

The Binding Sites of Antibodies Are Especially Versatile

All proteins must bind to particular ligands to carry out their various functions. But this binding capacity seems to have been most highly developed for proteins in the antibody family: our bodies have the capacity to produce a unique antibody that is capable of recognizing and binding tightly to the structure of any molecule imaginable.

Figure 4–30 The binding of a protein to another molecule is highly selective. Many weak bonds are needed to enable a protein to bind tightly to a second molecule (a *ligand*). The ligand must therefore fit precisely into the protein's binding site, like a hand into a glove, so that a large number of noncovalent bonds can be formed between the protein and the ligand.

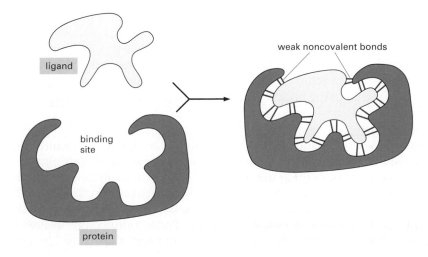

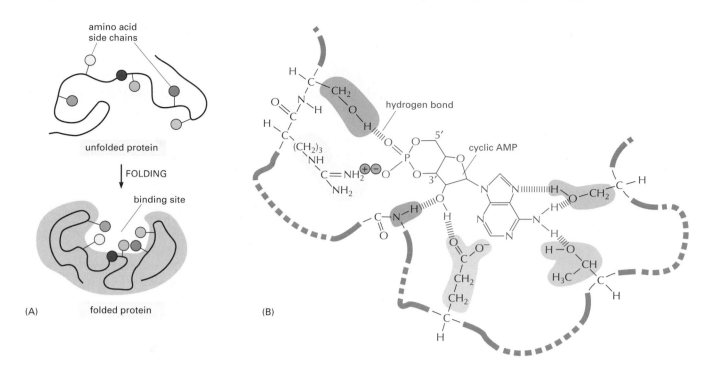

unfolded protein

amino acid side chains

FOLDING

binding site

folded protein

(A)

hydrogen bond

cyclic AMP

(B)

Antibodies, or immunoglobulins, are proteins produced by the immune system in response to foreign molecules, such as those on the surface of an invading microorganism. Each antibody binds to a particular target molecule extremely tightly, either inactivating the target directly or marking it for destruction. An antibody recognizes its target (called an **antigen**) with remarkable specificity, and because there are potentially billions of different antigens that a person might encounter, we have to be able to produce billions of different antibodies. The properties of antibodies—how they are made, how they help fight infection, and how they can be used to purify and study other proteins in the laboratory—are summarized in Panel 4–6 (pp. 164–165).

Antibodies are Y-shaped molecules with two identical binding sites that are each complementary to a small portion of the surface of the antigen molecule. A detailed examination of the antigen-binding sites of antibodies reveals that they are formed from several loops of polypeptide chain that protrude from the ends of a pair of closely juxtaposed protein domains (Figure 4–32). The amino acid sequence in these loops can be changed by mutation without altering the basic structure of the antibody. An enormous diversity of antigen-binding sites can be generated by changing only the length and amino acid sequence of the loops, which is how the wide variety of different antibodies is formed.

Loops of this kind are ideal for grasping other molecules. They allow a large number of chemical groups to surround a ligand so that the protein can link to it with many weak bonds. For this reason, peptide loops are used to form the ligand-binding sites in many proteins.

Enzymes Are Powerful and Highly Specific Catalysts

For many proteins, binding to another molecule is their only function. An antibody molecule need only bind to its target molecule on the surface of a bacterium or a virus and its job is done; an actin molecule need only associate with other actin molecules to form a filament. There are other proteins, however, for which ligand binding is simply a necessary first step in their function. This is the case for the large and very important class of proteins called **enzymes**. These remarkable molecules determine all of the chemical transformations that occur in cells.

Figure 4–31 Binding sites allow a protein to interact with specific ligands. (A) The folding of the polypeptide chain typically creates a crevice or cavity on the protein surface. This crevice contains a set of amino acid side chains disposed in such a way that they can make noncovalent bonds only with certain ligands. (B) Close-up view of an actual binding site showing the hydrogen bonds and ionic interactions formed between a protein and its ligand (in this example, the bound ligand is cyclic AMP, shown in *pink).*

Question 4–6

Explain how an enzyme (such as hexokinase, mentioned in the text) can distinguish substrates (here D-glucose) from their optical isomers (here L-glucose). (Hint: remembering that a carbon atom forms four single bonds that are tetrahedrally arranged and that the optical isomers are mirror images of each other around such a bond, draw the substrate as a simple tetrahedron with four different corners and then draw its mirror image. Using this drawing, indicate why only one compound might bind to a schematic active site of an enzyme.)

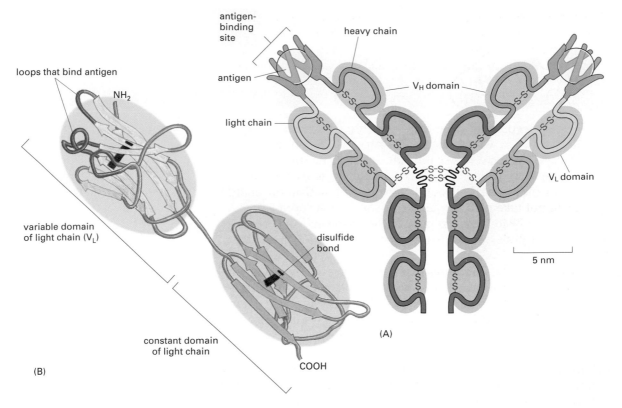

loops that bind antigen

NH$_2$

variable domain of light chain (V$_L$)

constant domain of light chain

(B)

antigen-binding site

antigen

light chain

heavy chain

V$_H$ domain

disulfide bond

V$_L$ domain

5 nm

(A)

COOH

Figure 4–32 An antibody is Y-shaped and has two identical binding sites for its antigen, one on each arm of the Y.
(A) Schematic drawing of a typical antibody molecule. The protein is composed of four polypeptide chains (two identical heavy chains and two identical and smaller light chains) held together by disulfide bonds. Each chain is made up of several different domains, here shaded either *blue* or *gray*. The antigen-binding site is formed where a heavy-chain variable domain (V$_H$) and a light-chain variable domain (V$_L$) come close together. These are the domains that differ most in their sequence and structure in different antibodies. (B) Ribbon drawing of a single light chain showing the parts of the V$_L$ domain most closely involved in binding to the antigen in *red*; these contribute half of the fingerlike loops that fold around each of the antigen molecules in (A).

Enzymes bind to one or more ligands, called **substrates**, and convert them into chemically modified products, doing this over and over again with amazing rapidity. They speed up reactions, often by a factor of a million or more, without themselves being changed—that is, enzymes act as *catalysts* that permit cells to make or break covalent bonds at will. This catalysis of organized sets of chemical reactions by enzymes creates and maintains the cell, making life possible.

Enzymes can be grouped into functional classes that carry out similar chemical reactions (Table 4–1). Each type of enzyme is highly specific, catalyzing only a single type of reaction. Thus, *hexokinase* adds a phosphate group to D-glucose but will ignore its optical isomer L-glucose; the blood-clotting enzyme *thrombin* cuts one type of blood protein between a particular arginine and its adjacent glycine and nowhere else. As discussed in detail in Chapter 3, enzymes often work in teams, with the product of one enzyme becoming the substrate for the next. The result is an elaborate network of metabolic pathways that provides the cell with energy and generates the many large and small molecules that the cell needs.

Lysozyme Illustrates How an Enzyme Works

To explain how enzymes catalyze chemical reactions, we will use the example of lysozyme—an enzyme that acts as a natural antibiotic in egg white, saliva, tears, and other secretions. Lysozyme severs the polysaccharide chains that form the cell walls of bacteria. Because the bacterial cell is under pressure due to osmotic forces, cutting even a small number of polysaccharide chains causes the cell wall to rupture and the bacterium to burst. Lysozyme is a relatively small and stable protein that can be isolated easily in large quantities. For these reasons, it has been intensively studied, and it was the first enzyme whose structure was worked out in atomic detail by X-ray crystallography.

The reaction catalyzed by lysozyme is a hydrolysis: the enzyme adds a molecule of water to a single bond between two adjacent sugar groups in the polysaccharide chain, thereby causing the bond to break. The reaction is energetically favorable because the free energy of the severed polysaccharide chain is lower than the free energy of the intact chain. However, the pure polysaccharide can sit for years in water without being hydrolyzed to any detectable degree. This is because there is an energy barrier to the reaction, as discussed in Chapter 3. For a colliding water molecule to break a bond linking two sugars, the polysaccharide molecule has to be distorted into a particular shape—the **transition state**—in which the atoms around the bond have an altered geometry and electron distribution. To generate this distortion, a large input of energy, called the *activation energy,* must be supplied through random molecular collisions. Without the activation energy, the reaction will not take place. In aqueous solution at room temperature, the energy of collisions almost never exceeds the activation energy. Consequently, hydrolysis occurs extremely slowly, if at all, under these conditions.

This is where the enzyme comes in. Like all enzymes, lysozyme has a special binding site on its surface, termed an **active site**, that cradles the contours of its substrate molecule. Here the catalysis of the chemical reaction occurs. Because its substrate is a polymer, lysozyme's active site is a long groove that holds six linked sugars at the same time. As soon as the polysaccharide binds to form an enzyme–substrate complex, the enzyme cuts the polysaccharide by adding a water molecule across one of its sugar–sugar bonds. The product chains are then quickly released, freeing the enzyme for further cycles of reaction (Figure 4–33).

The chemistry that underlies the binding of lysozyme to its substrate is the same as that for antibody binding—the formation of multiple non-covalent bonds. However, lysozyme holds its polysaccharide substrate in a particular way, so that one of the two sugars involved in the bond to be

Table 4–1 Some Common Functional Classes of Enzymes

ENZYME CLASS	BIOCHEMICAL FUNCTION
Hydrolase	General term for enzymes that catalyze a hydrolytic cleavage reaction.
Nuclease	Break down nucleic acids by hydrolyzing bonds between nucleotides.
Protease	Break down proteins by hydrolyzing peptide bonds between amino acids.
Synthase	General name used for enzymes that synthesize molecules in anabolic reactions by condensing two molecules together.
Isomerase	Catalyze the rearrangement of bonds within a single molecule.
Polymerase	Catalyze polymerization reactions such as the synthesis of DNA and RNA.
Kinase	Catalyze the addition of phosphate groups to molecules. Protein kinases are an important group of kinases that attach phosphate groups to proteins.
Phosphatase	Catalyze the hydrolytic removal of a phosphate group from a molecule.
Oxido-reductase	General name for enzymes that catalyze reactions in which one molecule is oxidized while the other is reduced. Enzymes of this type are often called oxidases, reductases, and dehydrogenases.
ATPase	Hydrolyze ATP. Many proteins with a wide range of roles have an energy-harnessing ATPase activity as part of their function, for example, motor proteins such as myosin and membrane transport proteins such as the sodium–potassium pump.

Enzyme names typically end in "-ase," with the exception of some enzymes, such as pepsin, trypsin, thrombin, lysozyme, and so on, which were discovered and named before the convention became generally accepted at the end of the nineteenth century. The common name of an enzyme usually indicates the substrate and the nature of the reaction catalyzed. For example, citrate synthase catalyzes the synthesis of citrate by a reaction between acetyl CoA and oxaloacetate.

broken is distorted from its normal, most stable conformation. The bond to be broken is also held close to two amino acids with acidic side chains: a glutamic acid and an aspartic acid within the active site.

Conditions are thereby created in the microenvironment of the lysozyme active site that greatly reduce the activation energy necessary for the hydrolysis to take place. Figure 4–34 shows the main intermediates in this enzymatically catalyzed reaction.

1. The enzyme stresses its bound substrate by bending some critical chemical bonds in one sugar, so that the shape of this sugar more closely resembles the shape of high-energy transition states formed during the reaction.

2. The negatively charged aspartic acid reacts with the C1 carbon atom on the distorted sugar, breaking this sugar–sugar bond and leaving the aspartic acid covalently linked to the site of bond cleavage.

3. Aided by the negatively charged glutamic acid, a water molecule reacts with the C1 carbon atom, displacing the aspartic acid and completing the process of hydrolysis.

The overall chemical reaction, from the initial binding of the polysaccharide on the surface of the enzyme through the final release of the severed chains, occurs many millions of times faster than it would in the absence of enzyme.

Other enzymes use similar mechanisms to lower activation energies and speed up the reactions they catalyze. In reactions involving two or more substrates, the active site also acts like a template or mold that brings the reactants together in the proper orientation for chemistry to occur between them (Figure 4–35A). As we saw for lysozyme, the active site of an enzyme contains precisely positioned atoms that speed up a reaction by using charged groups to alter the distribution of electrons in the substrates (Figure 4–35B). As we likewise saw, the binding to the enzyme will also change substrate shapes, bending bonds so as to drive a substrate toward a particular transition state (Figure 4–35C). Finally, like lysozyme, many enzymes participate intimately in the reaction by briefly forming a covalent bond between the substrate and a side chain of the enzyme. Subsequent steps in the reaction restore the side chain to its original state, so that the enzyme remains unchanged after the reaction.

Figure 4–33 Lysozyme cleaves a polysaccharide chain.
(A) Schematic view of the enzyme lysozyme (denoted E), which catalyzes the cutting of a polysaccharide chain, which is its substrate (denoted S). The enzyme first binds to the chain to form an enzyme–substrate complex (ES) and then catalyzes the cleavage of a specific covalent bond in the backbone of the polysaccharide, forming an enzyme–product complex (EP) that rapidly dissociates. Release of the severed chain (the products P) leaves the enzyme free to act on another substrate molecule. (B) A space-filling model of the lysozyme molecule bound to a short length of polysaccharide chain prior to cleavage. (B, courtesy of Richard J. Feldmann.)

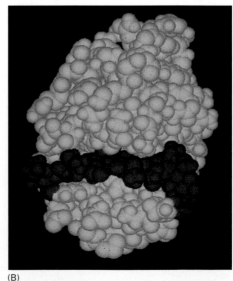

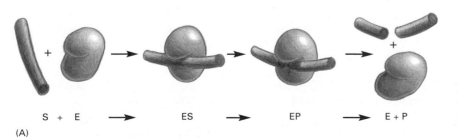

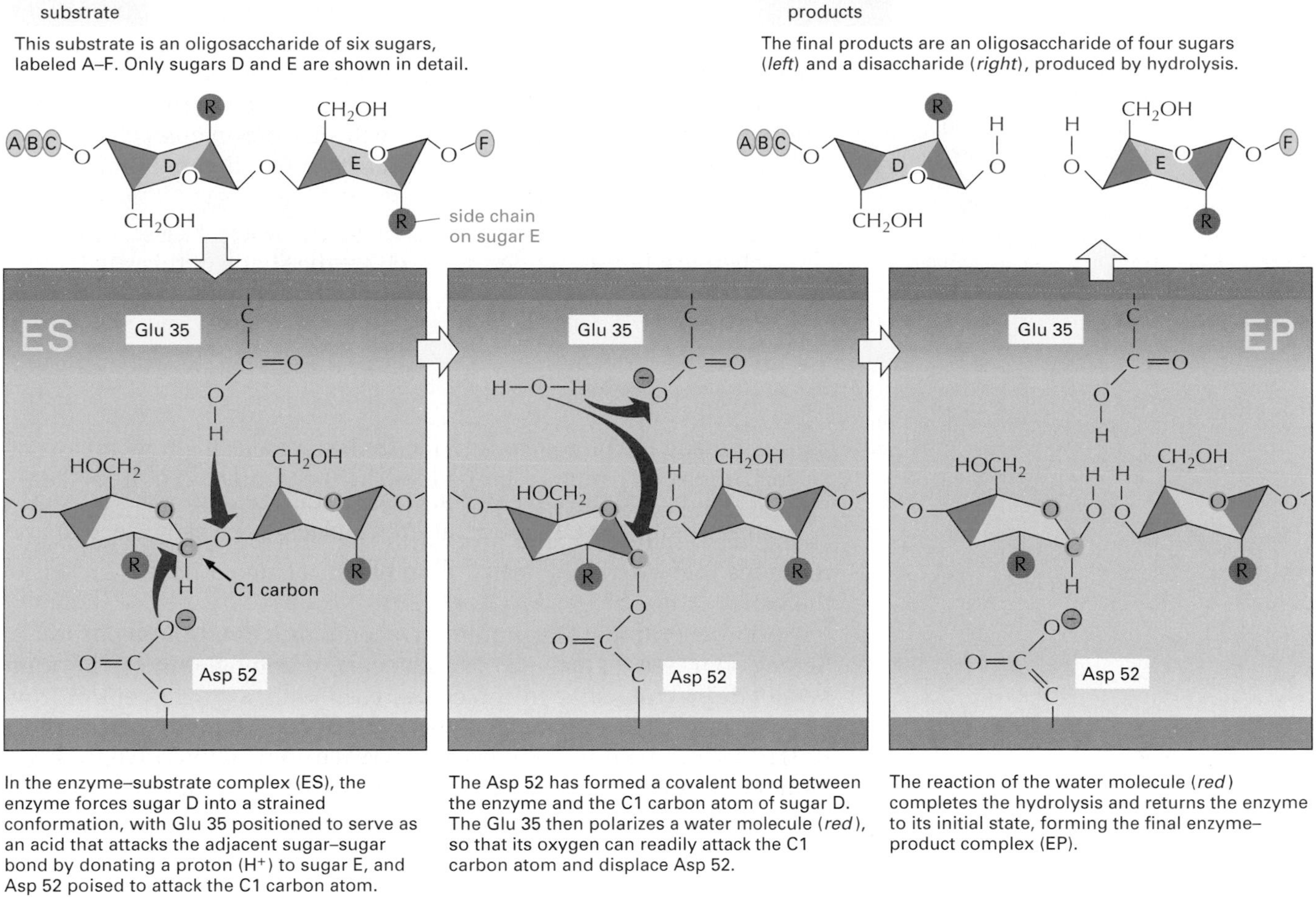

substrate

This substrate is an oligosaccharide of six sugars, labeled A–F. Only sugars D and E are shown in detail.

products

The final products are an oligosaccharide of four sugars (*left*) and a disaccharide (*right*), produced by hydrolysis.

ES Glu 35

C1 carbon

Asp 52

In the enzyme–substrate complex (ES), the enzyme forces sugar D into a strained conformation, with Glu 35 positioned to serve as an acid that attacks the adjacent sugar–sugar bond by donating a proton (H$^+$) to sugar E, and Asp 52 poised to attack the C1 carbon atom.

Glu 35

H–O–H

Asp 52

The Asp 52 has formed a covalent bond between the enzyme and the C1 carbon atom of sugar D. The Glu 35 then polarizes a water molecule (*red*), so that its oxygen can readily attack the C1 carbon atom and displace Asp 52.

Glu 35 **EP**

Asp 52

The reaction of the water molecule (*red*) completes the hydrolysis and returns the enzyme to its initial state, forming the final enzyme–product complex (EP).

Figure 4–34 In the active site of lysozyme, bonds are bent and broken. The top left and top right drawings depict the free substrate and the free products, respectively, whereas the other three drawings depict sequential events at the enzyme active site. Note the change in the conformation of sugar D in the enzyme–substrate complex; this is the sugar that is also distorted in unstable transition states. (Based on D.J. Vocadlo et al., *Nature* 412:835–838, 2001.)

Tightly Bound Small Molecules Add Extra Functions to Proteins

Although the order of amino acids in proteins gives these molecules their shape and the versatility to perform different functions, sometimes the amino acids by themselves are not enough. Just as we use tools to enhance and extend the capabilities of our hands, so proteins often employ small nonprotein molecules to perform functions that would be difficult or impossible using amino acids alone. Thus the signal receptor protein *rhodopsin*—which is the purple, light-sensitive

(A) enzyme binds to two substrate molecules and orients them precisely to encourage a reaction to occur between them

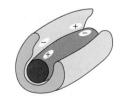

(B) binding of substrate to enzyme rearranges electrons in the substrate, creating partial negative and positive charges that favor a reaction

(C) enzyme strains the bound substrate molecule, forcing it toward a transition state to favor a reaction

Figure 4–35 Enzymes can encourage catalysis in several ways. (A) Holding substrates together in a precise alignment. (B) Charge stabilization of reaction intermediates. (C) Altering bond angles in the substrate to increase the rate of a particular reaction.

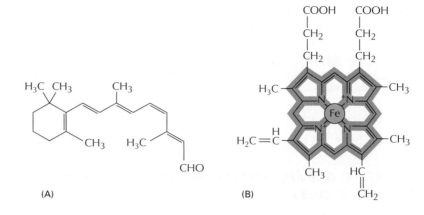

pigment made by the rod cells in the retina—detects light by means of a small molecule, *retinal,* embedded in the protein (Figure 4–36A). Retinal changes its shape when it absorbs a photon of light, and this change is amplified by the protein to trigger a cascade of enzymatic reactions that eventually leads to an electrical signal being carried to the brain.

Another example of a protein that contains a nonprotein portion is hemoglobin (see Figure 4–23). A molecule of hemoglobin carries four *heme* groups, ring-shaped molecules each with a single central iron atom (Figure 4–36B). Heme gives hemoglobin (and blood) its red color. By binding reversibly to oxygen gas through its iron atom, heme enables hemoglobin to pick up oxygen in the lungs and release it in the tissues.

Sometimes these small molecules are attached covalently and permanently to their protein, thereby becoming an integral part of the protein molecule itself. We will see in Chapter 11 that proteins are often anchored to cell membranes through covalently attached lipid molecules. And membrane proteins exposed on the surface of the cell, as well as proteins secreted outside the cell, are often modified by the covalent addition of sugars and oligosaccharides.

Enzymes frequently have a small molecule or metal atom tightly associated with their active site that assists with their catalytic function. *Carboxypeptidase,* an enzyme that cuts polypeptide chains, carries a tightly bound zinc ion in its active site. During the cleavage of a peptide bond by carboxypeptidase, the zinc ion forms a transient bond with one of the substrate atoms, thereby assisting the hydrolysis reaction. In other enzymes, a small organic molecule serves a similar purpose. *Biotin,* for example, is found in enzymes that transfer a carboxylate group (–COO⁻) from one molecule to another (see Figure 3–38). Biotin participates in these reactions by forming a transient covalent bond to the –COO⁻ group to be transferred; this small molecule is better suited for this function than any of the amino acids used to make proteins. Because biotin cannot be synthesized by humans, it must be provided by the diet; thus biotin is classified as a *vitamin.* Other vitamins are similarly needed to make small molecules that are essential components of our proteins; vitamin A, for example, is needed in the diet to make retinal, the light-sensitive part of rhodopsin.

How Proteins Are Controlled

Thus far we have examined how proteins do their jobs—how binding to other proteins or small molecules allows them to perform their specific functions. But inside the cell, most proteins and enzymes do not work continuously, or at full speed. Instead, their activity is regulated so that the cell can maintain itself in a state of equilibrium, generating only those molecules it requires to thrive under the current conditions. To achieve this balance, the activities of cellular proteins are controlled in an integrated fashion, with consideration of what reactions are occurring in other parts of the cell. By coordinating when—and how vigorously—proteins function, the cell ensures that it does not deplete its energy reserves by accumulating molecules it does not require while exhausting its stockpiles of critical substrates. We now consider how cells regulate the activity of proteins and enzymes. As we shall see, proteins can be switched on—or switched off—by a variety of mechanisms.

The Catalytic Activities of Enzymes Are Often Regulated by Other Molecules

A living cell contains thousands of enzymes, many of which operate at the same time and in the same small volume of the cytosol. By their catalytic action, enzymes generate a complex web of metabolic pathways, each composed of chains of chemical reactions in which the product of one enzyme becomes the substrate of the next. In this maze of pathways there are many branch points where different enzymes compete for the same substrate. The system is so complex (see Figure 3–2) that elaborate controls are required to regulate when and how rapidly each reaction occurs.

Regulation of enzyme activity occurs at many levels. At one level, the cell controls how many molecules of each enzyme it makes by regulating the expression of the gene that encodes that protein (discussed in Chapter 8). At another level, the cell controls enzymatic activities by confining sets of enzymes to particular subcellular compartments, enclosed by distinct membranes (discussed in Chapters 14 and 15). But the most rapid and general process used to adjust reaction rates operates at the level of the enzyme itself. In this case, an enzyme's activity changes in response to other specific molecules that it encounters.

The most common type of control occurs when a molecule other than a substrate binds to an enzyme at a special regulatory site outside of the active site, altering the rate at which the enzyme converts its substrates to products. In **feedback inhibition**, an enzyme acting early in a reaction pathway is inhibited by a late product of that pathway. Thus, whenever large quantities of the final product begin to accumulate, the product binds to the first enzyme and slows down its catalytic action, limiting further entry of substrates into that reaction pathway (Figure 4–37). Where pathways branch or intersect, there are usually multiple points of control by different final products, each of which works to regulate its own synthesis (Figure 4–38). Feedback inhibition can work almost instantaneously and is rapidly reversed when the product levels fall.

Feedback inhibition is *negative regulation:* it prevents an enzyme from acting. Enzymes can also be subject to *positive regulation,* in which the enzyme's activity is stimulated by a regulatory molecule rather than being shut down. Positive regulation occurs when a product in one branch of the metabolic maze stimulates the activity of an enzyme in another pathway. As one example, the accumulation of ADP activates several enzymes involved in the oxidation of sugar molecules, thereby stimulating the cell to convert more ADP to ATP.

Allosteric Enzymes Have Two Binding Sites That Influence One Another

There was one feature of feedback inhibition that was initially puzzling to those who discovered it: the regulatory molecule often has a shape

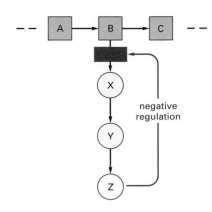

Figure 4–37 Feedback inhibition regulates the flow through biosynthetic pathways. The end product Z inhibits the first enzyme that is unique to its synthesis and thereby controls its own concentration in the cell. This is an example of negative regulation.

Question 4–7

Consider the drawing in Figure 4–37. What will happen if, instead of the indicated feedback,

A. Feedback inhibition from Z affects the step B → C only?

B. Feedback inhibition from Z affects the step Y → Z only?

C. Z is a positive regulator of the step B → X?

D. Z is a positive regulator of the step B → C?

For each case, discuss how useful these regulatory schemes would be for a cell.

Figure 4–38 Feedback inhibition at multiple sites regulates connected metabolic reactions. In this example, which shows the biosynthetic pathways for four different amino acids in bacteria, the *red arrows* indicate positions at which products feed back to inhibit enzymes. Each amino acid controls the first enzyme specific to its own synthesis, thereby controlling its own levels and avoiding a wasteful buildup of intermediates. The products can also separately inhibit the initial set of reactions common to all the syntheses; in this case, three different enzymes catalyze the initial reaction, each inhibited by a different product.

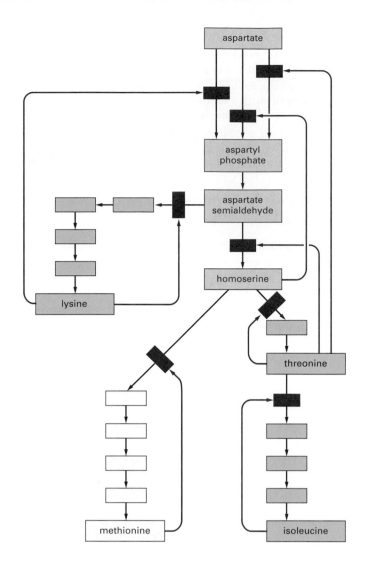

that is totally different from the shape of the enzyme's preferred substrate. Indeed, when this form of regulation was discovered in the 1960s, it was termed *allostery* (from the Greek *allo,* "other," and *stere,* "solid" or "shape"). As more was learned about feedback inhibition, researchers realized that many enzymes must have at least two different binding sites on their surface—the active site that recognizes the substrates and a second site that recognizes a regulatory molecule. Furthermore, these two sites must somehow "communicate" in a way that allows the catalytic events at the active site to be influenced by the binding of the regulatory molecule at its separate site on the protein's surface.

The interaction between sites that are located on separate regions of a protein molecule is now known to depend on a *conformational change* in the protein: binding at one of the sites causes a shift in the protein's structure from one folded shape to a slightly different folded shape. Many enzymes have two conformations that differ in activity, each stabilized by the binding of different ligands. During feedback inhibition, for example, the binding of an inhibitor at one site on the protein causes the protein to shift to a conformation in which its active site—located elsewhere in the protein—becomes less accommodating to the substrate molecule (Figure 4–39).

Many—if not most—protein molecules are **allosteric**: they can adopt two or more slightly different conformations, and by a shift from one to another, their activity can be regulated. This is true not only for

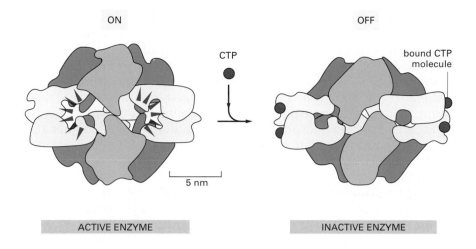

ON   OFF

CTP

bound CTP molecule

5 nm

ACTIVE ENZYME   INACTIVE ENZYME

Figure 4–39 Feedback inhibition triggers a conformational change. An enzyme used in early studies of allosteric regulation was aspartate transcarbamoylase from *E. coli*. This large multisubunit enzyme (see Figure 4–9) catalyzes an important reaction that begins the synthesis of the pyrimidine ring of C, U, and T nucleotides. One of the final products of this pathway, cytosine triphosphate (CTP), binds to the enzyme to turn it off whenever CTP is plentiful. This diagram shows the comformational change that occurs when the enzyme is turned off by CTP binding.

enzymes but for many other proteins—including receptors, structural proteins, and motor proteins. The chemistry involved here is extremely simple in concept: because each protein conformation will have somewhat different contours on its surface, the protein's binding site for ligands will be altered when the protein changes shape. Each ligand will stabilize the conformation that it binds to most strongly—and at high enough concentrations, the ligand will tend to "switch" the population of proteins to the conformation that it favors (Figure 4–40).

Phosphorylation Can Control Protein Activity by Triggering a Conformational Change

Enzymes are not only regulated by the binding of small molecules. A second method commonly used by eucaryotic cells to regulate protein activity involves attaching a phosphate group covalently to one of its amino acid side chains. Because each phosphate group carries two negative charges, the enzyme-catalyzed addition of a phosphate group to a protein can cause a major conformational change by, for example, attracting a cluster of positively charged amino acid side chains. This conformational change can, in turn, affect the binding of ligands elsewhere on the protein surface—thus altering the protein's activity. Removal of the phosphate group by a second enzyme returns the protein to its original conformation and restores its initial activity.

This reversible **protein phosphorylation** controls the activity of many different types of proteins in eucaryotic cells; in fact, this method is used so extensively that more than a third of the 10,000 or so proteins

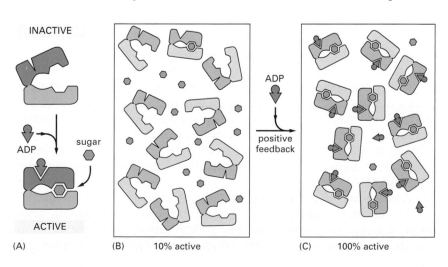

INACTIVE

ADP   sugar

ACTIVE

(A)

(B)   10% active

ADP

positive feedback

(C)   100% active

Figure 4–40 The equilibrium between two conformations of a protein is affected by ligand binding. This schematic diagram shows a hypothetical enzyme in which an increase in the concentration of ADP molecules (*green wedges*) acts as an activator to increase the rate at which sugar molecules (*orange hexagons*) are oxidized. (A) This hypothetical enzyme is allosterically regulated. It might, for example, catalyze a rate-limiting step in either glycolysis or the citric acid cycle, because when ADP accumulates it feeds back to such enzymes to accelerate sugar catabolism, thereby increasing the rate of production of ATP from ADP. (B) With no ADP present, only a small fraction of the molecules spontaneously adopt the active (closed) conformation; most are in the inactive (open) conformation. (C) Because ADP can bind only to the protein in its closed conformation, ADP addition lowers the energy of the closed conformation, locking nearly all of the enzyme molecules in the active form.

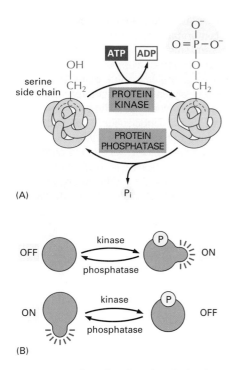

(A)

(B)

Figure 4–41 **Protein phosphorylation is a very common means of regulating protein activity.** Many thousands of proteins in a typical eucaryotic cell are modified by the covalent addition of a phosphate group. (A) The general reaction, shown here, entails transfer of a phosphate group from ATP to an amino acid side chain of the target protein by a protein kinase. Removal of the phosphate group is catalyzed by a second enzyme, a protein phosphatase. In this example, the phosphate is added to a serine side chain; in other cases, the phosphate is instead linked to the –OH group of a threonine or a tyrosine in the protein. (B) The phosphorylation of a protein by a protein kinase can either increase or decrease the protein's activity, depending on the site of phosphorylation and the structure of the protein.

Question 4–8

Explain how phosphorylation and the binding of a nucleotide can both be used to regulate protein activity. What do you suppose are advantages of either form of regulation?

in a typical mammalian cell appear to be phosphorylated at any one time. The addition and removal of phosphate groups from specific proteins often occurs in response to signals that specify some change in a cell's state. For example, the complicated series of events that takes place as a eucaryotic cell divides is timed in this way (discussed in Chapter 19). And many of the signals generated by hormones and neurotransmitters are carried from the plasma membrane to the nucleus by a cascade of protein phosphorylation events (discussed in Chapter 16).

Protein phosphorylation involves the enzyme-catalyzed transfer of the terminal phosphate group of ATP to the hydroxyl group on a serine, threonine, or tyrosine side chain of the protein. This reaction is catalyzed by a **protein kinase**. The reverse reaction—removal of the phosphate group, or *dephosphorylation*—is catalyzed by a **protein phosphatase** (Figure 4–41). Cells contain hundreds of different protein kinases, each responsible for phosphorylating a different protein or set of proteins. Cells also contain many different protein phosphatases; some of these are highly specific and remove phosphate groups from only one or a few proteins, whereas others act on a broad range of proteins. The state of phosphorylation of a protein at any moment in time, and thus its activity, will depend on the relative activities of the protein kinases and phosphatases that act on it.

For many proteins, a phosphate group is added to a particular side chain and then removed in a continuous cycle. Phosphorylation cycles of this kind allow proteins to switch rapidly from one state to another. The more rapidly the cycle is "turning," the faster the concentration of a phosphorylated protein can change in response to a sudden stimulus. The energy required to drive this cycle is derived from the free energy of hydrolysis of ATP, one molecule of which is consumed with each turn of the cycle.

GTP-Binding Proteins Are Also Regulated by the Cyclic Gain and Loss of a Phosphate Group

Eucaryotic cells have a second way to regulate protein activity by phosphate addition and removal. In this case, instead of being enzymatically transferred from ATP to the protein, the phosphate is part of a guanine nucleotide—either guanosine triphosphate (GTP) or guanosine diphosphate (GDP)—that is bound tightly to the protein. Such **GTP-binding proteins** are in their active conformations with GTP bound; the protein itself then hydrolyzes this GTP to GDP—releasing a phosphate—and flips to an inactive conformation. As with protein phosphorylation, this process is reversible. The active conformation is regained by dissociation of the GDP, followed by the binding of a fresh molecule of GTP (Figure 4–42).

There are a large number of related GTP-binding proteins that function as molecular switches in cells. The dissociation of GDP and its replacement by GTP, which turns the switch on, is often stimulated in response to a signal received by the cell. The GTP-binding proteins often bind to other proteins to control enzyme activities, and their crucial role in intracellular signaling pathways will be discussed in detail in Chapter 16. Here we shall look at their general mechanism of action by examining the bacterial elongation factor EF-Tu, a small GTP-binding protein that helps to load tRNA molecules onto ribosomes during protein synthesis.

Analysis of the three-dimensional structure of EF-Tu has revealed how an allosteric transition triggered by the gain or loss of a phosphate on the bound guanine nucleotide can cause a major shape change in a GTP-binding protein. Figure 4–43 shows how the loss of a

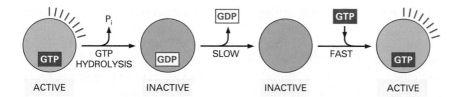

Figure 4–42 GTP-binding proteins form molecular switches. The activity of a GTP-binding protein generally requires the presence of a tightly bound GTP molecule (switch on). Hydrolysis of this GTP molecule produces GDP and inorganic phosphate (P_i), and it causes the protein to convert to a different, usually inactive, conformation (switch off). As shown here, resetting the switch requires that the tightly bound GDP dissociate, a slow step that is greatly accelerated by specific signals; once the GDP dissociates, a molecule of GTP is quickly re-bound.

single phosphate group, which initially causes only a tiny movement of 0.1 nm or so at the binding site, is magnified by the protein to create a movement 50 times larger. Dramatic shape changes of this type also underlie the very large movements created by the types of proteins that we consider next.

Nucleotide Hydrolysis Allows Motor Proteins to Produce Large Movements in Cells

We have seen how conformational changes in proteins play a central part in enzyme regulation and cell signaling. But conformational changes also play another important role in the operation of the cell: they enable proteins whose major function is to move other molecules, the **motor proteins**, to generate the forces responsible for muscle contraction and the dramatic movements of cells. Motor proteins also power smaller-scale intracellular movements: they help move chromosomes to

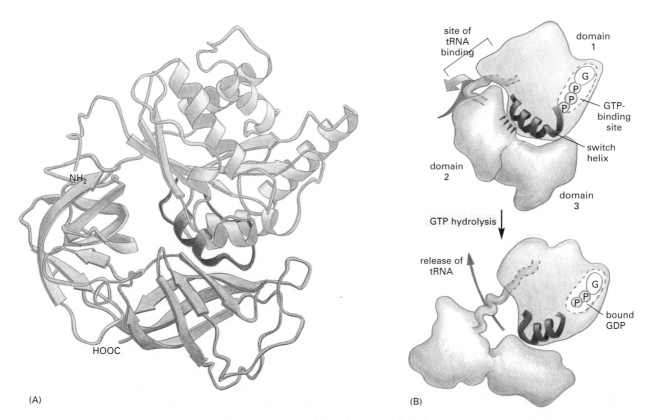

Figure 4–43 A large conformational change is produced in response to nucleotide hydrolysis. (A) The structure of the elongation factor Tu (EF-Tu), a GTP-binding protein that plays a role in the elongation of the polypeptide chain during protein synthesis. (B) The hydrolysis of bound GTP in EF-Tu causes only a minute change in the position (equivalent to a few times the diameter of a hydrogen atom) of amino acids at the nucleotide-binding site. But this small change is magnified by conformational changes within the protein to produce a much larger movement. The hydrolysis of GTP releases an intramolecular bond, like a "latch" (*red dashes* in the *upper right diagram*), which allows domains 2 and 3 to twist free and rotate by about 90° toward the viewer. This creates a major change in shape that releases the tRNA molecule that was initially tightly held by the protein, as required to allow protein synthesis to proceed on the ribosome. All of these structures were determined by X-ray crystallography; the structure at the *top* in (B) is the same as that in (A).

opposite ends of the cell during mitosis (discussed in Chapter 19), move organelles along molecular tracks within the cell (discussed in Chapter 17), and move enzymes along a DNA strand during the synthesis of a new DNA molecule (discussed in Chapter 6). An understanding of how proteins can operate as molecules with moving parts is therefore essential for understanding the molecular basis of cell behavior.

How are shape changes in proteins used to generate orderly movements in cells? If, for example, a protein is required to walk along a narrow thread such as a DNA molecule, it can do this by undergoing a series of conformational changes—as illustrated in Figure 4–44. However, with nothing to drive these changes in an orderly sequence—in one direction only—they will be perfectly reversible and the protein will wander randomly back and forth along the thread. We can look at this situation in another way. Because the directional movement of a protein does net work, the laws of thermodynamics demand that such movement utilize free energy from some other source, say, the hydrolysis of ATP. (Otherwise the protein could be used to make a perpetual motion machine!) Therefore, without an input of energy, the protein molecule can only wander aimlessly.

How, then, can one make the series of conformational changes unidirectional? To force the entire cycle to proceed in one direction, it is enough to make any one of the steps irreversible. For most proteins that are able to walk in a single direction for long distances, this irreversibility is achieved by coupling one of the conformational changes to the hydrolysis of an ATP molecule bound to the protein. The mechanism is similar to the one that drives allosteric shape changes by GTP hydrolysis. Because a great deal of free energy is released when ATP (or GTP) is hydrolyzed, it is very unlikely that the nucleotide-binding protein will undergo a reverse shape change—as required for moving backward—since this would require that it also reverse the ATP hydrolysis by adding a phosphate molecule to ADP to form ATP.

In the highly schematic model shown in Figure 4–45, ATP binding shifts a motor protein from conformation 1 to conformation 2. The bound ATP is then hydrolyzed to produce ADP and inorganic phosphate (P_i), causing a change from conformation 2 to conformation 3. Finally, the release of the bound ADP and P_i drives the protein back to conformation 1. Because the transition $2 \rightarrow 3$ is driven by the energy provided by ATP hydrolysis, this series of conformational changes will be effectively irreversible. Thus the entire cycle will go in only one direction, causing the protein molecule to walk continuously to the right in this example. Many motor proteins generate directional movement in this general way, including the muscle motor protein *myosin*—which "runs" along actin filaments to generate muscle contraction (discussed in Chapter 17)—and the *kinesin* protein involved in chromosome movements at mitosis (discussed in Chapter 19). Such movements can be rapid: some of the motor proteins involved in DNA replication propel themselves along a DNA strand at rates as high as 1000 nucleotides per second.

Proteins Often Form Large Complexes That Function as Protein Machines

As one progresses from small, single-domain proteins to large proteins formed from many domains, the functions that the proteins can perform become more elaborate. The most impressive tasks, however, are carried out by large protein assemblies formed from many protein molecules. Now that it is possible to reconstruct biological processes in cell-free systems in the laboratory, it is clear that each central process in a

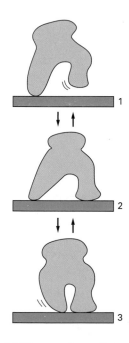

Figure 4–44 Changes in conformation allow a protein to "walk" along a filament or thread. This protein's three different conformations allow it to wander randomly back and forth while bound to a thread or a filament. But without an input of energy to drive its movement in a single direction, the protein will only shuffle aimlessly, getting nowhere.

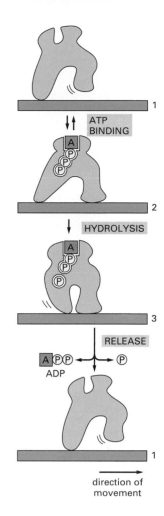

Figure 4–45 An allosteric motor protein, driven by ATP hydrolysis, moves in one direction. An orderly transition among three conformations is driven by the hydrolysis of a bound ATP molecule. Because one of these transitions is coupled to the hydrolysis of ATP, the entire cycle is essentially irreversible. By repeated cycles the protein moves continuously to the right along the thread.

direction of movement

cell—such as DNA replication, protein synthesis, vesicle budding, and transmembrane signaling—is catalyzed by a highly coordinated, linked set of 10 or more proteins. In most such **protein machines** the hydrolysis of bound nucleoside triphosphates (ATP or GTP) drives an ordered series of conformational changes in some of the individual protein subunits, enabling the ensemble of proteins to move coordinately. In this way, the appropriate enzymes can be moved directly into the positions where they are needed to carry out successive reactions in a series as, for example, in protein synthesis on a ribosome (discussed in Chapter 7), or in DNA replication—where a large multiprotein complex moves rapidly along the DNA. A simple mechanical analogy is illustrated in Figure 4–46.

Through evolution, cells have built protein machines that are capable of carrying out most biological reactions. Cells employ protein machines for the same reason that humans have invented mechanical and electronic machines: for almost any task, manipulations that are spatially and temporally coordinated through linked processes are much more efficient than is the sequential use of individual tools.

Large-Scale Studies of Protein Structure and Function Are Increasing the Pace of Discovery

We have made an enormous amount of progress in understanding the structure and function of proteins over the past 150 years (Table 4–2). These advances are the fruits of decades of painstaking research on isolated proteins, performed by individual scientists working tirelessly on single proteins or protein families, one by one, sometimes for their entire careers. But many future advances may come from **proteomics**, the large-scale study of cellular proteins in which the activities or structures of hundreds—even thousands—of proteins are analyzed at once. If scientists can perfect such methods, they might someday be able to

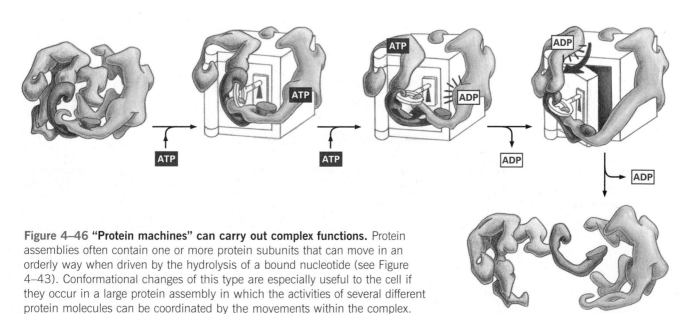

Figure 4–46 "Protein machines" can carry out complex functions. Protein assemblies often contain one or more protein subunits that can move in an orderly way when driven by the hydrolysis of a bound nucleotide (see Figure 4–43). Conformational changes of this type are especially useful to the cell if they occur in a large protein assembly in which the activities of several different protein molecules can be coordinated by the movements within the complex.

Table 4–2 Historical Landmarks in Our Understanding of Proteins

1838	The name **"protein"** (from the Greek proteios, "primary") was suggested by Berzelius for the complex nitrogen-rich substance found in the cells of all animals and plants.
1819–1904	Most of the 20 common **amino acids** found in proteins were discovered.
1864	Hoppe-Seyler crystallized, and named, the protein **hemoglobin**.
1894	Fischer proposed a **lock-and-key** analogy for enzyme–substrate interactions.
1897	Buchner and Buchner showed that cell-free extracts of yeast can ferment sucrose to form carbon dioxide and ethanol, thereby laying the foundations of **enzymology**.
1926	Sumner crystallized urease in pure form, demonstrating that proteins could possess the **catalytic activity** of enzymes; Svedberg developed the first **analytical ultracentrifuge** and used it to estimate the correct molecular weight of hemoglobin.
1933	Tiselius introduced **electrophoresis** for separating proteins in solution.
1934	Bernal and Crowfoot presented the first detailed **X-ray diffraction** patterns of a protein, obtained from crystals of the enzyme pepsin.
1942	Martin and Synge developed **chromatography**, a technique now widely used to separate proteins.
1951	Pauling and Corey proposed the structure of a helical conformation of a chain of amino acids—the **α helix**—and the structure of the **β sheet**, both of which were later found in many proteins.
1955	Sanger determined the analysis of the **amino acid sequence of insulin**, the first protein whose amino acid sequence was determined.
1956	Ingram produced the first **protein fingerprints**, showing that the difference between sickle-cell hemoglobin and normal hemoglobin is due to a change in a single amino acid.
1960	Kendrew described the first detailed **three-dimensional structure** of a protein (sperm whale myoglobin) to a resolution of 0.2 nm, and Perutz proposed a lower-resolution structure for hemoglobin.
1963	Monod, Jacob, and Changeux recognized that many enzymes are regulated through **allosteric changes** in their conformation.

Question 4–9

Explain why the enzymes in Figure 4–46 have a great advantage in opening the vault if they work as a protein complex, as opposed to working in an unlinked, sequential manner.

monitor all of the proteins that are present in a cell—assessing whether they are switched on (or off) and seeing which proteins they are partnered with—all in a single experiment.

Large-scale analyses of protein structures are already under way. Techniques are being scaled up and automated, allowing researchers to rapidly clone genes, produce proteins, grow crystals, and collect X-ray diffraction data for hundreds of proteins at a time. The goal is to catalog representative structures for every folding pattern that protein domains adopt in nature. The 2000 folding patterns already solved are estimated to comprise anywhere from one-fifth to one-half of the total. When the ultimate goal has been reached, one hopes to be able to take any amino acid sequence and predict the structure and function of the protein.

These powerful techniques should bring us closer to understanding the fundamental basis of living cells: how proteins work together to make it possible to create and maintain order in a universe that is always tending toward disorder.

Essential Concepts

- Living cells contain an enormously diverse set of protein molecules, each made as a linear chain of amino acids covalently linked together.
- Each type of protein has a unique amino acid sequence that determines both its three-dimensional shape and its biological activity.
- The folded structure of a protein is stabilized by noncovalent interactions between different parts of the polypeptide chain.
- Hydrogen bonds between neighboring regions of the polypeptide backbone can give rise to regular folding patterns, known as α helices and β sheets.

- The structure of many proteins can be subdivided into smaller globular regions of compact three-dimensional structure, known as protein domains.
- The biological function of a protein depends on the detailed chemical properties of its surface and how it binds to other molecules, called ligands.
- Enzymes are proteins that first bind tightly to specific molecules, called substrates, and then catalyze the formation or breakage of covalent bonds in these molecules.
- At the active site of an enzyme, the amino acid side chains of the folded protein are precisely positioned so that they favor the formation of the high-energy transition states that the substrates must pass through to be converted to product.
- The three-dimensional structure of many proteins has evolved so that the binding of a small ligand can induce a significant change in protein shape.
- Most enzymes are allosteric proteins that can exist in two conformations that differ in catalytic activity, and the enzyme can be turned on or off by ligands that bind to a distinct regulatory site to stabilize either the active or the inactive conformation.
- The activities of most enzymes within the cell are strictly regulated. One of the most common forms of regulation is feedback inhibition, in which an enzyme early in a metabolic pathway is inhibited by its binding to one of the pathway's end products.
- Many thousands of proteins in a typical eucaryotic cell are regulated either by cycles of phosphorylation and dephosphorylation, or by the binding and hydrolysis of GTP by a GTP-binding protein.
- The hydrolysis of ATP to ADP by motor proteins produces directed movements in the cell.
- Highly efficient protein machines are formed by assemblies of allosteric proteins in which conformational changes are coordinated to perform complex cellular functions.

Key Terms

active site	helix
allosteric	ligand
α helix	motor protein
amino acid sequence	polypeptide backbone
antibody	protein
antigen	protein domain
β sheet	protein family
binding site	protein kinase
coiled-coil	protein machine
conformation	protein phosphatase
disulfide bond	protein phosphorylation
enzyme	secondary structure
feedback inhibition	side chain
fibrous protein	substrate
globular protein	subunit
GTP-binding protein	transition state

Panel 4–3 Cell breakage and initial fractionation of cell extracts

BREAKING CELLS AND TISSUES

The first step in the purification of most proteins is to disrupt tissues and cells in a controlled fashion.

Using gentle mechanical procedures, called homogenization, the plasma membranes of cells can be ruptured so that the cell contents are released. Four commonly used procedures are shown here.

The resulting thick soup (called a homogenate or an extract) contains large and small molecules from the cytosol, such as enzymes, ribosomes, and metabolites, as well as all of the membrane-enclosed organelles.

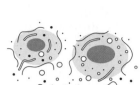

(1) Break cells with high-frequency sound.

(2) Use a mild detergent to make holes in the plasma membrane.

cell suspension or tissue

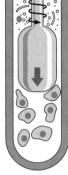

(3) Force cells through a small hole using high pressure.

(4) Shear cells between a close-fitting rotating plunger and the thick walls of a glass vessel.

When carefully applied, homogenization leaves most of the membrane-enclosed organelles intact.

THE CENTRIFUGE

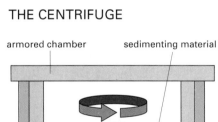

armored chamber   sedimenting material

fixed-angle rotor

refrigeration   vacuum

motor

swinging-arm rotor

centrifugal force

tube

metal bucket

CENTRIFUGATION

Many cell fractionations are done in a second type of rotor, a swinging-arm rotor.

The metal buckets that hold the tubes are free to swing outward as the rotor turns.

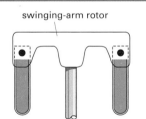

CELL HOMOGENATE before centrifugation

CENTRIFUGATION

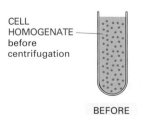

SUPERNATANT smaller and less dense components

PELLET larger and more dense components

BEFORE

AFTER

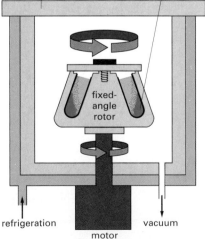
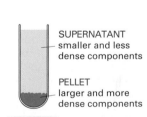

Centrifugation is the most widely used procedure to separate a homogenate into different parts, or fractions. The homogenate is placed in test tubes and rotated at high speed in a centrifuge or ultracentrifuge. Present-day ultracentrifuges rotate at speeds up to 100,000 revolutions per minute and produce enormous forces, as high as 600,000 times gravity.

Such speeds require centrifuge chambers to be refrigerated and evacuated so that friction does not heat up the homogenate. The centrifuge is surrounded by thick armor plating, because an unbalanced rotor can shatter with an explosive release of energy. A fixed-angle rotor can hold larger volumes than a swinging-arm rotor, but the pellet forms less evenly.

DIFFERENTIAL CENTRIFUGATION

Repeated centrifugation at progressively higher speeds will fractionate cell homogenates into their components.

Centrifugation separates cell components on the basis of size and density. The larger and denser components experience the greatest centrifugal force and move most rapidly. They sediment to form a pellet at the bottom of the tube, while smaller, less dense components remain in suspension above, a portion called the supernatant.

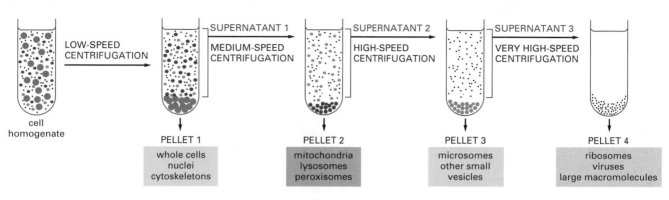

cell homogenate

LOW-SPEED CENTRIFUGATION

SUPERNATANT 1
MEDIUM-SPEED CENTRIFUGATION

SUPERNATANT 2
HIGH-SPEED CENTRIFUGATION

SUPERNATANT 3
VERY HIGH-SPEED CENTRIFUGATION

PELLET 1
whole cells
nuclei
cytoskeletons

PELLET 2
mitochondria
lysosomes
peroxisomes

PELLET 3
microsomes
other small
vesicles

PELLET 4
ribosomes
viruses
large macromolecules

VELOCITY SEDIMENTATION

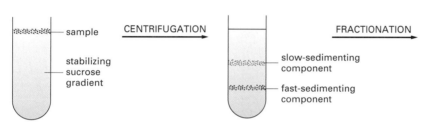

sample

stabilizing sucrose gradient

CENTRIFUGATION

slow-sedimenting component

fast-sedimenting component

FRACTIONATION

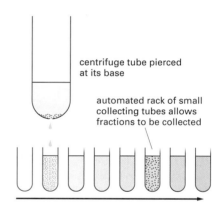

centrifuge tube pierced at its base

automated rack of small collecting tubes allows fractions to be collected

Subcellular components sediment at different rates according to their size after being carefully layered over a dilute salt solution and then centrifuged through it. In order to stabilize the sedimenting components against convective mixing in the tube, the solution contains a continuous shallow gradient of sucrose that increases in concentration toward the bottom of the tube. This is typically 5–20% sucrose. When sedimented through such a dilute sucrose gradient, different cell components separate into distinct bands that can be collected individually.

After an appropriate centrifugation time the bands may be collected, most simply by puncturing the plastic centrifuge tube and collecting drops from the bottom, as shown here.

EQUILIBRIUM SEDIMENTATION

The ultracentrifuge can also be used to separate cellular components on the basis of their buoyant density, independently of their size or shape. The sample is usually either layered on top of, or dispersed within, a steep density gradient that contains a very high concentration of sucrose or cesium chloride. Each subcellular component will move up or down when centrifuged until it reaches a position where its density matches its surroundings and then will move no further. A series of distinct bands will eventually be produced, with those nearest the bottom of the tube containing the components of highest buoyant density. The method is also called density gradient centrifugation.

The sample is distributed throughout the sucrose density gradient.

At equilibrium, components have migrated to a region in the gradient that matches their own density.

CENTRIFUGATION

CENTRIFUGATION

sample

steep sucrose gradient (e.g., 20–70%)

low–buoyant density component

high–buoyant density component

START

BEFORE EQUILIBRIUM

EQUILIBRIUM

A sucrose gradient is shown here, but denser gradients can be formed with cesium chloride that are particularly useful for separating the nucleic acids (DNA and RNA).

The final bands can be collected from the base of the tube, as shown above.

Panel 4–4 Protein separation by chromatography

PROTEIN SEPARATION

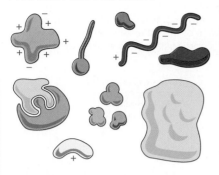

Proteins are very diverse. They differ by size, shape, charge, hydrophobicity, and their affinity for other molecules. All of these properties can be exploited to separate them from one another so that they can be studied individually.

THREE KINDS OF CHROMATOGRAPHY

Although the material used to form the matrix for column chromatography varies, it is usually packed in the column in the form of small beads. A typical protein purification strategy might employ in turn each of the three kinds of matrix described below, with a final protein purification of up to 10,000-fold.

Purity can easily be assessed by gel electrophoresis (see *opposite page*).

COLUMN CHROMATOGRAPHY

Proteins are often fractionated by column chromatography. A mixture of proteins in solution is applied to the top of a cylindrical column filled with a permeable solid matrix immersed in solvent. A large amount of solvent is then pumped through the column. Because different proteins are retarded to different extents by their interaction with the matrix, they can be collected separately as they flow out from the bottom. According to the choice of matrix, proteins can be separated according to their charge, hydrophobicity, size, or ability to bind to particular chemical groups (see *below*).

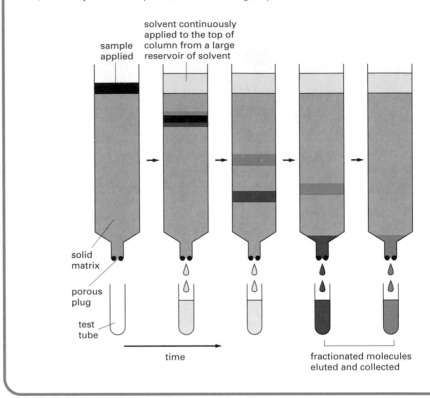

sample applied

solvent continuously applied to the top of column from a large reservoir of solvent

solid matrix

porous plug

test tube

time

fractionated molecules eluted and collected

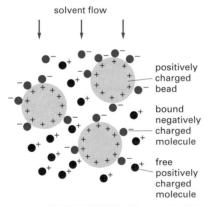

positively charged bead

bound negatively charged molecule

free positively charged molecule

(A) ION-EXCHANGE CHROMATOGRAPHY

Ion-exchange columns are packed with small beads carrying either positive or negative charges that retard proteins of the opposite charge. The association between a protein and the matrix depends on the pH and ionic strength of the solution passing down the column. These can be varied in a controlled way to achieve an effective separation.

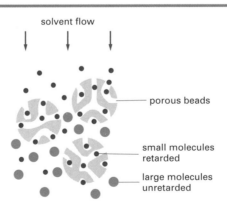

porous beads

small molecules retarded

large molecules unretarded

(B) GEL-FILTRATION CHROMATOGRAPHY

Gel-filtration columns separate proteins according to their size. The matrix consists of tiny porous beads. Protein molecules that are small enough to enter the holes in the beads are delayed and travel more slowly through the column. Proteins that cannot enter the beads are washed out of the column first. Such columns also allow an estimate of protein size.

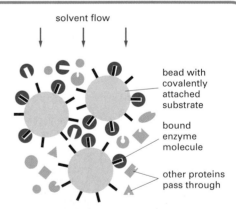

bead with covalently attached substrate

bound enzyme molecule

other proteins pass through

(C) AFFINITY CHROMATOGRAPHY

Affinity columns contain a matrix covalently coupled to a molecule that interacts specifically with the protein of interest (e.g., an antibody, or an enzyme substrate). Proteins that bind specifically to such a column can subsequently be released by a pH change or by concentrated salt solutions, and they emerge highly purified.

Panel 4–5 Protein separation by electrophoresis

GEL ELECTROPHORESIS

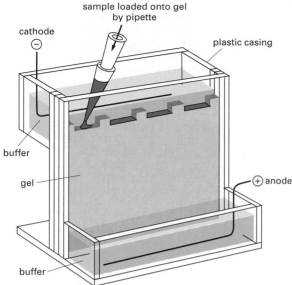

sample loaded onto gel by pipette

cathode

plastic casing

buffer

gel

(+) anode

buffer

When an electric field is applied to a solution containing protein molecules, the molecules will migrate in a direction and at a speed that reflects their size and net charge. This forms the basis of the technique called electrophoresis.

The detergent sodium dodecyl sulfate (SDS) is used to solubilize proteins for SDS polyacrylamide-gel electrophoresis (see *below*).

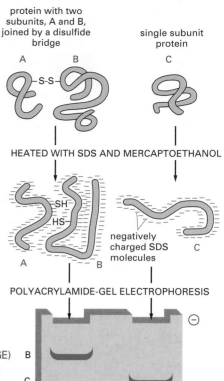

CH₃
|
CH₂
|
CH₂
|
CH₂
|
CH₂
|
CH₂
|
CH₂
|
CH₂
|
CH₂
|
CH₂
|
CH₂
|
CH₂
|
CH₂
|
CH₂
|
O
|
O=S=O
|
O⁻ Na⁺

SDS

protein with two subunits, A and B, joined by a disulfide bridge

single subunit protein

A   B   C

—S–S—

HEATED WITH SDS AND MERCAPTOETHANOL

SH
HS—

A   B

negatively charged SDS molecules

C

POLYACRYLAMIDE-GEL ELECTROPHORESIS

B

C

A

slab of polyacrylamide gel

SDS polyacrylamide-gel electrophoresis (SDS-PAGE) Individual polypeptide chains form a complex with negatively charged molecules of sodium dodecyl sulfate (SDS) and therefore migrate as a negatively charged SDS–protein complex through a slab of porous polyacrylamide gel. The apparatus used for this electrophoresis technique is shown above (*left*). A reducing agent (mercaptoethanol) is usually added to break any –S–S– linkages in or between proteins. Under these conditions, proteins migrate at a rate that reflects their molecular weight.

ISOELECTRIC FOCUSING

For any protein there is a characteristic pH, called the isoelectric point, at which the protein has no net charge and therefore will not move in an electric field. In isoelectric focusing, proteins are electrophoresed in a narrow tube of polyacrylamide gel in which a pH gradient is established by a mixture of special buffers. Each protein moves to a point in the gradient that corresponds to its isoelectric point and stays there.

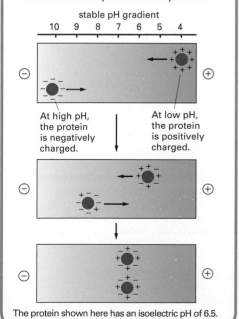

stable pH gradient

10 9 8 7 6 5 4

At high pH, the protein is negatively charged.

At low pH, the protein is positively charged.

The protein shown here has an isoelectric pH of 6.5.

TWO-DIMENSIONAL POLYACRYLAMIDE-GEL ELECTROPHORESIS

Complex mixtures of proteins cannot be resolved well on one-dimensional gels, but two-dimensional gel electrophoresis, combining two different separation methods, can be used to resolve more than 1000 proteins in a two-dimensional protein map. In the first step, native proteins are separated in a narrow gel on the basis of their intrinsic charge using isoelectric focusing (see *left*). In the second step, this gel is placed on top of a gel slab, and the proteins are subjected to SDS-PAGE (see *above*) in a direction perpendicular to that used in the first step. Each protein migrates to form a discrete spot.

All the proteins in an *E. coli* bacterial cell are separated in this 2-D gel, in which each spot corresponds to a different polypeptide chain. They are separated according to their isoelectric point from left to right and to their molecular weight from top to bottom. (Courtesy of Patrick O'Farrell.)

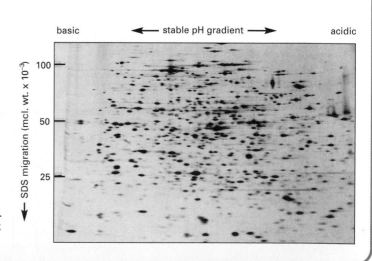

basic ← stable pH gradient → acidic

SDS migration (mol. wt. × 10⁻³)

100

50

25

Panel 4–6 Making and using antibodies

THE ANTIBODY MOLECULE

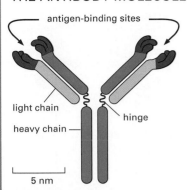

antigen-binding sites

light chain

heavy chain

hinge

5 nm

Antibodies are proteins that bind very tightly to their targets (antigens). They are produced in vertebrates as a defense against infection. Each antibody molecule is made of two identical light chains and two identical heavy chains, so the two antigen-binding sites are identical.

ANTIBODY SPECIFICITY

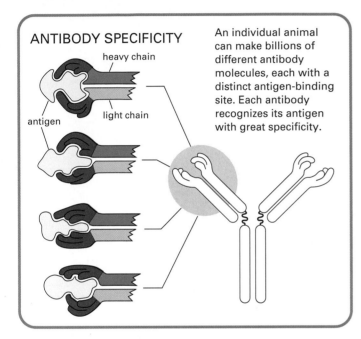

heavy chain

light chain

antigen

An individual animal can make billions of different antibody molecules, each with a distinct antigen-binding site. Each antibody recognizes its antigen with great specificity.

ANTIBODIES DEFEND US AGAINST INFECTION

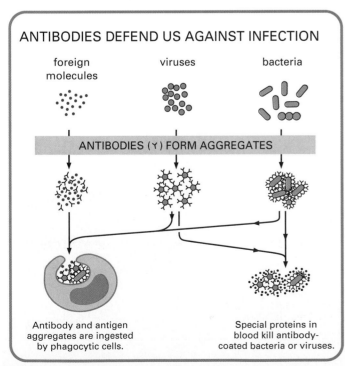

foreign molecules

viruses

bacteria

ANTIBODIES (Y) FORM AGGREGATES

Antibody and antigen aggregates are ingested by phagocytic cells.

Special proteins in blood kill antibody-coated bacteria or viruses.

B CELLS PRODUCE ANTIBODIES

Antibodies are made by a class of white blood cells called B lymphocytes, or B cells. Each resting B cell carries a different membrane-bound antibody molecule on its surface that serves as a receptor for recognizing a specific antigen. When antigen binds to this receptor, the B cell is stimulated to divide and to secrete large amounts of the same antibody in a soluble form.

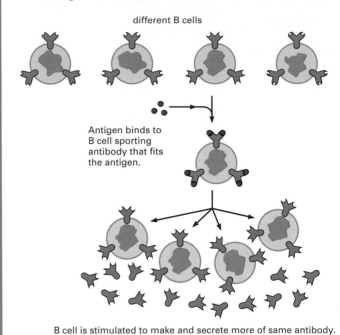

different B cells

Antigen binds to B cell sporting antibody that fits the antigen.

B cell is stimulated to make and secrete more of same antibody.

RAISING ANTIBODIES IN ANIMALS

Antibodies can be made in the laboratory by injecting an animal (usually a mouse, rabbit, sheep, or goat) with antigen A.

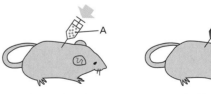

inject antigen A

take blood later

Repeated injections of the same antigen at intervals of several weeks stimulate specific B cells to secrete large amounts of anti-A antibodies into the bloodstream.

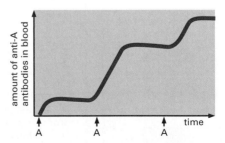

amount of anti-A antibodies in blood

time

A   A   A

Because many different B cells are stimulated by antigen A, the blood will contain a variety of anti-A antibodies, each of which binds A in a slightly different way.

USING ANTIBODIES TO PURIFY MOLECULES

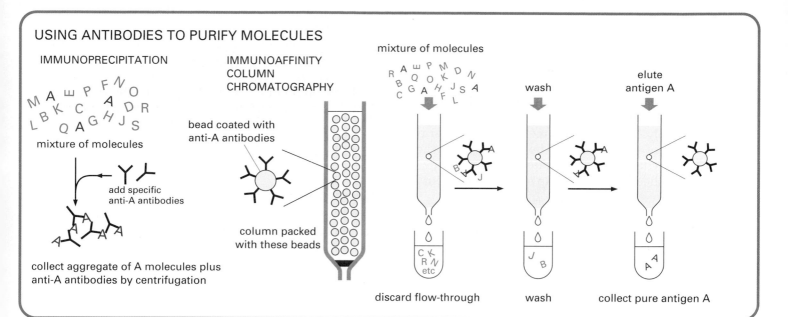

IMMUNOPRECIPITATION

mixture of molecules

add specific
anti-A antibodies

collect aggregate of A molecules plus
anti-A antibodies by centrifugation

IMMUNOAFFINITY COLUMN CHROMATOGRAPHY

bead coated with
anti-A antibodies

column packed
with these beads

mixture of molecules

wash

elute
antigen A

discard flow-through

wash

collect pure antigen A

MONOCLONAL ANTIBODIES

Large quantities of a single type of antibody
molecule can be obtained by fusing a B cell
(taken from an animal injected with antigen A)
with a tumor cell. The resulting hybrid cell divides
indefinitely and secretes anti-A antibodies of a
single (monoclonal) type.

B cell from animal
injected with antigen
A makes anti-A
antibody but does
not divide forever.

Tumor cell from
cell culture divides
indefinitely but
does not make
antibody.

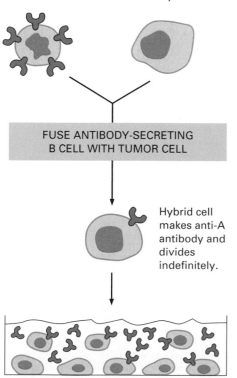

FUSE ANTIBODY-SECRETING
B CELL WITH TUMOR CELL

Hybrid cell
makes anti-A
antibody and
divides
indefinitely.

USING ANTIBODIES AS MOLECULAR TAGS

specific antibodies
against antigen A

couple to fluorescent dye,
colloidal gold particle, or
other special tag

labeled antibodies

cell
wall

MICROSCOPIC DETECTION

Fluorescent antibody binds to
antigen A in tissue and is
detected by fluorescence in a light
microscope. The antigen here is
pectin in the cell walls of a slice
of plant tissue.

Gold-labeled antibody binds
to antigen A in tissue and is
detected in an electron microscope. The antigen is pectin
in the cell wall of a single
plant cell.

BIOCHEMICAL DETECTION

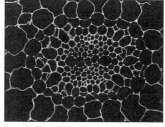

Antigen A is
separated from
other molecules
by electrophoresis.

Incubation with the
labeled antibodies
that bind to antigen A
allows the position of the
antigen to be determined.

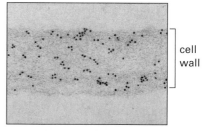

Labeled second antibody
(*blue*) binds to first
antibody (*black*).

antigen

Note: In all cases, the sensitivity can
be greatly increased by using multiple
layers of antibodies. This "sandwich"
method enables smaller numbers of
antigen molecules to be detected.

Questions

Question 4–10

Which of the following statements are correct? Explain your answers.

A. The active site of an enzyme usually occupies only a small fraction of its surface.

B. Catalysis by some enzymes involves the formation of a covalent bond between an amino acid side chain and a substrate molecule.

C. A β sheet can contain up to five strands, but no more.

D. The specificity of an antibody molecule is contained exclusively in loops on the surface of the folded light-chain domain.

E. The possible linear arrangements of amino acids are so vast that new proteins almost never evolve by alteration of old ones.

F. Allosteric enzymes have two or more binding sites.

G. Noncovalent bonds are too weak to influence the three-dimensional structure of macromolecules.

H. Affinity chromatography separates molecules according to their intrinsic charge.

I. Upon centrifugation of a cell homogenate, smaller organelles experience less friction and thereby sediment faster than larger ones.

Question 4–11

What common feature of α helices and β sheets makes them universal building blocks for proteins?

Question 4–12

Protein structure is determined solely by a protein's amino acid sequence. Should a genetically engineered protein in which the order of all amino acids is reversed therefore have the same structure as the original protein?

Question 4–13

Consider the following protein sequence as an α helix: Leu-Lys-Arg-Ile-Val-Asp-Ile-Leu-Ser-Arg-Leu-Phe-Lys-Val. How many turns does this helix make? Do you find anything remarkable about the arrangement of the amino acids in this sequence when folded into an α helix? (Hint: consult the properties of the amino acids in Figure 4–3.)

Question 4–14

Simple enzyme reactions often conform to the equation

$$E + S \rightleftharpoons ES \rightarrow EP \rightleftharpoons E + P$$

where E, S, and P are enzyme, substrate, and product, respectively.

A. What does ES represent in this equation?

B. Why is the first step shown with bidirectional arrows and the second step as a unidirectional arrow?

C. Why does E appear at both ends of the equation?

D. One often finds that high concentrations of P inhibit the enzyme. Suggest why this might occur.

E. Compound X resembles S and binds to the active site of the enzyme but cannot undergo the reaction catalyzed by it. What effects would you expect the addition of X to the reaction to have? Compare the effects of X and of accumulation of P.

Question 4–15

Which of the following amino acids would you expect to find more often near the center of a folded globular protein? Which ones would you expect to find more often exposed to the outside? Explain your answers. Ser, Ser-P (a Ser residue that is phosphorylated), Leu, Lys, Gln, His, Phe, Val, Ile, Met, Cys–S–S–Cys (two Cys residues that are disulfide-bonded), and Glu. Where would you expect to find the most N-terminal amino acid and the most C-terminal amino acid?

Question 4–16

Assume you want to make and study fragments of a protein. Would you expect that any fragment of the polypeptide chain would fold the same way as the corresponding sequence folds in the intact protein? Consider the protein shown in Figure 4–19. Which fragments do you suppose are most likely to fold correctly?

Question 4–17

An enzyme isolated from a mutant bacterium grown at 20°C works in a test tube at 20°C but not at 37°C (37°C is the temperature of the gut, where this bacterium normally lives). Furthermore, once the enzyme has been exposed to the higher temperature, it no longer works at the lower one. The same enzyme isolated from the normal bacterium works at both temperatures. Can you suggest what happens at the molecular level to the mutant enzyme as the temperature increases?

Question 4–18

A motor protein moves along filaments in the cell. Why are the elements shown in the illustration not sufficient to provide unidirectionality to the movement (Figure Q4–18)? With reference to Figure 4–45, modify the illustration shown here to include other elements that are required to create a unidirectional motor, and justify each modification you make to the illustration.

Figure Q4–18

Question 4–19

Gel-filtration chromatography separates molecules according to size (see Panel 4–4, p. 162). Smaller molecules diffuse faster in solution than larger ones, yet smaller molecules migrate more slowly through a gel-filtration column than larger ones. Explain this paradox. What should happen at very rapid flow rates?

Question 4–20

Both an α helix and the coiled-coil that forms from it are helical structures, but do they have the same handedness (refer to Figure 4–14)?

Highlights from *Essential Cell Biology 2 Interactive* CD-ROM

4.1   α Helix

4.2   β Sheet

4.11 The "Safe Crackers"

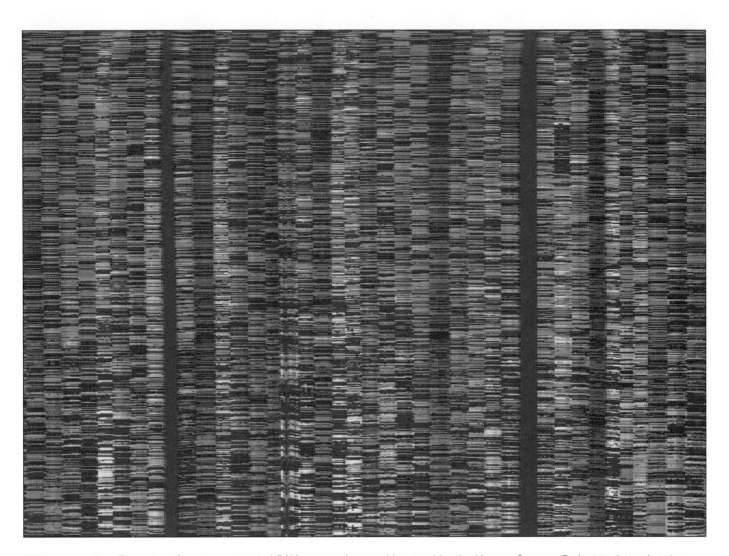

DNA sequencing. The output from an automated DNA sequencing machine used by the Human Genome Project to determine the complete human DNA sequence. Each vertical lane shows the sequence of nucleotides in a given stretch of DNA. Each of the four different nucleotides is labeled with one of the four colored dyes. The order of the nucleotides is analyzed by a computer and assembled to give the continuous nucleotide sequence of each chromosome. This image shows the sequence of only a tiny part of one chromosome. (Courtesy of Sanger Institute/Wellcome Photo Library.)

DNA and Chromosomes

Life depends on the ability of cells to store, retrieve, and translate the genetic instructions required to make and maintain a living organism. This hereditary information is passed on from a cell to its daughter cells at cell division, and from generation to generation of organisms through the reproductive cells. These instructions are stored within every living cell in its genes—the information-containing elements that determine the characteristics of a species as a whole and of the individuals within it.

At the beginning of the twentieth century, when genetics emerged as a science, scientists became intrigued by the chemical nature of genes. The information in genes is copied and transmitted from cell to daughter cell millions of times during the life of a multicellular organism, and it survives the process essentially unchanged. What kind of molecule could be capable of such accurate and almost unlimited replication, and also be able to direct the development of an organism and the daily life of a cell? What kind of instructions does the genetic information contain? How are these instructions physically organized so that the enormous amount of information required for the development and maintenance of even the simplest organism can be contained within the tiny space of a cell?

The answers to some of these questions began to emerge in the 1940s, when it was discovered from studies in simple fungi that genetic information consists primarily of instructions for making proteins. Proteins are the macromolecules that perform most cellular functions: they serve as building blocks for cellular structures, they form the enzymes that catalyze all of the cell's chemical reactions, they regulate gene expression, and they enable cells to move and to communicate with each other. With hindsight, it is hard to imagine what other type of instructions the genetic information could have contained.

The other crucial advance made in the 1940s was the recognition that **deoxyribonucleic acid** (**DNA**) was the likely carrier of this genetic information. But the mechanism whereby the hereditary information is copied for transmission from cell to cell, and how proteins are specified by the instructions in the DNA, remained mysterious until 1953, when the structure of DNA was determined by James Watson and Francis Crick. The structure immediately revealed how DNA might be copied, or replicated, and provided the first clues about how a molecule of DNA might encode the instructions for making proteins. Today, the fact that DNA is the genetic material is so fundamental to biological thought that it is difficult to appreciate what an enormous intellectual gap this discovery filled.

In this chapter we begin by describing the structure of DNA. We see how, despite its chemical simplicity, the structure and chemical

The Structure and Function of DNA

A DNA Molecule Consists of Two Complementary Chains of Nucleotides

The Structure of DNA Provides a Mechanism for Heredity

The Structure of Eucaryotic Chromosomes

Eucaryotic DNA Is Packaged into Chromosomes

Chromosomes Contain Long Strings of Genes

Chromosomes Exist in Different States Throughout the Life of a Cell

Interphase Chromosomes Are Organized Within the Nucleus

The DNA in Chromosomes Is Highly Condensed

Nucleosomes Are the Basic Units of Chromatin Structure

Chromosomes Have Several Levels of DNA Packing

Interphase Chromosomes Contain Both Condensed and More Extended Forms of Chromatin

Changes in Nucleosome Structure Allow Access to DNA

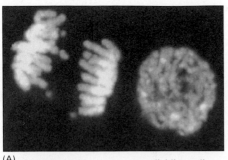

(A)
dividing cell   nondividing cell

(B)

10 µm

Figure 5–1 Chromosomes become visible as cells prepare to divide. (A) Two adjacent plant cells photographed through a light microscope. The DNA has been stained with a fluorescent dye (DAPI) that binds to it. The DNA is present in chromosomes, which become visible as distinct structures in the light microscope only when they become compact structures in preparation for cell division, as shown on the *left*. The cell on the *right*, which is not dividing, contains the identical chromosomes; they cannot be distinguished as individual chromosomes in the light microscope at this phase in the cell's life cycle because their DNA is in a much more extended conformation. (B) Schematic diagram of the outlines of the two cells along with their chromosomes. (A, courtesy of Peter Shaw.)

properties of DNA make it ideally suited as the raw material of genes. The genes of every cell on Earth are made of DNA, and insights into the relationship between DNA and genes have come from experiments in a wide variety of organisms. We then consider how genes and other important segments of DNA are arranged on the long molecules of DNA that are present in chromosomes. Finally, we discuss how eucaryotic cells fold these long DNA molecules into compact chromosomes. This packing has to be done in an orderly fashion so that the chromosomes can be replicated and apportioned correctly between the two daughter cells at each cell division. It must also allow access of chromosomal DNA to enzymes that repair it when it is damaged and to the specialized proteins that direct the expression of its many genes.

This is the first of five chapters that deal with basic genetic mechanisms—the ways in which the cell maintains, replicates, expresses, and occasionally improves the genetic information carried in its DNA. In the following chapter (Chapter 6) we discuss the mechanisms by which the cell accurately replicates and repairs DNA; we also describe how DNA sequences can be rearranged through the process of genetic recombination. Gene expression—the process through which the information encoded in DNA is interpreted by the cell to guide the synthesis of proteins—is the main topic of Chapter 7. In Chapter 8, we describe how gene expression is controlled by the cell to ensure that each of the many thousands of proteins encrypted in its DNA is manufactured at the proper time and place in the life of the cell. We turn in Chapter 9 to a discussion of how present-day genes and genomes evolved from distant ancestors. Following these five chapters on basic genetic mechanisms, we present an account of the experimental techniques used to study DNA and the role it plays in these fundamental cellular processes (Chapter 10).

The Structure and Function of DNA

Well before biologists understood the structure of DNA, they had recognized that genes are carried on *chromosomes,* which were discovered in the nineteenth century as threadlike structures in the nucleus of the eucaryotic cell that become visible as the cell begins to divide (Figure 5–1). As biochemical analysis became possible, researchers learned that chromosomes contain both DNA and protein. But which of these components encoded the organism's genetic information was not immediately clear.

We now know that the DNA carries the hereditary information of the cell, and that the protein components of chromosomes function largely to package and control the enormously long DNA molecules. But biologists in the 1940s had difficulty accepting DNA as the genetic material because of the apparent simplicity of its chemistry (see How We Know, pp. 172–174). DNA was thought of as simply a long polymer composed of only four types of subunits, which resemble one another chemically.

Then, early in the 1950s, DNA was examined by X-ray diffraction analysis, a technique for determining the three-dimensional atomic structure of a molecule (discussed in Chapter 4, pp. 129–131). The early X-ray diffraction results indicated that DNA was composed of two strands wound into a helix. The observation that DNA was double-stranded was of crucial significance. It provided one of the major clues that led, in 1953, to a correct model for the structure of DNA. Only when the Watson–Crick model of DNA structure was proposed did its potential for replication and information encoding become apparent.

In this section, we examine the structure of the DNA molecule and explain in general terms how it is able to store hereditary information.

A DNA Molecule Consists of Two Complementary Chains of Nucleotides

A DNA molecule consists of two long polynucleotide chains known as *DNA chains,* or *DNA strands.* Each of these chains is composed of four types of nucleotide subunits, and the two chains are held together by *hydrogen bonds* between the base portions of the nucleotides (Figure 5–2). As we saw in Chapter 2 (Panel 2–6, pp. 76–77), nucleotides are composed of a five-carbon sugar to which are attached one or more phosphate groups and a nitrogen-containing base. In the case of the nucleotides in DNA, the sugar is deoxyribose attached to a single phosphate group (hence the name deoxyribonucleic acid), and the base may be either *adenine (A), cytosine (C), guanine (G),* or *thymine (T).* The nucleotides are covalently linked together in a chain through the sugars and phosphates, which thus form a "backbone" of alternating sugar–phosphate–sugar–phosphate (see Figure 5–2). Because it is only the base that differs in each of the four types of subunits, each polynucleotide chain in DNA can be thought of as a necklace (the backbone) strung with four types of beads (the four bases A, C, G, and T). These same symbols (A, C, G, and T) are also commonly used to denote the four different nucleotides, that is, the bases with their attached sugar and phosphate groups.

The way in which the nucleotide subunits are linked together gives a DNA strand a chemical polarity. If we imagine that each nucleotide has a knob (the phosphate) and a hole (see Figure 5–2), each completed chain, formed by interlocking knobs with holes, will have all of its subunits lined up in the same orientation. Moreover, the two ends of the

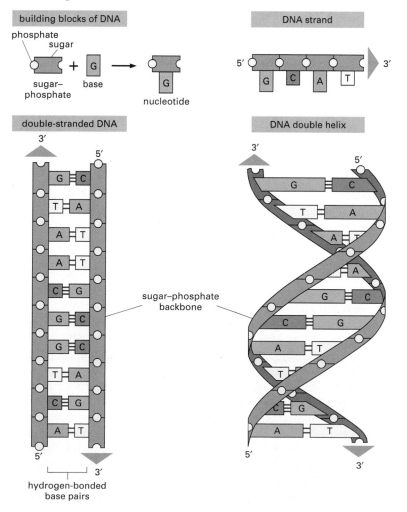

Figure 5–2 DNA is made of four nucleotide building blocks. The nucleotides are covalently linked together into polynucleotide chains with a sugar–phosphate backbone from which the bases (A, C, G, and T) extend. A DNA molecule is composed of two polynucleotide chains (DNA strands) held together by hydrogen bonds between the paired bases. The *arrows* on the DNA strands indicate the polarities of the two strands, which run antiparallel to each other in the DNA molecule. In the diagram at the bottom left of the figure, the DNA is shown straightened out; in reality, it is wound into a double helix, as shown on the right.

How We Know: Genes Are Made of DNA

By the 1920s, scientists generally agreed that genes reside on chromosomes, and they knew that chromosomes are composed of both DNA and proteins. But the connection between genes and DNA was not immediately accepted. Because DNA is so chemically simple, researchers naturally assumed that genes had to be made of proteins, which are much more chemically diverse. Even when the experimental evidence suggested otherwise, this assumption proved hard to shake.

Messages from the dead

The case for DNA began to take shape in the late 1920s, when a British medical officer named Fred Griffith made an astonishing discovery. He was studying *Streptococcus pneumoniae,* the bacterium that causes pneumonia. As antibiotics had not yet been discovered, infection with this organism was usually fatal. When grown in the laboratory, the bacteria come in two forms—a lethal form that causes disease when injected into animals, and a harmless form that is easily conquered by the animal's immune system and produces no infection.

In the course of his investigations, Griffith injected various preparations of these bacteria into mice. He confirmed the disease-causing tendencies of the two strains, and showed that pneumococci that had been killed by heating were unable to cause infection. The surprise came when Griffith injected both heat-killed pathogenic cells and live harmless cells into the same mouse. This mixture proved a lethal combination: not only did the animal die of pneumonia, but Griffith found that its blood was teeming with living bacteria of the pathogenic form (Figure 5–3). The heat-killed pneumococci had somehow converted the innocuous bacteria into the lethal form. What's more, Griffith found that the change was permanent: he could grow these transformed bacteria in culture and they remained virulent. But what was this mysterious material that turned harmless bacteria into killers? And how was this change passed on to progeny bacteria?

Blowing bubbles

Griffith's remarkable finding set the stage for the experiments that would provide the first strong evidence that genes are made of DNA. The American bacteriologist Oswald Avery, following up on Griffith's work, discovered that the harmless pneumococci could be transformed into a pathogenic strain in a culture tube by exposing it to an extract prepared from the disease-causing strain. It would take another 15 years, however, for Avery and his colleagues Colin MacLeod and Maclyn McCarty to successfully purify the "transforming principle" from this soluble extract and to demonstrate that the active ingredient was DNA. Because the transforming principle caused a heritable change in the bacteria that received it, DNA must be the very stuff of which genes are made.

The delay was in part a reflection of the academic climate—and the widespread supposition that the genetic material was likely to be made of protein. Because of the potential ramifications of their work, the researchers wanted to be absolutely certain that the transforming principle was DNA before they announced their findings. As Avery noted in a letter to his brother, also a bacteriologist, "It's lots of fun to blow bubbles, but it's wiser to prick them yourself before someone else tries to." So the researchers subjected the material to a battery of chemical tests (Figure 5–4). They found that the transforming principle exhibited all the chemical properties characteristic of DNA; furthermore, they showed that enzymes that destroy proteins and RNA did not affect its ability to transform bacteria, while enzymes that digested DNA inactivated it. And like Griffith before them, the investigators found that their purified preparation changed the bacteria permanently: DNA from the virulent species was taken up by the harmless species, and this change was faithfully passed on to subsequent generations of bacteria.

This landmark study offered rigorous proof that purified DNA can act as genetic material. But the resulting paper drew remarkably little attention. Despite the meticulous care with which these experiments were performed, geneticists were not immediately convinced that DNA is the hereditary material. Many argued that the transformation might have been caused by some trace protein contaminant in the preparations. Or that the extract might contain a mutagen that alters the genetic material of the harmless bacteria—converting it to the pathogenic form—rather than containing the genetic material itself.

Virus cocktails

The debate was not settled definitively until 1952, when Alfred Hershey and Martha Chase fired up their laboratory blender and demonstrated, once and for all, that genes are made of DNA. The researchers were studying T2—a virus that infects and eventually destroys the bacterium *E. coli.* These bacteria-killing viruses behave like little molecular syringes: they inject their genetic material into the host cell, while the empty virus heads remain outside the infected bacterium (Figure 5–5A). Once inside the cell, the viral genes direct the formation of new viral particles. Within

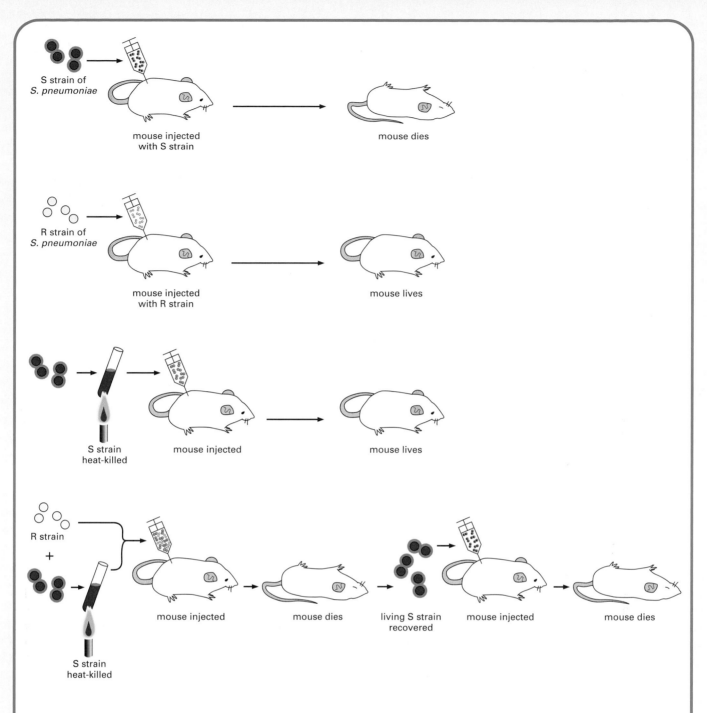

Figure 5–3 Griffith demonstrates that heat-killed bacteria can transform living cells. The bacterium *Streptococcus pneumoniae* comes in two forms that differ from one another in their microscopic appearance and in their ability to cause disease. Cells of the pathogenic strain, which are lethal when injected into mice, are encased in a slimy, glistening polysaccharide capsule. When grown on a plate of nutrients in the laboratory, this disease-causing bacterium forms colonies that look dome-shaped and smooth; hence it is designated the S form. The harmless strain of the pneumococcus, on the other hand, lacks this protective coat; it forms colonies that appear flat and rough—hence, it is referred to as the R form. Griffith found that a substance present in the virulent S strain could permanently change, or transform, the nonlethal R strain into the deadly S strain.

minutes, the infected cells explode, spewing thousands of new viruses into the medium. These then infect neighboring bacteria, and the process begins again.

The beauty of T2 is that these viruses contain only two kinds of molecules: DNA and protein. So the genetic material had to be one or the other. But which? The experiment was fairly

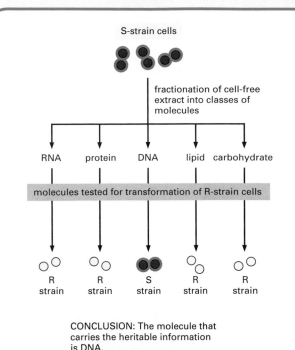

S-strain cells

fractionation of cell-free extract into classes of molecules

RNA   protein   DNA   lipid   carbohydrate

molecules tested for transformation of R-strain cells

R strain   R strain   S strain   R strain   R strain

CONCLUSION: The molecule that carries the heritable information is DNA.

Figure 5–4 Avery, MacLeod, and McCarty demonstrate that DNA is the genetic material. These researchers prepared an extract from the disease-causing S strain and identified the "transforming principle" that would permanently change R-strain pneumococci into the lethal S strain as DNA. This was the first evidence that DNA could serve as the genetic material.

straightforward. Because the viral genes enter the bacterial cell, while the rest of the viral particle remains outside, the researchers decided to radioactively label the protein in one batch of virus and the DNA in another. Then all they had to do was follow the radioactivity to see whether the DNA or the protein winds up inside the bacteria. To do this, the researchers incubated their labeled viruses with *E. coli;* after allowing a few minutes for infection to take place, they poured the mix into a Waring blender and hit "puree." The blender's spinning blades sheared the empty virus heads from the surfaces of the cells. The researchers then centrifuged the sample to separate the heavier, infected bacteria, which formed a pellet at the bottom of the centrifuge tube, from the empty viral coats, which remained in suspension (Figure 5–5B).

As you have probably surmised, Hershey and Chase found that the radioactive DNA entered the bacterial cells, while the labeled proteins remained with the empty virus heads. This radioactively labeled DNA, they found, was also incorporated into the next generation of virus particles. The experiment demonstrated conclusively that viral DNA enters bacterial host cells whereas viral protein does not. Thus the genetic material in this virus had to be made of DNA.

Together with the studies done by Avery, MacLeod, and McCarty, this evidence clinched the case for DNA as the agent of heredity.

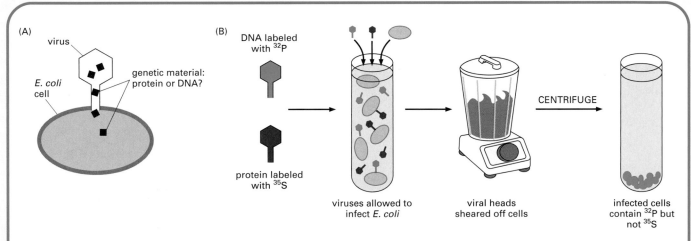

(A) virus

E. coli cell

genetic material: protein or DNA?

(B) DNA labeled with ^{32}P

protein labeled with ^{35}S

viruses allowed to infect *E. coli*

CENTRIFUGE

viral heads sheared off cells

infected cells contain ^{32}P but not ^{35}S

Figure 5–5 Hershey and Chase demonstrate definitively that genes are made of DNA. (A) The researchers worked with T2 viruses, which are made of protein and DNA. Each virus acts as a molecular syringe, injecting its genetic material into a bacterium; the empty viral capsule remains attached to the outside of the cell. (B) To determine whether the genetic material of the virus is protein or DNA, the researchers radioactively labeled the DNA in one batch of viruses with ^{32}P and the proteins in a second batch of viruses with ^{35}S. Because DNA lacks sulfur and proteins lack phosphorus, these radioactive isotopes provided a handy way for the researchers to distinguish these two types of molecules. These labeled viruses were then allowed to infect *E. coli*, and the mixture was disrupted by brief pulsing in a Waring blender to separate the infected bacteria from the empty viral heads. When the researchers measured the radioactivity, they found that most of the ^{32}P-labeled DNA had entered the bacterial cells, while most of the ^{35}S-labeled proteins remained in solution with the spent viral particles.

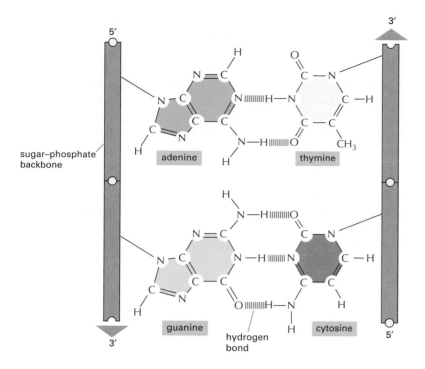

Figure 5–6 **Complementary base pairs are formed in the DNA double helix.** The shapes and chemical structure of the bases allow hydrogen bonds to form efficiently only between A and T and between G and C, where atoms that are able to form hydrogen bonds (see Panel 2–2, pp. 68–69) can be brought close together without perturbing the double helix. Two hydrogen bonds form between A and T, while three form between G and C. The bases can pair in this way only if the two polynucleotide chains that contain them are antiparallel, that is, oriented in opposite polarities.

chain will be easily distinguishable, as one has a hole (the 3′ hydroxyl) and the other a knob (the 5′ phosphate) at its terminus. This polarity in a DNA chain is indicated by referring to one end as the *3′ end* and the other as the *5′ end*. This convention is based on the details of the chemical linkage between the nucleotide subunits.

The two polynucleotide chains in the DNA **double helix** are held together by hydrogen-bonding between the bases on the different strands. All the bases are therefore on the inside of the helix, with the sugar–phosphate backbones on the outside (see Figure 5–2). The bases do not pair at random, however: A always pairs with T, and G with C (Figure 5–6). In each case, a bulkier two-ring base (a purine; discussed in Chapter 2) is paired with a single-ring base (a pyrimidine). This *complementary base-pairing* enables the **base pairs** to be packed in the energetically most favorable arrangement in the interior of the double helix. In this arrangement, each base pair is of similar width, thus holding the sugar–phosphate backbones an equal distance apart along the DNA molecule (Figure 5–7). The two sugar–phosphate backbones twist

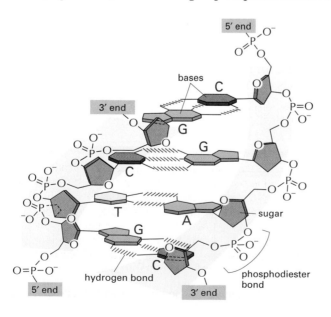

Figure 5–7 **The two strands of the DNA double helix are held together by base pairing.** A short section of the double helix viewed from its side. Four base pairs are shown. The nucleotides are linked together covalently by phosphodiester bonds through the 3′-hydroxyl (–OH) group of one sugar and the 5′-phosphate (PO_4) of the next. Thus, each polynucleotide strand has a chemical polarity; that is, its two ends are chemically different. The 3′ end carries an unlinked –OH group attached to the 3′ position on the sugar ring; the 5′ end carries a free phosphate group attached to the 5′ position on the sugar ring.

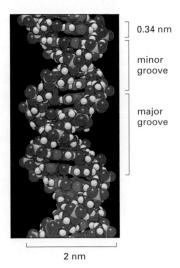

0.34 nm

minor groove

major groove

2 nm

Figure 5–8 The DNA double helix has a major and a minor groove. This space-filling model shows 1.5 turns of the DNA double helix. The coiling of the two strands around each other creates two grooves in the double helix. As indicated in the figure, the wider groove is called the major groove, and the smaller, the minor groove.

Question 5–1

Which of the following statements are correct? Explain your answers.

A. A DNA strand has a polarity because the bases contain hydrophilic amino groups.

B. G-C base pairs are more stable than A-T base pairs.

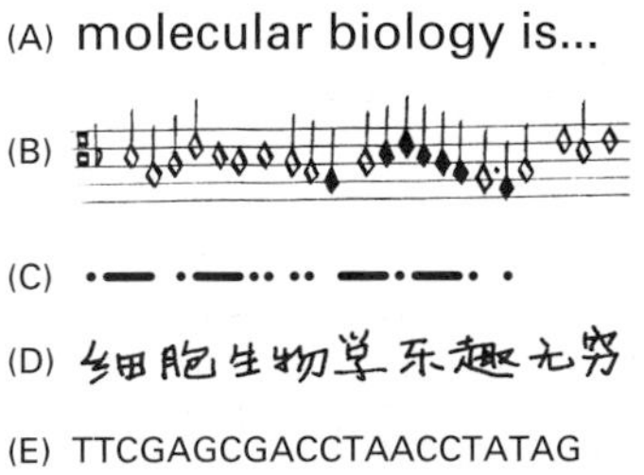

(A) molecular biology is...

(B) [musical score]

(C) •– –••• •–•• ••• • ••

(D) 细胞生物学乐趣无穷

(E) TTCGAGCGACCTAACCTATAG

Figure 5–9 Linear messages come in many forms. The languages are, (A) English, (B) a musical score, (C) Morse code, (D) Chinese, and (E) DNA.

around one another to form a double helix containing 10 bases per helical turn (Figure 5–8). This winding also contributes to the energetically favorable conformation of the DNA double helix.

The members of each base pair can fit together within the double helix only if the two strands of the helix are **antiparallel**, that is, only if the polarity of one strand is oriented opposite to that of the other strand (see Figure 5–2). A consequence of these base-pairing requirements is that each strand of a DNA molecule contains a sequence of nucleotides that is exactly **complementary** to the nucleotide sequence of its partner strand. This is of crucial importance for the copying of DNA, as we shall see in Chapter 6.

The Structure of DNA Provides a Mechanism for Heredity

Genes carry biological information that must be copied and transmitted accurately when a cell divides to form two daughter cells. This situation poses two central biological problems: how can the information for specifying an organism be carried in chemical form, and how is it accurately copied? The discovery of the structure of the DNA double helix was a landmark in twentieth-century biology because it immediately suggested answers to these two questions, and thereby resolved at the molecular level the problem of heredity. In this chapter we outline the answer to the first question, and in the next chapter we address in detail the answer to the second.

DNA encodes information in the order, or sequence, of the nucleotides along each strand. Each base—A, C, T, or G—can be considered as a letter in a four-letter alphabet that is used to spell out biological messages in the chemical structure of the DNA (Figure 5–9). Organisms differ from one another because their respective DNA molecules have different nucleotide sequences and, consequently, carry different biological messages. But how is the nucleotide alphabet used to make up messages, and what do they spell out?

It had already been established some time before the structure of DNA was determined that genes contain the instructions for producing proteins (Figure 5–10). The DNA messages, therefore, must somehow encode proteins. Consideration of the chemical character of proteins makes the problem easier to define. As discussed in Chapter 4, the function of a protein is determined by its three-dimensional structure, and its structure in turn is determined by the sequence of the amino acids of which it is composed. The linear sequence of nucleotides in a gene must therefore somehow spell out the linear sequence of amino acids in a protein.

The exact correspondence between the 4-letter nucleotide alphabet of DNA and the 20-letter amino acid alphabet of proteins—the genetic code—is not obvious from the structure of the DNA molecule,

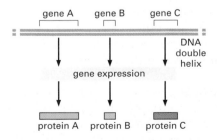

gene A   gene B   gene C

DNA double helix

gene expression

protein A   protein B   protein C

Figure 5–10 Genes contain information to make proteins.

Figure 5–11 Gene sequences can be written down and read like any text. Presented here is the sequence of nucleotides in the human β-globin gene. This gene carries the information for the amino acid sequence of one of the two types of subunits of the hemoglobin molecule, which carries oxygen in the blood. A different gene, the α-globin gene, carries the information for the other type of hemoglobin subunit (a hemoglobin molecule has four subunits, two of each type). Only one of the two strands of the DNA double helix containing the β-globin gene is shown; the other strand has the exact complementary sequence. The sequence should be read left to right in successive lines down the page as if it were normal English text. The DNA sequences highlighted in color show the three regions of the gene that specify the amino sequence for the β-globin protein. We will see in Chapter 7 how the cell connects these three sequences together in order to synthesize a full-length β-globin protein.

```
CCCTGTGGAGCCACACCCTAGGGTTGGCCA
ATCTACTCCCAGGAGCAGGGAGGGCAGGAG
CCAGGGCTGGGCATAAAAGTCAGGGCAGAG
CCATCTATTGCTTACATTTGCTTCTGACAC
AACTGTGTTCACTAGCAACTCAAACAGACA
CCATGGTGCACCTGACTCCTGAGGAGAAGT
CTGCCGTTACTGCCCTGTGGGGCAAGGTGA
ACGTGGATGAAGTTGGTGGTGAGGCCCTGG
GCAGGTTGGTATCAAGGTTACAAGACAGGT
TTAAGGAGACCAATAGAAACTGGGCATGTG
GAGACAGAGAAGACTCTTGGGTTTCTGATA
GGCACTGACTCTCTCTGCCTATTGGTCTAT
TTTCCCACCCTTAGGCTGCTGGTGGTCTAC
CCTTGGACCCAGAGGTTCTTTGAGTCCTTT
GGGGATCTGTCCACTCCTGATGCTGTTATG
GGCAACCCTAAGGTGAAGGCTCATGGCAAG
AAAGTGCTCGGTGCCTTTAGTGATGGCCTG
GCTCACCTGGACAACCTCAAGGGCACCTTT
GCCACACTGAGTGAGCTGCACTGTGACAAG
CTGCACGTGGATCCTGAGAACTTCAGGGTG
AGTCTATGGGACCCTTGATGTTTTCTTTCC
CCTTCTTTTCTATGGTTAAGTTCATGTCAT
AGGAAGGGGAGAAGTAACAGGGTACAGTTT
AGAATGGGAAACAGACGAATGATTGCATCA
GTGTGGAAGTCTCAGGATCGTTTTAGTTTC
TTTTATTTGCTGTTCATAACAATTGTTTTC
TTTTGTTTAATTCTTGCTTTCTTTTTTTTT
CTTCTCCGCAATTTTTACTATTATACTTAA
TGCCTTAACATTGTGTATAACAAAAGGAAA
TATCTCTGAGATACATTAAGTAACTTAAAA
AAAAACTTTACACAGTCTGCCTAGTACATT
ACTATTTGGAATATATGTGTGCTTATTTGC
ATATTCATAATCTCCCTACTTTATTTTCTT
TTATTTTTAATTGATACATAATCATTATAC
ATATTTATGGGTTAAAGTGTAATGTTTTAA
TATGTGTACACATATTGACCAAATCAGGGT
AATTTTGCATTTGTAATTTTAAAAAATGCT
TTCTTCTTTTAATATACTTTTTTGTTTATC
TTATTTCTAATACTTTCCCTAATCTCTTTC
TTTCAGGGCAATAATGATACAATGTATCAT
GCCTCTTTGCACCATTCTAAAGAATAACAG
TGATAATTTCTGGGTTAAGGCAATAGCAAT
ATTTCTGCATATAAATATTTCTGCATATAA
ATTGTAACTGATGTAAGAGGTTTCATATTG
CTAATAGCAGCTACAATCCAGCTACCATTC
TGCTTTTATTTTATGGTTGGGATAAGGCTG
GATTATTCTGAGTCCAAGCTAGGCCCTTTT
GCTAATCATGTTCATACCTCTTATCTTCCT
CCCACAGCTCCTGGGCAACGTGCTGGTCTG
TGTGCTGGCCCATCACTTTGGCAAAGAATT
CACCCCACCAGTGCAGGCTGCCTATCAGAA
AGTGGTGGCTGGTGTGGCTAATGCCCTGGC
CCACAAGTATCACTAAGCTCGCTTTCTTGC
TGTCCAATTTCTATTAAAGGTTCCTTTGTT
CCCTAAGTCCAACTACTAAACTGGGGGATA
TTATGAAGGGCCTTGAGCATCTGGATTCTG
CCTAATAAAAAACATTTATTTTCATTGCAA
TGATGTATTTAAATTATTTCTGAATATTTT
ACTAAAAAGGGAATGTGGGAGGTCAGTGCA
TTTAAAACATAAAGAAATGATGAGCTGTTC
AAACCTTGGGAAAATACACTATATCTTAAA
CTCCATGAAAGAAGGTGAGGCTGCAACCAG
CTAATGCACATTGGCAACAGCCCCTGATGC
CTATGCCTTATTCATCCCTCAGAAAAGGAT
TCTTGTAGAGGCTTGATTTGCAGGTTAAAG
TTTTGCTATGCTGTATTTTACATTACTTAT
TGTTTTAGCTGTCCTCATGAATGTCTTTTC
```

and it took more than a decade after the discovery of the double helix to work it out. In Chapter 7, we describe this code in detail in the course of elaborating the process, known as *gene expression,* through which a cell converts the nucleotide sequence of a gene into the amino acid sequence of a protein.

The complete set of information in an organism's DNA is called its **genome** (the term is also used to refer to the DNA that carries this information). The total amount of this information is staggering: written out in the four-letter nucleotide alphabet, the nucleotide sequence of a small gene from humans occupies a quarter of a page of text (Figure 5–11), while the complete sequence of the human genome would fill more than 1000 books the size of this one. Herein lies a problem that affects the architecture of all eucaryotic chromosomes: how can all this information be packed neatly into every cell nucleus? In the remainder of this chapter we discuss the answer to this fundamental question.

The Structure of Eucaryotic Chromosomes

Large amounts of DNA are required to encode all the information needed to make just a single-celled bacterium, and far more DNA is needed to encode the instructions for the development of multicellular organisms like ourselves. Each human cell contains about 2 m of DNA; yet the cell nucleus is only 5 to 8 μm in diameter. Tucking all this material into such a small space is the equivalent of trying to fold 40 km (24 miles) of extremely fine thread into a tennis ball.

In eucaryotic cells, enormously long double-stranded DNA molecules are packaged into *chromosomes* that not only fit readily inside the nucleus but can be easily apportioned between the two daughter cells at each cell division. As we see in this section, the complex task of packaging DNA is accomplished by specialized proteins that bind to and fold the DNA, generating a series of coils and loops that provide increasingly higher levels of organization and that prevent the DNA from becoming an unmanageable tangle. Amazingly, the DNA is compacted in such an orderly fashion that it can become accessible to all of the enzymes and other proteins that replicate it, repair it, and use its genes to produce proteins.

Bacteria typically carry their genes on a single, circular DNA molecule. This DNA is also associated with proteins that condense DNA, but they differ from the proteins that package eucaryotic DNA. Although it is often called a bacterial "chromosome," this procaryotic DNA does not have the same structure as eucaryotic chromosomes, and less is known about how it is packaged. Our discussion of chromosome structure in this chapter will therefore focus entirely on eucaryotic chromosomes.

Eucaryotic DNA Is Packaged into Chromosomes

In eucaryotes, such as ourselves, the DNA in the nucleus is distributed among a set of different **chromosomes**. The human genome, for example, contains approximately 3.2×10^9 nucleotides distributed over 24 chromosomes. Each chromosome consists of a single, enormously long, linear DNA molecule associated with proteins that fold and pack the fine thread of DNA into a more compact structure. The complex of DNA and protein is called *chromatin* (from the Greek *chroma,* "color," because of its staining properties). In addition to the proteins involved in packaging the DNA, chromosomes are also associated with other proteins involved in gene expression, DNA replication, and DNA repair.

With the exception of the germ cells (sperm and eggs) and highly specialized cells that lack DNA entirely (such as the red blood cell), human cells each contain two copies of each chromosome, one inherited from the mother and one from the father; the maternal and paternal chromosomes of a pair are called *homologous chromosomes (homologs).* The only nonhomologous chromosome pairs are the sex chromosomes in males, where a *Y chromosome* is inherited from the father and an *X chromosome* from the mother.

Human chromosomes can be distinguished by *DNA hybridization* (described in detail in Chapter 10); the technique uses a set of DNA molecules coupled to fluorescent molecules to "paint" each chromosome a different color (Figure 5–12). But the more traditional way of distinguishing one chromosome from another is to stain the chromosomes with dyes that bind to certain types of DNA sequences. These dyes mainly distinguish between DNA that is rich in A-T nucleotide pairs and DNA that is G-C rich, and they produce a striking and reliable pattern of bands along each chromosome (Figure 5–13). The pattern of bands on each type of chromosome is unique, allowing each chromosome to be identified and numbered.

A display of the full set of 46 human chromosomes is called the human **karyotype**. If parts of a chromosome are lost, or switched between chromosomes, these changes can be detected by changes in the banding patterns. Cytogeneticists use alterations in banding patterns to detect chromosomal abnormalities that are associated with some inherited defects (Figure 5–14) and with certain types of cancer.

Figure 5–12 Each human chromosome can be "painted" a different color to allow its unambiguous identification under the light microscope. The chromosomes, from a male, were isolated from a cell undergoing nuclear division (mitosis) and are therefore in a highly compact state. Chromosome painting is carried out by exposing the chromosomes to a collection of human DNA molecules that have been coupled to a combination of fluorescent dyes. For example, DNA molecules derived from Chromosome 1 are labeled with one specific dye combination, those from Chromosome 2 with another, and so on. Because the labeled DNA can form base pairs, or hybridize, only to its chromosome of origin (discussed in Chapter 10), each chromosome is differently labeled. For such experiments, the chromosomes are subjected to treatments that separate the double-helical DNA into individual strands, to enable base-pairing with the single-stranded labeled DNA while keeping the chromosome structure relatively intact. (A) The chromosomes as visualized as they originally spilled from the lysed cell. (B) The same chromosomes artificially lined up in order. In this karyotype, the homologous chromosomes are numbered and arranged in pairs; the presence of a Y chromosome indicates that the DNA was isolated from a male. (From E. Schröck et al., *Science* 273:494–497, 1996. © AAAS.)

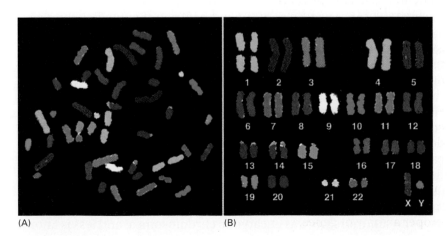

(A)　　　　(B)

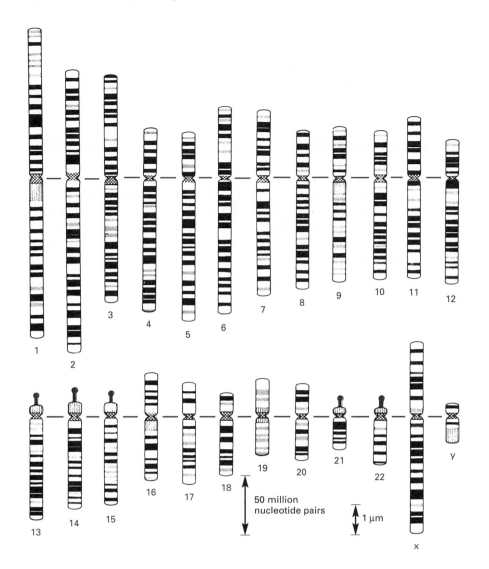

Figure 5–13 Unique banding patterns allow the identification of each human chromosome. Chromosomes 1 through 22 are numbered in approximate order of size. A typical human somatic (that is, non-germ) cell contains two of each of these chromosomes plus two sex chromosomes—two X chromosomes in a female, one X and one Y chromosome in a male. The chromosomes used to make these maps were stained at an early stage in mitosis, when the DNA is compacted. The *horizontal line* represents the position of the centromere, which appears as a constriction on mitotic chromosomes; the *knobs* on Chromosomes 13, 14, 15, 21, and 22 indicate the positions of genes that code for the large ribosomal RNAs (discussed in Chapter 7). These patterns are obtained by staining chromosomes with Giemsa stain, which produces dark bands in regions rich in A-T nucleotide pairs. (Adapted from U. Franke, *Cytogenet. Cell Genet.* 31:24–32, 1981.)

Chromosomes Contain Long Strings of Genes

The most important function of chromosomes is to carry genes—the functional units of heredity (Figure 5–15). A **gene** is usually defined as a segment of DNA that contains the instructions for making a particular protein (or, in some cases, a set of closely related proteins). Although this definition fits the majority of genes, some genes direct the production of an RNA molecule, instead of a protein, as their final product. Like proteins, RNA molecules perform a diverse set of structural and catalytic functions in the cell, as we will see in later chapters.

As might be expected, a correlation exists between the complexity of an organism and the number of genes in its genome. For example, the total number of genes range from less than 500 for a simple bacterium to about 30,000 for humans. Bacteria and some single-celled eucaryotes have especially compact genomes: the DNA molecules that make up their chromosomes are little more than strings of closely packed genes. However, chromosomes from many eucaryotes (including humans) contain, in addition to genes, a large excess of interspersed DNA that does not seem to carry critical information. This DNA is sometimes called junk DNA, as its usefulness to the cell has not yet been clearly demonstrated. Although the particular nucleotide sequence of this DNA may not be important, the DNA itself—acting as spacer material—may be crucial for the long-term evolution of the species and for the proper activity of genes.

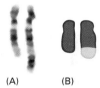

(A)   (B)

Figure 5–14 Abnormal chromosomes are associated with some inherited genetic defects. (A) A pair of chromosomes from a patient with inherited ataxia, a disease characterized by progressive deterioration of motor skills. The patient has one normal Chromosome 12 *(left)* and one aberrant Chromosome 12, as seen by its greater length. The additional material contained on the aberrant Chromosome 12 was deduced, from its pattern of bands, as a piece of Chromosome 4 that had become inappropriately attached to Chromosome 12. (B) The chromosomes are shown here with the segment corresponding to Chromosome 4 DNA "painted" *blue* and the parts corresponding to Chromosome 12 DNA painted *purple*. (From E. Schröck et al., *Science* 273:494–497, 1996. © AAAS.)

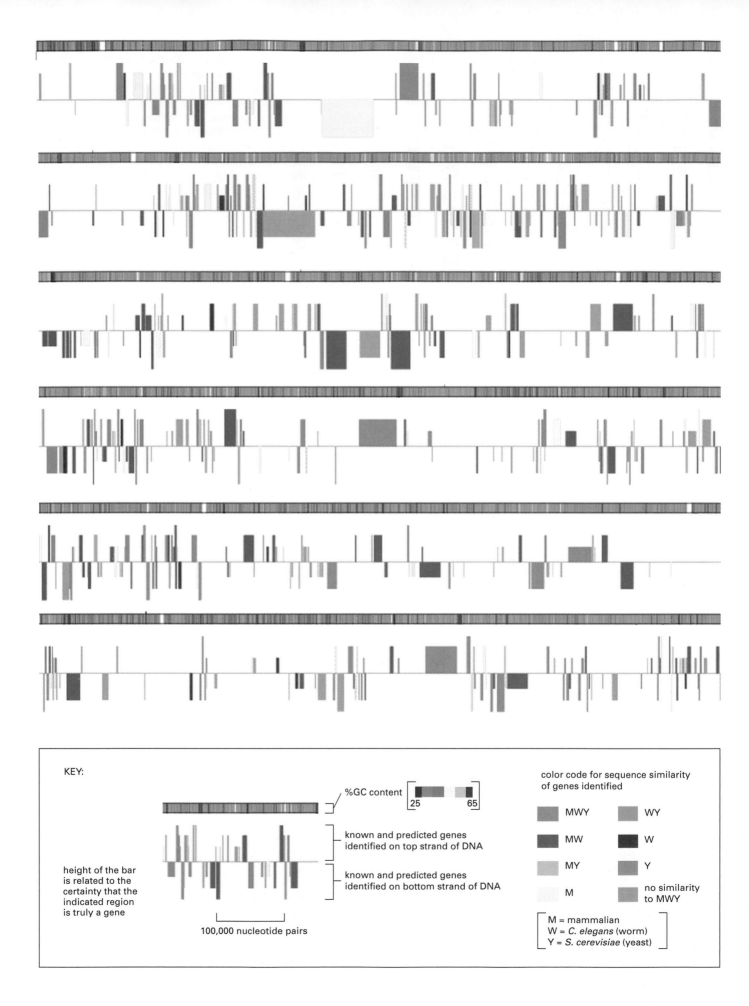

KEY:

%GC content

25   65

height of the bar is related to the certainty that the indicated region is truly a gene

known and predicted genes identified on top strand of DNA

known and predicted genes identified on bottom strand of DNA

100,000 nucleotide pairs

color code for sequence similarity of genes identified

MWY   WY

MW   W

MY   Y

M   no similarity to MWY

M = mammalian
W = *C. elegans* (worm)
Y = *S. cerevisiae* (yeast)

Figure 5–15 (opposite page) The genes carried on a portion of chromosome 2 from the fruit fly *Drosophila melanogaster* **are depicted.** This figure represents approximately 3% of the total *Drosophila* genome, arranged as six contiguous segments. As summarized in the key, the symbolic representations are *rainbow-colored horizontal bar,* G–C base-pair content; vertical colored bars, genes (both known and predicted) coded on one strand of DNA (boxes *above* the midline) and genes coded on the other strand (bars *below* the midline). As indicated in the key, the height of each gene bar is related to the certainty with which it represents a true gene: the higher the bar, the higher the confidence. The color of each gene bar (see *color code* in the key) indicates whether a closely related gene is found in other organisms. For example, MWY means the gene has close relatives in mammals, in the nematode worm *Caenorhabditis elegans,* and in the yeast *Saccharomyces cerevisiae.* MW indicates the gene has close relatives in mammals and the worm but not in yeast. (From Mark D. Adams et al., *Science* 287:2185–2195, 2000. © AAAS.)

In general, the more complex an organism is, the larger its genome. But this relationship does not always hold true. The human genome, for example, is 200 times larger than that of the yeast *S. cerevisiae*, but 30 times smaller than that of some plants and 200 times smaller than that of a species of amoeba. Furthermore, how the DNA is apportioned over chromosomes also differs from one species to another. Humans have 46 chromosomes, but a species of small deer has only 6, while certain carp species have more than 100. Even closely related species with similar genome sizes can have very different numbers and sizes of chromosomes (Figure 5–16). Thus, although gene number is roughly correlated with species complexity, there is no simple relationship between gene number, chromosome number, and total genome size. The genomes and chromosomes of modern species have each been shaped by a unique history of seemingly random genetic events, acted on by selection pressures.

Chromosomes Exist in Different States Throughout the Life of a Cell

To form a functional chromosome, a DNA molecule must do more than simply carry genes: it must be able to replicate, and the replicated copies must be separated and partitioned reliably into daughter cells at each cell division. These processes occur through an ordered series of stages, known collectively as the **cell cycle**. This cycle of cell growth is briefly summarized in Figure 5–17 and will be discussed in detail in Chapter 19. Only two of these stages need concern us in this chapter: *interphase*, when chromosomes are duplicated; and *mitosis*, when they are distributed to the two daughter nuclei.

During interphase, the chromosomes are extended as long, thin, tangled threads of DNA in the nucleus and cannot be easily distinguished in

Question 5–2

In a DNA double helix, adjacent nucleotide pairs are 0.34 nm apart. Use Figure 5–13 to estimate the length of the DNA in human Chromosome 1 if it were unraveled and stretched out. If the actual length of Chromosome 1 at this stage of mitosis is approximately 10 μm, what is the degree of compaction of the DNA in this state?

Figure 5–16 Closely related species can have very different chromosome numbers. In the evolution of the Indian muntjac deer, chromosomes that were initially separate fused without having a major effect on the animal. The two species shown have roughly the same number of genes. (Adapted from M.W. Strickberger, *Evolution*, 3rd edn. Sudbury, MA: Jones & Bartlett Publishers, 2000.)

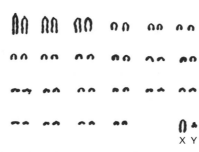

Chinese muntjac

Indian muntjac

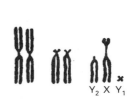

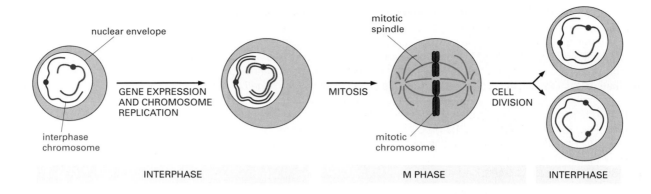

INTERPHASE M PHASE INTERPHASE

Figure 5–17 Replication and segregation of chromosomes occur through an ordered series of stages, called the cell cycle. During interphase, the cell is actively expressing its genes. Still during interphase and before cell division, the DNA is replicated and the chromosomes are duplicated. Once DNA replication is complete, the cell can enter *M phase,* when mitosis occurs. Mitosis is the division of the nucleus. During this stage, the chromosomes condense, gene expression largely ceases, the nuclear envelope breaks down, and the mitotic spindle forms from microtubules and other proteins. The condensed chromosomes are captured by the mitotic spindle, and one complete set of chromosomes is pulled to each end of the cell. A nuclear envelope forms around each chromosome set, and in the final step of M phase, the cell divides to produce two daughter cells.

the light microscope. We refer to chromosomes in this extended state as *interphase chromosomes.* Specialized DNA sequences found in all eucaryotes ensure that the interphase chromosomes replicate efficiently (Figure 5–18). One type of nucleotide sequence acts as a **replication origin**, the location at which duplication of the DNA begins, as we discuss in Chapter 6. Eucaryotic chromosomes contain many replication origins to ensure that the entire chromosome can be replicated rapidly. Another DNA sequence forms the **telomeres** found at each of the two ends of a chromosome. Telomeres contain repeated nucleotide sequences that enable the ends of chromosomes to be replicated. They also protect the end of the chromosome from being mistaken by the cell as a broken DNA molecule in need of repair. We discuss telomere replication and function further in the following chapters.

As the cell cycle proceeds, the DNA coils up, adopting a more and more compact structure, until the highly condensed *mitotic chromosomes* have been formed. This is the form in which chromosomes are most easily visualized; in fact, all of the images of chromosomes shown so far in the chapter are of mitotic chromosomes. In this highly condensed state, duplicated chromosomes can be easily separated when the cell divides. It is the presence of the third specialized DNA sequence, the **centromere**, that allows one copy of each duplicated chromosome to be apportioned to each daughter cell (see Figure 5–17).

Figure 5–18 Three DNA sequence elements are needed to produce a eucaryotic chromosome that can be replicated and then segregated at mitosis. Each chromosome has multiple origins of replication, one centromere, and two telomeres. Shown schematically is the sequence of events that a typical chromosome follows during the cell cycle. The DNA replicates in interphase beginning at the origins of replication and proceeding bidirectionally from the origins across the chromosome. In M phase, the centromere attaches the duplicated chromosomes to the mitotic spindle so that one copy is distributed to each daughter cell during mitosis. The centromere also helps to hold the duplicated chromosomes together until they are ready to be moved apart. The telomeres form special caps at each chromosome end.

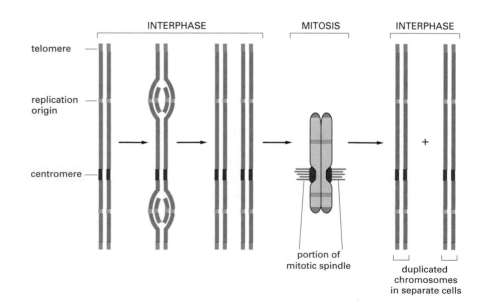

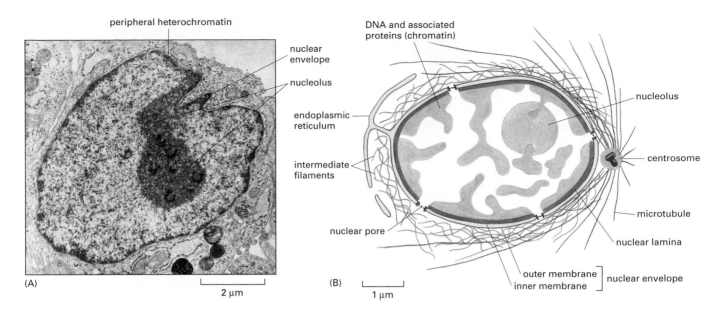

peripheral heterochromatin

nuclear envelope

nucleolus

endoplasmic reticulum

intermediate filaments

nuclear pore

DNA and associated proteins (chromatin)

nucleolus

centrosome

microtubule

nuclear lamina

outer membrane
inner membrane } nuclear envelope

(A) 2 μm

(B) 1 μm

Interphase Chromosomes Are Organized Within the Nucleus

Despite the fact that the chromosomes of cells in interphase are very much longer and finer than mitotic chromosomes, they are thought to be well organized within the nucleus. The nucleus is delimited by a *nuclear envelope* formed by two concentric membranes. The nuclear envelope is supported by two networks of protein filaments (discussed in Chapter 17): one, called the *nuclear lamina,* forms a thin layer underlying and supporting the inner nuclear membrane; while the other, less regularly organized, surrounds the outer nuclear membrane (Figure 5–19). The two membranes are punctured at intervals by nuclear pores, which actively transport selected molecules to and from the cytosol.

The interior of the nucleus is not a random jumble of its many DNA, RNA, and protein components. Each interphase chromosome probably occupies a particular region of the nucleus so that different chromosomes do not become entangled with each other. This organization is thought to be achieved at least in part by attachment of parts of the chromosomes to sites on the nuclear envelope or the nuclear lamina.

The most obvious example of chromosome organization in the interphase nucleus is the **nucleolus**, which is the most prominent structure evident in the interphase nucleus under the light microscope. This is a region where the parts of the different chromosomes carrying genes for ribosomal RNA cluster together (see Figures 5–13 and 5–19). Here, ribosomal RNAs are synthesized and combined with proteins to form ribosomes, the cell's protein-synthesizing machines (discussed in Chapter 7).

The DNA in Chromosomes Is Highly Condensed

As we have seen, all eucaryotic cells, whether in interphase or mitosis, package their DNA tightly into chromosomes. Human Chromosome 22, for example, contains about 48 million nucleotide pairs; stretched out end-to-end, its DNA would extend about 1.5 cm. Yet, during mitosis, Chromosome 22 measures only about 2 μm in length—that is nearly 10,000 times more compact than its extended form. This remarkable feat of compression is performed by proteins that coil and fold the DNA into higher and higher levels of organization. The DNA of interphase chromosomes, although less condensed than that of mitotic

Figure 5–19 Interphase chromosomes are organized within the nucleus. (A) Electron micrograph of a thin section through the nucleus of a human fibroblast. The nucleus is surrounded by a nuclear envelope composed of a double membrane, perforated by nuclear pores. Inside the nucleus, the chromatin appears as a diffuse speckled mass, with especially dense chromosomal regions, called heterochromatin (*dark staining*), located mainly around the periphery, immediately under the nuclear envelope. The large dark region is the nucleolus. (B) Diagrammatic view of a cross section of a typical cell nucleus. The nuclear envelope consists of two membranes, the outer one being continuous with the endoplasmic reticulum. Two networks of cytoskeletal filaments (*green*) provide mechanical support for the nuclear envelope; the one inside the nucleus forms a sheetlike *nuclear lamina* lining the internal face of the inner nuclear membrane. The nucleolus (*gray*) is the site of ribosomal RNA synthesis. (A, courtesy of E.G. Jordan and J. McGovern.)

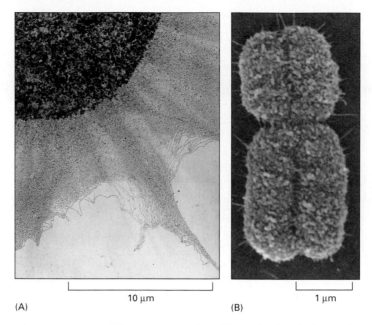

Figure 5–20 DNA in interphase chromosomes is less compact than that in mitotic chromosomes. (A) An electron micrograph showing an enormous tangle of chromosomal DNA spilling out of a lysed interphase nucleus. (B) A scanning electron micrograph of a mitotic chromosome, which is a duplicated chromosome in which the two new chromosomes are still linked together (see Figure 5–18). The constricted region indicates the position of the centromere. Note the difference in scales. (A, courtesy of Victoria Foe; B, courtesy of Terry D. Allen.)

(A) 10 µm   (B) 1 µm

chromosomes (Figure 5–20), is still packed tightly, with a compaction ratio of about 1000-fold.

In the next sections, we introduce the specialized proteins that make this compression possible. Bear in mind, though, that chromosome structure is dynamic. Not only do chromosomes condense and relax in concert with the cell cycle, but different regions of the interphase chromosome must unpack to allow cells to access specific DNA sequences for replication, repair, or gene expression. Chromosome packaging must therefore be flexible enough to allow rapid, localized, on-demand access to DNA.

Nucleosomes Are the Basic Units of Chromatin Structure

The proteins that bind to the DNA to form eucaryotic chromosomes are traditionally divided into two general classes: the **histones**, and the nonhistone chromosomal proteins. Histones are present in enormous quantities (about 60 million molecules of several different types in each cell), and their total mass in chromosomes is about equal to that of the DNA itself. The complex of both classes of protein with nuclear DNA is called **chromatin**.

Histones are responsible for the first and most fundamental level of chromatin packing, the **nucleosome**, which was discovered in 1974. When interphase nuclei are broken open very gently and their contents examined under the electron microscope, most of the chromatin is in the form of a fiber with a diameter of about 30 nm (Figure 5–21A). If this

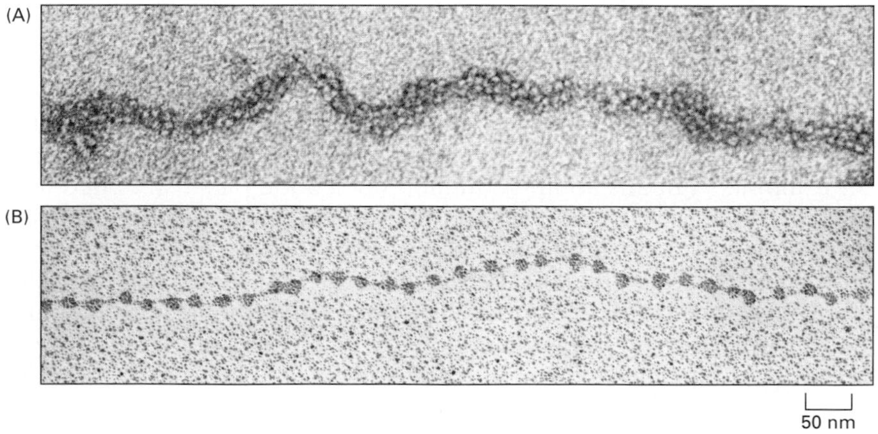

(A)

(B)

Figure 5–21 Nucleosomes can be seen in the electron microscope. (A) Chromatin isolated directly from an interphase nucleus appears in the electron microscope as a thread 30 nm thick. (B) This electron micrograph shows a length of chromatin that has been experimentally unpacked, or decondensed, after isolation to show the nucleosomes. (A, courtesy of Barbara Hamkalo; B, courtesy of Victoria Foe.)

50 nm

Figure 5–22 Nucleosomes contain DNA wrapped around a protein core of eight histone molecules. As indicated, the nucleosome core particle is released from chromatin by digestion of the linker DNA with a nuclease, an enzyme that breaks down DNA. (The nuclease can degrade the exposed DNA but cannot attack the DNA wound tightly around the nucleosome core.) After dissociation of the isolated nucleosome into its protein core and DNA, the length of the DNA that was wound around the core can be determined. Its length of 146 nucleotide pairs is sufficient to wrap almost twice around the histone core.

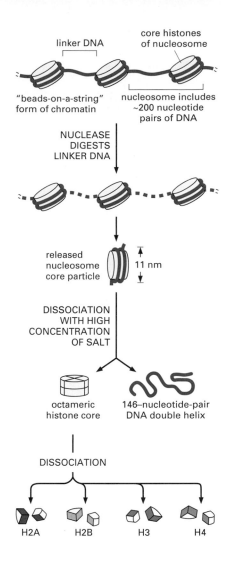

chromatin is subjected to treatments that cause it to unfold partially, it can then be seen under the electron microscope as a series of "beads on a string" (Figure 5–21B). The string is DNA, and each bead is a *nucleosome core particle* that consists of DNA wound around a core of proteins formed from histones.

The structure of nucleosomes was determined after first isolating them from the unfolded chromatin by digestion with particular enzymes (called nucleases) that break down DNA by cutting between the nucleotides. After digestion for a short period, the exposed DNA between the nucleosome core particles, the *linker DNA,* is degraded. An individual nucleosome core particle consists of a complex of eight histone proteins—two molecules each of histones H2A, H2B, H3, and H4—and a double-stranded DNA of about 146 nucleotide pairs that winds around this *histone octamer* (Figure 5–22). The high-resolution structure of the nucleosome core particle was solved in 1997, revealing in atomic detail the disc-shaped histone complex around which the DNA is tightly wrapped, making 1.65 turns in a left-handed coil (Figure 5–23).

Each nucleosome core particle is separated from the next by a region of linker DNA, which can vary in length from a few nucleotide pairs up to about 80. (The term "nucleosome" technically refers to a nucleosome core particle plus one of its adjacent DNA linkers, but is often used synonymously with "nucleosome core particle.") Formation of nucleosomes converts a DNA molecule into a chromatin thread approximately one-third of its initial length, and provides the first level of DNA packing.

All four of the histones that make up the nucleosome core are relatively small proteins with a high proportion of positively charged amino acids (lysine and arginine). The positive charges help the histones bind tightly to the negatively charged sugar–phosphate backbone of DNA. These numerous interactions explain in part why DNA of virtually any

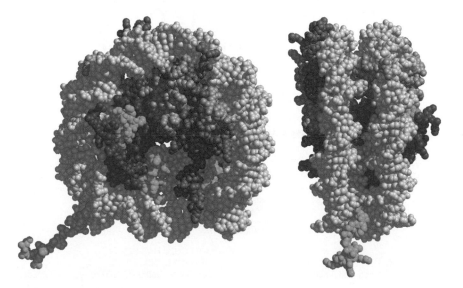

Figure 5–23 The structure of the nucleosome core particle, as determined by X-ray diffraction analysis, reveals how DNA is tightly wrapped around a disc-shaped histone core. Each histone is colored according to the scheme shown in Figure 5–22; the DNA helix is *gray*. A portion of the H3 histone tail can be seen extending from the nucleosome, but the tails of the other histone proteins are not shown. (Reprinted by permission from K. Luger et al., *Nature* 389:251–260, 1997. © Macmillan Magazines Ltd.)

sequence can bind to a histone core. Each of the core histones also has a long N-terminal amino acid "tail," which extends out from the DNA histone core. These histone tails are subject to several types of covalent modification that control many aspects of chromatin structure, as we will see shortly.

The histones that form the nucleosome core are among the most highly conserved of all known eucaryotic proteins: there are only two differences between the amino acid sequences of histone H4 from peas and cows, for example. Recently, histones have been found in archaea— procaryotes that form a phylogenetic kingdom distinct from eucaryotes and eubacteria (discussed in Chapter 1). This extreme evolutionary conservation reflects the vital structural role of histones in forming chromatin.

Chromosomes Have Several Levels of DNA Packing

Although long strings of nucleosomes form on most chromosomal DNA, chromatin in the living cell rarely adopts the extended beads-on-a-string form seen in Figure 5–21B. Instead, the nucleosomes are further packed upon one another to generate a more compact structure, the *30-nm fiber* (see Figure 5–21A). Packing of nucleosomes into the 30-nm fiber depends on a fifth histone called histone H1, which is thought to pull the nucleosomes together into a regular repeating array. The structure that results is illustrated, as part of a larger schematic of the various levels of chromosome packing, in Figure 5–24.

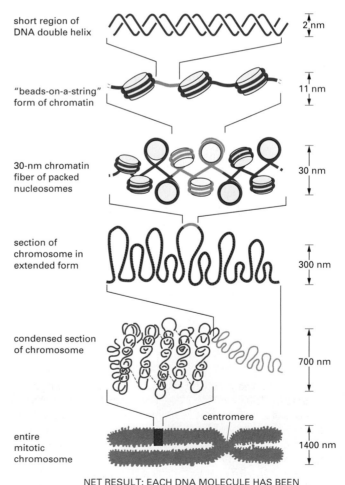

short region of
DNA double helix — 2 nm

"beads-on-a-string"
form of chromatin — 11 nm

30-nm chromatin
fiber of packed
nucleosomes — 30 nm

section of
chromosome in
extended form — 300 nm

condensed section
of chromosome — 700 nm

centromere

entire
mitotic
chromosome — 1400 nm

NET RESULT: EACH DNA MOLECULE HAS BEEN
PACKAGED INTO A MITOTIC CHROMOSOME THAT
IS 10,000-FOLD SHORTER THAN ITS EXTENDED LENGTH

Figure 5–24 Chromatin packing occurs on several levels. This schematic drawing shows some of the orders of chromatin packing thought to give rise to the highly condensed mitotic chromosome.

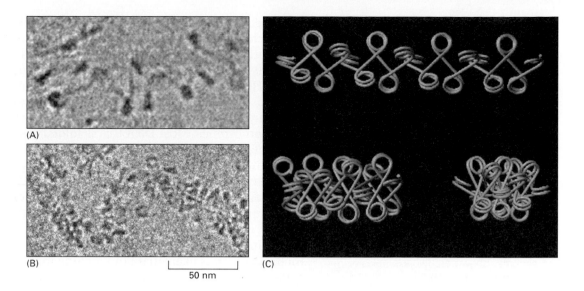

(A)

(B)

50 nm

(C)

Several models have been proposed to explain how nucleosomes are packed in the 30-nm chromatin fiber; the one most consistent with the available data is a series of structural variations known collectively as the *zigzag model* (Figure 5–25). The 30-nm structure found in chromosomes is probably a fluid mosaic of the different zigzag variations.

We know that the 30-nm chromatin fiber can be compacted still further. We saw earlier in this chapter that during mitosis chromatin becomes so highly condensed that the chromosomes can be seen under the light microscope. How is the 30-nm fiber folded to produce mitotic chromosomes? The answer to this question is not yet known in detail, but it is thought that the 30-nm fiber is further organized into loops emanating from a central axis (Figures 5–24 and 5–26). Finally, this string of loops is thought to undergo at least one more level of packing to form the mitotic chromosome (see Figure 5–24).

Figure 5–25 Chromatin fibers may be packed according to a zigzag model. (A and B) Electron microscopic evidence for the top and bottom-left model structures shown in (C). The structure of the 30-nm chromatin fiber may be a combination of these zigzag variations. An interconversion between these three variations may occur through an accordionlike expansion and contraction of the fiber. Note that the histone cores are omitted from the diagrams in (C). (From J. Bednar et al., *Proc. Natl. Acad. Sci. U.S.A.* 95:14173–14178, 1998. © National Academy of Sciences.)

Interphase Chromosomes Contain Both Condensed and More Extended Forms of Chromatin

As the daughter cells complete their separation following mitosis, the mitotic chromosomes unfold into a more extended form—the interphase chromosomes (see Figure 5–18). However, the chromatin in an interphase chromosome is not in the same packing state throughout the chromosome. In general, regions of the chromosome that contain genes that are being expressed are more extended, while those that contain quiescent genes are more compact. Thus, the detailed structure of an interphase chromosome can differ from one cell type to the next, depending on which genes are being expressed.

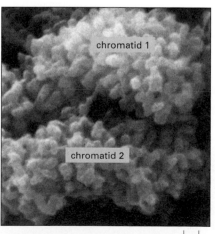

chromatid 1

chromatid 2

0.1 μm

Figure 5–26 The mitotic chromosome is formed from tightly packed chromatin. This scanning electron micrograph shows a region near one end of a typical mitotic chromosome. Each knoblike projection is believed to represent the tip of a separate loop of chromatin. The chromosome in this picture has duplicated, but the two new chromosomes (also called chromatids) are still held together (see Figure 5–20B). The ends of the two chromosomes can be easily distinguished in this micrograph. (From M.P. Marsden and U.K. Laemmli, *Cell* 17:849–858, 1989. © Elsevier.)

The most highly condensed form of interphase chromatin is called **heterochromatin** (from the Greek *heteros,* meaning "different," plus chromatin). It was first observed under the light microscope in the 1930s as discrete, strongly staining regions within the mass of chromatin. Heterochromatin typically makes up about 10% of an interphase chromosome, and in mammalian chromosomes, it is typically concentrated around the centromere region and in the telomeres at the ends of the chromosomes. Most DNA that is folded into heterochromatin does not contain genes. However, genes that do become packaged into heterochromatin usually become resistant to being expressed because heterochromatin is unusually compact (Figure 5–27). The rest of the interphase chromatin, which is in a variety of more extended states, is called *euchromatin* (from the Greek *eu,* meaning "true" or "normal," plus chromatin). It is not yet understood which of the levels of chromosome packaging shown in Figure 5–24 best describe euchromatin and heterochromatin; it is likely that both include mixtures of several packaging states.

Although most eucaryotic chromosomes contain regions of both euchromatin and heterochromatin, some important exceptions exist. Perhaps the most striking example is found in the interphase X chromosomes of female mammals. Female cells contain two X chromosomes, while male cells contain one X and one Y. Because a double dose of X-chromosome products would be lethal, female mammals have evolved a mechanism to permanently inactivate one of the two X chromosomes in each cell: at random, one or the other of the two X

Figure 5–27 Expression of a gene can be altered by moving it to another location in the genome. Shown are two examples of *position effects,* in which the activity of a gene depends on its position along a chromosome. (A) The yeast *ADE2* gene at its normal chromosomal location is expressed in all cells. When moved near the end of a yeast chromosome, which has been shown to be folded into a particularly compact form of chromatin, the gene is no longer expressed in most cells of the population. *ADE2* encodes one of the enzymes of adenine biosynthesis, and the absence of the *ADE2* gene product leads to the accumulation of a red pigment. Therefore, a colony of cells that expresses *ADE2* is *white,* and one composed of cells where the *ADE2* gene is not expressed is *red.* The white sectors around the red colony represent cells where the *ADE2* gene has spontaneously become active. This results from a heritable change in the packing state of chromatin near the *ADE2* gene in these cells. (B) Position effects can also be observed for the *white* gene in the fruit fly *Drosophila.* The *white* gene controls eye pigment production and is named after the mutation that first identified it. Wild-type flies with a normal *white* gene (*white⁺*) have normal pigment production, which gives them red eyes, but if the *white* gene is mutated and inactivated, the mutant flies (*white⁻*) make no pigment and have white eyes. In flies in which a normal *white⁺* gene has been moved near a region of heterochromatin, the eyes are mottled, with both red and white patches. The white patches represent cells where the *white⁺* gene is silenced by the effects of the heterochromatin and red patches represent cells that express the *white⁺* gene. The mottling occurs because the silencing of the *white⁺* gene by the heterochromatin is not complete. As for the yeast, the presence of large patches of red and white cells indicates that the state of the gene (either active or silenced) is inherited.

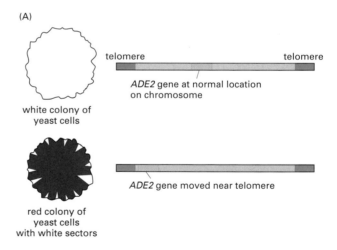

(A)

white colony of
yeast cells

red colony of
yeast cells
with white sectors

telomere .... telomere
ADE2 gene at normal location
on chromosome

ADE2 gene moved near telomere

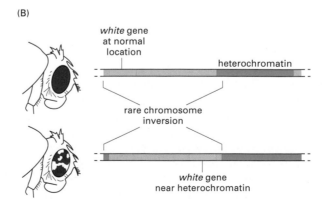

(B)

white gene
at normal
location

heterochromatin

rare chromosome
inversion

white gene
near heterochromatin

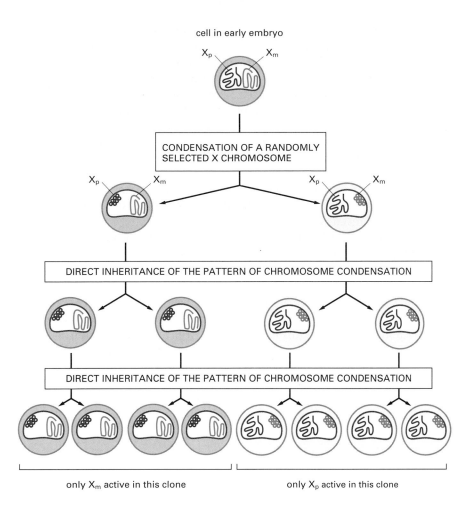

cell in early embryo

X_p X_m

CONDENSATION OF A RANDOMLY
SELECTED X CHROMOSOME

X_p X_m X_p X_m

DIRECT INHERITANCE OF THE PATTERN OF CHROMOSOME CONDENSATION

DIRECT INHERITANCE OF THE PATTERN OF CHROMOSOME CONDENSATION

only X_m active in this clone   only X_p active in this clone

Figure 5–28 An individual X chromosome can be completely inactivated by heterochromatin formation. Cells in the early female mammalian embryo contain two X chromosomes, one from the mother (X_m) and the other from the father (X_p). At an early stage of development, one of these two chromosomes in each cell becomes condensed into heterochromatin, apparently at random. At each cell division after this stage, the same chromosome becomes condensed in all the descendants of that original cell. In mice, X-chromosome inactivation occurs between the third and sixth days of development. In humans, too, X-inactivation occurs very early in development, before cells have been allocated to any particular developmental pathway. Thus the female ends up as a mosaic of cells bearing maternal or paternal inactivated X chromosomes. In most tissues and organs about half the cells will be of one type and the other half will be of the other.

chromosomes in each cell becomes highly condensed into heterochromatin early in embryonic development. Thereafter, in all of the many progeny of the cell, the condensed and inactive state of that X chromosome is inherited (Figure 5–28).

Changes in Nucleosome Structure Allow Access to DNA

So far we have discussed how DNA is packed carefully and tightly into chromatin; we now turn to the question of how this packaging can be dynamic, allowing rapid access to the underlying DNA. We have seen that the DNA in cells carries enormous amounts of coded information, and it is especially important that cells have on-demand access to it. As we shall see, direct access to DNA is typically localized, exposing only those regions of the genome that the cell needs at any particular time.

Eucaryotic cells have several ways to rapidly adjust the local structure of their chromatin. One approach takes advantage of **chromatin remodeling complexes**, protein machines that use the energy of ATP hydrolysis to change the structure of nucleosomes (Figure 5–29). These complexes can make the underlying DNA more accessible to other proteins in the cell, especially those involved in DNA replication, repair, and gene expression. During mitosis, at least some of the chromatin remodeling complexes are inactivated, which may help mitotic chromosomes maintain their tightly packed structure.

Another approach to the changing chromatin structure relies on the reversible modification of histone tails. The N-terminal tails of each of the four core histone proteins perform crucial functions in regulating

Question 5–3

Mutations in a particular gene on the X chromosome result in color blindness. All men carrying a mutant gene are color-blind. Most women carrying a mutant gene have proper color vision but see color images with reduced resolution, as though functional cone cells (the cells that contain the color photoreceptors) are spaced farther apart than normal in the retina. Can you give a plausible explanation for this observation? If a woman is color-blind, what could you say about her father? About her mother? Explain your answers.

chromatin structure. Each tail is subject to several types of covalent modification, which are added and removed after the nucleosome has been assembled by enzymes that reside in the nucleus. Although modifications of the histone tails have little direct effect on the stability of an individual nucleosome, some seem to directly affect the stability of the 30-nm chromatin fiber and of some of the higher-order structures discussed earlier.

However, the most profound effect of modified histone tails seems to be their ability to bind to and thereby attract specific proteins to stretches of chromatin. Different patterns of histone tail modifications attract different proteins; some of these proteins cause further condensation of the chromatin, others facilitate access to the DNA. Together, different combinations of tail modifications and different sets of histone-binding proteins can signal different things to the cell: one pattern might indicate that a particular stretch of chromatin has been newly replicated, another that gene expression should take place (Figure 5–30).

As is true for the chromatin remodeling complexes, the enzymes that modify histone tails are tightly regulated. They are brought to a particular region of chromatin by other cues, particularly by their interactions with proteins that bind to specific sequences in DNA, a subject we discuss in Chapter 8. It is likely that these histone-modifying enzymes work in concert with the chromatin remodeling complexes to condense and relax stretches of chromatin, allowing local chromatin structure to rapidly change according to the needs of the cell.

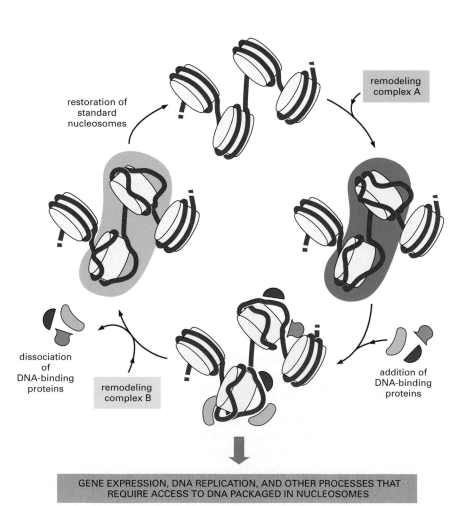

Figure 5–29 Chromatin remodeling complexes alter nucleosome structure. According to this model, different chromatin remodeling complexes disrupt and re-form nucleosomes, although, in principle, the same complex might catalyze both reactions. The DNA-binding proteins could be involved in gene expression, DNA replication, or DNA repair.

restoration of standard nucleosomes

remodeling complex A

dissociation of DNA-binding proteins

remodeling complex B

addition of DNA-binding proteins

GENE EXPRESSION, DNA REPLICATION, AND OTHER PROCESSES THAT REQUIRE ACCESS TO DNA PACKAGED IN NUCLEOSOMES

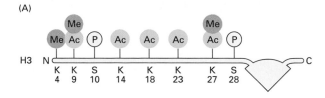

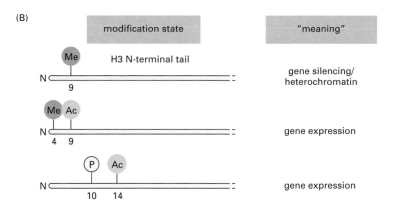

(A)

H3   N

Me
Me Ac P Ac Ac Ac Me Ac P C
K K S K K K Ac K S
4 9 10 14 18 23 27 28

(B)

modification state	"meaning"

H3 N-terminal tail

Me
N ━━━━━━━━━━ ═   gene silencing/
  9   heterochromatin

Me Ac
N ━━━━━━━━━━ ═   gene expression
  4 9

P Ac
N ━━━━━━━━━━ ═   gene expression
  10 14

Figure 5–30 The pattern of modification of histone tails may dictate how a stretch of chromatin is treated by the cell. (A) Each histone can be modified by the covalent attachment of a number of different molecules. Histone H3, for example, can receive an acetyl group (Ac), a methyl group (Me), or a phosphate (P). Note that some positions (e.g., lysine 9 and 27) can be modified in more than one way. (B) Different combinations of histone tail modifications may constitute a type of "histone code." According to the histone code hypothesis, each marking conveys a specific meaning to the stretch of chromatin on which it occurs. Only a few of the meanings of the modifications are known.

Essential Concepts

- Life depends on stable and compact storage of genetic information.
- Genetic information is carried by very long DNA molecules and encoded in the linear sequence of nucleotides A, T, G, and C.
- A molecule of DNA is in the form of a double helix composed of a pair of complementary strands of nucleotides held together by hydrogen bonds between G-C and A-T base pairs.
- Each strand of DNA has a chemical polarity due to the linkage of alternating sugars and phosphates in its backbone. The two strands of the DNA double helix run antiparallel—that is, in opposite orientations.
- The genetic material of a eucaryotic cell is contained within one or more chromosomes, each formed from a single, enormously long DNA molecule that contains many genes.
- The DNA in a eucaryotic chromosome contains, in addition to genes, many replication origins, one centromere, and two telomeres. These sequences ensure that the chromosome can be replicated efficiently and passed on to daughter cells.
- Chromosomes in eucaryotic cells consist of DNA tightly bound to a roughly equal mass of specialized proteins. These proteins fold the DNA into a more compact form so that it can fit into a cell nucleus. The complex of DNA and protein in chromosomes is called chromatin.
- Chromosomal proteins include the histones, which pack DNA into a repeating array of DNA–protein particles called nucleosomes.
- Nucleosomes pack together, with the aid of histone H1 molecules, to form a 30-nm fiber. This fiber can be further coiled and folded, producing more compact chromatin structures.
- Some forms of chromatin are so highly compacted that the packaged genes cannot be expressed into protein.
- Chromatin structure is dynamic: by temporarily altering its structure—using chromatin remodeling complexes and enzymes that modify histone tails—the cell can ensure that proteins involved in gene expression, replication, and repair have rapid, localized access to the necessary DNA sequences.

Key Terms

antiparallel	gene
base pair	genome
chromatin	heterochromatin
chromatin remodeling complex	histone
chromosome	karyotype
complementary	nucleosome
deoxyribonucleic acid (DNA)	replication origin
double helix	telomere

Questions

Question 5–4

A. The nucleotide sequence of one DNA strand of a DNA double helix is
5′-GGATTTTTGTCCACAATCA-3′.
What is the sequence of the complementary strand?

B. In the DNA of certain bacterial cells, 13% of the nucleotides are adenine. What are the percentages of the other nucleotides?

C. How many possible nucleotide sequences are there for a stretch of DNA that is N nucleotides long, if it is (a) single-stranded or (b) double-stranded?

D. Suppose you had a method of cutting DNA at specific sequences of nucleotides. How many nucleotides long (on average) would such a sequence have to be in order to make just one cut in a bacterial genome of 3×10^6 nucleotide pairs? How would the answer differ for the genome of an animal cell that contains 3×10^9 nucleotide pairs?

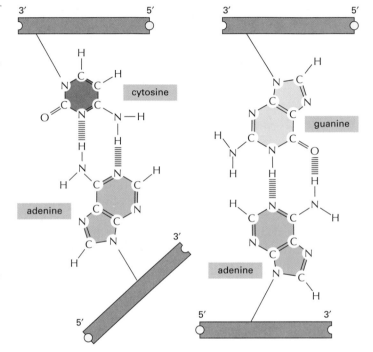

Figure Q5–5

Question 5–5

An A-T base pair is stabilized by only two hydrogen bonds. Hydrogen-bonding schemes of very similar strengths can also be drawn between other base combinations, such as the A-C and the A-G pairs shown in Figure Q5–5. What would happen if these pairs formed during DNA replication and the inappropriate bases were incorporated? Discuss why this does not happen often. (Hint: see Figure 5–6.)

Question 5–6

A. A macromolecule isolated from an extraterrestrial source superficially resembles DNA but upon closer analysis reveals quite different base structures (Figure Q5–6) in place of A, T, G, and C. Look at these structures closely. Could these DNA-like molecules have been derived from a living organism that uses principles of genetic inheritance similar to those used by cells on Earth? If so, what can you say about its properties?

B. Simply judged by their potential for hydrogen-bonding, could any of these extraterrestrial bases replace terrestrial A, T, G, or C in terrestrial DNA? Explain your answers.

Question 5–7

The two strands of DNA double helix can be separated by heating. If you raised the temperature of a solution containing the following three DNA molecules, in what order do you suppose they would "melt"? Explain your answer.

A. 5′-GCGGGCCAGCCCGAGTGGGTAGCCCAGG-3′
3′-CGCCCGGTCGGGCTCACCCATCGGGTCC-5′

B. 5′-ATTATAAAATATTTAGATACTATATTTACAA-3′
3′-TAATATTTTATAAATCTATGATATAAATGTT-5′

C. 5′-AGAGCTAGATCGAT-3′
3′-TCTCGATCTAGCTA-5′

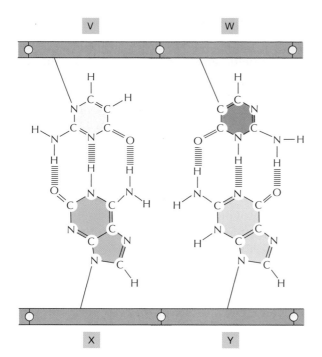

Figure Q5–6

Question 5–8

The total length of DNA in the human genome is about 1 m, and the diameter of the double helix is about 2 nm. Nucleotides in a DNA double helix are stacked at an interval of 0.34 nm. If the DNA were enlarged so that its diameter equaled that of an electrical extension cord (5 mm), how long would the extension cord be from one end to the other (assuming that it is completely stretched out)? How close would the bases be to each other? How long would a gene of 1000 nucleotide pairs be?

Question 5–9

A compact disc (CD) stores about 4.8×10^9 bits of information in a 96 cm^2 area. This information is stored as a binary code—that is, every bit is either a 0 or a 1.

    A.   How many bits would it take to specify each nucleotide pair in a DNA sequence?

    B.   How many CDs would it take to store the information contained in the human genome?

Question 5–10

Which of the following statements are correct? Explain your answers.

    A.   Each eucaryotic chromosome must contain the following DNA sequence elements: multiple origins of replication, two telomeres, and one centromere.

    B.   Nucleosome core particles are 30 nm in diameter and, when lined up, form 30-nm filaments.

Question 5–11

Define the following terms and their relationships to one another:

    A.   Interphase chromosome

    B.   Mitotic chromosome

    C.   Chromatin

    D.   Heterochromatin

    E.   Histones

    F.   Nucleosome

Question 5–12

Carefully consider the result shown in Figure 5–27A. Each of the two colonies shown is a clump of approximately 100,000 yeast cells that has grown up from a single cell that is now somewhere in the middle of the colony. As explained in the figure caption, the lower colony contains mostly red cells, because the *ADE2* gene is inactivated when it is positioned near the telomere. Explain why the white sectors have formed near the rim of the colony. Based on the existence of these sectors, what can you conclude about the propagation of the transcriptional state of the *ADE2* gene from mother to daughter cells?

Question 5–13

The two electron micrographs in Figure Q5–13 show nuclei of two different cell types. Can you tell from these pictures which of the two cells is transcribing more of its genes? Explain how you arrived at your answer.

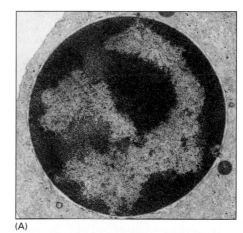

(A)

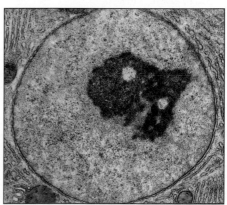

Figure Q5–13   (B)

Question 5–14

DNA forms a right-handed helix. Pick out the right-handed helix from those shown in Figure Q5–14.

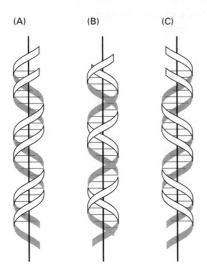

(A)   (B)   (C)

Figure Q5–14

Question 5–15

A single nucleosome is 11 nm long and contains 146 bp of DNA (0.34 nm/bp). What packing ratio (DNA length to nucleosome length) has been achieved by wrapping DNA around the histone octamer? Assuming that there are an additional 54 bp of extended DNA in the linker between nucleosomes, how condensed is "beads-on-a-string" DNA relative to fully extended DNA? What fraction of the 10,000-fold condensation that occurs at mitosis does this first level of packing represent?

Question 5–16

Assuming that the histone octamer forms a cylinder 9 nm in diameter and 5 nm in height and that the human genome forms 32 million nucleosomes, what volume of the nucleus (6 μm in diameter) is occupied by histone octamers? (Volume of a cylinder is $\pi r^2 h$; volume of a sphere is $4/3\ \pi r^3$.) What fraction of the total volume of the nucleus does the DNA and the histone octamers occupy?

Question 5–17

Histone proteins are among the most highly conserved proteins in eucaryotes. Histone H4 proteins from a pea and a cow, for example, differ in only 2 of 102 amino acids. Comparison of the gene sequences shows many more differences, but only two that change encoded amino acids. These observations indicate that mutations that change amino acids must be selected against. Why do you suppose that amino acid-altering mutations in histone genes are deleterious?

Highlights from *Essential Cell Biology 2 Interactive* CD-ROM

5.1   DNA Structure

5.2   Chromosome Coiling

5.3   Liver Cell: View 6

DNA Replication, Repair, and Recombination

The ability of a cell to maintain order in a chaotic environment depends on the accurate duplication of the vast quantity of genetic information carried in its DNA. This duplication process, called **DNA replication**, must occur before a cell can produce two genetically identical daughter cells. Maintaining order in a cell also requires the continual surveillance and repair of its genetic information, as DNA is subject to damage by chemicals and radiation from the environment, and by accidents and reactive molecules that occur inside the cell. As we shall see in this chapter, each cell contains elaborate machinery for accurately copying its store of genetic information, as well as specialized enzymes for repairing DNA when it is damaged. These enzymes catalyze some of the most rapid and accurate processes that take place within cells, and their actions reflect the elegance and efficiency of cellular chemistry.

Despite these systems for protecting the genetic instructions from copying errors and accidental damage, permanent changes, or *mutations,* sometimes do occur. Mutations in the DNA often affect the information it encodes. Occasionally, this can benefit the organism in which a mutation occurs: for example, mutations can make bacteria resistant to antibiotics that are used to kill them. Indeed, the accumulation of changes in DNA over millions of years provides the variety in genetic material that makes one species distinct from another, as we discuss in Chapter 9. Mutations also produce the smaller variations that underlie the differences between individuals of the same species that we can easily see in humans and other animals (Figure 6–1).

However, mutations are often detrimental: in humans, mutations are responsible for thousands of inherited diseases, and mutations that arise in the cells of the body throughout the lifetime of an individual may also cause disease, most notably the many types of cancer. Thus survival of a cell or organism can depend on preventing changes to its DNA. Without the cellular systems that are continually monitoring and repairing damage to DNA, it is questionable whether life could exist at all.

In this chapter, we begin by reviewing the cellular mechanisms—DNA replication and repair—that are responsible for keeping mutations to a minimum. We next consider some of the intriguing ways in which cells alter their genetic information, including *DNA recombination* and the movement of the special DNA sequences in our chromosomes called *transposable elements.* Finally, we consider viruses—little more than genes protected by a protein coat—which can move from cell to cell.

DNA Replication
Base-Pairing Enables DNA Replication
DNA Synthesis Begins at Replication Origins
New DNA Synthesis Occurs at Replication Forks
The Replication Fork Is Asymmetrical
DNA Polymerase Is Self-correcting
Short Lengths of RNA Act as Primers for DNA Synthesis
Proteins at a Replication Fork Cooperate to Form a Replication Machine
Telomerase Replicates the Ends of Eucaryotic Chromosomes
DNA Replication Is Relatively Well Understood

DNA Repair
Mutations Can Have Severe Consequences for an Organism
A DNA Mismatch Repair System Removes Replication Errors That Escape the Replication Machine
DNA Is Continually Suffering Damage in Cells
The Stability of Genes Depends on DNA Repair
The High Fidelity of DNA Maintenance Allows Closely Related Species to Have Proteins with Very Similar Sequences

DNA Recombination
Homologous Recombination Results in an Exact Exchange of Genetic Information
Recombination Can Also Occur Between Nonhomologous DNA Sequences
Mobile Genetic Elements Encode the Components They Need for Movement
A Large Fraction of the Human Genome Is Composed of Two Families of Transposable Sequences
Viruses Are Fully Mobile Genetic Elements That Can Escape from Cells
Retroviruses Reverse the Normal Flow of Genetic Information

Figure 6–1 **Hereditary information is passed faithfully from one generation to the next.** Changes in the DNA, however, can produce the variations that underlie the differences between individuals of the same species—or, over time, the differences between one species and another. In this family photo, the children resemble one another and their parents more closely than they resemble other people because they inherit their particular genes from their parents. The cat shares many features with humans, but during the millions of years of evolution that have separated humans and cats, we both have accumulated many hereditary changes that now make us quite different species. The chicken is an even more distant relative.

DNA Replication

At each cell division, a cell must copy its genome with extraordinary accuracy. In this section, we explore how the cell achieves this precision, while duplicating DNA at rates as high as 1000 nucleotides per second.

Base-Pairing Enables DNA Replication

In the preceding chapter, we saw that each strand of the DNA double helix contains a sequence of nucleotides that is exactly complementary to the nucleotide sequence of its partner strand. Each strand can therefore act as a **template**, or mold, for the synthesis of a new complementary strand (Figure 6–2). In other words, if we designate the two DNA strands as S and S′, strand S can serve as a template for making a new strand S′, while strand S′ can serve as a template for making a new strand S (Figure 6–3). Thus, the genetic information in DNA can be accurately copied by the beautifully simple process in which strand S separates from strand S′, and each separated strand then serves as a template for the production of a new complementary partner strand that is identical to its former partner.

The ability of each strand of a DNA molecule to act as a template for producing a complementary strand enables a cell to copy, or *replicate,* its genes before passing them on to its descendants. But the task is awe-inspiring, as it can involve copying billions of nucleotide pairs every time a cell divides. The copying must be carried out with speed and accuracy: in about 8 hours, a dividing animal cell will copy the

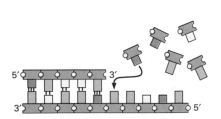

Figure 6–2 A DNA strand can serve as a template. Preferential binding occurs between pairs of nucleotides (A with T, and G with C) that can form base pairs. This enables each strand to act as a template for forming its complementary strand.

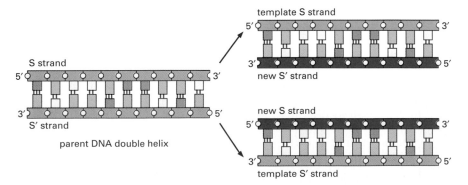

Figure 6–3 DNA acts as a template for its own duplication. Because the nucleotide A will successfully pair only with T, and G with C, each strand of DNA can serve as a template to specify the sequence of nucleotides in its complementary strand. In this way, double-helical DNA can be copied precisely. Keep in mind that although they are colored differently here, the template strands *(orange)* and the new strands *(red)* are chemically identical.

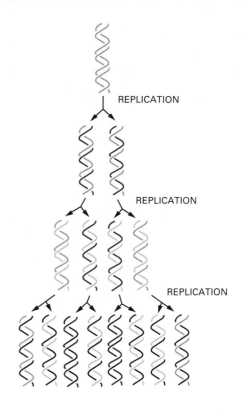

Figure 6–4 In each round of replication, each of the two strands of DNA is used as a template for the formation of a complementary DNA strand. The original strands, therefore, remain intact through many cell generations. DNA replication is "semiconservative" because each daughter DNA double helix is composed of one conserved strand and one newly synthesized strand.

equivalent of 1000 books like this one and, on average, get no more than a single letter or two wrong. This feat is performed by a cluster of proteins that together form a "replication machine." DNA replication produces two complete double helices from the original DNA molecule, each new DNA helix identical (except for rare errors) in nucleotide sequence to the parental DNA double helix (see Figure 6–3). Because each parental strand serves as the template for one new strand, each of the daughter DNA double helices ends up with one of the original (old) strands plus one strand that is completely new; this style of replication is said to be semiconservative (Figure 6–4).

DNA Synthesis Begins at Replication Origins

The DNA double helix is normally very stable: the two DNA strands are locked together firmly by the large numbers of hydrogen bonds between the bases on both strands (see Figure 5–2). As a result, only temperatures approaching those of boiling water provide enough thermal energy to separate these strands. To be used as a template, however, the double helix must first be opened up and the two strands separated to expose unpaired bases. How does this occur at the temperatures found in living cells?

The process of DNA replication is begun by initiator proteins that bind to the DNA and pry the two strands apart, breaking the hydrogen bonds between the bases (Figure 6–5). Although the hydrogen bonds collectively make the DNA helix very stable, individually each hydrogen bond is weak (discussed in Chapter 2). Separating a short length of DNA does not therefore require a large energy input and can occur with the assistance of these proteins at normal temperatures.

The positions at which the DNA is first opened are called **replication origins**, and they are usually marked by a particular sequence of nucleotides. In simple cells like those of bacteria or yeast, replication origins span approximately 100 base pairs; they are composed of DNA sequences that attract the initiator proteins, as well as stretches of DNA that are especially easy to open. We saw in Chapter 5 that an A-T base pair is held together by fewer hydrogen bonds than is a G-C base pair. Therefore, DNA rich in A-T base pairs is relatively easy to pull apart, and A-T-rich stretches of DNA are typically found at replication origins.

A bacterial genome, which is typically contained in a circular DNA molecule of several million nucleotide pairs, has a single origin of replication. The human genome, which is very much larger, has approximately 10,000 such origins. In humans, beginning DNA replication at many places at once allows a cell to replicate its DNA relatively quickly. In How We Know, pp. 198–200, we discuss experiments that reveal the locations of the replication origins in an organism's genome.

Once an initiator protein binds to DNA at the replication origin and locally opens up the double helix, it attracts a group of proteins that carry out DNA replication. This group operates as a *protein machine*, with each member carrying out a specific function. We will introduce each of these proteins shortly, after we consider the overall process of DNA replication.

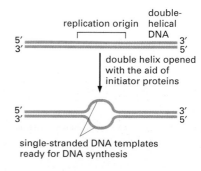

Figure 6–5 A DNA double helix is opened at its replication origin. Replication initiator proteins recognize sequences of DNA at replication origins and locally pry apart the two strands of the double helix. The exposed single strands can then serve as templates for copying the DNA.

How We Know: Finding Replication Origins

For eucaryotic cells, DNA replication is a monumental task. To transmit its genetic information to its daughters, a cell must copy its entire genome quickly, accurately, and completely. And it must copy its genetic material once and only once during each round of cell division. Failure to perform these tasks properly—essentially perfectly—can have catastrophic consequences. If the whole genome is not replicated, critical information may be lost; copying some information more than once can be equally disastrous. Both situations can cause mutations, duplications, rearrangements, even massive chromosomal breakage—defects that in a multicellular organism can cause cancer or other diseases.

But how do cells control this complicated molecular maneuver? We know quite a bit about the control of replication in procaryotes, where the situation is simpler: the relatively small, circular bacterial genome has a single origin of replication, so all a bacterial cell has to do is to start replication and make sure that it finishes.

Eucaryotic genomes, on the other hand, contain many origins of replication—hundreds in simple yeasts, thousands in humans—and not all of these origins are equal. Some are used faithfully in each round of replication; others are used periodically, occasionally, or even rarely. Furthermore, different origins are called into service at different times during S phase, the stage in the cell cycle when DNA is replicated. Some "fire" early in S phase, others launch replication later.

Their activity depends, at least in part, on where in the genome they are located.

With so many replication origins all over the genome, firing at different times, how does a cell coordinate the overall process of genome replication with such extraordinary accuracy. Current studies aimed at identifying and locating replication origins, and determining when they are used, are beginning to reveal some answers as to how origins are chosen, how the timing of their activity is determined, and how the whole process is controlled.

Origins at work

Unlike the situation in procaryotes, origins of replication in eucaryotes do not contain an easily identifiable DNA sequence. Most replication origins in the yeast genome contain variations of an 11-nucleotide sequence, but not all DNA fragments with this sequence can act as origins. In other eucaryotes, many different sequences seem capable of initiating replication. So how can researchers identify an origin?

The simplest approach to determining whether a piece of DNA contains an origin is to see whether the fragment can direct the replication of a plasmid. **Plasmids** are small loops of DNA that can exist inside a yeast or bacterial cell, separate from the cell's own genome. These circular DNA molecules have their own replication origins, which allow them

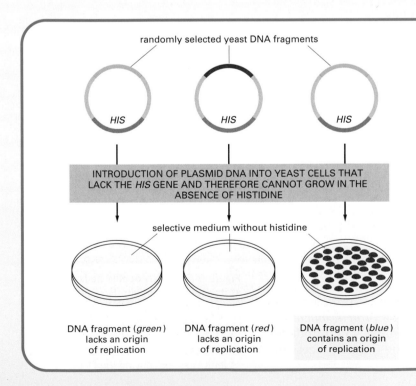

Figure 6–6 Segments of DNA can be analyzed for their potential to serve as a replication origin by inserting them into a plasmid. Random DNA fragments are inserted into a plasmid whose origin of replication has been removed. This plasmid, which contains a gene that allows yeast to grow in the absence of histidine, is then introduced into yeast cells, which are grown in medium lacking histidine. To survive and divide in this medium, yeast cells must contain a replicating plasmid. Thus any cells that multiply to form colonies will contain a plasmid that bears a DNA fragment that can drive replication. If the fragment does not support plasmid replication, the cells will not survive.

randomly selected yeast DNA fragments

HIS   HIS   HIS

INTRODUCTION OF PLASMID DNA INTO YEAST CELLS THAT LACK THE *HIS* GENE AND THEREFORE CANNOT GROW IN THE ABSENCE OF HISTIDINE

selective medium without histidine

DNA fragment (*green*) lacks an origin of replication

DNA fragment (*red*) lacks an origin of replication

DNA fragment (*blue*) contains an origin of replication

to replicate independently of the host cell's genome; they are described in greater detail in Chapter 10.

Using recombinant DNA techniques (which we also discuss in Chapter 10), an investigator can remove a plasmid's origin of replication and replace it with a DNA fragment that might contain a yeast replication origin. The modified plasmids are then introduced into a yeast cell and the cells are incubated to allow replication to occur: replication of the plasmid can only occur if its inserted DNA fragment contains an origin of replication. In this way, yeast origins of replication can be identified (Figure 6–6).

Arrays to the rescue

The plasmid maintenance method works well when one has just one piece of DNA—or a small collection of DNA fragments—that potentially contain a replication origin. But what if a researcher has no idea where the origins lie, or wishes to identify and study all of the origins in a eucaryote such as yeast? Powerful new techniques are now allowing investigators to locate the replication origins across a whole genome—and determine when they fire—in one fell swoop.

To map the positions and timing of all the replication origins in the yeast genome, researchers have turned to DNA microarrays—grids studded with thousands of DNA fragments of known sequence. As we will see in Chapter 10, these microarrays are used to determine whether nucleotide fragments of a particular sequence are present in a sample. They are widely used, for example, to monitor which genes are being expressed in a cell. Because the exact sequence—and position—of every DNA fragment on the microarray is known, the exact sequence of any nucleotide fragment that binds to its complementary sequence in the array can be ascertained simply by determining to which position on the array it is bound.

For the replication origin experiment, researchers use a microarray that contains DNA fragments that cover the entire yeast genome—tens of thousands of probes that represent unique sequences found at consecutive intervals of, say, 500 nucleotides along the genome. The yeast are allowed to begin replication and their DNA is collected, broken into fragments, and labeled with a fluorescent marker (Figure 6–7). These fragments are washed over the whole-genome

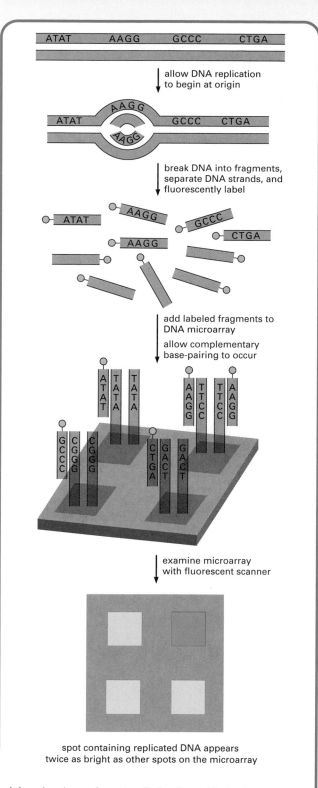

Figure 6–7 DNA microarrays can be used to locate DNA replication origins. A culture of yeast cells is allowed to begin replication, and the partially replicated DNA is collected. This DNA is broken into fragments, separated into single strands, and then labeled with a fluorescent marker. The labeled fragments are washed over a microarray composed of genomic DNA fragments and are allowed to bind to complementary fragments on the array. When this microarray is examined under a fluorescence scanner, the positions to which labeled DNA has bound will light up. The more DNA bound to a spot, the brighter that spot will appear. Spots representing replicated DNA will appear twice as bright, as they will bind twice as much DNA as spots representing nonreplicated DNA. Shown here is a grid spotted with four different DNA fragments only four nucleotides in length. Actual microarrays contain thousands of DNA fragments, typically tens to hundreds of nucleotides in length. For simplicity, each spot on the microarray drawn here shows only two DNA fragments bound to it; in reality, each spot on a microarray contains millions of DNA fragments of identical sequence.

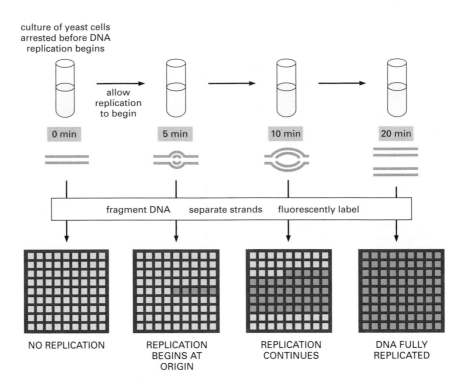

culture of yeast cells
arrested before DNA
replication begins

allow
replication
to begin

0 min   5 min   10 min   20 min

fragment DNA   separate strands   fluorescently label

NO REPLICATION   REPLICATION BEGINS AT ORIGIN   REPLICATION CONTINUES   DNA FULLY REPLICATED

Figure 6–8 **Collecting DNA at different times after replication begins allows an investigator to monitor the progress of replication through the genome.** Cells are synchronized so that they begin replication at the same time. DNA is collected and applied to the microarray as shown in Figure 6–7. Replication begins at an origin and proceeds, bidirectionally, until the entire genome has been copied. For simplicity only one origin is shown here. In yeast cells, replication begins at hundreds of origins located throughout the genome. The spots on these microarrays represent consecutive sequences along the yeast genome. Only 81 spots are shown here, but the actual arrays contain tens of thousands of sequences that span the entire yeast genome. Because the sequence of the DNA at each spot on the microarray is known, the location of every replication origin can be determined by monitoring which spots light up first.

microarray and allowed to bind to their complementary sequences fixed to the surface of the array. Researchers can then locate the replication origins by seeing which DNA segments are duplicated first. As shown in Figure 6–7, DNA segments that have been replicated will be present at twice the concentration of nonreplicated segments, generating a fluorescent spot with twice the intensity on the DNA microarray.

Ready, set, replicate

The situation outlined in Figure 6–7 is an overly simple one: it shows replication beginning at a single origin and proceeding for only a short distance. In reality, replication in yeast cells is initiated at hundreds of origins that fire at different times. Fortunately, the same experimental protocol can be adapted to determine which regions of DNA replicate first, which start later, and how quickly replication sweeps through different segments of the genome.

To follow replication over time, researchers collect DNA from yeast cells at different times after the start of replication (Figure 6–8). They begin by synchronizing the yeast cells so that each cell in the population will begin to repli-

cate its DNA at exactly the same time. This can be done by exposing the yeasts to a molecule that will cause them to arrest in G_1, the phase of the cell cycle in which the replication machinery assembles on the origins but in which DNA synthesis has yet to begin. When the drug is removed, the yeast cells will all enter S phase together.

DNA samples are collected, fluorescently labeled, and washed over the microarray as before. Now, however, researchers can monitor the progression of replication throughout the genome. Using such microarray-based techniques, researchers have identified some 300 origins of replication in the yeast genome; some of these had previously been identified using the plasmid maintenance methods described above, but many are new. They also found that replication origins are activated throughout S phase, with most firings occurring near mid-S phase.

Although the experiment was performed with yeast DNA, this method can be adapted to look for replication origins in human DNA. Such studies should enhance our understanding of how all cells achieve such exquisite control over a process as critical and complex as DNA replication.

New DNA Synthesis Occurs at Replication Forks

DNA molecules in the process of being replicated can be observed in the electron microscope (Figure 6–9), where it is possible to see Y-shaped junctions in the DNA, called **replication forks**. At these forks, the replication machine is moving along the DNA, opening up the two strands of the double helix and using each strand as a template to make a new daughter strand. Two replication forks are formed starting from each replication origin, and they move away from the origin in both directions, unzipping the DNA as they go. DNA replication in bacterial and eucaryotic chromosomes is therefore termed *bidirectional*. The forks move very rapidly—at about 1000 nucleotide pairs per second in bacteria and 100 nucleotide pairs per second in humans. The slower rate of fork movement in humans (indeed, in all eucaryotes) may be due to the difficulties in replicating through the more complex chromatin structure found in these higher organisms.

At the heart of the replication machine is an enzyme called **DNA polymerase**, which synthesizes new DNA using one of the old strands as a template. This enzyme catalyzes the addition of nucleotides to the 3' end of a growing DNA strand by forming a phosphodiester bond between this end and the 5'-phosphate group of the incoming nucleotide (Figure 6–10). The nucleotides enter the reaction initially as energy-rich nucleoside triphosphates, which provide the energy for the polymerization reaction. The hydrolysis of one phosphoanhydride bond in the nucleoside triphosphate provides the energy for the condensation reaction that links the nucleotide monomer to the chain and releases pyrophosphate (PP_i). The DNA polymerase couples the release of this energy to the polymerization reaction. Pyrophosphate is further hydrolyzed to inorganic phosphate (P_i), which makes the polymerization reaction effectively irreversible (see Figure 3–41).

DNA polymerase does not dissociate from the DNA each time it adds a new nucleotide to the growing chain; rather, it stays associated with the DNA and moves along the template strand stepwise for many cycles of the polymerization reaction. We will see, later in this chapter, how a special protein keeps the polymerase attached in this way.

Question 6–1

Look carefully at the micrograph in Figure 6–9.

A. Using the scale bar, estimate the lengths of the DNA strands between the replication forks. Numbering the replication forks sequentially from the left, how long will it take until forks 4 and 5, and forks 6 and 7, respectively, collide with each other? (Recall that the distance between the bases in DNA is 0.34 nm, and eucaryotic replication forks move at about 100 nucleotides per second.) For this question disregard the nucleosomes seen in the micrograph and assume that the DNA is fully extended.

B. The fly genome is about 1.8×10^8 nucleotide pairs in size. How much of the total fly DNA is shown in the micrograph?

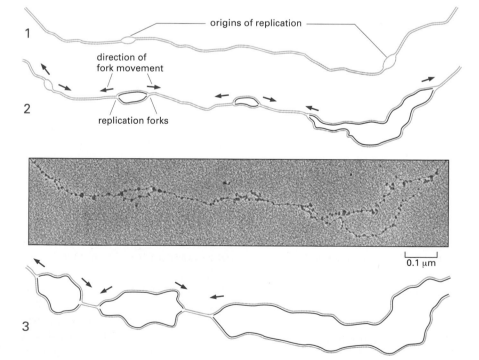

Figure 6–9 Replication forks move away in both directions from multiple replication origins in a eucaryotic chromosome. The electron micrograph shows DNA replicating in the early embryo of a fly. The particles visible along the DNA are nucleosomes, protein complexes present in eucaryotic chromosomes around which the DNA is wrapped. (1), (2), and (3) are drawings of the same portion of a DNA molecule as it might appear at successive stages of replication, drawn from electron micrographs. (2) is drawn from the electron micrograph shown here. The *orange lines* represent the parental DNA strands; the *solid red lines* represent the newly synthesized DNA. (Electron micrograph courtesy of Victoria Foe.)

Figure 6–10 DNA is synthesized in the 5′-to-3′ direction. Addition of a deoxyribonucleotide to the 3′-hydroxyl end of a polynucleotide chain is the fundamental reaction by which DNA is synthesized; the new DNA chain is therefore synthesized in the 5′-to-3′ direction. Base-pairing between the incoming deoxyribonucleotide and the template strand guides the formation of a new strand of DNA that is complementary in nucleotide sequence to the template chain (see Figure 6–2). The enzyme DNA polymerase catalyzes the addition of nucleotides to the free 3′ hydroxyl on the growing DNA strand. The nucleotides enter the reaction as nucleoside triphosphates. Breakage of a phosphoanhydride bond (indicated by the *asterisk*) in the incoming nucleoside triphosphate releases a large amount of free energy and thus provides the energy for the polymerization reaction.

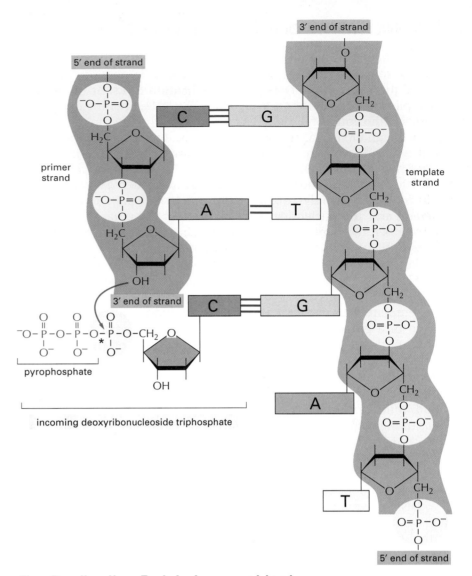

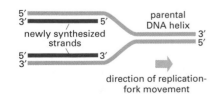

Figure 6–11 At a replication fork, the two newly synthesized DNA strands are of opposite polarity.

The Replication Fork Is Asymmetrical

The 5′-to-3′ direction of the DNA polymerization mechanism poses a problem at the replication fork. We saw in Figure 5–2 that the sugar–phosphate backbone of each strand of a DNA double helix has a unique chemical direction, or polarity, determined by the way each sugar residue is linked to the next, and that the two strands in the double helix run in opposite orientations. As a consequence, at the replication fork, one new DNA strand is being made on a template that runs in one direction (3′ to 5′), whereas the other new strand is being made on a template that runs in the opposite direction (5′ to 3′) (Figure 6–11). The replication fork is therefore asymmetrical. Both of the new DNA strands appear to be growing in the same direction, that is, the direction in which the replication fork is moving. On the face of it this suggests that one strand is being synthesized in the 3′-to-5′ direction and one is being synthesized in the 5′-to-3′ direction.

DNA polymerase, however, can catalyze the growth of the DNA chain in only one direction; it can add new subunits only to the 3′ end of the chain (see Figure 6–10). As a result, a new DNA chain can be synthesized only in a 5′-to-3′ direction. This can easily account for the synthesis of one of the two strands of DNA at the replication fork, but not the other. One might have expected a second type of DNA polymerase to synthesize the other DNA strand—one that works by adding subunits to the 5′ end of a DNA chain. However, no such enzyme exists. Instead,

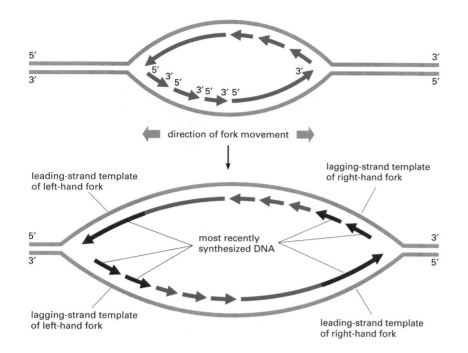

direction of fork movement

leading-strand template of left-hand fork

lagging-strand template of right-hand fork

most recently synthesized DNA

lagging-strand template of left-hand fork

leading-strand template of right-hand fork

Figure 6–12 DNA replication forks are asymmetrical. Because both of the new strands are synthesized in the 5'-to-3' direction, the lagging strand of DNA must be made initially as a series of short DNA strands that are later joined together. The upper diagram shows two replication forks moving in opposite directions; the lower diagram shows the same forks a short time later. To synthesize the lagging strand, DNA polymerase must "backstitch": it synthesizes short fragments (called Okazaki fragments) in the 5'-to-3' direction, and then moves in the opposite direction along the template strand (toward the fork) before synthesizing the next fragment.

the problem is solved by the use of a "backstitching" maneuver. The DNA strand whose 5' end must grow is made *discontinuously*, in successive separate small pieces, with the DNA polymerase working backward from the replication fork in the 5'-to-3' direction for each new piece. These pieces—called **Okazaki fragments** after the biochemist who discovered them—are later "stitched" together to form a continuous new strand (Figure 6–12). The DNA strand that is synthesized discontinuously in this way is called the **lagging strand**; the strand that is synthesized continuously is called the **leading strand**.

Although they differ in subtle details, the replication forks of all cells, procaryotic and eucaryotic, have leading and lagging strands. This common feature arises from the fact that all of the DNA polymerases used to replicate DNA polymerize in the 5'-to-3' direction only. We shall look at events on the lagging strand in more detail later in this chapter; first, we consider another feature of DNA polymerase that is common to all cells.

DNA Polymerase Is Self-correcting

DNA polymerase is so accurate that it makes only about one error in every 10^7 nucleotide pairs it copies. This error rate is much lower than can be accounted for simply by the accuracy of complementary base-pairing. Although A-T and C-G are by far the most stable base pairs, other, less stable base pairs—for example, G-T and C-A—can also be formed. Such incorrect base pairs are formed much less frequently than correct ones, but they occur often enough that they would kill the cell through an accumulation of mistakes in the DNA if they were allowed to remain. This catastrophe is avoided because DNA polymerase can correct its mistakes. As well as catalyzing the polymerization reaction, DNA polymerase has an error-correcting activity called **proofreading**. Before the enzyme adds a nucleotide to a growing DNA chain, it checks whether the previous nucleotide added is correctly base-paired to the template strand. If so, the polymerase adds the next nucleotide; if not, the polymerase removes the mispaired nucleotide by cutting the phosphodiester bond it has just made, releases the nucleotide, and tries again (Figure 6–13). Thus, DNA polymerase possesses both a 5'-to-3'

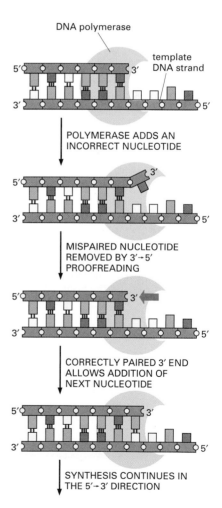

DNA polymerase

template DNA strand

POLYMERASE ADDS AN INCORRECT NUCLEOTIDE

MISPAIRED NUCLEOTIDE REMOVED BY 3'→5' PROOFREADING

CORRECTLY PAIRED 3' END ALLOWS ADDITION OF NEXT NUCLEOTIDE

SYNTHESIS CONTINUES IN THE 5'→3' DIRECTION

Figure 6–13 During DNA synthesis, DNA polymerase proofreads its own work. If an incorrect nucleotide is added to a growing strand, the DNA polymerase will cleave it from the strand and replace it with the correct nucleotide before continuing.

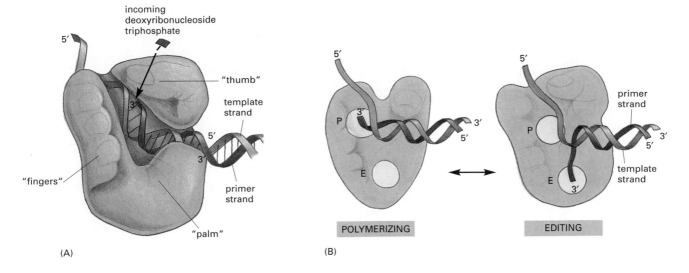

(A)

POLYMERIZING   EDITING

(B)

Figure 6–14 DNA polymerase contains separate sites for DNA synthesis and editing. (A) The structure of an *E. coli* DNA polymerase molecule, as determined by X-ray crystallography. Roughly speaking, the enzyme resembles a right hand in which the palm, fingers, and thumb grasp the DNA. (B) A cutaway outline of the structures of DNA polymerase complexed with the DNA template in the polymerizing mode *(left)* and the editing mode *(right)*. The catalytic site for the error-correcting exonuclease activity (E) and the polymerization activity (P) are indicated. To determine these structures by X-ray crystallography, researchers "froze" the polymerases in these two states, by using a mutant polymerase defective in the exonuclease domain *(right)* or by withholding the Mg^{2+} required for polymerization *(left)*. These drawings illustrate a DNA polymerase that functions during DNA repair, but the enzymes that replicate DNA have similar features. (A, adapted from L.S. Beese, V. Derbyshire, and T.A. Steitz, *Science* 260:352–355, 1993.)

polymerization activity and a 3′-to-5′ *exonuclease* (nucleic acid–degrading) activity. These activities are carried out by different domains within the polymerase molecule (Figure 6–14).

This proofreading mechanism explains why DNA polymerases synthesize DNA only in the 5′-to-3′ direction, despite the need this imposes for a cumbersome backstitching mechanism at the replication fork. As shown in Figure 6–15, a hypothetical DNA polymerase that synthesized in the 3′-to-5′ direction (and would thereby circumvent the need for backstitching) would be unable to proofread: if it removed an incorrectly paired nucleotide, the polymerase would create a chain end that is chemically dead, in the sense that it would no longer be able to elongate. Thus, for a DNA polymerase to function as a self-correcting enzyme that removes its own polymerization errors as it moves along the DNA, it must proceed only in the 5′-to-3′ direction.

Short Lengths of RNA Act as Primers for DNA Synthesis

We have seen that the accuracy of DNA replication depends on the requirement of the DNA polymerase for a correctly base-paired end before it can add more nucleotides. But because the polymerase can join a nucleotide only to a base-paired nucleotide in a DNA double helix, it cannot start a completely new DNA strand. A different enzyme is needed to begin a new DNA strand, an enzyme that can begin a new polynucleotide chain simply by joining two nucleotides together without the need for a base-paired end. This enzyme does not, however, synthesize DNA. It makes a short length of a closely related type of nucleic acid—**RNA (ribonucleic acid)**—using the DNA strand as a template. This short length of RNA, about 10 nucleotides long, is base-paired to the template strand and provides a base-paired 3′ end as a starting point for DNA polymerase. It thus serves as a *primer* for DNA synthesis, and the enzyme that synthesizes the RNA primer is known as *primase*. A strand of RNA is very similar chemically to a single strand of DNA except that it is made of ribonucleotide subunits, in which the sugar is ribose, not deoxyribose; RNA also differs from DNA in that it contains the base uracil (U) instead of thymine (T) (see Panel 2–6, pp. 76–77). However, because U can form a base pair with A, the RNA primer is synthesized on the DNA strand by complementary base-pairing in exactly the same way as is DNA.

For the leading strand, an RNA primer is needed only to start replication at a replication origin; once a replication fork has been

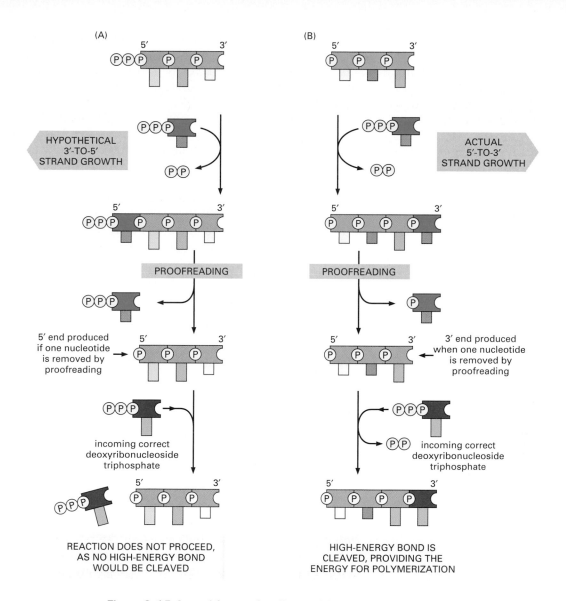

(A)

HYPOTHETICAL
3'-TO-5'
STRAND GROWTH

PROOFREADING

5' end produced
if one nucleotide
is removed by
proofreading

incoming correct
deoxyribonucleoside
triphosphate

REACTION DOES NOT PROCEED,
AS NO HIGH-ENERGY BOND
WOULD BE CLEAVED

(B)

ACTUAL
5'-TO-3'
STRAND GROWTH

PROOFREADING

3' end produced
when one nucleotide
is removed by
proofreading

incoming correct
deoxyribonucleoside
triphosphate

HIGH-ENERGY BOND IS
CLEAVED, PROVIDING THE
ENERGY FOR POLYMERIZATION

Figure 6–15 A need for proofreading explains why DNA chains are synthesized only in the 5′ to 3′ direction. (A) Proofreading in the hypothetical 3′-to-5′ polymerization scheme would allow the removal of an incorrect nucleotide (*dark green*), but would block addition of the correct nucleotide (*red*) and thereby prevent further chain elongation. (B) Growth in the 5′-to-3′ direction allows the chain to continue to be elongated when an incorrect nucleotide has been added and then removed by proofreading (see Figure 6–13).

established, the DNA polymerase is continuously presented with a base-paired 3′ end as it tracks along the template strand. But on the lagging strand, where DNA synthesis is discontinuous, new primers are needed continually, as one can see from Figure 6–12. As the movement of the replication fork exposes a new stretch of unpaired bases, a new RNA primer is made at intervals along the lagging strand. DNA polymerase adds a deoxyribonucleotide to the 3′ end of this primer to start a DNA strand, and it will continue to elongate this strand until it runs into the next RNA primer (Figure 6–16).

To produce a continuous new DNA strand from the many separate pieces of RNA and DNA made on the lagging strand, three additional enzymes are needed. These act quickly to remove the RNA primer, replace it with DNA, and join the DNA fragments together; a *nuclease* breaks apart the RNA primer, a DNA polymerase called a *repair polymerase* replaces the RNA with DNA (using the adjacent

Question 6–2

Discuss the following statement: "Primase is a sloppy enzyme that makes many mistakes. Eventually, the RNA primers it makes are disposed of and replaced with DNA by a polymerase with higher fidelity. This is wasteful. It would be more energy-efficient if a DNA polymerase made an accurate copy in the first place."

Okazaki fragment as a primer), and the enzyme *DNA ligase* joins the 5′-phosphate end of one new DNA fragment to the 3′-hydroxyl end of the next (see Figure 6–16). ATP or NADH is required for ligase activity. We will discuss these three enzymes in more detail in the section on DNA repair later in this chapter.

Primase can begin new polynucleotide chains, but this activity is possible because the enzyme does not proofread its work. As a result, primers contain a high frequency of mistakes. But because primers are made of RNA instead of DNA, they stand out as "suspect copy" to be automatically removed and replaced by DNA. This DNA is put in by DNA repair polymerases, which, like the replicative polymerases, proofread as they synthesize. In this way, the cell's replication machinery is able to begin new DNA chains and, at the same time, ensure that all of the DNA is copied faithfully.

Proteins at a Replication Fork Cooperate to Form a Replication Machine

As mentioned earlier, DNA replication requires a variety of proteins that act in concert with DNA polymerase. Here, we will discuss the additional proteins that, together with DNA polymerase and primase, form the protein machine that powers the replication fork forward and synthesizes new DNA behind it. Although it would make good sense for the three proteins that replace RNA primers with DNA—the nuclease, repair polymerase, and ligase—to also be a part of the replication machine, it is not yet known whether this is the case.

At the head of the replication machine is a *helicase*, a protein that uses the energy of ATP hydrolysis to speed along DNA, unzipping the double helix as it moves (Figure 6–17). We saw earlier in this chapter that the DNA double helix must be opened to begin DNA replication, and it must also be opened continuously as the replication fork progresses, in order to provide exposed templates for the polymerase. Another component of the replication machine—*single-strand binding protein*—clings to the single-stranded DNA exposed by the helicase and transiently prevents it from re-forming base pairs. Yet another protein, called a *sliding clamp*, keeps the DNA polymerase firmly attached to the DNA template; on the lagging strand, the sliding clamp releases the polymerase from the DNA each time an Okazaki fragment is completed. This clamp protein forms a ring around the DNA helix and binds polymerase, allowing it to slide along a template strand as it synthesizes new DNA (see Figure 6–17).

Most of the proteins involved in DNA replication are thought to be held together in a large multienzyme complex that moves as a unit

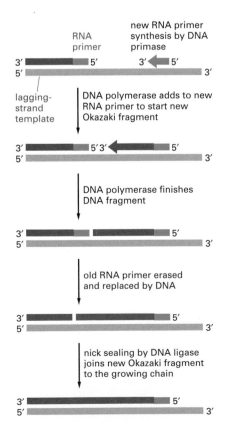

Figure 6–16 On the lagging strand, DNA is synthesized in fragments. In eucaryotes, RNA primers are made at intervals of about 200 nucleotides on the lagging strand, and each RNA primer is approximately 10 nucleotides long. In the bacterium *E. coli*, the primers and Okazaki fragments are about 5 and 1000 nucleotides long, respectively. Primers are removed by nucleases that recognize an RNA strand in an RNA/DNA helix and degrade it; this leaves gaps that are filled in by a DNA repair polymerase that can proofread as it fills in the gaps. The completed fragments are finally joined together by an enzyme called DNA ligase, which catalyzes the formation of a phosphodiester bond between the 3′-OH end of one fragment and the 5′-P end of the next, thus linking up the sugar–phosphate backbones.

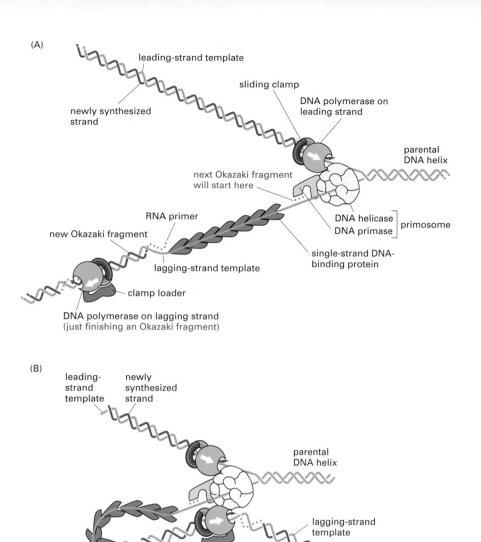

(A)

leading-strand template

sliding clamp

newly synthesized strand

DNA polymerase on leading strand

parental DNA helix

next Okazaki fragment will start here

RNA primer

new Okazaki fragment

DNA helicase
DNA primase } primosome

lagging-strand template

single-strand DNA-binding protein

clamp loader

DNA polymerase on lagging strand
(just finishing an Okazaki fragment)

(B)

leading-strand template

newly synthesized strand

parental DNA helix

lagging-strand template

new Okazaki fragment

newly synthesized strand

Figure 6–17 A group of proteins act together at a replication fork. (A) Two molecules of DNA polymerase are shown, one on the leading strand and one on the lagging strand. Both are held on to the DNA by a circular protein clamp that allows the polymerase to slide. DNA helicase uses the energy of ATP hydrolysis to propel itself forward and thereby separate the strands of the parental DNA double helix ahead of the polymerase. Single-stranded DNA-binding proteins maintain these separated strands as single-stranded DNA to provide access for the primase and polymerase. For simplicity, this figure shows the proteins working independently; in the cell they are held together in a large replication machine, as shown in (B). (B) This diagram shows a current view of how the replication proteins are arranged at the replication fork when the fork is moving. The structure in (A) has been altered by folding the DNA on the lagging strand to bring the lagging-strand DNA polymerase molecule in contact with the leading-strand DNA polymerase molecule. This folding process also brings the 3′ end of each completed Okazaki fragment close to the start site for the next Okazaki fragment. Because the lagging-strand DNA polymerase molecule is held to the rest of the replication proteins, it can be reused to synthesize successive Okazaki fragments; in this diagram, it is about to let go of its completed DNA fragment and move to the RNA primer that will be synthesized nearby, as required to start the next DNA fragment on the lagging strand.

along the DNA, enabling DNA to be synthesized on both strands in a coordinated manner. This complex can be likened to a tiny sewing machine composed of protein parts and powered by nucleoside triphosphate hydrolysis. Although the structures of the individual protein components of the replication machine have been determined, how these components fit together is not known in detail. Some ideas about the general appearance of the complex, however, have been proposed (Figure 6–17B).

Telomerase Replicates the Ends of Eucaryotic Chromosomes

Having discussed how DNA replication begins at origins and how movement of the replication fork proceeds, we now turn to the special problem of replicating the very ends of eucaryotic chromosomes. As we discussed previously, the fact that DNA is synthesized only in the 5′-to-3′ direction means that the lagging strand of the replication fork is synthesized in the form of discontinuous DNA fragments, each of which is

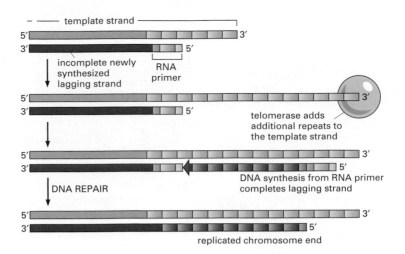

Figure 6–18 Telomeres allow the completion of DNA synthesis at the ends of eucaryotic chromosomes. To synthesize the lagging strand at the very end of a eucaryotic chromosome, the machinery of DNA replication requires a length of template DNA extending beyond the DNA that is to be copied. In a linear DNA molecule, synthesis of the lagging strand thus stops short just before the end of the template. But the enzyme telomerase adds a series of repeats of a DNA sequence to the template strand, which allows the lagging strand to be completed by DNA polymerase, as shown. In humans, the nucleotide sequence of the repeat is GGGGTTA. The telomerase enzyme contains within it a short piece of RNA of complementary sequence to the DNA repeat sequence; this RNA acts as the template for the telomerase DNA synthesis.

(Labels in figure: template strand; incomplete newly synthesized lagging strand; RNA primer; telomerase adds additional repeats to the template strand; DNA REPAIR; DNA synthesis from RNA primer completes lagging strand; replicated chromosome end)

primed with an RNA primer laid down by a separate enzyme (see Figure 6–16). When the replication fork approaches the end of a chromosome, however, the replication machinery encounters a serious problem: there is no place to lay down the RNA primer needed to start the Okazaki fragment at the very tip of the linear DNA molecule. Therefore, some DNA could easily be lost from the ends of a DNA molecule each time it is replicated.

Bacteria solve this "end-replication" problem by having circular DNA molecules as chromosomes. Eucaryotes solve it by having special nucleotide sequences at the ends of their chromosomes which are incorporated into *telomeres*. These repetitive telomeric DNA sequences attract an enzyme called **telomerase** to the chromosome. Telomerase adds multiple copies of the same telomere DNA sequence to the ends of the chromosomes, thereby producing a template that allows replication of the lagging strand to be completed (Figure 6–18).

In addition to allowing replication of chromosome ends, telomeres serve additional functions: for example, the repeated telomere DNA sequences, together with the regions adjoining them, form structures that are recognized by the cell as the true ends of chromosomes rather than breaks that sometimes occur in the middle of chromosomes and must be repaired.

DNA Replication Is Relatively Well Understood

Our current understanding of DNA replication is much more complete than that of many other aspects of cell biology, yet many mysteries still remain. For example, it is not yet understood how the polymerase on the leading strand is connected with that on the lagging strand in order to allow replication to proceed synchronously on both strands. Moreover, although we know in some detail how DNA replication begins at replication origins in bacteria, our understanding of this process in eucaryotes—including humans—is only just beginning.

Given the demands for accuracy during DNA replication, and the lengths to which cells go to achieve this precision, it is not surprising, as we shall see shortly, that cells have also evolved elaborate protein machines to scan the finished product for mistakes. These protein machines then correct any errors made during DNA replication (rare as they are) and repair any nucleotides that may have been accidentally damaged by light, by chemicals in the cell, or by other mutation-causing agents.

Question 6–3

A gene encoding one of the proteins involved in DNA replication has been inactivated by a mutation in a cell. In the absence of this protein the cell attempts to replicate its DNA for the very last time. What DNA products would be generated in each case if the following protein were missing?

A. DNA polymerase

B. DNA ligase

C. Sliding clamp for DNA polymerase

D. Nuclease that removes RNA primers

E. DNA helicase

F. Primase

DNA Repair

The diversity of living organisms and their success in colonizing almost every part of the Earth's surface depend on genetic changes accumulated gradually over millions of years. These changes allow organisms to adapt to changing conditions and to colonize new habitats. However, in the short term, and from the perspective of an individual organism, genetic change is often detrimental, especially in multicellular organisms, where a genetic change can upset an organism's extremely complex and finely tuned development and physiology. To survive and reproduce, individuals must be genetically stable. This stability is achieved not only through the extremely accurate mechanism for replicating DNA that we have just discussed, but also through mechanisms for correcting the rare copying mistakes made by the replication machinery and for repairing the accidental damage that continually occurs to the DNA. Most of these changes in DNA are only temporary because they are immediately corrected by processes collectively called **DNA repair**.

Mutations Can Have Severe Consequences for an Organism

Only rarely do the cell's DNA replication and repair processes fail and allow a permanent change in the DNA. Such a permanent change is called a **mutation**, and it can have profound consequences. A mutation affecting just a single nucleotide pair can severely compromise an organism's fitness if the change occurs in a vital position in the DNA sequence. Because the structure and activity of each protein depend on its amino acid sequence, a protein with an altered sequence may function poorly or not at all. For example, humans use the protein hemoglobin to transport oxygen in the blood; the sequence of nucleotides that encodes the amino acid sequence of one of the two types of protein chains (the β-globin chain) of the hemoglobin molecule is shown in Figure 5–11. A permanent change in a single nucleotide in this sequence can cause cells to make a β-globin chain with an incorrect sequence of amino acids. Such a mutation causes the disease *sickle-cell anemia* (Figure 6–19). The sickle-cell hemoglobin is less soluble than normal hemoglobin and forms fibrous precipitates, which lead to the

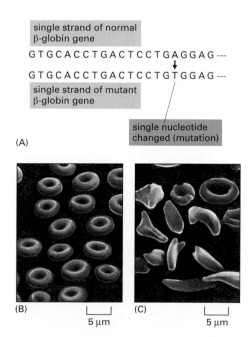

Figure 6–19 **A single nucleotide change causes the disease sickle-cell anemia.** The complete nucleotide sequence of the β-globin gene is given in Figure 5–11. Only a small portion of the sequence near the beginning of the gene is shown here (A). A single nucleotide change (mutation) in the sickle-cell gene produces a β-globin that differs from normal β-globin only by a change from glutamic acid to valine at the sixth amino acid position. (The β-globin molecule contains a total of 146 amino acids.) Humans carry two copies of each gene (one inherited from each parent); a sickle-cell mutation in one of the two β-globin genes generally causes no harm to the individual, as it is compensated for by the normal gene. However, an individual who inherits two copies of the mutant β-globin gene displays the symptoms of sickle-cell anemia. Normal red blood cells are shown in (B), and those from an individual suffering from sickle-cell anemia in (C). Although sickle-cell anemia can be a life-threatening disease, the mutation responsible can also be beneficial: patients with the disease, or who are heterozygous carriers of the mutation, are more resistant to malaria than unaffected individuals. The parasite that causes malaria grows poorly in red blood cells from homozygous sickle-cell patients or from heterozygous carriers.

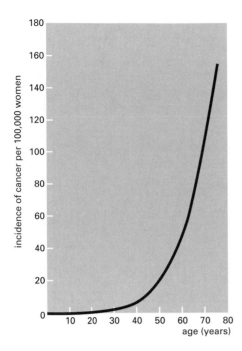

Figure 6–20 Cancer incidence increases dramatically as a function of age. The number of newly diagnosed cases of cancer of the colon in women in England and Wales in one year is plotted as a function of age at diagnosis. Because cells are continually experiencing accidental changes to their DNA that accumulate and are passed on to progeny cells, the chance that a cell will become cancerous increases greatly with age. (Data from C. Muir et al., Cancer Incidence in Five Continents, Vol. V. Lyon: International Agency for Research on Cancer, 1987.)

characteristic sickle shape of affected red blood cells. Because these cells are more fragile and frequently break in the bloodstream, patients with this potentially life-threatening disease have a reduced number of red blood cells (Figure 6–19C), a deficiency that can cause weakness, dizziness, headaches, pain, and total organ failure.

The example of sickle-cell anemia, which is an inherited disease, illustrates the importance of protecting reproductive cells (*germ cells*) against mutation. A mutation in one of these will be passed on to all the cells in the body of the multicellular organism that develops from it, including the germ cells for production of the next generation. However, the many other cells in a multicellular organism (its *somatic cells*) must also be protected from genetic change to safeguard the health and well-being of the individual. Nucleotide changes that occur in somatic cells can give rise to variant cells, some of which grow in an uncontrolled fashion at the expense of the other cells in the organism. In the extreme case, an uncontrolled cell proliferation known as cancer results. This disease, which is responsible for about 30% of the deaths that occur in Europe and North America, is due largely to a gradual accumulation of changes in the DNA sequences of somatic cells that is caused by random mutation (Figure 6–20). Increasing the mutation frequency even two- or threefold would cause a disastrous increase in the incidence of cancer by accelerating the rate at which somatic cell variants arise.

Thus the high fidelity with which DNA sequences are replicated and maintained is important both for the reproductive cells, which transmit the genes to the next generation, and for the somatic cells, which normally function as carefully regulated members of the complex community of cells in a multicellular organism. We should therefore not be surprised to find that all cells have acquired an elegant set of mechanisms to reduce the number of mutations that occur in their DNA.

A DNA Mismatch Repair System Removes Replication Errors That Escape the Replication Machine

In the first part of this chapter, we saw that the high fidelity of the cell's replication machinery generally prevents copying mistakes. Despite these safeguards, however, such errors do occur. Fortunately, the cell has a backup system—called *DNA mismatch repair*—which is dedicated to correcting these rare mistakes. The replication machine itself makes approximately one error per 10^7 nucleotides copied; DNA mismatch repair corrects 99% of these errors, increasing the overall accuracy to one mistake in 10^9 nucleotides copied. This level of accuracy is much higher than that generally encountered in the visible world around us (Table 6–1).

Table 6–1 Error Rates	
US Postal Service on-time delivery of local first-class mail	13 late deliveries per 100 parcels
Airline luggage system	1 lost bag per 200
A professional typist typing at 120 words per minute	1 mistake per 250 characters
Driving a car in the United States	1 death per 10^4 people per year
DNA replication (without mismatch repair)	1 mistake per 10^7 nucleotides copied
DNA replication (including mismatch repair)	1 mistake per 10^9 nucleotides copied

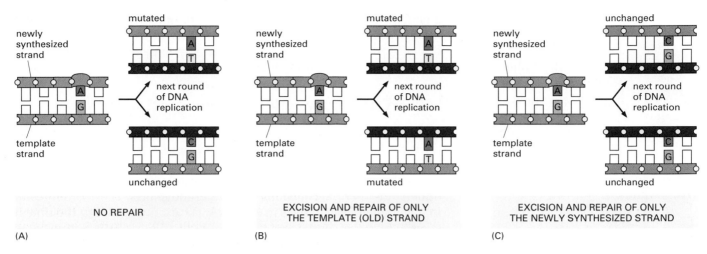

NO REPAIR

EXCISION AND REPAIR OF ONLY THE TEMPLATE (OLD) STRAND

EXCISION AND REPAIR OF ONLY THE NEWLY SYNTHESIZED STRAND

(A)　　　　　　　　　　(B)　　　　　　　　　　(C)

Whenever the replication machinery makes a copying mistake, it leaves a mispaired nucleotide (commonly called a mismatch) behind. If left uncorrected, the mismatch will result in a permanent mutation in the next round of DNA replication (Figure 6–21A). A complex of mismatch repair proteins recognizes these DNA mismatches, removes (excises) one of the two strands of DNA involved in the mismatch, and resynthesizes the missing strand (Figure 6–22). To be effective in correcting replication mistakes, this mismatch repair system must always excise only the newly synthesized DNA strand: excising the other strand (the old strand) would preserve the mistake instead of correcting it (see Figure 6–21).

In eucaryotes, it is not yet known for certain how the mismatch repair machinery distinguishes the two DNA strands. However, there is evidence that newly replicated DNA strands—both leading and lagging—are preferentially nicked; it is these nicks (single-stranded breaks) that appear to provide the signal that directs the mismatch repair machinery to the appropriate strand (see Figure 6–22).

The importance of mismatch repair in humans was recognized recently when it was discovered that an inherited predisposition to certain cancers (especially some types of colon cancer) is caused by a mutation in the gene responsible for producing one of the mismatch repair proteins. Humans inherit two copies of this gene (one from each parent), and individuals who inherit one damaged mismatch repair gene show no symptoms until the undamaged copy of the gene is accidentally mutated in a somatic cell. This gives rise to a clone of somatic cells that, because they are deficient in mismatch repair, accumulate mutations more rapidly than do normal cells. Because most cancers

Figure 6–21 Errors made during DNA replication are corrected by the cell. (A) If uncorrected, the mismatch will lead to a permanent mutation in one of the two DNA molecules produced by the next round of DNA replication. (B) If the mismatch is "repaired" using the newly synthesized DNA strand as the template, both DNA molecules produced by the next round of DNA replication will contain a mutation. (C) If the mismatch is corrected using the original template (old) strand as the template, the possibility of a mutation is eliminated. The scheme shown in (C) is used by cells to repair mismatches, as shown in Figure 6–22.

Figure 6–22 DNA mismatch repair proteins correct errors that occur during DNA replication. A DNA mismatch, formed when an incorrectly matched base is incorporated into a newly synthesized DNA chain, distorts the geometry of the double helix. This distortion is subsequently recognized by the DNA mismatch repair proteins, which then remove the newly synthesized DNA. The gap in the newly synthesized DNA is replaced by a DNA polymerase that proofreads as it synthesizes and is sealed by DNA ligase. As shown in the figure, a nick in the DNA has been proposed as the signal that allows the mismatch repair proteins to distinguish the newly synthesized DNA (which contains the mistake) from the old DNA. Such nicks are known to occur in the lagging strands (see Figure 6–12) and are observed also to occur, although less frequently, in the leading strands. These nicks remain for only a short period after a replication fork passes (see Figure 6–16 or 6–17), so that mismatch repair must occur quickly.

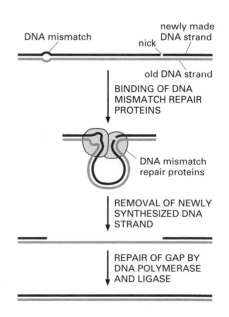

DNA mismatch — newly made DNA strand — nick — old DNA strand

BINDING OF DNA MISMATCH REPAIR PROTEINS

DNA mismatch repair proteins

REMOVAL OF NEWLY SYNTHESIZED DNA STRAND

REPAIR OF GAP BY DNA POLYMERASE AND LIGASE

Question 6–4

Discuss the following statement: "The DNA repair enzymes that correct defects introduced by deamination and depurination reactions must preferentially recognize such defects on newly synthesized DNA strands."

arise from cells that have accumulated multiple mutations (see Figure 6–20), a cell deficient in mismatch repair has a greatly enhanced chance of becoming cancerous. Thus, inheriting a damaged mismatch repair gene predisposes an individual to cancer.

DNA Is Continually Suffering Damage in Cells

Rare mistakes in DNA replication, as we have seen, can be corrected by the mismatch repair mechanism. There are also other ways in which the DNA can be damaged, and these require other mechanisms for their repair. Just like any other molecule in the cell, DNA is continually undergoing thermal collisions with other molecules. These often result in major chemical changes in the DNA. For example, during the time it takes to read this sentence, a total of about a trillion (10^{12}) purine bases (A and G) will be lost from the DNA of your cells by a spontaneous reaction called *depurination* (Figure 6–23). Depurination does not break the phosphodiester backbone but, instead, gives rise to lesions that resemble missing teeth. Another major change is the spontaneous loss of an amino group (*deamination*) from cytosine in DNA to produce the base uracil (see Figure 6–23). Some chemically reactive by-products of metabolism also occasionally react with the bases in DNA, altering them in such a way that their base-pairing properties are changed. The ultraviolet radiation in sunlight is also damaging to DNA; it promotes covalent linkage between two adjacent pyrimidine bases, forming, for example, the *thymine dimer* shown in Figure 6–24.

These are only a few of many chemical changes that can occur in our DNA. If left unrepaired, many of them would lead either to the substitution of one nucleotide pair by another as a result of incorrect base-pairing during replication or to deletion of one or more nucleotide pairs in the daughter DNA strand after DNA replication (Figure 6–25). Some types of DNA damage (thymine dimers, for example) often stall the DNA replication machinery at the site of the damage. All of these types of damage, if unrepaired, would have disastrous consequences for an organism.

Figure 6–23 Depurination and deamination are the most frequent chemical reactions known to create serious DNA damage in cells. (A) Depurination can release guanine as well as adenine from DNA. (B) The major type of deamination reaction converts cytosine to an altered DNA base, uracil, but deamination can occur on other bases as well. Both of these reactions take place on double-helical DNA; for convenience, only one strand is shown.

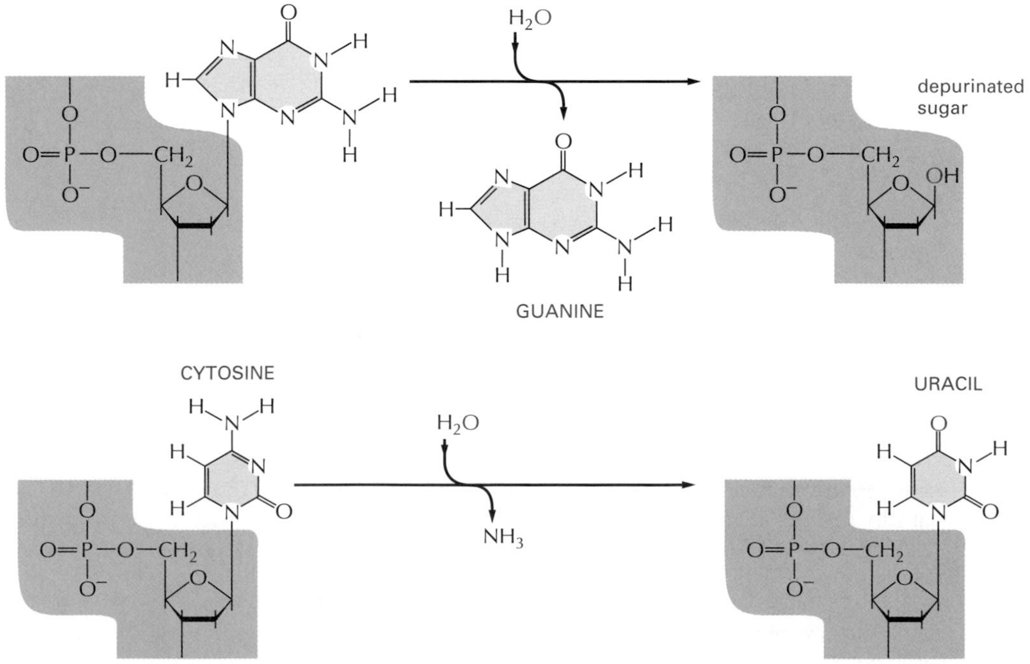

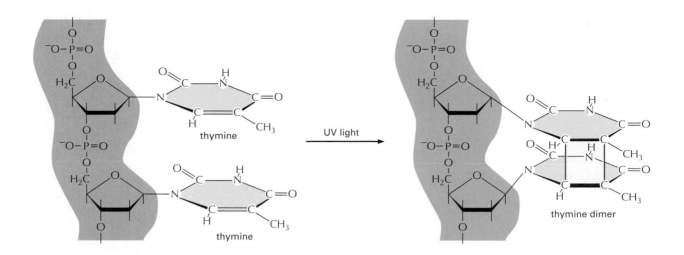

thymine

The Stability of Genes Depends on DNA Repair

The thousands of random chemical changes that occur every day in the DNA of a human cell, through metabolic accidents or exposure to DNA-damaging chemicals, are repaired by a variety of mechanisms, each catalyzed by a different set of enzymes. Nearly all these mechanisms depend on the existence of two copies of the genetic information, one in each strand of the DNA double helix: if the sequence in one strand is accidentally damaged, information is not lost irretrievably, because a backup version of the altered strand remains in the complementary sequence of nucleotides in the other strand. Most damage creates structures that are never encountered in an undamaged DNA strand; thus the good strand is easily distinguished from the bad. The basic pathway for repairing damage to DNA is illustrated schematically in Figure 6–26.

Figure 6–24 The ultraviolet radiation in sunlight causes DNA damage. Two adjacent thymine bases have become covalently attached to one another to form a thymine dimer. Skin cells that are exposed to sunlight are especially susceptible to this type of DNA damage.

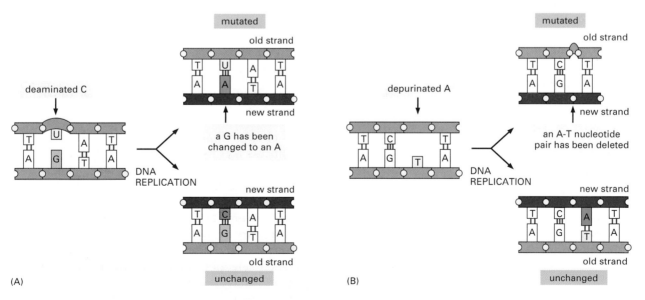

Figure 6–25 Chemical modifications of nucleotides, if left unrepaired, produce mutations. (A) Deamination of cytosine, if uncorrected, results in the substitution of one base for another when the DNA is replicated. As shown in Figure 6–23, deamination of cytosine produces uracil. Uracil differs from cytosine in its base-pairing properties and preferentially base-pairs with adenine. The DNA replication machinery therefore inserts an adenine when it encounters a uracil on the template strand. (B) Depurination, if uncorrected, can lead to the loss of a nucleotide pair. When the replication machinery encounters a missing purine on the template strand, it can skip to the next complete nucleotide, as shown, thus producing a nucleotide deletion in the newly synthesized strand. In other cases (not shown), the replication machinery places an incorrect nucleotide across from the missing base, again resulting in a mutation.

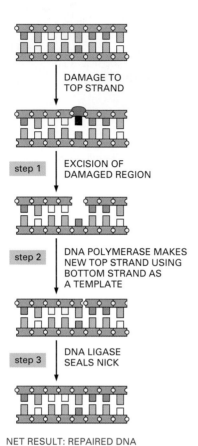

DAMAGE TO
TOP STRAND

step 1   EXCISION OF
          DAMAGED REGION

step 2   DNA POLYMERASE MAKES
          NEW TOP STRAND USING
          BOTTOM STRAND AS
          A TEMPLATE

step 3   DNA LIGASE
          SEALS NICK

NET RESULT: REPAIRED DNA

Figure 6–26 The basic mechanism of DNA repair involves three steps: excision, resynthesis, and ligation. In step 1 (excision), the damage is cut out by one of a series of nucleases, each specialized for a type of DNA damage. In step 2 (resynthesis), the original DNA sequence is restored by a repair DNA polymerase, which fills in the gap created by the excision events. In step 3 (ligation), DNA ligase seals the nick left in the sugar–phosphate backbone of the repaired strand. Nick sealing, which requires energy from ATP hydrolysis, remakes the broken phosphodiester bond between the adjacent nucleotides. Some types of DNA damage (the deamination of cytosine [Figure 6–23], for example) involve the replacement of a single nucleotide, as shown in the figure. For the repair of other kinds of DNA damage, such as thymine dimers (see Figure 6–24), a longer stretch of 10 to 20 nucleotides is removed from the damaged strand.

As indicated, it involves three steps:

1. The damaged DNA is recognized and removed by one of a variety of different nucleases, which cleave the covalent bonds that join the damaged nucleotides to the rest of the DNA molecule, leaving a small gap on one strand of the DNA double helix in this region.

2. A repair DNA polymerase binds to the 3′-hydroxyl end of the cut DNA strand. It then fills in the gap by making a complementary copy of the information stored in the undamaged strand. Although a different enzyme from the DNA polymerase that replicates DNA, a repair DNA polymerase synthesizes DNA strands in the same way. For example, it synthesizes chains in the 5′-to-3′ direction and has the same type of proofreading activity to ensure that the template strand is accurately copied. In many cells, this is the same enzyme that fills in the gap left after the RNA primers are removed in normal DNA replication (see Figure 6–16).

3. When the repair DNA polymerase has filled in the gap, a break remains in the sugar–phosphate backbone of the repaired strand. This nick in the helix is sealed by DNA ligase, the same enzyme that joins the lagging-strand DNA fragments during DNA replication.

Steps 2 and 3 are nearly the same for most types of DNA repair, including mismatch repair. However, step 1 uses a series of different enzymes, each specialized for removing different types of DNA damage.

The importance of these repair processes is indicated by the large investment that cells make in DNA repair enzymes. Single-celled organisms such as yeasts contain more than 50 different proteins that function in DNA repair, and DNA repair pathways are likely to be even more complex in humans. The importance of these DNA repair processes is also evident from the consequences of their malfunction. Humans with the genetic disease *xeroderma pigmentosum,* for example, cannot repair thymine dimers (see Figure 6–24) because they have inherited a defective gene for one of the proteins involved in this repair process. Such individuals develop severe skin lesions, including skin cancer, because of the accumulation of thymine dimers in cells that are exposed to sunlight and the consequent mutations that arise in the cells that contain them.

The High Fidelity of DNA Maintenance Allows Closely Related Species to Have Proteins with Very Similar Sequences

We have seen in this chapter that DNA is replicated and maintained with remarkable fidelity. As a consequence, changes in the DNA accumulate remarkably slowly in the course of evolution. Of course, the rate of evolutionary change in the DNA of a species depends also on the effects of natural selection: DNA copying errors that have harmful consequences for the organism are eliminated from the population through the death or reduced fertility of individuals carrying the misreplicated DNA. But the mechanisms of DNA replication and repair are so accurate that even where no such selection operates—at the many sites in the DNA where a change of nucleotide has no effect on the fitness of the organism—the genetic message is faithfully preserved over tens of millions of years. Thus humans and chimpanzees, after about 5 million years of divergent evolution, still have DNA sequences that are at least 98% identical. Even humans and whales, after 10 or 20 times this period, still have chromosomes that are unmistakably similar in their DNA sequence and many proteins with amino acid sequences that are almost identical (Figure 6–27). Thus, in our genomes, we and our relatives receive a message from the distant past—a message that is longer

whale   GTGTGGTCTCGTGATCAAAGGCGAAAGGTGGCTCTAGAGAATCCC
        |||||||||||| ||||| |||||| ||| |||||||||||||||||
human   GTGTGGTCTCGCGATCAGAGGCGCAAGATGGCTCTAGAGAATCCC

Figure 6–27 The sex-determination genes from humans and whales are unmistakably similar. Although their body plans are strikingly different, humans and whales are built from the same proteins. Despite the length of time since humans and whales diverged, the nucleotide sequences of many of their genes are still closely similar. The sequences of a part of the gene encoding the protein that determines maleness in humans and in whales are shown one above the other, and the positions where the two are identical are *shaded*.

and more detailed than any book. Thanks to the faithfulness of DNA replication and repair, 100 million years have scarcely changed its essential content.

DNA Recombination

Thus far we have discussed how the DNA sequences in cells can be maintained from generation to generation with very little change. However, it is also clear that the DNA sequence in chromosomes does change with time and can, as we discuss in this section, even be rearranged. The particular combination of genes present in any individual genome is often altered by such DNA rearrangements. In a population, this sort of genetic variation is important to allow organisms to evolve in response to a changing environment. These DNA rearrangements are caused by a class of mechanisms called **genetic recombination**.

We begin by reviewing several types of recombination mechanisms, each of which can introduce change into the genome of an individual cell. We discuss later in Chapter 9 how such rearrangements contribute to the genetic diversity and evolution of entire species.

Homologous Recombination Results in an Exact Exchange of Genetic Information

Of all the recombination mechanisms that exist, perhaps the most fundamental is **homologous recombination**. The central features that lie at the heart of homologous recombination seem to be the same in all organisms on Earth. Although the mechanism is not completely understood, the following characteristics are probably common to homologous recombination in all cells:

1. Two double-stranded DNA molecules that have regions of very similar (homologous) DNA sequence align so that their homologous sequences are in register. The DNA molecules can then "cross over": in a complex reaction, both strands of each double helix are broken and the broken ends are rejoined to the ends of the opposite DNA molecule to re-form two intact double helices, each made up of parts of the two different DNA molecules (Figure 6–28).

2. The site of exchange (that is, where a red double helix is joined to a green double helix in Figure 6–28) can occur anywhere in the homologous nucleotide sequences of the two participating DNA molecules.

3. No nucleotide sequences are altered at the site of exchange; the cleavage and rejoining events occur so precisely that not a single nucleotide is lost or gained.

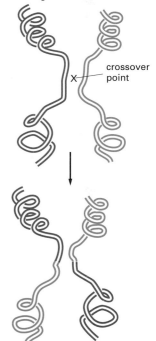

two homologous DNA double helices

crossover point

DNA molecules that have crossed over

Figure 6–28 Homologous recombination takes place between DNA molecules with similar nucleotide sequences. The breaking and rejoining of two homologous DNA double helices creates two DNA molecules that have "crossed over." Although the two original DNA molecules must have similar nucleotide sequences in order to cross over, they do not have to be identical; thus a crossover can create DNA molecules of novel nucleotide sequence.

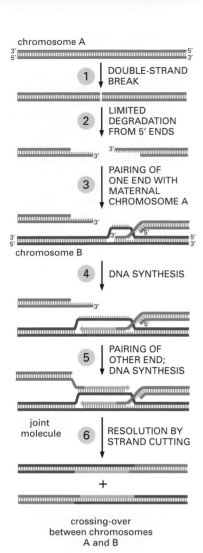

chromosome A

1 DOUBLE-STRAND BREAK

2 LIMITED DEGRADATION FROM 5′ ENDS

3 PAIRING OF ONE END WITH MATERNAL CHROMOSOME A

chromosome B

4 DNA SYNTHESIS

5 PAIRING OF OTHER END; DNA SYNTHESIS

joint molecule

6 RESOLUTION BY STRAND CUTTING

+

crossing-over between chromosomes A and B

Figure 6–29 Homologous recombination begins with a double-strand break in a chromosome. A DNA-digesting enzyme then creates protruding 3′ ends, which find the homologous region of a second chromosome. The joint molecule formed can eventually be resolved by selective strand cuts to produce two chromosomes that have crossed over, as shown.

Homologous recombination begins with a bold stroke: a special enzyme simultaneously cuts both strands of the double helix, creating a complete break in the DNA molecule (Figure 6–29). The 5′ ends at the break are then chewed back by a DNA-digesting enzyme, creating protruding single-stranded 3′ ends. Each of these single strands then searches for a homologous, complementary DNA helix with which to pair—leading to the formation of a "joint molecule" between the two chromosomes. The nicks in the DNA strands are then sealed so that the two DNA molecules are now held together physically by a crossing-over of one of each of their strands. This crucial intermediate in homologous recombination is known as a *cross-strand exchange* or *Holliday junction* (Figure 6–30).

To regenerate two separate DNA molecules, the two crossing strands must be cut. But if they are cut while the structure is still in the form shown in Figure 6–30A, the two original DNA molecules will separate from each other almost unaltered (Figure 6–30D). The structure can, however, undergo a series of rotational movements so that the two original noncrossing strands become crossing strands and vice versa (Figure 6–30B and C, and Figure 6–31). If the crossing strands are cut after rotation, one section of each original DNA helix is joined to a section of the other DNA helix; in other words, the two DNA molecules have crossed over, and two molecules of novel DNA sequence have been produced (Figure 6–30E).

As might be expected, cells use a set of specialized proteins to facilitate homologous recombination; these proteins break the DNA, catalyze strand exchange, and cleave Holliday structures. Because the essential features of homologous recombination are highly conserved, the proteins that carry out this process in different organisms are often very similar to one another in amino acid sequence.

Homologous recombination provides many advantages to cells and organisms. The process allows an organism to repair DNA that is damaged on both strands of the double helix, and it can fix other genetic accidents that occur during nearly every round of DNA replication. It is also essential for the accurate chromosome segregation that occurs during meiosis in fungi, plants, and animals, as we shall see in Chapter 20. The chromosomal "crossing-over" that occurs when homologous chromosomes come together causes bits of genetic information to be exchanged, generating new combinations of DNA sequences in each chromosome. The benefit of such gene mixing for the progeny organisms is apparently so great that the reassortment of genes by homologous recombination is not confined to sexually reproducing organisms; it is also widespread in asexually reproducing organisms, such as bacteria.

Recombination Can Also Occur Between Nonhomologous DNA Sequences

In homologous recombination, DNA rearrangements occur between DNA segments that are very similar in sequence. A second, more specialized type of recombination, called **site-specific recombination**,

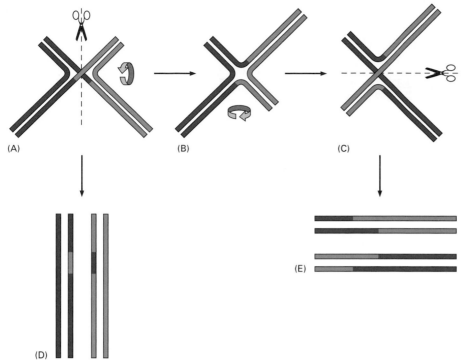

(A) (B) (C)

(D)

(E)

Figure 6–30 **The rotation of a Holliday junction allows recombination to occur.** In step A, a cross-strand exchange has formed, as shown in Figure 6–29. (A) Without a rotation (also called isomerization), cutting of the two crossing strands would terminate the exchange and homologous recombination would not occur (D). After rotation (B and C), cutting the two crossing strands creates two DNA molecules that have exchanged pieces of DNA (E). In the DNA strands shown here, the entire structure is rotated to yield the cross-strand exchange shown in C; to form B, the chromosome strands on the left-hand side of A are held steady while the strands on the right-hand side of the structure are rotated as indicated; to form C, the strands in the upper portion of B are held steady while the lower strands are rotated as indicated. In reality, however, only the regions immediately surrounding the site of crossing over need to rotate relative to one another to achieve the same result.

allows DNA exchanges to occur between DNA double helices that are dissimilar in nucleotide sequence. Although site-specific recombination performs a variety of tasks in the cell, perhaps its most prevalent function is to shuffle specialized bits of DNA called *mobile genetic elements*. These elements, found in the genomes of nearly all organisms, are short sequences of DNA that can move from one position in the genome to another through site-specific recombination.

Some of these mobile genetic elements are viruses that take advantage of site-directed recombination to move their genomes into and out of the chromosomes of their host cell. A virus can package its nucleic acid into viral particles that can move from one cell to another through the extracellular environment. However, most mobile elements can move only within a single cell and its descendants, as they lack any intrinsic ability to leave the cell in which they reside.

Mobile genetic elements often comprise a sizable fraction of an organism's DNA. For example, approximately 45% of the human genome is made up of mobile genetic elements; most of these elements, however, are fossils that—because they have been accumulating random mutations throughout the course of human evolution—have lost the ability to move within the genome.

Because they have a tendency to multiply, mobile DNA elements are sometimes called parasitic DNA. However, as we discuss in Chapter 9, mobile genetic elements also provide some advantages to their host genomes by generating the genetic variation upon which evolution depends.

Mobile Genetic Elements Encode the Components They Need for Movement

Unlike homologous recombination, site-specific recombination is guided by recombination enzymes that recognize short, specific nucleotide sequences present on one or both of the recombining DNA molecules; extensive DNA homology is not required. Each type of mobile genetic element generally encodes the enzyme that mediates its own movement and contains special sites upon which the enzyme acts

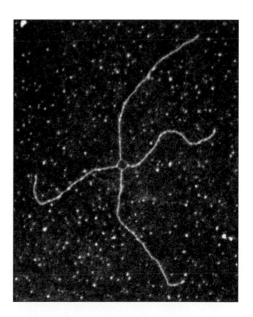

Figure 6–31 **A cross-strand exchange (Holliday junction) can be seen in this electron micrograph.** This view of the molecule corresponds to the open structure illustrated in Figure 6–30B. (Courtesy of Huntington Potter and David Dressler.)

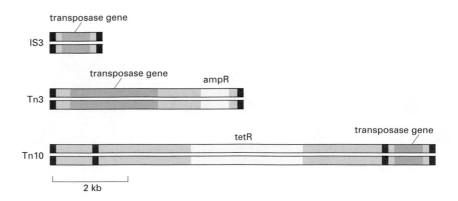

Figure 6–32 Bacteria contain many types of mobile genetic elements, three of which are shown. Each of these DNA-only transposons contains a gene that encodes a transposase *(blue)*, an enzyme that carries out some of the DNA breaking and joining reactions needed for the transposon to move. Each transposon also carries DNA sequences (indicated in *red)* that are recognized only by the transposase encoded by that element and are necessary for movement of the transposon. Some transposons carry, in addition, genes that encode enzymes that inactivate antibiotics such as ampicillin (ampR) and tetracycline (tetR). Movement of these genes presents a growing problem in medicine, as many disease-causing bacteria have become resistant to many of the antibiotics developed during the twentieth century.

(Figure 6–32). Many elements also carry other genes. For example, viruses encode coat proteins that enable them to exist outside cells, in addition to essential viral enzymes. The spread of mobile genetic elements that carry antibiotic resistance genes is a major factor underlying the widespread dissemination of antibiotic resistance in bacterial populations.

In bacteria, the most common genetic elements are called *DNA-only transposons*; these generally have only modest selectivity for their target sites and can thus insert themselves into many different DNA sequences. These transposons move from place to place within the genome by means of specialized recombination enzymes, called *transposases*, that are encoded by the transposable elements themselves (see Figure 6–32). The transposase first disconnects the transposon from the flanking DNA and then inserts it into a new target DNA site. Again, there is no requirement for homology between the ends of the element and the insertion site. Some bacterial transposons move to the target site using a cut-and-paste mechanism; others replicate before inserting into the new chromosomal site, leaving the original copy intact at its previous location (Figure 6–33).

A Large Fraction of the Human Genome Is Composed of Two Families of Transposable Sequences

As discussed in Chapter 5, a significant fraction of many vertebrate chromosomes is made up of repeated DNA sequences. Many of these

Figure 6–33 Bacterial transposons move by two mechanisms. (A) In cut-and-paste nonreplicative transposition, the transposon is cut out of the donor DNA and inserted into the target DNA, leaving behind a broken donor DNA molecule. (The donor can be repaired in a variety of ways, but this sometimes results in deletions or rearrangements of the donor molecule.) (B) In the course of replicative transposition, the transposon is copied by DNA replication. The end products are a molecule that appears identical to the original donor and a target molecule that has a transposon inserted into it. In general, a particular type of transposon moves by only one of these mechanisms. However, the two mechanisms have many enzymatic similarities, and a few transposons can move by either mechanism. The donor and target DNAs can be part of the same DNA molecule or reside on different molecules.

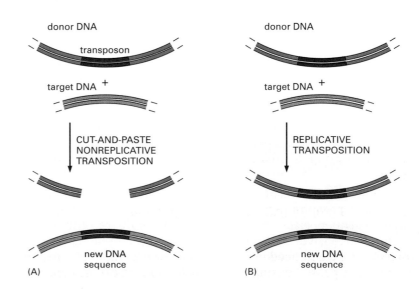

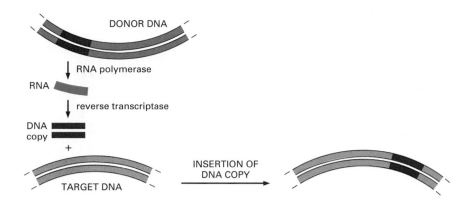

Figure 6–34 Retrotransposons move via an RNA intermediate. These transposable elements are first transcribed into an RNA intermediate. A DNA copy of this RNA is made by reverse transcriptase. The DNA copy of the transposon is then inserted into the target location. The target can be on the same or a different DNA molecule from the donor. These transposable elements are called retrotransposons because at one stage in their transposition their genetic information flows backward, from RNA to DNA.

repeated sequences are mobile DNA elements that have proliferated over evolutionary time-scales, although most can no longer move, as they have accumulated deleterious mutations. Some vertebrate mobile elements have moved from place to place within their host chromosomes using the cut-and-paste mechanism discussed above for bacterial transposons (see Figure 6–33A). However, many others have moved not as DNA but via an RNA intermediate. These are called **retrotransposons** and are, as far as is known, unique to eucaryotes.

One type of retrotransposon, the *L1 element* (sometimes referred to as *LINE-1*), is a highly repeated sequence that constitutes about 15% of the total mass of the human genome. Although most copies of the *L1* element are immobile, a few still retain the ability to move. Translocation of *L1* can sometimes result in human disease: a particular form of hemophilia, for example, is caused by insertion of an *L1* element into the gene that encodes Factor VIII, a protein essential for proper blood clotting.

The *L1* element transposes by first being transcribed into RNA by cellular RNA polymerases. A DNA copy of this RNA is then made using the enzyme *reverse transcriptase*, an unusual DNA polymerase that can use RNA as a template. The reverse transcriptase is encoded by the *L1* element itself. The DNA copy can then reintegrate into another site in the genome (Figure 6–34).

Almost as abundant in the human genome is the *Alu sequence*, which is unusually short (about 300 nucleotide pairs). *Alu* is present in about 1 million copies in the genome and constitutes about 11% of human DNA; thus it appears on average about once every 5000 nucleotide pairs. Only some of the *Alu* sequences in the genome can still be copied into RNA. Because *Alu* elements do not encode their own reverse transcriptase, they depend on enzymes already present in the cell to help them move.

Comparisons of the sequence and locations of the *L1* and *Alu*-like elements in different mammals suggest that these sequences have multiplied to high copy numbers in primates relatively recently in evolutionary time (Figure 6–35). These highly abundant sequences, scattered throughout our genome, must have had major effects on the expression of many of our genes. It is perhaps humbling to contemplate how many of our uniquely human qualities we might owe to these parasitic genetic elements.

Viruses Are Fully Mobile Genetic Elements That Can Escape from Cells

Viruses were first noticed as disease-causing agents that, by virtue of their tiny size, passed through ultrafine filters that can hold back even

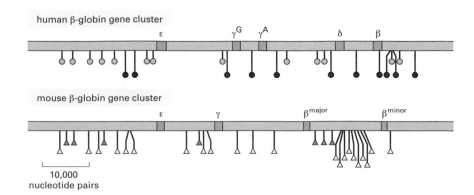

Figure 6–35 *L1* and *Alu*-like elements have multiplied to high copy numbers relatively recently in evolutionary time. The human genome contains a cluster of five globin genes *(top)*. Each gene (shown in *orange* and designated by a Greek letter) encodes a protein that carries oxygen in the blood. The comparable region from the mouse genome *(bottom)* contains only four globin genes. The positions of the human *Alu* sequences are indicated by *green circles,* and the human *L1* elements by *red circles.* The mouse genome contains different transposable elements: the positions of *B1* elements (which are related to the human *Alu* sequences) are indicated by *blue triangles,* and the positions of the mouse *L1* elements (which are related to the human *L1*) are indicated by *yellow triangles.* Because the DNA sequences of the mouse and human transposable elements are distinct and because the positions of these transposable elements on the β-globin gene cluster are very different between human and mouse, it is believed that they accumulated in mammalian genomes relatively recently in evolutionary time. (Courtesy of Ross Hardison and Webb Miller.)

the smallest cell. We now know that viruses are essentially genes enclosed by a protective protein coat. However, these genes must enter a cell and utilize the cell's machinery to express their genes as proteins and to replicate their chromosomes, so as to package themselves into newly made protective coats. Virus reproduction per se is often lethal to the cells in which it occurs; in many cases the infected cell breaks open (lyses) and thereby releases the progeny viruses and allows them access to nearby cells. Many of the medical symptoms of viral infection reflect the lytic effect of the virus. The cold sores formed by herpes simplex virus and the blisters caused by the chicken pox virus, for example, both reflect the localized killing of skin cells. Although the first viruses that were discovered attack mammalian cells, it is now recognized that many types of viruses exist. Some of these infect plant cells, while others use bacterial cells as their hosts.

Viral genomes can be made of DNA or RNA and can be single-stranded or double-stranded (Table 6–2 and Figure 6–36). The amount of DNA or RNA that can be packaged inside a protein shell is limited,

Table 6–2 Viruses That Cause Human Disease

VIRUS	GENOME TYPE	DISEASE
Herpes simplex virus	double-stranded DNA	recurrent cold sores
Epstein-Barr virus (EBV)	double-stranded DNA	infectious mononucleosis
Varicella-zoster virus	double-stranded DNA	chicken pox and shingles
Smallpox virus	double-stranded DNA	smallpox
Hepatitis B virus	part single-, part double-stranded DNA	serum hepatitis
Human immuno-deficiency virus (HIV)	single-stranded RNA	acquired immunodeficiency syndrome (AIDS)
Influenza virus type A	single-stranded RNA	respiratory disease (flu)
Poliovirus	single-stranded RNA	infantile paralysis
Rhinovirus	single-stranded RNA	common cold
Hepatitis A virus	single-stranded RNA	infectious hepatitis
Hepatitis C virus	single-stranded RNA	non-A, non-B type hepatitis
Yellow fever virus	single-stranded RNA	yellow fever
Rabies virus	single-stranded RNA	rabies
Mumps virus	single-stranded RNA	mumps
Measles virus	single-stranded RNA	measles

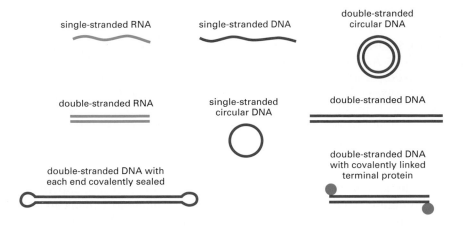

single-stranded RNA

single-stranded DNA

double-stranded circular DNA

double-stranded RNA

single-stranded circular DNA

double-stranded DNA

double-stranded DNA with each end covalently sealed

double-stranded DNA with covalently linked terminal protein

Figure 6–36 Viral genomes differ in structure. The smallest viruses contain only a few genes and can have an RNA or a DNA genome; the largest viruses contain hundreds of genes and have a double-stranded DNA genome. Some examples of these types of viruses are as follows: *single-stranded RNA*—tobacco mosaic virus, bacteriophage R17, poliovirus; *double-stranded RNA*—reovirus; *single-stranded DNA*—parvovirus; *single-stranded circular DNA*—M13 and φX174 bacteriophages; *double stranded circular DNA*—SV40 and polyoma-viruses; *double-stranded DNA*—T4 bacteriophage, herpesvirus; *double-stranded DNA with covalently linked terminal protein*—adenovirus; *double-stranded DNA with covalently sealed ends*—poxvirus. The peculiar ends (as well as the circular forms) present in some viral genomes overcome the difficulty of replicating the last few nucleotides at the end of a DNA chain (see Figure 6–18).

and is too small to encode the many different enzymes and other proteins that are required to replicate even the simplest virus. Viruses are therefore parasites that can reproduce themselves only inside a living cell, where they are able to hijack the cell's own biochemical machinery. Viral genomes typically encode the viral coat proteins as well as proteins that attract host enzymes to replicate their genome (Figure 6–37).

The simplest viruses consist of a protein coat made up primarily of many copies of a single polypeptide chain surrounding a small genome composed of as few as three genes. More complex viruses have larger genomes of up to several hundred genes, surrounded by an elaborate shell composed of many different proteins (Figure 6–38).

Even the largest viruses depend heavily on their host cells for biosynthesis; no known virus makes its own ribosomes or generates the ATP needed for nucleic acid replication, for example. Clearly, cells must have evolved before viruses.

Retroviruses Reverse the Normal Flow of Genetic Information

Although there are many similarities between bacterial and eucaryotic viruses, one important type of virus—the **retrovirus**—is found only in eucaryotic cells. In many respects, retroviruses resemble the retrotransposons we discussed earlier. A key feature found in both these genetic elements is a step where DNA is synthesized using RNA as a template (the term "retro" refers to the reversal of the usual flow of DNA information to RNA (see Figure 7–2), which is described in detail in the next chapter. The enzyme that carries out this step is *reverse transcriptase;* the retroviral genome (which is single-stranded RNA) encodes this enzyme, and a few molecules of the enzyme are packaged along with the RNA genome in each individual virus.

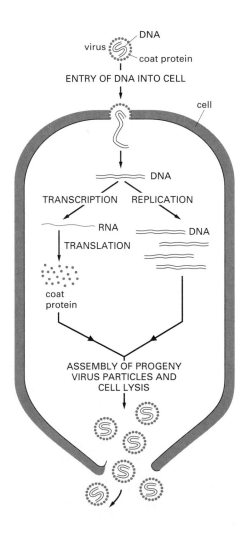

Figure 6–37 Viruses commandeer the host cell's machinery to reproduce. The simple, hypothetical virus illustrated consists of a small double-stranded DNA molecule that encodes just a single type of viral coat protein. In order to replicate, the viral genome must enter the cell. This is followed by replication of the viral DNA to form many copies. At the same time, the viral genes are expressed into protein through the sequential steps of *transcription* and *translation*, which will be described in the next chapter (see also Figure 7–2). The viral genomes assemble spontaneously with the coat proteins to form new virus particles.

The life cycle of a retrovirus is shown in Figure 6–39. When the single-stranded RNA genome of the retrovirus enters a cell, the reverse transcriptase brought in with it makes a complementary DNA strand to form a DNA/RNA hybrid double helix. The RNA strand is removed, and the reverse transcriptase (which can use either DNA or RNA as a template) now synthesizes a complementary strand to produce a DNA double helix. This DNA is then integrated into a randomly selected site in the host genome by a virally encoded integrase enzyme. In this state, the virus is latent: each time the host cell divides, it passes on a copy of the integrated viral genome to its progeny cells.

The next step in the replication of a retrovirus—which can take place long after its integration into the host genome—is the copying of the integrated viral DNA into RNA by a host cell enzyme, which can produce large numbers of single-stranded RNAs identical to the original infecting genome. (This enzyme, RNA polymerase, is discussed in Chapter 7). The viral genes are then expressed by the host cell machinery to produce the protein shell, the envelope proteins, and reverse transcriptase—all of which are assembled with the RNA genome into new virus particles.

The human immunodeficiency virus (HIV), which is the cause of AIDS, is a retrovirus. As with other retroviruses, the HIV genome can persist in the latent state as a DNA provirus embedded in the chromosomes of an infected cell. This ability of the virus to hide within host cells complicates any attempt to treat the infection with antiviral drugs. But because the HIV reverse transcriptase is not used by cells for any purpose of their own, one of the prime targets of drug development against AIDS is the viral reverse transcriptase.

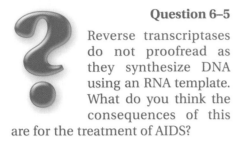

Question 6–5

Reverse transcriptases do not proofread as they synthesize DNA using an RNA template. What do you think the consequences of this are for the treatment of AIDS?

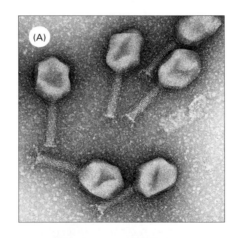

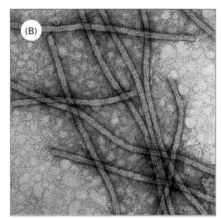

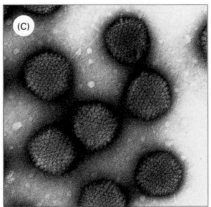

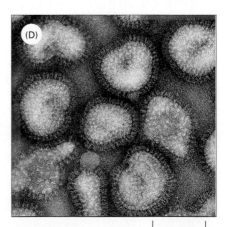

100 nm

Figure 6–38 Viral coats come in different shapes and sizes. These electron micrographs of virus particles are all shown at the same scale. (A) *T4,* a large DNA-containing virus that infects *E. coli* cells. The DNA is stored in the bacteriophage head and injected into the bacterium through the cylindrical tail. (B) *Potato virus X,* a plant virus that contains an RNA genome. (C) *Adenovirus,* a DNA-containing virus that can infect human cells. (D) *Influenza virus,* a large RNA-containing animal virus whose protein capsid is further enclosed in a lipid-bilayer-based envelope. The spikes protruding from the envelope are viral proteins embedded in the membrane bilayer (see Figure 6–39). (A, courtesy of James Paulson; B, courtesy of Graham Hills; C, courtesy of Mei Lie Wong; D, courtesy of R.C. Williams and H.W. Fisher.)

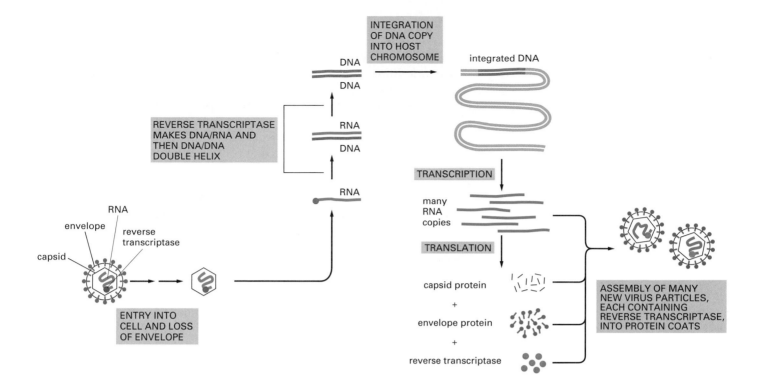

INTEGRATION OF DNA COPY INTO HOST CHROMOSOME

integrated DNA

DNA
DNA

REVERSE TRANSCRIPTASE MAKES DNA/RNA AND THEN DNA/DNA DOUBLE HELIX

RNA
DNA

RNA

RNA
envelope
reverse transcriptase
capsid

ENTRY INTO CELL AND LOSS OF ENVELOPE

TRANSCRIPTION

many RNA copies

TRANSLATION

capsid protein
+
envelope protein
+
reverse transcriptase

ASSEMBLY OF MANY NEW VIRUS PARTICLES, EACH CONTAINING REVERSE TRANSCRIPTASE, INTO PROTEIN COATS

Essential Concepts

- The ability of a cell to maintain order in a chaotic environment depends on the accurate duplication of the vast quantity of genetic information carried in its DNA.

- Each of the two DNA strands can act as a template for the synthesis of the other strand. A DNA double helix thus carries the same information in each of its strands.

- A DNA molecule is duplicated (replicated) by the polymerization of new complementary strands onto each of the old strands of the DNA double helix. This process of DNA replication, in which two identical DNA molecules are formed from the original molecule, enables the genetic information to be copied and passed on from cell to daughter cell and from parent to offspring.

- As a DNA molecule replicates, its two strands are pulled apart to form one or more Y-shaped replication forks. The enzyme DNA polymerase, situated in the fork, lays down a new complementary DNA strand on each parental strand, thereby making two new double-helical molecules.

- DNA polymerase replicates a DNA template with remarkable fidelity, making less than one error in every 10^7 bases read. This is possible because the enzyme removes its own polymerization errors as it moves along the DNA (proofreading).

- Because DNA polymerase can synthesize new DNA in only one direction, only one of the strands in the replication fork, the leading strand, can be replicated in a continuous fashion. On the lagging strand DNA is synthesized by the polymerase in a discontinuous "backstitching" process, making short fragments of DNA that are later joined up by the enzyme DNA ligase to make a single continuous DNA strand.

- The proofreading feature of DNA polymerase makes it incapable of starting a new DNA chain. DNA synthesis is primed by an RNA polymerase, called primase, that makes short lengths of RNA, called primers, that are subsequently erased and replaced with DNA.

Figure 6–39 The life cycle of a retrovirus includes reverse transcription and integration into the host genome. The retrovirus genome consists of an RNA molecule of about 8500 nucleotides; two such molecules are packaged into each viral particle. The enzyme reverse transcriptase first makes a DNA copy of the viral RNA molecule and then a second DNA strand, generating a double-stranded DNA copy of the RNA genome. The integration of this DNA double helix into the host chromosome is then catalyzed by a virus-encoded *integrase* enzyme. This integration is required for the synthesis of new viral RNA molecules by the host cell RNA polymerase, the enzyme that transcribes DNA into RNA (discussed in Chapter 7).

- DNA replication requires the cooperation of many proteins, which form a multienzyme replication machine, situated at the replication fork, that catalyzes DNA synthesis.
- Genetic information can be stored stably in DNA sequences only because a variety of DNA repair enzymes continuously scan the DNA and correct replication mistakes and replace damaged nucleotides. DNA can be repaired easily because one strand can be corrected using the other strand as a template.
- In eucaryotes, a special enzyme called telomerase replicates the DNA at the ends of the chromosomes.
- The rare copying mistakes that slip through the DNA replication machinery are dealt with by the mismatch repair proteins, which monitor newly replicated DNA and repair copying mistakes. The overall accuracy of DNA replication, including mismatch repair, is one mistake per 10^9 nucleotides copied.
- DNA damage caused by chemical reactions and ultraviolet irradiation is corrected by a variety of enzymes that recognize damaged DNA and excise a short stretch of the DNA strand that contains it. The missing DNA is resynthesized by a repair DNA polymerase that uses the undamaged strand as a template. DNA ligase reseals the DNA to complete the repair process.
- Homologous recombination is the process by which two double-stranded DNA molecules of similar nucleotide sequence can cross over to create DNA molecules of novel sequence.
- Mobile genetic elements are DNA sequences that can move from place to place in the genomes of their hosts. This movement creates change in the host genomes and provides a source of genetic variation.
- More than 50% of the human genome consists of DNA that is repeated many times in the genome. Approximately two-thirds of this repeated DNA (about 34% of the total genome) consists of two classes of transposons that have multiplied to especially high copy numbers in the genome.
- Viruses are little more than genes packaged in protective protein coats. They require host cells in order to reproduce themselves.
- Some viruses have RNA instead of DNA as their genomes. One group of RNA viruses—the retroviruses—must copy their RNA genomes into DNA in order to replicate.

Key Terms

DNA polymerase	proofreading
DNA repair	replication fork
DNA replication	retrotransposon
genetic recombination	retrovirus
homologous recombination	RNA (ribonucleic acid)
lagging strand	site-specific recombination
leading strand	telomerase
mobile genetic element	template
mutation	transposon
Okazaki fragment	virus
plasmid	

Questions

Question 6–6

DNA repair enzymes preferentially repair mismatched bases on the newly synthesized DNA strand, using the old DNA strand as a template. If mismatches were simply repaired without regard for which strand served as template, would this reduce replication errors? Explain your answer.

Question 6–7

Suppose a mutation affects an enzyme that is required to repair the damage to DNA caused by the loss of purine bases. This mutation causes the accumulation of about 5000 mutations in the DNA of each of your cells per day. As the average difference in DNA sequence between humans and chimpanzees is about 1%, how long will it take you to turn into an ape? What is wrong with this argument?

Question 6–8

Which of the following statements are correct? Explain your answers.
  A. The replication fork is asymmetrical because it contains two DNA polymerase molecules that are structurally distinct.
  B. Okazaki fragments are removed by an RNA nuclease.
  C. The error rate of DNA replication is reduced both by proofreading of the DNA polymerase and by DNA repair enzymes.
  D. In the absence of DNA repair, genes are unstable.
  E. None of the aberrant bases formed by deamination occur naturally in DNA.
  F. Cancer results from uncorrected mutations in somatic cells.

Question 6–9

Being a born skeptic, you plan to confirm for yourself the results of a classic experiment originally performed in the 1960s by Meselson and Stahl from which they concluded that each daughter cell inherits one and only one strand of its mother's DNA. To do so, you "synchronize" (using established methods that need not concern us here) a culture of growing cells, so that virtually all cells in your flask begin and then complete DNA synthesis at the same time. Your cells are first grown in a normal growth medium and then, after one round of DNA synthesis, grown further in a specially concocted (and very expensive) growth medium that contains nutrients highly enriched in heavy isotopes of nitrogen and carbon (^{15}N and ^{13}C in place of the naturally abundant ^{14}N and ^{12}C). Cells growing on this medium use the heavy isotopes to build all of their macromolecules, including nucleotides and nucleic acids. You then isolate DNA from cells that have grown for a different number of generations in the heavy-isotope medium and analyze the DNA for its density

using a gradient centrifugation technique (see Panel 4–3, pp. 160–161). The more heavy isotopes have been built into the DNA, the heavier it appears in this analysis. Your data, plotting the amount of DNA isolated over its density, are shown in Figure Q6–9. Are these results in agreement with your expectations? Explain the results.

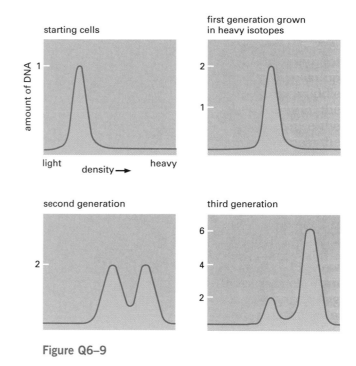

Figure Q6–9

Question 6–10

The speed of DNA replication at a replication fork is about 100 nucleotides per second in human cells. What is the minimum number of origins of replication that a human cell must have in order to replicate its DNA once every 24 hours? Recall that a human cell contains two copies of the human genome, one inherited from the mother, the other from the father, each consisting of 3×10^9 nucleotide pairs.

Question 6–11

Look carefully at the structures of the compounds shown in Figure Q6–11. One or the other of the two compounds is added to a DNA replication reaction.
  A. What would you expect if compound A were added in large excess over the concentration of the available deoxycytosine triphosphate (dCTP)?
  B. What would happen if it were added at 10% of the concentration of the available dCTP?
  C. What effects would you expect if compound B were added under the same conditions?

Question 6–12

The genetic material of a hypothetical organism is structurally indistinguishable from DNA of normal

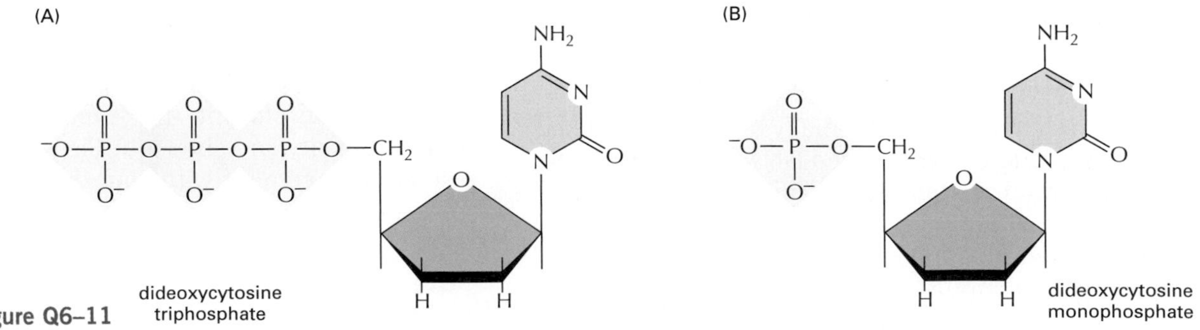

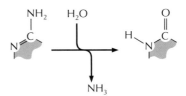

(A) dideoxycytosine triphosphate

Figure Q6–11

(B) dideoxycytosine monophosphate

cells. Surprisingly, analyses reveal that the DNA is synthesized from nucleoside triphosphates that contain free 5′-hydroxyl groups and triphosphate groups at the 3′ position. In what way must this organism's DNA polymerase differ from that of normal cells? Could it still proofread?

Question 6–13

Figure Q6–13 shows a snapshot of a replication fork in which the RNA primer has just been added to the lagging strand. Using this diagram as a guide, sketch the path of the DNA as the next Okazaki fragment is synthesized. Indicate the sliding clamp and the single-strand binding protein as appropriate.

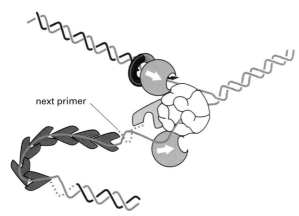

next primer

Figure Q6–13

Question 6–14

Approximately how many high-energy bonds are used to replicate a bacterial chromosome? How much glucose (compared with its own weight of 10^{-12} g) does a bacterium need to consume to provide enough energy to copy its DNA once? The number of base pairs in the bacterial chromosome is 3×10^6. Oxidation of one glucose molecule yields about 30 high-energy phosphate bonds. The molecular weight of glucose is 180 g/mole. (Recall that there are 6×10^{23} molecules in a mole; discussed in Chapter 2.)

Question 6–15

What, if anything, is wrong with the following statement: "Both reproductive-cell DNA stability and somatic-cell DNA stability are essential for the survival of a species." Explain your answer.

Question 6–16

A common type of error in DNA is produced by a spontaneous reaction termed *deamination* in which a nucleotide base loses an amino group (NH_2), which is replaced by a keto group (C=O) by the general reaction shown in Figure Q6–16. Write the structures of the bases A, G, C, T, and U and predict the products that will be produced by deamination. By looking at the products of this reaction—and remembering that, in the cell, these will need to be recognized and repaired—can you propose an explanation why DNA cannot contain uracil?

NH_2 H_2O

NH_3

Figure Q6–16

Question 6–17

A. Explain why telomeres and telomerase are needed for replication of eucaryotic chromosomes but not for replication of a circular bacterial chromosome. Draw a diagram to illustrate your explanation.

B. Would you still need telomeres and telomerase to complete eucaryotic chromosome replication if DNA primase always laid down the RNA primer at the very 3′ end of the template for the lagging strand?

Question 6–18

Discuss the following statement: "Viruses exist in the twilight zone of life: outside cells they are simply dead assemblies of molecules; inside cells, however, they are alive."

Question 6–19

Many transposons move within a genome by replicative mechanisms (such as those shown in Figures 6–33 and 6–34) and so increase the number of copies in the genome at each transposition. Although individual transposition events are rare, many transposons are found in multiple copies in genomes. What do you suppose keeps the transposons from completely overrunning their hosts' genomes?

Question 6–20

Describe the consequences that would arise if a eucaryotic chromosome

    A. Contained only one origin of replication:
        (i) At the exact center of the chromosome
        (ii) At one end of the chromosome
    B. Lacked one or both telomeres
    C. Had no centromere

Assume that the chromosome is 150 million nucleotide pairs in length, a typical size for an animal chromosome, and that DNA replication in animal cells proceeds at about 100 nucleotides per second.

Highlights from *Essential Cell Biology 2 Interactive* CD-ROM

6.4 Replication I
6.5 Replication II

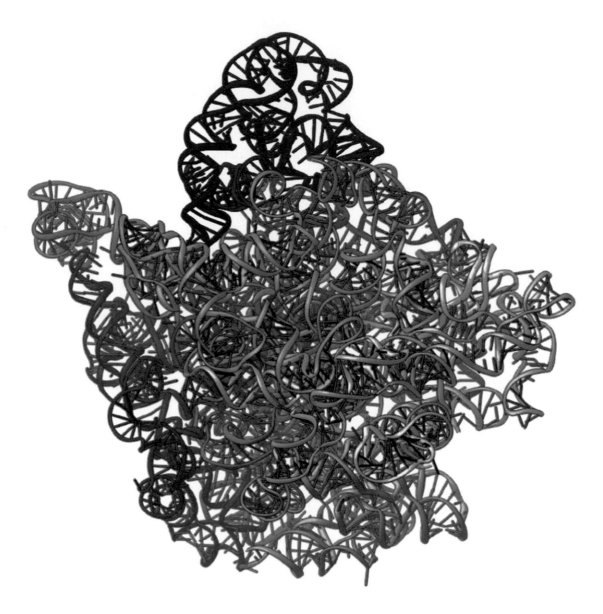

The Ribosome. Ribosomes are the complex macromolecular machines that catalyze the formation of the peptide bonds in proteins when messenger RNA sequences are translated into amino acid sequences during protein synthesis. Two-thirds of the mass of a ribosome is composed of RNA, which forms both its structural core and its catalytic site for peptide bond formation. Shown here is the complex of the two ribosomal RNA molecules that forms the core of the larger of the two ribosomal subunits in bacteria. (Adapted from N.Ban et al., *Science* 289:905–920, 2000.)

From DNA to Protein: How Cells Read the Genome

Once the structure of DNA (deoxyribonucleic acid) had been determined in the early 1950s, it became clear that the hereditary information in cells is encoded in DNA's sequence of nucleotides. We saw in Chapter 6 how this information can be passed on unchanged from a cell to its descendants through the process of DNA replication. But how does the cell decode and use the information? How do genetic instructions written in an alphabet of just four "letters"—the four different nucleotides in DNA—direct the formation of a bacterium, a fruit fly, or a human? We still have a lot to learn about how the information stored in an organism's genes produces even the simplest unicellular bacterium, let alone how it directs the development of complex multicellular organisms like ourselves. But the DNA code itself has been deciphered, and the language of genes can be read.

Even before the DNA code had been broken, it was known that the information contained in genes somehow directed the synthesis of proteins. Proteins are the principal constituents of cells and determine not only their structure but also their functions. In previous chapters, we have encountered some of the thousands of different kinds of proteins that cells can make. We saw in Chapter 4 that the properties and function of a protein molecule are determined by the linear order—the *sequence*—of the different amino acid subunits in its polypeptide chain: each type of protein has its own unique amino acid sequence, and this sequence dictates how the chain will fold to give a molecule with a distinctive shape and chemistry. The genetic instructions carried by DNA must therefore specify the amino acid sequences of proteins. We shall see in this chapter exactly how this is done.

DNA does not direct protein synthesis itself, but acts rather like a manager, delegating the various tasks required to a team of workers. When a particular protein is needed by the cell, the nucleotide sequence of the appropriate portion of the immensely long DNA molecule in a chromosome is first copied into another type of nucleic acid—*RNA (ribonucleic acid)*. It is these RNA copies of short segments of the DNA that are used as templates to direct the synthesis of the protein. Many thousands of these conversions from DNA to protein are occurring each second in every cell in our bodies. The flow of genetic information in cells is therefore from DNA to RNA to protein (Figure 7–1). All cells, from bacteria to humans, express their genetic information in this way—a principle so fundamental that it has been termed the *central dogma* of molecular biology.

The primary task of this chapter is to explain the mechanisms by which cells copy DNA into RNA (a process called *transcription*) and then use the information in RNA to make protein (a process called

From DNA to RNA
Portions of DNA Sequence Are Transcribed into RNA

Transcription Produces RNA Complementary to One Strand of DNA

Several Types of RNA Are Produced in Cells

Signals in DNA Tell RNA Polymerase Where to Start and Finish

Eucaryotic RNAs Are Transcribed and Processed Simultaneously in the Nucleus

Eucaryotic Genes Are Interrupted by Noncoding Sequences

Introns Are Removed by RNA Splicing

Mature Eucaryotic mRNAs Are Selectively Exported from the Nucleus

mRNA Molecules Are Eventually Degraded by the Cell

The Earliest Cells May Have Had Introns in Their Genes

From RNA to Protein
An mRNA Sequence Is Decoded in Sets of Three Nucleotides

tRNA Molecules Match Amino Acids to Codons in mRNA

Specific Enzymes Couple tRNAs to the Correct Amino Acid

The RNA Message Is Decoded on Ribosomes

The Ribosome Is a Ribozyme

Codons in mRNA Signal Where to Start and to Stop Protein Synthesis

Proteins Are Made on Polyribosomes

Inhibitors of Procaryotic Protein Synthesis Are Used as Antibiotics

Carefully Controlled Protein Breakdown Helps Regulate the Amount of Each Protein in a Cell

There Are Many Steps Between DNA and Protein

RNA and the Origins of Life
Life Requires Autocatalysis

RNA Can Both Store Information and Catalyze Chemical Reactions

RNA Is Thought to Predate DNA in Evolution

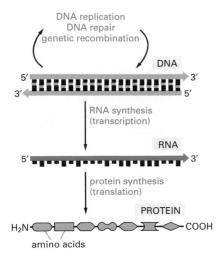

DNA replication
DNA repair
genetic recombination

DNA

5′ 3′
3′ 5′

RNA synthesis
(transcription)

RNA

5′ 3′

protein synthesis
(translation)

PROTEIN

H₂N— —COOH

amino acids

Figure 7–1 Genetic information directs the synthesis of protein. The flow of genetic information from DNA to RNA (transcription) and from RNA to protein (translation) occurs in all living cells.

Question 7–1

Consider the expression "central dogma," referring to the proposition that genetic information flows from DNA to RNA to protein. Is the word "dogma" appropriate in this scientific context?

translation). We shall also see that there are several variations on this basic scheme. Principal among these is *RNA splicing,* a process whereby RNA transcripts are rearranged before eucaryotic cells translate them into proteins. These alterations change the "meaning" of RNA molecules and therefore are crucial for understanding how cells decode the genome. In the final section of this chapter, we shall consider how the present scheme of information storage, transcription, and translation might have arisen from simpler systems in the earliest stages of cellular evolution.

From DNA to RNA

Transcription and translation are the means by which cells read out, or express, their genetic instructions—their *genes.* Many identical RNA copies can be made from the same gene, and each RNA molecule can direct the synthesis of many identical protein molecules. Because each cell contains only one or two copies of any particular gene, this successive amplification enables cells to synthesize the required amount of a protein much more rapidly than if the DNA itself were acting as the direct template for protein synthesis. Each gene can be transcribed and translated with a different efficiency, and this provides the cell with a way to make vast quantities of some proteins and tiny quantities of others (Figure 7–2). Moreover, as we shall see in Chapter 8, a cell can change (or regulate) the expression of each of its genes according to the needs of the moment. In this section, we shall discuss the production of RNA, the first step in gene expression.

Portions of DNA Sequence Are Transcribed into RNA

The first step a cell takes in reading out part of its genetic instructions is to copy the required portion of the nucleotide sequence of DNA—the gene—into a nucleotide sequence of RNA. The process is called **transcription** because the information, though copied into another chemical form, is still written in essentially the same language—the language of nucleotides. Like DNA, RNA is a linear polymer made of four different types of nucleotide subunits linked together by phosphodiester bonds (Figure 7–3). It differs from DNA chemically in two respects: (1) the nucleotides in RNA are *ribonucleotides*—that is, they contain the sugar ribose (hence the name *ribo*nucleic acid) rather than deoxyribose; (2) although, like DNA, RNA contains the bases adenine (A), guanine

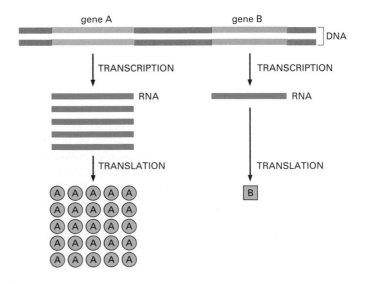

Figure 7–2 Genes can be expressed with different efficiencies. Gene A is transcribed and translated much more efficiently than is gene B. This allows the amount of protein A in the cell to be much higher than that of protein B. In this and later figures, the untranscribed portions of the DNA are shown in *gray.*

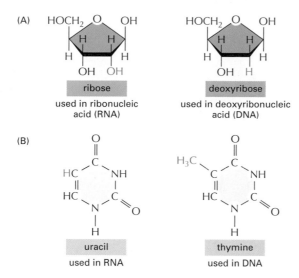

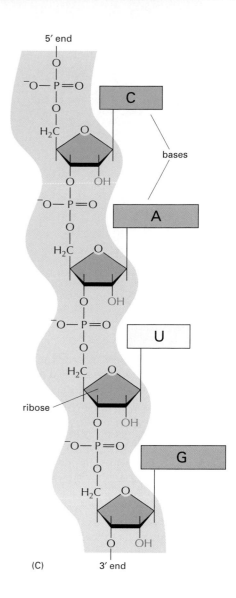

Figure 7–3 The chemical structure of RNA differs slightly from that of DNA. (A) RNA contains the sugar ribose, which differs from deoxyribose, the sugar used in DNA, by the presence of an additional –OH group. (B) RNA contains the base uracil, which differs from thymine, the equivalent base in DNA, by the absence of a –CH$_3$ group. (C) A short length of RNA. The chemical linkage between nucleotides in RNA is the same as that in DNA.

(G), and cytosine (C), it contains uracil (U) instead of the thymine (T) found in DNA. Because U, like T, can base-pair by hydrogen-bonding with A (Figure 7–4), the complementary base-pairing properties described for DNA in Chapter 5 apply also to RNA.

Although their chemical differences are small, DNA and RNA differ quite dramatically in overall structure. Whereas DNA always occurs in cells as a double-stranded helix, RNA is single-stranded. This difference has important functional consequences. Because an RNA chain is single-stranded, it can fold up into a variety of shapes, just as a polypeptide chain folds up to form the final shape of a protein (Figure 7–5); double-stranded DNA cannot fold in this fashion. As we shall see later in this chapter, the ability to fold into a complex three-dimensional shape allows RNA to carry out functions in cells in addition to conveying information between DNA and protein. Whereas DNA functions solely as an information store, RNA comes in different varieties, some having structural and even catalytic functions.

Transcription Produces RNA Complementary to One Strand of DNA

All of the RNA in a cell is made by transcription, a process that has certain similarities to DNA replication (discussed in Chapter 6). Transcription begins with the opening and unwinding of a small portion of the DNA double helix to expose the bases on each DNA strand. One of the two strands of the DNA double helix then acts as a template for the synthesis of RNA. Ribonucleotides are added, one-by-one, to the growing RNA chain, and as in DNA replication, the nucleotide sequence

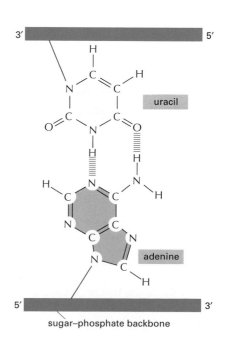

Figure 7–4 Uracil forms a base pair with adenine. Despite the absence of a methyl group, uracil has the same base-pairing properties as thymine. Thus, U-A base pairs closely resemble T-A base pairs (see Figure 5–6).

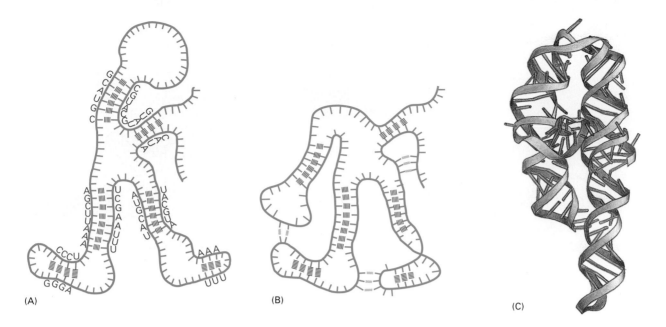

(A)   (B)   (C)

Figure 7–5 RNA molecules can form intramolecular base pairs and fold into specific structures. RNA is single-stranded, but it often contains short stretches of nucleotides that can base-pair with complementary sequences found elsewhere on the same molecule. These interactions, along with "nonconventional" base-pair interactions, allow an RNA molecule to fold into a three-dimensional structure that is determined by its sequence of nucleotides. (A) A diagram of a folded RNA structure showing only conventional (that is, Watson–Crick) base-pairing interactions; (B) structure with both conventional (red) and nonconventional (e.g., A-G) base-pair interactions (green); (C) structure of an actual RNA, a molecule involved in RNA splicing. Each conventional base-pair interaction is indicated by a "rung" in the double helix. Bases in other configurations are indicated by broken rungs.

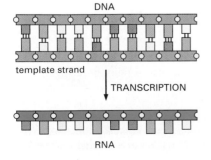

DNA

template strand

↓ TRANSCRIPTION

RNA

of the RNA chain is determined by complementary base-pairing with the DNA template. When a good match is made, the incoming ribonucleotide is covalently linked to the growing RNA chain in an enzymatically catalyzed reaction. The RNA chain produced by transcription—the *transcript*—is therefore elongated one nucleotide at a time and has a nucleotide sequence exactly complementary to the strand of DNA used as the template (Figure 7–6).

Transcription, however, differs from DNA replication in several crucial features. Unlike a newly formed DNA strand, the RNA strand does not remain hydrogen-bonded to the DNA template strand. Instead, just behind the region where the ribonucleotides are being added, the DNA helix re-forms and displaces the RNA chain. For this reason—and because only one strand of the DNA molecule is transcribed—RNA molecules are single-stranded. Further, as RNAs are copied from only a limited region of DNA, these molecules are much shorter than DNA molecules; DNA molecules in a human chromosome can be up to 250 million nucleotide pairs long, whereas most RNAs are no more than a few thousand nucleotides long, and many are much shorter than that.

The enzymes that carry out transcription are called **RNA polymerases**. Like the DNA polymerase that catalyzes DNA replication (discussed in Chapter 6), RNA polymerases catalyze the formation of the phosphodiester bonds that link the nucleotides together and form the sugar–phosphate backbone of the RNA chain. The RNA polymerase moves stepwise along the DNA, unwinding the DNA helix just ahead to expose a new region of the template strand for complementary base-pairing. In this way, the growing RNA chain is extended by one nucleotide at a time in the 5′-to-3′ direction (Figure 7–7), using ribonucleoside triphosphates (ATP, CTP, UTP, and GTP), whose high-energy bonds provide the energy that drives the reaction forward (see Figure 6–10).

The almost immediate release of the RNA strand from the DNA as it is synthesized means that many RNA copies can be made from the same gene in a relatively short time, the synthesis of the next RNA usually being started before the first RNA is completed (Figure 7–8). A medium-sized gene (say, 1500 nucleotide pairs) requires approximately

Figure 7–6 Transcription produces an RNA complementary to one strand of DNA. The top strand is sometimes called the *coding strand* because its sequence is equivalent to the RNA product.

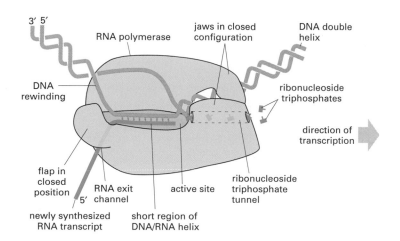

3' 5'

RNA polymerase

jaws in closed configuration

DNA double helix

DNA rewinding

ribonucleoside triphosphates

direction of transcription

flap in closed position

5'

RNA exit channel

active site

ribonucleoside triphosphate tunnel

newly synthesized RNA transcript

short region of DNA/RNA helix

Figure 7–7 DNA is transcribed by the enzyme RNA polymerase. The RNA polymerase *(pale blue)* moves stepwise along the DNA, unwinding the DNA helix in front of it. As it progresses, the polymerase adds nucleotides *(small "T" shapes)* one-by-one to the RNA chain at the polymerization site using an exposed DNA strand as a template. The resulting RNA transcript is thus a single-stranded complementary copy of one of the two DNA strands. The polymerase has a rudder that displaces the newly formed RNA, allowing the two strands of DNA behind the polymerase to rewind. A short region of DNA/RNA helix (approximately nine nucleotides in length) therefore forms only transiently, causing a "window" of DNA/RNA helix to move along the DNA with the polymerase. The incoming nucleotides are in the form of ribonucleoside triphosphates (ATP, UTP, CTP, and GTP), and the energy stored in their phosphate–phosphate bonds provides the driving force for the polymerization reaction (see Figure 6–10).

50 seconds for a molecule of RNA polymerase to transcribe it. In some cases, there may be 15 polymerases speeding along a single stretch of DNA, hard on one another's heels, allowing more than 1000 transcripts to be synthesized in an hour. For most genes, however, the amount of transcription is much less than this.

Although RNA polymerase catalyzes essentially the same chemical reaction as DNA polymerase, there are some important differences between the two enzymes. First, and most obvious, RNA polymerase catalyzes the linkage of ribonucleotides, not deoxyribonucleotides. Second, unlike the DNA polymerase involved in DNA replication, RNA polymerases can start an RNA chain without a primer. This difference may exist because transcription need not be as accurate as DNA replication; unlike DNA, RNA is not used as the permanent storage form of genetic information in cells, so mistakes in RNA transcripts have relatively minor consequences. RNA polymerases make about one mistake for every 10^4 nucleotides copied into RNA, compared with an error rate for DNA polymerase of about one in 10^7 nucleotides.

Several Types of RNA Are Produced in Cells

The vast majority of genes carried in a cell's DNA specify the amino acid sequence of proteins, and the RNA molecules that are copied from these genes (and that ultimately direct the synthesis of proteins) are collectively called **messenger RNA** (**mRNA**). The final product of other genes, however, is the RNA itself (Table 7–1). As we shall see in later sections of this chapter, these nonmessenger RNAs, like proteins, serve as structural and enzymatic components of cells, and they play key parts translating the genetic message into protein. *Ribosomal RNA (rRNA)* forms the core of the ribosomes, on which mRNA is translated into protein, and *transfer RNA (tRNA)* forms the adaptors that select amino acids and hold them in place on a ribosome for their incorporation into protein.

Question 7–2

In the electron micrograph in Figure 7–8, are the RNA polymerase molecules moving from right to left or from left to right? Why are the RNA transcripts so much shorter than the length of the DNA that encodes them?

Figure 7–8 Transcription can be visualized in the electron microscope. The micrograph shows many molecules of RNA polymerase simultaneously transcribing each of two adjacent genes. Molecules of RNA polymerase are visible as a series of dots along the DNA with the transcripts (fine threads) attached to them. The RNA molecules (called rRNAs) transcribed from the genes shown in this example are not translated into protein but are instead used directly as components of ribosomes, the machines on which translation takes place. The particles at the 5′ end (the free end) of each rRNA transcript are believed to be ribosomal proteins that have assembled on the rRNA. (Courtesy of Ulrich Scheer.)

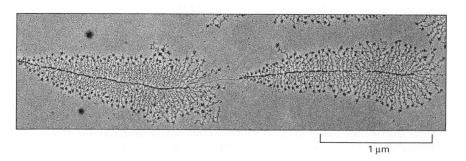

1 μm

Table 7–1 Types of RNA Produced in Cells

TYPE OF RNA	FUNCTION
mRNAs	code for proteins
rRNAs	form part of the structure of the ribosome and participate in protein synthesis
tRNAs	used in protein synthesis as adaptors between mRNA and amino acids
Small RNAs	used in pre-mRNA splicing, transport of proteins to the ER, and other cellular processes

While DNA molecules are typically very long and carry the instructions for thousands of different proteins, an RNA molecule is very much shorter, as it carries the information from just one portion of the DNA. In eucaryotes, each mRNA typically carries information transcribed from just one gene, coding for a single protein; in bacteria a set of adjacent genes is often transcribed as a single mRNA that therefore carries the information for several different proteins.

Signals in DNA Tell RNA Polymerase Where to Start and Finish

To begin transcription, RNA polymerase must be able to recognize the start of a gene and bind firmly to the DNA at this site. The way in which RNA polymerases recognize the transcription start site differs somewhat between bacteria and eucaryotes. Because the situation in bacteria is simpler, we shall focus on the procaryotic system first, and defer our discussion of transcription initiation in eucaryotes to the next chapter. This is an important topic, because the initiation of transcription is the main point at which the cell regulates which proteins are to be produced and at what rate.

In both procaryotes and eucaryotes, RNA polymerase molecules tend to stick weakly to the DNA when they randomly collide with it; the polymerase molecule then slides rapidly along the DNA. The enzyme latches tightly onto the DNA once it encounters a region called a **promoter**, which contains a sequence of nucleotides indicating the starting point for RNA synthesis. These nucleotide sequences are conserved, meaning that they occur, with some minor variations, in all promoters. The polymerase protein can recognize this DNA sequence, even though the DNA is in its double-helical form, by making specific contacts with the portions of the bases that are exposed on the outside of the helix. In eucaryotes, the binding of RNA polymerase to the promoter involves the participation of additional proteins; we shall discuss this more complicated situation in Chapter 8.

After the RNA polymerase makes contact with the promoter DNA and binds to it tightly, the enzyme opens up the double helix immediately in front of it to expose the nucleotides on a short stretch of DNA on each strand (Figure 7–9A). One of the two exposed DNA strands then acts as a template for complementary base-pairing with incoming ribonucleotides, two of which are joined together by the polymerase to begin the RNA chain. Chain elongation then continues until the enzyme encounters a second signal in the DNA, the terminator (or stop site), where the polymerase halts and releases both the DNA template and the newly made RNA chain (Figure 7–9B).

A subunit of bacterial polymerase, called *sigma (σ) factor*, is primarily responsible for recognizing the promoter sequence on DNA. Once the polymerase has latched onto the DNA at this site and has synthe-sized approximately 10 nucleotides of RNA, the sigma factor is released, enabling the polymerase to move forward and continue transcribing without it. After the polymerase is released at a terminator, it reassociates with a free sigma factor and searches for a promoter, where it can begin the process of transcription again.

Because DNA is double-stranded, two different RNA molecules could in principle be transcribed from any gene, using each of the two DNA strands as a template. However, the promoter is asymmetrical and binds the polymerase in only one orientation; thus, once properly positioned on a promoter, the RNA polymerase has no option but to

Figure 7–9 RNA polymerase transcribes a bacterial gene. (A) Production of an RNA molecule in bacteria. Bacterial RNA polymerase *(light blue)* contains a subunit called the sigma factor *(yellow)* that recognizes the promoter *(green)* on the DNA. Once transcription has begun, the sigma factor is released and the polymerase continues synthesizing the RNA without it. Chain elongation continues until the polymerase encounters a termination signal *(red)* in the DNA. There the enzyme halts and releases both the DNA template and the newly made transcript. Polymerase then reassociates with a free sigma factor and searches for another promoter to begin the process again. (B) The nucleotide sequences that signal to a bacterial RNA polymerase where to begin transcribing and where to stop. The *green-shaded* regions in the top diagram in (B) represent the DNA sequences that are required to create a promoter. The numbers represent the positions of nucleotides counting from the first nucleotide transcribed, which is designated +1. The *red-shaded* regions in the bottom diagram in (B) represent sequences that signal to RNA polymerase to terminate transcription.

(A)

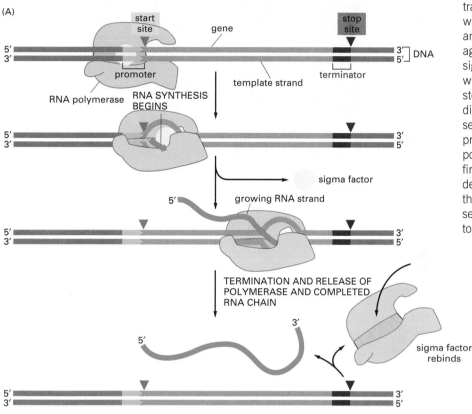

(B)

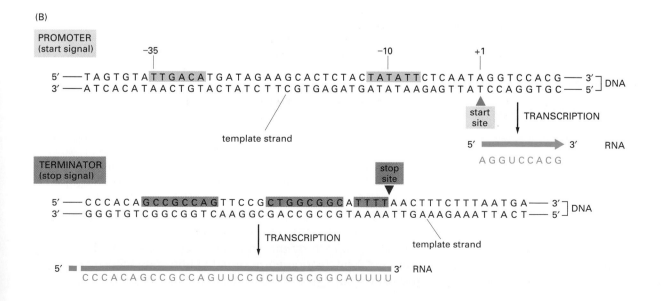

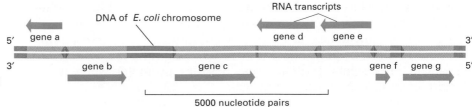

Figure 7–10 **Some genes are transcribed using one DNA strand as a template, while others are transcribed using the other DNA strand.** The direction of transcription is determined by the orientation of the promoter at the beginning of each gene (*green arrowheads*). Approximately 0.2% (10,000 base pairs) of the *E. coli* chromosome is depicted here. The genes transcribed from left to right use the bottom DNA strand as the template; those transcribed from right to left use the top strand as the template (see Figure 7–9).

Question 7–3

Could the RNA polymerase used for transcription be used as the polymerase that makes the RNA primer required for replication (discussed in Chapter 6)?

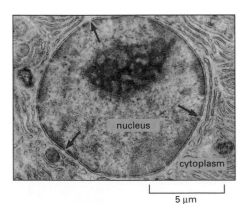

Figure 7–11 **Before they can be translated, mRNA molecules made in the nucleus move out into the cytoplasm via pores in the nuclear envelope** (*arrows*). Shown here is a section of a liver cell nucleus. (From D.W. Fawcett, A Textbook of Histology, 11th edn. Philadelphia: Saunders, 1986.)

transcribe the appropriate DNA strand, since transcription can proceed only in the 5′-to-3′ direction. The direction of transcription with respect to the chromosome as a whole will vary from gene to gene (Figure 7–10).

The RNA polymerase's requirement for binding tightly to DNA before it can start transcription means that a portion of DNA can be transcribed only if it is preceded by a promoter sequence. This ensures that only those parts of a DNA molecule that contain a gene will be transcribed into RNA. In bacteria, genes tend to lie very close to one another in the DNA, with only very short lengths of nontranscribed DNA between them. But in plant and animal DNA, including that of humans, individual genes are widely dispersed, with stretches of DNA up to 100,000 nucleotide pairs long between one gene and the next. Unnecessary transcription of these spacer-DNA regions, which—as far as is known—encode no genetic instructions, would waste a cell's valuable resources.

Eucaryotic RNAs Are Transcribed and Processed Simultaneously in the Nucleus

Although the templating principle by which DNA is transcribed into RNA is the same in all organisms, the way in which the RNA transcripts are handled before they can be used by the cell differs a great deal between bacteria and eucaryotes. Bacterial DNA lies directly exposed to the cytoplasm, which contains the *ribosomes* on which protein synthesis takes place. As mRNA molecules in bacteria are transcribed, ribosomes immediately attach to the free 5′ end of the RNA transcript and protein synthesis starts.

In eucaryotic cells, by contrast, DNA is enclosed within the *nucleus* (see Figure 5–19). Transcription takes place in the nucleus, but protein synthesis takes place on ribosomes in the cytoplasm. So, before a eucaryotic mRNA can be translated, it must be transported out of the nucleus through small pores in the nuclear envelope (Figure 7–11). In addition, before a eucaryotic RNA exits the nucleus, it must go through several different **RNA processing** steps. These reactions are tightly coupled to transcription and take place as the RNA is being transcribed. The enzymes responsible for RNA processing actually ride on the "tail" of the eucaryotic RNA polymerase as it transcribes an RNA, and then hop on to the nascent RNA molecule to begin processing as the transcript emerges from the RNA polymerase.

Depending on which type of RNA is being produced—mRNA or some other type—the transcripts are processed in various ways before leaving the nucleus. Two processing steps that occur only on transcripts destined to become mRNA molecules are *RNA capping* and *polyadenylation* (Figure 7–12):

1. RNA capping involves a modification of the 5′ end of the mRNA transcript, the end that is synthesized first during transcription. The RNA is capped by the addition of an atypical nucleotide—a guanine (G) nucleotide with a methyl group attached. This capping occurs after the RNA polymerase has produced about 25

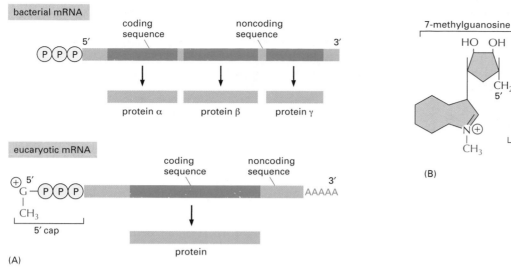

Figure 7–12 Unlike bacterial mRNAs, eucaryotic mRNA molecules are modified by capping and polyadenylation. (A) The 5′ and 3′ ends of a bacterial mRNA are the unmodified ends of the chain synthesized by the RNA polymerase, which initiates and terminates transcription at those points, respectively. The corresponding ends of a eucaryotic mRNA are modified by the addition of a 5′ cap and by cleavage of the primary transcript and the addition of a poly-A tail, respectively. The figure also illustrates another difference between the procaryotic and eucaryotic mRNAs: bacterial mRNAs can contain the instructions for several different proteins, whereas eucaryotic mRNAs nearly always contain the information for only a single protein. (B) The structure of the cap at the 5′ end of eucaryotic mRNA molecules. Note the unusual 5′-to-5′ linkage of the 7-methyl G to the remainder of the RNA. Many eucaryotic mRNA caps carry an additional modification: the 2′-hydroxyl group on the second ribose sugar in the mRNA is methylated (not shown).

nucleotides of RNA, long before it has completed transcribing the whole gene.

2. Polyadenylation provides most newly transcribed mRNAs with a special structure at their 3′, or tail, ends. In contrast with bacteria, where the 3′ end of an mRNA is simply the end of the chain synthesized by the RNA polymerase, the 3′ ends of eucaryotic RNAs are first trimmed by an enzyme that cuts the RNA chain at a particular sequence of nucleotides and are then finished off by a second enzyme that adds a series of repeated adenine (A) nucleotides (a *poly-A tail*) onto the cut end. The poly-A tail is generally a few hundred nucleotides long.

These two modifications—capping and polyadenylation—are thought to increase the stability of the eucaryotic mRNA molecule, to aid its export from the nucleus to the cytoplasm, and to generally identify the RNA molecule as an mRNA. They are also used by the protein-synthesis machinery as an indication that both ends of the mRNA are present and that the message is therefore complete.

Eucaryotic Genes Are Interrupted by Noncoding Sequences

Most eucaryotic RNAs have to undergo an additional processing step before they are functional. This step involves a far more radical modification of the primary RNA transcript than capping or polyadenylation,

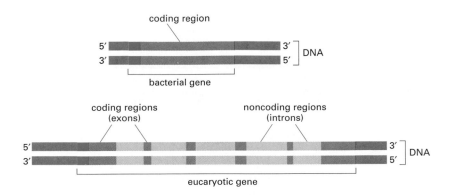

Figure 7–13 Eucaryotic and bacterial genes are organized differently. A bacterial gene consists of a single stretch of uninterrupted nucleotide sequence that encodes the amino acid sequence of a protein. In contrast, the coding sequences of most eucaryotic genes *(exons)* are interrupted by noncoding sequences *(introns)*. Promoters for transcription are indicated in *green*.

and it is the consequence of a surprising and critical feature of eucaryotic gene arrangement. In the 1970s, cell biologists studying transcription in eucaryotic cells were puzzled by the behavior of the RNA in the nucleus, which seemed to be quite different from that of the better-studied and more familiar bacterial mRNAs. They found that some nuclear RNAs that could be identified as prospective mRNAs by their G nucleotide caps and their poly-A tails became progressively shorter while in the nucleus, although they retained both their caps and their tails. In all, only about 5% of the RNA initially transcribed in the nucleus ever reached the cytoplasm. This seemed not only strangely wasteful but also deeply mysterious: how could the middle part of an RNA molecule shrink?

The answer to this mystery came in 1977 with the unexpected discovery that the organization of eucaryotic genes is fundamentally different from that of bacterial genes. In bacteria, most proteins are encoded by an uninterrupted stretch of DNA sequence that is transcribed into an RNA that, without any further processing, can serve as an mRNA. Most eucaryotic genes, in contrast, have their coding sequences interrupted by long, noncoding *intervening sequences*, called **introns** (Figure 7–13). The scattered pieces of coding sequences, or *expressed sequences*, called **exons**, are usually shorter than the introns, and the coding portion of a eucaryotic gene is often only a small fraction of the total length of the gene. Most introns range in length from about 80 nucleotides to 10,000 nucleotides, although even longer introns exist (Figure 7–14).

Introns Are Removed by RNA Splicing

To produce an mRNA in a eucaryotic cell, the entire length of the gene, introns as well as exons, is transcribed into a RNA. After capping, as the RNA polymerase continues to transcribe the gene, the process of **RNA splicing** begins, in which the intron sequences are removed from the newly synthesized RNA and the exons are stitched together. Each transcript ultimately receives a poly-A tail; in some cases, this happens after splicing, and in other cases it occurs before the final splicing reactions have been completed. Once a transcript has been spliced and its 5′ and

Figure 7–14 Most human genes are broken into exons and introns. (A) The nucleotide sequence of the β-globin gene, which encodes one of the subunits of the oxygen-carrying protein hemoglobin, was given in Figure 5–11. As indicated, it contains 3 exons. (B) The Factor VIII gene codes for a protein (Factor VIII) that functions in the blood-clotting pathway. Mutations in this large gene are responsible for the most prevalent form of hemophilia. As indicated, it contains 26 exons.

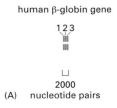

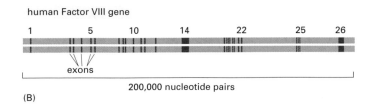

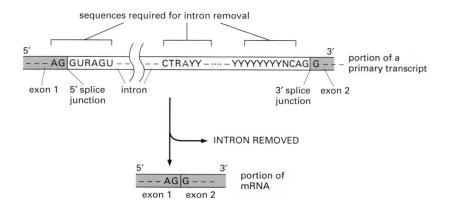

sequences required for intron removal

5′ --- AG|GURAGU -- ⁓ -- CTRAYY - ····· - YYYYYYYYYNCAG|G -- 3′ portion of a primary transcript

exon 1 | 5′ splice junction | intron | 3′ splice junction | exon 2

INTRON REMOVED

5′ --- AG|G --- 3′ portion of mRNA

exon 1 | exon 2

Figure 7–15 Special nucleotide sequences signal the beginning and the end of an intron. Only the three nucleotide sequences shown are required to remove an intron. The other positions in an intron can be occupied by any nucleotide. The special sequences are recognized by small nuclear ribonucleoproteins (snRNPs), which cleave the RNA at the intron–exon borders and covalently link the exons together. Here A, G, U, and C are the standard RNA nucleotides; R stands for either A or G; Y stands for either C or U. The A highlighted in *red* forms the branch point of the lariat produced in the splicing reaction (see Figure 7–16). The distances along the RNA between the three splicing sequences are highly variable; however, the distance between the branchpoint and the 3′ splice junction is typically much shorter than that between the 5′ splice junction and the branchpoint. The splicing sequences shown are from humans; other eukaryotic organisms have similar sequences that direct RNA splicing.

3′ ends have been modified, the RNA is a functional mRNA molecule that can now leave the nucleus and be translated into protein.

How does the cell determine which parts of the primary transcript to remove during splicing? Unlike the coding sequence of an exon, the exact nucleotide sequence of most of an intron seems to be unimportant. Although there is little general resemblance between the nucleotide sequences of different introns, each intron contains a few short nucleotide sequences that act as cues for its removal. These sequences are found at or near each end of the intron and are the same or very similar in all introns (Figure 7–15). The intron is cut out in the form of a "lariat" structure formed by the reaction of the "A" highlighted in red in this figure with the 5′ splice junction of the intron (Figure 7–16).

Unlike the other steps of mRNA production we have discussed, RNA splicing is performed largely by RNA molecules instead of proteins. RNA molecules recognize intron–exon boundaries and participate in the chemistry of splicing. These RNA molecules, called **small nuclear RNAs** (**snRNAs**), bind with additional proteins to form **small nuclear ribonucleoprotein particles** (**snRNPs**, pronounced "snurps"). These snRNPs form the core of the **spliceosome**, the large assembly of RNA and protein molecules that performs RNA splicing in the cell.

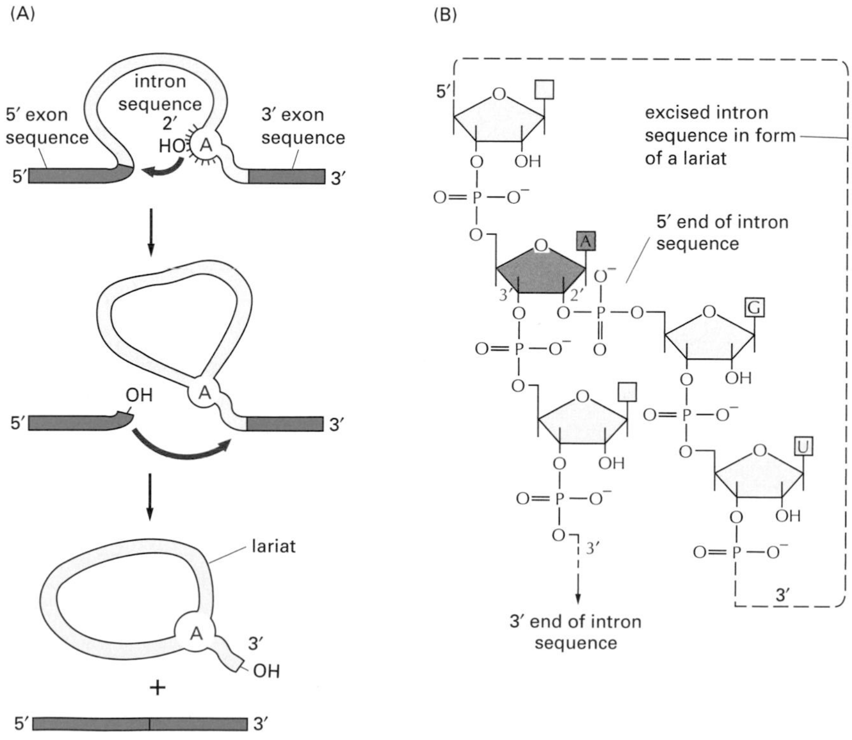

(A)

5′ exon sequence — intron sequence — 3′ exon sequence

intron sequence 2′ HO A

5′ ... 3′

OH A

5′ ... 3′

lariat

A 3′ OH

+

5′ ▬▬▬ 3′

(B)

5′

excised intron sequence in form of a lariat

5′ end of intron sequence

3′ end of intron sequence

Figure 7–16 An RNA chain forms a branched structure during splicing. (A) In the first step, a particular adenine nucleotide *(red)* in the intron sequence attacks the 5′ splice site and cuts the sugar–phosphate backbone of the RNA at this point. This adenine is the same one that is highlighted in Figure 7–15. The cut 5′ end of the intron becomes covalently linked to the adenine nucleotide to form a branched structure. The free 3′-OH end of the exon sequence then reacts with the start of the next exon sequence, joining the two exons together into a continuous coding sequence and releasing the intron in the form of a lariat. The lariat containing the intron is eventually degraded. (B) Details of the structure of the lariat branch. The cut 5′ end of the intron is linked to the 2′-OH group of the ribose of the branchpoint adenine nucleotide.

To splice an RNA, a group of snRNPs assemble at an intron–exon boundary, cut out the intron, and rejoin the RNA chain—releasing the excised intron as a lariat structure (Figure 7–17). One role of the snRNAs in the spliceosome is to recognize and—by using complementary base-pairing—to pair with the nucleotide sequences that mark the beginning and the branch point of each intron. The snRNPs thereby bring the two ends of the intron together so that splicing can take place. Although the snRNPs are central to the splicing reaction, additional proteins are also required.

The intron–exon type of gene arrangement in eucaryotes at first seems wasteful, but it does have positive consequences. This organization is likely to have been profoundly important in the early evolutionary

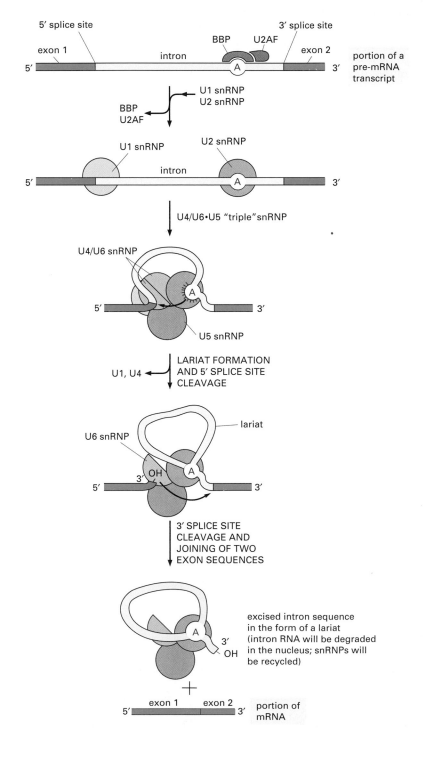

Figure 7–17 RNA splicing is catalyzed by an assembly of snRNPs (shown as *colored circles*) plus other proteins (most of which are not shown), which together form the spliceosome. The spliceosome recognizes the splicing signals on a pre-mRNA molecule, brings the two ends of the intron together, and provides the enzymatic activity for the two reactions. The branch-point site is first recognized by the BBP (branch-point–binding protein) and U2AF, a helper protein. In the next steps, the U2 snRNP displaces BBP and U2AF and forms base pairs with the branch-point site consensus sequence, and the U1 snRNP forms base pairs with the 5′ splice junction. At this point, the U4/U6•U5 "triple" snRNP enters the spliceosome. In this triple snRNP, the U4 and U6 snRNAs are held firmly together by base-pair interactions and the U5 snRNP is more loosely associated. Several RNA–RNA rearrangements then occur that break apart the U4/U6 base pairs and allow the U6 snRNP to displace U1 at the 5′ splice junction. Subsequent rearrangements create the active site of the spliceosome and position the appropriate portions of the pre-mRNA substrate for the splicing reaction to occur. The A highlighted in red is the same as that highlighted in Figures 7–15 and 7–16.

history of genes, where it is thought to have speeded up the emergence of new and useful proteins, as we shall discuss in Chapter 9. The presence of numerous introns in DNA makes genetic recombination between exons of different genes more likely. This means that genes for new proteins could have evolved quite rapidly by the combination of parts of preexisting genes, a mechanism resembling the assembly of a new type of machine from a kit of preexisting functional components. Indeed, many proteins in present-day cells resemble patchworks composed from a common set of protein pieces, called protein *domains* (see Figure 4–19).

RNA splicing also allows present-day eucaryotes to pack more information into every gene. The transcripts of many eucaryotic genes can be spliced in various ways to produce different mRNAs, depending on the cell type in which the gene is being expressed, or the stage of development of the organism. This allows different proteins to be produced from the same gene (Figure 7–18). An estimated 60% of human genes likely undergo such alternative splicing.

In sum, rather than being the wasteful process it seemed at first sight, RNA splicing enables eucaryotes to increase the already enormous coding potential of their genomes.

Mature Eucaryotic mRNAs Are Selectively Exported from the Nucleus

We have now seen how eucaryotic mRNA synthesis and processing takes place in an orderly fashion within the cell nucleus. However, these events create a special problem for eucaryotic cells: of the total mRNA that is synthesized, only a small fraction—the mature mRNA—is useful to the cell. The remaining RNA fragments—excised introns, broken RNAs, and aberrantly spliced transcripts—are not only useless, but could be dangerous to the cell if not destroyed. How, then, does the cell distinguish between the relatively rare mature mRNA molecules it needs to keep and the overwhelming amount of debris generated by RNA processing?

The answer is that the transport of mRNA from the nucleus to the cytoplasm, where it is translated into protein, is highly selective, as it is closely coupled to correct RNA processing. This coupling is achieved by the *nuclear pore complex*, which recognizes and transports only completed mRNAs. These aqueous pores connect the nucleoplasm with the cytosol, and as we discuss in Chapter 15, they act as gates that control which macromolecules can enter or exit the nucleus. To be "export ready," it seems that an mRNA molecule must be bound to an

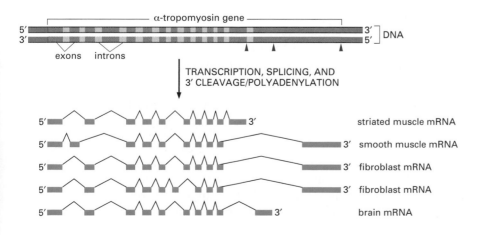

Figure 7–18 The α-tropomyosin gene can be spliced in different ways.
α-Tropomyosin is a coiled-coil protein (see Figure 4–16) that regulates contraction in muscle cells. The primary transcript can be spliced in different ways, as indicated in the figure, to produce distinct mRNAs that then give rise to variant proteins. Some of the splicing patterns are specific for certain types of cells. For example, the α-tropomyosin made in striated muscle is different from that made from the same gene in smooth muscle. The *arrowheads* in the top part of the figure represent sites where poly-A addition can occur.

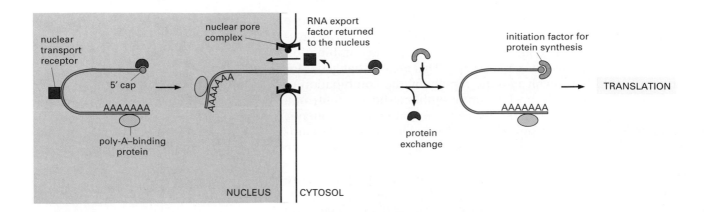

Figure 7–19 A specialized set of RNA-binding proteins mark a mature mRNA for export to the cytoplasm. As indicated in the schematic illustration, some of these proteins travel with the mRNA as it moves through the nuclear pore. One type of protein, a *nuclear transport receptor* dedicated to mRNA export, associates with mature mRNAs and actively guides them through the nuclear pore (discussed in Chapter 15). Once in the cytoplasm, the mRNA continues to shed previously bound proteins and acquire new ones; these substitutions affect the subsequent translation of the message. Because some are transported with the RNA, the proteins that become bound to an mRNA in the nucleus can influence its subsequent stability and translation in the cytosol.

appropriate set of proteins, each of which signals that the mRNA has been correctly processed. These proteins include poly-A–binding proteins, a cap-binding complex, and proteins that mark completed RNA splices (Figure 7–19). It is presumably the entire set of bound proteins, rather than any single protein, that ultimately determines whether an RNA molecule will leave the nucleus. The "waste RNAs" that remain behind in the nucleus are degraded, and the building blocks are reused for transcription.

mRNA Molecules Are Eventually Degraded by the Cell

Because the same mRNA molecule can be translated many times (see Figure 7–2), the length of time that a mature mRNA molecule persists in the cell affects the amount of protein it produces. Each mRNA molecule is eventually degraded into nucleotides by cellular RNases, but the lifetimes of mRNA molecules differ considerably—depending on the nucleotide sequence of the mRNA and the type of cell in which the mRNA is produced. Most mRNAs produced in bacteria are degraded rapidly, having a typical lifetime of about 3 minutes. The mRNAs in eucaryotic cells usually persist for longer amounts of time. Some transcripts, such as the one encoding β-globin, have lifetimes of more than 10 hours, whereas other eucaryotic mRNAs have lifetimes of less than 30 minutes.

These different lifetimes are in part controlled by nucleotide sequences that are present in the mRNA itself, most often in the portion of RNA called the 3′ untranslated region, that lies between the 3′ end of the coding sequence and the poly-A tail. The different lifetimes of mRNA help the cell specify the level of each protein that it synthesizes. In general, proteins made at high levels, such as β-globin, are translated from mRNAs that have long lifetimes, whereas those proteins present at low levels, or those whose levels must change rapidly in response to signals, are typically synthesized from short-lived mRNAs. These different lifetimes are the outcome of evolutionary fine-tuning in which the stability of mRNAs is tied to the needs of the cell.

The Earliest Cells May Have Had Introns in Their Genes

The process of transcription is universal: all cells use RNA polymerase, coupled with complementary base-pairing, to convert DNA into RNA. It may therefore seem puzzling that the resulting transcript is handled very differently in eucaryotes and procaryotes (Figure 7–20). In particular, RNA splicing seems to mark a fundamental difference between those two types of cells. But how did this dramatic difference arise?

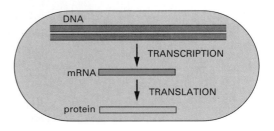

Figure 7–20 Procaryotes and eucaryotes handle their RNA transcripts somewhat differently. (A) In eucaryotic cells, the initial RNA molecule produced by transcription contains both intron and exon sequences. Its two ends are modified, and the introns are removed by an enzymatically catalyzed RNA splicing reaction. The resulting mRNA is then transported from the nucleus to the cytoplasm, where it is translated into protein. Although these steps are depicted as occurring one at a time, in a sequence, in reality they occur simultaneously. For example, the RNA cap is typically added and splicing typically begins before the transcript has been completed. Because of this coupling, complete primary transcripts do not typically exist in the cell. (B) In procaryotes, the production of mRNA molecules is simpler. The 5′ end of an mRNA molecule is produced by the initiation of transcription by RNA polymerase, and the 3′ end is produced by the termination of transcription. Because procaryotic cells lack a nucleus, transcription and translation take place in a common compartment. In fact, translation of a bacterial mRNA often begins before its synthesis has been completed. The amount of protein in a cell depends on the efficiency of each of these steps and on the rates of degradation of the RNA and protein molecules.

As we have seen, RNA splicing does provide eucaryotes with some degree of evolutionary flexibility, allowing them to produce a large variety of proteins and to develop systems for carefully controlling gene expression. However, these advantages come with a cost: the cell has to maintain a larger genome and to throw out a large fraction of the RNA it synthesizes. According to one school of thought, early cells—the common ancestors of procaryotes and eucaryotes—contained introns that were lost in procaryotes during subsequent evolution. By shedding their introns and adopting a smaller, more streamlined genome, procaryotes would have been able to reproduce more rapidly and efficiently. Consistent with this idea, simple eucaryotes that reproduce rapidly (some yeasts, for example) have relatively few introns, and these introns are usually much shorter than those found in higher eucaryotes.

On the other hand, some argue that introns were originally parasitic mobile genetic elements (discussed in Chapter 6) that happened to invade an early eucaryotic ancestor, colonizing its genome. These host cells then unwittingly replicated the selfish nucleotide snippets along with their own DNA, and modern eucaryotes have never bothered to sweep away the genetic clutter left from that ancient infection. The issue, however, is far from settled; whether introns evolved early—and were lost in procaryotes—or evolved later in eucaryotes is still a topic of scientific debate, and we return to it in Chapter 9.

From RNA to Protein

By the end of the 1950s, biologists had demonstrated that the information encoded in DNA is copied first into RNA and then into protein. The debate then centered on the "coding problem": how is the information in a linear sequence of nucleotides in RNA translated into the linear sequence of a chemically quite different set of subunits—the amino acids in proteins? This fascinating question stimulated much excitement among scientists at the time. Here was a cryptogram set up by nature that, after more than 3 billion years of evolution, could finally be solved by one of the products of evolution—human beings! And indeed, not only was the code eventually deciphered step-by-step (see How We Know, pp. 246–247), but the main features of the machinery by which cells make protein were pieced together.

An mRNA Sequence Is Decoded in Sets of Three Nucleotides

Transcription as a means of information transfer is simple to understand, since DNA and RNA are chemically and structurally similar, and DNA can act as a direct template for the synthesis of RNA through complementary base-pairing. As the term *transcription* signifies, it is as if a message written out by hand is being converted, say, into a typewritten text. The language itself and the form of the message do not change, and the symbols used are closely related.

In contrast, the conversion of the information in RNA into protein represents a **translation** of the information into another language that uses quite different symbols. Because there are only 4 different nucleotides in mRNA and 20 different types of amino acids in a protein, this translation cannot be accounted for by a direct one-to-one correspondence between a nucleotide in RNA and an amino acid in protein. The rules by which the nucleotide sequence of a gene, through the medium of mRNA, is translated into the amino acid sequence of a protein are known as the **genetic code**.

The sequence of nucleotides in the mRNA molecule is read consecutively in groups of three. Because RNA is a linear polymer made of four different nucleotides, there are thus $4 \times 4 \times 4 = 64$ possible combinations of three nucleotides: AAA, AUA, AUG, and so on. However, only 20 different amino acids are commonly found in proteins. Either some nucleotide triplets are never used, or the code is redundant and some amino acids are specified by more than one triplet. The second possibility is, in fact, correct, as shown by the completely deciphered genetic code in Figure 7–21. Each group of three consecutive nucleotides in RNA is called a **codon**, and each specifies one amino acid.

This code is used universally in all present-day organisms. Although a few slight differences in the code have been found, these occur chiefly in the DNA of mitochondria. Mitochondria have their own transcription and protein synthesis machinery that operate quite independently from those of the rest of the cell (discussed in Chapter 14), and they have been able to accommodate minor changes to the otherwise universal code.

In principle, an RNA sequence can be translated in any one of three different, nonoverlapping **reading frames**, depending on where the decoding process begins (Figure 7–22). However, only one of the three possible reading frames in an mRNA encodes the required protein. We shall see in a later section how a special punctuation signal at the beginning of each RNA message sets the correct reading frame.

Figure 7–21 The nucleotide sequence of an mRNA is translated into the amino acid sequence of a protein via the genetic code. All the three-nucleotide codons that specify a given amino acid are listed above that amino acid, which is given in both its three-letter and one-letter abbreviations (see Panel 2–5 for the full name of each amino acid and its structure). By convention, codons are always written with the 5'-terminal nucleotide to the left. Note that most amino acids are represented by more than one codon, and that there are some regularities in the set of codons that specify each amino acid. Codons for the same amino acid tend to contain the same nucleotides at the first and second positions and to vary at the third position. Three codons do not specify any amino acid but act as termination sites (stop codons), signaling the end of the protein-coding sequence. One codon—AUG—acts both as an initiation codon, signaling the start of a protein-coding message, and as the codon that specifies methionine.

	AGA									UUA					AGC				GUA	
	AGG									UUG					AGU					
GCA	CGA					GGA			CUA			CCA	UCA	ACA			GUA	UAA
GCC	CGC				AUA	GGC		AUA	CUC			CCC	UCC	ACC			GUC	UAG
GCG	CGG	GAC	AAC	UGC	GAA	CAA	GGG	CAC	AUC	CUG	AAA	AUG	UUC	CCG	UCG	ACG		UAC	GUG	UAG
GCU	CGU	GAU	AAU	UGU	GAG	CAG	GGU	CAU	AUU	CUU	AAG		UUU	CCU	UCU	ACU	UGG	UAU	GUU	UGA
Ala	Arg	Asp	Asn	Cys	Glu	Gln	Gly	His	Ile	Leu	Lys	Met	Phe	Pro	Ser	Thr	Trp	Tyr	Val	stop
A	R	D	N	C	E	Q	G	H	I	L	K	M	F	P	S	T	W	Y	V	

tRNA Molecules Match Amino Acids to Codons in mRNA

The codons in an mRNA molecule do not directly recognize the amino acids they specify: the group of three nucleotides does not, for example, bind directly to the amino acid. Rather, the translation of mRNA into protein depends on adaptor molecules that can recognize and bind both to the codon and, at another site on their surface, to the amino acid. These adaptors consist of a set of small RNA molecules known as **transfer RNAs** (**tRNAs**), each about 80 nucleotides in length.

We saw earlier that an RNA molecule will generally fold into a three-dimensional structure by forming base pairs between different regions of the molecule. If the base-paired regions are sufficiently extensive, they will form a double-helical structure, like that of double-stranded DNA. The tRNA molecules provide a striking example of this. Four short segments of the folded tRNA are double-helical, producing a molecule that looks like a cloverleaf when drawn schematically (Figure 7–23A). For example, a 5'-GCUC-3' sequence in one part of a polynucleotide chain can form a relatively strong association with a 5'-GAGC-3' sequence in another region of the same molecule. The cloverleaf undergoes further folding to form a compact L-shaped structure that is held together by additional hydrogen bonds between different regions of the molecule (see Figure 7–23).

Two regions of unpaired nucleotides situated at either end of the L-shaped molecule are crucial to the function of tRNA in protein synthesis. One of these regions forms the **anticodon**, a set of three consecutive nucleotides that pairs with the complementary codon in an mRNA molecule. The other is a short single-stranded region at the 3' end of the

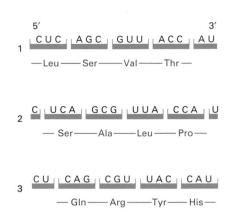

Figure 7–22 An RNA molecule can be translated in three possible reading frames. In the process of translating a nucleotide sequence (blue) into an amino acid sequence (red), the sequence of nucleotides in an mRNA molecule is read from the 5' to the 3' end in sequential sets of three nucleotides. In principle, therefore, the same RNA sequence can specify three completely different amino acid sequences, depending on the reading frame. In reality, however, only one of these reading frames encodes the actual message.

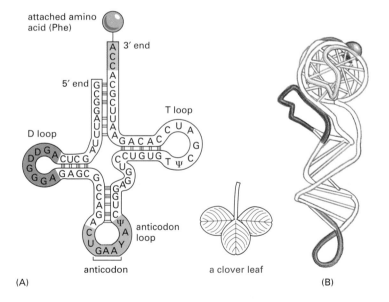
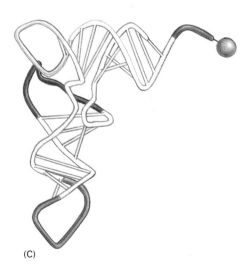

5' GCGGAUUUAGCUCAGDDGGGAGAGCGCCAGACUGAAYAΨCUGGAGGUCCUGUGTΨCGAUCCACAGAAUUCGCACCA 3'

(D)

anticodon

Figure 7–23 tRNA molecules are molecular adaptors, linking amino acids with codons. In this series of diagrams, the same tRNA molecule—in this case a tRNA specific for the amino acid phenylalanine (Phe)—is depicted in various ways. (A) The cloverleaf structure, a convention used to show the complementary base-pairing (red lines) that creates the double-helical regions of the molecule. The anticodon is the sequence of three nucleotides that base-pairs with a codon in mRNA. The amino acid matching the codon/anticodon pair is attached at the 3' end of the tRNA. tRNAs contain some unusual bases, which are produced by chemical modification after the tRNA has been synthesized. The bases denoted Ψ (for pseudouridine) and D (for dihydrouridine) are derived from uracil. (B and C) Views of the actual L-shaped molecule, based on X-ray diffraction analysis. These images are rotated 90° with respect to one another. (D) The linear nucleotide sequence of the molecule, color-coded to match A, B, and C.

How We Know: Cracking the Genetic Code

By the beginning of the 1960s, the "central dogma" had been accepted as the pathway along which information flows from gene to protein. It was clear that genes encode proteins, that genes are made of DNA, and that mRNA serves as an intermediary, carrying the information from the nucleus—where DNA is stored—to the cytoplasm, where protein translation takes place.

Even the general format of the genetic code had been worked out: each of the 20 amino acids found in proteins is represented by a triplet codon in an mRNA molecule. But an even greater challenge lay ahead: biologists, chemists, and even physicists set their sights on breaking the genetic code—attempting to figure out which amino acid each of the 64 possible nucleotide triplets designates. The most straightforward path to the solution would have been to compare the sequence of a segment of DNA or mRNA with its corresponding polypeptide product. Techniques for sequencing nucleic acids, however, would not be devised until the late 1960s.

So researchers decided that, to crack the genetic code, they would have to synthesize their own simple messages. If they could use these molecules to direct the synthesis of polypeptides, they would be on their way to deciphering which triplets encode which amino acids.

Losing the cells

Before researchers began preparing their synthetic mRNAs, they wanted to perfect a cell-free system for protein synthesis. This would allow them to translate their messages into polypeptides in a test tube. (Generally speaking, when working in the laboratory, the simpler the system, the cleaner the results.) To isolate the molecular machinery they needed for such a cell-free translation system, researchers broke open E. coli cells and loaded their contents into a centrifuge. Spinning these samples at high speed caused the membranes and other large chunks of cellular debris to be dragged to the bottom of the tube; the lighter cellular components required for protein synthesis—including mRNA, tRNA, ribosomes, enzymes, and other small molecules— were left floating in the supernatant. Researchers found that simply adding radioactive amino acids to this cell "soup" would trigger the production of radiolabeled proteins. By centrifuging this supernatant again, at a somewhat higher speed, the researchers could force the ribosomes and any newly synthesized peptides to the bottom of the tube; the labeled polypeptides could then be detected by measuring the radioactivity in the sediment remaining in the tube after the top layer had been decanted.

The trouble with this particular system was that it produced proteins encoded by cellular mRNAs already present in the extract. But, researchers wanted to use their own synthetic messages to direct protein synthesis. This problem was solved when Marshall Nirenberg discovered that he could destroy the cellular mRNA in the extract by adding the enzyme ribonuclease. Now all he needed to do was prepare his own message, load this synthetic mRNA into the cell-free system, and see what peptides came out.

Faking the message

Producing a polynucleotide with a defined sequence was not as simple as it sounds. Again, it would be years before chemists developed techniques that could be used to synthesize any given string of nucleic acids. Nirenberg decided to use polynucleotide phosphorylase, an enzyme that would join ribonucleotides together in the absence of a template. The sequence of the resulting RNA, then, depended entirely on which nucleotides were presented to the enzyme. A mixture of nucleotides would be sewn into an entirely random sequence; but a single nucleotide would yield a homogeneous polymer string containing only that one nucleotide. Thus Nirenberg, working with his collaborator Heinrich Matthaei, first produced an mRNA made entirely of uracil— poly U.

Together the researchers fed this poly U to their cell-free translation system. They then added radioactively labeled amino acids to the mix. After testing each amino acid—one at a time, in 20 different experiments—they determined that poly U directs the synthesis of a peptide containing only phenylalanine (Figure 7–24). Because UUU is the only triplet codon in this message, UUU, they reasoned, encodes phenylalanine. The first word in the genetic code had been deciphered.

Nirenberg and Matthaei then repeated the experiment with poly A and poly C and determined that AAA codes for lysine and CCC for proline. The meaning of poly G could not be ascertained by this method because this polynucleotide forms an odd triple-stranded helix and would not cooperate as a template in the cell-free system.

Feeding ribosomes synthetic mRNA seemed a fruitful technique. But with the single-nucleotide possibilities exhausted, researchers had three codons down, and 61 still to go. The other codon combinations, however, were harder to construct, and a new synthetic approach was needed. In the 1950s, the organic chemist Gobind Khorana had been developing methods for preparing mixed polynucleotides of defined sequence—but his techniques worked only for DNA. When he learned of Nirenberg's work with synthetic messages, Khorana directed his energies and skills to producing RNAs. He found that if he made DNAs of a defined sequence, he could then use RNA polymerase to produce

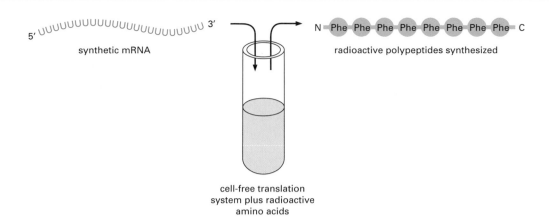

Figure 7–24 UUU codes for phenylalanine. Synthetic mRNAs are fed into a cell-free translation system containing ribosomes, tRNAs, enzymes, and other small molecules. Radioactive amino acids are added to this mix and the resulting polypeptides analyzed. In this case, poly U is shown to encode a polypeptide containing only phenylalanine.

RNAs from those. In this way, Khorana prepared a collection of different mRNAs of defined repeating sequence. He generated sequences of repeating dinucleotides (such as poly UC), trinucleotides (such as poly UUC), or tetranucleotides (such as poly UAUC).

These mixed polynucleotides, however, yielded results that were much more difficult to decipher than the mononucleotide messages that Nirenberg had used. Take poly UG, for example. When this repeating dinucleotide is added to the translation system, researchers discovered that it codes for a polypeptide of alternating cysteine and valine residues. This RNA, of course, contains two different alternating codons: UGU and GUG. So researchers could say that UGU and GUG code for cysteine and valine, although they could not tell which was which. Thus these mixed messages provided useful information, but did not definitively reveal which codons specified which amino acids (Figure 7–25).

Trapping the triplets

These final ambiguities in the code were resolved when Nirenberg and a young medical graduate named Phil Leder discovered that RNA fragments that were only three nucleotides in length—the size of a single codon—could bind to a ribosome and attract the appropriate charged tRNA molecule to the protein-making machinery. These complexes—containing one ribosome, one mRNA codon, and one radiolabeled aminoacyl-tRNA—could then be captured on a piece of filter paper and the attached amino acid identified.

Their trial run with UUU—the first word—worked splendidly. Leder and Nirenberg primed the usual cell-free translation system with snippets of UUU. These trinucleotides bound to the ribosomes, and Phe-tRNAs bound to the UUU. The new system was up and running, and the researchers had confirmed that UUU codes for phenylalanine.

All that remained was for researchers to produce all 64 possible codons—a task that was quickly accomplished in both the Nirenberg and Khorana laboratories. Because these small trinucleotides were much simpler to synthesize chemically, and the triplet-trapping tests were easier to perform and analyze than the previous decoding experiments, the researchers were able to work out the entire genetic code within the next year.

MESSAGE	PEPTIDES PRODUCED	CODON ASSIGNMENTS	
poly UG	...Cys–Val–Cys–Val...	UGU GUG	Cys, Val*
poly AG	...Arg–Glu–Arg–Glu...	AGA GAG	Arg, Glu
poly UUC	...Phe–Phe–Phe... + ...Ser–Ser–Ser... + ...Leu–Leu–Leu...	UUC UCU CUU	Phe, Ser, Leu
poly UAUC	...Tyr–Leu–Ser–Ile...	UAU CUA UCU AUC	Tyr, Leu, Ser, Ile

* One codon specifies Cys, the other Val. The same ambiguity exists for the other codon assignments shown here.

Figure 7–25 Messages of mixed repeating sequences further narrowed the coding possibilities. Although these mixed messages reveal the composition of the encoded peptides, they did not permit the unambiguous assignment of a single codon to a specific amino acid. For example, for poly UG it is still not clear which of the resulting codons, UGU or GUG, encodes cysteine and which encodes valine. The same ambiguity exists for experiments using di-, tri-, and tetranucleotides.

molecule; this is the site where the amino acid that matches the codon is attached to the tRNA.

We saw in the previous section that the genetic code is redundant; that is, several different codons can specify a single amino acid (see Figure 7–21). This redundancy implies either that there is more than one tRNA for many of the amino acids or that some tRNA molecules can base-pair with more than one codon. In fact, both situations occur. Some amino acids have more than one tRNA and some tRNAs are constructed so that they require accurate base-pairing only at the first two positions of the codon and can tolerate a mismatch (or *wobble*) at the third position. This wobble base-pairing explains why so many of the alternative codons for an amino acid differ only in their third nucleotide (see Figure 7–21). Wobble base-pairings make it possible to fit the 20 amino acids to their 61 codons with as few as 31 kinds of tRNA molecules. The exact number of different kinds of tRNAs, however, differs from one species to the next. For example, humans have 497 different tRNA genes, but among them only 48 anticodons are represented.

Specific Enzymes Couple tRNAs to the Correct Amino Acid

We have seen that in order to read the genetic code in DNA, cells make many different tRNAs. We now must consider how each tRNA molecule becomes *charged*—linked to the one amino acid in 20 that is its appropriate partner. Recognition and attachment of the correct amino acid depends on enzymes called **aminoacyl-tRNA synthetases**, which covalently couple each amino acid to its appropriate set of tRNA molecules. There is a different synthetase enzyme for each amino acid (that is, there are 20 synthetases in all); one attaches glycine to all tRNAs that recognize codons for glycine, another attaches alanine to all tRNAs that recognize codons for alanine, and so on. Specific nucleotides in both the anticodon and the amino acid–accepting arm allow the correct tRNA to be recognized by the synthetase enzyme. The synthetases are equal in importance to the tRNAs in the decoding process, because it is the combined action of synthetases and tRNAs that allows each codon in the mRNA molecule to associate with its proper amino acid (Figure 7–26).

The synthetase-catalyzed reaction that attaches the amino acid to the 3′ end of the tRNA is one of many cellular reactions coupled to the energy-releasing hydrolysis of ATP (see Figure 3–34), and it produces a high-energy bond between the charged tRNA and the amino acid. The energy of this bond is used at a later stage in protein synthesis to covalently link the amino acid to the growing polypeptide chain.

The RNA Message Is Decoded on Ribosomes

The recognition of a codon by the anticodon on a tRNA molecule depends on the same type of complementary base-pairing used in DNA replication and transcription. However, accurate and rapid translation of mRNA into protein requires a large molecular machine, which travels along the mRNA chain, capturing complementary tRNA molecules, holding them in position, and covalently linking the amino acids that they carry so as to form a protein chain. This protein-manufacturing machine is the **ribosome**—a large complex made from more than 50 different proteins (the *ribosomal proteins*) and several RNA molecules called **ribosomal RNAs** (**rRNAs**). A typical living cell contains millions of ribosomes in its cytoplasm (Figure 7–27). In a eucaryote, the ribosomal subunits are made in the nucleus, by the association of newly

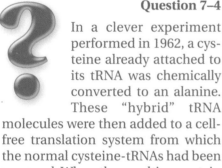

Question 7–4

In a clever experiment performed in 1962, a cysteine already attached to its tRNA was chemically converted to an alanine. These "hybrid" tRNA molecules were then added to a cell-free translation system from which the normal cysteine-tRNAs had been removed. When the resulting protein was analyzed, it was found that alanine had been inserted at every point in the protein chain where cysteine was supposed to be. Discuss what this experiment tells you about the role of aminoacyl-tRNA synthetases during the normal translation of the genetic code.

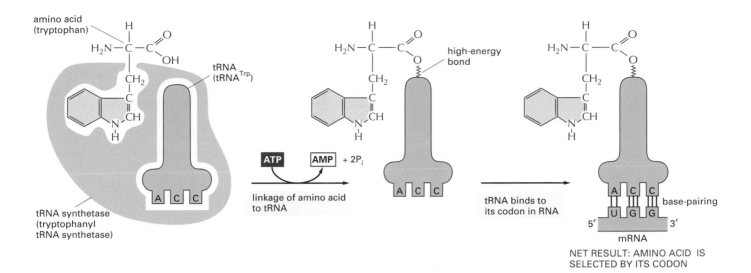

NET RESULT: AMINO ACID IS
SELECTED BY ITS CODON

transcribed rRNAs with ribosomal proteins, which have been transported into the nucleus after their synthesis in the cytoplasm. The individual ribosomal subunits are then exported to the cytoplasm to take part in protein synthesis.

Eucaryotic and procaryotic ribosomes are very similar in design and function. Both are composed of one large and one small subunit that fit together to form a complete ribosome with a mass of several million daltons (Figure 7–28); for comparison, an average-sized protein has a mass of 40,000 daltons. The small subunit matches the tRNAs to the codons of the mRNA, while the large subunit catalyzes the formation of the **peptide bonds** that link the amino acids together into a polypeptide chain (discussed in Chapter 2). The two subunits come together on an mRNA molecule, usually near its beginning (5′ end), to begin the synthesis of a protein. The ribosome then moves along the mRNA, translating the nucleotide sequence into an amino acid sequence one codon at a time, using the tRNAs as adaptors to add each amino acid in the correct sequence to the end of the growing polypeptide chain. Finally, the two subunits of the ribosome separate when synthesis of the protein is finished. Ribosomes operate with remarkable efficiency; in one second a ribosome in a eucaryotic cell adds about 2 amino acids to a polypeptide chain; bacterial ribosomes operate even faster, at a rate of about 20 amino acids per second.

Figure 7–26 The genetic code is translated by means of two adaptors that act one after another. The first adaptor is the aminoacyl-tRNA synthetase, which couples a particular amino acid to its corresponding tRNA; this coupling process is called charging. The second adaptor is the tRNA molecule itself, whose anticodon forms base pairs with the appropriate codon on the mRNA. A tRNA coupled with its amino acid is also called a charged tRNA. An error in either step—the charging or the binding of the charged tRNA to its codon—will cause the wrong amino acid to be incorporated into a protein chain. In the sequence of events shown, the amino acid tryptophan (Trp) is selected by the codon UGG on the mRNA.

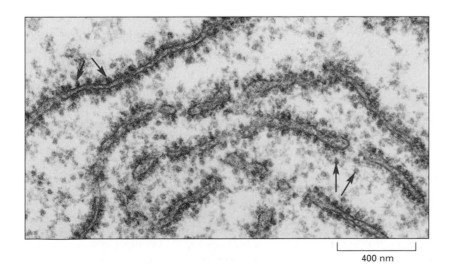

400 nm

Figure 7–27 Ribosomes are found in the cytoplasm of a eucaryotic cell. This electron micrograph shows a thin section of a small region of cytoplasm. The ribosomes appear as black dots (red arrows). Some are free in the cytosol; others are attached to membranes of the endoplasmic reticulum. (Courtesy of George Palade.)

How does the ribosome choreograph the movements required for translation? Each ribosome contains three binding sites for tRNA molecules, called the A-site, the P-site, and the E-site. Because tRNA molecules have the same basic shape, they are all capable of fitting into these sites. A tRNA molecule is held tightly at the A- and P-sites, however, only if its anticodon forms base pairs (allowing for wobble) with a complementary codon on the mRNA molecule that is bound to the ribosome. The A- and P-sites are sufficiently close together that their two tRNA molecules are forced to form base pairs with adjacent codons on the mRNA molecule (Figure 7–29).

Once protein synthesis has been initiated, each new amino acid is added to the elongating chain in a cycle of reactions. We shall step into the chain elongation process at a point where some amino acids have already been linked together and there is a tRNA joined to the growing polypeptide at the P-site on the ribosome (Figure 7–30, step 1). A tRNA carrying the next amino acid in the chain has bound to the vacant ribosomal A-site by forming base pairs with the codon in mRNA exposed at the A-site. In step 2, the carboxyl end of the polypeptide chain is uncoupled from the tRNA at the P-site (by breakage of the high-energy bond between the tRNA and its amino acid) and joined by a peptide bond to the free amino group of the amino acid linked to the tRNA at the A-site. This central reaction of protein synthesis is catalyzed by the activity of a *peptidyl transferase* enzyme, which is part of the ribosome. As shown in the figure, the peptidyl transferase reaction is thought to be accompanied by a shift of the large subunit relative to the small subunit, which actually holds on to the mRNA. This shift moves the two tRNAs into the E- and P-sites of the large subunit. In step 3, the small subunit moves

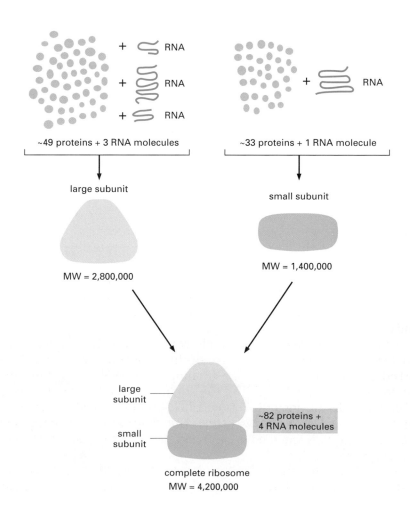

Figure 7–28 **A ribosome is a large complex of four RNAs and more than 80 proteins.** Shown here are the components of eucaryotic ribosomes. Procaryotic ribosomes are very similar.

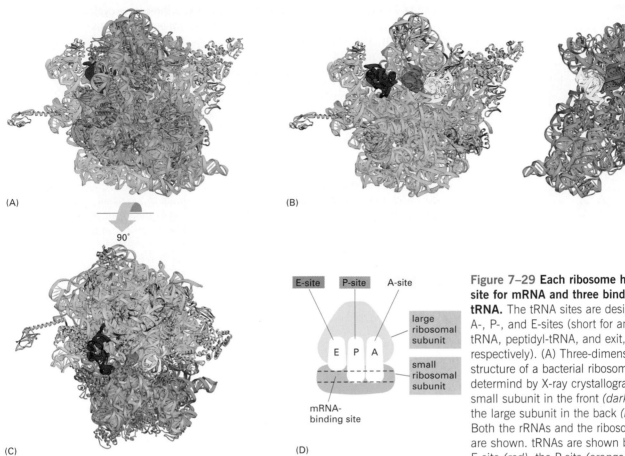

(A)

90°

(B)

(C)

(D)

E-site   P-site   A-site

large
ribosomal
subunit

E   P   A

small
ribosomal
subunit

mRNA-
binding site

Figure 7–29 Each ribosome has a binding site for mRNA and three binding sites for tRNA. The tRNA sites are designated the A-, P-, and E-sites (short for aminoacyl-tRNA, peptidyl-tRNA, and exit, respectively). (A) Three-dimensional structure of a bacterial ribosome, as determind by X-ray crystallography with the small subunit in the front *(dark green)* and the large subunit in the back *(light green)*. Both the rRNAs and the ribosomal proteins are shown. tRNAs are shown bound in the E-site *(red)*, the P-site *(orange)*, and the A-site *(yellow)*. Although all three tRNA sites are shown occupied here, during the process of protein synthesis not more than two of these sites are thought to contain tRNA molecules at any one time (see Figure 7–30). (B) Structure of the large *(left)* and small *(right)* ribosomal subunits arranged as though the ribosome in (A) were opened like a book. (C) Structure of the ribosome in (A) viewed from the top. (D) Highly schematized representation of a ribosome (in the same orientation as C), which will be used in subsequent figures. (A, B, and C, adapted from M.M. Yusupov et al., *Science* 292:883–896, 2001, courtesy of Albion Bausom and Harry Noller.)

exactly three nucleotides along the mRNA molecule, bringing it back to its original position relative to the large subunit, and the tRNA occupying the E-site dissociates. The entire cycle of three steps is repeated each time an amino acid is added to the polypeptide chain, with the chain growing from its amino to its carboxyl end until a stop codon is encountered.

The Ribosome Is a Ribozyme

The ribosome is one of the largest and most complex structures in the cell, composed of two-thirds RNA and one-third protein. The determination, in 2000, of the entire three-dimensional structure of its large and small subunits was a major triumph of modern biology. The structure strongly confirms earlier evidence that the rRNAs—not the proteins—are responsible for the ribosome's overall structure, its ability to position tRNAs on the mRNA, and even its catalytic function, forming peptide bonds.

The rRNAs are folded into highly compact, precise three-dimensional structures that form the core of the ribosome (Figure 7–31). In marked contrast to the central positioning of the rRNA, the ribosomal proteins are generally located on the surface, where they fill the gaps and crevices of the folded RNA. The main role of the ribosomal proteins seems to be to stabilize the RNA core, while permitting the changes in rRNA conformation that are necessary for this RNA to catalyze efficient protein synthesis.

Not only are the three binding sites for tRNAs (the A-, P-, and E-sites) on the ribosome formed primarily by the rRNAs, but the catalytic

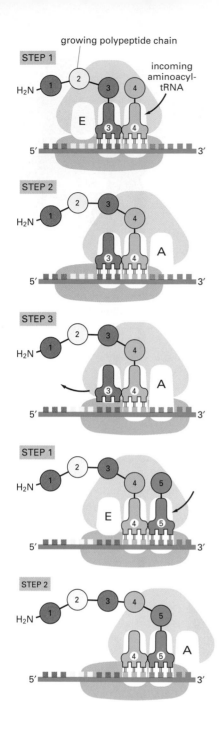

STEP 1

growing polypeptide chain

H₂N — 1 — 2 — 3 — 4 ← incoming aminoacyl-tRNA

E

5' — 3'

STEP 2

H₂N — 1 — 2 — 3 — 4

A

5' — 3'

STEP 3

H₂N — 1 — 2 — 3 — 4

A

5' — 3'

STEP 1

H₂N — 1 — 2 — 3 — 4 — 5

E

5' — 3'

STEP 2

H₂N — 1 — 2 — 3 — 4 — 5

A

5' — 3'

Figure 7–30 An mRNA molecule is translated in a three-step cycle. Each amino acid added to the growing end of a polypeptide chain is selected by complementary base-pairing between the anticodon on its attached tRNA molecule and the next codon on the mRNA chain. Because only one of the many types of tRNA molecules in a cell can base-pair with each codon, the codon determines the specific amino acid to be added to the growing polypeptide chain. The three-step cycle shown is repeated over and over during the synthesis of a protein chain. An aminoacyl-tRNA molecule binds to a vacant A-site on the ribosome in step 1, a new peptide bond is formed in step 2, and the mRNA moves a distance of three nucleotides through the small subunit in step 3, ejecting the spent tRNA molecule and "resetting" the ribosome so that the next aminoacyl-tRNA molecule can bind. Although the figure shows a large movement of the small ribosome subunit relative to the large subunit, the conformational changes that actually take place in the ribosome during translation are probably much more subtle. As indicated, the mRNA is translated in the 5'-to-3' direction, and the N-terminal end of a protein is made first, with each cycle adding one amino acid to the C-terminus of the polypeptide chain. The position at which the growing peptide chain is attached to a tRNA does not change during the elongation cycle: it is always linked to the tRNA present at the P-site of the large subunit.

site for peptide-bond formation appears to be formed by the 23S RNA of the large subunit; the nearest amino acid is located too far away to make contact with the incoming aminoacyl-tRNA or the growing polypeptide chain. This RNA-based catalytic site for peptidyl transferase is similar in many respects to those found in some proteins: it is a highly structured pocket that precisely orients the two reactants—the elongating peptide and the charged tRNA—and it provides a chemical group to act as a catalyst.

RNA molecules that possess catalytic activity are called **ribozymes**. In the final section of this chapter, we consider other ribozymes and discuss what RNA-based catalysis might mean for the early evolution of life on Earth. Here we need only note that there is good reason to suspect that RNA rather than protein molecules served

Figure 7–31 Ribosomal RNAs give the ribosome its overall shape. Shown are the detailed structures of the two rRNAs, the 23S rRNA *(blue)* and the 5S rRNA *(purple),* that form the core of the large subunit of bacterial ribosomes. One of the protein subunits of the ribosome (L1) is included as a reference point, as this protein forms a characteristic protrusion on the ribosome. Ribosomal components are commonly designated by their "S values," which refer to their rate of sedimentation in an ultracentrifuge. (Adapted from N. Ban et al., *Science* 289:905–920, 2000.)

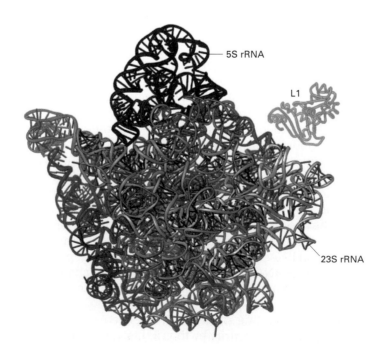

5S rRNA

L1

23S rRNA

Question 7–5

The following sequence of nucleotides in a DNA strand was used as a template to synthesize an mRNA that was then translated into protein: 5'-TTAACGGCTTTTTTC-3'. Predict the C-terminal amino acid and the N-terminal amino acid of the resulting polypeptide. Assume that the mRNA is translated without the need for a start codon.

as the first catalysts for living cells. If so, the ribosome, with its RNA core, might be viewed as a relic of an earlier time in life's history, when protein synthesis evolved in cells that were run almost entirely by ribozymes.

Codons in mRNA Signal Where to Start and to Stop Protein Synthesis

The site where protein synthesis begins on the mRNA is crucial, because it sets the reading frame for the whole length of the message. An error of one nucleotide either way at this stage will cause every subsequent codon in the message to be misread, so that a nonfunctional protein with a garbled sequence of amino acids will result (see Figure 7–22). The initiation step is also of great importance in another respect, because it is the last point at which the cell can decide whether the mRNA is to be translated and the protein synthesized; the rate of initiation thus determines the rate at which the protein is synthesized.

The translation of an mRNA begins with the codon AUG, and a special tRNA is required to initiate translation. This **initiator tRNA** always carries the amino acid methionine (or a modified form of methionine, formylmethionine, in bacteria) so that newly made proteins all have methionine as the first amino acid at their N-terminal end, the end of a protein that is synthesized first. This methionine is usually removed later by a specific protease. The initiator tRNA is distinct from the tRNA that normally carries methionine.

In eucaryotes, the initiator tRNA, which is coupled to methionine, is first loaded into the small ribosomal subunit along with additional proteins called **translation initiation factors** (Figure 7–32). Of all the charged tRNAs in the cell, only the charged initiator tRNA is capable of tightly binding to the P-site of the small ribosome subunit. Next, the loaded ribosomal subunit binds to the 5' end of an mRNA molecule, which is recognized in part by the cap that is present on eucaryotic mRNA (see Figure 7–12). The small ribosomal subunit then moves forward (5' to 3') along the mRNA searching for the first AUG. When this AUG is encountered, several initiation factors dissociate from the small ribosomal subunit to make way for the large ribosomal subunit to assemble and complete the ribosome. Because the initiator tRNA is bound to the P-site, protein synthesis is ready to begin with the addition of the next tRNA coupled to its amino acid (see Figure 7–32).

The mechanism for selecting a start codon is different in bacteria. Bacterial mRNAs have no 5' caps to tell the ribosome where to begin searching for the start of translation. Instead, they contain specific ribosome-binding sequences, up to six nucleotides long, that are located a few nucleotides upstream of the AUGs at which translation is to begin.

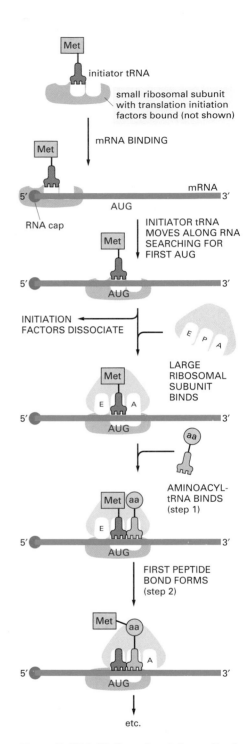

Figure 7–32 Initiation of protein synthesis in eucaryotes involves initiation factors and a special initiator tRNA. Although not shown here, efficient translation initiation also requires additional proteins (shown in Figure 7–19) bound at the 5' cap and poly-A tail of the mRNA. In this way, the translation apparatus can ascertain that both ends of the mRNA are intact before initiating translation. Following initiation the protein is elongated using the reactions outlined in Figure 7–30.

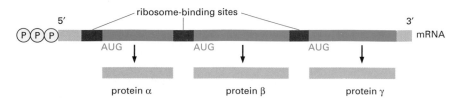

Figure 7–33 **A single procaryotic mRNA molecule can encode several different proteins.** Unlike eucaryotic ribosomes, which recognize a 5' cap, procaryotic ribosomes initiate transcription at ribosome-binding sites, which can be located in the interior of an mRNA molecule. This feature permits procaryotes to synthesize more than one type of protein from a single mRNA species.

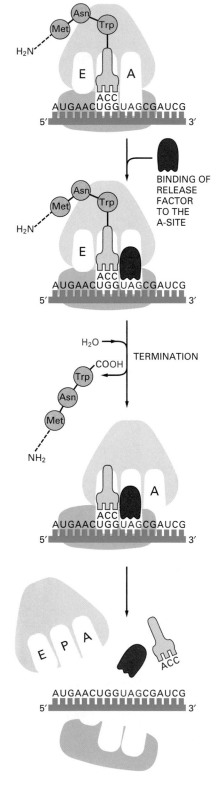

Figure 7–34 **In the final phase of protein synthesis, the binding of release factor to an A-site bearing a stop codon terminates translation.** The completed polypeptide is released, and the ribosome dissociates into its two separate subunits.

Unlike a eucaryotic ribosome, a procaryotic ribosome can readily bind directly to a start codon that lies in the interior of an mRNA, as long as a ribosome-binding site precedes it by several nucleotides. Such ribosome-binding sequences are necessary in bacteria, as procaryotic mRNAs are often *polycistronic*—that is, they encode several different proteins, each of which is translated from the same mRNA molecule (Figure 7–33). In contrast, a eucaryotic mRNA usually carries the information for a single protein.

The end of the protein-coding message is signaled by the presence of one of several codons (UAA, UAG, or UGA) called *stop codons* (see Figure 7–21). These are not recognized by a tRNA and do not specify an amino acid, but instead signal to the ribosome to stop translation. Proteins known as *release factors* bind to any stop codon that reaches the A-site on the ribosome, and this binding alters the activity of the peptidyl transferase in the ribosome, causing it to catalyze the addition of a water molecule instead of an amino acid to the peptidyl-tRNA (Figure 7–34). This reaction frees the carboxyl end of the growing polypeptide chain from its attachment to a tRNA molecule, and because this is the only attachment that holds the growing polypeptide to the ribosome, the completed protein chain is immediately released into the cytoplasm. The ribosome releases the mRNA and dissociates into its two separate subunits, which can then assemble on another mRNA molecule to begin a new round of protein synthesis.

We saw in Chapter 4 that most proteins can fold into their three-dimensional shape spontaneously; most do so as they are synthesized on the ribosome. Some proteins, however, require *molecular chaperones* (see p. 125) to help them fold correctly. Such proteins are often met by their chaperones as they emerge from the ribosome, and they are folded properly as they elongate.

Proteins Are Made on Polyribosomes

The synthesis of most protein molecules takes between 20 seconds and several minutes. But even during this very short period, multiple initiations usually take place on each mRNA molecule being translated. A new ribosome hops onto the 5' end of the mRNA molecule almost as soon as the preceding ribosome has translated enough of the nucleotide sequence to move out of the way. The mRNA molecules

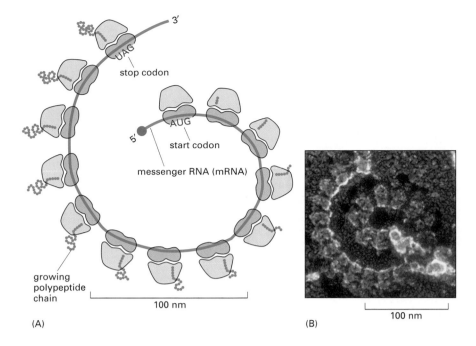

being translated are therefore usually found in the form of *polyribosomes* (also known as *polysomes),* large cytoplasmic assemblies made up of several ribosomes spaced as close as 80 nucleotides apart along a single mRNA molecule (Figure 7–35). These multiple initiations mean that many more protein molecules can be made in a given time than would be possible if each had to be completed before the next one could start.

Both bacteria and eucaryotes make use of polysomes, but bacteria can speed up the rate of protein synthesis even further. Because bacterial mRNA does not need to be processed and is also physically accessible to ribosomes while it is being made, ribosomes will attach to the free end of a bacterial mRNA molecule and start translating it even before the transcription of that RNA is complete; these ribosomes follow closely behind the RNA polymerase as it moves along DNA.

Inhibitors of Procaryotic Protein Synthesis Are Used as Antibiotics

The ability to accurately translate mRNAs into proteins is a fundamental feature of all life on Earth. Although the ribosome and other molecules that carry out this enormous task are very similar among organisms, there are some subtle differences, as we have seen, in the way bacteria and eucaryotes synthesize proteins. Through a quirk of evolution, these differences form the basis of one of the most important advances in modern medicine.

Many of our most effective antibiotics are compounds that act by inhibiting bacterial, but not eucaryotic, protein synthesis. Some of these drugs exploit the small structural and functional differences between bacterial and eucaryotic ribosomes, so they interfere preferentially with bacterial protein synthesis. These compounds can thus be taken in fairly high doses without being toxic to humans. Because different antibiotics bind to different regions of the bacterial ribosome, these drugs often inhibit different steps in the synthetic process. A few of the more common antibiotics of this kind are listed in Table 7–2.

It may seem odd, but many of the most common antibiotics were first isolated from fungi. Fungi and bacteria often occupy the same

Table 7–2 Antibiotics that Inhibit Protein or RNA Synthesis

ANTIBIOTIC	SPECIFIC EFFECT
Tetracycline	blocks binding of aminoacyl-tRNA to A-site of ribosome (step 1 in Figure 7–30)
Streptomycin	prevents the transition from initiation complex to chain-elongating ribosome; also causes miscoding
Chloramphenicol	blocks the peptidyl transferase reaction on ribosomes (step 2 in Figure 7–30)
Erythromycin	blocks the translocation reaction on ribosomes (step 3 in Figure 7–30)
Rifamycin	blocks initiation of RNA chains by binding to RNA polymerase

ecological niches; to gain a competitive edge, fungi have evolved, over time, potent toxins that kill bacteria but are harmless to the fungus. Because fungi and humans are more closely related to one another than either is to bacteria (see Figure 1–29), we can borrow these compounds to combat our own bacterial enemies.

Carefully Controlled Protein Breakdown Helps Regulate the Amount of Each Protein in a Cell

After a protein is released from the ribosome, it becomes subject to a number of controls by the cell. The number of copies of a protein in a cell depends, like the human population, not only on how quickly new individuals are made but also on how long they survive. So the breakdown of proteins into their constituent amino acids by cells is a way of regulating the amount of a particular protein present at a given time. Proteins vary enormously in their life span. Structural proteins that become part of a fairly permanent tissue such as bone or muscle may last for months or even years, whereas other proteins, such as metabolic enzymes and those that regulate the cycle of cell growth, mitosis, and division (discussed in Chapter 18), last only for days, hours, or even seconds. How does the cell control these lifetimes?

Cells possess specialized pathways to enzymatically break proteins down into their constituent amino acids (a process termed *proteolysis*). The enzymes that degrade proteins, first to short peptides and finally to individual amino acids, are known collectively as **proteases**. Proteases act by cutting (hydrolyzing) the peptide bonds between amino acids (see Panel 2–5, pp. 74–75). One function of proteolytic pathways is to rapidly degrade those proteins whose lifetimes must be short. Another is to recognize and eliminate proteins that are damaged or misfolded. Eliminating improperly folded proteins is critical for an organism, as neurodegenerative disorders such as Huntington's, Alzheimer's, and Creutzfeldt–Jacob diseases are caused by the aggregation of misfolded proteins. These protein aggregates can severely damage cells and tissues and even trigger cell death.

Although most damaged proteins are broken down in the cytosol, important protein degradation pathways also operate in other compartments in eucaryotic cells, such as lysosomes (discussed in Chapter 15). Most proteins degraded in the cytosol of eucaryotic cells are broken down by large complexes of proteolytic enzymes, called **proteasomes**. A proteasome contains a central cylinder formed from proteases whose active sites are thought to face into an inner chamber. Each end of the cylinder is stoppered by a large protein complex formed from at least 10 different types of protein subunits (Figure 7–36). These protein stoppers are thought to bind the proteins destined for digestion and then feed them into the inner chamber of the cylinder; there the proteases chop the proteins into short peptides, which are then released from the end of the proteasome. Housing the proteases inside this molecular chamber of destruction makes sense, as it prevents them from running rampant throughout the cell.

How does the proteasome select which proteins in the cell should enter the cylinder and be degraded? Proteasomes act primarily on proteins that have been marked for destruction by the covalent attachment of a small protein called *ubiquitin*. Specialized enzymes tag selected proteins with one or more ubiquitin molecules; these ubiquitinated proteins are then recognized and drawn into the proteasome, probably by proteins in the stopper.

Proteins that are meant to be short-lived often sport a short amino acid sequence that identifies the protein as one to be ubiquitinated and ultimately degraded by the proteasome. Denatured or misfolded proteins, as well as proteins containing oxidized or otherwise abnormal amino acids, are also recognized and degraded by this ubiquitin-dependent proteolytic system. The enzymes that add ubiquitin to such proteins presumably recognize signals that become exposed on these proteins as a result of the misfolding or chemical damage—for example, amino acid sequences or conformational motifs that are normally buried and inaccessible.

There Are Many Steps Between DNA and Protein

We have seen so far in this chapter that many different types of chemical reactions are required to produce a protein from the information contained in a gene (Figure 7–37). The final concentration of a protein in a cell, therefore, depends on the efficiency with which each of the many steps is carried out. We will see in the next chapter that cells have the ability to change the levels of most of their proteins according to their needs. In principle, all of the steps in Figure 7–37 can be regulated by the cell. However, as we shall see in the next chapter, the initiation of transcription is the most common point for a cell to regulate the expression of each of its genes.

Transcription and translation are universal processes that lie at the heart of life. However, when scientists came to consider how the flow of information from DNA to protein might have originated, they came to some unexpected conclusions.

(A)

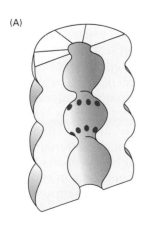

(B)

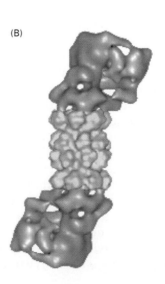

Figure 7–36 **The proteasome degrades short-lived and unwanted proteins in eucaryotic cells.** (A) A cutaway view of the structure of the central cylinder of the proteasome, determined by X-ray crystallography, with the active sites of the proteases indicated by *red dots*. (B) The structure of the entire proteasome, in which the central cylinder *(yellow)* is topped by a cap *(purple)* at each end. The complex cap structure selectively binds those proteins that have been marked for destruction; it then uses ATP hydrolysis to unfold their polypeptide chains and feed them into the inner chamber of the cylinder for digestion to short peptides. The cap structures were determined by computer processing of electron microscope images. (B, from W. Baumeister et al., *Cell* 92:367–380, 1998. © Elsevier Science.)

Figure 7–37 Protein production in a eucaryotic cell requires many steps. The final concentration of each protein in a eucaryotic cell depends upon the efficiency of each step depicted. Even after a protein has been translated, its activity can be regulated by degradation (shown), by the binding of small molecules, or by other posttranslational modifications such as phosphorylation (discussed in Chapter 4).

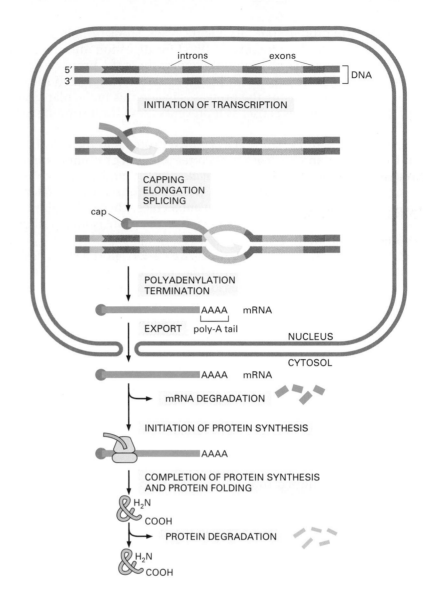

RNA and the Origins of Life

To understand fully the processes occurring in present-day cells, we need to consider how cells evolved. The most challenging and fundamental of all such problems is that of the expression of hereditary information, which today requires incredibly complex machinery and proceeds from DNA to protein through an RNA intermediate. How did this machinery arise? One view is that an *RNA world* existed on Earth before modern cells arose (Figure 7–38). According to this hypothesis, RNA both stored genetic information and catalyzed chemical reactions in primitive cells. Only later in evolutionary time did DNA take over as the genetic material and proteins become the major catalysts and structural components of cells. If this idea is correct, then the transition out of the RNA world was never completed; as we have seen in this chapter, RNA still catalyzes several fundamental reactions in modern-day cells. These RNA catalysts are essentially molecular fossils of an earlier world.

In this section we outline some of the arguments in support of the RNA world hypothesis. We will see that several of the more surprising features of modern cells, such as the ribosome and the RNA splicing machinery, are most easily explained by viewing them as descendants of a complex network of RNA-mediated interactions that dominated cell metabolism in the RNA world. We also discuss how DNA might have

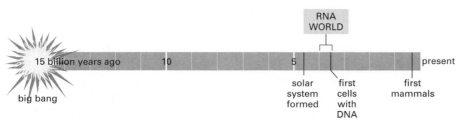

taken over as the genetic material and how proteins might have eclipsed RNA to perform the bulky biochemical catalysis in modern-day cells.

Life Requires Autocatalysis

The origin of life requires molecules possessing, if only to a small extent, one crucial property: the ability to catalyze reactions that lead—directly or indirectly—to production of more molecules like themselves. Catalysts with this special self-promoting property, once they had arisen by chance, would reproduce themselves and would therefore divert raw materials from the production of other substances. In this way one can envisage the gradual development of an increasingly complex chemical system of organic monomers and polymers that function together to generate more molecules of the same types, fueled by a supply of simple raw materials in the environment. Such an *autocatalytic* system would have many of the properties we think of as characteristic of living matter: the system would contain a far from random selection of interacting molecules; it would tend to reproduce itself; it would compete with other systems dependent on the same raw materials; and, if deprived of its raw materials or maintained at a temperature that upset the balance of reaction rates, it would decay toward chemical equilibrium and "die."

But what molecules could have had such autocatalytic properties? In present-day living cells the most versatile catalysts are polypeptides, which are able to adopt diverse three-dimensional forms that bristle with chemically reactive sites. Their versatility as catalysts notwithstanding, there is no known way in which a polypeptide can reproduce itself directly. Polynucleotides, however, can do all of these things.

RNA Can Both Store Information and Catalyze Chemical Reactions

We have seen that complementary base-pairing enables one polynucleotide to act as a template for the formation of another. Thus a single strand of RNA or DNA can specify the sequence of a complementary polynucleotide, which, in turn, can specify the sequence of the original molecule, allowing the original polynucleotide to be replicated (Figure 7–39). Such complementary templating mechanisms lie at the heart of DNA replication and transcription in modern-day cells.

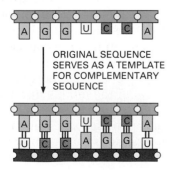

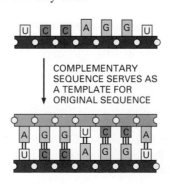

Figure 7–39 **An RNA molecule can in principle guide the formation of an exact copy of itself.** In the first step the original RNA molecule acts as a template to form an RNA molecule of complementary sequence. In the second step this complementary RNA molecule itself acts as a template, forming RNA molecules of the original sequence. Since each templating molecule can produce many copies of the complementary strand, these reactions can result in the "multiplication" of the original sequence.

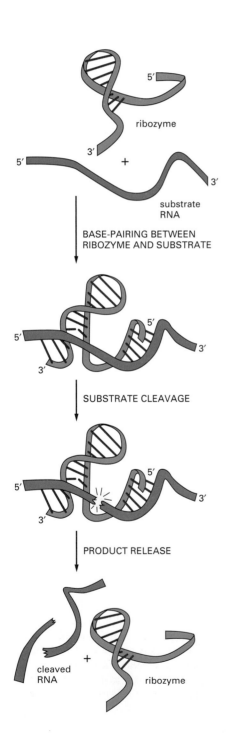

But the efficient synthesis of polynucleotides by such complementary templating mechanisms also requires catalysts to promote the polymerization reaction: without catalysts, polymer formation is slow, error-prone, and inefficient. Today, nucleotide polymerization is rapidly catalyzed by protein enzymes—such as the DNA and RNA polymerases. But how could this reaction be catalyzed before proteins with the appropriate catalytic specificity existed? The beginnings of an answer to this question were obtained in 1982, when it was discovered that RNA molecules themselves can act as catalysts. We have already seen in this chapter, for example, that a molecule of RNA appears to be the catalyst for the peptidyl transferase reaction that takes place on the ribosome. The unique potential of RNA molecules to act both as information carriers and as catalysts is thought to have enabled them to play the central role in the origin of life.

In present-day cells, RNA is synthesized as a single-stranded molecule, and we have seen that complementary base-pairing can occur between nucleotides in the same chain (see Figure 7–5). This base-pairing, along with "nonconventional" hydrogen bonds, can cause each RNA molecule to fold up in a unique way that is determined by its nucleotide sequence. Such associations produce complex three-dimensional patterns of folding, where the molecule as a whole adopts a unique shape.

We have seen that a protein enzyme is able to catalyze a biochemical reaction because it has a surface with unique contours and chemical properties on which a given substrate can react (discussed in Chapter 4). In the same way, RNA molecules, with their unique folded shapes, can serve as enzymes (Figure 7–40), although the fact that they are constructed of only four different subunits limits their catalytic efficiency and the range of chemical reactions they can catalyze compared with proteins. Nonetheless, ribozymes that catalyze quite a wide range of chemical reactions have now been found in nature or constructed in laboratories (Table 7–3). Relatively few such catalytic RNAs exist in present-day cells; most catalytic functions in present-day cells have been taken over by proteins. But the processes in which catalytic RNAs still seem to have major roles include some of the most fundamental steps in the expression of genetic information—especially those steps where RNA molecules themselves are spliced or translated into protein.

RNA, therefore, has all the properties required of a molecule that could catalyze its own synthesis (Figure 7–41), and such molecules could have been instrumental in the formation of early cells. Although self-replicating systems of RNA molecules have not been found in nature, scientists are confident that they can be constructed in the laboratory. While this demonstration would not prove that self-replicating RNA molecules were essential in the origin of life on Earth, it would certainly suggest that such a scenario is possible.

Figure 7–40 A ribozyme is an RNA molecule that possesses catalytic activity. The simple RNA molecule shown catalyzes the cleavage of a second RNA at a specific site. This ribozyme is found embedded in larger RNA genomes—called viroids—that infect plants. The cleavage, which occurs in nature at a distant location on the same RNA molecule that contains the ribozyme, is a step in the replication of the RNA genome. This reaction requires a molecule of magnesium (not shown), which is brought next to the site of cleavage on the substrate. (Adapted from T.R. Cech and O.C. Uhlenbeck, *Nature* 372:39–40, 1994.)

Table 7–3 Biochemical Reactions that Can Be Catalyzed by Ribozymes

ACTIVITY	RIBOZYMES
RNA cleavage, RNA ligation	self-splicing RNAs
DNA cleavage	self-splicing RNAs
Peptide bond formation in protein synthesis	ribosomal RNA
DNA ligation	*in vitro* selected RNA
RNA splicing	RNAs of the spliceosome (?), self-splicing RNAs
RNA polymerization	*in vitro* selected RNA
RNA phosphorylation	*in vitro* selected RNA
RNA aminoacylation	*in vitro* selected RNA
RNA alkylation	*in vitro* selected RNA
Isomerization (C–C bond rotation)	*in vitro* selected RNA

RNA Is Thought to Predate DNA in Evolution

The first cells on Earth would presumably have been much less complex and less efficient in reproducing themselves than even the simplest present-day cells, because catalysis by RNA molecules is less efficient than that by proteins. They would have consisted of little more than a simple membrane enclosing a set of self-replicating molecules and a few other components required to provide the materials and energy for their replication. If the evolutionary speculations about RNA outlined above are correct, these earliest cells would also have differed fundamentally from the cells we know today in having their hereditary information stored in RNA rather than DNA.

Evidence that RNA arose before DNA in evolution can be found in the chemical differences between them. Ribose (see Figure 7–3A), like glucose and other simple carbohydrates, is readily formed from formaldehyde (HCHO), which is one of the principal products of experiments simulating conditions on the primitive Earth. The sugar deoxyribose is harder to make, and in present-day cells it is produced from ribose in a reaction catalyzed by a protein enzyme, suggesting that ribose predates deoxyribose in cells. Presumably, DNA appeared on the scene later, and then proved more suited than RNA as a permanent repository of genetic information. In particular, the deoxyribose in its sugar–phosphate backbone makes chains of DNA chemically much more stable than chains of RNA, so that greater lengths of DNA can be maintained without breakage.

The other differences between RNA and DNA—the double-helical structure of DNA and the use of thymine rather than uracil—further enhance DNA stability by making the molecule easier to repair. We saw in the previous chapter that a damaged nucleotide on one strand of the double helix can be repaired using the other strand as a template. Furthermore, deamination, one of the most common unwanted chemical changes occurring in polynucleotides, is easier to detect and repair in DNA than in RNA (see Figures 6–23 and 6–25). This is because the product of the deamination of cytosine is, by chance, uracil, which already exists in RNA, so that such changes would be impossible for repair enzymes to detect in an RNA molecule. However, in DNA, which has thymine rather than uracil, any uracil produced by the accidental degradation of cytosine is easily detected and repaired.

Taken together, the evidence we have discussed points to the notion that RNA, having both genetic and catalytic properties, preceded

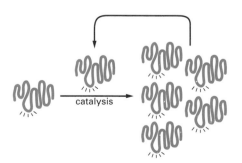

Figure 7–41 Could an RNA molecule catalyze its own synthesis? This hypothetical process would require the catalysis of both steps shown in Figure 7–39. The *red rays* represent the active site of this RNA enzyme.

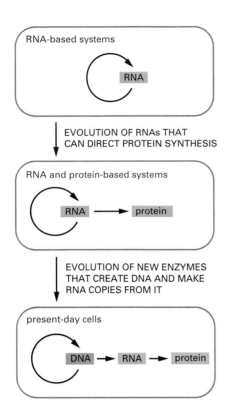

RNA-based systems

RNA

↓ EVOLUTION OF RNAs THAT
CAN DIRECT PROTEIN SYNTHESIS

RNA and protein-based systems

RNA ⟶ protein

↓ EVOLUTION OF NEW ENZYMES
THAT CREATE DNA AND MAKE
RNA COPIES FROM IT

present-day cells

DNA ⟶ RNA ⟶ protein

Figure 7–42 RNA may have preceded DNA and proteins in evolution. According to this idea, RNA molecules combined genetic, structural, and catalytic functions in the earliest cells. DNA is now the repository of genetic information, and proteins carry out almost all catalytic functions in cells. RNA now functions mainly as a go-between in protein synthesis, while remaining a catalyst for a few crucial reactions.

DNA in evolution. As cells more closely resembling present-day cells appeared, it is believed that many of the functions originally performed by RNA were taken over by molecules more specifically fitted to the tasks required. Eventually DNA took over the primary genetic function and proteins became the major catalysts, while RNA remained primarily as the intermediary connecting the two (Figure 7–42). With the advent of DNA, cells were able to become more complex, for they could then carry and transmit more genetic information than could be stably maintained in an RNA molecule. Because of the greater chemical complexity of proteins and the variety of chemical reactions they can catalyze, the shift (albeit incomplete) from RNA to proteins also provided a much richer source of structural components and enzymes. This enabled cells to evolve the great diversity of structure and function that we see in life today.

Essential Concepts

- The flow of genetic information in all living cells is DNA → RNA → protein. The conversion of the genetic instructions in DNA into RNAs and proteins is termed gene expression.
- To express the genetic information carried in DNA, the nucleotide sequence of a gene is first transcribed into RNA. Transcription is catalyzed by the enzyme RNA polymerase. Nucleotide sequences in the DNA molecule indicate to the RNA polymerase where to start and stop transcribing.
- RNA differs in several respects from DNA. It contains the sugar ribose instead of deoxyribose and the base uracil (U) instead of thymine (T). RNAs in cells are synthesized as single-stranded molecules, which often fold up into precise three-dimensional shapes.
- Cells make several different functional types of RNA, including messenger RNA (mRNA), which carries the instructions for making proteins; ribosomal RNA (rRNA), which is a component of ribosomes; and transfer RNA (tRNA), which acts as an adaptor molecule in protein synthesis.
- In eucaryotic DNA, most genes are composed of a number of smaller coding regions (exons) interspersed with noncoding regions (introns). When a eucaryotic gene is transcribed from DNA into RNA, both the exons and introns are copied.
- Introns are removed from the RNA transcripts in the nucleus by the process of RNA splicing. In a reaction catalyzed by small ribonucleoprotein complexes known as snRNPs, the introns are excised from the primary transcript and the exons are joined together.
- Eucaryotic mRNAs go through several additional RNA processing steps before they leave the nucleus, including RNA capping and polyadenylation. These reactions, along with splicing, are tightly coupled to transcription and take place as the RNA is being transcribed. The mature mRNA then moves to the cytoplasm.
- Translation of the nucleotide sequence of mRNA into a protein takes place in the cytoplasm on large ribonucleoprotein assemblies called

Question 7–6

Discuss the following: "During the evolution of life on Earth, RNA has been demoted from its glorious position as the first self-replicating catalyst. Its role now is as a mere messenger in the information flow from DNA to protein."

ribosomes. These attach to the mRNA and move stepwise along the mRNA chain, translating the message into protein.

- The nucleotide sequence in mRNA is read in sets of three nucleotides (codons), each codon corresponding to one amino acid.
- The correspondence between amino acids and codons is specified by the genetic code. The possible combinations of the 4 different nucleotides in RNA give 64 different codons in the genetic code. Most amino acids are specified by more than one codon.
- tRNA acts as an adaptor molecule in protein synthesis. Enzymes called aminoacyl-tRNA synthetases link amino acids to their appropriate tRNAs. Each tRNA contains a sequence of three nucleotides, the anticodon, which matches a codon in mRNA by complementary base-pairing between codon and anticodon.
- Protein synthesis begins when a ribosome assembles at an initiation codon (AUG) in mRNA, a process that is regulated by proteins called translation initiation factors. The completed protein chain is released from the ribosome when a stop codon (UAA, UAG, or UGA) is reached.
- The stepwise linking of amino acids into a polypeptide chain is catalyzed by an rRNA molecule in the large ribosomal subunit. Thus the ribosome is an example of a ribozyme, an RNA molecule that can catalyze a chemical reaction.
- The degradation of proteins in the cell is carefully controlled. Some proteins are degraded in the cytosol by large protein complexes called proteasomes.
- From our knowledge of present-day organisms and the molecules they contain, it seems likely that living systems began with the evolution of RNA molecules that could catalyze their own replication.
- It has been proposed that, as cells evolved, the DNA double helix replaced RNA as a more stable molecule for storing increased amounts of genetic information, and proteins replaced RNAs as major catalytic and structural components.
- The flow of information in present-day living cells is DNA → RNA → protein, with RNA serving primarily as a go-between. Some important reactions, however, are still catalyzed by RNA; these are thought to provide a glimpse into the ancient, RNA-based world.

Key Terms

aminoacyl-tRNA synthetase	ribosome
anticodon	ribozyme
codon	RNA polymerase
exon	RNA processing
genetic code	RNA splicing
initiator tRNA	small nuclear ribonucleoprotein
intron	particle (snRNP)
messenger RNA (mRNA)	small nuclear RNA (snRNA)
peptide bond	spliceosome
promoter	transcription
proteasome	transfer RNA (tRNA)
reading frame	translation
ribosomal RNA (rRNA)	translation initiation factor

Questions

Question 7–7

Which of the following statements are correct? Explain your answers.

A. An individual ribosome can make only one type of protein.

B. All mRNAs fold into particular three-dimensional structures that are required for their translation.

C. The large and small subunits of an individual ribosome always stay together and never exchange partners.

D. Ribosomes are cytoplasmic organelles that are encapsulated by a single membrane.

E. Because the two strands of DNA are complementary, the mRNA of a given gene can be synthesized using either strand as a template.

F. An mRNA may contain the sequence ATTGAC-CCCGGTCAA.

G. The amount of a protein present in the cell at a steady state depends on its rate of synthesis, its catalytic activity, and its rate of degradation.

Question 7–8

The Lacheinmal protein is a hypothetical protein that causes people to smile more often. It is inactive in many chronically unhappy people. The mRNA isolated from a number of different unhappy persons in the same family was found to lack an internal stretch of 173 nucleotides that is present in the Lacheinmal mRNA isolated from a control group of generally happy people. The DNA sequences of the *Lacheinmal* genes from the happy and unhappy families were determined and compared. They differed by just one nucleotide change—and no nucleotides were deleted. Moreover, the change was found in an intron. What can you say about the molecular basis of unhappiness in this family?

(Hints: consider the following two questions independently. [1] Can you hypothesize a molecular mechanism by which a single nucleotide change in a gene could cause the observed deletion in the mRNA? Note that it is an *internal* deletion. [2] What consequences for the Lacheinmal protein would result from removing a 173-ribonucleotide-long internal stretch from its mRNA? Assume that the 173 bases are deleted inside the coding region of the Lacheinmal mRNA.)

Question 7–9

Use the genetic code shown in Figure 7–21 to identify which of the following nucleotide sequences would code for the polypeptide sequence arginine-glycine-aspartate:

1. 5′-AGA-GGA-GAU-3′
2. 5′-ACA-CCC-ACU-3′
3. 5′-GGG-AAA-UUU-3′
4. 5′-CGG-GGU-GAC-3′

Question 7–10

"The bonds that form between the anticodon of a tRNA molecule and the three nucleotides of a codon in mRNA are _____." Complete this sentence with each of the following options and explain why the resulting statements are correct or incorrect.

A. Covalent bonds formed by GTP hydrolysis

B. Hydrogen bonds that form when the tRNA is at the A-site

C. Broken by the translocation of the ribosome along the mRNA

Question 7–11

List the ordinary, dictionary definitions of the terms *replication, transcription,* and *translation.* By their side, list the special meaning each term has when applied to the living cell.

Question 7–12

In an alien world, the genetic code is written in pairs of nucleotides. How many amino acids can it specify? In a different world a triplet code is used but the sequence of nucleotides is not important. It only matters which nucleotides are present. How many amino acids can this code specify? Would you expect to encounter any problems translating these codes?

Question 7–13

One remarkable regular feature of the genetic code is that amino acids with similar chemical properties often have similar codons. Thus, codons with U or C as the second nucleotide tend to specify hydrophobic amino acids. Can you suggest a possible explanation for this phenomenon in terms of the early evolution of the protein synthesis machinery?

Question 7–14

A mutation in DNA generates a UGA stop codon in the middle of the RNA coding for a particular protein. A second mutation in the cell leads to a single nucleotide change in a tRNA that allows the correct translation of the protein; that is, the second mutation "suppresses" the defect caused by the first. The altered tRNA translates the UGA as tryptophan. What nucleotide change has probably occurred in the mutant tRNA molecule? What consequences would the presence of such a mutant tRNA have for the translation of the normal genes in this cell?

Question 7–15

The charging of a tRNA with an amino acid can be represented by the following equation:

amino acid + tRNA + ATP → aminoacyl-tRNA + AMP + PP$_i$

where PP$_i$ is pyrophosphate (see Figure 3–41). In the aminoacyl-tRNA, the amino acid and tRNA are linked with a high-energy bond (discussed in Chapter 3); a large portion of the energy derived from the hydrolysis of ATP is thus stored in this bond and is available to drive peptide-bond formation at the later stages of protein synthesis. The free-energy change of the charging reaction shown in the equation is close to zero and therefore would not be expected to favor attachment of the amino acid to tRNA. Can you suggest a further step that could drive the reaction to completion?

Question 7–16

A. The average molecular weight of proteins in the cell is about 30,000 daltons. A few proteins, however, are very much larger. The largest known polypeptide chain made by any cell is a protein called titin (made by muscle cells), and it has a molecular weight of 3,000,000 daltons. Estimate how long it will take a muscle cell to translate an mRNA coding for titin (assume the average molecular weight of an amino acid to be 120, and a translation rate of two amino acids per second for eucaryotic cells).

B. Protein synthesis is very accurate: for every 10,000 amino acids joined together, only one mistake is made. What is the fraction of average-sized protein molecules and of titin molecules that are synthesized without any errors? (Hint: the probability P of obtaining an error-free protein is given by $P = (1 - E)^n$, where E is the error frequency and n the number of amino acids.)

C. The molecular weight of all eucaryotic ribosomal proteins combined is about 2.5×10^6 daltons. Would it be advantageous to synthesize them as a single protein?

D. Transcription occurs at a rate of about 30 nucleotides per second. Is it possible to calculate the time required to synthesize a titin mRNA from the information given here?

Question 7–17

Which of the following mutational changes would be predicted to harm an organism? Explain your answers.

A. Insertion of a single nucleotide near the end of the coding sequence
B. Removal of a single nucleotide near the beginning of the coding sequence
C. Deletion of three consecutive nucleotides in the middle of the coding sequence
D. Deletion of four consecutive nucleotides in the middle of the coding sequence
E. Substitution of one nucleotide for another in the middle of the coding sequence

Question 7–18

Without looking back at the figures, draw a diagram depicting the essential steps of the splicing reaction. Indicate the intron–exon junctions in the primary transcript, the snRNPs, the excised intron, and the mature mRNA. During splicing, the same complex of snRNPs catalyzes both the cleavage of the RNA at the first intron–exon junction and the subsequent joining of the two exons. What might happen if the two reactions were catalyzed by separate enzymes that were not associated in a complex?

Highlights from *Essential Cell Biology 2 Interactive* CD-ROM

7.3 RNA Splicing
7.6 Translation II
7.8 Ribosome Elongation Cycle

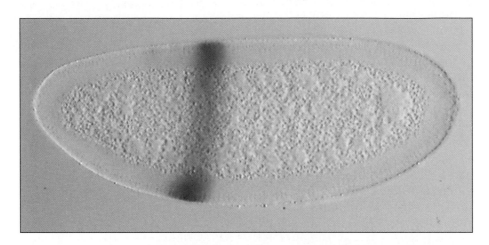

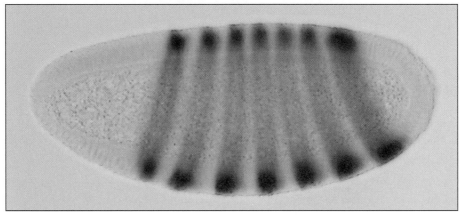

Differential gene expression. The patterning and development of multicellular organisms depends in part on the differential expression of genes. Gene regulatory proteins switch on the genes that are appropriate for each cell at each stage of development. One such regulatory protein involved in fruit fly development is called Eve. The images here show the seven characteristic stripes of Eve protein in the fly embryo (bottom photograph), with the position of the second stripe revealed by a reporter gene controlled by a small region of the regulatory region of the *eve* gene at the top. The story of the *eve* gene is told in the How We Know box on pp. 282–284. (Courtesy of Stephen Small and Michael Levine.)

Control of Gene Expression

An organism's DNA encodes all of the RNA and protein molecules that are needed to make its cells. But a complete description of the DNA sequence of an organism—be it the few million nucleotides of a bacterium or the few billion nucleotides in each human cell—would no more enable us to reconstruct the organism than a list of English words in a dictionary would enable us to reconstruct a play by Shakespeare. We need to know how the elements in the DNA sequence—the genes—are used. Even the simplest single-celled bacterium can use its genes selectively, switching genes on and off so that it makes different metabolic enzymes depending on the food sources available.

In multicellular plants and animals, **gene expression** is under even more elaborate control. In the course of embryonic development, a fertilized egg cell gives rise to many cell types that differ dramatically in both structure and function. The differences between a mammalian neuron and a lymphocyte, for example, are so extreme that it is difficult to imagine that the two cells contain the same DNA (Figure 8–1). For this reason, and because cells in an adult organism rarely lose their distinctive characteristics, biologists originally suspected that genes might be selectively lost when a cell becomes specialized. We now know, however, that nearly all the cells of a multicellular organism contain the same genome. Cell differentiation is instead achieved by changes in gene expression.

Hundreds of different cell types carry out a range of specialized functions that depend upon genes that are only switched on in that cell type: for example, the β cells of the pancreas make the protein hormone insulin, while the α cells of the pancreas make the hormone glucagon; the lymphocytes of the immune system are the only cells in the body to make antibodies, while developing red blood cells are the only cells that make the oxygen-transport protein hemoglobin. The differences between a neuron, a lymphocyte, a pancreas cell, and a red blood cell depend upon the precise control of gene expression. In each case the cell is using only some of the genes in its total repertoire.

In this chapter, we shall discuss the main ways in which gene expression is controlled in bacterial and eucaryotic cells. Although some mechanisms of control apply to both sorts of cells, eucaryotic cells, through their more complex chromosomal structure, have ways of controlling gene expression that are not available to bacteria.

An Overview of Gene Expression

The Different Cell Types of a Multicellular Organism Contain the Same DNA

Different Cell Types Produce Different Sets of Proteins

A Cell Can Change the Expression of Its Genes in Response to External Signals

Gene Expression Can Be Regulated at Many of the Steps in the Pathway from DNA to RNA to Protein

How Transcriptional Switches Work

Transcription Is Controlled by Proteins Binding to Regulatory DNA Sequences

Repressors Turn Genes Off, Activators Turn Them On

An Activator and a Repressor Control the *lac* Operon

Initiation of Eucaryotic Gene Transcription Is a Complex Process

Eucaryotic RNA Polymerase Requires General Transcription Factors

Eucaryotic Gene Regulatory Proteins Control Gene Expression from a Distance

Packing of Promoter DNA into Nucleosomes Can Affect Initiation of Transcription

The Molecular Mechanisms That Create Specialized Cell Types

Eucaryotic Genes Are Regulated by Combinations of Proteins

The Expression of Different Genes Can Be Coordinated by a Single Protein

Combinatorial Control Can Create Different Cell Types

Stable Patterns of Gene Expression Can Be Transmitted to Daughter Cells

The Formation of an Entire Organ Can Be Triggered by a Single Gene Regulatory Protein

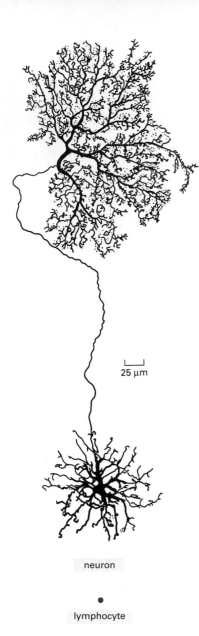

25 μm

neuron

lymphocyte

Figure 8–1 A neuron and a lymphocyte share the same genome.
The long branching processes of this neuron from the retina enable it to receive electrical signals from many cells and carry them to many neighboring cells. The lymphocyte is a white blood cell involved in the immune response to infection and moves freely through the body. Both of these mammalian cells contain the same genome, but they express different RNAs and proteins.

An Overview of Gene Expression

How does an individual cell specify which of its many thousands of genes to express as proteins? Deciding which genes are to be expressed is an especially important problem for multicellular organisms because, as the animal develops, cell types such as muscle, nerve, and blood cells become different from one another, eventually leading to the wide variety of cell types seen in the adult. This **differentiation** arises because cells make and accumulate different sets of RNA and protein molecules; that is, they express different genes.

The Different Cell Types of a Multicellular Organism Contain the Same DNA

As discussed above, cells have the ability to change which genes they express without altering the nucleotide sequence of their DNA. But how do we know this? If DNA were altered irreversibly during development, the chromosomes of a differentiated cell would be incapable of guiding the development of the whole organism. To test this idea, a nucleus from a skin cell of an adult frog was injected into a frog egg whose own nucleus had been removed. In at least some cases the egg developed normally, indicating that the transplanted skin cell nucleus cannot have lost any critical DNA sequences (Figure 8–2). Such nuclear transplantation experiments have also been carried out successfully using differentiated cells taken from adult mammals, including sheep, cows, pigs, goats, and mice. And in plants, individual cells removed from a carrot, for example, can be shown to regenerate an entire adult carrot plant. These experiments all show that the DNA in specialized cell types still contains the entire set of instructions needed to form a whole organism. The cells of an organism therefore differ not because they contain different genes, but because they express them differently.

Different Cell Types Produce Different Sets of Proteins

The extent of the differences in gene expression between different cell types may be roughly gauged by comparing the protein composition of cells in liver, heart, brain, and so on using the technique of two-dimensional gel electrophoresis (see Panel 4–5, p. 163). Experiments of this kind reveal that many proteins are common to all the cells of a multicellular organism. These *housekeeping* proteins include the structural proteins of chromosomes, RNA polymerases, DNA repair enzymes, ribosomal proteins, enzymes involved in glycolysis and other basic metabolic processes, and many of the proteins that form the cytoskeleton. Each different cell type also produces specialized proteins that are responsible for the cell's distinctive properties. In mammals, for example, hemoglobin is made in reticulocytes, the cells that develop into red blood cells, but it cannot be detected in any other cell type.

Many proteins in a cell are produced in such small numbers that they cannot be detected by the technique of gel electrophoresis. The mRNAs that encode such proteins, however, can be detected by more

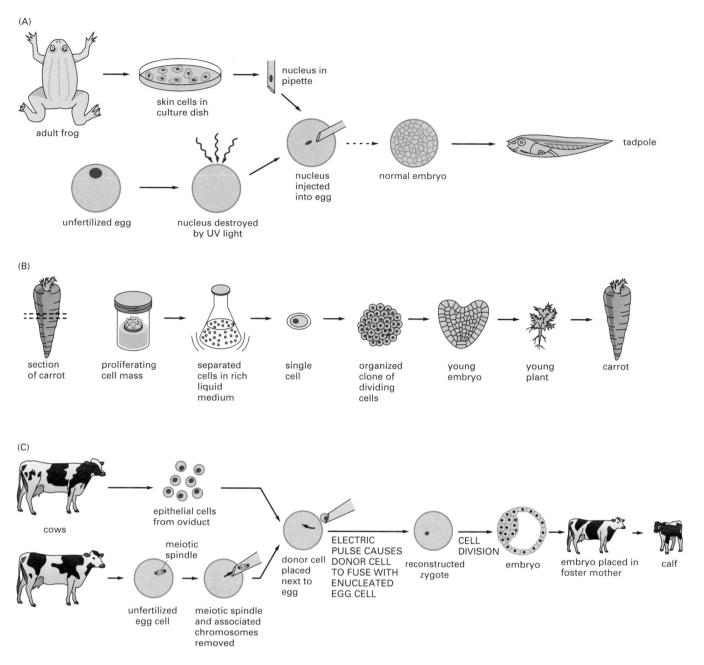

Figure 8–2 Differentiated cells contain all the genetic instructions necessary to direct the formation of a complete organism. (A) The nucleus of a skin cell from an adult frog transplanted into an egg whose nucleus has been removed can give rise to an entire tadpole. The broken arrow indicates that to give the transplanted genome time to adjust to an embryonic environment, a further transfer step is required in which one of the nuclei is taken from the early embryo that begins to develop and is put back into a second enucleated egg. (B) In many types of plants, differentiated cells retain the ability to "dedifferentiate," so that a single cell can form a clone of progeny cells that later give rise to an entire plant. (C) A differentiated cell from an adult cow introduced into an enucleated egg from a different cow can give rise to a calf. Different calves produced from the same differentiated cell donor are genetically identical and are therefore clones of one another. (A, modified from J.B. Gurdon, *Sci. Am.* 219(6):24–35, 1968.)

sensitive techniques (discussed in Chapter 10). Estimates of the number of different mRNA sequences in human cells suggest that, at any one time, a typical differentiated human cell expresses perhaps 10,000–20,000 genes from a repertoire of about 30,000. It is the expression of a different collection of genes in each cell type that causes the large variations seen in the size, shape, behavior, and function of differentiated cells.

A Cell Can Change the Expression of Its Genes in Response to External Signals

Most of the specialized cells in a multicellular organism are capable of altering their patterns of gene expression in response to extracellular cues. If a liver cell is exposed to a glucocorticoid hormone, for example, the production of several specific proteins is dramatically increased. Glucocorticoids are released in the body during periods of starvation or intense exercise and signal the liver to increase the production of glucose from amino acids and other small molecules; the set of proteins whose production is induced includes enzymes such as tyrosine aminotransferase, which helps to convert tyrosine to glucose. When the hormone is no longer present, the production of these proteins drops to its normal level.

Other cell types respond to glucocorticoids differently. In fat cells, for example, the production of tyrosine aminotransferase is reduced, while some other cell types do not respond to glucocorticoids at all. These examples illustrate a general feature of cell specialization: different cell types often respond in different ways to the same extracellular signal. Underlying such adjustments that occur in response to extracellular signals, there are features of the gene expression pattern that do not change and give each cell type its permanently distinctive character.

Gene Expression Can Be Regulated at Many of the Steps in the Pathway from DNA to RNA to Protein

If differences among the various cell types of an organism depend on the particular genes that the cells express, at what level is the control of gene expression exercised? As we saw in the last chapter, there are many steps in the pathway leading from DNA to protein, and all of them can in principle be regulated. Thus a cell can control the proteins it makes by (1) controlling when and how often a given gene is transcribed, (2) controlling how the primary RNA transcript is spliced or otherwise processed, (3) selecting which mRNAs are translated by ribosomes, or (4) selectively activating or inactivating proteins after they have been made (Figure 8–3).

Although examples of regulation at each of the steps shown in Figure 8–3 are known, for most genes the control of transcription (step number 1) is paramount. This makes sense because only transcriptional control can ensure that no unnecessary intermediates are synthesized. For the remainder of this chapter, therefore, we concentrate on the DNA and protein components that determine which genes a cell transcribes into RNA.

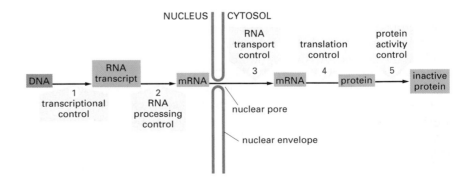

Figure 8–3 Eucaryotic gene expression can be controlled at several different steps. Examples of regulation at each of the steps are known, although for most genes the main site of control is step 1: transcription of a DNA sequence into RNA.

How Transcriptional Switches Work

Only 40 years ago the idea that genes could be switched on and off was revolutionary. This concept was a major advance, and it came originally from the study of how *E. coli* bacteria adapt to changes in the composition of their growth medium. Many of the same principles apply to eucaryotic cells. However, the enormous complexity of gene regulation in higher organisms, combined with the packaging of their DNA into chromatin, creates special challenges and some novel opportunities for control—as we shall see. We begin with a discussion of *gene regulatory proteins,* the proteins specialized for switching genes on and off.

Transcription Is Controlled by Proteins Binding to Regulatory DNA Sequences

Control of transcription is usually exerted at the step at which the process is initiated. In Chapter 7, we saw that the promoter region of a gene attracts the enzyme RNA polymerase and correctly orients the enzyme to begin its task of making an RNA copy of the gene. The promoters of both bacterial and eucaryotic genes include an *initiation site,* where transcription actually begins, and a sequence of approximately 50 nucleotides that extends "upstream" from the initiation site (if one likens the direction of transcription to the flow of a river). This region contains sites that are required for the RNA polymerase to bind to the promoter. In addition to the promoter, nearly all genes, whether bacterial or eucaryotic, have **regulatory DNA sequences** that are used to switch the gene on or off. Whether a gene is expressed or not depends on a variety of factors, including the type of cell, its surroundings, its age, and the extracellular signals that impinge on it.

Some regulatory DNA sequences are as short as 10 nucleotide pairs and act as simple gene switches that respond to a single signal. Such simple switches predominate in bacteria. Other regulatory DNA sequences, especially those in eucaryotes, are very long (sometimes more than 10,000 base pairs) and act as molecular microprocessors, responding to a variety of signals that they integrate into an instruction that determines how often transcription is initiated.

Regulatory DNA sequences do not work by themselves. To have any effect these sequences must be recognized by proteins called **gene regulatory proteins** that bind to the DNA. It is the combination of a DNA sequence and its associated protein molecules that acts as the switch to control transcription. Hundreds of regulatory DNA sequences have been identified, and each is recognized by one or more gene regulatory proteins.

Proteins that recognize a specific DNA sequence do so because the surface of the protein fits tightly against the special surface features of the double helix in that region. These features will vary depending on the nucleotide sequence, and thus different proteins will recognize different nucleotide sequences. In most cases, the protein inserts into the major groove of the DNA helix (see Figure 5–8) and makes a series of molecular contacts with the base pairs. The protein forms hydrogen bonds, ionic bonds, and hydrophobic interactions with the edges of the bases, usually without disrupting the hydrogen bonds that hold the base pairs together (Figure 8–4). Although each individual contact is weak, the 20 or so contacts that are typically formed at the protein–DNA interface combine to ensure that the interaction is both highly specific and very strong; in fact, protein–DNA interactions are among the tightest and most specific molecular interactions known in biology.

Although each example of protein–DNA recognition is unique in

Figure 8–4 A gene regulatory protein binds to the major groove of a DNA helix.

Only a single contact between the protein and one base pair in DNA is shown. Typically, the protein–DNA interface would consist of 10 to 20 such contacts, each involving a different amino acid and each contributing to the strength of the protein–DNA interaction.

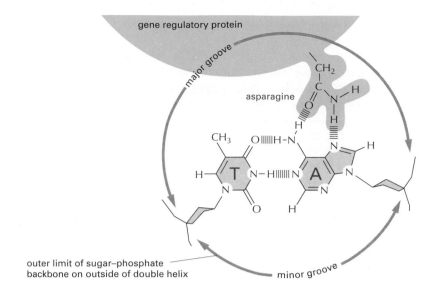

detail, many of the proteins responsible for gene regulation contain one of several particularly stable folding patterns. These fit into the major groove of the DNA double helix and form tight associations with a short stretch of DNA base pairs. The *DNA-binding motifs* shown in Figure 8–5—the homeodomain, the zinc finger, and the leucine zipper—are

Figure 8–5 Gene regulatory proteins contain a variety of DNA-binding motifs.

(A and B) Front and side views of the *homeodomain*—a structural motif in many eucaryotic DNA-binding proteins. It consists of three linked α helices, which are shown as *cylinders* in this figure. Most of the contacts with the DNA bases are made by helix 3 (which is seen end-on in B). The asparagine (Asn) in this helix contacts an adenine in the manner shown in Figure 8–4. This protein is a member of the helix–turn–helix family of DNA-binding proteins. (C) The *zinc finger* motif is built from an α helix and a β sheet (the latter shown as a *twisted arrow*) held together by a molecule of zinc (indicated by a *sphere*). Zinc fingers are often found in clusters covalently joined together to allow the α helix of each finger to contact the DNA bases in the major groove. The illustration here shows a cluster of three zinc fingers. (D) A *leucine zipper* motif. This DNA-binding motif is formed by two α helices, each contributed by a different protein molecule. Leucine zipper proteins thus bind to DNA as dimers, gripping the double helix like a clothespin on a clothesline. Each motif makes many contacts with DNA. For simplicity, only the hydrogen-bond contacts are shown in (B), and none of the individual protein–DNA contacts is shown in (C) and (D).

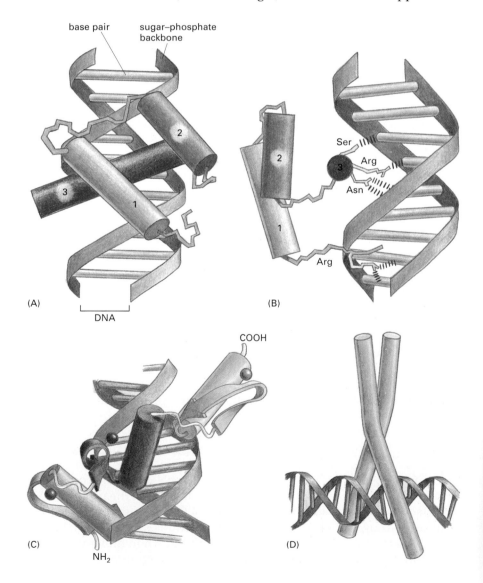

found in gene regulatory proteins that control the expression of thousands of different genes in virtually all eucaryotic organisms. In each example, an α helix of the protein contacts the major groove of DNA. Frequently, DNA-binding proteins bind in pairs (dimers) to the DNA helix. Dimerization roughly doubles the area of contact with the DNA, thereby increasing the strength and specificity of the protein–DNA interaction. Because two different proteins can be combined in pairs, dimerization also makes it possible for many different DNA sequences to be recognized by a limited number of proteins.

Repressors Turn Genes Off, Activators Turn Them On

The simplest and most completely understood examples of gene regulation occur in bacteria and in the viruses that infect them. The genome of the bacterium *E. coli* consists of a single circular DNA molecule of about 4.6×10^6 nucleotide pairs. This DNA encodes approximately 4300 proteins, although only a fraction of these are made at any one time. Bacteria regulate the expression of many of their genes according to the food sources that are available in the environment. For example, in *E. coli*, five genes code for enzymes that manufacture the amino acid tryptophan. These genes are arranged in a cluster on the chromosome and are transcribed from a single promoter as one long mRNA molecule from which the five proteins are translated (Figure 8–6). When tryptophan is present in the surroundings and enters the bacterial cell, these enzymes are no longer needed and their production is shut off. This situation arises, for example, when the bacterium is in the gut of a mammal that has just eaten a meal rich in protein. These five coordinately expressed genes are part of an *operon*—a set of genes that is transcribed into a single mRNA. Operons are common in bacteria but are not found in eucaryotes, where genes are regulated individually.

We now understand in considerable detail how the tryptophan operon functions. Within the promoter is a short DNA sequence (15 nucleotides in length) that is recognized by a gene regulatory protein. When the regulatory protein binds to this nucleotide sequence, termed the *operator*, it blocks access of RNA polymerase to the promoter; this prevents transcription of the operon and production of the tryptophan-producing enzymes. The gene regulatory protein is known as the **tryptophan repressor**, and it is regulated in an ingenious way: the repressor can bind to DNA only if it has also bound several molecules of the amino acid tryptophan (Figure 8–7).

The tryptophan repressor is an allosteric protein: the binding of tryptophan causes a subtle change in its three-dimensional structure so that it can now bind to the operator DNA. When the concentration of free tryptophan in the cell drops, the repressor no longer binds tryptophan and thus no longer binds to DNA, and the tryptophan operon is transcribed. The repressor is thus a simple device that switches production of a set of biosynthetic enzymes on and off according to the availability of the end product of the pathway that the enzymes catalyze.

Question 8–1

Explain how DNA-binding proteins can make sequence-specific contacts to a double-stranded DNA molecule without breaking the hydrogen bonds that hold the bases together. Indicate how, by making such contacts, a protein can distinguish a T-A from a C-G pair. Give your answer in a form similar to Figure 8–4, and indicate what sorts of noncovalent bonds (hydrogen bonds, ionic bonds, or hydrophobic interactions; see Panel 2–1, pp. 66–67) would be made. There is no need to specify any particular amino acid on the protein. The structures of all the base pairs in DNA were given in Figure 5–6.

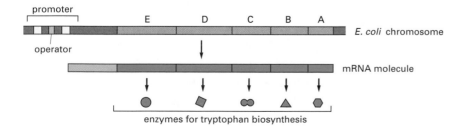

Figure 8–6 A cluster of bacterial genes can be transcribed from a single promoter. Each of these five genes encodes a different enzyme; all of the enzymes are needed to synthesize the amino acid tryptophan. The genes are transcribed as a single mRNA molecule, a feature that allows their expression to be coordinated. Clusters of genes transcribed as a single mRNA molecule are common in bacteria. Each such cluster is called an *operon*; expression of the tryptophan operon shown here is regulated at a regulatory sequence called the *operator*, situated within the promoter.

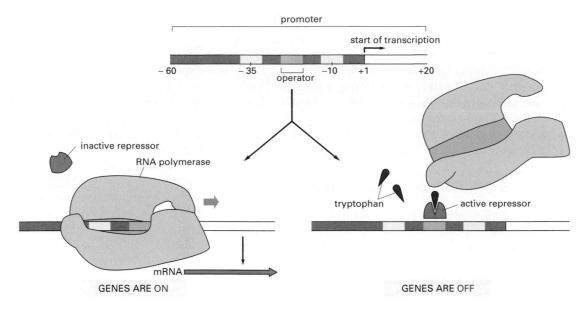

promoter

start of transcription

−60   −35   −10   +1   +20

operator

inactive repressor

RNA polymerase

tryptophan

active repressor

mRNA

GENES ARE ON

GENES ARE OFF

Figure 8–7 Genes can be switched on and off with repressor proteins. If the concentration of tryptophan inside the cell is low, RNA polymerase *(blue)* binds to the promoter and transcribes the five genes of the tryptophan operon *(left).* If the concentration of tryptophan is high, however, the repressor protein *(dark green)* becomes active and binds to the operator *(light green),* where it blocks the binding of RNA polymerase to the promoter *(right).* Whenever the concentration of intracellular tryptophan drops, the repressor releases its tryptophan and is released from the DNA, allowing the polymerase to again transcribe the operon (see also Figure 8–6). The promoter is marked by two key blocks of DNA sequence information, the −35 and −10 regions, highlighted in *yellow*.

The bacterium can respond very rapidly to the rise in tryptophan concentration because the tryptophan repressor protein itself is always present in the cell. The gene that encodes it is continuously transcribed at a low level, so that a small amount of the repressor protein is always being made. Such unregulated gene expression is known as *constitutive* gene expression.

The tryptophan repressor, as its name suggests, is a **repressor** protein that switches genes off, or *represses* them. Other bacterial gene regulatory proteins operate in the opposite manner by switching genes on, or *activating* them. These **activator** proteins act on promoters that—in contrast to the promoter for the tryptophan operator—are, on their own, only marginally functional in binding and positioning RNA polymerase; they may, for example, be recognized only poorly by the polymerase. However, these poorly functioning promoters can be made fully functional by activator proteins that bind to a nearby site on the DNA and contact the RNA polymerase in a way that helps it initiate transcription (Figure 8–8).

Like a repressor, an activator protein's ability to bind to DNA is often affected by its interaction with a second molecule. For example, the bacterial activator protein CAP has to bind cyclic AMP before it can bind to DNA. Genes activated by CAP are switched on in response to an

Figure 8–8 Gene expression can also be controlled with activator proteins. An activator protein binds to a regulatory sequence on the DNA and then interacts with the RNA polymerase to help it initiate transcription. Without the activator, the promoter fails to initiate transcription efficiently. In bacteria, the binding of the activator to DNA is often controlled by the interaction of a metabolite or other small molecule *(red triangle)* with the activator protein. For example, the bacterial activator protein CAP has to bind cyclic AMP (cAMP) before it can bind to DNA; thus CAP allows genes to be switched on in response to increases in intracellular cAMP concentration.

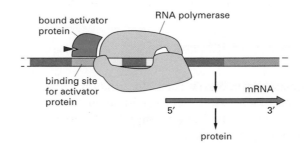

bound activator protein

RNA polymerase

binding site for activator protein

5′   3′

mRNA

protein

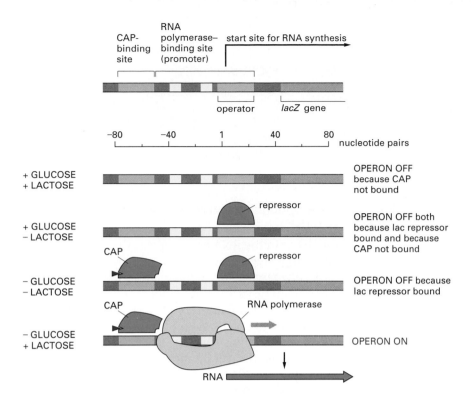

Figure 8–9 The *lac* operon is controlled by two signals. Glucose and lactose levels control the initiation of transcription of the *lac* operon through their effects on the *lac* repressor protein and CAP. Addition of lactose increases the concentration of allolactose, which binds to the repressor protein and removes it from the DNA. Addition of glucose decreases the concentration of cyclic AMP *(red triangle)*; when cyclic AMP is no longer bound to CAP, this gene activator protein dissociates from the DNA, turning off the operon. *lacZ*, the first gene of the operon, encodes the enzyme β-galactosidase, which breaks down lactose to galactose and glucose.

increase in intracellular cAMP concentration, which signals to the bacterium that glucose, its preferred carbon source, is no longer available; as a result, enzymes capable of degrading other sugars are made.

An Activator and a Repressor Control the *lac* Operon

In many instances, the activity of a single promoter can be controlled by two different signals. The *lac operon* in *E. coli*, for example, is controlled by both the *lac* repressor and the activator protein CAP. The *lac* operon encodes proteins required to import and digest the disaccharide lactose. In the absence of glucose, the cell's preferred carbon source, CAP switches on genes that allow the cell to utilize alternative sources of carbon—including lactose. It would be wasteful, however, for CAP to induce expression of the *lac* operon when lactose is not present. Thus the *lac* repressor ensures that the operon is shut off in the absence of lactose. This arrangement enables the control region of the *lac* operon to respond to and integrate two different signals, so that the operon is highly expressed only when two conditions are met: lactose must be present and glucose must be absent (Figure 8–9). Any of the other three possible signal combinations shuts down the cluster of genes.

The simple logic of this genetic switch first attracted the attention of biologists more than 50 years ago. The molecular basis of the switch was uncovered by a combination of genetics and biochemistry, providing the first insight into how gene expression is controlled. Although the same basic strategies are used to control gene expression in higher organisms, the genetic switches are usually much more complex.

Initiation of Eucaryotic Gene Transcription Is a Complex Process

The genetic switches present in bacteria are telling examples of the economy and simplicity often observed in biology. In eucaryotes, however, a typical gene responds to many different signals, and its regulation is consequently more complex.

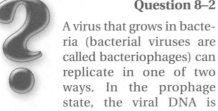

A virus that grows in bacteria (bacterial viruses are called bacteriophages) can replicate in one of two ways. In the prophage state, the viral DNA is inserted into the bacterial chromosome and is copied along with the bacterial genome with which it is passed from generation to generation. In the lytic state the viral DNA is released from the bacterial chromosome and replicates many times in the cell. This viral DNA then produces viral coat proteins that together with the replicated viral DNA form many new virus particles that burst out of the bacterial cell. These two stable states are controlled by two gene regulatory proteins, called cI ("c one") and cro, that are encoded by the virus. In the prophage state, cI is expressed; in the lytic state, cro is expressed. In addition to regulating the expression of other genes, cI is a repressor of transcription of the gene that encodes cro, and cro is a repressor of the gene that encodes cI (Figure Q8–2). When bacteria containing a phage in the prophage state are briefly irradiated with UV light, cI protein is degraded.

A. What will happen next?

B. Will the change in (A) be reversed when the UV light is switched off?

C. Why might this mechanism have evolved?

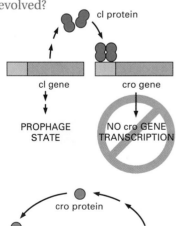

cI protein

cI gene   cro gene

PROPHAGE STATE   NO cro GENE TRANSCRIPTION

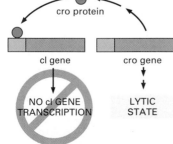

cro protein

cI gene   cro gene

NO cI GENE TRANSCRIPTION   LYTIC STATE

Figure Q8–2

Table 8–1 The Three RNA Polymerases in Eucaryotic Cells	
TYPE OF POLYMERASE	GENES TRANSCRIBED
RNA polymerase I	most rRNA genes
RNA polymerase II	all protein-coding genes, plus some genes for small RNAs (e.g., those in spliceosomes)
RNA polymerase III	tRNA genes 5S rRNA gene genes for some small structural RNAs

Regulation of transcription in eucaryotes differs in four important ways from that in bacteria:

- The first difference lies in the RNA polymerases themselves. While bacteria contain a single type of RNA polymerase, eucaryotic cells have three—*RNA polymerase I, RNA polymerase II,* and *RNA polymerase III.* These polymerases are responsible for transcribing different types of genes. RNA polymerases I and III transcribe the genes encoding transfer RNA, ribosomal RNA, and small RNAs that play structural and catalytic roles in the cell (Table 8–1). RNA polymerase II transcribes the vast majority of eucaryotic genes, including all those that encode proteins, and our subsequent discussion will therefore focus on this enzyme.

- A second difference is that while bacterial RNA polymerase is able to initiate transcription without the help of additional proteins, as we saw for the tryptophan operon (see Figure 8–7), eucaryotic RNA polymerases cannot. They require the help of a large set of proteins called *general transcription factors,* which must assemble at each promoter along with the polymerase before the polymerase can begin transcription.

- A third distinctive feature of gene regulation in eucaryotes is that gene regulatory proteins (repressors and activators) can influence the initiation of transcription even when they are bound to DNA thousands of nucleotide pairs away from the promoter. This feature allows a single promoter to be controlled by an almost unlimited number of regulatory sequences scattered along the DNA. In bacteria, in contrast, genes are often controlled by a single regulatory sequence that is typically located quite near the promoter (see Figures 8–7 and 8–8).

- Last but not least, eucaryotic transcription initiation must take into account the packing of DNA into nucleosomes and more compact forms of chromatin structure.

We now turn our attention to the latter three features and discuss how they are used to control eucaryotic gene regulation.

Eucaryotic RNA Polymerase Requires General Transcription Factors

The initial finding that, unlike bacterial RNA polymerase, purified eucaryotic RNA polymerase II could not on its own initiate transcription *in vitro* led to the discovery and purification of the **general transcription factors** required for this process. The general transcription factors are thought to position the RNA polymerase correctly at the promoter, to aid in pulling apart the two strands of DNA to allow transcription to begin, and to allow RNA polymerase to leave the promoter as transcription begins.

The term "general" refers to the fact that these proteins assemble on all promoters transcribed by RNA polymerase II. In this way they differ from the repressors and activators that we have just described in bacteria, which act only at particular genes or operons, and from the eucaryotic gene regulatory proteins we discuss below, which also act only at particular genes.

Figure 8–10 shows one of the ways that general transcription factors can assemble at promoters used by RNA polymerase II. The assembly process starts with the binding of the general transcription factor TFIID to a short double-helical DNA sequence primarily composed of T and A nucleotides; because of its composition, this sequence is known as the TATA sequence, or **TATA box**. Upon binding to DNA, TFIID causes a dramatic local distortion in the DNA (Figure 8–11), which helps to serve as a landmark for the subsequent assembly of other proteins at the promoter. The TATA box is a key component of nearly all promoters used by RNA polymerase II, and it is typically located 25 nucleotides upstream from the transcription start site. Once the first general transcription factor is bound to this DNA site, then the other factors are assembled, along with RNA polymerase II, to form a complete *transcription initiation complex*. The precise order of assembly of the general transcription factors likely differs from one promoter to the next.

After RNA polymerase II has been tethered to the promoter DNA in the transcription initiation complex, it must be released from the complex of transcription factors in order to begin its task of making an RNA molecule. A key step in this release is the addition of phosphate groups to the RNA polymerase, performed by the general transcription factor TFIIH, which contains a protein kinase enzyme as one of its subunits (see Figure 8–10E). This phosphorylation is thought to help the polymerase disengage from the cluster of transcription factors, allowing transcription to begin. Phosphorylation of the RNA polymerase II tail also enables proteins that process the nascent RNA transcript (discussed in Chapter 7) to assemble on the polymerase and move along with it (Figure 8–12). Once transcription has begun, most of the general

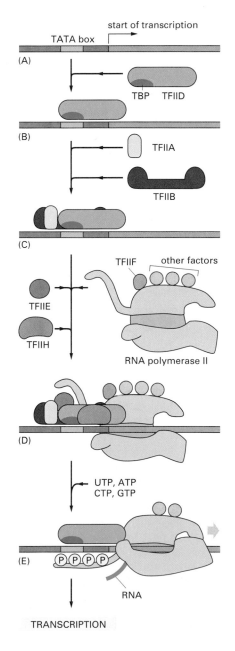

Figure 8–10 To begin transcription, eucaryotic RNA polymerase II requires the general transcription factors. These transcription factors are called TFIIA, TFIIB, and so on. (A) The promoter contains a DNA sequence called the TATA box, which is located 25 nucleotides away from the site where transcription is initiated. (B) The TATA box is recognized and bound by transcription factor TFIID, which then enables the adjacent binding of TFIIB (C). For simplicity the DNA distortion produced by the binding of TFIID (see Figure 8–11) is not shown. (D) The rest of the general transcription factors as well as the RNA polymerase itself assemble at the promoter. (E) TFIIH then uses ATP to pry apart the double helix at the transcription start point, allowing transcription to begin. TFIIH also phosphorylates RNA polymerase II, releasing it from the general factors so it can begin the elongation phase of transcription. As shown, the site of phosphorylation is a long polypeptide tail that extends from the polymerase molecule.

The exact order in which the general transcription factors assemble on each promoter is not known with certainty. In some cases, most of the general factors are thought to first assemble with the polymerase independent of the DNA, with this whole assembly then binding to the DNA in a single step. The general transcription factors have been highly conserved in evolution; some of those from human cells can be replaced in biochemical experiments by the corresponding factors from simple yeasts.

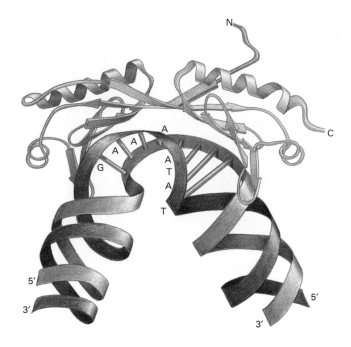

Figure 8–11 TATA-binding protein (TBP) binds to TATA box sequences and distorts the DNA. TBP is the subunit of the general transcription factor TFIID that is responsible for recognizing and binding to the TATA box sequence in the DNA *(red)*. The unique DNA bending caused by TBP—two kinks in the double helix separated by partially unwound DNA—may serve as a landmark that helps attract the other general transcription factors. TBP is a single polypeptide chain that is folded into two very similar domains *(blue and green)*. (Adapted from J.L. Kim et al., *Nature* 365:520–527, 1993.)

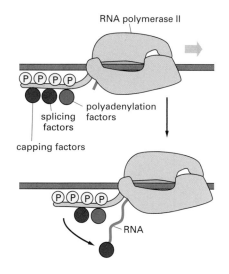

Figure 8–12 Phosphorylation of RNA polymerase II allows RNA-processing proteins to assemble on its tail. Not only does the polymerase transcribe DNA into RNA, it also carries RNA-processing proteins which then act on the nascent RNA at the appropriate time. The RNA-processing proteins first bind to the RNA polymerase tail when it is phosphorylated late in the process of transcription initiation (see Figure 8–10). Once RNA polymerase II finishes transcribing, it is released from the DNA, the phosphates on its tail are stripped off by phosphatases, and it can reinitiate transcription. Only the dephosphorylated form of RNA polymerase II can start RNA synthesis at a promoter.

transcription factors are released from the DNA so that they are available to initiate another round of transcription with a new RNA polymerase molecule.

Eucaryotic Gene Regulatory Proteins Control Gene Expression from a Distance

We have seen that bacteria use gene regulatory proteins—activators and repressors—to regulate the expression of their genes. Eucaryotic cells use the same basic strategy. Although the eucaryotic general transcription factors and RNA polymerase together can initiate transcription *in vitro* (see Figure 8–10), inside the cell these proteins on their own are not sufficient to initiate transcription. Nearly all eucaryotic promoters also require activator proteins to aid the assembly of the general transcription factors and RNA polymerase onto chromosomal DNA.

The DNA sites to which the eucaryotic gene activators bound were originally termed *enhancers,* because their presence "enhanced," or increased, the rate of transcription dramatically. It was surprising to biologists when, in 1979, it was discovered that these activator proteins could be bound thousands of nucleotide pairs away from the promoter. Moreover, eucaryotic activators could influence transcription of a gene when bound either upstream or downstream from it. How do enhancer sequences and the proteins bound to them function over these long distances? How do they communicate with the promoter?

Many models for "action at a distance" have been proposed, but the simplest of these seems to apply in most cases. The DNA between the enhancer and the promoter loops out to allow the activator proteins to directly influence events that take place at the promoter (Figure 8–13). The DNA thus acts as a tether, causing a protein bound to an enhancer even thousands of nucleotide pairs away to interact with the proteins in the vicinity of the promoter. Often, additional proteins serve to "link" the distantly bound gene regulatory proteins to the RNA polymerase and general transcription factors; the most important is a large complex of proteins known as the *mediator* (see Figure 8–13). One of the ways that activator proteins function is by aiding in the assembly of the general transcription factors and RNA polymerase at the promoter.

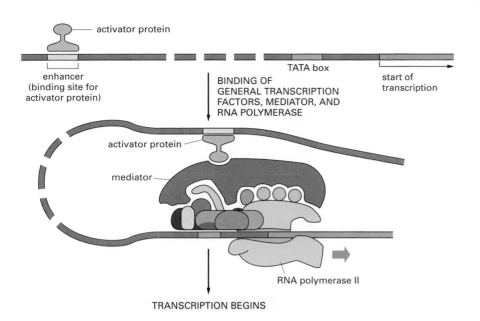

activator protein

enhancer
(binding site for
activator protein)

TATA box

BINDING OF
GENERAL TRANSCRIPTION
FACTORS, MEDIATOR, AND
RNA POLYMERASE

start of
transcription

activator protein

mediator

RNA polymerase II

TRANSCRIPTION BEGINS

Figure 8–13 In eucaryotes, gene activation occurs at a distance. An activator protein bound near a promoter attracts RNA polymerase complex and general transcription factors (see Figure 8–10) to the promoter. Looping of the DNA permits contact between the activator protein bound to the enhancer and the transcription complex bound to the promoter. In the case shown here, a large protein complex called the mediator serves as a go-between. The broken stretch of DNA signifies that the length of DNA between the enhancer and the start of transcription varies, sometimes reaching tens of thousands of nucleotide pairs in length.

Eucaryotic repressor proteins can do the opposite—they can decrease transcription by preventing or sabotaging the assembly of the same protein complex.

Eucaryotic activator and repressor proteins have an additional mechanism of action: they attract proteins that modulate chromatin structure and thereby affect the accessibility of the promoter to the general transcription factors and RNA polymerase, as we discuss next.

Packing of Promoter DNA into Nucleosomes Can Affect Initiation of Transcription

Initiation of transcription in eucaryotic cells must also take into account the packing of the DNA into chromatin. As we saw in Chapter 5, the genetic material in eucaryotic cells is packed into nucleosomes, which, in turn, are folded into higher-order structures. How do gene regulatory proteins, general transcription factors, and RNA polymerase gain access to such DNA? Nucleosomes can inhibit the initiation of transcription if they are positioned over a promoter, probably because they simply block the assembly of the general transcription factors or RNA polymerase on the promoter. In fact, such chromatin packaging may have evolved in part to prevent leaky gene expression—initiation of transcription in the absence of the proper activator proteins.

In eucaryotic cells, activator and repressor proteins exploit chromatin structure to help turn genes on and off. As we saw in Chapter 5, cells can alter local chromatin structure by enlisting the aid of chromatin remodeling complexes or by modifying the histone proteins that form the core of the nucleosome (see Figures 5–29 and 5–30). Many gene activators take advantage of these mechanisms by locally recruiting chromatin remodeling complexes or histone-modifying proteins, such as the histone acetylases that attach an acetyl group to selected lysines in the tail of histone proteins (Figure 8–14). The resulting alterations in chromatin structure allow greater accessibility to the underlying DNA. This enhanced availability facilitates the assembly of the general transcription factors and the RNA polymerase at the promoter.

Likewise, gene repressor proteins can modify chromatin in ways that reduce the efficiency of transcription initiation. Repressors attract, for example, **histone deacetylases**—enzymes that remove the acetyl groups from histone tails, thereby rendering the underlying DNA less

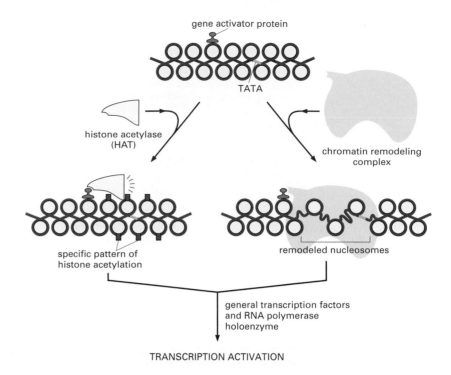

Figure 8–14 Eucaryotic gene activator proteins can direct local alterations in chromatin structure. Activator proteins can recruit histone acetylases or chromatin remodeling complexes to the promoter region of a gene. The action of these proteins renders the DNA packaged in chromatin more accessible to other proteins in the cell, including those required for transcription initiation.

accessible to the general transcription factors. Although some eucaryotic repressor proteins work on a gene-by-gene basis, others can orchestrate the formation of large swaths of condensed chromatin containing many genes. As we discussed in Chapter 5, these transcription-resistant regions of DNA include the heterochromatin found in interphase chromosomes and the entire X chromosome in female mammals. The proteins responsible for forming these large regions of condensed chromatin structure are only beginning to be understood.

The Molecular Mechanisms That Create Specialized Cell Types

Although all cells must be able to switch genes on and off in response to changes in their environments, the cells of multicellular organisms have evolved this capacity to an extreme degree and in highly specialized ways to form an organized array of differentiated cell types. In particular, once a cell in a multicellular organism becomes committed to differentiate into a specific cell type, the choice of fate is generally maintained through many subsequent cell generations, which means that the changes in gene expression involved in the choice must be remembered. This phenomenon of *cell memory* is a prerequisite for the creation of organized tissues and for the maintenance of stably differentiated cell types. In contrast, the simplest changes in gene expression in both eucaryotes and bacteria are only transient; the tryptophan repressor, for example, switches off the tryptophan genes in bacteria only in the presence of tryptophan; as soon as tryptophan is removed from the medium, the genes are switched back on, and the descendants of the cell will have no memory that their ancestors had been exposed to tryptophan.

In this section, we discuss some of the special features of transcriptional regulation that are found in multicellular organisms. Our focus will be on how these mechanisms create and maintain the specialized cell types that give a worm, a fly, or a human its distinctive characteristics.

Eucaryotic Genes Are Regulated by Combinations of Proteins

Because eucaryotic gene regulatory proteins can control transcription when bound to DNA many base pairs away from the promoter, the nucleotide sequences that control the expression of a gene can be spread over long stretches of DNA. In animals and plants it is not unusual to find the regulatory sequences of a gene dotted over distances as great as 50,000 nucleotide pairs, although much of this DNA serves as "spacer" sequence and is not recognized by gene regulatory proteins.

So far in this chapter we have treated gene regulatory proteins as though each functioned individually to turn a gene on or off. While this idea holds true for many bacterial activators and repressors, most eucaryotic gene regulatory proteins work as part of a "committee" of regulatory proteins, all of which are necessary to express the gene in the right cell, in response to the right conditions, at the right time, and at the required amount.

The term **combinatorial control** refers to the way groups of proteins work together to determine the expression of a single gene. We saw a simple example of such regulation by multiple signals when we discussed the bacterial *lac* operon (see Figure 8–9). In eucaryotes, the regulatory inputs have been amplified: a typical human gene, for example, is controlled by dozens of gene regulatory proteins. As shown in the simple cartoon in Figure 8–15, many different proteins bind to regulatory sequences to influence whether transcription is initiated in eucaryotes. Often, some of these regulatory proteins are repressors and some are activators; the molecular mechanisms by which the effects of all of these proteins are added up to determine the final level of expression for a gene are only now beginning to be understood. An example of such a complex regulatory system—one that participates in the development of a fruit fly from a fertilized egg—is described in How We Know, pp. 282–284.

The Expression of Different Genes Can Be Coordinated by a Single Protein

Both bacteria and eucaryotes need to be able not only to switch genes on and off individually, but also to coordinate the expression of different genes. When a eucaryotic cell receives a signal to divide, for example, a

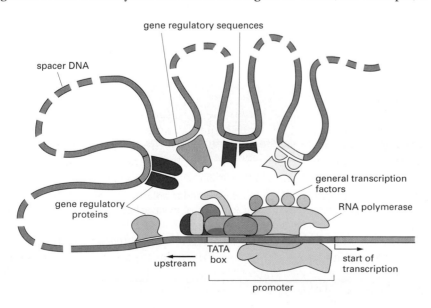

Figure 8–15 **Regulatory proteins work together as a "committee" to control the expression of a eucaryotic gene.** Whereas the general transcription factors that assemble at the promoter are the same for all genes transcribed by polymerase II, the gene regulatory proteins and the locations of their binding sites relative to the promoters are different for different genes. The effects of multiple gene regulatory proteins combine to determine the rate of transcription initiation. However, it is not yet understood in detail for any gene how these effects are combined.

How We Know: Gene Regulation—The Story of eve

The ability to regulate the activity of genes is key to the proper development of a multicellular organism from a fertilized egg to a fertile adult. Beginning at the earliest moments in development, a succession of programs controls the differential expression of genes that allows an animal to form a proper body plan—helping to distinguish its back from its belly, its head from its tail. These cues ultimately direct the correct placement of a wing or a leg, a mouth or an anus, a neuron or a sex cell.

A central problem in development, then, is understanding how an organism generates these patterns of gene expression, which are laid down within hours of fertilization. A large part of the story rests on the action of gene regulatory proteins. By interacting with different regulatory DNA sequences, these proteins instruct every cell in the embryo to switch on the genes that are appropriate for that cell at each timepoint during development. How can a protein binding to a piece of DNA help direct the development of a complex multicellular organism? To see how we can address that large question, we now review the story of eve.

In the Big Egg

Even-skipped—eve, for short—is a gene whose expression plays an important role in the development of the *Drosophila* embryo. If this gene is inactivated by mutation, many parts of the embryo fail to form and the fly larva dies early in development. At the stage of development when *eve* is first switched on, the embryo developing within the egg is still a single, giant cell containing multiple nuclei afloat in a common cytoplasm. This embryo, which is some 400 μm long and 160 μm in diameter, is formed from the fertilized egg through a series of rapid nuclear divisions that occur without cell division. Eventually each nucleus will be enclosed in a plasma membrane and become a cell; however, the events that concern us happen before such cellularization.

The common cytoplasm is far from uniform: the anterior (head) end of the embryo contains different proteins from those in the posterior (tail) end. The presence of these asymmetries in the fertilized egg and the early embryo had first been demonstrated by experiments in which *Drosophila* eggs were made to leak. If the front end of an egg is punctured carefully and a small amount of the anterior cytoplasm is allowed to ooze out, the embryo will fail to develop head segments. Further, if cytoplasm taken from the posterior end of another egg is then injected into this somewhat depleted anterior area, the animal will develop a second set of abdominal segments where its head parts should have been (Figure 8–16).

Finding the Proteins

This egg-draining experiment shows that the normal head-to-tail pattern of development is controlled by substances

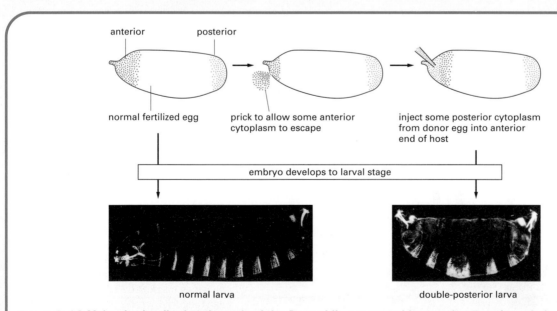

Figure 8–16 Molecules localized at the ends of the *Drosophila* egg control its anterior-posterior polarity. A small amount of cytoplasm is allowed to leak out of the anterior end of the egg and is replaced by an injection of posterior cytoplasm. The resulting double-tailed embryo *(right)* shows a duplication of the last three abdominal segments. A normal embryo *(left)* is shown for comparison. (From H.G. Frohnhöfer, R. Lehmann, and C. Nüsslein-Volhard, *J. Embryol. Exp. Morphol.* 97[suppl]:169–179, 1986. © The Company of Biologists.)

located at either end of the embryo. To begin to identify these molecules researchers turned to genetics and searched for mutations that mimic the effects of losing the anterior or posterior cytoplasm. As we will see in Chapter 10, such mutations offer a powerful means for identifying genes and learning what their protein products do in the cell or organism.

In this case, researchers determined that these genes encoded regulatory proteins; in particular, four regulatory proteins were important for setting up the anterior–posterior polarity in the *Drosophila* embryo: Bicoid, Hunchback, Krüppel, and Giant. (*Drosophila* genes are often given colorful or whimsical names that reflect the appearance of flies in which the gene is inactivated by mutation.) Once these proteins had been identified, researchers could prepare antibodies that would recognize each. These antibodies, coupled to fluorescent markers, were then used to determine where in the early embryo each protein is localized (see Panel 1–1, pp. 8–9).

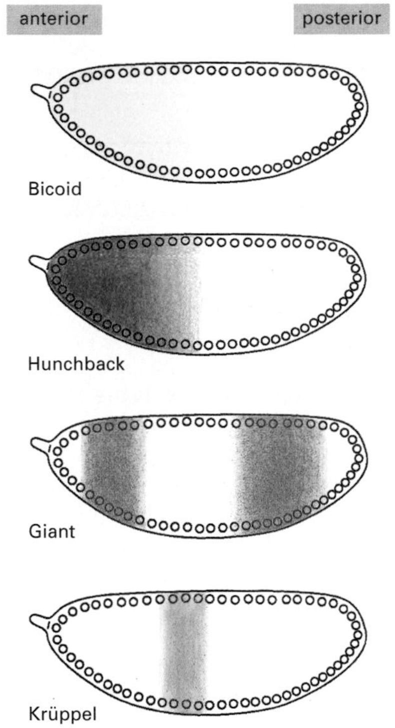

The results of these antibody-staining experiments are quite striking. The cytoplasm of the early embryo, it turns out, contains a mixture of these gene regulatory proteins, each distributed unevenly along the length of the embryo (Figure 8–17). As a result, the nuclei inside this giant, multinucleate cell begin to express different genes depending on which gene regulatory proteins they encounter, which in turn, depends on their location in the embryo. The nuclei near the anterior end of the embryo, for example, are exposed to a set of gene regulatory proteins that is distinct from the set that influences nuclei at its posterior end. Thus the differing amounts of these proteins provide nuclei in the developing embryo with positional information that allows them to determine where they are located along the anterior-posterior axis of the embryo.

This is where *eve* comes in. The expression of the *eve* gene is controlled by these four regulatory proteins. The regulatory DNA sequences of *eve* are designed to read the concentrations of the gene regulatory proteins at each position along the length of the embryo and to interpret this information in such a way that the *eve* gene is expressed in seven stripes, each positioned precisely along the anteroposterior axis of the embryo. To find out how these regulatory proteins control the expression of *eve* with such precision, researchers next set their sights on the regulatory region of the *eve* gene.

Dissecting the DNA

As we have seen in this chapter, regulatory sequences control which cells in an organism will express a particular gene, and at what point that gene will be turned on. One way to learn when and where a regulatory DNA sequence will promote gene activation is to hook the sequence up to a reporter gene—a gene encoding a protein whose activity is easy to monitor experimentally. The regulatory sequences will now drive the activity of the reporter gene. This artificial

Figure 8–17 **The early *Drosophila* embryo shows a nonuniform distribution of four gene regulatory proteins.**

DNA construct is then reintroduced into a cell or an organism and the activity of the reporter protein is measured.

By coupling various portions of the regulatory sequence of *eve* to a reporter gene, researchers discovered that the *eve* gene contains a series of seven regulatory modules, each of which is responsible for specifying a single stripe of *eve* expression along the embryo. So, for example, researchers could remove the regulatory module that specifies stripe 2 from its normal setting upstream of *eve*, place it in front of a reporter gene, and reintroduce this engineered DNA sequence into the *Drosophila* genome (Figure 8–18A). When embryos derived from flies carrying this genetic construct are examined, the reporter gene is found to be expressed in precisely the position of stripe 2 (Figure 8–18B). Similar experiments revealed the existence of other regulatory modules, one for each of the other six stripes.

The question then becomes: how does each of these modules direct the formation of a single stripe in a specific position? The answer, researchers found, is that each module contains a unique combination of regulatory sequences that bind different combinations of the four gene regulatory proteins that are present in gradients in the early embryo. The stripe 2 unit, for example, contains recognition sequences for all four regulatory proteins—two that activate *eve* transcription, Bicoid and Hunchback; and two that repress it, Krüppel and Giant (Figure 8–19). The relative concentrations of these four proteins determine whether protein complexes that form

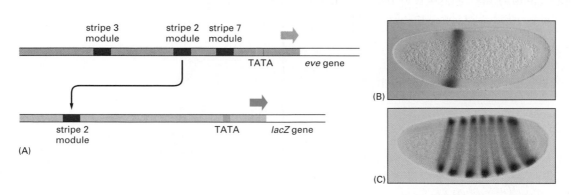

(A)

(B)

(C)

Figure 8–18 A reporter gene reveals the modular construction of the *eve* gene regulatory region. (A) A 480-nucleotide piece of the *eve* regulatory region was removed and inserted upstream of the *E. coli lacZ* gene, which encodes the enzyme β-galactosidase. Enzyme activity is assayed by the addition of X-gal, a modified sugar that when cleaved by β-galactosidase generates an insoluble blue product. (B) When this engineered DNA construct is reintroduced into the genome of a *Drosophila* embryo, the resulting embryo expresses β-galactosidase precisely in the position of the second of the seven *eve* stripes. (C) Embryos stained with antibodies to the Eve protein show the seven characteristic Eve stripes. (B and C, courtesy of Stephen Small and Michael Levine.)

at the stripe 2 module will turn on transcription of the *eve* gene in that segment of the embryo.

The other stripe regulatory modules are thought to be constructed along similar lines; each module is designed to read positional information provided by some unique combination of gene regulatory proteins. In the case of *eve,* the entire gene control region is strung out over 20,000 nucleotide pairs of DNA and binds more than 20 different proteins including the four we discussed. A large and complex control region is thereby built from a series of smaller modules, each of which consists of a unique arrangement of short DNA sequences recognized by specific gene regulatory proteins. In this way, a single gene can respond to an enormous number of combinatorial inputs. Eve itself is a gene regulatory protein and it—in combination with many other regulatory proteins—controls key events later in development. This organization begins to explain how the development of a complex organism can be orchestrated by repeated applications of a few basic principles.

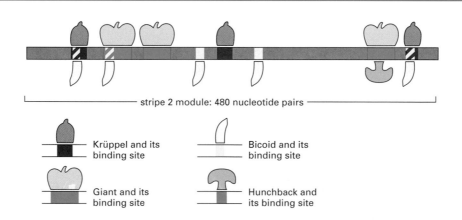

stripe 2 module: 480 nucleotide pairs

Krüppel and its binding site

Bicoid and its binding site

Giant and its binding site

Hunchback and its binding site

Figure 8–19 The regulatory module for *eve* stripe 2 contains binding sites for four different gene regulatory proteins. All four regulatory proteins are responsible for the proper expression of *eve* in stripe 2. Flies that are deficient in the two activators, Bicoid and Hunchback, fail to form stripe 2 efficiently; in flies deficient in either of the two repressors, Giant or Krüppel, stripe 2 expands and covers an abnormally broad region of the embryo. As indicated in the top diagram, in some cases the binding sites for the gene regulatory proteins overlap and the proteins compete for binding to the DNA. For example, the binding of Bicoid and Krüppel to the site at the far right is thought to be mutually exclusive.

number of hitherto unexpressed genes are turned on together to set in motion the events that lead eventually to cell division (discussed in Chapter 19). One way bacteria coordinate the expression of a set of genes is by having them clustered together in an operon under the control of a single promoter (see Figure 8–6). This is not the case in eucaryotes, in which each gene is regulated individually. So how do eucaryotes coordinate gene expression? In particular, given that a eucaryotic cell uses a committee of regulatory proteins to control each of its genes, how can it rapidly and decisively switch whole groups of genes on or off? The answer is that even though control of gene expression is combinatorial, the effect of a single gene regulatory protein can still be decisive in switching any particular gene on or off, simply by completing the combination needed to activate or repress that gene. This is like dialing in the final number of a combination lock: the lock will spring open if the other numbers have been previously entered. Just as the same number can complete the combination for different locks, the same protein can complete the combination for several different genes. If a number of different genes contain the regulatory site for the same gene regulatory protein, it can be used to regulate the expression of all of them.

An example of this style of regulation in humans is seen with the *glucocorticoid receptor protein*. In order to bind to regulatory sites in DNA this gene regulatory protein must first form a complex with a molecule of a glucocorticoid steroid hormone (for example, cortisol; see Table 16–1, p. 537). This hormone is released in the body during times of starvation and intense physical activity, and among its other activities, it stimulates cells in the liver to increase the production of glucose from amino acids and other small molecules. In response to glucocorticoid hormones, liver cells increase the expression of many different genes, one of which encodes the enzyme tyrosine aminotransferase, as discussed earlier. These genes are all regulated by the binding of the hormone–glucocorticoid receptor complex to a regulatory site in the DNA of the gene. When the body has recovered and the hormone is no longer present, the expression of all of these genes drops to its normal level. In this way a single gene regulatory protein can control the expression of many different genes (Figure 8–20).

Combinatorial Control Can Create Different Cell Types

The ability to switch many different genes on or off using just one protein is not only useful in the day-to-day regulation of cell function. It is also one of the means by which eucaryotic cells differentiate into particular types of cells during embryonic development.

A striking example of the effect of a single gene regulatory protein on differentiation comes from studying the development of muscle cells. A mammalian skeletal muscle cell is a highly distinctive cell type. It is typically an extremely large cell that is formed by the fusion of many muscle precursor cells called *myoblasts* (and therefore contains many nuclei). The mature muscle cell is distinguished from other cells by the production of a large number of characteristic proteins, such as the actin and myosin that make up the contractile apparatus (discussed in Chapter 17) as well as the receptor proteins and ion channel proteins in the cell membranes that make the muscle cell sensitive to nerve stimulation. Genes encoding these muscle-specific proteins are all switched on coordinately as the myoblasts begin to fuse. Studies of muscle cells differentiating in culture have identified key gene regulatory proteins, expressed only in potential muscle cells, that coordinate the gene expression and thus are crucial for muscle cell differentiation. These gene regulatory proteins activate the transcription of the genes that

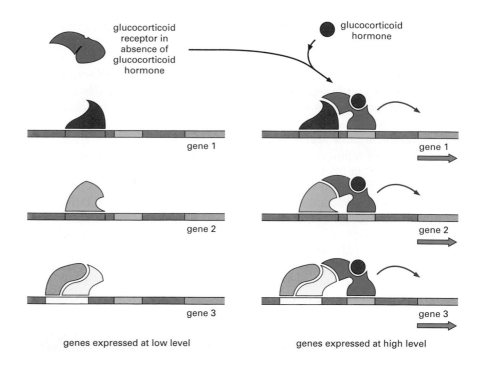

Figure 8–20 A single gene regulatory protein can coordinate the expression of many different genes. The action of the glucocorticoid receptor is illustrated. On the left is shown a series of genes, each of which has various gene activator proteins bound to its regulatory region. However, these bound proteins are not sufficient on their own to activate transcription efficiently. On the right is shown the effect of adding an additional gene regulatory protein—the glucocorticoid receptor in a complex with glucocorticoid hormone—that can bind to the regulatory region of each gene. The glucocorticoid receptor completes the combination of gene regulatory proteins required for efficient initiation of transcription, and the genes are now switched on as a set.

20 μm

Figure 8–21 Fibroblasts can be converted to muscle cells by a single gene regulatory protein. As shown in this immunofluorescence micrograph, fibroblasts from the skin of a chick embryo have been converted to muscle cells by the experimentally induced expression of the *myoD* gene. The fibroblasts that have received the MyoD gene regulatory protein have fused to form elongated multinucleate musclelike cells, which are stained *green* with an antibody that detects a muscle-specific protein. Fibroblasts that did not receive the *myoD* gene are barely visible in the background. (Courtesy of Stephen Tapscott and Harold Weintraub.)

code for the muscle-specific proteins by binding to sites present in their regulatory regions.

These key regulatory proteins can convert nonmuscle cells to myoblasts by activating the changes in gene expression typical of differentiating muscle cells. When the gene for one of these gene regulatory proteins, MyoD, is introduced into fibroblasts cultured from skin connective tissue, the fibroblasts start to behave like myoblasts and fuse to form musclelike cells. The dramatic effect of expressing the *myoD* gene in fibroblasts is shown in Figure 8–21. It appears that the fibroblasts, which are derived from the same broad class of embryonic cells as muscle cells, have already accumulated all of the other necessary gene regulatory proteins required for the combinatorial control of the muscle-specific genes, and that addition of MyoD completes the unique combination that directs the cells to become muscle. Some other cell types fail to be converted to muscle by the addition of MyoD; these cells presumably have not accumulated the other required gene regulatory proteins during their developmental history.

How the accumulation of different gene regulatory proteins can lead to the generation of different cell types is illustrated schematically in Figure 8–22. This figure also illustrates how, thanks to the possibilities of combinatorial control and shared regulatory sequences, a limited set of gene regulatory proteins can control the expression of a much larger number of genes.

The conversion of one cell type (fibroblast) to another (muscle) by a single gene regulatory protein emphasizes one of the most important principles discussed in this chapter: the dramatic differences between cell types—such as size, shape, and function—are produced by differences in gene expression.

Stable Patterns of Gene Expression Can Be Transmitted to Daughter Cells

Although all cells, whether bacterial or eucaryotic, must be able to switch genes on and off, multicellular organisms require special gene-switching mechanisms for generating and maintaining their different types of cells. In particular, once a cell in a multicellular organism has become differentiated into a particular cell type, it will generally remain differentiated, and if it is able to divide, all its progeny cells will be of that same cell type. Some highly specialized cells never divide again once they have differentiated, for example, skeletal muscle cells and neurons. But many other differentiated cells, such as fibroblasts, smooth muscle cells, and liver cells (hepatocytes), will divide many times in the life of an individual. All of these cell types give rise only to cells like themselves when they divide: smooth muscle does not give rise to liver cells, nor liver cells to fibroblasts.

This means that the changes in gene expression that give rise to a differentiated cell must be remembered and passed on to its daughter cells through all subsequent cell divisions, unlike the temporary changes in gene expression that can occur in both bacterial and eucaryotic cells. For example, in the cells illustrated in Figure 8–22, the production of each gene regulatory protein, once begun, has to be perpetuated in the daughter cells of each cell division. How might this be accomplished?

Cells have several ways of ensuring that daughter cells "remember" what kind of cells they are supposed to be. One of the simplest is through a **positive feedback loop**, where a key gene regulatory protein activates transcription of its own gene in addition to that of other cell-type–specific genes (Figure 8–23). For example, the MyoD gene regulatory protein discussed earlier functions in such a positive feedback loop. Another way of maintaining cell type is through the faithful propagation of a condensed chromatin structure from parent to daughter cell even though DNA replication intervenes. We saw an example of this in Figure 5–28, where the same X chromosome is inactive through many

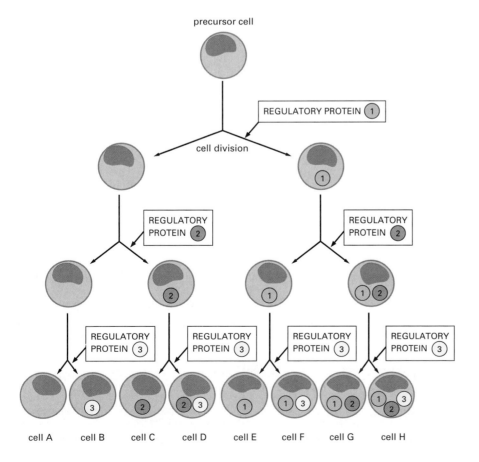

Figure 8–22 Combinations of a few gene regulatory proteins can generate many different cell types during development. In this simple scheme a "decision" to make a new gene regulatory protein (shown as a *numbered circle*) is made after each cell division. Repetition of this simple rule enables eight cell types (A through H) to be created using only three different regulatory proteins. Each of these hypothetical cell types would then express different genes, as dictated by the combination of gene regulatory proteins that are present within it.

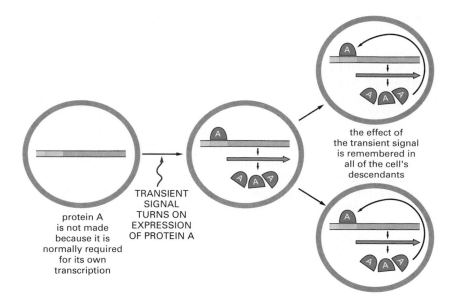

Figure 8–23 A positive feedback loop can create cell memory. Protein A is a gene regulatory protein that activates its own transcription. All of the descendants of the original cell will therefore "remember" that the progenitor cell had experienced a transient signal that initiated the production of the protein.

the effect of the transient signal is remembered in all of the cell's descendants

TRANSIENT SIGNAL TURNS ON EXPRESSION OF PROTEIN A

protein A is not made because it is normally required for its own transcription

cell generations. The molecular mechanism through which the chromatin state is passed on is not understood in detail, but a general hypothesis is shown in Figure 8–24.

The Formation of an Entire Organ Can Be Triggered by a Single Gene Regulatory Protein

We have seen that even though combinatorial control is the norm for eucaryotic genes, a single gene regulatory protein, if it completes the appropriate combination, can be decisive in switching a whole set of genes on or off, and we have seen how this can convert one cell type into another. A dramatic extension of this principle comes from studies of eye development in *Drosophila,* mice, and humans. Here, a single gene regulatory protein (called Ey in flies and Pax-6 in vertebrates) is crucial for eye development. When expressed in the proper type of cell, Ey can trigger the formation of not just a single cell type but a whole organ—the eye—composed of different types of cells all properly organized in three-dimensional space.

The best evidence for the action of Ey comes from experiments in fruit flies in which the *ey* gene is artificially expressed early in development in cells that normally go on to form legs. This abnormal gene expression causes eyes to develop in the middle of the legs (Figure 8–25). The *Drosophila* eye is composed of thousands of cells, and how the Ey protein coordinates the specification of each cell in the eye is an actively studied topic in *developmental biology*. Here, we note that Ey

Figure 8–24 States of chromatin structure can be inherited directly during DNA replication. According to this model, portions of a cluster of chromatin proteins bound to the DNA are transferred directly from the parental DNA helix *(top left)* to both daughter helices. The inherited cluster then causes each of the daughter DNA helices to bind additional copies of the same proteins. Because the binding to DNA is cooperative—i.e., the binding of one molecule enhances the binding of additional molecules—the DNA synthesized from an identical parental DNA helix that lacks the bound proteins *(top right)* will remain free of them.

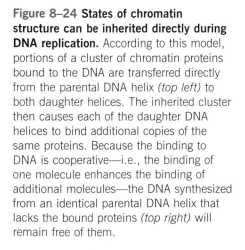
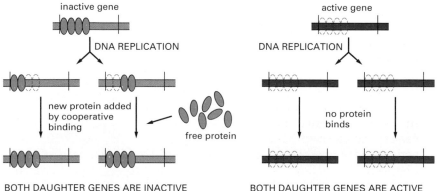

inactive gene

DNA REPLICATION

new protein added by cooperative binding

free protein

BOTH DAUGHTER GENES ARE INACTIVE

active gene

DNA REPLICATION

no protein binds

BOTH DAUGHTER GENES ARE ACTIVE

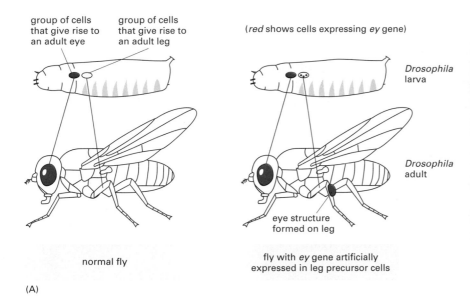

group of cells that give rise to an adult eye

group of cells that give rise to an adult leg

(*red* shows cells expressing *ey* gene)

Drosophila larva

Drosophila adult

eye structure formed on leg

normal fly

fly with *ey* gene artificially expressed in leg precursor cells

(A)

(B)

Figure 8–25 Expression of the *Drosophila* *ey* gene in the precursor cells of the leg triggers the development of an eye on the leg. (A) Simplified diagrams showing the result when a fruit fly larva contains either the normally expressed *ey* gene *(left)* or an *ey* gene that is additionally expressed artificially in cells that will give rise to legs *(right)*. (B) Photograph of an abnormal leg that contains a misplaced eye. (B, courtesy of Walter Gehring.)

directly controls the expression of many genes by binding to their regulatory regions. Some of the genes controlled by Ey encode additional gene regulatory proteins that, in turn, control the expression of other genes. Moreover, some of these regulatory proteins act back on *ey* itself to create a positive feedback loop that ensures the continued production of the Ey protein (see Figure 8–23). So the action of just one regulatory protein can turn on a cascade of gene regulatory proteins whose actions result in forming an organized group of many different types of cells. One can begin to imagine how, by repeated applications of this principle, a complex organism is built up piece-by-piece.

All of the instructions needed to form these regulatory networks are encoded in each organism's genome. One of the great challenges of biology in this century is to decipher this information and to determine how these networks specify the development of complex organisms—including ourselves.

Essential Concepts

- A typical eucaryotic cell expresses only a fraction of its genes, and the distinct types of cells in multicellular organisms arise because different sets of genes are expressed as a cell differentiates.
- Although all of the steps involved in expressing a gene can in principle be regulated, for most genes the initiation of transcription is the most important point of control.
- The transcription of individual genes is switched on and off in cells by gene regulatory proteins. These act by binding to short stretches of DNA called regulatory DNA sequences.
- Although each gene regulatory protein has unique features, most bind to DNA using one of a small number of protein structure motifs. The precise amino acid sequence that is folded into the DNA-binding motif determines the particular DNA sequence that is recognized.
- RNA polymerase binds to the DNA and initiates transcription at a site called the promoter.
- In bacteria, regulatory proteins usually bind to regulatory DNA sequences close to where RNA polymerase binds and then either activate or repress transcription of the gene. In eukaryotes, these regulatory DNA sequences are often separated from the promoter by many thousands of nucleotide pairs.

- To initiate transcription, eucaryotic RNA polymerases require the assembly of a complex of general transcription factors at the promoter.
- Eucaryotic gene regulatory proteins act in two fundamental ways: (1) they can directly affect the assembly process of RNA polymerase and the general transcription factors at the promoter, and (2) they can locally modify the chromatin structure of promoter regions.
- In eucaryotes, the expression of a gene is generally controlled by a combination of gene regulatory proteins.
- In multicellular plants and animals, the production of different gene regulatory proteins in different cell types ensures the expression of only those genes appropriate to the particular type of cell.
- A single gene regulatory protein, if expressed in the appropriate precursor cell, can trigger the formation of a specialized cell type or even an entire organ.

Key Terms

activator	positive feedback loop
combinatorial control	regulatory DNA sequence
differentiation	reporter gene
gene expression	repressor
gene regulatory protein	TATA box
general transcription factor	tryptophan repressor
histone deacetylase	

Questions

Question 8–4

(True/False) When the nucleus of a fully differentiated carrot cell is injected into a frog egg whose nucleus has been removed, the injected donor nucleus is capable of programming the recipient egg to produce a normal carrot. Explain your answer.

Question 8–5

Which of the following statements are correct? Explain your answers.
A. In bacteria the genes encoding ribosomal RNA, tRNA, and mRNA are transcribed by different RNA polymerases.
B. In bacteria, but not in eucaryotes, most mRNAs encode more than one protein.
C. Most DNA-binding proteins bind to the major groove of the double helix.
D. Of the major control points in gene expression (transcription, RNA processing, RNA transport, translation, and control of a protein's activity), transcription initiation is used for the vast majority of gene regulation events.
E. The zinc atoms in DNA-binding proteins that contain zinc finger domains contribute to the binding specificity through sequence-specific interactions, that they form with the bases.

Question 8–6

Your task in the laboratory of Professor Quasimodo is to determine how far an enhancer (a binding site for an activator protein) could be moved from the promoter of the *straightspine* gene and still activate transcription. You systematically vary the number of nucleotide pairs between these two sites and then determine the amount of transcription by measuring the production of Straightspine mRNA. At first glance, your data look confusing (Figure Q8–6). What would

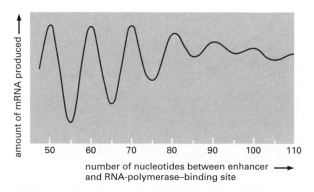

Figure Q8–6

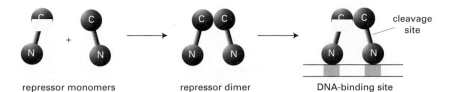

Figure Q8–9 repressor monomers   repressor dimer   DNA-binding site

you have expected for the results of this experiment? Can you save your reputation and explain these results to Professor Quasimodo?

Question 8–7

Many gene regulatory proteins form dimers of identical or slightly different subunits on the DNA. Why is this advantageous? Describe three structural motifs that are often used to contact DNA. What are the particular features that suit them for this purpose?

Question 8–8

Bacterial cells can take up the amino acid tryptophan (Trp) from their surroundings, or if there is an insufficient external supply, they can synthesize tryptophan from other small molecules. The Trp repressor is a bacterial gene regulatory protein that shuts off the transcription of genes that code for the enzymes required for the synthesis of tryptophan. Trp repressor binds to a site in the promoter of these genes only when molecules of tryptophan are bound to it (see Figure 8–7).

A. Why is this a useful property of the Trp repressor?

B. What would happen to the regulation of the tryptophan biosynthesis enzymes in cells that express a mutant form of Trp repressor that (1) cannot bind to DNA or (2) binds to DNA even if no tryptophan is bound to it?

C. What would happen in scenarios (1) and (2) if the cells, in addition, produced normal Trp repressor protein from a second, normal gene?

Question 8–9

The λ repressor binds as a dimer to critical sites on the λ genome to keep the lytic genes turned off, which stabilizes the prophage (integrated) state. Each molecule of the repressor consists of an N-terminal DNA-binding domain and a C-terminal dimerization domain (Figure Q8–9). Upon induction (for example, by irradiation with UV light), the genes for lytic growth are expressed, λ progeny are produced, and the bacterial cell lyses. Induction is initiated by cleavage of the λ repressor at a site between the DNA-binding domain and the dimerization domain. In the absence of bound repressor, RNA polymerase binds and initiates lytic growth. Given that the number (concentration) of DNA-binding domains is unchanged by cleavage of the repressor, why do you suppose its cleavage results in its removal from the DNA?

Question 8–10

The enzymes for arginine biosynthesis are located at several positions around the genome of *E. coli*, and they are regulated coordinately by a gene regulatory protein encoded by the *argR* gene. The activity of ArgR is modulated by arginine. Upon binding arginine, ArgR alters its conformation, dramatically changing its affinity for the regulatory sequences in the promoters of the genes for the arginine biosynthetic enzymes. Given that ArgR is a gene repressor, would you expect that ArgR would bind more tightly or less tightly to the regulatory sequences when arginine is abundant? If ArgR functioned instead as a gene activator, would you expect the binding of arginine to increase or to decrease its affinity for its regulatory sequences? Explain your answers.

Question 8–11

When enhancers were initially found to influence activity at remote promoters, two principal models were invoked to explain this action at a distance. In the "DNA looping" model, direct interactions between proteins bound at enhancers and promoters were proposed to stimulate RNA polymerase. In the "scanning" or "entry-site" model, RNA polymerase (or a transcription factor) was proposed to bind at the enhancer and then scan along the DNA until it reached the promoter. These two models were distinguished using an enhancer on one piece of DNA and a β-globin gene and promoter on a separate piece of DNA (Figure Q8–11). The β-globin gene was not expressed from the mixture of pieces. When the two segments of DNA were joined via a protein linker, the β-globin gene was expressed.

How does this experiment distinguish between the DNA looping model and the scanning model? Explain your answer.

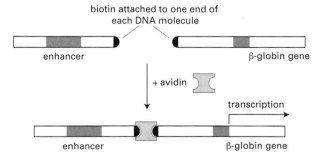

Figure Q8–11

Question 8–12

All differentiated cells in an organism contain the same genes. (Among the few exceptions to this rule are the cells of the mammalian immune system, where the formation of specialized cells is based on small

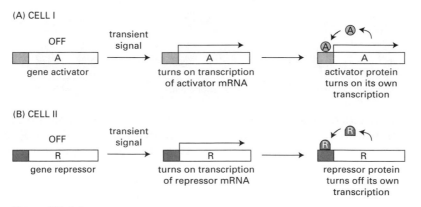

(A) CELL I

OFF — gene activator → transient signal → turns on transcription of activator mRNA → activator protein turns on its own transcription

(B) CELL II

OFF — gene repressor → transient signal → turns on transcription of repressor mRNA → repressor protein turns off its own transcription

Figure Q8–14

rearrangements of the genome.) Describe an experiment that substantiates the first sentence of this question, and explain why it does.

Question 8–13

Figure 8–22 shows a simple scheme by which three gene regulatory proteins might be used during development to create eight different cell types. How many cell types could you create, using the same rules, with four different gene regulatory proteins? MyoD is a regulatory protein that by itself is sufficient to induce muscle-specific gene expression in fibroblasts. How does this observation fit the scheme in Figure 8–22?

Question 8–14

Imagine the two situations shown in Figure Q8–14. In cell I, a transient signal induces the synthesis of protein A, which is a gene activator that turns on many genes including its own. In cell II, a transient signal induces the synthesis of protein R, which is a gene repressor that turns off many genes including its own. In which, if either, of these situations will the descendants of the original cell "remember" that the progenitor cell had experienced the transient signal? Explain your reasoning.

Question 8–15

Discuss the following argument: "If the expression of every gene depends on a set of gene regulatory proteins, then the expression of these gene regulatory proteins must also depend on the expression of other gene regulatory proteins, and their expression must depend on the expression of still other gene regulatory proteins, and so on. Cells would therefore need an infinite number of genes, most of which would code for gene regulatory proteins." How does the cell get by without having to achieve the impossible?

Highlights from *Essential Cell Biology 2 Interactive* CD-ROM

8.1   Homeodomain
8.2   Zinc Finger Domain
8.3   Leucine Zipper

How Genes and Genomes Evolve

In Chapter 6 we saw that organisms go to great lengths to ensure that their genetic material is accurately maintained and replicated so that the information it contains can be passed on unchanged to their progeny. However, one individual of a species is far from being an exact replica of the next. Look at any crowd of people: each individual differs in a host of heritable characteristics—in eye color, skin color, hair color, height, build, and so on. Plainly their genomes do not contain exactly the same nucleotide sequences.

Differences in nucleotide sequences provide the raw material upon which evolution works. Sculpted by selective pressures over billions of cell generations, since the beginning of life on Earth, these changes have engendered the whole spectacular menagerie of modern-day life forms, from bacteria to whales. The diversity of species thus depends on a delicate balance between the conservative accuracy of genome replication that enables progeny to inherit the virtues of their parents, and the creative errors of genome replication and maintenance that enable the progeny to acquire novel features and evolve new capabilities. If this balance were struck differently, the whole history of life on Earth would be different.

In this chapter, we discuss how genes and genomes change over time. We examine the molecular mechanisms by which genetic changes occur, and we consider how the information in present-day genomes can be deciphered to yield a historical record of the evolutionary processes that have shaped them. Genome sequencing has revolutionized our understanding of evolution, providing a wealth of new insights into the origins and relationships of genes and of living species. We shall end the chapter by taking a closer look at the human genome to see what our own DNA sequence tells us about who we are and where we come from.

Generating Genetic Variation

In discussing evolution, we think of living things in terms of their pedigrees—the series of cell divisions that give rise to the family tree that links each individual to its ancestors. For a unicellular organism that reproduces by copying its genome and splitting in two, such as a bacterium, the family tree is a simple branching diagram of cell divisions. But, for a multicellular organism that reproduces sexually, the family tree of cell divisions is more complicated because only a subset of its cells will ferry its genome to the next generation. As we discuss in Chapter 20, these specialized reproductive cells or *germ cells* come together during fertilization to give rise to a new individual. The other cells of the body—the *somatic cells*—are doomed to die without leaving

Generating Genetic Variation

Five Main Types of Genetic Change Contribute to Evolution

Genome Alterations Are Caused by Failures of the Normal Mechanisms for Copying and Maintaining DNA

DNA Duplications Give Rise to Families of Related Genes Within a Single Cell

The Evolution of the Globin Gene Family Shows How DNA Duplications Contribute to the Evolution of Organisms

Gene Duplication and Divergence Provide a Critical Source of Genetic Novelty for Evolving Organisms

New Genes Can Be Generated by Repeating the Same Exon

Novel Genes Can Also Be Created by Exon Shuffling

The Evolution of Genomes Has Been Accelerated by the Movement of Transposable Elements

Genes Can Be Exchanged Between Organisms by Horizontal Gene Transfer

Reconstructing Life's Family Tree

Genetic Changes That Offer an Organism a Selective Advantage Are the Most Likely to Be Preserved

The Genome Sequences of Two Species Differ in Proportion to the Length of Time That They Have Evolved Separately

Humans and Chimpanzee Genomes Are Similar in Organization As Well As Detailed Sequence

Functionally Important Sequences Show Up As Islands of DNA Sequence Conservation

Genome Comparisons Suggest That "Junk DNA" Is Dispensable

Sequence Conservation Allows Us to Trace Even the Most Distant Evolutionary Relationships

Examining the Human Genome

The Nucleotide Sequence of the Human Genome Shows How Our Genes Are Arranged

Genetic Variation Within the Human Genome Contributes to Our Individuality

Comparing Our DNA with That of Related Organisms Helps Us to Interpret the Human Genome

The Human Genome Contains Copious Information Yet to Be Deciphered

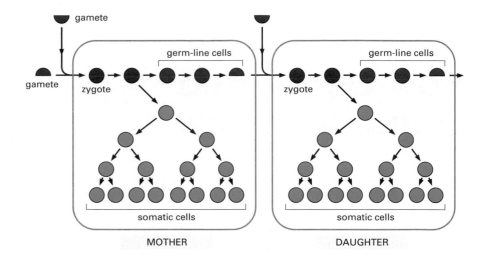

Figure 9–1 Germ-line cells and somatic cells carry out fundamentally different functions. In sexually reproducing organisms, the germ-line cells *(red)* propagate genetic information into the next generation. Somatic cells *(blue),* which form the body of the organism, ultimately leave no progeny of their own. In a sense, they exist only to help cells of the germ line survive and propagate.

descendants of their own (Figure 9–1). The cell lineage that gives rise to the germ cells is called the *germ line,* and it is through sequences of germ-line cell divisions that every individual traces its descent back to its ancestors and, ultimately, back to the ancestors of us all—the first cells that existed, at the origin of life more than 3.5 billion years ago. In this sense, the somatic cells can be considered to exist only to help cells of the germ line survive and propagate.

A mutation that occurs in the germ line, therefore, is passed on to the next generation. A mutation in a somatic cell, though it might have unfortunate consequences—such as cancer—for the individual in whom it occurs, is not transmitted to the organism's offspring (Figure 9–2). Thus in tracking the genetic changes that accumulate during evolution, we concentrate on events that occur in the germ line.

Tracing the family tree of genetic change in sexually reproducing organisms is further complicated because sex itself generates genetic variation. When the germ-line cells from two individuals fuse, they generate an offspring with a genome that is genetically distinct from that of either parent (see Figure 20–11).

Although sexual reproduction involves the mixing of genomes—which influences how genetic variations are propagated—the basic mechanisms that generate genetic change are the same for both sexually and asexually reproducing organisms, as we now discuss.

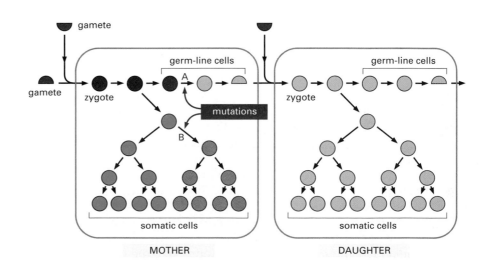

Figure 9–2 Germ-line and somatic mutations have different consequences. Mutations that occur in germ-line cells (A) are passed on to progeny *(green)*; those that arise in somatic cells (B) affect only the individual in which they occur *(orange)*. Somatic mutations are responsible for most human cancers (see Figure 6–20).

Five Main Types of Genetic Change Contribute to Evolution

Evolution works on the DNA sequences each organism inherits from its ancestors: there is no natural mechanism for making long stretches of new random sequence. In this sense, no gene—or genome—is ever entirely new. Evolution is more a tinkerer than an inventor: the astonishing diversity in form and function that we see in living systems is all the result of variations on preexisting themes. And yet, as variation is piled on variation over millions of generations, the cummulative effect is radical change.

So how are the elementary variations produced? At least five basic types of genetic change are crucial in evolution (Figure 9–3):

- *Mutation within a gene*: an existing gene can be modified by mutations that change a single nucleotide or that delete or duplicate one or more nucleotides in its DNA sequence. These so called point mutations are usually a result of rare "mistakes" made during DNA replication or to a failure in DNA repair following DNA damage.

- *Gene duplication*: an existing gene, a larger segment of DNA, or even a whole genome can be duplicated, creating a set of closely related genes within a single cell. As this cell and its progeny divide, these duplicated genes can then undergo additional mutations and assume functions distinct from those of the original genes.

- *Gene deletion*: individual genes, or whole blocks of genes can be lost through chromosome breakage and failures of repair.

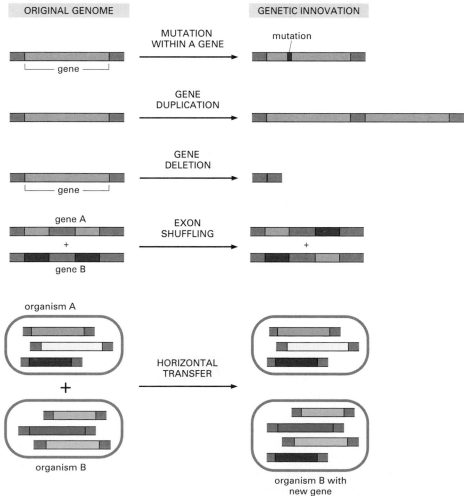

Figure 9–3 **Genes and genomes can be altered by several different mechanisms.**

- *Exon shuffling*: two or more existing genes can be broken and rejoined to make a hybrid gene containing DNA segments that originally belonged to separate genes. Because the breaking and rejoining often occurs within intron sequences in eucaryotic cells, the process does not have to be precise to result in a functional gene.

- *Horizontal (intercellular) gene transfer*: a piece of DNA can be transferred from the genome of one cell to that of another—even to that of another species. This process, rare among eucaryotes but common among procaryotes, differs from the usual vertical transfer of genetic information from parent to progeny.

Each of these forms of genetic variation—from the simple mutations that occur within a gene to the more extensive duplications, deletions, rearrangements, and additions that occur within a genome—has played an important part in the evolution of modern organisms. We now discuss these basic mechanisms of genetic change in more detail, and we consider their consequences for genome evolution.

Genome Alterations Are Caused by Failures of the Normal Mechanisms for Copying and Maintaining DNA

Despite the elaborate mechanisms that cells employ to maintain their DNA sequences, each nucleotide pair in the human genome runs a certain small risk—about 1 in 10^{10}—of changing each time a human cell divides. Such minor changes, called *point mutations*, typically arise from small errors in DNA replication or repair (discussed in Chapter 6).

The nucleotide mutation rate has been determined directly in experiments with bacteria such as *Escherichia coli*—a resident of the human gut. Under laboratory conditions, *E. coli* divides about once every 20–25 minutes; in less than a day, a single *E. coli* can produce more descendants than there are humans on Earth—enough to provide a good chance for almost any conceivable point mutation to occur. A culture containing 10^9 *E. coli* cells thus harbors millions of mutant cells whose genomes differ from the ancestor cell. Some of these mutations may confer a selective advantage on individual cells—resistance to a poison, for example, or the ability to survive when deprived of a standard nutrient. By exposing the culture to the selective condition—adding an antibiotic or removing an essential nutrient—one can find the needles in the haystack: the cells that have undergone a specific mutation enabling them to survive in conditions where the original cells cannot. In this way, one can discover how frequently the specific mutations occur, and from this one can estimate the overall mutation frequency in *E. coli*: it is about 1 nucleotide change per 10^9 nucleotide pairs per cell generation.

For humans, the point mutation rate can be estimated in other ways—for example, from the frequency with which babies are born with abnormalities that are identifiable as the result of harmful mutations in specific genes. Such genetic defects are very rare individually, but in the aggregate quite common. For a given gene, the frequency of such mutations is typically of the order of 1 in 100,000 births. When one takes account of the number of critical nucleotides in a gene and the number of germ-line cell generations in a single human generation, one arrives at an estimate for the human point mutation rate of about 0.1 nucleotide change per 10^9 nucleotide pairs each time the DNA is copied—about tenfold lower than that for *E. coli*.

Point mutations provide a way of fine tuning the function of a gene by making small adjustments to its sequence. They can also eliminate

Question 9–1

In this chapter it is argued that genetic variability is beneficial for a species because it enhances its ability to adapt to changing conditions. Why, then, does a cell go to great lengths to assure the fidelity of DNA replication?

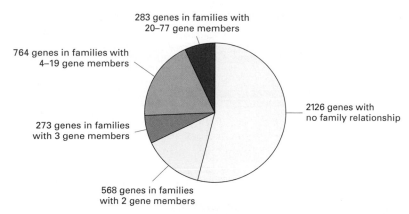

283 genes in families with 20–77 gene members

764 genes in families with 4–19 gene members

273 genes in families with 3 gene members

2126 genes with no family relationship

568 genes in families with 2 gene members

Figure 9–4 The *Bacillus subtilis* genome contains many families of evolutionarily related genes. The largest gene family in the *B. subtilis* genome contains 77 genes coding for a variety of ABC transporters, a class of proteins—found in bacteria, archaea, and eucaryotes—that transport various materials across the cell membrane. (After F. Kunst et al., *Nature* 390:253–256, 1997. © Macmillan Magazines Ltd.)

the activity of genes. Very often, however, they do neither of these things. At many sites in the genome, a point mutation has absolutely no effect on the appearance or viability of the organism—because the change leads to no alteration in the amino acid sequence of any protein, or in the function of any regulatory piece of DNA. Such silent, *selectively neutral mutations* accumulate steadily in the genome of a species over evolutionary time. As we shall discuss later, they can be used like the ticks of an evolutionary clock to estimate how many generations separate two individuals, or two closely related species, from their last common ancestor.

DNA Duplications Give Rise to Families of Related Genes Within a Single Cell

Point mutations can adjust the activity of a gene. But how do new genes come into being? Gene duplication is perhaps the most important mechanism for generating new genes. Once a gene has been duplicated, one of the two gene copies is free to mutate and become specialized to perform a different function. This specialization generally occurs gradually, as mutations accumulate in the descendants of the original cell in which gene duplication occurred. Repeated rounds of this process of **duplication** and **divergence**, over many millions of years, can allow one gene to give rise to a whole family of genes within a single genome. Analysis of genome sequences reveals many examples of such gene families: in *Bacillus subtilis*, for example, nearly half of the genes have one or more obvious relatives elsewhere in the genome (Figure 9–4).

Eucaryotic genomes also contain many families of related genes. For example, different opsins—proteins that detect light of different wavelengths—are expressed in different retinal cells, and different collagen genes are expressed in the various types of connective tissue. A great deal of evidence indicates that these gene families arose by successive duplication and divergence from a single primordial gene. For example, the globin gene family in humans, discussed in detail in the next section, clearly arose this way. But what is the mechanism that gives rise to gene duplication in the first place?

It is thought that gene duplications often result from a type of recombination event that occurs between two homologous segments of DNA—chromosomes or portions of chromosomes that are identical or very similar in sequence. This type of recombination normally takes place only when two long stretches of DNA that are nearly identical become paired, most often exactly the same region of DNA on two homologous chromosomes (discussed in Chapter 6). But on rare occasions, the recombination between two chromosomes will instead occur between two short repeated DNA sequences on the opposite sides of a

Figure 9–5 Gene duplication can occur by unequal crossing-over between short repeated sequences on homologous segments of DNA. A pair of homologous chromosomes undergoes a crossover (recombination) at a short sequence *(red)*, which is repeated on each chromosome, with a gene (or set of genes) bracketed between copies of this repeated sequence. In some cases, these repeated sequences are remnants of transposons, which are present in many copies in the human genome (see Figure 6–35). After the crossover, the long chromosome has two copies of the intervening gene, while the short chromosome has the gene deleted. If this process occurs in the germ line, some progeny will inherit the long chromosome, while others will inherit the short one.

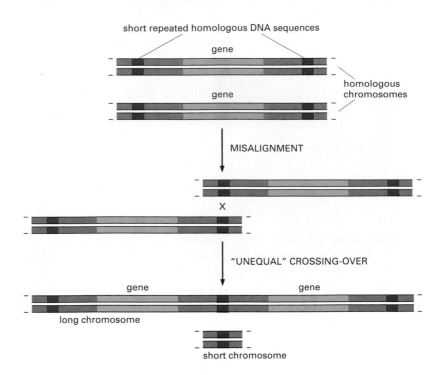

gene. One of the two resulting chromosomes will then end up with an extra copy of the DNA containing that gene (Figure 9–5). Once a gene has been duplicated in this way, subsequent unequal crossover events can readily add extra copies to the duplicated set by pairing a gene on one chromosome with its adjacent copy on a second chromosome. As a result, it is not surprising to find an entire set of closely related genes arranged in series, as observed in the globin gene family.

The Evolution of the Globin Gene Family Shows How DNA Duplications Contribute to the Evolution of Organisms

The globin gene family provides a particularly good example of how DNA duplication generates new proteins, because its evolutionary history has been worked out particularly well. The unmistakable homologies in amino acid sequence and structure among the present-day globins indicate that they all must derive from a common ancestral gene, even though the genes that encode some of them are now widely separated in the mammalian genome.

We can reconstruct some of the past events that produced the various types of oxygen-carrying hemoglobin molecules by considering the different forms of the protein in organisms at different positions on the phylogenetic tree of life. Oxygen-binding proteins related to hemoglobin are found even in bacteria, but they are especially important in multicellular animals that grow to a large size and cannot rely on the simple diffusion of oxygen through the body surface to oxygenate their tissues adequately. The most primitive oxygen-carrying molecule in animals is a globin polypeptide chain of about 150 amino acids, which is found in many marine worms, insects, and primitive fish. The hemoglobin molecule in higher vertebrates, however, is composed of two kinds of globin chains. It appears that about 500 million years ago, during the evolution of higher fish, a series of gene mutations and duplications occurred. These events established two slightly different globin genes, coding for the α- and β-globin chains in the genome of each individual. In modern higher vertebrates each hemoglobin molecule is a complex of two α chains and two β chains (Figure 9–6). The four oxygen-binding sites in the $\alpha_2\beta_2$ molecule interact, allowing a cooperative allosteric change in the molecule as it binds and releases oxygen, which enables hemoglobin to take up and to release oxygen more efficiently than the single-chain version.

Still later, during the evolution of mammals, the β-chain gene apparently underwent duplication and mutation to give rise to a second β-like chain that is synthesized specifically in the fetus (Figure 9–7). The resulting hemoglobin molecule has a higher affinity for oxygen than adult hemoglobin and thus helps in the transfer of oxygen from the mother to the fetus. The gene for the new β-like chain subsequently mutated and duplicated again to produce two new genes, ε and γ, the ε chain being produced earlier in development (to form $\alpha_2\varepsilon_2$) than the fetal γ chain, which forms $\alpha_2\gamma_2$. A duplication of the adult β-chain gene occurred still later, during primate evolution, to give rise to a δ-globin gene and thus to a minor form of hemoglobin ($\alpha_2\delta_2$) found only in adult primates.

Each of these duplicated genes has been modified by point mutations that affect the properties of the final hemoglobin molecule, as well as by changes in regulatory regions that determine the timing and level of expression of the gene. As a result, each globin is made in different amounts at different times of human development.

The end result of the gene duplication processes that have given rise to the diversity of globin chains is seen clearly in the human genes that arose from the original β gene; these genes are arranged as a series of homologous DNA sequences located within 50,000 nucleotide pairs of one another. A similar cluster of α-globin genes is located on a separate human chromosome. Because the α- and β-globin gene clusters are on separate chromosomes in birds and mammals but are together in the frog *Xenopus,* it is believed that a chromosome breakage event separated the two gene clusters about 300 million years ago (see Figure 9–7).

There are several duplicated globin DNA sequences in the α- and β-globin gene clusters that are not functional genes, but *pseudogenes*. These have a close homology to the functional genes but have been disabled by mutations that prevent their expression. The existence of such pseudogenes make it clear that, as expected, not every DNA duplication leads to a new functional gene. We also know that nonfunctional DNA sequences are not rapidly discarded, as indicated by the large excess of noncoding DNA that is found in mammalian genomes.

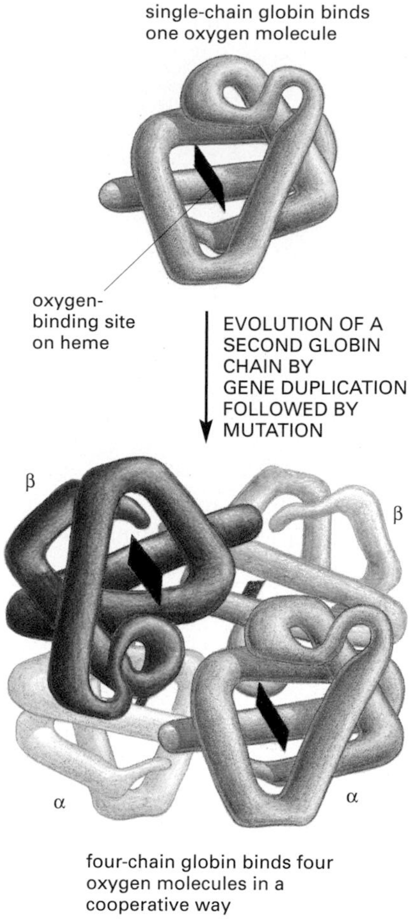

Gene Duplication and Divergence Provide a Critical Source of Genetic Novelty for Evolving Organisms

Gene duplication has clearly allowed the development of more complex life forms; as we have seen, it provides an organism with a cornucopia of spare gene copies, which are free to mutate to serve divergent purposes. Almost every gene in the genome of a vertebrate exists in multiple versions, indicating that the entire vertebrate genome has undergone successive duplications during evolution either piecemeal, with different portions of the genome undergoing duplication at different

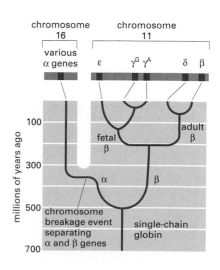

single-chain globin binds one oxygen molecule

oxygen-binding site on heme

EVOLUTION OF A SECOND GLOBIN CHAIN BY GENE DUPLICATION FOLLOWED BY MUTATION

β   β

α   α

four-chain globin binds four oxygen molecules in a cooperative way

Figure 9–6 A single-chain globin molecule gave rise to the four-chain hemoglobin used by humans and other mammals. The mammalian hemoglobin molecule is a complex of two α- and two β-globin chains. The single-chain globin, which is found in some primitive vertebrates, forms a dimer that dissociates when it binds oxygen, representing an intermediate stage in the evolution of the four-chain molecule.

Figure 9–7 Repeated rounds of duplication and mutation generated the globin gene family in humans. An ancestral globin gene duplicated and gave rise to the β-globin family and to the related α family. In vertebrates, a molecule of hemoglobin is formed from two α chains and two β chains, as shown in Figure 9–6. The scheme shown was worked out by comparing globin genes from many different organisms. For example, the nucleotide sequences of the γ^G and γ^A genes, which produce globin chains that form fetal hemoglobin, are much more similar to each other than either of them is to the adult β gene. The location of the globin genes in the human genome is shown at the top of the figure.

times, or as a whole through duplication of the entire chromosome set at one blow. According to one hypothesis, at an early stage in the evolution of vertebrates, the entire genome underwent duplication twice in succession, giving rise to four copies of every gene. In some groups of vertebrates, such as fish in the salmon and carp families (including the zebrafish, a popular laboratory animal), there may have been yet another duplication, creating an eightfold multiplicity of genes.

The precise course of vertebrate genome evolution remains uncertain, however, because many other changes have occurred since these ancient evolutionary events. Genes that were once identical have diverged, and many of the gene copies have been lost through deleterious mutations. Moreover, in each branch of the vertebrate family tree, the genome has suffered repeated rearrangements, breaking up most of the original gene order. Indeed, it is entirely possible that the present state of affairs is the result of many separate duplications of fragments of the genome, rather than duplications of the genome as a whole.

There is, however, no doubt that whole-genome duplications do occur from time to time. The frog genus *Xenopus*, for example, comprises a set of closely similar species related to one another by repeated duplications or triplications of the whole genome (Figure 9–8). Such large-scale duplications can happen if cell division fails to occur following a round of genome replication in the germ line of a particular individual. Once an accidental doubling of the genome occurs, it will be faithfully passed on to other germ cells in the individual and, ultimately, to its progeny.

New Genes Can Be Generated by Repeating the Same Exon

The role of DNA duplication in evolution is not confined to the expansion of gene families. It can also act on a smaller scale to modify single genes by creating internal duplications. As we discussed in Chapter 4, many proteins in eucaryotes are composed of a series of repeating protein *domains*. Such proteins include the albumins and immunoglobulins (see Figure 4–32), as well as most fibrous proteins such as the collagens. These proteins are encoded by genes that have evolved by repeated duplications of a single DNA segment within a gene.

In such genes, each individual protein domain is often encoded by a separate exon. The duplications necessary to form a single gene encoding a protein with repeating domains can then occur by breaking and rejoining the DNA anywhere in the long introns on either side of the exon that codes for the protein domain (Figure 9–9). Without introns there would be very few sites in the original gene at which a recombinational exchange between homologous chromosomes could duplicate the domain without damaging it. In contrast, the introns can be snipped at arbitrary sites and rejoined, like strings linking jewels in a necklace. The evolution of new proteins is therefore thought to have been greatly facilitated by the organization of eucaryotic DNA coding sequences as a series of relatively short exons separated by long, noncoding introns (see Figure 7–13).

Novel Genes Can Also Be Created by Exon Shuffling

The type of recombination that allows exons to be duplicated within a gene can also occur between two different genes, joining together two initially separate exons that code for quite different protein domains—an important process called **exon shuffling**. The lack of precision that can be tolerated in a recombination event that breaks and rejoins two

Figure 9–8 Different species of the frog *Xenopus* have different DNA contents. *X. tropicalis (above)* has an ordinary diploid genome; *X. laevis (below)* has a duplicated genome with twice as much DNA per cell. (Courtesy of Enrique Amaya.)

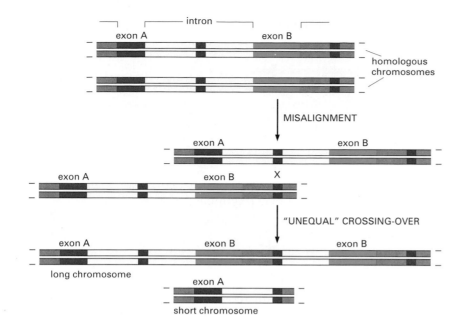

Figure 9–9 **An exon can be duplicated by unequal crossing-over.** The general scheme is the same as that of Figure 9–5, with a short repeated sequence indicated in *red,* however, here an exon within a gene, rather than an entire gene, is being duplicated. The mRNA from the original gene contains two exons, A and B, whereas the long chromosome will produce an mRNA with three exons (A, B, and B). Because the duplicated exons are joined by an intron with its splicing sequences intact, the modified nucleotide sequence can be readily spliced after transcription to produce a functional mRNA.

introns greatly increases the probability that a chance recombination event will generate a hybrid gene that joins exons in this way. The presumed results of such recombinations are seen in many present-day proteins, which are a patchwork of many different protein domains (Figure 9–10).

It has been proposed that all the proteins encoded by the human genome (approximately 30,000) arose from the duplication and shuffling of a few thousand distinct exons, each encoding a protein domain of approximately 30–50 amino acids. This remarkable idea suggests that the great diversity of protein structures is generated from a quite small universal "list of parts" pieced together in different combinations.

The Evolution of Genomes Has Been Accelerated by the Movement of Transposable Elements

The mobile DNA elements described in Chapter 6 are another important source of genomic change. In particular, transposable elements (transposons) have profoundly affected the structure of genomes. These parasitic DNA sequences can colonize a genome and can spread within it. During this process, they often disrupt the function or alter the regulation of existing genes; sometimes they even create altogether novel genes through fusions between transposon sequences and segments of existing genes.

The insertion of a transposable element into the coding sequence of a gene or into its regulatory region is a frequent cause of the "spontaneous" mutations that are observed in many organisms. Transposable elements can severely disrupt a coding sequence if they land directly in it, producing an insertion mutation that destroys the gene's capacity to encode a useful protein. For example, a number of the mutations in the Factor VIII gene that cause hemophilia in humans result from insertion of transposable elements in the gene.

Transposable elements also provide opportunities for genome rearrangements by serving as targets of homologous recombination. For example, the duplications that gave rise to the β-globin gene cluster are thought to have occurred by homologous recombination between *Alu*-like sequences (see Figures 9–5 and 9–7). However, transposons also have more direct roles in the evolution of genomes. In addition to moving themselves, transposable elements occasionally move or

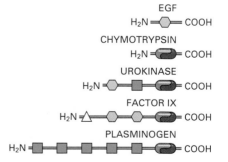

Figure 9–10 **Exon shuffling can generate proteins with new combinations of protein domains.** Each type of symbol represents a different type of protein domain. During evolution, these have been joined together end-to-end, as shown, to create the modern-day proteins identified by name.

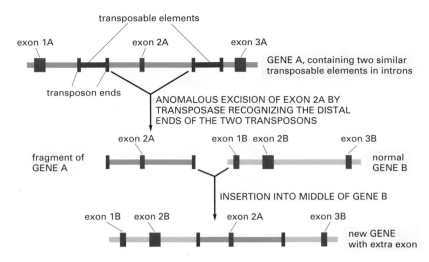

Figure 9–11 Transposable elements can create new exon arrangements. When two transposable elements of the same type (*red*) happen to insert near each other in a chromosome, the transposition mechanism may occasionally use the ends of two different elements (instead of the two ends of the same element) and thereby move the chromosomal DNA between them to a new chromosomal site. Because introns are very large relative to exons, the illustrated insertion of a new exon into a preexisting intron is a frequent outcome of such a transposition event.

Question 9–2

Why do you suppose that horizontal gene transfer is more prevalent in single-celled organisms than multicellular organisms?

rearrange neighboring DNA sequences of the host genome. When two transposable elements that are recognized by the same transposase integrate into neighboring chromosomal sites, the DNA between them can itself be transposed. In eucaryotic genomes, this provides a particularly effective pathway for the movement of exons, generating new genes by creating novel arrangements of existing exons (Figure 9–11).

The activity of transposons can also change the way that existing genes are expressed. For example, an insertion of a transposable element in the regulatory region of a gene will often have a dramatic effect on where and when that gene is turned on (Figure 9–12). Transposable elements are a significant source of developmental changes, and they are thought to have been particularly important in the evolution of the body plans of multicellular plants and animals.

Genes Can Be Exchanged Between Organisms by Horizontal Gene Transfer

So far we have considered genetic changes that take place within the genome of an individual organism. However, genes and other portions of genomes can also be exchanged between individuals of different species. This mechanism, known as **horizontal gene transfer**, is rare among eucaryotes, but common among bacteria (Figure 9–13).

E. coli, for example, appears to have acquired at least 18% of its genome from other species within the past 100 million years. And such genetic exchanges are currently responsible for the rise of new and potentially dangerous strains of drug-resistant bacteria. For example, genes that confer resistance to an antibiotic can be transferred from species to species. This DNA exchange provides the recipient bacterium with an enormous selective advantage in evading the antimicrobial compounds that constitute modern medicine's frontline attack against

Figure 9–12 Mutation due to a transposable element can induce dramatic alterations in the body plan of an organism. (A) A normal fruit fly *(Drosophila melanogaster)*. (B) The fly's antennae have been transformed into legs because of a mutation in regulatory DNA that allows genes for the construction of a leg to be activated in circumstances that normally activate genes for construction of an antenna. Although this particular change is not advantageous to the fly, it does illustrate how a DNA rearrangement caused by a transposable element can produce a dramatic change in the organism. (A, courtesy of E.B. Lewis; B, courtesy of Matthew Scott.)

(A)

(B)

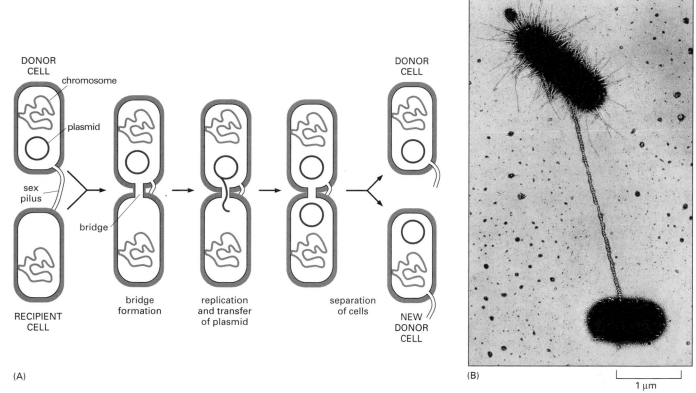

Figure 9–13 **Bacterial cells can horizontally transfer DNA by a process called conjugation.** (A) In this case, the DNA to be transferred is a small self-replicating circle of double-stranded DNA *(red)* called a plasmid. Plasmids, which often contain genes for antibiotic resistance, will be described in Chapter 10. Conjugation begins when the donor cell attaches to the recipient by a fine appendage, called a sex pilus. A cytoplasmic bridge then forms between the cells and a copy of the plasmid is transferred across it into the recipient cell. Following transfer, the bacteria separate and both partners can independently seek out new recipient cells. (B) In this electron micrograph, the cell donating its DNA *(top)* is seen linked to the recipient cell by a sex pilus. To make it more visible, the pilus has been labeled along its length by viruses that specifically adhere to it. (B, Courtesy of Charles C. Brinton Jr. and Judith Carnahan.)

bacterial infection. As a result, many antibiotics are no longer effective against the common bacterial infections for which they were originally used. For example, most strains of *Neisseria gonorrhoeae*, the bacterium that causes gonorrhea, are now resistant to penicillin; this antibiotic is therefore no longer the primary treatment for this disease.

Gene swapping may have been even more rampant in the early days of life on Earth. Indeed, it has been suggested that the genomes of present-day bacteria, archaea, and eucaryotes originated not by divergent lines of descent from a single genome in a single ancestral cell, but from independent anthologies of genes that arose from a pool of genes that were shared by a primordial community of promiscuous cells (Figure 9–14). This genetic promiscuity in ancestral cells could explain the otherwise puzzling observation that eucaryotes seem similar to one class of procaryotes—the archaea—in their genes for replication, transcription, and translation, but more similar to another class—the eubacteria—in their genes for metabolic processes.

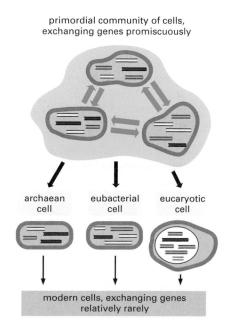

Figure 9–14 **Primordial cells may have been genetically promiscuous.** In the early days of life on Earth, cells may have been less strict about maintaining their separate identities and may have exchanged genes through horizontal transfer much more readily than cells do now. In this way, the archaean, bacterial, and eucaryotic lineages may have inherited different but overlapping subsets of genes from a primordial community of cells that readily exchanged DNA.

Reconstructing Life's Family Tree

Given an understanding of the basic molecular mechanisms by which genomes change, we can begin to decipher the clues to our evolutionary history that come from comparing and analyzing genome sequences. Perhaps the most astonishing revelation of such genome analyses has been that **homologous genes**—genes that are similar in their nucleotide sequence because of a common ancestry—can often be recognized across vast phylogenetic distances. Unmistakable homologs of many human genes are easy to detect in such organisms as worms, fruit flies, yeasts, and even bacteria. The nematode worm *Caenorhabditis elegans*, the fly *Drosophila melanogaster*, and the vertebrate *Homo sapiens*—the first three animals for which a complete genome sequence was obtained—are very distant relatives: the lineage leading to vertebrates is thought to have diverged from that leading to nematodes and insects more than 600 million years ago. Yet when the 19,000 genes of *C. elegans*, the 14,000 genes of *Drosophila*, and the 30,000 or so genes of *Homo sapiens* are systematically compared, we find that about 50% of the genes in each of these species have homologs in one or both of the other two species. In other words, clearly recognizable versions of at least half of all human genes were already present in the common ancestor of worms, flies, and humans.

By tracing such relationships among genes, we can begin to define the evolutionary relationships among different species, placing each bacterium, protist, animal, plant, or fungus in a single vast family tree of life. In this section, we discuss how these relationships are determined and what they tell us about our genetic heritage.

Genetic Changes That Offer an Organism a Selective Advantage Are the Most Likely to Be Preserved

Evolution is commonly thought of as progressive, but a large part of the process, at a molecular level, is random. Consider the fate of a point mutation. As we discussed earlier, on rare occasions such a mutation may represent a change for the better; very often, it will cause no significant difference in the organism's prospects; and sometimes it will cause serious damage—by disrupting the coding sequence for a key protein, for example. Changes due to mutations of the first type will tend to be perpetuated, because the organism that inherits them will have an increased likelihood of reproducing itself. Changes due to mutations of the second type—*selectively neutral* changes—may be perpetuated or not: in the competition for limited resources, it is a matter of chance whether the mutant organism or its cousins will succeed. In contrast, mutations that cause serious damage lead nowhere: the organism that inherits them dies, leaving no progeny. Through endless repetition of this cycle of error and trial—of mutation and natural selection—organisms evolve: their genetic specifications change, and they develop new ways to exploit the environment more effectively, to survive in competition with others, and to reproduce successfully.

Clearly, some parts of the genome can accumulate mutations more easily than others in the course of evolution. A segment of DNA that does not code for protein and has no significant regulatory role is free to change at a rate limited only by the frequency of random mutation. In contrast, a gene that codes for a highly optimized essential protein or RNA molecule cannot be altered so easily: when mutations occur, the faulty organism is almost always eliminated. Genes of this latter sort are therefore *highly conserved*; that is, the proteins they encode are very similar from organism to organism. Throughout 3.5 billion years or

Question 9–3

Highly conserved genes such as those for ribosomal RNA are present as clearly recognizable relatives in all organisms on Earth; thus, they have evolved very slowly over time. Were such genes "born" perfect?

more of evolutionary history, many features that have become part of the human genome have changed beyond all recognition, but the most highly conserved genes remain perfectly recognizable in all living species. These latter genes are the ones that must be examined if we wish to trace family relationships among the most distantly related organisms in the tree of life.

On the other hand, if we are interested in more closely related species, it is more informative for many purposes to focus on the selectively neutral changes. These accumulate steadily at a rate that is independent of selection pressures, and thus provide us with a simple and easily readable evolutionary clock. The evolution of primates provides a good example.

The Genome Sequences of Two Species Differ in Proportion to the Length of Time That They Have Evolved Separately

We saw earlier that errors occur every time our genome is copied: about 1 nucleotide in 10^{10} is altered at random in each cell division cycle, and the alterations are passed on to the next cell generation, where more errors creep in, and so on. Not more than 2% of the human genome codes for protein, with perhaps a similar fraction having important functions in gene regulation. This leaves at least 95% of the DNA free to change its sequence without any significant effect on the viability of the cell or the fitness of the person who inherits it. Thus most of the randomly occurring point mutations are selectively neutral: there is no selection pressure to eliminate them.

It is estimated that about 200 cell divisions occur in the germ line on average from the time of conception to the time of production of the eggs and sperm that go to make the next generation, and the total amount of human DNA in a cell (including maternally and paternally derived sets of chromosomes) is about 6 billion nucleotide pairs. From this it follows that in each human generation random mutations will contribute, in round figures, 100 new differences between the DNA sequences of the children and those of the parents. In another family, the same will be true, but the new mutations will be at different sites in the genome. Thus the total genetic difference between the two families will increase by about 200 nucleotide differences per generation. This type of evolutionary divergence is due to random mutation largely unconstrained by selection pressures.

After 150 generations, the genetic difference between two families that trace their origins back to a common ancestor in the time of the early Pharaohs but have lived apart ever since will have increased by about 30,000 nucleotide pairs. After about 10 million years, this difference will have increased to about 1% of the genome. This value is roughly the proportion of nucleotides by which a human genome differs from that of a chimpanzee: to be precise, the human and chimpanzee genomes are only 1.2% different, or in other words 98.8% identical. If humans and chimpanzees were all descended from a single pair of genetically identical individuals, we could infer that these last common ancestors must have lived about 10 million years ago. From more realistic (and complicated) calculations along these lines, allowing for a population of last common ancestors that already contained some genetic diversity, it is estimated that the two species diverged from one another about 5 million years ago.

Of course, there are uncertainties in this estimate even though it is in reasonable agreement with the fossil record. We do not know, for example, the precise value of the mutation rate, in nucleotide changes

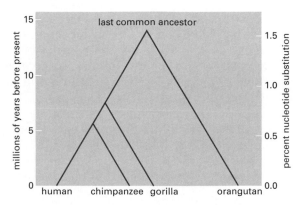

Figure 9–15 Phylogenetic trees display the relationships among modern life forms. In the higher-primate family tree, humans fall closer to chimpanzees than to gorillas or orangutans. This phylogenetic tree is based on a comparison of nucleotide sequence data. As indicated, the genome sequences of all four species are estimated to differ from the sequence of their last common ancestor by about 1.5%. Because changes occur independently on both diverging lineages, the divergence between two species will be double the amount of change that takes place between each species and the last common ancestor. For example, humans and orangutans typically show sequence divergences of a bit more than 3%, while human and chimp genomes differ by about 1.2%. Although this phylogenetic tree is based solely on nucleotide sequences, the estimated times of divergence derive, ultimately, from data obtained from the fossil record. (Modified from F.C. Chen and W.H. Li, *Am. J. Hum. Genet.* 68:444–456, 2001.)

per year, for ancestral humans and ancestral chimpanzees, and it might be different from that today. On the other hand, we do know that the number of differences between the human genome and that of the gorilla is substantially greater than between human and chimpanzee, and that the difference between the human and the orangutan is greater still. In this way, by comparing genome sequences, we can deduce the evolutionary relationships of these primate species and their relative times of divergence from common ancestors, even if there are uncertainties as to the exact mutation rates. The conclusions are shown in a **phylogenetic tree**—a diagram of the lines of evolutionary descent (Figure 9–15).

Humans and Chimpanzee Genomes Are Similar in Organization As Well As Detailed Sequence

Humans and chimpanzees are so closely related that it is possible to reconstruct the gene sequences of the extinct, common ancestor of the two species (Figure 9–16). Not only do humans and chimpanzees appear to have essentially the same set of 30,000 genes, but these genes are arranged in nearly the same way along the chromosomes of the two species (Figure 9–17). The only substantial exception is human Chromosome 2, which arose from a fusion of two chromosomes that remain separate in the chimpanzee, the gorilla, and the orangutan.

Even the massive resculpting of genomes that can be produced by the activities of transposable elements (discussed in Chapter 6) has had only minor effects on the overall structure of human and chimp genomes. More than 99% of the million copies of the *Alu* retrotransposon sequences that are present in both genomes are in corresponding positions. This observation indicates that most of the *Alu* sequences in our genome underwent duplication and transposition before humans and chimpanzees diverged. However, members of the *Alu* family are still capable of transposing, as is evident from a small number of observed cases in which new *Alu* insertions have caused human genetic disease. These cases involve transposition of this DNA into sites that were unoccupied in the genomes of the patients' parents.

Functionally Important Sequences Show Up As Islands of DNA Sequence Conservation

As we delve back further into our evolutionary history and compare our genomes with those of more distant relatives, the picture changes. The lineages of humans and mice, for example, diverged about 75 million years ago. We both have genomes of practically the same size, with practically the same numbers of genes, and both genomes are peppered

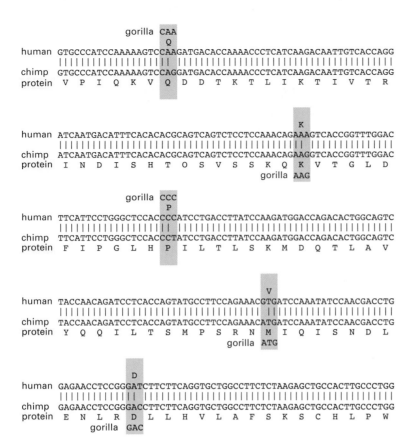

Figure 9–16 Ancestral gene sequences can be reconstructed by comparing the sequences of closely related modern-day species. Shown here in five contiguous sections is the sequence from the coding region of the leptin gene from humans and chimpanzees. Leptin is a hormone that regulates food intake and energy utilization. As indicated by the codons boxed in *green*, only 5 (of 441 nucleotides total) differ between the two sequences. Only one of these changes results in a change in the amino acid sequence. The sequence of the last common ancestor was likely the same as the human and chimp sequences where they agree; in the few places where they disagree, the gorilla sequence can be used as a "tiebreaker." This strategy is based on the relationship shown in Figure 9–15: differences between humans and chimpanzees reflect relatively recent events in evolutionary history, and the gorilla sequence reveals the most likely precursor sequence. For convenience, only the first 300 nucleotides of the leptin coding sequences are shown. The last 141 are identical between humans and chimpanzees.

with transposons. But the detailed distribution of the transposons is quite different, implying that the transposons have been independently proliferating and moving around the genome in each lineage since we diverged (Figure 9–18). Moreover, the large-scale organization of the genomes has become scrambled by many episodes of chromosome breakage and recombination—a total of about 180 "break-and-join" events, it is estimated. As a result, the overall structures of the chromosomes have changed dramatically. For example, in humans most of the centromeres lie near the middle of a chromosome, and none are found directly at a chromosome end, where the centromeres are located on each mouse chromosome (see Figure 9–28, below). Nevertheless, in spite of the genetic shuffling, one can still recognize many blocks of **conserved synteny**—regions where corresponding genes that began as neighbors have remained neighbors, strung together in the same sequence in both species. More than 90% of the mouse and human genomes can be partitioned into such corresponding regions of conserved synteny. Within these regions, we can align the DNA of the mouse with that of the human and compare the nucleotide sequences in detail.

Such genome-wide sequence comparisons reveal that in the roughly 75 million years since humans and mice diverged from their

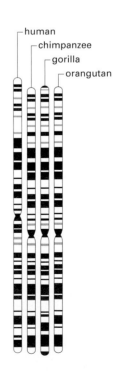

Figure 9–17 Human chromosomes are very similar in organization to those of the chimpanzee. Comparison of chromosomal staining patterns indicates that human chromosomes are more closely related to those of the chimpanzee than to those of the gorilla or orangutan. Chromosome 1 from each species is shown here. As discussed in Chapter 5, this banding pattern results from staining that distinguishes A-T–rich regions from G-C–rich regions. (Adapted from M.W. Strickberger, Evolution, 3rd edn., Sudbury, MA: Jones & Bartlett Publishers, 2000.)

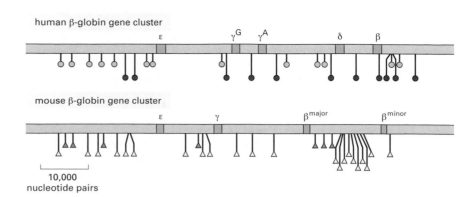

Figure 9–18 The positions of transposons in the human and the mouse genomes provide additional evidence of the long divergence time separating the two species. This stretch of human genome contains five functional β-globin–like genes *(orange);* the comparable region from the mouse genome has only four. The positions of the human *Alu* sequence *(green circles)* and the human *L1* sequences *(red circles)* are shown. The mouse genome contains related transposable elements: the positions of *B1* elements (a relative of the human *Alu* sequences; *blue triangles)* and the mouse *L1* elements (a relative of the human *L1* sequences; *yellow triangles)* are shown. The absence of transposable elements within the globin structural genes can be attributed to natural selection, which would have eliminated any insertion that compromised gene function. (Courtesy of Ross Hardison and Webb Miller.)

common ancestor, about 50% of the nucleotides have changed. Against this background of dissimilarity, however, one can now begin to see very clearly the regions where changes are not tolerated and the human and mouse sequences have remained much more closely similar (Figure 9–19). Here, the sequences have been conserved by **purifying selection**—that is, the elimination of individuals carrying mutations that interfere with important functions.

The selectively conserved DNA segments are of two main sorts. Some are exons coding for protein domains whose amino acid sequence must not be changed if the protein is to do its job correctly, or are genes coding for functionally important RNA molecules such as the ribosomal RNAs. Others are pieces of regulatory DNA—mostly binding sites for proteins that control the expression of adjacent genes. As we discuss below, comparisons between the DNA sequences of pairs of species such as mouse and human are an important tool for molecular biologists trying to pinpoint these functionally important segments of the genome.

Genome Comparisons Suggest That "Junk DNA" Is Dispensable

Going back still further in evolution, we can compare our own genome with that of a fish. The fish and mammal lineages diverged about 400 million years ago. This is long enough for random sequence changes and differing selection pressures to have obliterated almost every trace of similarity in nucleotide sequence except where purifying selection has operated to prevent change. The conserved regions thus stand out even more strikingly. One can still recognize most of the same genes and even many of the same segments of regulatory DNA. Blocks of conserved synteny can still be identified, though not so easily as in comparisons of human with mouse. On the other hand, the extent of duplication of any given gene is often different, resulting in different numbers of members of gene families such as the globin genes. The total size of the genome may also be different—in some cases dramatically so.

An extreme example is the puffer fish, *Fugu rubripes* (Figure 9–20), whose genome has been chosen for sequencing for the very reason that

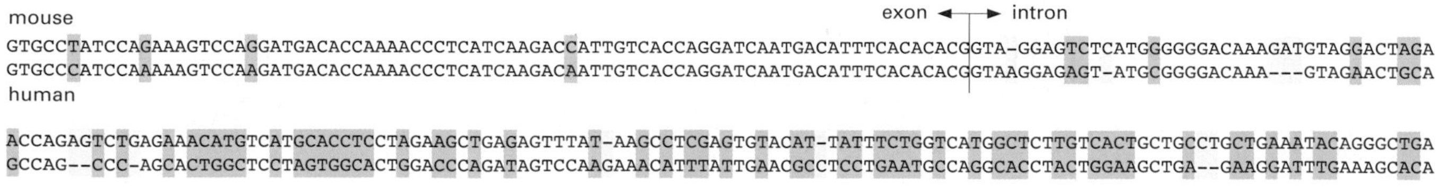

Figure 9–19 Accumulated mutations have resulted in considerable divergence in the nucleotide sequences of the human and the mouse genomes. Shown here in two contiguous sections are portions of the human and mouse leptin gene sequences. Positions where the sequences differ by a single nucleotide substitution are boxed in *green,* and positions where they differ by the addition or deletion of nucleotides are boxed in *yellow.* Note that the coding sequence of the exon is much more conserved than the adjacent intron sequence.

```
mouse                                                                    exon ◄──► intron
GTGCCTATCCAGAAAGTCCAGGATGACACCAAAACCCTCATCAAGACCATTGTCACCAGGATCAATGACATTTCACACACGGTA-GGAGTCTCATGGGGGGACAAAGATGTAGGACTAGA
GTGCCCATCCAAAAAGTCCAAGATGACACCAAAACCCTCATCAAGACAATTGTCACCAGGATCAATGACATTTCACACACGGTAAGGAGAGT-ATGCGGGGACAAA---GTAGAACTGCA
human

ACCAGAGTCTGAGAAACATGTCATGCACCTCCTAGAAGCTGAGAGTTTAT-AAGCCTCGAGTGTACAT-TATTTCTGGTCATGGCTCTTGTCACTGCTGCCTGCTGAAATACAGGGCTGA
GCCAG--CCC-AGCACTGGCTCCTAGTGGCACTGGACCCAGATAGTCCAAGAAACATTTATTGAACGCCTCCTGAATGCCAGGCACCTACTGGAAGCTGA--GAAGGATTTGAAAGCACA
```

Figure 9–20 The puffer fish, *Fugu rubripes*, has a remarkably compact genome. With only 400 million nucleotide pairs, the *Fugu* genome is only one-quarter the size of the zebrafish genome, even though the two species have a similar number of genes. (From a woodcut by Hiroshige, courtesy of Arts and Designs of Japan.)

it is small—only about one eighth of the number of nucleotide pairs in a mammalian genome. This small size is not because the number of genes is small: *Fugu* has just as many genes as a mammal. Conserved exons and conserved segments of regulatory DNA are still seen. Gene structure has also been well conserved between *Fugu* and human, in the sense that the numbers and positions of introns are almost identical and therefore presumably identical to those of our common ancestor. The dramatic difference of genome size is due almost entirely to a reduction, not in the numbers, but in the sizes of the introns, and in the sizes of the regions of noncoding DNA that lie between genes (Figure 9–21). Somehow or other, *Fugu* has either managed to rid itself of most of the noncoding DNA that clutters the genome of other species, or has managed to avoid accumulating this DNA in the first place. Nevertheless, it is a perfectly viable and vigorous fish. Many biologists take this as strong evidence that most of the intronic and intergenic DNA that is so plentiful in other vertebrates—including humans, as we shall discuss—is truly superfluous—that the bulk of our "junk DNA" is actually junk.

Sequence Conservation Allows Us to Trace Even the Most Distant Evolutionary Relationships

As we go back further still to the genomes of even more distant relatives, beyond apes, mice, fish, flies, worms, plants, and yeasts, all the way to bacteria, we find fewer and fewer resemblances that we can recognize in the genomic landscape. Yet even across this enormous evolutionary divide, purifying selection has kept some things almost constant. As we saw in Chapter 1, a core set of a few hundred fundamentally important genes have recognizable homologs in all domains of the living world. By comparing the sequences of these genes in different organisms and seeing how far they have diverged, we can attempt to construct a phylogenetic tree that goes all the way back to the ultimate ancestors—the cells at origins of life from which we all derive.

Figure 9–21 The positions of *Fugu* introns are conserved relative to their positions in mammalian genomes. Comparison of the nucleotide sequences of the human and *Fugu* genes encoding the huntingtin protein. Both genes *(red)* contain 67 short exons that align in 1:1 correspondence with one another; these exons are connected by curved lines. The human gene is 7.5 times larger than the *Fugu* gene (180,000 versus 24,000 nucleotide pairs), due entirely to the presence of larger introns in the human sequence. The larger size of the human introns is a result in part of the presence of transposable elements whose positions are represented by *blue* vertical bars. In humans, mutation of the huntingtin gene causes Huntington's disease, an inherited neurodegenerative disorder. (Adapted from S. Baxendale et al., *Nat. Genet.* 10:67–76, 1995.)

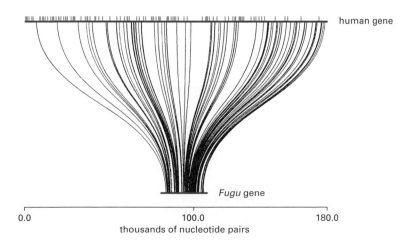

human gene

Fugu gene

| 0.0 | 100.0 | 180.0 |

thousands of nucleotide pairs

```
GTTCCGGGGGGAGTATGGTTGCAAAGCTGAAACTTAAAGGAATTGACGGAAGGGCACCACCAGGAGTGGAGCCTGCGGCTTAATTTGACTCAACACGGGAAACCTCACCC   human
 I  IIIIIIII III  IIII   IIIIIIIIIIIII III  I IIII II I IIIIIII  IIIIIII IIIIII IIII I II III  I I  III III
GCCGCCTGGGGAGTACGGTCGCAAGACTGAAACTTAAAGGAATTGGCGGGGGGAGCACTACAACGGGTGGAGCCTGCGGTTTAATTGGATTCAACGCCGGGCATCTTACCA   Methanococcus
 IIIIIIIIIII IIIIIIIIIIII IIIIIIIIIII IIIIIII IIIIIII I IIII II  IIIIII IIIIIIIIII IIIII  IIIIIII I III IIIIIIII
ACCGCCTGGGGAGTACGGCCGCAAGGTTAAAACTCAAATGAATTGACGGGGGCCCGC.ACAAGCGGTGGAGCATGTGGTTTAATTCGATGCAACGCGAAGAACCTTACCT   E. coli
 I I IIIIIIII III IIIIII  I IIII  IIIII III  II IIIIIIII  I I IIIIIII IIII I IIIIIII  IIIII  II III  IIIII  IIII
GTTCCGGGGGGAGTATGGTTGCAAAGCTGAAACTTAAAGGAATTGACGGAAGGGCACCACCAGGAGTGGAGCCTGCGGCTTAATTTGACTCAACACGGGAAACCTCACCC   human
```

Figure 9–22 Some genetic information has been conserved since the beginnings of life. A part of the gene for the small subunit ribosomal RNA is shown. Corresponding segments of nucleotide sequence from three distantly related species (*Methanococcus jannaschii,* an archaea; *Escherichia coli,* a eubacterium; and *Homo sapiens,* a eucaryote) are aligned in parallel. Sites where the nucleotides are identical between species are indicated by a vertical line; the human sequence is repeated at the bottom of the alignment so that all three two-way comparisons can be seen. A dot halfway along the *E. coli* sequence denotes a site where a nucleotide has been either deleted from the eubacterial lineage in the course of evolution, or inserted in the other two lineages. Note that the sequences from these three distantly related organisms all differ from one another to a roughly similar degree, while still retaining unmistakable similarities.

To construct such a tree, biologists have focused on one particular gene that is conserved in all living species: the gene that codes for one of the ribosomal RNAs that is found in the smaller of the two ribosomal subunits (see Figure 7–28). Because the process of translation is fundamental to all living cells, this component of the ribosome has been well conserved since early in the history of life on Earth (Figure 9–22).

By applying the same principles used to construct the primate family tree (see Figure 9–15), the small subunit rRNA nucleotide sequences have been used to create a single all-encompassing tree of life (Figure 9–23). Although many aspects of this phylogenetic tree were anticipated by classical taxonomy (which is based on the outward appearance of organisms), there were also many surprises. Perhaps the most important was the realization that some of the organisms that were traditionally classed as "bacteria" are as widely divergent in their evolutionary origins as is any procaryote from any eucaryote. As discussed in Chapter 1, it now appears that the procaryotes comprise two distinct groups—the *bacteria* (or eubacteria) and the *archaea* (or archaebacteria)—that diverged early in the history of life on Earth, either before the ancestors of the eucaryotes diverged as a separate group or at about the same time. The living world therefore has three major divisions or *domains:* bacteria, archaea, and eucaryotes (see Figure 9–23).

Although we humans have been classifying the visible world since antiquity, we now realize that most of life's genetic diversity lies in the world of invisible microbes. Traditionally, these organisms have tended to go unnoticed, unless they cause disease or rot the timbers of our houses. Yet microorganisms make up most of the total mass of living matter on our planet. Only now—through the analysis of DNA sequences—are we beginning to get a picture of life on Earth that is not grossly distorted by our biased perspective as large animals living on dry land.

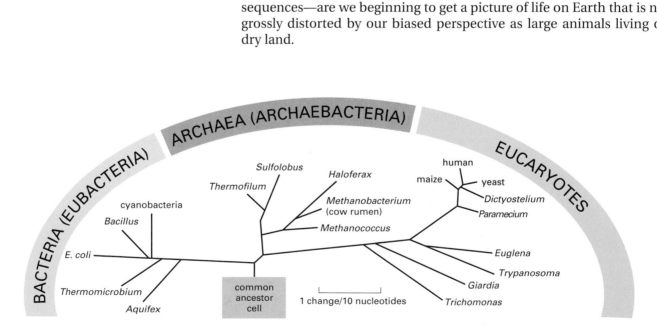

Figure 9–23 The tree of life has three major divisions. Each branch is labeled with the name of a representative member of that group, and the length of the branches corresponds to the degree of difference in their small subunit rRNA sequences (see Figure 9–22). Note that all the organisms we can see with the unaided eye (highlighted in *yellow)*—animals, plants, and fungi—represent only a small subset of the diversity of life.

Examining the Human Genome

We have seen how genomes change gradually over time, and how comparing the genomes of different species can reveal clues to their evolutionary histories. We shall now examine our own genome, which contains numerous clues about where we came from and who we are.

The entire human genome—all 3.2×10^9 nucleotide pairs—is distributed over 22 autosomes and 2 sex chromosomes. The *human genome sequence* refers to the complete nucleotide sequence of the DNA contained in these 24 chromosomes. A wide variety of humans contributed DNA for the genome-sequencing project, and because individual humans differ from one another by an average of 1 nucleotide in 1000, the published human genome sequence is a composite of many individual sequences. Thus the genome sequence, annotated with all of its variations, represents at once both our unity and our diversity as a species.

The sheer quantity of information provided by the Human Genome Project is unprecedented in biology. The human genome is 25 times larger than any other genome sequenced previously (Figure 9–24). At its peak, the Human Genome Project generated raw nucleotide sequences at a rate of 1000 nucleotides per second, around the clock.

It will be many decades before the information contained in our genome is fully analyzed, and the human genome sequence will continue to stimulate many new experiments; it has already affected the content of every chapter in this book. In this section, we describe some of its most striking features.

The Nucleotide Sequence of the Human Genome Shows How Our Genes Are Arranged

When the DNA sequence of human Chromosome 22, one of the smallest human chromosomes, was completed in 1999, it became possible for the first time to see exactly how genes are arranged along an entire vertebrate chromosome (Figure 9–25). With the publication of the "first draft" of the entire human genome in 2001, we got our first panoramic view of the genetic landscape of all human chromosomes (Table 9–1).

The first striking feature of the human genome is how little of it—only a few percent—codes for proteins, structural RNAs, and catalytic RNAs (Figure 9–26). Much of the remaining DNA is made up of transposable elements that have gradually colonized our genome over evolutionary time (discussed in Chapter 6). Although all multicellular eucaryote genomes analyzed to date contain transposable elements, in humans (and in other primates) these parasitic DNA sequences have proliferated extensively, nearly overrunning our chromosomes. Because they have gradually accumulated mutations, most of these transposable elements can no longer move; rather, they are relics from an earlier evolutionary era when the movement of transposons was rampant in our genome.

A second notable feature of the human genome is the very large average gene size of 27,000 nucleotide pairs. Only about 1300 nucleotide pairs are required to encode a protein of average size (about 430 amino acids, in humans). Most of the remaining DNA in a gene consists of long stretches of noncoding DNA that interrupt the relatively short protein-coding exons (see Figure 9–25D).

In addition to the introns and exons, each gene is associated with regulatory DNA sequences that ensure that the gene is expressed at the proper level and time, and in the proper type of cell. In humans, these regulatory sequences are typically spread out over tens of thousands of

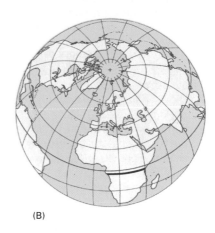

(A)

(B)

Figure 9–24 The human genome is vast. If each nucleotide pair is drawn to span 1 mm as in (A), then the human genome would extend 3200 km (approximately 2000 miles)—far enough to stretch across the center of Africa, the site of our human origins (*red line* in B). At this scale, there would be, on average, a protein-coding gene every 300 m. An average gene would extend for 30 m, but the coding sequences in this gene would add up to only just over a meter.

Question 9–4

Transposable DNA sequences, such as *Alu* sequences, are found in many copies in human DNA. In what ways could the presence of an *Alu* sequence affect a nearby gene?

Figure 9–25 Chromosome 22 was the first human chromosome to have its nucleotide sequence determined.
(A) Chromosome 22 contains 48×10^6 nucleotide pairs and makes up approximately 1.5% of the entire human genome. Most of its left arm consists of short repeated sequences packaged tightly into heterochromatin, as discussed in Chapter 5. (B) A tenfold expansion of a portion of Chromosome 22 shows about 40 genes. Known genes are shown in *dark brown*; predicted genes in *light brown*. (C) An expanded portion of (B) shows the entire length of several genes. (D) The intron–exon arrangement of a typical gene is shown after an additional tenfold expansion. Each exon *(red)* codes for a portion of a protein, while the nucleotide sequence of the introns *(gray)* is relatively unimportant. (Adapted from The International Human Genome Sequencing Consortium, *Nature* 409:860–921, 2001.)

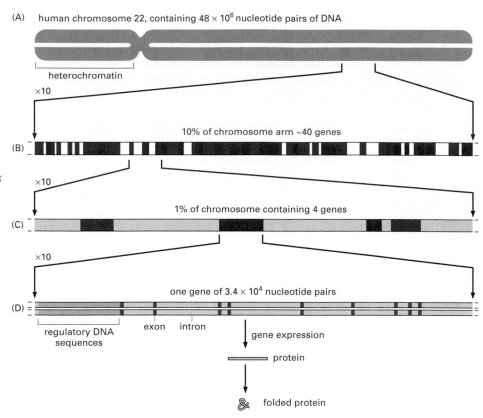

(A) human chromosome 22, containing 48×10^6 nucleotide pairs of DNA

heterochromatin

×10

(B) 10% of chromosome arm ~40 genes

×10

(C) 1% of chromosome containing 4 genes

×10

(D) one gene of 3.4×10^4 nucleotide pairs

regulatory DNA sequences   exon   intron

gene expression

protein

folded protein

Table 9–1 Vital Statistics of Human Chromosome 22 and the Entire Human Genome

	CHROMOSOME 22	HUMAN GENOME
DNA length	48×10^6 nucleotide pairs*	3.2×10^9
Number of genes	approximately 700	approximately 30,000
Smallest protein-coding gene	1000 nucleotide pairs	not analyzed
Largest gene	583,000 nucleotide pairs	2.4×10^6 nucleotide pairs
Mean gene size	19,000 nucleotide pairs	27,000 nucleotide pairs
Smallest number of exons per gene	1	1
Largest number of exons per gene	54	178
Mean number of exons per gene	5.4	8.8
Smallest exon size	8 nucleotide pairs	not analyzed
Largest exon size	7600 nucleotide pairs	17,106 nucleotide pairs
Mean exon size	266 nucleotide pairs	145 nucleotide pairs
Number of pseudogenes**	more than 134	not analyzed
Percentage of DNA sequence in exons (protein-coding sequences)	3%	1.5%
Percentage of DNA in high-copy repetitive elements	42%	approximately 50%
Percentage of total human genome	1.5%	100%

*The nucleotide sequence of 33.8×10^6 nucleotides is known; the rest of the chromosome consists primarily of very short repeated sequences that do not code for proteins or RNA.

**A pseudogene is a nucleotide sequence of DNA closely resembling that of a functional gene, but containing numerous mutations that prevent its proper expression. Most pseudogenes arise from the duplication of a functional gene followed by the accumulation of damaging mutations in one copy.

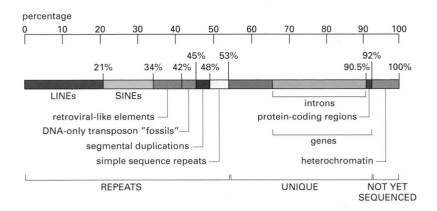

percentage

Figure 9–26 The bulk of the human genome is built of repetitive, noncoding nucleotide sequences. LINEs, SINEs, retroviral-like elements, and DNA-only transposons are all mobile genetic elements that have multiplied in our genome by replicating themselves and inserting the new copies in different positions on our chromosomes. Simple sequence repeats are short nucleotide sequences (less than 14 nucleotide pairs) that are repeated again and again for long stretches. Segmental duplications are large blocks of the genome (1000–200,000 nucleotide pairs) that are present at two or more locations in the genome. More than half of the unique sequence consists of the introns and exons of genes, and a substantial portion of the remainder is probably involved in gene regulation. Most of the DNA present in heterochromatin, which contains relatively few genes, has not yet been sequenced. (Adapted from *Unveiling the Human Genome, Supplement to the Wellcome Trust Newsletter.* London: Wellcome Trust, February 2001.)

nucleotide pairs, most of which is "spacer" DNA. In contrast, organisms with more compact genomes, such as the yeast *S. cerevisiae* (see Figure 10–6), have more compressed regulatory sequences, and most of their genes lack introns altogether.

Perhaps most surprising is the number of genes encoded in the human genome, as judged from analysis of the genome sequence. Earlier estimates had been in the neighborhood of 100,000 (see How We Know, pp. 314–315). Although the exact number is still not certain, revised estimates place the number of human genes at about 30,000, bringing us much closer to the gene numbers for simpler multicellular animals, such as *Drosophila* (14,000) and *C. elegans* (19,000).

Finally, the nucleotide sequence of the human genome has revealed that the critical information it carries seems to be in an alarming state of disarray. As one commentator described our genome, "In some ways it may resemble your garage/bedroom/refrigerator/life: highly individualistic, but unkempt; little evidence of organization; much accumulated clutter (referred to by the uninitiated as 'junk'); virtually nothing ever discarded; and the few patently valuable items indiscriminately, apparently carelessly, scattered throughout."

Genetic Variation Within the Human Genome Contributes to Our Individuality

As we have already emphasized, with the exception of identical twins, no two people have the exact same genome. When the same region of the genome from two different humans is compared, the nucleotide sequences typically differ by about 0.1%. That might seem an insignificant degree of variation, but considering the size of the human genome, that amounts to some 3 million genetic differences in each maternal or paternal chromosome set between one person and the next. Detailed analysis of the data on human genetic variation suggests that the bulk of this variation was already present early in our evolution, perhaps 100,000 years ago, when the human population was still small. This means that a great deal of the genetic variation we possess today was inherited from our early human ancestors.

Most of the genetic variation in the human genome takes the form of single base changes called **single-nucleotide polymorphisms** (SNPs, pronounced "snips"). These polymorphisms are simply points in the genome that differ in nucleotide sequence between one portion of the population and another—positions where one person has an A-T nucleotide pair, for example, while another has a G-C. By the time the final version of the human genome sequence was published, more than 3 million SNPs had been located. These polymorphisms are scattered

How We Know: Counting Genes

How many genes does it take to make a human? It seems a natural thing to wonder. If 6000 genes can produce a yeast, and 14,000 a fly, how many are needed to encode a human being—a complex creature, curious and clever enough to study its own genome? Until researchers completed the first draft of the human genome sequence, the most frequently cited estimate was 100,000. But where did that figure come from? And how was the revised estimate of only 30,000 genes derived?

Walter Gilbert, a physicist-turned-biologist who won a Nobel Prize for developing techniques for sequencing DNA, was one of the first to throw out a ballpark estimate of the number of human genes. In the mid-1980s, Gilbert suggested that humans could have 100,000 genes, an estimate based on the average size of the few human genes known at the time (about 3×10^4 nucleotide pairs) and the size of our genome (3×10^9 nucleotide pairs). This back-of-the-envelope calculation yielded a number with such a pleasing roundness that it wound up being quoted widely in articles and textbooks.

The calculation provides an estimate of the number of genes a human could have in principle, but it does not address the question of how many genes we actually do have. As it turns out, that question is not so easy to answer, even with the complete human genome sequence in hand. The problem is, how does one identify a gene? We learned in Chapter 5 that genes are defined as regions of DNA that determine the characteristics of a cell or organism, and that these DNA segments usually encode a protein or a functional RNA. We now know that such coding segments comprise only a few percent of the human genome. So, looking at a given piece of raw DNA sequence—an apparently random string of As, Ts, Gs, and Cs—how can one tell which parts are genes and which parts are "junk"? Being able to accurately and reliably distinguish the coding sequences from the noncoding sequences in a genome is necessary before one can hope to locate and count its genes.

Signals and chunks

As always, the situation is simplest in procaryotes and simple eucaryotes such as yeasts. To identify genes in such a genome, one essentially searches through the entire DNA sequence looking for open reading frames (ORFs). These are long sequences—say, 100 codons or more—that lack stop codons. A random sequence of nucleotides will by chance encode a stop signal for protein synthesis about once every 20 codons (as there are three stop codons in the set of 64 possible codons). So finding an ORF, a continuous nucleotide sequence that encodes more than 100 amino acids, is the first step in identifying a good candidate for a presumptive gene. Today computer programs are used to search for such ORFs, which begin with an initiation codon, usually ATG, and end with a termination codon, TAA, TAG, or TGA (Figure 9–27).

In animals and plants, the process of identifying ORFs is complicated by the presence of large intron sequences that interrupt the coding portions of genes. As we have seen, these sequences are generally much larger than the exons themselves, which might represent only a few percent of the gene. In human DNA, exons sometimes contain as few as 50 codons (150 nucleotide pairs), while introns may exceed 10,000 nucleotide pairs in length. Fifty codons is too short to generate a statistically significant "ORF signal," as it is not all that unusual for 50 random codons to lack a stop signal. Moreover, introns are so long that they are likely to contain by chance quite a bit of "ORF noise," numerous stretches of sequence lacking stop signals. Finding the true ORFs in this sea of information in which the "noise" often outweighs the "signal" is quite difficult. Thus, to identify genes in eucaryotic DNA, it is also necessary to search for other distinctive features that mark the presence of a gene: splicing sequences that signal an intron–exon boundary or distinctive DNA regulatory sequences that lie upstream of a gene.

One of the most powerful approaches to identify genes is through homology with genes from other organisms. For example, even a very short ORF is likely to be an exon if the amino acid sequence it encodes matches up with a known protein from another organism. In addition, if a presumptive ORF is highly conserved in several different genomes, it is very likely to code for protein, even though the gene that contains it may not yet have been identified or studied in any organism. Through such comparisons (for example, human versus mouse versus zebrafish) it is possible to identify short ORFs of unknown function and, with more work, to piece them together into whole genes.

In 1992, Craig Venter and his colleagues used a computer program to predict protein-coding regions in preliminary human sequence data. The researchers found two genes in a 58,000–nucleotide-pair segment of Chromosome 4, and five genes in a 106,000–nucleotide-pair segment of Chromosome 19—or an average of 1 gene every 23,000 nucleotide pairs. Extrapolating from that density to the whole genome would give humans nearly 130,000 genes. It turned out, however, that the chromosomes the researchers analyzed had been chosen for sequencing precisely because they appeared to be gene-rich. When the estimate was adjusted to take into account the gene-poor regions of the human genome—guessing that half of the human genome had maybe one-tenth that gene density—the number dropped to 71,000.

Matching tags

Of course, these estimates are based on what we think genes look like; however, we are still learning how to recognize them. A different but complementary approach to counting the coding regions in a genome involves determining experimentally how many genes are actually expressed.

To determine which genes are expressed in a particular cell type or tissue, mRNAs from a variety of different tissues are isolated and converted into **complementary DNAs**, or cDNAs (discussed in Chapter 10). Because mRNAs are produced by transcription and splicing from protein-coding genes, this collection of cDNAs represents the sequences of all genes that were being expressed in the cells from which the mRNA was prepared. The cDNAs are prepared from a variety of tissues because the goal of this approach is to identify every gene, and different tissues express different genes. An additional benefit to working with cDNAs stems from the fact that mRNAs lack the introns and the nonessential "spacer" DNA that lies between genes; thus cDNA sequences correspond directly to the coding sequences in the genome.

Short fragments of these cDNAs, called **expressed sequence tags**, or ESTs, are then sequenced; the resulting EST sequences are compared with the nucleotide sequence of the entire genome to locate the gene that contains each. By carefully analyzing how ESTs map onto the human genome, researchers arrived at an estimate of about 30,000 genes.

Human gene countdown

Right now, the most accurate approach to predicting genes combines different types of data, including (1) EST analyses, (2) computerized searches of the genome for ORFs and for sequences that signal the splice sites at the ends of every exon, and (3) comparisons with genome sequences from other organisms, especially the mouse. This last approach is particularly powerful because, as we have seen in this chapter, the mouse and human genomes are sufficiently divergent that only the most crucial portions of their genomes—the exons and regulatory sequences, for example—are similar.

Although the estimates are all converging around 30,000, it could be many years before we have the final answer to how many genes it takes to make a human. As computational methods are refined, and as we collect more data on the human genome and on the genome sequences of other organisms, our ability to predict where genes reside within a particular DNA sequence can only improve. In the end, however, knowing the exact number of genes is not nearly as important as understanding the functions of each gene and how it interacts with other genes to build a human being. These are central questions that are likely to occupy biologists for at least the next century.

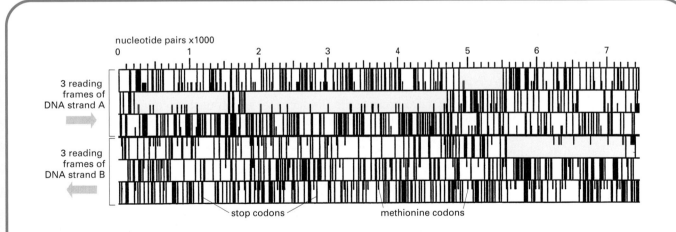

Figure 9–27 **Computer programs are used to find genes.** In this simple example, a DNA sequence of 7500 nucleotide pairs from the pathogenic yeast *Candida albicans* was fed into a computer which then translated the entire sequence in all six possible open reading frames, three from each strand (see Figure 7–22). The output displays each reading frame as a horizontal column, with the stop codons (TGA, TAA, and TAG) marked by the longer vertical lines and methionine codons (ATG) marked by the shorter lines. Four open reading frames (*yellow*) can be clearly identified by the statistically significant absence of stop codons. For each ORF, the presumptive initiation codon (ATG) is indicated in *red*. The additional ATG codons in the ORFs code for methionine.

throughout the genome and more than 90% of all human genes contain at least one SNP. Because SNPs are present at such a high density, they provide useful markers for conducting genetic analyses in which one attempts to link a specific trait (such as disease susceptibility) with a particular pattern of SNPs (see Figure 20–30). This type of analysis may lead to improvements in health care by allowing doctors to determine whether an individual is susceptible to a disease, such as heart disease, long before he or she shows any symptoms. The person can then change his or her behavior to help prevent the disease before it arises.

In addition to the SNPs that we inherited from our ancestors, humans also possess repetitive nucleotide sequences that are particularly prone to new mutations. A dramatic example can be found in the CA repeats, sequences that are ubiquitous in the human genome and in the genomes of other eucaryotes. Nucleotide sequences containing large numbers of CA repeats are replicated poorly because of a slippage that occurs between the template and the newly synthesized DNA strands during replication; hence the precise number of repeats can vary widely from one genome to the next. Because they show such exceptional variability (and because this variability has arisen so late in evolution), these repeats, and others like them, make ideal genetic markers. They can provide the basis for identifying individuals by DNA analysis in crime investigations, paternity suits, and other forensic applications (see Figure 10–30).

Most of the SNPs and other common variations in the human genome sequence are genetically silent. Many, for example, affect the DNA sequences in noncritical regions of the genome. These SNPs therefore have no effect on how we look or how our cells function. However, a tiny subset of SNPs are presumably responsible for nearly all of the heritable aspects of human individuality. As we discuss in Chapter 20, a major challenge in human genetics is to learn to recognize those relatively few variations that are functionally important against the large background of neutral variation that distinguishes the genomes of any two human beings.

Comparing Our DNA with That of Related Organisms Helps Us to Interpret the Human Genome

A major obstacle in interpreting the nucleotide sequences of human chromosomes is the fact that much of the sequence appears unimportant. Moreover, the coding regions of the genome (the exons) are typically found in short segments (average size about 145 nucleotide pairs) floating in a sea of DNA whose exact nucleotide sequence is of little consequence. This arrangement makes it very difficult to identify all of the exons in a genome, and to determine where all of its genes begin and end.

Comparative genome analyses provide a valuable tool for identifying genes as well as functionally important regulatory sequences. The approach works because, as we have seen, sequences that have a function are conserved during evolution, whereas those without a function are free to mutate randomly. One way to identify human genes, therefore, is to compare the human sequence with that of the corresponding regions of a related genome, such as that of the mouse. The common sequences, which have survived over evolutionary timescales (see Figure 9–19), must be critical to both organisms. Such studies have also highlighted regions of conserved synteny, where the mouse and human genomes contain corresponding genes in the same order (Figure 9–28). Thus, knowing the location, and possibly the function, of a gene in one genome consequently makes it easier to identify and predict the function of the corresponding gene in the other genome.

Question 9–5

You are interested in finding out the function of a particular gene in the mouse genome. You have sequenced the gene, defined the portion that codes for its protein product, and searched the database; however, neither the gene nor the encoded protein resembles anything in the database. What (various) types of information about the gene or the encoded protein would you like to know in order to narrow down the possible functions, and why? Focus on the information you want, rather than on the techniques you might use to get that information.

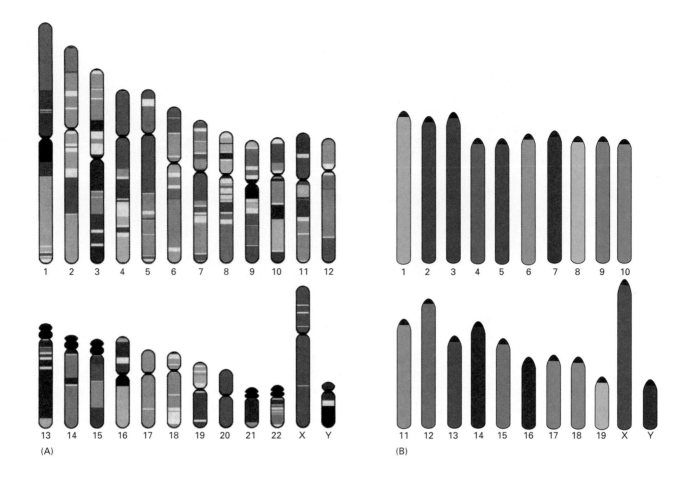

The Human Genome Contains Copious Information Yet to Be Deciphered

Even with the human genome sequence in hand, many questions will continue to challenge cell biologists throughout the next century. Perhaps the most perplexing one is this: given that a human, a chimp, and a mouse contain essentially the same genes, and are therefore formed from the same set of proteins, what makes these creatures so different?

The answer, it seems, will come in large part from studies of gene regulation. The proteins encoded in the genome are like the components of a construction kit. By assembling the components in different combinations, many different things can be built with the same kit. In the end, however, the overall shape of the final object is determined by the instructions that prescribe how the components are to be put together.

To a large extent, the instructions needed to produce a multicellular animal are contained in the noncoding, regulatory DNA that is associated with each gene. As discussed in Chapter 8, this DNA contains, scattered within it, dozens of separate regulatory elements—short DNA segments that serve as binding sites for specific gene regulatory proteins. This regulatory DNA can be said to define the sequential program of development: the rules that cells follow as they proliferate, assess their positions in the embryo, and switch on new sets of genes accordingly (Figure 9–29). If the regulatory DNA sequences are altered, the control proteins will not bind correctly and the adjacent genes will be switched on or off inappropriately. An extreme example that can result from this type of change was shown in Figure 9–12B.

Figure 9–28 The human and mouse genomes contain many regions where the order of genes has been preserved. Regions from different mouse chromosomes (B) show conserved gene order with the indicated regions of the human genome (A). For example, the genes present in the upper portion of human Chromosome 1 (*orange* in A) are present in the same order in a portion of mouse Chromosome 4. Regions of human chromosomes that contain primarily short repeated sequences are shown in *black*. Mouse centromeres (*black* in B) are located at the ends of the chromosomes; human centromeres (indicated by constrictions) tend to occupy more internal positions on the chromosomes. (Adapted from The International Human Genome Sequencing Consortium, *Nature* 409:860–921, 2001.)

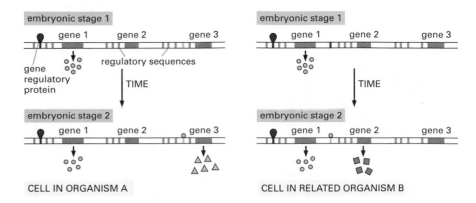

Figure 9–29 Regulatory genes and proteins define an organism's developmental program. The genomes of organisms A and B code for the same set of regulatory proteins, but the regulatory DNA controlling expression of the proteins is different. Both organisms express the same proteins at stage 1, but because of the differences in their regulatory DNA they follow different pathways in stage 2.

Figure 9–30 Alternative splicing of RNA transcripts can produce many distinct proteins. The *Drosophila* DSCAM proteins are receptors that help nerve cells to make the appropriate connections. The final mRNA transcript contains 24 exons, four of which (denoted A, B, C, and D) are present in the *DSCAM* gene as arrays of alternative exons. Each mature mRNA contains 1 of 12 alternatives for exon A *(red)*, 1 of 48 alternatives for exon B *(green)*, 1 of 33 alternatives for exon C *(blue)*, 1 of 2 alternatives for exon D *(yellow)*, and all of the 19 invariant exons *(gray)*. If all possible splicing combinations are used, 38,016 different proteins could in principle be produced from the *DSCAM* gene. Only one of the many possible splicing patterns (indicated by the *red line* and by the mature mRNA below it) is shown. (Adapted from D.L. Black, *Cell* 103:367–370, 2000.)

We are only beginning to be able to recognize and interpret the regulatory elements in our own genome. Their short lengths make them difficult to pick out from the vast excess of "junk" DNA, and the fact that control of gene expression occurs in complex and combinatorial ways (see Figure 8–15) complicates our attempts to decipher when in development and in which type of cell each gene is expressed.

Another challenge in interpreting the information encoded in the human genome comes from the prevalence of *alternative splicing*. We know that most human genes (an estimated 60%) undergo alternative splicing, allowing cells to produce a range of related but distinct proteins from a single gene (see Figure 7–18). Often this splicing is regulated so that one form of the protein is produced in one type of tissue, while other forms are produced preferentially in other tissues. In one extreme case, from *Drosophila*, a single gene may produce as many as 38,000 different protein variants through alternative splicing (Figure 9–30). Thus an organism can produce far more protein products than it has genes. We do not yet know enough about the biology of alternative splicing to predict exactly which human genes are subject to this process and when and where during development such regulation might occur. We also do not understand the extent to which alternative splicing may contribute to the marked differences between a mouse and a human.

The human body is formed as the result of complex sequences of decisions that cells make as they proliferate and specialize during our development: which RNA molecules and which proteins are to be made

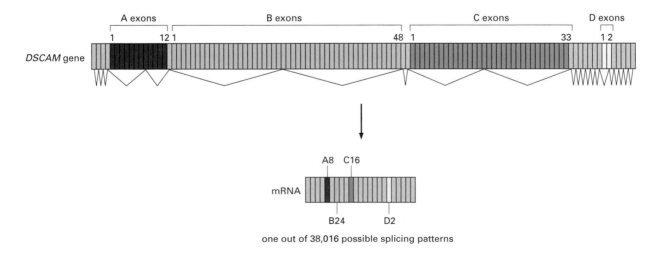

one out of 38,016 possible splicing patterns

where, and exactly when and how much of each is to be produced. The information for all these decisions is contained within the human genome sequence, but we do not yet know how to read this type of code. Deciphering this information is one of the great challenges for the next generation of cell biologists.

Essential Concepts

- The vast diversity of life we see around us has arisen through changes in DNA sequences that have accumulated since the first cells on Earth arose some 3.5 billion years ago.
- Genetic changes that offer an organism a selective advantage or those that are selectively neutral are the most likely to be perpetuated. Changes that seriously compromise an organism's fitness are eliminated through natural selection.
- Genetic variation—the raw material for evolutionary change—occurs by a variety of mechanisms that alter nucleotide sequences, producing simple point mutations or larger-scale duplications or rearrangements in a genome.
- Gene duplication is one of the most important sources of genetic diversity. Once a gene has been duplicated, one of the two gene copies is free to mutate and become specialized to perform a different function. Repeated rounds of such duplication and divergence, over evolutionary time, can result in the creation of large gene families.
- The evolution of new proteins is thought to have been greatly facilitated by the organization of eucaryotic genes as relatively short exons separated by long, noncoding introns. The presence of introns greatly increases the probability that a chance recombination event can generate a functional hybrid gene by joining together two initially separate exons coding for quite different protein domains—a process called exon shuffling.
- By comparing the nucleotide or amino acid sequences of contemporary organisms, we are beginning to be able to reconstruct how genomes have evolved in the billions of years that have elapsed since the appearance of the first cells.
- The human genome contains 3.2×10^9 nucleotide pairs divided among 22 autosomes and 2 sex chromosomes. Only a few percent of that DNA codes for proteins or for structural or catalytic RNAs. Human genes tend to be large, and consist of small, protein-coding exons surrounded by vast noncoding intron sequences.
- Individual humans differ from one another by an average of 1 nucleotide pair in every 1000; this variation underlies our individuality and provides the basis for identifying individuals by DNA analysis.
- Comparative genome analyses provide a valuable tool for identifying genes as well as functionally important regulatory sequences. Such comparisons have revealed that mice and humans share most of the same genes, and that large blocks of the mouse and human genomes contain these genes in the same order.
- One of the great challenges facing biologists is to determine how organisms built from essentially the same set of proteins can be so different. This will require understanding how genes are regulated and alternatively spliced to define each organism's developmental programs.

Questions

Question 9–6

Discuss the following statement: "Transposable elements are parasites. They are harmful to the host organism, and therefore place it at an evolutionary disadvantage."

Question 9–7

On chromosome 22 (48 Mb in length) there are about 700 genes, averaging 19,000 nucleotide pairs in length and containing an average of 5.4 exons, each of which averages 266 nucleotide pairs. On average what fraction of a gene is present in mRNA? What fraction of the chromosome do genes occupy?

Question 9–8

(True/False) The majority of human DNA is thought to be unimportant junk. Explain your answer.

Question 9–9

Transposable elements of four types—long interspersed elements (LINES), short interspersed elements (SINES), retrotransposons, and DNA transposons—are inserted more or less randomly throughout the human genome. However, at the four homeobox gene clusters, HoxA, HoxB, HoxC, and HoxD, these elements are rare, as illustrated for HoxD on chromosome 2 (Figure Q9–9). Each Hox cluster is about 100 kb in length and contains 9 to 11 genes, whose differential expression along the antero-posterior axis of the developing embryo establishes the basic body plan for humans (and for other animals). Why do you suppose that transposable elements are so rare in the Hox clusters? In Figure Q9–9, lines that project *upward* indicate exons of known genes. Lines that project *downward* indicate transposable elements; they are so numerous they merge into nearly a solid block outside of the Hox clusters. For comparison an equivalent region of chromosome 22, which lacks a Hox cluster, is shown.

Question 9–10

The earliest graphical method for comparing nucleotide sequences—the so-called diagon plot—still yields one of the best visual comparisons of sequence relatedness. An example is illustrated in Figure Q9–10, where the human β-globin gene is compared to the human cDNA for β globin (which contains only the coding portion of the gene; Figure Q9–10A) and to the mouse β-globin gene (Figure Q9–10B). Diagon plots are generated by comparing blocks of sequence, in this case blocks of 11 nucleotides at a time. If 9 or more of the nucleotides match, a dot is placed on the diagram at the coordinates corresponding to the blocks being compared. A comparison of all possible blocks generates diagrams such as the ones shown in Figure Q9–10, in which sequence homologies show up as diagonal lines.

- A. From the comparison of the human β-globin gene with the human β-globin cDNA (Figure Q9–10A), deduce the positions of exons and introns in the β-globin gene.
- B. Are the entire exons of the human β-globin gene (indicated by shading in Figure Q9–10B) homologous to the mouse β-globin gene? Identify and explain any discrepancies.
- C. Is there any homology between the human and mouse β-globin genes that lies outside the exons? If so, identify its location and offer an explanation for its preservation during evolution.
- D. Have either of the genes undergone a change of intron length during their evolutionary divergence? How can you tell?

chromosome 22

chromosome 2

100 kb

HoxD cluster

Figure Q9–9

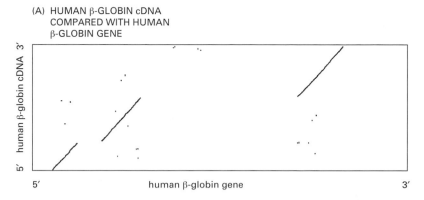

(A) HUMAN β-GLOBIN cDNA COMPARED WITH HUMAN β-GLOBIN GENE

human β-globin cDNA (3′ to 5′)

human β-globin gene (5′ to 3′)

(B) MOUSE β-GLOBIN GENE COMPARED WITH HUMAN β-GLOBIN GENE

mouse β-globin gene (3′ to 5′)

5′ human β-globin gene 3′

Figure Q9–10

Question 9–11

Your advisor, the brilliant bioinformatician, has a high regard for your intellect and industry. She suggests that you write a computer program that will identify the exons of protein-encoding genes directly from the sequence of the human genome. In preparation for that task, you decide to write down a list of the features that might distinguish coding sequences from intronic DNA and sequences outside of genes. What features would you list? (You may wish to review basic aspects of gene expression in Chapter 7.)

Question 9–12

Why do you expect to encounter a stop codon about every 20 codons, or so, on average in a random sequence of DNA?

Question 9–13

Define a "gene."

Question 9–14

The genetic code (see Figure 7–21) specifies the entire set of codons that relate the nucleotide sequence of mRNA to the amino acid sequence of encoded proteins. Ever since the code was deciphered four decades ago, some have claimed it must be a frozen accident, while others have argued that it was shaped by natural selection.

A striking feature of the genetic code is its inherent resistance to the effects of mutation. For example, a change in the third position of a codon often specifies the same amino acid or one with similar chemical properties. But is the natural code more resistant to mutation (less susceptible to error) than other possible versions? The answer is an emphatic "Yes," as illustrated in Figure Q9–14. Only one in a million computer-generated "random" codes is more error-resistant than the natural genetic code.

Does the extraordinary mutation resistance of the genetic code argue in favor of its origin as a frozen accident or as a result of natural selection? Explain your reasoning.

Question 9–15

Which one of the processes listed below is NOT thought to contribute significantly to the evolution of new genes?

A. Duplication of genes to create extra copies that can acquire new function.
B. Formation of new genes *de novo* from noncoding DNA in the genome.
C. Horizontal transfer of DNA between cells of different species.
D. Mutation of existing genes to create new functions.
E. Shuffling of domains of genes by gene rearrangement.

Question 9–16

Some genes evolve rapidly, whereas others are highly conserved. How can we tell whether a gene has evolved rapidly, or simply had a long time to diverge from it relatives? The most reliable approach is to compare several genes from the same two species, as shown for rat and human in the table on page 322. Two measures of rates of nucleotide substitution are indicated in the table. Nonsynonymous changes refer to single nucleotide changes in the DNA sequence that alter the encoded amino acid (ATC → TTC, which gives I → F, for example). Synonymous changes refer to those that do not alter the encoded amino acid (ATC → ATT, which gives I → I, for example). (As is apparent in the genetic code, listed inside the front cover, there are many cases where several codons correspond to the same amino acid.)

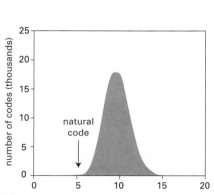

Figure Q9–14

number of codes (thousands)

natural code

susceptibility to mutation

Gene	Amino Acids	Rates of Change	
		Nonsynonymous	Synonymous
Histone H3	135	0.0	4.5
Hemoglobin α	141	0.6	4.4
Interferon γ	136	3.1	5.5

Rates are expressed as nucleotide changes per site per 10^9 years. The average rate of nonsynonymous changes for several dozen rat and human genes is about 0.8.

A. Why are there such large differences between the synonymous and nonsynonymous rates of nucleotide substitution?

B. Considering that the rates of synonymous changes are about the same for all three genes, how is it possible for the histone H3 gene to resist so effectively those nucleotide changes that alter the amino acid sequence?

C. In principle, a gene might be highly conserved because it exists in a 'privileged' site in the genome that is subject to very low mutation rates. What feature of the data in the table argues against this possibility for the histone H3 gene?

Question 9–17

Plant hemoglobins were found initially in legumes, where they function in root nodules to lower the oxygen concentration so that the resident bacteria can fix nitrogen. These hemoglobins impart a characteristic pink color to the root nodules. When these genes were first discovered, it was so surprising to find a gene typical of animal blood that it was hypothesized that the plant gene arose by horizontal transfer from some animal. Many more hemoglobin genes have now been sequenced, and a phylogenetic tree based on some of these sequences is shown in Figure Q9–17.

A. Does this tree support or refute the hypothesis that the plant hemoglobins arose by horizontal gene transfer?

B. Supposing that the plant hemoglobin genes were originally derived from a parasitic nematode, for example, what would you expect the phylogenetic tree to look like?

Question 9–18

The accuracy of DNA replication is such that on average only about 0.6 out of the 6 billion nucleotides in a human cell is altered at each cell division. Because most of our DNA does not code for protein and is not subject to any precise constraint on its sequence, most of these changes are selectively neutral and are as likely as not to be conserved by evolution. Any two modern humans picked at random will show about 1 difference of nucleotide sequence in every 1000 nucleotides. Suppose we are all descended from a single pair of ancestors—Adam and Eve—who were genetically identical and homozygous (had inherited identical maternal and paternal gene copies). How many cell generations must have elapsed since their days for this number of differences to have accumulated in the lineages leading to the modern descendants? Assuming that each human generation corresponds on average to 200 cell-division cycles in the germ-cell lineage; allowing 30 years per human generation, how many years ago would this ancestral couple have lived?

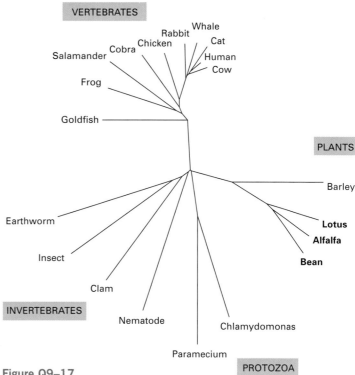

Figure Q9–17

Highlight from *Essential Cell Biology 2 Interactive* CD-ROM

9.1   Conjugation

Manipulating Genes and Cells

The twenty-first century promises to be a particularly exciting time for cell biology. New methods for analyzing and manipulating DNA, RNA, and proteins are fueling an information explosion and allowing us to study genes and cells in previously unimagined ways. We now have access to the sequences of many billions of nucleotides, providing the molecular blueprints for dozens of organisms—from microbes and plants to insects, humans, and other mammals. And powerful new techniques are helping us to decipher this information, allowing us not only to compile huge, detailed catalogs of genes and proteins, but to begin to unravel how these components work together to form functional cells and organisms. The goal is nothing short of obtaining a complete understanding of what takes place inside cells as they respond to their environment and interact with their neighbors.

This technological revolution has been powered, in large part, by the development of methods that have dramatically increased our ability to handle DNA. In the early 1970s, it became possible, for the first time, to isolate a given piece of DNA out of the many millions of nucleotide pairs in a typical chromosome. This in turn made it possible to generate new DNA molecules in the test tube and to introduce this custom-made genetic material back into living organisms. These developments, called variously "recombinant DNA," "gene splicing," or "genetic engineering," make it possible to create chromosomes with combinations of genes that could never have formed naturally—or combinations that could conceivably occur in nature but might take thousands of years of chance events to come together.

Of course, humans have been experimenting with DNA, albeit without realizing it, for millennia. Modern garden-rose varieties, for example, are the product of centuries of selective breeding between strains of wild roses from China and Europe (Figure 10–1A). Similarly, the enormous variation in the sizes, colors, shapes, and even behaviors of different breeds of dogs is the result of deliberate breeding experiments—with selection for desired traits—carried out since the gray wolf, the ancestor of the modern dog, was first domesticated some 10,000–15,000 years ago (Figure 10–1B).

Modern genetic engineering techniques, however, allow us to alter DNA with much greater precision and speed. Now even a beginning student can isolate a region of DNA containing a specific gene from a genome, produce a virtually unlimited number of exact copies of this DNA, and determine its nucleotide sequence with ease. Using variations of these techniques, the isolated gene can be redesigned in the laboratory and then transferred back into cells in culture to elucidate its function. With more sophisticated techniques, the redesigned genes

Isolating Cells and Growing Them in Culture
A Uniform Population of Cells Can Be Obtained from a Tissue
Cells Can Be Grown in a Culture Dish
Maintaining Eucaryotic Cells in Culture Poses Special Challenges

How DNA Molecules Are Analyzed
Restriction Nucleases Cut DNA Molecules at Specific Sites
Gel Electrophoresis Separates DNA Fragments of Different Sizes
The Nucleotide Sequence of DNA Fragments Can Be Determined
Genome Sequences Are Searched to Identify Genes

Nucleic Acid Hybridization
DNA Hybridization Facilitates the Diagnosis of Genetic Diseases
Hybridization on DNA Microarrays Monitors the Expression of Thousands of Genes at Once
In Situ Hybridization Locates Nucleic Acid Sequences in Cells or on Chromosomes

DNA Cloning
DNA Ligase Joins DNA Fragments Together to Produce a Recombinant DNA Molecule
Recombinant DNA Can Be Copied Inside Bacterial Cells
Specialized Plasmid Vectors Are Used to Clone DNA
Human Genes Are Isolated by DNA Cloning
cDNA Libraries Represent the mRNA Produced by a Particular Tissue
The Polymerase Chain Reaction Amplifies Selected DNA Sequences

DNA Engineering
Completely Novel DNA Molecules Can Be Constructed
Rare Cellular Proteins Can Be Made in Large Amounts Using Cloned DNA
Engineered Genes Can Reveal When and Where a Gene Is Expressed
Mutant Organisms Best Reveal the Function of a Gene
Animals Can Be Genetically Altered
Transgenic Plants Are Important for Both Cell Biology and Agriculture

Figure 10–1 Humans have been experimenting with DNA for millennia. (A) The oldest known depiction of a rose in Western art, from the palace of Knossos in Crete, around 2000 BC. Modern roses are the result of centuries of breeding between such wild roses. (B) A poodle and a pug illustrate the range of dog breeds. All dogs, regardless of breed, belong to a single species. (B, courtesy of Heather Angel.)

(A) (B)

can be inserted into animals and plants, so that they become a functional and heritable part of the organism's genome.

These technical breakthroughs have had a dramatic impact on all aspects of cell biology. They have made possible our present knowledge of the organization and evolutionary history of the complex genomes of eucaryotes (as discussed in Chapter 9) and led to the discovery of whole new classes of genes and proteins. They provide new means to determine the functions of proteins and of individual domains within proteins, revealing a host of unexpected relationships among them. And they give biologists an important set of tools for unraveling the mechanisms by which a whole animal or plant can develop from a single cell.

Recombinant DNA technology has also had a profound influence on many aspects of human life outside of scientific research: it is used to detect the mutations in DNA that are responsible for inherited diseases and to diagnose an individual's predisposition to genetic diseases, such as cancer; it is used in forensic science to identify or acquit possible suspects in a crime; it is used to produce an increasing number of human pharmaceuticals, including insulin for diabetics and the blood-clotting protein Factor VIII for hemophiliacs. Even our laundry detergents, which contain heat-stable proteases that digest food spills and blood spots, make use of the products of DNA technology. Of all the discoveries described in this book, those of DNA technology are likely to have the greatest impact on our everyday lives.

In this chapter we describe the methods commonly used to manipulate genes, proteins, and cells. We begin with a presentation of the techniques used for isolating cells and keeping them alive outside the body. These cells then serve as experimental subjects and as sources of the DNA, proteins, and other macromolecules that we wish to analyze and understand. We then survey the principal methods of the revolutionary field of **recombinant DNA technology,** beginning with a discussion of the basic techniques of DNA analysis. Next we describe how DNA sequences can be isolated and produced in large numbers by the techniques of *DNA cloning* and the *polymerase chain reaction (PCR)*, and how these sequences can be used to produce and study proteins. In the final section of the chapter we look at the application of DNA technology to *genetic engineering,* the genetic manipulation of cells and whole organisms.

Isolating Cells and Growing Them in Culture

Cells are small but complex. It is hard to see their structure, hard to discover their molecular composition, and harder still to find out how their various components work together to allow them to survive, grow, and divide. As we saw in Chapter 1, the organelles and the largest molecules in a single cell can be visualized with microscopes; however, truly understanding how cells work requires detailed biochemical studies of their individual components.

Most of these biochemical procedures require large numbers of cells, which are then physically disrupted to isolate their components (as illustrated in Panel 4–3, pp. 160–161). If the sample is a piece of tissue, composed of different types of cells, components of all of its different cell populations will be mixed together. The resulting sample will contain a confusion of macromolecules from a variety of different cell types. To preserve the character of each individual cell type, biologists have developed ways of dissociating cells from tissues and separating the various types that are present. This manipulation results in relatively homogeneous populations of cells that can then be analyzed directly. In many cases, the number of cells can be greatly increased by allowing them to proliferate in the laboratory as pure cultures. We begin this section by describing a few common cell-separation techniques and then discuss how cells can be coaxed to proliferate in culture.

A Uniform Population of Cells Can Be Obtained from a Tissue

Several approaches can be used to separate a particular type of cell from the cells that surround it in the body. If the cells are part of a compact tissue, they must first be dissociated from each other. This is often accomplished using proteolytic enzymes and other agents that disrupt the adhesive bonds between cells. The tissue can then be teased apart into single living cells by gentle agitation.

Next, the different types of cells in the tissue must be isolated from each other. One of the most sophisticated methods makes use of antibodies that bind specifically to the surface of only one cell type in the tissue. These antibodies might recognize, for example, a specific protein that is expressed on the surface of only one particular type of cell. If such an antibody is chemically coupled to a fluorescent dye, the fluorescently labeled cells can be separated from the unlabeled ones in an electronic *fluorescence-activated cell sorter*. In this remarkable machine, individual cells traveling in single file in a fine stream pass through a laser beam and the fluorescence of each cell is measured. Slightly farther downstream, tiny droplets, most containing either one cell or no cells, are formed by a vibrating nozzle. The droplets containing a single cell are automatically given a positive or a negative charge at the moment of formation, depending on whether the cell they contain is fluorescent; they are then deflected by a strong electric field into an appropriate container. Such machines can select 1 cell in 1000 and sort several thousand cells each second (Figure 10–2).

Cells Can Be Grown in a Culture Dish

Given the appropriate surroundings, most plant and animal cells will live, proliferate, and even express specialized properties in a tissue-culture dish. The cells can be watched continuously under the microscope or analyzed biochemically, and the effects of adding or removing specific molecules, such as hormones or growth factors, can be explored. In addition, by mixing two cell types, the interactions between one cell type and another can be studied. Experiments performed using cultured cells are sometimes said to be carried out *in vitro* (literally, "in glass") to contrast them with experiments on intact organisms, which are said to be carried out *in vivo* (literally, "in the living"). These terms can be confusing, however, because they are often used in a very different sense by biochemists. In the biochemistry laboratory, *in vitro* refers to reactions carried out in a test tube in the absence of cells, whereas *in vivo* refers to any reaction taking place inside a living cell, even cells that are growing in culture.

Although not true for all types of cells, most cells grown in culture display the differentiated properties appropriate to their origin: fibroblasts, the precursor cells that give rise to connective tissue, continue to secrete collagen; cells derived from embryonic skeletal muscle fuse to form muscle fibers that spontaneously contract in the culture dish; nerve cells extend axons that are electrically excitable and make synapses with other nerve cells; and epithelial cells form extensive sheets with many of the properties of an intact epithelium (Figure 10–3). Because these phenomena occur in culture, they are accessible to study in ways that are often not possible in intact tissues.

Maintaining Eucaryotic Cells in Culture Poses Special Challenges

Many vertebrate cells stop proliferating after a finite number of cell divisions in culture. Normal human fibroblasts, for example, typically divide only 25–40 times in culture. Like most human somatic cells, these cells do not express the enzyme telomerase, whose job it is to renew the ends of chromosomes at each cell division (see Figure 6–18). As a result, the chromosomes of human fibroblasts progressively shrink at each cell division, and cell division stops when critical information is lost from the ends of chromosomes. This feature ensures that, in the body, fibroblasts and other somatic cells do not divide indiscriminately and develop into cancerous cells. These cells, however, can be coaxed to proliferate indefinitely in the laboratory by providing them with the gene that encodes the catalytic subunit of telomerase. Cells that can divide indefinitely as the result of a genetic change are said to be **immortalized** and can be propagated in culture as a **cell line**. Such cell

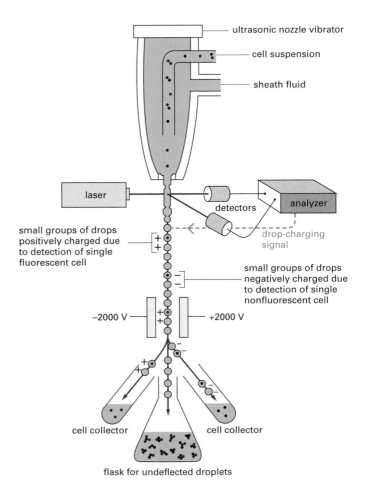

Figure 10–2 A fluorescence-activated cell sorter allows the isolation of specific types of cells. A cell passing through the laser beam is monitored for fluorescence. Droplets containing single cells are given a negative or positive charge, depending on whether the cell is fluorescent or not. The droplets are then deflected by an electric field into collection tubes according to their charge. Note that the cell concentration must be adjusted so that most droplets contain no cells and flow to a waste container. Occasional clumps of cells, detected by their increased light scattering, are left uncharged and are also discarded. The same apparatus can also be used to separate fluorescently labeled chromosomes from one another, providing valuable starting material for the isolation and mapping of genes.

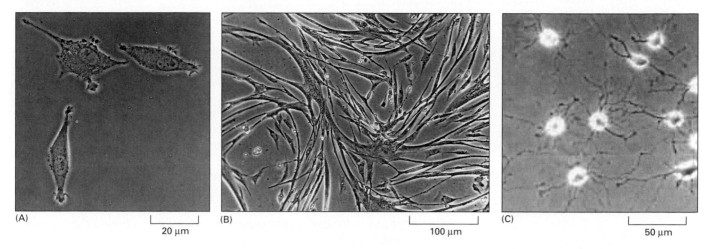

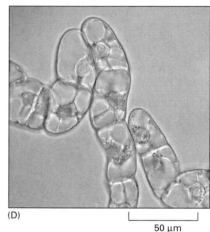

(A)   20 μm   (B)   100 μm   (C)   50 μm

(D)   50 μm

Figure 10–3 Cells in culture often display properties that reflect their origin. (A) Phase-contrast micrograph of fibroblasts in culture. (B) Micrograph of cultured myoblasts, the precursor cells that give rise to muscle, shows cells fusing to form multinucleate muscle cells. (C) Cultured precursor cells that give rise to oligodendrocytes, the glial cells that support and nurture neurons in the brain. (D) Tobacco cells, from an immortal cell line, grown in liquid culture. (A, courtesy of Daniel Zicha; B, courtesy of Rosalind Zalin; C, from Tang et al., *J. Cell Biol.* 148:971–984, 2000; D, courtesy of Gethin Roberts.)

lines can be maintained for many years, and they provide a convenient and widely used source of homogeneous cells.

Among the most promising cell lines to be developed—particularly from a medical point of view—are the human *embryonic stem (ES) cell lines*. These cells, harvested from the inner cell mass of the early embryo, can proliferate indefinitely in culture. The critical importance of these cell lines lies in the fact that the cells are undifferentiated; yet, given the appropriate treatment, they can give rise to any tissue in the body. As discussed in Chapter 21, ES cells could potentially revolutionize medicine by providing a source of cells capable of replacing or repairing tissues that have been damaged by injury or disease.

How DNA Molecules Are Analyzed

Until the development of recombinant DNA techniques, crucial clues for understanding how cells work remained locked in the genome. Once scientists realized that genetic information was encoded in the sequence of nucleotides in DNA, they wanted to closely examine this DNA in order to discover what genes look like and how they function. Before the revolution in DNA technology that took place in the 1970s, this task was almost impossible. Important advances in understanding gene structure and regulation had been made by indirect genetic means in "model" organisms such as *E. coli* and *Drosophila,* but the goal of isolating and closely examining a single gene from a large chromosome seemed unattainable. Unlike a protein, a gene does not exist as a discrete entity in cells, but rather as a part of a much larger DNA molecule. Even bacterial genomes, which are much less complex than the chromosomes of eucaryotes, are enormously long. The *E. coli* genome, for example, contains 4.6 million nucleotides.

Large pieces of DNA can be broken into small pieces by mechanical shear; however, the fragment containing a particular gene will still be only one among a hundred thousand or more DNA fragments that would be obtained, for example, from a mammalian genome by these means. And in a sample containing many identical copies of the same large DNA molecule, each molecule would be broken up differently by shear, producing a confusing set of random fragments. How then, can a gene be isolated and purified?

The solution to this problem emerged with the discovery of a class of bacterial enzymes known as *restriction nucleases*. A nuclease catalyzes the hydrolysis of a phosphodiester bond in a nucleic acid. But these enzymes have a property that is distinct from other nucleases: they cut double-stranded DNA only at particular sites, determined by a short sequence of nucleotide pairs. Restriction nucleases can therefore be used to produce a reproducible set of specific DNA fragments from any genome. We begin this section by describing how these enzymes work and how the DNA fragments produced by them can be separated from each other. We then explain how the order of nucleotides (the DNA sequence) in a DNA fragment that has been isolated in this way can be determined.

Restriction Nucleases Cut DNA Molecules at Specific Sites

Like most of the tools of DNA technology, restriction nucleases were discovered by researchers studying a specialized biological problem that had gripped their interest. Certain bacteria, it was noticed, always degraded "foreign" DNA that was introduced into them experimentally. A search for the cause of this degradation revealed a novel class of nucleases present inside the host bacterium. The most important feature of these nucleases is that they cleave DNA only at certain nucleotide sequences. The bacterium's own DNA is protected from cleavage by chemical modification of these same sequences. Because these enzymes restricted the transfer of DNA between certain strains of bacteria, the name **restriction nuclease** was given to them. Different bacterial species contain different restriction nucleases, each cutting at a different, specific sequence of nucleotides.

The restriction nucleases used in DNA technology come mainly from bacteria, and because their target sequences are short—generally 4–8 nucleotide pairs—sites of cleavage will occur, purely by chance, in any long DNA molecule. Thus restriction nucleases can be used to analyze DNA from any source. The main reason they are so useful is that a given enzyme will always cut a given DNA molecule at the same sites. Thus, for a sample of DNA from a human, treatment with a given restriction nuclease will always produce the same set of DNA fragments. Restriction nucleases are now a hot commodity in DNA technology and are typically ordered through the mail; one supply catalog lists hundreds of such enzymes, each able to cut a different DNA sequence. A few examples are shown in Figure 10–4.

The target sequences of restriction nucleases vary in the frequency with which they will occur in DNA. As shown in Figure 10–4, the enzyme Hae III cuts at a sequence of four nucleotide pairs; this sequence would be expected to occur purely by chance approximately once every 256 nucleotide pairs (1 in 4^4). By similar reasoning, the enzyme Not I, which has a target sequence of eight nucleotides, would be expected to cleave DNA on average once every 65,536 nucleotide pairs (1 in 4^8). The average sizes of the DNA fragments produced by different restriction nucleases can thus be very different. This feature makes it possible to cleave a long DNA molecule into the fragment sizes that are most suitable for a given application.

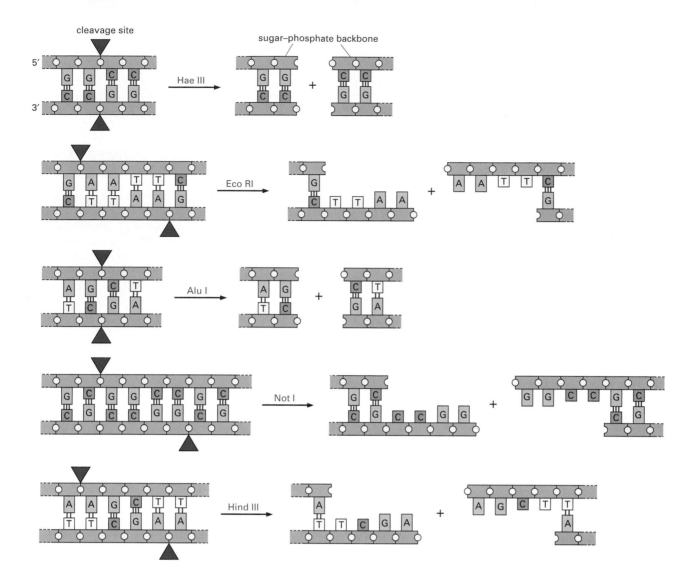

Gel Electrophoresis Separates DNA Fragments of Different Sizes

After a large DNA molecule is cleaved into smaller pieces using a restriction nuclease, it is often desirable to separate the DNA fragments from one another. This is usually accomplished using gel electrophoresis, which separates the fragments on the basis of their length. The mixture of DNA fragments is loaded at one end of a slab of agarose or polyacrylamide gel, which contains a microscopic network of pores. A voltage is then applied across the gel slab. Because DNA is negatively charged, the fragments migrate toward the positive electrode; the larger fragments migrate more slowly because their progress is more impeded by the agarose matrix. Over several hours, the DNA fragments become spread out across the gel according to size, forming a ladder of discrete bands, each composed of a collection of DNA molecules of identical length (Figure 10–5A). Isolating a particular DNA fragment is fairly simple: a small section of the gel containing the band can be cut out using a scalpel or a razor blade.

DNA bands on agarose or polyacrylamide gels are invisible unless the DNA is labeled or stained in some way. One sensitive method of staining DNA is to expose it to a dye that fluoresces under ultraviolet light when it is bound to DNA (Figure 10–5B). An even more sensitive detection method involves incorporating a radioisotope into the DNA

Figure 10–4 Restriction nucleases cleave DNA at specific nucleotide sequences.
Target sequences are often palindromic (that is, the nucleotide sequence is symmetrical around a central point). In these examples, both strands of DNA are cut at specific points within the target sequence. Some enzymes, such as Hae III and Alu I, cut straight across the DNA double helix and leave two blunt-ended DNA molecules; for others, such as Eco RI, Not I, and Hind III, the cuts on each strand are staggered. These staggered cuts generate "sticky ends," short, single-stranded overhangs that help the cut DNA molecules join back together through complementary base-pairing. This rejoining of DNA molecules becomes important for DNA cloning, as we shall discuss later in the chapter. Restriction nucleases are usually obtained from bacteria, and their names reflect their origins: for example, the enzyme Eco RI comes from *Escherichia coli*.

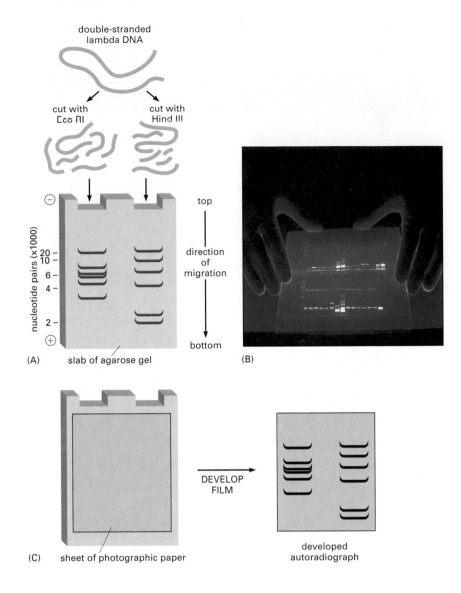

Figure 10–5 DNA molecules can be separated by size using gel electrophoresis. (A) This schematic illustration compares the results of cutting the same DNA molecule (in this case the genome of a bacteria-infecting virus called lambda) with two different restriction nucleases—Eco RI *(left)* and Hind III *(right)*. The fragments are then separated by gel electrophoresis. The mixture of DNA fragments obtained from treatment with each enzyme is placed at the top of a thin gel slab, and under the influence of an electric field, the fragments move through the gel toward the positive electrode. Larger fragments migrate more slowly than smaller fragments, and thus the fragments in the mixture become separated by size. For example, the two lowermost bands in the lane on the right correspond to the two smallest DNA fragments produced by Hind III digestion. To visualize the DNA bands, the gel is soaked in a dye that binds to DNA and fluoresces brightly under ultraviolet light (B). (C) An alternative method for visualizing the DNA bands is autoradiography. Prior to cleavage with restriction enzymes, the DNA can be "labeled" with the radioisotope ^{32}P by substituting ^{32}P for some of the nonradioactive phosphorus atoms. This could be done, for example, by replicating the virus in the presence of ^{32}P. Because the β particles emitted from ^{32}P will expose photographic film, a sheet of film placed flat on top of the agarose gel will, when developed, show the position of all the DNA bands. (B, courtesy of Science Photo Library.)

Question 10–1

Which products are produced when the following piece of double-stranded DNA is digested with (A) Eco RI, (B) Alu I, (C) Not I, or (D) all three of these enzymes together? (See Figure 10–4 for the target sequences of these enzymes.)

molecules before electrophoresis; ^{32}P is often used, as it can be incorporated into the phosphates of DNA and emits an energetic β particle that is easily detected by the technique of autoradiography (Figure 10–5C).

One of the earliest applications of restriction nuclease cleavage followed by the separation of individual DNA fragments was in the construction of physical maps of small regions of DNA. A physical map characterizes a stretch of DNA by charting the position of various landmarks present along it; restriction nuclease cleavage sites are one type of landmark. By comparing the sizes of the DNA fragments produced from a particular region of DNA after treatment with different combinations of restriction nucleases, a physical map of the region can be constructed showing the location of each cutting site along the DNA molecule. Such a map is known as a **restriction map**.

Of course, the ultimate physical map of DNA is its complete nucleotide sequence, and in the next section we see how this goal is achieved.

5′-AAGAATTGCGGAATTCGAGCTTAAGGGCCGCGCCGAAGCTTTAAA-3′
5′-TTCTTAACGCCTTAAGCTCGAATTCCCGGCGCGGCTTCGAAATTT-3′

The Nucleotide Sequence of DNA Fragments Can Be Determined

In the late 1970s, researchers developed methods that allow the nucleotide sequence of any purified DNA fragment to be determined simply and quickly. These techniques have made it possible to determine the complete nucleotide sequences of hundreds of thousands of genes and the complete genome sequences of many organisms, including the budding yeast *Saccharomyces cerevisiae* (Figure 10–6), the nematode worm *Caenorhabditis elegans*, the fruit fly *Drosophila melanogaster* (see Figure 5–15), the model plant *Arabidopsis thaliana*, and the human (discussed in Chapter 9). The volume of DNA sequence information is now so large (many tens of billions of nucleotides) that powerful computers and highly sophisticated software are used to organize and analyze it. Because each organism's genome sequence specifies all of the possible RNA and protein molecules used to construct that organism, having this information in hand greatly expedites future studies of these organisms.

Several schemes for sequencing DNA have been developed, but the most widely used is the **dideoxy method**, which is based on DNA synthesis carried out *in vitro* in the presence of chain-terminating dideoxyribonucleoside triphosphates (Figure 10–7A). In this technique, DNA polymerase is used to make partial copies of the DNA fragment to be sequenced. These DNA replication reactions are performed under conditions that ensure that the new DNA strands terminate when a given nucleotide (A, G, C, or T) is reached (Figure 10–7B). As illustrated in Figure 10–7C, this method produces a collection of different DNA copies that terminate at every position in the original DNA, and thus differ in length by a single nucleotide. These DNA copies can be separated on the basis of their length by gel electrophoresis, and the nucleotide sequence of the original DNA can be determined from the order of these DNA fragments in the gel.

Although the same basic method is still used today, many improvements have been made. DNA sequencing is now completely automated: robotic devices mix the reagents and then load, run, and read the order of the nucleotide bases from the gel. This process is facilitated by the use of chain-terminating nucleotides that are each tagged with a different-colored fluorescent dye; all four synthesis reactions can thus be performed in the same tube, and the products can be separated in a single lane on a gel. A detector positioned near the bottom of the gel reads and records the color of the fluorescent label on each band as it moves past, and a computer stores the sequence for subsequent analysis (Figure 10–8). The complete genome sequence is then assembled from this raw

Figure 10–6 The complete nucleotide sequence of a simple eucaryote, the budding yeast *S. cerevisiae*, was finished in 1996. The complete genome sequence, written out start to finish, would occupy about 4000 pages of text. Shown here are two representations of this vast amount of information. (A) The genome is distributed over 16 chromosomes, and its nucleotide sequence was determined by a cooperative effort involving scientists working in many different locations, as indicated (*gray*, Canada; *orange*, European Union [including scientists from many different European countries]; *yellow*, United Kingdom; *blue*, Japan; *light green*, St. Louis, Missouri; *dark green*, Palo Alto, California). The constriction present on each chromosome represents the position of its centromere. The small region of Chromosome 11 highlighted in *red* is magnified in (B) to show the high density of genes characteristic of this species. As indicated, some genes are transcribed from the lower strand, while others are transcribed from the upper strand. There are about 6200 genes contained in the complete genome, which is 12,147,813 nucleotide pairs long.

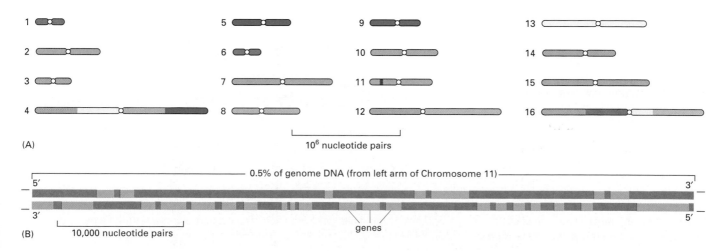

(A)

10^6 nucleotide pairs

0.5% of genome DNA (from left arm of Chromosome 11)

5′   3′

3′   5′

(B)   10,000 nucleotide pairs   genes

Figure 10–7 The enzymatic or dideoxy method is the most commonly used technique for sequencing DNA.

(A) This method relies on the use of dideoxyribonucleoside triphosphates, derivatives of the normal deoxyribonucleoside triphosphates that lack the 3′ hydroxyl group. (B) Purified DNA is synthesized *in vitro* in a mixture that contains single-stranded molecules of the DNA to be sequenced *(gray)*, the enzyme DNA polymerase, a short primer DNA *(orange)* to enable the polymerase to start replication, and the four deoxyribonucleoside triphosphates (dATP, dCTP, dGTP, dTTP: *blue* A, C, G, and T). If a dideoxyribonucleoside analog *(red)* of one of these nucleotides is also present in the nucleotide mixture, it becomes incorporated into a growing DNA chain. The chain now lacks a 3′-OH group, the addition of the next nucleotide is blocked, and the DNA chain terminates at that point. In the example illustrated, a small amount of dideoxyATP (ddATP, symbolized here as a *red* A) has been included in the nucleotide mixture. It competes with an excess of the normal deoxyATP (dATP, *blue* A), so that ddATP is occasionally incorporated, at random, into a growing DNA strand. This reaction mixture will eventually produce a set of DNAs of different lengths complementary to the template DNA that is being sequenced and terminating at each of the different As. (C) To determine the complete sequence of a DNA fragment, the double-stranded DNA is first separated into its single strands and one of the strands is used as the template for sequencing. Four different chain-terminating dideoxyribonucleoside triphosphates (ddATP, ddCTP, ddGTP, ddTTP, again shown in *red*) are used in four separate DNA synthesis reactions on copies of the same single-stranded DNA template *(gray)*. Each reaction produces a set of DNA copies that terminate at different points in the sequence. The products of these four reactions are separated by electrophoresis in four parallel lanes of a polyacrylamide gel (labeled here A, T, C, and G). The newly synthesized fragments are detected by a label (either radioactive or fluorescent) that has been incorporated either into the primer or into one of the deoxyribonucleoside triphosphates used to extend the DNA chain. In each lane, the bands represent fragments that have terminated at a given nucleotide (e.g., A in the leftmost lane) but at different positions in the DNA. By reading off the bands in order, starting at the bottom of the gel and working across all lanes, the DNA sequence of the newly synthesized strand can be determined. The sequence is given in the *green arrow* to the right of the gel. This sequence is identical to that of the 5′ → 3′ strand *(green)* of the original double-stranded DNA.

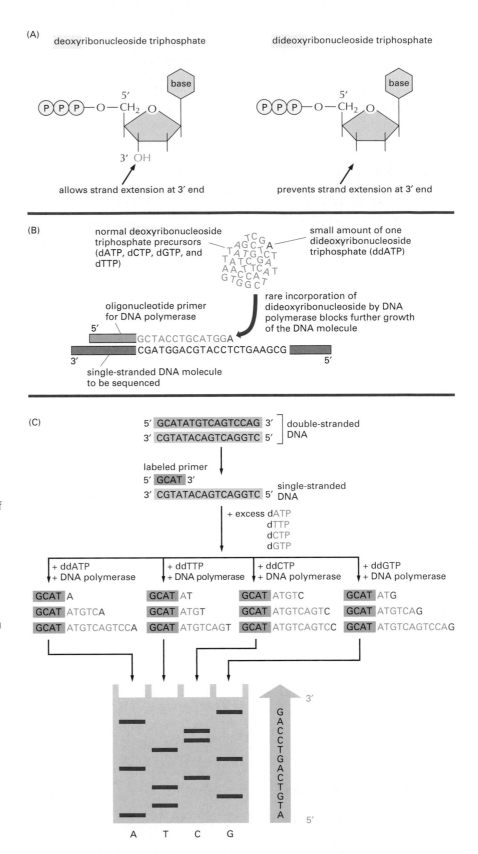

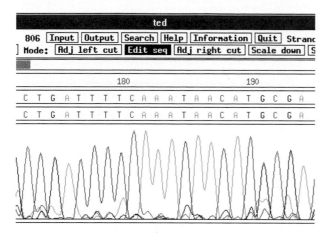

Figure 10–8 DNA sequencing is now fully automated. Shown here is a tiny part of the data from an automated sequencing run as it appears on the computer screen. Each colored peak represents a nucleotide of the DNA sequence; a clear stretch of nucleotide sequence can be read between positions 173 and 194 relative to the start of the sequence. This example is taken from the international project that determined the complete nucleotide sequence of the genome of the plant *Arabidopsis*. (Courtesy of George Murphy.)

sequence by piecing together large stretches of sequence from a set of smaller fragments. These fragments typically have some overlapping sequence, which allows them to be linked up with their neighbors. This vast assembly process is described in How We Know, pp. 334–335.

Genome Sequences Are Searched to Identify Genes

A genome sequence is not an end, but a beginning. Once the complete nucleotide sequence of a genome has been obtained, the challenge becomes identifying its resident genes and determining how they function. The difficult process of interpreting a genome sequence by locating its genes and assigning functions to them is called **annotation**.

Identifying genes is easiest when the DNA sequence is from a simple genome, such as a bacterial chromosome, that lacks introns and other nonessential DNA. The location of genes in these nucleotide sequences can be predicted by examining the DNA for certain telltale features of genes. For example, genes that encode proteins can be identified by searching the nucleotide sequence for open reading frames (ORFs) that begin with an initiation codon (usually ATG) and end with a termination codon (TAA, TAG, or TGA). To minimize errors, computers searching for ORFs are often directed to count as genes only those sequences that are more than, say, 100 codons in length (see How We Know, pp. 314–315).

For more complex genomes, such as those of plants and vertebrates, the annotation process is severely complicated by the presence of large introns embedded within the coding portion of genes and by large stretches of DNA that lie between genes. In many multicellular organisms, including humans, the coding regions of the genome—the exons—are relatively short and are typically embedded in vast stretches of DNA whose exact nucleotide sequence is not critical. This arrangement makes it difficult to identify all of the exons in a stretch of DNA sequence; determining where a gene begins and ends and how many exons it has is even harder. To locate genes in eucaryotic DNA, one must also search for additional features that signal the presence of a gene, for example, sequences that mark an intron–exon boundary (see Figure 7–15) or distinctive gene regulatory sequences of the sort discussed in Chapter 8.

As we saw in Chapter 9, comparison of DNA sequences from related organisms also helps to uncover coding sequences and other sequences important for gene regulation. Nucleotide sequences that have a function are conserved during evolution, whereas those without a function are free to mutate randomly. Consequently, chromosomal regions that contain functionally important exons and regulatory sequences tend to be conserved throughout evolution; nonconserved

Question 10–2

What are the consequences for a DNA sequencing reaction if the ratio of dideoxyribonucleoside triphosphates to deoxyribonucleoside triphosphates is increased? What happens if this ratio is decreased?

How We Know: Sequencing the Human Genome

When DNA sequencing techniques became fully automated, determining the order of the nucleotides in a piece of DNA went from being an elaborate Ph.D. thesis project to a routine laboratory chore. Feed DNA into the sequencing machine, add the necessary reagents, and out comes the sought-after result: the order of As, Ts, Gs, and Cs. Nothing could be simpler.

So why was sequencing the human genome such a formidable task? Largely because of its size. Today's DNA sequencing methods are limited by the physical size of the gel that is used to separate the labeled fragments (see Figure 10–7C). At most, only a few hundred nucleotides can be read from a single gel. How, then, do you handle a genome that contains billions of nucleotides?

The solution is to break the genome into fragments and sequence those smaller pieces. The main challenge then comes in piecing the short fragments together in the correct order to yield a comprehensive sequence of a whole chromosome, and ultimately a whole genome. To accomplish this breakage and reassembly, researchers have generally adopted two different strategies for sequencing genomes: the shotgun method and the clone-by-clone approach.

Shotgun sequencing

The most straightforward approach to sequencing a genome is to break it into random fragments, sequence each of the

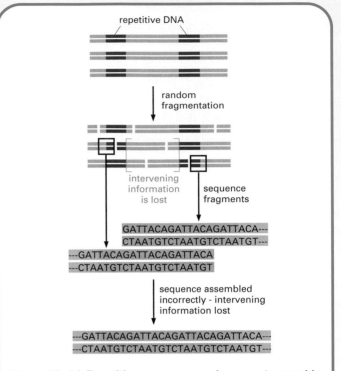

Figure 10–10 Repetitive sequences make correct assembly difficult. In this example, the DNA contains two segments of repetitive DNA, each made of many copies of the sequence GATTACA. When the resulting sequences are examined, two fragments from different parts of the DNA appear to overlap. Assembling these sequences incorrectly would result in a loss of the information (in brackets) that lies between the original repeats.

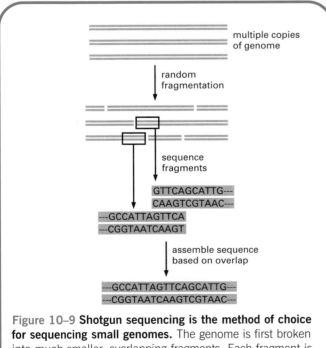

Figure 10–9 Shotgun sequencing is the method of choice for sequencing small genomes. The genome is first broken into much smaller, overlapping fragments. Each fragment is then sequenced, and the genome is assembled based on overlapping sequences.

fragments, and then use a powerful computer to order these pieces using sequence overlaps to guide the assembly (Figure 10–9). This approach is called the shotgun sequencing strategy. As an analogy, imagine shredding several copies of *Essential Cell Biology (ECB)* into small pieces, mixing the pieces up, and then trying to put one whole copy of the book back together again by matching up the words or phrases or sentences that appear on the different slips of paper. (Several copies would be needed to generate the overlap necessary for reassembly.) It could be done, but it would be much easier if the book were only, say, 2 pages long.

For this reason, a straight-out shotgun approach is the strategy of choice only for sequencing small genomes. The method proved its worth in 1995, when it was used to sequence the genome of the infectious bacterium *Haemophilus influenzae,* the first organism to have its complete genome sequence determined. The trouble with shotgun sequencing is that the reassembly process can be flummoxed by repetitive nucleotide sequences (Figure 10–10). Although rare in bacteria, these sequences make up a large fraction of vertebrate genomes (see Figure 9–26). Highly repetitive DNA segments make it difficult to piece DNA sequences back together accurately. Returning to the *ECB* analogy, this chapter alone contains more than a dozen instances of the phrase "the human genome." Imagine that one slip of paper from the shredded *ECBs* contains the information "So why was sequencing the human genome" (which appears at the start of this section); another contains the information "the human genome sequence consortium combined shotgun sequencing with a BAC-based, clone-by-clone approach" (which appears in the next paragraph). You might be tempted to join these two segments together based on the overlapping phrase "the human genome." But you would wind up with the nonsensical statement: "So why was sequencing the human genome sequence consortium combined shotgun sequencing with a BAC-based, clone-by-clone approach." Not only that, but this *ECB* assembly would eliminate the several paragraphs of important text that originally appeared between these two instances of "the human genome."

And that's just in this section. The phrase "the human genome" appears in nearly every chapter of this book. Such repetition compounds the problem of placing each fragment in its correct context. To circumvent these assembly problems, researchers in the human genome sequence consortium combined shotgun sequencing with a BAC-based, clone-by-clone approach.

Clone-by-clone

In this approach, researchers started by breaking the genome into overlapping fragments, 100 to 200 kilobase pairs in size. They then plugged these segments into BACs (bacterial artificial chromosomes) and inserted them into *E. coli*. As the bacteria divided, they copied the BACs, thus producing a collection of overlapping cloned fragments. This procedure is described in detail later in this chapter (see Figure 10–23).

The researchers then mapped each of these DNA fragments to its correct position in the genome. To do this, investigators used restriction enzymes to generate a "fingerprint" of each clone (Figure 10–11). The locations of the restriction enzyme sites in each fragment allowed researchers to map each BAC clone onto a previously generated restriction map of the entire human genome.

Knowing the relative positions of the cloned fragments, the researchers then selected some 30,000 BACs, sheared each into smaller fragments, and determined the nucleotide sequence of each BAC separately using the shotgun method. They could then assemble the whole genome sequence by stitching together the sequences of thousands of individual BACs that span the length of the genome.

The beauty of this approach is that it is relatively easy to accurately determine where the BAC fragments belong in the genome. This mapping step reduces the likelihood that regions containing repetitive sequences will be assembled incorrectly, and it virtually eliminates the possibility that sequences from different chromosomes will be mistakenly joined together. Returning to the textbook analogy, the BAC-based approach is akin to first separating your copies of *ECB* into individual pages and then shredding each page into its own separate pile. It should be much easier to put the book back together when one pile of fragments contains words from page 1, a second pile from page 2, and so on. And there's virtually no chance of mistakenly sticking a sentence from page 40 into the middle of a paragraph on page 412.

All together now

The clone-by-clone approach produced the first draft of the human genome sequence that we examined in Chapter 9. This draft sequence continues to be updated as additional clones are sequenced (to complete the coverage of the genome) and the original clones are resequenced to ensure the accuracy of the data.

The completion of the first draft of the human genome sequence is a landmark achievement in cell biology. As the set of instructions that specify all of the RNA and protein molecules needed to build a human being, this string of genetic bits holds the secrets to human development, physiology, and medicine. But the human sequence will also be of great value to researchers interested in comparative genomics or in the physiology of other organisms; it will ease the assembly of nucleotide sequences from other mammalian genomes—mice, rats, dogs, and other primates. Thus the human sequence will likely be the only large genome to be completed in this detailed and methodical way. Thanks to the human genome project, sequencing can only get easier from here.

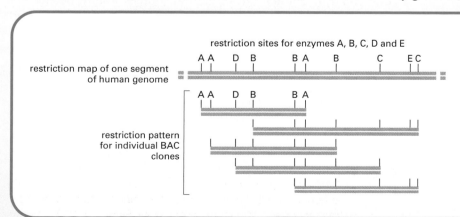

restriction sites for enzymes A, B, C, D and E

restriction map of one segment of human genome

restriction pattern for individual BAC clones

Figure 10–11 Individual BAC clones are positioned on the physical map of the human genome sequence on the basis of their restriction digest "fingerprints." Clones are digested with restriction endonucleases, and the sites at which the different enzymes cut each clone are recorded. The distinctive pattern of restriction sites allows investigators to order the fragments and place them on a previously generated restriction map of the human genome.

regions, in contrast, represent DNA whose sequence is generally not critical for function. By reviewing the results of the longest natural "experiment" ever conducted, comparative DNA sequencing can highlight the most biologically interesting regions in any genome.

Nucleic Acid Hybridization

Once a gene has been identified, we might want to know at what stages in development and in what tissues the gene is transcribed. We might also want to know whether an organism whose genome has not been sequenced contains a related gene. If the gene came from humans, we might want to determine if mutations present in it cause any human diseases. Or we might want to closely examine the same gene in a wide variety of individuals to see if it is responsible for any heritable traits. All of these issues can be addressed in the laboratory by taking advantage of a fundamental property of DNA: through the formation of Watson-Crick base pairs, a strand of DNA can pair in a highly selective manner with a second strand of *complementary* nucleotide sequence. The two strands of a DNA double helix are held together by relatively weak hydrogen bonds that can be broken by heating the DNA to around 90°C or by subjecting it to extremes of pH. These treatments release the two strands from each other but do not break the covalent bonds that link nucleotides together within each strand. If this process is slowly reversed (that is, by slowly lowering the temperature to normal body temperature or by bringing the pH back to neutral), the complementary strands will readily re-form double helices. This process is called **hybridization** or *renaturation,* and it results from a restoration of the complementary hydrogen bonds (Figure 10–12).

A similar hybridization reaction will occur between any two single-stranded nucleic acid chains (DNA/DNA, RNA/RNA, or RNA/DNA), provided they have complementary nucleotide sequences. The fundamental capacity of a single-stranded nucleic acid molecule to form a double helix only with a molecule complementary to it provides a powerful technique to detect specific nucleotide sequences in both DNA and RNA.

DNA Hybridization Facilitates the Diagnosis of Genetic Diseases

To search for a nucleotide sequence by hybridization, one first needs a piece of nucleic acid with which to search. This *DNA probe* is a single-stranded DNA molecule, typically 10–1000 nucleotides long, that is used in hybridization reactions to detect nucleic acid molecules containing a complementary sequence. At first, scientists were limited to using probes that could be obtained from natural sources. Today, because of advances in nucleotide chemistry, short lengths of DNA of any desired sequence can be synthesized nonenzymatically in the laboratory.

Figure 10–12 A molecule of DNA can undergo denaturation and renaturation (hybridization). For hybridization to occur, the two single strands must have complementary nucleotide sequences that allow base-pairing. In this example, the *red* and *orange* strands are complementary to each other, and the *blue* and *green* strands are complementary to each other.

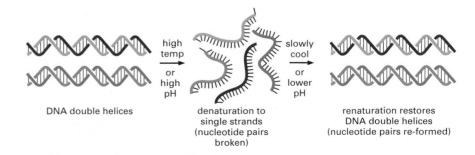

DNA double helices → high temp or high pH → denaturation to single strands (nucleotide pairs broken) → slowly cool or lower pH → renaturation restores DNA double helices (nucleotide pairs re-formed)

Machines the size of a microwave oven can be programmed to string nucleotides together by chemical synthesis to produce single-stranded DNA chains of any sequence up to several hundred nucleotides in length.

Of the many uses of DNA probes, one of the most important is in identifying carriers of genetic diseases. More than 3000 different human genetic diseases are caused by mutations in single genes. In many of these cases, the mutation is *recessive*—that is, it shows its effect only when an individual inherits two defective copies of the gene, one from each parent. For some of these diseases, it is now possible to identify early in a pregnancy fetuses that carry two copies of a defective gene; this information may be a factor in decisions relating to possible termination of the pregnancy.

Examining a single gene in the human genome requires searching through a total genome of over 3 billion nucleotides. However, the incredible specificity of DNA hybridization makes this relatively simple. For the recessive genetic disease sickle-cell anemia, for example, the exact nucleotide change in the mutant gene is known; the sequence GAG is changed to GTG at a certain position in the DNA strand that codes for the β-globin chain of hemoglobin (Figure 10–13A). This causes a change in the amino acid encoded by that sequence from a glutamic acid to a valine; this small change is sufficient to alter the properties of the resulting hemoglobin molecules to produce the disease (see Figure 6–19). For prenatal diagnosis of sickle-cell anemia, DNA is extracted from fetal cells. Two DNA probes are used to test the fetal DNA—one corresponding to the normal β-globin gene sequence in the region of the mutation and the other corresponding to the mutant gene sequence. If the probes are short (about 20 nucleotides), they can be hybridized with DNA at a temperature at which only the perfectly matched helices will be stable. Using this technique, it is possible to distinguish whether DNA isolated from the fetus contains one, two, or no defective β-globin genes (Figure 10–13B). For example, a fetus carrying two copies of the mutant β-globin gene (which will result in the disease) can be recognized because its DNA will hybridize only with the probe

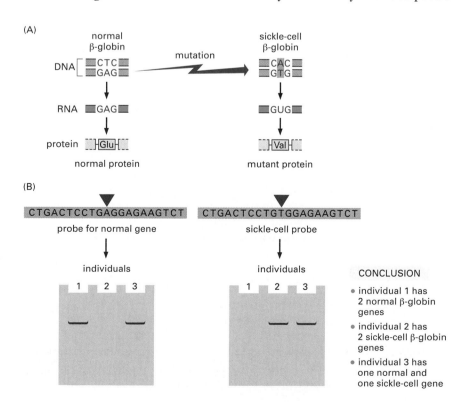

Figure 10–13 **The mutation that causes sickle-cell anemia can be detected using DNA hybridization.** (A) Sickle-cell anemia is caused by a mutation in the gene for β-globin that results in a change of a single amino acid from glutamic acid to valine in the β-globin protein. Individuals carrying two defective β-globin genes inherit the disease; individuals carrying one mutant gene and one normal gene usually do not develop symptoms. (B) The mutant gene can be detected in fetal DNA by DNA hybridization (see Figure 10–12). DNA samples from the fetus are first treated with restriction nucleases, and all the resulting DNA fragments (including those that contain the relevant portion of the β-globin gene) are electrophoresed through the gel. The gel is then treated with a DNA probe that detects only the restriction fragment that carries the β-globin gene (see Figure 10–14). Two different synthetic DNA probes are used, one corresponding to the normal sequence and one corresponding to the mutant sequence. The probes are labeled either with radioactive isotopes or with a fluorescent dye.

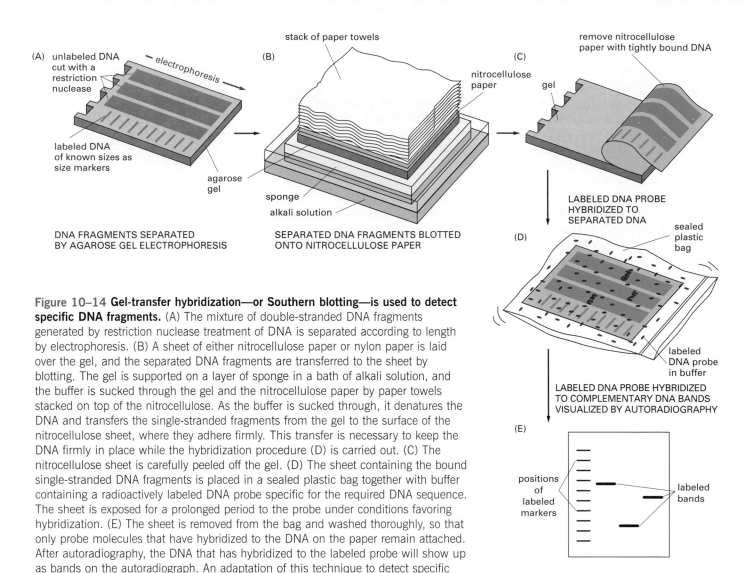

(A) unlabeled DNA cut with a restriction nuclease

~ electrophoresis

labeled DNA of known sizes as size markers

agarose gel

DNA FRAGMENTS SEPARATED BY AGAROSE GEL ELECTROPHORESIS

(B) stack of paper towels

nitrocellulose paper

sponge

alkali solution

SEPARATED DNA FRAGMENTS BLOTTED ONTO NITROCELLULOSE PAPER

(C) remove nitrocellulose paper with tightly bound DNA

gel

LABELED DNA PROBE HYBRIDIZED TO SEPARATED DNA

(D) sealed plastic bag

labeled DNA probe in buffer

LABELED DNA PROBE HYBRIDIZED TO COMPLEMENTARY DNA BANDS VISUALIZED BY AUTORADIOGRAPHY

(E) positions of labeled markers

labeled bands

Figure 10–14 Gel-transfer hybridization—or Southern blotting—is used to detect specific DNA fragments. (A) The mixture of double-stranded DNA fragments generated by restriction nuclease treatment of DNA is separated according to length by electrophoresis. (B) A sheet of either nitrocellulose paper or nylon paper is laid over the gel, and the separated DNA fragments are transferred to the sheet by blotting. The gel is supported on a layer of sponge in a bath of alkali solution, and the buffer is sucked through the gel and the nitrocellulose paper by paper towels stacked on top of the nitrocellulose. As the buffer is sucked through, it denatures the DNA and transfers the single-stranded fragments from the gel to the surface of the nitrocellulose sheet, where they adhere firmly. This transfer is necessary to keep the DNA firmly in place while the hybridization procedure (D) is carried out. (C) The nitrocellulose sheet is carefully peeled off the gel. (D) The sheet containing the bound single-stranded DNA fragments is placed in a sealed plastic bag together with buffer containing a radioactively labeled DNA probe specific for the required DNA sequence. The sheet is exposed for a prolonged period to the probe under conditions favoring hybridization. (E) The sheet is removed from the bag and washed thoroughly, so that only probe molecules that have hybridized to the DNA on the paper remain attached. After autoradiography, the DNA that has hybridized to the labeled probe will show up as bands on the autoradiograph. An adaptation of this technique to detect specific sequences in RNA is called *Northern blotting.* In this case mRNA molecules are electrophoresed through the gel and the probe is usually a single-stranded DNA molecule.

Question 10–3

DNA sequencing of your own two β-globin genes (one from each of your two Chromosome 11s) reveals a mutation in one of the genes. Given this information alone, how much should you worry about being a carrier of an inherited disease that could be passed on to your children? What other information would you like to have to assess your risk?

that is exactly complementary to the mutant DNA sequence. A common laboratory procedure used to visualize the hybridization is called *Southern blotting,* as shown in Figure 10–14.

The same techniques can also be used to ascertain an individual's susceptibility to future disease. They can, for example, identify individuals who have inherited abnormal copies of a DNA mismatch repair gene (see Figure 6–22). Because they cannot repair mistakes in DNA replication efficiently and because most cancers are caused by accumulation of mutations (see Figure 6–20), such individuals (estimated at 1 in 200 North Americans) have a greatly increased risk of cancer, especially a certain type of colon cancer. They need to take protective measures, such as receiving frequent checkups, to improve their prospects for remaining healthy.

Hybridization on DNA Microarrays Monitors the Expression of Thousands of Genes at Once

We saw in Chapter 8 that a cell expresses only a subset of the genes available in its genome. One of the most important uses of nucleic acid hybridization is to determine, for a population of cells, exactly which genes are being transcribed into mRNA and which genes are transcriptionally silent. For many years, scientists have been able to monitor the expression of genes, one at a time, but **DNA microarrays**, developed in the last decade, have revolutionized the way we analyze genes by allowing the RNA products of thousands of genes to be monitored at the same time. By examining the expression of so many genes simultaneously, we can begin to identify and study the complex gene expression patterns that underlie cellular physiology: we can see which genes are switched on (or off) as cells grow, divide, or respond to hormones, toxins, or infection.

DNA microarrays are simply glass microscope slides studded with a large number of DNA fragments, each containing a nucleotide sequence that serves as a probe for a specific gene. The most dense arrays contain tens of thousands of these fragments in an area smaller than a postage stamp, allowing thousands of hybridization reactions to be performed in parallel (Figure 10–15). Some types of microarrays carry DNA fragments corresponding to entire genes that are spotted onto the slides by a robot. Other types contain short oligonucleotides that are synthesized on the surface of the wafer using techniques similar to those that are used to etch circuits onto computer chips. In either case, the exact sequence—and position—of every DNA probe on the chip is known.

To use a DNA microarray to simultaneously monitor the expression of every gene in a cell, mRNA from the cells being studied is extracted and copied into a complementary DNA form, as we will describe in the next section. This complementary DNA, or cDNA, which is easier to manipulate than the original RNA, is then labeled with a fluorescent probe. The microarray is incubated with the labeled cDNA sample, and hybridization is allowed to occur (see Figure 10–15). The array is then washed to remove unbound molecules, and the positions to which labeled DNA fragments have hybridized are identified as fluorescent spots by an automated scanning-laser microscope. The array positions are then matched to the particular genes whose DNA was originally spotted at each location.

DNA microarrays have been used to examine everything from the changes in gene expression that make strawberries ripen to the "signatures" of different types of human cancer cells. Comparisons of the gene expression profiles of human cancers, for example, can be used to readily distinguish one type of cancer cell from another. By relating these expression patterns to clinical data gathered for each cancer—including how rapidly it progresses and whether it responds to treatment—it may be possible to predict whether a particular patient will respond to a

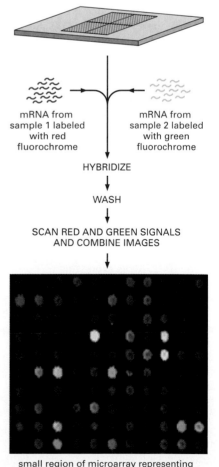

Figure 10–15 DNA microarrays are used to simultaneously monitor the expression of hundreds or thousands of genes. In this example, mRNA is collected from two different cell samples for direct comparison of their relative levels of gene expression. These samples are converted to cDNA and labeled, one with a red fluorochrome, the other with a green fluorochrome. The labeled samples are mixed and then allowed to hybridize to the microarray. After incubation, the array is washed and the fluorescence scanned. Only a small proportion of the microarray is shown; it represents 110 yeast genes. *Red spots* indicate that the gene in sample 1 is expressed at a higher level than the corresponding gene in sample 2, and *green spots* indicate that expression of the gene is more vigorous in sample 2 than in sample 1. *Yellow spots* reveal genes that are expressed at equal levels in both cell samples. Dark spots indicate little or no expression of the gene whose fragment is located at that position in the array. For another use of DNA microarrays, see How We Know (pp. 198–200).

mRNA from sample 1 labeled with red fluorochrome

mRNA from sample 2 labeled with green fluorochrome

HYBRIDIZE

WASH

SCAN RED AND GREEN SIGNALS AND COMBINE IMAGES

small region of microarray representing expression of 110 genes from yeast

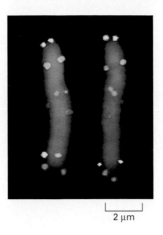

2 µm

Figure 10–16 *In situ* hybridization is used to locate genes on chromosomes. Here, six different DNA probes have been used to mark the location of their respective nucleotide sequences on human Chromosome 5 isolated in the metaphase stage of mitosis (see Figure 5–18 and Panel 19–1, pp. 642–643). The probes have been chemically labeled and are detected using fluorescent antibodies specific for the chemical label. Both the maternal and paternal copies of Chromosome 5 are shown, aligned side-by-side. Each probe produces two dots on each chromosome because chromosomes undergoing mitosis have already replicated their DNA and therefore each chromosome contains two identical DNA helices. (Courtesy of David C. Ward.)

specific type of therapy. Thus microarray-based "profiles" of cancer cells will likely lead to much more precise treatments for this often fatal disease.

In Situ Hybridization Locates Nucleic Acid Sequences in Cells or on Chromosomes

Nucleic acids, like other macromolecules, occupy precise positions in cells and tissues, and a great deal of potential information is lost when these molecules are extracted from cells. For this reason, techniques have been developed whereby nucleic acid probes are used to locate specific nucleic acid sequences while they are still in place within cells or are still part of chromosomes. This procedure is called *in situ* **hybridization** (from the Latin *in situ*, "in place"), and can be applied to detect either DNA sequences in chromosomes or RNA sequences in cells. For the former application, nucleic acid probes labeled with fluorescent dyes or radioactive isotopes are hybridized to whole chromosomes that have been exposed briefly to a very high pH to separate the two DNA strands. The chromosomal regions that bind the labeled probe can then be visualized (Figure 10–16). This type of experiment reveals the position of a gene along a whole chromosome; it has also been used to study the folded structures of chromosomes and their positions within the cell.

 In situ hybridization can also be used to reveal the distribution of gene expression patterns in cells that make up tissues or organs (Figure 10–17). This technique in particular has led to great advances in our

Figure 10–17 **Cells expressing a particular mRNA can be visualized by** *in situ* **hybridization.** This example shows a group of cells in a growing shoot tip of a snapdragon. Only a few of the cells (stained *dark blue*) are expressing an mRNA for a cyclin protein that triggers the cells to divide. The cells have been treated with a DNA probe that hybridizes to the cyclin mRNA. This probe was then specifically linked to an enzyme that produces a dark blue reaction product. The nuclei in other cells appear *light blue* because their DNA has been stained with the dye DAPI. (Courtesy of John Doonan.)

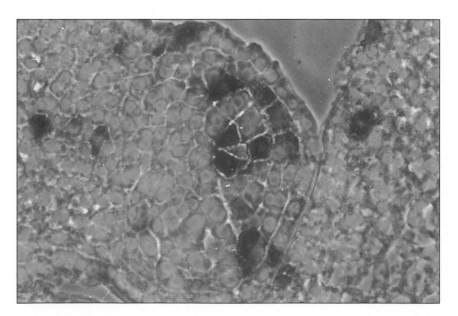

understanding of embryonic development, by making easily visible the many changes in patterns of gene expression that occur in different cells of the developing embryo.

DNA Cloning

We have seen that DNA molecules can be cleaved into shorter fragments using restriction nucleases and that these fragments can be separated from one another by gel electrophoresis. We have discussed how the nucleotide sequence of DNA can be determined and how important regions of a sequenced genome can be identified. We have also seen that hybridization can be used to pick out a match with a DNA probe of known sequence. In this section of the chapter, we shall see how these procedures are combined in order to obtain and work with a physical piece of DNA from a genome, rather than just the sequence information. In other words, we will discuss how a particular piece of any genome can be cloned. In cell biology, the term **DNA cloning** is used in two senses: in one it literally refers to the act of making many identical copies of a DNA molecule. However, it is also used to describe the separation of a particular stretch of DNA (often a particular gene) from the rest of a cell's DNA, because this isolation is facilitated by steps in which many identical copies of the DNA of interest are made. Obtaining any defined segment of DNA from a genome is one of the most important feats of recombinant DNA technology, as it is the starting point for understanding the function of each stretch of DNA within the genome.

DNA Ligase Joins DNA Fragments Together to Produce a Recombinant DNA Molecule

Modern DNA technology depends both on the ability to break long DNA molecules into conveniently sized fragments and on the ability to join these fragments back together in new combinations. The cell itself has provided the means to perform these molecular manipulations. As discussed in Chapter 6, the enzyme **DNA ligase** reseals the nicks in the DNA backbone that arise during DNA replication and DNA repair (see Figures 6–16 and 6–26). This enzyme has become one of the most important tools of recombinant DNA technology, as it allows scientists to join together any two DNA fragments (Figure 10–18). Because DNA has the same chemical structure in all organisms, this simple maneuver allows DNAs from any source to be united. In this way, isolated DNA fragments can be recombined in the test tube to produce DNA molecules not found in nature. Once two DNA molecules have been joined by ligase, the cell cannot tell that the two DNAs were originally separate and will treat the resulting DNA as a single molecule. If such a foreign stretch of DNA is appropriately introduced into the DNA of a host cell, it will be replicated and transcribed as if it were a normal part of the cell's own DNA.

Recombinant DNA Can Be Copied Inside Bacterial Cells

For many applications of DNA technology, it is desirable to make identical copies of (that is, clone) a defined stretch of DNA, often a gene. As we shall see, this replication can be accomplished in several ways. One way is to introduce the DNA to be copied into a rapidly dividing bacterium; each time the bacterium replicates its own DNA it also copies the introduced DNA.

 DNA can be introduced into bacteria by a mechanism called **transformation**. Some bacteria naturally take up DNA molecules present in

Almost all the cells in an individual animal contain identical genomes. In an experiment, a tissue composed of multiple different cell types is fixed and subjected to *in situ* hybridization with a DNA probe to a particular gene. To your surprise, the hybridization signal is much stronger in some cells than in others. Explain this result.

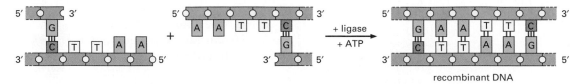

(A) JOINING TWO COMPLEMENTARY STAGGERED ENDS

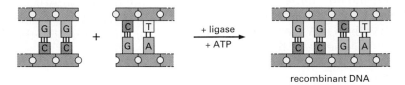

(B) JOINING TWO BLUNT ENDS

(C) JOINING A BLUNT END WITH A STAGGERED END

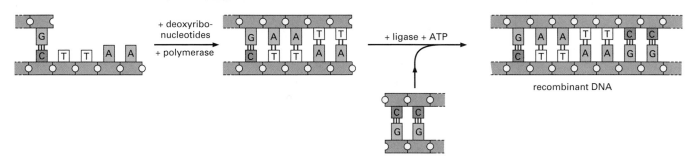

Figure 10–18 Recombinant DNA molecules can be formed *in vitro*. The enzyme DNA ligase can join any two DNA fragments together regardless of the source of the DNA. ATP provides the energy necessary for the ligase to reseal the sugar–phosphate backbone of DNA. (A) The joining of two DNA fragments produced by the restriction nuclease Eco RI. Note that the staggered ends produced by this enzyme enable the ends of the two fragments to base-pair correctly with each other, greatly facilitating their rejoining. This ligation reaction also reconstructs the original restriction nuclease-cutting site, which allows DNA fragments to be easily inserted or removed. (B) The joining of a DNA fragment produced by the restriction nuclease Hae III to one produced by Alu I. (C) The joining of DNA fragments produced by Eco RI and Hae III, respectively, using DNA polymerase to fill in the staggered cut produced by Eco RI. Each DNA fragment shown in the figure is oriented so that its 5′ ends are the left end of the upper strand and the right end of the lower strand, as indicated in (A).

their surroundings by pulling the DNA through their cell membrane to the inside of the cell (Figure 10–19). The incoming DNA is often then incorporated into the genome by recombination. The term "transformation" originated from early observations of this phenomenon in which it appeared that one bacterial strain had become transformed into another. The transformation of one strain of bacterium with purified DNA derived from another strain provided one of the first proofs that DNA is indeed the genetic material (see How We Know, pp. 172–174).

In a natural bacterial population, a source of DNA for transformation is provided by bacteria that have died and released their contents (including DNA) into the environment. In the laboratory, bacteria such as *E. coli* can be coaxed to take up recombinant DNA that has been created in the laboratory. A great advantage to the experimenter is that naked DNA from any source, not just the DNA of the same bacterial species, can be taken up by this route. Bacterial transformation thus allows DNA from complex organisms, such as humans, to be studied easily in the laboratory.

Specialized Plasmid Vectors Are Used to Clone DNA

As mentioned previously, DNA introduced into bacteria is often incorporated into the bacterial genome. Investigators interested in cloning, however, find it easier to manipulate, copy, and purify their recombinant DNA when it is maintained as an independent molecule, separate from the bacterial chromosome. To maintain foreign DNA in a bacterial cell, a bacterial plasmid is used as a carrier, or *vector*. The **plasmids** typically used for gene cloning are relatively small circular DNA molecules of several thousand nucleotide pairs that can replicate inside a bacterium (Figure 10–20). A plasmid vector contains a replication origin,

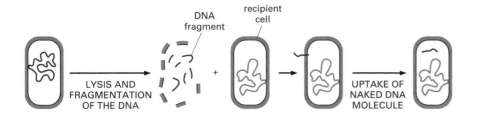

LYSIS AND
FRAGMENTATION
OF THE DNA

DNA
fragment

recipient
cell

UPTAKE OF
NAKED DNA
MOLECULE

Figure 10–19 Some bacteria can efficiently take up DNA from their surroundings. In nature, the source of this DNA is often other bacterial cells that have died. Once inside the recipient cell, the donor DNA can become a part of the recipient genome (through the process of homologous recombination) or—in special cases—can be maintained as a piece of DNA independent of the bacterial chromosome.

which enables it to replicate in the bacterial cell independently of the bacterial chromosome. It also has a cutting site for a convenient restriction nuclease, so that the plasmid can be opened and a foreign DNA fragment can be inserted. Plasmids also usually contain a gene for some selectable property, such as antibiotic resistance, which enables bacteria that take up the recombinant DNA to be easily identified.

Plasmids occur naturally in many different species of bacteria. They were first recognized by physicians and scientists because they often carry genes that render their host bacteria resistant to one or more antibiotics. Indeed, historically potent antibiotics (for example, penicillin) are no longer effective against many of today's bacterial infections because plasmids have spread among bacterial species by horizontal transfer (see Figure 9–13). The plasmids used for recombinant DNA research are derived from such naturally occurring plasmids; however, they are typically much simpler, containing only the DNA elements discussed above.

To insert a piece of DNA to be cloned into a laboratory plasmid, the purified plasmid DNA is exposed to a restriction nuclease that cleaves it in just one place, and the DNA fragment to be cloned is covalently inserted into it using DNA ligase (Figure 10–21). This recombinant DNA molecule is then introduced into a bacterium (usually *E. coli*) by transformation, and the bacterium is allowed to grow in nutrient broth, where it doubles in number every 30 minutes. Each time it doubles, the number of copies of the recombinant DNA molecule also doubles, and after just a day, hundreds of millions of copies of the plasmid will have been produced. The bacteria are then lysed, and the plasmid DNA is purified (by virtue of its small size) from the rest of the cell contents, including the large bacterial chromosome. The purified preparation of plasmid DNA will contain millions of copies of the original DNA fragment (Figure 10–22). This DNA fragment can be recovered by cutting it cleanly out of the plasmid DNA using the appropriate restriction enzyme and separating it from the plasmid DNA by gel electrophoresis (see Figure 10–5). These steps effectively allow the purification of a given stretch of DNA from the rest of the genome.

Human Genes Are Isolated by DNA Cloning

We have seen how any DNA fragment can be produced in large numbers by inserting it into a bacterium. But how are these DNA fragments identified and chosen in the first place? In particular, how is an individual human gene (one of around 30,000 genes in the human genome) first isolated by cloning? As an example, let us imagine we are going to clone the gene for the human blood-clotting protein Factor VIII. Although the exact methods by which human genes have been isolated and identified differ from one case to the next, this example will illustrate many of the general features of DNA cloning.

Defects in the gene for Factor VIII are the cause of the most common type of hemophilia—hemophilia A. This genetically determined disease has been recognized for over a thousand years and affects

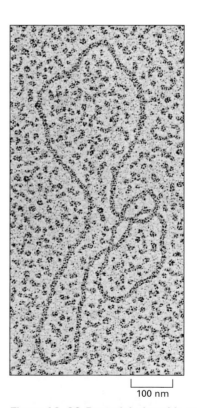

100 nm

Figure 10–20 Bacterial plasmids are commonly used as cloning vectors. This circular, double-stranded DNA molecule consists of several thousand nucleotide pairs. The staining required to make the DNA visible in this electron micrograph makes the DNA appear much thicker than it actually is. (Courtesy of Brian Wells.)

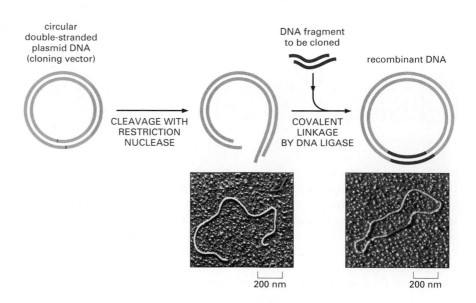

Figure 10–21 A DNA fragment is inserted into a bacterial plasmid using the enzyme DNA ligase. The plasmid is cut open with a restriction nuclease (in this case one that produces staggered ends) and is mixed with the DNA fragment to be cloned (which has been prepared using the same restriction nuclease), DNA ligase, and ATP. The staggered ends base-pair, and the nicks in the DNA backbone are sealed by the DNA ligase to produce a complete recombinant DNA molecule. (Micrographs courtesy of Huntington Potter and David Dressler.)

approximately one in 10,000 males. People with hemophilia A fail to produce fully active Factor VIII, and thus have repeated episodes of uncontrolled bleeding. Until recently, the standard treatment for this disease had been the injection of concentrated Factor VIII protein, pooled from many blood samples. Tragically, this treatment has exposed hemophiliacs to risk of infection by viruses, including HIV (the AIDS virus). The commercial production of pure Factor VIII using recombinant DNA technology offers a significant improvement in the treatment of this disease. This achievement required the cloning of the normal human gene that codes for Factor VIII and the piecing together of its coding sequence. This coding sequence was then used to produce large amounts of the purified protein, as described later in this chapter.

Dealing with the 3×10^9 nucleotide pairs of the complete human genome is a daunting task, and the first step in cloning any human gene is to break up the total genomic DNA into smaller, more manageable pieces to make it easier to work with. One general procedure for doing this is outlined in Figure 10–23. Human DNA is first extracted from a tissue sample or cell culture and cleaved with a restriction nuclease, which produces millions of different DNA fragments. The mixture of DNA fragments is then ligated into plasmid vectors under conditions that favor the insertion of one DNA fragment for each plasmid molecule. The plasmid most commonly used for human DNA libraries is the bacterial artificial chromosome (BAC), a vector derived from a plasmid that occurs naturally in *E. coli*. These recombinant plasmids are mixed with a culture of *E. coli* at a concentration that ensures that no more than one plasmid molecule is taken up by each bacterium. The collection of cloned DNA fragments in the resulting bacterial culture is known as a **DNA library**. In this case it is called a *genomic library*, as the

Figure 10–22 A cloned DNA fragment can be replicated inside a bacterial cell. To produce many copies of a particular fragment of DNA, it is first inserted into a plasmid vector, as shown in Figure 10–21. The resulting recombinant plasmid DNA is then introduced into a bacterium where it can be replicated many millions of times as the bacterium multiplies.

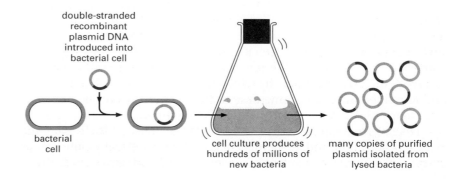

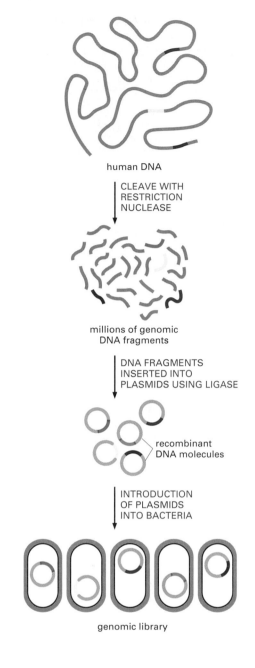

Figure 10–23 Human genomic libraries can be constructed using restriction nucleases and ligase. A genomic library comprises a set of bacteria, each carrying a different small fragment of human DNA. For simplicity, cloning of just a few representative fragments *(colored)* is shown. In reality, all of the *gray* DNA fragments will also be cloned.

human DNA

CLEAVE WITH
RESTRICTION
NUCLEASE

millions of genomic
DNA fragments

DNA FRAGMENTS
INSERTED INTO
PLASMIDS USING LIGASE

recombinant
DNA molecules

INTRODUCTION
OF PLASMIDS
INTO BACTERIA

genomic library

DNA fragments are derived directly from the chromosomal DNA. As we see later, there are other types of DNA libraries. If colonies derived from a single bacterium are isolated on Petri dishes, each bacterial colony will represent a clone of one particular stretch of human DNA. A collection of several million colonies in this library should thus represent all of the human genome.

To find a particular gene, we now face a problem analogous to that of entering a real library, wishing to find a book, and realizing that there is no card catalog or computerized listing of the millions of books in the library. How do we find a particular stretch of DNA (in our case the Factor VIII gene) in the vast human DNA library? The key is to exploit the hybridization properties of nucleic acids discussed earlier in this chapter. If we had a DNA probe for the Factor VIII gene, we could use it to find the matching clone in the library. But where does such a probe come from before the gene itself has been identified?

In the case of Factor VIII, a small amount of the protein was purified from human blood donors, using blood clotting as the biochemical assay. The partial amino acid sequence of the protein was deduced (today this would be done using mass spectrometry; see How We Know, pp. 129–131). Applying the genetic code in reverse, this amino acid sequence information was then used to deduce the partial nucleotide sequence of the gene. This nucleotide sequence was then chemically synthesized to create a DNA probe. Using this probe, the rare bacterial clones in the DNA library containing the complementary Factor VIII fragment were identified by hybridization (Figure 10–24).

When such a Factor VIII probe was first used on a human genomic library, a single complementary clone from the DNA library was identified. The nucleotide sequence of this cloned DNA showed that it contained only a small portion of the Factor VIII gene, and the entire gene had to be laboriously pieced together. We now know that the Factor VIII gene is 180,000 nucleotide pairs long and contains many introns (see Figure 7–14B), so it is hardly surprising that no single genomic clone contained the entire gene.

Many human genes were originally identified and cloned using variations on the procedure described for Factor VIII. However, now that the complete human genome sequence is known, cloning a particular gene often is much easier. For example, once the partial amino acid sequence of a protein of interest is known, it can, with the aid of computers, be used to directly search the human genome sequence for the matching gene.

For many applications of DNA technology, it is advantageous to obtain a gene in a form that contains only the coding sequence, that is, a form that lacks the intron DNA. For example, in the case of the Factor VIII gene the complete genomic clone—introns and exons—is so large and unwieldy that it is necessary to work with the gene in pieces. Moreover, if one wanted to deduce the complete amino acid sequence of the Factor VIII protein solely from the nucleotide sequence of its gene, it would be difficult to figure out where each exon begins and ends; after all, the great majority of the gene sequence is relatively useless intron sequences (see Figure 7–14B). However, as we shall see in the next section, it is relatively simple to isolate a gene free of all its introns. For this purpose, a different type of library, called a *cDNA library,* is used.

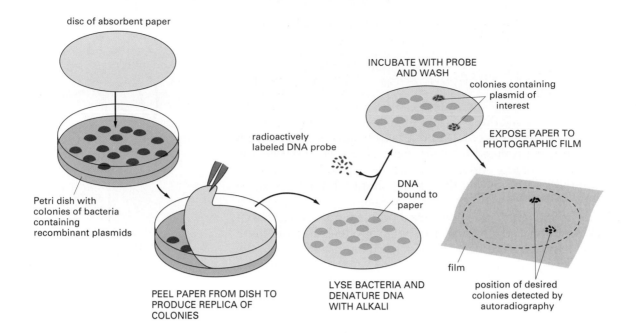

disc of absorbent paper

Petri dish with colonies of bacteria containing recombinant plasmids

PEEL PAPER FROM DISH TO PRODUCE REPLICA OF COLONIES

radioactively labeled DNA probe

LYSE BACTERIA AND DENATURE DNA WITH ALKALI

DNA bound to paper

INCUBATE WITH PROBE AND WASH

colonies containing plasmid of interest

EXPOSE PAPER TO PHOTOGRAPHIC FILM

film

position of desired colonies detected by autoradiography

Figure 10–24 A bacterial colony carrying a particular DNA clone can be identified by hybridization. A replica of the arrangement of colonies on the Petri dish is made by pressing a piece of absorbent paper against the surface of the dish. This replica is treated with alkali (to lyse the cells and separate the plasmid DNA into single strands), and the paper is then hybridized to a highly radioactive DNA probe. Those bacterial colonies that have bound the probe are identified by autoradiography. Living cells containing the plasmid can then be isolated from the original dish.

cDNA Libraries Represent the mRNA Produced by a Particular Tissue

A human cDNA library is similar to the genomic library described earlier in that it also contains numerous clones containing many different human DNA sequences. But it differs in one important respect. The DNA that goes into a cDNA library is not genomic DNA (chromosomal DNA), but is instead DNA copied from the mRNAs present in a particular tissue or cell culture. To prepare a **cDNA** library, the total mRNA is extracted from the cells, and DNA copies (cDNA, short for complementary DNA) of the mRNA molecules are produced by the enzyme reverse transcriptase (Figure 10–25). The cDNA molecules are then cloned, just like the genomic DNA fragments described earlier, to produce the cDNA library. For example, using such a cDNA library prepared from liver, the organ that normally produces Factor VIII, it was possible to isolate the complete coding sequence of the Factor VIII gene, devoid of introns, and present on one piece of DNA. The Factor VIII cDNA was isolated from a cDNA library by using a portion of the genomic Factor VIII DNA as a probe and employing the procedure shown previously in Figure 10–24. We will see in the final part of this chapter how the coding sequence was used to produce purified human Factor VIII protein on a commercial scale.

There are several important differences between genomic DNA clones and cDNA clones, as illustrated in Figure 10–26. Genomic clones represent a random sample of all of the DNA sequences found in an organism's genome and, with very rare exceptions, will contain the same sequences regardless of the cell type from which the DNA came. Also, genomic clones from eucaryotes contain large amounts of repetitive DNA sequences, introns, gene regulatory regions, and spacer DNA, as well as sequences that code for proteins (see Figure 9–26). By contrast, cDNA clones contain only coding sequences, and only those for genes that have been transcribed into mRNA in the tissue from which the RNA came. As the cells of different tissues produce distinct sets of mRNA molecules, a different cDNA library will be obtained for each type of tissue. Patterns of gene expression change during development, so cDNA libraries will also reflect the genes expressed by cells at different stages in their development.

By far the most important advantage of cDNA clones is that they contain the uninterrupted coding sequence of the gene. Thus, if the aim of cloning the gene is either to deduce the amino acid sequence of the protein from the DNA or to produce the protein in bulk by expressing the cloned gene in a bacterial or yeast cell (neither of which can remove introns from mammalian RNA transcripts), it is essential to start with cDNA.

The main advantage of genomic clones, on the other hand, is that they contain introns as well as exons, and they include the regulatory sequences that determine when genes are expressed, in which tissues, and how much protein will be made. For this reason, genomic clones are used to determine the complete nucleotide sequences of genomes. As described in How We Know (pp. 334–335), a complete genome sequence is pieced together from the individual nucleotide sequences of an enormous number of genomic clones.

The Polymerase Chain Reaction Amplifies Selected DNA Sequences

Cloning via DNA libraries was once the only route to gene isolation, and it is still used in sequencing whole genomes and in dealing with very large genes. However, a method known as the **polymerase chain reaction (PCR)** provides a quicker and less expensive alternative for many cloning applications, particularly for those organisms whose complete genome sequence is known. Invented in the 1980s, PCR can be carried out entirely *in vitro* without the use of cells. Using this technique, a given nucleotide sequence can be selectively and rapidly replicated in large amounts from any DNA sample that contains it. For example, the

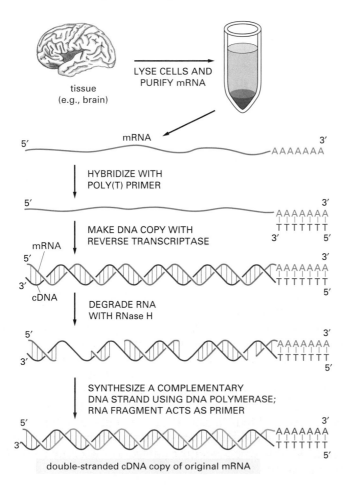

Figure 10–25 Complementary DNA (cDNA) is synthesized from mRNA. Total mRNA is extracted from a particular tissue, and DNA copies (cDNA) of the mRNA molecules are produced by the enzyme reverse transcriptase (see Figure 6–39). For simplicity, the copying of just one of these mRNAs into cDNA is illustrated here. A short oligonucleotide complementary to the poly-A tail at the 3′ end of the mRNA (discussed in Chapter 7) is first hybridized to the RNA to act as a primer for the reverse transcriptase, which then copies the RNA into a complementary DNA chain, thereby forming a DNA/RNA hybrid helix. Treating the DNA/RNA hybrid with the enzyme RNase H creates nicks and gaps in the RNA strand. The remaining single-stranded cDNA is then copied into double-stranded cDNA by the enzyme DNA polymerase. The primer for this synthesis reaction is provided by a fragment of the original mRNA, as shown.

Figure 10–26 Genomic DNA clones and cDNA clones derived from the same region of DNA are different. In this example, gene A is infrequently transcribed, whereas gene B is frequently transcribed, and both genes contain introns (green). In the genomic DNA library, both introns and the nontranscribed DNA (pink) are included in the clones, and most clones will contain only part of the coding sequence of a gene (red). In the cDNA clones the intron sequences (yellow) have been removed by RNA splicing during the formation of the mRNA (blue), and a continuous coding sequence is therefore present in each clone. Because gene B is transcribed more frequently than is A in the cells from which the cDNA library was made, it will be represented much more often than A in the cDNA library. In contrast, A and B should be represented equally in the genomic library.

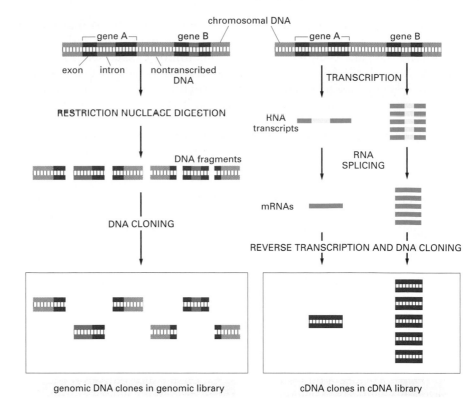

genomic DNA clones in genomic library

cDNA clones in cDNA library

Question 10–5

A. If the PCR reaction shown in Figure 10–27 is carried through an additional two rounds of amplification, how many of the DNA fragments labeled in gray, green, or red or outlined in yellow are produced? If many additional cycles are carried out, which fragments will predominate?

B. Assume you start with one double-stranded DNA molecule and amplify a 500–nucleotide-pair sequence contained within it. Approximately how many cycles of PCR amplification will you need to produce 100 ng of this DNA? 100 ng is an amount that can be easily detected after staining with a fluorescent dye. (Hint: for this calculation, you need to know that each nucleotide has an average molecular weight of 330 g/mole.)

polymerase chain reaction is now widely used to provide large amounts of any gene from a small sample of human DNA. It also has many other applications, including amplifying DNA for use in diagnostic tests for disease genes and in forensic medicine, as we discuss shortly.

PCR is based on the use of DNA polymerase to copy a DNA template in repeated rounds of replication. The polymerase is guided to the sequence to be copied by short primer oligonucleotides that are hybridized to the DNA template at the beginning and end of the desired DNA sequence. These oligonucleotides are designed by the experimenter so that they provide a primer for DNA replication on each strand of the original double-stranded DNA. Because the primers have to be chemically synthesized, PCR can be used only to clone DNA whose beginning and end sequences are known. Guided by these primers, DNA polymerase is then used to make many copies (billions are typical) of the sequence required. The method is outlined in Figure 10–27. PCR is extremely sensitive; it can detect a single copy of a DNA sequence in a sample by amplifying it so much that it becomes detectable by, for example, staining after separation by gel electrophoresis (see Figure 10–5).

There are several especially useful applications of PCR. First, PCR is now the method of choice for cloning relatively short DNA fragments (say, under 10,000 nucleotide pairs) from a cell. The original template for the reaction can be either DNA or RNA, so PCR can be used to obtain either a full genomic copy or a cDNA copy of the gene (Figure 10–28).

Another application of PCR, which relies on its extraordinary sensitivity, is to detect infections by pathogens at very early stages. Here, short sequences complementary to the pathogen's genome are used as primers, and following many cycles of amplification, the presence or absence of even a few copies of an invading genome in a sample of blood can be ascertained (Figure 10–29). For many infections, PCR is the most sensitive method of detection; already it is replacing the use of antibodies against surface proteins to detect the presence of pathogens in human samples.

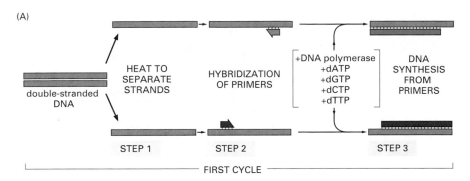

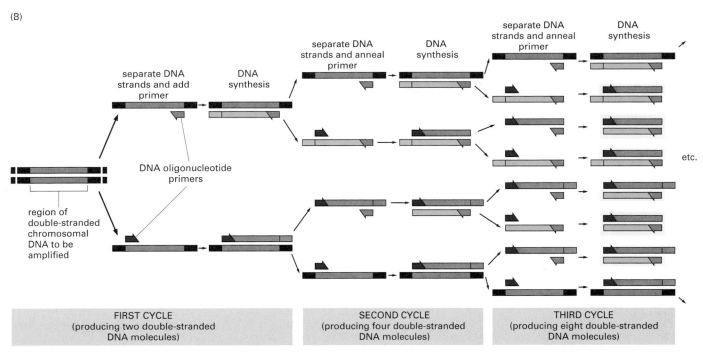

Figure 10–27 **PCR is used to amplify a DNA sequence.** Knowledge of the DNA sequence to be amplified is used to design two synthetic DNA oligonucleotides, each complementary to the sequence on one strand of the DNA double helix at opposite ends of the region to be amplified. These oligonucleotides serve as primers for *in vitro* DNA synthesis, which is carried out by a DNA polymerase, and they determine the segment of the DNA that is amplified. (A) PCR starts with a double-stranded DNA, and each cycle of the reaction begins with a brief heat treatment to separate the two strands (step 1). After strand separation, cooling of the DNA in the presence of a large excess of the two primer DNA oligonucleotides allows these primers to hybridize to complementary sequences in the two DNA strands (step 2). This mixture is then incubated with DNA polymerase and the four deoxyribonucleoside triphosphates so that DNA is synthesized, starting from the two primers (step 3). The cycle is then begun again by a heat treatment to separate the newly synthesized DNA strands. The technique depends on the use of a special DNA polymerase isolated from a thermophilic bacterium; this polymerase is stable at much higher temperatures than eucaryotic DNA polymerases, so it is not denatured by the heat treatment shown in step 1. It therefore does not have to be added again after each cycle of reaction. (B) As the procedure is carried out over and over again, the newly synthesized fragments serve as templates in their turn, and within a few cycles the predominant DNA is identical to the sequence bracketed by and including the two primers in the original template. In practice, 20–30 cycles are required for useful DNA amplification. Each cycle doubles the amount of DNA synthesized in the previous cycle. Each cycle takes only about 5 minutes, and automation of the whole procedure now enables cell-free cloning of a DNA fragment in a few hours, compared with the several days required for standard cloning procedures. Of the DNA put into the original reaction, only the sequence bracketed by the two primers is amplified because there are no primers attached anywhere else. In the example illustrated in (B), three cycles of reaction produce 16 DNA chains, 8 of which *(boxed in yellow)* are the same length as and correspond exactly to one or the other strand of the original bracketed sequence shown at the far left; the other strands contain extra DNA downstream of the original sequence, which is replicated in the first few cycles. After four more cycles, 240 of the 256 DNA chains will correspond exactly to the original sequence, and after several more cycles, essentially all of the DNA strands will have this unique length.

Figure 10–28 PCR can be used to obtain genomic or cDNA clones. (A) To obtain a genomic clone using PCR, chromosomal DNA is first purified from cells. PCR primers that flank the stretch of DNA to be cloned are added, and many cycles of the PCR reaction are completed (see Figure 10–27). Because only the DNA between (and including) the primers is amplified, PCR provides a way to obtain selectively a short stretch of chromosomal DNA in an effectively pure form. (B) To use PCR to obtain a cDNA clone of a gene, mRNA is first purified from cells. The first primer is then added to the population of mRNAs, and reverse transcriptase is used to make a complementary DNA strand. The second primer is then added, and the single-stranded DNA molecule is amplified through many cycles of PCR, as shown in Figure 10–27.

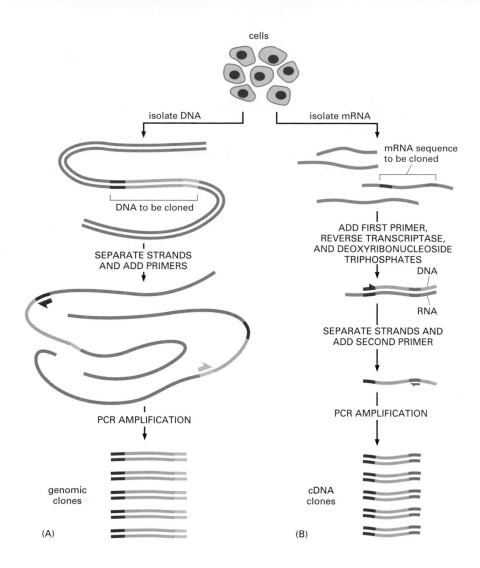

Finally, PCR has great potential in forensic medicine. Its extreme sensitivity makes it possible to work with a very small sample—minute traces of blood and tissue that may contain the remnants of only a single cell—and still obtain a *DNA fingerprint* of the person from whom it came. The genome of each human (with the exception of identical twins) differs in DNA sequence from the genome of every other human; the DNA amplified by PCR using a particular primer pair is therefore quite likely to differ in sequence from one individual to another. Using a carefully selected set of primer pairs that cover the known highly variable parts of the human genome, PCR can generate a distinctive DNA fingerprint for each individual (Figure 10–30).

Figure 10–29 PCR can be used to detect the presence of a viral genome in a sample of blood. The genome of the human immunodeficiency virus (HIV), the cause of AIDS, is a single-stranded RNA molecule. Because of its ability to enormously amplify the signal from every single molecule of nucleic acid, PCR is an extraordinarily sensitive method for detecting trace amounts of virus in a sample of blood or tissue without the need to purify the virus. In addition to HIV, many viruses that infect humans are now monitored in this way.

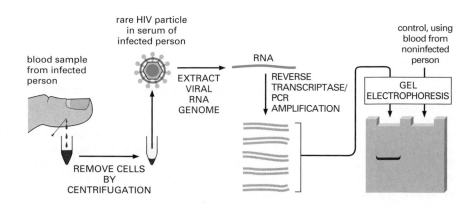

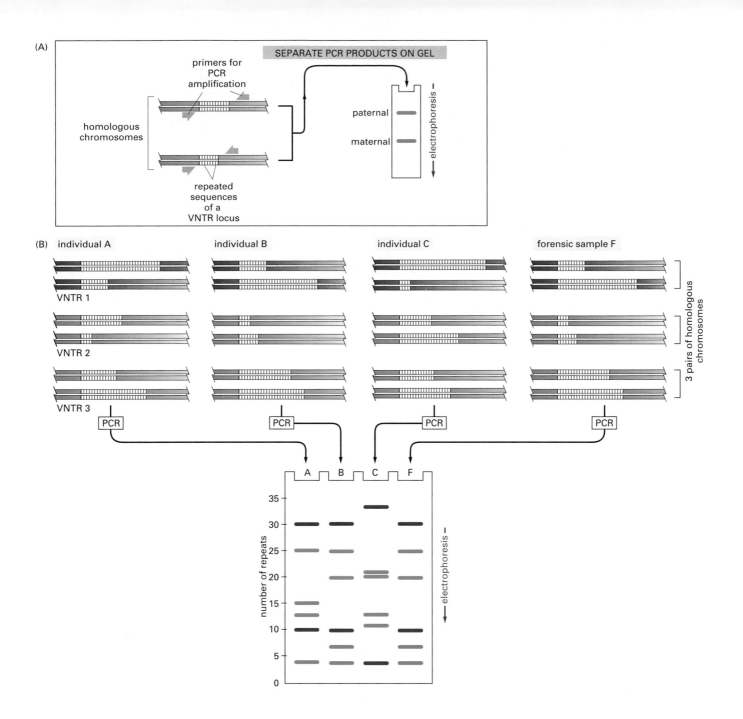

Figure 10–30 PCR is used in forensic science. (A) The DNA sequences that create the variability used in this analysis contain runs of short, repeated sequences, such as GTGTGT..., which are found in various positions (loci) in the human genome. The number of repeats in each run is highly variable in the population, ranging from 4 to 40 in different individuals. A run of repeated nucleotides of this type is commonly referred to as a *hypervariable microsatellite* sequence, also known as a VNTR *(variable number of tandem repeats)* sequence. Because of the variability in these sequences, individuals will usually inherit a different variant of each VNTR locus from their mother and from their father; two unrelated individuals therefore do not usually contain the same pair of sequences. A PCR reaction using primers that bracket the locus produces a pair of bands of amplified DNA from each individual, one band representing the maternal variant and the other representing the paternal variant. The length of the amplified DNA, and thus its position after electrophoresis, will depend on the exact number of repeats at the locus. (B) In the schematic example shown here, the same three VNTR loci are analyzed from three suspects (individuals A, B, and C), producing six bands for each person after polyacrylamide gel electrophoresis. Although some individuals have several bands in common, the overall pattern is quite distinctive for each. The band pattern can therefore serve as a "fingerprint" to identify an individual nearly uniquely. The fourth lane (F) contains the products of the same PCR reactions carried out on a forensic sample. The starting material for such a PCR reaction can be a single hair or a tiny sample of blood that was left at the crime scene. The more loci that are examined, the more confident we can be about the results. When examining the variability at 5–10 different VNTR loci, the odds that two random individuals would share the same fingerprint by chance are approximately one in 10 billion. In the case shown here, individuals A and C can be eliminated from inquiries, while B remains a clear suspect. A similar approach is now used routinely for paternity testing.

DNA Engineering

In this section we describe how extensions of the methods we have introduced so far in this chapter have revolutionized all other aspects of cell biology by providing new ways to study the functions of genes, RNA molecules, and proteins.

Completely Novel DNA Molecules Can Be Constructed

We have seen that recombinant DNA molecules are generally made by using DNA ligase to join together two DNA molecules (see Figure 10–18). For production of a DNA library, one of the DNA molecules is the vector, derived from a bacterial plasmid or virus, while the other is either a cDNA molecule or fragment of a chromosome (see Figure 10–23). Such a recombinant DNA molecule can in turn serve as a vector for the insertion and cloning of additional DNA, and by a repetition of this procedure, a cloned DNA can be generated that is different from any DNA that occurs naturally (Figure 10–31). This new DNA can be made from the combination of naturally occurring DNA sequences, or its sequence can be determined entirely by the experimenter by using chemically synthesized DNA. By repeated DNA cloning steps, synthetic and naturally occurring DNAs can be joined together in various combinations to produce longer DNA molecules of any desired sequence. One of the most important uses of such custom-designed DNA sequences is for the high-level production of what are normally rare cellular proteins, as we shall see next.

Rare Cellular Proteins Can Be Made in Large Amounts Using Cloned DNA

Until recently, the only proteins that could be studied easily were those produced in relative abundance by cells. Starting with several hundred grams of cells, a major protein—one that constitutes 1% or more of the total cellular protein—can be purified by sequential chromatography steps (see Panel 4–4, p. 162) to yield perhaps 0.1 g (100 mg) of pure protein. This amount is sufficient for conventional amino acid sequencing, for analysis of the protein's biological activity (for example, what biochemical reaction it might catalyze), and for the production of antibodies against the protein, which can then be used to detect it within the cell. Moreover, if suitable crystals of the protein can be grown (often a difficult task), its three-dimensional structure can be determined by X-ray crystallography (see How We Know, pp. 129–131). The structure and function of many abundant proteins, including hemoglobin isolated from red blood cells, have been analyzed in this way.

But most of the thousands of different proteins in a eucaryotic cell, including many with crucially important functions, are present in very small amounts. For these, it used to be extremely difficult, if not impossible, to obtain more than a few micrograms of pure material. One of the most important contributions of DNA cloning and genetic engineering to cell biology is that they make it possible to produce any protein, including the rare ones, in large amounts.

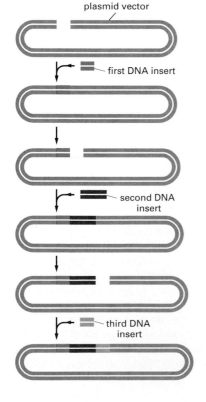

plasmid vector

first DNA insert

second DNA insert

third DNA insert

Figure 10–31 Serial DNA cloning can be used to splice together a set of DNA fragments derived from different sources. After each DNA insertion step, the recombinant plasmid is cloned to purify and amplify the new DNA (see Figure 10–22). The recombinant molecule is then cut once with a restriction nuclease, as indicated, and used as a cloning vector for the next DNA fragment.

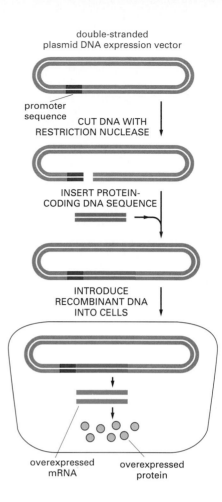

Figure 10–32 Large amounts of a protein can be produced from a protein-coding DNA sequence cloned into an expression vector and introduced into cells. A plasmid vector has been engineered to contain a highly active promoter, which causes unusually large amounts of mRNA to be produced from an adjacent protein-coding gene inserted into the plasmid vector. Depending on the characteristics of the cloning vector, the plasmid is introduced into bacterial, yeast, insect, or mammalian cells, where the inserted gene is efficiently transcribed and translated into protein.

This high-level production is usually accomplished by using specially designed vectors known as *expression vectors*. Unlike the cloning vectors discussed earlier, these vectors also include appropriate gene regulatory and promoter DNA sequences necessary to enable an adjacent protein-coding DNA insert to be efficiently transcribed in cells. These vectors can thus direct the production of large amounts of mRNA, which can be translated into protein within the cell (Figure 10–32). Different expression vectors are designed for use in bacterial, yeast, insect, or mammalian cells, each containing the appropriate regulatory sequences for transcription and translation in these cells. The expression vector is replicated at each round of cell division, giving rise to a cell culture able to synthesize very large amounts of the protein of interest. Because the protein encoded by the expression vector is typically produced inside the cell, it must be purified from the host cell proteins after cell lysis by chromatography; but because it is so plentiful in the cell lysate (often comprising 1–10% of the total cell protein), purification is usually easy to accomplish in only a few steps.

The experimental techniques which take us from protein to gene and back again are summarized in Figure 10–33. They have been used to produce many proteins of biological interest in large enough amounts for the detailed structural and functional studies that were previously possible only for a few (Figure 10–33). This technology is now also used to make large amounts of many medically useful proteins. The Factor VIII protein, for example, is now made commercially from cultures of genetically engineered mammalian cells and is thus free of viral contamination. Many other useful proteins, including hormones (such as insulin), growth factors, and viral coat proteins for use in vaccines, are also produced in this fashion.

Engineered Genes Can Reveal When and Where a Gene Is Expressed

So far in this chapter we have assumed that the function of a cloned gene is known. But suppose that you have discovered a gene that codes

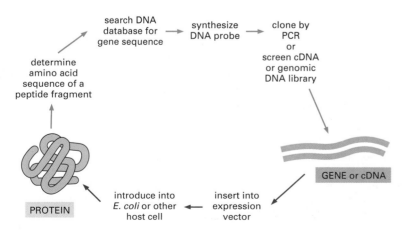

Figure 10–33 Knowledge of the molecular biology of cells makes it possible to experimentally move from gene to protein and from protein to gene. A small quantity of a purified protein is used to obtain a partial amino acid sequence. This provides sequence information that enables the corresponding gene to be cloned from a DNA library (see Figure 10–24) or to be identified in a sequenced genome and amplified by PCR (see Figure 10–27). Once the gene has been cloned, its protein-coding sequence can be used to design a DNA that can then be used to produce large quantities of the protein from genetically engineered cells (see Figure 10–32).

for a protein of unknown function. How do you determine what the protein does? Now that genome-sequencing projects are rapidly identifying new genes from the DNA sequence alone, this has become a common question in cell biology. For example, there are more than 10,000 human genes whose functions are unknown. Neither the complete nucleotide sequence of a gene nor even the three-dimensional structure of the protein is sufficient to deduce a protein's function unless it is closely related to a protein that has already been studied. Many proteins—such as those that have a structural role in the cell or normally form part of a large multienzyme complex—will have no obvious biochemical activity on their own. Even those that do have known activities (motor proteins and protein kinases, for example) could in principle participate in any number of different pathways in the cell; in other words, it is not always clear from their biochemical activities how the proteins are actually used by the cell.

In many cases, clues to a protein's function can be obtained by examining when and where its gene is expressed in the cell or in the organism as a whole. Determining the pattern and timing of a gene's expression can be accomplished by joining the regulatory region of the gene under study to a *reporter gene*—one whose activity can be easily monitored. As we discussed in detail in Chapter 8, gene expression is controlled by regulatory DNA sequences, usually located upstream of the coding region, that are not transcribed themselves. These regulatory sequences, which control which cells will express a gene and under what conditions, can also be made to control the expression of a reporter gene. The level, timing, and cell specificity of reporter protein production will reflect the function of the original gene as well as the action of the regulatory sequences that belong to it (Figure 10–34). In most cases, the expression of the reporter gene is monitored by tracking the fluorescence or enzymatic activity of its protein product.

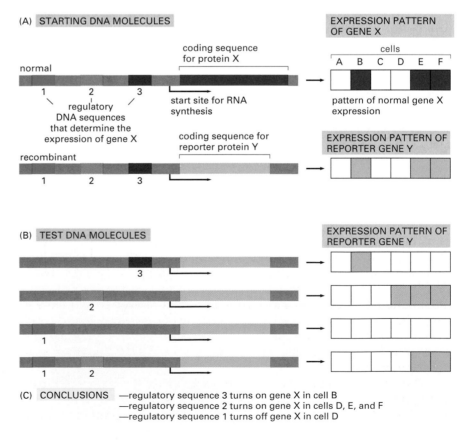

Figure 10–34 Reporter genes can be used to determine the pattern of a gene's expression. (A) In this example the coding sequence for protein X is replaced by the coding sequence for reporter protein Y. (B) Various fragments of DNA containing candidate regulatory sequences are added in combinations. The recombinant DNA molecules are then tested for expression after their transfection into a variety of different types of mammalian cells, and the results are summarized in (C). For experiments in eucaryotic cells, two commonly used reporter proteins are the enzymes β-galactosidase (see Figure 8–18B) and green fluorescent protein, or GFP (see Figure 10–35).

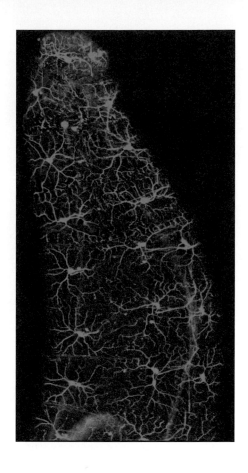

Figure 10–35 Green fluorescent protein (GFP) can be used to identify specific cells in a living animal. For this experiment, carried out in the fruit fly, the GFP gene was joined (using recombinant DNA techniques) to a fly promoter that is active only in a specialized set of neurons. This image of a live fly embryo was captured by a fluorescence microscope and shows approximately 20 neurons, each with long projections (axons and dendrites) that communicate with other (nonfluorescent) cells. These neurons are located just under the surface of the animal and allow it to sense its immediate environment. (From W.B. Grueber et al., *Curr. Biol.* 13:618–626, 2003.)

One of the most popular reporter proteins used today is **green fluorescent protein** (**GFP**), the molecule that gives luminescent jellyfish their greenish glow. In many cases, the GFP gene can simply be attached to one end of the gene that encodes a protein of interest. Often, the resulting GFP fusion protein behaves in the same way as the original protein, and its distribution in the cell or in the organism can easily be monitored simply by following its fluorescence by microscopy (Figure 10–35). The GFP fusion protein strategy has become a standard way to determine the distribution and to track the movement of any protein of interest in a living organism. From this information, many insights about a protein's function in the organism can be obtained.

Mutant Organisms Best Reveal the Function of a Gene

Ultimately, the cell biologist wishes to determine how genes, and the proteins they encode, function in an intact organism. As discussed in Chapter 20, classical genetics provides a powerful way to answer these questions. Using this approach, one begins by isolating mutants that have an interesting or unusual appearance: fruit flies with white eyes or stumpy or curly wings, for example. Working backward from this **phenotype**—the appearance or behavior of the individual—one then determines the organism's **genotype**, the form of the gene responsible for that characteristic. Before the advent of gene cloning, the functions of most known genes were identified in this way, according to the cellular or physiological processes that were altered by specific mutations. The classical genetic approach is most easily applicable to organisms that reproduce rapidly and can be easily mutated in the laboratory—such as bacteria, yeasts, nematode worms, and fruit flies.

Recombinant DNA technology has made possible a different type of genetic approach. Instead of beginning with a randomly generated mutant and using it to identify a gene and its protein, one can start with a cloned gene and proceed to make mutations in it *in vitro*. Then, by reintroducing the altered gene back into the organism from which it originally came, one can produce a mutant organism in which the gene's function may be revealed. Using the techniques to be discussed shortly, the coding sequence of a cloned gene can be altered to change the functional properties of the protein product or even to eliminate it altogether. Alternatively, the regulatory region of the gene can be altered so that the amount of protein made is altered, or so that the gene is expressed in a different cell type from normal or at a different time during development.

The ability to manipulate DNA *in vitro* makes it possible to introduce precise mutations, and genes can thus be altered in very subtle ways. It is often desirable, for example, to change the protein that the

Question 10–6

After decades of work, Dr. Ricky M. isolated a small amount of attractase, an enzyme producing a powerful human pheromone, from hair samples of Hollywood celebrities. To produce attractase for his personal use, he obtained a complete genomic clone of the attractase gene, connected it to a strong bacterial promoter on an expression plasmid, and introduced the plasmid into *E. coli* cells. He was devastated to find that no attractase was produced in the cells. What is a likely explanation for the failure?

Figure 10–36 A synthetic oligonucleotide can be used to modify the protein-coding region of a gene by site-directed mutagenesis. (A) A recombinant plasmid containing a gene insert is separated into its two DNA strands. A synthetic oligonucleotide primer corresponding to part of the gene sequence but containing a single altered nucleotide at a predetermined point is added to the single-stranded DNA under conditions that permit less than perfect DNA hybridization. (B) The primer hybridizes to the DNA, forming a single mismatched nucleotide pair. (C) The recombinant plasmid is made double-stranded by *in vitro* DNA synthesis starting from the primer and sealing by DNA ligase. (D) The double-stranded DNA is introduced into a cell, where it is replicated. Replication of one strand produces a normal DNA molecule, but replication of the other (the strand that contains the primer) produces a DNA molecule carrying the desired mutation. Only half of the progeny cells will end up with a plasmid that contains the desired mutant gene; however, a progeny cell that contains the mutated gene can be identified, separated from other cells, and cultured to produce a pure population of cells, all of which carry the mutated gene. Only one of the many changes that can be engineered in this way is shown here. With an oligonucleotide of the appropriate sequence, more than one amino acid substitution can be made at a time, or one or more amino acids can be added or deleted.

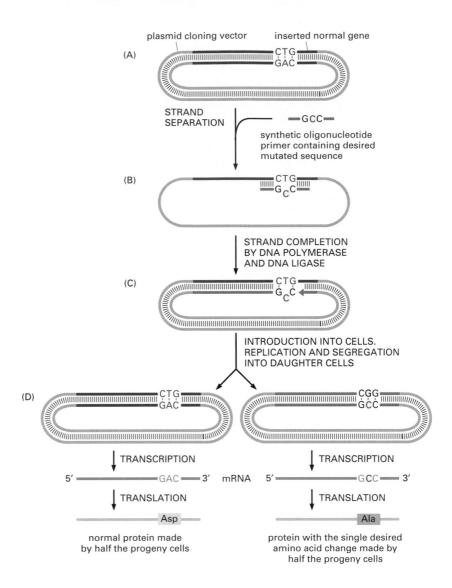

gene encodes by just one or a few amino acids. The use of the technique of **site-directed mutagenesis** to achieve this is outlined in Figure 10–36. By changing selected amino acids in this way, one can determine which parts of the polypeptide chain are crucial to such fundamental processes as protein folding, protein–ligand interactions, and enzymatic catalysis. Moreover, site-directed mutagenesis allows one to determine the biological roles of each part of a given protein.

Animals Can be Genetically Altered

The ultimate test of the function of a gene mutated as described in Figure 10–36 is to insert it into the genome of an organism and see what effect it has. Organisms into which a new gene has been introduced, or those whose genomes have been altered in other ways using recombinant DNA techniques, are known as **transgenic organisms**.

To study the function of a gene mutated *in vitro,* ideally one would like simply to replace the normal gene with the altered one so that the function of the mutant protein can be analyzed in the absence of the normal protein. Such gene replacement can be accomplished quite easily by homologous recombination between the introduced mutant DNA and the chromosomal DNA in some simple haploid organisms such as bacteria or yeasts (Figure 10–37A). Homologous recombination with an introduced DNA that contains a deletion in the gene of interest can lead to a large deletion in the normal gene and the gene's complete inactivation, or *knockout* (Figure 10–37B). Another possibility is that the mutant gene is added to the genome without any alteration made to the normal gene in the process (Figure 10–37C).

It is more difficult, but still possible, to alter genes in diploid organisms with large and complex genomes, such as mice (Figure 10–38). If the altered gene can be introduced into cells that will form part of the germ line (that is, the reproductive cells), such transgenic animals will be able to pass the altered gene on to at least some of their progeny as a permanent part of their genome (see Figure 9–1). Technically, even the human germ line could now be altered in this way, although this is unlawful for a variety of ethical reasons. Similar techniques are, however, being explored to correct genetic defects in human somatic cells. Somatic cells, such as the cells that form the liver, pancreas, bone, or skin, reproduce within an individual human being but are not passed on to progeny (see Figure 9–1). Some genetic diseases could be alleviated or even cured by the introduction of genetically corrected somatic cells into the tissue most affected by the disease; however, the alterations would not be passed on to descendants.

If the normal gene is replaced by a deletion or by an otherwise inactive version of the gene (see Figure 10–37B), transgenic techniques make it possible to produce complex organisms that are missing certain gene products entirely. For example, many *knockout mice*—strains of mice that have a particular gene permanently inactivated—have now been produced. The study of mice with gene knockouts has led to considerable advances in identifying the roles that genes play in the complex physiology of the organism (Figure 10–39).

Recently, a much easier way to inactivate genes has been discovered. Called *RNA interference* (*RNAi*, for short), this method exploits a recently discovered natural mechanism used in a wide variety of plants and animals to regulate selected genes. The technique relies on introducing into a cell or organism a double-stranded RNA molecule whose nucleotide sequence matches that of the gene to be inactivated. The RNA molecule hybridizes with the mRNA produced by the target gene and directs its degradation. Small fragments of this degraded RNA are subsequently used by the cell to produce more double-stranded RNA which directs the continued elimination of the target mRNA. Because these short RNA fragments can be passed on to progeny cells, RNAi can cause heritable changes in gene expression. But there is a second mechanism through which RNAi can stably inactivate genes. RNA fragments, produced by degradation in the cytosol can enter the nucleus and interact with the target gene itself, directing its packaging into a transcriptionally repressive form of chromatin (discussed in Chapter 8, pp. 279–280). Although the mechanism is not understood in detail, the RNA is presumed to "find" its target gene through complementary base

Figure 10–37 Several types of gene alterations can be made in genetically engineered organisms. (A) The normal gene can be completely replaced by a mutant copy of the gene, a process called gene replacement. This will provide information on the activity of the mutant gene, without interference from the normal gene, and thus the effects of small and subtle mutations can be determined. (B) The normal gene can be inactivated completely, for example, by making a large deletion in it; the gene is said to have suffered a knockout. This type of alteration is very widely used to obtain information on the possible function of the normal gene in the whole animal. (C) A mutant gene can simply be added to the genome. In some organisms this is the easiest type of genetic engineering to perform. Even this alteration can still provide useful information when the introduced mutant gene overrides the function of the normal gene.

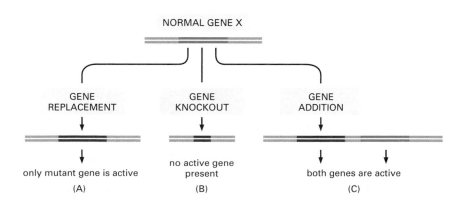

NORMAL GENE X

GENE REPLACEMENT

GENE KNOCKOUT

GENE ADDITION

only mutant gene is active

(A)

no active gene present

(B)

both genes are active

(C)

pairing. This dual mode of controlling gene expression makes RNAi an especially effective tool for shutting down genes individually.

RNAi is frequently used to inactivate genes in mammalian cell culture lines. It has also been widely used to study gene function in the nematode, *C. elegans*. When working with worms, introducing the double-stranded RNA is quite simple: the RNA can be injected directly into the intestine of the animal, or the worm can be fed with *E.coli* engineered to produce the RNA. The RNA is distributed throughout the body of the worm, where it inhibits expression of the target gene in different tissue types. Because the entire genome of *C. elegans* has been sequenced, RNAi is being used to help in assigning functions to the entire complement of worm genes. More recently, a related technique has also been widely applied to mice.

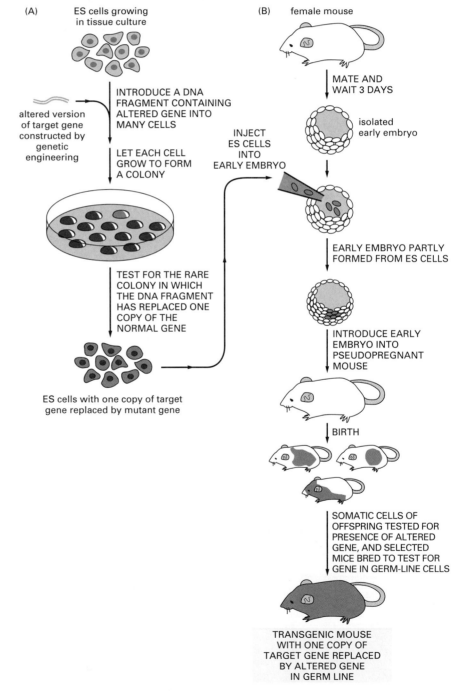

Figure 10–38 Gene replacement in mice utilizes embryonic stem (ES) cells. In the first step (A), an altered version of the gene is introduced into cultured ES cells. In a few rare ES cells, the corresponding normal genes will be replaced by the altered gene through homologous recombination. Although the procedure is often laborious, these rare cells can be identified and cultured to produce many descendants, each of which carries an altered gene in place of one of its two normal corresponding genes. In the next step of the procedure (B), these altered ES cells are injected into a very early mouse embryo; the cells are incorporated into the growing embryo, and a mouse produced by such an embryo will contain some somatic cells (indicated by *orange*) that carry the altered gene. Some of these mice will also contain germ-line cells that contain the altered gene. When bred with a normal mouse, some of the progeny of these mice will contain the altered gene in all of their cells. If two such mice are in turn bred, some of the progeny will contain two altered genes (one on each chromosome) in all of their cells. If the original gene alteration completely inactivates the function of the gene, these mice are known as "knockout" mice. It is often the case that mice missing genes that function during development die long before they reach adulthood.

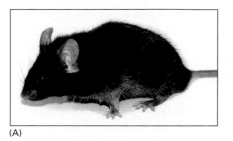 

(A)   (B)

Figure 10–39 Transgenic mice engineered to express a mutant DNA helicase show premature aging. The helicase, encoded by the *XPD* gene, is involved in both transcription and DNA repair. Compared with a wild-type mouse (A), a transgenic mouse that expresses a defective version of *XPD* (B) exhibits many of the symptoms of premature aging, including osteoporosis, emaciation, early graying, infertility, and reduced life span. The mutation in *XPD* used here impairs the activity of the helicase and mimics a mutation that in humans causes trichothiodystrophy, a disorder characterized by brittle hair, skeletal abnormalities, and a strongly reduced life expectancy. These results support the hypothesis that an accumulation of DNA damage contributes to the aging process in both humans and mice. (From J. de Boer et al., *Science* 296:1276–1279, 2002. © AAAS.)

Transgenic Plants Are Important for Both Cell Biology and Agriculture

Although we tend to think of recombinant DNA research in terms of animal biology, these techniques have also had a profound impact on our study of plants. In fact, certain features of plants make them especially amenable to recombinant DNA methods.

When a piece of plant tissue is cultured in a sterile medium containing nutrients and appropriate growth regulators, many of the cells are stimulated to proliferate indefinitely in a disorganized manner, producing a mass of relatively undifferentiated cells called a *callus*. If the nutrients and growth regulators are carefully manipulated, one can induce the formation of a shoot within the callus, and in many species, a whole new plant can be regenerated from such shoots. In a number of plants—including tobacco, petunia, carrot, potato, and *Arabidopsis*—a single cell from such a callus can be grown into a small clump of cells from which a whole plant can be regenerated (see Figure 8–2B). Just as mutant mice can be derived by genetic manipulation of embryonic stem cells in culture, so transgenic plants can be created from plant cells transfected with DNA in culture (Figure 10–40).

The ability to produce transgenic plants has greatly accelerated progress in many areas of plant cell biology. It has played an important part, for example, in isolating receptors for growth regulators and in

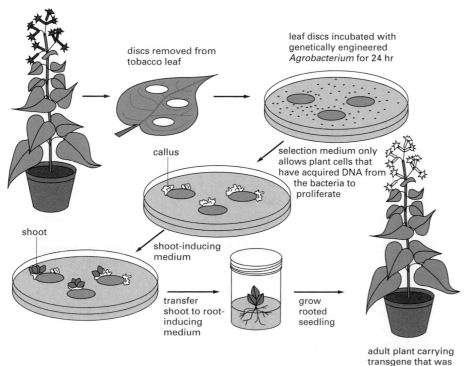

Figure 10–40 Transgenic plants can be made using recombinant DNA techniques. A disc is cut out of a leaf and incubated in an *Agrobacterium* culture in which the bacterial cells carry a recombinant plasmid with both a selectable marker and a desired transgene. The wounded cells at the edge of the disc release substances that attract the bacteria and cause them to inject DNA into these cells. Only those plant cells that take up the appropriate DNA and express the selectable marker gene survive to proliferate and form a callus. The manipulation of growth factors supplied to the callus induces it to form shoots that subsequently root and grow into adult plants carrying the transgene.

discs removed from tobacco leaf

leaf discs incubated with genetically engineered *Agrobacterium* for 24 hr

callus

selection medium only allows plant cells that have acquired DNA from the bacteria to proliferate

shoot

shoot-inducing medium

transfer shoot to root-inducing medium

grow rooted seedling

adult plant carrying transgene that was originally present in the bacteria

analyzing the mechanisms of morphogenesis and of gene expression in plants. These techniques have also opened up many new possibilities in agriculture that could benefit both the farmer and the consumer. They have made it possible, for example, to modify the ratio of lipid, starch, and protein in seeds, to impart pest and virus resistance to plants, and to create modified plants that tolerate extreme habitats such as salt marshes or water-stressed soil. One variety of rice has been genetically engineered to produce β-carotene, the precursor to vitamin A. This "golden rice"—so called because of its faint yellow color—could help to alleviate severe vitamin A deficiency, which causes blindness in hundreds of thousands of children in the developing world each year.

Essential Concepts

- Individual cells can be isolated from their resident tissues; these purified cells can be used for biochemical analysis or for establishing cell cultures.
- Many animal and plant cells survive and proliferate in culture provided they have a suitable medium containing nutrients and the necessary growth factor proteins.
- Most vertebrate cells cease to proliferate after a finite number of cell divisions, but in many cases immortalized cell lines can be generated by providing the cells with the gene that encodes the catalytic subunit of telomerase. These cell lines provide a convenient source of homogeneous cells.
- Recombinant DNA technology has revolutionized the study of the cell, making it possible for researchers to pick out any gene at will from the thousands of genes in a cell and, after an amplification step, to determine the exact molecular structure of the gene.
- A crucial element in this technology is the ability to cut a large DNA molecule into a specific and reproducible set of DNA fragments using restriction nucleases, each of which cuts the DNA double helix only at a particular nucleotide sequence.
- DNA fragments can be separated from one another on the basis of size using gel electrophoresis.
- Techniques are now available for rapidly determining the nucleotide sequence of any isolated DNA fragment.
- The complete nucleotide sequences of the genomes of dozens of single-celled organisms (including bacteria, archaea, and yeasts), as well as several more complex organisms (*Caenorhabditis elegans*, *Drosophila*, *Arabidopsis*, mice, and humans), are now known.
- Nucleic acid hybridization can detect any given DNA or RNA sequence in a mixture of nucleic acid fragments. This technique relies on the fact that a single strand of DNA or RNA will form a double helix only with another nucleic acid strand of the complementary nucleotide sequence.
- Single-stranded DNAs of known sequence and labeled with fluorescent dyes or radioisotopes are used as probes in hybridization reactions. Nucleic acid hybridization can be used to detect the precise location of genes in chromosomes, or RNAs in cells and tissues.
- By presenting a platform for performing a large number of simultaneous hybridization reactions, DNA microarrays can be used to monitor the expression of thousands of genes at once.
- Short DNA strands of any sequence can be made by chemical (nonenzymatic) synthesis in the laboratory.
- DNA cloning techniques enable a DNA sequence to be selected from millions of other sequences and produced in unlimited amounts in pure form.

- DNA fragments can be joined together *in vitro* using DNA ligase to form recombinant DNA molecules not found in nature.
- The first step in a typical cloning procedure is to insert the DNA fragment to be cloned into a DNA molecule capable of replication, such as a plasmid or a viral genome. This recombinant DNA molecule is then introduced into a rapidly dividing host cell, usually a bacterium, so that the DNA is replicated at each cell division.
- A collection of cloned fragments of chromosomal DNA representing the complete genome of an organism is known as a genomic library. The library is often maintained as clones of bacteria, each clone carrying a different DNA fragment.
- cDNA libraries contain cloned DNA copies of the total mRNA of a particular cell type or tissue. Unlike genomic DNA clones, cloned cDNAs contain only protein-coding sequences; they lack introns, gene regulatory sequences, and promoters. They are thus most suitable for use when the cloned gene is to be expressed to make a protein.
- The polymerase chain reaction (PCR) is a powerful form of DNA amplification that is carried out *in vitro* using a purified DNA polymerase. PCR requires a prior knowledge of the sequence to be amplified, because two synthetic oligonucleotide primers must be synthesized that bracket the portion of DNA to be replicated.
- Genetic engineering has far-reaching consequences. Bacteria, yeasts, and mammalian cells can be engineered to synthesize a particular protein from any organism in large quantities, thus making it possible to study proteins that are otherwise rare or difficult to isolate.
- Using DNA engineering techniques, a protein can be joined to a molecular tag, such as the green fluorescent protein (GFP), which allows the tracking of its movement inside the cell. In the case of GFP, the protein can be monitored over time in living organisms.
- Cloned genes can be permanently inserted into the genome of a cell or an organism by the techniques of genetic engineering. Cloned DNA can be altered *in vitro* to create mutant genes which can then be reinserted into a cell or an organism to study gene function.

Key Terms

annotation	hybridization
cDNA	immortalize
cell line	*in situ* hybridization
dideoxy method	phenotype
DNA cloning	plasmid
DNA library	polymerase chain reaction (PCR)
DNA ligase	recombinant DNA technology
DNA microarray	restriction map
embryonic stem cell line	restriction nuclease
genome	site-directed mutagenesis
genotype	transformation
green fluorescent protein (GFP)	transgenic organism

Questions

Question 10–7

Fluorescence-activated cell sorting is just one of the ways for generating homogeneous cell populations. Why do you suppose it is important to have a homogeneous cell population for many experiments?

Question 10–8

Which of the following statements are correct? Explain your answers.

A. Restriction nucleases cut DNA at specific sites that are always located between genes.

B. DNA migrates toward the positive electrode during electrophoresis.

C. Clones isolated from cDNA libraries contain promoter sequences.

D. PCR utilizes a heat-stable DNA polymerase because for each amplification step, double-stranded DNA must be heat-denatured.

E. Digestion of genomic DNA with Alu I, a restriction enzyme that recognizes a four-nucleotide sequence, produces fragments that are all exactly 256 nucleotides in length.

F. To make a cDNA library, both a DNA polymerase and a reverse transcriptase must be used.

G. DNA fingerprinting by PCR relies on the fact that different individuals have different numbers of repeats in VNTR regions in their genome.

H. It is possible for a coding region of a gene to be present in a genomic library prepared from a particular tissue, but not to be represented in a cDNA library prepared from the same tissue.

Question 10–9

Why are human embryonic stem cell lines thought to be especially promising from a medical perspective?

Question 10–10

A. Determine the sequence of the DNA that was used in the sequencing reaction shown in Figure Q10–10. The four lanes show the products of sequencing reactions that contained ddG (lane 1), ddA (lane 2), ddT (lane 3), and ddC (lane 4). The numbers to the right of the autoradiograph represent the

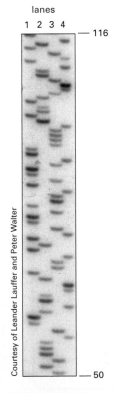

lanes
1 2 3 4

— 116

— 50

Courtesy of Leander Lauffer and Peter Walter

Figure Q10–10

positions of DNA fragments of 50 and 116 nucleotides.

B. The DNA was derived from the middle of a cDNA clone of a mammalian protein. What is the amino acid sequence of this portion of the protein?

Question 10–11

A. How many different DNA fragments would you expect to obtain if you cleaved human genomic DNA with Hae III? (Recall that there are 3×10^9 nucleotide pairs per haploid genome.) How many fragments would you expect with Eco RI, or with Not I?

B. Human genomic libraries are often made from fragments obtained by cleaving human DNA with Hae III in such a way that the DNA is only partially digested, that is, not all the possible Hae III sites have been cleaved. What is a possible reason for doing this?

Question 10–12

A molecule of double-stranded DNA was cleaved with three different restriction nucleases, and the resulting products were separated by electrophoresis (Figure Q10–12). DNA fragments of known sizes were electrophoresed on the same gel for use as size markers. The size of the DNA markers is given in kilobase pairs (kb), where 1 kb = a length of 1000 nucleotide pairs. Using the size markers as a guide, estimate the size of each restriction fragment obtained. From this information, deduce a map of the original DNA molecule that indicates the relative positions of all the restriction sites.

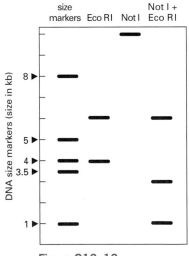

Figure Q10–12

Question 10–13

A mutation engineered *in vitro* as shown in Figure 10–36C introduces a mismatch into the DNA. Would you expect this mismatch to be recognized and repaired by DNA mismatch repair enzymes (see Figure 6–22) when the plasmid that contains the mismatch is introduced into cells? Explain your answer.

Question 10–14

You have isolated a small amount of a rare protein. You cleaved the protein into fragments using proteases, separated some of the fragments by chromatography, and determined their amino acid sequence. Unfortunately, as is often the case when only small amounts of protein are available to start with, you obtained only three short stretches of amino acid sequence from the protein:

1. Trp-Met-His-His-Lys
2. Leu-Ser-Arg-Leu-Arg
3. Tyr-Phe-Gly-Met-Gln

A. Using the genetic code (see Figure 7–21), design oligonucleotide probes that could be used to detect the gene in a cDNA library by hybridization. Which of the three sets of oligonucleotide probes would it be preferable to use first? Explain your answer.

B. You have also been able to determine that the Gln of your peptide #3 is the C-terminal (i.e., the final) amino acid of your protein. How would you go about designing oligonucleotide primers that could be used to amplify a portion of the gene from a cDNA library using PCR?

C. Suppose the PCR amplification in (B) yields a DNA that is precisely 300 nucleotides long. Upon determining the nucleotide sequence of this DNA, you find the sequence CTAT-CACGCTTTAGG approximately in its middle. What would you conclude from these observations?

Question 10–15

Discuss the following statement: "From the nucleotide sequence of a cDNA clone, the complete amino acid sequence of a protein can be deduced by applying the genetic code. Thus, protein biochemistry has become superfluous because there is nothing more that can be learned by studying the protein."

Question 10–16

Assume that a DNA sequencing reaction is carried out as shown in Figure 10–7, except that the four different dideoxyribonucleoside triphosphates are modified so that each contains a covalently attached dye of a different color (which does not interfere with its incorporation into the DNA chain). What would the products be if you added a mixture of all four of these labeled dideoxyribonucleoside triphosphates along with the four unlabeled deoxyribonucleoside triphosphates into a single sequencing reaction? What would the results look like if you electrophoresed these products in a single lane of a gel?

Question 10–17

As described in the answer to Question 10–10B, genomic DNA clones are often used to "walk" along a chromosome. In this approach, one cloned DNA is used to isolate other clones that contain overlapping DNA sequences (Figure Q10–17). Using this method, it is possible to build up a stretch of DNA sequence and thus identify new genes in near proximity to a previously cloned gene.

A. Would it be faster to use cDNA clones in this method, because they do not contain any intron sequences?

B. What would the consequences be if you encountered a repetitive DNA sequence, like the L1 transposon (see Figure 6–35), which is found in many copies and in many different places in the genome?

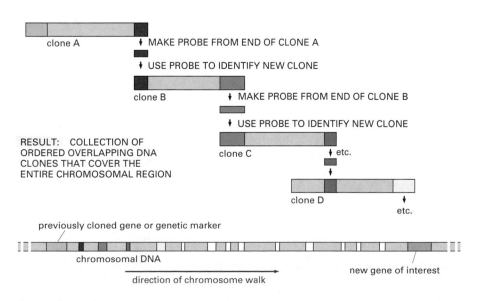

Figure Q10–17

Question 10–18

There has been a colossal snafu in the maternity ward at your local hospital. Four sets of male twins, born within an hour of each other, were inadvertently shuffled in the excitement occasioned by that unlikely event. You have been called in to set things right. As a first step, you want to get the twins matched up. To that end you analyze a small blood sample from each infant using a hybridization probe that detects variable number tandem repeat (VNTR) polymorphisms located in widely scattered regions of the genome. The results are shown in Figure Q10–18.

A. Which infants are brothers?

B. How will you match brothers to the correct parents?

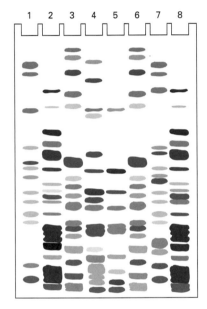

Figure Q10–18

Question 10–19

One of the first organisms that was genetically engineered using modern DNA technology was a bacterium that normally lives on the surface of strawberry plants. This bacterium makes a protein, called ice-protein, that causes the efficient formation of ice crystals around it when the temperature drops to just below freezing. Thus, strawberries harboring this bacterium are particularly susceptible to frost damage because their cells are destroyed by the ice crystals. Consequently, strawberry farmers have a considerable interest in preventing ice crystallization.

Figure Q10–19

A genetically engineered version of this bacterium was constructed in which the ice-protein gene was knocked out. The mutant bacteria were then introduced in large numbers into strawberry fields, where they displaced the normal bacteria by competition for their ecological niche. This approach has been successful: strawberries bearing the mutant bacteria show a much reduced susceptibility to frost damage.

Nevertheless, at the time they were first carried out, the initial open-field trials triggered an intense debate because they represented the first release into the environment of an organism that had been genetically engineered using recombinant DNA techniques. Indeed, all preliminary experiments were carried out with extreme caution and in strict containment. The photograph gives an idea of the containment conditions during the initial applications of the bacteria to strawberry plants (Figure Q10–19).

Discuss some of the issues that arise from such applications of DNA technology. Do you think that bacteria lacking the ice-protein could be isolated without the use of modern DNA technology? Is it likely that such mutations have already occurred in nature? Would the use of a mutant bacterial strain isolated from nature be of lesser concern? Should we be concerned about the risks posed by the application of genetic engineering techniques in agriculture, medicine, and technology? Explain your answers.

Highlight from *Essential Cell Biology 2 Interactive* CD-ROM

10.1   Polymerase Chain Reaction

Membrane Structure

A living cell is a self-reproducing system of molecules held inside a container. That container is the **plasma membrane**—a fatty film so thin and transparent that it cannot be seen directly in the light microscope. Every cell on Earth uses a membrane to separate and protect its chemical components from the outside environment. Without membranes there would be no cells, thus no life.

The cellular membrane is simple in form: its structure is based on a two-ply sheet of lipid molecules about 5 nm—or 50 atoms—thick. Its properties, however, are unlike those of any sheet of material that we are familiar with in the everyday world. Although it serves as a barrier to prevent the contents of the cell from escaping and mixing with the surrounding medium (Figure 11–1A), the plasma membrane does much more than that. If a cell is to survive and grow, nutrients must pass inward, across the plasma membrane; and waste products must pass out. To facilitate this exchange, the membrane is penetrated by highly selective channels and pumps—protein molecules that allow specific substances to be imported while others are exported. Other protein molecules in the membrane act as sensors that enable the cell to respond to changes in its environment. The mechanical properties of the membrane are equally remarkable. When a cell grows or changes shape, so does its membrane: it enlarges in area by adding new membrane without ever losing its continuity, and it can deform without tearing (Figure 11–2). If the membrane is pierced, it neither collapses like a balloon nor remains torn; instead, it quickly reseals.

The simplest bacteria have only a single membrane—the plasma membrane. Eucaryotic cells, however, contain in addition a profusion of *internal membranes* that enclose intracellular compartments. These other membranes are constructed on the same principles as the plasma membrane, and they, too, serve as highly selective barriers between spaces containing distinct collections of molecules (as shown in Figure 11–1B). Thus the membranes of the endoplasmic reticulum, Golgi apparatus, mitochondria, and other membrane-enclosed organelles (Figure 11–3) maintain the characteristic differences in composition and function among these discrete organelles. These internal membranes act as more than just barriers: subtle differences among them, especially differences in the resident membrane proteins, give each organelle its distinctive character.

The Lipid Bilayer

Membrane Lipids Form Bilayers in Water

The Lipid Bilayer Is a Two-dimensional Fluid

The Fluidity of a Lipid Bilayer Depends on Its Composition

The Lipid Bilayer Is Asymmetrical

Lipid Asymmetry Is Generated Inside the Cell

Membrane Proteins

Membrane Proteins Associate with the Lipid Bilayer in Various Ways

A Polypeptide Chain Usually Crosses the Bilayer as an α Helix

Membrane Proteins Can Be Solubilized in Detergents and Purified

The Complete Structure Is Known for a Few Membrane Proteins

The Plasma Membrane Is Reinforced by the Cell Cortex

The Cell Surface Is Coated with Carbohydrate

Cells Can Restrict the Movement of Membrane Proteins

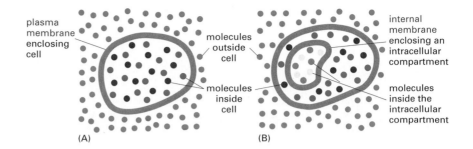

Figure 11–1 Cell membranes act as selective barriers. **Figure 11–1 Cell membranes act as selective barriers.** Membranes serve as barriers between two compartments—either between the inside and the outside of the cell (A) or between two intracellular compartments (B). In either case the membrane prevents molecules on one side from mixing with those on the other.

Regardless of their location, all cell membranes are composed of lipids and proteins and share a common general structure (Figure 11–4). The lipid component consists of many millions of lipid molecules arranged in two closely apposed sheets, forming a *lipid bilayer* (see Figure 11–4B and C). This lipid bilayer gives the membrane its basic structure and serves as a permeability barrier. The protein molecules mediate most of the other functions of the membrane and give different membranes their individual characteristics.

In this chapter we consider the structure and organization of the two main constituents of biological membranes—the lipids and the membrane proteins. Although we focus mainly on the plasma membrane, most of the concepts we discuss also apply to the various internal cellular membranes. The functions of cell membranes, including their role in the transport of small molecules and in energy generation, are considered in later chapters.

The Lipid Bilayer

The **lipid bilayer** has been firmly established as the universal basis of cell-membrane structure. The properties of this bilayer are responsible for the general properties of cell membranes. We begin this section by considering how the structure of the lipid bilayer is a consequence of the way these lipid molecules behave in a watery (aqueous) environment.

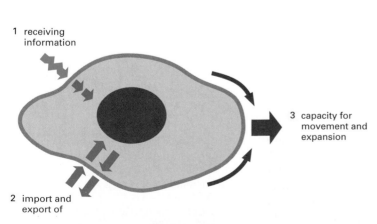

Figure 11–2 The plasma membrane is involved in cell signaling, the transport of small molecules, and cellular growth and motility.

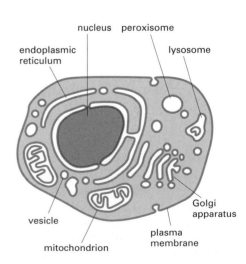

Figure 11–3 Membranes form the many different compartments in a eucaryotic cell. The membrane-enclosed organelles in a typical animal cell are shown here. Note that the nucleus and mitochondria are each enclosed by two membranes.

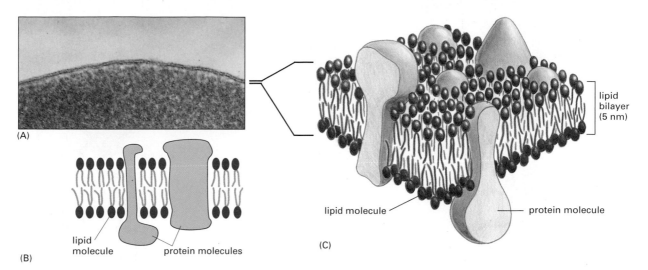

(A)

(B)

(C)

lipid bilayer (5 nm)

lipid molecule

protein molecule

lipid molecule

protein molecules

Figure 11–4 A cell membrane can be viewed in a number of ways. (A) An electron micrograph of a plasma membrane (of a human red blood cell) seen in cross section. (B and C) Schematic drawings showing two-dimensional and three-dimensional views of a cell membrane. (A, courtesy of Daniel S. Friend.)

Membrane Lipids Form Bilayers in Water

The lipids in cell membranes combine two very different properties in a single molecule: each lipid has a hydrophilic ("water-loving") *head* and one or two hydrophobic ("water-hating") *hydrocarbon tails* (Figure 11–5). The most abundant lipids in cell membranes are the **phospholipids**, molecules in which the hydrophilic head is linked to the rest of the lipid through a phosphate group. The most common type of phospholipid in most cell membranes is **phosphatidylcholine**, which has the small molecule choline attached to a phosphate as its hydrophilic head and two long hydrocarbon chains as its hydrophobic tails (Figure 11–6).

Molecules with both hydrophilic and hydrophobic properties are termed **amphipathic**. This chemical property is also shared by other types of membrane lipids—the *sterols* (such as the cholesterol found in animal cell membranes) and the *glycolipids,* which have sugars as part of their hydrophilic head (Figure 11–7)—and it plays a crucial part in driving these lipid molecules to assemble into bilayers.

As discussed in Chapter 2, hydrophilic molecules dissolve readily in water because they contain charged atoms or polar groups, that is, groups with an uneven distribution of positive and negative charges; these charged atoms can form electrostatic bonds or hydrogen bonds with water molecules, which are themselves polar (Figure 11–8). Hydrophobic molecules, by contrast, are insoluble in water because all—or almost all—of their atoms are uncharged and nonpolar; they therefore cannot form bonds with water molecules. Instead, these nonpolar atoms force adjacent water molecules to reorganize themselves into a cagelike structure around the hydrophobic molecule (Figure 11–9). Because the cagelike structure is more highly ordered than the surrounding water, its formation requires energy. The energy cost is minimized, however, if the hydrophobic molecules cluster together, limiting their contact with water to the smallest possible number of water molecules. Thus, purely hydrophobic molecules, like the fats found in animal fat cells and the oils found in plant seeds (Figure 11–10A), coalesce into a single large drop when dispersed in water.

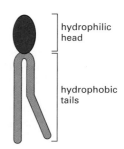

hydrophilic head

hydrophobic tails

Figure 11–5 A typical membrane lipid molecule has a hydrophilic head and hydrophobic tails.

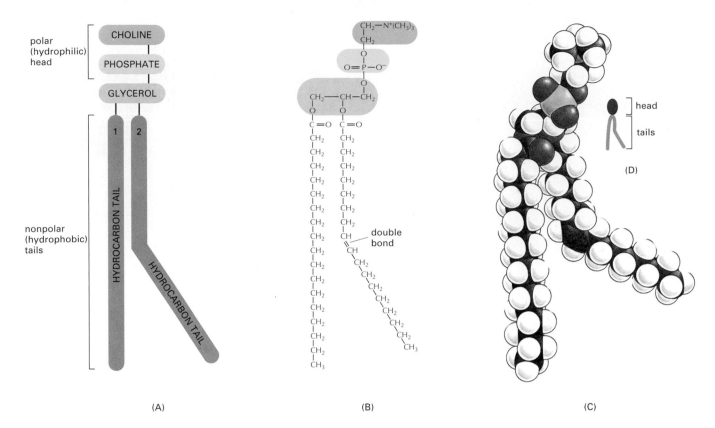

Figure 11–6 Phosphatidylcholine is the most common phospholipid in cell membranes. It is represented (A) schematically, (B) in formula, (C) as a space-filling model, and (D) as a symbol. This particular phospholipid is built from five parts: the hydrophilic head, *choline,* is linked via a *phosphate* to *glycerol,* which in turn is linked to two *hydrocarbon chains,* forming the hydrophobic tail. The two hydrocarbon chains originate as *fatty acids*—that is, hydrocarbon chains with a –COOH group at one end—which become attached to glycerol via their –COOH groups. A kink in one of the hydrocarbon chains occurs where there is a double bond between two carbon atoms; it is exaggerated in these drawings for emphasis. The "phosphatidyl-" part of the name of phospholipids refers to the phosphate–glycerol–fatty acid portion of the molecule.

Figure 11–7 Different types of membrane lipids are all amphipathic. Each of the three types shown here has a hydrophilic head and one or two hydrophobic tails. The hydrophilic head (shaded *blue* and *green*) is serine phosphate in phosphatidylserine, an –OH group in cholesterol, and a sugar (galactose) and an –OH group in galactocerebroside. See also Panel 2–4, pp. 72–73.

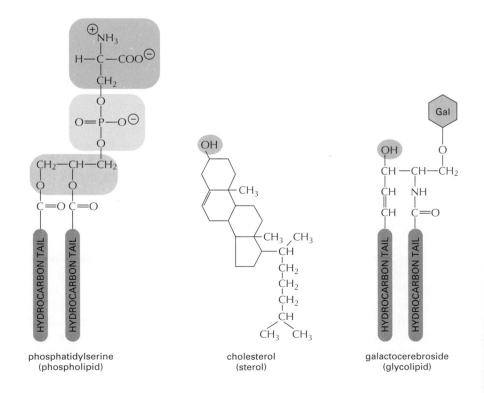

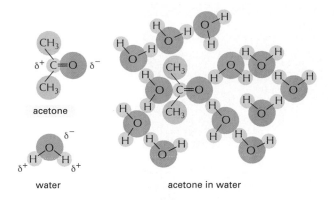

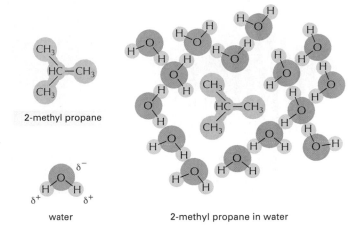

Figure 11–8 A hydrophilic molecule attracts water molecules. Because acetone is polar, it can form favorable interactions with water molecules, which are also polar. Thus acetone readily dissolves in water. δ^- indicates a partial negative charge, and δ^+ indicates a partial positive charge. Polar atoms are shown in color *(red and blue)*; nonpolar groups are shown in *gray*.

Figure 11–9 A hydrophobic molecule tends to avoid water. Because the 2-methyl propane molecule is entirely hydrophobic, it cannot form favorable interactions with water and forces adjacent water molecules to reorganize into a cagelike structure around it.

In contrast, amphipathic molecules, such as phospholipids (Figure 11–10B), are subject to two conflicting forces: the hydrophilic head is attracted to water, while the hydrophobic tail shuns water and seeks to aggregate with other hydrophobic molecules. This conflict is beautifully resolved by the formation of a lipid bilayer—an arrangement that satisfies all parties and is energetically most favorable. The hydrophilic heads face the water at each of the two surfaces of the sheet of molecules; the hydrophobic tails are all shielded from the water and lie next to one another in the interior of the sandwich (Figure 11–11).

Question 11–1

Water molecules are said "to arrange into a cagelike structure" around hydrophobic compounds (e.g., see Figure 11–9). This seems paradoxical because water molecules do not interact with the hydrophobic compound. So how could they "know" about its presence and change their behavior to interact differently with one another? Discuss this argument and in doing so develop a clear concept of what is meant by a "cagelike" structure. How does it compare to ice? Why would this cagelike structure be energetically unfavorable?

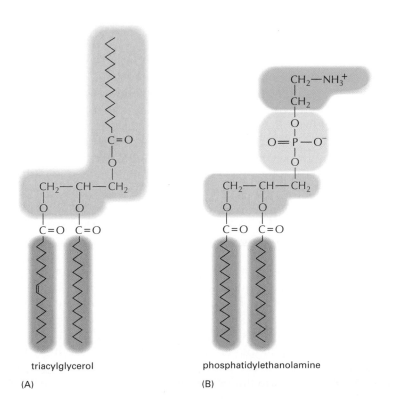

triacylglycerol

(A)

phosphatidylethanolamine

(B)

Figure 11–10 Fat molecules are hydrophobic, whereas phospholipids are amphipathic. (A) Triacylglycerol, a fat molecule, is entirely hydrophobic. (B) Phospholipids such as phosphatidylethanolamine are amphipathic, containing both hydrophobic and hydrophilic portions. The hydrophobic parts are shaded *red,* and the hydrophilic parts are shaded *blue* and *green.* (The third hydrophobic tail of the triacylglycerol molecule is drawn here facing upward for comparison with the phospholipid, although normally it is depicted facing down.)

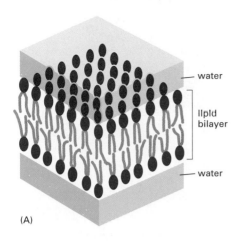

water

llpld
bilayer

water

(A)

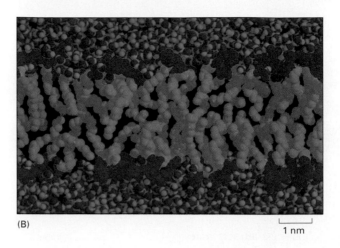

(B)

1 nm

Figure 11–11 Amphipathic phospholipids form a bilayer in water. (A) Schematic drawing of a phospholipid bilayer in water. (B) Computer simulation showing the phospholipid molecules *(red heads* and *orange tails)* and the surrounding water molecules *(blue)* in a cross section of a lipid bilayer. (B, adapted from *Science* 262:223–228 (1993), courtesy of R. Venable and R. Pastor.)

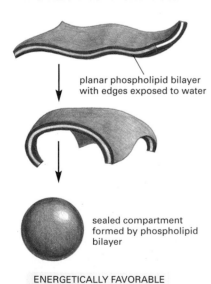

ENERGETICALLY UNFAVORABLE

planar phospholipid bilayer
with edges exposed to water

sealed compartment
formed by phospholipid
bilayer

ENERGETICALLY FAVORABLE

Figure 11–12 Phospholipid bilayers spontaneously close in on themselves to form sealed compartments. The closed structure is stable because it avoids the exposure of the hydrophobic hydrocarbon tails to water, which would be energetically unfavorable.

The same forces that drive the amphipathic molecules to form a bilayer confer on that bilayer a self-sealing property. Any tear in the bilayer will create a free edge that is exposed to water. Because this is energetically unfavorable, the molecules of the bilayer will spontaneously rearrange to eliminate the free edge. If the tear is small, this spontaneous rearrangement will exclude the water molecules and lead to repair of the bilayer, restoring a single continuous sheet. If the tear is large, the sheet may begin to fold in on itself and break up into separate closed vesicles. In either case, the overriding principle is that free edges are quickly eliminated.

The prohibition on free edges has a profound consequence: the only way a finite sheet can avoid having free edges is to form a boundary around a closed space (Figure 11–12). Therefore, amphipathic molecules such as phospholipids necessarily assemble to form self-sealing containers that define closed compartments. This remarkable behavior, fundamental to the creation of a living cell, is in essence simply a result of the property that each molecule is hydrophilic at one end and hydrophobic at the other.

The Lipid Bilayer Is a Two-dimensional Fluid

The aqueous environment inside and outside a cell prevents membrane lipids from escaping from the bilayer, but nothing stops these molecules from moving about and changing places with one another within the plane of the bilayer. The membrane therefore behaves as a two-dimensional fluid, which is crucial for membrane function. This property is distinct from *flexibility,* which is the ability of the membrane to bend. Membrane flexibility is also important, and it sets a lower limit of about 25 nm to the size of vesicle that cell membranes can form.

The fluidity of lipid bilayers can be studied using synthetic lipid bilayers, which are easily produced by the spontaneous aggregation of amphipathic lipid molecules in water. Two types of synthetic lipid bilayers are commonly used in experiments. Closed spherical vesicles, called *liposomes,* form if pure phospholipids are added to water; they vary in size from about 25 nm to 1 mm in diameter (Figure 11–13). Alternatively, flat phospholipid bilayers can be formed across a hole in a partition between two aqueous compartments (Figure 11–14).

These simple artificial bilayers allow delicate measurements of the movements of the lipid molecules, revealing that some types of movement are rare while others are frequent and rapid. Thus, in synthetic lipid bilayers, phospholipid molecules very rarely tumble from one monolayer (one half of the bilayer) to the other. Without proteins to facilitate the process and under conditions similar to those in a cell, it is

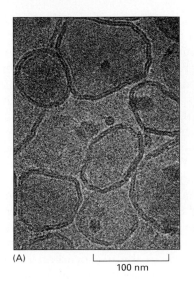 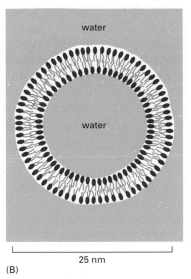

(A)

100 nm

(B)

25 nm

Figure 11–13 **Pure phospholipids can form closed, spherical liposomes.** (A) An electron micrograph of phospholipid vesicles (liposomes) showing the bilayer structure of the membrane. (B) A drawing of a small spherical liposome seen in cross section. (A, courtesy of Jean Lepault.)

estimated that this event, called "flip-flop," occurs less than once a month for any individual lipid molecule. On the other hand, as the result of thermal motions, lipid molecules within a monolayer continuously exchange places with their neighbors (Figure 11–15). This exchange leads to rapid diffusion in the plane of the membrane so that, for example, a lipid molecule in an artificial bilayer may diffuse a length equal to that of a large bacterial cell (~2 μm) in about one second. If the temperature is decreased, the drop in thermal energy decreases the rate of lipid movement, making the bilayer less fluid.

Similar observations are made when one examines isolated cell membranes and whole cells, indicating that the lipid bilayer of a cell membrane also behaves as a two-dimensional fluid in which the constituent lipid molecules are free to move within their own layer in any direction in the plane of the membrane. These studies also show that lipid hydrocarbon chains are flexible and that individual lipid molecules within a monolayer rotate very rapidly about their long axis—some reaching speeds of 30,000 rpm (see Figure 11–15). In cells, as in synthetic bilayers, individual phospholipid molecules are normally confined to their own monolayer and do not flip-flop spontaneously.

The Fluidity of a Lipid Bilayer Depends on Its Composition

The fluidity of a cell membrane—the ease with which its lipid molecules move within the plane of the bilayer—is important for membrane function and has to be maintained within certain limits. Just how fluid a lipid bilayer is at a given temperature depends on its phospholipid composition and, in particular, on the nature of the hydrocarbon tails: the closer and more regular the packing of the tails, the more viscous and less fluid

Question 11–2

Five students in your class always sit together in the front row. This could be because (A) they really like each other or (B) nobody else in your class wants to sit next to them. Which explanation holds for the assembly of a lipid bilayer? Explain. Suppose instead that the other explanation held for lipid molecules. How would the properties of the lipid bilayer be different?

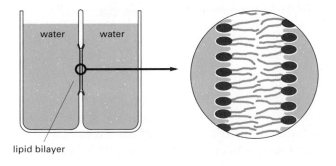

Figure 11–14 **A synthetic phospholipid bilayer can be formed across a small hole (about 1 mm in diameter) in a partition that separates two aqueous compartments.** To make the planar bilayer, the partition is submerged in an aqueous solution and a phospholipid solution (in a nonaqueous solvent) is painted across the hole with a paintbrush.

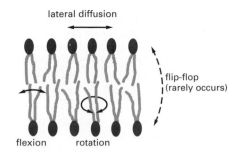

lateral diffusion

flip-flop
(rarely occurs)

flexion   rotation

Figure 11–15 Phospholipids can move within the plane of the membrane. The drawing shows the types of movement possible for phospholipid molecules in a lipid bilayer.

the bilayer will be. Two major properties of hydrocarbon tails affect how tightly they pack in the bilayer—their length, and their degree of *unsaturation* (that is, the number of double bonds they contain). A shorter chain length reduces the tendency of the hydrocarbon tails to interact with one another and therefore increases the fluidity of the bilayer. The hydrocarbon tails of membrane phospholipid molecules vary in length between 14 and 24 carbon atoms, with 18–20 atoms being most usual.

As for unsaturation, most phospholipids usually contain one hydrocarbon tail that has one or more double bonds between adjacent carbon atoms, and a second tail with single bonds only (see Figure 11–6). The chain that harbors a double bond does not contain the maximum number of hydrogen atoms that could, in principle, be attached to its carbon backbone; it is thus said to be **unsaturated** with respect to hydrogens. The fatty acid tail with no double bonds has a full complement of hydrogen atoms; it is said to be **saturated**. Each double bond in an unsaturated tail creates a small kink in the hydrocarbon tail (see Figure 11–6) that makes it more difficult for the tails to pack against one another. For this reason, lipid bilayers that contain a large proportion of unsaturated hydrocarbon tails are more fluid than those with lower proportions.

In bacterial and yeast cells, which have to adapt to varying temperatures, both the lengths and the unsaturation of the hydrocarbon tails in the bilayer are constantly adjusted to maintain the membrane at a relatively constant fluidity: at higher temperatures, for example, the cell makes membrane lipids with tails that are longer and that contain fewer double bonds. A related trick is used in the manufacture of margarine from vegetable oils. The fats produced by plants are generally unsaturated and therefore liquid at room temperatures, unlike animal fats such as butter or lard, which are generally saturated and therefore solid at room temperature. Margarine is made of hydrogenated vegetable oils, whose double bonds have been removed by the addition of hydrogen, so that they are more solid and butterlike at room temperature.

In animal cells, membrane fluidity is modulated by the inclusion of the sterol **cholesterol** (Figure 11–16A). These short, rigid molecules are present in especially large amounts in the plasma membrane, where they fill the spaces between neighboring phospholipid molecules left by the kinks in their unsaturated hydrocarbon tails (Figure 11–16B). In this way cholesterol tends to stiffen the bilayer, making it more rigid and less permeable.

For all cells, membrane fluidity is important for many reasons. It enables membrane proteins to diffuse rapidly in the plane of the bilayer and to interact with one another, as is crucial, for example, in

Figure 11–16 Cholesterol stiffens cell membranes. (A) The structure of cholesterol. (B) How cholesterol fits into the gaps between phospholipid molecules in a lipid bilayer. The chemical formula of cholesterol is shown in Figure 11–7.

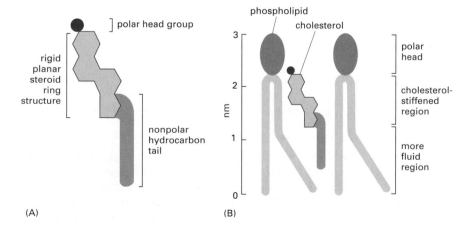

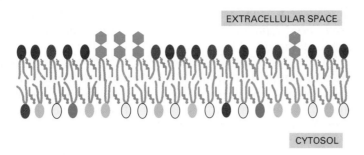

Figure 11–17 Phospholipids and glycolipids are distributed asymmetrically in the plasma membrane lipid bilayer. Five types of phospholipid molecules are shown in different colors: phosphatidylcholine *(red)*, sphingomyelin *(brown)*, phosphatidylserine *(light green)*, phosphatidylinositol *(dark green)*, and phosphatidylethanolamine *(yellow)*. The glycolipids are drawn with *blue* hexagonal head groups to represent sugars. All of the glycolipid molecules are in the external monolayer of the membrane, while cholesterol *(gray)* is distributed almost equally in both monolayers.

cell signaling (discussed in Chapter 16). It permits membrane lipids and proteins to diffuse from sites where they are inserted into the bilayer after their synthesis to other regions of the cell. It allows membranes to fuse with one another and mix their molecules, and it ensures that membrane molecules are distributed evenly between daughter cells when a cell divides. If biological membranes were not fluid, it is hard to imagine how cells could live, grow, and reproduce.

The Lipid Bilayer Is Asymmetrical

Cell membranes are generally asymmetrical: they present a very different face to the interior of the cell or organelle than they show to the exterior. The two halves of the bilayer often include strikingly different selections of phospholipids and glycolipids (Figure 11–17). Moreover, the proteins are embedded in the bilayer with a specific orientation, which is crucial for their function.

The lipid asymmetry is established at the point of manufacture. In eucaryotic cells, new phospholipid molecules are synthesized by enzymes bound to the part of the ER membrane that faces the cytosol; these enzymes use as substrates fatty acids available in the cytosolic half of the bilayer—that is, the cytosolic monolayer—and they release the newly made phospholipid into the same monolayer. To enable the membrane as a whole to grow evenly, a proportion of the lipid molecules then have to be transferred to the opposite monolayer. This transfer is catalyzed by enzymes called *flippases* (Figure 11–18). Some flippases transfer specific phospholipid molecules selectively, so that different types become concentrated in each monolayer.

One-sided insertion and selective flippases are not the only ways of producing asymmetry in lipid bilayers, however. In particular, a different mechanism operates for glycolipids—the class of lipid molecules that shows the most striking and consistent asymmetric distribution in animal cells. To explain their distribution, it is necessary to look more carefully at the route by which new membrane is produced in eucaryotic cells.

Lipid Asymmetry Is Generated Inside the Cell

In eucaryotic cells nearly all new membrane synthesis occurs in one intracellular compartment—the *endoplasmic reticulum*, or *ER* (discussed in more detail in Chapter 15). The new membrane assembled there is exported to the other membranes of the cell through a cycle of vesicle budding and fusion: bits of membrane pinch off from the ER to form small vesicles, which then become incorporated into another membrane by fusing with it. Because the orientation of the bilayer relative to the cytosol is preserved during this vesicular transport process, all cell membranes, whether the external plasma membrane or an intracellular membrane surrounding an organelle, have distinct "inside" and

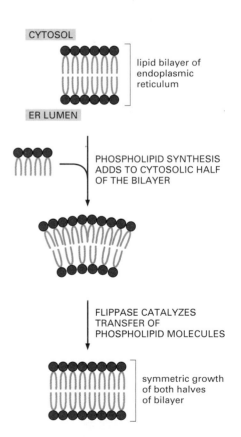

CYTOSOL

ER LUMEN

lipid bilayer of endoplasmic reticulum

PHOSPHOLIPID SYNTHESIS ADDS TO CYTOSOLIC HALF OF THE BILAYER

FLIPPASE CATALYZES TRANSFER OF PHOSPHOLIPID MOLECULES

symmetric growth of both halves of bilayer

Figure 11–18 Flippases play a role in synthesizing the lipid bilayer. Although the newly synthesized phospholipid molecules are all added to one side of the bilayer, flippases transfer some of these to the opposite monolayer so that the entire bilayer expands.

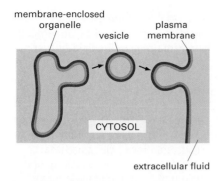

Figure 11–19 Membrane vesicles are generated by budding and fusing. A membrane vesicle is shown budding from a membrane-enclosed organelle and fusing with the plasma membrane. Note that the orientation of the membrane is preserved during the process of vesicular budding and fusion, so that the cytosolic surface remains the cytosolic surface.

Question 11–3

It seems paradoxical that a lipid bilayer can be fluid yet asymmetrical. Explain.

"outside" faces: the *cytosolic* face is always adjacent to the cytosol, while the *noncytosolic* face is exposed to either the cell exterior or the interior space of an organelle (Figure 11–19).

Glycolipids are located mainly in the plasma membrane, and they are found only in the noncytosolic half of the bilayer. Their sugar groups therefore are exposed to the exterior of the cell (see Figure 11–17), where they form part of a continuous protective coat of carbohydrate that surrounds most animal cells. The glycolipid molecules acquire their sugar groups in the Golgi apparatus (discussed in Chapter 15). The enzymes that add the sugar groups are confined to the inside of the Golgi apparatus so that the sugars are added only to lipid molecules in the noncytosolic half of the lipid bilayer. Once a glycolipid molecule has been created in this way, it remains trapped in this monolayer, as there are no flippases to transfer it to the cytosolic monolayer. Thus, when it is finally delivered to the plasma membrane, the glycolipid molecule faces away from the cytosol and displays its sugar on the exterior of the cell (see Figure 11–19).

Other lipid molecules show different types of asymmetric distributions, related to other functions. The *inositol phospholipids,* for example, are minor components of the plasma membrane, but they play a special role in relaying signals from the cell surface to the intracellular components that respond to those signals (discussed in Chapter 16). They act only after the signal has been transmitted across the plasma membrane; thus they are concentrated in the cytosolic half of this lipid bilayer (see Figure 11–17).

Membrane Proteins

Although the lipid bilayer provides the basic structure of all cell membranes and serves as a permeability barrier to the molecules on either side of it, most membrane functions are carried out by **membrane proteins.** In animals, proteins constitute about 50% of the mass of most plasma membranes, the remainder being lipid plus relatively small amounts of carbohydrate. Because lipid molecules are much smaller than protein molecules, however, a cell membrane typically contains about 50 times more lipid molecules than protein (see Figure 11–4).

Membrane proteins not only transport particular nutrients, metabolites, and ions across the lipid bilayer; they serve many other functions. Some anchor the membrane to macromolecules on either side. Others function as receptors that detect chemical signals in the cell's environment and relay them to the cell's interior, and still others work as enzymes to catalyze specific reactions (Figure 11–20; Table 11–1). Each type of cell membrane contains a different set of proteins, reflecting the specialized functions of the particular membrane. In this section we discuss the structure of membrane proteins and illustrate the different ways that they can be associated with the lipid bilayer.

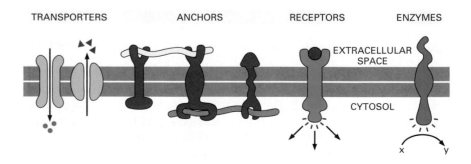

Figure 11–20 Plasma membrane proteins have a variety of functions.

Table 11–1 Some Examples of Plasma Membrane Proteins and Their Functions

FUNCTIONAL CLASS	PROTEIN EXAMPLE	SPECIFIC FUNCTION
Transporters	Na+ pump	actively pumps Na+ out of cells and K+ in
Anchors	integrins	link intracellular actin filaments to extracellular matrix proteins
Receptors	platelet-derived growth factor (PDGF) receptor	binds extracellular PDGF and, as a consequence, generates intracellular signals that cause the cell to grow and divide
Enzymes	adenylyl cyclase	catalyzes the production of intracellular cyclic AMP in response to extracellular signals

Membrane Proteins Associate with the Lipid Bilayer in Various Ways

Proteins can be associated with the lipid bilayer of a cell membrane in several ways (Figure 11–21).

1. Many membrane proteins extend through the bilayer, with part of their mass on either side (Figure 11–21A). Like their lipid neighbors, these *transmembrane proteins* have both hydrophobic and hydrophilic regions. Their hydrophobic regions lie in the interior of the bilayer, nestled against the hydrophobic tails of the lipid molecules. Their hydrophilic regions are exposed to the aqueous environment on either side of the membrane.

2. Other membrane proteins are located entirely in the cytosol, associated with the inner leaflet of the lipid bilayer by an amphipathic α helix exposed on the surface of the protein (Figure 11–21B).

3. Some proteins lie entirely outside the bilayer, on one side or the other, attached to the membrane only by one or more covalently attached lipid groups (Figure 11–21C).

4. Yet other proteins are bound indirectly to one or the other face of the membrane, held in place only by their interactions with other membrane proteins (Figure 11–21D).

Proteins that are directly attached to membranes—whether they are transmembrane, monolayer-associated, or lipid-linked—can be removed only by disrupting the lipid bilayer with detergents, as discussed shortly. Such proteins are known as *integral membrane proteins.* The remaining membrane proteins are known as *peripheral membrane proteins;* they can be released from the membrane by relatively gentle

Figure 11–21 Membrane proteins can associate with the lipid bilayer in several different ways. (A) Transmembrane proteins can extend across the bilayer as a single α helix, as multiple α helices, or as a rolled-up β sheet (called a β barrel). (B) Some membrane proteins are anchored to the cytosolic surface by an amphipathic α helix. (C) Others are attached to either side of the bilayer solely by a covalent attachment to a lipid molecule *(red zigzag lines).* (D) Finally, many proteins are attached to the membrane only by relatively weak, noncovalent interactions with other membrane proteins.

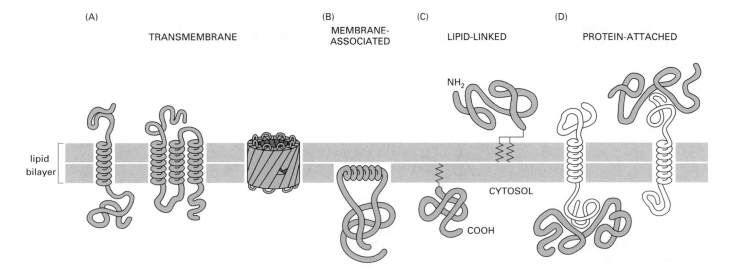

(A) TRANSMEMBRANE (B) MEMBRANE-ASSOCIATED (C) LIPID-LINKED (D) PROTEIN-ATTACHED

lipid bilayer

NH2

CYTOSOL

COOH

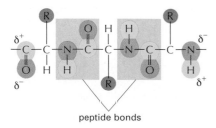

Figure 11–22 **The peptide bonds (shaded in *gray*) that join adjacent amino acids together in a polypeptide chain are polar and therefore hydrophilic.** δ^- indicates a partial negative charge, and δ^+ indicates a partial positive charge.

peptide bonds

extraction procedures that interfere with protein–protein interactions but leave the lipid bilayer intact.

A Polypeptide Chain Usually Crosses the Bilayer as an α Helix

All membrane proteins have a unique orientation in the lipid bilayer: a transmembrane receptor protein, for example, always displays the same region to the cytosol. This orientation is a consequence of the way in which the protein is synthesized (as discussed in Chapter 15). The portions of a transmembrane protein that exist outside the lipid bilayer are connected by specialized membrane-spanning segments of the polypeptide chain. These segments, which run through the hydrophobic environment of the interior of the lipid bilayer, are composed largely of amino acids with hydrophobic side chains. Because these side chains cannot form favorable interactions with water molecules, they prefer the lipid environment, where no water is present.

In contrast to the hydrophobic side chains, however, the peptide bonds that join the successive amino acids in a protein are normally polar, making the polypeptide backbone hydrophilic (Figure 11–22). Because water is absent from the bilayer, atoms forming the backbone are driven to form hydrogen bonds with one another. Hydrogen bonding is maximized if the polypeptide chain forms a regular α helix, and so the great majority of the membrane-spanning segments of polypeptide chains traverse the bilayer as α helices. In these membrane-spanning α helices, the hydrophobic amino acid side chains are exposed on the outside of the helix, where they contact the hydrophobic lipid tails, while atoms in the polypeptide backbone form hydrogen bonds with one another on the inside of the helix (Figure 11–23).

In many transmembrane proteins the polypeptide chain crosses the membrane only once (see Figure 11–21A). Many of these proteins are receptors for extracellular signals: their extracellular part binds the signal molecule, while their cytoplasmic part signals to the cell's interior (see Figure 11–20).

Other transmembrane proteins form aqueous pores that allow water-soluble molecules to cross the membrane. Such pores cannot be formed by proteins with a single, uniformly hydrophobic transmembrane α helix. These more complicated transmembrane proteins usually possess a series of α helices that snake across the bilayer a number of times (see Figure 11–21A). In many of these proteins, one or more of the transmembrane regions are formed from α helices that contain both hydrophobic and hydrophilic amino acid side chains. The hydrophobic side chains lie on one side of the helix, exposed to the lipids of the membrane. The hydrophilic side chains are concentrated on the other side, where they form part of the lining of a hydrophilic pore created by packing several helices side-by-side in a ring within the hydrophobic environment of the lipid bilayer (Figure 11–24). How such pores function in the selective transport of small water-soluble molecules across membranes is discussed in Chapter 12.

Although the α helix is by far the most common form in which a polypeptide chain crosses a lipid bilayer, the polypeptide chain of some transmembrane proteins crosses the lipid bilayer as a β sheet that is curved into a cylinder, forming an open-ended keglike structure called

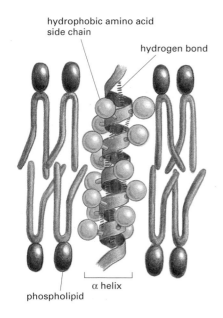

hydrophobic amino acid side chain

hydrogen bond

α helix

phospholipid

Figure 11–23 A segment of α helix crosses a lipid bilayer. The hydrophobic side chains of the amino acids forming the α helix contact the hydrophobic hydrocarbon tails of the phospholipid molecules, while the hydrophilic parts of the polypeptide backbone form hydrogen bonds with one another in the interior of the helix. About 20 amino acids are required to completely traverse a membrane in this way.

a *β barrel*. As expected, the amino acid side chains that face the inside of the barrel, and therefore line the aqueous channel, are mostly hydrophilic, while those on the outside of the barrel, which contact the hydrophobic core of the lipid bilayer, are exclusively hydrophobic. The most striking example of a β barrel structure is found in the *porin* proteins, which form large, water-filled pores in mitochondrial and bacterial membranes (Figure 11–25). Mitochondria and some bacteria are surrounded by a double membrane, and porins allow the passage of nutrients and small ions across their outer membranes while preventing the entry of large molecules such as antibiotics and toxins. Unlike α helices, β barrels can form only wide channels, because there is a limit to how tightly the β sheet can be curved to form the barrel. In this respect a β barrel is less versatile than a collection of α helices.

Membrane Proteins Can Be Solubilized in Detergents and Purified

To understand a protein fully one needs to know its structure in detail, and for membrane proteins this presents special problems. Most biochemical procedures are designed for studying molecules dissolved in water or a simple solvent; membrane proteins, however, are built to operate in an environment that is partly aqueous and partly fatty, and taking them out of this environment and purifying them while preserving their essential structure is no easy task.

Before an individual protein can be studied in detail, it must be separated from all the other cellular proteins. For most membrane proteins, the first step in this separation process involves solubilizing the membrane with agents that destroy the lipid bilayer by disrupting hydrophobic associations. The most widely used disruptive agents are **detergents**, which are small, amphipathic, lipidlike molecules that have both a hydrophilic and a hydrophobic region (Figure 11–26). Detergents differ from membrane phospholipids in that they have only a single hydrophobic tail and, consequently, behave in a significantly different way. Because of their single hydrophobic tail, detergent molecules are shaped more like cones; in water, they tend to aggregate into small clusters called *micelles*, rather than forming a bilayer as do the phospholipids, which are more cylindrical in shape.

When mixed in great excess with membranes, the hydrophobic ends of detergent molecules bind to the membrane-spanning hydrophobic region of the transmembrane proteins, as well as to the hydrophobic tails of the phospholipid molecules, thereby separating the proteins from the phospholipids. Because the other end of the detergent molecule is hydrophilic, this association tends to bring the membrane proteins into solution as protein–detergent complexes (Figure 11–27). At the same time, the detergent solubilizes the phospholipids. The protein–detergent complexes can then be separated from one another and from the lipid–detergent complexes by a technique such as SDS polyacrylamide-gel electrophoresis (discussed in Chapter 4).

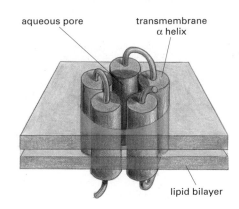

Figure 11–24 **A transmembrane hydrophilic pore can be formed by multiple α helices.** In this example, five transmembrane α helices form a water-filled channel across the lipid bilayer. The hydrophobic amino acid side chains *(green)* on one side of each helix contact the hydrophobic hydrocarbon tails, while the hydrophilic side chains *(red)* on the opposite side of the helices form a water-filled pore.

Question 11–4

Explain why the polypeptide chain of most transmembrane proteins crosses the lipid bilayer as an α helix or a β barrel.

Figure 11–25 **Porin proteins form water-filled channels in the outer membrane of a bacterium *(Rhodobacter capsulatus).*** The protein consists of a 16-stranded β sheet curved around on itself to form a transmembrane water-filled channel, as shown in this three-dimensional structure, determined by X-ray crystallography. Although not shown in the drawing, three porin proteins associate to form a trimer, which has three separate channels. (From S.W. Cowan, *Curr. Opin. Struct. Biol.* 3:501–507, 1993.)

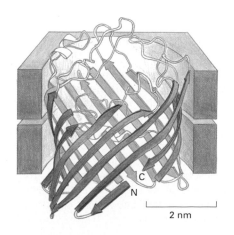

Figure 11–26 SDS and Triton X-100 are two commonly used detergents.

Sodium dodecyl sulfate (SDS) is a strong ionic detergent (that is, it has an ionized group at its hydrophilic end), and Triton X-100 is a mild nonionic detergent (that is, it has a nonionized but polar structure at its hydrophilic end). The hydrophobic portion of each detergent is shown in *blue,* and the hydrophilic portion is shown in *red.* The bracketed portion of Triton X-100 is repeated about eight times. Strong ionic detergents like SDS not only displace lipid molecules from proteins but also unfold the proteins as well (see Panel 4–5, p. 163).

sodium dodecyl sulfate
(SDS)

Triton X-100

The Complete Structure Is Known for a Few Membrane Proteins

Much of what we know about the structure of membrane proteins has been learned by indirect means. The standard direct method for determining protein structure is X-ray crystallography (discussed in Chapter 4), but this requires ordered crystalline arrays of the molecule, and membrane proteins have proved harder to crystallize than the soluble proteins that inhabit the cell cytosol. Among those membrane proteins whose structures have been determined to high resolution are *bacteriorhodopsin* and a *photosynthetic reaction center,* both of which are bacterial membrane proteins with important roles in the capture and use of energy from sunlight. The structures of these proteins have revealed exactly how α helices cross the lipid bilayer, as well as how a set of different protein molecules can assemble to form functional complexes in a membrane.

The structure of **bacteriorhodopsin** is shown in Figure 11–28. This small protein (about 250 amino acids) is found in large amounts in the plasma membrane of an archaebacterium, *Halobacterium halobium,* that lives in salt marshes. Bacteriorhodopsin acts as a membrane transport protein that pumps H$^+$ (protons) out of the bacterium. Pumping requires energy, and bacteriorhodopsin gets its energy directly from sunlight. Each bacteriorhodopsin molecule contains a single light-absorbing nonprotein molecule—called *retinal*—that gives the protein (and the bacterium) a deep purple color. This small

Question 11–5

For the two detergents shown in Figure 11–26, explain why the red portions of the molecules are hydrophilic and the blue portions hydrophobic. Draw a short stretch of a polypeptide chain made up of three amino acids with hydrophobic side chains (see Panel 2–5, pp. 74–75) and apply a similar color scheme.

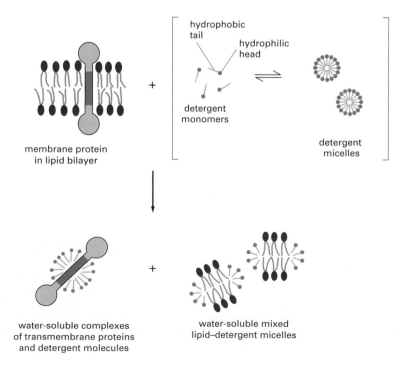

Figure 11–27 Membrane proteins can be solubilized with a mild detergent such as Triton X-100.

The detergent disrupts the lipid bilayer and brings the proteins into solution as protein–detergent complexes. The phospholipids in the membrane are also solubilized by the detergents. As illustrated, detergent molecules in water tend to aggregate into clusters called micelles.

membrane protein
in lipid bilayer

hydrophobic tail

hydrophilic head

detergent monomers

detergent micelles

water-soluble complexes
of transmembrane proteins
and detergent molecules

water-soluble mixed
lipid–detergent micelles

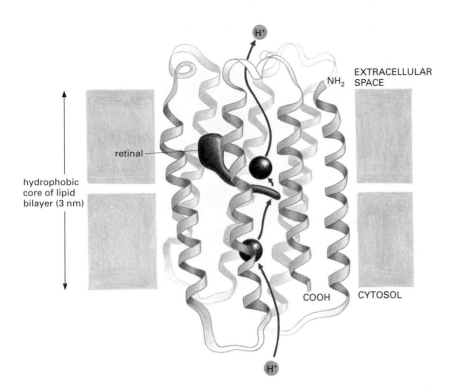

Figure 11–28 Bacteriorhodopsin acts as a proton pump. The polypeptide chain crosses the lipid bilayer as seven α helices. The location of the retinal and the probable pathway taken by protons during the light-activated pumping cycle are shown; two polar amino acid side chains thought to be involved in the H$^+$ transfer process are shown in *black*. Note that the pathway taken by the protons (*red* arrows) enables them to avoid contact with the lipid bilayer. Retinal is also used to detect light in our own eyes, where it is attached to a protein with a structure very similar to bacteriorhodopsin. (Adapted from R. Henderson et al., *J. Mol. Biol.* 213:899–929, 1990.)

hydrophobic molecule is covalently attached to one of bacteriorhodopsin's seven transmembrane α helices and lies in the plane of the lipid bilayer, entirely surrounded by the seven α helices (see Figure 11–28). When retinal absorbs a photon of light, it changes its shape, and in doing so it causes the protein embedded in the lipid bilayer to undergo a series of small conformational changes. These changes result in the transfer of one H$^+$ from the retinal to the outside of the bacterium: the H$^+$ moves across the bilayer along a pathway of strategically placed polar amino acid side chains (see Figure 11–28). The retinal is then regenerated by taking up a H$^+$ from the cytosol, returning the protein to its original conformation so that it can repeat the cycle. The overall outcome is the transfer of one H$^+$ out of the bacterium, which lowers the H$^+$ concentration inside the cell.

In the presence of sunlight, thousands of bacteriorhodopsin molecules pump H$^+$ out of the cell, generating a concentration gradient of H$^+$ across the bacterial membrane. This H$^+$ gradient serves as an energy store, like water behind a dam. And just as dammed water can be used to generate electricity when it flows downhill through a turbine, so an H$^+$ gradient can be used to generate ATP when H$^+$ flows back into the bacterium through a second membrane protein, called *ATP synthase*. The same type of ATP synthase generates much of the ATP in plant and animal cells, as discussed in Chapter 14.

The structure of a bacterial photosynthetic reaction center is shown in Figure 11–29. It is a large complex composed of four protein molecules. Three are transmembrane proteins; two of these (M and L) have multiple α helices passing through the lipid bilayer, while the other (H) has only one. The fourth protein (cytochrome) is associated with the outer surface of the membrane, bound to the transmembrane proteins. The entire protein complex serves as a molecular machine, taking in light energy absorbed by chlorophyll and producing high-energy electrons required for photosynthetic reactions (discussed in Chapter 14). Many membrane proteins are arranged in large complexes, and the structure of the photosynthetic reaction center is a good model for thousands of other membrane proteins whose structures are not yet known.

Figure 11–29 **The photosynthetic reaction center of the bacterium *Rhodopseudomonas viridis* captures energy from sunlight.** The three-dimensional structure was determined by X-ray diffraction analysis of crystals of this transmembrane protein complex. The complex consists of four subunits—L, M, H, and a cytochrome. The L and M subunits form the core of the reaction center, and each contains five α helices that span the lipid bilayer. All the α helices are shown as cylinders. The locations of the various electron carrier groups, which are covalently bound to the protein subunits, are shown in *black,* except for the special pair of chlorophyll molecules that are excited by light, which are shown in *dark green.* Note that the cytochrome is bound to the outer surface of the membrane solely by its attachment to the transmembrane subunits (see Figure 11–21D). (Adapted from a drawing by J. Richardson based on data from J. Deisenhofer, O. Epp, K. Miki, R. Huber, and H. Michel, *Nature* 318:618–624, 1985.)

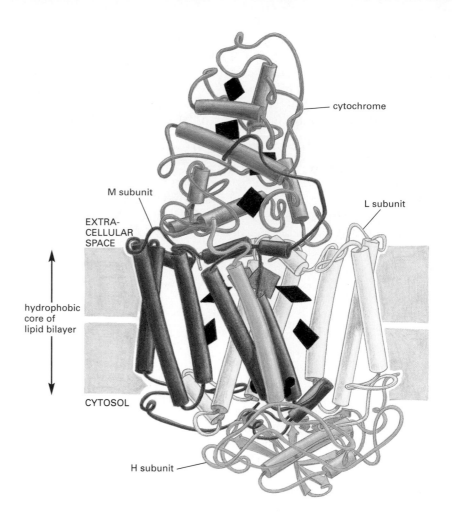

cytochrome

M subunit

L subunit

EXTRA-
CELLULAR
SPACE

hydrophobic
core of
lipid bilayer

CYTOSOL

H subunit

Question 11–6

Look at the structure of the photosynthetic reaction center in Figure 11–29. As you would expect, many α helices span the membrane. At the lower right-hand corner, however, there is a stretch of the polypeptide chain of the L subunit that forms a disordered loop within the hydrophobic core of the lipid bilayer. Does this invalidate the general rule that transmembrane proteins span the lipid bilayer as α helices or β sheets?

The Plasma Membrane Is Reinforced by the Cell Cortex

A cell membrane by itself is extremely thin and fragile. It would require nearly 10,000 cell membranes laid on top of one another to achieve the thickness of this paper. Most cell membranes are therefore strengthened and supported by a framework of proteins, attached to the membrane via transmembrane proteins. In particular, the shape of the cell and the mechanical properties of the plasma membrane are determined by a meshwork of fibrous proteins, called the *cell cortex,* that is attached to the cytosolic surface of the membrane.

The cell cortex of human red blood cells is a relatively simple and regular structure and is by far the best understood. Red blood cells are small and have a distinctive flattened shape (Figure 11–30). The main

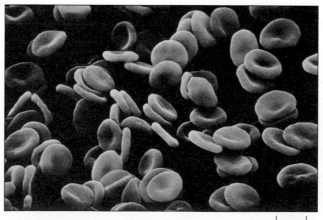

Figure 11–30 **Human red blood cells have a distinctive flattened shape, as seen in this scanning electron micrograph.** These cells lack a nucleus and other intracellular organelles. (Courtesy of Bernadette Chailley.)

5 μm

component of their cortex is the protein *spectrin,* a long, thin, flexible rod about 100 nm in length. It forms a meshwork that provides support for the plasma membrane and maintains cell shape. The spectrin meshwork is connected to the membrane through intracellular attachment proteins that link the spectrin to specific transmembrane proteins (Figure 11–31). The importance of this meshwork is seen in mice and humans that have genetic abnormalities in spectrin structure. These individuals are anemic: they have fewer red blood cells than normal, and the red cells they do have are spherical instead of flattened and are abnormally fragile.

Proteins similar to spectrin and to its associated attachment proteins are present in the cortex of most of our cells, but the cortex in these cells is much more complex than that in red blood cells. While red blood cells need their cortex mainly to provide mechanical strength as they are pumped through blood vessels, other cells also need their cortex to allow them to change their shape actively and to move, as we discuss in Chapter 17.

The Cell Surface Is Coated with Carbohydrate

We saw earlier that in eucaryotic cells many of the lipids in the outer layer of the plasma membrane have sugars covalently attached to them. The same is true for most of the proteins in the plasma membrane. The great majority of these proteins have short chains of sugars, called *oligosaccharides,* linked to them; they are called *glycoproteins.* Other membrane proteins have one or more long polysaccharide chains attached to them; they are called *proteoglycans.* All of the carbohydrate on the glycoproteins, proteoglycans, and glycolipids is located on one side of the membrane, the noncytosolic side, where it forms a sugar coating called the **carbohydrate layer** (Figure 11–32).

By forming a physical layer on top of the lipid bilayer, the carbohydrate layer helps to protect the cell surface from mechanical and chemical damage. As the oligosaccharides and polysaccharides in the carbohydrate layer adsorb water, they give the cell a slimy surface. This coating helps motile cells such as white blood cells to squeeze through narrow spaces, and it prevents blood cells from sticking to one another or to the walls of blood vessels.

Cell-surface carbohydrates do more than just protect and lubricate the cell, however. They have an important role in cell–cell recognition and adhesion. Just as many proteins will recognize and bind to a particular site on another protein, some proteins (called *lectins*) are

Question 11–7

Look carefully at the transmembrane proteins shown in Figure 11–31. What can you say about their mobility in the membrane?

Figure 11–31 A spectrin meshwork forms the cell cortex in human red blood cells. (A) Spectrin dimers, together with a smaller number of actin molecules, are linked together into a netlike meshwork that is attached to the plasma membrane by the binding of at least two types of attachment proteins (shown here in *yellow* and *blue*) to two kinds of transmembrane proteins (shown here in *green* and *brown*). (B) Electron micrograph showing the spectrin meshwork on the cytoplasmic side of a red blood cell membrane. The meshwork has been stretched out to show the details of its structure; in the normal cell the meshwork shown would be much more crowded and would occupy only about one-tenth of this area. (B, courtesy of T. Byers and D. Branton, *Proc. Natl. Acad. Sci. U.S.A.* 82:6153–6157, 1985. © National Academy of Sciences.)

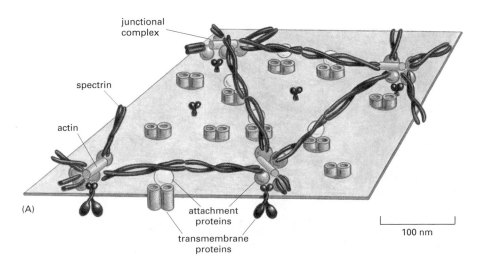

(A)

junctional complex

spectrin

actin

attachment proteins

transmembrane proteins

100 nm

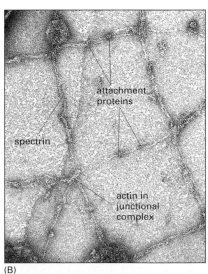

attachment proteins

spectrin

actin in junctional complex

(B)

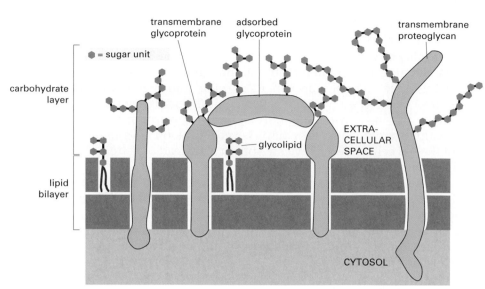

Figure 11–32 Eucaryotic cells are coated with sugars. The carbohydrate layer is made of the oligosaccharide side chains attached to membrane glycolipids and glycoproteins, and of the polysaccharide chains on membrane proteoglycans. Glycoproteins and proteoglycans that have been secreted by the cell and then adsorbed back onto its surface can also contribute to the carbohydrate layer. Note that all the carbohydrate is on the external (noncytosolic) surface of the plasma membrane.

specialized to recognize particular oligosaccharide side chains and bind to them. The oligosaccharide side chains of glycoproteins and glycolipids, although short (typically fewer than 15 sugar units), are enormously diverse. Unlike polypeptide (protein) chains, in which the amino acids are all joined together linearly by identical peptide bonds (see Figure 11–22), sugars can be joined together in different ways and in varied sequences, often forming branched oligosaccharide chains (see Panel 2–3, pp. 70–71). Even three sugar groups can be put together in enough different combinations of covalent linkages that they can form hundreds of different trisaccharides.

In a multicellular organism the carbohydrate layer can thus serve as a kind of distinctive clothing, like a police officer's uniform, that is characteristic of cells specialized for a particular function and that is recognized by other cells with which each must interact. Specific oligosaccharides in the carbohydrate layer are involved, for example, in the recognition of an egg by a sperm (discussed in Chapter 20). They are also involved in inflammatory responses. In the early stages of a bacterial infection, for instance, the carbohydrate on the surface of white blood cells called *neutrophils* is recognized by a lectin on the cells lining the blood vessels at the site of infection. This recognition process causes the neutrophils to adhere to the blood vessels and then migrate from the bloodstream into the infected tissues, where they help to remove the invading bacteria (Figure 11–33).

Figure 11–33 The recognition of the cell-surface carbohydrate on neutrophils is the first stage of their migration out of the blood at sites of infection. Specialized transmembrane proteins (called lectins) are made by the cells lining the blood vessel (called endothelial cells) in response to chemical signals emanating from the site of infection. These proteins recognize particular groups of sugars carried by glycolipids and glycoproteins on the surface of neutrophils circulating in the blood. The neutrophils consequently stick to the blood vessel wall. This association is not very strong, but it leads to another, much stronger protein–protein interaction (not shown) that helps the neutrophil migrate out of the bloodstream between the endothelial cells into the tissue at the site of infection.

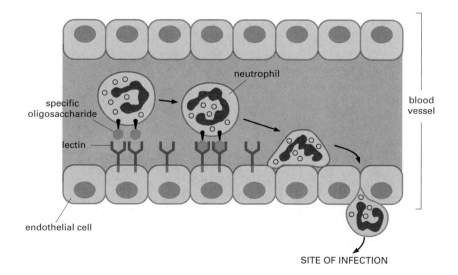

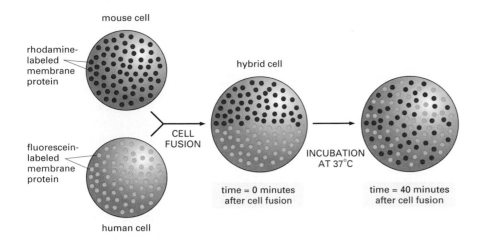

mouse cell

rhodamine-labeled membrane protein

fluorescein-labeled membrane protein

human cell

CELL FUSION

hybrid cell

time = 0 minutes after cell fusion

INCUBATION AT 37°C

time = 40 minutes after cell fusion

Figure 11–34 Formation of mouse–human hybrid cells shows that plasma membrane proteins can move. The mouse and human proteins are initially confined to their own halves of the newly formed hybrid-cell plasma membrane, but they intermix within a short time. To reveal the proteins, two antibodies that bind to human and mouse proteins, respectively, are labeled with different fluorescent tags (rhodamine or fluorescein) and added to the cells. The two fluorescent antibodies can be distinguished in a fluorescence microscope because fluorescein is *green*, whereas rhodamine is *red*. (Based on observations of L.D. Frye and M. Edidin, *J. Cell Sci.* 7:319–335, 1970. © The Company of Biologists.)

Cells Can Restrict the Movement of Membrane Proteins

Because a membrane is a two-dimensional fluid, many of its proteins, like its lipids, can move freely within the plane of the lipid bilayer. This diffusion can be neatly demonstrated by fusing a mouse cell with a human cell to form a double-size hybrid cell and then monitoring the distribution of mouse and human plasma membrane proteins. Although at first the mouse and human proteins are confined to their own halves of the newly formed hybrid cell, within half an hour or so the two sets of proteins become evenly mixed over the entire cell surface (Figure 11–34).

The picture of a membrane as a sea of lipid in which all proteins float freely is too simple, however. Cells have ways of confining particular plasma membrane proteins to localized areas within the bilayer, thereby creating functionally specialized regions, or **membrane domains**, on the cell or organelle surface (Figure 11–35). We describe some of the modern techniques for studying the movement of membrane proteins in How We Know, pp. 384–385.

Proteins can be linked to fixed structures outside the cell—for example, to molecules in the extracellular matrix (discussed in Chapter 21). Membrane proteins also can be anchored to relatively immobile structures inside the cell, especially to the cell cortex (see Figure 11–31). Finally, cells can create barriers that restrict particular membrane components to one membrane domain. In epithelial cells that line the gut, for example, it is important that transport proteins involved in the

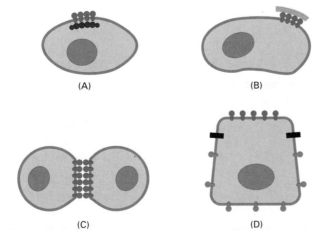

(A)

(B)

(C)

(D)

Figure 11–35 The lateral mobility of plasma membrane proteins can be restricted in several ways. Proteins can be tethered to the cell cortex inside the cell (A), to extracellular matrix molecules outside the cell (B), or to proteins on the surface of another cell (C). Finally, diffusion barriers (shown as *black* bars) can restrict proteins to a particular membrane domain (D).

How We Know: Measuring Membrane Flow

An essential feature of the lipid bilayer is its fluidity. This vital molecular flow gives biological membranes both their flexibility and integrity. It allows the resident proteins to float about the bilayer, coupling and uncoupling, engaging in the molecular interactions that support cellular life. The dynamic nature of cellular membranes is so central to their proper function that our working model of membrane structure is commonly called the *fluid-mosaic model.*

Given its importance for membrane structure and function, how do we measure and study the fluidity of cellular membranes? The most common methods are visual: simply label some of the molecules native to the bilayer and then watch them move. Such an approach first demonstrated the diffusion of membrane proteins that had been tagged with labeled antibodies (see Figure 11–34). This experiment, however, left researchers with the impression that membrane proteins drift freely, without restriction, in an open sea of lipids. We now know that this image is not entirely accurate. To probe membrane dynamics more thoroughly, researchers had to invent more precise methods for tracking the movement of proteins within a membrane such as the plasma membrane of a living cell.

The FRAP attack

One such technique, called *fluorescence recovery after photobleaching (FRAP),* involves uniformly labeling proteins across the cell surface, bleaching the label from a small region in this fluorescent sea, and then seeing how quickly the surrounding labeled proteins seep into this bleached patch of membrane. To start, the membrane protein of interest is tagged with a specific fluorescent group. This labeling can be done either with a fluorescent antibody or by fusing the membrane protein with a fluorescent protein such as green fluorescent protein (GFP) using recombinant DNA techniques (discussed in Chapter 10).

Once the cell has been labeled, it is placed under a microscope and a small patch of its membrane is irradiated with an intense pulse from a sharply focused laser beam. This treatment irreversibly bleaches the fluorescent groups in a spot, typically 1 μm square, on the cell surface (Figure 11–36). The time it takes for fluorescent proteins to migrate from the adjacent areas into the bleached region of the membrane can then be measured. The rate of this "fluorescence recovery" is a direct measure of the rate at which the surrounding protein molecules can diffuse within the membrane. Such experiments reveal that, generally speaking, the cell membrane is about as viscous as olive oil.

One-by-one

One drawback to the FRAP approach is that the technique monitors the movement of fairly large populations of proteins—hundreds or thousands—across relatively large areas of the membrane. With this technique it is impossible to see

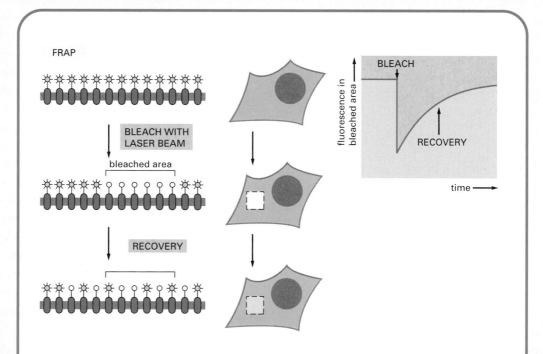

Figure 11–36 Photobleaching techniques can be used to measure the rate of lateral diffusion of a membrane protein. A specific protein of interest can be labeled with a fluorescent antibody (as shown here) or can be expressed as a fusion to green fluorescent protein (GFP), which is intrinsically fluorescent. In the FRAP technique, fluorescent molecules are bleached in a small area using a laser beam. The fluorescence intensity recovers as the bleached molecules diffuse away and unbleached molecules diffuse into the irradiated area (shown here in side and top views; the bleached area is halfway recovered). The diffusion coefficient is calculated from a graph of the rate of recovery: the greater the diffusion coefficient of the membrane protein, the faster the recovery.

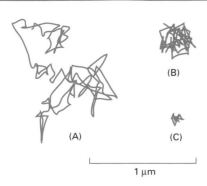

Figure 11–37 Proteins show different patterns of motion.
Single-particle tracking studies reveal some of the pathways that real proteins follow on the surface of a living cell. Shown here are some trajectories representative of different kinds of proteins in the plasma membrane.
(A) Tracks made by a protein that is free to diffuse randomly around the plasma membrane. (B) Tracks made by a protein that is corralled by, or tethered to, other proteins in the plasma membrane. (C) Tracks made by a protein that is tethered to the cytoskeleton and hence is essentially immobile. The movement of the proteins is monitored over a period of seconds.

what individual molecules are doing. If a protein fails to migrate into the bleached zone over the course of a FRAP study, for example, is it because the molecule is immobile, essentially anchored in one place in the membrane? Or is it restricted to movement within a very small region—fenced in by cytoskeletal proteins—and thus it only appears motionless?

To get around this problem, researchers have developed methods for labeling and tracking the movement of individual molecules or small clusters of molecules. One such technique, dubbed *single-particle tracking (SPT) microscopy,* relies on labeling protein molecules with antibodies coated with gold particles. The gold spheres appear as tiny black dots under the microscope, and their movement, and thus the movement of the tagged proteins, can be tracked using video microscopy.

From the studies carried out to date, it appears that membrane proteins can make a variety of patterns of movement, from random diffusion to complete immobility (Figure 11–37). Some proteins show combinations of these different kinds of motion.

Sans cells
In the end, researchers often wish to study the behavior of a particular protein in isolation, in the absence of molecules that might restrain its movement or activity. For such studies, membrane proteins can be removed from cells and reconstituted in artificial phospholipid vesicles (Figure 11–38). The lipids allow the isolated proteins to maintain their proper structure, and the activity and behavior of these purified proteins can then be analyzed in detail.

It is apparent from these studies that membrane proteins diffuse more freely and rapidly in artificial lipid bilayers than in cell membranes. The fact that proteins show restricted mobility in the cell membrane makes sense, as these membranes are crowded with proteins and contain a greater variety of lipids than an artificial lipid bilayer. Furthermore, many membrane proteins are tethered to proteins in the extracellular matrix, anchored to cytoskeletal elements tucked just under the plasma membrane, or both (as illustrated in Figure 11–35).

Taken together, such studies of the movement of molecules throughout the lipid bilayer reveal information about the architecture and organization of the cell membrane, allowing us to paint a more accurate portrait of the membrane as a dynamic fluid mosaic.

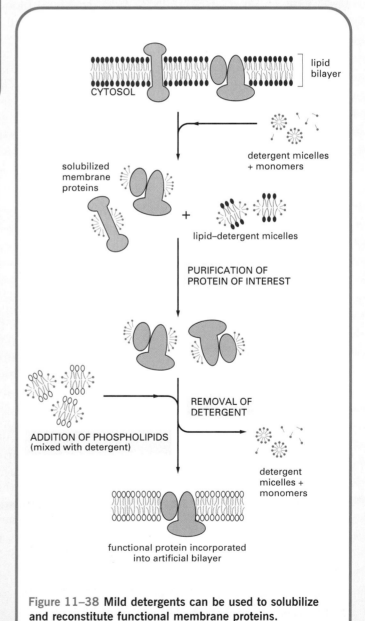

Figure 11–38 Mild detergents can be used to solubilize and reconstitute functional membrane proteins.

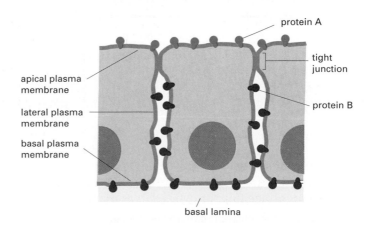

Figure 11–39 A membrane protein is restricted to a particular domain of the plasma membrane of an epithelial cell in the gut. Protein A (in the apical membrane) and protein B (in the basal and lateral membranes) can diffuse laterally in their own membrane domains but are prevented from entering the other domain by a specialized cell junction called a tight junction.

Question 11–8

Describe the different methods that cells use to restrict proteins to specific regions of the plasma membrane. Is a membrane with many anchored proteins still fluid?

uptake of nutrients from the gut be confined to the *apical* surface of the cells (the surface that faces the gut contents) and that other proteins involved in the transport of solutes out of the epithelial cell into the tissues and bloodstream be confined to the *basal* and *lateral* surfaces (Figure 11–39). This asymmetric distribution of membrane proteins is maintained by a barrier formed along the line where the cell is sealed to adjacent epithelial cells by a so-called *tight junction*. At this site, specialized junctional proteins form a continuous belt around the cell where it contacts its neighbors, creating a seal between adjacent cell membranes (see Figure 21–22). Membrane proteins cannot diffuse past the junction.

In the next chapter we examine the individual functions of the protein molecules that the cell takes such care to localize on its surface. In particular, we discuss how these proteins transport molecules into and out of the cell.

Essential Concepts

- Cell membranes enable a cell to create barriers that confine particular molecules to specific compartments.
- Cell membranes consist of a continuous double layer—a bilayer—of lipid molecules in which proteins are embedded.
- The lipid bilayer provides the basic structure and barrier function of all cell membranes.
- Membrane lipid molecules have both hydrophobic and hydrophilic regions. They assemble spontaneously into bilayers when placed in water, forming closed compartments that reseal if torn.
- There are three major classes of membrane lipid molecules: phospholipids, sterols, and glycolipids.
- The lipid bilayer is fluid, and individual lipid molecules are able to diffuse within their own monolayer; they do not, however, spontaneously flip from one monolayer to the other.
- The two layers of the lipid bilayer have different lipid compositions, reflecting the different functions of the two faces of a cell membrane.
- Cells adjust their membrane fluidity by modifying the lipid composition of their membranes.
- Membrane proteins are responsible for most of the functions of a membrane, such as the transport of small water-soluble molecules across the lipid bilayer.
- Transmembrane proteins extend across the lipid bilayer, usually as one or more α helices but sometimes as a β sheet curved into the form of a barrel.
- Other membrane proteins do not extend across the lipid bilayer but are attached to one or the other side of the membrane either by non-

covalent association with other membrane proteins or by covalent attachment to lipids.

- Many of the proteins and some of the lipids exposed on the surface of cells have attached chains of sugars, which help protect and lubricate the cell surface and are involved in cell–cell recognition.
- Most cell membranes are supported by an attached framework of proteins. An example is the meshwork of fibrous proteins forming the cell cortex underneath the plasma membrane.
- Although many membrane proteins can diffuse rapidly in the plane of the membrane, cells have ways of confining proteins to specific membrane domains and of immobilizing particular proteins by attaching them to intracellular or extracellular macromolecules.

Key Terms

amphipathic	membrane protein
bacteriorhodopsin	phosphatidylcholine
carbohydrate layer	phospholipid
cholesterol	plasma membrane
detergent	saturated
lipid bilayer	unsaturated
membrane domain	

Questions

Question 11–9

Which of the following statements are correct? Explain your answers.
- A. Lipids in a lipid bilayer rotate rapidly around their long axis.
- B. Lipids in a lipid bilayer rapidly exchange positions with one another in the plane of the membrane.
- C. Lipids in a lipid bilayer do not flip-flop readily from one lipid monolayer to the other.
- D. Hydrogen bonds that form between lipid head groups and water molecules are continually broken and re-formed.
- E. Glycolipids move through different membrane-enclosed compartments during their synthesis but remain restricted to one side of the lipid bilayer.
- F. Margarine contains more saturated lipids than the vegetable oil from which it is made.
- G. Some membrane proteins are enzymes.
- H. The sugar coat that surrounds all cells is called a carbohydrate layer and makes cells more slippery.

Question 11–10

What is meant by the term "two-dimensional fluid"?

Question 11–11

The structure of a lipid bilayer is determined by the particular properties of its lipid molecules. What would happen if
- A. Phospholipids had only one hydrocarbon chain instead of two?
- B. The hydrocarbon chains were shorter than normal, say, about 10 carbon atoms long?
- C. All of the hydrocarbon chains were saturated?
- D. All of the hydrocarbon chains were unsaturated?
- E. The bilayer contained a mixture of two kinds of lipid molecules, one with two saturated hydrocarbon tails and the other with two unsaturated hydrocarbon tails?
- F. Each lipid molecule were covalently linked through the end carbon atom of one of its hydrocarbon chains to a lipid molecule in the opposite monolayer?

Question 11–12

What are the differences between a lipid molecule and a detergent molecule? How would the structure of a lipid molecule need to change to make it a detergent?

Question 11–13

- A. Lipid molecules exchange places with their lipid neighbors every 10^{-7} second. A lipid molecule diffuses from one end of a 2-μm-long bacterial cell to the other in about 1 second. Are these two numbers in agreement (assume that the diameter of a lipid head group is about 0.5 nm)? If not, can you think of a reason for the difference?

B. To get an appreciation for the great speed of molecular motions, assume that a lipid head group is about the size of a Ping-Pong ball (4 cm diameter) and that the floor of your living room (6 m × 6 m) is covered wall-to-wall with these balls. If two neighboring balls exchanged positions once every 10^{-7} second, what would their speed be in kilometers per hour? How long would it take for a ball to move from one end of the room to the other?

Question 11–14

Why does a red blood cell membrane need proteins?

Question 11–15

Consider a transmembrane protein that forms a hydrophilic pore across the plasma membrane of a eucaryotic cell, allowing Na^+ to enter the cell when it is activated upon binding a specific ligand on its extracellular side. It is made of five similar transmembrane subunits, each containing a membrane-spanning α helix with hydrophilic amino acid side chains on one surface of the helix and hydrophobic amino acid side chains on the opposite surface. Considering the function of the protein as an ion channel, propose a possible arrangement of the five membrane-spanning α helices in the membrane.

Question 11–16

In the membrane of a human red blood cell the ratio of the mass of protein (average molecular weight 50,000) to phospholipid (molecular weight 800) to cholesterol (molecular weight 386) is about 2:1:1. How many lipid molecules are there for every protein molecule?

Question 11–17

Draw a schematic diagram that shows a close-up view of two plasma membranes as they come together during cell fusion, as shown in Figure 11–34. Show the membrane proteins in both cells that were labeled from the outside by the binding of differently colored fluorescent antibody molecules. Indicate in your drawing the fates of these color tags as the cells fuse. Will they still be only on the outside of the hybrid cell (A) after cell fusion and (B) after the mixing of membrane proteins that occurs during the incubation at 37°C? How would the experimental outcome be different if the incubation were done at 0°C?

Question 11–18

Compare the hydrophobic forces that hold a membrane protein in the lipid bilayer to those that help proteins fold into a unique three-dimensional structure.

Question 11–19

Predict which of the following organisms will have the highest percentage of unsaturated phospholipids in their membranes. Explain your answer.

A. Antarctic fish
B. Desert snake
C. Human being
D. Polar bear
E. Thermophilic bacterium

Question 11–20

Which of the three 20-amino acid sequences listed below in the single letter amino acid code is the most likely candidate to form a transmembrane region (α-helix) of a transmembrane protein? Explain your answer.

A. I T L I Y F G N M S S V T Q T I L L I S
B. L L L I F F G V M A L V I V V I L L I A
C. L L K K F F R D M A A V H E T I L E E S

Highlights from *Essential Cell Biology 2 Interactive* CD-ROM

11.2 Fluidity of the Lipid Bilayer
11.6 Rolling Leukocytes
11.7 FRAP

Membrane Transport

12

Principles of Membrane Transport
The Ion Concentrations Inside a Cell Are Very Different from Those Outside
Lipid Bilayers Are Impermeable to Solutes and Ions
Membrane Transport Proteins Fall into Two Classes: Carriers and Channels
Solutes Cross Membranes by Passive or Active Transport

Carrier Proteins and Their Functions
Concentration Gradients and Electrical Forces Drive Passive Transport
Active Transport Moves Solutes Against Their Electrochemical Gradients
Animal Cells Use the Energy of ATP Hydrolysis to Pump Out Na^+
The Na^+-K^+ Pump Is Driven by the Transient Addition of a Phosphate Group
Animal Cells Use the Na^+ Gradient to Take Up Nutrients Actively
The Na^+-K^+ Pump Helps Maintain the Osmotic Balance of Animal Cells
Intracellular Ca^{2+} Concentrations Are Kept Low by Ca^{2+} Pumps
H^+ Gradients Are Used to Drive Membrane Transport in Plants, Fungi, and Bacteria

Ion Channels and the Membrane Potential
Ion Channels Are Ion-Selective and Gated
Ion Channels Randomly Snap Between Open and Closed States
Different Types of Stimuli Influence the Opening and Closing of Ion Channels
Voltage-gated Ion Channels Respond to the Membrane Potential
Membrane Potential Is Governed by Membrane Permeability to Specific Ions

Ion Channels and Signaling in Nerve Cells
Action Potentials Provide for Rapid Long-Distance Communication
Action Potentials Are Usually Mediated by Voltage-gated Na^+ Channels
Voltage-gated Ca^{2+} Channels Convert Electrical Signals into Chemical Signals at Nerve Terminals
Transmitter-gated Channels in Target Cells Convert Chemical Signals Back into Electrical Signals
Neurons Receive Both Excitatory and Inhibitory Inputs
Transmitter-gated Ion Channels Are Major Targets for Psychoactive Drugs
Synaptic Connections Enable You to Think, Act, and Remember

Cells live and grow by exchanging molecules with their environment, and the plasma membrane acts as a barrier that controls the transit of molecules into and out of the cell. Because the interior of the lipid bilayer is hydrophobic, as we saw in Chapter 11, the plasma membrane tends to block the passage of almost all water-soluble molecules. But various water-soluble molecules must be able to cross the plasma membrane: cells must import nutrients such as sugars and amino acids, eliminate metabolic waste products such as CO_2, and regulate the intracellular concentrations of a variety of inorganic ions. A few of these solutes, CO_2 and O_2 for example, can simply diffuse across the lipid bilayer, but the vast majority cannot. Instead, their transfer depends on **membrane transport proteins** that span the membrane, providing private passageways across the bilayer for select substances (Figure 12–1).

In this chapter we consider how membranes control the traffic of small molecules into and out of cells. Cells can also selectively transfer macromolecules such as proteins across their membranes, but this transport requires more elaborate machinery, as we discuss in Chapter 15. Here we begin by outlining some of the general principles that guide the passage of small, water-soluble molecules through the cell membrane. We then examine, in turn, the two main classes of membrane proteins that mediate this transfer. *Carrier proteins*, which have moving parts, can shift small molecules from one side of the membrane to the other by changing their shape. Solutes transported in this way can be either small organic molecules or inorganic ions. *Channel proteins*, in contrast, form tiny hydrophilic pores in the membrane through which solutes can pass by diffusion. Most channel proteins let through inorganic ions only and are therefore called *ion channels*. Because these ions are electrically charged, their movements can create powerful electric forces across the membrane. In the final part of the chapter we will see how these forces enable nerve cells to communicate, ultimately carrying out the astonishing range of behaviors of which the human brain is capable.

Principles of Membrane Transport

To provide a foundation for discussing membrane transport, we first consider the differences in ion composition between a cell's interior and its environment. This will help make it clear why the transport of ions by both carrier proteins and ion channels is of such fundamental importance to cells.

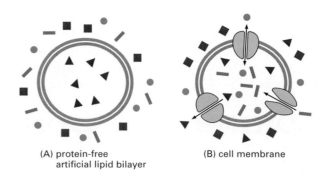

Figure 12–1 Membrane transport proteins are responsible for transferring small water-soluble molecules across cell membranes. Whereas protein-free artificial lipid bilayers are impermeable to most water-soluble molecules (A), cell membranes are not (B). Note that each type of transport protein in a cell membrane transfers a particular type of molecule, causing a selective set of solutes to end up inside the membrane-enclosed compartment.

(A) protein-free
artificial lipid bilayer

(B) cell membrane

The Ion Concentrations Inside a Cell Are Very Different from Those Outside

Living cells maintain an internal ion composition that is very different from the ion composition in the fluid around them, and these differences are crucial for a cell's survival and function. Inorganic ions such as Na^+, K^+, Ca^{2+}, Cl^-, and H^+ (protons) are the most plentiful of all the solutes in a cell's environment, and their movements across cell membranes play an essential part in many biological processes, including the activity of nerve cells, as we discuss later, and the production of ATP by all cells, as we discuss in Chapter 14.

Na^+ is the most plentiful positively charged ion (cation) outside the cell, while K^+ is the most plentiful inside (Table 12–1). For a cell to avoid being torn apart by electrical forces, the quantity of positive charge inside the cell must be balanced by an almost exactly equal quantity of negative charge, and the same is true for the charge in the surrounding fluid. However, tiny excesses of positive or negative charge, concentrated in the neighborhood of the plasma membrane, do occur, and they have important electrical effects, as we discuss later. The high concentration of Na^+ outside the cell is balanced chiefly by extracellular Cl^-.

Table 12–1 A Comparison of Ion Concentrations Inside and Outside a Typical Mammalian Cell

COMPONENT	INTRACELLULAR CONCENTRATION (mM)	EXTRACELLULAR CONCENTRATION (mM)
Cations		
Na^+	5–15	145
K^+	140	5
Mg^{2+}	0.5	1–2
Ca^{2+}	10^{-4}	1–2
H^+	7×10^{-5} ($10^{-7.2}$ M or pH 7.2)	4×10^{-5} ($10^{-7.4}$ M or pH 7.4)
Anions*		
Cl^-	5–15	110

* The cell must contain equal quantities of positive and negative charges (that is, be electrically neutral). Thus, in addition to Cl^-, the cell contains many other anions not listed in this table; in fact, most cellular constituents are negatively charged (HCO_3^-, PO_4^{3-}, proteins, nucleic acids, metabolites carrying phosphate and carboxyl groups, etc.). The concentrations of Ca^{2+} and Mg^{2+} given are for the free ions. There is a total of about 20 mM Mg^{2+} and 1–2 mM Ca^{2+} in cells, but this is mostly bound to proteins and other substances and, for Ca^{2+}, stored within various organelles.

The high concentration of K$^+$ inside is balanced by a variety of negatively charged intracellular ions (anions).

This differential distribution of ions inside and outside the cell is controlled in part by the activity of membrane transport proteins, and in part by the permeability characteristics of the lipid bilayer itself.

Lipid Bilayers Are Impermeable to Solutes and Ions

The hydrophobic interior of the lipid bilayer creates a barrier to the passage of most hydrophilic molecules, including ions. They are as reluctant to enter a fatty environment as hydrophobic molecules are reluctant to enter water. But given enough time, virtually any molecule will diffuse across a lipid bilayer. The *rate* at which it diffuses, however, varies enormously depending on the size of the molecule and its solubility properties. In general, the smaller the molecule and the more soluble it is in oil (that is, the more hydrophobic, or nonpolar, it is), the more rapidly it will diffuse across. Thus,

1. *Small nonpolar molecules,* such as molecular oxygen (O$_2$, molecular mass 32 daltons) and carbon dioxide (44 daltons), readily dissolve in lipid bilayers and therefore rapidly diffuse across them; indeed, cells require this permeability to gases for the cellular respiration processes discussed in Chapter 14.

2. *Uncharged polar molecules* (molecules with an uneven distribution of electric charge) also diffuse rapidly across a bilayer, if they are small enough. Water (18 daltons) and ethanol (46 daltons), for example, cross fairly rapidly; glycerol (92 daltons) diffuses less rapidly; and glucose (180 daltons) diffuses hardly at all (Figure 12–2).

3. In contrast, lipid bilayers are highly impermeable to all *ions and charged molecules,* no matter how small. The molecules' charge and their strong electrical attraction to water molecules inhibit them from entering the hydrocarbon phase of the bilayer. Thus synthetic bilayers are a billion (10^9) times more permeable to water than they are to even such small ions as Na$^+$ or K$^+$.

Cell membranes allow water and small nonpolar molecules to permeate by simple diffusion. But for cells to take up nutrients and release wastes, membranes must also allow the passage of many other molecules, such as ions, sugars, amino acids, nucleotides, and many cell metabolites. These molecules cross lipid bilayers far too slowly by simple diffusion; thus, specialized transport proteins are required to transfer them efficiently across cell membranes.

Membrane Transport Proteins Fall into Two Classes: Carriers and Channels

Membrane transport proteins occur in many forms and in all types of biological membranes. Each protein provides a private passageway across the membrane for a particular class of molecule—ions, sugars, or amino acids, for example. Most of these protein portals are even more exclusive, allowing entrance of only select members of a particular molecular class: some, for example, are open to Na$^+$ but not K$^+$, others to K$^+$ but not Na$^+$. The set of transport proteins present in the plasma membrane, or in the membrane of an intracellular organelle, determines exactly which solutes can pass into and out of that cell or organelle. Each type of membrane therefore has its own characteristic set of transport proteins.

As discussed in Chapter 11, the membrane transport proteins that have been studied in detail have polypeptide chains that traverse the

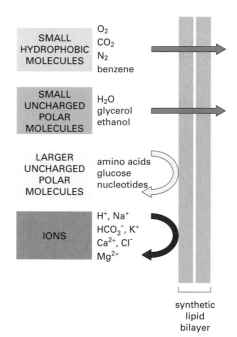

Figure 12–2 The rate at which a molecule diffuses across a synthetic lipid bilayer depends on its size and solubility. The smaller the molecule and, more important, the fewer its favorable interactions with water (that is, the less polar it is), the more rapidly the molecule diffuses across the bilayer. Note that many of the molecules that the cell uses as nutrients are too large and polar to pass through a pure lipid bilayer.

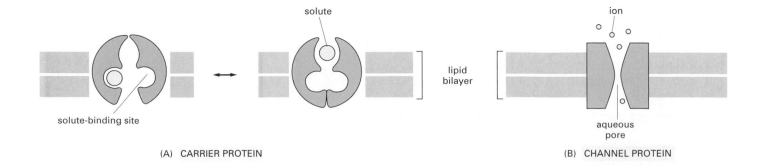

solute

solute-binding site

(A) CARRIER PROTEIN

ion

lipid
bilayer

aqueous
pore

(B) CHANNEL PROTEIN

Figure 12–3 Small molecules can enter the cell through a carrier or a channel. (A) A carrier protein undergoes a series of conformational changes to transfer small water-soluble molecules across the lipid bilayer. (B) A channel protein, in contrast, forms a hydrophilic pore across the bilayer through which specific inorganic ions can diffuse. As would be expected, channel proteins transport molecules at a much greater rate than carrier proteins. Ion channels can exist in either an open or a closed conformation, and they transport only in the open conformation, which is shown. Channel opening and closing is usually controlled by an external stimulus or by conditions within the cell.

lipid bilayer multiple times—that is, they are multipass transmembrane proteins (see Figure 11–24). By crisscrossing back and forth across the bilayer, the polypeptide chain is thought to form a continuous protein-lined pathway that allows selected small hydrophilic molecules to cross the membrane without coming into direct contact with the hydrophobic interior of the lipid bilayer.

Membrane transport proteins can be divided into two main classes: carriers and channels. The basic difference between **carrier proteins** and **channel proteins** is the way they discriminate between solutes, transporting some solutes but not others (Figure 12–3). A channel protein discriminates mainly on the basis of size and electric charge: if the channel is open, molecules small enough and carrying the appropriate charge can slip through, as through a narrow trapdoor. A carrier protein, on the other hand, allows passage only to solute molecules that fit into a binding site on the protein; it then transfers these molecules across the membrane one at a time by changing its own conformation, acting more like a turnstile than an open door. Carrier proteins bind their solutes with great specificity in the same way that an enzyme binds its substrate, and it is this requirement for specific binding that makes the transport selective.

Solutes Cross Membranes by Passive or Active Transport

Transport proteins allow small molecules to cross the cell membrane, but what controls whether these solutes move into or out of the cell? The direction of transport depends, in large part, on the relative concentrations of the solute. Molecules will flow "downhill" from a region of high concentration to a region of low concentration spontaneously, provided a pathway exists. Such movements are called *passive,* because they need no other driving force. If, for example, a solute is present at a higher concentration outside the cell than inside and an appropriate channel or carrier protein is present in the plasma membrane, the solute will move spontaneously across the membrane into the cell by **passive transport** (sometimes called *facilitated diffusion*), without expenditure of energy by the transport protein. All channel proteins and many carrier proteins can act as a conduit for such passive transport.

To move a solute against its concentration gradient, however, a transport protein must do work: it has to drive the "uphill" flow by coupling it to some other process that provides energy (as discussed in Chapter 3 for enzyme reactions). Transmembrane solute movement driven in this way is termed **active transport**, and it is carried out only by special types of carrier proteins that can harness some energy source to the transport process (Figure 12–4). We now examine a variety of carrier proteins and see how they function to move molecules across cell membranes.

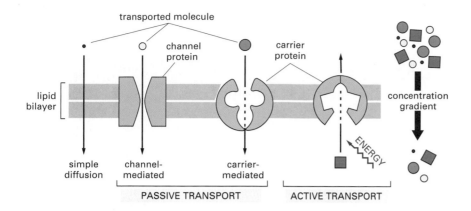

Figure 12–4 **Solutes cross cell membranes by passive or active transport.** If uncharged solutes are small enough, they can move down their concentration gradients directly across the lipid bilayer itself by simple diffusion. Examples of such solutes are ethanol, carbon dioxide, and oxygen. Most solutes, however, can cross the membrane only if there is a membrane transport protein (a carrier protein or a channel protein) to transfer them. As indicated, passive transport, in the same direction as a concentration gradient, occurs spontaneously, whereas transport against a concentration gradient (active transport) requires an input of energy. Only carrier proteins can carry out active transport, but both carrier proteins and channel proteins can carry out passive transport.

Carrier Proteins and Their Functions

Carrier proteins are required for the transport of almost all small organic molecules across cell membranes, with the exception of fat-soluble molecules and small uncharged molecules that can pass directly through the lipid bilayer by *simple diffusion*. Each carrier protein is highly selective, often transporting just one type of molecule. To guide and propel the complex traffic of small molecules into and out of the cell and between the cytosol and the different membrane-enclosed organelles, each cell membrane contains a set of different carrier proteins appropriate to that particular membrane. Thus the plasma membrane contains carriers to import nutrients such as sugars, amino acids, and nucleotides; the inner membrane of mitochondria contains carriers for importing pyruvate and ADP and for exporting ATP (Figure 12–5).

To understand fully how a carrier protein transfers solutes across a membrane, we would need to know its three-dimensional structure in detail. As yet this information is available for only a very few membrane transport proteins. One is bacteriorhodopsin, which functions as a light-activated H$^+$ pump (see Figure 11–28). Another is the protein pump that moves Ca^{2+} from the cytosol into the sarcoplasmic reticulum—a specialized form of endoplasmic reticulum found in skeletal muscle cells (Figure 12–6).

Although the detailed molecular mechanisms that underlie transport are known for only a few carrier proteins, the general principles that govern the function of these proteins are well understood.

Concentration Gradients and Electrical Forces Drive Passive Transport

Solutes can cross the membrane by passive or active transport—and carrier proteins are capable of facilitating both types of movement. A simple example of a carrier protein that mediates passive transport is

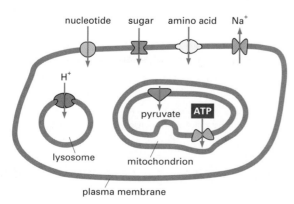

Figure 12–5 **Each cell membrane has its own characteristic set of carrier proteins.**

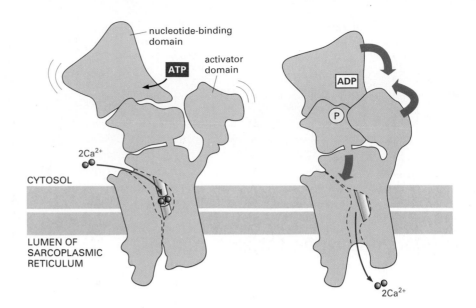

Figure 12–6 A calcium pump returns Ca²⁺ to the sarcoplasmic reticulum in a skeletal muscle cell. The three-dimensional structure of this membrane transport protein has been determined by X-ray crystallography and electron microscopy. When a muscle cell is stimulated, calcium floods from the sarcoplasmic reticulum into the cytosol, allowing the cell to contract; to recover from the contraction, calcium is returned to the sarcoplasmic reticulum by this calcium pump. The polypeptide chain of the protein crosses the membrane as 10 α helices. Phosphorylation of the pump causes a rearrangement of the nucleotide-binding and activator domains in the protein. This in turn leads to a rearrangement of the transmembrane helices, which forces bound Ca²⁺ ions that entered a binding site from the cytoplasmic side of the membrane to be released into the lumen of the sarcoplasmic reticulum. Note that the pathway taken by Ca²⁺ ions leads through the protein, thereby allowing the ions to avoid contact with the lipid bilayer. (Adapted from C. Toyoshima et al., *Nature* 405:647–655, 2000.)

Figure 12–7 A conformational change in a carrier protein could mediate the passive transport of a solute like glucose. In this hypothetical model, the carrier protein can exist in two conformational states: in state A the binding sites for the solute are exposed on the outside of the membrane; in state B the same sites are exposed on the other side of the membrane. The transition between the two states is proposed to occur randomly and independently of whether the solute is bound and to be completely reversible. If the concentration of the solute is higher on the outside of the membrane, it will be more often caught up in A → B transitions that carry it into the cell than in B → A transitions that carry it out, and there will be a net transport of the solute down its concentration gradient.

the *glucose carrier* found in the plasma membrane of mammalian liver cells (and of many other types of cells). It consists of a protein chain that crosses the membrane at least 12 times. It is thought that the protein can adopt at least two conformations and switches reversibly and randomly between them. In one conformation the carrier exposes binding sites for glucose to the exterior of the cell; in the other it exposes this site to the interior of the cell (Figure 12–7).

When sugar is plentiful outside the liver cell (after a meal), glucose molecules bind to the externally displayed binding sites; when the protein switches conformation, it carries these molecules inward and releases them into the cytosol, where the glucose concentration is low. Conversely, when blood sugar levels are low (when you are hungry), the hormone glucagon stimulates the liver cell to produce large amounts of glucose by the breakdown of glycogen. As a result, the glucose concentration is higher inside the cell than outside, and glucose binds to any internally displayed binding sites on the carrier protein; when the protein switches conformation in the opposite direction, the glucose is transported out of the cell. The flow of glucose can thus go either way, according to the direction of the glucose *concentration gradient* across the membrane—inward if glucose is more concentrated outside the cell than inside, and outward if the opposite is true. Transport proteins of this type, which permit a flux of solute but play no part in determining its direction, carry out passive transport. Although passive, the transport is highly selective: the binding sites in the glucose transporter bind only D-glucose and not, for example, its mirror image L-glucose, which the cell cannot use for glycolysis.

For glucose, which is an uncharged molecule, the direction of passive transport is determined solely by its concentration gradient. For

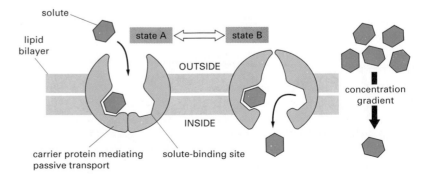

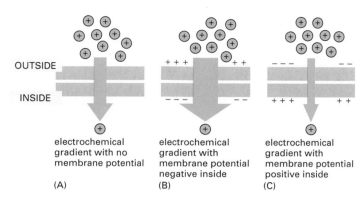

(A) electrochemical gradient with no membrane potential

(B) electrochemical gradient with membrane potential negative inside

(C) electrochemical gradient with membrane potential positive inside

Figure 12–8 An electrochemical gradient has two components. The net driving force (the electrochemical gradient) tending to move a charged solute (ion) across the membrane is the sum of the concentration gradient of the solute and the voltage across the membrane (the membrane potential). The width of the *green arrow* represents the magnitude of the electrochemical gradient for the same positively charged solute in three different situations. In (A), there is only a concentration gradient. In (B), the concentration gradient is supplemented by a membrane potential that increases the driving force. In (C), the membrane potential decreases the driving force that is caused by the concentration gradient.

electrically charged molecules, either small organic ions or inorganic ions, an additional force comes into play. For reasons we explain later, most cell membranes have a voltage across them, a difference in the electrical potential on each side of the membrane, which is referred to as the *membrane potential*. This difference in potential exerts a force on any molecule that carries an electric charge. The cytoplasmic side of the plasma membrane is usually at a negative potential relative to the outside, and this tends to pull positively charged solutes into the cell and drive negatively charged ones out. At the same time, a charged solute will also tend to move down its concentration gradient.

The net force driving a charged solute across the membrane is therefore a composite of two forces, one due to the concentration gradient and the other due to the voltage across the membrane. This net driving force is called the **electrochemical gradient** for the given solute. This gradient determines the direction of passive transport across the membrane. For some ions, the voltage and concentration gradient work in the same direction, creating a relatively steep electrochemical gradient (Figure 12–8B). This is the case for Na^+, which is positively charged and at a higher concentration outside cells than inside. Na^+ therefore tends to enter cells if given an opportunity. If, however, the voltage and concentration gradients have opposing effects, the resulting electrochemical gradient can be small (Figure 12–8C). This is the case for K^+, a positively charged ion that is present at a much higher concentration inside cells than outside. Because of these opposing effects, K^+ has a small electrochemical gradient across the membrane, despite its large concentration gradient, and therefore there is little net movement of K^+ across the membrane.

Active Transport Moves Solutes Against Their Electrochemical Gradients

Of course cells cannot rely solely on passive transport. Active transport of solutes against their electrochemical gradient is essential to maintain the intracellular ionic composition of cells and to import solutes that are at a lower concentration outside the cell than inside. Cells carry out active transport in three main ways (Figure 12–9). (1) *Coupled transporters* couple the uphill transport of one solute across the membrane to the downhill transport of another. (2) *ATP-driven pumps* couple uphill transport to the hydrolysis of ATP. (3) *Light-driven pumps*, which are found mainly in bacterial cells, couple uphill transport to an input of energy from light, as discussed in Chapter 11 for bacteriorhodopsin.

Because a substance has to be carried uphill before it can flow downhill, the different forms of active transport are necessarily linked. Thus in the plasma membrane of an animal cell, an ATP-driven pump transports Na^+ out of the cell against its electrochemical gradient, and the Na^+ then flows back in, down its electrochemical gradient. Because

Question 12–1

A simple enzyme reaction can be described by the equation E + S $\rightleftharpoons$ ES $\rightarrow$ E + P, where E is the enzyme, S the substrate, P the product, and ES the enzyme–substrate complex.

A. Write a corresponding equation describing the workings of a carrier protein (CP) that mediates the transport of a solute (S) down its concentration gradient.

B. What does this equation tell you about carrier protein function?

C. Why would this equation be an inappropriate description of channel protein function?

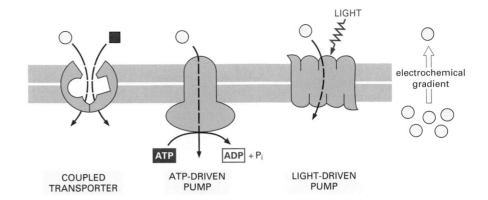

Figure 12–9 Cells drive active transport in three main ways. The actively transported molecule is shown in *yellow,* and the energy source is shown in *red.*

COUPLED TRANSPORTER

ATP-DRIVEN PUMP

LIGHT-DRIVEN PUMP

the ion flows through Na^+-coupled transporters, the influx of Na^+ drives the active movement of many other substances into the cell against their electrochemical gradients. If the Na^+ pump ceased operating, the Na^+ gradient would soon run down, and transport through Na^+-coupled transporters would come to a halt. The ATP-driven Na^+ pump, therefore, has a central role in membrane transport in animal cells. In plant cells, fungi, and many bacteria, a similar role is played by ATP-driven pumps that create an electrochemical gradient of H^+ ions (protons) by pumping H^+ out of the cell, as we discuss later.

Animal Cells Use the Energy of ATP Hydrolysis to Pump Out Na^+

The ATP-driven Na^+ pump in animal cells hydrolyzes ATP to ADP to transport Na^+ out of the cell; this pump is therefore not only a carrier protein, but also an enzyme—an ATPase. At the same time, the protein couples the outward transport of Na^+ to an inward transport of K^+. The pump is therefore known as the *Na^+-K^+ ATPase,* or the **Na^+-K^+ pump** (Figure 12–10).

This carrier protein plays a central part in the energy economy of animal cells, typically accounting for 30% or more of their total ATP consumption. Like a bilge pump in a leaky ship, it operates ceaselessly to expel the Na^+ that is constantly entering through other carrier proteins and ion channels. In this way, the pump keeps the Na^+ concentration in the cytosol about 10–30 times lower than in the extracellular fluid and the K^+ concentration about 10–30 times higher (see Table 12–1, p. 390). Under normal conditions, the interior of most cells is at a negative electric potential compared with the exterior, so that positive ions tend to be pulled into the cell; thus the inward electrochemical driving force for Na^+ is large, as it includes the driving force due to the

Figure 12–10 The Na^+-K^+ pump plays a central role in membrane transport in animal cells. This carrier protein uses the energy of ATP hydrolysis to pump Na^+ out of the cell and K^+ in, both against their electrochemical gradients. Ouabain is a drug that binds to the pump and inhibits its activity by preventing K^+ binding.

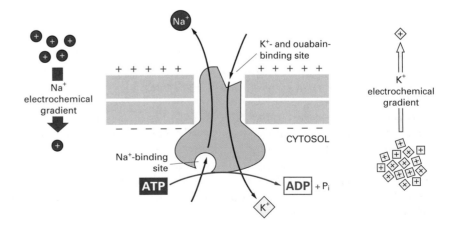

concentration gradient and a driving force in the same direction due to the voltage gradient (see Figure 12–8B).

The Na$^+$ outside the cell, on the uphill side of its electrochemical gradient, is like a large volume of water behind a high dam: it represents a very large store of energy (Figure 12–11). Even if one artificially halts the operation of the Na$^+$-K$^+$ pump with a toxin such as the plant glycoside *ouabain*, which binds to the pump and prevents K$^+$ binding (see Figure 12–10), the energy in this store is sufficient to sustain for many minutes the other transport processes that are driven by the downhill flow of Na$^+$.

For K$^+$ the situation is different. The electric force is the same as for Na$^+$, because it depends only on the charge carried by the ion. The concentration gradient, however, is in the opposite direction. The result, under normal conditions, is that the net driving force for movement of K$^+$ across the membrane is close to zero: the electric force pulling K$^+$ into the cell is almost exactly balanced against the concentration gradient tending to drive it out.

The Na$^+$-K$^+$ Pump Is Driven by the Transient Addition of a Phosphate Group

The Na$^+$-K$^+$ pump provides a beautiful illustration of how a protein couples one reaction to another, following the principles discussed in Chapter 3. The pump works in a cycle, as illustrated schematically in Figure 12–12. Na$^+$ binds to the pump at sites exposed inside the cell (stage 1), activating the ATPase activity. ATP is split, with the release of ADP and the transfer of a phosphate group into a high-energy linkage to the pump itself—that is, the pump phosphorylates itself (stage 2). Phosphorylation causes the pump to switch its conformation so as to release Na$^+$ at the exterior surface of the cell and, at the same time, to expose a binding site for K$^+$ at the same surface (stage 3). The binding of extracellular K$^+$ triggers the removal of the phosphate group (dephosphorylation; stages 4 and 5), causing the pump to switch back to its original conformation, discharging the K$^+$ into the cell interior (stage 6). The whole cycle, which takes about 10 milliseconds, can then be repeated. Each step in the cycle depends on the one before, so that if any of the individual steps is prevented from occurring, all the functions of the pump are halted. This tight coupling ensures that the pump operates only when the appropriate ions are available to be transported, thereby avoiding useless ATP hydrolysis.

Animal Cells Use the Na$^+$ Gradient to Take Up Nutrients Actively

A gradient of any solute across a membrane, like the Na$^+$ gradient generated by the Na$^+$-K$^+$ pump, can be used to fuel the active transport of a second molecule. The downhill movement of the first solute down its gradient provides the energy to drive the uphill transport of the second. The carrier proteins that do this are called **coupled transporters** (see Figure 12–9). They may couple the movement of one inorganic ion to that of another, the movement of an inorganic ion to that of an organic molecule, or the movement of one organic molecule to that of another. If the transporter moves both solutes in the same direction across the membrane, it is called a *symport* (Figure 12–13). If it moves them in opposite directions, it is called an *antiport*. A carrier protein, like the passive glucose transporter mentioned earlier, that ferries only one type of solute across the membrane (and is therefore not a coupled transporter) is called a *uniport*.

Figure 12–11 Na$^+$ outside the cell is like water behind a high dam. The water in the dam has potential energy, which can be used to drive energy-requiring processes. In the same way, an ion gradient across a membrane can be used to drive active processes in a cell, including the active transport of other molecules. Shown here is the Blyde River dam in South Africa. (Courtesy of Paul Franklin. © Oxford Scientific Films.)

Question 12–2

A transmembrane protein has the following properties: it has two binding sites, one for solute A and one for solute B. The protein can undergo a conformational change to switch between two states: either both binding sites are exposed exclusively on one side of the membrane or both binding sites are exposed exclusively on the other side of the membrane. The protein can switch between the two conformational states only if both binding sites are occupied or if both binding sites are empty, but cannot switch if only one binding site is occupied.

A. What kind of protein do these properties define?

B. Do you need to specify any additional properties to turn this protein into a symport that couples the movement of solute A up its concentration gradient to the movement of solute B down its electrochemical gradient?

C. Write a set of rules that defines an antiport.

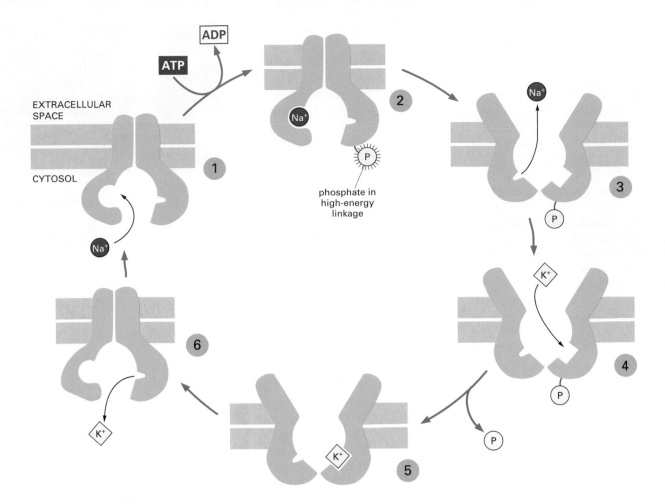

Figure 12–12 The Na⁺-K⁺ pump transports ions in a cyclic manner. The binding of Na⁺ (1) and the subsequent phosphorylation by ATP of the cytosolic face of the pump (2) induce the protein to undergo a conformational change that transfers the Na⁺ across the membrane and releases it on the outside (3). The high-energy linkage of the phosphate to the protein provides the energy to drive the conformational change. The binding of K⁺ on the extracellular surface (4) and the subsequent dephosphorylation (5) return the protein to its original conformation, which transfers the K⁺ across the membrane and releases it into the cytosol (6). These changes in conformation are analogous to the A ⇌ B transitions shown in Figure 12–7 except that here the Na⁺-dependent phosphorylation and the K⁺-dependent dephosphorylation of the protein cause the conformational transitions to occur in an orderly manner, enabling the protein to do useful work. For simplicity, only one Na⁺- and one K⁺-binding site are shown. In the real pump in mammalian cells there are thought to be three Na⁺- and two K⁺-binding sites. The net result of one cycle of the pump is therefore to transport three Na⁺ out of the cell and two K⁺ in.

Although we do not yet know the three-dimensional structure of any coupled transporter in detail, it is possible to imagine how the molecular mechanism of coupled transport might work, as outlined in Figure 12–14. This model is based on a simple modification of the mechanism proposed in Figure 12–7. Regardless of uncertainties about details, one principle of coupled transport is fundamental: if one of the co-transported solutes is absent, transport of its companion cannot occur.

In animal cells an especially important role is played by symports that use the inward flow of Na⁺ down its steep electrochemical gradient to drive the import of other solutes into the cell. The epithelial cells that line the gut, for example, transfer glucose from the gut across the gut epithelium. If these cells had only the passive glucose transporters described earlier, they would release glucose into the gut after a sugar-free meal as freely as they take it up from the gut after a sugar-rich meal. But they also possess a glucose–Na⁺ symport, with which they can take up glucose actively from the gut lumen even when the concentration of glucose is higher inside the cell than in the gut. If the gut epithelial cells had *only* this symport, however, they could never release glucose for use by the other cells of the body. These cells, therefore, have two types of glucose carriers. In the apical domain of the plasma membrane, which faces the lumen of the gut, they have glucose–Na⁺ symports that take up glucose actively, creating a high glucose concentration in the cytosol. In the basal and lateral domains of the plasma membrane, they have passive glucose uniports that release the glucose down its concentration gradient for use by other tissues (Figure 12–15). The two types of glucose carriers are kept segregated in their proper domains of the plasma membrane by a diffusion barrier formed by a tight junction around the

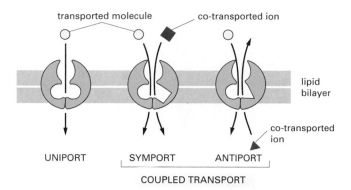

transported molecule   co-transported ion

lipid
bilayer

co-transported
ion

UNIPORT   SYMPORT   ANTIPORT

COUPLED TRANSPORT

Figure 12–13 Carrier proteins can transport solutes in several different ways. Some carrier proteins transport a single solute across the membrane (uniports). In coupled transport, by contrast, the transfer of one solute depends on the simultaneous or sequential transfer of another solute, either in the same direction (symports) or in the opposite direction (antiports). Uniports, symports, and antiports are used for both passive and active transport.

apex of the cell, which prevents mixing of membrane components between the apical and the basal and lateral domains, as discussed in Chapter 11 (see Figure 11–39).

Cells in the lining of the gut and in many other organs, such as the kidney, contain a variety of symports in their plasma membrane that are similarly driven by the electrochemical gradient of Na^+; each of these carrier proteins specifically imports a small group of related sugars or amino acids into the cell. Na^+-driven antiports are important for cell function, too. For example, the *Na^+-H^+ exchanger* in the plasma membranes of many animal cells uses the downhill influx of Na^+ to pump H^+ out of the cell and is one of the main devices that animal cells use to control the pH in their cytosol.

The Na^+-K^+ Pump Helps Maintain the Osmotic Balance of Animal Cells

The plasma membrane is permeable to water (see Figure 12–2), and if the total concentration of solutes is low on one side of it and high on the other, water will tend to move across it so as to make the solute concentrations equal. Such movement of water from a region of low solute concentration (high water concentration) to a region of high solute concentration (low water concentration) is called **osmosis**. The driving force for the water movement is equivalent to a difference in water pressure and is called the **osmotic pressure**. In the absence of any counteracting pressure, the osmotic movement of water into a cell will cause it

Figure 12–14 The glucose–Na^+ symport protein uses the electrochemical Na^+ gradient to drive the import of glucose. Glucose can be moved across epithelial cell membranes using both active and passive transporters. Shown here is one way in which the glucose–Na^+ symport protein could actively pump glucose across the membrane using the influx of Na^+ down its gradient to drive glucose transport. The pump oscillates randomly between two alternate states, A and B. In the A state the protein is open to the extracellular space; in the B state it is open to the cytosol. Although Na^+ and glucose bind equally well to the protein in either state, they bind effectively only if both are present together: the binding of Na^+ induces a conformational change in the protein that greatly increases the protein's affinity for glucose and vice versa. Because the Na^+ concentration is much higher in the extracellular space than in the cytosol, glucose is more likely to bind to the pump in the A state; therefore, both Na^+ and glucose enter the cell (via an A → B transition) much more often than they leave it (via a B → A transition). The overall result is the net transport of both glucose and Na^+ into the cell. Note that, because the binding is cooperative, if one of the two solutes is missing, the other will fail to bind to the pump, and it will not be transported. An alternative way in which coupled transport may work is considered in Question 12–3.

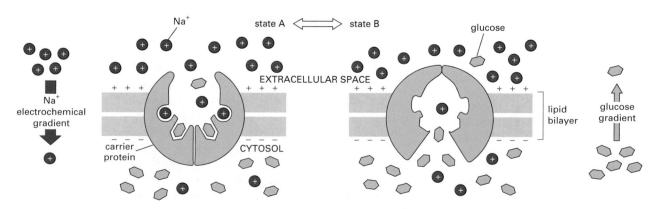

Na^+ state A ⟺ state B   glucose

EXTRACELLULAR SPACE

Na^+
electrochemical
gradient

carrier
protein   CYTOSOL

lipid
bilayer

glucose
gradient

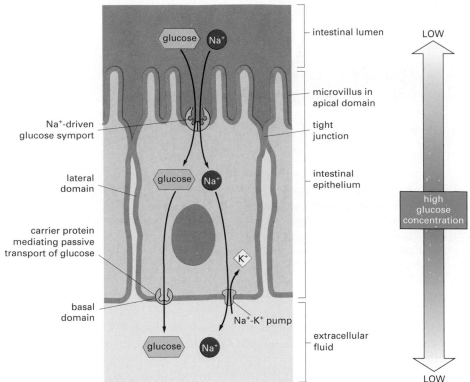

Figure 12–15 Two types of glucose carriers enable gut epithelial cells to transfer glucose across the gut lining. Glucose is actively transported into the cell by Na⁺-driven glucose symports at the apical surface, and it is released from the cell down its concentration gradient by passive glucose uniports at the basal and lateral surfaces. The two types of glucose carriers are kept segregated in the plasma membrane by the tight junction.

to swell. Such effects are a severe problem for animal cells, which have no rigid external wall to prevent them from swelling. Placed in pure water, such cells will generally swell until they burst (Figure 12–16).

In the tissues of the animal body, cells are bathed by a fluid that is rich in solutes, especially Na⁺ and Cl⁻. This balances the concentration of organic and inorganic solutes confined inside the cell and so prevents osmotic disaster. But the osmotic balance is always in danger of being upset, as the external solutes are constantly leaking into the cell down their individual electrochemical gradients. The cells thus have to do continuous work, pumping out unwanted solutes to maintain the osmotic equilibrium (Figure 12–17A). This function is performed mainly by the Na⁺-K⁺ pump, which pumps out the Na⁺ that leaks in. At the same time, by helping to maintain a membrane potential (as we explain later), the Na⁺-K⁺ pump also tends to prevent the entry of Cl⁻, which is negatively charged. If the pump is halted with an inhibitor such as ouabain, or if the cell simply runs out of the ATP needed to fuel the pump, Na⁺ and Cl⁻ ions enter through carrier proteins and open ion channels, upsetting the osmotic balance so that the cell swells and eventually bursts.

Other cells cope with their osmotic problems in different ways. Plant cells are prevented from swelling and bursting by their tough cell walls and so can tolerate a large osmotic difference across their plasma

Figure 12–16 The diffusion of water is known as osmosis. If the concentration of solutes inside a cell is higher than that outside, water will move in by osmosis, causing the cell to swell. If the difference in solute concentration is great enough, the cell will burst.

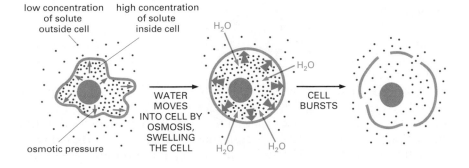

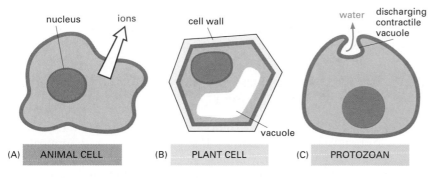

Figure 12–17 Cells use different tactics to avoid osmotic swelling. The animal cell keeps the intracellular solute concentration low by pumping out ions (A). The plant cell is saved from swelling and bursting by its tough wall (B). The protozoan avoids swelling by periodically ejecting the water that moves into the cell (C).

membrane (Figure 12–17B). The cell wall exerts a counteracting pressure that tends to balance the osmotic pressure created by the solutes in the cell and thereby limits the movement of water into the cell. Osmosis, together with the active transport of ions into the cell, results in a *turgor pressure* that keeps plant cells distended with water, with their cell wall tense. Thus plant cells are like footballs whose leather outer case is held taut by the pressure in the pumped-up rubber bladder inside; the cell wall acts like the leather outer case, and the plasma membrane acts like the rubber bladder. The turgor pressure serves various functions. It holds plant stems rigid and leaves extended. It also plays a part in regulating gas exchange through the stomata—the microscopic "mouths" in the surface of a leaf; these pores are opened and closed by the guard cells that surround them. Guard cells control their own turgor pressure by regulating the movement of K^+ across their plasma membranes.

In some protozoans living in fresh water, such as amoebae, the excess water that continually flows into the cell by osmosis is collected in contractile vacuoles that periodically discharge their contents to the exterior (Figure 12–17C). The cell first allows the vacuole to fill with a solution rich in solutes, which causes water to follow by osmosis. The cell then retrieves the solutes by actively pumping them back into the cytosol before emptying the vacuole to the exterior. But for most animal cells, the Na^+-K^+ pump is crucial for maintaining osmotic balance.

Intracellular Ca^{2+} Concentrations Are Kept Low by Ca^{2+} Pumps

Ca^{2+}, like Na^+, is also kept at a low concentration in the cytosol compared with its concentration in the extracellular fluid, but it is much less plentiful than Na^+, both inside and outside cells. The movement of Ca^{2+} across cell membranes, however, is crucially important because Ca^{2+} can bind tightly to many other molecules in the cell, altering their activities. An influx of Ca^{2+} into the cytosol through Ca^{2+} channels, for example, is often used as a signal to trigger other intracellular events, such as the secretion of signaling molecules and the contraction of muscle cells.

The lower the background concentration of free Ca^{2+} in the cytosol, the more sensitive the cell is to an increase in cytosolic Ca^{2+}. Thus eucaryotic cells in general maintain very low concentrations of free Ca^{2+} in their cytosol (about 10^{-4} mM) in the face of very much higher extracellular Ca^{2+} concentrations (typically 1–2 mM). This is achieved mainly by means of ATP-driven Ca^{2+} pumps in both the plasma membrane and the endoplasmic reticulum membrane, which actively pump Ca^{2+} out of the cytosol.

Like the Na^+-K^+ pump, the Ca^{2+} pump is an ATPase that is phosphorylated and dephosphorylated during its pumping cycle (see Figure 12–6). It is thought to work in much the same way as depicted for the Na^+-K^+ pump in Figure 12–12 except that it returns to its original conformation without binding and transporting a second ion. These two

Question 12–3

A rise in the intracellular Ca^{2+} concentration causes muscle cells to contract. In addition to an ATP-driven Ca^{2+} pump, muscle cells that contract quickly and regularly, such as those of the heart, have an antiporter that exchanges Ca^{2+} for extracellular Na^+ across the plasma membrane. The majority of the Ca^{2+} ions that have entered the cell during contraction are rapidly pumped back out of the cell by this antiporter, thus allowing the cell to relax. Ouabain and digitalis are important drugs for treating patients with heart disease because they make the heart muscle contract more strongly. Both drugs function by partially inhibiting the Na^+-K^+ pump in the membrane of the heart muscle cell. Can you propose an explanation for the effects of the drugs in the patients? What will happen if too much of either drug is taken?

ATP-driven pumps have similar amino acid sequences and structures, with about 10 membrane-spanning α helices in each subunit, and it is likely that they have a common evolutionary origin.

H⁺ Gradients Are Used to Drive Membrane Transport in Plants, Fungi, and Bacteria

Plant cells, fungi (including yeasts), and bacteria do not have Na^+-K^+ pumps in their plasma membranes. Instead of an electrochemical gradient of Na^+, they rely mainly on an electrochemical gradient of H^+ to drive the transport of solutes into the cell. The gradient is created by H^+ pumps in the plasma membrane, which pump H^+ out of the cell, thus setting up an electrochemical proton gradient, with H^+ higher outside than inside; in the process, the H^+ pump also creates an acid pH in the medium surrounding the cell. The uptake of many sugars and amino acids into bacterial cells, then, is driven by H^+ symports, which use the electrochemical gradient of H^+ across the plasma membrane in much the same way that animal cells use the electrochemical gradient of Na^+.

In some photosynthetic bacteria the H^+ gradient is created by the activity of light-driven H^+ pumps such as bacteriorhodopsin. In other bacteria the gradient is created by the activities of plasma membrane proteins that carry out the final stages of cellular respiration leading up to ATP synthesis, as discussed in Chapter 14. But plants, fungi, and many other bacteria set up their H^+ gradient by means of ATPases in their plasma membranes that use the energy of ATP hydrolysis to pump H^+ out of the cell; these ATPases resemble the Na^+-K^+ pumps and Ca^{2+} pumps in mammalian cells that we discussed earlier.

A different type of H^+ ATPase is found in the membranes of some intracellular organelles, such as the lysosomes of animal cells and the central vacuole of plant and fungal cells. Their function is to pump H^+ out of the cytosol into the organelle, thereby helping to keep the pH of the cytosol neutral and the pH of the interior of the organelle acidic. The acid environment in many organelles is crucial to their function, as we discuss in Chapter 15.

Some of the carrier proteins present in animal and plant cells that we have discussed are shown in Figure 12–18. These and some other carrier proteins considered in this chapter are listed in Table 12–2.

We now turn to the transport of ions through channel proteins, and we see how this ion flow can generate a membrane potential.

Figure 12–18 There are similarities and differences in carrier-mediated solute transport in animal and plant cells. Whereas an electrochemical gradient of Na^+, generated by the Na^+-K^+ pump (Na^+-K^+ ATPase), is often used to drive the active transport of solutes across the animal cell plasma membrane (A), an electrochemical gradient of H^+, usually set up by an H^+ ATPase, is often used for this purpose in plant cells (B), as well as in bacteria and fungi (including yeasts). The lysosomes in animal cells and the vacuoles in plant and fungal cells contain an H^+ ATPase in their membrane that pumps in H^+, helping to keep the internal environment of these organelles acidic. (C) An electron micrograph showing the vacuole in plant cells in a young tobacco leaf. (C, courtesy of J. Burgess.)

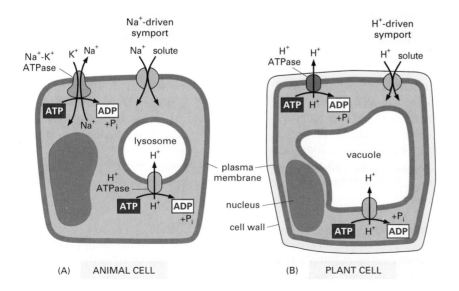

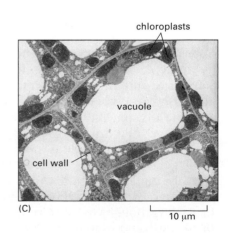

Table 12–2 Some Examples of Carrier Proteins

CARRIER PROTEIN	LOCATION	ENERGY SOURCE	FUNCTION
Glucose carrier	plasma membrane of most animal cells	none	passive import of glucose
Na^+-driven glucose pump	apical plasma membrane of kidney and intestinal cells	Na^+ gradient	active import of glucose
Na^+-H^+ exchanger	plasma membrane of animal cells	Na^+ gradient	active export of H^+ ions, pH regulation
Na^+-K^+ pump (Na^+-K^+ ATPase)	plasma membrane of most animal cells	ATP hydrolysis	active export of Na^+ and import of K^+
Ca^{2+} pump (Ca^{2+} ATPase)	plasma membrane of eucaryotic cells	ATP hydrolysis	active export of Ca^{2+}
H^+ pump (H^+ ATPase)	plasma membrane of plant cells, fungi, and some bacteria	ATP hydrolysis	active export of H^+ from cell
H^+ pump (H^+ ATPase)	membranes of lysosomes in animal cells and of vacuoles in plant and fungal cells	ATP hydrolysis	active export of H^+ from cytosol into vacuole
Bacteriorhodopsin	plasma membrane of some bacteria	light	active export of H^+ out of the cell

Ion Channels and the Membrane Potential

In principle, the simplest way to allow a small water-soluble molecule to cross from one side of a membrane to the other is to create a hydrophilic channel through which the molecule can pass. Channel proteins perform this function in cell membranes, forming transmembrane aqueous pores that allow the passive movement of small water-soluble molecules into or out of the cell or organelle.

A few channel proteins form relatively large pores: examples are the proteins that form *gap junctions* between two adjacent cells (see Figure 21–28) and the *porins* that form channels in the outer membrane of mitochondria and some bacteria (see Figure 11–25). But such large, permissive channels would lead to disastrous leaks if they directly connected the cytosol of a cell to the extracellular space. Most of the channel proteins in the plasma membrane of animal and plant cells are therefore quite different and have narrow, highly selective pores. Almost all of these proteins are **ion channels**, concerned exclusively with the transport of inorganic ions, mainly Na^+, K^+, Cl^-, and Ca^{2+}.

Ion Channels Are Ion-selective and Gated

Two important properties distinguish ion channels from simple aqueous pores. First, they show *ion selectivity,* permitting some inorganic ions to pass but not others. Ion selectivity depends on the diameter and shape of the ion channel and on the distribution of charged amino acids in its lining: the channel is narrow enough in places to force ions into contact with the wall of the channel so that only those of appropriate size and charge are able to pass (Figure 12–19). Narrow channels, for example, will not pass large ions, and channels with a negatively charged lining will deter negative ions from entering because of the mutual electrostatic repulsion between like charges. In this way channels have evolved that are selective for just one type of ion, such as Cl^- or K^+. Each ion in aqueous solution is surrounded by a small shell of water molecules, and the ions have to shed most of their associated

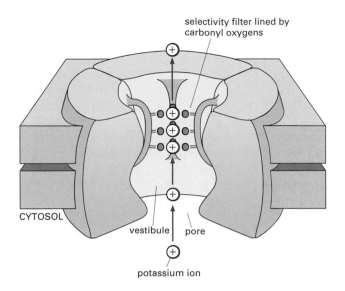

Figure 12–19 A K+ channel protein possesses a selectivity filter that controls which ion it will transport across the membrane. Shown here is a portion of a bacterial K+ channel protein. The fourth subunit has been omitted from the drawing to expose the interior structure of the pore. From the cytosolic side, the pore opens into a vestibule that sits in the middle of the membrane. K+ ions in the vestibule are still cloaked in their associated water molecules. The narrow selectivity filter, which links the vestibule with the outside of the cell, is lined with carbonyl oxygen atoms that bear a partial negative charge and form transient binding sites for the K+ ions that have shed their watery shells. (Adapted from D.A. Doyle et al., *Science* 280:69–77, 1998.)

water molecules in order to pass, single file, through the selectivity filter in the narrowest part of the channel. There, ions make important but very transient contacts with atoms that line the walls of the selectivity filter (see Figure 12–19). These precisely positioned atoms allow the channel to discriminate between ions that differ only minutely in size. This step in the transport process also limits the maximum rate of passage of ions through the channel. Thus, as ion concentrations are increased, the flow of ions through a channel at first increases proportionally but then levels off (saturates) at a maximum rate.

The second important distinction between simple aqueous pores and ion channels is that ion channels are not continuously open. Ion transport would be of no value to the cell if there were no means of controlling the flow and if all of the many thousands of ion channels in a cell membrane were open all of the time. Instead ion channels open briefly and then close again (Figure 12–20). As we discuss later, most ion channels are *gated:* a specific stimulus triggers them to switch between an open and a closed state by a change in their conformation.

Ion channels have a large advantage over carrier proteins with respect to their maximum rate of transport. More than a million ions can pass through one channel each second, which is a rate 1000 times greater than the fastest rate of transport known for any carrier protein. On the other hand, channels cannot couple the ion flow to an energy source to carry out active transport. The function of most ion channels is simply to make the membrane transiently permeable to selected inorganic ions, mainly Na+, K+, Ca2+, or Cl−, allowing these to diffuse rapidly down their electrochemical gradients across the membrane when the channel gates are open.

Thanks to active transport by pumps and other carrier proteins, most ion concentrations are far from equilibrium across the membrane.

Figure 12–20 A typical ion channel fluctuates between closed and open conformations. The channel shown here in cross section forms a hydrophilic pore across the lipid bilayer only in the "open" conformation. Polar groups are thought to line the wall of the pore, while hydrophobic amino acid side chains interact with the lipid bilayer (not shown). The pore narrows to atomic dimensions in the selectivity filter, where the ion selectivity of the channel is largely determined (see Figure 12–19).

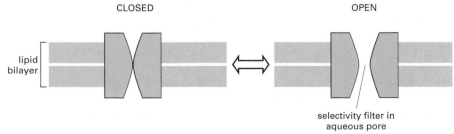

When a channel opens, therefore, ions rush through it. The rush of ions amounts to a pulse of electric charge delivered either into the cell (as ions flow in) or out of the cell (as ions flow out). The ion flow changes the voltage across the membrane—the *membrane potential*—thus altering the electrochemical driving forces for transmembrane movements of all the other ions. It also forces other ion channels, which are specifically sensitive to changes in the membrane potential, to open or close in a matter of milliseconds. The resulting flurry of electrical activity can spread rapidly from one region of the cell membrane to another, conveying an electrical signal, as we discuss later in the context of nerve cells. This type of electrical signaling is not restricted to animals; it also occurs in protozoans and plants. Carnivorous plants like the Venus flytrap, for example, use electrical signaling to sense and trap insects (Figure 12–21).

The membrane potential is the basis of all electrical activity in cells, whether they are plant cells, animal cells, or protozoans. Before we discuss how the membrane potential is generated, however, we look at how ion-channel activity is measured.

Ion Channels Randomly Snap Between Open and Closed States

Measuring changes in electrical current is the main method used to study ion movements and ion channels in living cells. Amazingly, electrical recording techniques have been refined to the point where it is now possible to detect and measure the electric current flowing through a single channel molecule. The procedure for doing this is known as **patch-clamp recording**, and it provides a direct and surprising picture of how individual ion channels behave.

In patch-clamp recording, a fine glass tube is used as a *microelectrode* to make electrical contact with the surface of the cell. The microelectrode is made by heating the glass tube and pulling it to create an extremely fine tip with a diameter of no more than a few micrometers; the tube is then filled with an aqueous conducting solution, and the tip is pressed against the cell surface. With gentle suction, a tight electrical seal is formed where the cell membrane contacts the mouth of the microelectrode (Figure 12–22A). If one wishes to expose the cytosolic face of the membrane, the patch of membrane held in the microelectrode is gently detached from the cell (Figure 12–22B). At the other, open end of the microelectrode, a metal wire is inserted. Current that enters the microelectrode through ion channels in the small patch of membrane covering its tip passes via the wire, through measuring instruments, back into the bath of medium in which the cell or the detached patch is sitting (Figure 12–22D). Patch-clamp recording makes it possible to record from ion channels in all sorts of cell types—not only in large nerve cells, which are famous for their electrical activities, but also in cells such as yeasts that are too small for the electrical events in them to be detected by other means.

By varying the concentrations of ions in the medium on either side of the membrane patch, one can test which ions will go through its channels. With the appropriate electronic circuitry, the voltage across the membrane patch—that is, the membrane potential—can also be set and held clamped at any chosen value (hence the term "patch-clamp"). This makes it possible to see how changes in membrane potential affect the opening and closing of the channels in the membrane.

With a sufficiently small area of membrane in the patch, sometimes only a single ion channel will be present. Modern electrical instruments are sensitive enough to reveal the ion flow through a single channel,

Figure 12–21 A Venus flytrap uses electrical signaling to capture its prey. The leaves snap shut in less than half a second when an insect moves on them. The response is triggered by touching any two of the three trigger hairs in succession in the center of each leaf. This mechanical stimulation opens ion channels and thereby sets off an electrical signal, which, by an unknown mechanism, leads to a rapid change in turgor pressure that closes the leaf. (Courtesy of J.S. Sira, Garden Picture Library.)

Figure 12–22 The technique of patch-clamp recording is used to monitor ion channel activity. Because of the extremely tight seal between the mouth of the microelectrode and the membrane, current can enter or leave the microelectrode only by passing through the channels in the patch of membrane covering its tip. The term "clamp" is used because an electronic device is employed to maintain, or "clamp," the membrane potential at a set value while recording the ionic current through individual channels. Recordings of the current through these channels can be made with the patch still attached to the rest of the cell, as in (A), or detached, as in (B). The advantage of the detached patch is that it is easy to alter the composition of the solution on either side of the membrane to test the effect of various solutes on channel behavior. The micrograph (C) shows a nerve cell from the eye held in a suction pipette (the tip of which is shown on the left) while a microelectrode *(upper right)* is being used for patch-clamp recording. (D) The circuitry for patch-clamp recording. (C, from T.D. Lamb, H.R. Mathews, and V. Torre, *J. Physiol.* 37: 315–349, 1986.)

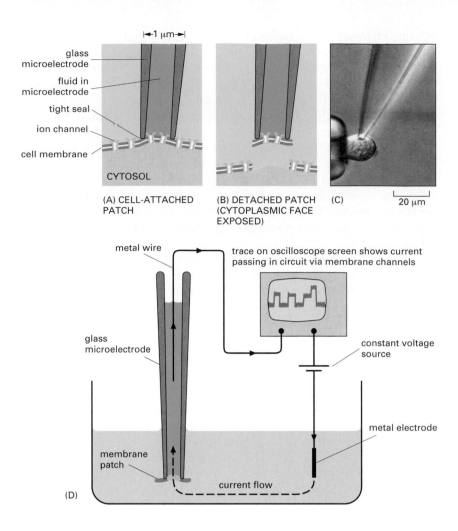

(A) CELL-ATTACHED PATCH

(B) DETACHED PATCH (CYTOPLASMIC FACE EXPOSED)

(C)

(D)

Figure 12–23 The current through a single ion channel can be measured by the patch-clamp technique. The voltage (the membrane potential) across the isolated patch of membrane is held constant during the recording. In this example the membrane is from a muscle cell and contains a single channel protein that is responsive to the neurotransmitter acetylcholine. This ion channel opens to allow passage of positive ions when acetylcholine binds to the exterior face of the channel, as is the case here. Even when acetylcholine is bound to the channel as is the case during the three bursts of channel opening shown here, it does not remain open all the time. Instead the channel flickers between open and closed states, as seen in the brief transient drops in current in each of the three bursts. If acetylcholine were not bound to the channel, the channel would rarely open. (Courtesy of David Colquhoun.)

detected as a minute electric current (of the order of 10^{-12} ampere). These currents typically behave in a surprising way: even when conditions are held constant, the currents switch abruptly on and abruptly off again, as though an on/off switch were being jiggled randomly (Figure 12–23). This behavior implies that the channel has moving parts and is snapping back and forth between open and closed conformations (see Figure 12–20). As this behavior is seen even when conditions are constant, it presumably indicates that the channel protein is being knocked from one conformation to the other by the random thermal movements of the molecules in its environment. Single-channel recording is one of a very few techniques that can be used to monitor the conformational changes of a single protein molecule. The picture it paints, of a jerky piece of machinery subjected to a constant, violent buffeting, is certain to apply also to other proteins with moving parts.

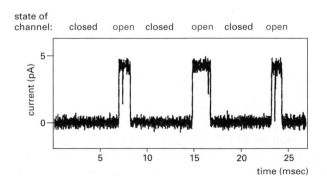

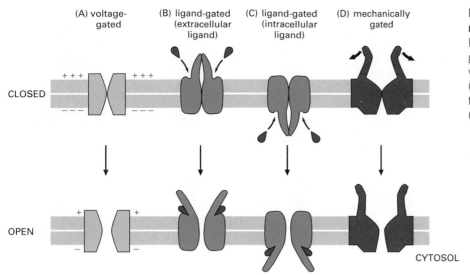

| (A) voltage-gated | (B) ligand-gated (extracellular ligand) | (C) ligand-gated (intracellular ligand) | (D) mechanically gated |

CLOSED

OPEN

CYTOSOL

Figure 12–24 Gated ion channels respond to different types of stimuli. Depending on the type of ion channel, the gates open in response to a change in the voltage difference across the membrane (A), to the binding of a chemical ligand to the channel, outside (B) or inside the cell (C), or to mechanical stimulation (D).

If ion channels randomly snap between open and closed conformations even when conditions on each side of the membrane are held constant, how can their state be regulated by conditions inside or outside the cell? The answer is that when the appropriate conditions change, the random behavior continues but with a greatly changed probability: if the altered conditions tend to open the channel, for example, the channel will now spend a much greater proportion of its time in the open conformation, although it will not remain open continuously (see Figure 12–23). When an ion channel is open, it is fully open, and when it is closed, it is fully closed.

Different Types of Stimuli Influence the Opening and Closing of Ion Channels

There are more than a hundred types of ion channels, and even simple organisms can possess many different channels. The nematode worm *C. elegans*, for example, has genes that encode 68 different but related K+ channels alone. Ion channels differ from one another primarily with respect to their (1) *ion selectivity*—the type of ions they allow to pass; and (2) *gating*—the conditions that influence their opening and closing. For a **voltage-gated channel**, the probability of being open is controlled by the membrane potential (Figure 12–24A). For a **ligand-gated channel**, it is controlled by the binding of some molecule (the ligand) to the channel protein (Figure 12–24B and C). For a **stress-activated channel**, opening is controlled by a mechanical force applied to the channel (Figure 12–24D). The *auditory hair cells* in the ear are an important example of cells that depend on this type of channel. Sound vibrations pull stress-activated channels open, causing ions to flow into the hair cells; this sets up an electrical signal that is transmitted from the hair cell to the auditory nerve, which conveys the signal to the brain (Figure 12–25).

Voltage-gated Ion Channels Respond to the Membrane Potential

Voltage-gated ion channels play the major role in propagating electrical signals in nerve cells. They are present in many other cells, too, including muscle cells, egg cells, protozoans, and even plant cells, where they enable electrical signals to travel from one part of the plant to another,

Question 12–4

Figure Q12–4 shows a recording from a patch-clamp experiment in which the electrical current passing across a patch of membrane is measured as a function of time. The membrane patch was plucked from the plasma membrane of a muscle cell by the technique shown in Figure 12–22 and contains molecules of the acetylcholine receptor, which is a ligand-gated cation channel that is opened by the binding of acetylcholine to the extracellular face of the channel protein. To obtain a recording, acetylcholine was added to the solution in the microelectrode. Describe what you can learn about the channels from this recording. How would the recording differ if acetylcholine were (i) omitted or (ii) added to the solution outside the microelectrode only?

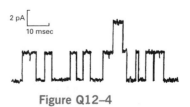

Figure Q12–4

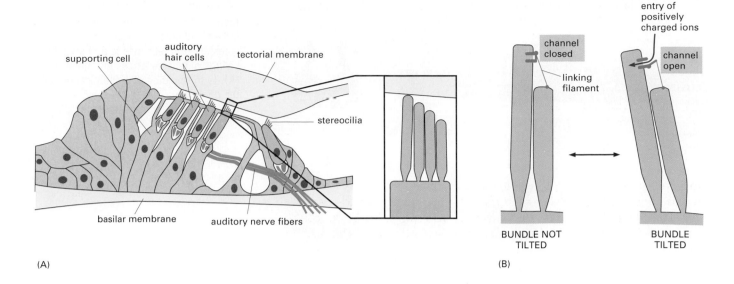

(A)

(B)

channel closed

linking filament

entry of positively charged ions

channel open

BUNDLE NOT TILTED

BUNDLE TILTED

Figure 12–25 Stress-activated ion channels allow us to hear. (A) A section through the organ of Corti, which runs the length of the cochlea of the inner ear. Each auditory hair cell has a tuft of spiky extensions called stereocilia projecting from its upper surface. The hair cells are embedded in a sheet of supporting cells, which is sandwiched between the *basilar membrane* below and the *tectorial membrane* above. (These are not lipid bilayer membranes but sheets of extracellular matrix.) (B) Sound vibrations cause the basilar membrane to vibrate up and down, causing the stereocilia to tilt. Each stereocilium in the staggered array on each hair cell is attached to the next shorter stereocilium by a fine filament. The tilting stretches the filaments, which pull open stress-activated ion channels in the stereocilium membrane, allowing positively charged ions to enter from the surrounding fluid. The influx of ions activates the hair cells, which stimulate underlying nerve cells that convey the auditory signal to the brain. The hair-cell mechanism is astonishingly sensitive: the force required to open a single channel is estimated to be about 2×10^{-13} newton, and the faintest sounds we can hear have been estimated to stretch the filaments by an average of about 0.04 nm, which is less than the diameter of a hydrogen ion.

as in the leaf-closing response of the mimosa tree (Figure 12–26). Voltage-gated ion channels have specialized charged protein domains called *voltage sensors* that are extremely sensitive to changes in the membrane potential: changes above a certain threshold value exert sufficient electrical force on these domains to encourage the channel to switch from its closed to its open conformation, or vice versa. A change in the membrane potential does not affect how wide the channel is open but alters the probability that it will be found in its open conformation. Thus in a large patch of membrane, containing many molecules of the channel protein, one might find that on average 10% of them are open at any instant when the membrane is held at one potential, while 90% are open when it is held at another potential.

To appreciate the function of voltage-gated ion channels in a living cell, we have to consider what controls the membrane potential. The simple answer is that ion channels themselves control it, and the opening and closing of these channels is what makes it change. This control loop, from ion channels → membrane potential → ion channels, is fundamental to all electrical signaling in cells. Having seen how the membrane potential can regulate ion channels, we now discuss how ion channels can control the membrane potential. In the last part of the chapter we consider how this control loop works as a whole.

Membrane Potential Is Governed by Membrane Permeability to Specific Ions

All cells have an electrical potential difference, or **membrane potential**, across their plasma membranes. To understand how this potential arises, it is helpful to recall some basic principles of electricity. While electricity in metals is carried by electrons, electricity in aqueous solutions is carried by ions, which are either positively (cations) or negatively (anions) charged. An ion flow across a cell membrane is detectable as an electric current, and an accumulation of ions, if not exactly balanced by an accumulation of oppositely charged ions, is detectable as an accumulation of electric charge, or a membrane potential (Figure 12–27).

To see how the membrane potential is generated and maintained, consider the ion movements into and out of a typical animal cell in an unstimulated "resting" state. The negative charges on the organic molecules confined within the cell are largely balanced by K^+, the predominant

  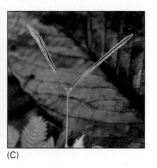

(A)   (B)   (C)

Figure 12–26 **Voltage-gated ion channels underlie the leaf-closing response in mimosa.** (A) Resting leaf. (B and C) Successive responses to touch. A few seconds after the leaf is touched, the leaflets snap shut. The response involves the opening of voltage-gated ion channels, generating an electric impulse. When the impulse reaches specialized hinge cells at the base of each leaflet, a rapid loss of water by these cells occurs, causing the leaflets to fold closed suddenly and progressively down the leaf stalk. (Courtesy of G.I. Bernard. © Oxford Scientific Films.)

positive ion inside the cell. The high intracellular concentration of K^+ is in part generated by the Na^+-K^+ pump, which actively pumps K^+ into the cell. This leads to a large concentration difference for K^+ across the plasma membrane, with the concentration of K^+ being much higher inside the cell than outside. The plasma membrane, however, also contains a set of K^+ channels known as *K^+ leak channels*. These channels randomly flicker between open and closed states no matter what the conditions are inside or outside the cell, and when they are open, they allow K^+ to move freely. In a resting cell, these are the main ion channels open in the plasma membrane, thus making the resting plasma membrane much more permeable to K^+ than to other ions.

K^+ will have a tendency to flow out of the cell through these channels down its steep concentration gradient. But any transfer of positive charge to the exterior leaves behind unbalanced negative charge within the cell, thereby creating an electrical field, or membrane potential, which will oppose any further movement of K^+ out of the cell. Within a millisecond or so, an equilibrium condition is established in which the membrane potential is just strong enough to counterbalance the tendency of K^+ to move out down its concentration gradient—that is, in which the electrochemical gradient for K^+ is zero, even though there is still a much higher concentration of K^+ inside the cell than out (Figure 12–28).

The *resting membrane potential* is the membrane potential in such steady-state conditions, in which the flow of positive and negative ions across the plasma membrane is precisely balanced, so that no further difference in charge accumulates across the membrane. The membrane potential is measured as a voltage difference across the membrane. In animal cells, the resting membrane potential varies between –20 and –200 millivolts (mV), depending on the organism and cell type. It is expressed as a negative value because the interior of the cell is negative

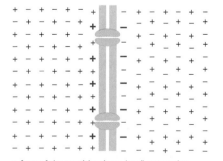

lipid bilayer   ion channel

exact balance of charges on each side of the membrane so that each positive ion is balanced by a negative counterion; membrane potential = 0

a few of the positive ions (*red*) cross the membrane from right to left, leaving their negative counterions (*red*) behind; this sets up a nonzero membrane potential

Figure 12–27 **The distribution of ions on either side of the bilayer gives rise to the membrane potential.** The membrane potential results from a thin (<1 nm) layer of ions close to the membrane, held in place by their electrical attraction to oppositely charged counterparts on the other side of the membrane. The number of ions that must move across the membrane to set up a membrane potential is a tiny fraction of the ions present. (6000 K^+ ions crossing 1 μm^2 of cell membrane are enough to shift the membrane potential by about 100 mV; the number of K^+ ions in 1 μm^3 of bulk cytoplasm is 70,000 times larger than this.)

Figure 12–28 K⁺ leak channels plays a major role in generating the membrane potential across the plasma membrane. Starting from a hypothetical situation where the membrane potential is zero, K⁺ will tend to leave the cell, moving down its concentration gradient through K⁺ leak channels. Assuming the membrane contains no open channels permeable to other ions, K⁺ ions will cross the membrane, but negative ions will be unable to follow them. The result will be an excess of positive charge on the outside of the membrane and of negative charge on the inside. This gives rise to a membrane potential that tends to drive K⁺ back in. At equilibrium, the effect of the K⁺ concentration gradient is exactly balanced by the effect of the membrane potential, and there is no net movement of K⁺.

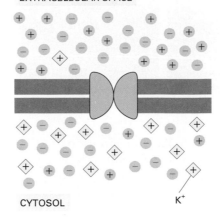

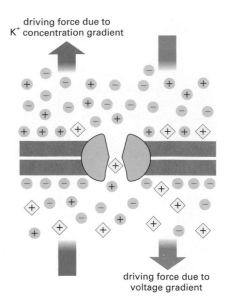

(A) K⁺ channel closed, membrane potential = 0; more K⁺ inside the cell than outside, but zero *net* charge on each side (positive and negative charges balanced exactly)

(B) K⁺ channel open; K⁺ moves out, leaving negative ions behind, and this charge distribution creates a membrane potential that balances the tendency of K⁺ to move out

The force tending to drive an ion across a membrane is made up of two components: one due to the electrical membrane potential and one due to the concentration gradient. At equilibrium, the two forces are balanced and satisfy a simple mathematical relationship given by the

Nernst equation

$$V = 62 \log_{10}(C_o/C_i)$$

where V is the membrane potential in millivolts, and C_o and C_i are the outside and inside concentrations of the ion, respectively. This form of the equation assumes that the ion carries a single positive charge and that the temperature is 37°C.

Figure 12–29 The Nernst equation can be used to calculate the resting potential of a membrane.

with respect to the exterior, as the negative charges inside the cell are in slight excess over positive charges. The actual value of the resting membrane potential in animal cells is chiefly a reflection of the K⁺ concentration gradient across the plasma membrane because, at rest, this membrane is chiefly permeable to K⁺ and K⁺ is the main positive ion inside the cell. A simple formula called the **Nernst equation** (Figure 12–29) expresses the equilibrium quantitatively and makes it possible to calculate the theoretical resting membrane potential if the ratio of internal to external ion concentrations is known.

Suppose now that other channels permeable to some other ion—say, Na⁺—are suddenly opened in the resting plasma membrane. Because Na⁺ is at a higher concentration outside the cell than inside, Na⁺ will move into the cell through these channels and the membrane potential will become less negative, maybe even reversing sign to become positive (so that the interior of the cell is positive with respect to the exterior). The membrane potential will shift toward a new value that is a compromise between the negative value that would correspond to equilibrium for K⁺ and the positive value that would correspond to equilibrium for Na⁺. Any change in the membrane's permeability to specific ions—that is, any change in the numbers of ion channels of different sorts that are open—thus causes a change in the membrane potential. The membrane potential, therefore, is determined by both the state of the ion channels and the ion concentrations in the cytosol and extracellular medium. Because the electrical processes at the plasma membrane occur very quickly compared with changes in the bulk ion concentrations, however, over the short term—milliseconds to seconds or minutes—it is the ion channels that are most important in controlling the membrane potential.

To see how the interplay between the membrane potential and ion channels is used for electrical signaling, we now turn from the behavior of ions and ion channels to the behavior of entire cells. We take nerve cells as our prime example, for they, more than any other cell type, have made a profession of electrical signaling and employ ion channels in the most sophisticated ways.

Ion Channels and Signaling in Nerve Cells

The fundamental task of a nerve cell, or **neuron**, is to receive, conduct, and transmit signals. Neurons carry signals inward from sense organs to the central nervous system—the brain and spinal cord. In the central nervous system, neurons signal from one to another through networks of enormous complexity, allowing the brain and spinal cord to analyze, interpret, and respond to the signals coming in from the sense organs. From the central nervous system, neurons extend processes outward to convey signals for action to muscles and glands. To perform these functions, neurons are often extremely elongated: the motor neurons in a human that carry signals from the spinal cord to a muscle in the foot, for example, may be a meter long.

Every neuron consists of a *cell body* (containing the nucleus) that has a number of long, thin extensions radiating outward from it. Usually, a neuron has one long **axon**, which conducts signals away from the cell body toward distant target cells; and several shorter, branching *dendrites,* which extend from the cell body like antennae and provide an enlarged surface area to receive signals from the axons of other neurons (Figure 12–30). The axon commonly divides at its far end into many branches, each of which ends in a **nerve terminal**, so that the neuron's message can be passed simultaneously to many target cells—either other neurons or muscle or gland cells. Likewise, the branching of the dendrites can be extensive, in some cases sufficient to receive as many as 100,000 inputs on a single neuron.

No matter what the meaning of the signal a neuron carries—whether it is visual information from the eye, a motor command to a muscle, or one step in a complex network of neural communication in the brain—the form of the signal is always the same: it consists of changes in the electrical potential across the neuron's plasma membrane.

Action Potentials Provide for Rapid Long-Distance Communication

A neuron is stimulated by a signal—typically from another neuron—delivered to a localized site on its surface. This signal initiates a change in the membrane potential at that site. To transmit the signal onward, however, the change in membrane potential has to spread from this point, which is usually on a dendrite or the cell body, to the axon terminals, which relay the signal to the next cells in the pathway. Although a local change in membrane potential will spread passively along an axon or a dendrite to adjacent regions of the plasma membrane, it rapidly becomes weaker with increasing distance from the source. Over short distances this weakening is unimportant, but for long-distance

Question 12–5

From the concentrations given in Table 12–1 (p. 390), calculate the equilibrium membrane potential of K^+ and Na^+ (assume the concentration of intracellular Na^+ is 10 mM). What membrane potential would you predict in a resting cell? Explain your answer. What would happen if a large number of Na^+ channels suddenly opened, making the membrane much more permeable to Na^+ than to K^+? (Note that because few ions need to move across the membrane to change the charge distribution across the membrane drastically, you should assume that the ion concentrations on either side of the membrane do not change significantly.) What would you predict would happen next if the Na^+ channels closed again?

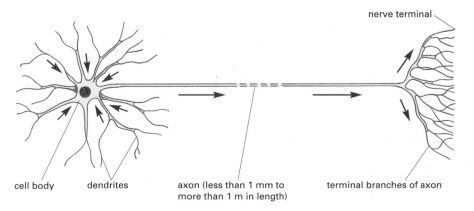

cell body   dendrites   axon (less than 1 mm to more than 1 m in length)   terminal branches of axon

Figure 12–30 A typical neuron has a cell body, a single axon, and multiple dendrites. The axon conducts signals away from the cell body, while the multiple dendrites receive signals from the axons of other neurons. The *red arrows* indicate the direction in which signals are conveyed.

Figure 12–31 The squid *Loligo* has a large nerve cell with a giant axon that allows the animal to respond rapidly to threats in its environment. Long before patch clamping allowed recordings from single channels in small cells, scientists were able to record action potentials in the squid giant axon and deduce the existence of ion channels in membranes. (Courtesy of Howard Hall. © Oxford Scientific Films.)

Figure 12–32 An action potential is triggered by a rapid change in membrane potential. The resting membrane potential in this neuron is –60 mV. The action potential is triggered when a stimulus depolarizes the plasma membrane by about 20 mV, making the membrane potential –40 mV, which is the threshold value in this cell for initiating an action potential. Once an action potential is triggered, the membrane rapidly depolarizes further: the membrane potential swings past zero and reaches +40 mV before it returns to its resting negative value, as the action potential terminates. The *green curve* shows how the membrane potential simply would have relaxed back to the resting value, after the initial depolarizing stimulus if there had been no voltage-gated ion channels in the plasma membrane.

communication such *passive spread* is inadequate. In the same way, a telephone signal can be transmitted without amplification the short distances through the wires in your hometown, but for transmission across an ocean by an undersea cable, the strength of the signal has to be boosted at intervals.

Neurons solve this long-distance communication problem by employing an active signaling mechanism: a local electrical stimulus of sufficient strength triggers an explosion of electrical activity in the plasma membrane that is propagated rapidly along the membrane of the axon and sustained by automatic renewal all along the way. This traveling wave of electrical excitation, known as an **action potential**, or a *nerve impulse*, can carry a message, without the signal weakening, from one end of a neuron to the other at speeds of up to 100 meters per second.

All of the early research that established the mechanism of electrical signaling along nerve axons was done on the giant axon of the squid (Figure 12–31). This axon has such a large diameter that it is possible to record its electrical activity from an electrode inserted directly into it (How We Know, pp. 414–415). From such studies it was deduced that action potentials are the direct consequence of the properties of voltage-gated ion channels (see Figure 12–24A) in the nerve cell membrane, as we now explain.

Action Potentials Are Usually Mediated by Voltage-gated Na⁺ Channels

An action potential in a neuron is typically triggered by a sudden local *depolarization* of the plasma membrane—that is, by a shift in the membrane potential to a less negative value. We discuss later how such a depolarization is caused by the action of signaling molecules—*neurotransmitters*—released by another neuron. A stimulus that causes a sufficiently large depolarization to pass a certain threshold value promptly causes **voltage-gated Na⁺ channels** to open temporarily at that site, allowing a small amount of Na⁺ to enter the cell down its electrochemical gradient. The influx of positive charge depolarizes the membrane further (that is, it makes the membrane potential even less negative), thereby opening more voltage-gated Na⁺ channels, which admit more Na⁺ ions and cause still further depolarization. This process continues in a self-amplifying fashion until, within about a millisecond, the membrane potential in the local region of membrane has shifted from its resting value of about –60 mV to about +40 mV (Figure 12–32). This voltage is close to the membrane potential at which the electrochemical driving force for movement of Na⁺ across the membrane is zero—that

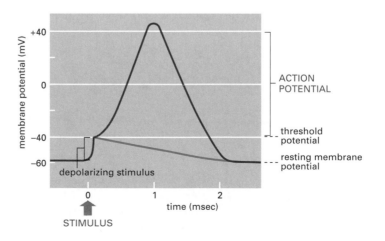

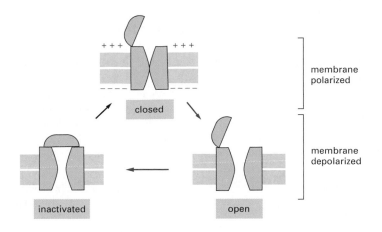

membrane
polarized

closed

membrane
depolarized

inactivated

open

Figure 12–33 A voltage-gated Na⁺ channel can adopt at least three conformations. The channel can flip from one conformation (state) to another, depending on the membrane potential. When the membrane is at rest (highly polarized), the closed conformation is the most stable. When the membrane is depolarized, however, the open conformation is more stable, and so the channel has a high probability of opening; but in the depolarized membrane the inactivated conformation is more stable still, and so, after a brief period spent in the open conformation, the channel becomes inactivated and cannot open. The *red arrows* indicate the sequence that follows a sudden depolarization, and the *black arrow* indicates the return to the original conformation after the membrane is repolarized.

is, at which the effects of the membrane potential and the concentration gradient for Na⁺ are equal and opposite and Na⁺ has no further tendency to enter or leave the cell. At this point the cell would get stuck with all of its voltage-gated Na⁺ channels predominantly open if the channels continued indefinitely to respond to the altered membrane potential in the same way.

The cell is saved from this fate because the Na⁺ channels have an automatic inactivating mechanism, which causes them to rapidly adopt (within a millisecond or so) a special inactive conformation, where the channel is unable to open again: even though the membrane is still depolarized, the Na⁺ channels will remain in this *inactivated state* until a few milliseconds after the membrane potential returns to its initial negative value. A schematic illustration of these three distinct states of the voltage-gated Na⁺ channel—*closed, open,* and *inactivated*—is shown in Figure 12–33. How they contribute to the rise and fall of the action potential is shown in Figure 12–34.

The membrane is also helped to return to its resting value by the opening of *voltage-gated K⁺ channels*. These also open in response to depolarization of the membrane, but not as promptly as the Na⁺ channels, and they then stay open as long as the membrane remains depolarized. As the action potential reaches its peak, K⁺ ions (carrying positive charge) therefore start to flow out of the cell through these K⁺ channels down their electrochemical gradient, temporarily unhindered by

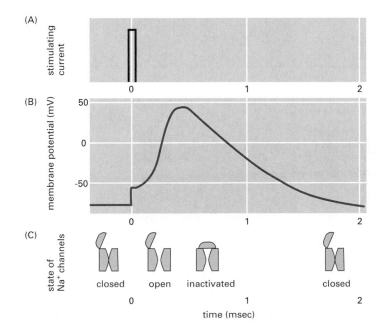

(A) stimulating current

(B)

membrane potential (mV)

50

0

-50

0   1   2

(C) state of Na⁺ channels

closed   open   inactivated   closed

0   1   2

time (msec)

Figure 12–34 Ion flows dictate the rise and fall of an action potential. In this example the action potential is triggered by a brief pulse of electric current (A), which partially depolarizes the membrane, as shown in the plot of membrane potential versus time (B). (B) shows the course of the action potential that is caused by the opening and subsequent inactivation of voltage-gated Na⁺ channels, whose state is shown in (C). Even if restimulated, the membrane cannot produce a second action potential until the Na⁺ channels have returned from the inactivated to the simply closed conformation (see Figure 12–33); until then the membrane is resistant, or refractory, to stimulation.

How We Know: Squid Reveal Secrets of Membrane Excitability

Each spring, *Loligo pealei* migrate to the shallow waters off Cape Cod on the eastern coast of the United States. There they spawn, launching the next generation of squid. But more than just meeting and breeding, these animals provide neuroscientists summering at the Marine Biological Laboratories in Woods Hole, Massachusetts, with an excellent system for studying the mechanism of electrical signaling along nerve axons.

Like most animals, squid survive by catching prey and escaping predators. Fast reflexes and an ability to accelerate rapidly and make sudden changes in swimming direction help them avoid danger while chasing down a decent meal. Squid derive their speed and agility from a specialized biological jet propulsion system: they draw water into their mantle cavity and then contract their muscular body wall to rapidly expel the collected water through a tubular siphon, thus propelling themselves through the water.

Controlling such quick and coordinated muscle contraction requires a nervous system that can convey signals with great speed down the length of the animal's body. Indeed, *Loligo pealei* possess some of the largest nerve fibers found in nature. Squid giant axons can reach 10 cm in length and are over 100 times the diameter of a mammalian axon—about the width of a pencil lead. Generally speaking, the larger the diameter of an axon, the more rapidly signals can travel along its length.

In the 1930s, scientists first started to take advantage of the squid giant axon for studying the electrophysiology of the nerve cell. Because of its relatively large size, researchers can insert electrodes into the axon to measure its electrical activity and monitor its action potentials. This experimental system allowed researchers to address a variety of questions about membrane conductance in neurons, including which ions are important for initiating and propagating an action potential, how membrane permeability changes as an action potential sweeps by, and how these changes in membrane potential control the opening and closing of ion channels.

Setup for action

Because the squid axon is so long and wide, an electrode made from a glass capillary tube containing a conducting solution can be thrust down the axis of the axon so that its tip lies deep in the cytoplasm (Figure 12–35A). This setup then allows one to measure the voltage difference between the inside and the outside of the axon—that is, the membrane potential—as an action potential sweeps past the end of the electrode (Figure 12–35B). The action potential itself is triggered by applying a brief electrical stimulus to the end of the axon. It does not really matter which end is stimulated, as the excitation can travel in either direction; it also does not matter how big the stimulus is, as long as it exceeds a certain threshold: an action potential is all or nothing.

Once researchers could reliably generate and measure an action potential, they could use the squid axon system to answer other questions about membrane excitability. For example, which ions are critical for an action potential? The three most plentiful ions, both inside and outside the axon, are Na^+, K^+, and Cl^-. Do they have equal importance when it comes to the action potential? Because the squid axon is so large and robust, it is possible to extrude the cytoplasm from the axon like toothpaste from a tube (Figure 12–36A).

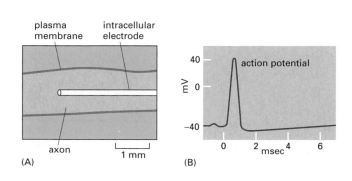

Figure 12–35 An electrode inserted into the squid giant axon (A) can be used to measure action potentials (B).

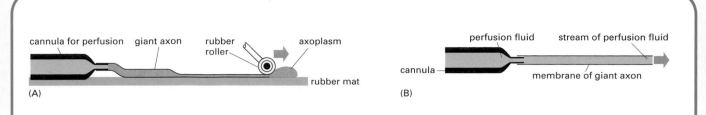

Figure 12–36 The cytoplasm in an axon can be removed and replaced with an artificial solution of pure ions.
(A) The axon cytoplasm is extruded using a rubber roller. (B) A perfusion fluid containing the desired concentration of ions is pumped gently through the axon.

The axon can then be perfused internally with a pure solution of Na⁺, K⁺, Cl⁻, or SO_4^{2-} (Figure 12–36B). Remarkably, researchers performing this experiment discovered that the axon will generate a normal action potential if and only if the concentrations of Na⁺ and K⁺ approximate the natural concentrations found inside and outside the cell. Thus, the cellular components key to the action potential are the plasma membrane, Na⁺ and K⁺ ions, and the energy provided by the concentration gradients of these ions across the membrane; all other components, including other sources of metabolic energy, were presumably removed by the perfusion.

Channel traffic

Once Na⁺ and K⁺ had been singled out as critical to an action potential, the question then became, What does each of these ions contribute to the action potential? How permeable is the membrane to each, and how does the membrane permeability change as an action potential sweeps by? Again, the squid giant axon provided some answers. The concentrations of Na⁺ and K⁺ outside the membrane could be altered, and the effects that these changes have on membrane potential could be measured directly. From such studies it was determined that, at rest, the membrane potential of an axon is close to the equilibrium potential for K⁺. When the external concentration of K⁺ is varied, the resting potential of the axon changes roughly in accordance with the Nernst equation (see Figure 12–29). At rest, therefore, the membrane is chiefly permeable to K⁺; as we now know, K⁺ leak channels provide the main pathway these ions can take through the cell membrane.

The situation for Na⁺ is very different. When the external concentration of Na⁺ is varied, there is no effect on the resting potential of the axon. However, the height of the peak of the action potential varies with the concentration of Na⁺ outside the membrane (Figure 12–37). During the action potential, therefore, the membrane appears to be chiefly permeable to Na⁺, as the result of the opening of Na⁺ channels. In the aftermath of the action potential, the sodium channels close and the membrane potential reverts to a negative value that depends on the external concentration of K⁺. As the membrane loses its permeability to Na⁺, it becomes even more permeable to K⁺ than before. The opening of additional K⁺ channels helps speed this resetting of the membrane potential to the resting state. This readies the membrane for the next action potential.

These studies on the squid giant axon made an enormous contribution to our understanding of neuronal excitability, and the researchers who set up the system were rewarded with a Nobel Prize in 1963.

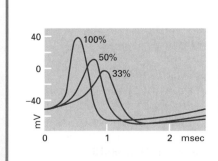

the form of the action potential when the external medium contains 100%, 50%, or 33% of the normal concentration of Na⁺

Figure 12–37 The shape of the action potential depends on the concentration of Na⁺ outside the membrane.
Shown here are action potentials recorded when the external medium contains 100%, 50%, or 33% of the normal concentration of Na⁺.

Figure 12–38 An action potential can be propagated along the length of an axon.
(A) The voltages that would be recorded from a set of intracellular electrodes placed at intervals along the axon, whose width is greatly exaggerated in this schematic figure. Note that the action potential does not weaken as it travels. (B) The changes in the Na$^+$ channels and the current flows *(orange arrows)* that give rise to the traveling disturbance of the membrane potential. The region of the axon with a depolarized membrane is shaded in *blue*. Note that an action potential can travel only away from the site of depolarization because Na$^+$-channel inactivation prevents the depolarization from spreading backward (see also Figure 12–34). In myelinated axons, clusters of Na$^+$ channels can be millimeters apart from each other.

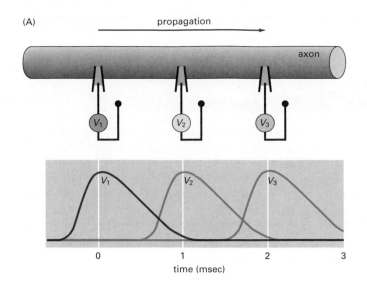

(B)   instantaneous view at $t = 0$

instantaneous view at $t = 1$ msec

Question 12–6

Explain in no more than 100 words how an action potential is passed along an axon.

the negative membrane potential that restrains them in the resting cell. The rapid outflow of K$^+$ through the voltage-gated K$^+$ channels brings the membrane back to its resting state more quickly than could be achieved by K$^+$ outflow through the K$^+$ leak channels alone.

The description just given of an action potential concerns only a small patch of plasma membrane. The self-amplifying depolarization of the patch, however, is sufficient to depolarize neighboring regions of membrane, which then go through the same self-amplifying cycle. In this way the action potential spreads outward as a traveling wave from the initial site of depolarization, eventually reaching the extremities of the axon (Figure 12–38).

Voltage-gated Ca²⁺ Channels Convert Electrical Signals into Chemical Signals at Nerve Terminals

When an action potential reaches the ends of the axon—the *nerve terminals*—the signal must somehow be relayed to the *target cells* that the nerve terminals contact, which are usually neurons or muscle cells. The signal is transmitted at specialized junctions known as **synapses**. At most synapses the plasma membranes of the transmitting and receiving cells—the *presynaptic* and the *postsynaptic* cells, respectively—are separated from each other by a narrow *synaptic cleft* (typically 20 nm across), which the electrical signal cannot cross (Figure 12–39). For the message to be transmitted from one neuron to another, the electrical signal is converted into a chemical signal, in the form of a small signaling molecule known as a **neurotransmitter**.

Neurotransmitters are stored ready-made in the nerve terminals, packaged in membrane-enclosed **synaptic vesicles** (see Figure 12–39). When the action potential reaches the terminal, the neurotransmitters are released from the nerve ending by exocytosis (discussed in Chapter 15). This link between the action potential and secretion involves the activation of yet another type of voltage-gated cation channel. The depolarization of the nerve-terminal plasma membrane caused by the arrival of the action potential transiently opens *voltage-gated Ca²⁺ channels,* which are concentrated in the plasma membrane of the presynaptic nerve terminal. Because the Ca²⁺ concentration outside the cell is more than 1000 times greater than the free Ca²⁺ concentration in the cytosol, Ca²⁺ rushes into the nerve terminal through the open channels. The resulting increase in Ca²⁺ concentration in the cytosol of the nerve terminal triggers the fusion of the synaptic vesicles with the presynaptic plasma membrane, releasing the neurotransmitter into the synaptic cleft. Thanks to the voltage-gated Ca²⁺ channels, the electrical signal has now been converted into a chemical signal (Figure 12–40).

Transmitter-gated Channels in Target Cells Convert Chemical Signals Back into Electrical Signals

The released neurotransmitter rapidly diffuses across the synaptic cleft and binds to *neurotransmitter receptors* concentrated in the postsynaptic membrane on the target cell. The binding of neurotransmitter to its receptors causes a change in the membrane potential of the target cell, which can trigger the cell to fire an action potential. The

Figure 12–39 Neurons transmit chemical signals across synapses. An electron micrograph (A) and drawing (B) of a cross section of two nerve terminals *(yellow)* forming synapses on a single nerve cell dendrite *(blue)* in the mammalian brain. Note that both the presynaptic and postsynaptic membranes are thickened at the synapse. (A, courtesy of Cedric Raine.)

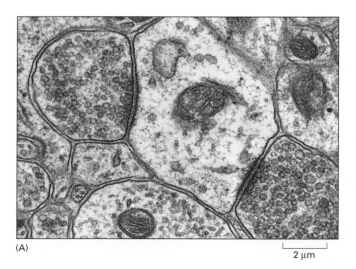

(A) 2 µm

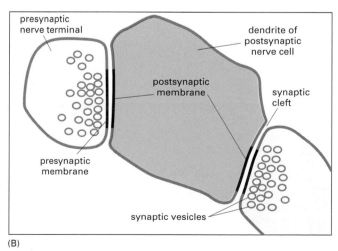

(B)

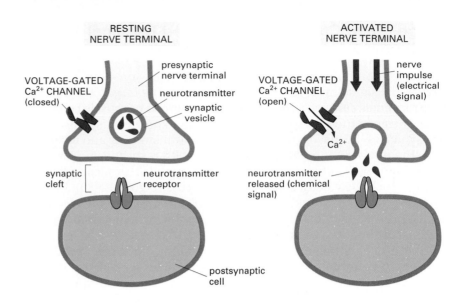

Figure 12–40 An electrical signal is converted into a chemical signal at a nerve terminal. When an action potential reaches a nerve terminal, it opens voltage-gated Ca^{2+} channels in the plasma membrane, allowing Ca^{2+} to flow into the terminal. The increased Ca^{2+} in the nerve terminal stimulates the synaptic vesicles to fuse with the plasma membrane, releasing their neurotransmitter into the synaptic cleft.

Question 12–7

In the disease myasthenia gravis, the human body makes—by mistake—antibodies to its own acetylcholine receptor molecules. These antibodies bind to and inactivate acetylcholine receptors on the plasma membrane of muscle cells. The disease leads to a devastating progressive weakening of the patients. Early on, they may have difficulty opening their eyelids, for example, and, in an animal model of the disease, rabbits have difficulty holding their ears up. As the disease progresses, most muscles weaken, and patients have difficulty speaking and swallowing. Eventually, impaired breathing can cause death. Explain which step of muscle function is affected.

neurotransmitter is then quickly removed from the synaptic cleft—either by enzymes that destroy it, or by reuptake into the nerve terminals that released it or into neighboring cells. This rapid removal of the neurotransmitter ensures that when the presynaptic cell falls quiet, the postsynaptic cell will fall quiet as well.

Neurotransmitter receptors can be of various types; some mediate relatively slow effects in the target cell, others trigger more rapid responses. Rapid responses—on a time scale of milliseconds—depend on receptors that are *transmitter-gated ion channels*. These constitute a subclass of ligand-gated ion channels (see Figure 12–24B), and their function is to convert the chemical signal carried by a neurotransmitter back into an electrical signal. The channels open transiently in response to the binding of the neurotransmitter, thus changing the permeability of the postsynaptic membrane to ions. This in turn causes a change in the membrane potential (Figure 12–41); if the change is big enough, it can trigger an action potential in the postsynaptic cell. A well-studied example of a transmitter-gated ion channel is found at the *neuromuscular junction*—the specialized type of synapse formed between a neuron and a muscle cell. In vertebrates the neurotransmitter here is *acetylcholine,* and the transmitter-gated ion channel is the *acetylcholine receptor* (Figure 12–42).

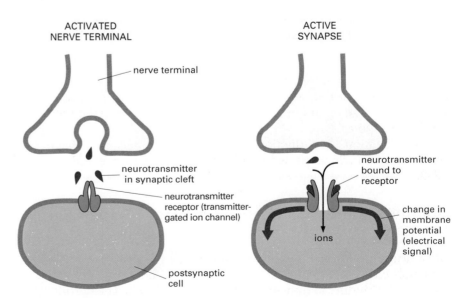

Figure 12–41 A chemical signal is converted into an electrical signal by transmitter-gated ion channels at a synapse. The released neurotransmitter binds to and opens the transmitter-gated ion channels in the plasma membrane of the postsynaptic cell. The resulting ion flows alter the membrane potential of the postsynaptic cell, thereby converting the chemical signal back into an electrical one.

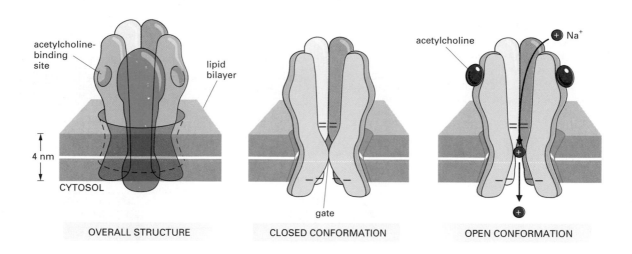

acetylcholine-binding site

lipid bilayer

4 nm

CYTOSOL

OVERALL STRUCTURE

gate

CLOSED CONFORMATION

acetylcholine

Na⁺

OPEN CONFORMATION

Neurons Receive Both Excitatory and Inhibitory Inputs

The response produced by a neurotransmitter at a synapse can be either excitatory or inhibitory. Some neurotransmitters (delivered by axon terminals of *excitatory neurons)* cause the postsynaptic cell to fire an action potential, whereas others (delivered by axon terminals of *inhibitory neurons)* prevent the postsynaptic cell from firing. The drug curare, which is used by surgeons to relax muscles during an operation, causes paralysis by blocking the delivery of excitatory signals at neuromuscular junctions, while the poison strychnine causes muscle spasms, convulsions, and death by blocking the delivery of inhibitory signals.

Excitatory and inhibitory neurotransmitters bind to different receptors, and it is the character of the receptor that makes the difference between excitation and inhibition. The chief receptors for excitatory neurotransmitters, mainly *acetylcholine* and *glutamate,* are ion channels that allow the passage of Na^+ and Ca^{2+}, respectively. When the neurotransmitter binds, the channels open to allow an influx mainly of Na^+, which depolarizes the plasma membrane toward the threshold potential required for triggering an action potential. Stimulation of these receptors thus tends to activate the postsynaptic cell. The receptors for inhibitory neurotransmitters, mainly *γ-aminobutyric acid (GABA)* and *glycine,* by contrast, are usually channels for Cl^-. When the neurotransmitter binds, the channels open. Very little Cl^- enters the cell at this point because the driving force for movement of Cl^- across the membrane is close to zero at the resting membrane potential. If Na^+ channels are also opened, however, Na^+ will rush into the cell, causing the membrane potential to shift away from its resting value. This shift causes Cl^- to move into the cell, neutralizing the effect of the Na^+ influx (Figure 12–43). In this way inhibitory neurotransmitters suppress the production of an action potential by making the target cell membrane harder to depolarize.

The locations and functions of these ion channels, and of some of the other channels discussed in this chapter, are summarized in Table 12–3.

Transmitter-gated Ion Channels Are Major Targets for Psychoactive Drugs

Most drugs used in the treatment of insomnia, anxiety, depression, and schizophrenia exert their effects at synapses in the brain, and many of them act by binding to transmitter-gated ion channels. The barbiturates

Figure 12–42 The acetylcholine receptor, present in the plasma membrane of muscle cells, opens when it binds to the neurotransmitter acetylcholine, released by a nerve. This transmitter-gated ion channel is composed of five transmembrane protein subunits that combine to form an aqueous pore across the lipid bilayer. The pore is lined by five transmembrane α helices, one contributed by each subunit. Negatively charged amino acid side chains at either end of the pore ensure that only positively charged ions, mainly Na^+ and K^+, can pass. When the channel is in its closed conformation, the pore is occluded by hydrophobic amino acid side chains in the region called the *gate;* when acetylcholine binds, the protein undergoes a conformational change in which these side chains move apart and the gate opens, allowing Na^+ and K^+ to flow across the membrane, down their electrochemical gradients. Even with acetylcholine bound, the channel flickers randomly between the open and closed states (see Figure 12–23); without acetylcholine bound, however, it rarely opens.

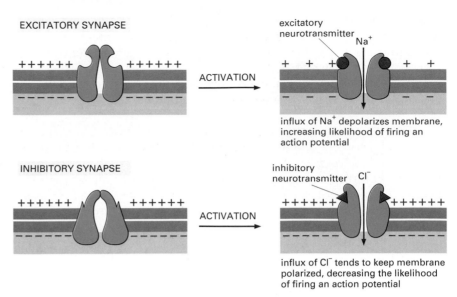

Figure 12–43 Synapses can be excitatory or inhibitory. Excitatory neurotransmitters activate ion channels that allow the passage of Na$^+$ and Ca^{2+}, whereas inhibitory neurotransmitters activate ion channels that allow the passage of Cl$^-$.

EXCITATORY SYNAPSE

ACTIVATION

excitatory neurotransmitter

Na$^+$

influx of Na$^+$ depolarizes membrane, increasing likelihood of firing an action potential

INHIBITORY SYNAPSE

ACTIVATION

inhibitory neurotransmitter

Cl$^-$

influx of Cl$^-$ tends to keep membrane polarized, decreasing the likelihood of firing an action potential

and tranquilizers such as Valium, Halcion, and temazepam, for example, bind to GABA-gated Cl$^-$ channels. Their binding makes the channels easier to open by GABA, thus making the cell more sensitive to GABA's inhibitory action. By contrast, the antidepressant Prozac blocks the reuptake of an excitatory neurotransmitter, *serotonin*, increasing the amount of serotonin available at those synapses that use this transmitter. Why this should relieve depression is still a mystery.

The number of distinct types of neurotransmitter receptors is very large, although they fall into a small number of families. There are, for example, many subtypes of acetylcholine, glutamate, GABA, glycine, and serotonin receptors; they are usually located in different neurons and often differ only subtly in their properties. With such a large variety of receptors, it may be possible to design a new generation of psychoactive drugs that will act more selectively on specific sets of neurons to alleviate the mental illnesses that devastate so many people's lives. One percent of the human population, for example, has schizophrenia, and another 1 percent suffers from manic–depressive disease.

Synaptic Connections Enable You to Think, Act, and Remember

At a chemical synapse, the nerve terminal of the presynaptic cell converts an electrical signal into a chemical one, and the postsynaptic cell

Table 12–3 Some Examples of Ion Channels

ION CHANNEL	TYPICAL LOCATION	FUNCTION
K$^+$ leak channel	plasma membrane of most animal cells	maintenance of resting membrane potential
Voltage-gated Na$^+$ channel	plasma membrane of nerve cell axon	generation of action potentials
Voltage-gated K$^+$ channel	plasma membrane of nerve cell axon	return of membrane to resting potential after initiation of an action potential
Voltage-gated Ca^{2+} channel	plasma membrane of nerve terminal	stimulation of neurotransmitter release
Acetylcholine receptor (acetylcholine-gated Na$^+$ and Ca^{2+} channel)	plasma membrane of muscle cell (at neuromuscular junction)	excitatory synaptic signaling
GABA receptor (GABA-gated Cl$^-$ channel)	plasma membrane of many neurons (at synapses)	inhibitory synaptic signaling
Stress-activated cation channel	auditory hair cell in inner ear	detection of sound vibrations

converts the chemical signal back into an electrical one. Interference with these processes, for good or ill, is of enormous practical importance to us. But why has evolution favored such an apparently inefficient way to pass on an electrical signal? It would seem more efficient to have a direct electrical connection between the pre- and postsynaptic cells, or to do away with the synapse altogether and use a single continuous cell.

The value of chemical synapses becomes clear when we consider them in the context of a functioning nervous system—a huge network of neurons, interconnected by many branching pathways, performing complex computations, storing memories, and generating plans for action. To carry out these functions, neurons have to do more than merely generate and relay signals: they must also combine them, interpret them, and record them. Chemical synapses make these activities possible. A motor neuron in the spinal cord, for example, receives inputs from hundreds or thousands of other neurons that make synapses on it (Figure 12–44). Some of these signals tend to stimulate the neuron, while others tend to inhibit it. The motor neuron has to combine all of the information it receives and react either by firing action potentials along its axon to stimulate a muscle or by remaining quiet. This task of computing an appropriate output from the babble of inputs is achieved by a complicated interplay between different types of ion channels in the neuron's plasma membrane. Each of the hundreds of types of neurons in your brain has its own characteristic set of receptors and ion channels that enables the cell to respond in a particular way to a certain set of inputs and thus to perform its specialized task. Moreover, the ion channels and other components at a synapse can undergo lasting modifications according to the usage they have experienced, thereby preserving traces of past events. In this way, memories are stored. Ion channels, therefore, are at the heart of the machinery that enables you to act, think, feel, speak, and—perhaps most important of all—to remember everything you read in this book.

Figure 12–44 Thousands of synapses form on the cell body and dendrites of a motor neuron in the spinal cord. (A) Many thousands of nerve terminals synapse on the neuron, delivering signals from other parts of the animal to control the firing of action potentials along the neuron's axon. (B) A rat nerve cell in culture. Its cell body and dendrites *(green)* are stained with a fluorescent antibody that recognizes a cytoskeletal protein. Thousands of axon terminals *(red)* from other nerve cells (not visible) make synapses on the cell's surface; they are stained with a fluorescent antibody that recognizes a protein in synaptic vesicles. Electrical signals are sent out along axons, relayed across synapses, and passed in along dendrites toward the nerve cell body. The signaling depends on movements of ions across the plasma membranes of the nerve cells. (B, courtesy of Olaf Mundigl and Pietro de Camilli.)

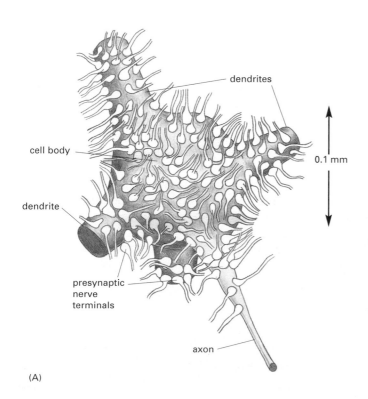

(A)

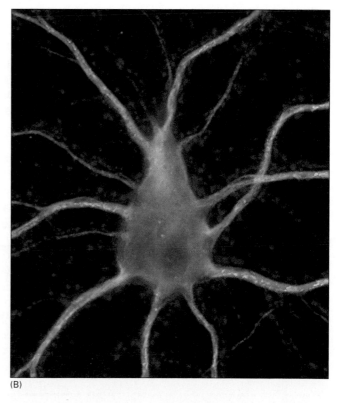

(B)

Essential Concepts

- The lipid bilayer of cell membranes is permeable to small nonpolar molecules such as oxygen and carbon dioxide and to very small polar molecules such as water. It is highly impermeable to most large, water-soluble molecules and all ions. Transfer of nutrients, metabolites, and ions across the plasma membrane and internal cell membranes is carried out by membrane transport proteins.

- Cell membranes contain a variety of transport proteins, each of which is responsible for transferring a particular type of solute across the membrane. There are two classes of membrane transport proteins—carrier proteins and channel proteins.

- The electrochemical gradient represents the net driving force on an ion due to its concentration gradient and the electric field.

- In passive transport an uncharged solute moves spontaneously down its concentration gradient, and a charged solute (an ion) moves spontaneously down its electrochemical gradient. In active transport an uncharged solute or an ion is transported against its concentration or electrochemical gradient in an energy-requiring process.

- Carrier proteins bind specific solutes (inorganic ions, small organic molecules, or both) and transfer them across the lipid bilayer by undergoing conformational changes that expose the solute-binding site first on one side of the membrane and then on the other.

- Carrier proteins can act as pumps to transport a solute uphill against its electrochemical gradient, using energy provided by ATP hydrolysis, by a downhill flow of Na^+ or H^+ ions, or by light.

- The Na^+-K^+ pump in the plasma membrane of animal cells is an ATPase that actively transports Na^+ out of the cell and K^+ in, maintaining the steep Na^+ gradient across the plasma membrane that is used to drive other active transport processes and to convey electrical signals.

- Channel proteins form aqueous pores across the lipid bilayer through which solutes can diffuse. Whereas transport by carrier proteins can be active or passive, transport by channel proteins is always passive.

- Most channel proteins are selective ion channels that allow inorganic ions of appropriate size and charge to cross the membrane down their electrochemical gradients. Transport through ion channels is at least 1000 times faster than transport through any known carrier protein.

- Most ion channels are gated; they open transiently in response to a specific stimulus, such as a change in membrane potential (voltage-gated channels) or the binding of a ligand (ligand-gated channels).

- Even when opened by their specific stimulus, ion channels do not remain continuously open: they flicker randomly between open and closed conformations. An activating stimulus increases the proportion of time the channel spends in the open state.

- The membrane potential is determined by the unequal distribution of electric charge on the two sides of the plasma membrane and is altered when ions flow through open channels. In most animal cells, K^+-selective leak channels hold the resting membrane potential at a negative value, close to the value where the driving force for movement of K^+ across the membrane is almost zero.

- Neurons propagate signals in the form of action potentials, which can travel long distances along an axon without weakening. Action potentials are usually mediated by voltage-gated Na^+ channels that open in response to depolarization of the plasma membrane.

- Voltage-gated Ca^{2+} channels in nerve terminals couple electrical signals to transmitter release at synapses. Transmitter-gated ion channels convert these chemical signals back into electrical signals in the postsynaptic target cell.
- Excitatory neurotransmitters open transmitter-gated channels that are permeable to Na^+ and thereby depolarize the postsynaptic cell membrane toward the threshold potential for firing an action potential. Inhibitory neurotransmitters open transmitter-gated Cl^- channels and thereby suppress firing by keeping the postsynaptic cell membrane polarized.

Key Terms

action potential	nerve terminal
active transport	neuron
axon	neurotransmitter
carrier protein	osmosis
channel protein	osmotic pressure
coupled transporter	passive transport
electrochemical gradient	patch-clamp recording
ion channel	stress-activated channel
ligand-gated channel	synapses
membrane potential	synaptic vesicles
membrane transport protein	voltage-gated channel
Na^+-K^+ pump	voltage-gated Na^+ channel
Nernst equation	

Questions

Question 12–8

The diagram in Figure 12–7 shows a passive carrier protein that mediates the transfer of a solute down its concentration gradient across the membrane. How would you need to change the diagram to convert the carrier protein into a pump that transports the solute up its electrochemical gradient by hydrolyzing ATP? Explain the need for each of the steps in your new illustration.

Question 12–9

Which of the following statements are correct? Explain your answers.

A. The plasma membrane is highly impermeable to all charged molecules.

B. Channel proteins must first bind to solute molecules before they can select those that they allow to pass.

C. Without a continual input of energy, cells will burst.

D. Carrier proteins allow solutes to cross a membrane at much faster rates than do channel proteins.

E. Certain H^+ pumps are fueled by light energy.

F. The plasma membrane of many animal cells contains open K^+ channels, yet the K^+ concentration in the cytosol is much higher than outside the cell.

G. A symport would function as an antiport if its orientation in the membrane were reversed (i.e., if the portion of the molecule normally exposed to the cytosol faced the outside of the cell instead).

H. The membrane potential of an axon temporarily becomes more negative when an action potential excites it.

Question 12–10

List the following compounds in order of increasing membrane permeability: RNA, Ca^{2+}, glucose, ethanol, N_2, water.

Question 12–11

Name at least one similarity and at least one difference between the following (it may help to review the definitions of the terms using the Glossary):
A. Symport and antiport
B. Active transport and passive transport
C. Membrane potential and electrochemical gradient
D. Pump and carrier protein
E. Axon and telephone wire
F. Solute and ion

Question 12–12

Discuss the following statement: "The differences between a channel and a carrier protein are like the differences between a bridge and a ferry."

Question 12–13

The neurotransmitter acetylcholine is made in the cytosol and then transported into synaptic vesicles, where its concentration is more than 100-fold higher than in the cytosol. When synaptic vesicles are isolated from neurons, they can take up additional acetylcholine added to the solution in which they are suspended, but only when ATP is present. Na^+ ions are not required for acetylcholine uptake, but, curiously, raising the pH of the solution in which the synaptic vesicles are suspended increases the rate of acetylcholine uptake. Furthermore, transport is inhibited when drugs are added that make the membrane permeable to H^+ ions. Suggest a mechanism that is consistent with all of these observations.

Question 12–14

The resting membrane potential of a cell is about –70 mV, and the thickness of a lipid bilayer is about 4.5 nm. What is the strength of the electric field across the membrane in V/cm? What do you suppose would happen if you applied this voltage to two metal electrodes separated by a 1-cm air gap?

Question 12–15

Phospholipid bilayers form sealed spherical vesicles in water (discussed in Chapter 11). Assume you have constructed lipid vesicles that contain Na^+-K^+ pumps as the sole membrane protein, and assume for the sake of simplicity that each pump transports one Na^+ one way and one K^+ the other way in each pumping cycle. All the Na^+-K^+ pumps have the portion of the molecule that normally faces the cytosol oriented toward the outside of the vesicles. With the help of Figure 12–12, determine what would happen if
A. Your vesicles were suspended in a solution containing both Na^+ and K^+ ions and had a solution with the same ionic composition inside them.
B. You add ATP to the suspension described in (A).
C. You add ATP, but the solution—outside as well as inside the vesicles—contains only Na^+ ions and no K^+ ions.
D. Half of the pump molecules embedded in the membrane of each vesicle were oriented the other way around so that the normally cytosolic portions of these molecules faced the inside of the vesicles. You then add ATP to the suspension.
E. You add ATP to the suspension described in (A), but in addition to Na^+-K^+ pumps, the membrane of your vesicles also contains K^+ leak channels.

Question 12–16

Name the three ways in which an ion channel can be gated.

Question 12–17

One thousand Ca^{2+} channels open in the plasma membrane of a cell that is 1000 μm^3 in size and has a cytosolic Ca^{2+} concentration of 100 nM. For how long would the channels need to stay open in order for the cytosolic Ca^{2+} concentration to rise to 5 μM? There is virtually unlimited Ca^{2+} available in the outside medium (the extracellular Ca^{2+} concentration in which most animal cells live is a few millimolar), and each channel passes 10^6 calcium ions per second.

Question 12–18

Amino acids are taken up by animal cells using a symport in the plasma membrane. What is the most likely ion whose electrochemical gradient drives the import? Is ATP consumed in the process? If so, how?

Question 12–19

We shall see in Chapter 15 that an acidic pH inside endosomes, which are membrane-enclosed intracellular organelles, is required for their function. Acidification is achieved by an H^+ pump in the endosomal membrane. The endosomal membrane also contains Cl^- channels. If the channels do not function properly (e.g., because of a mutation in the genes encoding the channel proteins), acidification is also impaired.
A. Can you explain how Cl^- channels might help acidification?
B. According to your explanation, would the Cl^- channels be absolutely required to lower the pH inside the endosome?

Concentration of	Rate of Transport (μmol/min)	
Carbon Source (mM)	Compound A	Compound B
0.1	2.0	18
0.3	6.0	46
1.0	20	100
3.0	60	150
10.0	200	182

Question 12–20

Some bacterial cells can grow on either ethanol
(CH_3CH_2OH) or acetate (CH_3COO^-) as their only carbon
source. Dr. Schwips measured the rate at which the
two compounds traverse the bacterial plasma membrane but, due to excessive inhalation of one of the
compounds (which one?), failed to label his data accurately.
  A. Plot the data from the table above.
  B. Determine from your graph whether the data
     describing compound A correspond to the
     uptake of ethanol or acetate.
  C. Determine the rates of transport for com-
     pounds A and B at 0.5 mM and 100 mM. (This
     part of the question requires that you be famil-
     iar with the principles of enzyme kinetics dis-
     cussed in Chapter 3.)
Explain your answers.

Question 12–21

Acetylcholine-gated cation channels do not discriminate among Na^+, K^+ and Ca^{2+} ions, allowing all to pass
through them freely. So why is it that when acetylcholine binds to this protein complex in muscle cells,
the channel opens and there is a large net influx of primarily Na^+ ions?

Question 12–22

The ion channels that are regulated by binding of neurotransmitters, such as acetylcholine, glutamate,
GABA, or glycine, have a similar overall structure. Yet,
each class of these channels consists of a very diverse
set of subtypes with different ligand affinities, different
channel conductances, and different rates of opening
and closing. Do you suppose that such extreme diversity is a good or a bad thing from the standpoint of the
pharmaceutical industry?

Highlights from *Essential Cell Biology 2 Interactive* CD-ROM

12.2	Na^+-K^+ Pump
12.5	Potassium Channel
12.8	Action Potentials

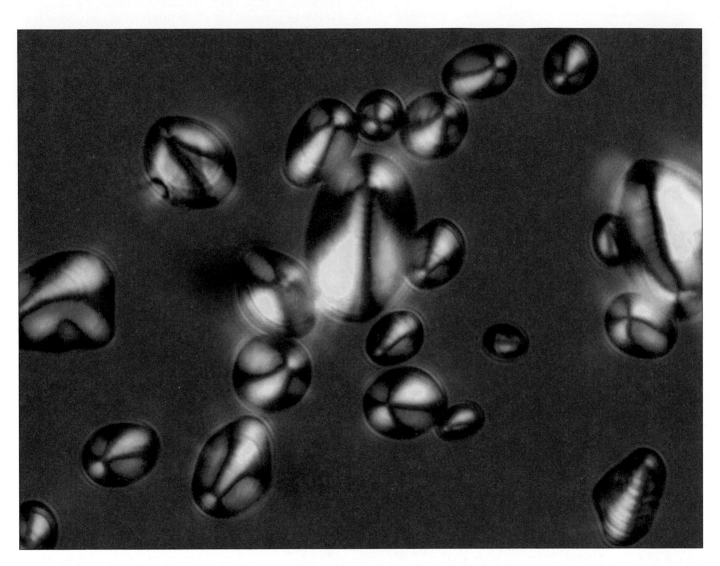

Starch. Sugar production in plants is driven by the photosynthetic activity of the plant chloroplasts. While some of this sugar is used directly by the plant, the rest can be converted into a storage polysaccharide called starch. Much of the energy for human existence ultimately derives from the starch stored in our staple food crops, which include rice, maize, wheat, and potatoes. The large starch grains shown here, viewed in polarized light, have been isolated from a raw potato. (Courtesy of Tatiana Bogracheva.)

How Cells Obtain Energy from Food

As we discussed in Chapter 3, cells require a constant supply of energy to generate and maintain the biological order that keeps them alive. This energy comes from the chemical bond energy in food molecules, which thereby serve as fuel for cells.

Perhaps the most important fuel molecules are the sugars. Plants make their own sugars by photosynthesis, whereas animals obtain sugars—and other molecules, such as starch, that are easily converted to sugars—by eating other organisms. Nevertheless, the process whereby these sugars are oxidized to generate energy is very similar in both animals and plants. In both cases, the cells that form the organism harvest useful energy from the chemical bond energy locked in sugars as the sugar molecule is broken down and oxidized to CO_2 and H_2O. This energy is stored as high-energy chemical bonds in activated carrier molecules, such as ATP and NADPH, which in turn serve as portable sources of the chemical groups and electrons needed for biosynthesis (discussed in Chapter 3).

If a fuel molecule such as glucose were oxidized to CO_2 and H_2O in a single step (as happens in nonliving systems), it would release an amount of energy many times larger than any carrier molecule could capture. Instead, living cells use enzymes to carry out the oxidation of sugars in a tightly controlled series of reactions, as illustrated in Figure 13–1A: the glucose molecule is degraded step-by-step, paying out energy in small packets to activated carrier molecules by means of coupled reactions. In this way, much of the energy released by oxidizing glucose is saved and made available to do useful work for the cell.

In the first part of this chapter we trace the major steps in the breakdown—or catabolism—of sugars and show how this oxidation produces ATP, NADH, and other activated carrier molecules in cells. We concentrate on the breakdown of glucose because these reactions dominate energy production in most animal cells. A very similar pathway operates in plants, fungi, and many bacteria. Other molecules, such as fatty acids and proteins, can also serve as energy sources if they are funneled through appropriate enzymatic pathways.

In the second part of the chapter, we examine how cells store food molecules for their future metabolic needs, and see how many of the molecules generated from the breakdown of sugars and fats can be used to build macromolecules. We will save our discussion of how cells produce most of the ATP that they need for Chapter 14.

The Breakdown of Sugars and Fats

Food Molecules Are Broken Down in Three Stages

Glycolysis Is a Central ATP-producing Pathway

Fermentations Allow ATP to Be Produced in the Absence of Oxygen

Glycolysis Illustrates How Enzymes Couple Oxidation to Energy Storage

Sugars and Fats Are Both Degraded to Acetyl CoA in Mitochondria

The Citric Acid Cycle Generates NADH by Oxidizing Acetyl Groups to CO_2

Electron Transport Drives the Synthesis of the Majority of the ATP in Most Cells

Storing and Utilizing Food

Organisms Store Food Molecules in Special Reservoirs

Chloroplasts and Mitochondria Collaborate in Plant Cells

Many Biosynthetic Pathways Begin with Glycolysis or the Citric Acid Cycle

Metabolism Is Organized and Regulated

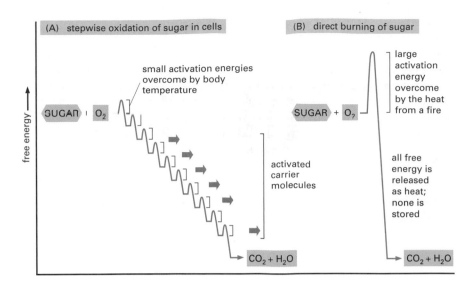

Figure 13–1 The controlled stepwise oxidation of sugar that occurs in the cell preserves useful energy, unlike the simple burning of the same fuel molecule. In the cell, enzymes catalyze oxidations via a series of small steps in which free energy is transferred in conveniently sized packets to carrier molecules—most often ATP and NADH. At each step, an enzyme controls the reaction by reducing the activation energy barrier that has to be surmounted before the specific reaction can occur. The total free energy released is exactly the same in (A) and (B).

The Breakdown of Sugars and Fats

Animal cells make ATP in two ways. In one, specific steps in a series of enzyme-catalyzed reactions are directly coupled to the energetically unfavorable reaction ADP + P$_i$ → ATP. The reactions required to drive this process oxidize food molecules. The second process takes place in mitochondria and uses the energy from activated carrier molecules to drive ATP production; this process involves membranes, and it is described in detail in Chapter 14. Here we focus on the sequence of reactions by which food molecules are oxidized—both in the cytosol and inside the mitochondria. These reactions produce both ATP and the activated carrier molecules that will drive the production of much larger amounts of ATP in the membranes of mitochondria.

Food Molecules Are Broken Down in Three Stages

The proteins, lipids, and polysaccharides that make up most of the food we eat must be broken down into smaller molecules before our cells can use them—either as a source of energy or as building blocks for other molecules. The breakdown processes must act on food taken in from outside, but not on the macromolecules inside our own cells. Therefore stage 1 in the enzymatic breakdown of food molecules—*digestion*—occurs either outside of cells (in our intestine) or in a specialized organelle within cells called the lysosome. A membrane that surrounds the lysosome keeps its digestive enzymes separated from the cytosol (discussed in Chapter 15).

In either case, digestive enzymes reduce the large polymeric molecules in food into their monomeric subunits—proteins into amino acids, polysaccharides into sugars, and fats into fatty acids and glycerol. After digestion, the small organic molecules derived from food enter the cytosol of a cell, where their gradual oxidation begins. As illustrated in Figure 13–2, this oxidation occurs in two further stages: stage 2 starts in the cytosol and ends in mitochondria, while stage 3 is confined to the mitochondria.

In stage 2 of cellular catabolism, a chain of reactions called *glycolysis* converts each molecule of glucose into two smaller molecules of pyruvate. Sugars other than glucose are also converted to pyruvate, after they are first converted to one of the sugar intermediates in this glycolytic pathway. During the formation of pyruvate, two types of activated carrier molecules are produced—ATP and NADH. The pyruvate then passes from the cytosol into mitochondria. There, each pyruvate molecule is converted into CO_2 plus a two-carbon acetyl group; this acetyl group then gets attached to coenzyme A (CoA), forming acetyl CoA, another of the activated carrier molecules discussed in Chapter 3 (see Figure 3–37). Large amounts of acetyl CoA are also produced by the stepwise breakdown and oxidation of fatty acids derived from fats. Fatty acids are carried in the bloodstream, imported into cells, and then moved into mitochondria for acetyl CoA production.

Stage 3 of the oxidative breakdown of food molecules takes place entirely in mitochondria. Because the acetyl group in acetyl CoA is linked to coenzyme A through a high-energy bond, it is easily transferable to other molecules. After its transfer to the four-carbon molecule

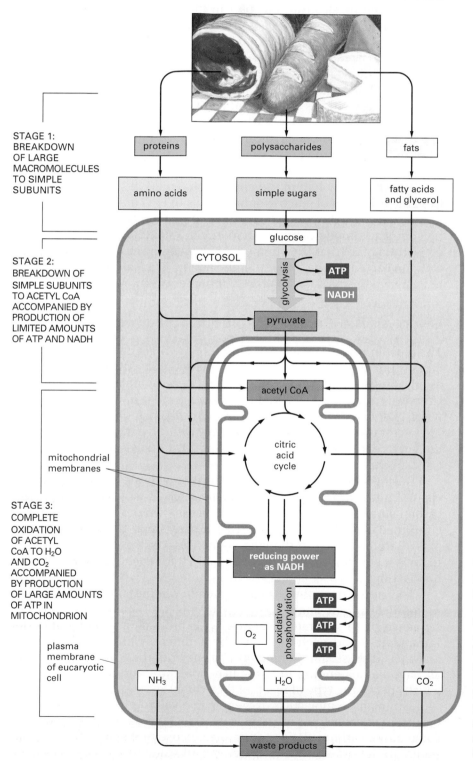

STAGE 1:
BREAKDOWN
OF LARGE
MACROMOLECULES
TO SIMPLE
SUBUNITS

STAGE 2:
BREAKDOWN OF
SIMPLE SUBUNITS
TO ACETYL CoA
ACCOMPANIED BY
PRODUCTION OF
LIMITED AMOUNTS
OF ATP AND NADH

mitochondrial
membranes

STAGE 3:
COMPLETE
OXIDATION
OF ACETYL
CoA TO H_2O
AND CO_2
ACCOMPANIED
BY PRODUCTION
OF LARGE AMOUNTS
OF ATP IN
MITOCHONDRION

plasma
membrane
of eucaryotic
cell

Figure 13–2 **The three stages of cellular metabolism lead from food to waste products in animal cells.** This series of reactions produces ATP, which is then used to drive biosynthetic reactions and other energy-requiring processes in the cell. Stage 1 mostly occurs outside cells—although special organelles called lysosomes can digest large molecules in the cell interior. Stage 2 occurs mainly in the cytosol, except for the final step of conversion of pyruvate to acetyl groups on acetyl CoA, which occurs in mitochondria. Stage 3 occurs in mitochondria.

oxaloacetate, the acetyl group enters a series of reactions called the *citric acid cycle*. As we discuss shortly, the acetyl group is oxidized to CO_2 in these reactions, and large amounts of the electron carrier NADH are generated. Finally, the high-energy electrons from NADH are passed along an electron-transport chain within the mitochondrial inner membrane, where the energy released by their transfer is used to drive a process that produces ATP and consumes molecular oxygen (O_2). It is in these final steps that most of the energy released by oxidation is harnessed to produce most of the cell's ATP.

Because the energy for it ultimately derives from the oxidative breakdown of food molecules, the phosphorylation of ADP to form ATP that is driven by electron transport in the mitochondrial inner membrane is known as *oxidative phosphorylation*. The ATP thus generated is then moved out of the mitochondrion and into the cytosol, to be used by the cell as needed. The fascinating events that occur during oxidative phosphorylation are the focus of Chapter 14.

Through the production of ATP, the energy derived from the breakdown of sugars and fats is redistributed as packets of chemical energy in a form convenient for use elsewhere in the cell. Roughly 10^9 molecules of ATP are in solution in a typical cell at any instant, and in many cells, all of this ATP is turned over (that is, used up and replaced) every 1–2 minutes.

In total, nearly half of the energy that could in theory be derived from the oxidation of glucose or fatty acids to H_2O and CO_2 is captured and used to drive the energetically unfavorable reaction $P_i + ADP \rightarrow ATP$. By contrast, a modern combustion engine, such as a car engine, can convert no more than 20% of the available energy in its fuel into useful work. In each instance, the rest of the energy is released as heat, which in living organisms helps to make our bodies warm.

Glycolysis Is a Central ATP-producing Pathway

The most important process in stage 2 of the breakdown of food molecules is the degradation of glucose in the sequence of reactions known as **glycolysis**—from the Greek *glykys*, "sweet," and *lysis*, "splitting." Glycolysis produces ATP without the involvement of molecular oxygen (O_2 gas). It occurs in the cytosol of most cells, including many anaerobic microorganisms. Glycolysis probably evolved early in the history of life, before photosynthetic organisms introduced oxygen into the atmosphere.

During glycolysis, a glucose molecule, with six carbon atoms, is cleaved into two molecules of *pyruvate*, each of which contains three carbon atoms. For each molecule of glucose, two molecules of ATP are consumed to provide energy to drive the early steps, but four molecules of ATP are produced in the later steps. Thus, at the end of glycolysis, there is a net gain of two molecules of ATP for each glucose molecule broken down.

The glycolytic pathway is presented in outline in Figure 13–3, and in more detail in Panel 13–1 (pp. 432–433). Glycolysis involves a sequence of 10 separate reactions, each producing a different sugar intermediate and each catalyzed by a different enzyme. Like most enzymes (discussed in Chapter 4), the enzymes that catalyze glycolysis all have names ending in *-ase*—like *isomerase* and *dehydrogenase*—which specify the type of reaction they catalyze.

Although no molecular oxygen is involved in glycolysis, oxidation occurs, in that electrons are removed by NAD^+ (producing NADH) from some of the carbons derived from the glucose molecule. The stepwise nature of the process allows the energy of oxidation to be released in

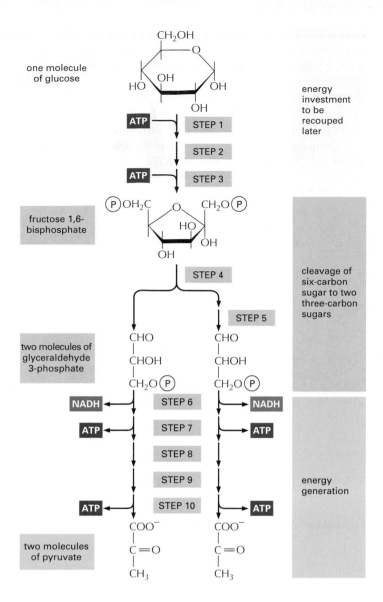

one molecule
of glucose

fructose 1,6-
bisphosphate

two molecules of
glyceraldehyde
3-phosphate

two molecules
of pyruvate

ATP STEP 1
STEP 2
ATP STEP 3

energy
investment
to be
recouped
later

STEP 4

STEP 5

cleavage of
six-carbon
sugar to two
three-carbon
sugars

NADH STEP 6 NADH
ATP STEP 7 ATP
STEP 8
STEP 9
ATP STEP 10 ATP

energy
generation

Figure 13–3 The stepwise oxidation of sugars begins with glycolysis. Each of the 10 steps of glycolysis is catalyzed by a different enzyme. Note that step 4 cleaves a six-carbon sugar into two three-carbon sugars, so that the number of molecules at every stage after this doubles. As indicated, step 6 begins the energy-generation phase of glycolysis, which causes the net synthesis of ATP and NADH molecules (see also Panel 13–1). Glycolysis is also sometimes referred to as the Embden–Meyerhof pathway.

small packets, so that much of it can be stored in carrier molecules rather than all of it being released as heat (see Figure 13–1). Some of the energy released by oxidation drives the direct synthesis of ATP molecules from ADP and P_i, while much remains with the electrons in the high-energy electron carrier NADH.

Two molecules of NADH are formed per molecule of glucose in the course of glycolysis. In aerobic organisms these NADH molecules donate their electrons to the electron-transport chain, as described in detail in Chapter 14. The electrons are passed along this chain to molecular oxygen (O_2), forming water, and the NAD^+ formed from the NADH is used again for glycolysis (see step 6 in Panel 13–1, p. 433).

Fermentations Allow ATP to Be Produced in the Absence of Oxygen

For most animal and plant cells, glycolysis is only a prelude to the third and final stage of the breakdown of food molecules. In these cells, the pyruvate formed at the end of glycolysis is rapidly transported into the mitochondria, where it is converted into CO_2 plus acetyl CoA, which is then completely oxidized to CO_2 and H_2O. But for many anaerobic organisms, which do not use molecular oxygen and can grow and divide in its absence, glycolysis is the principal source of the cell's ATP. The

Panel 13–1 Details of the 10 steps of glycolysis

For each step, the part of the molecule that undergoes a change is shadowed in blue,
and the name of the enzyme that catalyzes the reaction is in a yellow box.

Step 1 Glucose is phosphorylated by ATP to form a sugar phosphate. The negative charge of the phosphate prevents passage of the sugar phosphate through the plasma membrane, trapping glucose inside the cell.

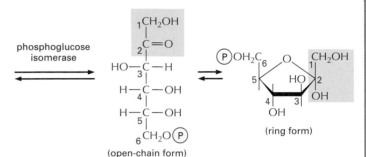

glucose + ATP → glucose 6-phosphate + ADP + H⁺

Step 2 A readily reversible rearrangement of the chemical structure (isomerization) moves the carbonyl oxygen from carbon 1 to carbon 2, forming a ketose from an aldose sugar. (See Panel 2–3, pp. 70–71.)

glucose 6-phosphate (ring form / open-chain form) phosphoglucose isomerase fructose 6-phosphate (open-chain form / ring form)

Step 3 The new hydroxyl group on carbon 1 is phosphorylated by ATP, in preparation for the formation of two three-carbon sugar phosphates. The entry of sugars into glycolysis is controlled at this step, through regulation of the enzyme *phosphofructokinase*.

fructose 6-phosphate + ATP →(phosphofructokinase)→ fructose 1,6-bisphosphate + ADP + H⁺

Step 4 The six-carbon sugar is cleaved to produce two three-carbon molecules. Only the glyceraldehyde 3-phosphate can proceed immediately through glycolysis.

fructose 1,6-bisphosphate (ring form / open-chain form) aldolase dihydroxyacetone phosphate + glyceraldehyde 3-phosphate

Step 5 The other product of step 4, dihydroxyacetone phosphate, is isomerized to form glyceraldehyde 3-phosphate.

dihydroxyacetone phosphate triose phosphate isomerase glyceraldehyde 3-phosphate

Step 6 The two molecules of glyceraldehyde 3-phosphate are oxidized. The energy-generation phase of glycolysis begins, as NADH and a new high-energy anhydride linkage to phosphate are formed (see Figure 13–5).

glyceraldehyde
3-phosphate
+ NAD⁺ + P$_i$

glyceraldehyde 3-phosphate
dehydrogenase

1,3-bisphosphoglycerate
+ NADH + H⁺

Step 7 The transfer to ADP of the high-energy phosphate group that was generated in step 6 forms ATP.

1,3-bisphosphoglycerate
+ ADP

phosphoglycerate kinase

3-phosphoglycerate
+ ATP

Step 8 The remaining phosphate ester linkage in 3-phosphoglycerate, which has a relatively low free energy of hydrolysis, is moved from carbon 3 to carbon 2 to form 2-phosphoglycerate.

3-phosphoglycerate

phosphoglycerate mutase

2-phosphoglycerate

Step 9 The removal of water from 2-phosphoglycerate creates a high-energy enol phosphate linkage.

2-phosphoglycerate

enolase

phosphoenolpyruvate
+ H$_2$O

Step 10 The transfer to ADP of the high-energy phosphate group that was generated in step 9 forms ATP, completing glycolysis.

phosphoenolpyruvate
+ ADP + H⁺

pyruvate kinase

pyruvate
+ ATP

NET RESULT OF GLYCOLYSIS

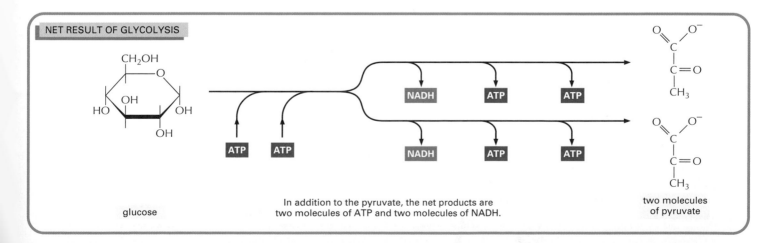

glucose

In addition to the pyruvate, the net products are two molecules of ATP and two molecules of NADH.

two molecules
of pyruvate

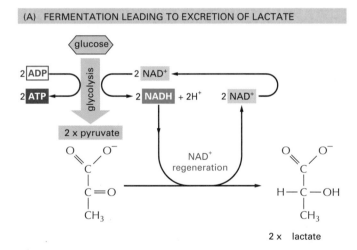

Question 13–1

At first glance, the final steps in fermentation appear to be optional add-on reactions to glycolysis. Explain why cells growing in the absence of oxygen could not simply discard pyruvate as a waste product. Which products derived from glucose would accumulate in cells unable to generate either lactate or ethanol by fermentation?

same is true in certain animal tissues, such as skeletal muscle, that can continue to function at low levels of molecular oxygen. In these anaerobic conditions, the pyruvate and the NADH electrons stay in the cytosol. The pyruvate is converted into products that are excreted from the cell: into lactate in muscle, for example, or into ethanol and CO_2 in the yeasts used in brewing and breadmaking. In this process, the NADH gives up its electrons and is converted back into NAD$^+$. This regeneration of NAD$^+$ is required to maintain the reactions of glycolysis (Figure 13–4).

Anaerobic energy-yielding pathways like these are called **fermentations**. Studies of the commercially important fermentations carried out by yeasts inspired much of early biochemistry. Work in the nineteenth century led in 1896 to the then startling recognition that these processes could be studied outside living organisms, in cell extracts. This revolutionary discovery eventually made it possible to dissect out and study each of the individual reactions in the fermentation process. The piecing together of the complete glycolytic pathway in the 1930s was a major triumph of biochemistry, and it was quickly followed by the recognition of the central role of ATP in cellular processes.

Glycolysis Illustrates How Enzymes Couple Oxidation to Energy Storage

We have previously used a "paddle wheel" analogy to explain how cells harvest useful energy from the oxidation of organic molecules by using

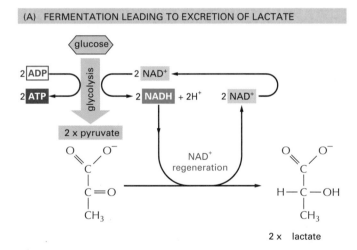

(A) FERMENTATION LEADING TO EXCRETION OF LACTATE

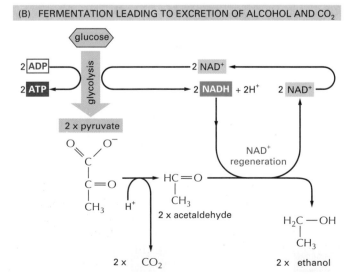

(B) FERMENTATION LEADING TO EXCRETION OF ALCOHOL AND CO_2

Figure 13–4 Pyruvate can be broken down by fermentation in the absence of oxygen. (A) When inadequate oxygen is present, for example, in a muscle cell undergoing vigorous contraction, the pyruvate produced by glycolysis is converted to lactate as shown. This reaction restores the NAD$^+$ consumed in step 6 of glycolysis, but the whole pathway yields much less energy overall than complete oxidation. (B) In some organisms that can grow anaerobically, such as yeasts, pyruvate is converted via acetaldehyde into carbon dioxide and ethanol. Again, this pathway regenerates NAD$^+$ from NADH, as required to enable glycolysis to continue. Both (A) and (B) are examples of fermentations.

enzymes to couple an energetically unfavorable reaction to an energetically favorable one (see Figure 3–31). Enzymes play the part of the paddle wheel in our analogy, and we now return to a step in glycolysis that we have previously discussed, in order to illustrate exactly how coupled reactions occur.

Two central reactions in glycolysis (steps 6 and 7 in Panel 13–1, p. 433) convert the three-carbon sugar intermediate glyceraldehyde 3-phosphate (an aldehyde) into 3-phosphoglycerate (a carboxylic acid). This conversion entails the oxidation of an aldehyde group to a carboxylic acid group, which occurs in two steps. The overall reaction releases enough free energy to convert a molecule of ADP to ATP and to transfer two electrons from the aldehyde to NAD^+ to form NADH, while still releasing enough heat to the environment to make the overall reaction energetically favorable ($\Delta G°$ for the overall reaction is –3.0 kcal/mole).

The means by which this remarkable feat of energy harvesting is accomplished is outlined in Figure 13–5. The indicated chemical reactions are precisely guided by two enzymes to which the sugar intermediates are tightly bound. In fact, as detailed in Figure 13–5, the first enzyme (glyceraldehyde 3-phosphate dehydrogenase) forms a short-lived covalent bond to the aldehyde through a reactive –SH group on the enzyme, and catalyzes its oxidation in this attached state. The reactive enzyme–substrate bond is then displaced by an inorganic phosphate ion to produce a high-energy phosphate intermediate, which is released from the enzyme. This intermediate binds to the second enzyme (phosphoglycerate kinase), which catalyzes the energetically favorable transfer of the high-energy phosphate just created to ADP, forming ATP and completing the process of oxidizing an aldehyde to a carboxylic acid.

We have shown this particular oxidation process in some detail because it provides a clear example of enzyme-mediated energy storage through coupled reactions (Figure 13–6). These reactions (steps 6 and 7) are the only ones in glycolysis that create a high-energy phosphate linkage directly from inorganic phosphate. As such, they account for the net yield of two ATP molecules and two NADH molecules per molecule of glucose.

As we have just seen, ATP can be formed readily from ADP when reaction intermediates are formed with phosphate bonds of higher energy than those in ATP. The energy of phosphate bonds can be ordered by determining the standard free-energy change ($\Delta G°$) for the breakage of each bond by hydrolysis; Figure 13–7 compares the high-energy phosphoanhydride bonds in ATP with some other phosphate bonds that are generated during glycolysis.

Sugars and Fats Are Both Degraded to Acetyl CoA in Mitochondria

We now move on to consider stage 3 of catabolism, a process that requires abundant molecular oxygen (O_2 gas). The Earth is thought to have developed an atmosphere containing O_2 gas between 1 and 2 billion years ago, whereas abundant life forms are known to have existed on the Earth for 3.5 billion years. Therefore the use of O_2 in the reactions that we discuss next is thought to be of relatively recent origin. In contrast, the mechanism used to produce ATP in Figure 13–5 does not require oxygen, and relatives of that elegant pair of coupled reactions could have arisen very early in the history of life on Earth.

In aerobic metabolism, the pyruvate produced by glycolysis is rapidly decarboxylated by a giant complex of three enzymes, called the pyruvate dehydrogenase complex. The products of pyruvate

Question 13–2

Arsenate (AsO_4^{3-}) is chemically very similar to phosphate (PO_4^{3-}) and is used as an alternative substrate by many phosphate-requiring enzymes. In contrast to phosphate, however, an anhydride bond between arsenate and carbon is very quickly hydrolyzed in water. Knowing this, suggest why arsenate is a compound of choice for murderers but not for cells. Formulate your explanation in the context of Figure 13–6.

**Figure 13–5 Energy is harvested in steps
6 and 7 of glycolysis.** In these steps the
oxidation of an aldehyde to a carboxylic
acid is coupled to the formation of ATP and
NADH. (A) Step 6 begins with the
formation of a covalent bond between the
substrate (glyceraldehyde 3-phosphate)
and an –SH group exposed on the surface
of the enzyme (glyceraldehyde 3-phosphate
dehydrogenase). The enzyme then
catalyzes transfer of hydrogen (as a hydride
ion—a proton plus two electrons) from the
bound glyceraldehyde 3-phosphate to a
molecule of NAD$^+$. Part of the energy
released in this oxidation is used to form a
molecule of NADH, and part is used to
convert the original linkage between the
enzyme and its substrate to a high-energy
thioester bond (shown in *red*). A molecule
of inorganic phosphate then displaces this
high-energy bond on the enzyme, creating
a high-energy sugar–phosphate bond
instead *(red)*. At this point the enzyme has
not only stored energy (in NADH), but also
coupled the energetically favorable
oxidation of an aldehyde to the
energetically unfavorable formation of a
high-energy phosphate bond. The second
reaction has been driven by the first,
thereby acting like the "paddle wheel"
coupler in Figure 3–31. Note that the
portion of the glyceraldehyde 3-phosphate
molecule in the shaded box remains
unchanged.

In reaction step 7, the high-energy
molecule that was just produced,
1,3-bisphosphoglycerate, binds to a second
enzyme, phosphoglycerate kinase. The
reactive phosphate is transferred to ADP,
forming a molecule of ATP and leaving a
free carboxylic acid group on the oxidized
sugar. (B) Summary of the overall chemical
change produced by reactions 6 and 7.

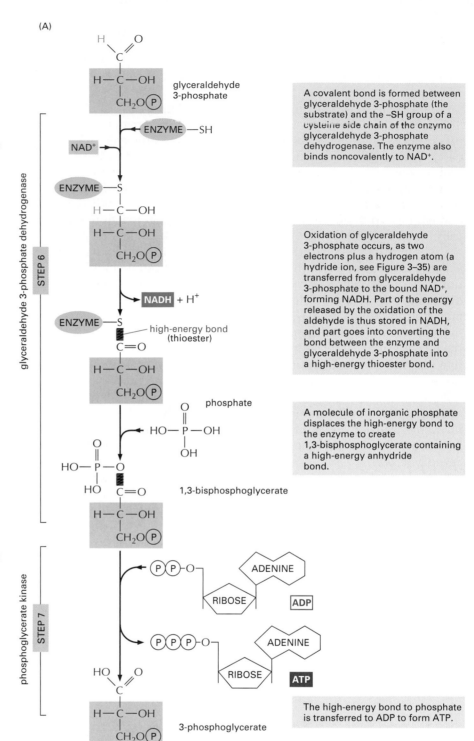

A covalent bond is formed between
glyceraldehyde 3-phosphate (the
substrate) and the –SH group of a
cysteine side chain of the enzyme
glyceraldehyde 3-phosphate
dehydrogenase. The enzyme also
binds noncovalently to NAD$^+$.

Oxidation of glyceraldehyde
3-phosphate occurs, as two
electrons plus a hydrogen atom (a
hydride ion, see Figure 3–35) are
transferred from glyceraldehyde
3-phosphate to the bound NAD$^+$,
forming NADH. Part of the energy
released by the oxidation of the
aldehyde is thus stored in NADH,
and part goes into converting the
bond between the enzyme and
glyceraldehyde 3-phosphate into
a high-energy thioester bond.

A molecule of inorganic phosphate
displaces the high-energy bond to
the enzyme to create
1,3-bisphosphoglycerate containing
a high-energy anhydride
bond.

The high-energy bond to phosphate
is transferred to ADP to form ATP.

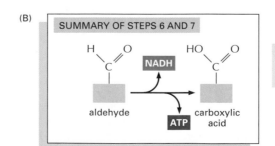

Much of the energy of oxidation
has been stored in the activated
carriers ATP and NADH.

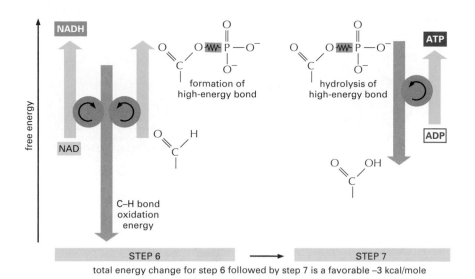

Figure 13–6 Coupled reactions form NADH and ATP in steps 6 and 7 of glycolysis. The C–H bond oxidation energy drives the formation of both NADH and a high-energy phosphate bond. The breakage of the high-energy bond in step 7 then drives ATP formation.

total energy change for step 6 followed by step 7 is a favorable –3 kcal/mole

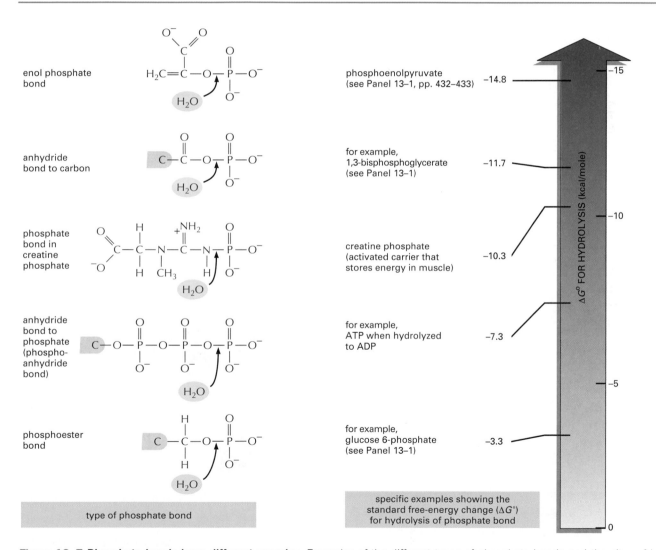

Figure 13–7 Phosphate bonds have different energies. Examples of the different types of phosphate bonds and the sites of hydrolysis are shown in the molecules depicted on the left. Those starting with a *gray* carbon atom show only part of a molecule. Examples of molecules containing such bonds are given on the right, with the free-energy change for hydrolysis. The transfer of a phosphate group from one molecule to another is energetically favorable if the standard free-energy change (ΔG^o) for hydrolysis of the phosphate bond of the first molecule is more negative than that for hydrolysis of the phosphate bond (once present) in the second. Thus, a phosphate group is readily transferred from 1,3-bisphosphoglycerate to ADP to form ATP. The hydrolysis reaction can be viewed as the transfer of the phosphate group to water.

Figure 13–8 Pyruvate is oxidized to acetyl CoA and CO₂ by pyruvate dehydrogenase. (A) General structure of the pyruvate dehydrogenase complex, which contains three different enzymes and some 60 polypeptide chains. In this large multienzyme complex, located in the mitochondrion of eucaryotic cells, reaction intermediates are passed directly from one enzyme to another. (B) The reactions carried out by the pyruvate dehydrogenase complex. The complex converts pyruvate to acetyl CoA in the mitochondrial matrix; NADH is also produced in this reaction. A, B, and C are the three enzymes *pyruvate decarboxylase, lipoamide reductase-transacetylase,* and *dihydrolipoyl dehydrogenase,* illustrated in (A); their activities are linked as shown.

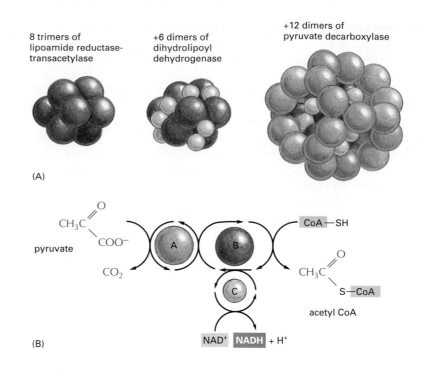

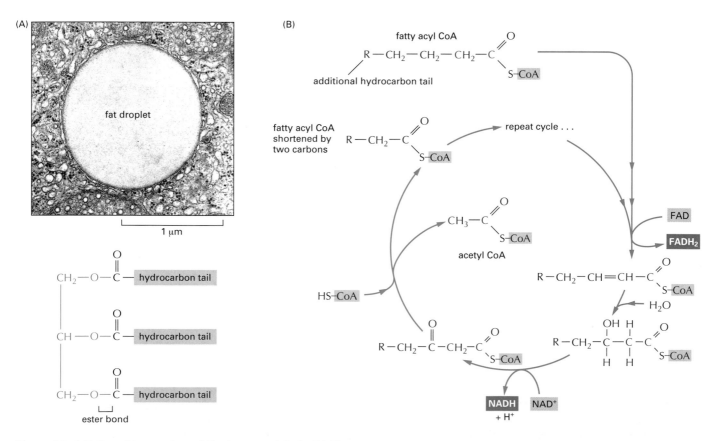

Figure 13–9 Fatty acids are also oxidized to acetyl CoA. (A) Electron micrograph of a lipid droplet in the cytoplasm *(top),* and the structure of fats *(bottom).* Fats are triacylglycerols. The glycerol portion, to which three fatty acids are linked through ester bonds, is shown here in *blue.* Fats are insoluble in water and form large lipid droplets in the specialized fat cells (called adipocytes) in which they are stored. (B) The fatty acid oxidation cycle. The cycle is catalyzed by a series of four enzymes in the mitochondrion. Each turn of the cycle shortens the fatty acid chain by two carbons (shown in *red)* and generates one molecule of acetyl CoA and one molecule each of NADH and FADH₂. (A, courtesy of Daniel S. Friend.)

decarboxylation are a molecule of CO_2 (a waste product), a molecule of NADH, and acetyl CoA. The three-enzyme complex is located in the mitochondria of eucaryotic cells; its structure and mode of action are outlined in Figure 13–8.

The enzymes that degrade the fatty acids derived from fats likewise produce acetyl CoA in mitochondria. Each molecule of fatty acid (in the form of the activated molecule fatty acyl CoA) is broken down completely by a cycle of reactions that trims two carbons at a time from its carboxyl end, generating one molecule of acetyl CoA for each turn of the cycle. A molecule of NADH and a molecule of $FADH_2$ are also produced in this process (Figure 13–9).

Sugars and fats provide the major energy sources for most nonphotosynthetic organisms, including humans. However, the majority of the useful energy that can be extracted from the oxidation of both types of foodstuffs remains stored in the acetyl CoA molecules that are produced by the two types of reactions just described. The citric acid cycle of reactions, in which the acetyl group in acetyl CoA is oxidized to CO_2 and H_2O, is therefore central to the energy metabolism of aerobic organisms. In eucaryotes these reactions all take place in mitochondria, the organelles to which pyruvate and fatty acids are directed for acetyl CoA production (Figure 13–10). We should therefore not be surprised to discover that the mitochondria are the place where most of the ATP is produced in animal cells. In contrast, aerobic bacteria carry out all of their reactions in a single compartment, the cytosol, and it is here that the citric acid cycle takes place in these cells.

The Citric Acid Cycle Generates NADH by Oxidizing Acetyl Groups to CO_2

In the nineteenth century, biologists noticed that in the absence of air (anaerobic conditions) cells produce lactic acid (for example, in muscle) or ethanol (for example, in yeast), while in the presence of air (aerobic conditions) cells consume O_2 and produce CO_2 and H_2O. Intensive efforts to define the pathways of aerobic metabolism eventually focused on the oxidation of pyruvate and led in 1937 to the discovery of the **citric acid cycle**, also known as the *tricarboxylic acid cycle* or the *Krebs cycle* (see How We Know, pp. 442–443). The citric acid cycle accounts for about two-thirds of the total oxidation of carbon compounds in most cells, and its major end products are CO_2 and high-energy electrons in the form of NADH. The CO_2 is released as a waste product, while the high-energy electrons from NADH are passed to a membrane-bound electron-transport chain, eventually combining with O_2 to produce H_2O. Although the citric acid cycle itself does not use O_2, it requires O_2 to proceed because there is no other efficient way for the NADH to get rid of its electrons and thus regenerate the NAD^+ that is needed to keep the cycle going.

Question 13–3

Many catabolic and anabolic reactions are based on reactions that are similar but work in opposite directions, such as the hydrolysis and condensation reactions described in Figure 3–39. This is true for fatty acid breakdown and fatty acid synthesis. From what you know about the mechanism of fatty acid breakdown outlined in Figure 13–9, would you expect the fatty acids found in cells to most commonly have an even or an odd number of carbon atoms?

Figure 13–10 In eucaryotic cells, acetyl CoA is produced in the mitochondria from molecules derived from sugars and fats. Most of the cell's oxidation reactions occur in these organelles, and most of its ATP is made here.

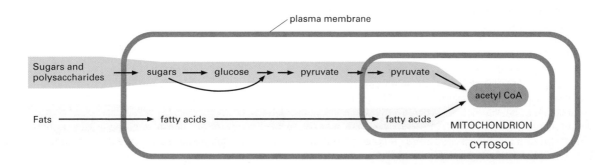

The citric acid cycle, which takes place inside mitochondria in eucaryotic cells, catalyzes the complete oxidation of the carbon atoms of the acetyl groups in acetyl CoA, converting them into CO_2. But the acetyl group is not oxidized directly. Instead, it is transferred from acetyl CoA to a larger, four-carbon molecule, *oxaloacetate*, to form the six-carbon tricarboxylic acid—*citric acid*, for which the subsequent cycle of reactions is named. The citric acid molecule is then gradually oxidized, and the energy of this oxidation is harnessed to produce energy-rich carrier molecules, in much the same manner as was described for glycolysis. The chain of eight reactions forms a cycle, because the oxaloacetate that began the process is regenerated at the end, as shown in outline in Figure 13–11.

We have thus far discussed only one of the three types of activated carrier molecules that are produced by the citric acid cycle, NADH. But in addition to three molecules of NADH, each turn of the cycle also produces one molecule of **FADH$_2$** (reduced flavin adenine dinucleotide) from FAD and one molecule of the ribonucleotide **GTP** (guanosine triphosphate) from GDP (see Figure 13–11). The structures of these two activated carrier molecules are illustrated in Figure 13–12. GTP is a close relative of ATP, and the transfer of its terminal phosphate group to ADP produces one ATP molecule in each cycle. Like NADH, FADH$_2$ is a carrier of high-energy electrons and hydrogen. As we discuss shortly, the energy that is stored in the readily transferred high-energy electrons of NADH and FADH$_2$ will subsequently be used to produce ATP through the process of *oxidative phosphorylation*, the only step in the oxidative catabolism of foodstuffs that directly requires gaseous oxygen (O_2) from the atmosphere.

The complete citric acid cycle is presented in Panel 13–2 (pp. 450–451). Note that the oxygen atoms required to make CO_2 from the acetyl groups entering the citric acid cycle are supplied not by molecular oxygen but by water. As illustrated in the panel, three molecules of water are split in each cycle, and the oxygen atoms of some of them are ultimately

Figure 13–11 The citric acid cycle catalyzes the complete oxidation of the carbon atoms in acetyl CoA. The reaction of acetyl CoA with oxaloacetate starts the cycle by producing citrate (citric acid). Each turn of the cycle produces two molecules of CO_2 (as waste products), three molecules of NADH, one molecule of GTP, and one molecule of FADH$_2$. The number of carbon atoms in each intermediate is shown in a *yellow box*. (See also Panel 13–2, pp. 450–451.)

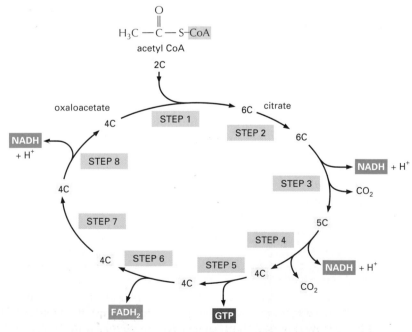

NET RESULT: ONE TURN OF THE CYCLE PRODUCES THREE NADH, ONE GTP, AND ONE FADH$_2$, AND RELEASES TWO MOLECULES OF CO_2

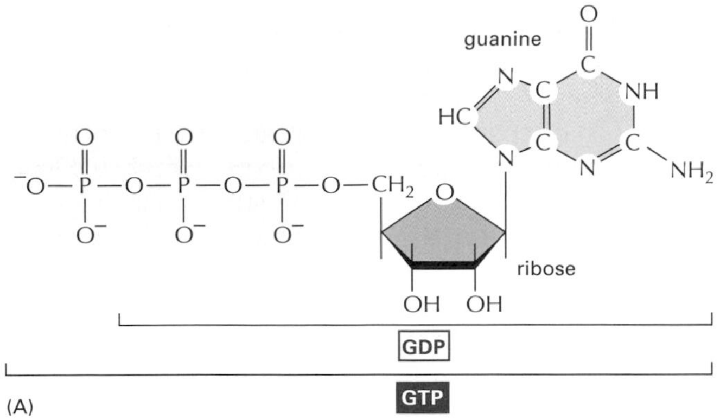

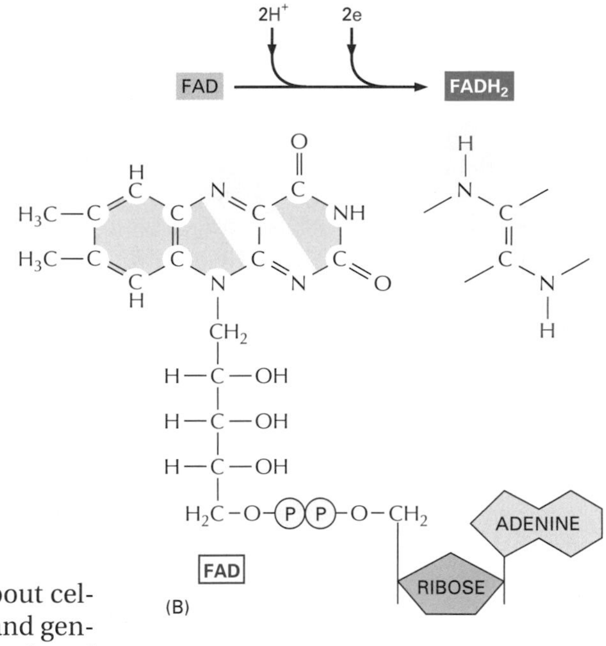

(A)

(B)

used to make CO_2. This brings up a common misconception about cellular respiration: although the process requires an input of O_2, and generates CO_2 as a product, the oxygen molecules of the O_2 gas are reduced to water, not incorporated directly into CO_2.

In addition to pyruvate and fatty acids, some amino acids pass from the cytosol into mitochondria, where they are also converted into acetyl CoA or one of the other intermediates of the citric acid cycle (see Figure 13–2). Thus, in the eucaryotic cell, the mitochondrion is the center toward which all energy-yielding processes lead, whether they begin with sugars, fats, or proteins.

As will be discussed later, the citric acid cycle also functions as a starting point for important biosynthetic reactions by producing vital carbon-containing intermediates, such as *oxaloacetate* and *α-ketoglutarate*. These substances produced by catabolism are transferred back from the mitochondrion to the cytosol, where they serve in anabolic reactions as precursors for the synthesis of many essential molecules, such as amino acids.

Figure 13–12 **Each turn of the citric acid cycle produces one molecule of GTP and one molecule of FADH$_2$, whose structures are shown here.** (A) GTP and GDP are close relatives of ATP and ADP, respectively, the only difference being the substitution of the base guanine for adenine. (B) Despite its very different structure, FADH$_2$, like NADH and NADPH, is a carrier of hydrogens and high-energy electrons. It is shown here in its oxidized form (FAD) with the hydrogen-carrying atoms highlighted in *yellow*.

Electron Transport Drives the Synthesis of the Majority of the ATP in Most Cells

It is in the last step in the oxidation of a food molecule that the major portion of its chemical energy is released. In this final process, the electron carriers NADH and FADH$_2$ transfer the electrons that they have gained when oxidizing other molecules to the **electron-transport chain**, which is embedded in the inner membrane of the mitochondrion. As the electrons pass along this long chain of specialized electron acceptor and donor molecules, they fall to successively lower energy states. The energy they release in this process is used to drive H$^+$ ions (protons) across the membrane—from the inner mitochondrial compartment to the outside. A gradient of H$^+$ ions is thereby generated. This gradient serves as a source of energy, like a battery, that is tapped to drive a variety of energy-requiring reactions. The most prominent of these reactions is the generation of ATP by the phosphorylation of ADP.

At the end of this series of transfers, the electrons are passed to molecules of oxygen gas (O_2) that have diffused into the mitochondrion; these reduced oxygen molecules simultaneously combine with protons (H$^+$) from the surrounding solution to produce molecules of water. The electrons have now reached their lowest energy levels, and therefore all the available energy has been extracted from the food molecule being

Question 13–4

Looking at the chemistry detailed in Panel 13–2 (pp. 450–451), why do you suppose it is useful to link the acetyl group first to another carbon skeleton, oxaloacetate, before completely oxidizing both carbons to CO_2?

"I have often been asked how the work on the citric acid cycle arose and developed," stated the biological chemist Hans Krebs in a lecture and review article in which he described his Nobel Prize-winning discovery of the cycle of reactions that lies at the center of cell metabolism. Did the concept stem from a sudden inspiration, a revelatory vision? "It was nothing of the kind," answered Krebs. Instead his realization that these reactions occur in a cycle—rather than a set of linear pathways—arose from a "very slow evolutionary process" that occurred over a five-year period, during which Krebs coupled insight and reasoning to careful experimentation to discover one of the central pathways that underlies energy metabolism in cells.

Minced tissues, curious catalysis

By the early 1930s, Krebs and other investigators had discovered that a select set of molecules are oxidized extraordinarily rapidly in various types of tissue preparations—slices of kidney or liver, or suspensions of minced pigeon muscle. Because these reactions depend on the presence of oxygen, the researchers surmised that this set of compounds might include intermediates that are important in cellular respiration—the consumption of O_2 and production of CO_2 that accompanies the metabolism of foodstuffs.

Using the minced-tissue preparations, Krebs and others made the following observations. First, that in the presence of oxygen certain organic acids—citrate, succinate, fumarate, and malate—are readily oxidized to carbon dioxide. These reactions depend on a continuous supply of oxygen.

Second, that the oxidation of these compounds falls into a pair of linear, sequential pathways:

$$citrate \rightarrow \alpha\text{-ketoglutarate} \rightarrow succinate$$
and
$$succinate \rightarrow fumarate \rightarrow malate \rightarrow oxaloacetate$$

Third, that small amounts of several of these compounds, when added to minced-muscle suspensions, stimulated an unusually large uptake of oxygen—far greater than that needed to oxidize only the added molecules. To explain this surprising catalytic observation, Albert Szent-Györgyi (the Nobelist who worked out the second pathway above) suggested that a single molecule of each compound must somehow act catalytically to stimulate the oxidation of many molecules of some endogenous substance in the muscle.

At this point, most of the reactions central to the citric acid cycle were known. What was not yet clear—and caused great confusion, even to future Nobel laureates—is how these apparently linear reactions could drive such a catalytic consumption of oxygen, where each molecule of metabolite fuels the oxidation of many more molecules. To simplify the

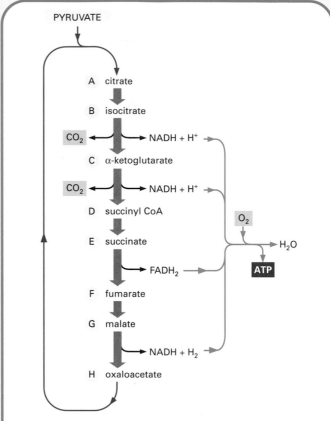

Figure 13–13 In this simplified representation of the citric acid cycle, O_2 is consumed and CO_2 is liberated as the molecular intermediates become oxidized. Krebs and others did not initially realize that these oxidation reactions occur in a cycle, as shown here.

discussion of how Krebs ultimately solved this puzzle—by linking these linear reactions together into a circle—we will now refer to the molecules involved by a sequence of designated letters, A through H (Figure 13–13).

A poison suggests a cycle

Many of the clues that Krebs used to formulate the citric acid cycle came from experiments using malonate—a poisonous compound that specifically inhibits the enzyme succinate dehydrogenase, which converts E to F. Malonate closely resembles succinate in its structure (Figure 13–14), and it serves as a competitive inhibitor of the enzyme. Because the

Figure 13–14 The structure of malonate closely resembles that of succinate.

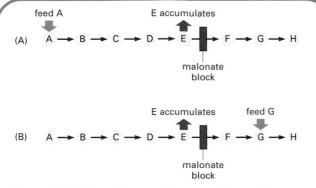

Figure 13–15 Poisoning muscle preparations with malonate provided clues to the cyclic nature of these oxidative reactions. (A) Adding A, B, or C to malonate-poisoned muscle causes an accumulation of E. (B) Addition of F, G, or H to a malonate-poisoned preparation also causes an accumulation of E, suggesting that enzymatic reactions can convert these molecules into E. The discovery that citrate (A) can be formed from oxaloacetate (H) and pyruvate allowed Krebs to join these two reaction pathways into a complete circle.

addition of malonate poisons respiration in tissues, Krebs concluded that succinate dehydrogenase (and the entire pathway linked to it) must play a critical role in cellular respiration.

Krebs then discovered that when A, B, or C is added to malonate-poisoned tissue suspensions, E accumulates (Figure 13–15A). This observation reinforces the importance of succinate dehydrogenase for successful respiration. However, he found that E also accumulates when F, G, or H is added to malonate-poisoned muscle (Figure 13–15B). The latter result suggests that an additional set of reactions must exist that can convert F, G, and H molecules into E, since E was previously shown to be a precursor for F, G, and H, rather than a product of their reactions.

At about this time Krebs also determined that when muscle suspensions are incubated with pyruvate and oxaloacetate, citrate forms:

pyruvate + oxaloacetate → citrate
or,   pyruvate + H → A

This observation led Krebs to postulate that when oxygen is present, pyruvate and H condense to form A, converting the previously delineated string of linear reactions into a cyclic sequence (see Figure 13–13). The discovery that acetyl CoA acts as an intermediary between pyruvate and oxaloacetate in this reaction, however, did not come for another decade.

Explaining the mysterious stimulatory effects

The cycle of reactions that was proposed by Krebs clearly explained how the addition of small amounts of any of the intermediates A through H could cause the large increase in the uptake of molecular oxygen that had been observed. Pyruvate is an abundant substance in minced tissues, and can be readily produced by glycolysis from the glucose that can be generated from glycogen stores (see Figure 13–3). Its oxidation requires a functioning citric acid cycle, in which each turn of the cycle results in the oxidation of one molecule of pyruvate. If the intermediates A through H are in small enough supply, the rate at which the entire cycle turns will be restricted. Adding a supply of any one of these intermediates will then have a dramatic effect on the rate at which the entire citric acid cycle operates (Figure 13–16). Thus it is easy to see how a large number of pyruvate molecules can be oxidized, and a great deal of oxygen consumed, for every molecule that is added of a citric acid cycle intermediate.

Krebs went on to demonstrate that all of the individual enzymatic reactions in his postulated cycle take place in tissue preparations. Furthermore, they occur at rates high enough to account for the rate of pyruvate and oxygen consumption in these tissues. Krebs therefore concluded that this series of reactions is the major, if not the sole, pathway for the oxidation of pyruvate—at least in muscle. By fitting together pieces of information like a jigsaw puzzle, and by searching for missing links, Krebs arrived at a coherent picture of the metabolic processes that underlie the oxidation of foodstuffs. Remarkably, he worked out this intricate metabolic pathway without the aid of reagents and techniques considered essential by modern biochemists: radioactive markers that allow one to trace labeled compounds through these reaction pathways—or mass spectrometry, a powerful method for rapidly identifying the various chemical intermediates that occur along the way.

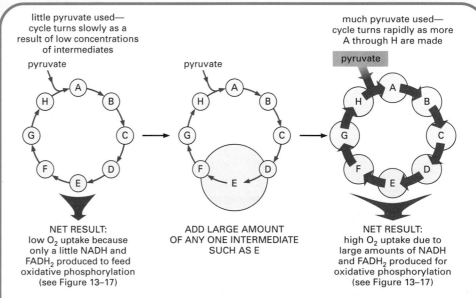

Figure 13–16 Replenishing the supply of any single intermediate has a dramatic effect on the rate at which the entire citric acid cycle operates.

Figure 13–17 In the final stages of oxidation of food molecules, NADH (and FADH$_2$, not shown) produced by the citric acid cycle donate high-energy electrons that are eventually used to reduce oxygen gas to water. A major portion of the energy released during an elaborate series of electron-transfers in the mitochondrial inner membrane (or in the plasma membrane of bacteria) is harnessed to drive the synthesis of ATP through the process of oxidative phosphorylation.

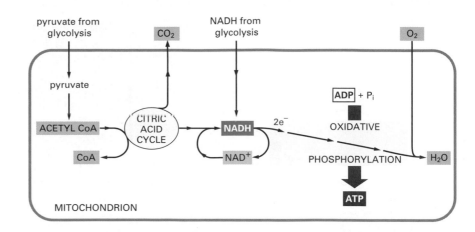

Question 13–5

What, if anything, is wrong with the following statement: "The oxygen consumed during the oxidation of glucose in animal cells is returned as part of CO_2 to the atmosphere." How could you support your answer experimentally?

oxidized. This process, termed *oxidative phosphorylation* (Figure 13–17), also occurs in the plasma membrane of bacteria. One of the most remarkable achievements of cellular evolution, oxidative phosphorylation will be a central topic of Chapter 14.

In total, the complete oxidation of a molecule of glucose to H_2O and CO_2 produces about 30 molecules of ATP. In contrast, only 2 molecules of ATP are produced per molecule of glucose by glycolysis alone.

Storing and Utilizing Food

All organisms need to restore their ATP pools constantly in order for biological order to be maintained in their cells. Yet animals have only periodic access to food, and plants need to survive overnight without sunlight, without the possibility of sugar production from photosynthesis. Thus animals and plants have evolved the means to store food molecules for consumption whenever these energy sources are scarce.

Of course, food provides cells with more than energy; these molecules also serve as the building blocks for many other cellular molecules. In this section, we discuss how organisms store and use the molecules derived from food.

Organisms Store Food Molecules in Special Reservoirs

To compensate for long periods of fasting, animals store food within their cells. Fatty acids are stored as fat droplets composed of water-insoluble triacylglycerols, largely in specialized fat cells (see Figure 13–9A). Sugar is stored as glucose subunits in the large, branched polysaccharide **glycogen**, which is present as small granules in the cytoplasm of many cells, including liver and muscle (Figure 13–18). The synthesis and degradation of glycogen are rapidly regulated according to need. When more ATP is needed than can be generated from the food molecules taken in from the bloodstream, cells break down glycogen in a reaction that produces glucose 1-phosphate, which enters glycolysis.

Quantitatively, **fat** is a far more important storage form than glycogen, in part because the oxidation of a gram of fat releases about twice as much energy as the oxidation of a gram of glycogen. Moreover, glycogen differs from fat in binding a great deal of water, producing a sixfold difference in the actual mass of glycogen required to store the same amount of energy as fat. An average adult human stores enough glycogen for only about a day of normal activities, but enough fat to last for nearly a month. If our main fuel reservoir had to be carried as glycogen instead of fat, body weight would need to be increased by an average of about 60 pounds.

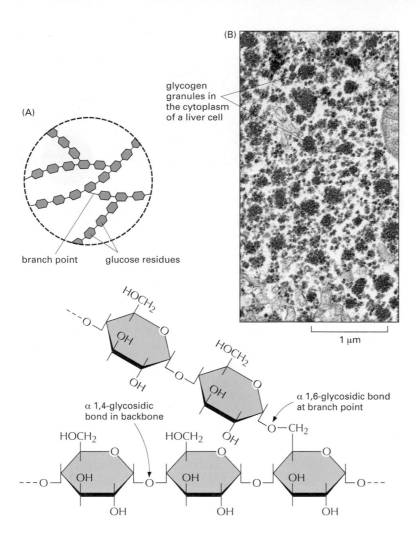

(A)

glycogen granules in the cytoplasm of a liver cell

branch point glucose residues

α 1,4-glycosidic bond in backbone

α 1,6-glycosidic bond at branch point

1 μm

Figure 13–18 Animal cells store glycogen to provide energy in times of fasting.
(A) The structure of glycogen (and of starch). Both are polymers of the sugar glucose and differ only in the frequency of branch points (the region in *yellow* that is shown enlarged below). Glycogen has many more branches than starch.
(B) An electron micrograph shows glycogen granules in the cytoplasm of a liver cell.
(B, courtesy of Robert Fletterick and Daniel S. Friend.)

Most of our fat is stored in adipose tissue (Figure 13–19), from which it is released into the bloodstream for other cells to use as needed. The need arises after a period of not eating; even a normal overnight fast results in the mobilization of fat, so that in the morning most of the acetyl CoA entering the citric acid cycle is derived from fatty acids rather than from glucose. After a meal, however, most of the acetyl CoA entering the citric acid cycle comes from glucose derived from food, and any excess glucose is used to replenish depleted glycogen stores or to synthesize fats. (While animal cells readily convert sugars to fats, they cannot convert fatty acids to sugars.)

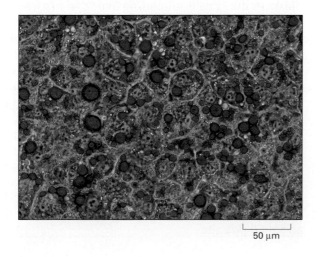

50 μm

Figure 13–19 Animal cells also store fats.
Fat droplets (stained *red)* beginning to accumulate in developing fat cells.
(Courtesy of P. Tontonoz and Ronald M. Evans.)

Figure 13–20 Plant cells store both starch and fat in their chloroplasts. A thin section of a single chloroplast from a plant cell, showing the starch granules and lipid droplets that have accumulated as a result of the biosyntheses occurring there. (Courtesy of K. Plaskitt.)

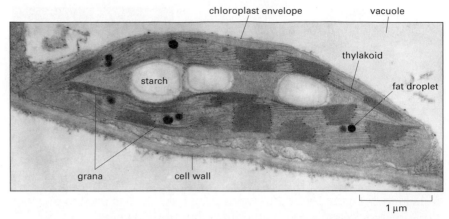

chloroplast envelope · vacuole

thylakoid

starch

fat droplet

grana · cell wall

1 μm

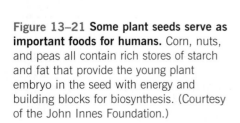

Question 13–6

After looking at the structures of sugars and fatty acids (discussed in Chapter 2), give an intuitive explanation as to why oxidation of a sugar yields only about half as much energy as the oxidation of an equivalent dry weight of a fatty acid.

Plants convert some of the sugars that they make through photosynthesis during the daylight into fats and into starch, a polymer of glucose analogous to the glycogen of animals. The fats in plants are triacylglycerols, just like the fats in animals, from which they differ only in the types of fatty acids that predominate. Fat and starch are both stored in the chloroplast—the specialized organelle that carries out photosynthesis in plant cells—where they serve as food reservoirs that can be mobilized to produce ATP during periods of darkness (Figure 13–20).

The embryos inside plant seeds must live on stored sources of energy for a prolonged period, until they germinate to produce leaves that can harvest the energy in sunlight. They also require these materials to build their cell walls and synthesize many other biological molecules as they grow. For this reason plant seeds often contain especially large amounts of fats and starch—which make them a major food source for animals, including ourselves (Figure 13–21). Germinating seeds can convert the stored fat into glucose as needed.

Chloroplasts and Mitochondria Collaborate in Plant Cells

Chloroplasts play a central role in energy metabolism in plant cells, using the energy of sunlight to produce the activated carriers ATP and NADPH through the process of photosynthesis. In this specialized organelle, sugars are produced by a carbon-fixation process that requires these activated carriers. The inner membrane of the chloroplast is impermeable to both NADPH and ATP, which therefore cannot be exported directly to the rest of the cell. These activated carriers are instead used to produce sugars in the chloroplast that are exported to the cytosol. Much of this sugar ends up fueling mitochondria, where most of the ATP needed by the plant is synthesized using exactly the same pathways for the oxidative breakdown of sugars that are used by animal cells and other nonphotosynthetic organisms (Figure 13–22). As

Figure 13–21 Some plant seeds serve as important foods for humans. Corn, nuts, and peas all contain rich stores of starch and fat that provide the young plant embryo in the seed with energy and building blocks for biosynthesis. (Courtesy of the John Innes Foundation.)

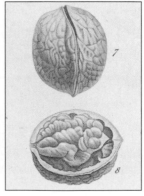

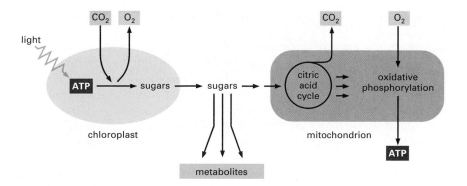

Figure 13–22 **In plants, the chloroplasts and mitochondria collaborate to supply cells with metabolites and ATP.** The chloroplast's inner membrane is impermeable to the ATP and NADPH produced by photosynthesis. Hence this organelle exports sugars to the rest of the plant cell. In mitochondria, the sugars are oxidized and ATP is generated. The mitochondrial membranes are permeable to ATP, as indicated.

indicated in Figure 13–22, the exported sugars are also used to produce building blocks and other metabolites that the cell needs.

Many Biosynthetic Pathways Begin with Glycolysis or the Citric Acid Cycle

Catabolism produces both energy for the cell and the building blocks from which many other molecules of the cell are made (see Figure 3–3). Thus far, we have emphasized energy production rather than the provision of the starting materials for biosynthesis. But many of the intermediates formed in glycolysis and the citric acid cycle are siphoned off by other biosynthetic pathways, where enzymes use them to produce the amino acids, nucleotides, lipids, and other small organic molecules that the cell needs. An idea of the complexity of this process can be gathered from Figure 13–23, which illustrates some of the branches from the central catabolic reactions that lead to biosyntheses.

The existence of so many branching pathways in the cell requires that the choices at each branch be carefully regulated, as we discuss next.

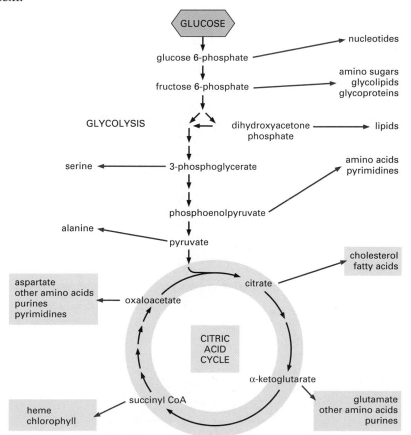

Figure 13–23 **Glycolysis and the citric acid cycle provide the precursors needed to synthesize many important biological molecules.** The amino acids, nucleotides, lipids, sugars, and other molecules—shown here as products—in turn serve as the precursors for the many macromolecules of the cell. Each *black arrow* in this diagram denotes a single enzyme-catalyzed reaction; the *red arrows* generally represent pathways with many steps that are required to produce the indicated products.

Metabolism Is Organized and Regulated

One can get a sense of the intricacy of a cell as a chemical machine from Figure 13–24, which charts a selection of the enzymatic pathways in a cell. Highlighted in red on that diagram are glycolysis and the citric acid cycle. It is obvious that our discussion of cell metabolism has dealt with only a tiny fraction of cellular chemistry.

All of these reactions occur in a cell that is less than 0.1 mm in diameter, and each step requires a different enzyme. As is clear from Figure 13–24, the same molecule can often be part of many different pathways. Pyruvate, for example, is a substrate for half a dozen or more different enzymes, each of which modifies it chemically in a different way. One enzyme converts pyruvate to acetyl CoA, another to oxaloacetate; a third enzyme changes pyruvate to the amino acid alanine, a fourth to lactate, and so on. All of these different pathways compete for the same pyruvate molecule, and similar competitions for thousands of other small molecules go on at the same time. One might think that the whole system would be so finely balanced that any minor upset, such as a temporary change in dietary intake, would be disastrous.

In fact, the metabolic balance of a cell is amazingly stable. Whenever the balance is perturbed, the cell reacts so as to restore the initial state. The cell can adapt and continue to function during starvation or disease.

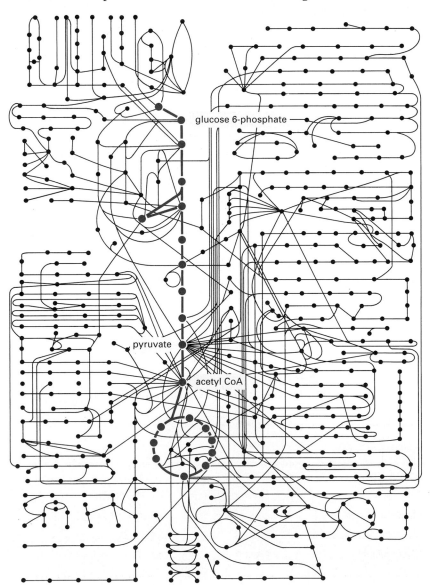

Figure 13–24 Glycolysis and the citric acid cycle lie at the center of metabolism. Some 500 metabolic reactions of a typical cell are shown schematically with the reactions of glycolysis and the citric acid cycle in *red*. Other reactions either lead into these two central pathways—delivering small molecules to be catabolized with production of energy—or they lead outward and thereby supply carbon compounds for the purpose of biosynthesis.

Mutations of many kinds can damage or even eliminate particular reaction pathways, and yet—provided that certain minimum requirements are met—the cell survives. As was previously explained in Chapter 4, this resilience is made possible by an elaborate network of *control mechanisms* that act on enzymes; these regulate and coordinate the rates of the many metabolic reactions in a cell.

Essential Concepts

- Glucose and other food molecules are broken down by controlled stepwise oxidation to provide useful chemical energy in the form of the activated carriers ATP and NADH.
- Sugars derived from food are broken down by distinct sets of reactions: glycolysis (which occurs in the cytosol), the citric acid cycle (in the mitochondrial matrix), and oxidative phosphorylation (in the inner mitochondrial membrane).
- The reactions of glycolysis degrade the six-carbon sugar glucose to two molecules of the three-carbon sugar pyruvate, producing a relatively small amount of ATP and NADH.
- In the presence of oxygen, pyruvate is converted to acetyl CoA plus CO_2. The citric acid cycle then converts the acetyl group in acetyl CoA to CO_2 and H_2O. Much of the energy released in these oxidation reactions is stored as high-energy electrons in the activated carriers NADH and $FADH_2$. In eucaryotic cells these reactions occur in mitochondria.
- The other major energy source in foods is fat. The fatty acids produced from fats are imported into mitochondria and converted to acetyl CoA molecules. These acetyl CoA molecules are then further oxidized through the citric acid cycle, producing NADH and $FADH_2$, just like the acetyl CoA derived from pyruvate.
- The NADH and $FADH_2$ pass their high-energy electrons to an electron-transport chain in the inner mitochondrial membrane, where a series of electron transfers is used to drive the formation of ATP. Most of the energy captured during the breakdown of food molecules is harvested during this process of oxidative phosphorylation (described in detail in Chapter 14).
- Cells store food molecules in special reservoirs. Glucose subunits are stored as glycogen in animals and as starch in plants; both animals and plants store food as fats. The food reservoirs produced by plants are major sources of food for animals, including humans.
- Molecules ingested as food are used not only as sources of metabolic energy but also as raw materials for biosynthesis. Thus many intermediates of glycolysis and the citric acid cycle are starting points for pathways that lead to the synthesis of proteins, nucleic acids, and the many other specialized molecules of the cell.
- The thousands of different reactions carried out simultaneously by a cell are closely coordinated, enabling the cell to adapt and continue to function under a wide range of external conditions.

Key Terms

acetyl CoA	fat	glycolysis
ADP, ATP	fermentation	NAD^+, NADH
citric acid cycle	GDP, GTP	oxidative phosphorylation
electron-transport chain	glucose	pyruvate
FAD, $FADH_2$	glycogen	starch

Question 13–7

A cyclic reaction pathway requires that the starting material be regenerated and available at the end of each cycle. If compounds of the citric acid cycle are siphoned off as building blocks used in a variety of metabolic reactions, why does the citric acid cycle not quickly cease to exist?

Panel 13-2 The complete citric acid cycle

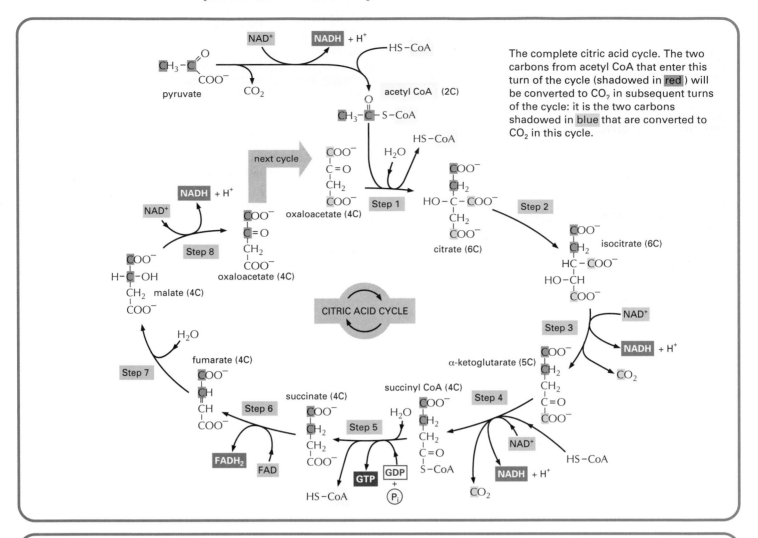

The complete citric acid cycle. The two carbons from acetyl CoA that enter this turn of the cycle (shadowed in red) will be converted to CO_2 in subsequent turns of the cycle: it is the two carbons shadowed in blue that are converted to CO_2 in this cycle.

Details of the eight steps are shown below. For each step, the part of the molecule that undergoes a change is shadowed in blue, and the name of the enzyme that catalyzes the reaction is in a yellow box.

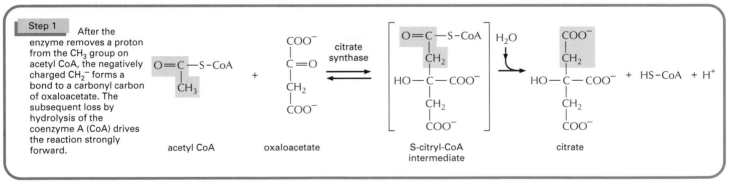

Step 1 After the enzyme removes a proton from the CH_3 group on acetyl CoA, the negatively charged CH_2^- forms a bond to a carbonyl carbon of oxaloacetate. The subsequent loss by hydrolysis of the coenzyme A (CoA) drives the reaction strongly forward.

Step 2 An isomerization reaction, in which water is first removed and then added back, moves the hydroxyl group from one carbon atom to its neighbor.

Step 3 In the first of four oxidation steps in the cycle, the carbon carrying the hydroxyl group is converted to a carbonyl group. The immediate product is unstable, losing CO_2 while still bound to the enzyme.

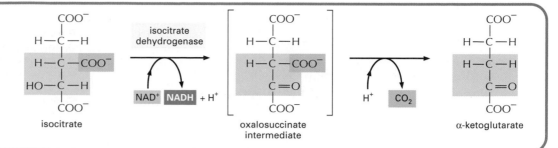

isocitrate

oxalosuccinate intermediate

α-ketoglutarate

Step 4 The *α-ketoglutarate dehydrogenase complex* closely resembles the large enzyme complex that converts pyruvate to acetyl CoA (*pyruvate dehydrogenase*). It likewise catalyzes an oxidation that produces NADH, CO_2, and a high-energy thioester bond to coenzyme A (CoA).

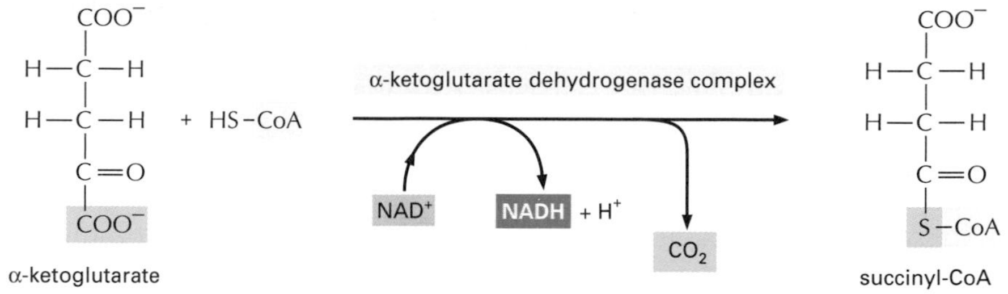

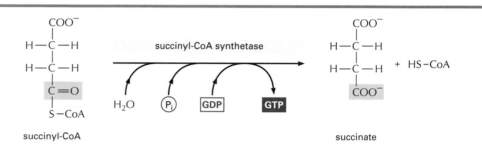

α-ketoglutarate

succinyl-CoA

Step 5 A phosphate molecule from solution displaces the CoA, forming a high-energy phosphate linkage to succinate. This phosphate is then passed to GDP to form GTP. (In bacteria and plants, ATP is formed instead.)

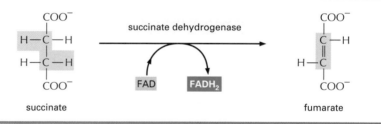

succinyl-CoA

succinate

Step 6 In the third oxidation step in the cycle, FAD removes two hydrogen atoms from succinate.

succinate dehydrogenase

FAD   FADH₂

succinate

fumarate

Step 7 The addition of water to fumarate places a hydroxyl group next to a carbonyl carbon.

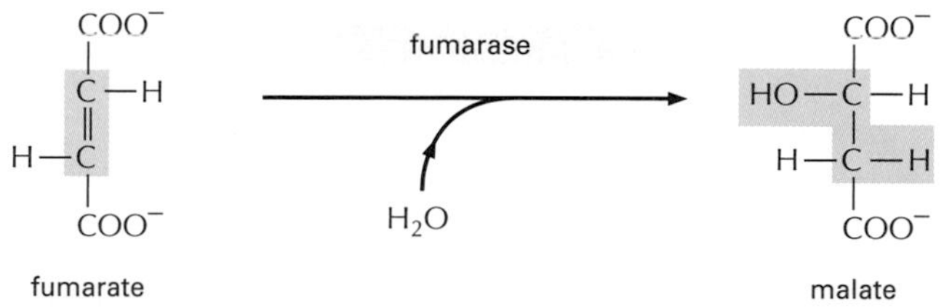

fumarase

H_2O

fumarate

malate

Step 8 In the last of four oxidation steps in the cycle, the carbon carrying the hydroxyl group is converted to a carbonyl group, regenerating the oxaloacetate needed for step 1.

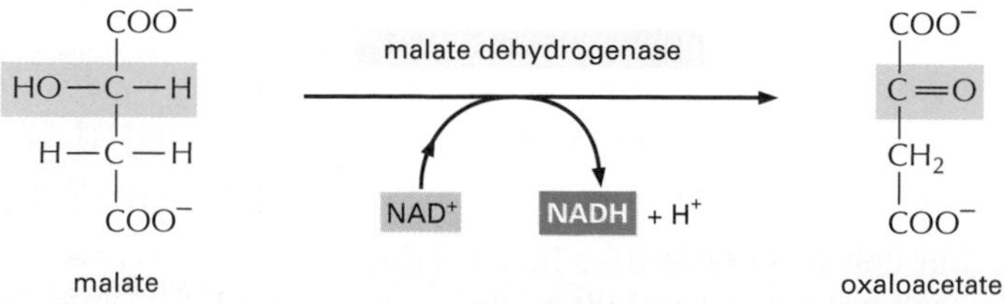

malate dehydrogenase

NAD⁺   NADH + H⁺

malate

oxaloacetate

Questions

Question 13–8

The oxidation of sugar molecules by the cell takes place according to the general reaction $C_6H_{12}O_6$ (glucose) $+ 6O_2 \rightarrow 6CO_2 + 6H_2O + $ energy. Which of the following statements are correct? Explain your answers.

    A. All of the energy produced is in the form of heat.

    B. None of the produced energy is in the form of heat.

    C. The energy is produced by a process that involves the oxidation of carbon atoms.

    D. The reaction supplies the cell with essential water.

    E. In cells the reaction takes place in more than one step.

    F. Many steps in the oxidation of sugar molecules involve reaction with oxygen gas.

    G. Some organisms carry out the reverse reaction.

    H. Some cells that grow in the absence of O_2 produce CO_2.

Question 13–9

An exceedingly sensitive instrument (yet to be devised) shows that one of the carbon atoms in Charles Darwin's last breath is resident in your bloodstream, where it forms part of a hemoglobin molecule. Suggest how this carbon atom might have traveled from Darwin to you, and list some of the molecules it could have entered en route.

Question 13–10

Yeast cells can grow both in the presence of molecular oxygen (aerobically) and in its absence (anaerobically). Under which of the two conditions could you expect the cells to grow better? Explain your answer.

Question 13–11

During movement, muscle cells require large amounts of ATP to fuel their contractile apparatus. These cells contain high levels of creatine phosphate (shown in Figure 13–7). Why is this a useful compound to store energy? Justify your answer with the information shown in Figure 13–7.

Question 13–12

Identical pathways that make up the complicated sequence of reactions of glycolysis, shown in Panel 13–1 (pp. 432–433), are found in most living cells, from bacteria to humans. One could envision, however, countless alternative chemical reaction mechanisms that would allow the oxidation of sugar molecules and that could, in principle, have evolved to take the place of glycolysis. Discuss this fact in the context of evolution.

Question 13–13

Assume that an animal cell is a cube that has a side length of 10 μm. The cell contains 10^9 ATP molecules that it uses up every minute. ATP is regenerated by oxidizing glucose molecules. After what amount of time will the cell have used up an amount of oxygen gas that is equal to its own volume? (Recall that one mole contains 6×10^{23} molecules. One mole of a gas has a volume of 22.4 liters.)

Question 13–14

Under the conditions existing in the cell, the free energies of the first few reactions in glycolysis (in Panel 13–1, pp. 432–433) are:

step 1	$\Delta G = -8.0$ kcal/mole
step 2	$\Delta G = -0.6$ kcal/mole
step 3	$\Delta G = -5.3$ kcal/mole
step 4	$\Delta G = -0.3$ kcal/mole

Are these reactions energetically favorable? Using these values, draw to scale an energy diagram (A) for the overall reaction and (B) for the pathway composed of the four individual reactions.

Question 13–15

The chemistry of most metabolic reactions was deciphered by synthesizing metabolites containing atoms that are different isotopes from those occurring naturally. The products of reactions starting with isotopically labeled metabolites can be analyzed to determine precisely which atoms in the products are derived from which atoms in the starting material. The methods of detection exploit, for example, the fact that different isotopes have different masses that can be distinguished using biophysical techniques such as mass spectrometry. Moreover, some isotopes are radioactive and can therefore be readily recognized with Geiger counters or photographic film that becomes exposed by radiation.

    A. Assume pyruvate containing radioactive ^{14}C in the carboxyl group is added to a cell extract that can support oxidative phosphorylation. What compound is produced that will contain the vast majority of the ^{14}C added?

    B. Assume oxaloacetate containing radioactive ^{14}C in its keto group (refer to Panel 13–2, pp. 450–451) is added to the extract. Where would the ^{14}C atom be after precisely one turn of the cycle?

Question 13–16

In cells that can grow both aerobically and anaerobically, fermentation is inhibited in the presence of molecular oxygen. Suggest a reason for this observation.

Highlight from *Essential Cell Biology 2 Interactive* CD-ROM

13.1   Citric Acid Cycle

Energy Generation in Mitochondria and Chloroplasts

The fundamental need to generate energy efficiently has had a profound influence on the history of life on Earth. Much of the structure, function, and evolution of cells and organisms can be related to their need for energy. The earliest cells may have produced ATP by breaking down organic molecules, left by earlier geochemical processes, using some form of fermentation. Fermentation reactions occur in the cytosol of present-day cells; these reactions use the energy derived from the partial oxidation of energy-rich food molecules to form ATP, the chemical energy currency of cells.

But very early in the history of life, a much more efficient method for generating energy and synthesizing ATP appeared. This process is based on the transport of electrons along membranes. Billions of years later, it is so central to the survival of life on Earth that we devote this entire chapter to it. As we shall see, this membrane-based mechanism is used by cells to acquire energy from a wide variety of sources: for example, it is central to the conversion of light energy into chemical bond energy in photosynthesis, and to the aerobic respiration that enables us to use oxygen to produce large amounts of ATP from food molecules. The mechanism we will describe first appeared in bacteria some 3.5 billion years ago. The descendants of these pioneering cells crowd every corner and crevice of the land and the oceans with a wild menagerie of living forms, and they survive within eucaryotic cells in the form of chloroplasts and mitochondria.

Where we come from and how we are related to other living things are puzzles that have fascinated humans since the beginning of recorded time. The story that we can tell now, worked out through a long chain of scientific investigation, is one of the most dramatic and exciting histories ever told. And we are not yet done. Each year, further discoveries in cell biology enable us to add more details through molecular detective work of dramatically increasing power.

Absolutely central to life's progression was the provision of an abundant source of energy for cells. In this chapter, we discuss the remarkable mechanism that made this all possible.

Cells Obtain Most of Their Energy by a Membrane-based Mechanism

The main chemical energy currency in cells is ATP (see Figure 3–32). In eucaryotic cells, small amounts of ATP are generated during glycolysis in the cytosol (as discussed in Chapter 13), but most ATP is produced by membrane-based processes in mitochondria (and also in chloroplasts in plant and algal cells). Very similar processes also occur in the cell

Cells Obtain Most of Their Energy by a Membrane-based Mechanism

Mitochondria and Oxidative Phosphorylation

A Mitochondrion Contains an Outer Membrane, an Inner Membrane, and Two Internal Compartments

High-Energy Electrons Are Generated via the Citric Acid Cycle

A Chemiosmotic Process Converts Oxidation Energy into ATP

Electrons Are Transferred Along a Chain of Proteins in the Inner Mitochondrial Membrane

Electron Transport Generates a Proton Gradient Across the Membrane

The Proton Gradient Drives ATP Synthesis

Coupled Transport Across the Inner Mitochondrial Membrane Is Driven by the Electrochemical Proton Gradient

Proton Gradients Produce Most of the Cell's ATP

The Rapid Conversion of ADP to ATP in Mitochondria Maintains a High ATP/ADP Ratio in Cells

Electron-Transport Chains and Proton Pumping

Protons Are Readily Moved by the Transfer of Electrons

The Redox Potential Is a Measure of Electron Affinities

Electron Transfers Release Large Amounts of Energy

Metals Tightly Bound to Proteins Form Versatile Electron Carriers

Cytochrome Oxidase Catalyzes Oxygen Reduction

The Mechanism of H^+ Pumping Will Soon Be Understood in Atomic Detail

Respiration Is Amazingly Efficient

Chloroplasts and Photosynthesis

Chloroplasts Resemble Mitochondria but Have an Extra Compartment

Chloroplasts Capture Energy from Sunlight and Use It to Fix Carbon

Excited Chlorophyll Molecules Funnel Energy into a Reaction Center

Light Energy Drives the Synthesis of ATP and NADPH

Carbon Fixation Is Catalyzed by Ribulose Bisphosphate Carboxylase

Carbon Fixation in Chloroplasts Generates Sucrose and Starch

The Origins of Chloroplasts and Mitochondria

Oxidative Phosphorylation Gave Ancient Bacteria an Evolutionary Advantage

Photosynthetic Bacteria Made Even Fewer Demands on Their Environment

The Lifestyle of *Methanococcus* Suggests That Chemiosmotic Coupling Is an Ancient Process

membranes of many bacteria. The fundamental mechanism for making all of this ATP arose very early in life's history, and it was so successful that its essential features have been retained in the long evolutionary journey from early procaryotes to modern cells. The process consists of two linked stages, both of which are carried out by protein complexes embedded in a membrane.

Stage 1. Electrons (derived from the oxidation of food molecules or from other sources discussed later) are transferred along a series of electron carriers—called an *electron-transport chain*—embedded in the membrane. These electron transfers release energy that is used to pump protons (H+, derived from the water that is ubiquitous in cells) across the membrane and thus generate an electrochemical proton gradient (Figure 14–1A). As discussed in Chapter 12, an ion gradient across a membrane is a form of stored energy; this energy can be harnessed to do useful work when the ions are allowed to flow back across the membrane down their electrochemical gradient.

Stage 2. H+ flows back down its electrochemical gradient through a protein complex called *ATP synthase*, which catalyzes the energy-requiring synthesis of ATP from ADP and inorganic phosphate (P_i). This ubiquitous enzyme serves the role of a turbine, permitting the proton gradient to drive the production of ATP (Figure 14–1B).

The linkage of electron transport, proton pumping, and ATP synthesis was called the *chemiosmotic hypothesis* when it was first proposed in the 1960s, because of the link between the chemical bond–forming reactions that synthesize ATP ("chemi-") and the membrane transport processes ("osmotic," from the Greek *osmos*, "to push"). It is now known as **chemiosmotic coupling**.

Chemiosmotic coupling first evolved in bacteria. It is perhaps not surprising, therefore, that aerobic eucaryotic cells appear to have adopted the bacterial chemiosmotic mechanisms intact, first by engulfing aerobic bacteria to form mitochondria, and somewhat later—in the lineages leading to algae and plants—by engulfing cyanobacteria to form chloroplasts, as described in Chapter 1.

In this chapter we shall consider energy generation in both mitochondria and chloroplasts, emphasizing the common principles by which proton gradients are created and used in these organelles, as well as in bacteria. We start by describing the structure and function of mitochondria, looking in detail at the events that occur in the mitochondrial membrane to create the proton gradient and generate ATP. We next

Figure 14–1 Cells have evolved systems for harnessing the energy required for life. (A) The essential requirements for chemiosmosis are a membrane, in which are embedded a pump protein and an ATP synthase, and sources of high-energy electrons (e^-) and of protons (H+). The pump harnesses the energy of electron transfer (details not shown here) to pump protons derived from water, creating a proton gradient across the membrane. The source of the high-energy electrons can be sunlight or organic or inorganic molecules. (B) The gradient produced in (A) serves as a versatile energy store. It is used to drive a variety of energy-requiring reactions in mitochondria, chloroplasts, and bacteria—including the synthesis of ATP by the ATP synthase. The *red arrow* shows the direction of proton movement at each stage.

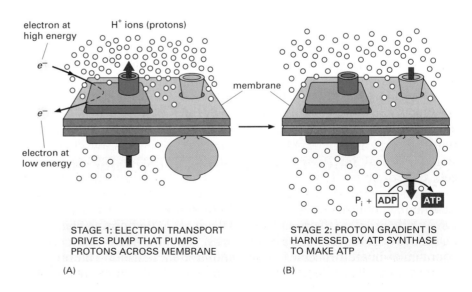

STAGE 1: ELECTRON TRANSPORT DRIVES PUMP THAT PUMPS PROTONS ACROSS MEMBRANE

(A)

STAGE 2: PROTON GRADIENT IS HARNESSED BY ATP SYNTHASE TO MAKE ATP

(B)

consider photosynthesis as it occurs in the chloroplasts of plant cells. Finally, we trace the evolutionary pathways that gave rise to these mechanisms of energy generation. By examining the lifestyles of a variety of single-celled organisms—including those that might resemble our early ancestors—we can begin to see the role that chemiosmotic coupling has played in the rise of complex eucaryotes and in the development of all life on Earth.

Mitochondria and Oxidative Phosphorylation

Mitochondria are present in nearly all eucaryotic cells—in plants, animals, and most eucaryotic microorganisms—and it is in these organelles that most of a cell's ATP is produced. Without them, present-day eucaryotes would be dependent on the relatively inefficient process of glycolysis (described in Chapter 13) for all of their ATP production, and it seems unlikely that complex multicellular organisms could have been supported in this way. When glucose is converted to pyruvate by glycolysis, less than 10% of the total free energy potentially available from the glucose is released. In the mitochondria, the metabolism of sugars is completed, and the energy released is harnessed so efficiently that about 30 molecules of ATP are produced for each molecule of glucose oxidized. By contrast, only two molecules of ATP are produced per glucose molecule by glycolysis alone.

Defects in mitochondrial function can have serious repercussions for an organism. Consider, for example, an inherited disorder called *myoclonic epilepsy and ragged red fiber disease (MERRF)*. This disease, caused by a mutation in one of the mitochondrial transfer RNA genes, is characterized by a decrease in synthesis of the mitochondrial proteins required for electron transport and ATP production. As a result, patients with this disorder typically experience muscle weakness or heart problems (from effects on cardiac muscle) and epilepsy or dementia (from effects on nerve cells). Muscle and nervous tissues suffer most when mitochondria are defective because they need particularly large amounts of ATP to function optimally.

The same metabolic reactions that occur in mitochondria also take place in aerobic bacteria, which do not possess these organelles; in these organisms the plasma membrane carries out the chemiosmotic processes. But unlike a bacterial cell, which also has to carry out many other functions, the mitochondrion has become highly specialized for energy generation, as we see next.

A Mitochondrion Contains an Outer Membrane, an Inner Membrane, and Two Internal Compartments

Mitochondria are generally similar in size and shape to bacteria, although these attributes can vary depending on the cell type. They contain their own DNA and RNA, and a complete transcription and translation system including ribosomes, which allows them to synthesize some of their own proteins. Time-lapse movies of living cells reveal mitochondria as remarkably mobile organelles, constantly changing shape and position. Present in large numbers—1000 to 2000 in a liver cell, for example—these organelles can form long, moving chains in association with microtubules of the cytoskeleton (discussed in Chapter 17). In other cells, they remain fixed in one cellular location to target ATP directly to a site of unusually high ATP consumption. In a heart muscle cell, for example, mitochondria are located close to the contractile apparatus, while in a sperm they are wrapped tightly around the motile flagellum (Figure 14–2). The number of mitochondria present

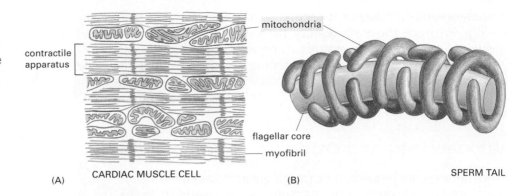

Figure 14–2 Mitochondria are located near sites of high ATP utilization. (A) In a cardiac muscle cell, mitochondria are located close to the contractile apparatus, in which ATP hydrolysis provides the energy for contraction. (B) In a sperm, mitochondria are located in the tail, around the core of the motile flagellum, which requires ATP for its movement.

contractile apparatus

mitochondria

flagellar core

myofibril

(A) CARDIAC MUSCLE CELL

(B) SPERM TAIL

in different cell types varies dramatically, and can change with the energy needs of the cell. In skeletal muscle cell, for example, the number of mitochondria may increase five- to tenfold due to mitochondrial growth and division that occurs if the muscle has been repeatedly stimulated to contract.

Each mitochondrion is bounded by two highly specialized membranes—one wrapped around the other—that play a crucial part in its activities. The outer and inner mitochondrial membranes create two mitochondrial compartments: a large internal space called the **matrix** and the much narrower *intermembrane space* (Figure 14–3). If purified mitochondria are gently processed and fractionated into separate components by differential centrifugation (see Panel 4–3, pp. 160–161), the biochemical composition of each of the two membranes and of the spaces enclosed by them can be determined. Each contains a unique collection of proteins.

The *outer membrane* contains many molecules of a transport protein called porin, which, as described in Chapter 11, forms wide aqueous channels through the lipid bilayer. As a result, the outer membrane is like a sieve that is permeable to all molecules of 5000 daltons or less, including small proteins. This makes the intermembrane space chemically equivalent to the cytosol with respect to the small molecules it contains. In contrast, the *inner membrane,* like other membranes in the cell, is impermeable to the passage of ions and most small molecules, except where a path is provided by membrane transport proteins. The mitochondrial matrix therefore contains only molecules that can be selectively transported into the matrix across the inner membrane, and its contents are highly specialized.

The inner mitochondrial membrane is the site of electron transport and proton pumping, and it contains the ATP synthase. Most of the proteins embedded in the inner mitochondrial membrane are components of the electron-transport chains required for oxidative phosphorylation. This membrane has a distinctive lipid composition and also contains a variety of transport proteins that allow the entry of selected small molecules, such as pyruvate and fatty acids, into the matrix.

The inner membrane is usually highly convoluted, forming a series of infoldings, known as *cristae,* that project into the matrix space to greatly increase the surface area of the inner membrane (see Figure 14–3). These folds provide a large surface on which ATP synthesis can take place; in a liver cell, for example, the inner mitochondrial membranes of all the mitochondria constitute about a third of the total membranes of the cell. And the number of cristae is three times greater in a mitochondrion of a cardiac muscle cell than in a mitochondrion of a liver cell, presumably because of the greater demand for ATP in heart cells.

Question 14–2

Electron micrographs show that mitochondria in heart muscle have a much higher density of cristae than mitochondria in skin cells. Suggest an explanation for this observation.

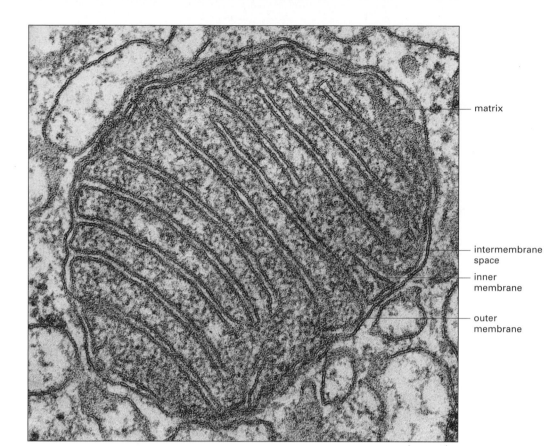

Figure 14–3 A mitochondrion is organized into four separate compartments. Two of these compartments are membranes and two are internal spaces. Each of these compartments contains a unique set of proteins that enables it to perform its distinct functions. In liver mitochondria, an estimated 67% of the total mitochondrial protein is located in the matrix, 21% is located in the inner membrane, 6% in the outer membrane, and 6% in the intermembrane space. (Micrograph courtesy of Daniel S. Friend.)

matrix

intermembrane space

inner membrane

outer membrane

100 nm

Matrix. This large internal space contains a highly concentrated mixture of hundreds of enzymes, including those required for the oxidation of pyruvate and fatty acids and for the citric acid cycle. The matrix also contains several identical copies of the mitochondrial DNA genome, special mitochondrial ribosomes, tRNAs, and various enzymes required for expression of the mitochondrial genes.

Inner membrane. The inner membrane (*red*) is folded into numerous cristae, which greatly increase its total surface area. It contains proteins with three types of functions: (1) those that carry out the oxidation reactions of the electron-transport chain, (2) the ATP synthase that makes ATP in the matrix, and (3) transport proteins that allow the passage of metabolites into and out of the matrix. An electrochemical gradient of H+, which drives the ATP synthase, is established across this membrane, and so it must be impermeable to ions and most small charged molecules.

Outer membrane. Because it contains a large channel-forming protein (called porin), the outer membrane is permeable to all molecules of 5000 daltons or less. Other proteins in this membrane include enzymes involved in mitochondrial lipid synthesis and enzymes that convert lipid substrates into forms that are subsequently metabolized in the matrix.

Intermembrane space. This space (*white*) contains several enzymes that use the ATP passing out of the matrix to phosphorylate other nucleotides.

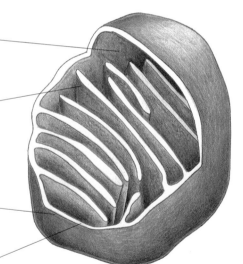

High-Energy Electrons Are Generated via the Citric Acid Cycle

In the mitochondria the metabolism of food molecules is completed. Mitochondria can use both pyruvate and fatty acids as fuel. Pyruvate comes mainly from glucose and other sugars, and fatty acids come from fats. Both of these fuel molecules are transported across the inner mitochondrial membrane and then converted to the crucial metabolic intermediate *acetyl CoA* by enzymes located in the mitochondrial matrix (see Figure 13–10). The acetyl groups in acetyl CoA are then oxidized in the matrix via the citric acid cycle, as described in Chapter 13. The cycle converts the carbon atoms in acetyl CoA to CO_2, which is released from

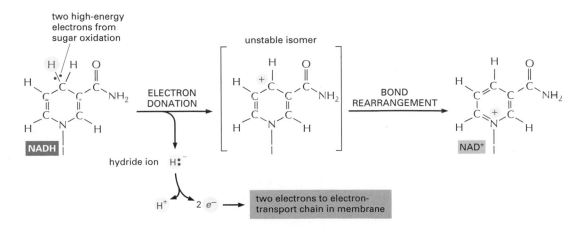

two high-energy electrons from sugar oxidation

unstable isomer

ELECTRON DONATION

BOND REARRANGEMENT

NADH

NAD⁺

hydride ion H:⁻

H⁺   2 e⁻ → two electrons to electron-transport chain in membrane

Figure 14–4 NADH donates its electrons to the electron-transport chain. In this drawing, the high-energy electrons are shown as two *red dots* on a *yellow* hydrogen atom. A hydride ion ($H^{:-}$, a hydrogen atom with an extra electron) is removed from NADH and is converted into a proton and two high-energy electrons: $H^{:-} \rightarrow H^+ + 2e^-$. Only the ring that carries the electrons in a high-energy linkage is shown; for the complete structure and the conversion of NAD^+ back to NADH, see the structure of the closely related NADPH in Figure 3–35. Electrons are also carried in a similar way by $FADH_2$, whose structure is shown in Figure 13–12B.

the cell as a waste product. In addition, the cycle generates high-energy electrons, carried by the activated carrier molecules NADH and $FADH_2$ (Figure 14–4). These high-energy electrons are then transferred to the inner mitochondrial membrane, where they enter the electron-transport chain; the loss of electrons regenerates the NAD^+ and FAD that are needed for continued oxidative metabolism. Electron transport along the chain now begins. The entire sequence of reactions is outlined in Figure 14–5.

A Chemiosmotic Process Converts Oxidation Energy into ATP

Although the citric acid cycle is considered to be part of aerobic metabolism, it does not itself use molecular oxygen. Only in the final catabolic reactions that take place on the inner mitochondrial membrane is O_2 directly consumed. Nearly all the energy available from burning carbohydrates, fats, and other foodstuffs in the earlier stages of their

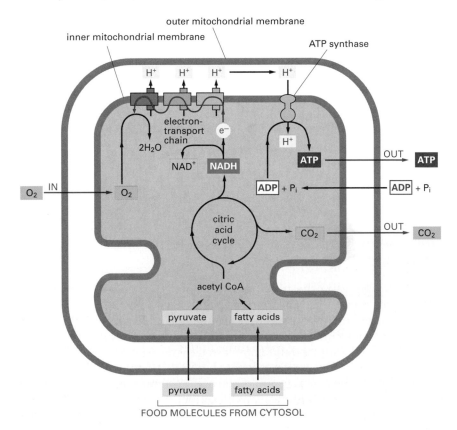

Figure 14–5 Mitochondria can use pyruvate or fatty acids to generate energy. Pyruvate and fatty acids enter the mitochondrion *(bottom),* are converted to acetyl CoA, and are then metabolized by the citric acid cycle, which reduces NAD^+ to NADH (and FAD to $FADH_2$, not shown). In the process of oxidative phosphorylation, high-energy electrons from NADH (and $FADH_2$) are then passed along the electron-transport chain in the inner membrane to oxygen (O_2). This electron transport generates a proton gradient across the inner membrane, which is used to drive the production of ATP by ATP synthase.

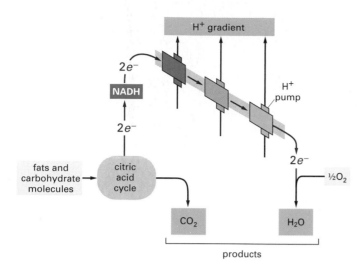

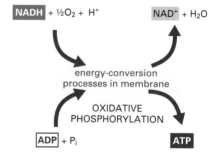

Figure 14–6 **Protons are pumped across the inner mitochondrial membrane.** Only stage 1 of chemiosmotic coupling is shown. Inputs are *yellow*, products are *blue*, and the path of electron flow is indicated by *red arrows*.

oxidation is initially saved in the form of the activated carrier molecules generated during glycolysis and the citric acid cycle—NADH and FADH$_2$. These carrier molecules donate their high-energy electrons to the electron-transport chain in the mitochondrial membrane, and thus become oxidized to NAD$^+$ and FAD. The electrons are quickly passed along the chain to molecular oxygen (O$_2$) to form water (H$_2$O).

The energy released during passage of the electrons along the electron-transport chain is harnessed to pump protons across the inner mitochondrial membrane (Figure 14–6), and this proton gradient in turn drives the synthesis of ATP to complete the chemiosmotic mechanism. The process involves both the consumption of O$_2$ and the synthesis of ATP through the addition of a phosphate group to ADP, and so is called **oxidative phosphorylation** (Figure 14–7).

Even though the mechanism escaped detection for many years (see How We Know, pp. 460–461), the vast majority of living organisms use chemiosmotic coupling to generate ATP. The source of the electrons that power proton pumping differs widely. As we have seen, in the aerobic respiration that produces ATP in mitochondria and aerobic bacteria, the electrons are derived ultimately from the oxidation of glucose or fatty acids, and molecular oxygen (O$_2$) acts as the final electron acceptor, producing water as a waste product (see Figure 14–7). In chemiosmotic coupling in photosynthesis, the required electrons are derived from the action of light on the green pigment chlorophyll. And many bacteria use inorganic substances such as hydrogen, iron, and sulfur as the source of the high-energy electrons that they need to make ATP.

We now outline the types of reactions that make oxidative phosphorylation possible.

Electrons Are Transferred Along a Chain of Proteins in the Inner Mitochondrial Membrane

The **electron-transport chain** that carries out oxidative phosphorylation is present in many copies in the inner mitochondrial membrane. Also known as the *respiratory chain*, it contains over 40 proteins, of which about 15 are directly involved in electron transport. Most of these proteins are embedded in the lipid bilayer and function only in an intact membrane, making them difficult to study. However, the components of the electron-transport chain, like other membrane proteins, can be solubilized using nonionic detergents (see Figure 11–27), purified, and then reconstituted in operational form in small membrane vesicles. Such studies reveal that most of the proteins involved in the

Figure 14–7 **Mitochondria catalyze a major conversion of energy.** In oxidative phosphorylation, the energy released by the oxidation of NADH to NAD$^+$ is harnessed—through energy-conversion processes in the membrane (electron transfer, proton pumping, and the flow of protons through ATP synthase)—to the energy-requiring phosphorylation of ADP to form ATP. The high-energy electrons lost from NADH move along an electron-transport chain in the membrane and eventually combine with molecular oxygen and H$^+$ to form water. The net equation for this electron-transfer process, in which two electrons pass from NADH to oxygen, is NADH $+$ ½O$_2$ $+$ H$^+$ $\rightarrow$ NAD$^+$ $+$ H$_2$O (see Figure 14–6).

How We Know: How Chemiosmotic Coupling Drives ATP Synthesis

In 1861 Louis Pasteur discovered that yeast cells grow and divide more vigorously when air is present, the first demonstration that aerobic metabolism is more efficient than anaerobic metabolism. His observations make sense in that we now know that oxidative phosphorylation is a much more efficient means of generating ATP than is glycolysis: electron-transport systems produce about 30 molecules of ATP for each molecule of glucose oxidized, compared with the two molecules of ATP generated by glycolysis alone. But it took another hundred years for researchers to determine that the process of chemiosmotic coupling—using proton pumping to power ATP synthesis—allows cells to generate energy with such efficiency.

Imaginary intermediates

In the 1950s, many researchers believed that the oxidative phosphorylation that takes place in mitochondria generates ATP via a mechanism similar to that used in glycolysis. During glycolysis, ATP is produced when a molecule of ADP receives a phosphate group directly from a high-energy intermediate. Such "substrate-level" phosphorylation occurs in steps 7 and 10 of glycolysis, where the high-energy phosphate groups from 1,3-bisphosphoglycerate and phosphoenolpyruvate are transferred to ADP to form ATP (see Panel 13–1, pp. 432–433). It was assumed that the electron-transport chain in mitochondria would similarly generate some high-energy intermediate that could then donate its phosphate group directly to ADP. This model inspired a frustrating search for this mysterious intermediate that lasted for years. Investigators occasionally claimed to discover the missing intermediate, but the compounds turned out to be either unrelated to electron transport or, as one researcher put it in a review of the history of bioenergetics, "products of high-energy imagination."

Harnessing the force

It wasn't until 1961 that Peter Mitchell suggested that the "high-energy intermediate" his colleagues were seeking was, in fact, the electrochemical proton gradient generated by the electron-transport system. His proposal, dubbed the chemiosmotic hypothesis, stated that the energy of a H^+ gradient formed during the transfer of electrons through the transport chain could be tapped to drive ATP synthesis.

Several lines of evidence offered support for such chemiosmotic coupling. First, mitochondria do generate a proton gradient across their inner membrane. But what does this gradient do? If the H^+ electrochemical gradient (also called the proton-motive force) is required to drive ATP synthesis, as the chemiosmotic hypothesis posits, then destruction of that gradient—or of the membrane itself—should inhibit energy generation. In fact researchers found this to be true. Physical disruption of the inner mitochondrial membrane halts ATP synthesis. Similarly, dissipation of the proton gradient by chemical "uncoupling" agents such as 2,4-dinitrophenol also prevents ATP from being made. These gradient-busting chemicals insert into the inner mitochondrial membrane, where they act as H^+ carriers, providing a pathway for the flow of H^+ that bypasses the ATP synthase (Figure 14–8). In this way they uncouple electron transport from ATP synthesis. As a result of this short-circuiting, the proton-motive force is dissipated completely and ATP can no longer be made.

Such uncoupling occurs naturally in some specialized fat cells. In these cells, called *brown fat cells,* most of the energy from oxidation is dissipated as heat rather than converted into ATP. The inner membranes of the large mitochondria in these cells contain a special transport protein

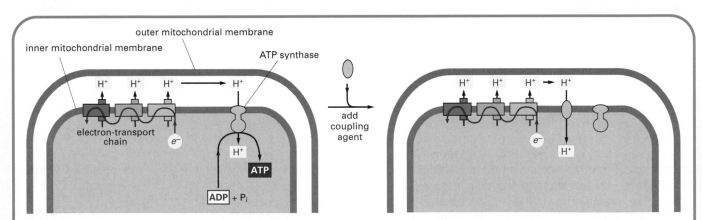

Figure 14–8 Uncoupling agents are H^+ carriers that can insert into the mitochondrial inner membrane. They render the membrane permeable to protons, allowing H^+ to flow into the mitochondrion without passing through ATP synthase. This short circuit effectively uncouples electron transport from ATP synthesis.

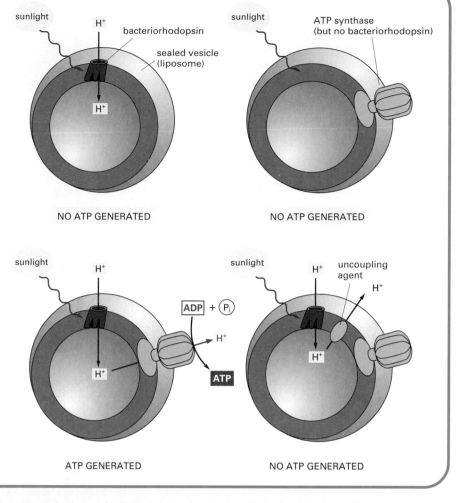

Figure 14–9 Experiments with bacteriorhodopsin and an ATP synthase from cow-heart mitochondria provided strong evidence that proton gradients can power ATP production. When bacteriorhodopsin is added to artificial vesicles, the protein generates a proton gradient in response to light. In artificial vesicles containing both bacteriorhodopsin and an ATP synthase, this proton gradient drives the formation of ATP. Uncoupling agents that abolish the gradient eliminate the ATP synthesis.

sunlight
H^+
bacteriorhodopsin
sealed vesicle (liposome)
H^+
NO ATP GENERATED

sunlight
ATP synthase (but no bacteriorhodopsin)
NO ATP GENERATED

sunlight
H^+
H^+
ADP + P_i
H^+
ATP
ATP GENERATED

sunlight
H^+
uncoupling agent
H^+
H^+
NO ATP GENERATED

that allows protons to move down their electrochemical gradient, circumventing ATP synthase. As a result, the cells oxidize their fat stores at a rapid rate and produce more heat than ATP. Tissues containing brown fat serve as biological heating pads, helping to revive hibernating animals and to protect sensitive areas of newborn human babies (such as the backs of their necks) from the cold.

Artificial ATP generation

If disrupting the proton gradient across the mitochondrial membrane terminates ATP synthesis, then, conversely, generating an artificial proton gradient should stimulate the production of ATP. Again, this is exactly what happens. When a H^+ gradient is imposed artificially by lowering the pH on the cytoplasmic side of the mitochondrial membrane, ATP is synthesized, even in the absence of an oxidizable substrate.

How does this proton gradient drive ATP production? That's where the ATP synthase comes in. In 1974 Efraim Racker and Walther Stoeckenius demonstrated elegantly that the combination of an ATP synthase plus a proton gradient will produce ATP. These researchers found that they could reconstitute a complete artificial energy-generating system by combining an ATPase from cow-heart mitochondria with a protein from the purple membrane of *Halobacterium halobium*. As discussed in Chapter 11, the plasma membrane

of this archaebacterium is packed with bacteriorhodopsin, a protein that pumps H^+ out of the cell in response to sunlight (see Figure 11–28). Thus the purple membrane protein generates a proton gradient when exposed to light.

When bacteriorhodopsin is reconstituted into artificial lipid vesicles, Racker and Stoeckenius showed that in the presence of light it pumps H^+ into the vesicles, generating a proton gradient. (For some reason the orientation of the protein is reversed in these membranes, so that H^+ ions are transported into the vesicles; in the bacterium protons are pumped out.) And when an ATPase purified from mitochondria is incorporated into these vesicles, the system catalyzes ATP synthesis in response to light. This ATP formation requires the H^+ gradient, as the researchers found that eliminating bacteriorhodopsin from the system or adding uncoupling agents abolished ATP synthesis (Figure 14–9).

Thus, although Mitchell's hypothesis initially met with considerable resistance—biochemists had hoped to discover a high-energy intermediate rather than having to settle for an elusive electrochemical force—the experimental evidence that eventually accumulated to support the importance of chemiosmotic coupling in cellular energy generation could not be ignored, and Mitchell was awarded a Nobel Prize in 1978.

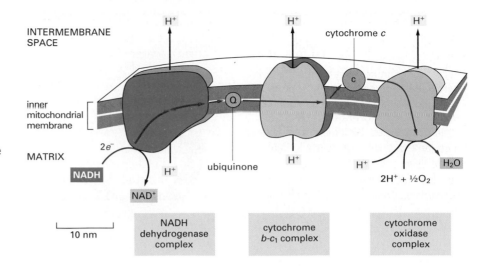

Figure 14–10 Electrons are transferred through three respiratory enzyme complexes in the inner mitochondrial membrane. The relative size and shape of each complex is indicated. During the transfer of electrons from NADH to oxygen (*red lines*), protons derived from water are pumped across the membrane by each of the respiratory enzyme complexes. The ubiquinone (Q) and cytochrome *c* (c) serve as mobile carriers that ferry electrons from one complex to the next. The structures and roles of these two molecules will be discussed later (see Figures 14–20 and 14–23).

mitochondrial electron-transport chain are grouped into three large *respiratory enzyme complexes,* each containing multiple individual proteins. Each complex includes transmembrane proteins that hold the entire protein complex firmly in the inner mitochondrial membrane.

The three respiratory enzyme complexes are (1) the *NADH dehydrogenase complex,* (2) the *cytochrome b-c₁ complex,* and (3) the *cytochrome oxidase complex.* Each contains metal ions and other chemical groups that form a pathway for the passage of electrons through the complex. The respiratory complexes are the sites of proton pumping, and each can be thought of as a protein machine that pumps protons across the membrane as electrons are transferred through it.

Electron transport begins when a hydride ion (H^-) is removed from NADH and is converted into a proton and two high-energy electrons: $H^- \rightarrow H^+ + 2e^-$, as previously explained in Figure 14–4. As illustrated in Figure 14–10, this reaction is catalyzed by the first of the respiratory enzyme complexes, the NADH dehydrogenase, which accepts the electrons. The electrons are then passed along the chain to each of the other enzyme complexes in turn, using mobile electron carriers that will be described later. The transfer of electrons along the chain is energetically favorable: the electrons start out at very high energy and lose energy at each step as they pass along the chain, eventually entering the cytochrome oxidase, where they combine with a molecule of O_2 to form water. This is the oxygen-requiring step of cellular respiration, and it consumes nearly all of the oxygen that we breathe.

Electron Transport Generates a Proton Gradient Across the Membrane

The proteins of the respiratory chain guide the electrons so that they move sequentially from one enzyme complex to another—with no short circuits that skip a complex. Each electron transfer is an oxidation–reduction reaction: as described in Chapter 3, the molecule or atom donating the electron becomes oxidized, while the receiving molecule or atom becomes reduced (see p. 90). Electrons will pass spontaneously from molecules that have a relatively low affinity for their available electrons, and thus lose them easily, to molecules with a higher electron affinity. For example, NADH with its high-energy electrons has a low electron affinity, so that its electrons are readily passed to the NADH dehydrogenase. The electrical batteries of our common experience are based on similar electron transfers between two chemical substances with different electron affinities.

In the absence of any means for harnessing the energy released by electron transfers, this energy is simply liberated as heat. But just as a battery can be connected to a device that pumps water (Figure 14–11), cells harness much of the energy of electron transfer by having the electron transfers take place within proteins that can pump protons (H⁺). Each of the respiratory enzyme complexes couples the energy released by electron transfer across it to an uptake of protons from water (H₂O → H⁺ + OH⁻) in the mitochondrial matrix, accompanied by the release of protons on the other side of the membrane into the intermembrane space. As a result, the energetically favorable flow of electrons along the electron-transport chain pumps protons across the membrane out of the matrix, creating an electrochemical proton gradient across the inner mitochondrial membrane (see Figure 14–10).

The active pumping of protons thus has two major consequences:

1. It generates a gradient of proton (H⁺) concentration (a pH gradient) across the inner mitochondrial membrane, with the pH about one unit higher in the matrix (around pH 8) than in the intermembrane space, where the pH is generally close 7. (Because H⁺, like any small molecule, equilibrates freely across the outer membrane of the mitochondrion, the pH in the intermembrane space is the same as in the cytosol.) This is a tenfold drop in H⁺ concentration in the matrix compared with the intermembrane space.

2. It generates a membrane potential (discussed in Chapter 12) across the inner mitochondrial membrane, with the inside (the matrix side) negative and the outside positive as a result of a net outflow of protons, which are positive ions.

As discussed in Chapter 12, the force driving the passive flow of ions such as Na⁺ and K⁺ across a membrane is proportional to the electrochemical gradient for the ion across the membrane. This in turn depends on the voltage across the membrane, which is measured as the membrane potential, and on the concentration gradient of the ion (see Figure 12–8).

Because protons are positively charged, they will move more readily across a membrane if the membrane has an excess of negative electrical charges on the other side. In the case of the inner mitochondrial membrane, the pH gradient and membrane potential work together to create a steep electrochemical proton gradient that makes it energetically

(A)

(B)

(C)

positive electrode (metal-capped carbon rod)

insulator

manganese dioxide and powdered carbon in fine porous sac

electrolyte gel

zinc negative electrode

negative ions

electron flow from zinc to manganese dioxide

electron flow in wire

all chemical energy from electron transfer converted to heat energy

pump

chemical energy from electron transfer converted to the potential energy stored in a difference in water levels; less energy is therefore lost as heat energy

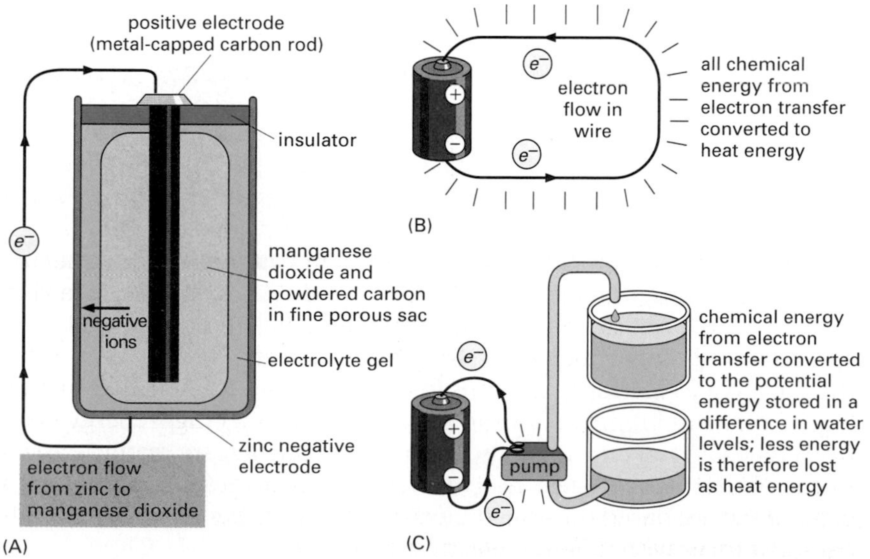

Figure 14–11 Electrical batteries are powered by chemical reactions based on electron transfers. (A) A standard flashlight battery, in which electrons are passed from zinc metal (Zn) to the manganese atom in manganese dioxide (MnO₂), forming Zn²⁺ and manganous oxide (MnO) as products. (The carbon in the battery simply serves to conduct electrons.) (B) If the battery terminals are directly connected to each other, the energy released by electron transfer is all converted into heat. (C) If the battery is connected to a pump, much of the energy released by electron transfers can be harnessed to do work instead (in this case, to pump water). Cells can similarly harness the energy of electron transfer to a pumping mechanism, as illustrated in Figure 14–1.

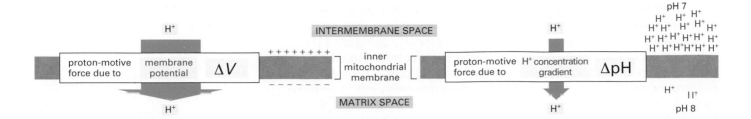

Figure 14–12 The total electrochemical

Figure 14–12 The total electrochemical gradient of H$^+$ across the inner mitochondrial membrane consists of a large force due to the membrane potential (ΔV) and a smaller force due to the H$^+$ concentration gradient (ΔpH). Both forces combine to produce the total *proton-motive force* that drives H$^+$ into the matrix space.

very favorable for H$^+$ to flow back into the mitochondrial matrix. In all the energy-producing membranes that we will discuss in this chapter, the membrane potential adds to the driving force pulling H$^+$ back across the membrane; hence this potential increases the amount of energy stored in the proton gradient (Figure 14–12).

The Proton Gradient Drives ATP Synthesis

As explained previously, the electrochemical proton gradient across the inner mitochondrial membrane is used to drive ATP synthesis in the process of oxidative phosphorylation (Figure 14–13). The device that makes this possible is a large membrane-bound enzyme called **ATP synthase**. This enzyme creates a hydrophilic pathway across the inner mitochondrial membrane that allows protons to flow down their electrochemical gradient. As these ions thread their way through the ATP synthase, they are used to drive the energetically unfavorable reaction between ADP and P$_i$ that makes ATP (see Figure 2–24). The ATP synthase is of ancient origin; the same enzyme occurs in the mitochondria of animal cells, the chloroplasts of plants and algae, and in the plasma membrane of bacteria.

The structure of ATP synthase is shown in Figure 14–14. It is a large, multisubunit protein. A large enzymatic portion, shaped like a lollipop head, projects on the matrix side of the inner mitochondrial membrane

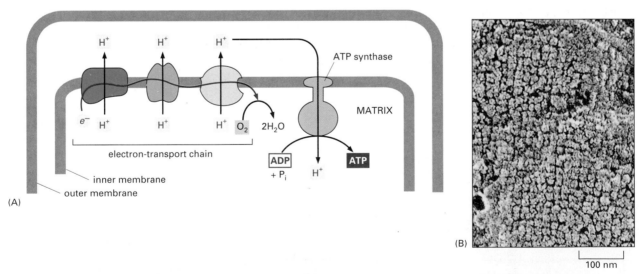

Figure 14–13 The electrochemical gradient across the inner mitochondrial membrane drives ATP synthesis by oxidative phosphorylation. (A) As a high-energy electron is passed along the electron-transport chain, some of the energy released is used to drive the three respiratory enzyme complexes that pump H$^+$ out of the matrix space. The resulting electrochemical proton gradient across the inner membrane drives H$^+$ back through the ATP synthase, a transmembrane protein complex that uses the energy of the H$^+$ flow to synthesize ATP from ADP and P$_i$ in the matrix.
(B) Electron micrograph of the inside surface of the inner mitochondrial membrane in a plant cell. The densely packed particles are the protruding portions of the ATP synthases and the respiratory enzyme complexes.

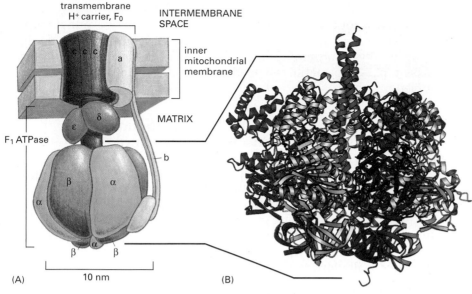

transmembrane
H⁺ carrier, F₀

INTERMEMBRANE
SPACE

inner
mitochondrial
membrane

MATRIX

F₁ ATPase

(A) 10 nm

(B)

Figure 14–14 ATP synthase is embedded in the inner mitochondrial membrane. (A) The enzyme is composed of a head portion, called the F_1 ATPase, and a transmembrane H^+ carrier, called F_0. Both F_1 and F_0 are formed from multiple subunits, as indicated. A rotating stalk turns with a rotor formed by a ring of 10 to 14 c subunits *(red)* in the membrane. The stator *(light green)* is formed from transmembrane a subunits, tied to other subunits that create an elongated arm. This arm fixes the stator to a ring of three α and three β subunits that forms the head of the F_1 ATPase. (B) The three-dimensional structure of the F_1 ATPase, as determined by X-ray crystallography. This part of the ATP synthase derives its name from its ability to carry out the reverse of the ATP synthesis reaction, namely, the hydrolysis of ATP to ADP and P_i, when detached from the transmembrane portion. (B, courtesy of John Walker, from J.P. Abrahams et al., *Nature* 370:621–628, 1994. © Macmillan Magazines Ltd.)

and is attached through a thinner multisubunit "stalk" to a transmembrane proton carrier. As protons pass through a narrow channel within the transmembrane carrier, their movement causes the stalk to spin rapidly within the head, inducing the head to make ATP. The synthase essentially acts as an energy-generating molecular motor, converting the energy of proton flow down a gradient into the mechanical energy of two sets of proteins rubbing against one another—rotating stalk proteins pushing against stationary head proteins. The changes in protein conformation driven by the rotating stalk then convert this mechanical energy into the chemical bond energy needed to generate ATP. This marvelous device is capable of producing more than 100 molecules of ATP per second, and about three protons need to pass through the synthase to make each molecule of ATP.

The ATP synthase is a reversible coupling device. It can either harness the flow of protons down their electrochemical gradient to make ATP (its normal role in mitochondria and the plasma membrane of bacteria growing aerobically) or use the energy of ATP hydrolysis to pump protons across a membrane, like the H^+ pumps described in Chapter 12 (Figure 14–15). Whether the ATP synthase primarily makes or consumes ATP depends on the magnitude of the electrochemical proton gradient across the membrane in which it sits. In many bacteria that can grow

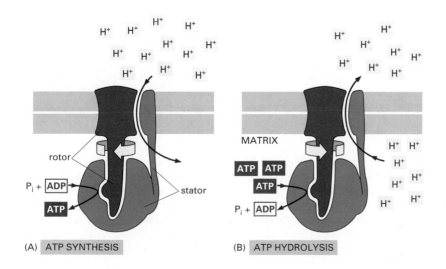

(A) ATP SYNTHESIS

rotor

P_i + ADP

ATP

stator

(B) ATP HYDROLYSIS

MATRIX

ATP ATP

ATP

P_i + ADP

Figure 14–15 ATP synthase is a reversible coupling device that can convert the energy of the electrochemical proton gradient into chemical bond energy or vice versa. The ATP synthase can either synthesize ATP by harnessing the H^+ gradient (A) or pump protons against their electrochemical gradient by hydrolyzing ATP (B). The direction of operation at any given instant depends on the net free-energy change (ΔG, discussed in Chapter 3) for the coupled processes of H^+ translocation across the membrane and the synthesis of ATP from ADP and P_i. Thus, for example, if the electrochemical proton gradient falls below a certain level, the ΔG for the H^+ transport into the matrix space will no longer be large enough to drive ATP production. Instead, ATP will be hydrolyzed by the ATP synthase to rebuild the gradient.

Question 14–4

The remarkable properties that allow ATP synthase to run in either direction allow the interconversion of energy stored in the H⁺ gradient and energy stored in ATP in either direction. (A) If ATP synthase making ATP can be likened to a water-driven turbine producing electricity, what would be an appropriate analogy when it works in the opposite direction? (B) Under what conditions would one expect the ATP synthase to stall, running neither forward nor backward? (C) What determines the direction in which the ATP synthase operates?

either aerobically or anaerobically, the direction in which the ATP synthase works is routinely reversed when the bacterium runs out of O_2. At this point, the ATP synthase uses some of the ATP generated inside the cell by glycolysis to pump protons out of the cell, creating the proton gradient that the bacterial cell needs to import its essential nutrients by coupled transport.

Coupled Transport Across the Inner Mitochondrial Membrane Is Driven by the Electrochemical Proton Gradient

The synthesis of ATP is not the only process driven by the electrochemical proton gradient. In mitochondria, many charged molecules, such as pyruvate, ADP, and P_i, are pumped into the matrix from the cytosol, while others, such as ATP, must be moved in the opposite direction. Carrier proteins that bind these molecules can couple their transport to the energetically favorable flow of H⁺ into the mitochondrial matrix. Thus, for example, pyruvate and inorganic phosphate (P_i) are co-transported inward with H⁺ as the latter moves into the matrix.

In contrast, ADP is co-transported with ATP in opposite directions by a single carrier protein. Because an ATP molecule has one more negative charge than ADP, each nucleotide exchange results in the moving of a total of one negative charge out of the mitochondrion. The ADP–ATP co-transport is therefore driven by the charge difference across the membrane (Figure 14–16).

In eucaryotic cells, the proton gradient is consequently used to drive both the formation of ATP and the transport of certain metabolites across the inner mitochondrial membrane. In bacteria, the proton gradient across the bacterial plasma membrane serves all of these functions, but, in addition, is itself an important source of directly usable energy: in motile bacteria, the gradient drives the rapid rotation of the bacterial flagellum, which propels the bacterium along (Figure 14–17).

Proton Gradients Produce Most of the Cell's ATP

As discussed in Chapter 13, glycolysis alone produces a net yield of two molecules of ATP for every molecule of glucose, which is the total energy yield for the fermentation processes that occur in the absence of

Figure 14–16 The electrochemical proton gradient across the inner mitochondrial membrane is also used to drive some coupled transport processes. Pyruvate, inorganic phosphate (P_i), and ADP are moved into the matrix, while ATP is pumped out. The charge on each of the transported molecules is indicated for comparison with the membrane potential, which is negative inside, as shown. The outer membrane is freely permeable to all of these compounds. The active transport of molecules across membranes by carrier proteins is discussed in Chapter 12.

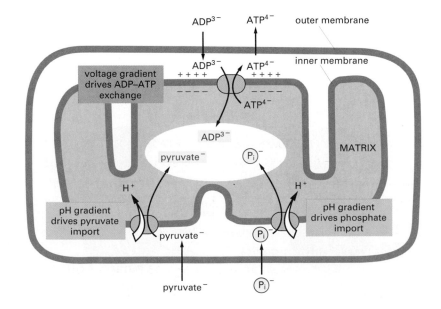

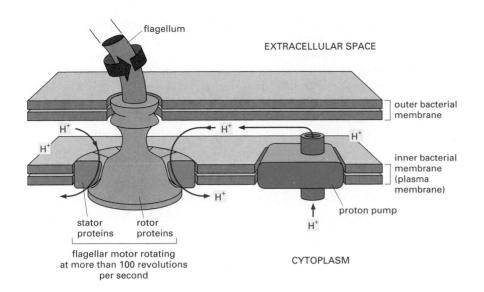

flagellum

EXTRACELLULAR SPACE

outer bacterial membrane

H^+

H^+

H^+

H^+

H^+

inner bacterial membrane (plasma membrane)

proton pump

H^+

stator proteins   rotor proteins

flagellar motor rotating at more than 100 revolutions per second

CYTOPLASM

Figure 14–17 The rotation of the bacterial flagellum is driven by H^+ flow. The flagellum is attached to a series of protein rings (shown in *orange*), which are embedded in the outer and inner (plasma) membranes and rotate with the flagellum (see also Panel 1–2, p. 25). The rotation is driven by a flow of protons through an outer ring of proteins (the *stator*) by mechanisms that may resemble those used by the ATP synthase, although they are not yet understood in detail.

O_2. In contrast, during oxidative phosphorylation each pair of electrons donated by the NADH produced in mitochondria is thought to provide energy for the formation of about 2.5 molecules of ATP, once one includes the energy needed for transporting this ATP to the cytosol. Oxidative phosphorylation also produces 1.5 ATP molecules per electron pair from $FADH_2$, or from the NADH molecules produced by glycolysis in the cytosol. From the product yields of glycolysis and the citric acid cycle summarized in Table 14–1, one can readily calculate that the complete oxidation of one molecule of glucose—starting with glycolysis and ending with oxidative phosphorylation—gives a net yield of about 30 ATPs (see Question 14–5 and its solution). Thus, the vast majority of the ATP produced from the oxidation of glucose in an animal cell is produced by chemiosmotic mechanisms on the mitochondrial membrane. Oxidative phosphorylation in the mitochondrion also produces a large amount of ATP from the NADH and the $FADH_2$ derived from the oxidation of fats (see also Figures 13–9 and 13–10).

Question 14–5

A. Calculate the number of ATP molecules produced per pair of electrons transferred from NADH to oxygen, if (i) five protons are pumped across the inner mitochondrial membrane for each electron passed through the three respiratory enzyme complexes, (ii) three protons must pass through the ATP synthase for each ATP molecule that it produces from ADP and inorganic phosphate inside the mitochondrion, and (iii) one proton is used to produce the voltage gradient needed to transport each ATP molecule out of the mitochondrion to the cytosol where it is used.

B. Use the information in Table 14–1 together with your ATP yield per electron pair from (A) to calculate (i) the number of ATP molecules produced per molecule of glucose as a result of the citric acid cycle alone (assume that the oxidation which results from the transfer of a pair of electrons from each $FADH_2$ molecule results in pumping of six protons, and that GTP and ATP are freely interconvertible according to the reaction GTP + ADP $\rightleftharpoons$ GDP + ATP), and (ii) the total number of ATP molecules produced by the complete oxidation of glucose (assume that electrons from each NADH molecule produced in the cytosol by glycolysis are also fed, indirectly, into the electron-transport chain; however, the yield of ATP is only 1.5 ATP molecules for each of these NADH molecules).

Table 14–1 Summary of Product Yields from the Oxidation of Sugars and Fats

A.   Net products from oxidation of one molecule of glucose

In cytosol (glycolysis)

glucose → 2 pyruvate + 2 NADH + 2 ATP

In mitochondrion (pyruvate dehydrogenase and citric acid cycle)

2 pyruvate → 2 acetyl CoA + 2 NADH

2 acetyl CoA → 6 NADH + 2 $FADH_2$ + 2 GTP

Net result in mitochondrion

2 pyruvate → 8 NADH + 2 $FADH_2$ + 2 GTP

B.   Net products from oxidation of one molecule of palmitoyl CoA (activated form of palmitate, a fatty acid)

In mitochondrion (fatty acid oxidation and citric acid cycle)

palmitoyl CoA → 8 acetyl CoA + 7 NADH + 7 $FADH_2$

8 acetyl CoA → 24 NADH + 8 $FADH_2$ + 8 GTP

Net result in mitochondrion

palmitoyl CoA → 31 NADH + 15 $FADH_2$ + 8 GTP

The Rapid Conversion of ADP to ATP in Mitochondria Maintains a High ATP/ADP Ratio in Cells

Using the transport proteins in the inner mitochondrial membrane discussed earlier, the ADP molecules produced by ATP hydrolysis in the cytosol rapidly enter mitochondria for recharging, while the bulk of the ATP molecules formed in the mitochondrial matrix by oxidative phosphorylation are rapidly pumped into the cytosol, where they are needed. Some of this ATP is used by the mitochondrion itself to power its replication, protein synthesis, and other energy-consuming reactions. All in all, a typical ATP molecule in the human body shuttles out of a mitochondrion and back into it (as ADP) for recharging more than once per minute, keeping the concentration of ATP in the cell about 10 times higher than that of ADP.

As discussed in Chapter 3, biosynthetic enzymes often drive energetically unfavorable reactions by coupling them to the energetically favorable hydrolysis of ATP (see Figure 3–34). The ATP pool is used to drive cellular processes in much the same way that a battery can be used to drive electric engines: if the activity of the mitochondria were halted, ATP levels would fall and the cell's battery would run down; eventually, energetically unfavorable reactions would no longer be driven and the cell would die. The poison cyanide, which blocks electron transport in the inner mitochondrial membrane, causes death in exactly this way.

Electron-Transport Chains and Proton Pumping

Having considered in general terms how a mitochondrion functions, we need to examine in more detail the mechanisms that underlie its membrane-based energy-conversion processes. In doing so, we will also be accomplishing a larger purpose. As emphasized at the beginning of this chapter, very similar energy-conversion devices are used by mitochondria, chloroplasts, and bacteria, and the basic principles that we shall discuss next therefore underlie the function of nearly all living things. To emphasize this central point, we shall end this chapter by describing how electron transport provides energy for one of the many types of bacteria that flourish in huge numbers, in total darkness and without oxygen, miles below the surface of the ocean.

For many years, the reason that electron-transport chains were embedded in membranes eluded the biochemists who were struggling to understand them. The puzzle was solved as soon as the fundamental role of transmembrane proton gradients in energy generation was proposed in the early 1960s (see How We Know, pp. 460–461). But the process of chemiosmotic coupling entails an interplay between chemical and electrical forces that is not easy to decipher at a molecular level; indeed, the idea was so novel that it was not widely accepted until many years later, after additional supporting evidence had accumulated from experiments designed as tests of the hypothesis.

Although investigators today are still unraveling many of the details of chemiosmotic coupling, the fundamentals are now clear. In this part of the chapter we shall look at some of the principles that underlie the electron-transport process and explain how it can generate a proton gradient.

Protons Are Readily Moved by the Transfer of Electrons

Although protons resemble other positive ions such as Na^+ and K^+ in their movement across membranes, in some respects they are unique.

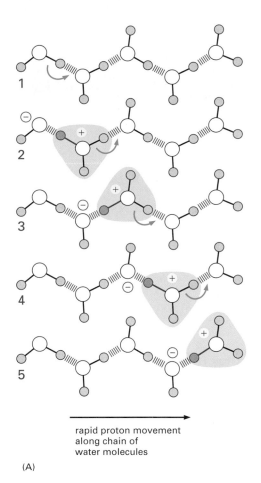

Figure 14–18 The protons in water are highly mobile. (A) Protons move very rapidly along a chain of hydrogen-bonded water molecules. In this diagram, proton jumps are indicated by arrows, and hydronium ions are indicated by *green shading*. As discussed in Chapter 2, naked protons rarely exist as such, and are instead associated with a water molecule in the form of a hydronium ion, H_3O^+. At a neutral pH (pH 7.0), the hydronium ions are present at a concentration of 10^{-7} M; however, for simplicity one usually refers to this as a H^+ (proton) concentration of 10^{-7} M (see Panel 2–2). (B) Electron transfer can cause the transfer of entire hydrogen atoms, because protons are readily accepted from or donated to water inside cells. In this example, A picks up an electron plus a proton when it is reduced, and B loses an electron plus a proton when it is oxidized.

rapid proton movement along chain of water molecules

(A)

Hydrogen atoms are by far the most abundant type of atom in living organisms and are plentiful not only in all carbon-containing biological molecules, but also in the water molecules that surround them. The protons in water are highly mobile, flickering through the hydrogen-bonded network of water molecules by rapidly dissociating from one water molecule in order to associate with its neighbor, as illustrated in Figure 14–18A. Thus, water, which is everywhere in cells, serves as a ready reservoir for donating and accepting protons.

Whenever a molecule is reduced by acquiring an electron, the electron (e^-) brings with it a negative charge. In many cases, this charge is rapidly neutralized by the addition of a proton (H^+) from water, so that the net effect of the reduction is to transfer an entire hydrogen atom, H^+ + e^- (Figure 14–18B). Similarly, when a molecule is oxidized, the hydrogen atom can be readily dissociated into its constituent electron and proton, allowing the electron to be transferred separately to a molecule that accepts electrons, while the proton is passed to the water. Therefore, in a membrane in which electrons are being passed along an electron-transport chain, it is a relatively simple matter, in principle, to pump protons from one side of the membrane to another. All that is required is that the electron carrier be arranged in the membrane in a way that causes it to pick up a proton from one side of the membrane when it accepts an electron, while releasing the proton on the other side of the membrane as the electron is passed on to the next carrier molecule in the chain (Figure 14–19).

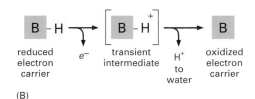

(B)

The Redox Potential Is a Measure of Electron Affinities

In biochemical reactions, any electrons removed from one molecule are always passed to another, so that whenever one molecule is oxidized, another is reduced. Like any other chemical reaction, the tendency of such oxidation–reduction reactions, or **redox reactions**, to proceed spontaneously depends on the free-energy change (ΔG) for the electron transfer, which in turn depends on the relative affinities of the two molecules for electrons. (The role of free energy in chemical reactions is discussed in Chapter 3, pp. 93–94.)

Because electron transfers provide most of the energy for living things, it is worth spending a little time to understand them. Many readers are already familiar with acids and bases, which donate and accept protons (see Panel 2–2, pp. 68–69). Acids and bases exist in conjugate acid–base pairs, where the acid is readily converted into the base by the loss of a proton. For example, acetic acid (CH_3COOH) is converted into its conjugate base (CH_3COO^-) in the reaction

$$CH_3COOH \rightleftharpoons CH_3COO^- + H^+$$

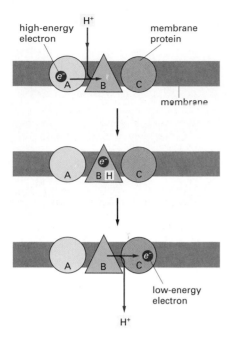

Figure 14–19 Protons can be pumped across membranes by the transfer of electrons. As an electron passes along an electron-transport chain, it can bind and release a proton at each step. In this schematic diagram, electron carrier B picks up a proton (H^+) from one side of the membrane when it accepts an electron (e^-) from carrier A; it releases the proton to the other side of the membrane when it donates its electron to carrier C. As we shall see, other mechanisms can also be employed for this proton pumping.

In exactly the same way, pairs of compounds such as NADH and NAD^+ are called **redox pairs**, because NADH is converted to NAD^+ by the loss of electrons in the reaction

$$NADH \rightleftharpoons NAD^+ + H^+ + 2e^-$$

NADH is a strong electron donor: because its electrons are held in a high-energy linkage, the free-energy change for passing its electrons to many other molecules is favorable (see Figure 14–4). Conversely, it is difficult to form the high-energy linkage in NADH, so its partner, NAD^+, is of necessity a weak electron acceptor.

The tendency to transfer electrons from any redox pair can be measured experimentally. All that is required is to form an electrical circuit that links a 1:1 (equimolar) mixture of the redox pair to a second redox pair that has been arbitrarily selected as a reference standard, so that the voltage difference between them can be measured (Panel 14–1, p. 471). This voltage difference is defined as the **redox potential**; as defined, electrons will move spontaneously from a redox pair like $NADH/NAD^+$ with a low redox potential (a low affinity for electrons) to a redox pair like O_2/H_2O with a high redox potential (a high affinity for electrons). Thus NADH is a good molecule to donate electrons to the respiratory chain, while O_2 is well suited to act as the "sink" for electrons at the end of the pathway. As explained in Panel 14–1, the difference in redox potential, $\Delta E_0'$, is a direct measure of the standard free-energy change ($\Delta G°$) for the transfer of an electron from one molecule to another.

Electron Transfers Release Large Amounts of Energy

As just discussed, those pairs of compounds that have the most negative redox potentials have the weakest affinity for electrons and therefore the strongest tendency to donate electrons. Conversely, those pairs that have the most positive redox potentials have the strongest affinity for electrons and therefore the strongest tendency to accept electrons. A 1:1 mixture of NADH and NAD^+ has a redox potential of –320 mV, indicating that NADH has a strong tendency to donate electrons; a 1:1 mixture of H_2O and $\frac{1}{2}O_2$ has a redox potential of +820 mV, indicating that O_2 has a strong tendency to accept electrons. The difference in redox potential is 1.14 volts (1140 mV), which means that the transfer of each electron from NADH to O_2 under these standard conditions is enormously favorable, where $\Delta G° = -26.2$ kcal/mole (–52.4 kcal/mole for the two electrons transferred per NADH molecule; see Panel 14–1). If we compare this free-energy change with that for the formation of the phosphoanhydride bonds in ATP ($\Delta G° = +7.3$ kcal/mole; see Figure 13–7), we see that more than enough energy is released by the oxidization of one NADH molecule to synthesize several molecules of ATP from ADP and P_i.

Living systems could certainly have evolved enzymes that would allow NADH to donate electrons directly to O_2 to make water in the reaction

$$2H^+ + 2e^- + \frac{1}{2}O_2 \rightarrow H_2O$$

But because of the huge free-energy drop, this reaction would proceed with almost explosive force and nearly all of the energy would be released as heat. Cells do perform this reaction, but they make it proceed much more gradually by passing the high-energy electrons from NADH to O_2 via the many electron carriers in the electron-transport chain. Because each successive carrier in the chain holds its electrons more tightly than the last, the energetically favorable reaction $2H^+ + 2e^- + \frac{1}{2}O_2 \rightarrow H_2O$ is made to occur in many small steps. This enables nearly half of the released energy to be stored, instead of being lost to the environment as heat.

Panel 14–1 Redox potentials

HOW REDOX POTENTIALS ARE MEASURED

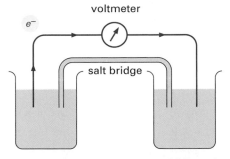

voltmeter

e^-

salt bridge

$A_{reduced}$ and $A_{oxidized}$ in equimolar amounts

1 M H+ and 1 atmosphere H_2 gas

One beaker *(left)* contains substance A with an equimolar mixture of the reduced ($A_{reduced}$) and oxidized ($A_{oxidized}$) members of its redox pair. The other beaker contains the hydrogen reference standard ($2H^+ + 2e^- \rightleftharpoons H_2$), whose redox potential is arbitrarily assigned as zero by international agreement. (A salt bridge formed from a concentrated KCl solution allows K^+ and Cl^- to move between the beakers and neutralize the charges when electrons flow between the beakers.) The metal wire *(red)* provides a resistance-free path for electrons, and a voltmeter then measures the redox potential of substance A. If electrons flow from $A_{reduced}$ to H^+, as indicated here, the redox pair formed by substance A is said to have a negative redox potential. If they instead flow from H_2 to $A_{oxidized}$, this redox pair is said to have a postive redox potential.

By convention, the redox potential for a redox pair is designated as *E*. Since biological reactions occur at pH 7, biologists define the standard state as $A_{reduced}$ = $A_{oxidized}$ and $H^+ = 10^{-7}$ M and use it to determine the standard redox potential E_0'.

examples of redox potentials	redox potential E_0'
$NADH \rightleftharpoons NAD^+ + H^+ + 2e^-$	–320 mV
reduced ubiquinone $\rightleftharpoons$ oxidized ubiquinone $+ 2H^+ + 2e^-$	+30 mV
reduced cytochrome c $\rightleftharpoons$ oxidized cytochrome c $+ e^-$	+230 mV
$H_2O \rightleftharpoons \tfrac{1}{2}O_2 + 2H^+ + 2e^-$	+820 mV

CALCULATION OF $\Delta G°$ FROM REDOX POTENTIALS

$\Delta E_0' = E_0'(acceptor) - E_0'(donor) = +350 mV$

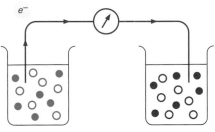

e^-

1:1 mixture of NADH and NAD+

1:1 mixture of oxidized and reduced ubiquinone

$\Delta G° = -n(0.023)\Delta E_0'$ where *n* is the number of electrons transferred across a redox potential change of $\Delta E_0'$ millivolts (mV)

Example: The transfer of one electron from NADH to ubiquinone has a favorable $\Delta G°$ of –8.0 kcal/mole, calculated as follows:

$\Delta G° = -n(0.023)\Delta E_0' = -1(0.023)(350) = -8.0$ kcal/mole

The same calculation reveals that the transfer of one electron from ubiquinone to oxygen has an even more favorable $\Delta G°$ of –18.2 kcal/mole. The $\Delta G°$ value for the transfer of one electron from NADH to oxygen is the sum of these two values, –26.2 kcal/mole.

EFFECT OF CONCENTRATION CHANGES

As explained in Chapter 3 (see pp. 94–95), the actual free-energy change for a reaction, ΔG, depends on the concentration of the reactants and generally will be different from the standard free-energy change, $\Delta G°$. The standard redox potentials are for a 1:1 mixture of the redox pair. For example, the standard redox potential of –320 mV is for a 1:1 mixture of NADH and NAD^+. But when there is an excess of NADH over NAD^+, electron transfer from NADH to an electron acceptor becomes more favorable. This is reflected by a more negative redox potential and a more negative ΔG for electron transfer.

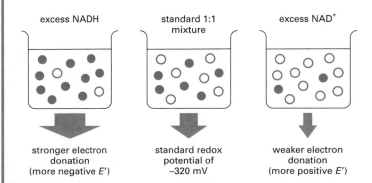

excess NADH

standard 1:1 mixture

excess NAD+

stronger electron donation (more negative *E'*)

standard redox potential of –320 mV

weaker electron donation (more positive *E'*)

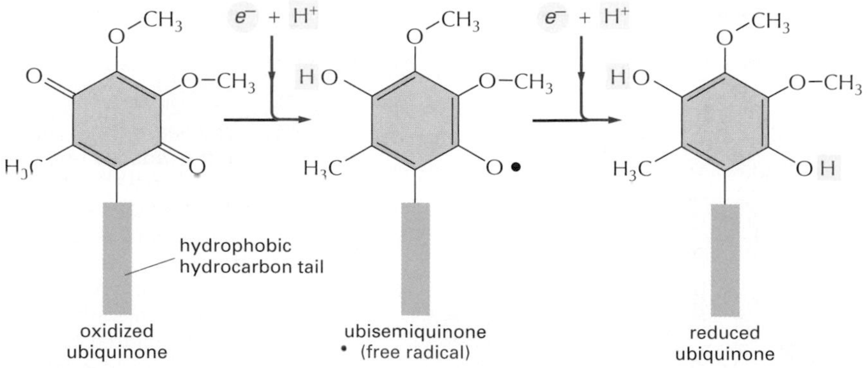

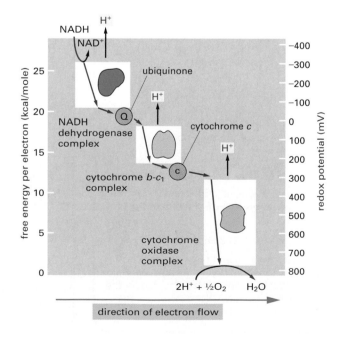

Figure 14–20 Quinones carry electrons within the lipid bilayer. Ubiquinone in the mitochondrial electron-transport chain picks up one H⁺ from the aqueous environment for every electron it accepts, and it can carry either one or two electrons as part of a hydrogen atom *(yellow)*. When reduced ubiquinone donates its electrons to the next carrier in the chain, the protons are released. The long hydrophobic tail confines ubiquinone to the membrane and consists of 6–10 five-carbon isoprene units, the number depending on the organism. The corresponding quinone electron carrier in the photosynthetic membranes of chloroplasts is plastoquinone, which is almost identical in structure. For simplicity, both ubiquinone and plastoquinone will be referred to as *quinone* and abbreviated as Q.

Metals Tightly Bound to Proteins Form Versatile Electron Carriers

In the respiratory chain, the electron carriers are arranged in order of increasing redox potential, thus making possible the gradual release of the energy stored in NADH electrons. Within each of the three respiratory enzyme complexes, the electrons move mainly between metal atoms that are tightly bound to the proteins, travelling by skipping from one metal ion to the next.

In contrast, electrons are carried between the different respiratory complexes by molecules that diffuse along the lipid bilayer, picking up electrons from one complex and delivering them to another in an orderly sequence. *Ubiquinone*, a small hydrophobic molecule that dissolves in the lipid bilayer, is the only carrier that is not part of a protein. In the mitochondrial respiratory chain, ubiquinone picks up electrons from the NADH dehydrogenase complex and delivers them to the cytochrome b-c_1 complex (see Figure 14–10). As shown in Figure 14–20, a **quinone** (like ubiquinone) can pick up or donate either one or two electrons, and it picks up one H⁺ from the surroundings with each electron that it carries. Its redox potential of +30 mV places ubiquinone about one-quarter of the way down the chain from NADH in terms of energy loss (Figure 14–21).

All the rest of the electron carriers in the electron-transport chain are small molecules that are tightly bound to proteins. To get from

Figure 14–21 Redox potential increases along the mitochondrial electron-transport chain. For the symbols used here see Figure 14–10. Note that the big increases in redox potential occur across each of the three respiratory enzyme complexes, as required for each of them to pump protons.

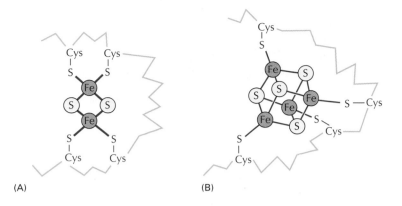

(A) (B)

NADH to ubiquinone, the electrons are passed inside the NADH dehydrogenase complex between a flavin group (see Figure 13–12 for structure) bound to the protein complex and a set of **iron–sulfur centers** of increasing redox potentials. Iron–sulfur centers have structures resembling those shown in Figure 14–22, and they carry one electron at a time. The final iron–sulfur center in the dehydrogenase donates its electrons to ubiquinone.

Iron–sulfur centers have relatively low affinities for electrons and thus would be less useful in the later part of the electron-transport chain, in the pathway from ubiquinone to O_2. In this part of the pathway, iron atoms in heme groups that are tightly bound to cytochrome proteins are commonly used as electron carriers, as in the cytochrome b-c_1 and cytochrome oxidase complexes. The **cytochromes** constitute a family of colored proteins (hence their name, from the Greek *chroma*, "color"); each contains one or more heme groups whose iron atom changes from the ferric (Fe^{3+}) to the ferrous (Fe^{2+}) state whenever it accepts an electron. As one would expect, the various cytochromes increase in redox potential as one progresses down the mitochondrial electron-transport chain toward oxygen. The structure of *cytochrome c,* the small protein that shuttles electrons between the cytochrome b-c_1 complex and the cytochrome oxidase complex, is shown in Figure 14–23: its redox potential is +230 mV.

Question 14–6

At many steps in the electron-transport chain Fe ions are used as part of heme or FeS clusters to bind the electrons in transit. Why do these functional groups that carry out the chemistry of electron transfer need to be bound to proteins? Provide several different reasons why this is necessary.

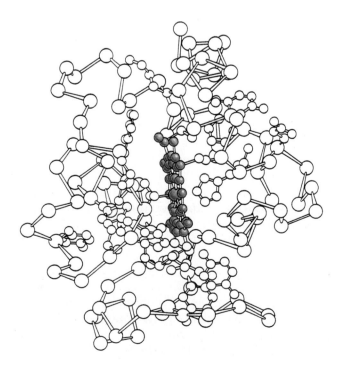

Figure 14–23 Cytochrome *c* is an electron carrier in the electron-transport chain. This small protein contains just over 100 amino acids and is held loosely on the outer face of the inner membrane by ionic interactions (see Figure 14–10). The iron atom *(orange)* in the bound heme *(blue)* can carry a single electron. The structure of the heme group in hemoglobin, which reversibly binds O_2 rather than an electron, was shown in Figure 4–36.

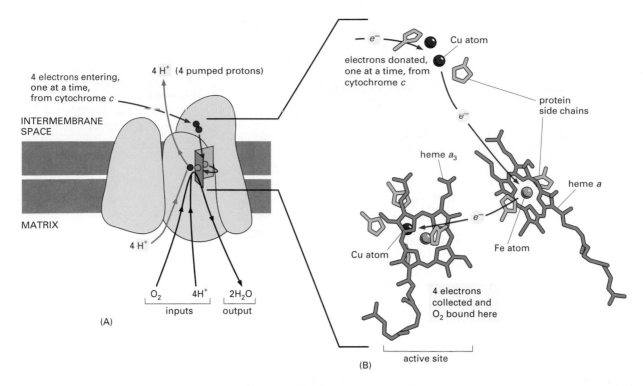

Four electrons entering,
one at a time,
from cytochrome c

4 H⁺ (4 pumped protons)

Cu atom

electrons donated,
one at a time, from
cytochrome c

INTERMEMBRANE
SPACE

protein
side chains

heme a_3

heme a

MATRIX

Cu atom

Fe atom

4 H⁺

O_2 4H⁺ $2H_2O$

inputs   output

4 electrons
collected and
O_2 bound here

(A)

(B)

active site

Figure 14–24 Cytochrome oxidase, the final enzyme complex in the electron transport chain, consumes nearly all the oxygen we breathe. (A) Spatial orientation of the reactions catalyzed by cytochrome oxidase. Four electrons from cytochrome c and four protons from the aqueous environment are added to each O_2 molecule to produce two molecules of H_2O. In addition, four protons are pumped across the membrane. The shapes and arrangement of the three subunits shown are based on the complete structure of cytochrome oxidase determined by X-ray diffraction. The inside of the membrane is the matrix space in mitochondria, and the cytoplasm in aerobic bacteria. (B) Details of the active site in (A), showing the location of the metal ions within cytochrome oxidase that serve as temporary resting places for the electrons as they hop from cytochrome c—through the cytochrome oxidase protein—and finally to oxygen (O_2). As in (A), the two iron (Fe) atoms are *orange,* the three copper (Cu) atoms are *red*, and the two hemes that carry the iron atoms are *blue*. All of these metal ions are buried within the protein, and the electrons jump between them by a phenomenon called electron tunneling.

At the very end of the respiratory chain, just before oxygen, the electron carriers are those in the cytochrome oxidase complex. The carriers here are either iron atoms in heme groups or copper atoms that are tightly bound to the complex in specific ways that give them a high redox potential. Throughout the chain, the redox potential of each of the electron carriers has been fine-tuned to facilitate efficient electron transfer by binding the carrier atom or molecule to a particular protein, which adjusts its normal affinity for electrons to fit its position in the respiratory chain.

Cytochrome Oxidase Catalyzes Oxygen Reduction

Cytochrome oxidase receives electrons from cytochrome c, thus oxidizing it (hence its name), and donates these electrons to oxygen. A schematic view of the reaction catalyzed by cytochrome oxidase is presented in Figure 14–24A. In brief, four electrons from cytochrome c and four protons from the aqueous environment are added to each O_2 molecule in the reaction $4e^- + 4H^+ + O_2 \rightarrow 2H_2O$. In addition, four more protons are pumped across the membrane during electron transfer, building up the electrochemical proton gradient. Proton pumping is caused by allosteric changes in the conformation of the protein, which are driven by energy derived from electron transport.

At its active site, cytochrome oxidase contains a complex of a heme iron atom juxtaposed with a tightly bound copper atom (Figure 14–24B). It is here that nearly all of the oxygen we breathe is used, serving as the final repository for the electrons that NADH donated at the start of the electron-transport chain. Oxygen is useful for this purpose because of its very high affinity for electrons. Once O_2 picks up one electron, however, it forms the superoxide radical O_2^-; this radical is dangerously reactive and will avidly take up another three electrons wherever it can find them, a tendency that can cause serious damage to nearby DNA, proteins, and lipid membranes. One of the roles of cytochrome oxidase is to hold on tightly to its oxygen molecule until all four of the electrons needed to convert it to two H_2O molecules are in hand, thereby preventing a random attack on cellular macromolecules by superoxide radicals—damage that is postulated to be a major cause of human aging. The invention of cytochrome oxidase was therefore crucial to the evolution of cells that are able to use O_2 as an electron acceptor.

The cytochrome oxidase reaction is estimated to account for 90% of the total oxygen uptake in most cells. This protein complex is therefore crucial for all aerobic life. Cyanide and azide are extremely toxic because they bind tightly to the cell's cytochrome oxidase complexes to stop electron transport, thereby greatly reducing ATP production.

For a protein to pump protons actively across a membrane, the conformational change in the protein pump must be coupled to an energetically favorable reaction. For cytochrome oxidase, the energetically favorable reaction is the addition of electrons to O_2 at the enzyme's active site. Although the cytochrome oxidase in mammals contains 13 different protein subunits, most of these seem to have a subsidiary role, helping to regulate either the activity or the assembly of the three subunits that form the core of the enzyme. The complete structure of this large enzyme complex has recently been determined by X-ray crystallography, as illustrated in Figure 14–25. Such atomic-resolution structures have helped to reveal the detailed mechanisms of this finely tuned protein machine.

The Mechanism of H⁺ Pumping Will Soon Be Understood in Atomic Detail

Some respiratory enzyme complexes pump one H^+ per electron across the inner mitochondrial membrane, whereas others pump two. The detailed mechanism by which electron transport is coupled to H^+ pumping is different for the three different respiratory enzyme complexes. In the cytochrome b-c_1 complex, the quinones clearly have a role. As mentioned previously, a quinone picks up a H^+ from the aqueous medium along with each electron it carries and liberates it when it

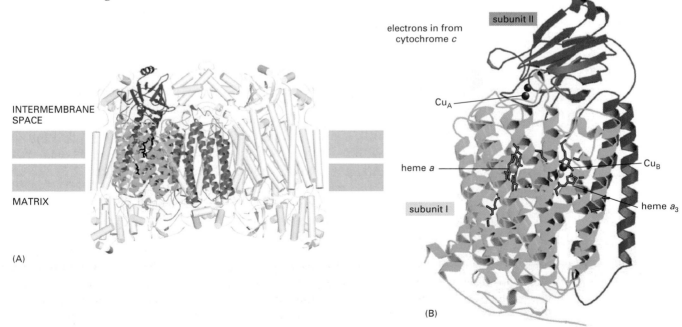

Figure 14–25 Cytochrome oxidase is a finely tuned protein machine. The protein is a dimer formed from a monomer with 13 different protein subunits. The three colored subunits that form the functional core of the complex are encoded by the mitochondrial genome. As electrons pass through this protein on the way to its bound O_2 molecule, they cause the protein to pump protons across the membrane. (A) The entire protein is shown, positioned in the inner mitochondrial membrane. (B) The electron carriers are located in subunits I and II, as indicated.

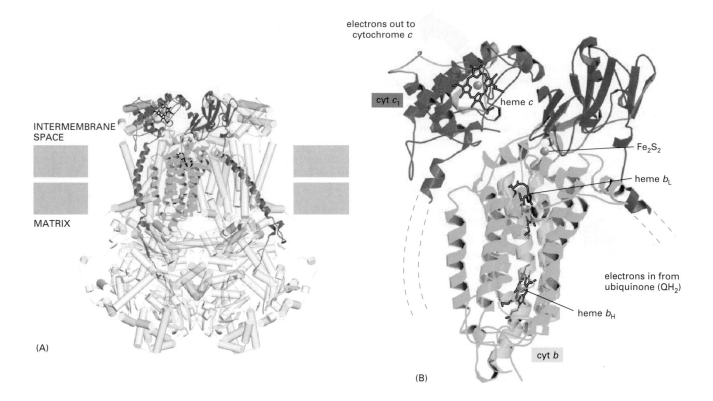

electrons out to
cytochrome c

INTERMEMBRANE
SPACE

MATRIX

(A)

cyt c_1

heme c

Fe$_2$S$_2$

heme b_L

electrons in from
ubiquinone (QH$_2$)

heme b_H

cyt b

(B)

Figure 14–26 The atomic structure of cytochrome b-c_1 shows where electrons enter and exit the complex. This protein is a dimer. In mammals, the monomer is composed of 11 different protein subunits. The three colored proteins form the functional core of the enzyme: cytochrome b *(green)*, cytochrome c_1 *(blue)*, and the protein that contains an iron–sulfur center *(purple)*. (A) The interaction of these proteins across the two monomers. (B) Their electron carriers, along with the entrance and exit sites for electrons.

releases the electron (see Figure 14–20). Because ubiquinone is freely mobile in the lipid bilayer, it can accept electrons near the inside surface of the membrane and donate them to the cytochrome b-c_1 complex near the outside surface. Thus ubiquinone transfers one H$^+$ across the bilayer for every electron it transports. However, two protons are pumped per electron in the cytochrome b-c_1 complex, and there is good evidence for a so-called *Q-cycle*, in which ubiquinone is recycled through the complex in an ordered way that makes this two-for-one transfer possible. Exactly how this occurs is being worked out at the atomic level, as the complete structure of the cytochrome b-c_1 complex has been determined by X-ray crystallography (Figure 14–26).

Allosteric changes in protein conformations driven by electron transport can also pump H$^+$, just as H$^+$ is pumped when ATP is hydrolyzed by the ATP synthase running in reverse (see Figure 14–15). For both the NADH dehydrogenase complex and the cytochrome oxidase complex, it seems likely that electron transport drives sequential allosteric changes in protein conformation that casue a portion of the protein to pump H$^+$ across the mitochondrial inner membrane. A general mechanism for this type of H$^+$ pumping is presented in Figure 14–27.

Question 14–7

Two different diffusible electron carriers, ubiquinone and cytochrome c, shuttle electrons between the three protein complexes of the electron-transport chain. Could, in principle, the same diffusible carrier be used for both steps? Explain your answer.

Respiration Is Amazingly Efficient

The free-energy changes for burning fats and carbohydrates directly into CO$_2$ and H$_2$O can be compared with the total amount of energy generated and stored in the phosphate bonds of ATP during the corresponding biological oxidations. When this is done, it is seen that the efficiency with which oxidation energy is converted into ATP bond energy is often greater than 40%. This is considerably better than the efficiency of most nonbiological energy-conversion devices. If cells worked only with the efficiency of an electric motor or a gasoline engine (10–20%), an organism would have to eat voraciously in order to maintain itself. Moreover, because wasted energy is liberated as heat, large organisms (including ourselves) would need more efficient mechanisms than they presently have for giving up heat to the environment.

Students sometimes wonder why the chemical interconversions in cells follow such complex pathways. The oxidation of sugars to CO_2 plus H_2O could certainly be accomplished more directly, eliminating the citric acid cycle and many of the steps in the respiratory chain. This would make respiration easier for students to learn, but it would be a disaster for the cell. Oxidation produces huge amounts of free energy, which can be utilized efficiently only in small bits. Biological oxidative pathways involve many intermediates, each differing only slightly from its predecessor. The energy released is thereby parceled out into small packets that can be efficiently converted to high-energy bonds in useful molecules, such as ATP and NADH, by means of coupled reactions (see Figure 13–1).

Having seen how chemiosmotic coupling is used to generate ATP in mitochondria, we now look at how it harnesses light energy for the generation of ATP in chloroplasts.

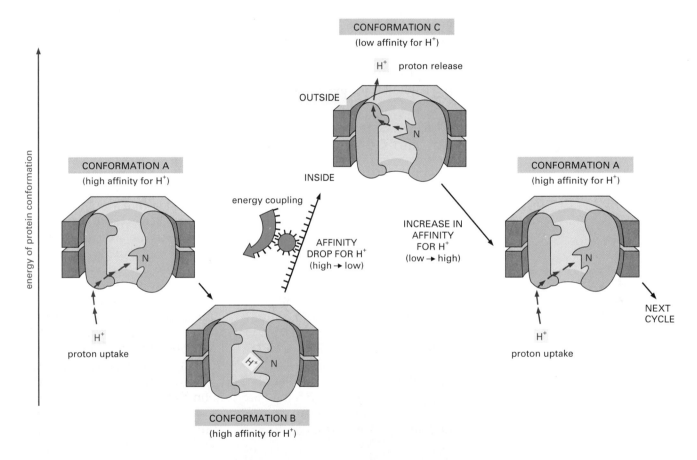

Figure 14–27 H$^+$ pumping can be caused by a conformational change in a protein pump, driven by an energetically favorable reaction. This model for H$^+$ pumping by a transmembrane protein is based on mechanisms that are thought to be used by NADH dehydrogenase and cytochrome oxidase, as well as by the light-driven procaryotic proton pump, bacteriorhodopsin. The protein is driven through a cycle of three conformations: A, B, and C. As indicated by their vertical spacing, these protein conformations have different energies. In conformation A, the protein has a high affinity for H$^+$, causing it to pick up a H$^+$ on the inside of the membrane. In conformation C, the protein has a low affinity for H$^+$, causing it to release a H$^+$ on the outside of the membrane. The transition from conformation B to conformation C that releases the H$^+$ is energetically unfavorable, and it occurs only because it is driven by being allosterically coupled to an energetically favorable reaction occurring elsewhere on the protein *(blue arrow)*. The other two conformational changes, A → B and C → A, lead to states of lower energy, and they proceed spontaneously. Because the overall cycle A → B → C → A → B → C releases free energy, H$^+$ is pumped from the inside (the matrix in mitochondria) to the outside (the intermembrane space in mitochondria). For cytochrome oxidase, the energy required for the transition B → C is provided by electron transport, whereas for bacteriorhodopsin this energy is provided by light (see Figure 11–28). For yet other proton pumps, the energy is derived from ATP hydrolysis.

Chloroplasts and Photosynthesis

Virtually all of the organic materials required by present-day living cells are produced by **photosynthesis**—the series of light-driven reactions that creates organic molecules from atmospheric carbon dioxide (CO_2). Plants, algae, and the most advanced photosynthetic bacteria, such as the cyanobacteria, use electrons from water and the energy of sunlight to convert atmospheric CO_2 into organic compounds. In the course of splitting water they liberate into the atmosphere vast quantities of oxygen gas. This oxygen is in turn required for cellular respiration—not only in animals but also in plants and many bacteria. Thus the activity of early photosynthetic bacteria, which filled the atmosphere with oxygen, enabled the evolution of life forms that use aerobic metabolism to make their ATP (Figure 14–28).

In plants, photosynthesis is carried out in a specialized intracellular organelle—the **chloroplast**. Chloroplasts perform photosynthesis during the daylight hours and thereby produce ATP and NADPH, which in turn are used to convert CO_2 into sugars inside the chloroplast. The sugars produced are exported to the surrounding cytosol, where they are used as fuel to make ATP and as starting materials for many of the other organic molecules that the plant cell needs. Sugar is also exported to all those cells in the plant that lack chloroplasts. As is the case for animal cells, most of the ATP present in the cytosol of plant cells is made by the oxidation of sugars and fats in mitochondria.

Chloroplasts Resemble Mitochondria but Have an Extra Compartment

Chloroplasts carry out their energy interconversions by means of proton gradients in much the same way that mitochondria do. Although larger (Figure 14–29A), they are organized on the same principles. They have a highly permeable outer membrane, a much less permeable inner membrane—in which membrane transport proteins are embedded—

Figure 14–28 Microorganisms that carry out oxygen-producing photosynthesis changed the Earth's atmosphere.
(A) Living stromatolites from a lagoon in western Australia. These structures are produced by large colonies of oxygen-producing photosynthetic cyanobacteria, which lay down successive layers of material, and are formed in specialized environments. (B) Cross section of a modern stromatolite, showing its layered structure. (C) Cross section through a fossil stromatolite in a rock 3.5 billion years old. Note the layered structure similar to that in (B). Fossil stromatolites are thought to have been formed by photosynthetic bacteria very similar to modern cyanobacteria. The activities of bacteria like these, which liberate O_2 gas as a waste product of photosynthesis, would have slowly changed the Earth's atmosphere. (A, courtesy of Sally Birch, © Oxford Scientific Films; B and C, courtesy of S.M. Awramik, University of California/Biological Photo Service.)

(A)

(B)

(C)

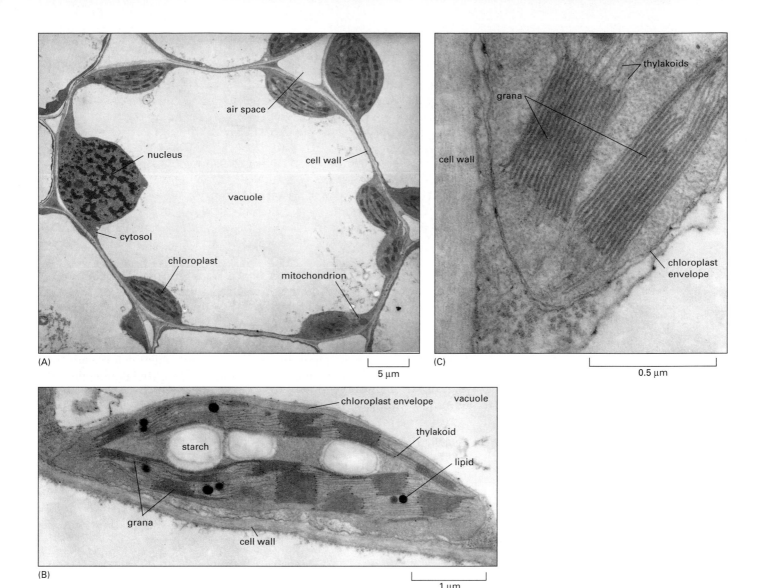

(A)

5 µm

(B)

1 µm

(C)

0.5 µm

Figure 14–29 Photosynthesis takes place in chloroplasts. Electron micrographs show structures of chloroplasts. (A) A wheat leaf cell in which a thin rim of cytoplasm containing nucleus, chloroplasts, and mitochondria surrounds a large vacuole. (B) A thin section of a single chloroplast, showing the chloroplast envelope, starch granules, and lipid (fat) droplets that have accumulated in the stroma as a result of the biosyntheses occurring there. (C) A high-magnification view of two *grana*, which is the name given to a stack of thylakoids. (Courtesy of K. Plaskitt.)

Question 14–8

Chloroplasts have a third internal compartment, the thylakoid space, bounded by the thylakoid membrane. This membrane contains the photosystems, reaction centers, electron-transport chain, and ATP synthase. In contrast, mitochondria use their inner membrane for electron transport and ATP synthesis. In both organelles, protons are pumped out of the largest internal compartment (the matrix in mitochondria and the stroma in chloroplasts). The thylakoid space is completely sealed off from the rest of the cell. Why does this arrangement allow a larger H+ gradient in chloroplasts than can be achieved for mitochondria?

and a narrow intermembrane space in between. Together these membranes form the chloroplast envelope (Figure 14–29B). The inner membrane surrounds a large space called the **stroma**, which is analogous to the mitochondrial matrix and contains many metabolic enzymes. Like the mitochondrion, the chloroplast evolved from an engulfed bacterium, and it still contains its own genome and genetic system. The stroma, therefore, like the mitochondrial matrix, also contains a special set of ribosomes, RNA, and DNA.

There is, however, an important difference between the organization of mitochondria and that of chloroplasts. The inner membrane of the chloroplast does not contain the electron-transport chains. Instead, the light-capturing systems, the electron-transport chains, and ATP synthase are all contained in the *thylakoid membrane*, a third membrane

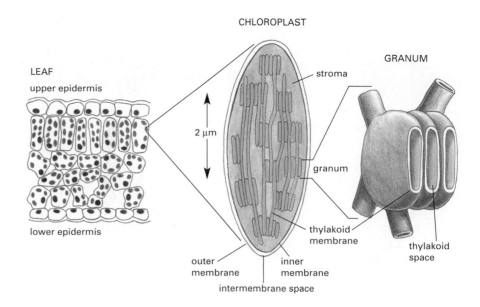

Figure 14–30 A chloroplast contains a third internal compartment. This photosynthetic organelle contains three distinct membranes (the outer membrane, the inner membrane, and the thylakoid membrane) that define three separate internal compartments (the intermembrane space, the stroma, and the thylakoid space). The thylakoid membrane contains all of the energy-generating systems of the chloroplast, including its chlorophyll. In electron micrographs this membrane appears to be broken up into separate units that enclose individual flattened vesicles (see Figure 14–29C), but these are probably joined into a single, highly folded membrane in each chloroplast. As indicated, the individual thylakoids are interconnected, and they tend to stack to form grana.

that forms a set of flattened disclike sacs, the *thylakoids* (Figure 14–29C). These are arranged in stacks, and the space inside each thylakoid is thought to be connected with that of other thylakoids, thereby defining a continuous third internal compartment that is separated from the stroma by the thylakoid membrane (Figure 14–30). The structural similarities and differences between mitochondria and chloroplasts are illustrated in Figure 14–31.

Chloroplasts Capture Energy from Sunlight and Use It to Fix Carbon

The many reactions that occur during photosynthesis in plants can be grouped into two broad categories (Figure 14–32):

1. In the *photosynthetic electron-transfer reactions* (also called the "light reactions"), energy derived from sunlight energizes an electron in the green organic pigment *chlorophyll,* enabling the electron to move along an electron-transport chain in the thylakoid membrane in much the same way that an electron moves along the respiratory chain in mitochondria. The chlorophyll obtains its electrons from water (H_2O), producing O_2 as a by-product. During the electron-transport process, H^+ is pumped across the thylakoid membrane, and the resulting electrochemical proton gradient drives the synthesis of ATP in the stroma. As the final step in this

Figure 14–31 Compared with mitochondria, chloroplasts are larger and have an extra compartment. A chloroplast contains, in addition to an inner and an outer membrane, a thylakoid membrane enclosing a thylakoid space. The thylakoid membrane contains the light-capturing systems, the electron-transport chains, and ATP synthase. Unlike the chloroplast inner membrane, the mitochondrial inner membrane is folded into cristae in order to increase its surface area.

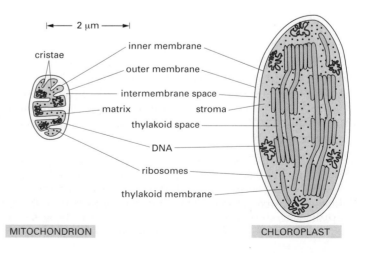

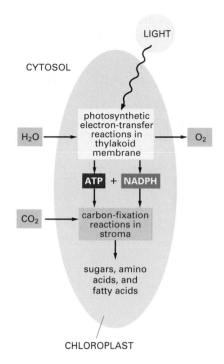

Figure 14–32 **Both categories of photosynthetic reactions take place in the chloroplast.** Water is oxidized and oxygen is released in the photosynthetic electron-transfer reactions, while carbon dioxide is assimilated (fixed) to produce sugars and a variety of other organic molecules in the carbon-fixation reactions.

series of reactions, high-energy electrons are loaded (together with H^+) onto $NADP^+$, converting it to NADPH. All of these reactions are confined to the chloroplast.

2. In the *carbon-fixation reactions* (also called the "dark reactions"), the ATP and the NADPH produced by the photosynthetic electron-transfer reactions serve as the source of energy and reducing power, respectively, to drive the conversion of CO_2 to carbohydrate. The carbon-fixation reactions, which begin in the chloroplast stroma and continue in the cytosol, produce sucrose and many other organic molecules in the leaves of the plant. The sucrose is exported to other tissues as a source of both organic molecules and energy for growth.

Thus the formation of ATP, NADPH, and O_2 (which requires light energy directly) and the conversion of CO_2 to carbohydrate (which requires light energy only indirectly) are separate processes, although elaborate feedback mechanisms interconnect the two. Several of the chloroplast enzymes required for carbon fixation, for example, are inactivated in the dark and reactivated by light-stimulated electron-transport processes.

Excited Chlorophyll Molecules Funnel Energy into a Reaction Center

Visible light is a form of electromagnetic radiation composed of many different wavelengths, ranging from violet (wavelength 400 nm) to deep red (700 nm). When we consider events at the level of a single molecule—such as the absorption of light by a molecule of chlorophyll, we have to picture light as being composed of discrete packets of energy called *photons*. Light of different colors is distinguished by photons of different energy, with longer wavelengths corresponding to lower energies. Thus photons of red light have a lower energy than photons of green light.

When sunlight is absorbed by a molecule of the green pigment **chlorophyll**, electrons in the molecule interact with photons of light and are raised to a higher energy level. The electrons in the extensive network of alternating single and double bonds in the chlorophyll molecule (Figure 14–33) absorb red light most strongly.

An isolated molecule of chlorophyll is incapable of converting the light it absorbs to a form of energy useful to living systems. It can do this only when it is associated with the appropriate proteins and embedded in a membrane. In plant thylakoid membranes and in the membranes of photosynthetic bacteria, the light-absorbing chlorophylls are held in large multiprotein complexes called **photosystems**. The *antenna* portion of a photosystem consists of hundreds of chlorophyll molecules

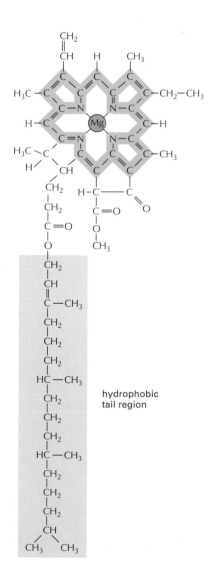

Figure 14–33 **Chlorophyll is a green pigment that absorbs energy from light photons.** A magnesium atom (*orange*) is held in the center of a porphyrin ring, which is structurally similar to the porphyrin ring that binds iron in heme. Light is absorbed by electrons within the bond network shown in *blue*, while the long hydrophobic tail (*green*) helps to hold the chlorophyll in the thylakoid membrane.

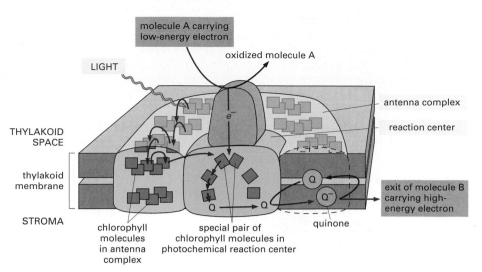

Figure 14–34 A photosystem contains a reaction center and an antenna. The antenna collects the energy of electrons that have been excited by light and funnels this energy to a special pair of chlorophyll molecules in the reaction center. The reaction center thereby acquires a high-energy electron that can be passed rapidly to the electron-transport chain in the thylakoid membrane, via the quinone (Q).

that capture light energy in the form of excited (high-energy) electrons (Figure 14–34). These chlorophylls are arranged so that the energy of an excited electron can be passed from one chlorophyll molecule to another, funneling the energy into an adjacent protein complex in the membrane—the **reaction center**. There the energy is trapped and used to energize one electron in a *special pair* of chlorophyll molecules.

The reaction center is a transmembrane complex of proteins and organic pigments that lies at the heart of photosynthesis. It is thought to have first evolved more than 3.5 billion years ago in primitive photosynthetic bacteria. Detailed structural and functional studies have revealed how it functions at an atomic level of detail. The special pair of chlorophyll molecules in the reaction center acts as an irreversible trap for an excited electron, because these chlorophylls are poised to pass a high-energy electron to a precisely positioned neighboring molecule in the same protein complex, creating a charge separation by moving the energized electron rapidly away from the chlorophylls, the reaction center transfers it to an environment where it is much more stable.

Because the chlorophyll molecule in the reaction center loses an electron, it becomes positively charged. As illustrated in Figure 14–35A, this chlorophyll rapidly regains an electron from an adjacent electron donor (*orange*) to return to its unexcited, uncharged state. Then, in slower reactions, the electron donor has its missing electron replaced with an electron removed from water, and the high-energy electron that was generated by the excited chlorophyll is transferred to the electron-transport chain (Figure 14–35B).

Light Energy Drives the Synthesis of ATP and NADPH

As discussed in Chapter 3, biosynthesis is, in a sense, the opposite of oxidative breakdown. To build its molecular components, the cell needs not only energy in the form of ATP, but also reducing power in the form of the hydrogen carrier NADPH (see Figure 3–35). Because a primary function of photosynthesis is to synthesize organic molecules from CO_2, the process has a huge requirement for both ATP and reducing power. The need for reducing power is met by making NADPH from $NADP^+$, using the energy captured from sunlight to convert low-energy electrons in water into the high-energy electrons in NADPH.

Photosynthesis in plants and cyanobacteria produces ATP and NADPH by a process requiring two photons of light. ATP is made after the first photon is absorbed, NADPH after the second. To perform these

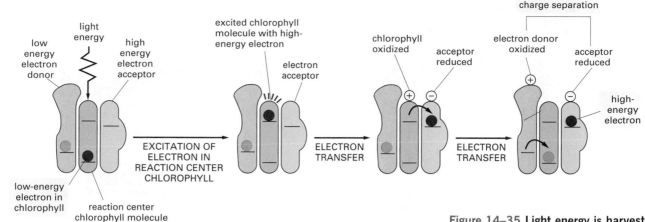

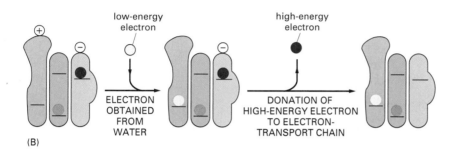

Figure 14–35 Light energy is harvested by a reaction center chlorophyll molecule. (A) The initial events in a reaction center create a charge separation. A chlorophyll molecule of the special pair (*green*) is tightly held in a pigment-protein complex, positioned so that both a potential low-energy electron donor (*gray*) and a potential high-energy electron acceptor (*blue*) are available. As soon as the *red* electron in the chlorophyll is excited by light energy, it is passed to the electron acceptor and stabilized as a high-energy electron. The positively charged chlorophyll molecule quickly attracts the low-energy *orange* electron and returns to its resting state. These reactions require less than 10^{-6} second to complete. (B) The final production of a high-energy electron from a low-energy electron. In this process, which follows that in (A), the entire reaction center is restored to its resting state. As a result, a high-energy electron in the thylakoid membrane has been produced from a low-energy electron obtained from water.

reactions, two different photosystems work in series. Together these photosystems impart to an electron a high enough energy to produce NADPH. Along the way, a proton gradient is also generated, allowing ATP to be made.

In outline, light energy is absorbed initially by one photosystem (paradoxically called *photosystem II* for historical reasons), where it is used to produce a high-energy electron that is propelled via an electron-transport chain toward the second photosystem. While traveling down the electron-transport chain, the electron drives an H^+ pump in the thylakoid membrane and creates a proton gradient in the manner described previously for oxidative phosphorylation (Figure 14–36). An ATP synthase in the thylakoid membrane then uses this proton gradient to drive the synthesis of ATP on the stromal side of the membrane.

The electron then arrives at the second photosystem in the pathway (*photosystem I*), where it fills a positively charged "hole" left in the reaction center of this photosystem when it absorbs a second photon of light. Because photosystem I is designed to start at a higher energy level than photosystem II, it is able to boost electrons to the very high energy level needed to make NADPH from $NADP^+$ (see Figure 14–36). The redox potentials of the components along this electron-transport chain are shown in Figure 14–37.

In the overall process described thus far, we have seen that an electron that is removed from a chlorophyll molecule at the reaction center of photosystem II travels all the way through the electron-transport chain in the thylakoid membrane until it winds up being donated to NADPH. This initial electron must be replaced to return the system to its unexcited state. The replacement electron comes from a low-energy electron donor, which, in plants and many photosynthetic bacteria, is water (see Figure 14–35B). The reaction center of photosystem II includes a water-splitting enzyme that holds the oxygen atoms of two water molecules bound to a cluster of manganese atoms in the protein

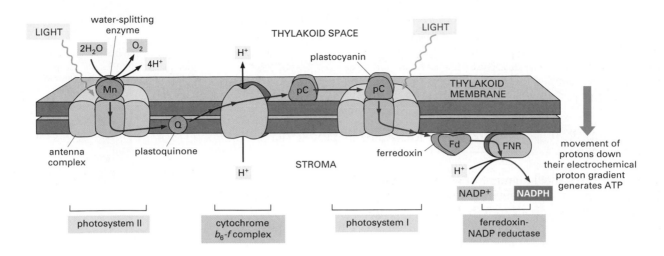

Figure 14–36 During photosynthesis electrons travel down an electron transport chain in the thylakoid membrane. Light is harvested by the antenna complexes in both membrane-embedded photosystems and funneled to chlorophyll molecules in their reaction centers (as shown in Figure 14–34). These electrons are then passed to the electron-transport chain via the mobile electron carriers plastoquinone (which closely resembles the ubiquinone of mitochondria), plastocyanin (a small, copper-containing protein), and ferredoxin (a small protein containing an iron–sulfur center). The cytochrome b_6-f complex resembles the cytochrome b-c_1 complex of mitochondria, and it is the sole site of active H^+-pumping in the chloroplast electron-transport chain. The H^+ released by water oxidation and the H^+ taken up during NADPH formation also contribute to generating the electrochemical proton gradient. As indicated by the *blue* arrow, the proton gradient drives an ATP synthase located in the same membrane to generate ATP (the ATP synthase is not shown here).

Figure 14–37 The components in the electron-transport chain have different redox potentials. The redox potential for each molecule is indicated by its position along the vertical axis. Photosystem II passes electrons from its excited chlorophyll through an electron-transport chain in the thylakoid membrane that leads to photosystem I. The net electron flow through the two photosystems linked in series is from water to $NADP^+$, and it produces NADPH as well as ATP. The ATP is synthesized by an ATP synthase (not shown) that harnesses the electrochemical proton gradient produced by electron transport.

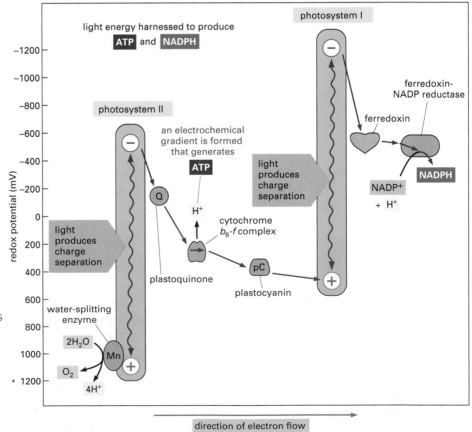

(see Figures 14–36 and 14–37). This enzyme removes electrons one at a time from the water to fill the holes created by light in the chlorophyll molecules of the reaction center. When four electrons have been removed from two water molecules (which requires four photons of light), O_2 is released. It is this critical process, occurring over billions of years, that has generated all of the O_2 in the Earth's atmosphere.

Carbon Fixation Is Catalyzed by Ribulose Bisphosphate Carboxylase

Having seen how the light reactions of photosynthesis generate ATP and NADPH, we must now consider how these compounds are used in the reactions of **carbon fixation**. When carbohydrates are oxidized to CO_2 and H_2O, a large amount of free energy is released. Clearly, the reverse reaction, in which CO_2 and H_2O combine to make carbohydrate, must therefore be energetically very unfavorable. In order to occur, it must be coupled to an energetically favorable reaction that drives it.

The central reaction of photosynthetic carbon fixation, in which an atom of inorganic carbon (as CO_2) is converted to organic carbon, is illustrated in Figure 14–38. CO_2 from the atmosphere combines with the five-carbon sugar derivative *ribulose 1,5-bisphosphate* plus water to give two molecules of the three-carbon compound 3-phosphoglycerate. This reaction, which was discovered in 1948, is catalyzed in the chloroplast stroma by a large enzyme called *ribulose bisphosphate carboxylase (rubisco)*. Since this enzyme works extremely sluggishly compared with most other enzymes (processing about three molecules of substrate per second compared with 1000 molecules per second for a typical enzyme), many enzyme molecules are needed. Ribulose bisphosphate carboxylase often represents more than 50% of the total chloroplast protein, and it is widely claimed to be the most abundant protein on Earth.

The reaction in which CO_2 is initially fixed is energetically favorable, but only because it receives a continuous supply of the energy-rich compound ribulose 1,5-bisphosphate, to which each molecule of CO_2 is added (see Figure 14–38). The elaborate metabolic pathway by which this compound is regenerated requires both ATP and NADPH; it was worked out in one of the first successful applications of radioisotopes as tracers in biochemistry. This *carbon-fixation cycle* (or Calvin cycle) is outlined in Figure 14–39; it is a cyclic process, beginning and ending with ribulose 1,5-bisphosphate. However, for every three molecules of carbon dioxide that enter the cycle, one new molecule of *glyceraldehyde 3-phosphate* is produced—the three-carbon sugar that is the net product of the cycle. This sugar then provides the starting material for the synthesis of many other sugars and organic molecules.

Question 14–9

Which of the following statements are correct? Explain your answers.

A. After an electron has been removed by light, the affinity for electrons of the positively charged chlorophyll in the reaction center of the first photosystem (photosystem II) is even greater than the electron affinity of O_2.

B. Photosynthesis is the light-driven transfer of an electron from chlorophyll to a second molecule with a much lower affinity for electrons.

C. Because it requires the absorption of four photons to release one O_2 molecule from two H_2O molecules, the water-splitting enzyme has to keep the reaction intermediates tightly bound so as to prevent partly reduced, and therefore hazardous, superoxide radicals from escaping.

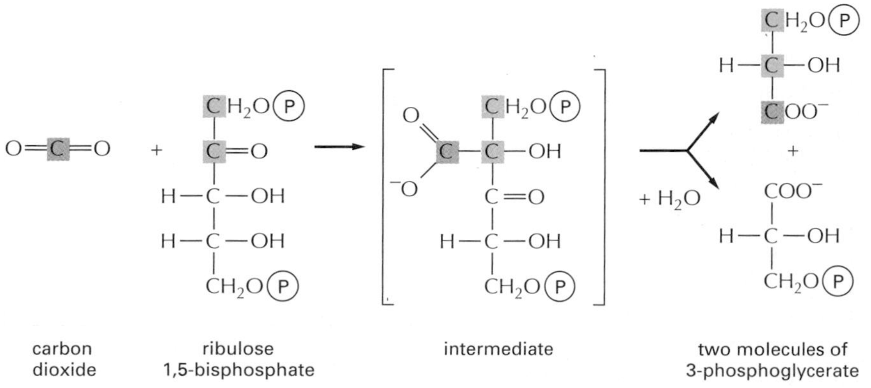

Figure 14–38 **The initial reaction in carbon fixation, in which carbon dioxide is converted into organic carbon, is catalyzed in the chloroplast stroma by the abundant enzyme ribulose bisphosphate carboxylase.** The product is 3-phosphoglycerate.

Figure 14–39 The carbon-fixation cycle forms organic molecules from CO_2 and H_2O. The number of carbon atoms in each type of molecule is indicated in the *white box*. There are many intermediates between glyceraldehyde 3-phosphate and ribulose 5-phosphate, but they have been omitted here for clarity. The entry of water into the cycle is also not shown.

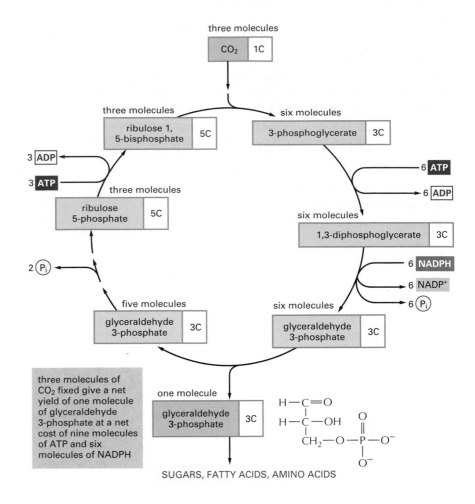

SUGARS, FATTY ACIDS, AMINO ACIDS

In the carbon-fixation cycle, three molecules of ATP and two molecules of NADPH are consumed for each CO_2 molecule converted into carbohydrate. Thus both phosphate bond energy (as ATP) and reducing power (as NADPH) are required for the formation of sugar molecules from CO_2 and H_2O.

Carbon Fixation in Chloroplasts Generates Sucrose and Starch

Much of the glyceraldehyde 3-phosphate produced in chloroplasts is moved out of the chloroplast into the cytosol. Some of it enters the glycolytic pathway (see Figure 13–3), where it is converted to pyruvate that is then used to produce ATP by oxidative phosphorylation in plant cell mitochondria. The glyceraldehyde 3-phosphate is also converted into many other metabolites, including the disaccharide sucrose. *Sucrose* is the major form in which sugar is transported between plant cells: just as glucose is transported in the blood of animals, sucrose is exported from the leaves via the vascular bundle to provide carbohydrate to the rest of the plant.

The glyceraldehyde 3-phosphate that remains in the chloroplast is mainly converted to *starch* in the stroma. Like glycogen in animal cells, starch is a large polymer of glucose that serves as a carbohydrate reserve (see Figure 13–18). The production of starch is regulated so that it is synthesized and stored as large grains in the chloroplast stroma during periods of excess photosynthetic capacity (see Figure 14–29B). At night, starch is broken down to sugars to help support the metabolic needs of the plant. Starch forms an important part of the diet of all animals that eat plants.

The Origins of Chloroplasts and Mitochondria

It is now widely accepted that chloroplasts and mitochondria evolved from bacteria that were engulfed by ancestral eucaryotic cells more than a billion years ago (see Figures 1–19 and 1–21). As a relic of this evolutionary past, both types of organelles contain their own genomes, as well as their own biosynthetic machinery for making RNA and organelle proteins. The way that mitochondria and chloroplasts reproduce—through the growth and division of preexisting organelles—provides additional evidence of their bacterial ancestry (Figure 14–40).

The growth and proliferation of mitochondria and chloroplasts is complicated, however, by the fact that their component proteins are encoded by two separate genetic systems—one in the organelle and one in the cell nucleus. In the case of the mitochondrion, most of the original bacterial genes have become transposed to the cell nucleus, leaving only relatively few genes inside the organelle itself. Animal mitochondria in fact contain a uniquely simple genetic system: the human mitochondrial genome, for example, contains only 16,569 nucleotide pairs of DNA encoding 37 genes. The vast majority of mitochondrial proteins—including those needed to make the mitochondrion's RNA polymerase and ribosomal proteins, and all of the enzymes of its citric acid cycle—are instead produced from nuclear genes, and these proteins must therefore be imported into the mitochondria from the cytosol, where they are made (discussed in Chapter 15).

Like the mitochondrion, the chloroplast contains many of its own genes, as well as a complete transcription and translation system for producing proteins from these genes. Chloroplast genomes are considerably larger than mitochondrial genomes, in higher plants, for example, the chloroplast genome contains about 120 genes in 120,000 nucleotide pairs. These genes are strikingly similar to the genes of cyanobacteria, the photosynthetic bacteria from which chloroplasts are thought to have been derived. Even so, many chloroplast proteins are now encoded by nuclear genes and must be imported from the cytosol.

The same techniques that have allowed us to analyze the genomes of mitochondria and chloroplasts have also permitted us to identify and explore the molecular biology of many microorganisms on the Earth. Some of these organisms thrive in the most inhospitable habitats on the planet. These include sulfurous hot springs or hydrothermal vents deep on the ocean floor. In these seemingly odd, modern microbes, we can readily find clues to life's history—in the form of the many molecules from which they are made. Like the fingerprints left at the scene of a

Figure 14–40 A mitochondrion divides like a bacterium. (A) Both mitochondrial fission and mitochondrial fusion are observed to occur. The fission process is conceptually similar to bacterial division processes. (B) An electron micrograph of a dividing mitochondrion in a liver cell. (B, Courtesy of Daniel S. Friend.)

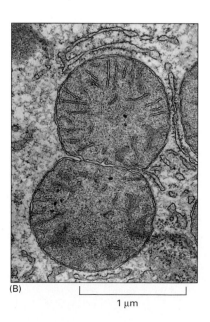

FISSION

mitochondrial DNA

FUSION

(A)

(B)

1 µm

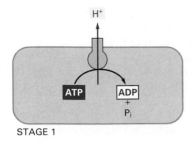

STAGE 1

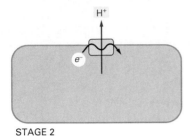

STAGE 2

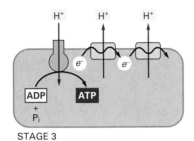

STAGE 3

Figure 14–41 Oxidative phosphorylation might have developed in stages, because bacteria that could generate ATP using an ATP synthase driven by the protons pumped by an electron transport chain would have had a selective advantage.

crime, these molecules provide powerful evidence that allows us to trace the history of ancient events, permitting speculations on the origin of the ATP-generating systems that are found in today's mitochondria and chloroplasts. We therefore end this chapter with a discussion of the evolution of the energy-harvesting systems that we have discussed in detail previously.

Oxidative Phosphorylation Gave Ancient Bacteria an Evolutionary Advantage

As we have already mentioned, the first living cells on Earth—both procaryotes and primitive eucaryotes—most likely consumed geochemically produced organic molecules, and generated ATP by fermentation. In the absence of oxygen, which was not yet present in the atmosphere, these cells most likely removed electrons from a hydrogen-rich organic molecule, such as glucose, and then transferred these electrons (via NADH or NADPH) to another organic molecule, which consequently became reduced. The resulting waste products—reduced organic acids such as lactic or formic acids, for example—would then be excreted into the environment (see Figure 13–4A).

This excretion of organic acids probably lowered the pH of the environment, favoring the survival of cells that evolved transmembrane proteins that could pump H+ out of the cytosol, keeping the cell from becoming too acidified (stage 1 in Figure 14–41). One of these pumps may have used the energy available from ATP hydrolysis to eject H+ from the cell; such a protein pump could have been the ancestor of the present-day ATP synthase.

As the Earth's supply of fermentable nutrients began to dwindle, organisms that could find a way to pump H+ without consuming ATP would have been at an advantage: they could save the small amounts of ATP they derived from the fermentation of foodstuffs to fuel other important cellular activities. Selective pressures such as the scarcity of nutrients might therefore have led to the evolution of the first electron-transport proteins; these carrier proteins allowed cells to use the movement of electrons between molecules of different redox potentials as the energy source for transporting H+ across the plasma membrane (stage 2 in Figure 14–41). Some of these cells might have used the nonfermentable organic acids that neighboring cells had excreted as waste to provide the electrons needed to feed the system. Some present-day bacteria grow on formic acid, for example, using the small amount of redox energy derived from the transfer of electrons from formic acid to fumarate to pump H+ (Figure 14–42).

Eventually some bacteria would have developed H+-pumping electron-transport systems that were so efficient that they could harvest more redox energy than they needed to maintain their internal pH. These cells most likely generated large electrochemical proton gradients, which they could then use to produce ATP. Protons could leak back into the cell through the ATP-driven H+ pumps, essentially running them in reverse so that they synthesized ATP (stage 3 in Figure 14–41). Because such cells required much less of the increasingly scarce supply of fermentable nutrients, they would have proliferated at the expense of their neighbors.

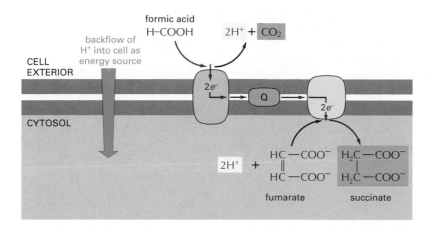

Figure 14–42 **Some present-day anaerobic bacteria can oxidize formic acid using an electron-transport chain in their plasma membrane.** In such anaerobic bacteria, including *E. coli*, the oxidation is mediated by an energy-conserving electron-transport chain in the plasma membrane. As indicated, the starting materials are formic acid and fumarate, and the products are succinate and CO_2. Note that H^+ is consumed inside the cell and generated outside the cell, which is equivalent to pumping H^+ to the cell exterior. Thus, this membrane-bound electron-transport system can generate an electrochemical proton gradient across the plasma membrane. The redox potential of the formic acid–CO_2 pair is –420 mV, while that of the fumarate–succinate pair is +30 mV.

Photosynthetic Bacteria Made Even Fewer Demands on Their Environment

The major evolutionary breakthrough in energy metabolism, however, was almost certainly the formation of photochemical reaction centers that could use the energy of sunlight to produce molecules such as NADH. It is thought that this development occurred early in the process of cellular evolution—more than 3 billion years ago, in the ancestors of the green sulfur bacteria. Present-day green sulfur bacteria use light energy to transfer hydrogen atoms (as an electron plus a proton) from H_2S to NADPH, thereby creating the strong reducing power required for carbon fixation (Figure 14–43).

The next step, which is thought to have occurred with the rise of cyanobacteria more than 3 billion years ago, was the evolution of organisms capable of using water as the electron source for photosynthesis. This entailed the evolution of a water-splitting enzyme and the addition of a second photosystem, acting in tandem with the first, to bridge the enormous gap in redox potential between H_2O and NADPH (see Figure 14–37). The biological consequences of this evolutionary step were far-reaching. For the first time, there were organisms that made only very minimal chemical demands on their environment. These cells could spread and evolve in ways denied to the earlier photosynthetic bacteria, which needed H_2S or organic acids as a source of electrons. Consequently, large amounts of biologically synthesized, fermentable

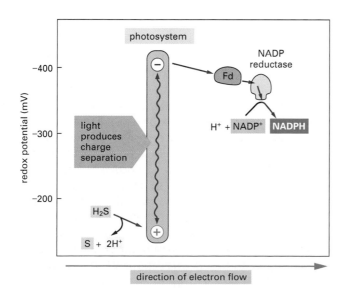

Figure 14–43 **Modern green sulfur bacteria engage in a form of photosynthesis that uses H_2S as an electron source.** The photosystem in green sulfur bacteria resembles photosystem I in plants and cyanobacteria. Both photosystems use a series of iron–sulfur centers as the electron acceptors that eventually donate their high-energy electrons to ferredoxin (Fd). An example of a bacterium of this type is *Chlorobium tepidum*, which can thrive at high temperatures and low light intensities in hot springs.

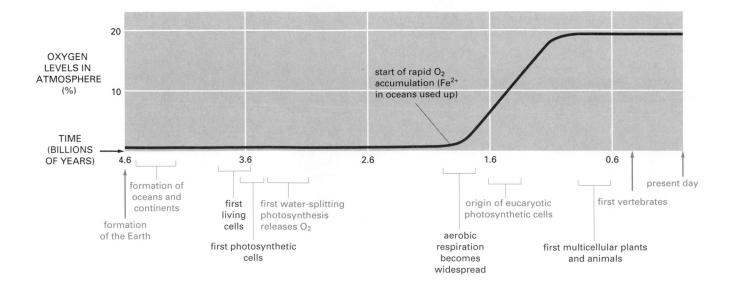

Figure 14–44 **Life on Earth has evolved over billions of years.** With the evolution of the membrane-based process of photosynthesis, organisms were no longer dependent on preformed organic chemicals. They could now make their own organic molecules from CO_2 gas. The delay of more than a billion years between the appearance of bacteria that split water and released O_2 during photosynthesis and the accumulation of high levels of O_2 in the atmosphere is thought to be due to the initial reaction of the oxygen with abundant ferrous iron (Fe^{2+}) dissolved in the early oceans. This iron would have removed oxygen from the atmosphere, and formed the enormous deposits of iron oxide found in some rocks of this age. Only when the iron was used up would oxygen have started to accumulate in the atmosphere. Membrane-based aerobic respiration presumably arose in response to the rising amount of oxygen in the atmosphere. As nonphotosynthetic oxygen-using organisms appeared, the concentration of oxygen in the atmosphere leveled out.

organic materials accumulated. Moreover, oxygen entered the atmosphere for the first time (Figure 14–44).

The availability of O_2 made possible the development of bacteria that relied on aerobic metabolism to make their ATP. As explained previously, these organisms could harness the large amount of energy released by breaking down carbohydrates and other reduced organic molecules all the way to CO_2 and H_2O.

As organic materials accumulated as a by-product of photosynthesis, some photosynthetic bacteria—including the ancestors of *E. coli*—lost their ability to survive on light energy alone and came to rely entirely on cellular respiration. Mitochondria arose when a primitive eucaryotic cell engulfed such a respiration-dependent bacterium. And plants arose somewhat later when a descendant of this early aerobic eucaryote captured a photosynthetic bacterium, which became the precursor of the chloroplast. Once eucaryotes had acquired the bacterial symbionts that became mitochondria and chloroplasts, they could then embark on the amazing pathway of evolution that eventually led to complex multicellular organisms.

The Lifestyle of *Methanococcus* Suggests That Chemiosmotic Coupling Is an Ancient Process

The conditions today that most resemble those under which cells are thought to have lived 3.5–3.8 billion years ago may be those near deep-ocean hydrothermal vents. These vents represent places where the Earth's molten mantle is breaking through the crust, expanding the width of the ocean floor. Indeed, the modern organisms that appear to be most closely related to the hypothetical cells from which all life evolved live at high temperatures (75°C to 95°C, close to the temperature of boiling water). This ability to thrive at such extreme temperatures suggests that life's common ancestor—the cell that gave rise to bacteria, archaea, and eucaryotes—lived under very hot, anaerobic conditions.

One of the archaea that live in this environment today is *Methanococcus jannaschii*. Originally isolated from a hydrothermal vent more than a mile beneath the ocean surface, the organism grows entirely on inorganic nutrients in the complete absence of light and gaseous oxygen, utilizing as nutrients hydrogen gas (H_2), CO_2, and nitrogen gas (N_2) that bubble up from the vent. Its mode of existence gives us a hint of how early cells might have used electron transport to derive their energy and their carbon molecules from inorganic materials that were freely available on the hot early Earth.

Methanococcus relies on N_2 gas as its source of nitrogen for organic molecules such as amino acids. The organism reduces N_2 to ammonia (NH_3) by the addition of hydrogen, a process called **nitrogen fixation.** Nitrogen fixation requires a large amount of energy, as does the carbon-fixation process that the bacterium needs to convert CO_2 into sugars. Much of the energy required for both processes is derived from the transfer of electrons from H_2 to CO_2, with the release of large amounts of methane (CH_4) as a waste product (thus producing natural gas and giving the organism its name; Figure 14–45A). Part of this electron transfer occurs in the membrane and results in the pumping of protons (H^+) across it. The resulting electrochemical proton gradient drives an ATP synthase in the same membrane to make ATP. The fact that such chemiosmotic coupling exists in an organism as primitive as *Methanococcus* suggests that the storage of energy derived from electron transport in an H^+ gradient is an extremely ancient process.

Methanococcus also engages in a process of carbon fixation, but the mechanism this organism uses to fix carbon is completely different from the pathway found in plants, algae, and cyanobacteria that we discussed previously. As indicated in Figure 14–45B, in addition to being the source of the high-energy electrons for the membrane-based process that generates ATP, hydrogen gas (H_2) is used to reduce CO_2 to an enzyme-bound molecule of carbon monoxide (CO). The CO then reacts with a methyl group, produced as an intermediate in the process of methanogenesis, to form acetyl CoA. The acetyl CoA is then converted to sugars, amino acids, nucleotides, and many other small and large molecules through the familiar enzyme-catalyzed pathways that require ATP.

Methanococcus is by no means a simple organism; for example, among its 1800 genes are more than 60 that encode enzymes that function in the pathway from CO_2 to CH_4 alone. The earliest cells that contained proteins must have been much simpler; as a guess, they might have required less than a total of 100 genes. The simplest known living cell today, the tiny bacterium *Mycoplasma genitalium*, possesses only 500 genes.

The genomes of scores of diverse single-celled organisms have been sequenced, giving us ever more powerful tools for reconstructing

Figure 14–45 *Methanococcus* uses chemiosmotic coupling to generate energy, but fixes carbon using a pathway that differs from that used by plants, algae, and cyanobacteria. This deep-sea archean uses hydrogen gas (H_2) as the source of reducing power in both of the pathways shown. (A) Energy generation. The production of methane (CH_4) from CO_2 occurs in several stages. The initial reduction steps take place via enzyme-catalyzed reactions in the cytoplasm. In contrast, the final reduction step involves a membrane-based electron transfer that generates a proton gradient that drives ATP synthesis, while producing methane as a waste product. (B) Carbon fixation. The main pathway for the fixation of carbon dioxide results in the production of acetyl CoA. This is the source for the sugars, fatty acids, and nucleotides that the cell needs for biosynthesis. The circles and squares in the diagram represent a series of special coenzymes to which the indicated metabolic intermediates are bound.

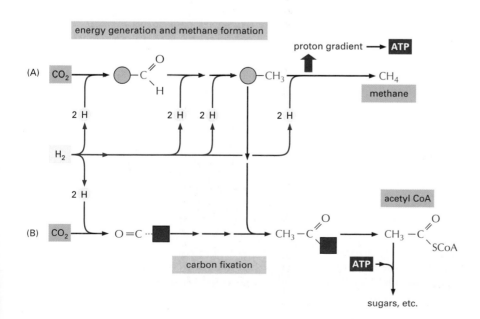

the past. However, it will be much easier to pin down relatively "recent" events, such as the adoption of specific bacteria by our eucaryotic ancestors to produce the first mitochondria, than to push back into cell origins, where major conceptual challenges remain. For example, we presently have no convincing way of describing how RNA-based cells, which are postulated to have existed very early in cellular evolution (see Figure 7–38), might have evolved to exploit the energy-yielding processes of fermentation or of membrane-based electron transport—the mechanisms that power all of the cells that we know about today.

Essential Concepts

- Mitochondria, chloroplasts, and many bacteria produce ATP by a membrane-based mechanism known as chemiosmotic coupling.
- Mitochondria produce most of an animal cell's ATP, using energy derived from oxidation of sugars and fatty acids.
- Mitochondria are enclosed by two concentric membranes, the innermost of which encloses the mitochondrial matrix. The matrix space contains many enzymes, including those of the citric acid cycle. These enzymes produce large amounts of NADH and $FADH_2$ from the oxidation of acetyl CoA.
- In the inner mitochondrial membrane, high-energy electrons donated by NADH and $FADH_2$ pass along an electron-transport chain—the respiratory chain—eventually combining with molecular oxygen (O_2) in an energetically favorable reaction.
- Much of the energy released by electron transfers along the respiratory chain is harnessed to pump H^+ out of the matrix, thereby creating a transmembrane electrochemical proton (H^+) gradient. The proton pumping is carried out by three large respiratory enzyme complexes embedded in the membrane.
- The resulting electrochemical proton gradient across the inner mitochondrial membrane is harnessed to make ATP when H^+ ions flow back into the matrix through ATP synthase, an enzyme located in the inner mitochondrial membrane.
- The electrochemical proton gradient also drives the active transport of metabolites into and out of the mitochondrion.
- In photosynthesis in chloroplasts and photosynthetic bacteria, high-energy electrons are generated when sunlight is absorbed by chlorophyll; this energy is captured by protein complexes known as photosystems, which are located in the thylakoid membranes of chloroplasts.
- Electron-transport chains associated with photosystems transfer electrons from water to $NADP^+$ to form NADPH, with the concomitant production of an electrochemical proton gradient across the thylakoid membrane. Molecular oxygen (O_2) is generated as a by-product.
- As in mitochondria, the proton gradient across the thylakoid membrane is used by an ATP synthase embedded in the membrane to generate ATP.
- The ATP and the NADPH made by photosynthesis are used within the chloroplast to drive the carbon-fixation cycle in the chloroplast stroma, thereby producing carbohydrate from CO_2.
- Carbohydrate is exported to the cell cytosol, where it is metabolized to provide organic carbon, ATP (mostly via mitochondria), and reducing power for the rest of the cell.
- Both mitochondria and chloroplasts are thought to have evolved from bacteria that were endocytosed by primitive eucaryotic cells.

Each retains its own genome and divides by processes that resemble a bacterial cell division.

- Chemiosmotic coupling mechanisms are widespread and of ancient origin. Modern microorganisms that live in environments similar to those thought to have been present on the early Earth also use chemiosmotic coupling to produce ATP.

Key Terms

ATP synthase	nitrogen fixation
carbon fixation	oxidative phosphorylation
chemiosmotic coupling	photosynthesis
chlorophyll	photosystem
chloroplast	quinone
cytochrome	reaction center
electron-transport chain	redox pair
iron–sulfur center	redox potential
matrix	redox reaction
mitochondria	stroma

Questions

Question 14–11

Which of the following statements are correct? Explain your answers.

A. Many, but not all, electron-transfer reactions involve metal ions.

B. The electron-transport chain generates an electrical potential across the membrane because it moves electrons from the intermembrane space into the matrix.

C. The electrochemical proton gradient consists of two components: a pH difference and an electrical potential.

D. Ubiquinone and cytochrome c are both diffusible electron carriers.

E. Plants have chloroplasts and therefore can live without mitochondria.

F. Both chlorophyll and heme contain an extensive system of double bonds that allows them to absorb visible light.

G. The role of chlorophyll in photosynthesis is equivalent to that of heme in mitochondrial electron transport.

H. Most of the dry weight of a tree comes from the minerals that are taken up by the roots.

Question 14–12

A single proton moving down its electrochemical gradient into the mitochondrial matrix space liberates 4.6 kcal/mole of free energy. How many protons have to flow across the inner mitochondrial membrane to synthesize one molecule of ATP if the ΔG for ATP synthesis under intracellular conditions is between 11 and 13 kcal/mole? (ΔG is discussed in Chapter 3, pp. 93–99.) Why is a range given for this latter value, and not a precise number? Under which conditions would the lower value apply?

Question 14–13

In the following statement, choose the correct one of the alternatives in italics and justify your answer. "If no O_2 is available, all components of the mitochondrial electron-transport chain will accumulate in their *reduced/oxidized* form. If O_2 is suddenly added again, the electron carriers in cytochrome oxidase will become *reduced/oxidized before/after* those in NADH dehydrogenase."

Question 14–14

Assume that the conversion of oxidized ubiquinone to reduced ubiquinone by NADH dehydrogenase occurs on the matrix side of the inner mitochondrial membrane and that its oxidation by cytochrome b-c_1 occurs on the intermembrane space side of the membrane (see Figures 14–10 and 14–20). What are the consequences of this arrangement for the generation of the H⁺ gradient across the membrane?

Question 14–15

If a voltage is applied to two platinum wires (electrodes) immersed in water, then water molecules become split into H_2 and O_2 gas. At the negative electrode, electrons are donated and H_2 gas is released; at the positive electrode, electrons are accepted and O_2 gas is produced. When photosynthetic bacteria and plant cells split water, they produce O_2, but no H_2. Why?

Question 14–16

In an insightful experiment performed in the 1960s, chloroplasts were first soaked in an acidic solution at pH 4, so that the stroma and thylakoid space became acidified (Figure Q14–16). They were then transferred to a basic solution (pH 8). This quickly increased the pH of the stroma to 8, while the thylakoid space temporarily remained at pH 4. A burst of ATP synthesis was observed, and the pH difference between the thylakoid and the stroma then disappeared.

A. Explain why these conditions lead to ATP synthesis.
B. Is light needed for the experiment to work?
C. What would happen if the solutions were switched so that the first incubation is in the pH 8 solution and the second one in the pH 4 solution?
D. Does the experiment support or question the chemiosmotic model?

Explain your answers.

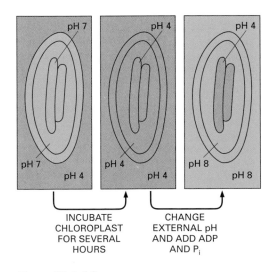

Figure Q14–16

pH 7 / pH 7 / pH 4
pH 4 / pH 4 / pH 4
pH 4 / pH 8 / pH 8

INCUBATE CHLOROPLAST FOR SEVERAL HOURS

CHANGE EXTERNAL pH AND ADD ADP AND P_i

Question 14–17

As your first experiment in the laboratory, your adviser asks you to reconstitute purified bacteriorhodopsin, a light-driven H⁺ pump from the plasma membrane of photosynthetic bacteria, and purified ATP synthase

from ox-heart mitochondria together into the same membrane vesicles—as shown in Figure Q14–17. You are then asked to add ADP and P_i to the external medium and shine light into the suspension of vesicles.

A. What do you observe?
B. What do you observe if not all the detergent is removed and the vesicle membrane therefore remains leaky to ions?
C. You tell a friend over dinner about your new experiments, and he questions the validity of an approach that utilizes components from so widely divergent, unrelated organisms: "Why would anybody want to mix vanilla pudding with brake fluid?" Defend your approach against his critique.

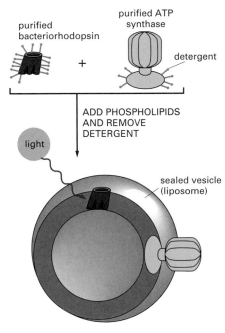

purified bacteriorhodopsin + purified ATP synthase

detergent

ADD PHOSPHOLIPIDS AND REMOVE DETERGENT

light

sealed vesicle (liposome)

Figure Q14–17

Question 14–18

FADH₂ is produced in the citric acid cycle by a membrane-embedded enzyme complex, called succinate dehydrogenase, that contains bound FAD and carries out the reactions

$$\text{succinate} + \text{FAD} \rightarrow \text{fumarate} + \text{FADH}_2$$

and

$$\text{FADH}_2 \rightarrow \text{FAD} + 2H^+ + 2e^-$$

The redox potential of FADH₂, however, is only –220 mV. Referring to Panel 14–1 (p. 471) and Figure 14–21, suggest a plausible mechanism by which its electrons could be fed into the electron-transport chain. Draw a diagram to illustrate your proposed mechanism.

Question 14–19

Some bacteria have become specialized to live in an environment of high pH (pH ~10). Do you suppose that these bacteria use a proton gradient across their plasma membrane to produce their ATP? (Hint: all cells must maintain their cytoplasm at a pH close to neutrality.)

MITOCHONDRION

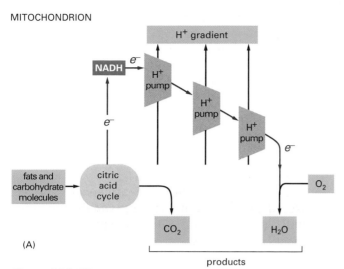

(A)

CHLOROPLAST

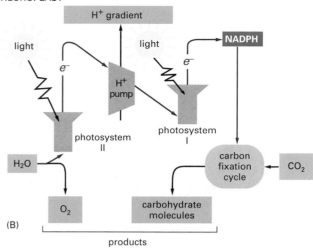

(B)

Figure Q14–20

Question 14–20

Figure Q14–20 summarizes the circuitry used by mitochondria and chloroplasts to interconvert different forms of energy. Is it accurate to say
  A. that the products of chloroplasts are the substrates for mitochondria?
  B. that the activation of electrons by the photosystems enables chloroplasts to drive electron transfer from H_2O to carbohydrate, which is opposite to the direction of electron transfer in the mitochondrion?
  C. that the citric acid cycle is the reverse of the normal carbon-fixation cycle?

Question 14–21

A manuscript has been submitted for publication to a prestigious scientific journal. In the paper the authors describe an experiment in which they have succeeded in trapping an individual ATP synthase molecule and then mechanically rotating its head by applying a force to it. The authors show that upon rotating the head of the ATP synthase, ATP is produced, in the absence of an H^+ gradient. What might this mean about the mechanism whereby ATP synthase functions? Should this manuscript be considered for publication in one of the best journals?

Question 14–22

You mix the following components in a solution. Assuming that the electrons must follow the path specified in Figure 14–10, in which experiments would you expect a net transfer of electrons to cytochrome c? Discuss why no electron transfer occurs in the other experiments.
  A. reduced ubiquinone and oxidized cytochrome c
  B. oxidized ubiquinone and oxidized cytochrome c
  C. reduced ubiquinone and reduced cytochrome c
  D. oxidized ubiquinone and reduced cytochrome c
  E. reduced ubiquinone, oxidized cytochrome c, and cytochrome b-c_1 complex
  F. oxidized ubiquinone, oxidized cytochrome c, and cytochrome b-c_1 complex
  G. reduced ubiquinone, reduced cytochrome c, and cytochrome b-c_1 complex
  H. oxidized ubiquinone, reduced cytochrome c, and cytochrome b-c_1 complex

Highlights from *Essential Cell Biology 2 Interactive* CD-ROM

14.1   Tomograph of Mitochondrion
14.3   ATP Synthase—A Molecular Turbine
14.6   Light Harvesting

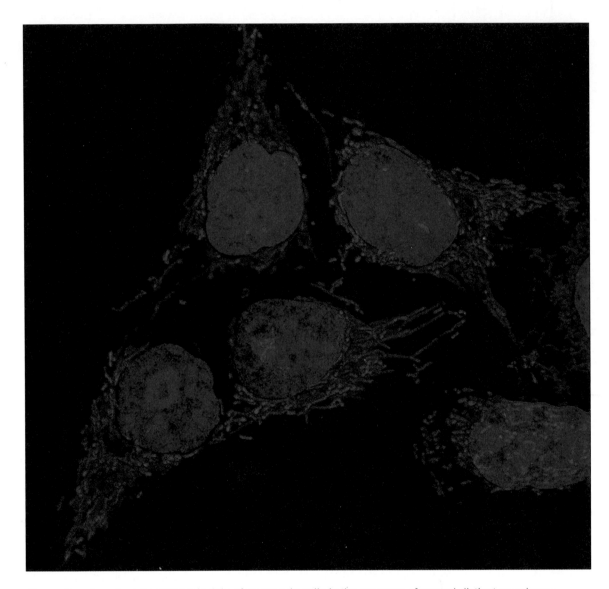

Organelles. A major organizing principle of eucaryotic cells is the presence of several distinct membrane-enclosed compartments. Most cells contain a single nucleus and multiple mitochondria, and in this chapter we discuss the different strategies whereby proteins, made in the cytosol, enter these organelles. This image is of a collection of cells growing in culture with their nuclei shown in *blue* and their mitochondria shown in *red*. (Courtesy of Paul J. Smith and Rachel Errington/Wellcome Photo Library.)

Intracellular Compartments and Transport

15

At any one time, a typical eucaryotic cell carries out thousands of different chemical reactions, many of which are mutually incompatible. One series of reactions makes glucose, for example, while another breaks down glucose; some enzymes synthesize peptide bonds, whereas others hydrolyze them, and so on. Indeed, if the cells of an organ such as the liver are broken apart and their contents mixed together in a test tube, chemical chaos results, and the cells' enzymes and other proteins are quickly degraded by their own proteolytic enzymes. For a cell to operate effectively, the different intracellular processes that occur simultaneously must somehow be segregated.

Cells have evolved several strategies for isolating and organizing their chemical reactions. One strategy used by both procaryotic and eucaryotic cells is to aggregate the different enzymes required to catalyze a particular sequence of reactions into a single large protein complex. Such multiprotein complexes are used, for example, in the synthesis of DNA, RNA, and proteins. A second strategy, which is most highly developed in eucaryotic cells, is to confine different metabolic processes, and the proteins required to perform them, within different membrane-enclosed compartments. As discussed in Chapters 11 and 12, cell membranes provide selectively permeable barriers through which the transport of most molecules can be controlled. In this chapter we consider the strategy of compartmentalization and some of its consequences.

In the first section we describe the principal membrane-enclosed compartments, or *membrane-enclosed organelles,* of eucaryotic cells and briefly consider their main functions. In the second section we discuss how the protein composition of the different compartments is set up and maintained. Each compartment contains a unique set of proteins that have to be transferred selectively from the cytosol, where they are made, to the compartment in which they are used. This transfer process, called *protein sorting,* depends on signals built into the amino acid sequence of the proteins. In the third section we describe how certain membrane-enclosed compartments in a eucaryotic cell communicate with one another by forming small membranous sacs, or *vesicles,* that pinch off from one compartment, move through the cytosol, and fuse with another compartment in a process called *vesicular transport.* In the last two sections we discuss how this constant vesicular traffic also provides the main routes for releasing proteins from the cell by the process of *exocytosis* and for importing them by the process of *endocytosis.*

Membrane-enclosed Organelles

Eucaryotic Cells Contain a Basic Set of Membrane-enclosed Organelles

Membrane-enclosed Organelles Evolved in Different Ways

Protein Sorting

Proteins Are Imported into Organelles by Three Mechanisms

Signal Sequences Direct Proteins to the Correct Compartment

Proteins Enter the Nucleus Through Nuclear Pores

Proteins Unfold to Enter Mitochondria and Chloroplasts

Proteins Enter the Endoplasmic Reticulum While Being Synthesized

Soluble Proteins Are Released into the ER Lumen

Start and Stop Signals Determine the Arrangement of a Transmembrane Protein in the Lipid Bilayer

Vesicular Transport

Transport Vesicles Carry Soluble Proteins and Membrane Between Compartments

Vesicle Budding Is Driven by the Assembly of a Protein Coat

The Specificity of Vesicle Docking Depends on SNAREs

Secretory Pathways

Most Proteins Are Covalently Modified in the ER

Exit from the ER Is Controlled to Ensure Protein Quality

Proteins Are Further Modified and Sorted in the Golgi Apparatus

Secretory Proteins Are Released from the Cell by Exocytosis

Endocytic Pathways

Specialized Phagocytic Cells Ingest Large Particles

Fluid and Macromolecules Are Taken Up by Pinocytosis

Receptor-mediated Endocytosis Provides a Specific Route into Animal Cells

Endocytosed Macromolecules Are Sorted in Endosomes

Lysosomes Are the Principal Sites of Intracellular Digestion

Membrane-enclosed Organelles

Whereas a procaryotic cell consists of a single compartment, the **cytosol**, enclosed by the plasma membrane, a eucaryotic cell is elaborately subdivided by internal membranes. These membranes create enclosed compartments in which sets of enzymes can operate without interference from reactions occurring in other compartments. When a cross section through a plant or an animal cell is examined in the electron microscope, numerous small, membrane-enclosed sacs, tubes, spheres, and irregularly shaped structures can be seen, often arranged without much apparent order (Figure 15–1). These structures are all distinct membrane-enclosed organelles, or parts of such organelles, each of which contains a unique set of large and small molecules and carries out a specialized function. In this section, we review these functions and discuss how different membrane-enclosed organelles may have evolved.

Eucaryotic Cells Contain a Basic Set of Membrane-enclosed Organelles

The major **membrane-enclosed organelles** of an animal cell are illustrated in Figure 15–2, and their functions are summarized in Table 15–1. These organelles are surrounded by the cytosol, which is enclosed by the plasma membrane. The *nucleus* is generally the most prominent organelle in eucaryotic cells. It is surrounded by a double membrane, known as the *nuclear envelope*, and communicates with the cytosol via *nuclear pores* that perforate the envelope. The outer nuclear membrane is continuous with the membrane of the *endoplasmic reticulum (ER)*, a system of interconnected sacs and tubes of membrane that often extends throughout most of the cell. The ER is the major site of new-membrane synthesis in the cell. Large areas of the system have ribosomes attached to the cytosolic surface and are designated *rough endoplasmic reticulum*. The ribosomes are actively synthesizing proteins that are delivered into the ER lumen or ER membrane. The *smooth ER* is scanty in most cells but is highly developed for performing particular functions in others: it is the site of steroid hormone synthesis in cells of

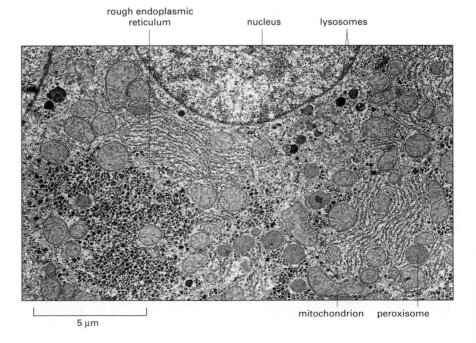

Figure 15–1 In eucaryotic cells, internal membranes create enclosed compartments and organelles in which different metabolic processes are segregated. Examples of many of the major membrane-enclosed organelles can be identified in this electron micrograph of part of a liver cell, seen in cross section. The small black granules between the membrane-enclosed compartments are glycogen granules (glycosomes), which are aggregates of glycogen and the enzymes that control its synthesis and breakdown. (Courtesy of Daniel S. Friend.)

rough endoplasmic reticulum   nucleus   lysosomes

mitochondrion   peroxisome

5 μm

the adrenal gland, for example, and the site where a variety of organic molecules, including alcohol, are detoxified in liver cells. In many eucaryotic cells the smooth ER also sequesters Ca^{2+} from the cytosol; the release and reuptake of Ca^{2+} from the ER is involved in the rapid response to many extracellular signals, as discussed in Chapter 12 and 16.

The *Golgi apparatus,* which is usually situated near the nucleus, receives proteins and lipids from the ER, modifies them, and then dispatches them to other destinations in the cell. Small sacs of digestive enzymes called *lysosomes* degrade worn-out organelles, as well as macromolecules and particles taken into the cell by endocytosis. On their way to lysosomes, endocytosed materials must first pass through a series of compartments called *endosomes,* which sort the ingested molecules and recycle some back to the plasma membrane. *Peroxisomes* are small organelles enclosed by a single membrane. They contain enzymes used in a variety of oxidative reactions that break down lipids and destroy toxic molecules. *Mitochondria* and (in plant cells) *chloroplasts* are each surrounded by a double membrane and are the sites of oxidative phosphorylation and photosynthesis, respectively (discussed in Chapter 14); both contain membranes that are highly specialized for the production of ATP.

Many of the membrane-enclosed organelles, including the ER, Golgi apparatus, mitochondria, and chloroplasts, are held in their relative locations in the cell by attachment to the cytoskeleton, especially to microtubules. Cytoskeletal filaments provide tracks for moving the organelles around and for directing the traffic of vesicles between them. These movements are driven by motor proteins that use the energy of ATP hydrolysis to propel the organelles and vesicles along the filaments, as discussed in Chapter 17.

On average, the membrane-enclosed organelles together occupy nearly half the volume of a eucaryotic cell (Table 15–2), and the total amount of membrane associated with them is enormous: in a typical mammalian cell, for example, the area of the endoplasmic reticulum membrane is 20–30 times greater than that of the plasma membrane. In terms of its area and mass, the plasma membrane is only a minor membrane in most eucaryotic cells.

Much can be learned about the composition and function of an organelle once it has been isolated from other cell structures. For the most part, organelles are far too small to be isolated by hand, but it is possible to separate one type of organelle from another by differential

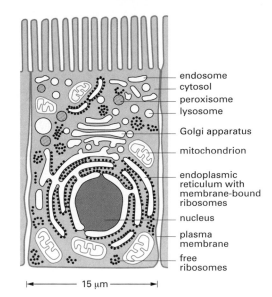

Figure 15–2 A cell from the lining of the intestine contains the basic set of organelles found in most animal cells. The nucleus, endoplasmic reticulum (ER), Golgi apparatus, lysosomes, endosomes, mitochondria, and peroxisomes are distinct compartments separated from the cytosol *(gray)* by at least one selectively permeable membrane. Ribosomes are also shown, even though they are not enclosed by a membrane and are too small to be seen in a light microscope and therefore do not fit the original definition of an organelle. Some ribosomes are found free in the cytosol, while others are bound to the cytosolic surface of the ER.

Table 15–1 The Main Functions of the Membrane-enclosed Compartments of a Eucaryotic Cell

COMPARTMENT	MAIN FUNCTION
Cytosol	contains many metabolic pathways (Chapters 3 and 13); protein synthesis (Chapter 7)
Nucleus	contains main genome (Chapter 5); DNA and RNA synthesis (Chapters 6 and 7)
Endoplasmic reticulum (ER)	synthesis of most lipids (Chapter 11); synthesis of proteins for distribution to many organelles and to the plasma membrane (this chapter)
Golgi apparatus	modification, sorting, and packaging of proteins and lipids for either secretion or delivery to another organelle (this chapter)
Lysosomes	intracellular degradation (this chapter)
Endosomes	sorting of endocytosed material (this chapter)
Mitochondria	ATP synthesis by oxidative phosphorylation (Chapter 14)
Chloroplasts (in plant cells)	ATP synthesis and carbon fixation by photosynthesis (Chapter 14)
Peroxisomes	oxidation of toxic molecules

Table 15–2 The Relative Volumes Occupied by the Major Membrane-enclosed Organelles in a Liver Cell (Hepatocyte)		
INTRACELLULAR COMPARTMENT	PERCENT OF TOTAL CELL VOLUME	APPROXIMATE NUMBER PER CELL
Cytosol	54	1
Mitochondria	22	1700
Endoplasmic reticulum	12	1
Nucleus	6	1
Golgi apparatus	3	1
Peroxisomes	1	400
Lysosomes	1	300
Endosomes	1	200

centrifugation (described in Panel 4–3, pp. 160–161). Once a purified sample of one type of organelle has been obtained, the organelle's proteins can be identified. In many cases the organelle itself can be incubated in a test tube under conditions that allow its functions to be studied. Isolated mitochondria, for example, can produce ATP from the oxidation of pyruvate to CO_2 and water, provided they are adequately supplied with ADP and O_2.

Membrane-enclosed Organelles Evolved in Different Ways

In trying to understand the relationships between the different compartments of a modern eucaryotic cell, it is helpful to consider how they might have evolved. The compartments probably evolved in stages. The precursors of the first eucaryotic cells are thought to have been simple microorganisms, resembling bacteria, which had a plasma membrane but no internal membranes. The plasma membrane in such cells would have provided all membrane-dependent functions, including ATP synthesis and lipid synthesis, as does the plasma membrane in most modern bacteria. Bacteria can get by with this arrangement because of their small size and thus their high surface-to-volume ratio: their plasma membrane area is sufficient to sustain all the vital functions for which membranes are required. Present-day eucaryotic cells, however, have volumes 1000 to 10,000 times greater than that of a typical bacterium such as *E. coli*. Such a large cell has a small surface-to-volume ratio, and presumably could not survive with a plasma membrane as its only membrane. Thus the increase in size typical of eucaryotic cells probably could not have occurred without the development of internal membranes.

Membrane-enclosed organelles are thought to have arisen in evolution in at least two ways. The nuclear membranes and the membranes of the ER, Golgi apparatus, endosomes, and lysosomes are believed to have originated by invagination of the plasma membrane (Figure 15–3). These membranes, and the organelles they enclose, are all part of what is collectively called the *endomembrane system*. As we discuss later, the interiors of these organelles (with the exception of the nucleus) communicate extensively with one another and with the outside of the cell by means of small vesicles that bud off from one of these organelles and fuse with another. Consistent with this proposed evolutionary origin, the interiors of these organelles are treated by the cell in many ways as "extracellular," as we shall see. The hypothetical scheme shown in

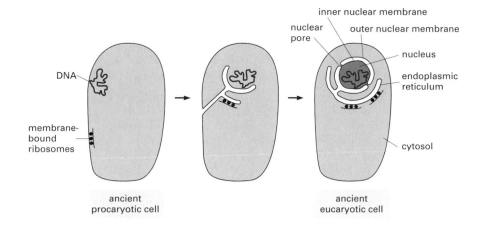

inner nuclear membrane
nuclear pore
outer nuclear membrane
nucleus
endoplasmic reticulum
cytosol
DNA
membrane-bound ribosomes
ancient procaryotic cell
ancient eucaryotic cell

Figure 15–3 Nuclear membranes and the ER may have evolved through invagination of the plasma membrane. In bacteria, the single DNA molecule is typically attached to the plasma membrane. It is possible that in a very ancient procaryotic cell, the plasma membrane, with its attached DNA, could have invaginated and eventually formed a two-layered envelope of membrane completely surrounding the DNA. This envelope is presumed to have eventually pinched off completely from the plasma membrane, producing a nuclear compartment surrounded by a double membrane. This nuclear envelope is penetrated by channels called nuclear pores, which enable it to communicate directly with the cytosol. Other portions of the same membrane formed the ER, to which some of the ribosomes became attached. This hypothetical scheme would explain why the space between the inner and outer nuclear membranes is continuous with the lumen of the ER.

Figure 15–3 would also explain why the nucleus is surrounded by two membranes. Although membrane invagination is rare in present-day bacteria, it does occur in some photosynthetic bacteria in which the regions of the plasma membrane containing the photosynthetic apparatus are internalized, forming intracellular vesicles.

Mitochondria and chloroplasts are thought to have originated in a different way. They differ from all other organelles in that they possess their own small genomes and can make some of their own proteins, as discussed in Chapter 14. The similarity of these genomes to those of bacteria and the close resemblance of some of their proteins to bacterial proteins strongly suggest that mitochondria and chloroplasts evolved from bacteria that were engulfed by primitive eucaryotic cells with which they initially lived in symbiosis. This theory would also explain why these organelles are enclosed by two membranes (Figure 15–4). As might be expected from their origins, mitochondria and chloroplasts remain isolated from the extensive vesicular traffic that connects the interiors of most of the other membrane-enclosed organelles to one another and to the outside of the cell.

Having briefly reviewed the main membrane-enclosed organelles of the eucaryotic cell, we turn now to the question of how each organelle acquires its unique set of proteins.

Question 15–1

As shown in the drawings in Figure 15–3, the lipid bilayer of the inner and outer nuclear membranes forms a continuous sheet, joined around the nuclear pores. As membranes are two-dimensional fluids, this would imply that membrane proteins can diffuse freely between the two nuclear membranes. Yet each of these two nuclear membranes has a different protein composition, reflecting different functions. How could you reconcile this apparent paradox?

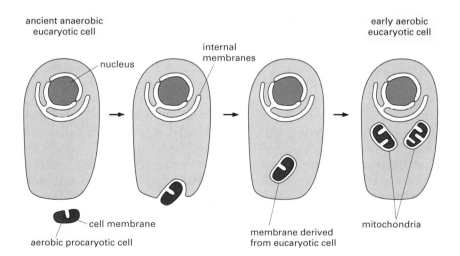

ancient anaerobic eucaryotic cell
early aerobic eucaryotic cell
internal membranes
nucleus
cell membrane
aerobic procaryotic cell
membrane derived from eucaryotic cell
mitochondria

Figure 15–4 Mitochondria are thought to have originated when an aerobic procaryote was engulfed by a larger anaerobic eucaryotic cell. Chloroplasts are thought to have originated later in a similar way, when an aerobic eucaryotic cell engulfed a photosynthetic procaryote. This theory would explain why these organelles have two membranes and why they do not participate in the vesicular traffic that connects many other intracellular compartments.

Protein Sorting

Before a eucaryotic cell reproduces by dividing in two, it has to duplicate its membrane-enclosed organelles. A cell cannot make these organelles from scratch: it requires information and materials contained in the organelle itself. Thus most of the organelles are formed from preexisting organelles, which grow and then divide. As cells grow, membrane-enclosed organelles enlarge by incorporation of new molecules; the organelles then divide and at cell division are distributed between the two daughter cells. The nuclear envelope, ER, and Golgi apparatus break up into small vesicles, which then coalesce again as the two daughter cells are formed (discussed in Chapter 19). Organelle growth requires a supply of new lipids to make more membrane and a supply of the appropriate proteins—both membrane proteins and the soluble proteins that will occupy the interior of the organelle. Even in cells that are not dividing, proteins are being produced continually. These newly synthesized proteins must be accurately delivered to organelles—some for eventual secretion from the cell and some to replace organelle proteins that have been degraded. Directing newly made proteins to their correct organelle is therefore necessary for a cell to be able to grow, divide, and function properly.

For some organelles, including the mitochondria, chloroplasts, peroxisomes, and the interior of the nucleus, proteins are delivered directly from the cytosol. For others, including the Golgi apparatus, lysosomes, endosomes, and the nuclear membranes, proteins and lipids are delivered indirectly via the ER, which is itself a major site of lipid and protein synthesis. Proteins enter the ER directly from the cytosol: some are retained there, but most are transported by vesicles to the Golgi apparatus and then onward to other organelles or the plasma membrane.

In this section we discuss the mechanisms by which proteins directly enter membrane-enclosed organelles from the cytosol. Proteins made in the cytosol are dispatched to different locations in the cell according to specific address labels that they contain in their amino acid sequences. Once at the correct address, the protein enters the organelle.

Proteins Are Imported into Organelles by Three Mechanisms

The synthesis of virtually all proteins in the cell begins on ribosomes in the cytosol. The exceptions are the few mitochondrial and chloroplast proteins that are synthesized on ribosomes inside these organelles; most mitochondrial and chloroplast proteins, however, are made in the cytosol and subsequently imported. The fate of any protein molecule synthesized in the cytosol depends on its amino acid sequence, which can contain a *sorting signal* that directs the protein to the organelle in which it is required. Proteins that lack such signals remain as permanent residents in the cytosol; those that possess a sorting signal move from the cytosol to the appropriate organelle. Different sorting signals direct proteins into the nucleus, mitochondria, chloroplasts (in plants), peroxisomes, and the ER.

When a membrane-enclosed organelle imports proteins from the cytosol or from another organelle it faces a problem: how can it draw the protein across membranes that are normally impermeable to hydrophilic macromolecules? This task is accomplished in different ways for different organelles, but each mechanism requires an input of energy.

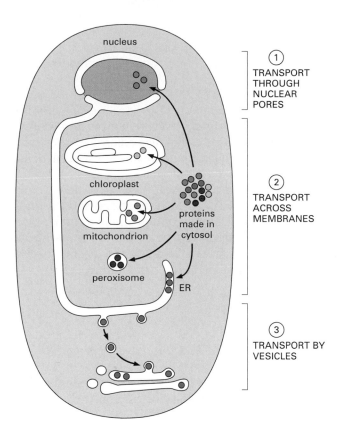

Figure 15–5 **Membrane-enclosed organelles import proteins by one of three mechanisms.** All of these processes require energy. The protein remains folded during the transport steps in mechanisms 1 and 3 but usually has to be unfolded in mechanism 2.

Figure labels:
- ① TRANSPORT THROUGH NUCLEAR PORES
- nucleus
- chloroplast
- mitochondrion
- proteins made in cytosol
- peroxisome
- ER
- ② TRANSPORT ACROSS MEMBRANES
- ③ TRANSPORT BY VESICLES

1. Proteins moving from the cytosol into the nucleus are transported through the nuclear pores that penetrate the inner and outer nuclear membranes; the pores function as selective gates that actively transport specific macromolecules but also allow free diffusion of smaller molecules (mechanism 1 in Figure 15–5).

2. Proteins moving from the cytosol into the ER, mitochondria, chloroplasts, or peroxisomes are transported across the organelle membrane by *protein translocators* located in the membrane; unlike transport through nuclear pores, the transported protein molecule must usually unfold in order to snake through the membrane (mechanism 2 in Figure 15–5). Bacteria have similar protein translocators in their plasma membrane.

3. Proteins moving from the ER onward and from one compartment of the endomembrane system to another are transported by a mechanism that is fundamentally different from the other two. These proteins are ferried by *transport vesicles*, which become loaded with a cargo of proteins from the interior space, or *lumen*, of one compartment, as they pinch off from its membrane. The vesicles subsequently discharge their cargo into a second compartment by fusing with its membrane (mechanism 3 in Figure 15–5). In the process, membrane lipids and membrane proteins are also delivered from the first compartment to the second.

Signal Sequences Direct Proteins to the Correct Compartment

The typical sorting signal on proteins is a continuous stretch of amino acid sequence, typically 15–60 amino acids long. This **signal sequence** is often (but not always) removed from the finished protein once the sorting decision has been executed. Some of the signal sequences used to specify different destinations in the cell are shown in Table 15–3.

Table 15–3 Some Typical Signal Sequences

FUNCTION OF SIGNAL	EXAMPLE OF SIGNAL SEQUENCE
Import into ER	^+H_3N-Met-Met-Ser-Phe-Val-Ser-Leu-Leu-Leu-Val-Gly-Ile-Leu-Phe-Trp-Ala-Thr-Glu-Ala-Glu-Gln-Leu-Thr-Lys-Cys-Glu-Val-Phe-Gln-
Retention in lumen of ER	-Lys-Asp-Glu-Leu-COO^-
Import into mitochondria	^+H_3N-Met-Leu-Ser-Leu-Arg-Gln-Ser-Ile-Arg-Phe-Phe-Lys-Pro-Ala-Thr-Arg-Thr-Leu-Cys-Ser-Ser-Arg-Tyr-Leu-Leu-
Import into nucleus	-Pro-Pro-Lys-Lys-Lys-Arg-Lys-Val-
Import into peroxisomes	-Ser-Lys-Leu-

Positively charged amino acids are shown in *red,* and negatively charged amino acids in *blue.* An extended block of hydrophobic amino acids is enclosed in a *yellow* box. ^+H_3N indicates the N-terminus of a protein; COO^- indicates the C-terminus. The ER retention signal is commonly referred to by its single-letter amino acid abbreviation, KDEL.

Signal sequences are both necessary and sufficient to direct a protein to a particular organelle. This has been shown by experiments in which the sequence is either deleted or transferred from one protein to another by genetic engineering techniques (discussed in Chapter 10). Deleting a signal sequence from an ER protein, for example, converts it into a cytosolic protein, while placing an ER signal sequence at the beginning of a cytosolic protein redirects the protein to the ER (Figure 15–6). The signal sequences specifying the same destination can vary greatly even though they have the same function: physical properties, such as hydrophobicity or the placement of charged amino acids, often appear to be more important for the function of these signals than the exact amino acid sequence.

Proteins Enter the Nucleus Through Nuclear Pores

The **nuclear envelope** encloses the nuclear DNA and defines the nuclear compartment. It is formed from two concentric membranes. The inner nuclear membrane contains proteins that act as binding sites for the chromosomes (discussed in Chapter 5) and for the *nuclear lamina,* a finely woven meshwork of protein filaments that lines the inner

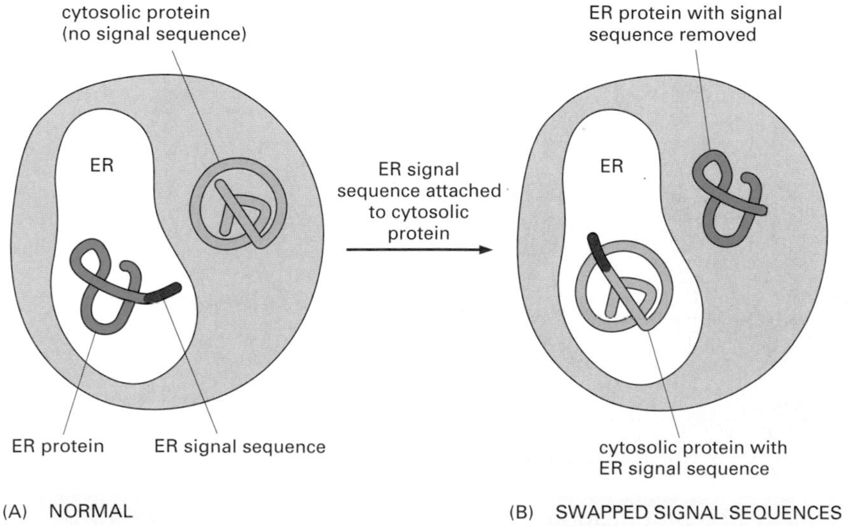

Figure 15–6 Signal sequences direct proteins to the correct organelle.
(A) Proteins destined for the ER possess an N-terminal signal sequence that directs them to that organelle, whereas those destined to remain in the cytosol lack this sequence. (B) In one experiment, recombinant DNA techniques are used to change the location of the two proteins: the signal sequence is removed from the ER protein and attached to the cytosolic protein. The result, illustrated here, is that the altered proteins are redirected, each ending up in an abnormal location in the cell. Such experiments indicate that the ER signal sequence is both necessary and sufficient to direct a protein to the ER.

face of this membrane and provides a structural support for the nuclear envelope (discussed in Chapter 17). The composition of the outer nuclear membrane closely resembles the membrane of the ER, with which it is continuous (Figure 15–7).

The nuclear envelope in all eucaryotic cells is perforated by **nuclear pores** that form the gates through which all molecules enter or leave the nucleus. Traffic occurs in both directions through the pores: newly made proteins destined for the nucleus enter from the cytosol; RNA molecules, which are synthesized in the nucleus, and ribosomal subunits, which are assembled in the nucleus, are exported. Messenger RNA molecules that are incompletely spliced are not exported from the nucleus, suggesting that nuclear transport serves as a final quality-control step in mRNA synthesis and processing (discussed in Chapter 7).

A nuclear pore is a large, elaborate structure composed of about 100 different proteins (Figure 15–8). Each pore contains one or more water-filled channels through which small water-soluble molecules can pass freely and nonselectively between the nucleus and the cytosol. Larger molecules (such as RNAs and proteins) and macromolecular complexes, however, cannot pass through the pores unless they carry an appropriate sorting signal. The signal sequence that directs a protein

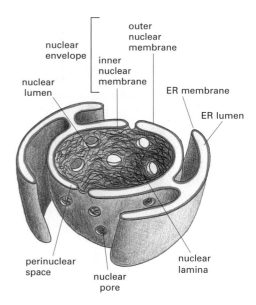

Figure 15–7 The outer nuclear membrane is continuous with the ER. The double membrane of the nuclear envelope is penetrated by nuclear pores. The ribosomes that are normally bound to the cytosolic surface of the ER membrane and outer nuclear membrane are not shown.

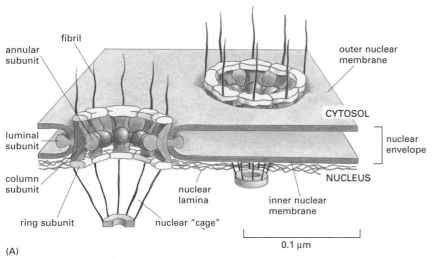

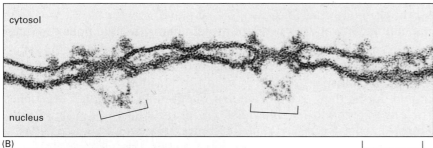

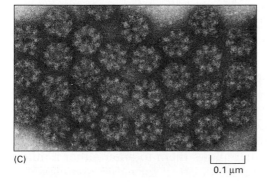

Figure 15–8 The nuclear pore complex forms a gate through which molecules enter or exit the nucleus. (A) Drawing of a small region of the nuclear envelope showing two pore complexes. Each complex is composed of a large number of distinct protein subunits. Protein fibrils protrude from both sides of the complex; on the nuclear side they converge to form a cagelike structure. The spacing between the fibrils is wide enough that the fibrils do not obstruct access to the pores. (B) Electron micrograph of a region of nuclear envelope showing a side view of two nuclear pore complexes *(brackets)*. (C) Electron micrograph showing a face-on view of nuclear pore complexes; the membranes have been extracted with detergent. (B, courtesy of Werner W. Franke; C, courtesy of Ron Milligan.)

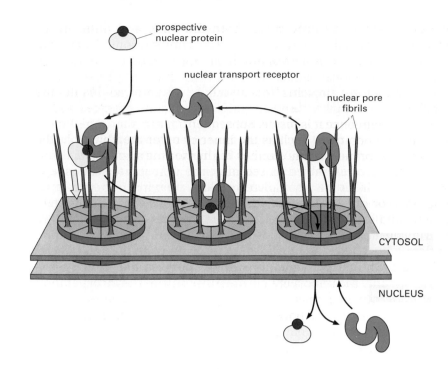

Figure 15–9 Proteins bound for the nucleus are actively transported through nuclear pores. First, specialized cytosolic proteins called nuclear transport receptors bind to the prospective nuclear protein. The resulting complex is guided to a nuclear pore by fibrils that extend from the pore into the cytosol. The binding of the nuclear protein to the pore opens the pore, and the nuclear protein, with its bound receptors, is actively transported into the nucleus. The receptors are then exported back through the pores into the cytosol for reuse. A similar type of transport receptor, operating in the reverse direction, exports mRNAs from the nucleus (discussed in Chapter 7; see Figure 7–19); both groups of receptors have a similar basic structure.

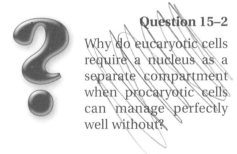

Question 15–2

Why do eucaryotic cells require a nucleus as a separate compartment when procaryotic cells can manage perfectly well without?

from the cytosol into the nucleus, called a *nuclear localization signal*, typically consists of one or two short sequences containing several positively charged lysines or arginines (see Table 15–3, p. 504).

The initial interaction of a newly synthesized prospective nuclear protein with a nuclear pore requires the aid of other proteins in the cytosol. These cytosolic proteins, called *nuclear transport receptors*, bind to the nuclear localization signal and help direct the new protein to the pore by interacting with the nuclear pore fibrils (Figure 15–9). The prospective nuclear protein is then actively transported into the nucleus by a process that uses the energy provided by GTP hydrolysis. A structure in the center of the nuclear pore functions like a close-fitting diaphragm: it opens just the right amount to allow the protein complex to pass through. The nuclear transport receptors are then returned to the cytosol via the nuclear pore for reuse (see Figure 15–9). How nuclear pore proteins operate this molecular gate, which can pump macromolecules in both directions through the pores while avoiding congestion and head-on collisions, is currently unknown.

The nuclear pores transport proteins in their fully folded conformation and transfer ribosomal components as assembled particles. This distinguishes the nuclear transport mechanism from the mechanisms that transport proteins into other organelles. Proteins have to unfold during their transport across membranes into other organelles such as mitochondria, chloroplasts, and the ER, as we discuss next. Little is known about how proteins are transported into peroxisomes.

Proteins Unfold to Enter Mitochondria and Chloroplasts

Both mitochondria and chloroplasts are surrounded by inner and outer membranes, and both organelles specialize in the synthesis of ATP. Chloroplasts also contain a third membrane system, the thylakoid membrane (discussed in Chapter 14). Although both organelles contain their own genomes and make some of their own proteins, most mitochondrial and chloroplast proteins are encoded by genes in the nucleus and are imported from the cytosol. These proteins usually have a signal sequence at their N-terminus that allows them to enter either a mitochondrion or a chloroplast. Proteins destined for either organelle are translocated simultaneously across both the inner and outer membranes at specialized sites where the two membranes are in contact with each other. Each protein is unfolded as it is transported, and the signal sequence is cleaved off after translocation is completed (Figure 15–10). Chaperone proteins (discussed in Chapter 4) inside the organelles help to pull the protein across the two membranes and to refold the protein once it is inside. Subsequent transport to a particular site within the organelle, such as the inner or outer membrane or the thylakoid membrane, usually requires further sorting signals in the protein, which are often only exposed after the first signal sequence is removed. The insertion of transmembrane proteins into the inner membrane, for example, is guided by signal sequences in the protein that start and stop the transfer process across the membrane, as we describe later for the insertion of transmembrane proteins in the ER membrane.

The growth and maintenance of mitochondria and chloroplasts requires not only the import of new proteins but also the incorporation of new lipids into their membranes. Most of their membrane phospholipids are thought to be imported from the ER, which is the main site of lipid synthesis in the cell. Phospholipids are transported individually to these organelles by water-soluble lipid-carrying proteins that extract a phospholipid molecule from one membrane and deliver it into another. These proteins ensure that the different cellular membranes retain their characteristic lipid composition.

Proteins Enter the Endoplasmic Reticulum While Being Synthesized

The **endoplasmic reticulum (ER)** is the most extensive membrane system in a eucaryotic cell (Figure 15–11A), and, unlike the organelles discussed so far, it serves as an entry point for proteins destined for other organelles, as well as for the ER itself. Proteins destined for the Golgi apparatus, endosomes, and lysosomes, as well as proteins destined for the cell surface, all first enter the ER from the cytosol. Once inside the ER or in the ER membrane, individual proteins will not reenter the cytosol during their onward journey. They will be ferried by transport vesicles from organelle to organelle and, in some cases, from organelle to plasma membrane or the cell exterior.

Two kinds of proteins are transferred from the cytosol to the ER: (1) water-soluble proteins are completely translocated across the ER membrane and are released into the ER lumen; (2) prospective transmembrane proteins are only partly translocated across the ER membrane and become embedded in it. The water-soluble proteins are destined either for secretion (by release at the cell surface) or for the lumen of an

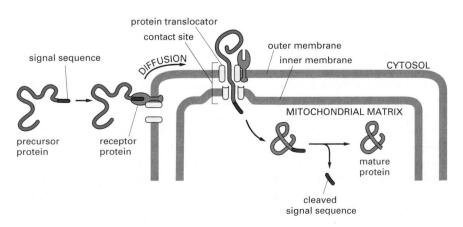

Figure 15–10 An unfolded protein is imported into a mitochondrion. The signal sequence is recognized by a receptor in the outer mitochondrial membrane. The complex of receptor and attached protein diffuses laterally in the membrane to a contact site, where the protein is translocated across both the outer and inner membranes by a protein translocator. The signal sequence is cleaved off by a signal peptidase inside the organelle. Proteins are imported into chloroplasts by a similar mechanism. The chaperone proteins that help to pull the protein across the membranes and help it to refold are not shown.

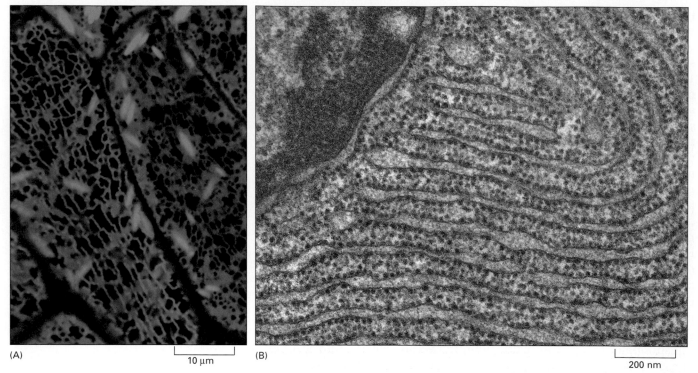

(A) 10 µm   (B) 200 nm

Figure 15–11 The endoplasmic reticulum is the most extensive membrane network in eucaryotic cells.
(A) Fluorescence micrograph of living plant cells showing the ER as a complex network of sheets and tubes. The cells shown here have been genetically engineered so that they contain a fluorescent protein in their ER. (B) An electron micrograph showing the rough ER in a cell from a dog's pancreas that makes and secretes large amounts of digestive enzymes. The cytosol is filled with closely packed sheets of ER studded with ribosomes. At the top left is a portion of the nucleus and its nuclear envelope; note that the outer nuclear membrane, which is continuous with the ER, is also studded with ribosomes. (A, courtesy of Jim Haseloff; B, courtesy of Lelio Orci.)

organelle; the transmembrane proteins are destined to reside in either the ER membrane, the membrane of another organelle, or the plasma membrane. All of these proteins are initially directed to the ER by an *ER signal sequence,* a segment of eight or more hydrophobic amino acids (see Table 15–3, p. 504) that is also involved in the process of translocation across the membrane.

Unlike the proteins that enter the nucleus, mitochondria, chloroplasts, and peroxisomes, most of the proteins that enter the ER begin to be threaded across the ER membrane before the polypeptide chain is completely synthesized. This requires that the ribosome synthesizing the protein be attached to the ER membrane. These membrane-bound ribosomes coat the surface of the ER, creating regions termed **rough endoplasmic reticulum** because of the characteristic beaded appearance when viewed in an electron microscope (Figure 15–11B).

There are, therefore, two separate populations of ribosomes in the cytosol. *Membrane-bound ribosomes* are attached to the cytosolic side of the ER membrane (and outer nuclear membrane) and are making proteins that are being translocated into the ER. *Free ribosomes* are unattached to any membrane and are making all of the other proteins encoded by the nuclear DNA. Membrane-bound ribosomes and free ribosomes are structurally and functionally identical; they differ only in the proteins they are making at any given time. When a ribosome happens to be making a protein with an ER signal sequence, the signal sequence directs the ribosome to the ER membrane. As an mRNA molecule is translated, many ribosomes bind to it, forming a *polyribosome* (discussed in Chapter 7). In the case of an mRNA molecule encoding a protein with an ER signal sequence, the polyribosome becomes riveted to the ER membrane by the growing polypeptide chains, which have become inserted into the membrane (Figure 15–12).

Soluble Proteins Are Released into the ER Lumen

The ER signal sequence is guided to the ER membrane with the aid of at least two components: (1) a *signal-recognition particle (SRP)*, present in the cytosol, which binds to the ER signal sequence when it is exposed on the ribosome, and (2) an *SRP receptor*, embedded in the membrane of the ER, which recognizes the SRP. Binding of an SRP to a signal sequence causes protein synthesis by the ribosome to slow down, until the ribosome and its bound SRP bind to an SRP receptor. After binding to its receptor, the SRP is released and protein synthesis recommences, with the polypeptide now being threaded into the lumen of the ER through a *translocation channel* in the ER membrane (Figure 15–13). Thus the SRP and SRP receptor function as molecular matchmakers, connecting ribosomes that are synthesizing proteins containing ER signal sequences to available ER translocation channels.

In addition to directing proteins to the ER, the signal sequence—which for soluble proteins is almost always at the N-terminus—functions to open the translocation channel. The signal peptide remains bound to the channel while the rest of the protein chain is threaded through the membrane as a large loop. At some stage during translocation, the signal sequence is cleaved off by a signal peptidase located on the luminal side of the ER membrane; the signal peptide is then released from the translocation channel and rapidly degraded to amino acids. Once the C-terminus of the protein has passed through the membrane, the protein is released into the ER lumen (Figure 15–14).

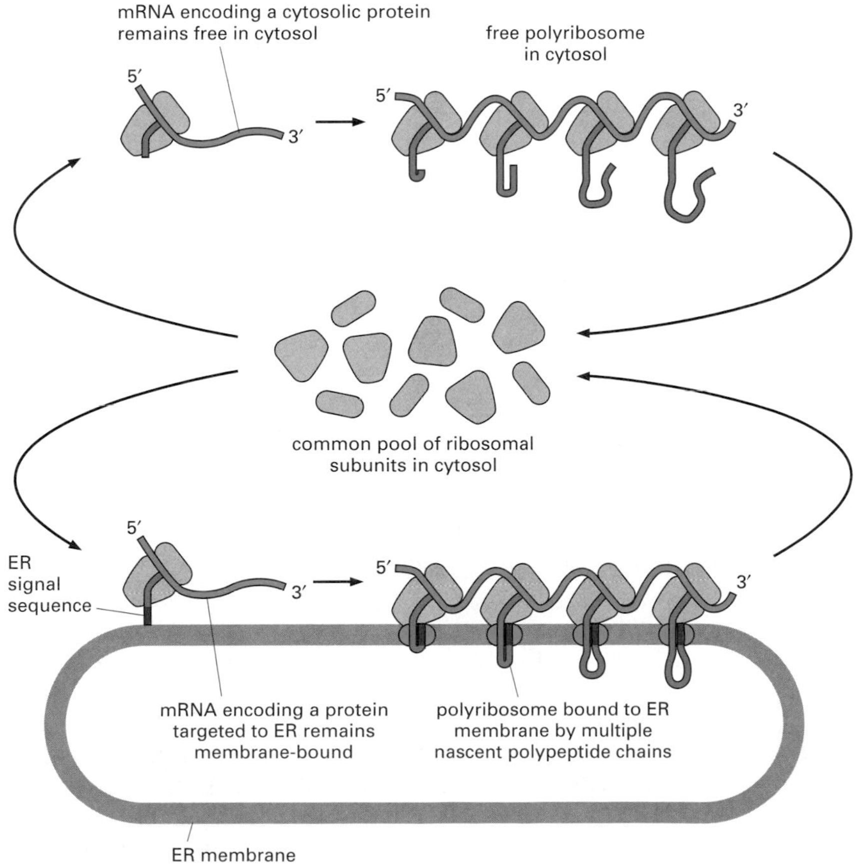

mRNA encoding a cytosolic protein remains free in cytosol

free polyribosome in cytosol

common pool of ribosomal subunits in cytosol

ER signal sequence

mRNA encoding a protein targeted to ER remains membrane-bound

polyribosome bound to ER membrane by multiple nascent polypeptide chains

ER membrane

Figure 15–12 A common pool of ribosomes is used to synthesize both the proteins that stay in the cytosol and those that are transported into membrane-enclosed organelles, including the ER. Ribosomes that are translating cytosolic proteins remain free in the cytosol. For proteins that are bound for the ER, a signal sequence *(red)* on the growing polypeptide chain directs the ribosome to the ER membrane. Many ribosomes bind to each mRNA molecule, forming a polyribosome. At the end of each round of protein synthesis, the ribosomal subunits are released and rejoin the common pool in the cytosol.

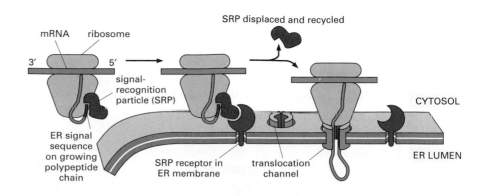

Figure 15–13 An ER signal sequence and an SRP direct a ribosome to the ER membrane. The SRP binds to the exposed ER signal sequence and to the ribosome, thereby slowing protein synthesis by the ribosome. The SRP-ribosome complex then binds to an SRP receptor in the ER membrane. The SRP is then released, passing the ribosome to a protein translocation channel in the ER membrane. The protein translocation channel then inserts the polypeptide chain into the membrane and starts to transfer it across the lipid bilayer.

Start and Stop Signals Determine the Arrangement of a Transmembrane Protein in the Lipid Bilayer

Not all proteins that enter the ER are released into the ER lumen. Some remain embedded in the ER membrane as transmembrane proteins. The translocation process for such proteins is more complicated than it is for soluble proteins, as some parts of the polypeptide chain must be translocated clear across the lipid bilayer while others remain fixed in the membrane.

In the simplest case, that of a transmembrane protein with a single membrane-spanning segment, the N-terminal signal sequence initiates translocation, just as for a soluble protein. But the transfer process is halted by an additional sequence of hydrophobic amino acids, a *stop-transfer sequence*, further into the polypeptide chain (Figure 15–15). This second sequence is released from the translocation channel and drifts into the plane of the lipid bilayer, where it forms an α-helical membrane-spanning segment that anchors the protein in the membrane. Simultaneously, the N-terminal signal sequence is also released from the channel into the lipid bilayer and is cleaved off. As a result, the translocated protein ends up as a transmembrane protein inserted in the membrane with a defined orientation—the N-terminus on the luminal side of the lipid bilayer and the C-terminus on the cytosolic side (see Figure 15–15). As discussed in Chapter 11, once inserted into the membrane, a transmembrane protein does not change its orientation, which is retained throughout any subsequent vesicle budding and fusion events.

Figure 15–14 A soluble protein crosses the ER membrane and enters the lumen. A protein translocation channel binds the signal sequence and actively transfers the rest of the polypeptide across the lipid bilayer as a loop. At some point during the translocation process, the signal peptide is cleaved from the growing protein by a signal peptidase. The translocation channel then opens and ejects the signal sequence into the bilayer, where it is degraded. The translocated polypeptide is released as a soluble protein into the ER lumen. It is thought that protein serving as a plug then binds from the ER lumen to close the inactive channel. The membrane-bound ribosome is omitted from this and the following two figures for clarity.

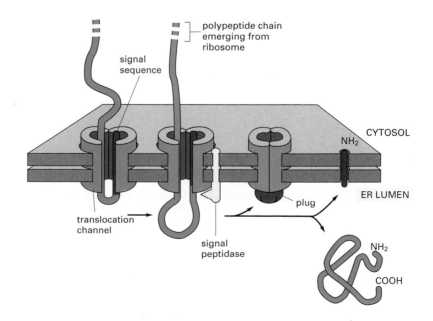
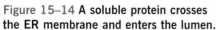

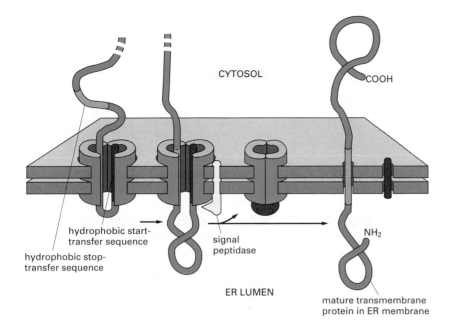

CYTOSOL

COOH

hydrophobic start-
transfer sequence

hydrophobic stop-
transfer sequence

signal
peptidase

NH₂

ER LUMEN

mature transmembrane
protein in ER membrane

Figure 15–15 A transmembrane protein is integrated into the ER membrane. An N-terminal ER signal sequence *(red)* initiates transfer as in Figure 15–14. In addition, the protein also contains a second hydrophobic sequence, a stop-transfer sequence *(orange)*. When this sequence enters the translocation channel, the channel discharges the protein sideways into the lipid bilayer. The N-terminal signal sequence is cleaved off, leaving the transmembrane protein anchored in the membrane. Protein synthesis on the cytosolic side continues to completion.

In some transmembrane proteins, an internal, rather than an N-terminal, signal sequence is used to start the protein transfer; this internal signal sequence, called a *start-transfer sequence*, is never removed from the polypeptide. This arrangement occurs in some transmembrane proteins in which the polypeptide chain passes back and forth across the lipid bilayer. In these cases hydrophobic signal sequences are thought to work in pairs: an internal start-transfer sequence serves to initiate translocation, which continues until a stop-transfer sequence is reached; the two hydrophobic sequences are then released into the bilayer, where they remain as membrane-spanning α helices (Figure 15–16). In complex multipass proteins, in which many hydrophobic α helices span the bilayer, additional pairs of stop and start sequences come into play: one sequence reinitiates translocation further down the polypeptide chain, and the other stops translocation and causes polypeptide release, and so on for subsequent starts and stops. Thus, multipass membrane proteins are stitched into the lipid bilayer as they

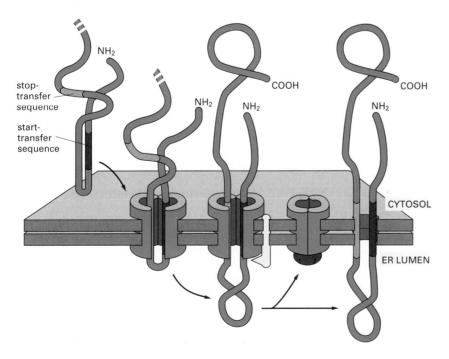

NH₂

stop-
transfer
sequence

start-
transfer
sequence

COOH

NH₂

NH₂

COOH

NH₂

CYTOSOL

ER LUMEN

Figure 15–16 A double-pass transmembrane protein uses an internal start-transfer sequence to integrate into the ER membrane. An internal ER signal sequence *(red)* acts as a start-transfer signal and initiates the transfer of the polypeptide chain. Like the N-terminal ER signal sequence, the internal start-transfer signal is recognized by an SRP that brings the ribosome to the ER membrane (not shown). When a stop-transfer sequence *(orange)* enters the translocation channel, the channel discharges both sequences into the plane of the lipid bilayer. Neither the start-transfer nor the stop-transfer sequence is cleaved off, and the entire polypeptide chain remains anchored in the membrane as a double-pass transmembrane protein. Proteins that span the membrane more times contain further pairs of stop and start sequences, and the same process is repeated for each pair.

Protein Sorting **511**

A. Predict the membrane orientation of a protein that is synthesized with an uncleaved, internal signal sequence (shown as the *red* start-transfer sequence in Figure 15–16) but does not contain a stop-transfer peptide.

B. Similarly, predict the membrane orientation of a protein that is synthesized with an N-terminal cleaved signal sequence followed by a stop-transfer sequence, followed by a start-transfer sequence.

C. What arrangement of signal sequences would enable the insertion of a multipass protein with an odd number of transmembrane segments?

are being synthesized, by a mechanism resembling the workings of a sewing machine.

Having considered how proteins enter the ER lumen or become embedded in the ER membrane, we now discuss how they are carried onward by vesicular transport.

Vesicular Transport

Entry into the ER is usually only the first step on a pathway to another destination. That destination, initially at least, is the Golgi apparatus. Transport from the ER to the Golgi apparatus and from the Golgi apparatus to other compartments of the endomembrane system is carried out by the continual budding and fusion of **transport vesicles**. The transport pathways mediated by these vesicles extend outward from the ER to the plasma membrane, and inward from the plasma membrane to lysosomes, and thus provide routes of communication between the interior of the cell and its surroundings. As proteins and lipids are transported outward along these pathways, many of them undergo various types of chemical modification, such as the addition of carbohydrate side chains (to both proteins and lipids) and the formation of disulfide bonds (in polypeptides) that stabilize protein structure.

In this section we discuss how vesicles shuttle proteins and membranes between cellular compartments, allowing cells to eat and secrete. We also consider how these transport vesicles are directed to their proper destination, be it the ER, Golgi apparatus, plasma membrane, or some other membrane-enclosed compartment.

Transport Vesicles Carry Soluble Proteins and Membrane Between Compartments

The vesicular traffic between membrane-enclosed compartments of the endomembrane system is highly organized. A major outward *secretory pathway* starts with the biosynthesis of proteins on the ER membrane and their entry into the ER, and leads through the Golgi apparatus to the cell surface; at the Golgi apparatus a side branch leads off through endosomes to lysosomes (Figure 15–17). A major

Figure 15–17 Vesicles bud from one membrane and fuse with another, carrying membrane components and soluble proteins between cellular compartments. Each compartment encloses a space, or lumen, that is topologically equivalent to the outside of the cell (see Figure 11–19). The extracellular space and each of the membrane-enclosed compartments (*shaded gray*) communicate with one another by means of transport vesicles, as shown. In the outward secretory pathway (*red arrows*) protein molecules are transported from the ER, through the Golgi apparatus, to the plasma membrane or (via late endosomes) to lysosomes. In the inward endocytic pathway (*green arrows*) extracellular molecules are ingested in vesicles derived from the plasma membrane and are delivered to early endosomes and then (via late endosomes) to lysosomes.

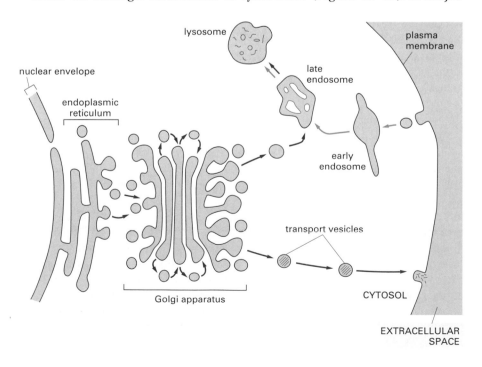

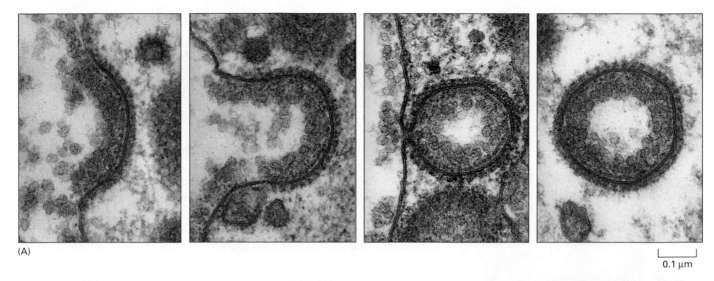

0.1 µm

inward *endocytic pathway*, which is responsible for the ingestion and degradation of extracellular molecules, moves materials from the plasma membrane, through endosomes, to lysosomes.

To function correctly, each transport vesicle that buds off from a compartment must take with it only the proteins appropriate to its destination and must fuse only with the appropriate target membrane. A vesicle carrying cargo from the Golgi apparatus to the plasma membrane, for example, must exclude proteins that are to stay in the Golgi apparatus, and it must fuse only with the plasma membrane and not with any other organelle. While participating in this constant flow of membrane components, each organelle must maintain its own distinct identity, that is, its own distinctive protein and lipid composition. All of these recognition events depend on proteins associated with the transport vesicle membrane. As we shall see, different types of transport vesicles shuttle between the various organelles, each carrying a distinct set of molecules.

Vesicle Budding Is Driven by the Assembly of a Protein Coat

Vesicles that bud from membranes usually have a distinctive protein coat on their cytosolic surface and are therefore called **coated vesicles**. After budding from its parent organelle, the vesicle sheds its coat, allowing its membrane to interact directly with the membrane to which it will fuse. Cells produce several kinds of coated vesicles, each with a distinctive protein coat. The coat is thought to serve at least two functions: it shapes the membrane into a bud, and it helps to capture molecules for onward transport.

The best-studied vesicles are those that have coats made largely of the protein **clathrin**. These *clathrin-coated vesicles* bud from the Golgi apparatus on the outward secretory pathway and from the plasma membrane on the inward endocytic pathway. At the plasma membrane, for example, each vesicle starts off as a *clathrin-coated pit*. Clathrin molecules assemble into a basketlike network on the cytosolic surface of the membrane, and it is this assembly process that starts shaping the membrane into a vesicle (Figure 15–18). A small GTP-binding protein called *dynamin* assembles as a ring around the neck of each deeply invaginated coated pit. Together with other proteins recruited to the neck of the vesicle, the dynamin causes the ring to constrict, thereby pinching off the vesicle from the membrane. Other kinds of transport vesicles, with different coat proteins, are also involved in vesicular

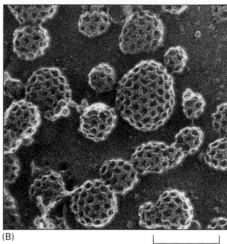

(B)

0.2 µm

Figure 15–18 Clathrin molecules form basketlike cages that help shape membranes into vesicles. (A) Electron micrographs showing the sequence of events in the formation of a clathrin-coated vesicle from a clathrin-coated pit. The clathrin-coated pits and vesicles shown here are unusually large and are being formed at the plasma membrane of a hen oocyte. They are involved in taking up particles made of lipid and protein into the oocyte to form yolk. (B) Electron micrograph showing numerous clathrin-coated pits and vesicles budding from the inner surface of the plasma membrane of cultured skin cells. (A, courtesy of M.M. Perry and A.B. Gilbert, *J. Cell Sci.* 39:257–272, 1979. © The Company of Biologists; B, from J. Heuser, *J. Cell Biol.* 84:560–583, 1980. © Rockefeller University Press.)

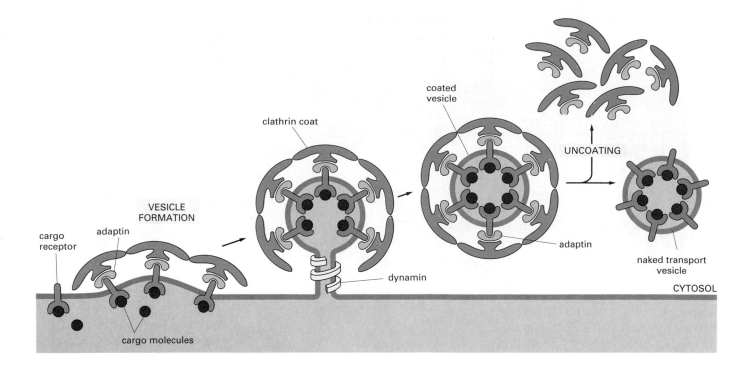

Figure 15–19 Clathrin-coated vesicles transport selected cargo molecules. Cargo receptors, with their bound cargo molecules, are captured by adaptins, which also bind clathrin molecules to the cytosolic surface of the budding vesicle. Dynamin proteins assemble around the neck of budding vesicles; once assembled, the dynamin molecules hydrolyze their bound GTP and, with the help of other proteins recruited to the area, pinch off the vesicle. After budding is complete, the coat proteins are removed and the naked vesicle can fuse with its target membrane. Functionally similar coat proteins are found in other types of coated vesicles.

transport. They form in a similar way and carry their own characteristic sets of molecules between the endoplasmic reticulum, the Golgi apparatus, and the plasma membrane. But how does a transport vesicle select its particular cargo? The mechanism is best understood for clathrin-coated vesicles.

Clathrin itself plays no part in capturing specific molecules for transport. This is the function of a second class of coat proteins called *adaptins,* which both secure the clathrin coat to the vesicle membrane and help select cargo molecules for transport. Molecules for onward transport carry specific *transport signals* that are recognized by *cargo receptors* in the compartment membrane. Adaptins help capture specific cargo molecules by trapping the cargo receptors that bind them. In this way a selected set of cargo molecules, bound to their specific receptors, is incorporated into the lumen of each newly formed clathrin-coated vesicle (Figure 15–19). There are at least two types of adaptins: the adaptins that bind cargo receptors in the plasma membrane are different from those that bind cargo receptors in the Golgi apparatus, reflecting the differences in the cargo molecules from each of these sources.

A different class of coated vesicles, called *COP-coated vesicles* (COP is shorthand for "coat protein"), is involved in transporting molecules between the ER and the Golgi apparatus and from one part of the Golgi apparatus to another (Table 15–4).

Table 15–4 Some Types of Coated Vesicles

TYPE OF COATED VESICLE	COAT PROTEINS	ORIGIN	DESTINATION
Clathrin-coated	clathrin + adaptin 1	Golgi apparatus	lysosome (via endosomes)
Clathrin-coated	clathrin + adaptin 2	plasma membrane	endosomes
COP-coated	COP proteins	ER Golgi cisterna Golgi apparatus	Golgi apparatus Golgi cisterna ER

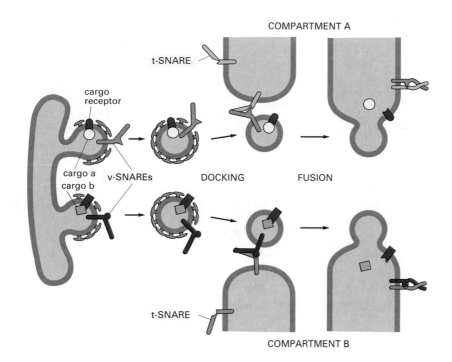

COMPARTMENT A

t-SNARE

cargo receptor

cargo a
cargo b   v-SNAREs   DOCKING   FUSION

t-SNARE

COMPARTMENT B

Figure 15–20 SNAREs help direct transport vesicles to their target membranes. Vesicles that bud from a membrane carry specific marker proteins called vesicle SNAREs (v-SNAREs) on their surfaces, which bind to complementary target SNAREs (t-SNAREs) on the target membrane. Many different complementary pairs of v-SNAREs and t-SNAREs are thought to play a crucial role in guiding transport vesicles to their appropriate target membranes.

The Specificity of Vesicle Docking Depends on SNAREs

After a transport vesicle buds from a membrane, it must find its way to its correct destination to deliver its contents. In most cases, the vesicle is actively transported by motor proteins that move along cytoskeletal fibers, as discussed in Chapter 17.

Once a transport vesicle has reached its target it must recognize and dock with the organelle. Only then can the vesicle membrane fuse with the target membrane and unload the vesicle's cargo. The impressive specificity of vesicular transport suggests that each type of transport vesicle in the cell displays on its surface molecular markers that identify the vesicle according to its origin and cargo. These markers must be recognized by complementary receptors on the appropriate target membrane, including the plasma membrane. Although the mechanism of this recognition process is not known for certain, it is thought to involve a family of related transmembrane proteins called **SNAREs**. SNAREs on the vesicle (called v-SNAREs) are recognized specifically by complementary SNAREs on the cytosolic surface of the target membrane (called t-SNAREs; Figure 15–20). Each organelle and each type of transport vesicle is believed to carry a unique SNARE, and the interactions between complementary SNAREs help ensure that transport vesicles fuse only with the correct membrane.

Once a transport vesicle has recognized its target membrane and docked there, the vesicle has to fuse with the membrane to deliver its cargo. Fusion not only delivers the contents of the vesicle into the interior of the target organelle, it also adds the vesicle membrane to the membrane of the organelle. Membrane fusion does not always follow immediately after docking, however; it often awaits a specific molecular signal. Whereas docking requires only that the two membranes come close enough for proteins protruding from the two lipid bilayers to interact, fusion requires a much closer approach: the two bilayers must come within 1.5 nm of each other so that their lipids can intermix. For this close approach, water must be displaced from the hydrophilic surface of the membrane—a process that is energetically highly unfavorable. It is likely, therefore, that all membrane fusions in cells are catalyzed by specialized proteins that assemble at the fusion site to form a

Question 15–5

The budding of clathrin-coated vesicles from eucaryotic plasma membrane fragments can be observed when adaptins, clathrin, and dynamin-GTP are added to the membrane preparation. What would you observe if you omitted (A) adaptins, (B) clathrin, or (C) dynamin? (D) What would you observe if the plasma membrane fragments were from a procaryotic cell?

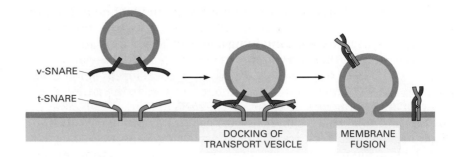

Figure 15–21 SNARE proteins play a central role in membrane fusion. Pairing of v-SNAREs and t-SNAREs forces the two lipid bilayers into close apposition. Lipids then flow between the two bilayers and the membranes fuse. In a cell, other proteins recruited to the fusion site presumably cooperate with SNAREs to initiate fusion. Additional proteins help to pry the SNAREs apart.

v-SNARE

t-SNARE

DOCKING OF TRANSPORT VESICLE

MEMBRANE FUSION

fusion complex that provides the means to cross this energy barrier. The SNARE proteins themselves are thought to play a central role in the fusion process: after pairing, v-SNAREs and t-SNAREs wrap around each other, thereby acting like a winch that pulls the two membranes into close proximity (Figure 15–21).

Secretory Pathways

Vesicular traffic is not confined to the interior of the cell. It extends to and from the plasma membrane. Newly made proteins, lipids, and carbohydrates are delivered from the ER, via the Golgi apparatus, to the cell surface by transport vesicles that fuse with the plasma membrane in a process called **exocytosis**. Each molecule that travels along this route passes through a fixed sequence of membrane-enclosed compartments and is often chemically modified en route.

In this section we follow the outward path of proteins as they travel from the ER, where they are made and modified, through the Golgi apparatus, where they are further modified and sorted, to the plasma membrane. As a protein passes from one compartment to another, it is monitored to check that it has folded properly and assembled with its appropriate partners, so that only correctly built proteins are released at the cell surface, while all of the others are degraded in the cell.

Most Proteins Are Covalently Modified in the ER

Most proteins that enter the ER are chemically modified there. Disulfide bonds are formed by the oxidation of pairs of cysteine side chains (see Figure 4–29), a reaction catalyzed by an enzyme that resides in the ER lumen. The disulfide bonds help to stabilize the structure of those proteins that may encounter changes in pH and degradative enzymes outside the cell—either after they are secreted or after they are incorporated into the plasma membrane. Disulfide bonds do not form in the cytosol, because of the reducing environment there.

Many of the proteins that enter the ER lumen or ER membrane are converted to glycoproteins in the ER by the covalent attachment of short oligosaccharide side chains. This process of *glycosylation* is carried out by glycosylating enzymes found in the ER but not in the cytosol. Very few proteins in the cytosol are glycosylated, and those that are have only a single sugar residue attached to them. The oligosaccharides on proteins serve various functions, depending on the protein. They can protect the protein from degradation, hold it in the ER until it is properly folded, or help guide it to the appropriate organelle by serving as a transport signal for packaging the protein into appropriate transport vesicles (as in the case of lysosomal proteins discussed later). When displayed on the cell surface, oligosaccharides form part of the cell's carbohydrate layer (see Figure 11–32) and can function in the recognition of one cell by another.

In the ER, individual sugars are not added one-by-one to the protein to create the oligosaccharide side chain. Instead, a preformed, branched oligosaccharide containing a total of 14 sugars is attached en bloc to all proteins that carry the appropriate site for glycosylation. The oligosaccharide is originally attached to a specialized lipid, called *dolichol,* in the ER membrane; it is then transferred to the amino (NH_2) group of an asparagine side chain on the protein immediately after the target asparagine emerges in the ER lumen during protein translocation (Figure 15–22). The addition takes place in a single enzymatic step catalyzed by a membrane-bound enzyme (an oligosaccharide protein transferase) that has its active site exposed on the luminal side of the ER membrane, which explains why cytosolic proteins are not glycosylated in this way. A simple sequence of three amino acids, of which the asparagine is one, defines which asparagine residues in a protein receive the oligosaccharide. Oligosaccharide side chains linked to an asparagine NH_2 group in a protein are said to be *N-linked* and are by far the most common type of linkage found on glycoproteins.

The addition of the 14-sugar oligosaccharide in the ER is only the first step in a series of further modifications before the mature glycoprotein emerges at the other end of the outward pathway. Despite their initial similarity, the *N*-linked oligosaccharides on mature glycoproteins are remarkably diverse. All of the diversity results from extensive modification of the original precursor structure shown in Figure 15–22. This *oligosaccharide processing* begins in the ER and continues in the Golgi apparatus.

Exit from the ER Is Controlled to Ensure Protein Quality

Some proteins made in the ER are destined to function there. They are retained in the ER (and are returned to the ER when they escape to the Golgi apparatus) by a C-terminal sequence of four amino acids called an *ER retention signal* (see Table 15–3, p. 504), which is recognized by a membrane-bound receptor protein in the ER and Golgi apparatus. Most proteins that enter the ER, however, are destined for other locations;

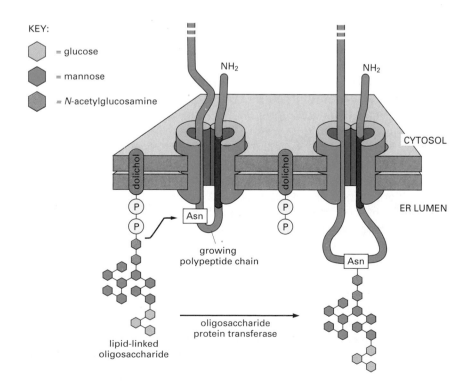

KEY:

= glucose

= mannose

= *N*-acetylglucosamine

Figure 15–22 Many proteins are glycosylated in the ER. Almost as soon as the polypeptide chain enters the ER lumen, it is glycosylated by addition of oligosaccharide side chains to particular asparagines in the polypeptide. Each oligosaccharide chain is transferred as an intact unit to the asparagine from a lipid called dolichol. Asparagines that are glycosylated are always present in the tripeptide sequences asparagine-X-serine or asparagine-X-threonine, where X can be any amino acid.

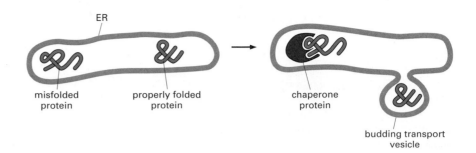

ER

misfolded protein

properly folded protein

chaperone protein

budding transport vesicle

Figure 15–23 Chaperones prevent misfolded or partially assembled proteins from leaving the ER. Misfolded proteins bind to chaperone proteins in the ER lumen and are thereby retained, whereas normally folded proteins are transported in transport vesicles to the Golgi apparatus. If the misfolded proteins fail to refold normally, they are transported into the cytosol, where they are degraded.

they are packaged into transport vesicles that bud from the ER and fuse with the Golgi apparatus. Exit from the ER, however, is highly selective. Proteins that fold up incorrectly, and dimeric or multimeric proteins that fail to assemble properly, are actively retained in the ER by binding to chaperone proteins that reside there. Interaction with chaperones holds the proteins in the ER until proper folding occurs; otherwise, the proteins are ultimately degraded (Figure 15–23). Antibody molecules, for example, are composed of four polypeptide chains (see Figure 4–32) that assemble into the complete antibody molecule in the ER. Partially assembled antibodies are retained in the ER until all four polypeptide chains have assembled; any antibody molecule that fails to assemble properly is ultimately degraded. In this way the ER controls the quality of the proteins that it exports to the Golgi apparatus.

Sometimes, however, this quality-control mechanism can be detrimental to the organism. The predominant mutation that causes the common genetic disease *cystic fibrosis,* which causes severe degeneration of the lung, for example, produces a plasma-membrane transport protein that is slightly misfolded; even though the mutant protein could function normally as a chloride channel if it reached the plasma membrane, it is retained in the ER, with dire consequences. The devastating disease results not because the mutation inactivates an important protein but because the active protein is discarded by the cells before it is given an opportunity to function.

Proteins Are Further Modified and Sorted in the Golgi Apparatus

The **Golgi apparatus** is usually located near the cell nucleus, and in animal cells it is often close to the centrosome, a small structure near the cell center. This organelle consists of a collection of flattened, membrane-enclosed sacs *(cisternae;* singular, *cisterna),* which are piled like stacks of plates. Each stack contains 3–20 cisternae (Figure 15–24). The number of Golgi stacks per cell varies greatly depending on the cell type: some cells contain one large stack, while others contain hundreds of very small ones.

Each Golgi stack has two distinct faces: an entry, or *cis,* face and an exit, or *trans,* face. The *cis* face is adjacent to the ER, while the *trans* face points toward the plasma membrane. The outermost cisterna at each face is connected to a network of interconnected membranous tubes and vesicles (see Figure 15–24A). Soluble proteins and membrane enter the *cis Golgi network* via transport vesicles derived from the ER. The proteins travel through the cisternae in sequence by means of transport vesicles that bud from one cisterna and fuse with the next. Proteins exit from the *trans* Golgi network in transport vesicles destined for either the cell surface or another compartment (see Figure 15–17). Both the *cis* and *trans* Golgi networks are thought to be important for protein sorting: proteins entering the *cis* Golgi network can either move onward through the Golgi stack or, if they contain an ER retention signal, be

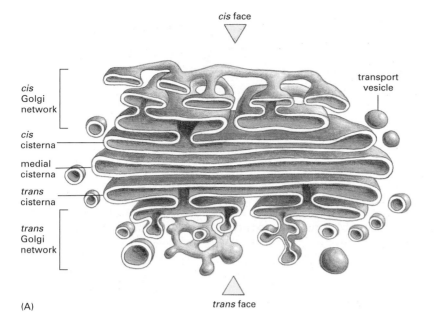

cis face

cis Golgi network

cis cisterna

medial cisterna

trans cisterna

trans Golgi network

transport vesicle

(A)

trans face

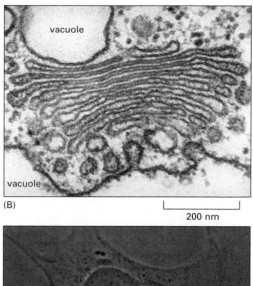

vacuole

vacuole

(B)

200 nm

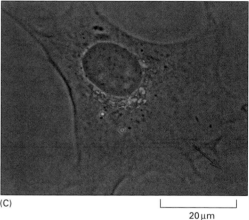

(C)

20 μm

Figure 15–24 The Golgi apparatus is made of a stack of flattened, membrane-enclosed sacs. (A) Three-dimensional reconstruction of a Golgi stack. It was reconstructed from electron micrographs of the Golgi apparatus in a secretory animal cell. (B) Electron micrograph of a Golgi stack from a plant cell, where the Golgi apparatus is especially distinct. The Golgi apparatus is oriented as in (A). (C) The Golgi apparatus in a cultured fibroblast stained with a fluorescent antibody that labels the Golgi apparatus specifically. A *red arrow* indicates the direction of the cell's movement. The *cis* face of the Golgi apparatus is close to the nucleus, and its *trans* face is oriented toward the direction of movement. (A, redrawn from A. Rambourg and Y. Clermont, *Eur. J. Cell Biol.* 51:189–200, 1990; B, courtesy of George Palade; C, courtesy of John Henley and Mark McNiven.)

returned to the ER; proteins exiting the *trans* Golgi network are sorted according to whether they are destined for lysosomes or for the cell surface. We discuss some examples of sorting by the *trans* Golgi network later, and we present some of the methods for tracking proteins through the secretory pathways of the cell in How We Know, pp. 520–521.

Many of the oligosaccharide groups that are added to proteins in the ER undergo further modifications in the Golgi apparatus. On some proteins, for example, complex oligosaccharide chains are created by a highly ordered process in which sugars are added and removed by a series of enzymes that act in a rigidly determined sequence as the protein passes through the Golgi stack. There is a clear correlation between the position of an enzyme in the chain of processing events and its localization in the Golgi stack: enzymes that act early are found in cisternae close to the *cis* face, while enzymes that act late are found in cisternae near the *trans* face.

Secretory Proteins Are Released from the Cell by Exocytosis

In all eucaryotic cells there is a steady stream of vesicles that bud from the *trans* Golgi network and fuse with the plasma membrane. This *constitutive exocytosis pathway* operates continually and supplies newly made lipids and proteins to the plasma membrane; it is the pathway for

How We Know: Tracking protein and vesicle transport

Over the years, biologists have taken advantage of a variety of techniques to untangle the pathways and mechanisms by which proteins are sorted and transported into and out of the cell and its resident organelles. As we saw earlier, transferring an ER signal sequence to a cytosolic protein allowed researchers to confirm that such signal peptides serve to target proteins to specific intracellular

compartments—in this example, the ER (see Figure 15–6). But such signal-swapping experiments are not the only way to track a protein's progress through the cell. Biochemical, genetic, and molecular biological and microscopic techniques also provide a means for studying how proteins shuttle from one cellular compartment to another. In some cases, these methods can be used to track the migration of proteins and transport vesicles in real time inside living cells.

In a tube
A protein bearing a signal sequence can be introduced to a preparation of isolated organelles in a test tube. This mixture can then be tested to see whether the protein will be taken up by the organelle being examined. The protein is usually produced *in vitro* by cell-free translation of a purified mRNA encoding the polypeptide; in the process, radioactive amino acids can be used to label the protein so that it will be easy to isolate and to follow. The labeled protein is incubated with a selected organelle and its translocation monitored by one of several different methods (Figure 15–25).

Ask a yeast
Movement of proteins between different cellular compartments via transport vesicles has been studied extensively using genetic techniques. Studies of mutant yeast cells that are defective for secretion at high temperatures have identified more than 25 genes that are involved in exocytosis. Many of these mutant genes encode temperature-sensitive proteins that are involved in transport and secretion. These mutant proteins may function normally at 25°C, but when the yeast cells are shifted to 35°C, they are inactivated. As a result, when researchers raise the temperature, proteins destined for secretion instead accumulate inappropriately in the ER, the Golgi apparatus, or transport vesicles (Figure 15–26).

At the movies
Perhaps the most dramatic method for tracking a protein as it moves throughout the cell involves tagging the polypeptide with green fluorescent protein (GFP). Using the genetic engineering techniques discussed in Chapter 10, this small protein can be fused to other cellular proteins. Fortunately, for most proteins studied, the addition of GFP does not perturb the molecule's normal function or transport. The movement of a GFP-tagged protein can then be monitored in a living cell with a fluorescent microscope.

GFP fusion proteins are widely used to study the location and movement of proteins in cells (Figure 15–27). GFP fused to

Figure 15–25 Several methods can be used to determine whether a protein bearing a particular signal sequence is transported into a preparation of isolated organelles. (A) The labeled protein with or without a signal sequence is incubated with the organelles and the preparation is centrifuged. Only those labeled proteins that contained a signal sequence will be transported and therefore should co-fractionate with the organelle. (B) The labeled proteins are incubated with the organelle and a protease is added to the preparation. A transported protein will be selectively protected from digestion by the protease; adding a detergent that disrupts the organelle membrane will eliminate that protection and the transported protein will also be degraded.

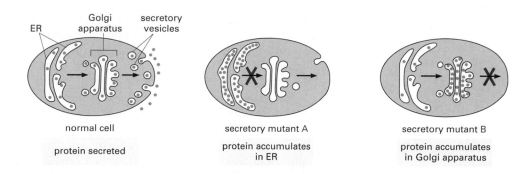

Figure 15–26 Temperature-sensitive mutants have been used to dissect the protein secretory pathway in yeast. Mutations in genes involved at different stages of the transport process result in the accumulation of proteins in the ER, the Golgi apparatus, or other transport vesicles. For example, a mutation A that blocks transport from the ER to the Golgi apparatus will cause a buildup of proteins in the ER. A mutation B that blocks exit of proteins from the Golgi apparatus will cause proteins to accumulate within that organelle.

proteins that shuttle in and out of the nucleus, for example, can be used to study nuclear transport events. GFP fused to plasma membrane proteins can be used to measure the kinetics of their movement through the secretory pathway. Movies demonstrating the power and beauty of this technique are included on the CD that accompanies this book.

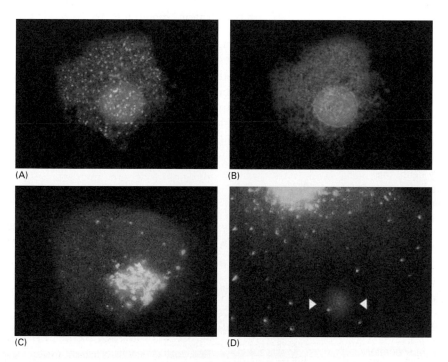

Figure 15–27 GFP fusion allows proteins to be tracked throughout the cell. In this experiment, GFP is fused to a viral coat protein and expressed in cultured cells. In an infected cell, the viral protein will move through the secretory pathway from the ER to the cell surface, where a virus particle would be assembled. The viral coat protein used in this experiment contains a mutation that allows export from the ER only at a low temperature. (A) At high temperatures, the fusion protein labels the ER. (B) As the temperature is lowered, the GFP fusion protein rapidly accumulates at ER exit sites. (C) The fusion protein then moves to the Golgi apparatus. (D) Finally, the fusion protein is delivered to the plasma membrane. The halo between the two arrowheads marks the spot where a single vesicle has fused expelling the viral coat protein into the plasma membrane. (A–D, courtesy of Jennifer Lippincott-Schwartz.)

plasma membrane growth when cells enlarge before dividing. The constitutive pathway also carries proteins to the cell surface to be released to the outside, a process called **secretion**. Some of the released proteins adhere to the cell surface, where they become peripheral proteins of the plasma membrane; some are incorporated into the extracellular matrix; still others diffuse into the extracellular fluid to nourish or to signal other cells. Because entry into this nonselective pathway does not require a particular signal sequence (like the ones that direct proteins to lysosomes or back to the ER), it is sometimes referred to as the *default pathway*.

In addition to the constitutive exocytosis pathway, which operates continually in all eucaryotic cells, there is a *regulated exocytosis pathway*, which operates only in cells that are specialized for secretion. Specialized *secretory cells* produce large quantities of particular products, such as hormones, mucus, or digestive enzymes, which are stored in **secretory vesicles** for later release. These vesicles bud off from the *trans* Golgi network and accumulate near the plasma membrane. There they wait for the extracellular signal that will stimulate them to fuse with the plasma membrane and release their contents to the cell exterior (Figure 15–28). An increase in blood glucose, for example, signals cells in the pancreas to secrete the hormone insulin (Figure 15–29).

Proteins destined for secretory vesicles are sorted and packaged in the *trans* Golgi network. Proteins that travel by this pathway have special surface properties that cause them to aggregate with one another under the ionic conditions (acidic pH and high Ca^{2+}) that prevail in the *trans* Golgi network. The aggregated proteins are recognized by an unknown mechanism and packaged into secretory vesicles, which pinch off from the network. Proteins secreted by the constitutive pathway do not aggregate and are therefore carried automatically to the plasma membrane by the transport vesicles of the constitutive pathway. Selective aggregation has another function: it allows secretory proteins to be packaged into secretory vesicles at concentrations much higher than the concentration of the unaggregated protein in the Golgi lumen. This increase in concentration can reach up to 200-fold, enabling secretory cells to release large amounts of the protein promptly when triggered to do so (see Figure 15–29).

Figure 15–28 In secretory cells, the regulated and constitutive pathways of exocytosis diverge in the *trans* Golgi network. Many soluble proteins are continually secreted from the cell by the constitutive secretory pathway, which operates in all cells. This pathway also continually supplies the plasma membrane with newly synthesized lipids and proteins. Specialized secretory cells have, in addition, a regulated exocytosis pathway, by which selected proteins in the *trans* Golgi network are diverted into secretory vesicles, where the proteins are concentrated and stored until an extracellular signal stimulates their secretion. It is unclear how aggregates of secretory proteins are segregated into secretory vesicles. Secretory vesicles have unique proteins in their membranes; perhaps some of these proteins act as receptors for secretory protein aggregates in the *trans* Golgi network.

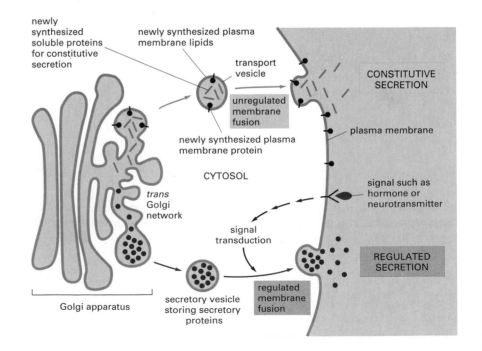

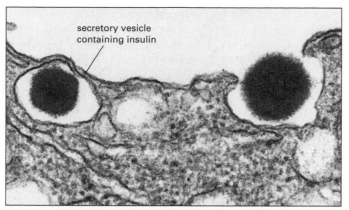

secretory vesicle
containing insulin

0.2 μm

Figure 15–29 **Secretory vesicles package and discharge concentrated aggregates of protein.** The electron micrograph shows the release of insulin into the extracellular space from a secretory vesicle of a pancreatic β cell. The insulin is stored in a highly concentrated form in each secretory vesicle and is released only when the cell is signaled to secrete by an increase in glucose levels in the blood. (Courtesy of Lelio Orci, from L. Orci, J.D. Vassali, and A. Perrelet, *Sci. Am.* 256:85–94, 1988.)

When a secretory vesicle or transport vesicle fuses with the plasma membrane and discharges its contents by exocytosis, its membrane becomes part of the plasma membrane. Although this should greatly increase the surface area of the plasma membrane, it does so only transiently because membrane components are removed from other regions of the surface by endocytosis almost as fast as they are added by exocytosis. This removal returns both the lipids and the proteins of the vesicle membrane to the Golgi network, where they can be used again.

Endocytic Pathways

Eucaryotic cells are continually taking up fluid, as well as large and small molecules, by the process of endocytosis. Specialized cells are also able to internalize large particles and even other cells. The material to be ingested is progressively enclosed by a small portion of the plasma membrane, which first buds inward and then pinches off to form an intracellular *endocytic vesicle*. The ingested material is ultimately delivered to lysosomes, where it is digested. The metabolites generated by digestion are transferred directly out of the lysosome into the cytosol, where they can be used by the cell.

Two main types of **endocytosis** are distinguished on the basis of the size of the endocytic vesicles formed. *Pinocytosis* ("cellular drinking") involves the ingestion of fluid and molecules via small vesicles (<150 nm in diameter). *Phagocytosis* ("cellular eating") involves the ingestion of large particles, such as microorganisms and cell debris, via large vesicles called *phagosomes* (generally >250 nm in diameter). Whereas all eucaryotic cells are continually ingesting fluid and molecules by pinocytosis, large particles are ingested mainly by specialized *phagocytic cells*.

In this final section we trace the endocytic pathway from the plasma membrane to lysosomes. We start by considering the uptake of large particles by phagocytosis.

Specialized Phagocytic Cells Ingest Large Particles

The most dramatic form of endocytosis, **phagocytosis**, was first observed more than a hundred years ago. In protozoa, phagocytosis is a form of feeding: microorganisms ingest large particles, such as bacteria, by taking them up into phagosomes; these phagosomes then fuse with lysosomes, where the food particles are digested. Few cells in multicellular organisms are able to ingest large particles efficiently. In the animal gut, for example, large particles of food have to be broken down to individual molecules by extracellular enzymes before they can be taken up by the absorptive cells lining the gut.

Question 15–7

What would you expect to happen in cells that secrete large amounts of protein through the regulated secretory pathway if the ionic conditions in the ER lumen could be changed to resemble those in the lumen of the *trans* Golgi network?

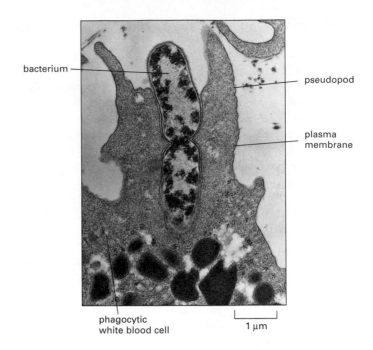

Figure 15–30 A white blood cell ingests a bacterium. Electron micrograph of a phagocytic white blood cell (a neutrophil) ingesting a bacterium, which is in the process of dividing. The white blood cell has extended surface projections called pseudopods, which progressively envelop the bacterium. (Courtesy of Dorothy F. Bainton.)

bacterium

pseudopod

plasma membrane

phagocytic white blood cell

1 μm

Nevertheless, phagocytosis is important in most animals for purposes other than nutrition. **Phagocytic cells**—including *macrophages,* which are widely distributed in tissues, and some other white blood cells—defend us against infection by ingesting invading microorganisms. To be taken up by a macrophage or other white blood cell, particles must first bind to the phagocytic cell surface and activate one of a variety of surface receptors. Some of these receptors recognize antibodies, the proteins that protect us against infection by binding to the surface of microorganisms. Binding of antibody-coated bacteria to these receptors induces the phagocytic cell to extend sheetlike projections of the plasma membrane, called *pseudopods,* that engulf the bacterium (Figure 15–30) and fuse at their tips to form a *phagosome.* This compartment then fuses with a lysosome and the microbe is digested. Some pathogenic bacteria have evolved tricks for subverting the system: for example, *Mycobacterium tuberculosis,* the agent responsible for tuberculosis, can inhibit the membrane fusion that unites the phagosome with a lysosome. Instead of being destroyed, the engulfed organism survives and multiplies within the macrophage. How the bacterium accomplishes this task is still unknown.

Phagocytic cells also play an important part in scavenging dead and damaged cells and cellular debris. Macrophages, for example, ingest more than 10^{11} of our worn-out red blood cells each day (Figure 15–31).

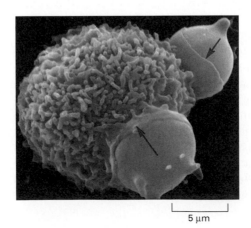

Figure 15–31 A macrophage scavenges a pair of red blood cells. The *red arrows* in this scanning electron micrograph point to the edges of the fine sheets of membrane—pseudopods—that the macrophage is extending like collars to engulf the red cells. The red blood cells are misshapen as they are being squeezed by the macrophage. (Courtesy of Jean Paul Revel.)

5 μm

Fluid and Macromolecules Are Taken Up by Pinocytosis

Eucaryotic cells continually ingest bits of their plasma membrane in the form of small pinocytic vesicles that are later returned to the cell surface. The rate at which plasma membrane is internalized by **pinocytosis** varies from cell type to cell type, but it is usually surprisingly large. A macrophage, for example, swallows 25% of its own volume of fluid each hour. This means that it removes 3% of its plasma membrane each minute, or 100% in about half an hour. Fibroblasts endocytose at a somewhat lower rate, whereas some phagocytic amoebae ingest their plasma membrane even more rapidly. Because a cell's total surface area and volume remain unchanged during this process, it is clear that as much membrane is being added to the cell surface by vesicle fusion (exocytosis) as is being removed by endocytosis.

Pinocytosis is mainly carried out by the clathrin-coated pits and vesicles that we discussed earlier (see Figures 15–18 and 15–19). After they pinch off from the plasma membrane, clathrin-coated vesicles rapidly shed their coat and fuse with an *endosome*. Extracellular fluid is trapped in the coated pit as it invaginates to form a coated vesicle, and so substances dissolved in the extracellular fluid are internalized and delivered to endosomes. This fluid intake is generally balanced by fluid loss during exocytosis.

Receptor-mediated Endocytosis Provides a Specific Route into Animal Cells

Pinocytosis, as just described, is indiscriminate. The endocytic vesicles simply trap any molecules that happen to be present in the extracellular fluid and carry them into the cell. In most animal cells, however, pinocytosis via clathrin-coated vesicles also provides an efficient pathway for taking up specific macromolecules from the extracellular fluid. The macromolecules bind to complementary receptors on the cell surface and enter the cell as receptor–macromolecule complexes in clathrin-coated vesicles. This process, called **receptor-mediated endocytosis**, provides a selective concentrating mechanism that increases the efficiency of internalization of particular macromolecules more than 1000-fold compared with ordinary pinocytosis, so that even minor components of the extracellular fluid can be taken up in large amounts without taking in a correspondingly large volume of extracellular fluid. An important example of receptor-mediated endocytosis is the ability of animal cells to take up the cholesterol they need to make new membrane.

Cholesterol is extremely insoluble and is transported in the bloodstream bound to protein in the form of particles called *low-density lipoproteins,* or *LDL.* The LDL binds to receptors located on cell surfaces, and the receptor–LDL complexes are ingested by receptor-mediated endocytosis and delivered to endosomes. The interior of endosomes is more acid than the surrounding cytosol or the extracellular fluid, and in this acidic environment the LDL dissociates from its receptor: the receptors are returned in transport vesicles to the plasma membrane for reuse, while the LDL is delivered to lysosomes. In the lysosomes the LDL is broken down by hydrolytic enzymes; the cholesterol is released and escapes into the cytosol, where it is available for new membrane synthesis. The LDL receptors on the cell surface are continually internalized and recycled, whether they are occupied by LDL or not (Figure 15–32).

This pathway for cholesterol uptake is disrupted in individuals who inherit a defective gene encoding the LDL receptor protein. In some

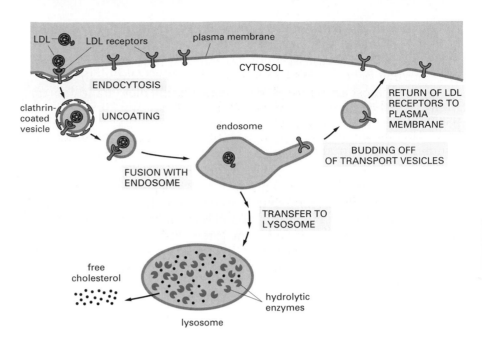

Figure 15–32 LDL enters cells via receptor-mediated endocytosis. LDL binds to receptors on the cell surface and is internalized in clathrin-coated vesicles. The vesicles lose their coat and then fuse with endosomes. In the acidic environment of the endosome, LDL dissociates from its receptors. Whereas the LDL ends up in lysosomes, where it is degraded to release free cholesterol, the LDL receptors are returned to the plasma membrane via transport vesicles to be used again. For simplicity, only one LDL receptor is shown entering the cell and returning to the plasma membrane. Whether it is occupied or not, an LDL receptor typically makes one round trip into the cell and back every 10 minutes, making a total of several hundred trips in its 20-hour life span.

cases the receptors are missing; in others they are present but nonfunctional. In either case, because the cells are deficient in taking up LDL, cholesterol accumulates in the blood and predisposes the individuals to develop atherosclerosis. Most die at an early age of heart attacks resulting from cholesterol clogging the arteries that supply the heart.

Receptor-mediated endocytosis is also used to take up many other essential metabolites, such as vitamin B_{12} and iron, that cells cannot take up by the processes of membrane transport discussed in Chapter 12. Vitamin B_{12} and iron are both required, for example, for the synthesis of hemoglobin, which is the major protein in red blood cells; these metabolites enter immature red blood cells as a complex with protein. Many cell-surface receptors that bind extracellular signaling molecules are also ingested by this pathway: some are recycled to the plasma membrane for reuse, whereas others are degraded in lysosomes. Unfortunately, receptor-mediated endocytosis can also be exploited by viruses: the influenza virus and HIV, which causes AIDS, gain entry into cells in this way.

Endocytosed Macromolecules Are Sorted in Endosomes

Because extracellular material taken up by pinocytosis is rapidly transferred to endosomes, it is possible to visualize the endosomal compartment by incubating living cells in fluid containing an electron-dense marker that will show up when viewed in an electron microscope. When examined in this way, the endosomal compartment reveals itself to be a complex set of connected membrane tubes and larger vesicles. Two sets of endosomes can be distinguished in such loading experiments: the marker molecules appear first in *early endosomes,* just beneath the plasma membrane; 5–15 minutes later they show up in *late endosomes,* near the nucleus. Early endosomes mature gradually into late endosomes as the vesicles within them fuse, either with one another or with a preexisting late endosome. The interior of the endosome compartment is kept acidic (pH 5–6) by an ATP-driven H$^+$ (proton) pump in the endosomal membrane that pumps H$^+$ into the endosome lumen from the cytosol.

The endosomal compartment acts as the main sorting station in the inward endocytic pathway, just as the *trans* Golgi network serves this function in the outward secretory pathway. The acidic environment of the endosome plays a crucial part in the sorting process by causing many receptors to release their bound cargo. The routes taken by receptors once they have entered an endosome differ according to the type of receptor: (1) most are returned to the same plasma membrane domain from which they came, as is the case for the LDL receptor discussed earlier; (2) some travel to lysosomes, where they are degraded; and (3) some proceed to a different domain of the plasma membrane, thereby transferring their bound cargo molecules from one extracellular space to another, a process called *transcytosis* (Figure 15–33).

Cargo molecules that remain bound to their receptors share the fate of their receptors. Those that dissociate from their receptors in the endosome are doomed to destruction in lysosomes along with most of the contents of the endosome lumen. It remains uncertain how molecules move from endosomes to lysosomes. One possibility is that they are carried in transport vesicles; another is that endosomes gradually convert into lysosomes.

Lysosomes Are the Principal Sites of Intracellular Digestion

Many extracellular particles and molecules ingested by cells end up in **lysosomes,** which are membranous sacs of hydrolytic enzymes that carry out the controlled intracellular digestion of both extracellular materials and worn-out organelles. They contain about 40 types of hydrolytic enzymes, including those that degrade proteins, nucleic acids, oligosaccharides, and phospholipids. All of these enzymes are optimally active in the acidic conditions (pH ~5) maintained within lysosomes. The membrane of the lysosome normally keeps these destructive enzymes out of the cytosol (whose pH is about 7.2), but the acid dependence of the enzymes protects the contents of the cytosol against damage even if some leakage should occur.

Like all other intracellular organelles, the lysosome not only contains a unique collection of enzymes but also has a unique surrounding membrane. The lysosomal membrane contains transport proteins that allow the final products of the digestion of macromolecules, such as amino acids, sugars, and nucleotides, to be transported to the cytosol, from where they can be either excreted or utilized by the cell. The membrane also contains an ATP-driven H^+ pump, which, like the ATPase in the endosome membrane, pumps H^+ (protons) into the lysosome, thereby maintaining its contents at an acidic pH (Figure 15–34). Most of the lysosomal membrane proteins are unusually highly glycosylated; the sugars, which cover much of the protein surfaces facing the lumen, protect the proteins from digestion by the lysosomal proteases.

The specialized digestive enzymes and membrane proteins of the lysosome are synthesized in the ER and transported through the Golgi apparatus to the *trans* Golgi network. While in the ER and the *cis* Golgi network, the enzymes are tagged with a specific phosphorylated sugar group (mannose 6-phosphate), so that when they arrive in the *trans* Golgi network they can be recognized by an appropriate receptor, the

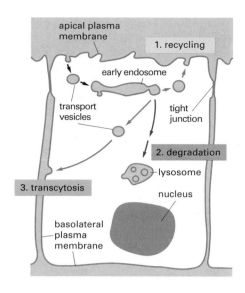

Figure 15–33 The fate of the receptor proteins involved in endocytosis depends on the type of receptor. Three pathways from the endosomal compartment in an epithelial cell are shown. Receptors that are not specifically retrieved from early endosomes follow the pathway from the endosomal compartment to lysosomes, where they are degraded. Retrieved receptors are returned either to the same plasma membrane domain from which they came (*recycling*) or to a different domain of the plasma membrane (*transcytosis*). If the ligand that is endocytosed with its receptor stays bound to the receptor in the acidic environment of the endosome, it will follow the same pathway as the receptor; otherwise it will be delivered to lysosomes for degradation.

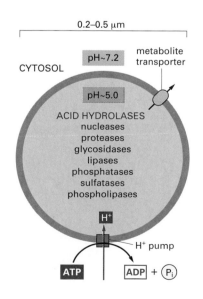

Figure 15–34 A lysosome contains hydrolytic enzymes and a H^+ pump. The acid hydrolases are hydrolytic enzymes that are active under acidic conditions. The lumen of the lysosome is maintained at an acidic pH by a H^+ ATPase in the membrane that pumps H^+ into the lumen.

Figure 15–35 Materials destined for degradation follow different pathways to the lysosome. Each pathway leads to the intracellular digestion of materials derived from a different source. The compartments resulting from the three pathways can sometimes be distinguished morphologically—hence the terms "autophagolysosome," "phagolysosome," and so on. Such lysosomes differ, however, only because of the different materials they are digesting.

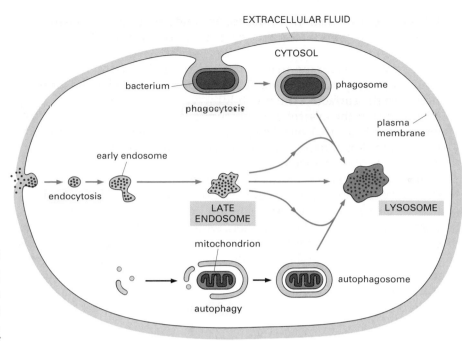

Question 15–8

Iron (Fe) is an essential trace metal that is needed by all cells. It is required, for example, for the synthesis of heme groups that are part of the active site of many enzymes involved in electron-transfer reactions; it is also required in hemoglobin, the main protein in red blood cells. Iron is taken up by cells by receptor-mediated endocytosis. The iron-uptake system has two components, a soluble protein called transferrin, which circulates in the bloodstream; and a transferrin receptor—a transmembrane protein that, like the LDL receptor in Figure 15–32, is continually endocytosed and recycled to the plasma membrane. Fe ions bind to transferrin at neutral pH but not at acidic pH. Transferrin binds to the transferrin receptor at neutral pH only when it has an Fe ion bound, but it binds to the receptor at acidic pH even in the absence of bound iron. From these properties, describe how iron is taken up, and discuss the advantages of this elaborate scheme.

mannose 6-phosphate receptor. This tagging permits the enzymes to be sorted and packaged into transport vesicles, which bud off and deliver their contents to lysosomes via late endosomes (see Figure 15–17).

Depending on their source, materials follow different paths to lysosomes. We have seen that extracellular particles are taken up into phagosomes, which fuse with lysosomes, and that extracellular fluid and macromolecules are taken up into smaller endocytic vesicles, which deliver their contents to lysosomes via endosomes. But cells have an additional pathway for supplying materials to lysosomes; this pathway is used for degrading obsolete parts of the cell itself. In electron micrographs of liver cells, for example, one often sees lysosomes digesting mitochondria, as well as other organelles. The process seems to begin with the enclosure of the organelle by a double membrane, creating an *autophagosome,* which then fuses with lysosomes (Figure 15–35). It is not known what marks an organelle for such destruction.

Essential Concepts

- Eucaryotic cells contain many membrane-enclosed organelles, including a nucleus, an endoplasmic reticulum (ER), a Golgi apparatus, lysosomes, endosomes, mitochondria, chloroplasts (in plant cells), and peroxisomes.
- Most organelle proteins are made in the cytosol and transported into the organelle where they function. Sorting signals in the amino acid sequence guide the proteins to the correct organelle; proteins that function in the cytosol have no signals and remain where they are made.
- Nuclear proteins contain nuclear localization signals that help direct their active transport from the cytosol into the nucleus through nuclear pores, which penetrate the double-membrane nuclear envelope. Proteins can enter the nucleus without being unfolded.
- Most mitochondrial and chloroplast proteins are made in the cytosol and are then actively transported into the organelles by protein

translocators in their membranes. Proteins must be unfolded to allow them to snake through the chloroplast or mitochondrial membrane.

- The ER is the membrane factory of the cell; it makes most of the cell's lipids and many of its proteins. The proteins are made by ribosomes bound to the surface of the rough ER.

- Ribosomes in the cytosol are directed to the ER if the protein they are making has an ER signal sequence, which is recognized by a signal-recognition particle (SRP) in the cytosol; the binding of the ribosome–SRP complex to a receptor on the ER membrane initiates the translocation process that threads the growing polypeptide across the ER membrane through a translocation channel.

- Soluble proteins destined for secretion or the lumen of an organelle pass completely into the ER lumen, while transmembrane proteins destined for the ER membrane or other cell membranes remain anchored in the lipid bilayer by one or more membrane-spanning α helices.

- In the ER lumen, proteins fold up, assemble with other proteins, form disulfide bonds, and become decorated with oligosaccharide chains.

- Exit from the ER is an important quality-control step; proteins that either fail to fold properly or fail to assemble with their normal partners are retained in the ER and are eventually degraded.

- Protein transport from the ER to the Golgi apparatus and from the Golgi apparatus to other destinations is mediated by transport vesicles that continually bud off from one membrane and fuse with another, a process called vesicular transport.

- Budding transport vesicles have distinctive coat proteins on their cytosolic surface; the assembly of the coat drives the budding process, and the coat proteins help incorporate receptors with their bound cargo molecules into the forming vesicle.

- Coated vesicles lose their protein coat soon after pinching off, enabling them to dock and then fuse with a particular target membrane; docking and fusion are thought to be mediated by proteins on the vesicle and on the target membranes, called v-SNAREs and t-SNAREs, respectively.

- The Golgi apparatus receives newly made proteins from the ER; it modifies their oligosaccharides, sorts the proteins, and dispatches them from the *trans* Golgi network to the plasma membrane, lysosomes, or secretory vesicles.

- In all eucaryotic cells, transport vesicles continually bud from the *trans* Golgi network and fuse with the plasma membrane, a process called constitutive exocytosis; the process delivers plasma membrane lipids and proteins to the cell surface and also releases molecules from the cell, a process called secretion.

- Specialized secretory cells also have a regulated exocytosis pathway, where molecules stored in secretory vesicles are released from the cell by exocytosis when the cell is signaled to secrete.

- Cells ingest fluid, molecules, and sometimes even particles, by endocytosis, in which regions of plasma membrane invaginate and pinch off to form endocytic vesicles.

- Much of the material that is endocytosed is delivered to endosomes and then to lysosomes, where it is degraded by hydrolytic enzymes; most of the components of the endocytic vesicle membrane, however, are recycled in transport vesicles back to the plasma membrane for reuse.

Questions

Question 15–9

Which of the following statements are correct? Explain your answers.

A. Ribosomes are cytoplasmic structures that, during protein synthesis, become linked by an mRNA molecule to form polyribosomes.

B. The amino acid sequence Leu-His-Arg-Leu-Asp-Ala-Gln-Ser-Lys-Leu-Ser-Ser is a signal sequence that directs proteins to the ER.

C. All transport vesicles in the cell must have a v-SNARE protein in their membrane.

D. Transport vesicles deliver proteins and lipids to the cell surface.

E. If the delivery of prospective lysosomal proteins from the *trans* Golgi network to the late endosomes were blocked, lysosomal proteins would be secreted by the constitutive secretion pathways shown in Figure 15–28.

F. Lysosomes digest only substances that have been taken up by cells by endocytosis.

G. *N*-linked sugar chains are found on glycoproteins that face the cell surface, as well as on glycoproteins that face the lumina of the ER, *trans* Golgi network, and mitochondria.

Question 15–10

How do you suppose that proteins with a nuclear export signal get into the nucleus?

Question 15–11

Influenza viruses are surrounded by a membrane that contains a fusion protein, which is activated by acidic pH. Upon activation, the protein causes the viral membrane to fuse with cell membranes. An old folk remedy against flu recommends that one should spend a night in a horse's stable. Odd as it may sound, there is a rational explanation for this advice. Air in stables contains ammonia (NH_3) generated by bacteria in the horse's urine. Sketch a diagram showing the pathway (in detail) by which flu virus enters cells, and speculate how NH_3 may protect cells from virus infection. (Hint: NH_3 can neutralize acidic solutions by the reaction $NH_3 + H^+ \rightarrow NH_4^+$.)

Question 15–12

Consider the v-SNAREs that direct transport vesicles from the *trans* Golgi network to the plasma membrane. They, like all other v-SNAREs, are membrane proteins that are integrated into the membrane of the ER during their biosynthesis and are then transported by transport vesicles to their destination. Thus, transport vesicles budding from the ER contain at least two kinds of v-SNAREs—those that target the vesicles to the *cis* Golgi cisternae, and those that are in transit to the *trans* Golgi network to be packaged in different transport vesicles destined for the plasma membrane. (A) Why might this be a problem? (B) Suggest possible ways in which the cell might solve it.

Question 15–13

A particular type of *Drosophila* mutant becomes paralyzed when the temperature is raised. The mutation affects the structure of dynamin, causing it to be inactivated at the higher temperature. Indeed, the function of dynamin was discovered by analyzing the defect in these mutant fruit flies. The complete paralysis at the elevated temperature suggests that synaptic transmission between nerve and muscle cells (discussed in Chapter 12) is blocked. Suggest why signal transmission at a synapse might require dynamin. On the basis of your hypothesis, what would you expect to see in electron micrographs of synapses of flies that were exposed to the elevated temperature?

Question 15–14

Edit the following statements, if required, to make them true: "Because nuclear localization sequences are not cleaved off by proteases following protein import into the nucleus, they can be reused to import nuclear proteins after mitosis, when cytosolic and nuclear proteins have become intermixed. This is in contrast to ER signal sequences, which are cleaved off by a signal peptidase once they reach the lumen of the ER. ER signal sequences cannot therefore be reused to import ER proteins after mitosis, when cytosolic and ER proteins have become intermixed; these ER proteins must therefore be degraded and resynthesized."

Question 15–15

Consider a protein that contains an ER signal sequence at its N-terminus and a nuclear localization sequence in its middle. What do you think the fate of this protein would be? Explain your answer.

Question 15–16

Compare and contrast protein import into the ER and into the nucleus. List at least two major differences in the mechanisms, and speculate why the ER mechanism might not work for nuclear import and vice versa.

Question 15–17

During mitosis, the nuclear envelope breaks down into small vesicles, and intranuclear proteins completely intermix with cytosolic proteins. Is this consistent with the evolutionary scheme proposed in Figure 15–3?

Question 15–18

A protein that inhibits certain proteolytic enzymes (proteases) is normally secreted into the bloodstream by liver cells. This inhibitor protein, antitrypsin, is absent from the bloodstream of patients who carry a mutation that results in a single amino acid change in the protein. Antitrypsin deficiency causes a variety of severe problems, particularly in lung tissue, because of the uncontrolled activity of proteases. Surprisingly, when the mutant antitrypsin is synthesized in the laboratory, it is as active as the normal antitrypsin at inhibiting proteases. Why, then, does the mutation cause the disease? Think of more than one possibility, and suggest ways in which you could distinguish between them.

Question 15–19

Dr. Outonalimb's claim to fame is her discovery of forgettin, a protein predominantly made by the pineal gland in human teenagers. The protein causes selec-

tive short-term unresponsiveness and memory loss when the auditory system receives statements like "Please take out the garbage!" Her hypothesis is that forgettin has a hydrophobic ER signal sequence at its C-terminus that is recognized by an SRP and causes it to be translocated across the ER membrane by the mechanism shown in Figure 15–13. She predicts that the protein is secreted from pineal cells into the bloodstream, from where it exerts its devastating systemic effects. You are a member of the committee deciding whether she should receive a grant for further work on her hypothesis. Consider that grant reviews should be polite and constructive.

Question 15–20

Taking the evolutionary scheme in Figure 15–3 one step further, suggest how the Golgi apparatus could have evolved. Sketch a simple diagram to illustrate your ideas. For the Golgi apparatus to be functional, what else would have to have evolved?

Question 15–21

If membrane proteins are integrated into the ER membrane by means of the ER protein translocation channel (which is itself composed of membrane proteins), how do the first protein translocation channels become incorporated into the ER membrane?

Question 15–22

The sketch in Figure Q15–22 is a schematic drawing of the electron micrograph shown in the third panel of Figure 15–18A. Name the structures that are labeled in the sketch.

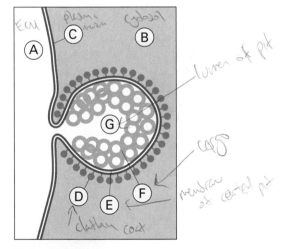

Figure Q15–22

Highlights from *Essential Cell Biology 2 Interactive* CD-ROM

15.2   Ribosome/ER Translocator

15.3   Freeze Fracture of Yeast Cell

15.8   Clathrin

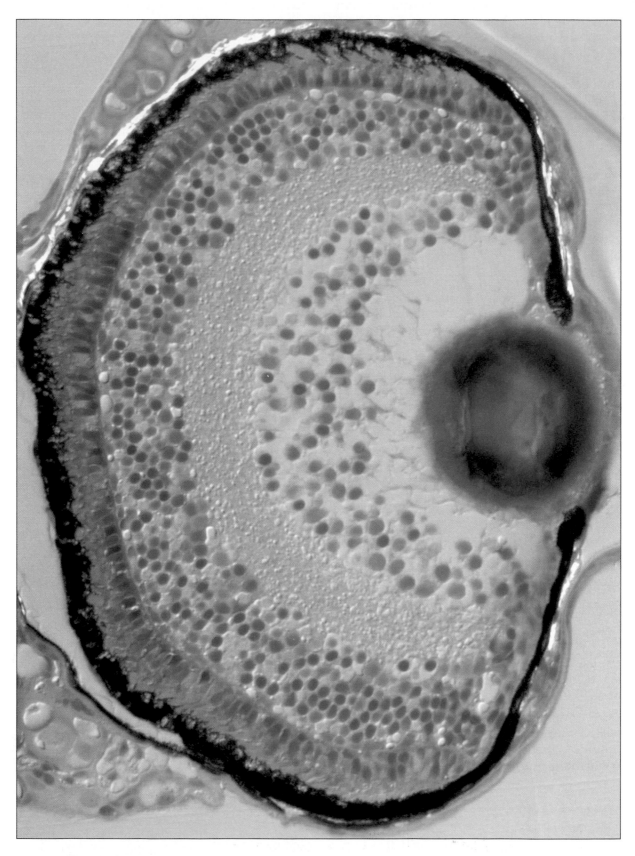

The eye. The act of seeing involves a relay chain of signals and cell communication. Light impinges on photoreceptor cells in the retina, activating receptor molecules, which generate intracellular signals that control release of a chemical messenger—a neurotransmitter—which acts on adjacent nerve cells in the retina and sends the message on through further relays to the brain—all in a few hundredths of a second. Shown here is a section through the developing eye of a 5-day-old zebrafish, with the lens on the right and the retina on the left.

Cell Communication

16

Individual cells, like multicellular organisms, need to sense and respond to their environment. A typical free-living cell—even a primitive bacterium—must be able to track down nutrients, tell the difference between light and dark, and avoid poisons and predators. And if such a cell is to have any kind of "social life," it must be able to communicate with other cells. When a yeast cell is ready to mate, for example, it secretes a small protein called a mating factor. Yeast cells of the opposite "sex" detect this chemical mating call and respond by halting their progress through the cell cycle and reaching out toward the cell that emitted the signal (Figure 16–1).

In a multicellular organism, things are much more complicated. Cells must interpret the multitude of signals they receive from other cells to help coordinate their behaviors. During animal development, for example, cells in the embryo exchange signals to determine which specialized role each cell will adopt, what position it will occupy in the animal, and whether it will survive, divide, or die; later, a large variety of signals coordinate the animal's growth and its day-to-day physiology and behavior. In plants, too, cells are in constant communication with one another. Their interactions allow the plant to respond to the conditions of light, dark, and temperature that guide the cycles of its growth, flowering, and fruiting, and to coordinate what happens in its roots, stems, and leaves.

In this chapter, we examine some of the most important means by which cells communicate, and we discuss how cells send signals and interpret the signals they receive. Although we concentrate on the mechanisms of signal reception and interpretation in animal cells, we also present a brief review of what is known about signaling pathways in plant cells. We begin our discussion with an overview of the general principles of cell signaling and then consider two of the main systems animal cells use to receive and interpret signals.

General Principles of Cell Signaling

Information can come in a variety of forms, and communication frequently involves converting information signals from one form to another. When you phone a friend, for instance, the sound waves of your voice are converted to electrical signals that travel over a telephone wire. The critical points in this relay occur where the message is converted from one form to another. This process of conversion is called **signal transduction** (Figure 16–2).

General Principles of Cell Signaling

Signals Can Act over Long or Short Range

Each Cell Responds to a Limited Set of Signals

Receptors Relay Signals via Intracellular Signaling Pathways

Nitric Oxide Crosses the Plasma Membrane and Activates Intracellular Enzymes Directly

Some Hormones Cross the Plasma Membrane and Bind to Intracellular Receptors

Cell-Surface Receptors Fall into Three Main Classes

Ion-channel–linked Receptors Convert Chemical Signals into Electrical Ones

Many Intracellular Signaling Proteins Act as Molecular Switches

G-protein–linked Receptors

Stimulation of G-protein–linked Receptors Activates G-Protein Subunits

Some G Proteins Regulate Ion Channels

Some G Proteins Activate Membrane-bound Enzymes

The Cyclic AMP Pathway Can Activate Enzymes and Turn On Genes

The Inositol Phospholipid Pathway Triggers a Rise in Intracellular Ca^{2+}

A Ca^{2+} Signal Triggers Many Biological Processes

Intracellular Signaling Cascades Can Achieve Astonishing Speed, Sensitivity, and Adaptability: A Look at Photoreceptors in the Eye

Enzyme-linked Receptors

Activated Receptor Tyrosine Kinases Assemble a Complex of Intracellular Signaling Proteins

Receptor Tyrosine Kinases Activate the GTP-binding Protein Ras

Some Enzyme-linked Receptors Activate a Fast Track to the Nucleus

Protein Kinase Networks Integrate Information to Control Complex Cell Behaviors

Multicellularity and Cell Communication Evolved Independently in Plants and Animals

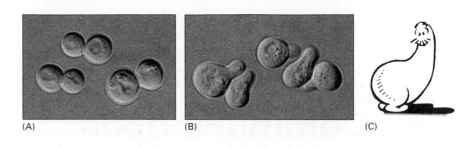

Figure 16–1 Yeast cells respond to mating factor. Budding yeast (*Saccharomyces cerevisiae*) cells are normally spherical (A), but when exposed to mating factor produced by neighboring yeast cells they extend a protrusion toward the source of the factor (B). Cells that adopt this shape in response to the mating signal are called "shmoos" after a classic 1940s cartoon character created by Al Capp (C). (A and B, courtesy of Michael Snyder; C, © Capp Enterprises, Inc., all rights reserved.)

The signals that pass between cells are far simpler than the sorts of messages that humans ordinarily exchange. In a typical communication between cells, the *signaling cell* produces a particular type of *signal molecule* that is detected by the *target cell*. The target cells possess *receptor proteins* that recognize and respond specifically to the signal molecule. Signal transduction begins when the receptor protein on the target cell receives an incoming extracellular signal and converts it to the intracellular signals that alter cell behavior. Most of this chapter will be concerned with signal reception and transduction—the events that cell biologists have in mind when they refer to **cell signaling**. First, however, we look briefly at the different types of signals that cells send to one another.

Signals Can Act over Long or Short Range

Single cells and cells in multicellular organisms use hundreds of kinds of extracellular molecules to send signals to one another—proteins, peptides, amino acids, nucleotides, steroids, fatty acid derivatives, and even dissolved gases—but they rely on only a handful of basic styles of communication for getting the message across (Figure 16–3).

In multicellular organisms, the most "public" style of communication involves broadcasting the signal throughout the whole body by secreting it into the bloodstream (in an animal) or the sap (in a plant). Signal molecules used in this way are called **hormones**, and in animals, the cells that produce hormones are called *endocrine* cells (Figure 16–3A). For example, part of the pancreas is an endocrine gland that produces the hormone insulin, which regulates glucose uptake in cells all over the body.

Somewhat less public is the process known as *paracrine signaling*. In this case, rather than entering the bloodstream, the signal molecules diffuse locally through the extracellular medium, remaining in the neighborhood of the cell that secretes them. Thus they act as **local mediators** on nearby cells (Figure 16–3B). Many of the signal molecules that regulate inflammation at the site of an infection or control cell proliferation in a healing wound function in this way.

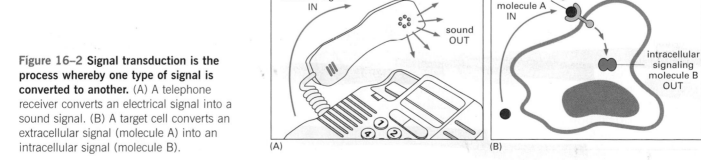

Figure 16–2 Signal transduction is the process whereby one type of signal is converted to another. (A) A telephone receiver converts an electrical signal into a sound signal. (B) A target cell converts an extracellular signal (molecule A) into an intracellular signal (molecule B).

Neuronal signaling constitutes a third form of cell communication. Like endocrine cells, neurons can deliver messages across long distances. In the case of neuronal signaling, however, a message is not broadcast widely but is delivered quickly and specifically to individual target cells through private lines (Figure 16–3C). As described in Chapter 12, the axon of a neuron terminates at specialized junctions (*synapses*) on target cells that can lie far from the neuronal cell body. The axons that connect a person's spinal cord and big toe, for example, can be more than 1 m in length. When activated by signals from the environment or from other nerve cells, a neuron sends electrical impulses racing along its axon at speeds of up to 100 m/sec. On reaching the axon terminal, these electrical signals are converted into a chemical form: each electrical impulse stimulates the nerve terminal to release a pulse of an extracellular chemical signal called a **neurotransmitter**. These neurotransmitters then diffuse across the narrow (< 100 nm) gap between the axon-terminal membrane and the membrane of the target cell in less than 1 msec.

A fourth style of signal-mediated cell–cell communication—the most intimate and short-range of all—does not require the release of a secreted molecule. Instead, the cells make direct contact through signaling molecules lodged in their plasma membranes. The message is delivered when a signal molecule anchored in the plasma membrane of the signaling cell binds to a receptor molecule embedded in the plasma membrane of the target cell (Figure 16–3D). In embryonic development, for example, such *contact-dependent signaling* plays an important part in tissues in which adjacent cells that are initially similar are destined to become specialized in different ways (Figure 16–4).

To relate these different signaling styles, imagine trying to advertise a potentially stimulating lecture—or a concert or football game. An endocrine signal would be akin to broadcasting the information over a radio station. A flyer posted on select notice boards would be the equivalent of a localized paracrine signal. Neuronal signals—long-distance

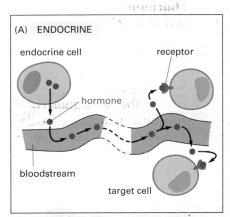

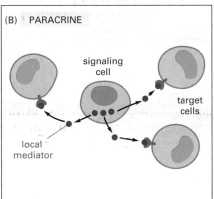

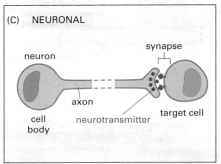

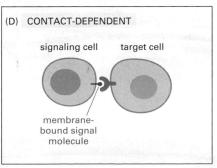

Figure 16–3 Animal cells can signal to one another in various ways.
(A) Hormones produced in endocrine glands are secreted into the bloodstream and are often distributed widely throughout the body. (B) Paracrine signals are released by cells into the extracellular fluid in their neighborhood and act locally.
(C) Neuronal signals are transmitted along axons to remote target cells. (D) Cells that maintain an intimate membrane-to-membrane interface can engage in contact-dependent signaling. Many of the same types of signal molecules are used for endocrine, paracrine, and neuronal signaling. The crucial differences lie in the speed and selectivity with which the signals are delivered to their targets.

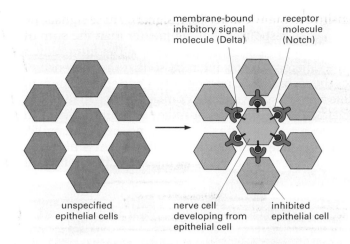

membrane-bound inhibitory signal molecule (Delta)

receptor molecule (Notch)

unspecified epithelial cells

nerve cell developing from epithelial cell

inhibited epithelial cell

Figure 16–4 Contact-dependent signaling controls nerve-cell production. The nervous system originates in the embryo from a sheet of epithelial cells. Isolated cells in this sheet begin to specialize as neurons, while their neighbors remain nonneuronal and maintain the epithelial structure of the sheet. The signals that control this process are transmitted via direct cell–cell contacts: each future neuron delivers an inhibitory signal to the cells next to it, deterring them from specializing as neurons too. Both the signal molecule (in this case, Delta) and the receptor molecule (called Notch) are transmembrane proteins. The same mechanism, mediated by essentially the same molecules, controls the detailed pattern of differentiated cell types in various other tissues, in both vertebrates and invertebrates. In mutants in which the mechanism fails, some cell types (such as neurons) are produced in great excess at the expense of others.

Question 16–1

To remain a local stimulus, paracrine signal molecules must be prevented from straying too far from their points of origin. Suggest different ways by which this could be accomplished. Explain your answers.

but personal—would be similar to a phone call or an e-mail, and contact-dependent signaling would be like a good, old-fashioned face-to-face conversation.

Table 16–1 lists some examples of hormones, local mediators, neurotransmitters, and contact-dependent signal molecules. The action of several of these is discussed in more detail later in this chapter.

Each Cell Responds to a Limited Set of Signals

A typical cell in a multicellular organism is exposed to hundreds of different signal molecules in its environment. These may be free in the extracellular fluid, embedded in the extracellular matrix in which cells rest, or bound to the surfaces of neighboring cells. Each cell must respond selectively to this mixture of signals, disregarding some and reacting to others, according to the cell's specialized function.

Whether a cell responds to a signal molecule depends first of all on whether it possesses a receptor for that signal. Without the appropriate receptor, a cell will be deaf to the signal and will not react. By producing only a limited set of receptors out of the thousands that are possible, the cell restricts the types of signals that can affect it. But this limited range of signals can still be used to control the behavior of the cell in complex ways. The complexity is of two sorts.

First, one signal, binding to one type of receptor protein, can cause a multitude of effects in the target cell: it can alter the cell's shape, movement, metabolism, and gene expression. As we shall see, the signal from a cell-surface receptor is generally conveyed into the cell interior via a set of interacting molecular mediators that are capable of producing widespread effects in the cell. This intracellular relay system and the intracellular targets on which it acts vary from one type of specialized cell to another, so that different types of cells respond to the same signal in different ways. For example, when a heart muscle cell is exposed to the neurotransmitter *acetylcholine*, the rate and force of its contractions decrease, but when a salivary gland is exposed to the same signal, it secretes components of saliva (Figure 16–5). These responses occur rapidly—within seconds to minutes—because the signal affects the activity of proteins and other molecules that are already present inside the cells.

The second kind of complexity arises because a typical cell possesses a collection of different receptors—tens to hundreds of thousands of receptors of a few dozen types. Such variety makes the cell simultaneously sensitive to many extracellular signals. These signals, by acting together, can evoke responses that are greater than the sum of the effects that each signal would evoke on its own. The intracellular relay systems for the different signals interact, so that the presence of one signal modifies the responses to another. Thus one combination of signals might enable a cell to survive; another might drive it to differentiate in some specialized way; and another might cause it to divide. In the absence of any signals, most animal cells are programmed to kill themselves (Figure 16–6). Because the execution of such a complex program often requires the synthesis of new proteins, it might take the cell

Table 16–1 Some Examples of Signal Molecules

SIGNAL MOLECULE	SITE OF ORIGIN	CHEMICAL NATURE	SOME ACTIONS
Hormones			
Adrenaline (epinephrine)	adrenal gland	derivative of the amino acid tyrosine	increases blood pressure, heart rate, and metabolism
Cortisol	adrenal gland	steroid (derivative of cholesterol)	affects metabolism of proteins, carbohydrates, and lipids in most tissues
Estradiol	ovary	steroid (derivative of cholesterol)	induces and maintains secondary female sexual characteristics
Glucagon	α cells of pancreas	peptide	stimulates glucose synthesis, glycogen breakdown, and lipid breakdown, e.g., in liver and fat cells
Insulin	β cells of pancreas	protein	stimulates glucose uptake, protein synthesis, and lipid synthesis, e.g., in liver cells
Testosterone	testis	steroid (derivative of cholesterol)	induces and maintains secondary male sexual characteristics
Thyroid hormone (thyroxine)	thyroid gland	derivative of the amino acid tyrosine	stimulates metabolism of many cell types
Local Mediators			
Epidermal growth factor (EGF)	various cells	protein	stimulates epidermal and many other cell types to proliferate
Platelet-derived growth factor (PDGF)	various cells, including blood platelets	protein	stimulates many cell types to proliferate
Nerve growth factor (NGF)	various innervated tissues	protein	promotes survival of certain classes of neurons; promotes growth of their axons
Transforming growth factor-β (TGF-β)	many cell types	protein	inhibits cell proliferation; stimulates extracellular matrix production
Histamine	mast cells	derivative of the amino acid histidine	causes blood vessels to dilate and become leaky, helping to cause inflammation
Nitric oxide (NO)	nerve cells; endothelial cells lining blood vessels	dissolved gas	causes smooth muscle cells to relax; regulates nerve cell activity
Neurotransmitters			
Acetylcholine	nerve terminals	derivative of choline	excitatory neurotransmitter at many nerve–muscle synapses and in central nervous system
γ-Aminobutyric acid (GABA)	nerve terminals	derivative of the amino acid glutamic acid	inhibitory neurotransmitter in central nervous system
Contact-dependent Signal Molecules			
Delta	prospective neurons; various other developing cell types	transmembrane protein	inhibits neighboring cells from becoming specialized in same way as the signaling cell

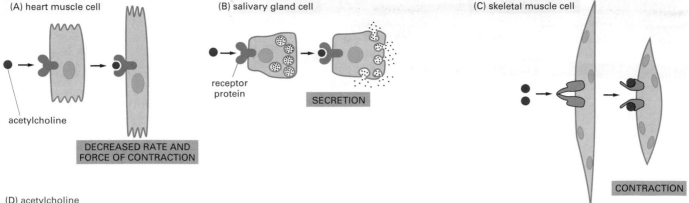

(A) heart muscle cell

acetylcholine

DECREASED RATE AND
FORCE OF CONTRACTION

(B) salivary gland cell

receptor
protein

SECRETION

(C) skeletal muscle cell

CONTRACTION

(D) acetylcholine

$$H_3C - \overset{\overset{\displaystyle O}{\|}}{C} - O - CH_2 - CH_2 - \overset{\overset{\displaystyle CH_3}{|}}{\underset{\underset{\displaystyle CH_3}{|}}{N^+}} - CH_3$$

Figure 16–5 The same signal molecule can induce different responses in different target cells. Different cell types are configured to respond to the neurotransmitter acetylcholine in different ways. Acetylcholine binds to similar receptor proteins on heart muscle cells (A) and salivary gland cells (B), but it evokes different responses in each cell type. Skeletal muscle cells (C) produce a different type of receptor protein for the same signal. As we shall see, the different receptor types generate different intracellular signals, thus enabling the different types of muscle cells to react differently to acetylcholine. (D) Chemical structure of acetylcholine. For such a versatile molecule, acetylcholine has a fairly simple structure.

hours to fully respond to the incoming signals. Overall, the integration of extracellular cues allows a relatively small number of signal molecules, used in different combinations, to exert subtle and complex control over cell behavior.

Receptors Relay Signals via Intracellular Signaling Pathways

Signal reception begins at the point where a signal originating outside the target cell encounters a target molecule belonging to the cell itself. In virtually every case, the target molecule is a **receptor protein** (also called a **receptor**), and each receptor is usually activated by only one type of signal. The receptor protein performs the primary transduction step: it receives an external signal and generates a new intracellular signal in response (see Figure 16–2B). As a rule, this is only the first event in a chain of intracellular signal transduction processes. In this game of molecular tag, the message is passed from one *intracellular signaling molecule* to another, each activating or generating the next signaling

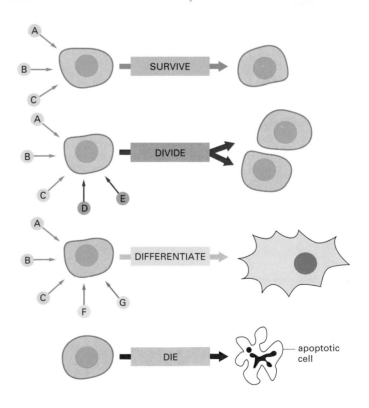

Figure 16–6 An animal cell depends on multiple extracellular signals. Every cell type displays a set of receptor proteins that enables it to respond to a specific set of signal molecules produced by other cells. These signal molecules work in combinations to regulate the behavior of the cell. As shown here, cells may require multiple signals (*blue arrows*) to survive, additional signals (*red arrows*) to divide, and still other signals (*green arrows*) to differentiate. If deprived of survival signals, most cells undergo a form of cell suicide known as programmed cell death, or apoptosis (discussed in Chapter 18).

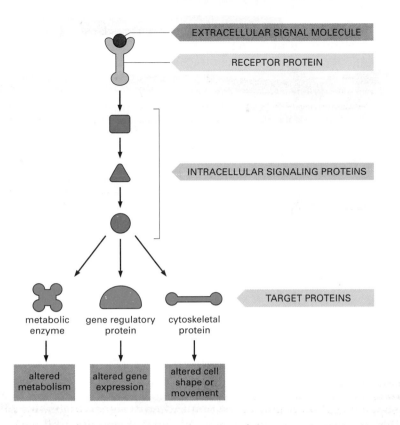

EXTRACELLULAR SIGNAL MOLECULE

RECEPTOR PROTEIN

INTRACELLULAR SIGNALING PROTEINS

TARGET PROTEINS

metabolic enzyme

gene regulatory protein

cytoskeletal protein

altered metabolism

altered gene expression

altered cell shape or movement

Figure 16–7 Extracellular signals alter the activity of a variety of cell proteins to change the behavior of the cell. In this case, the signal molecule binds to a cell-surface receptor protein. The receptor protein activates an intracellular signaling pathway that is mediated by a series of intracellular signaling proteins. Some of these signaling proteins interact with target proteins, altering them to change the behavior of the cell.

molecule in line, until, say, a metabolic enzyme is kicked into action, a gene is switched on, or the cytoskeleton is tweaked into a new configuration. This final outcome is called the *response* of the cell (Figure 16–7).

These relay chains, or **signaling cascades**, of intracellular signaling molecules have several crucial functions (Figure 16–8):

1. They *transform*, or *transduce*, the signal into a molecular form suitable for passing the signal along or stimulating a response.

2. They *relay* the signal from the point in the cell at which it is received to the point at which the response is produced.

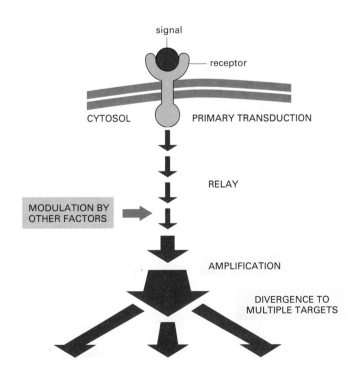

signal

receptor

CYTOSOL

PRIMARY TRANSDUCTION

RELAY

MODULATION BY OTHER FACTORS

AMPLIFICATION

DIVERGENCE TO MULTIPLE TARGETS

Question 16–2

When a single photon of light is absorbed by a rhodopsin photo-receptor, it activates about 500 individual molecules of an intracellular signaling protein called transducin. Each molecule of transducin, in turn, binds to and activates an enzyme, phosphodiesterase, that hydrolyzes about 4000 molecules of cyclic GMP per second. Cyclic GMP is a small molecule, similar to cyclic AMP, that in the cytosol of rod photoreceptor cells binds to the Na^+ channels in the plasma membrane and keeps them in their open conformation, as we discuss later (see Figure 16–28). If you only consider the decrease in cyclic GMP, what is the degree of signal amplification if each transducin molecule remains active for 100 milliseconds?

Figure 16–8 Cellular signaling cascades can follow a complex path. A receptor protein located on the cell surface transduces an extracellular signal into an intracellular signal, initiating a signaling cascade that transfers the signal into the cell interior, amplifying and distributing it en route. Many of the steps in the cascade can be modulated by other molecules or events in the cell.

3. In many cases, signaling cascades also *amplify* the signal received, making it stronger, so that a few extracellular signal molecules are enough to evoke a large intracellular response.

4. The signaling cascades can also *distribute* the signal so as to influence several processes in parallel: at any step in the pathway, the signal can *diverge* and be relayed to a number of different intracellular targets, creating branches in the information flow diagram and evoking a complex response.

5. Each step in this signaling cascade is open to *modulation* by other factors, including other external signals, so that the effects of the signal can be tailored to the conditions prevailing inside or outside the cell.

Most signaling pathways trace a long and branching route, and enlist many molecular players, as they relay information from receptors at the cell surface to appropriate machinery in the cell's interior. But some signaling pathways are simpler and more direct, as we discuss in the next two sections.

Nitric Oxide Crosses the Plasma Membrane and Activates Intracellular Enzymes Directly

Extracellular signal molecules in general fall into two classes. The first and largest class of signals consists of molecules that are too large or too hydrophilic to cross the plasma membrane of the target cell. They rely on receptors on the surface of the target cell to relay their message across the membrane (Figure 16–9A). The second, and smaller, class of signals consists of molecules that are small enough or hydrophobic enough to slip easily through the plasma membrane (Figure 16–9B). Once inside, these signal molecules either activate intracellular enzymes or bind to intracellular receptor proteins that regulate gene expression.

For an extracellular signal to alter a cell within a few seconds or minutes, direct activation of an enzyme is an effective strategy. **Nitric oxide (NO)** acts in this way. This dissolved gas diffuses readily out of the cell that generates it and enters neighboring cells. NO is made from the amino acid arginine and operates as a local mediator in many tissues. The gas acts only locally because it is quickly converted to nitrates and nitrites (with a half-life of about 5–10 seconds) by reaction with oxygen and water outside cells. Endothelial cells—the flattened cells that line every blood vessel—release NO in response to stimulation by nerve endings. This NO signal causes smooth muscle cells in the vessel wall to relax, allowing the vessel to dilate, so that blood flows through it more freely (Figure 16–10). The effect of NO on blood vessels accounts for the action of nitroglycerine, which has been used for almost 100 years to treat patients with angina (pain caused by inadequate blood flow to the heart muscle). In the body, nitroglycerine is converted to NO, which rapidly relaxes coronary blood vessels and increases blood flow to the heart. Many nerve cells also use NO to signal neighboring cells: NO released by nerve terminals in the penis, for instance, triggers the local blood-vessel dilation that is responsible for penile erection.

Inside many target cells, NO binds to the enzyme *guanylyl cyclase*, stimulating the formation of *cyclic GMP* from the nucleotide GTP. Cyclic GMP itself is a small intracellular signaling molecule that forms the next link in the signaling chain that leads to the cell's ultimate response. The impotence drug Viagra enhances penile erection by blocking the degradation of cyclic GMP, prolonging the NO signal. Cyclic GMP is very similar in its structure and mechanism of action to *cyclic AMP*, a much more

(A) CELL-SURFACE RECEPTORS

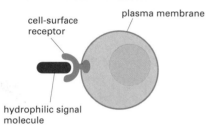

(B) INTRACELLULAR RECEPTORS

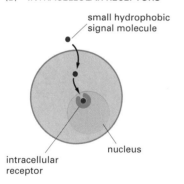

Figure 16–9 Extracellular signal molecules bind either to cell-surface receptors or to intracellular enzymes or receptors. (A) Most signal molecules are large and hydrophilic and are therefore unable to cross the plasma membrane directly; instead, they bind to cell-surface receptors, which in turn generate one or more signals inside the target cell (as shown in Figure 16–7). (B) Some small hydrophobic signal molecules, by contrast, diffuse across the target cell's plasma membrane and activate enzymes or bind to intracellular receptors—either in the cytosol or in the nucleus (as shown).

Figure 16–10 Nitric oxide (NO) triggers smooth muscle relaxation in a blood-vessel wall. (A) Drawing shows a nerve contacting a blood vessel. (B) Sequence of events leading to dilation of the blood vessel. Acetylcholine released by nerve terminals in the blood-vessel wall stimulates endothelial cells lining the blood vessel to make and release NO. The NO diffuses out of the endothelial cells and into adjacent smooth muscle cells, causing the muscle cells to relax. Note that NO gas is highly toxic when inhaled and should not be confused with nitrous oxide (N_2O), also known as laughing gas.

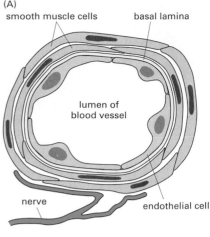

(A)

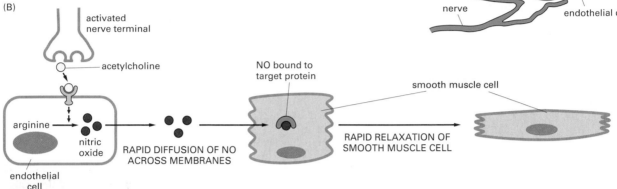

(B)

commonly used intracellular messenger molecule whose actions we discuss later.

Some Hormones Cross the Plasma Membrane and Bind to Intracellular Receptors

Gases such as NO are not the only signal molecules that can cross the plasma membrane. Hydrophobic signal molecules such as the **steroid hormones**—including *cortisol, estradiol,* and *testosterone*—and the *thyroid hormones* such as *thyroxine* (Figure 16–11) all pass through the plasma membrane of the target cell. Instead of activating intracellular enzymes, however, they bind to receptor proteins located in either the cytosol or the nucleus. These hormone receptors are proteins capable of regulating gene transcription, but they are typically present in an inactive form in unstimulated cells. When a hormone binds, the receptor protein undergoes a large conformational change that activates the

Figure 16–11 Some small hydrophobic hormones bind to intracellular receptors that act as gene regulatory proteins. Although these signal molecules differ in their chemical structure and function, they all act by binding to intracellular receptor proteins. Their receptors are not identical, but they are evolutionarily related, belonging to the *nuclear receptor superfamily* of gene regulatory proteins. The sites of origin and functions of these hormones are given in Table 16–1 (p. 537).

Consider the structure of cholesterol (Figure Q16–3), a small hydrophobic molecule with a sterol backbone similar to that of three of the hormones shown in Figure 16–11, but possessing fewer polar groups such as –OH, =O, and –COO⁻. If cholesterol were not normally found in cell membranes, could it be used effectively as a hormone if an appropriate intracellular receptor evolved?

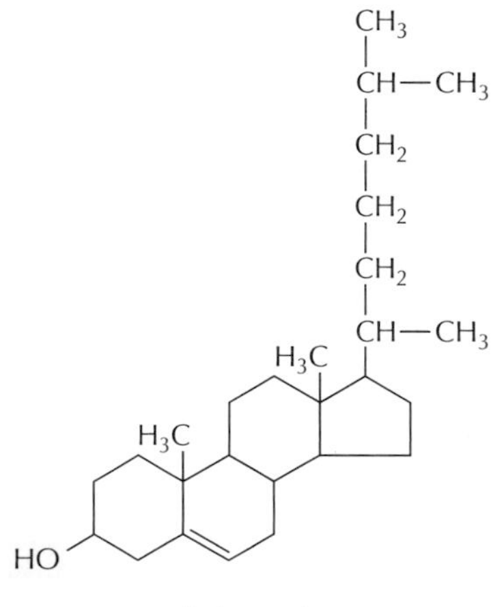

cholesterol

Figure Q16–3

protein, allowing it to promote or inhibit the transcription of a selected set of genes (Figure 16–12). Each hormone binds a different receptor protein, and each receptor acts at a different set of regulatory sites in DNA (discussed in Chapter 8). Because the hormones regulate different sets of genes, they evoke a variety of physiological responses (see also Table 16–1, p. 537).

Steroid hormone receptors play an essential role in human physiology, as illustrated by the dramatic consequences of a lack of the receptor for testosterone in humans. The male sex hormone testosterone shapes the formation of the external genitalia and influences brain development in the fetus; at puberty, it triggers the development of male secondary sexual characteristics. Some very rare individuals are genetically male (that is, they have both an X and a Y chromosome) but lack the testosterone receptor as a result of a mutation in the corresponding gene; thus they make the hormone, but their cells cannot respond to it. As a result, these individuals develop as females, which is the pathway of sexual and brain development that would occur if no male or female hormones were produced. This demonstrates the key role of the testosterone receptor in sexual development, and also shows that the receptor is required not just in one cell type to mediate one effect of testosterone, but in many cell types to help produce the whole range of features that distinguish men from women.

Cell-Surface Receptors Fall into Three Main Classes

In contrast to NO and the steroid and thyroid hormones, the vast majority of signal molecules are too large or hydrophilic to cross the plasma membrane of the target cell. These proteins, peptides, and other bulky, water-soluble molecules bind to receptor proteins that span the plasma membrane (Figure 16–13). The transmembrane receptors detect a signal on the outside and relay the message, in a new form, across the membrane into the interior of the cell.

Most cell-surface receptor proteins belong to one of three large families: *ion-channel–linked receptors, G-protein–linked receptors,* or *enzyme-linked receptors.* These families differ in the nature of the

Figure 16–12 The steroid hormone cortisol acts by activating a gene regulatory protein. Cortisol diffuses directly across the plasma membrane and binds to its receptor protein, which is located in the cytosol. The hormone–receptor complex is then transported into the nucleus via the nuclear pores. Cortisol binding activates the receptor protein, which is then able to bind to specific regulatory sequences in the DNA and activate (or represses, not shown) gene transcription. The receptors for cortisol and some other steroid hormones are located in the cytosol; those for the other signal molecules of this family are already bound to DNA in the nucleus.

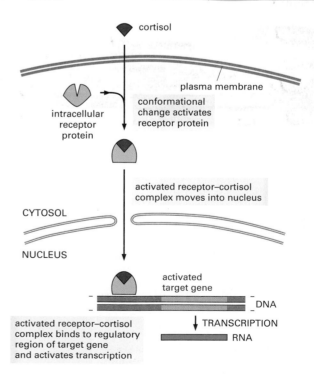

Figure 16–13 Most signal molecules bind to receptor proteins on the target cell surface. Shown here, the three-dimensional structure of human growth hormone *(red)* bound to its receptor. Binding of the hormone brings together two identical receptor proteins (one shown in *green*, the other in *blue*). The structures shown were determined by X-ray crystallographic studies of complexes formed between the hormone and extracellular receptor domains produced by recombinant DNA technology. Hormone binding activates cytoplasmic enzymes that are tightly bound to the cytosolic tails of the transmembrane receptors (not shown). (From A.M. deVos, M. Ultsch, and A.A. Kossiakoff, *Science* 255:306–312, 1992. © AAAS.)

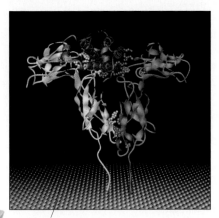

plasma membrane

intracellular signal that they generate when the extracellular signal molecule binds to them. For ion-channel–linked receptors, the resulting signal is a flow of ions across the membrane, which produces an electrical current (Figure 16–14A). G-protein–linked receptors activate a class of membrane-bound protein (a *trimeric GTP-binding protein* or *G protein*), which is then released to migrate in the plane of the plasma membrane, initiating a cascade of other effects (Figure 16–14B). And enzyme-linked receptors, when activated, act as enzymes or are associated with enzymes inside the cell (Figure 16–14C). Switching on this enzymatic activity then generates a host of additional signals, including small molecules that are released into the cytosol.

The number of different types of receptors in these three classes is even greater than the number of extracellular signals that act on them, because for many extracellular signal molecules there is more than one type of receptor. The neurotransmitter acetylcholine, for example, acts

(A) ION-CHANNEL–LINKED RECEPTOR

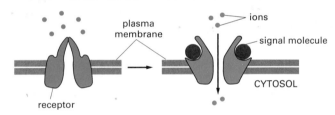

(B) G-PROTEIN–LINKED RECEPTORS

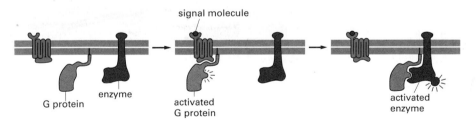

(C) ENZYME-LINKED RECEPTORS

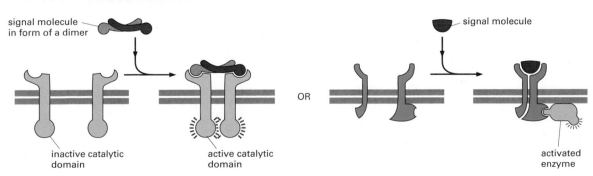

Figure 16–14 Cell-surface receptors fall into three basic classes. (A) An ion-channel–linked receptor opens (or closes, not shown) in response to binding of its signal molecule. (B) When a G-protein–linked receptor binds its extracellular signal molecule, the signal is passed first to a GTP-binding protein (a G protein) that is associated with the receptor. The activated G protein then leaves the receptor and turns on a target enzyme (or ion channel, not shown) in the plasma membrane. For simplicity, the G protein is shown here as a single molecule; as we shall see, it is in fact a complex of three subunits that can dissociate. (C) An enzyme-linked receptor binds its extracellular signal molecule, switching on an enzyme activity at the other end of the receptor, inside the cell. Although many enzyme-linked receptors have their own enzyme activity *(left)*, others rely on associated enzymes *(right)*.

Table 16–2 Some Substances That Mimic Natural Signal Molecules

MIMIC	SIGNAL MOLECULE	RECEPTOR ACTION	EFFECT
Valium and barbiturates	γ-aminobutyric acid (GABA)	stimulate GABA-activated ion-channel–linked receptors	relief of anxiety; sedation
Nicotine	acetylcholine	stimulates acetylcholine-activated ion-channel–linked receptors	constriction of blood vessels; elevation of blood pressure
Morphine and heroin	endorphins and enkephalins	stimulate G-protein–linked opiate receptors	analgesia (relief of pain); euphoria
Curare	acetylcholine	blocks acetylcholine-activated ion-channel–linked receptors	blockage of neuromuscular transmission, resulting in paralysis
Strychnine	glycine	blocks glycine-activated ion-channel–linked receptors	blockage of inhibitory synapses in spinal cord, resulting in seizures and muscle spasm

Question 16–4

The signaling mechanisms used by a steroid hormone receptor and by an ion-channel–linked receptor are relatively simple as they have few components. Can they lead to an amplification of the initial signal, and, if so, how?

on skeletal muscle cells via an ion-channel–linked receptor, whereas in heart muscle cells it acts through a G-protein–linked receptor (see Figure 16–5A and C). These two types of receptors generate different intracellular signals, and thus enable the two types of muscle cells to react to acetylcholine in different ways, increasing contraction in skeletal muscle and decreasing the frequency of contractions in heart.

The multitude of different cell-surface receptors that the body requires for signaling purposes are also targets for many foreign substances that interfere with our physiology and sensations, from heroin and nicotine to tranquilizers and chili peppers. These substances either mimic the natural ligand for a receptor, occupying the normal ligand-binding site, or bind to the receptor at some other site, blocking or overstimulating the receptor's natural activity. Many drugs and poisons act in this way (Table 16–2), and a large part of the pharmaceutical industry is devoted to the search for substances that will exert a precisely defined effect by binding to a specific type of cell-surface receptor.

Ion-channel–linked Receptors Convert Chemical Signals into Electrical Ones

Of all the cell-surface receptor types, *ion-channel–linked receptors* (also known as *transmitter-gated ion channels*) function in the simplest and most direct way. These receptors are responsible for the rapid transmission of signals across synapses in the nervous system. They transduce a chemical signal, in the form of a pulse of neurotransmitter delivered to the outside of the target cell, directly into an electrical signal, in the form of a change in voltage across the target cell's plasma membrane. When the neurotransmitter binds, this type of receptor alters its conformation so as to open or close a channel for the flow of specific types of ions—such as Na^+, K^+, Ca^{2+}, or Cl^-—across the plasma membrane (see Figure 16–14A). Driven by an electrochemical gradient, the ions rush into or out of the cell, creating a change in the membrane potential within a millisecond or so. This change in potential may trigger a nerve impulse, or alter the ability of other signals to do so. As we discuss later in this chapter, the opening of Ca^{2+} channels has special effects, as changes in the intracellular Ca^{2+} concentration can profoundly alter the activities of many proteins. The function of ion-channel–linked receptors is discussed in greater detail in Chapter 12.

Whereas ion-channel–linked receptors are a specialty of the nervous system and of other electrically excitable cells such as muscle,

G-protein–linked receptors and enzyme-linked receptors are used by practically every cell type of the body. Most of the remainder of this chapter will deal with these receptor families and with the signal transduction processes that they initiate.

Many Intracellular Signaling Proteins Act as Molecular Switches

Signals received via G-protein–linked or enzyme-linked receptors are transmitted to elaborate relay systems formed from cascades of intracellular signaling molecules. Apart from a few small molecules (such as cyclic GMP, cyclic AMP, and Ca^{2+}), these intracellular signaling molecules are proteins. Some serve as chemical transducers: in response to one type of chemical signal they generate another. Others serve as messengers, receiving a signal in one part of the cell and moving to another to exert an effect; and so on (see Figure 16–8).

Most of the key intracellular signaling proteins behave as **molecular switches**: receipt of a signal switches them from an inactive to an active state. Once activated, these proteins can turn on other proteins in the pathway. They then persist in an active state until some other process switches them off again. The importance of the switching-off process is often underappreciated. If a signaling pathway is to recover after transmitting a signal and make itself ready to transmit another, every molecular switch must be reset to its original, unstimulated state. Thus, at every step, for every activation mechanism there has to be an inactivation mechanism. The two are equally important for the function of the system.

Proteins that act as molecular switches mostly fall into one of two main classes. The first and by far the largest class consists of proteins whose activity is turned on or off by phosphorylation, as discussed in Chapter 4 (see Figure 4–41). For these, the switch is thrown in one direction by a protein kinase, which tacks a phosphate group onto the switch protein, and in the other direction by a protein phosphatase, which plucks the phosphate off the switch protein (Figure 16–15A). Many of the switch proteins controlled by phosphorylation are themselves protein kinases, and these are often organized into *phosphorylation cascades:* one protein kinase, activated by phosphorylation, phosphorylates the next protein kinase in the sequence, and so on, transmitting the signal onward and, in the process, amplifying, distributing, and modulating it.

The other main class of switch proteins involved in signaling consists of GTP-binding proteins. These switch between an active and an

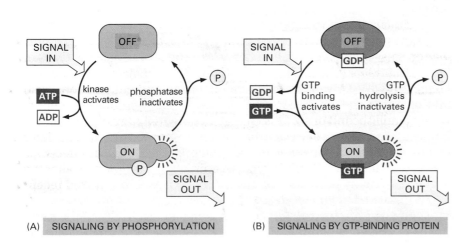

(A) **SIGNALING BY PHOSPHORYLATION**   (B) **SIGNALING BY GTP-BINDING PROTEIN**

Figure 16–15 Many Intracellular signaling proteins act as molecular switches. Intracellular signaling proteins can be activated by the addition of a phosphate group and inactivated by the removal of the phosphate. In some cases, the phosphate is added covalently to the protein by a protein kinase that transfers the terminal phosphate group from ATP to the signaling protein; the phosphate is then removed by a protein phosphatase (A). In other cases, a GTP-binding signaling protein is induced to exchange its bound GDP for GTP, which activates the protein; hydrolysis of the bound GTP to GDP then switches the protein off (B).

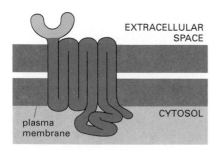

Figure 16–16 All G-protein–linked receptors possess a similar structure. The cytoplasmic portions of the receptor are responsible for binding to the G protein inside the cell. Receptors that bind to signal molecules that are proteins usually have a large extracellular ligand-binding domain *(light green)*. This domain, together with some of the transmembrane segments, binds the protein ligand. Receptors that recognize small signal molecules such as adrenaline, however, have small extracellular domains, and the ligand usually binds deep within the plane of the membrane to a site that is formed by amino acids from several transmembrane segments (not shown).

inactive state according to whether they have GTP or GDP bound to them (Figure 16–15B). The mechanisms that control the switch on and the switch off will be described in the next section. GTP-binding proteins are important in several signaling pathways. One class of GTP-binding protein, the G proteins, has a central role in signaling via G-protein–linked receptors, to which we now turn.

G-protein–linked Receptors

G-protein–linked receptors form the largest family of cell-surface receptors, with hundreds of members already identified in mammalian cells. They mediate responses to an enormous diversity of extracellular signal molecules, including hormones, local mediators, and neurotransmitters. These signal molecules are as varied in structure as they are in function: they can be proteins, small peptides, or derivatives of amino acids or fatty acids, and for each one of them there is a different receptor or set of receptors.

Despite the diversity of the signal molecules that bind to them, all G-protein–linked receptors that have been analyzed possess a similar structure: each is made of a single polypeptide chain that threads back and forth across the lipid bilayer seven times (Figure 16–16). This superfamily of *seven-pass transmembrane receptor proteins* includes rhodopsin (the light-activated photoreceptor protein in the vertebrate eye), the olfactory (smell) receptors in the vertebrate nose, and the receptors that participate in the mating rituals of single-celled yeasts. Evolutionarily speaking, G-protein–linked receptors are ancient: even bacteria possess structurally similar membrane proteins—such as the bacteriorhodopsin that functions as a light-driven H⁺ pump (discussed in Chapter 11). Although they resemble eucaryotic G-protein–linked receptors, these bacterial receptors do not act through G proteins; instead they are coupled to different signal transduction systems.

Stimulation of G-protein–linked Receptors Activates G-Protein Subunits

When an extracellular signal molecule binds to a seven-pass transmembrane receptor, the receptor protein undergoes a conformational change that enables it to activate a G protein located on the underside of the plasma membrane. To explain how this activation leads to the transmission of a signal, we must first consider how G proteins are constructed and how they function.

There are several varieties of G proteins. Each is specific for a particular set of receptors and a particular set of downstream target proteins, as we discuss shortly. All of these G proteins, however, have a similar general structure and operate in a similar way. They are composed of three protein subunits—α, β, and γ—two of which are tethered to the plasma membrane by short lipid tails. In the unstimulated state, the α subunit has GDP bound to it, and the G protein is idle (Figure 16–17A). When an extracellular ligand binds to its receptor, the altered receptor activates a G protein by causing the α subunit to lose some of its affinity for GDP, which it exchanges for a molecule of GTP. This activation breaks up the G protein subunits: the "switched-on" α subunit, clutching its GTP, detaches from the βγ complex, giving rise to two separate molecules that now roam independently along the plasma membrane (Figure 16–17B and C). The two activated parts of a G protein—the α subunit and the βγ complex—can both interact directly with target proteins located in the plasma membrane, which in turn may relay the signal to yet other destinations. The longer these target proteins have an α or a βγ subunit bound to them, the stronger and more prolonged the relayed signal will be.

The amount of time that the α and βγ subunits remain dissociated—and hence available to relay signals—is limited by the behavior of the α subunit. The α subunit has an intrinsic GTP-hydrolyzing (*GTPase*) activity, and it eventually hydrolyzes its bound GTP back to GDP; the α subunit then reassociates with a βγ complex and the signal is shut off (Figure 16–18). This reunion generally occurs within seconds after the G protein has been activated. The reconstituted G protein is now ready to be reactivated by another activated receptor.

Again, this system demonstrates a general principle of cell signaling: the mechanisms that shut a signal off are as important as the mechanisms that turn it on (see Figure 16–15B). They offer as many opportunities for control, and as many dangers of mishap. Take cholera, for example. The disease is caused by a bacterium that multiplies in the intestine, where it produces a protein called *cholera toxin*. This protein enters the cells that line the intestine and modifies the α subunit of a G protein (called G_S, because it *stimulates* the enzyme adenylyl cyclase, discussed later) in such a way that it can no longer hydrolyze its bound GTP. The altered α subunit thus remains in the active state indefinitely, continuously transmitting a signal to its target proteins. In intestinal cells, this causes a prolonged and excessive outflow of Cl⁻ and water into the gut, resulting in catastrophic diarrhea and dehydration. The condition often leads to death unless urgent steps are taken to replace the lost water and ions.

A similar situation occurs in whooping cough (pertussis), a common respiratory infection against which infants are now routinely vaccinated. In this case, the disease-causing bacterium colonizes the lung, where it produces a protein called *pertussis toxin*. This protein alters the α subunit of a different type of G protein (called G_i, because it *inhibits*

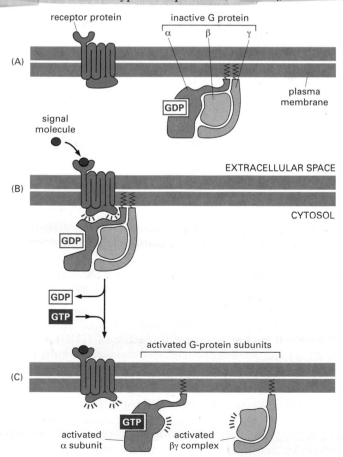

Figure 16–17 G proteins dissociate into two signaling proteins when activated.
(A) In the unstimulated state, the receptor and the G protein are both inactive. Although they are shown here as separate entities in the plasma membrane, in some cases, at least, they are associated in a preformed complex. (B) Binding of an extracellular signal to the receptor changes the conformation of the receptor, which in turn alters the conformation of the G protein that is bound to the receptor. (C) The alteration of the α subunit of the G protein allows it to exchange its GDP for GTP. This causes the G protein to break up into two active components—an α subunit and a βγ complex, both of which can regulate the activity of target proteins in the plasma membrane. The receptor stays active while the external signal molecule is bound to it, and it can therefore catalyze the activation of many molecules of G protein. Note that both the α and γ subunits of the G protein have covalently attached lipid molecules (*red*) that help anchor them to the plasma membrane.

Figure 16–18 The G-protein α subunit switches itself off by hydrolyzing its bound GTP. When an activated α subunit encounters and binds its target, it turns on its protein partner (or in some cases inactivates it, not shown) for as long as the two remain in touch. Within seconds, the GTP on the α subunit is hydrolyzed to GDP by the α subunit's intrinsic GTPase activity. This loss of GTP inactivates the α subunit, which dissociates from its target protein and reassociates with a βγ complex to re-form an inactive G protein. The G protein is now ready to couple to another receptor, as in Figure 16–17B. Both the activated α subunit and the free βγ complex can regulate target proteins.

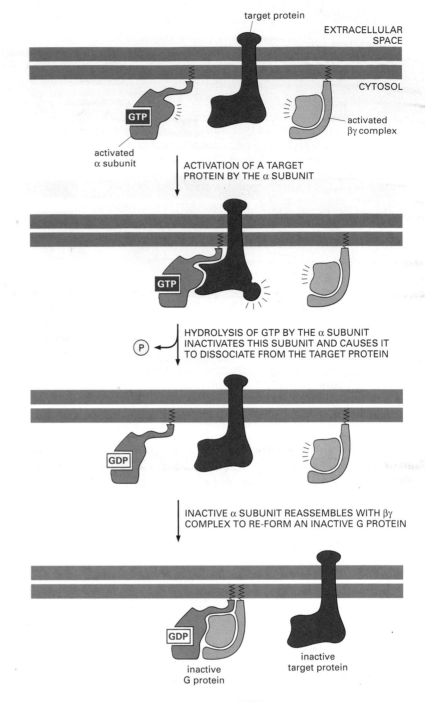

Question 16–5

G-protein–linked receptors activate G proteins by reducing the strength of GDP binding. This results in rapid dissociation of bound GDP, which is then replaced by GTP, which is present in the cytosol in much higher concentrations than GDP. What consequences would result from a mutation in the α subunit of a G protein that caused its affinity for GDP to be reduced without significantly changing its affinity for GTP? Compare the effects of this mutation with the effects of cholera toxin.

adenylyl cyclase). In this case, however, modification by the toxin disables the G protein by locking it into its inactive GDP-bound state. Knocking out G_i, like activating G_s, results in the generation of a prolonged, inappropriate signal. Oddly, although the biochemical effects of cholera and pertussis toxins are known in detail, it is not clear how the bacteria benefit from their actions. In any case, what cholera and pertussis toxins do show us is that, like a car accelerating out of control, intracellular signaling pathways can be rendered dangerously overactive either by gluing down the molecular gas pedal or by cutting the molecular brakes.

Some G Proteins Regulate Ion Channels

The target proteins for G-protein subunits are either ion channels or membrane-bound enzymes. Different targets are affected by different

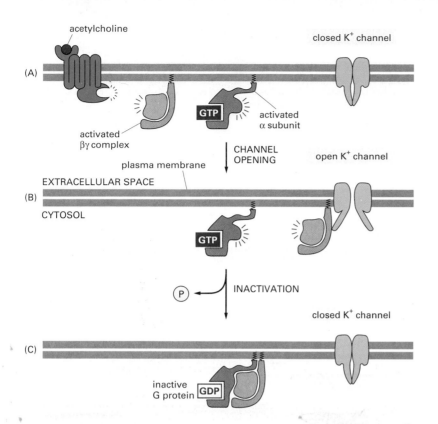

Figure 16–19 G proteins couple receptor activation to the opening of K$^+$ channels in the plasma membrane of heart muscle cells. (A) Binding of the neurotransmitter acetylcholine to its G-protein–linked receptor on heart muscle cells results in the dissociation of the G protein into an activated βγ complex and an activated α subunit. (B) The activated βγ complex binds to and opens a K$^+$ channel in the heart cell plasma membrane. (C) Inactivation of the α subunit by hydrolysis of bound GTP causes it to reassociate with the βγ complex to form an inactive G protein, allowing the K$^+$ channel to close.

types of G proteins (of which about 20 have so far been discovered in mammalian cells), and these various G proteins are themselves activated by different classes of cell-surface receptors. In this way, binding of an extracellular signal molecule to a G-protein–linked receptor leads to effects on a particular subset of the possible target proteins, eliciting a response that is appropriate for that signal and that type of cell.

We look first at an example of G-protein regulation of ion channels. The heartbeat in animals is controlled by two sets of nerve fibers: one speeds the heart up, the other slows it down. The nerves that signal a slowdown in heartbeat do so by releasing acetylcholine, which binds to a G-protein–linked receptor on the surface of the heart muscle cells. When acetylcholine binds to this receptor, a G protein (G$_i$) is activated— dissociating into an α subunit and a βγ complex (Figure 16–19A). In this particular example, the βγ complex is the active signaling component: it binds to the intracellular face of a K$^+$ channel in the heart muscle cell plasma membrane, forcing the ion channel into an open conformation and allowing K$^+$ to flow out of the cell (Figure 16–19B). This alters the electrical properties of the heart muscle cell, inhibiting its activity. The signal is shut down—and the K$^+$ channel recloses—when the α subunit inactivates itself by hydrolyzing its bound GTP and reassociates with the βγ complex to re-form an inactive G protein (Figure 16–19C).

Some G Proteins Activate Membrane-bound Enzymes

When G proteins interact with ion channels, they cause an immediate change in the state and behavior of the cell. Their interactions with enzyme targets have more complex consequences, leading to the production of additional intracellular signaling molecules. The most frequent target enzymes for G proteins are *adenylyl cyclase*, the enzyme responsible for production of the small intracellular signaling molecule *cyclic AMP*, and *phospholipase C*, the enzyme responsible for production of the small intracellular signaling molecules *inositol trisphosphate* and *diacylglycerol*. These two enzymes are activated by different types

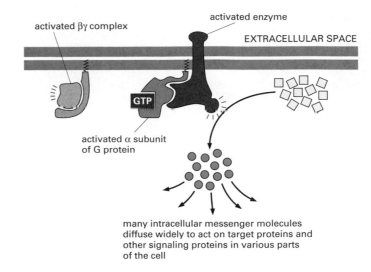

Figure 16–20 Enzymes activated by G proteins catalyze the synthesis of intracellular second-messenger molecules. Because each activated enzyme generates many second-messenger molecules, the signal is greatly amplified at this step in the pathway. The signal is passed on by the messenger molecules, which bind to target proteins and other signaling proteins in the cell and influence their activity.

activated βγ complex

activated enzyme

EXTRACELLULAR SPACE

GTP

activated α subunit of G protein

many intracellular messenger molecules diffuse widely to act on target proteins and other signaling proteins in various parts of the cell

of G proteins, so that cells are able to couple the production of the small intracellular signaling molecules to different extracellular signals. As we saw earlier, the coupling may be either stimulatory or inhibitory. We concentrate here on G proteins that stimulate enzyme activity. The small intracellular signaling molecules generated in these cascades are often called **second messengers** (the "first messengers" being the extracellular signals); they are produced in large numbers when a membrane-bound enzyme—such as adenylyl cyclase or phospholipase C—is activated, and they rapidly diffuse away from their source, spreading the signal throughout the cell (Figure 16–20).

Different second-messenger molecules, of course, produce different cellular responses. We will first examine the consequences of an increase in the intracellular concentration of cyclic AMP. This will take us along one of the main types of signaling pathways that lead from the activation of G-protein–linked receptors. We then discuss the actions of inositol trisphosphate and diacylglycerol, second-messenger molecules that will lead us along a different molecular route.

The Cyclic AMP Pathway Can Activate Enzymes and Turn On Genes

Many extracellular signals acting via G-protein–linked receptors affect the activity of **adenylyl cyclase** and thus alter the concentration of the messenger molecule **cyclic AMP** inside the cell. Most commonly, the activated G-protein α subunit switches on the adenylyl cyclase, causing a dramatic and sudden increase in the synthesis of cyclic AMP from ATP (which is always present in the cell). Because it stimulates the cyclase, this G protein is called G_s. To help eliminate the signal, a second enzyme, called *cyclic AMP phosphodiesterase,* rapidly converts cyclic AMP to ordinary AMP (Figure 16–21). One way that caffeine acts as a stimulant is by inhibiting this phosphodiesterase in the nervous system, blocking cyclic AMP degradation and keeping the concentration of this second-messenger high.

Cyclic AMP phosphodiesterase is continuously active inside the cell. Because it breaks cyclic AMP down so quickly, the concentrations

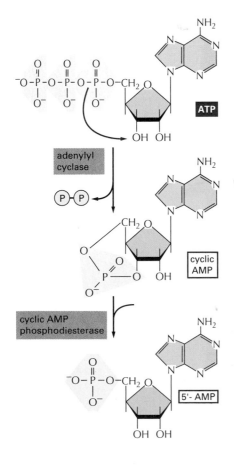

ATP

adenylyl cyclase

P P

cyclic AMP

cyclic AMP phosphodiesterase

5'- AMP

Figure 16–21 Cyclic AMP is synthesized by adenylyl cyclase and degraded by cyclic AMP phosphodiesterase. Cyclic AMP is formed from ATP by a cyclization reaction that removes two phosphate groups from ATP and joins the "free" end of the remaining phosphate group to the sugar part of the ATP molecule. The degradation reaction breaks this second bond, forming AMP.

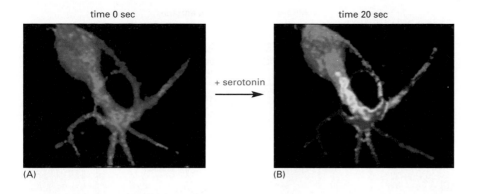

time 0 sec   + serotonin   time 20 sec

(A)   (B)

Figure 16–22 Cyclic AMP concentration rises rapidly in response to an extracellular signal. A nerve cell in culture responds to the binding of the neurotransmitter serotonin to a G-protein–linked receptor by synthesizing cyclic AMP. The concentration of intracellular cyclic AMP was monitored by injecting into the cell a fluorescent protein whose fluorescence changes when it binds cyclic AMP. *Blue* indicates a low level of cyclic AMP, *yellow* an intermediate level, and *red* a high level. (A) In the resting cell, the cyclic AMP concentration is about 5×10^{-8} M. (B) Twenty seconds after adding serotonin to the culture medium, the intracellular concentration of cyclic AMP has risen to more than 10^{-6} M, an increase of more than twentyfold. (Courtesy of Roger Tsien.)

of this second messenger can change rapidly in response to extracellular signals, rising or falling tenfold in a matter of seconds (Figure 16–22). Cyclic AMP is a water-soluble molecule, so it can carry its signal throughout the cell, traveling from the site on the membrane where it is synthesized to interact with proteins located in the cytosol, the nucleus, or other organelles.

Many cell responses are mediated by cyclic AMP; a few are listed in Table 16–3. As the table shows, different target cells respond very differently to extracellular signals that change intracellular cyclic AMP concentrations. In many types of animal cells, stimulating cyclic AMP production boosts the rate of consumption of metabolic fuel. When we are frightened or excited, for example, the adrenal gland releases the hormone *adrenaline*, which circulates in the bloodstream and binds to a class of G-protein–linked receptors (adrenergic receptors) that are present on many types of cells. The consequences vary from one cell type to another, but all of the cell responses help prepare the body for sudden action. In skeletal muscle, for example, adrenaline triggers a rise in the intracellular concentration of cyclic AMP, which causes the breakdown of glycogen (the polymerized storage form of glucose). This glycogen breakdown makes more glucose available as fuel for anticipated muscular activity. Adrenaline also acts on fat cells, stimulating the breakdown of triglyceride (the storage form of fat) to fatty acids—an immediately usable form of cell fuel (discussed in Chapter 13), which can also be exported to other cells.

Cyclic AMP exerts these various effects mainly by activating the enzyme **cyclic-AMP–dependent protein kinase (PKA)**. This enzyme is normally held inactive in a complex with another protein. The binding of cyclic AMP forces a conformational change that unleashes the active kinase. Activated PKA then catalyzes the phosphorylation of particular serines or threonines on certain intracellular proteins, thus altering

Table 16–3 Some Cell Responses Mediated by Cyclic AMP

EXTRACELLULAR SIGNAL MOLECULE*	TARGET TISSUE	MAJOR RESPONSE
Adrenaline	heart	increase in heart rate and force of contraction
Adrenaline	muscle	glycogen breakdown
Adrenaline, ACTH, glucagon	fat	fat breakdown
ACTH	adrenal gland	cortisol secretion

*Although all of the signal molecules listed here are hormones, some responses to local mediators and to neurotransmitters are also mediated by cyclic AMP.

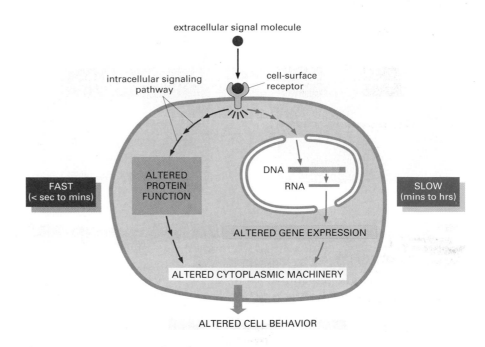

Figure 16–23 Extracellular signals can act slowly or rapidly. Certain types of altered cell behavior, such as increased cell growth and division, involve changes in gene expression and the synthesis of new proteins; they therefore occur relatively slowly. Other responses—such as changes in cell movement, secretion, or metabolism—need not involve the nuclear machinery and therefore occur more quickly; they may involve the rapid phosphorylation of target proteins in the cytoplasm, for example.

extracellular signal molecule

intracellular signaling pathway

cell-surface receptor

FAST (< sec to mins)

ALTERED PROTEIN FUNCTION

DNA

RNA

SLOW (mins to hrs)

ALTERED GENE EXPRESSION

ALTERED CYTOPLASMIC MACHINERY

ALTERED CELL BEHAVIOR

Question 16–6

Explain why cyclic AMP must be broken down rapidly in a cell to allow rapid signaling.

their activity. In different cell types, different sets of target proteins are available to be phosphorylated, which explains why the effects of cyclic AMP vary with the target cell.

In some cases the effects of activating a cyclic AMP cascade are rapid; in others the effects are slow (Figure 16–23). In skeletal muscle cells, for example, activated PKA phosphorylates enzymes involved in glycogen metabolism, triggering the mechanism that breaks down glycogen to glucose. This response occurs within seconds. At the other extreme, some cyclic AMP responses take minutes or hours to develop. Included in this slow class are responses that involve changes in gene expression, an important means of regulating cell behavior. Thus in some cells, the PKA phosphorylates gene regulatory proteins that then activate the transcription of selected genes, a process that requires minutes or hours. In endocrine cells in the hypothalamus, for example, a rise in the amount of intracellular cyclic AMP stimulates the production and secretion of a peptide hormone called somatostatin. Increases in cyclic AMP concentrations in neurons, by contrast, control the production of proteins involved in long-term memory. Figure 16–24 shows the extensive relay chain for such a pathway from the plasma membrane to the nucleus.

We now turn to the other enzyme-mediated signaling cascade that leads from G-protein–linked receptors—the pathway that begins with the activation of the membrane-bound enzyme *phospholipase C* and leads to the generation of the second messengers inositol trisphosphate and diacylglycerol.

The Inositol Phospholipid Pathway Triggers a Rise in Intracellular Ca²⁺

Some extracellular signal molecules exert their effects via a type of G protein that activates the membrane-bound enzyme **phospholipase C** instead of adenylyl cyclase. A few examples are given in Table 16–4.

Once activated, phospholipase C propagates its signal by cleaving a lipid molecule that is a component of the cell membrane. The molecule is an **inositol phospholipid** (a phospholipid that has the sugar inositol attached to its head) that is present in small quantities in the inner half of the plasma membrane lipid bilayer (see Figure 11–17). Because of the

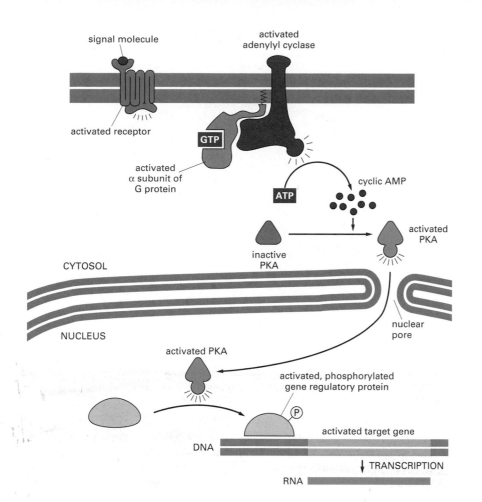

Figure 16–24 A rise in intracellular cyclic AMP can activate gene transcription.
Binding of a signal molecule to its G-protein–linked receptor can lead to the activation of adenylyl cyclase and a rise in the concentration of intracellular cyclic AMP. In the cytosol, cyclic AMP activates PKA, which then moves into the nucleus and phosphorylates specific gene regulatory proteins. Once phosphorylated, these proteins stimulate the transcription of a whole set of target genes. This type of signaling pathway controls many processes in cells, ranging from hormone synthesis in endocrine cells to the production of proteins involved in long-term memory in the brain. Activated PKA can also phosphorylate and thereby regulate other proteins and enzymes in the cytosol (not shown).

involvement of this phospholipid, the signaling pathway that begins with the activation of phospholipase C is often known as the *inositol phospholipid pathway*. This signaling cascade occurs in almost all eucaryotic cells and affects a host of different target proteins.

The cascade works in the following way. When phospholipase C chops the sugar-phosphate head off the inositol phospholipid, it generates two small messenger molecules—**inositol 1,4,5-trisphosphate** (**IP$_3$**) and **diacylglycerol** (**DAG**). IP$_3$, a hydrophilic sugar phosphate, diffuses into the cytosol, while the lipid DAG remains embedded in the plasma membrane. Both molecules play a crucial part in signaling inside the cell, and we will consider them in turn.

The IP$_3$ released into the cytosol will eventually encounter the endoplasmic reticulum; there it binds to and opens Ca^{2+} channels that are embedded in the endoplasmic reticulum membrane. Ca^{2+} stored

Table 16–4 Some Cell Responses Mediated by Phospholipase C Activation

SIGNAL MOLECULE	TARGET TISSUE	MAJOR RESPONSE
Vasopressin (a protein hormone)	liver	glycogen breakdown
Acetylcholine	pancreas	secretion of amylase (a digestive enzyme)
Acetylcholine	smooth muscle	contraction
Thrombin (a proteolytic enzyme)	blood platelets	aggregation

Figure 16–25 Phospholipase C activates two signaling pathways. Two intracellular messenger molecules are produced when a membrane inositol phospholipid is hydrolyzed by activated phospholipase C. Inositol 1,4,5-trisphosphate (IP_3) diffuses through the cytosol and triggers the release of Ca^{2+} from the endoplasmic reticulum by binding to and opening special Ca^{2+} channels in the endoplasmic reticulum membrane. The large electrochemical gradient for Ca^{2+} causes Ca^{2+} to rush out into the cytosol. Diacylglycerol remains in the plasma membrane and, together with Ca^{2+}, helps to activate the enzyme protein kinase C (PKC), which is recruited from the cytosol to the cytosolic face of the plasma membrane. PKC then phosphorylates its own set of intracellular target proteins, further propagating the signal.

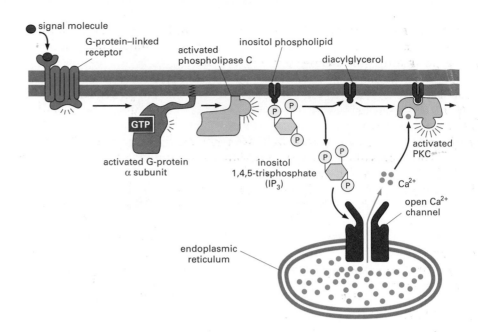

inside the endoplasmic reticulum rushes out into the cytosol through these open channels (Figure 16–25), causing a sharp rise in the cytosolic concentration of free Ca^{2+}, which is normally kept very low.

Together with Ca^{2+}, diacylglycerol helps recruit and activate a protein kinase, which translocates from the cytosol to the plasma membrane. This enzyme is called **protein kinase C (PKC)** because it also needs to bind Ca^{2+} to become active (see Figure 16–25). Once activated, PKC phosphorylates a set of intracellular proteins that varies depending on the cell type. PKC operates on the same principle as PKA, although most of its target proteins are different.

A Ca^{2+} Signal Triggers Many Biological Processes

Ca^{2+} has such an important and widespread role as an intracellular messenger that we must digress to consider its functions more generally. A surge in the cytosolic concentration of free Ca^{2+} is triggered by many different signals, not only those that act through G-protein–linked receptors. When a sperm fertilizes an egg cell, for example, Ca^{2+} channels open, and the resulting rise in cytosolic Ca^{2+} triggers the start of embryonic development (Figure 16–26); for skeletal muscle cells, a signal from a nerve triggers a rise in cytosolic Ca^{2+} that initiates contraction; and in many secretory cells, including nerve cells, Ca^{2+} triggers secretion. Ca^{2+} stimulates all these responses by binding to and influencing the activity of Ca^{2+}-sensitive proteins.

The concentration of free Ca^{2+} in the cytosol of an unstimulated cell is extremely low (10^{-7} M) compared with its concentration in the extracellular fluid and in the endoplasmic reticulum. These differences

Figure 16–26 Fertilization of an egg by a sperm triggers an increase in cytosolic Ca^{2+} in the egg. This starfish egg was injected with a Ca^{2+}-sensitive fluorescent dye before it was fertilized. When a sperm penetrates the egg, a wave of cytosolic Ca^{2+} *(red)*—released from the endoplasmic reticulum—sweeps across the egg from the site of sperm entry *(arrow)*. This Ca^{2+} wave provokes a change in the egg surface, preventing entry of other sperm, and initiates embryonic development. (Courtesy of Stephen A. Stricker.)

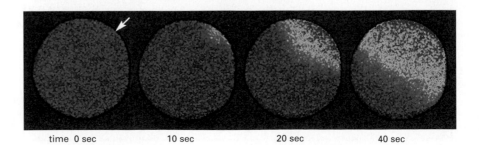

time 0 sec   10 sec   20 sec   40 sec

are maintained by membrane-embedded pumps that actively pump Ca^{2+} out of the cytosol—either into the endoplasmic reticulum or across the plasma membrane and out of the cell. As a result, a steep electrochemical gradient of Ca^{2+} exists across the endoplasmic reticulum membrane and across the plasma membrane (discussed in Chapter 12). When a signal transiently opens Ca^{2+} channels in either of these membranes, Ca^{2+} rushes into the cytosol down its electrochemical gradient, triggering changes in Ca^{2+}-responsive proteins in the cytosol.

The effects of Ca^{2+} in the cytosol are largely indirect: they are mediated through the interaction of Ca^{2+} with various transducer proteins, known collectively as *Ca²⁺-binding proteins.* The most widespread and common of these is the Ca^{2+}-responsive protein **calmodulin**. Calmodulin is present in the cytosol of all eucaryotic cells that have been examined, including those of plants, fungi, and protozoa. When calmodulin binds to Ca^{2+}, the protein undergoes a conformational change that enables it to wrap around a wide range of target proteins in the cell, altering their activities (Figure 16–27). One particularly important class of targets for calmodulin are the **Ca²⁺/calmodulin-dependent protein kinases (CaM-kinases)**. When these kinases are activated by binding to calmodulin complexed with Ca^{2+}, they influence other processes in the cell by phosphorylating selected proteins. In the mammalian brain, for example, a neuron-specific CaM-kinase is abundant at synapses, where it is thought to play a part in learning and memory. Some memories, it seems, depend on this CaM-kinase and the pulses of Ca^{2+} signals that occur during neural activity: mutant mice that lack the kinase show a marked inability to remember where things are.

Intracellular Signaling Cascades Can Achieve Astonishing Speed, Sensitivity, and Adaptability: A Look at Photoreceptors in the Eye

The steps in the signaling cascades associated with G-protein–linked receptors take a long time to describe, but they often take only seconds to execute. Consider how quickly a thrill can make your heart beat faster (when adrenaline stimulates the G-protein–linked receptors in your heart muscle cells, accelerating your heartbeat), or how fast the smell of food can make you salivate (through the G-protein–linked receptors for odors in your nose and the G-protein–linked receptors for acetylcholine

Question 16–7

Why do you suppose cells have evolved intracellular Ca^{2+} stores for signaling even though there is abundant extracellular Ca^{2+}?

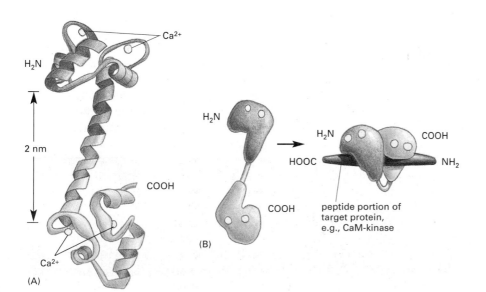

Figure 16–27 X-ray diffraction and NMR studies reveal the structure of Ca²⁺/calmodulin. (A) The calmodulin molecule has a dumbbell shape, with two globular ends connected by a long, flexible α helix. Each end has two Ca^{2+}-binding domains. (B) Simplified representation of the structure, showing the conformational changes in Ca^{2+}/calmodulin that occur when it binds to a target protein. Note that the α helix has jackknifed to surround the target protein. (A, based on X-ray crystallographic data from Y.S. Babu et al., *Nature* 315:37–40, 1985. © Macmillan Magazines Ltd.; B, based on X-ray crystallographic data from W.E. Meador, A.R. Means, and F.A. Quiocho, *Science* 257:1251–1255, 1992, and on NMR data from M. Ikura et al., *Science* 256:632–638, 1992. © AAAS.)

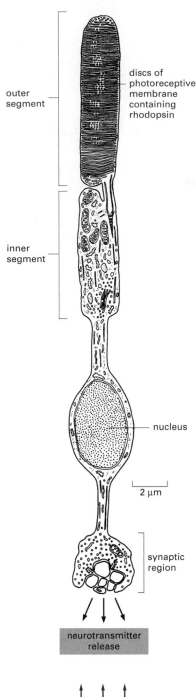

outer segment

discs of photoreceptive membrane containing rhodopsin

inner segment

nucleus

2 μm

synaptic region

neurotransmitter release

LIGHT

Figure 16–28 A rod photoreceptor cell from the retina is exquisitely sensitive to light. Drawing of a rod photoreceptor. The light-absorbing molecules of rhodopsin are embedded in many pancake-shaped vesicles *(discs)* of membrane inside the outer segment of the cell. Neurotransmitter is released from the opposite end of the cell to control firing of the retinal nerve cells that pass on the signal to the brain. When the rod cell is stimulated by light, a signal is relayed from the rhodopsin molecules in the discs, through the cytosol of the outer segment, to Na$^+$ channels in the plasma membrane of the outer segment. The Na$^+$ channels close in response to the signal, producing a change in the membrane potential of the rod cell. By mechanisms similar to those that control neurotransmitter release in ordinary nerve cells, the change in membrane potential alters the rate of neurotransmitter release from the synaptic region of the cell. (Adapted from T.L. Leutz, Cell Fine Structure. Philadelphia: Saunders, 1971.)

in salivary cells that stimulate secretion). Among the fastest of all responses mediated by a G-protein–linked receptor, however, is the response of the eye to bright light: it takes only 20 msec for the most quickly responding photoreceptor cells of the retina (the cone photoreceptors) to produce their electrical response to a sudden flash of light.

This speed is achieved in spite of the necessity to relay the signal over several steps of an intracellular signaling cascade. But photoreceptors also provide a beautiful illustration of the positive advantages of signaling cascades: in particular, such cascades allow spectacular amplification of the incoming signal and allow cells to adapt so as to be able to detect signals of widely varying intensity. The quantitative details have been most thoroughly analyzed for the rod photoreceptor cells in the eye (Figure 16–28). In this cell, light is harvested by *rhodopsin,* a G-protein–linked light receptor. Light-activated rhodopsin activates a G protein called *transducin.* The activated α subunit of transducin then activates an intracellular signaling cascade that causes Na$^+$ channels to close in the plasma membrane of the photoreceptor cell. This produces a change in the voltage across the cell membrane, with the ultimate consequence that a nerve impulse is sent to the brain.

The signal is repeatedly amplified as it is relayed along this pathway (Figure 16–29). When lighting conditions are dim (as on a moonless night), the amplification is enormous, and as few as a dozen photons absorbed in the entire retina will cause a perceptible signal to be delivered to the brain. In bright sunlight, when photons flood through each photoreceptor cell at a rate of billions per second, the signaling cascade *adapts,* stepping down the amplification more than 10,000-fold so that the photoreceptor cells are not overwhelmed and can still register increases and decreases in the strong light. The adaptation depends on negative feedback: an intense response in the photoreceptor cell generates an intracellular signal (a change in Ca^{2+} concentration) that inhibits the enzymes responsible for signal amplification.

Adaptation also occurs in signaling pathways that respond to chemical signals; again, it allows cells to remain sensitive to changes of signal intensity over a wide range of background levels of stimulation. Adaptation, in other words, allows a cell to respond to both messages that are whispered and those that are shouted.

In addition to vision, taste and smell also depend on G-protein–linked receptors. It seems likely that this mechanism of signal reception, invented early in the evolution of the eucaryotes, has its origins in the basic and universal need of cells to sense and respond to

Figure 16–29 The light-induced signaling cascade in rod photoreceptor cells greatly amplifies the light signal. When rod photoreceptors are adapted for dim light, signal amplification is enormous. The intracellular signaling pathway from the G protein transducin uses components that differ from the ones previously described. The cascade functions as follows. In the absence of a light signal, the messenger molecule cyclic GMP is continuously produced in the photoreceptor cell and binds to Na^+ channels in the photoreceptor cell plasma membrane, keeping them open. Activation of rhodopsin by light results in formation of activated transducin α subunits. These activate an enzyme called cyclic GMP phosphodiesterase, which breaks down cyclic GMP to GMP. The sharp fall in the intracellular concentration of cyclic GMP causes the bound cyclic GMP to dissociate from the Na^+ channels, which therefore close. The *red arrows* indicate the steps at which amplification occurs.

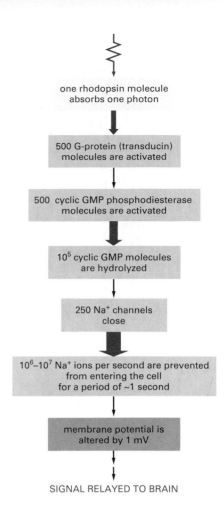

one rhodopsin molecule absorbs one photon

500 G-protein (transducin) molecules are activated

500 cyclic GMP phosphodiesterase molecules are activated

10^5 cyclic GMP molecules are hydrolyzed

250 Na^+ channels close

10^6–10^7 Na^+ ions per second are prevented from entering the cell for a period of ~1 second

membrane potential is altered by 1 mV

SIGNAL RELAYED TO BRAIN

their environment. Of course, G-protein–linked receptors are not the only receptors that activate intracellular signaling cascades. We now turn to another class of cell-surface receptors that play a key part in controlling cell numbers, cell differentiation, and cell movement in multicellular animals.

Enzyme-linked Receptors

Like G-protein–linked receptors, enzyme-linked receptors are transmembrane proteins that display their ligand-binding domains on the outer surface of the plasma membrane. Instead of associating with a G protein, however, the cytoplasmic domain of the receptor acts as an enzyme—or forms a complex with another protein that acts as an enzyme. Enzyme-linked receptors (see Figure 16–14C) came to light through their role in responses to extracellular signal proteins that regulate the growth, proliferation, differentiation, and survival of cells in animal tissues (see Table 16–1, p. 537, for examples). Most of these signal proteins act as local mediators and can act at very low concentrations (about 10^{-9} to 10^{-11} M). Responses to them are typically slow (on the order of hours) and require many intracellular transduction steps that eventually lead to changes in gene expression.

Enzyme-linked receptors also mediate direct, rapid reconfigurations of the cytoskeleton, controlling the way a cell moves and changes its shape. The extracellular signals for these architectural alterations are often not diffusible signal proteins, but proteins attached to the surfaces over which a cell is crawling. Disorders of cell growth, proliferation, differentiation, survival, and migration are fundamental to cancer, and abnormalities in signaling via enzyme-linked receptors play a major role in the initiation of this class of diseases.

The largest class of enzyme-linked receptors is made up of those with a cytoplasmic domain that functions as a tyrosine protein kinase, phosphorylating tyrosine side chains on selected intracellular proteins. Such receptors are called **receptor tyrosine kinases**, and we shall focus on these receptors here.

Activated Receptor Tyrosine Kinases Assemble a Complex of Intracellular Signaling Proteins

To do its job as a signal transducer, an enzyme-linked receptor has to switch on the enzyme activity of its intracellular domain (or of an

Question 16–8

One important feature of any signaling cascade is its ability to turn off. Consider the cascade shown in Figure 16–29. Where would off switches be required? Which ones do you suppose are the most important?

associated enzyme) when an external signal molecule binds to its extracellular domain. Unlike the seven-pass G-protein–linked receptors, enzyme-linked receptor proteins usually have only one transmembrane segment, which is thought to span the lipid bilayer as a single α helix. There is, it seems, no way to transmit a conformational change through a single α helix, and so enzyme-linked receptors have a different strategy for transducing the extracellular signal. In many cases, the binding of a signal molecule causes two receptor molecules to come together in the membrane, forming a dimer. Contact between the two adjacent intracellular receptor tails activates their kinase function, with the result that each receptor phosphorylates the other. In the case of receptor tyrosine kinases, the phosphorylations occur on specific tyrosines located on the cytosolic tail of the receptors.

This phosphorylation then triggers the assembly of an elaborate intracellular signaling complex on the receptor tails. The newly phosphorylated tyrosines serve as binding sites for a whole zoo of intracellular signaling proteins—perhaps as many as 10 or 20 different molecules—which themselves can become activated upon binding (Figure 16–30). While it lasts, this protein complex transmits its signal along several routes simultaneously to many destinations inside the cell, thus activating and coordinating the numerous biochemical changes that are required to trigger a complex response, such as cell proliferation. To terminate the activation of the receptor, the cell contains *protein tyrosine phosphatases*, which remove the phosphates that were added in response to the extracellular signal. In some cases, activated receptors are disposed of in a more brutal way: they are dragged into the interior of the cell by endocytosis and then destroyed by digestion in lysosomes.

Different receptor tyrosine kinases recruit different collections of intracellular signaling proteins, producing different effects; but certain components seem to be used quite widely. These include, for example, a phospholipase that functions in the same way as phospholipase C to activate the inositol phospholipid signaling pathway (see Figure 16–25). Receptor tyrosine kinases can also activate an important signaling enzyme called *phosphatidyl-inositol 3-kinase (PI 3-kinase)*, which phosphorylates inositol phospholipids in the plasma membrane. These then become docking sites for other intracellular signaling proteins. One of these signaling proteins is *protein kinase B (PKB)*, which phosphorylates target proteins on serines and threonines and is especially important in signaling cells to survive and grow.

The main signaling pathway from receptor tyrosine kinases to the nucleus, however, takes another route. This pathway has become well known for a sinister reason: mutations that cause a runaway activation

Figure 16–30 Activation of a receptor tyrosine kinase stimulates the assembly of an intracellular signaling complex. Typically, the binding of a signal molecule to the extracellular domain of a receptor tyrosine kinase causes two receptor molecules to associate into a dimer. The signal molecule shown here is itself a dimer and thus can physically cross-link two receptor molecules. In other cases, binding of the signal molecule changes the conformation of the receptor molecules in such a way that they dimerize. Dimer formation brings the kinase domains of each intracellular receptor tail into contact with the other; this activates the kinases and enables them to phosphorylate each other on several tyrosine side chains. Each phosphorylated tyrosine serves as a specific binding site for a different intracellular signaling protein, which then helps relay the signal to the cell's interior.

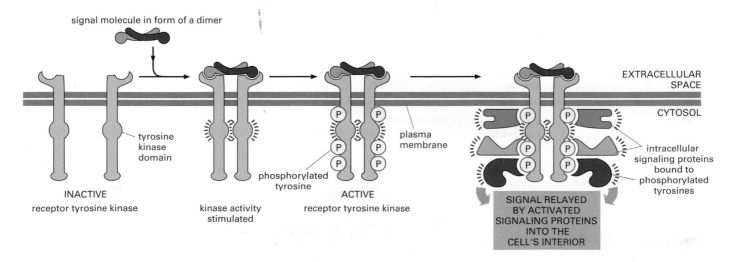

of this signaling cascade— thereby stimulating cell division inappropri- ately—help trigger many types of cancers. We shall conclude our dis- cussion of receptor tyrosine kinases by tracing this pathway from the receptor to the nucleus.

Receptor Tyrosine Kinases Activate the GTP-binding Protein Ras

As just discussed, activated receptor tyrosine kinases recruit many kinds of intracellular signaling proteins. Some of these proteins func- tion solely as physical *adaptors;* they help build a large signaling aggre- gate by coupling the receptor to other proteins, which in turn may bind to and activate yet other proteins that pass the message along. One of the key players in these adaptor-assembled signaling complexes is **Ras**—a small protein that is bound by a lipid tail to the cytoplasmic face of the plasma membrane (Figure 16–31). Virtually all receptor tyrosine kinases activate Ras, from the platelet-derived growth factor (PDGF) receptors that mediate cell proliferation in wound healing to the nerve growth factor (NGF) receptors that prevent certain neurons from dying in the developing nervous system.

The Ras protein is a member of a large family of small, single-sub- unit GTP-binding proteins, often called the *monomeric GTP-binding proteins* to distinguish them from the *trimeric G proteins* that we encountered earlier in this chapter. Ras resembles the α subunit of a G protein and functions as a molecular switch in much the same way. It cycles between two distinct conformational states—active when GTP is bound and inactive when GDP is bound (see Figure 16–15B). Interaction with an activating protein encourages Ras to exchange its GDP for GTP, thus switching Ras to its activated state. After a delay, Ras switches itself off again by hydrolyzing its GTP to GDP.

In its active state, Ras promotes the activation of a phosphorylation cascade in which a series of protein kinases phosphorylate and activate one another in sequence, like an intracellular game of dominoes (Figure 16–32). This relay system, which carries the signal from the plasma membrane to the nucleus, is called a **MAP-kinase cascade**, in honor of the final kinase in the chain **MAP-kinase** (*m*itogen-*a*ctivated *p*rotein kinase). In this cascade, MAP-kinase is phosphorylated and activated by an enzyme called, logically enough, *MAP-kinase-kinase*. And this pro- tein is itself switched on by a *MAP-kinase-kinase-kinase* (which is acti- vated by Ras). At the end of the signal cascade, MAP-kinase phosphory- lates certain gene regulatory proteins on serines and threonines, alter- ing their ability to control gene transcription and thereby causing a change in the pattern of gene expression. This shift may stimulate cell proliferation, promote cell survival, or induce cell differentiation: the precise outcome will depend on which other genes are active in the cell

Figure 16–31 Receptor tyrosine kinases activate Ras. An adaptor protein docks on a particular phosphotyrosine on the activated receptor (the other signaling proteins that are shown bound to the receptor in Figure 16–30 are omitted for simplicity). The adaptor recruits and stimulates a protein accomplice that functions as a Ras-activating protein. This protein in turn stimulates Ras to exchange its bound GDP for GTP. The activated Ras protein then stimulates the next steps in the signaling pathway, one of which is shown in Figure 16–32. Note that the Ras protein contains a covalently attached lipid group *(red)* that helps anchor the protein to the plasma membrane.

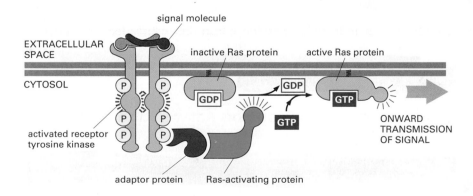

Figure 16–32 Ras activates a MAP-kinase phosphorylation cascade. A Ras protein activated by the process shown in Figure 16–31 triggers a phosphorylation cascade of three protein kinases, which relay and distribute the signal. The final kinase in the cascade, MAP-kinase, phosphorylates various downstream target proteins. These targets can include other protein kinases and, most important, gene regulatory proteins that control gene expression. Changes in gene expression and protein activity result in complex changes in cell behaviors such as proliferation and differentiation—typical outcomes of the Ras/MAP-kinase signaling pathway.

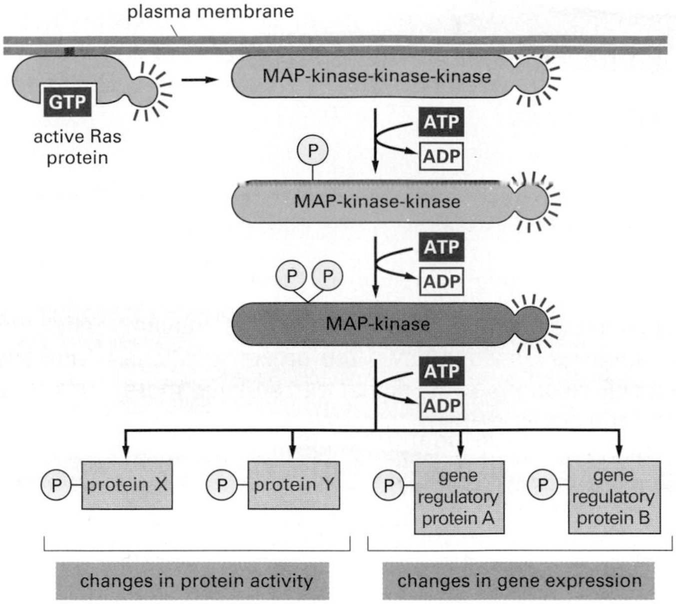

and what other signals the cell receives. How researchers unravel such complex signaling cascades is discussed in How We Know, pp. 561–563.

The importance of Ras has been demonstrated in various ways. If Ras is inhibited by an intracellular injection of Ras-inactivating antibodies, for example, a cell may no longer respond to some of the extracellular signals that it would normally recognize. Conversely, if Ras activity is permanently switched on, the cell may act as if it is being bombarded continuously by proliferation-stimulating extracellular signals (*mitogens*). Before it was discovered in normal cells, the Ras protein was found in human cancer cells, in which a mutation in the gene for Ras caused the production of such a hyperactive form of Ras. This mutant Ras protein helps stimulate the cells to divide even in the absence of mitogens. The resulting uncontrolled cell proliferation contributes to the formation of cancer.

About 30% of human cancers contain such activating mutations in *ras* genes, and many other cancers have mutations in genes whose products lie in the same signaling pathway as Ras. Many of the genes that encode these intracellular signaling proteins were identified in the quest for cancer-promoting *oncogenes*, which are discussed in Chapter 21. The normal versions of the genes—which encode the signaling proteins essential for proper cell function—are often known as *proto-oncogenes*, because they are capable of being converted into oncogenes by mutation.

Cancer is a disease in which cells in the body behave in a selfish and antisocial way—destroying the harmony of the multicellular organism by proliferating when they should not and invading tissues that they should not enter. The molecular derangements responsible for such unruly behavior are discussed more fully in Chapter 21. But it seems appropriate to note here that the common occurrence in cancer of mutations in genes for cell-signaling components reflects a familiar truth: maintaining order in a complex, integrated community depends above all on good communication.

Some Enzyme-linked Receptors Activate a Fast Track to the Nucleus

Not all enzyme-linked receptors trigger complex signaling cascades that require the cooperation of a sequence of protein kinases to carry a

Question 16–9

Would you expect to activate G-protein–linked receptors and receptor tyrosine kinases by exposing cells to antibodies that bind to the respective proteins? (Hint: review Panel 4–6, on pp. 164–165, regarding the properties of antibody molecules.)

How We Know: Untangling Cell Signaling Pathways

Intracellular signaling pathways are never mapped out in a single experiment. Instead, investigators figure out, piece by piece, how all the links in the chain fit together—and how each contributes to the cell's response to an extracellular signal such as the hormone insulin. The process involves breaking down the broad questions about how a cell responds to the signal into smaller, more manageable questions: Which protein is the insulin receptor? Which intracellular proteins become activated when insulin is present? With which proteins do these activated proteins interact? How does one protein activate another? Here we discuss the kinds of experiments that provide answers to such riddles.

Stimulated phosphorylation

When cells are exposed to an extracellular signal molecule, one result is that a number of proteins become phosphorylated. Some of these will be the intracellular signaling proteins responsible for propagating the message throughout the cell; others will be target proteins responsible for the cell's response. To determine which molecules have been activated by phosphorylation, researchers break open the cells, separate the proteins by size on a gel (discussed in Chapter 4, Panels 4–3 to 4–5), and then use antibodies to detect phosphorylated proteins.

Another common way to visualize newly phosphorylated proteins involves supplying cells with a radioactive version of ATP when they are exposed to an extracellular signal molecule. Protein kinases activated by the signal will transfer radioactive phosphate from the labeled ATP to their protein substrates. Again, the cell proteins are separated on a gel, but then the radiolabeled proteins are detected by exposing the gel to an X-ray film.

Close encounters

Once the activated proteins have been identified, one can determine which proteins interact with them. To identify interacting proteins, scientists often make use of *co-immunoprecipitation*. In this technique, antibodies are used to latch onto a specific protein, dragging it out of solution and down to the bottom of a test tube (discussed in Chapter 4, Panel 4–6). If the captured protein happens to be bound to other proteins, these will be dragged down as well. In this way, researchers can identify which proteins interact when cells are stimulated by a signal molecule.

Once two proteins are known to bind to each other, the experimenter can proceed to pinpoint which parts of the proteins are required for the interaction. This often involves using recombinant DNA technology to construct a set of mutant proteins, each of which differs slightly from the normal one. To determine which phosphorylated tyro-

sine on a receptor tyrosine kinase a certain intracellular signaling protein binds to, for example, a series of mutant receptors is used, each missing a different tyrosine from its cytoplasmic domain (Figure 16–33). In this way, the specific tyrosines required for binding can be determined. Similarly, one can determine whether this tyrosine docking site is required for the receptor to transmit a signal to the cell.

Jamming the pathway

Ultimately, one wants to assess how important a particular protein is for a signaling process. A first test involves using recombinant DNA technology to introduce into cells a gene encoding a constantly active form of the protein, to see if this mimics the effect of the extracellular signal. Take Ras, for example. The form of Ras involved in human cancers is constantly active because it has lost its ability to hydrolyze the GTP that keeps it switched on. This continuously active form of Ras can stimulate some cells to proliferate even in the absence of mitogens, which is one way it contributes to cancer (Figure 16–34).

The ultimate test of the importance of an intracellular protein in a signaling pathway is to inactivate the protein or its gene and see if the signaling pathway is affected. In the case of Ras, for example, one can introduce into cells a "dominant negative" mutant form of Ras. This disabled form of Ras clings too tightly to GDP, and therefore cannot be activated. Because it can still bind to other signaling partners in the pathway, it jams the pathway, preventing normal copies of Ras from doing their job. Such stalled cells do not proliferate in response to extracellular, division-stimulating signals, indicating the importance of normal Ras signaling in the proliferative response.

Ordering the pathway

Most signaling pathways take decades to untangle. Although insulin was first isolated from dog pancreas in the early 1920s, the molecular chain of events that links the binding of insulin to its receptor with the activation of the transporter proteins that take up glucose is still incompletely understood.

One powerful strategy that scientists use to identify proteins that participate in cell signaling involves screening a massive number of animals—usually tens of thousands of fruit flies or nematode worms that have been treated with a mutagen. They are looking for mutants in which a signaling pathway is not functioning properly. Flies and worms are useful because they reproduce rapidly and can be maintained in vast numbers in the laboratory. By examining enough mutant animals, many of the genes that encode the proteins involved in a signaling cascade can be identified—

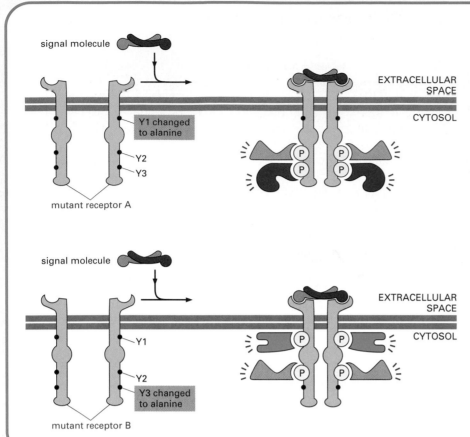

Figure 16–33 **Mutant proteins can help to determine exactly where an intracellular signaling molecule binds.** As shown in Figure 16–30, on binding their signal molecule, a pair of receptor tyrosine kinases come together and phosphorylate specific tyrosines on each other's cytoplasmic tails. These phosphorylated tyrosines attract different intracellular signaling molecules, which then become activated and pass on the signal. To determine which tyrosine binds to a specific intracellular signaling molecule, a series of mutant receptors are constructed. In the mutants shown, single tyrosines (Y1 or Y3) have been replaced by an alanine. As a result, the mutant receptors no longer bind to one of the intracellular signaling proteins. The effect on the cell's response to the signal can then be determined. It is important that the mutant receptor be tested in a cell that does not have its own normal receptors for the signal molecule.

including receptors, protein kinases, gene regulatory proteins, and so on.

Such genetic scans can also reveal the order in which intracellular signaling proteins act in a pathway. Imagine that a genetic screen reveals two new proteins X and Y in the Ras signaling pathway (Figure 16–35A). If insertion of a gene that encodes a continuously active version of Ras "rescues" the signaling pathway in cells in which a defective protein X was

blocking the pathway, then Ras must operate downstream of X in the signaling cascade (Figure 16–35B). If Ras operates upstream of protein Y in the pathway, a constantly active Ras would be unable to transmit a signal past the obstruction caused by the disabled protein Y (Figure 16–35C).

Used together, these biochemical and genetic techniques allow even the most complex intracellular signaling pathways to be dissected.

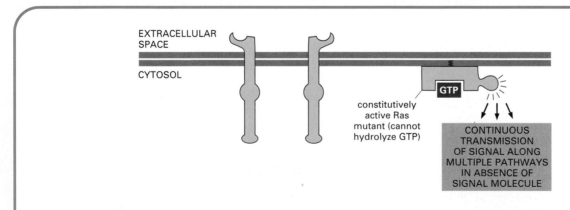

Figure 16–34 **A constitutively active form of Ras transmits a signal even in the absence of an extracellular signal molecule.** As shown in Figure 16–31, the normal Ras protein is activated in response to certain extracellular signals. The overactive form of Ras shown here has lost the ability to hydrolyze GTP. Thus, it cannot shut down its activity and, as a result, is constantly active.

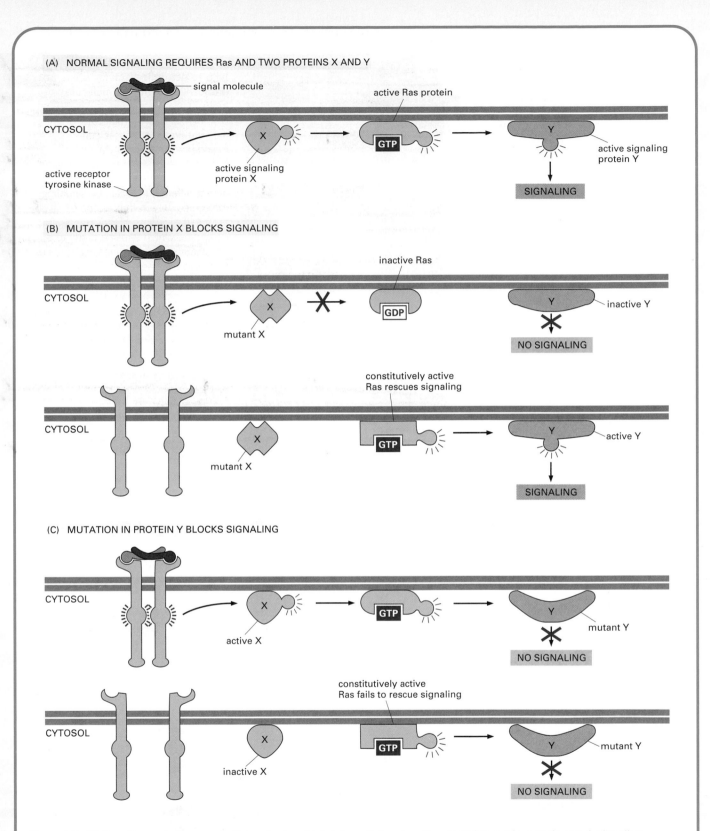

Figure 16–35 Genetic analysis reveals the order in which intracellular signaling proteins act in a pathway. A signaling pathway can be inactivated by mutations in any one of its components. Here, we show how a Ras signaling pathway (A) can be shut down by a mutation in either protein X (B) or protein Y (C). Addition of a constitutively active form of Ras to these cells can help to unravel where in the pathway the mutant proteins lie. Adding a continuously active Ras to cells with a mutation in X restores activity to the pathway, allowing the signal to be transmitted even in the absence of an extracellular signal molecule (B). An overactive Ras can rescue these cells because Ras lies downstream of the mutant protein X that is jamming the pathway. Adding a continuously active Ras to cells with a mutation in protein Y has no effect, as Ras lies upstream of the blockage (C).

message to the nucleus. Some receptors use a more direct route to control gene expression.

A few hormones and many local mediators called *cytokines* bind to receptors that can activate gene regulatory proteins that are held in a latent state at the plasma membrane. Once activated, these regulatory proteins head straight for the nucleus, where they stimulate the transcription of specific genes. This direct signaling pathway is used, for example, by *interferons*, which are cytokines that instruct cells to produce proteins that will make them more resistant to viral infection. Unlike the receptor tyrosine kinases that stimulate elaborate signaling cascades, **cytokine receptors** have no intrinsic enzyme activity. Instead, they are associated with cytoplasmic tyrosine kinases called *JAKs* that are activated when a cytokine binds to its receptor. Once activated, the JAKs phosphorylate and activate cytoplasmic gene regulatory proteins called STATs, which then migrate to the nucleus, where they stimulate transcription of specific target genes (Figure 16–36).

Different cytokine receptors evoke different cellular responses by activating different STATs. Like any pathway that is turned on by phosphorylation, the cytokine signal is shut off by protein phosphatases that remove the phosphate groups from the activated signaling proteins.

An even more direct signaling pathway is used by another class of enzyme-linked receptors that resemble receptor tyrosine kinases. These are receptor **serine/threonine kinases** that directly phosphorylate and activate cytoplasmic gene regulatory proteins (called SMADs) when stimulated by an extracellular signal molecule (Figure 16–37). The hormones and local mediators that activate these receptors belong to the *TGF-β superfamily* of extracellular proteins, which play an especially important role in animal development.

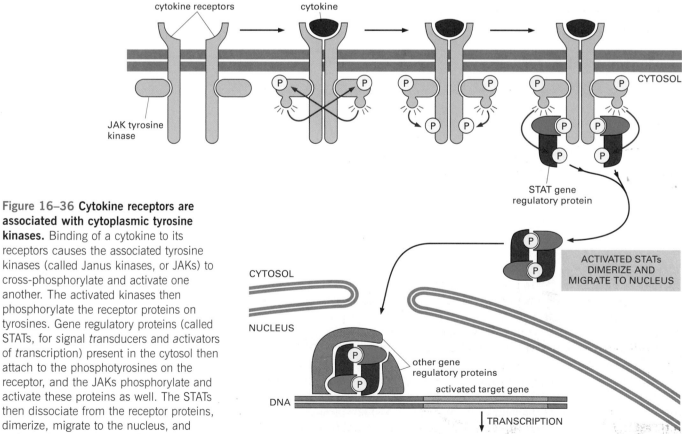

Figure 16–36 Cytokine receptors are associated with cytoplasmic tyrosine kinases. Binding of a cytokine to its receptors causes the associated tyrosine kinases (called Janus kinases, or JAKs) to cross-phosphorylate and activate one another. The activated kinases then phosphorylate the receptor proteins on tyrosines. Gene regulatory proteins (called STATs, for *signal transducers and activators of transcription*) present in the cytosol then attach to the phosphotyrosines on the receptor, and the JAKs phosphorylate and activate these proteins as well. The STATs then dissociate from the receptor proteins, dimerize, migrate to the nucleus, and activate the transcription of specific target genes.

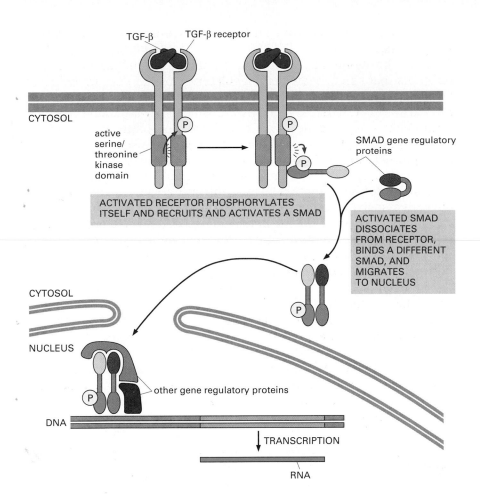

TGF-β

TGF-β receptor

CYTOSOL

active
serine/
threonine
kinase
domain

P

P

P

SMAD gene regulatory
proteins

**ACTIVATED RECEPTOR PHOSPHORYLATES
ITSELF AND RECRUITS AND ACTIVATES A SMAD**

**ACTIVATED SMAD
DISSOCIATES
FROM RECEPTOR,
BINDS A DIFFERENT
SMAD, AND
MIGRATES
TO NUCLEUS**

CYTOSOL

P

NUCLEUS

other gene regulatory proteins

P

DNA

TRANSCRIPTION

RNA

Figure 16–37 TGF-β receptors activate gene regulatory proteins directly at the plasma membrane. These receptor serine/threonine kinases phosphorylate themselves and then recruit and activate cytoplasmic gene regulatory proteins (called SMADs, after the related proteins *Sma* in nematodes and *Mad* in flies). The SMADs then dissociate from the receptors and bind to other SMADs, and the complexes then migrate to the nucleus, where they stimulate transcription of specific target genes. TGF-β stands for transforming growth factor-β.

Question 16–10

If cell-surface receptors can rapidly signal to the nucleus by activating latent gene regulatory proteins such as STATs and SMADs at the plasma membrane, why do most cell-surface receptors use long, indirect signaling cascades to influence gene transcription in the nucleus?

Protein Kinase Networks Integrate Information to Control Complex Cell Behaviors

In this chapter, we have outlined several major pathways for conveying a signal from the cell surface to the cell interior. Figure 16–38 compares four of these pathways: the routes from G-protein–linked receptors via adenylyl cyclase and via phospholipase C, and the routes from enzyme-linked receptors via phospholipase C and via Ras. Each pathway differs from the others, yet they use common components to transmit their signals. Because all these pathways eventually activate protein kinases, it seems that each is capable in principle of regulating practically any process in the cell.

In fact, the complexity of cell signaling is much greater than we have described. First, we have not presented every intracellular signaling pathway available to cells; second, the major signaling cascades we have introduced interact in ways that we have not described. They are connected by interactions of many sorts, but the most extensive links are those mediated by the protein kinases present in each of the pathways. These kinases often phosphorylate, and hence regulate, components in other signaling pathways in addition to components in the pathway to which they themselves belong. Thus, a certain amount of *cross talk* occurs among the different pathways (see Figure 16–38), and indeed between virtually all of the control systems of the cell. To give an idea of the scale of these regulatory systems, genome sequencing studies suggest that about 2% of our ~30,000 genes code for protein kinases; moreover, hundreds of distinct types of protein kinases are thought to be present in a single mammalian cell. How can we make sense of this tangled web of interacting signaling pathways, and what is the function of such complexity?

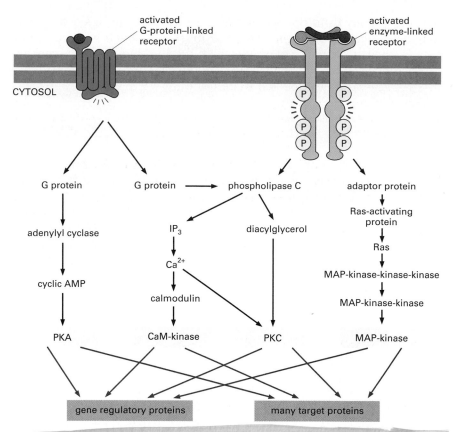

Figure 16–38 Signaling pathways can be highly interconnected. The diagram sketches the pathways from G-protein–linked receptors via adenylyl cyclase and via phospholipase C, and from enzyme-linked receptors via phospholipase C and via Ras. The protein kinases in these pathways phosphorylate many proteins, including proteins belonging to the other pathways. The resulting dense network of regulatory interconnections is symbolized by the *red arrows* radiating from each kinase shaded in *yellow;* some kinases phosphorylate some of the same target proteins.

A cell receives messages from many sources, and it must integrate this information to generate an appropriate response—to live or die, to divide or differentiate, to change shape, relocate, or send out a chemical message of its own. Through the cross talk between signaling pathways, the cell is able to put two or more bits of information together and react to the combination. Thus, some intracellular signaling proteins act as integrating devices, usually by having several potential phosphorylation sites, each of which can be phosphorylated by a different protein kinase. Information received from different sources can converge on such proteins, which then convert the input to a single outgoing signal (Figure 16–39). The integrating proteins in turn can deliver a signal to many downstream targets. In this way, the intracellular signaling system may act like a network of nerve cells in the brain—or like a collection of microprocessors in a computer—interpreting complex information and generating complex responses.

Our exploration of the pathways that cells use to process signals from their environment has led us from receptors on the cell surface to the proteins that form the elaborate control systems that operate deep within the cell's interior. We have examined a large array of signaling networks that allow cells to combine and process inputs from different sources, store information, and respond in an appropriate manner that benefits the organism. But our understanding of these intricate networks is still evolving: we are still discovering new links in the chains, new signaling partners, new connections, and even new pathways. And while we still have much to learn about signaling pathways in animal cells, we know even less about such pathways in plants.

Multicellularity and Cell Communication Evolved Independently in Plants and Animals

Plants and animals have been evolving independently for more than a billion years, the last common ancestor being a single-celled eucaryote that most likely lived on its own. Because these kingdoms diverged so long ago—when it was still "every cell for itself"—each has evolved its own molecular solutions to multicellular functioning. Thus, the mechanisms for cell–cell communication in plants and animals evolved separately and would be expected to be quite different. At the same time, however, plants and animals started with a common set of eucaryotic genes—including some used by single-cell organisms to communicate among themselves—and so their signaling systems should show some similarities.

A striking resemblance occurs at the cell surface. Like animals, plants make extensive use of membrane-embedded cell-surface receptors—especially enzyme-linked receptors. The spindly weed *Arabidopsis thaliana*—a plant studied by many present-day biologists—has hundreds of genes encoding receptor serine/threonine kinases, which are structurally distinct from those found in animal cells (see Figure 16–37). Such receptors are thought to play an important part in a large variety of plant cell signaling processes, including those governing growth, development, and disease resistance. In contrast to animal cells, plant cells appear not to use receptor tyrosine kinases, steroid-hormone–type nuclear receptors, or cyclic AMP, and they seem to use few G-protein–linked receptors.

One of the best-studied signaling systems in plants mediates the response of cells to ethylene—a gaseous hormone that regulates a diverse array of developmental processes, including seed germination and fruit ripening. Tomato growers use ethylene to ripen their fruit, even after it has been picked. Ethylene receptors are related to the proteins that bacteria use to locate nutrients or flee from poisons. Like the bacterial proteins, they function as histidine kinases and are unlike any receptor proteins yet found in animal cells. Activated ethylene receptors activate a MAP-kinase cascade that is similar to MAP-kinase cascades found in animal cells—presumably reflecting the common ancestry of plants and animals. For most of the receptor kinases in plants, however, the signal transduction pathways that link receptor activation to a cell response are not yet known.

Unraveling cell signaling pathways is an active area of research, and new discoveries are being made in both plant and animal systems every day. Genome sequencing projects are providing long lists of components involved in signal transduction in a large variety of organisms.

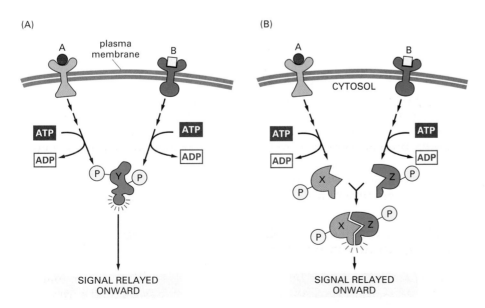

(A)　(B)

SIGNAL RELAYED ONWARD

Figure 16–39 Some intracellular signaling proteins serve to integrate incoming signals. (A) Signals A and B may activate different cascades of protein phosphorylations, each of which leads to the phosphorylation of protein Y but at different sites on the protein. Protein Y is activated only when both of these sites are phosphorylated, and therefore it is active only when signals A and B are simultaneously present. (B) Alternatively, signals A and B could lead to the phosphorylation of two proteins, X and Z, which then bind to each other to create the active protein XZ.

Even when we have identified all the pieces, it will remain a major challenge to figure out exactly how they fit together to allow cells to integrate the diverse array of signals in their environment and respond in appropriate ways.

Essential Concepts

- Cells in multicellular organisms communicate through a large variety of extracellular chemical signals.
- Hormones are carried in the blood to distant target cells, but most other extracellular signal molecules act over only a short range. Neighboring cells often communicate through direct cell-surface contacts.
- Cells are stimulated by an extracellular signal molecule when it binds to and activates a receptor protein. Each receptor protein recognizes a particular signal molecule.
- Receptor proteins act as transducers, converting signals from one physical form to another.
- Most extracellular signal molecules cannot pass through the plasma membrane; they bind to cell-surface receptor proteins that transduce the extracellular signal into different intracellular signals.
- Small hydrophobic extracellular signal molecules such as steroid hormones and nitric oxide can diffuse directly across the plasma membrane; they activate intracellular receptor proteins, which are either gene regulatory proteins or enzymes.
- There are three main classes of cell-surface receptors: (1) ion-channel–linked receptors, (2) G-protein–linked receptors, and (3) enzyme-linked receptors.
- G-protein–linked receptors and enzyme-linked receptors respond to extracellular signals by initiating cascades of intracellular signaling reactions that alter the behavior of the cell.
- G-protein–linked receptors activate a class of trimeric GTP-binding proteins called G proteins; these act as molecular switches, transmitting the signal onward for a short period and then switching themselves off by hydrolyzing their bound GTP to GDP.
- Some G proteins directly regulate ion channels in the plasma membrane. Others activate the enzyme adenylyl cyclase, increasing the intracellular concentration of cyclic AMP. Still other G proteins activate the enzyme phospholipase C, which generates the messenger molecules inositol trisphosphate (IP_3) and diacylglycerol.
- IP_3 opens ion channels in the membrane of the endoplasmic reticulum, releasing a flood of free Ca^{2+} ions into the cytosol. Ca^{2+} itself acts as an intracellular messenger, altering the activity of a wide range of proteins.
- A rise in cyclic AMP activates protein kinase A (PKA), while Ca^{2+} and diacylglycerol in combination activate protein kinase C (PKC).
- PKA and PKC phosphorylate selected target proteins on serines and threonines, thereby altering protein activity. Different cell types contain different sets of target proteins and are affected in different ways.
- In general, stimulation of G-protein–linked receptors produces rapid and reversible cell responses.

- Many enzyme-linked receptors have intracellular protein domains that function as enzymes; most are receptor tyrosine kinases, which phosphorylate tyrosines on selected intracellular proteins.
- Activated receptor tyrosine kinases cause the assembly of an intracellular signaling complex on the intracellular tail of the receptor; a part of this complex serves to activate Ras, a small GTP-binding protein, which activates a cascade of protein kinases that relay the signal from the plasma membrane to the nucleus.
- Mutations that stimulate cell proliferation by making Ras constantly active are a common feature of many cancers.
- Some enzyme-linked receptors activate a direct pathway to the nucleus. Instead of activating signaling cascades, they turn on gene regulatory proteins right at the plasma membrane.
- Different intracellular signaling pathways interact, enabling cells to produce an appropriate response to a complex combination of signals. Some combinations of signals enable a cell to survive; other combinations of signals will cause it to proliferate; and in the absence of any signals, most cells will kill themselves by undergoing apoptosis.
- Plants, like animals, use enzyme-linked cell-surface receptors to control their growth and development.

Key Terms

adaptation	molecular switch
adenylyl cyclase	neurotransmitter
Ca^{2+}/calmodulin-dependent protein kinases (CaM-kinases)	nitric oxide (NO)
	nuclear receptor
calmodulin	phospholipase C
cell signaling	protein kinase C (PKC)
cyclic AMP	Ras
cyclic-AMP–dependent protein kinase (PKA)	receptor
	receptor protein
cytokine receptor	receptor serine/threonine kinase
diacylglycerol (DAG)	receptor tyrosine kinase
G-protein–linked receptor	second messenger
hormone	serine/threonine kinase
inositol phospholipid	signal transduction
inositol 1,4,5-trisphosphate (IP_3)	signaling cascade
local mediator	steroid hormone
MAP-kinase	tyrosine kinase
MAP-kinase cascade	

Questions

Question 16–11

Which of the following statements are correct? Explain your answers.

A. The signal molecule acetylcholine has different effects on different cell types in an animal and binds to different receptor molecules on different cell types.

B. After acetylcholine is secreted from cells, it is long-lived, because it has to reach target cells all over the body.

C. Both the GTP-bound α subunits and nucleotide-free βγ complexes—but not GDP-bound, fully assembled G proteins—activate other molecules downstream of G-protein–linked receptors.

D. IP$_3$ is produced directly by cleavage of an inositol phospholipid without incorporation of an additional phosphate group.

E. Calmodulin regulates the intracellular Ca^{2+} concentration.

F. Different signals originating from the plasma membrane can be integrated by cross talk between different signaling pathways inside the cell.

G. *ras* is an oncogene.

H. Tyrosine phosphorylation serves to build binding sites for other proteins to bind to receptor tyrosine kinases.

Question 16–12

The Ras protein functions as a molecular switch that is set to its on state by other proteins that cause it to bind GTP. A GTPase-activating protein resets the switch to the off state by inducing Ras to hydrolyze its bound GTP to GDP much more rapidly than it would without this encouragement. Thus Ras works like a light switch that one person turns on and another turns off. You are given a mutant cell that lacks the GTPase-activating protein. What abnormalities would you expect to find in the way that Ras activity responds to extracellular signals?

Question 16–13

A. Compare and contrast signaling by neurons, which secrete neurotransmitters at synapses, to signaling carried out by endocrine cells, which secrete hormones into the blood.

B. Discuss the relative advantages of the two mechanisms.

Question 16–14

Two intracellular molecules, X and Y, are both normally synthesized at a constant rate of 1000 molecules per second per cell. Molecule X is broken down slowly: each molecule of X survives on average for 100 seconds. Molecule Y is broken down 10 times faster: each molecule of Y survives on average for 10 seconds.

A. Calculate how many molecules of X and Y the cell contains at any time.

B. If the rates of synthesis of both X and Y are suddenly increased tenfold to 10,000 molecules per second per cell—without any change in their degradation rates—how many molecules of X and Y will there be after one second?

C. Which molecule would be preferred for rapid signaling?

Question 16–15

"One of the great kings of the past ruled an enormous kingdom that was more beautiful than anywhere else in the world. Every plant glistened as brilliantly as polished jade, and the softly rolling hills were as sleek as the waves of the summer sea. The wisdom of all of his decisions relied on a constant flow of information brought to him daily by messengers who told him about every detail of his kingdom so that he could take quick, appropriate actions when needed. Despite the abundance of beauty and efficiency, his people felt doomed living under his rule, for he had an adviser who had studied cell signal transduction and accordingly administered the king's Department of Information. The adviser had implemented the policy that all messengers shall be immediately beheaded whenever spotted by the Royal Guard, because for rapid signaling the lifetime of messengers ought to be short. Their plea 'Don't hurt me, I'm only the messenger!' was to no avail, and the people of the kingdom suffered terribly because of the rapid loss of their sons and daughters." Why is the analogy on which the king's adviser based his policies inappropriate? Briefly discuss the features that set cell signaling pathways apart from the human communication pathway described in the story.

Question 16–16

In a series of experiments, genes that code for mutant forms of a receptor tyrosine kinase are introduced into cells. The cells also express their own normal form of the receptor from their normal gene, although the mutant genes are constructed so that they are expressed at considerably higher concentrations than the normal gene. What would be the consequences of introducing a mutant gene that codes for a receptor tyrosine kinase (A) lacking its extracellular domain, or (B) lacking its intracellular domain?

Question 16–17

Discuss the following statement: "Membrane proteins that span the membrane many times can undergo a conformational change upon ligand binding that can be sensed on the other side of the membrane. Thus individual protein molecules can transmit a signal across a membrane. In contrast, individual single-span membrane proteins cannot transmit a conformational change across the membrane but require oligomerization."

Question 16–18

What are the similarities and differences between the reactions that lead to the activation of G proteins and the reactions that lead to the activation of Ras?

Question 16–19

Why do you suppose cells use Ca^{2+} (which is kept by Ca^{2+} pumps at an intracellular concentration of 10^{-7} M) for intracellular signaling and not another ion such as Na^+ (which is kept by the Na^+ pump at an intracellular concentration of 10^{-3} M)?

Question 16–20

It seems counterintuitive that a cell, having a perfectly abundant supply of nutrients available, would commit suicide if not constantly stimulated by signals from other cells (see Figure 16–6). What do you suppose might be the advantages of such regulation?

Question 16–21

The contraction of the myosin–actin system in muscle cells is triggered by a rise in intracellular Ca^{2+}. Muscle cells have specialized Ca^{2+}-release channels—called ryanodine receptors because of their sensitivity to the drug ryanodine—that lie in the membrane of the sarcoplasmic reticulum, a specialized form of the endoplasmic reticulum. In contrast to the IP_3-gated Ca^{2+} channels in the endoplasmic reticulum shown in Figure 16–25, the signaling molecule that opens ryanodine receptors is Ca^{2+} itself. Discuss the consequences of ryanodine channels for muscle cell contraction.

Question 16–22

Two protein kinases, K1 and K2, function sequentially in an intracellular signaling cascade. If either kinase contains a mutation that permanently inactivates its function, no response is seen in cells when an extracellular signal is received. A different mutation in K1 makes it permanently active, so that in cells containing that mutation a response is observed even in the absence of an extracellular signal. You characterize a double mutant cell that contains K2 with the inactivating mutation and K1 with the activating mutation. You observe that the response is seen even when no signal is received by these cells. In the normal signaling pathway, does K1 activate K2 or does K2 activate K1? Explain your answer.

Question 16–23

A. Trace the steps of a long and indirect signaling pathway from a cell-surface receptor to a change in gene expression in the nucleus.

B. Compare this pathway with the shortest and most direct pathway from the cell surface to the nucleus.

Question 16–24

Why do you think animal cells and plant cells have such different intracellular signaling mechanisms and yet share some common mechanisms?

Highlights from *Essential Cell Biology 2 Interactive* CD-ROM

16.2   G-Protein Signaling

16.4   Inositol Phosphate Signaling

16.5   Calcium Signaling

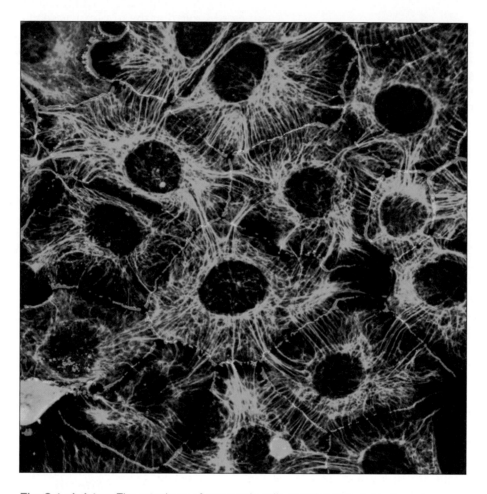

The Cytoskeleton. The cytoplasm of eucaryotic cells contains three intricate and dynamic networks of protein filaments, collectively called the cytoskeleton. These are responsible for the shaping, moving, dividing, and interactions of cells. One of these, the intermediate filament network, is shown here in a sheet of epithelial cells in culture. The filaments *(green)* connect, through junctions at the cell boundaries *(blue),* to the filament networks in adjacent cells, in this case helping to create a tough and durable cell sheet. (Courtesy of Kathleen Green and Evangeline Amargo.)

Cytoskeleton

17

The ability of eucaryotic cells to adopt a variety of shapes, organize the many components in their interior, interact mechanically with the environment, and carry out coordinated movements depends on the **cytoskeleton**—an intricate network of protein filaments that extends throughout the cytoplasm (Figure 17–1). This filamentous architecture helps to support the large volume of cytoplasm in a eucaryotic cell, a function that is particularly important in animal cells, which have no cell walls. Unlike our own bony skeleton, however, the cytoskeleton is a highly dynamic structure that is continuously reorganized as a cell changes shape, divides, and responds to its environment. The cytoskeleton is not only the "bones" of a cell but its "muscles" too, and it is directly responsible for large-scale movements such as the crawling of cells along a surface, contraction of muscle cells, and the changes in cell shape that take place as an embryo develops. Without the cytoskeleton, wounds would never heal, muscles would be useless, and sperm would never reach the egg.

The interior of the cell itself is also in constant motion, and the cytoskeleton provides the machinery for intracellular movements such as the transport of organelles from one place to another, the segregation of chromosomes into the two daughter cells at mitosis, and the pinching apart of animal cells at cell division. The cytoskeleton is especially prominent in the large and structurally complex eucaryotic cell, although some simple cytoskeletal components have been found in bacteria. The eucaryotic cell, like any factory making a complex product, has a highly organized interior in which specialized machines are concentrated in different areas but linked by transport systems (discussed in Chapter 15). The cytoskeleton controls the location of the organelles that conduct these specialized functions and also provides the transport between them.

The cytoskeleton is built on a framework of three types of protein filaments: *intermediate filaments, microtubules,* and *actin filaments.* Each type of filament has distinct mechanical properties and is formed from a different protein subunit (Figure 17–2). A family of fibrous proteins form the intermediate filaments; *tubulin* is the subunit in microtubules; and *actin* is the subunit in actin filaments. In each case, thousands of subunits assemble into a fine thread of protein that sometimes extends across the entire cell.

In this chapter we consider the structure and function of the three types of protein filament networks in turn. We begin with the

Intermediate Filaments

Intermediate Filaments Are Strong and Ropelike

Intermediate Filaments Strengthen Cells Against Mechanical Stress

The Nuclear Envelope Is Supported by a Meshwork of Intermediate Filaments

Microtubules

Microtubules Are Hollow Tubes with Structurally Distinct Ends

The Centrosome Is the Major Microtubule-organizing Center in Animal Cells

Growing Microtubules Show Dynamic Instability

Microtubules Are Maintained by a Balance of Assembly and Disassembly

Microtubules Organize the Interior of the Cell

Motor Proteins Drive Intracellular Transport

Organelles Move Along Microtubules

Cilia and Flagella Contain Stable Microtubules Moved by Dynein

Actin Filaments

Actin Filaments Are Thin and Flexible

Actin and Tubulin Polymerize by Similar Mechanisms

Many Proteins Bind to Actin and Modify Its Properties

An Actin-rich Cortex Underlies the Plasma Membrane of Most Eucaryotic Cells

Cell Crawling Depends on Actin

Actin Associates with Myosin to Form Contractile Structures

Extracellular Signals Control the Arrangement of Actin Filaments

Muscle Contraction

Muscle Contraction Depends on Bundles of Actin and Myosin

During Muscle Contraction Actin Filaments Slide Against Myosin Filaments

Muscle Contraction Is Triggered by a Sudden Rise in Ca^{2+}

Muscle Cells Perform Highly Specialized Functions in the Body

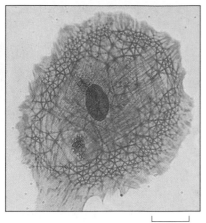

Figure 17–1 **The cytoskeleton gives a cell its shape and allows the cell to organize its internal components.** A skin cell (fibroblast) in culture has been fixed and stained with Coomassie Blue, a general stain for proteins. A variety of filamentous structures extend throughout the cell. The dark oval in the center is the nucleus. (Courtesy of Colin Smith.)

10 μm

intermediate filaments that provide cells with mechanical strength. We then see how cell appendages built from microtubules propel motile cells like protozoa and sperm, and how the actin cytoskeleton provides the motive force for a crawling fibroblast. Finally we discuss how the cytoskeleton drives one of the most obvious and best-studied forms of cell movement, the contraction of muscle.

Intermediate Filaments

Intermediate filaments have great tensile strength, and their main function is to enable cells to withstand the mechanical stress that occurs when cells are stretched. The filaments are called "intermediate" because their diameter (about 10 nm) is between that of the thin actin-containing filaments and the thicker myosin filaments of smooth muscle cells, the cells in which the intermediate filaments were first discovered. Intermediate filaments are the toughest and most durable of the three types of cytoskeletal filaments: when cells are treated with concentrated

Figure 17–2 **Three types of protein filaments form the cytoskeleton.** The cells illustrated are epithelial cells lining the gut.

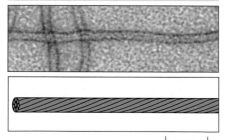

25 μm

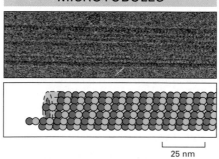

25 μm

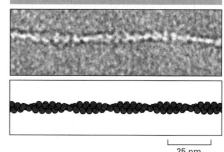

25 μm

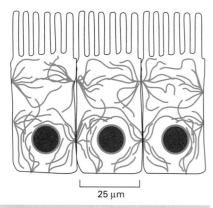

INTERMEDIATE FILAMENTS

25 nm

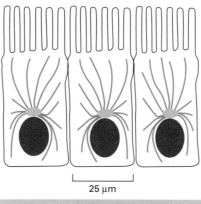

MICROTUBULES

25 nm

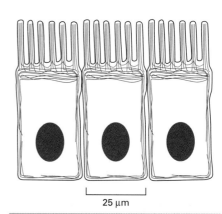

ACTIN FILAMENTS

25 nm

Intermediate filaments are ropelike fibers with a diameter of about 10 nm; they are made of intermediate filament proteins, which constitute a large and heterogeneous family. One type of intermediate filament forms a meshwork called the nuclear lamina just beneath the inner nuclear membrane. Other types extend across the cytoplasm, giving cells mechanical strength and distributing the mechanical stresses in an epithelial tissue by spanning the cytoplasm from one cell-cell junction to another. (Micrograph courtesy of Roy Quinlan.)

Microtubules are long, hollow cylinders made of the protein tubulin. With an outer diameter of 25 nm, they are more rigid than actin filaments or intermediate filaments. Microtubules are long and straight and typically have one end attached to a single microtubule-organizing center called a *centrosome*. (Micrograph courtesy of Richard Wade.)

Actin filaments (also known as *microfilaments*) are helical polymers of the protein actin. They are flexible structures, with a diameter of about 7 nm, that are organized into a variety of linear bundles, two-dimensional networks, and three-dimensional gels. Although actin filaments are dispersed throughout the cell, they are most highly concentrated in the *cortex*, the layer of cytoplasm just beneath the plasma membrane. (Micrograph courtesy of Roger Craig.)

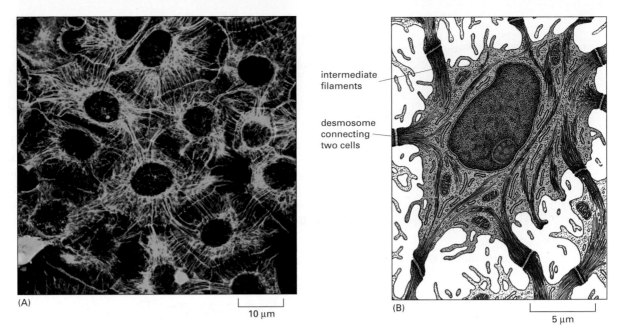

(A)

10 μm

(B)

intermediate filaments

desmosome connecting two cells

5 μm

Figure 17–3 Intermediate filaments form a strong, durable network in the cytoplasm of the cell. (A) Immunofluorescence micrograph of a sheet of epithelial cells in culture stained to show the lacelike network of intermediate keratin filaments *(green)* that surround the nuclei and extend through the cytoplasm of the cells. The filaments in each cell are indirectly connected to those of neighboring cells through the desmosomes (discussed in Chapter 21), establishing a continuous mechanical link from cell to cell throughout the tissue. This mechanical link strengthens the epithelium, allowing its cells to form a continuous sheet lining the tissue cavity. A second protein *(blue)* has been stained to show the locations of the cell boundaries. (B) Drawing from an electron micrograph of a section of epidermis showing the bundles of intermediate filaments that traverse the cytoplasm and that are inserted at desmosomes. (A, courtesy of Kathleen Green and Evangeline Amargo. B, from R. V. Krstić, Ultrastructure of the Mammalian Cell: An Atlas. Berlin: Springer, 1979.)

salt solutions and nonionic detergents, the intermediate filaments remain intact while most of the rest of the cytoskeleton is destroyed.

Intermediate filaments are found in the cytoplasm of most animal cells. They typically form a network throughout the cytoplasm, surrounding the nucleus and extending out to the cell periphery. There they are often anchored to the plasma membrane at cell–cell junctions such as desmosomes (discussed in Chaper 21), where the external face of the membrane is connected to that of another cell (Figure 17–3). Intermediate filaments are also found within the nucleus; a mesh of intermediate filaments, the *nuclear lamina,* underlies and strengthens the nuclear envelope in all eucaryotic cells. In this section, we see how the structure and assembly of intermediate filaments makes them particularly suited to strengthening cells and protecting them from mechanical stress.

Intermediate Filaments Are Strong and Ropelike

Intermediate filaments are like ropes with many long strands twisted together to provide tensile strength. The strands of this rope—the subunits of intermediate filaments—are elongated fibrous proteins, each composed of an N-terminal globular head, a C-terminal globular tail, and a central elongated rod domain (Figure 17–4A). The rod domain consists of an extended α-helical region that enables pairs of intermediate filament proteins to form stable dimers by wrapping around each other in a coiled-coil configuration (Figure 17–4B), as described in Chapter 4. Two of these coiled-coil dimers then associate by noncovalent bonding to form a tetramer (Figure 17–4C), and the tetramers then bind to one another end-to-end and side-by-side, also by noncovalent bonding, to generate the final ropelike intermediate filament (Figure 17–4D and E).

The central rod domains of different intermediate filament proteins are all similar in size and amino acid sequence, so that when they pack together they always form filaments of similar diameter and internal structure. By contrast, the globular head and tail regions, which are exposed on the surface of the filament, allow it to interact with other components of the cytoplasm. The globular domains vary greatly in both size and amino acid sequence from one intermediate filament protein to another.

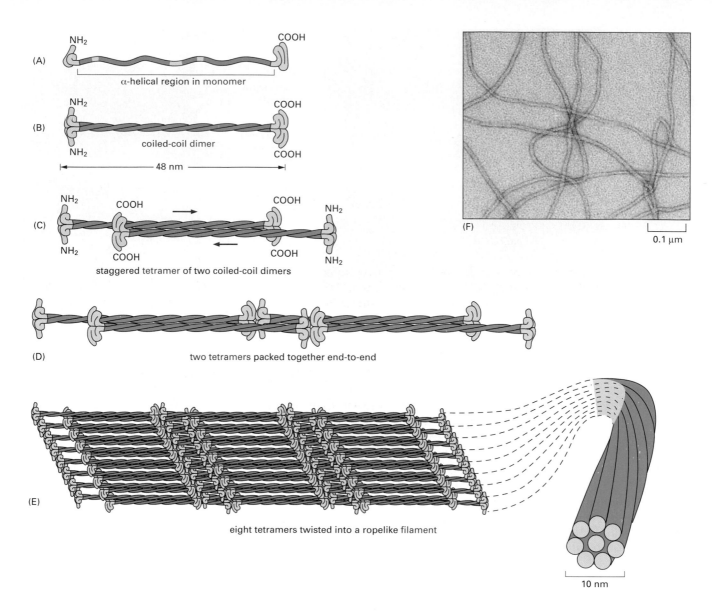

(A) α-helical region in monomer

(B) coiled-coil dimer
48 nm

(C) staggered tetramer of two coiled-coil dimers

(D) two tetramers packed together end-to-end

(E) eight tetramers twisted into a ropelike filament

(F) 0.1 µm

10 nm

Figure 17–4 Intermediate filaments are like ropes made of long, twisted strands of protein. The intermediate filament protein monomer shown in (A) consists of a central rod domain with globular regions at either end. Pairs of monomers associate to form a dimer (B), and two dimers then line up to form a staggered tetramer (C). Tetramers can pack together end-to-end as shown in (D) and assemble into a helical array containing eight strands of tetramers (shown here spread out flat for clarity) that generates the final ropelike intermediate filament (E). (F) An electron micrograph of the final 10-nm filament. (F, courtesy of Roy Quinlan.)

Intermediate Filaments Strengthen Cells Against Mechanical Stress

Intermediate filaments are particularly prominent in the cytoplasm of cells that are subject to mechanical stress. They are present in large numbers, for example, along the length of nerve cell axons, providing essential internal reinforcement to these extremely long and fine cell extensions. They are also abundant in muscle cells and in epithelial cells such as those of the skin. In all these cells, intermediate filaments, by stretching and distributing the effect of locally applied forces, keep cells and their membranes from breaking in response to mechanical shear (Figure 17–5). A similar principle is used to manufacture reinforced concrete and other composite materials in which tension-bearing linear elements such as steel bars, fiberglass, or carbon fibers are embedded in a space-filling matrix to give the material strength.

Intermediate filaments found in the cytoplasm can be grouped into four classes: (1) *keratin filaments* in epithelial cells; (2) *vimentin* and *vimentin-related filaments* in connective-tissue cells, muscle cells, and supporting cells of the nervous system (neuroglial cells); (3) *neurofilaments* in nerve cells; and (4) *nuclear lamins*, which strengthen the nuclear membrane of all animal cells (Figure 17–6). Filaments of each class are formed by polymerization of their corresponding protein subunits.

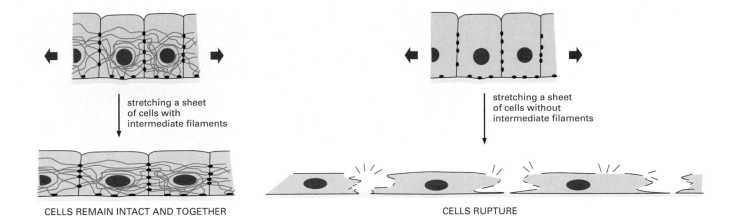

stretching a sheet
of cells with
intermediate filaments

CELLS REMAIN INTACT AND TOGETHER

stretching a sheet
of cells without
intermediate filaments

CELLS RUPTURE

The most diverse subunit family are the *keratins.* Different sets of keratins are found in different epithelia—one set in the lining of the gut, another in the epidermal layers of the skin. Specialized keratins also occur in hair, feathers, and claws. In each case, the keratin filaments are formed from a mixture of different keratin subunits. Keratin filaments typically span the interiors of epithelial cells from one side of the cell to the other, and filaments in adjacent epithelial cells are indirectly connected through cell–cell junctions called *desmosomes* (see Figure 17–3). The ends of the keratin filaments are anchored to the desmosomes, and they associate laterally with other cell components through their globular head and tail domains, which project from the surface of the assembled filament. This cabling of high tensile strength, formed by the filaments throughout the epithelial sheet, distributes the stress that occurs when the skin is stretched. The importance of this function is illustrated by the rare human genetic disease *epidermolysis bullosa simplex,* in which mutations in the keratin genes interfere with the formation of keratin filaments in the epidermis. As a result, the skin is highly vulnerable to mechanical injury, and even a gentle pressure can rupture its cells, causing the skin to blister.

Many of the intermediate filaments are further stabilized and reinforced by accessory proteins that cross-link the filament bundles into strong arrays. One particularly interesting cross-linking accessory protein is *plectin.* In addition to holding together bundles of intermediate filaments (particularly vimentin), these proteins link intermediate filaments to microtubules, to actin filaments, and to adhesive structures in the desmosomes (Figure 17–7). Mutations in the gene for plectin cause a devastating human disease that combines features of epidermolysis bullosa simplex (caused by disruption of skin keratin), muscular dystrophy (caused by disruption of intermediate filaments in muscle), and neurodegeneration (caused by disruption of neurofilaments). Mice

Figure 17–5 Intermediate filaments strengthen animal cells. If a sheet of epithelial cells is stretched by external forces (due to the growth or movements of the surrounding tissues, for example), then the network of intermediate filaments and desmosomal junctions that extends through the sheet develops tension and limits the extent of stretching. If the junctions alone were present, then the same forces would cause a major deformation of the cells, even to the extent of causing their plasma membranes to rupture.

INTERMEDIATE FILAMENTS			
CYTOPLASMIC			NUCLEAR
keratins	vimentin and vimentin-related	neurofilaments	nuclear lamins
in epithelia	in connective tissue, muscle cells, and neuroglial cells	in nerve cells	in all animal cells

Figure 17–6 Intermediate filaments can be divided into several different categories.

Figure 17–7 Plectin aids in the bundling of intermediate filaments and links these filaments to other cytoskeletal protein networks. Plectin (*green*) links intermediate filaments (*blue*) to other intermediate filaments, to microtubules (*red*), and to actin filaments (not shown). In this electron micrograph, the *yellow dots* are gold particles linked to antibodies that recognize plectin. The actin filament network has been removed to reveal these protein linkages. (From T.M. Svitkina and G.G. Borisy, *J. Cell Biol.* 135:991–1007, 1996. © The Rockefeller University Press.)

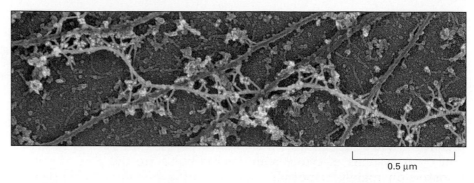

0.5 μm

Question 17–1

Which of the following types of cells would you expect to contain a high density of intermediate filaments in their cytoplasm? Explain your answers.

A. *Amoeba proteus* (a free-living amoeba)

B. Skin epithelial cell

C. Smooth muscle cell in the digestive tract

D. *Escherichia coli*

E. Nerve cell in the spinal cord

F. Sperm cell

G. Plant cell

lacking a functional plectin gene die within a few days of birth, with blistered skin and abnormal skeletal and heart muscles. Thus, although plectin may not be necessary for the initial formation of intermediate filaments, its cross-linking action is required to provide cells with the strength they need to withstand the mechanical stresses inherent to vertebrate life.

The Nuclear Envelope Is Supported by a Meshwork of Intermediate Filaments

While cytoplasmic intermediate filaments form ropelike structures, the intermediate filaments lining and strengthening the inside surface of the inner nuclear membrane are organized as a two-dimensional mesh (Figure 17–8). The intermediate filaments within this tough **nuclear lamina** are constructed from a class of intermediate filament proteins called *lamins* (not to be confused with *laminin,* which is an extracellular matrix protein). In contrast to the very stable cytoplasmic intermediate filaments found in many cells, the intermediate filaments of the nuclear lamina disassemble and re-form at each cell division, when the nuclear envelope breaks down during mitosis and then re-forms in each daughter cell (discussed in Chapter 19).

Disassembly and reassembly of the nuclear lamina is controlled by the phosphorylation and dephosphorylation (discussed in Chapter 4) of the lamins by protein kinases. When the lamins are phosphorylated, the consequent conformational change weakens the binding between the tetramers and causes the filament to fall apart. Dephosphorylation at the end of mitosis causes the lamins to reassemble (see Figure 19–18).

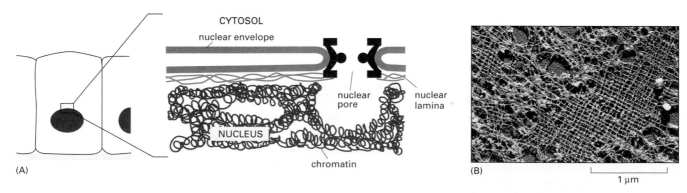

Figure 17–8 Intermediate filaments support and strengthen the nuclear envelope. (A) Schematic cross section through the nuclear envelope. The intermediate filaments of the nuclear lamina line the inner face of the nuclear envelope and are thought to provide attachment sites for the DNA-containing chromatin. (B) Electron micrograph of a portion of the nuclear lamina from a frog egg, or oocyte. The lamina is formed from a square lattice of intermediate filaments composed of lamins. (Nuclear laminae from other cell types are not always as regularly organized as the one shown here.) (B, courtesy of Ueli Aebi.)

Microtubules

Microtubules have a crucial organizing role in all eucaryotic cells. They are long and relatively stiff hollow tubes of protein that can rapidly disassemble in one location and reassemble in another. In a typical animal cell, microtubules grow out from a small structure near the center of the cell, called the *centrosome*. Extending out toward the cell periphery, they create a system of tracks within the cell, along which vesicles, organelles, and other cell components are moved (Figure 17–9A). These and other systems of cytoplasmic microtubules are the part of the cytoskeleton mainly responsible for anchoring membrane-enclosed organelles within the cell and for guiding intracellular transport.

When a cell enters mitosis, the cytoplasmic microtubules disassemble and then reassemble into an intricate structure called the *mitotic spindle.* As described in Chapter 19, the mitotic spindle provides the machinery that will segregate the chromosomes equally into the two daughter cells just before a cell divides (Figure 17–9B). Microtubules can also form permanent structures, as exemplified by the rhythmically beating hairlike structures called *cilia* and *flagella* (Figure 17–9C). These extend from the surface of many eucaryotic cells, which use them either as a means of propulsion or to sweep fluid over the cell surface. The core of a eucaryotic cilium or flagellum consists of a highly organized and stable bundle of microtubules. (Bacterial flagella have an entirely different structure and act as propulsive structures by a different mechanism.)

In this section we first look at the structure and assembly of microtubules and then discuss their role in organizing the cytoplasm. Their organizing function depends on the association of microtubules with accessory proteins, especially the *motor proteins* that propel organelles along cytoskeletal tracks. Finally, we discuss the structure and function of cilia and flagella, in which microtubules are permanently associated with motor proteins that power ciliary beating.

Microtubules Are Hollow Tubes with Structurally Distinct Ends

Microtubules are built from subunits—molecules of **tubulin**—each one of which is itself a dimer composed of two very similar globular proteins called *α-tubulin* and *β-tubulin,* bound tightly together by noncovalent bonding. The tubulin dimers stack together, again by noncovalent bonding, to form the wall of the hollow cylindrical microtubule. This tubelike structure is a cylinder made of 13 parallel *protofilaments,* each a linear chain of tubulin dimers with α- and β-tubulin alternating along its length (Figure 17–10). Each protofilament has a structural polarity, with α-tubulin exposed at one end and β-tubulin exposed at the other, and this **polarity**—the directional arrow embodied in the structure—is the same for all the protofilaments, giving a structural polarity to the microtubule as a whole. One end of the microtubule, thought to be the β-tubulin end, is called its *plus end,* and the other, the α-tubulin end, its *minus end.*

A microtubule grows from an initial ring of 13 tubulin molecules; tubulin dimers are added individually, gradually building up the structure of the hollow tube. *In vitro,* in a concentrated solution of pure tubulin, tubulin dimers will add to either end of a growing microtubule, although they add more rapidly to the plus end than the minus end (which is why the ends were originally named this way). The polarity of the microtubule—the fact that its structure has a definite direction, with the two ends being chemically different and behaving differently—is

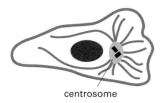

(A) INTERPHASE CELL

centrosome

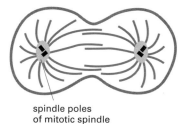

(B) DIVIDING CELL

spindle poles
of mitotic spindle

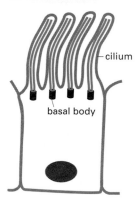

(C) CILIATED CELL

cilium

basal body

Figure 17–9 Microtubules usually grow out of an organizing structure. Unlike intermediate filaments, microtubules *(dark green)* extend from an organizing center such as (A) a centrosome, (B) a spindle pole, or (C) the basal body of a cilium.

Figure 17–10 Microtubules are hollow tubes of tubulin.

(A) One tubulin molecule (an αβ dimer) and one protofilament shown schematically, together with their location in the microtubule wall. Note that the tubulin molecules are all arranged in the protofilaments with the same orientation, so that the microtubule has a definite structural polarity. (B and C) Schematic diagrams of a microtubule showing how tubulin molecules pack together in the microtubule wall. At the top, the 13 molecules are shown in cross section. Below this, a side view of a short section of a microtubule shows how the tubulin molecules are aligned in rows, or protofilaments. (D) Cross section of a microtubule with its ring of 13 distinct subunits, each of which corresponds to a separate tubulin dimer. (E) Microtubule viewed lengthwise in an electron microscope. (D, courtesy of Richard Linck; E, courtesy of Richard Wade.)

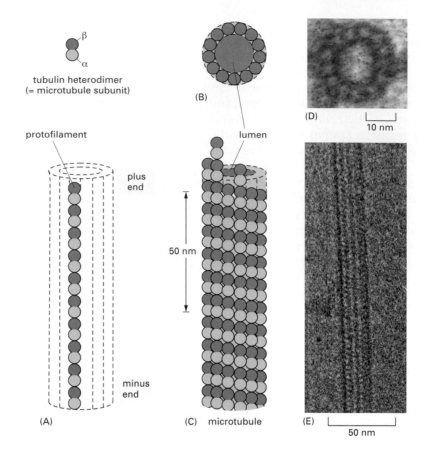

crucial, both for the assembly of microtubules and for their role once they are formed. If they had no polarity, they could not serve their function in defining a direction for intracellular transport, for example.

The Centrosome Is the Major Microtubule-organizing Center in Animal Cells

Microtubules in cells are formed by outgrowth from specialized organizing centers, which control the number of microtubules formed, their location, and their orientation in the cytoplasm. In animal cells, for example, the **centrosome**, which is typically present on one side of the cell nucleus when the cell is not in mitosis, organizes the array of microtubules that radiates outward from it through the cytoplasm (see Figure 17–9A). Centrosomes contain hundreds of ring-shaped structures formed from another type of tubulin, γ-tubulin, and each γ-tubulin ring serves as the starting point, or *nucleation site,* for the growth of one microtubule (Figure 17–11A). The αβ-tubulin dimers add to the γ-tubulin ring in a specific orientation, with the result that the minus end of each microtubule is embedded in the centrosome and growth occurs only at the plus end—that is, the outward-facing end (Figure 17–11B).

The γ-tubulin rings in the centrosome should not be confused with the **centrioles**, curious structures each made of a cylindrical array of short microtubules embedded in the centrosome of most animal cells. The centrioles have no role in the nucleation of microtubules in the centrosome (the γ-tubulin rings alone are sufficient), and their function there remains something of a mystery, especially as plant cells lack them. Centrioles are, however, similar, if not identical, to the *basal bodies* that form the organizing centers for the microtubules in cilia and flagella (see Figure 17–9C), as discussed later in this chapter.

Question 17–2

Why do you suppose it is much easier to add tubulin to existing microtubules than to start a new microtubule from scratch? Explain how γ-tubulin in the centrosome helps to overcome this hurdle.

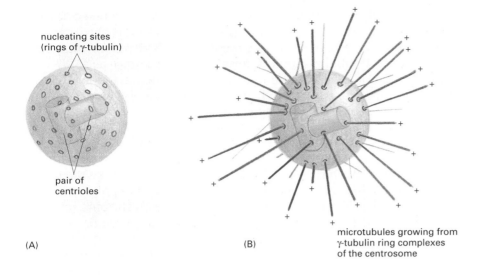

nucleating sites
(rings of γ-tubulin)

pair of
centrioles

(A)

(B)

microtubules growing from
γ-tubulin ring complexes
of the centrosome

Figure 17–11 Tubulin polymerizes from nucleation sites on a centrosome.
(A) Schematic drawing showing that a centrosome consists of an amorphous matrix of protein containing the γ-tubulin rings that nucleate microtubule growth. In animal cells, the centrosome contains a pair of centrioles, each made up of a cylindrical array of short microtubules. (B) A centrosome with attached microtubules. The minus end of each microtubule is embedded in the centrosome, having grown from a nucleating ring, whereas the plus end of each microtubule is free in the cytoplasm.

Microtubules need nucleating sites such as those provided by the γ-tubulin rings in the centrosome because it is much harder to start a new microtubule from scratch, by first assembling a ring of αβ-tubulin dimers, than to add such dimers to a preexisting microtubule structure. Purified free αβ-tubulin can polymerize spontaneously *in vitro* when at a high concentration, but in the living cell, the concentration of free αβ-tubulin is too low to drive the difficult first step of assembling the initial ring of a new microtubule. By providing organizing centers containing nucleation sites, and keeping the concentration of free αβ-tubulin dimers low, cells can thus control where microtubules form.

Growing Microtubules Show Dynamic Instability

Once a microtubule has been nucleated, its plus end typically grows outward from the organizing center by the addition of subunits for many minutes. Then, without warning, the microtubule suddenly undergoes a transition that causes it to shrink rapidly inward by losing subunits from its free end. It may shrink partially and then, no less suddenly, start growing again, or it may disappear completely, to be replaced by a new microtubule from the same γ-tubulin ring (Figure 17–12).

This remarkable behavior, known as **dynamic instability**, stems from the intrinsic capacity of tubulin molecules to hydrolyze GTP. Each free tubulin dimer contains one tightly bound GTP molecule that is hydrolyzed to GDP (still tightly bound) shortly after the subunit is added to a growing microtubule. The GTP-associated tubulin molecules pack efficiently together in the wall of the microtubule, whereas tubulin molecules carrying GDP have a different conformation and bind less strongly to each other.

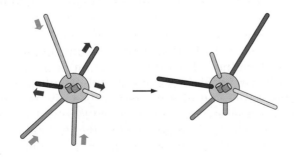

Figure 17–12 Each microtubule filament grows and shrinks independently of its neighbors. The array of microtubules anchored in a centrosome is continually changing as new microtubules grow (*red arrows*) and old microtubules shrink (*blue arrows*).

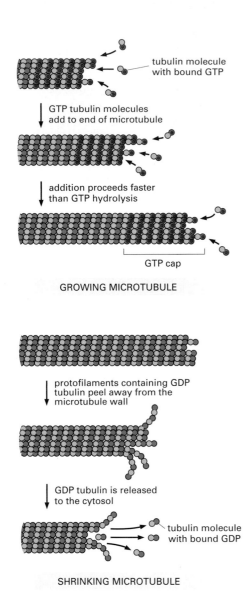

GROWING MICROTUBULE

SHRINKING MICROTUBULE

Figure 17–13 GTP hydrolysis controls the growth of microtubules. Tubulin dimers carrying GTP *(red)* bind more tightly to one another than do tubulin dimers carrying GDP *(dark green)*. Therefore, microtubules that have freshly added tubulin dimers at their end with GTP bound tend to keep growing. From time to time, however, especially when microtubule growth is slow, the subunits in this "GTP cap" will hydrolyze their GTP to GDP before fresh subunits loaded with GTP have time to bind. The GTP cap is thereby lost; the GDP-carrying subunits are less tightly bound in the polymer and are readily released from the free end, so that the microtubule begins to shrink continuously.

When polymerization is proceeding rapidly, tubulin molecules add to the end of the microtubule faster than the GTP they carry is hydrolyzed. The end of a growing microtubule is therefore composed entirely of GTP-tubulin subunits, forming what is known as a *GTP cap*. In this situation, because the microtubule can depolymerize only by losing subunits from its free end, the growing microtubule will continue to grow. Because of the randomness of chemical processes, however, it will occasionally happen that tubulin at the free end of the microtubule hydrolyzes its GTP before the next tubulin has been added, so that the free ends of protofilaments are now composed of GDP-tubulin subunits. This tips the balance in favor of disassembly. Because the rest of the microtubule is composed of GDP-tubulin, once depolymerization has started, it will tend to continue, often at a catastrophic rate; the microtubule starts to shrink rapidly and continuously and may even disappear (Figure 17–13).

The GDP-containing tubulin molecules that are freed as the microtubule depolymerizes join the unpolymerized tubulin molecules already in the cytosol. In a typical fibroblast, for example, at any one time about half of the tubulin in the cell is in microtubules, while the remainder is free in the cytosol, forming a pool of subunits available for microtubule growth. (This situation is quite unlike the arrangement with the more stable intermediate filaments, where the subunits are typically almost completely in the fully assembled form.) The tubulin molecules joining the pool then exchange their bound GDP for GTP, thereby becoming competent again to add to another microtubule that is in a growth phase.

Microtubules Are Maintained by a Balance of Assembly and Disassembly

The relative instability of microtubules allows them to undergo continual rapid remodeling, and this is crucial for microtubule function, as demonstrated by the effect of drugs that prevent polymerization or depolymerization of tubulin. Consider the mitotic spindle, the microtubule framework that guides the chromosomes during mitosis (see Figure 17–9B). If a cell in mitosis is exposed to the drug *colchicine*, which binds tightly to free tubulin and prevents its polymerization into microtubules, the mitotic spindle rapidly disappears and the cell stalls in the middle of mitosis, unable to partition its chromosomes into two groups. This shows that the mitotic spindle is normally maintained by a continuous balanced addition and loss of tubulin subunits: when tubulin addition is blocked by colchicine, tubulin loss continues until the spindle disappears.

The drug *taxol* has the opposite action at the molecular level. It binds tightly to microtubules and prevents them from losing subunits. Because new subunits can still be added, the microtubules can grow but cannot shrink. However, despite the differences in molecular detail, taxol has the same overall effect on the cell as colchicine: it also arrests dividing cells in mitosis. We learn from this that for the spindle to function, microtubules must be able not only to assemble but also to disassemble. The behavior of the spindle is discussed in more detail in Chapter 19, when we consider mitosis.

The inactivation or destruction of the mitotic spindle eventually kills dividing cells. Cancer cells, which are dividing with less control than most other cells of the body, can be killed preferentially by microtubule-stabilizing and microtubule-destabilizing *antimitotic drugs*. Thus drugs derived from both colchicine and taxol are used in the clinical treatment of cancer.

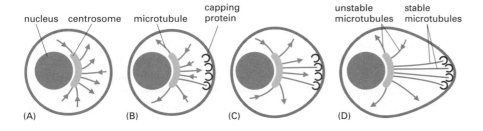

nucleus centrosome microtubule | capping protein | unstable microtubules | stable microtubules

(A)　(B)　(C)　(D)

Figure 17–14 The selective stabilization of microtubules can polarize a cell. A newly formed microtubule will persist only if both its ends are protected from depolymerization. In cells, the minus ends of microtubules are generally protected by the organizing centers from which these filaments grow. The plus ends are initially free but can be stabilized by other proteins. Here, for example, a nonpolarized cell is depicted in (A) with new microtubules growing from and shrinking back to a centrosome in many directions randomly. Some of these microtubules happen by chance to encounter proteins (capping proteins) in a specific region of the cell cortex that can bind to and stabilize the free plus ends of microtubules (B). This selective stabilization will lead to a rapid reorientation of the microtubule arrays (C) and convert the cell to a strongly polarized form (D).

In a normal cell, as a consequence of dynamic instability, the centrosome (or other organizing center) is continually shooting out new microtubules in an exploratory fashion in different directions and retracting them. A microtubule growing out from the centrosome can, however, be prevented from disassembling if its plus end is somehow permanently stabilized by attachment to another molecule or cell structure so as to prevent tubulin depolymerization. If stabilized by attachment to a structure in a more distant region of the cell, the microtubule will establish a relatively stable link between that structure and the centrosome (Figure 17–14). The centrosome can be compared to a fisherman casting a line: if there is no bite at the end of the line, the line is quickly withdrawn and a new cast is made; but if a fish bites, the line remains in place, tethering the fish to the fisherman. This simple strategy of random exploration and selective stabilization enables the centrosome and other nucleating centers to set up a highly organized system of microtubules linking selected parts of the cell. This system is used to position organelles relative to one another, as we now see.

Microtubules Organize the Interior of the Cell

Cells are able to modify the dynamic instability of their microtubules for particular purposes. As cells enter mitosis, for example, microtubules become initially more dynamic, switching between growing and shrinking much more frequently than cytoplasmic microtubules normally do. This enables them to disassemble rapidly and then reassemble into the mitotic spindle. On the other hand, when a cell has differentiated into a specialized cell type and taken on a definite fixed structure, the dynamic instability of its microtubules is often suppressed by proteins that bind to the ends of microtubules or along their length and stabilize them against disassembly. The stabilized microtubules then serve to maintain the organization of the cell.

Most differentiated animal cells are *polarized*; that is, one end of the cell is structurally or functionally different from the other. Nerve cells, for example, put out an axon from one end of the cell and dendrites from the other; cells specialized for secretion have their Golgi apparatus positioned toward the site of secretion, and so on. The cell's polarity is a reflection of the polarized systems of microtubules in its interior, which help to position organelles in their required location within the cell and to guide the streams of traffic moving between one part of the cell and another. In the nerve cell, for example, all the microtubules in the axon point in the same direction, with their plus ends toward the axon terminal (Figure 17–15). Along these oriented tracks the cell is able to send cargoes of materials, such as membrane vesicles and proteins for secretion, that are made in the cell body but required far away at the end of the axon.

Some of these materials move at speeds in excess of 10 cm a day, which means that they may still take a week or more to travel to the end of a long axon in larger animals. But movement along microtubules is

Question 17–3

Dynamic instability causes microtubules either to grow or to shrink rapidly. Consider an individual microtubule that is in its shrinking phase.

A. What must happen at the end of the microtubule in order for it to stop shrinking and to start growing?

B. How would a change in the tubulin concentration affect this switch?

C. What would happen if only GDP, but no GTP, were present in the solution?

D. What would happen if the solution contained an analog of GTP that cannot be hydrolyzed?

Figure 17–15 Microtubules transport cargo along a nerve cell axon. In nerve cells all the microtubules in the axon point in the same direction, with their plus ends toward the axon terminal. The oriented microtubules serve as tracks for the directional transport of materials synthesized in the cell body but required at the axon terminal (such as membrane proteins required for growth). For an axon passing from your spinal cord to a muscle in your shoulder, say, the journey takes about two days. In addition to this *outward* traffic of material *(red circles)* driven by one set of motor proteins, there is inward traffic *(blue circles)* in the reverse direction driven by another set of proteins. The *inward* traffic carries materials ingested by the tip of the axon or produced by the breakdown of proteins and other molecules back toward the cell body.

cell body

microtubule

axon

axon terminal

inward transport

outward transport

immeasurably faster and more efficient than free diffusion. A protein molecule traveling by free diffusion would take years to reach the end of a long axon, if it arrived at all (see Question 17–12).

It is important to realize, though, that the microtubules in living cells do not act alone. Their functions, like those of other cytoskeletal filaments, depend on a large variety of accessory proteins that bind to microtubules and serve various functions. Some microtubule-associated proteins stabilize microtubules against disassembly, for example, while others link microtubules to other cell components, including the other types of cytoskeletal filaments. Because the components of the cytoskeleton can interact with each other, their functions can be coordinated.

Microtubules influence the distribution of membranes in a eucaryotic cell by means of microtubule-associated motor proteins, which move along microtubules. As we discuss next, motor proteins use the energy of ATP hydrolysis to transport organelles, vesicles, and other cell materials along tracks in the cytoplasm provided by actin filaments and microtubules.

Motor Proteins Drive Intracellular Transport

If a living cell is observed in a light microscope, its cytoplasm is seen to be in continual motion (Figure 17–16). Mitochondria and the smaller membrane-enclosed organelles and vesicles move in small, jerky steps—that is, they move for a short period, stop, and then start again. This *saltatory movement* is much more sustained and directional than the continual, small Brownian movements caused by random thermal motions. Both microtubules and actin filaments are involved in saltatory and other directed intracellular movements in eucaryotic cells. In both cases the movements are generated by **motor proteins**, which bind to actin filaments or microtubules and use the energy derived from repeated cycles of ATP hydrolysis to travel steadily along the actin filament or the microtubule in a single direction (see Figure 4–45). At the same time, these motor proteins also attach to other cell components,

Figure 17–16 Organelles move along microtubules at different speeds. In this series of video-enhanced images of a flattened area of an invertebrate nerve cell, numerous membrane vesicles and mitochondria are present, many of which can be seen to move. The white circle provides a fixed frame of reference. These images were recorded at intervals of 400 milliseconds. (Courtesy of P. Forscher.)

 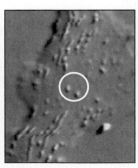 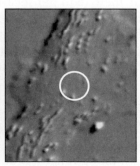 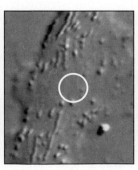 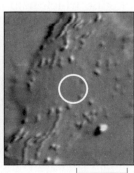

5 μm

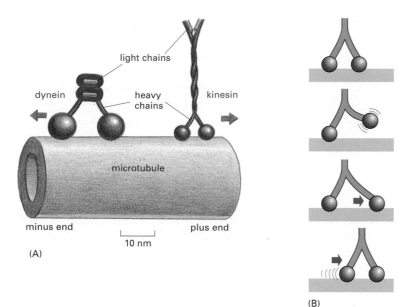

light chains

dynein

heavy chains

kinesin

microtubule

minus end

plus end

10 nm

(A)

(B)

Figure 17–17 Motor proteins move along microtubules using their globular heads. (A) Kinesins and cytoplasmic dyneins are microtubule motor proteins that generally move in opposite directions along a microtubule. Each of these proteins (drawn here to scale) has two heavy chains and several smaller light chains. Each heavy chain forms a globular head that interacts with microtubules. (B) Diagram of a motor protein showing ATP-dependent "walking" along a filament.

and thus transport this cargo along the filaments. Dozens of motor proteins have been identified. They differ in the type of filament they bind to, the direction in which they move along the filament, and the cargo they carry.

The motor proteins that move along cytoplasmic microtubules, such as those in the axon of a nerve cell, belong to two families: the **kinesins** generally move toward the plus end of a microtubule (away from the centrosome; outward in Figure 17–15), while the **dyneins** move toward the minus end (toward the centrosome; inward in Figure 17–15). Kinesins and dyneins both have two globular ATP-binding heads and a tail (Figure 17–17). The heads interact with microtubules in a stereospecific manner, so that kinesin, for example, will attach to a microtubule only if it is "pointing" in the correct direction. The tail of a motor protein generally binds stably to some cell component, such as a vesicle or an organelle, and thereby determines the type of cargo that the motor protein can transport (Figure 17–18). The globular heads of kinesin and dynein are enzymes with ATP-hydrolyzing (ATPase) activity. This reaction provides the energy for a cycle of conformational changes in the head that enable it to move along the microtubule by a cycle of binding, release, and rebinding to the microtubule (see Figure 4–45). For a discussion of the discovery and study of motor proteins, see How We Know, pp. 586–588.

Organelles Move Along Microtubules

Microtubules and their associated motor proteins play an important part in positioning membrane-enclosed organelles within a eucaryotic

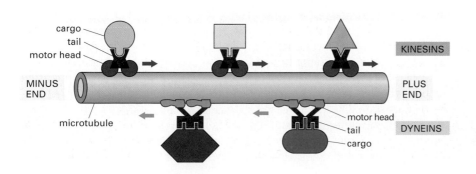

cargo

tail

motor head

MINUS END

microtubule

KINESINS

PLUS END

motor head

tail

cargo

DYNEINS

Figure 17–18 Different motor proteins transport cargo along microtubules. Most kinesins move toward the plus end of a microtubule, whereas dyneins move toward the minus end. Both types of microtubule motor proteins exist in many forms, each of which is thought to transport a different cargo. The tail of the motor protein determines what cargo the protein transports.

The movement of organelles throughout the cell cytoplasm has been observed, measured, and speculated about since the middle of the nineteenth century. But it wasn't until the middle of the 1980s that biologists could identify the molecules that drive this movement of organelles and vesicles from one part of the cell to another.

Why the lag between observation and understanding? The problem was in the proteins—or, more precisely, in the difficulty of studying them in isolation outside the cell. To investigate the activity of an enzyme, for example, biochemists first purify the polypeptide: they break open cells or tissues and separate the protein of interest from other molecular components (see Panels 4–3 to 4–5, pp. 160–163). They can then study the protein on its own, *in vitro,* controlling its exposure to substrates, inhibitors, ATP, and so on. Unfortunately, this approach did not seem to work for studies of the motile machinery that underlies intracellular transport. It is not possible to break open a cell and pull out a fully active transport system, free of extraneous material, that continues to carry mitochondria and vesicles from place to place.

The techniques needed to move the research forward came from two different sources. First, advances in microscopy allowed biologists to see that an operational transport system (with extraneous material still attached) could be squeezed from the right kind of living cell. At the same time, biochemists realized that they could assemble a working transport system from scratch—using purified cables, motors, and cargo—outside the cell. The breakthroughs start with a squid.

Teeming cytoplasm

As we saw in Chapter 12, neuroscientists interested in the electrical properties of nerve cell membranes have long studied the giant axon from squid (see How We Know, pp. 414–415). Because of its large size, researchers found that they could squeeze the cytoplasm from the axon like toothpaste, and then study how ions move back and forth through various channels in the empty, tubelike membrane. The physiologists simply discarded the cytoplasmic jelly, as it appeared to be inert (thus uninteresting) when examined under a standard light microscope.

Then along came video-enhanced microscopy. This type of microscopy, developed by Shinya Inoué, Robert Allen, and others, allows one to detect structures that are smaller than the resolution power of standard light microscopes, about 0.2 μm, or 200 nm (see Panel 1–1, pp. 8–9). Sample images are captured by a video camera, and then enhanced by computer processing to reduce the background and heighten contrast. When researchers in the early 1980s

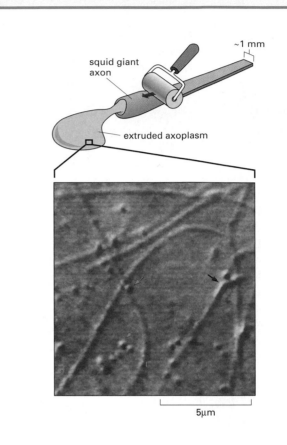

Figure 17–19 Video-enhanced microscopy of cytoplasm squeezed from a squid giant axon reveals the motion of organelles. In this micrograph numerous cytoskeletal filaments are visible, along with transported particles including a mitochondrion *(red arrow)* and smaller vesicles *(blue arrows).* (From R.D. Vale, B.J. Schnapp, T.S. Reese, and M.P. Sheetz, *Cell* 40:449–454, 1985. © Elsevier Science.)

applied this new technique to preparations of squid axon cytoplasm, they observed, for the first time, the motion of vesicles and membranous organelles along cytoskeletal filaments.

Under the video microscope, extruded axoplasm is seen to be teeming with tiny particles—from vesicles of 30 to 50 nm in diameter to mitochondria some 5000 nm in length, all moving to and fro along cytoskeletal filaments at speeds of up to 5 μm per second. If the axoplasm is spread thin enough, individual filaments can be seen (Figure 17–19).

The movement continues for hours, allowing researchers to manipulate the preparation and study the effects. Ray Lasek and Scott Brady discovered, for example, that organelle

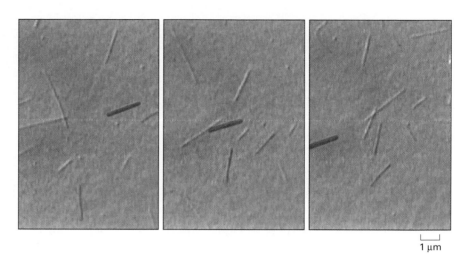

Figure 17–20 A motor protein causes microtubule gliding. In an active *in vitro* motility assay, purified kinesin is mixed with microtubules in a buffer containing ATP. When a drop of the mixture is placed on a glass slide and examined by video-enhanced microscopy, individual microtubules can be seen gliding over the slide driven by kinesin molecules. Images recorded at 1 second intervals. (Courtesy of Nick Carter and Rob Cross.)

1 µm

movement requires ATP. Substitution of ATP analogs, such as AMP-PNP, which bind to the enzyme active site but cannot be hydrolyzed (and thus provide no energy) inhibit the translocation.

Snaking tubes

More work was needed to identify the individual components that drive the transport system in squid axons: What are the filaments made of? What are the molecular machines that shuttle the vesicles and organelles along these filaments? Identifying the cables was relatively easy. Studies using antibodies to α-tubulin revealed that the filaments are microtubules. But what about the motor proteins? To find these, Ron Vale, Thomas Reese, and Michael Sheetz set up a system in which they could fish for proteins that power organelle movement.

Their strategy was simple yet elegant: add together purified cables and purified cargo and then look for molecules that induce motion. They took purified microtubules from squid optic lobe, added organelles isolated from squid axons, and showed that movement could be triggered by the addition of an extract from squid axon cytoplasm. In this preparation, the researchers could watch organelles travel along the microtubules, and microtubules glide snakelike over the surface of a glass coverslip (see Question 17–18). Their challenge was to isolate the protein responsible for movement in this reconstituted system.

To do that, Vale and his colleagues took advantage of Lasek and Brady's earlier work with the ATP analog AMP-PNP. Although this analog inhibits the movement of vesicles along microtubules, it still allows these components to attach to the microtubule filaments. So the researchers incubated the

cytoplasm extract with microtubules, in the presence of AMP-PNP; they then pulled out the microtubules with what they hoped were the motor proteins still attached. Vale and his team then added ATP to release the attached proteins, and they found a 110-kilodalton polypeptide that could bind to, and initiate movement of, microtubules *in vitro* (Figure 17–20). They dubbed the molecule kinesin (from the Greek *kinein,* "to move").

Such *in vitro* motility assays have been instrumental in the study of motor proteins and their activities. Subsequent studies showed that kinesin moves along microtubules from the minus end to the plus end, and also helped to identify many other kinesin-related motor proteins.

Lights, camera, action

Combining these *in vitro* assays with ever more refined microscopic techniques, researchers can now monitor the movement of individual motor proteins along single microtubules, even in living cells. In an assay developed by Steven Block and his colleagues in 1990, microscopic silica beads coated with low concentrations of kinesin (such that only one molecule of kinesin is present on each bead) can be monitored as they make their way down a microtubule (Figure 17–21). Other observations of single kinesin molecules are made possible by coupling the motor protein with a fluorescent marker protein such as GFP.

Such single-molecule studies have revealed that kinesin moves along microtubules *processively*—that is, each molecule takes 100 or so "steps" along the filament before falling off (Figure 17–22). The length of each step is 8 nm, which corresponds to the spacing of tubulin dimers along the microtubule. Combining these observations with assays of

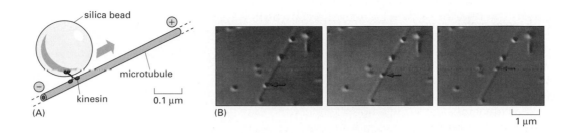

Figure 17–21 Video microscopy can be used to track the movement of a single kinesin molecule. (A) In this assay, silica beads are coated with kinesin molecules at a concentration such that each bead, on average, will have only one kinesin molecule attached to it. Kinesin is then allowed to walk along a microtubule, and its movement is monitored by tracking the movement of the bead. (B) In this series of images, the bead is captured by a laser-based optical tweezer, placed on a microtubule filament, and then allowed to move. Thirty seconds elapses between each frame. (From S. Block et al., *Nature* 348:348–352, 1990. © Macmillan Magazines Ltd.)

ATP hydrolysis, researchers have confirmed that one molecule of ATP is consumed per step. Kinesin can move in a processive manner because it has two heads (see Figure 17–22). The motor is thought to walk its way toward the plus end of the microtubule in a hand-over-hand fashion, each head binding and releasing the filament in turn. Further studies are required to confirm (or disprove) this model, and researchers are now actively working to improve their methods so that they can watch not only single molecules of kinesin, but each individual head as it moves, in relation to its partner, along the microtubule. The results will yield additional insights into the molecular movements that underlie the organization and activity of eucaryotic cells.

Figure 17–22 A single molecule of kinesin moves along a microtubule. (A) Electron micrograph of a single kinesin molecule showing the two head domains *(red arrows)*. (B) Three frames, separated by intervals of 1 second, record the movement of an individual kinesin-GFP molecule *(green)* along a microtubule *(red)* at a speed of 0.3 μm/sec. (C) Series of molecular models of the two heads of a kinesin molecule, showing how they are thought to processively walk their way along a microtubule in a series of 8 nm steps. (A, courtesy of John Heuser; B and C, courtesy of Ron Vale.)

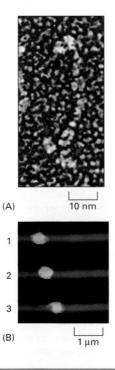

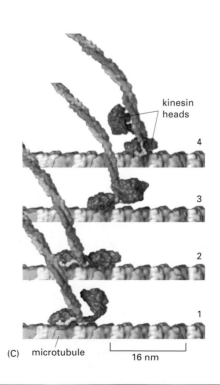

cell. In most animal cells, for example, the tubules of the endoplasmic reticulum reach almost to the edge of the cell, whereas the Golgi apparatus is located in the interior of the cell near the centrosome (Figure 17–23). Both the endoplasmic reticulum and the Golgi apparatus depend on microtubules for their alignment and positioning. The membranes of the endoplasmic reticulum extend out from their points of connection with the nuclear envelope (see Figure 1–22), aligning with microtubules that extend from the centrosome out to the plasma membrane. As the cell develops and the endoplasmic reticulum grows, kinesins attached to the outside of the endoplasmic reticulum membrane pull it outward along microtubules, stretching it like a net. Dyneins pull the Golgi apparatus the other way along microtubules, inward toward the cell center. In this way the regional differences in internal membranes, on which the successful function of the cell depends, are created and maintained.

When cells are treated with a drug such as colchicine that causes microtubules to disassemble, both of these organelles change their location dramatically. The endoplasmic reticulum, which has connections to the nuclear envelope, collapses to the center of the cell, while the Golgi apparatus, which is not attached to any other organelle, fragments into small vesicles, which disperse throughout the cytoplasm. When the drug is removed, the organelles return to their original positions, dragged by motor proteins moving along the re-formed microtubules. The normal positioning of these organelles is thought to be mediated by receptor proteins on their membranes that bind to motor proteins—to kinesins for the endoplasmic reticulum and to dyneins for the Golgi apparatus.

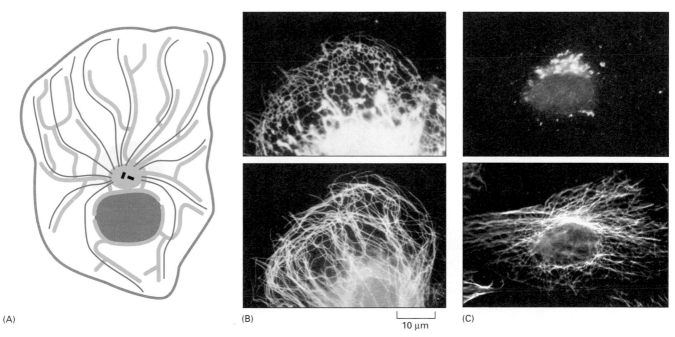

(A)   (B)   10 μm   (C)

Figure 17–23 Microtubules help to arrange the organelles in a eucaryotic cell. (A) Schematic diagram of a cell showing the typical arrangement of microtubules *(dark green)*, endoplasmic reticulum *(blue)*, and Golgi apparatus *(yellow)*. The nucleus is shown in *brown*, and the centrosome in *light green*. (B) Cell stained with antibodies to endoplasmic reticulum *(upper panel)* and to microtubules *(lower panel)*. Motor proteins pull the endoplasmic reticulum out along microtubules. (C) Cell stained with antibodies to the Golgi apparatus *(upper panel)* and to microtubules *(lower panel)*. In this case motor proteins move the Golgi apparatus inward to its position near the centrosome. (B, courtesy of Mark Terasaki, Lan Bo Chen, and Keigi Fujiwara; C, courtesy of Viki Allan and Thomas Kreis.)

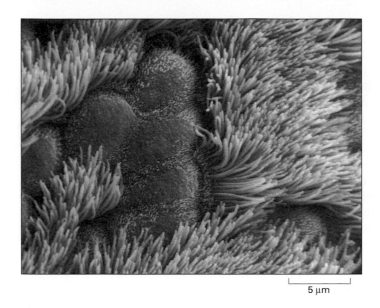

Figure 17–24 Hairlike cilia coat the surface of many eucaryotic cells. Scanning electron micrograph of the ciliated epithelium on the surface of the human respiratory tract. The thick tufts of cilia on the ciliated cells are interspersed with the dome-shaped surfaces of nonciliated epithelial cells. (Reproduced from R.G. Kessel and R.H. Karden, Tissues and Organs. San Francisco: W.H. Freeman & Co., 1979.)

5 μm

Cilia and Flagella Contain Stable Microtubules Moved by Dynein

Earlier in this chapter we mentioned that many microtubules in cells are stabilized through their association with other proteins, and therefore no longer show dynamic instability. Stable microtubules are employed by cells as stiff supports on which to construct a variety of polarized structures, including the remarkable cilia and flagella that allow eucaryotic cells to move water over their surface. **Cilia** (singular, **cilium**) are hairlike structures about 0.25 μm in diameter, covered by plasma membrane, that extend from the surface of many kinds of eucaryotic cells (see Figure 17–9C). A single cilium contains a core of stable microtubules, arranged in a bundle, that grow from a *basal body* in the cytoplasm; the basal body serves as the organizing center for the cilium.

The primary function of cilia is to move fluid over the surface of a cell, or to propel single cells through a fluid. Some protozoa, for example, use cilia to collect food particles, and others use them for locomotion. On the epithelial cells lining the human respiratory tract (Figure 17–24), huge numbers of cilia (more than a billion per square centimeter) sweep layers of mucus containing trapped dust particles and dead cells up toward the throat, to be swallowed and eventually eliminated from the body. Cilia on the cells of the oviduct wall create a current that helps to move eggs along the oviduct. Each cilium acts as a small oar, moving in a repeated cycle that generates the current that washes over the cell surface (Figure 17–25).

The **flagella** (singular, **flagellum**) that propel sperm and many protozoa are much like cilia in their internal structure, but usually very much longer. They are designed to move the entire cell, and instead of generating a current, they propagate regular waves along their length that drive the cell through liquid (Figure 17–26).

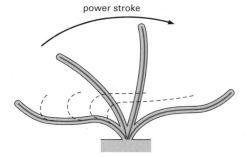

power stroke

Figure 17–25 A cilium beats by performing a repetitive cycle of movements consisting of a power stroke followed by a recovery stroke. In the fast power stroke, the cilium is fully extended and fluid is driven over the surface of the cell; in the slower recovery stroke the cilium curls back into position with minimal disturbance to the surrounding fluid. Each cycle typically requires 0.1–0.2 second and generates a force perpendicular to the axis of the cilium.

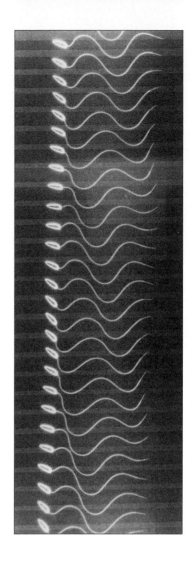

The microtubules in cilia and flagella are slightly different from the cytoplasmic microtubules; they are arranged in a curious and distinctive pattern that was one of the most striking revelations of early electron microscopy. A cross section through a cilium shows nine doublet microtubules arranged in a ring around a pair of single microtubules (Figure 17–27A). This "9 + 2" array is characteristic of almost all forms of eucaryotic cilia and flagella, from those of protozoa to those found in humans.

The movement of a cilium or a flagellum is produced by the bending of its core as the microtubules slide against each other. The microtubules are associated with numerous proteins (Figure 17–27B), which project at regular positions along the length of the microtubule bundle. Some serve as cross-links to hold the bundle of microtubules together, whereas others generate the force that causes the cilium to bend.

The most important of these accessory proteins is the motor protein *ciliary dynein,* which generates the bending motion of the core. It closely resembles cytoplasmic dynein and functions in much the same way. Ciliary dynein is attached by its tail to one microtubule, while its heads interact with an adjacent microtubule to generate a sliding force between the two filaments. Because of the multiple links that hold the adjacent microtubule doublets together, what would be a simple parallel sliding movement between free microtubules is converted to a bending motion in the cilium (Figure 17–28).

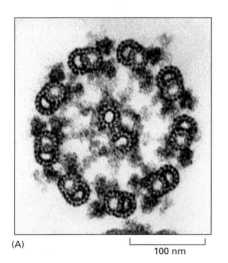

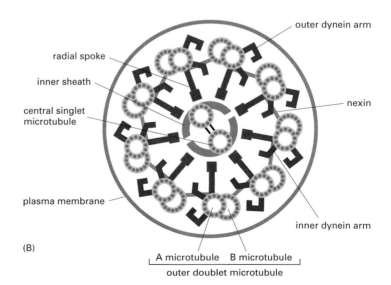

Figure 17–27 Microtubules in a cilium or flagellum are arranged in a "9 + 2" array. (A) Electron micrograph of a flagellum of *Chlamydomonas* shown in cross section, illustrating the distinctive 9 + 2 arrangement of microtubules. (B) Diagram of the flagellum in cross section. The nine outer microtubules (each a special paired structure) carry two rows of dynein molecules. The heads of these dyneins appear in this view like pairs of arms reaching toward the adjacent microtubule. In a living cilium, these dynein heads periodically make contact with the adjacent microtubule and move along it, thereby producing the force for ciliary beating. Various other links and projections shown are proteins that serve to hold the bundle of microtubules together and to convert the sliding motion produced by dyneins into bending, as illustrated in Figure 17–28. (A, courtesy of Lewis Tilney.)

Figure 17–28 The movement of dynein causes the flagellum to bend.
(A) If the outer doublet microtubules and their associated dynein molecules are freed from other components of a sperm flagellum and then exposed to ATP, the doublets slide against each other, telescope-fashion, due to the repetitive action of their associated dyneins.
(B) In an intact flagellum, however, the doublets are tied to each other by flexible protein links so that the action of the system produces bending rather than sliding.

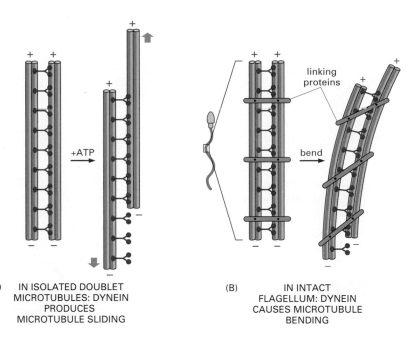

(A) IN ISOLATED DOUBLET MICROTUBULES: DYNEIN PRODUCES MICROTUBULE SLIDING

(B) IN INTACT FLAGELLUM: DYNEIN CAUSES MICROTUBULE BENDING

Question 17–4

Dynein arms in a cilium are arranged so that, when activated, the heads push their neighboring outer doublet outward toward the tip of the cilium. Consider a cross section of a cilium (see Figure 17–27). Why would no bending motion of the cilium result, if all dynein molecules were active at the same time? Suggest a pattern of dynein activity—consistent with the direction of pushing of dynein (see Figure 17–28) and the structure of the cilium—that can account for the bending of a cilium in one direction.

Actin Filaments

Actin filaments are found in all eucaryotic cells and are essential for many of their movements, especially those involving the cell surface. Without actin filaments, for example, an animal cell could not crawl along a surface, engulf a large particle by phagocytosis, or divide in two. Like microtubules, many actin filaments are unstable, but they can also form stable structures in cells, such as the contractile apparatus of muscle. Actin filaments are associated with a large number of *actin-binding proteins* that enable the filaments to serve a variety of functions in cells. Depending on their association with different proteins, actin filaments can form stiff and relatively permanent structures, such as the *microvilli* on the brush-border cells lining the intestine (Figure 17–29A) or small *contractile bundles* in the cytoplasm that can contract and act like the "muscles" of a cell (Figure 17–29B); they can also form temporary structures, such as the protrusions formed at the leading edge of a crawling fibroblast (Figure 17–29C) or the *contractile ring* that pinches the cytoplasm in two when an animal cell divides (Figure 17–29D). In this section, we see how the arrangements of actin filaments in a cell depend on the types of actin-binding proteins present. Even though actin filaments and microtubules are formed from unrelated types of proteins, we shall see that the principles according to which they assemble and disassemble, control cell structure, and bring about movement are strikingly similar.

Figure 17–29 Actin filaments allow eucaryotic cells to adopt a variety of shapes and perform a variety of functions. Various actin-containing structures are shown here in *red*:
(A) microvilli; (B) contractile bundles in the cytoplasm; (C) sheetlike (*lamellipodia*) and fingerlike (*filopodia*) protrusions from the leading edge of a moving cell;
(D) contractile ring during cell division.

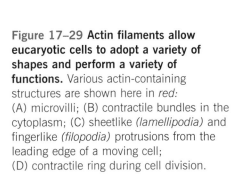
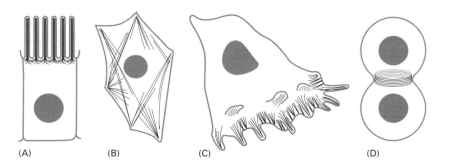

(A)   (B)   (C)   (D)

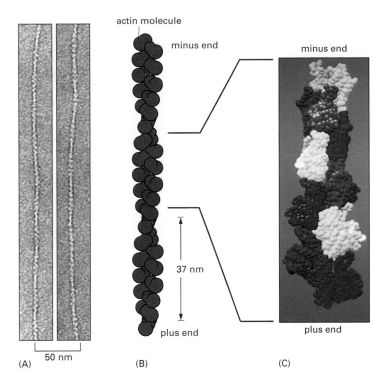

actin molecule

minus end

minus end

37 nm

plus end

plus end

50 nm

(A)   (B)   (C)

Figure 17–30 Actin filaments are thin, flexible protein threads. (A) Electron micrographs of negatively stained actin filaments. (B) Arrangement of actin molecules in an actin filament. Each filament may be thought of as a two-stranded helix with a twist repeating every 37 nm. Strong interactions between the two strands prevent the strands from separating. (C) The identical subunits of an actin filament are depicted in different colors to emphasize the close interaction between each actin molecule and its four nearest neighbors. (A, courtesy of Roger Craig; C, from K.C. Holmes et al., *Nature* 347:44–49, 1990. © Macmillan Magazines Ltd.)

Actin Filaments Are Thin and Flexible

Actin filaments appear in electron micrographs as threads about 7 nm in diameter. Each filament is a twisted chain of identical globular actin molecules, all of which "point" in the same direction along the axis of the chain. Like a microtubule, therefore, an actin filament has a structural polarity, with a plus end and a minus end (Figure 17–30).

Actin filaments are thinner, more flexible, and usually shorter than microtubules. There are, however, many more individual actin filaments in a cell than microtubules; thus the total length of all the actin filaments in a cell is at least 30 times greater than the total length of all of the microtubules. Actin filaments rarely occur in isolation in the cell; they are generally found in cross-linked bundles and networks, which are much stronger than the individual filaments.

Actin and Tubulin Polymerize by Similar Mechanisms

Actin filaments can grow by addition of actin monomers at either end, but the rate of growth is faster at the plus end than at the minus end. A naked actin filament, like a microtubule without associated proteins, is inherently unstable, and it can disassemble from both ends. Each free actin monomer carries a tightly bound nucleoside triphosphate, in this case ATP, which is hydrolyzed to ADP soon after the incorporation of the actin monomer into the filament. As with the GTP of tubulin, hydrolysis of bound ATP to ADP in an actin filament reduces the strength of binding between monomers and decreases the stability of the polymer. Nucleotide hydrolysis thereby promotes depolymerization, helping the cell to disassemble filaments after they have formed (Figure 17–31).

As for microtubules, the ability to assemble and disassemble is required for many of the functions performed by actin filaments, such as their role in cell locomotion. Actin filament function can be perturbed experimentally by certain toxins produced by fungi or marine sea sponges. Some prevent actin polymerization, such as the *cytochalasins;* others stabilize actin filaments against depolymerization, such

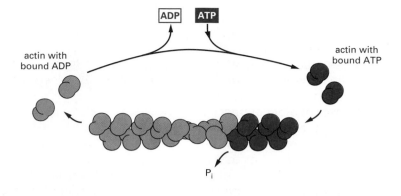

Figure 17–31 ATP hydrolysis decreases the stability of the actin polymer. Actin monomers in the cytosol carry ATP, which is hydrolyzed to ADP soon after assembly into a growing filament. The ADP molecules remain trapped within the actin filament, unable to exchange with ATP until the actin monomer that carries them dissociates from the filament.

Question 17–5

The formation of actin filaments in the cytosol is controlled by actin-binding proteins. Some actin-binding proteins significantly increase the rate at which formation of an actin filament is initiated. Suggest a mechanism by which they might do this.

as the *jasplakinolides.* Addition of these toxins to the medium bathing cells or tissues, even in low concentrations, instantaneously freezes cell movements such as the crawling motion of a fibroblast. Thus, the function of actin filaments depends on a dynamic equilibrium between the actin filaments and the pool of actin monomers, and many filaments persist for only a few minutes after they are formed.

Many Proteins Bind to Actin and Modify Its Properties

About 5% of the total protein in a typical animal cell is actin; about half of this actin is assembled into filaments, while the other half remains as actin monomers in the cytosol. The concentration of monomer is therefore high—much higher than the concentration required for purified actin monomers to polymerize *in vitro.* What, then, keeps the actin monomers in cells from polymerizing totally into filaments? The answer is that cells contain small proteins, such as *thymosin* and *profilin,* that bind to actin monomers in the cytosol, preventing them from adding to the ends of actin filaments. By keeping actin monomers in reserve until they are required, thymosin in particular plays a crucial role in regulating actin polymerization.

There are a great many other actin-binding proteins in cells. Most of these bind to assembled actin filaments rather than to actin monomers, and control the behavior of the intact filaments (Figure 17–32). Actin-bundling proteins, for example, hold actin filaments together in parallel bundles in microvilli; other cross-linking proteins hold actin filaments together in a gel-like meshwork within the *cell cortex*—the layer of cytoplasm just beneath the plasma membrane; filament-severing proteins, like *gelsolin,* fragment actin filaments into shorter lengths and thus can convert an actin gel to a more fluid state. Actin filaments can also be associated with motor proteins to form contractile bundles, as in muscle cells. And they often form tracks along which motor proteins transport organelles, a function that is especially conspicuous in plant cells.

In the remainder of this chapter, we consider some characteristic structures that actin filaments can form, and discuss how different types of actin-binding proteins are involved in their formation. We begin with the actin-rich *cell cortex* and its role in cell locomotion, and in the final section we consider the contractile apparatus of muscle cells as an example of a stable structure based on actin filaments.

An Actin-rich Cortex Underlies the Plasma Membrane of Most Eucaryotic Cells

Although actin is found throughout the cytoplasm of a eucaryotic cell, in most cells it is concentrated in a layer just beneath the plasma membrane. In this region, called the **cell cortex**, actin filaments are linked by actin-binding proteins into a meshwork that supports the outer surface of the cell and gives it mechanical strength. In red blood cells, as described in Chapter 11, a simple and regular network of fibrous proteins attached to the plasma membrane provides it with support necessary to maintain its simple discoid shape (see Figure 11–30). The cell cortex of other animal cells, however, is thicker and more complex and supports a far richer repertoire of shapes and movements. Like the red cell, it contains spectrin and ankyrin; but it includes a dense network of actin filaments that project into the cytoplasm, where they become cross-linked into a three-dimensional meshwork. This cortical actin mesh governs the shape and mechanical properties of the plasma membrane and the cell surface. As we shall see, actin rearrangements within the cortex provide the molecular basis for changes in cell shape and for cell locomotion.

Cell Crawling Depends on Actin

Many cells move by crawling over surfaces, rather than by swimming by means of cilia or flagella. Carnivorous amoebae crawl continually, in search of food. White blood cells known as *neutrophils* migrate from the

Figure 17–32 Actin-binding proteins control the behavior of actin filaments in vertebrate cells. Actin is shown in *red,* and the actin-binding proteins are shown in *green.*

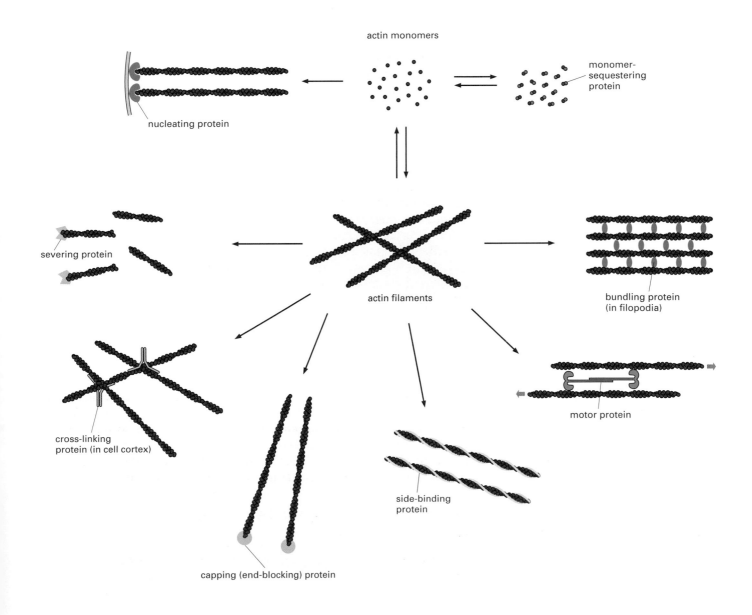

bloodstream into tissues when they "smell" small diffusing molecules released by bacteria, which the neutrophils eventually engulf and destroy. The advancing tip of a developing axon migrates in response to growth factors, following a path of substrate-bound and diffusible chemicals to its eventual synaptic target.

The molecular mechanisms of these and other forms of cell crawling are difficult to dissect. They entail coordinated changes of many molecules in different regions of the cell, and no single, easily identifiable locomotory organelle, such as a flagellum, is responsible. In broad terms, however, three interrelated processes are known to be essential: (1) the cell pushes out protrusions at its "front," or leading edge; (2) these protrusions adhere to the surface over which the cell is crawling; and (3) the rest of the cell drags itself forward by traction on these anchorage points (Figure 17–33).

All three processes involve actin, but in different ways. The first step, the pushing forward of the cell surface, is driven by actin polymerization. The leading edge of a crawling fibroblast in culture regularly extends thin, sheetlike **lamellipodia** (Figure 17–34), which contain a dense meshwork of actin filaments, oriented so that most of the filaments have their plus ends close to the plasma membrane. Many cells also extend thin, stiff protrusions called **filopodia**, both at the leading edge and elsewhere on their surface. These are about 0.1 μm wide and 5–10 μm long, and each contains a loose bundle of 10–20 actin filaments, again oriented with their plus ends pointing outward. The advancing tip (growth cone) of a developing nerve cell axon extends even longer filopodia, up to 50 μm long, which help it to probe its environment and find the correct path to its target. Both lamellipodia and filopodia are exploratory, motile structures that form and retract with great speed, moving at around 1 μm per second. Both are thought to be generated by rapid local growth of actin filaments, which are nucleated at the plasma membrane and push out the membrane without tearing it as they elongate.

The formation and growth of actin filaments at the leading edge of a cell is assisted by various actin-binding accessory proteins that help to nucleate actin polymers at the plasma membrane. One set of proteins, the *actin-related proteins,* or *ARP*s, promote the formation of branched

Figure 17–33 Forces generated in the actin-rich cortex move a cell forward. In this model, actin polymerization at the leading edge of the cell *pushes* the plasma membrane forward and forms new regions of actin cortex, shown here in *red*. New points of anchorage are made between the actin filaments and the surface on which the cell is crawling (the substratum). Contraction at the rear of the cell then draws the body of the cell forward. New anchorage points are established at the front, and old ones are released at the back as the cell crawls forward. The same cycle is repeated over and over again, moving the cell forward in a stepwise fashion.

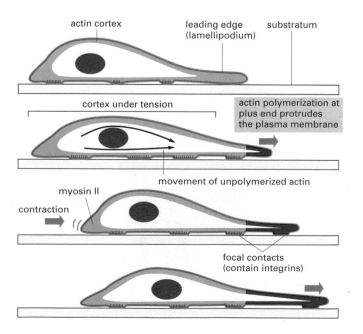

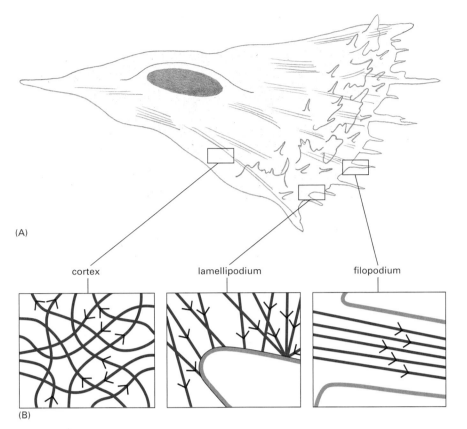

(A)

cortex   lamellipodium   filopodium

(B)

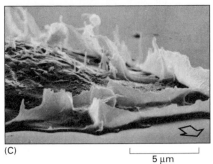

(C)

5 μm

Figure 17–34 Actin filaments allow animal cells to migrate.
(A) Schematic drawing of a fibroblast showing flattened lamellipodia and fine filopodia projecting from its surface, especially in the regions of the leading edge. (B) Details of the arrangement of actin filaments in three regions of the fibroblast are shown, with arrowheads pointing toward the plus end of the filaments. (C) Scanning electron micrograph showing lamellipodia and filopodia at the leading edge of a human fibroblast migrating in culture. An arrow shows the direction of migration. (C, courtesy of Julian Heath.)

actin filaments. These proteins form a complex that binds to the existing actin filaments, where they nucleate the growth of a new filament, which grows out at an angle to form a side branch (Figure 17–35). Thus, these complexes promote the formation of a two-dimensional, treelike web of actin. With the aid of additional actin-binding proteins, this web undergoes assembly at the front end and disassembly at the back, pushing the lamellipodia forward (Figure 17–36).

When the lamellipodia and filopodia touch down on a favorable patch of surface, they stick: transmembrane proteins in their plasma membrane, known as *integrins,* adhere to molecules in the extracellular matrix or on the surface of another cell over which the moving cell is crawling. Meanwhile, on the intracellular face of the membrane of the crawling cell, integrins capture actin filaments, thereby creating a robust anchorage for the system of actin filaments inside the crawling

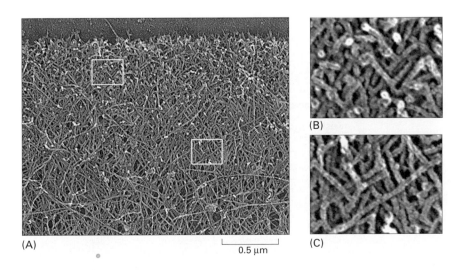

(A)   0.5 μm   (C)

(B)

Figure 17–35 Actin filament organization in a lamellipodium. Highly motile keratocytes from frog skin were fixed, dried, shadowed with platinum, and examined in an electron microscope. (A) Actin filaments form a dense network with filament fast-growing ends terminating at the leading margin of the lamellipodium (top of figure). (B and C) Boxed regions from (A) are enlarged to show Y-junctions between individual filaments. (Courtesy of Tatyana Svitkina and Gary Borisy.)

Figure 17–36 Assembly of an actin meshwork pushes forward the leading edge of a lamellipodium. Nucleation of new actin filaments (red) is mediated by ARP complexes (orange) at the front of the web. Newly formed filaments are thereby attached to the sides of preexisting filaments, as seen in Figure 17–35. As these filaments elongate, they push the plasma membrane forward. The actin filament plus ends will become protected by capping proteins (blue), preventing further assembly or disassembly from the old plus ends at the front of the array. Hydrolysis of ATP bound to the polymerized actin subunits promotes depolymerization at the rear end of the actin complex by a depolymerizing protein (green). The spatial separation of assembly and disassembly allows the network as a whole to move forward at a steady rate.

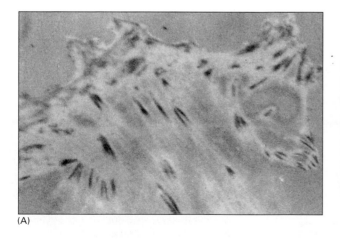

cell leading edge

newborn actin filament

ARP complex — depolymerizing protein — actin monomer — capping protein

Figure 17–37 Actin filaments attach to points where the crawling cell is anchored to the extracellular matrix. (A) Contacts made between a living fibroblast and a glass slide are revealed by reflection-interference microscopy. In this technique, light is reflected from the lower surface of a cell attached to a glass slide, and the focal contacts appear as dark patches. (B) Staining the same cell (after fixation) with fluorescent antibodies to actin shows that most of the cell's actin filament bundles terminate at, or close to, these sites of contact. (Courtesy of Grenham Ireland.)

cell (Figure 17–37). To use this anchorage to drag its body forward, the cell now makes use of internal contractions to exert a pulling force (see Figure 17–33). These too depend on actin, but in a different way—through the interaction of actin filaments with motor proteins known as *myosins.*

It is still not certain how this pulling force is produced: contraction of bundles of actin filaments in the cytoplasm or contraction of the actin meshwork in the cell cortex, or both, may be responsible. The general principles of how myosin motor proteins interact with actin filaments to cause movement, however, is clear, as we now discuss.

Actin Associates with Myosin to Form Contractile Structures

All actin-dependent motor proteins belong to the **myosin** family. They bind to and hydrolyze ATP, which provides the energy for their movement

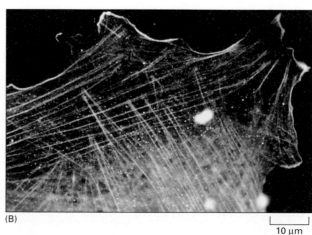

(A)

(B)

10 μm

along actin filaments from the minus end of the filament toward the plus end. Myosin, along with actin, was first discovered in skeletal muscle, and much of what we know about the interaction of these two proteins was learned from studies of muscle. There are several different types of myosins in cells, of which the *myosin-I* and *myosin-II* subfamilies are most abundant. Myosin-I is found in all types of cells, and as it is simpler in structure and mechanism of action, we shall discuss it first.

Myosin-I molecules have only one head domain and a tail (Figure 17–38A). The head domain interacts with actin filaments and has ATP-hydrolyzing motor activity that enables it to move along the filament in a cycle of binding, detachment, and rebinding. The tail varies among the different types of myosin-I, and it determines what cell components will be dragged along by the motor. For example, the tail may bind to a particular type of membrane vesicle and propel it through the cell along actin filament tracks (Figure 17–38B), or it may bind to the plasma membrane and move it relative to cortical actin filaments, thus pulling the membrane into a different shape (Figure 17–38C).

Extracellular Signals Control the Arrangement of Actin Filaments

In the preceding sections we have discussed how actin-binding proteins can regulate the length, location, organization, and dynamic behavior of actin filaments. The activity of these accessory proteins, in turn, can be regulated by extracellular signals, allowing the cell to rearrange its cytoskeleton in response to the environment.

For the actin cytoskeleton, such structural rearrangements are triggered by activation of a variety of receptor proteins embedded in the plasma membrane. All of these signals then seem to converge inside the cell on a group of closely related monomeric GTP-binding proteins called the **Rho protein family**. As we saw in Chapter 16, proteins of this kind behave as molecular switches that control cellular processes by cycling between an active, GTP-bound state and an inactive, GDP-bound state (see Figure 16–15B). In the case of the cytoskeleton, activation of different members of the Rho family affects the organization of

Question 17–6

Suppose that the actin molecules in a cultured skin cell have been randomly labeled in such a way that 1 in 10,000 molecules carries a fluorescent marker. What would you expect to see if you examined the lamellipodium (leading edge) of this cell through a fluorescence microscope? Assume that your microscope is sensitive enough to detect single fluorescent molecules.

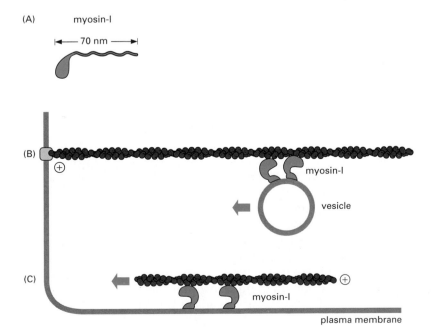

(A) myosin-I

|← 70 nm →|

(B)

myosin-I

vesicle

(C)

myosin-I

plasma membrane

Figure 17–38 The short tail of a myosin-I molecule contains sites that bind to various components of the cell, including membranes. (A) Myosin-I has a single globular head and a tail that attaches to another molecule or organelle in the cell. This arrangement allows the head domain to move a vesicle relative to an actin filament (B), or an actin filament and the plasma membrane relative to each other (C). Note that the head group of the myosin always walks toward the plus end of the actin filament it contacts.

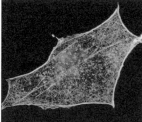
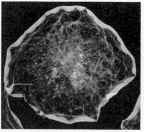
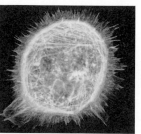

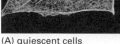

(A) quiescent cells   (B) Rac activation   (C) Cdc42 activation

Figure 17–39 Activation of GTP-binding proteins has a dramatic effect on the organization of actin filaments in fibroblasts. In these micrographs, actin has been stained with fluorescently labeled phalloidin, a molecule that binds to actin filaments. (A) Unstimulated fibroblasts have actin filaments primarily in the cortex. (B) Microinjection of an activated form of the small, monomeric GTP-binding protein Rac causes the formation of an enormous lamellipodium that extends from the entire circumference of the cell. (C) Microinjection of an activated form of the related protein, Cdc42, causes the protrusion of many long filopodia at the cell periphery. (From A. Hall, *Science* 279:509–514, 1988. © AAAS.)

Question 17–7

At the leading edge of a crawling cell, the plus ends of actin filaments are located close to the plasma membrane and actin monomers are added at these ends, pushing the membrane outward to form lamellipodia or filopodia. What do you suppose holds the filaments at their other ends to prevent them from just being pushed into the cell's interior?

actin filaments in different ways. For example, activation of a GTP-binding protein called Cdc42 triggers actin polymerization and bundling to form filopodia, whereas activation of the Rac GTP-binding protein promotes the formation of sheetlike lamellipodia and membrane ruffles (Figure 17–39).

These dramatic and complex structural changes occur because each of these molecular switches interacts with numerous target proteins, including protein kinases and the accessory proteins that control actin organization and dynamics. Activation of Cdc42, for example, enhances the actin-nucleating activities of ARP complexes. Rac also enhances actin nucleation at the ARP complex, but in addition promotes the uncapping of the plus ends of actin filaments. This uncapping provides additional sites for actin assembly at the plasma membrane, which promotes the formation of large lamellipodia.

One of the most tightly regulated rearrangements of cytoskeletal elements occurs when actin associates with myosin in muscle fibers in response to signals from the nervous system. We now discuss how this molecular interaction can generate the rapid, repetitive, forceful movements characteristic of the contraction of vertebrate muscles.

Muscle Contraction

Muscle contraction is the most familiar and the best understood of animal cell movements. In vertebrates, running, walking, swimming, and flying all depend on the ability of *skeletal muscle* to contract strongly and move various bones. Involuntary movements such as heart pumping and gut peristalsis depend on *cardiac muscle* and *smooth muscle,* respectively, which are formed from muscle cells that differ in structure from skeletal muscle, but which use actin and myosin in a similar way to contract. Although muscle cells are highly specialized, many cell movements—from the locomotion of whole cells down to the motion of components inside cells—depend on the interaction of actin and myosin. Much of our understanding of the mechanisms of cell movement comes from studies of muscle cell contraction. In this section, we review how actin and myosin interact to create motion.

Muscle Contraction Depends on Bundles of Actin and Myosin

Muscle myosin belongs to the myosin-II subfamily of myosins, all of which have two ATPase heads and a long, rodlike tail (Figure 17–40A). Myosin-II forms contractile structures with actin filaments; these are best known and most abundant in muscle, but they also occur in many other types of animal cells. Each myosin-II molecule is a dimer composed of a pair of identical myosin molecules held together by their tails; it has two globular ATPase heads at one end and a single coiled-coil tail at the other. Clusters of myosin-II molecules bind to each other

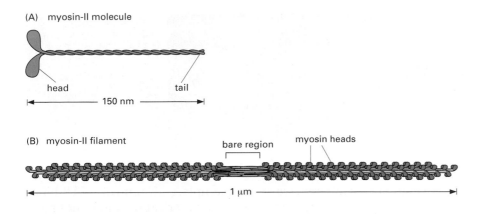

(A) myosin-II molecule

head   tail

|← 150 nm →|

(B) myosin-II filament

bare region   myosin heads

|← 1 μm →|

Figure 17–40 Myosin-II molecules can associate with one another to form myosin filaments. (A) A molecule of myosin-II has two globular heads and a coiled-coil tail. (B) The tails of myosin-II associate with one another to form a bipolar myosin filament in which the heads project outward from the middle in opposite directions. The bare region in the middle of the filament consists of tails only.

through their coiled-coil tails, forming a bipolar *myosin filament* in which the heads project from the sides (Figure 17–40B).

The myosin filament is like a double-headed arrow, with the two sets of heads pointing in opposite directions away from the middle. One set of heads binds to actin filaments in one orientation and moves them one way; the other set of heads binds to other actin filaments in the opposite orientation and moves them in the opposite direction (Figure 17–41). The overall effect is to slide sets of oppositely oriented actin filaments past one another. We can see how, therefore, if actin filaments and myosin filaments are organized together in a bundle, the bundle can generate a contractile force. This is seen most clearly in muscle contraction, but it also occurs in the *contractile bundles* of actin filaments and myosin-II filaments (see Figure 17–29B) that assemble transiently in nonmuscle cells, and in the *contractile ring* that pinches a dividing cell in two by contracting and pulling inward on the plasma membrane (discussed in Chapter 19).

During Muscle Contraction Actin Filaments Slide Against Myosin Filaments

The long fibers of skeletal muscle are huge single cells formed by the fusion of many separate smaller cells. The individual nuclei of the contributing cells are retained in the muscle fiber and lie just beneath the plasma membrane. The bulk of the cytoplasm is made up of **myofibrils**, the contractile elements of the muscle cell. These cylindrical structures are 1–2 μm in diameter and may be as long as the muscle cell itself (Figure 17–42A).

A myofibril consists of a chain of identical tiny contractile units, or **sarcomeres**. Each sarcomere is about 2.5 μm long, and the repeating pattern of sarcomeres gives the vertebrate myofibril a striped, or striated, appearance (Figure 17–42B). Sarcomeres are highly organized assemblies of two types of filaments—actin filaments and filaments of muscle-specific myosin-II. Myosin filaments (the *thick filaments*) are centrally positioned in each sarcomere, whereas the more slender actin

Question 17–8

If both the thick and thin filaments of muscle are made up of subunits held together by weak noncovalent bonds, how is it possible for a human being to lift heavy objects?

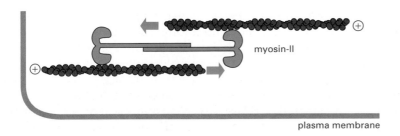

myosin-II

plasma membrane

Figure 17–41 Even small bipolar filaments composed of myosin-II molecules can slide actin filaments over each other, thus mediating local *shortening* of an actin filament bundle. As with myosin-I, the head group of myosin-II walks toward the plus end of the actin filament it contacts.

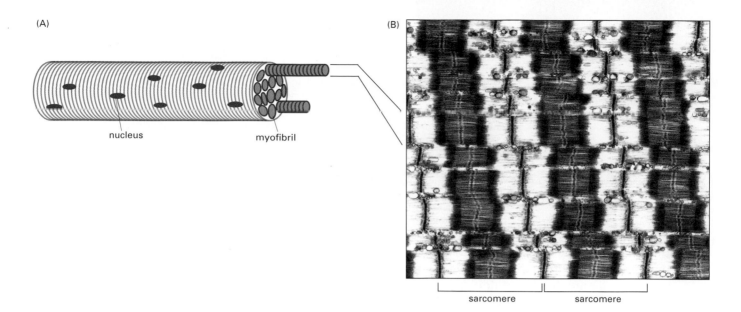

(A)

nucleus   myofibril

(B)

sarcomere   sarcomere

Figure 17–42 A skeletal muscle cell is packed with myofibrils, each of which consists of a repeating chain of sarcomeres. (A) In an adult human these huge multinucleated cells (also called muscle fibers) are typically 50 μm in diameter, and they can be several centimeters long. They contain numerous myofibrils in which actin filaments and myosin filaments are arranged in a highly organized structure, giving the myofibril a striated or striped appearance.
(B) Low-magnification electron micrograph of a longitudinal section through a skeletal muscle cell of a rabbit showing the regular organization of sarcomeres, the contractile units of the myofibrils. (B, courtesy of Roger Craig.)

Z disc

myofibril

(A)

sarcomere ~2.2 μm

Z disc   Z disc
thick filament (myosin) ——
thin filament (actin) ——

(B)

filaments (the *thin filaments)* extend inward from each end of the sarcomere (where they are anchored by their plus ends to a structure known as the *Z disc)* and overlap the ends of the myosin filaments (Figure 17–43).

The contraction of a muscle cell is caused by a simultaneous shortening of all the sarcomeres, which in turn is caused by the actin filaments sliding past the myosin filaments, with no change in the length of either type of filament (Figure 17–44). The sliding motion is generated by myosin heads that project from the sides of the myosin filament and interact with adjacent actin filaments. When a muscle is stimulated to contract, the myosin heads start to walk along the actin filament in repeated cycles of attachment and detachment. During each cycle, a myosin head binds and hydrolyzes one molecule of ATP. This is thought to cause a series of conformational changes in the myosin molecule that move the tip of the head by about 5 nm along the actin filament toward the plus end. This movement, repeated with each round of ATP hydrolysis, propels the myosin molecule unidirectionally along the actin filament (Figure 17–45). In so doing, the myosin heads pull against the actin filament, causing it to slide against the myosin filament. The concerted action of many myosin heads pulling the actin and myosin filaments past each other causes the sarcomere to contract. After a contraction is completed, the myosin heads lose contact with the actin filaments completely, and the muscle relaxes.

Each myosin filament has about 300 myosin heads. Each myosin head can attach and detach from actin about five times per second, allowing the myosin and actin filaments to slide past one another at speeds of up to 15 μm per second. This speed is sufficient to take a sarcomere from a fully extended state (3 μm) to a fully contracted state (2 μm) in less than a tenth of a second. All of the sarcomeres of a muscle

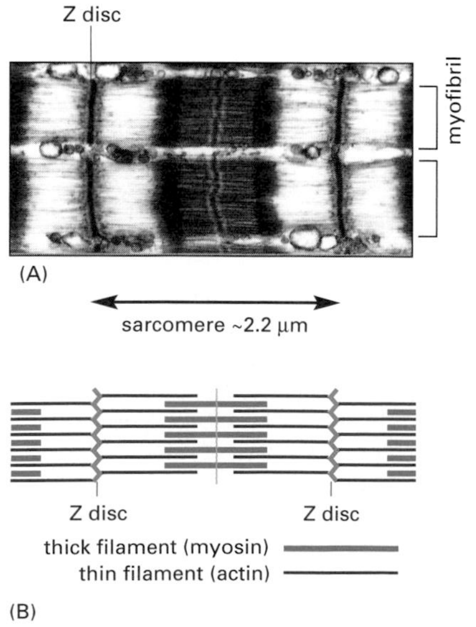

Figure 17–43 Sarcomeres are the contractile units of muscle.
(A) Detail of the skeletal muscle cell shown in Figure 17–42 showing two myofibrils, with the extent of one sarcomere marked.
(B) Schematic diagram of a single sarcomere showing the origin of the light and dark bands seen in the microscope. Z discs at either end of the sarcomere are attachment points for actin filaments; the centrally located thick filaments *(green)* are each composed of many myosin-II molecules. (A, courtesy of Roger Craig.)

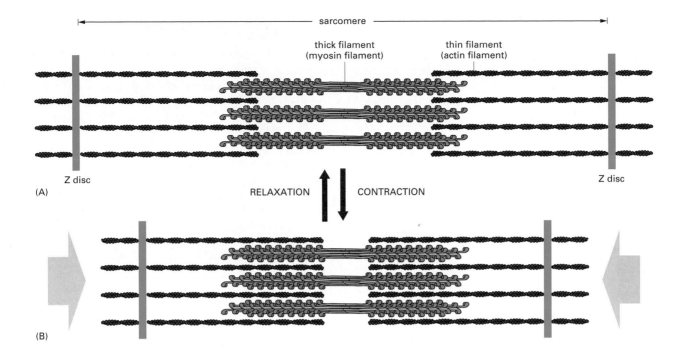

sarcomere

thick filament
(myosin filament)

thin filament
(actin filament)

Z disc

(A)

RELAXATION   CONTRACTION

(B)

Z disc

are coupled together and are triggered almost instantaneously by the system of signals described in the following section. Therefore, the entire muscle contracts extremely rapidly, usually within a tenth of a second.

Muscle Contraction Is Triggered by a Sudden Rise in Ca^{2+}

The force-generating molecular interaction between myosin and actin filaments takes place only when the skeletal muscle receives a signal from the nervous system. The signal from a nerve terminal triggers an action potential (discussed in Chapter 12) in the muscle cell plasma membrane. This electrical excitation spreads in a matter of milliseconds into a series of membranous tubes, called *transverse* (or *T*) *tubules*, that extend inward from the plasma membrane around each myofibril. The electrical signal is then relayed to the *sarcoplasmic reticulum*, an adjacent sheath of interconnected flattened vesicles that surrounds each myofibril like a net stocking (Figure 17–46).

The sarcoplasmic reticulum is a specialized region of the endoplasmic reticulum in muscle cells. It contains a very high concentration of Ca^{2+}, and in response to the incoming electrical excitation, much of this Ca^{2+} is released into the cytosol through ion channels that open in the sarcoplasmic reticulum membrane in response to the change in voltage across the plasma membrane (Figure 17–47). As discussed in Chapter 16, Ca^{2+} is widely used as an intracellular signal to relay a message from the exterior to the internal machinery of the cell. In the case of muscle, the Ca^{2+} interacts with a molecular switch made of specialized accessory proteins closely associated with the actin filaments (Figure 17–48A). One of these proteins is *tropomyosin*, a rigid, rod-shaped molecule that binds in the groove of the actin helix, overlapping seven actin monomers, and prevents the myosin heads from associating with the actin filament. The other is *troponin*, a protein complex that includes a Ca^{2+}-sensitive protein (*troponin-C*), which is associated with the end of a tropomyosin molecule. When the level of Ca^{2+} rises in the cytosol, Ca^{2+} binds to troponin and induces a change in its shape. This in turn causes the tropomyosin molecules to shift their position

Figure 17–44 Muscles contract by a sliding-filament mechanism. (A) The myosin and actin filaments of a sarcomere overlap with the same relative polarity on either side of the midline. Recall that actin filaments are anchored by their plus ends to the Z disc and that myosin filaments are bipolar. (B) During contraction, the actin and myosin filaments slide past each other without shortening. The sliding motion is driven by the myosin heads walking toward the plus end of the adjacent actin filament.

Question 17–9

Compare the structure of intermediate filaments with that of the myosin-II filaments in skeletal muscle cells. What are the major similarities? What are the major differences? How do the differences in structure relate to their function?

slightly, allowing myosin heads to bind to the actin filament and initiating contraction (Figure 17–48B).

Because the signal from the plasma membrane is passed within milliseconds (via the transverse tubules and sarcoplasmic reticulum) to every sarcomere in the cell, all the myofibrils in the cell contract at the same time. The increase in Ca^{2+} in the cytosol ceases as soon as the nerve signal stops because the Ca^{2+} is rapidly pumped back into the sarcoplasmic reticulum by abundant Ca^{2+} pumps in its membrane (discussed in Chapter 12). As soon as Ca^{2+} concentrations have returned to their resting level, troponin and tropomyosin molecules move back to their original positions, where they block myosin binding and thus end contraction.

Figure 17–45 A myosin molecule walks along an actin filament through a cycle of structural changes. (Based on I. Rayment et al., *Science* 261:50–58, 1993. © AAAS.)

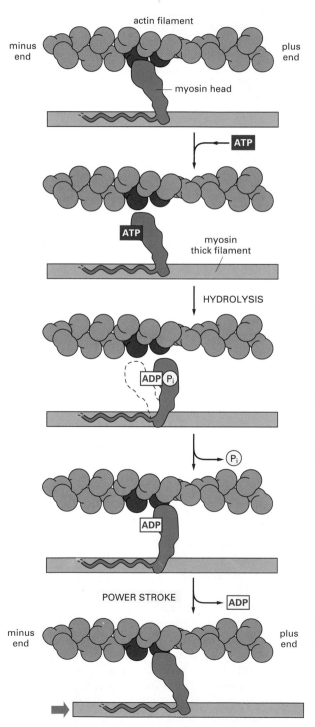

ATTACHED At the start of the cycle shown in this figure, a myosin head lacking a bound nucleotide is locked tightly onto an actin filament in a *rigor* configuration (so named because it is responsible for *rigor mortis*, the rigidity of death). In an actively contracting muscle, this state is very short-lived, being rapidly terminated by the binding of a molecule of ATP.

RELEASED A molecule of ATP binds to the large cleft on the "back" of the head (that is, on the side furthest from the actin filament) and immediately causes a slight change in the conformation of the domains that make up the actin-binding site. This reduces the affinity of the head for actin and allows it to move along the filament. (The space drawn here between the head and actin emphasizes this change, although in reality the head probably remains very close to the actin.)

COCKED The cleft closes like a clam shell around the ATP molecule, triggering a large shape change that causes the head to be displaced along the filament by a distance of about 5 nm. Hydrolysis of ATP occurs, but the ADP and inorganic phosphate (P_i) produced remain tightly bound to the protein.

FORCE-GENERATING A weak binding of the myosin head to a new site on the actin filament causes release of the inorganic phosphate produced by ATP hydrolysis, concomitantly with the tight binding of the head to actin. This release triggers the power stroke—the force-generating change in shape during which the head regains its original conformation. In the course of the power stroke, the head loses its bound ADP, thereby returning to the start of a new cycle.

ATTACHED At the end of the cycle, the myosin head is again locked tightly to the actin filament in a rigor configuration. Note that the head has moved to a new position on the actin filament.

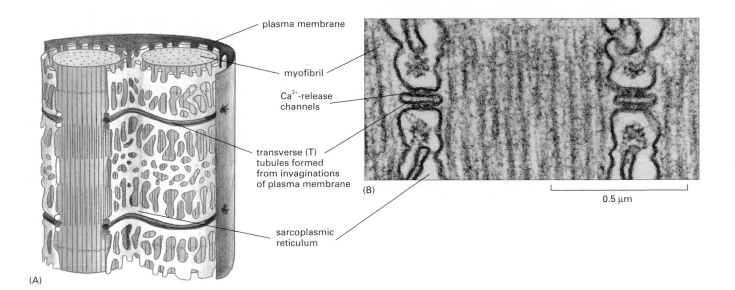

Figure 17–46 **T tubules and sarcoplasmic reticulum surround the myofibrils.** (A) Drawing of the two membrane systems that relay the signal to contract from the muscle cell plasma membrane to all of the myofibrils in the cell. (B) Electron micrograph showing a cross section of two T tubules and their adjacent sarcoplasmic reticulum compartments. (B, courtesy of Clara Franzini-Armstrong.)

Muscle Cells Perform Highly Specialized Functions in the Body

The highly specialized contractile machinery in muscle cells is thought to have evolved from the simpler contractile bundles of myosin and actin filaments found in all eucaryotic cells. The myosin-II in nonmuscle cells is also activated by a rise in cytosolic Ca^{2+}, but the mechanism of activation is quite different. An increase in Ca^{2+} leads to the phosphorylation of myosin-II, which alters the myosin conformation and enables it to interact with actin. A similar activation mechanism operates in *smooth muscle,* which lies in the walls of the stomach, intestine, uterus, and arteries, and in many other structures in which slow and sustained contractions are needed. Contractions produced by this second mode are slower because time is needed for enzyme molecules to diffuse to the myosin heads and to carry out the phosphorylation or dephosphorylation. However, this mechanism has the advantage that it is less specialized and can be driven by a variety of incoming signals: thus smooth muscle, for example, is triggered to contract by adrenaline, serotonin, prostaglandins, and a variety of other extracellular signals.

In addition to skeletal and smooth muscle, other forms of muscle each perform a specific mechanical function in the body. Perhaps the most familiar is the heart, or *cardiac*, muscle that drives the circulation of blood. This remarkable organ contracts autonomously for the lifetime

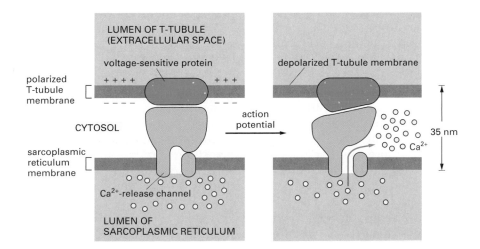

Figure 17–47 **A Ca^{2+} release channel in the sarcoplasmic reticulum membrane is thought to be opened by a voltage-sensitive transmembrane protein in the adjacent T tubule.**

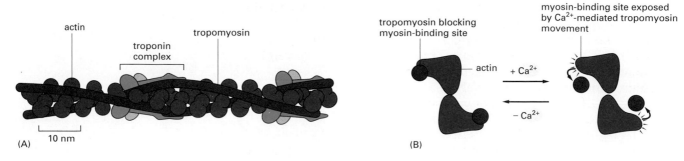

(A)

10 nm

(B)

Figure 17–48 Skeletal muscle contraction is controlled by troponin. (A) A muscle thin filament showing the positions of tropomyosin and troponin along the actin filament. Every tropomyosin molecule has seven evenly spaced regions of homologous sequence, each of which is thought to bind to an actin subunit in the filament. (B) When Ca^{2+} binds to troponin, the troponin moves the tropomyosin that otherwise blocks the interaction of actin with the myosin heads. In this diagram, the thin filament is shown end-on.

Question 17–10

A. Note that in Figure 17–48 troponin molecules are evenly spaced along an actin filament with one troponin found every seventh actin molecule. How do you suppose troponin molecules can be positioned this regularly? What does this tell you about the binding of troponin to actin filaments?

B. What do you suppose would happen if you mixed actin filaments with either (i) troponin alone, (ii) tropomyosin alone, or (iii) troponin plus tropomyosin, and then added myosin? Would the effects be dependent on Ca^{2+}?

of the organism—some 3 billion (3×10^9) times in a human. Even subtle changes in the actin and myosin of heart muscle can lead to serious heart disease.

The contraction of muscle cells represents a highly specialized use of the basic components of the eucaryotic cytoskeleton. In the following chapters, we see how the cytoskeleton participates in perhaps the most fundamental cell movement of all, the formation of two daughter cells during the process of cell division.

Essential Concepts

- The cytoplasm of a eucaryotic cell is supported and spatially organized by a cytoskeleton of intermediate filaments, microtubules, and actin filaments.
- Intermediate filaments are stable, ropelike polymers of fibrous proteins that give cells mechanical strength. Some types underlie the nuclear membrane to form the nuclear lamina; others are distributed throughout the cytoplasm.
- Microtubules are stiff, hollow tubes formed by polymerization of tubulin dimer subunits. They are polarized structures with a slow-growing "minus" end and a fast-growing "plus" end.
- Microtubules are nucleated in, and grow out from, organizing centers such as the centrosome. The minus ends of the microtubules are embedded in the organizing center.
- Many of the microtubules in a cell are in a labile, dynamic state in which they alternate between a growing state and a shrinking state. These transitions, known as dynamic instability, are controlled by the hydrolysis of GTP bound to tubulin dimers.
- Each tubulin dimer has a tightly bound GTP molecule that is hydrolyzed to GDP after the tubulin assembles into a microtubule. GTP hydrolysis reduces the affinity of the subunit for its neighbors and decreases the stability of the polymer, causing it to disassemble.
- Microtubules can be stabilized by proteins that capture the plus end—a process that influences the position of microtubule arrays in a cell. Cells contain many microtubule-associated proteins that stabilize microtubules, bind them to other cell components, and harness them for specific functions.
- Kinesins and dyneins are motor proteins that use the energy of ATP hydrolysis to move unidirectionally along microtubules. They carry

specific membrane vesicles and other cargoes and in this way maintain the spatial organization of the cytoplasm.

- Eucaryotic cilia and flagella contain a bundle of stable microtubules. Their beating is caused by bending of the microtubules, driven by a motor protein called ciliary dynein.
- Actin filaments are helical polymers of actin molecules. They are more flexible than microtubules and are often found in bundles or networks associated with the plasma membrane.
- Actin filaments are polarized structures with a fast- and a slow-growing end, and their assembly and disassembly are controlled by the hydrolysis of ATP tightly bound to each actin monomer.
- The varied forms and functions of actin filaments in cells depend on multiple actin-binding proteins. These control the polymerization of actin filaments, cross-link the filaments into loose networks or stiff bundles, attach them to membranes, or move them relative to one another.
- A network of actin filaments underneath the plasma membrane forms the cell cortex and is responsible for the shape and movement of the cell surface, including the movements involved when a cell crawls along a surface.
- A variety of actin-binding proteins are required to drive the leading edge of a migrating cell forward, to adhere to the substratum, and to retract its trailing edge. All of these processes are triggered by external stimuli working via small GTP-binding proteins.
- Myosins are motor proteins that use the energy of ATP hydrolysis to move along actin filaments: they can carry organelles along actin-filament tracks or cause adjacent actin filaments to slide past each other in contractile bundles.
- In muscle, huge arrays of overlapping actin filaments and myosin filaments generate contractions by sliding over one another.
- Muscle contraction is initiated by a sudden rise in cytosolic Ca^{2+}, which delivers a signal to the contractile apparatus via Ca^{2+}-binding proteins.

Key Terms

actin filament	kinesin
cell cortex	lamellipodium
centriole	microtubule
centrosome	motor protein
cilium	myofibril
cytoskeleton	myosin
dynamic instability	nuclear lamina
dynein	polarity
filopodium	Rho protein family
flagellum	sarcomere
intermediate filament	tubulin

Questions

Question 17–11

Which of the following statements are correct? Explain your answers.

  A. Kinesin moves endoplasmic-reticulum membranes along microtubules so that the network of ER tubules becomes stretched throughout the cell.
  B. Without actin, cells can form a functional mitotic spindle and pull their chromosomes apart but cannot divide.
  C. Lamellipodia and filopodia are "feelers" that a cell extends to find anchor points on the substratum that it will then crawl over.
  D. GTP is hydrolyzed by tubulin to cause the bending of flagella.
  E. Cells having an intermediate-filament network that cannot be depolymerized would die.
  F. The plus ends of microtubules grow faster because they have a larger GTP cap.
  G. The transverse tubules in muscle cells are an extension of the plasma membrane, with which they are continuous, and likewise, the sarcoplasmic reticulum is an extension of the endoplasmic reticulum.
  H. Activation of myosin movement on actin filaments is triggered by phosphorylation of troponin in some situations and by Ca^{2+} binding to troponin in others.

Question 17–12

The average time taken for a molecule or an organelle to diffuse a distance of x cm is given by the formula

$$t = x^2/2D$$

where t is the time in seconds and D is a constant called the diffusion coefficient for the molecule or particle. Using the above formula, calculate the time it would take for a small molecule, a protein, and a membrane vesicle to diffuse from one side to another of a cell 10 μm across. Typical diffusion coefficients in units of cm^2/sec are small molecule, 5×10^{-6}; protein molecule, 5×10^{-7}; vesicle, 5×10^{-8}. How long would a membrane vesicle take to reach the end of an axon 10 cm long by free diffusion?

Question 17–13

Why do eucaryotic cells, and especially animal cells, have such large and complex cytoskeletons? List the differences between animal cells and bacteria that depend on the eucaryotic cytoskeleton.

Question 17–14

Examine the structure of an intermediate filament shown in Figure 17–4. Does the filament have a unique polarity—that is, could you distinguish one end from the other by chemical or other means? Explain your answer.

Question 17–15

There are no known motor proteins that move on intermediate filaments. Suggest an explanation for this.

Question 17–16

When cells enter mitosis, their existing array of cytoplasmic microtubules has to be rapidly broken down and replaced with the mitotic spindle that forms to pull the chromosomes into the daughter cells. The enzyme katanin, named after Japanese samurai swords, is activated during the onset of mitosis, and chops microtubules into short pieces. What do you suppose is the fate of the microtubule fragments created by katanin? Explain your answer.

Question 17–17

The drug taxol, extracted from the bark of yew trees, has the opposite effect of the drug colchicine, an alkaloid from autumn crocus. Taxol binds tightly to microtubules and stabilizes them; when added to cells, it causes much of the free tubulin to assemble into microtubules. In contrast, colchicine prevents microtubule formation. Taxol is just as pernicious to dividing cells as colchicine, and both are used as anticancer drugs. Based on your knowledge of microtubule dynamics, suggest why both drugs are toxic to dividing cells despite their opposite actions.

Question 17–18

A useful technique for studying microtubule motors is to attach them by their tails to a glass coverslip (which can be accomplished quite easily because the tails stick avidly to a clean glass surface) and then to allow microtubules to settle onto them. The microtubules may then be viewed in a light microscope as they are propelled over the surface of the coverslip by the heads of the motor proteins. Since the motor proteins attach at random orientations to the coverslip, however, how can they generate coordinated movement of individual microtubules rather than engaging in a tug-of-war? In which direction will microtubules crawl on a "bed" of kinesin molecules (i.e., plus end first, or minus end first)?

Question 17–19

A typical time course of polymerization of purified tubulin to form microtubules is shown in Figure Q17–19.

  A. Explain the different parts of the curve (labeled A, B, and C). Draw a diagram that shows the behavior of tubulin molecules in each of the three phases.
  B. How would the curve in the figure change if centrosomes were added at the outset?

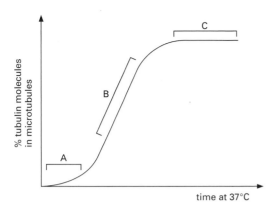

Figure Q17–19

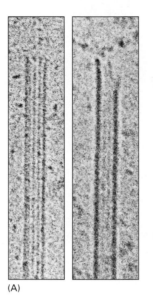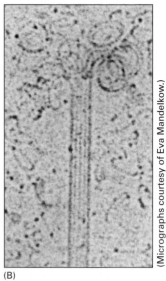

(A) (B)

(Micrographs courtesy of Eva Mandelkow.)

Figure Q17–20

Question 17–20

The electron micrograph shown in Figure Q17–20A was obtained from a population of microtubules that was growing rapidly. Figure Q17–20B was obtained from microtubules undergoing "catastrophic" shrinking. Comment on any differences between the two images, and suggest likely explanations for the differences that you observe.

Question 17–21

The locomotion of fibroblasts in culture is immediately halted by the drug cytochalasin, whereas colchicine causes fibroblasts to cease to move directionally and to begin extending lamellipodia in seemingly random directions. Injection of fibroblasts with antibodies to vimentin has no discernible effect on their migration. What do these observations suggest to you about the involvement of the three different cytoskeletal filaments in fibroblast locomotion?

Question 17–22

Complete the following sentence accurately, explaining your reason for accepting or rejecting each of the four phrases. The role of calcium in muscle contraction is

    A. To detach myosin heads from actin.

    B. To spread the action potential from the plasma membrane to the contractile machinery.

    C. To bind to troponin, cause it to move tropomyosin, and thereby expose actin filaments to myosin heads.

    D. To maintain the structure of the myosin filament.

Question 17–23

Which of the following changes takes place when a skeletal muscle contracts?

    A. Z discs move farther apart.

    B. Actin filaments contract.

    C. Myosin filaments contract.

    D. Sarcomeres become shorter.

Highlights from *Essential Cell Biology 2 Interactive* CD-ROM

17.2   Dynamic Instability of Microtubules

17.5   Kinesin

17.10 Beating Heart Cell

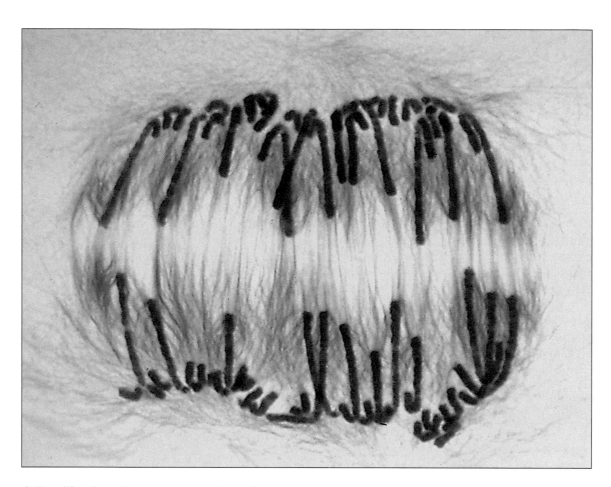

Cell proliferation. The various steps of the cell cycle include two key events: the replication of the chromosomes and their segregation into two daughter cells. Shown here is a dividing plant cell, photographed shortly after the replicated chromosomes *(dark blue)* were separated by the microtubules *(red)* of the mitotic spindle. (Courtesy of Andrew Bajer.)

Cell-Cycle Control and Cell Death

18

"Where a cell arises, there must be a previous cell, just as animals can only arise from animals and plants from plants." This *cell doctrine,* proposed by the German pathologist Rudolf Virchow in 1858, carried with it a profound message for the continuity of life. Cells are generated from cells, and the only way to make more cells is by the division of those that already exist. All living organisms, from the unicellular bacterium to the multicellular mammal, are products of repeated rounds of cell growth and division extending back in time to the beginnings of life over three billion years ago.

A cell reproduces by carrying out an orderly sequence of events in which it duplicates its contents and then divides in two. This cycle of duplication and division, known as the **cell cycle**, is the essential mechanism by which all living things reproduce. The details of the cell cycle vary from organism to organism and at different times in an individual organism's life. Certain characteristics, however, are universal, as the cycle must comprise, at a minimum, a set of processes that a cell has to perform to accomplish its most fundamental task—to copy and pass on its genetic information to the next generation of cells. To produce two genetically identical daughter cells, the DNA in each chromosome must be faithfully replicated, and the replicated chromosomes must then be accurately segregated into the two daughter cells, so that each cell receives a copy of the entire genome (Figure 18–1). Most cells must also duplicate their other organelles and macromolecules; otherwise, each time they divided they would get smaller and smaller. Thus, to maintain their size, dividing cells must also coordinate their growth with their division.

To explain how cells reproduce, we therefore have to consider three major questions: (1) How do cells duplicate their contents? (2) How do they partition the duplicated contents and split in two? (3) How do they coordinate all the machinery that is required for these two processes? The first two problems are discussed elsewhere in this book: in Chapter 6, we discuss how DNA is replicated, and in Chapters 7, 11, 15, and 17 we describe how the eucaryotic cell manufactures other components, such as proteins, membranes, organelles, and cytoskeletal filaments. The second problem—the physical process of cell division—is covered in detail in Chapter 19.

In this chapter, we tackle the third and most difficult problem—how the eucaryotic cell coordinates the various steps of its reproductive cycle. To ensure correct progression through the cell cycle, eucaryotic cells have evolved a complex network of regulatory proteins, known as the *cell-cycle control system*. The core of this system is an ordered series

Overview of the Cell Cycle
The Eucaryotic Cell Cycle Is Divided into Four Phases
A Central Control System Triggers the Major Processes of the Cell Cycle

The Cell-Cycle Control System
The Cell-Cycle Control System Depends on Cyclically Activated Protein Kinases
Cyclin-dependent Protein Kinases Are Regulated by the Accumulation and Destruction of Cyclins
The Activity of Cdks Is Also Regulated by Phosphorylation and Dephosphorylation
Different Cyclin–Cdk Complexes Trigger Different Steps in the Cell Cycle
S-Cdk Initiates DNA Replication and Helps Block Rereplication
Cdks Are Inactive Through Most of G_1
The Cell-Cycle Control System Can Arrest the Cycle at Specific Checkpoints
Cells Can Dismantle Their Control System and Withdraw from the Cell Cycle

Programmed Cell Death (Apoptosis)
Apoptosis Is Mediated by an Intracellular Proteolytic Cascade
The Death Program Is Regulated by the Bcl-2 Family of Intracellular Proteins

Extracellular Control of Cell Numbers and Cell Size
Animal Cells Require Extracellular Signals To Divide, Grow, and Survive
Mitogens Stimulate Cell Division
Extracellular Growth Factors Stimulate Cells to Grow
Animal Cells Require Survival Factors to Avoid Apoptosis
Some Extracellular Signal Proteins Inhibit Cell Growth, Division, or Survival

The division of a hypothetical eucaryotic cell with two chromosomes is shown to illustrate how two genetically identical daughter cells are produced in each cycle. Each of the daughter cells will often divide again by going through additional cell cycles.

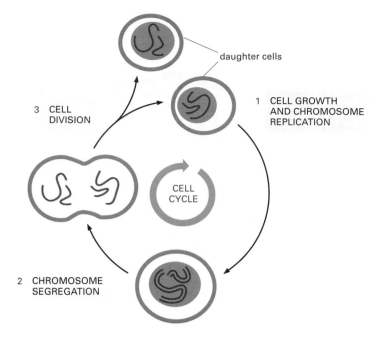

Question 18–1

Consider the following statement: "All present-day cells have arisen by an uninterrupted series of cell divisions extending back in time to the first cell division." Is this strictly true?

of biochemical switches that controls the main events of the cycle, including DNA replication and segregation of the duplicated chromosomes.

To coordinate these activities, the cell-cycle control system responds to various signals from inside and outside the cell. Inside the cell, the control system must monitor progression through the cell cycle to ensure, for example, that division does not begin before DNA replication is complete. It must also keep track of conditions outside the cell. In a multicellular animal, the control system is highly responsive to signals from other cells, stimulating cell division when more cells are needed and blocking it when they are not. The cell-cycle control system, therefore, plays a central part in the regulation of cell numbers in the tissues of the body; if the system malfunctions such that cell division is excessive, cancer can result.

We begin our discussion of how cell division is controlled and coordinated with an overview of the events that take place during the cell cycle. We then discuss how the cell-cycle control system coordinates these activities by responding to both intracellular and extracellular signals. We next describe how multicellular organisms eliminate unwanted cells by the process of *programmed cell death*, or *apoptosis*, in which a cell commits suicide when the interests of the organism demand it. Finally, we consider how animals regulate cell numbers and cell size, using extracellular signals to control cell division, cell survival, and cell growth.

Overview of the Cell Cycle

The most basic function of the cell cycle is to duplicate accurately the vast amount of DNA in the chromosomes and then precisely distribute the copies into genetically identical daughter cells. The duration of the cell cycle varies greatly from one cell type to another. A single-celled yeast can divide every 90–120 minutes in ideal conditions, whereas a mammalian liver cell divides on average less than once a year (Table 18–1). We focus here on the sequence of events that occur in a fairly rapidly dividing mammalian cell, with a cell-cycle time of about 24 hours, and we describe the cell-cycle control system that ensures that the various events of the cycle take place at the correct time.

Table 18–1 Some Eucaryotic Cell-Cycle Times	
CELL TYPE	CELL-CYCLE TIMES
Early frog embryo cells	30 minutes
Yeast cells	1.5–3 hours
Intestinal epithelial cells	~12 hours
Mammalian fibroblasts in culture	~20 hours
Human liver cells	~1 year

The Eucaryotic Cell Cycle Is Divided into Four Phases

The eucaryotic cell cycle is traditionally divided into four *phases*. Seen under a microscope, the two most dramatic events in the cycle are when the nucleus divides, a process called *mitosis,* and when the cell splits in two, a process called *cytokinesis.* These two processes together constitute the **M phase** of the cell cycle. In a typical mammalian cell, the whole of M phase takes about an hour, which is only a small fraction of the total cell-cycle time.

The period between one M phase and the next is called **interphase**. Under the microscope, it appears, deceptively, as an uneventful interlude during which the cell simply increases in size. Interphase, however, is a very busy time for the cell, and it encompasses the remaining three phases of the cell cycle. During **S phase** (S = synthesis), the cell replicates its nuclear DNA, an essential prerequisite for cell division. S phase is flanked by two phases in which the cell continues to grow. The **G_1 phase** (G = gap) is the interval between the completion of M phase and the beginning of S phase (DNA synthesis). The **G_2 phase** is the interval between the end of S phase and the beginning of M phase (Figure 18–2). During these gap phases, the cell monitors the internal and external environments to ensure that conditions are suitable and preparations are complete before it commits itself to the major upheavals of S phase and mitosis. At particular points in G_1 and G_2, the cell decides whether to proceed to the next phase or pause to allow more time to prepare.

During all of interphase, a cell continues to transcribe genes, synthesize proteins, and grow in mass. Together, G_1 and G_2 phases provide additional time for the cell to grow and duplicate its cytoplasmic organelles: if interphase lasted only long enough for DNA replication, the cell would not have time to double its mass before it divided and would consequently get smaller and smaller with each division. Indeed, in some special circumstances that is just what happens. In

Question 18–2

A population of proliferating cells is stained with a dye that becomes fluorescent when it binds to DNA, so that the amount of fluorescence is directly proportional to the amount of DNA in each cell. To measure the amount of DNA in each cell, the cells are then passed through a fluorescence-activated cell sorter (FACS), an instrument that measures the amount of fluorescence in individual cells. The number of cells with a given DNA content is plotted on the graph shown in Figure Q18–2. Indicate on

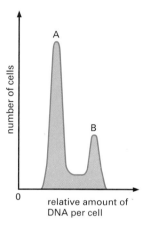

Figure Q18–2

the graph where you would expect to find cells that are in the following stages: G_1, S, G_2, and mitosis. Which is the longest phase of the cell cycle in this population of cells?

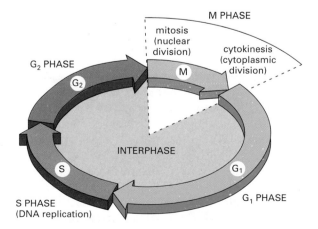

Figure 18–2 The cell cycle is divided into four phases. The cell grows continuously in interphase, which consists of three phases: DNA replication is confined to S phase; G_1 is the gap between M phase and S phase; G_2 is the gap between S phase and the M phase. During M phase, the nucleus and then the cytoplasm divide.

some animal embryos, for example, the first few cell divisions after fertilization (called *cleavage divisions*) serve to subdivide a giant egg cell into many smaller cells as quickly as possible. In these cell cycles, the G_1 and G_2 phases are drastically shortened, and the cells do not grow before they divide.

The first readily visible sign that a cell is about to enter M phase is the progressive *condensation* of its chromosomes, which were replicated earlier, during S phase. (Following replication, the two copies of each chromosome remain tightly bound together.) Chromosome condensation marks the end of the G_2 phase. At this stage in the cell cycle, the replicated chromosomes first become visible in the light microscope as long threads, which gradually get shorter and thicker by the process of compaction, described in Chapter 5. This condensation makes the chromosomes less likely to get entangled, so that they are easier to segregate to the two daughter cells during mitosis.

A Central Control System Triggers the Major Processes of the Cell Cycle

The cell-cycle control system operates much like the control system of an automatic clothes-washing machine. The washing machine functions in a series of stages: it takes in water, mixes it with detergent, washes the clothes, rinses them, and spins them dry. These essential processes of the wash cycle are analogous to the essential processes of the cell cycle—DNA replication, mitosis, and so on. In both cases, a central controller triggers each process in a set sequence (Figure 18–3). The controller is itself regulated at certain critical points of the cycle by feedback from the processes that are being performed. In the washtub, sensors monitor the water levels, for example, and send signals back to the controller to prevent the next process from beginning before the previous one has finished. Without such feedback, an interruption or a delay in any of the processes could be disastrous.

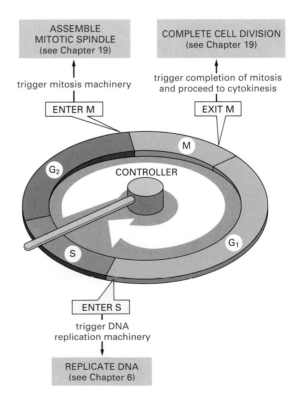

Figure 18–3 The essential processes of the cell cycle, such as DNA replication, mitosis, and cytokinesis, are triggered by a cell-cycle control system. By analogy with a washing machine, the cell-cycle control system is drawn as a controller arm that rotates clockwise, triggering essential processes when it reaches specific points on the outer dial.

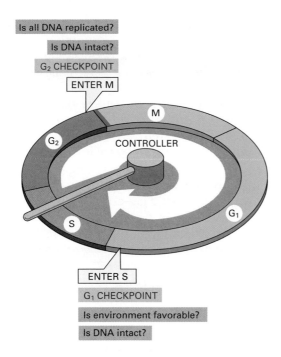

Is all DNA replicated?

Is DNA intact?

G$_2$ CHECKPOINT

ENTER M

M

G$_2$

CONTROLLER

S

G$_1$

ENTER S

G$_1$ CHECKPOINT

Is environment favorable?

Is DNA intact?

Figure 18–4 Feedback from the intracellular events of the cell cycle, as well as signals from the cell's environment, determine whether the control system will pass through certain checkpoints. Two prominent checkpoints occur at locations marked with *yellow* boxes: at the checkpoint in G$_1$, the control system determines whether the cell proceeds to S phase; at the one in G$_2$, the control system determines whether the cell proceeds to mitosis. There are other checkpoints in the cell cycle, as we discuss later in this chapter and in Chapter 19.

The events of the cell cycle must also occur in a particular sequence, and this sequence must be preserved even if one of the steps takes longer than usual. All of the nuclear DNA, for example, must be replicated before the nucleus begins to divide, which means that a complete S phase must precede M phase. If DNA synthesis is slowed down or stalled, mitosis and cell division must also be delayed. Similarly, if DNA is damaged, the cycle must arrest in either G$_1$ or G$_2$ so that the cell can repair the damage, either before the DNA is replicated or before the cell enters M phase. The cell-cycle control system achieves all of this by means of molecular brakes that can stop the cycle at various **checkpoints.** In this way, the control system does not trigger the next step in the cycle unless the cell is prepared.

Two important checkpoints occur in G$_1$ and G$_2$. The G$_1$ checkpoint allows the cell to confirm that the environment is favorable for cell proliferation and its DNA is intact before committing to S phase. Cell proliferation depends on nutrients and specific signal molecules in the extracellular environment; if extracellular conditions are unfavorable, cells can delay progress through G$_1$ and may even enter a specialized resting state known as G_0 (G zero). Many cells, including nerve cells and skeletal muscle cells, remain in G$_0$ for the lifetime of the organism. The G$_2$ checkpoint ensures that cells do not enter mitosis until damaged DNA is repaired and DNA replication is complete (Figure 18–4).

Checkpoints are especially important as points in the cell cycle where the control system can be regulated by signals from other cells. Some of these signals promote progress through the cell cycle, whereas others inhibit it. Later in this chapter, we consider the factors that influence the decisions made at these checkpoints, but first we discuss the proteins that form the core of the cell-cycle control system.

The Cell-Cycle Control System

Two types of machinery are involved in cell division: one manufactures the new components of the growing cell, and another hauls the components into their correct places and partitions them appropriately when the cell divides in two, as described in Chapter 19. No less important,

Question 18–3

What might be the consequences if a cell replicated damaged DNA before repairing it?

however, is the central control system that switches all this machinery on and off at the correct times and coordinates the activities that go into making the final product. It is the **cell-cycle control system** that ensures correct progression through the cell cycle by regulating the cell-cycle machinery.

In this section, we first consider some of the basic principles on which the cell-cycle control system operates. We then review the protein components of the control system and discuss how they work together to trigger the different phases of the cycle.

The Cell-Cycle Control System Depends on Cyclically Activated Protein Kinases

The cell-cycle control system governs the cell-cycle machinery by cyclically activating and then inactivating the key proteins and protein complexes that initiate or regulate DNA replication, mitosis, and cytokinesis. As discussed in Chapter 4, phosphorylation followed by dephosphorylation is one of the most common ways used by cells to switch the activity of a protein on and then off, and the cell-cycle control system uses this mechanism repeatedly. The phosphorylation reactions that control the cell cycle are carried out by a specific set of protein kinases, enzymes that transfer a phosphate group from ATP to a particular amino acid side chain on the target protein. The effects of phosphorylation can be rapidly reversed by removal of the phosphate group (dephosphorylation), a reaction carried out by another set of enzymes, called protein phosphatases.

The protein kinases that are part of the core of the cell-cycle control system are present in proliferating cells throughout the cell cycle. They are activated, however, only at appropriate times in the cycle, after which they quickly become deactivated again. Thus, the activity of each of these kinases rises and falls in a cyclical fashion. Some of the protein kinases, for example, become active toward the end of G_1 phase and are responsible for driving the cell into S phase; another kinase becomes active just before M phase and is responsible for driving the cell into mitosis.

Switching these kinases on and off at the appropriate times is partly the responsibility of a second set of protein components of the control system—the **cyclins**. Cyclins have no enzymatic activity themselves, but they have to bind to the cell-cycle kinases before the kinases can become enzymatically active. The kinases of the cell-cycle control system are therefore known as **cyclin-dependent protein kinases**, or **Cdks** (Figure 18–5). Cyclins are so-named because, unlike the Cdks, their concentrations vary in a cyclical fashion during the cell cycle. The cyclical changes in cyclin concentrations help drive the cyclic assembly and activation of the cyclin–Cdk complexes; activation of these complexes in turn triggers various cell-cycle events, such as entry into S phase or M phase. We discuss how these molecules were discovered later in the chapter (see How We Know, pp. 618–619).

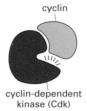

cyclin

cyclin-dependent
kinase (Cdk)

Figure 18–5 Progression through the cell cycle depends on cyclin-dependent protein kinases (Cdks). A Cdk depends on a regulatory subunit called a cyclin for the Cdk to be enzymatically active. The active complex phosphorylates key proteins in the cell that are required to initiate a particular step in the cell cycle. The cyclin also helps direct the Cdk to its target proteins.

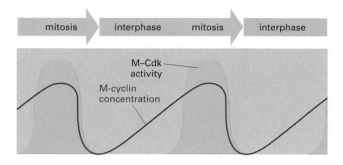

Figure 18–6 The rise and fall of M–Cdk activity during the cell cycle. The increase in M-cyclin concentration leads to the formation of the active M-cyclin–Cdk complex (M-Cdk), which drives entry into M phase. M-cyclin concentration increases steadily during interphase, peaks at mitosis, and falls rapidly as mitosis ends. Whereas M-Cdk activity rises and falls with each mitosis, the concentration of the Cdk does not change during the course of the cell cycle.

Cyclin-dependent Protein Kinases Are Regulated by the Accumulation and Destruction of Cyclins

The regulation of cyclin concentrations plays an important part in timing the events of the cell cycle, such as entry into M phase. The cyclin that helps drive cells into M phase is called **M-cyclin**, and the active complex it forms with its Cdk is called **M-Cdk**. Synthesis of M-cyclin starts immediately after cell division and continues steadily throughout interphase. The cyclin accumulates, so that its concentration rises gradually and helps time the onset of mitosis; its rapid elimination then helps initiate the exit from mitosis (Figure 18–6).

The sudden fall in M-cyclin concentration toward the end of mitosis is the result of its rapid destruction by the ubiquitin-dependent proteolytic system (discussed in Chapter 7). As mitosis nears completion, multiple molecules of the protein ubiquitin are covalently attached to the M-cyclin molecules. This ubiquitination marks the cyclin for degradation in proteasomes, large proteolytic machines found in all eucaryotic cells. Destruction of the cyclin inactivates the Cdk (Figure 18–7).

But what controls when cyclins are ubiquitinated and thus tagged for elimination? In the case of M-cyclin, a protein complex called the **anaphase promoting complex** (**APC**) adds ubiquitin to the cyclin and to other proteins involved in the regulation of mitosis. APC, however, is not active at all stages of the cell cycle. Its activity is switched on late in mitosis in a process that requires the activity of M-Cdk. Activation of M-Cdk initiates a process that—with a built-in delay—leads to the activation of APC. This, in turn, leads to the ubiquitination and degradation of M-cyclin, and thereby the inactivation of M-Cdk. In this way, M-Cdk contributes to its own eventual inactivation.

APC does more than trigger the degradation of M-cyclin. As we discuss in Chapter 19, the physical separation of the replicated chromosomes at a stage in mitosis called *anaphase* also depends on APC—hence the name "anaphase promoting complex."

The Activity of Cdks Is Also Regulated by Phosphorylation and Dephosphorylation

The rise and fall of cyclin levels plays an important part in regulating Cdk activity during the cell cycle, but there is more to the story. In the

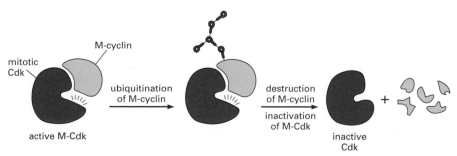

Figure 18–7 The activity of Cdks is regulated by cyclin degradation. Ubiquitination of a cyclin marks the protein for destruction. Loss of the cyclin leaves its Cdk partner inactive.

How We Know: Discovery of Cyclins and Cdks

For many years, cell biologists watched the "puppet show" of DNA synthesis, mitosis, and cytokinesis but had no idea what was behind the curtain controlling these events. The cell-cycle control system was simply a "black box" inside the cell. It was not even clear whether there was a separate control system, or whether the cell-cycle machinery somehow controlled itself. A breakthrough came with the identification of the key proteins of the control system and the realization that they are distinct from the components of the cell-cycle machinery—the enzymes and other proteins that perform the essential processes of DNA replication, chromosome segregation, and so on.

The first components of the cell-cycle control system to be discovered were the cyclins and cyclin-dependent kinases (Cdks) that drive cells into M phase. They were found in studies of cell division conducted in animal eggs.

Back to the egg

The fertilized eggs of many animals are especially suitable for biochemical studies of the cell cycle because they are exceptionally large and divide rapidly. An egg of the frog *Xenopus*, for example, is just over 1 mm in diameter (Figure 18–8). After fertilization, it divides rapidly, to partition the egg into many smaller cells. These rapid cell cycles mainly consist of repeated S and M phases, with little or no G_1 or G_2 phases between them. There is no new gene transcription: all of the mRNAs, as well as most of the proteins, required for this early stage of embryonic development are already packed into the very large egg during its develop-

ment as an oocyte in the ovary of the mother. In these early division cycles *(cleavage divisions)*, no cell growth occurs, and all the cells of the embryo divide synchronously.

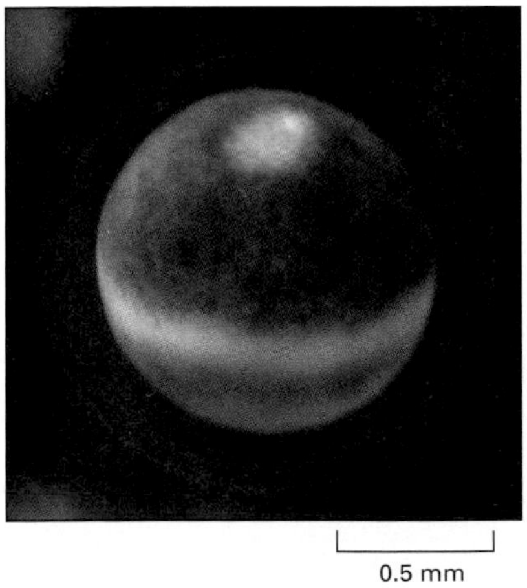

Because of the synchrony, it is possible to prepare an extract from frog eggs at a particular stage of the cell cycle that is representative of that cell-cycle stage. The biological activity of such an extract can then be tested by injecting it into a *Xenopus* oocyte (the immature precursor of the unfertilized egg) and observing, microscopically, its effects on cell-cycle behavior. The *Xenopus* oocyte is an especially convenient test system for detecting an activity that drives cells into M phase, as it has completed DNA replication and is arrested at a stage in the meiotic cell cycle (discussed in Chapter 20) that is equivalent to the G_2 phase of a mitotic cell cycle.

Give us an M

In such experiments, researchers found that an extract from an M-phase egg instantly drives the oocyte into M phase, whereas cytoplasm from a cleaving egg at other phases of the cycle does not. When first discovered, the biochemical identity and mechanism of action of the factor responsible for this activity were unknown, and the activity was simply called *maturation promoting factor*, or MPF (Figure 18–9). By testing cytoplasm from different stages of the cell cycle, MPF activity was found to oscillate dramatically during the course of each cell cycle: it increased rapidly just before the start of mitosis and fell rapidly to zero toward the end of mitosis (Figure 18–10). This oscillation made MPF a strong candidate for a component involved in cell-cycle control.

When MPF was finally purified, it was found to contain a protein kinase, which was required for its activity. But the kinase portion of MPF did not act alone. It had to have a specific protein (now known to be M-cyclin) bound to it in order to function. M-cyclin was discovered in a different type of experiment, involving clam eggs.

Fishing in clams

M-cyclin was initially identified as a protein whose concentration rose gradually during interphase and then fell rapidly to zero as cleaving clam eggs went through M phase (see Figure 18–6). The protein repeated this performance in each cell cycle. As cyclin itself has no enzymatic activity, its role in cell-cycle control was initially obscure. The breakthrough occurred when cyclin was found to be a component of MPF and to be required for MPF activity. Thus, MPF, which we now call M-Cdk, is a protein complex containing two subunits—a regulatory subunit, M-cyclin, and a catalytic subunit, the mitotic Cdk. After the components of M-Cdk were identified, other types of cyclins and Cdks were isolated, whose concentrations or activities, respectively, rise and fall at other stages in the cell cycle.

0.5 mm

Figure 18–8 A mature *Xenopus* egg provides a convenient system for studying cell division. (Courtesy of Tony Mills.)

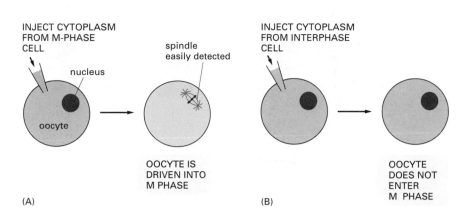

INJECT CYTOPLASM
FROM M-PHASE
CELL

nucleus

spindle
easily detected

oocyte

OOCYTE IS
DRIVEN INTO
M PHASE

(A)

INJECT CYTOPLASM
FROM INTERPHASE
CELL

OOCYTE
DOES NOT
ENTER
M PHASE

(B)

Figure 18–9 MPF activity was discovered by injecting *Xenopus* egg cytoplasm into *Xenopus* oocytes. (A) A *Xenopus* oocyte is injected with cytoplasm taken from a *Xenopus* egg in M phase. The cell extract drives the oocyte into M phase of the first meiotic division, causing the large nucleus to break down and a spindle to form. (B) When the cytoplasm is taken from a cleaving egg in interphase, it does not cause the oocyte to enter M phase. Thus, the extract in (A) must contain some activity—a maturation promoting factor (MPF)—that triggers entry into M phase.

All in the family

While biochemists were identifying the proteins that regulate the cell cycles of frog and clam embryos, yeast geneticists were taking a different approach to dissecting the cell-cycle control system. By studying mutants that get stuck or misbehave at specific points in the cell cycle, these researchers were able to identify many genes responsible for cell-cycle control. Some of these genes turned out to encode cyclin or Cdk proteins, which were unmistakably similar—in both amino acid sequence and function—to their counterparts in frogs and clams. Similar genes were soon identified in human cells.

Many of the cell-cycle control genes have changed so little during evolution that the human version of the gene will function perfectly well in a yeast cell. For example, a yeast with a defective copy of the gene encoding its only Cdk fails to divide; the mutant will divide normally, however, if a copy of the appropriate human gene is artificially introduced into the defective cell. Surely, even Darwin would have been astonished at such clear evidence that humans and yeasts are cousins. Despite a billion years of divergent evolution, all eucaryotic cells—whether yeast, animal, or plant—use essentially the same molecules to control the events of their cell cycle.

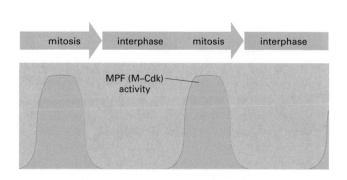

mitosis | interphase | mitosis | interphase

MPF (M–Cdk) activity

Figure 18–10 The activity of MPF oscillates during the cell cycle in *Xenopus* embryos. The activity assayed using the test outlined in Figure 18–9 rises rapidly just before the start of mitosis and falls rapidly to zero toward the end of mitosis.

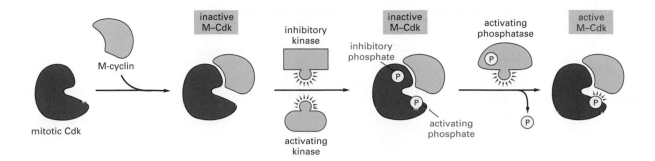

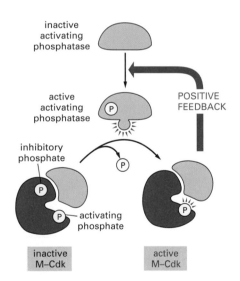

Figure 18–11 For M-Cdk to be active, it must be phosphorylated at some sites and dephosphorylated at others. The M-cyclin–Cdk complex is enzymatically inactive when first formed. Subsequently, the Cdk is phosphorylated at sites that are required for its activity and at other overriding sites that inhibit its activity. At this point, the M-Cdk remains inactive until it is finally activated by a phosphatase that removes the inhibitory phosphate groups. It is still not clear how the timing of this complex activation process is controlled.

Figure 18–12 Activated M-Cdk indirectly activates more M-Cdk. Once activated, M-Cdk phosphorylates, and thereby activates, more activating phosphatase. The phosphatase can now activate more M-Cdk by removing the inhibitory phosphate groups from the Cdk subunit. Although not shown, activated M-Cdk also inhibits the inhibitory kinase shown in Figure 18–11, further promoting the activation of M-Cdk. In these ways, activated M-Cdk indirectly activates more M-Cdk, so that the activation of M-Cdk occurs explosively.

case of M-Cdk, M-cyclin concentration increases gradually throughout interphase, but M-Cdk activity switches on abruptly at the end of interphase (see Figure 18–6). So what triggers this rapid activation of M-Cdk?

For M-Cdk to be maximally active it has to be phosphorylated at one or more sites by a specific protein kinase, and dephosphorylated at other sites by a specific protein phosphatase. The removal of the inhibitory phosphate groups by the phosphatase is the final step that activates the M-Cdk at the end of interphase (Figure 18–11). Once activated, each M-cyclin–Cdk complex can activate more of the same complexes, as illustrated in Figure 18–12. This positive feedback produces the sudden, explosive increase in M-Cdk activity that drives the cell abruptly into M phase.

Different Cyclin–Cdk Complexes Trigger Different Steps in the Cell Cycle

There are several types of cyclins and, in most eucaryotes, several types of Cdks involved in cell-cycle control. Different cyclin–Cdk complexes trigger different steps of the cell cycle. Whereas the M-Cdk complex acts in G_2 to trigger entry into M phase, distinct cyclins, called *S-cyclins* and *G_1/S-cyclins,* bind to a distinct Cdk protein late in G_1 to form S-Cdk and G_1/S-Cdk, respectively, which trigger entry into S phase. Other cyclins, called *G_1-cyclins,* act earlier in G_1 and bind to other Cdk proteins to form *G_1-Cdks,* which help drive the cell through G_1 toward S phase. We see later that the formation of these G_1–Cdks in animal cells usually depends on extracellular signal molecules that stimulate cells to divide. The names of the individual cyclins and their Cdks are listed in Table 18–2.

As discussed for M-cyclin, the concentration of each type of cyclin rises gradually, and then falls sharply, at a specific time in the cell cycle as a result of degradation by the ubiquitin pathway. The increase in concentration helps to activate the appropriate Cdk partner, while the rapid fall returns the Cdk to its inactive state (Figure 18–13). Thus, the slow accumulation of a cyclin to a critical level is one way that the cell-cycle control system measures the time intervals between one cell-cycle step and the next.

As is the case with M-Cdk, each of the different Cdks also has to be phosphorylated and dephosphorylated appropriately in order to act. Each of these activated cyclin–Cdk complexes act on a different set of target proteins in the cell. As a result, each type of complex triggers a different transition step in the cycle. M-Cdk, for example, phosphorylates key proteins that cause the chromosomes to condense, the nuclear envelope to break down, and the microtubules of the cytoskeleton to reorganize to form a mitotic spindle—events that herald the entry into mitosis, as we discuss in Chapter 19.

Thus far, we have mainly focused our attention on the activation of M-Cdk, which drives cells into mitosis. Equally important, however, is the process by which S-Cdk triggers entry into S phase, as we discuss next.

Table 18–2 The Major Cyclins and Cdks of Vertebrates		
Cyclin–Cdk Complex	Cyclin	Cdk Partner
G_1-Cdk	cyclin D*	Cdk4, Cdk6
G_1/S-Cdk	cyclin E	Cdk2
S-Cdk	cyclin A	Cdk2
M-Cdk	cyclin B	Cdk1**

*There are three D cyclins in mammals (cyclins D1, D2, and D3).
**The original name of Cdk1 was Cdc2 in vertebrates.

S-Cdk Initiates DNA Replication and Helps Block Rereplication

A cell must solve several problems in controlling the initiation and completion of DNA replication. Not only must replication occur with extreme accuracy to minimize the risk of mutations in the next cell generation, but every nucleotide in the genome must be copied once, and only once, to prevent the potentially damaging effects of gene amplification. It falls to the cell-cycle control system to see that DNA replication is initiated at the right time—and not more than once per cycle.

As we discuss in Chapter 6, DNA replication begins at *origins of replication*, nucleotide sequences that are scattered at various locations along each chromosome. These sequences recruit specific proteins that control the initiation and completion of DNA replication. One multiprotein complex, the **origin recognition complex (ORC),** remains bound to the origins of replication throughout the cell cycle, where it serves as a sort of landing pad for additional regulatory proteins that bind before the start of S phase.

One of these regulatory proteins is called Cdc6. It is present at low levels during most of the cell cycle, but its concentration increases transiently in early G_1. When Cdc6 binds to ORCs in G_1, it promotes the binding of additional proteins to form a *pre-replicative complex*. Once

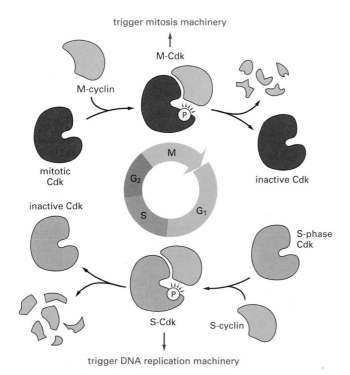

Figure 18–13 Distinct Cdks associate with different cyclins to trigger the different events of the cell cycle. For simplicity, only two types of cyclin–Cdk complexes are shown, one that triggers S phase and one that triggers M phase. In both cases, the activation of the Cdk requires cyclin binding (as well as phosphorylation and dephosphorylation, as shown in Figure 18–11), and its inactivation depends on cyclin degradation.

the pre-replicative complex is assembled, the replication origin is ready to "fire." The activation of S-Cdk in late G_1 then pulls the "trigger," initiating DNA replication (Figure 18–14).

S-Cdk not only initiates origin firing; it also helps prevent rereplication of the DNA. It helps phosphorylate Cdc6, causing it and the other proteins in the pre-replicative complex to dissociate from the ORC after an origin has fired. This disassembly prevents replication from occurring again at the same origin. In addition, phosphorylation by S-Cdk (and by M-Cdk, which becomes active at the start of M phase) marks Cdc6 for ubiquitination and degradation (see Figure 18–14), ensuring that DNA replication is not reinitiated later in the same cell cycle.

Cdks Are Inactive Through Most of G_1

We have discussed how activated Cdks trigger the transitions from one part of the cell cycle to the next. But how are these changes reversed so that the daughter cells can later go through another cycle?

At the end of mitosis, all Cdk activity in the cell is reduced to zero. S-Cdks are inactivated by the destruction of S-cyclin at the end of S phase, and M-Cdk is inactivated by the destruction of M-cyclin toward the end of mitosis (see Figure 18–13). The inactivation of M-Cdk leads to all of the other events that take the cell out of mitosis.

Cdks then remain inhibited throughout most of G_1. Several mechanisms prevent Cdk reactivation during G_1, including the binding of *Cdk inhibitory proteins* (discussed below). These mechanisms delay progression into the next S phase and allow the cell time to grow. Escape from this inhibitory state usually occurs through the accumulation of G_1-cyclins, which, in animal cells, is stimulated by extracellular signals that promote cell proliferation, as we see later.

The Cell-Cycle Control System Can Arrest the Cycle at Specific Checkpoints

We have seen that the cell-cycle control system triggers the events of the cycle in a specific order. It triggers mitosis, for example, only after all of

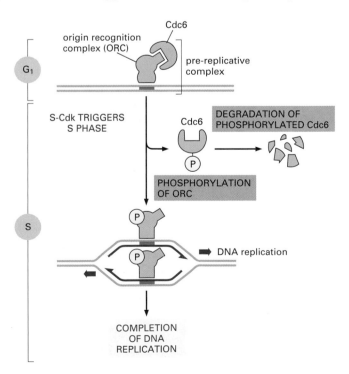

Figure 18–14 S-Cdk triggers DNA replication and ensures that DNA replication is initiated only once per cell cycle. The ORC remains associated with the replication origin throughout the cell cycle. In early G_1, the regulatory protein Cdc6 associates with ORC. Aided by Cdc6, additional proteins bind to the adjacent DNA (not shown), resulting in the formation of a pre-replicative complex. S-Cdk (with assistance from another protein kinase, not shown) then triggers origin firing by causing the assembly of DNA polymerase and the initiation of DNA synthesis (discussed in Chapter 6). S-Cdk also helps block rereplication by helping to phosphorylate Cdc6, which dissociates from the origin and is degraded. The protein complex that assembles at the replication origin to catalyze DNA replication when the origin fires is not shown.

the DNA has been replicated, and it permits the cell to divide in two only after mitosis has been completed. If one of the steps is delayed, the control system delays the activation of the next steps so that the normal sequence is maintained. This self-regulating property of the control system ensures, for example, that if DNA synthesis is halted for some reason during S phase, the cell will not proceed into M phase with its DNA only half replicated. As mentioned earlier, the control system accomplishes this feat through the action of molecular brakes that can stop the cell cycle at specific *checkpoints*, allowing the cell to monitor its internal state and its environment before continuing through the cycle (see Figure 18–4).

For the most part, the molecular mechanisms responsible for stopping cell-cycle progression at checkpoints are poorly understood. In some cases, however, specific **Cdk inhibitor proteins** come into play; these block the assembly or activity of one or more cyclin–Cdk complexes. One of the best understood checkpoint mechanisms halts the cell cycle in G_1 if DNA is damaged, helping to ensure that a cell does not replicate damaged DNA. DNA damage causes an increase in both the concentration and activity of a gene regulatory protein called *p53*, which activates the transcription of a gene encoding a Cdk inhibitor protein called *p21*. The p21 protein binds to G_1/S-Cdk and S-Cdk, preventing them from driving the cell into S phase (Figure 18–15). The

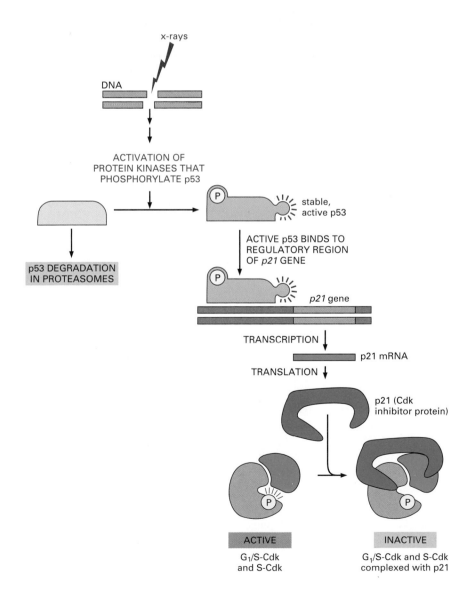

Figure 18–15 **DNA damage arrests the cell cycle in G_1.** When DNA is damaged, the p53 protein, which is normally rapidly degraded, is stabilized and activated. This is partly because p53 becomes phosphorylated by specific protein kinases that are activated in response to DNA damage. Activated p53 accumulates and stimulates the transcription of the gene that encodes the Cdk inhibitor protein, p21. The p21 protein binds to G_1/S-Cdk and S-Cdk and inactivates them, so that the cell cycle arrests in G_1.

arrest of the cell cycle in G_1 allows the cell time to repair the damaged DNA before replicating it. If p53 is missing or defective, the unrestrained replication of damaged DNA leads to a high rate of mutation and the production of cells that tend to become cancerous. In fact, mutations in the *p53* gene are found in about half of all human cancers.

Another important cell-cycle checkpoint occurs in mitosis. At this point, the cell determines whether all of its chromosomes are attached appropriately to the mitotic spindle. As we discuss in Chapter 19, the mitotic spindle is a cytoskeletal machine that physically pulls the duplicated chromosomes apart and segregates them into the two daughter cells. If the cell begins to segregate its chromosomes before all chromosomes are properly attached to the spindle, one daughter will receive an incomplete set of chromosomes, while the other daughter receives a surplus. Both situations can be lethal for the cell. Thus, a dividing cell must be sure that every last chromosome is attached properly to the spindle before it completes mitosis. To monitor chromosome attachment, the cell makes use of a negative signal: unattached chromosomes send a "stop" signal to the cell-cycle control system. Although the exact nature of the signal remains elusive, it inhibits further progress through mitosis by blocking the activation of APC. Without active APC, the duplicated chromosomes remain glued together. Thus, none of the duplicated chromosomes can be pulled apart until every chromosome is positioned correctly on the mitotic spindle.

Cells Can Dismantle Their Control System and Withdraw from the Cell Cycle

The most radical decision that the cell-cycle control system can make is to withdraw from the cell cycle entirely and stop the cell from dividing. This is a different matter from pausing in the middle of a cycle to cope with a temporary delay, and it has a special importance in multicellular organisms. In the human body, for example, nerve cells and skeletal muscle cells persist for a lifetime without dividing; they enter G_0, a modified G_1 state in which the cell-cycle control system is largely dismantled, in that many of the Cdks and cyclins disappear. Some cell types, such as liver cells, normally divide only once every year or two, while certain epithelial cells in the gut divide more than twice a day in order to renew the lining of the gut continually. Many of our cells fall somewhere in between: they can divide if the need arises but normally do so infrequently.

It seems to be a general rule that mammalian cells will multiply *(proliferate)* only if they are stimulated to do so by signals from other cells. If deprived of such signals, the cell cycle arrests at a G_1 checkpoint and enters the G_0 state, where the cells can remain for days or weeks, or for the lifetime of the organism. Most of the diversity in cell-division rates in the adult body thus lies in the variation in the time cells spend in G_0 or in G_1; once past the G_1 checkpoint (see Figure 18–4), a cell usually proceeds through the rest of the cell cycle quickly—typically within 12–24 hours in mammals. The G_1 checkpoint is therefore sometimes called *Start,* because passing it represents a commitment to complete a full division cycle, although a better name might be Stop (Figure 18–16). Some of the main checkpoints in the cell cycle are summarized in Figure 18–17.

Question 18–5

Why do you suppose cells have evolved a special G_0 state to exit the cell cycle, rather than just stopping in a G_1 state at a G_1 checkpoint?

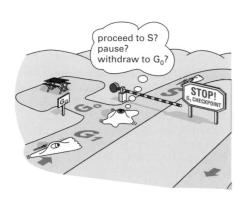

Figure 18–16 The G_1 checkpoint offers the cell a crossroad. The cell can commit to completing another cell cycle, pause transiently until conditions are right, or withdraw from the cell cycle altogether and enter G_0.

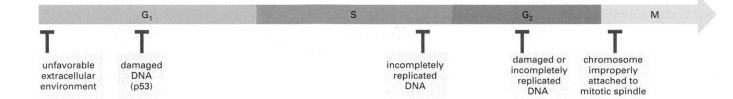

G_1	S	G_2	M

unfavorable extracellular environment

damaged DNA (p53)

incompletely replicated DNA

damaged or incompletely replicated DNA

chromosome improperly attached to mitotic spindle

The starting and stopping of cell proliferation are fundamentally important in controlling cell numbers and bodily proportions in a multicellular organism. But controls on cell division are only half the story. On the other side of the balance sheet lie other, equally important, controls that determine whether a cell lives, or whether it dies by suicide, as we discuss next.

Programmed Cell Death (Apoptosis)

The cells of a multicellular organism are members of a highly organized community. The number of cells in this community is tightly regulated—not simply by controlling the rate of cell division, but also by controlling the rate of cell death. If cells are no longer needed, they commit suicide by activating an intracellular death program. This process is therefore called **programmed cell death,** although it is more commonly called **apoptosis** (from a Greek word meaning "falling off," as leaves fall from a tree).

The amount of programmed cell death that occurs in both developing and adult tissues is astonishing. In the developing vertebrate nervous system, for example, more than half of the nerve cells produced normally die soon after they are formed. In a healthy adult human, billions of cells die in the bone marrow and intestine every hour. It seems remarkably wasteful for so many cells to die, especially as the vast majority are perfectly healthy at the time they kill themselves. What purposes does this massive cell death serve?

In some cases, the answers are clear. Mouse paws—and our own hands and feet—are sculpted by apoptosis during embryonic development: they start out as spadelike structures, and the individual fingers and toes separate only as the cells between them die (Figure 18–18). In other cases, cells die when the structure they form is no longer needed. When a tadpole changes into a frog at metamorphosis, the cells in the tail die, and the tail, which is not needed in the frog, disappears (Figure 18–19). In still other cases, cell death helps regulate cell numbers, as we discuss later. In all these cases, the unneeded cells die by apoptosis.

In adult tissues, cell death exactly balances cell division. If this were not so, the tissue would grow or shrink. If part of the liver is removed in an adult rat, for example, liver cell proliferation increases to make up

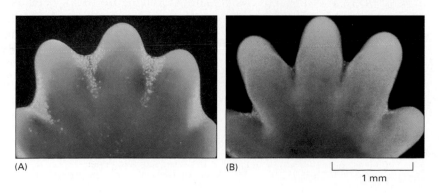

(A)

(B)

1 mm

Figure 18–17 The cell-cycle control system can arrest the cycle at various checkpoints. The *red "T"s* represent points in the cycle where the control system can apply molecular brakes to stop progression in response to DNA damage, intracellular processes that are not completed, or an unfavorable extracellular environment. The checkpoint indicated in M phase ensures that all of the chromosomes are appropriately attached to the mitotic spindle before progressing into anaphase, where the daughter chromosomes separate and move to opposite poles of the spindle (discussed in Chapter 19).

Figure 18–18 Apoptosis in the developing mouse paw sculpts the digits. (A) The paw in this mouse embryo has been stained with a dye that specifically labels cells that have undergone apoptosis. The apoptotic cells appear as *bright green dots* between the developing digits. (B) This interdigital cell death eliminates the tissue between the developing digits, as seen in the paw shown one day later. Here, few, if any, apoptotic cells can be seen. (From W. Wood et al., *Development* 127:5245–5252, 2000. © The Company of Biologists.)

Figure 18–19 Apoptosis helps eliminate the tail during the metamorphosis of a tadpole into a frog. As a tadpole changes into a frog, the cells in the tadpole tail are induced to undergo apoptosis. All of the changes that occur during metamorphosis, including the induction of apoptosis in the tail, are stimulated by an increase in thyroid hormone in the blood.

the loss. Conversely, if a rat is treated with the drug phenobarbital, which stimulates liver cell division, the liver enlarges. However, when the phenobarbital treatment is stopped, apoptosis in the liver greatly increases until the organ has returned to its original size, usually within a week or so. Thus, the liver is kept at a constant size through the regulation of both the cell death rate and the cell birth rate.

In this short section, we discuss the molecular mechanisms of apoptosis and its intracellular control. We discuss the control of apoptosis by extracellular signals in the final section.

Apoptosis Is Mediated by an Intracellular Proteolytic Cascade

Cells that die as a result of acute injury typically swell and burst, spilling their contents all over their neighbors, a process called *cell necrosis* (Figure 18–20A). This eruption triggers a potentially damaging inflammatory response. By contrast, a cell that undergoes apoptosis dies neatly, without damaging its neighbors. A cell in the throes of apoptosis shrinks and condenses (Figure 18–20B). The cytoskeleton collapses, the nuclear envelope disassembles, and the nuclear DNA breaks up into fragments. Most important, the cell surface is altered in such a manner that it immediately attracts phagocytic cells, usually specialized phagocytic cells called macrophages (discussed in Chapter 15). These cells engulf the apoptotic cell before it spills its contents (Figure 18–20C). This rapid removal of the dying cell avoids the damaging consequences of cell necrosis, and also allows the organic components of the apoptotic cell to be recycled by the cell that ingests it.

Figure 18–20 Apoptosis kills cells quickly and cleanly. Electron micrographs showing cells that have died by necrosis (A) or by apoptosis (B and C). The cells in (A) and (B) died in tissue culture, whereas the cell in (C) died in a developing tissue and has been engulfed by a phagocytic cell. Note that the cell in (A) seems to have exploded, while those in (B) and (C) have condensed but seem relatively intact. The large vacuoles seen in the cytoplasm of the cell in (B) are a variable feature of apoptosis. (Courtesy of Julia Burne.)

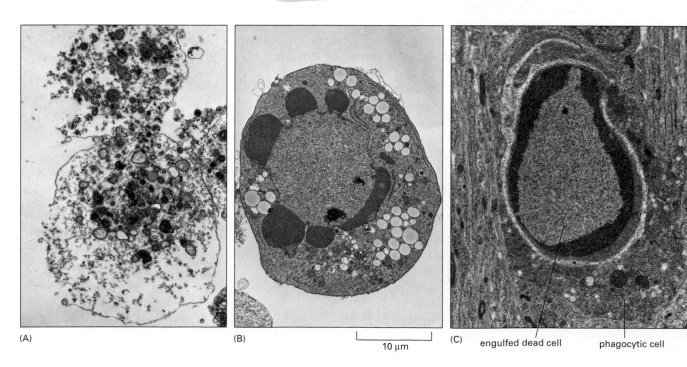

(A) (B) 10 μm (C) engulfed dead cell   phagocytic cell

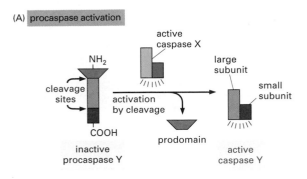

(A) procaspase activation

(B) caspase cascade

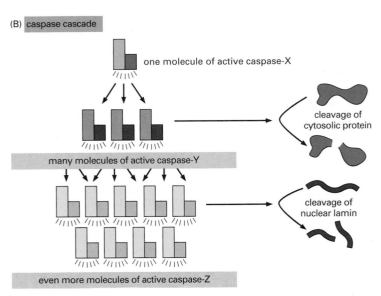

Figure 18–21 Apoptosis is mediated by an intracellular proteolytic cascade.
(A) Each suicide protease (caspase) is made as an inactive proenzyme, a procaspase, which is itself activated by proteolytic cleavage by another member of the same protease family. (B) Each activated caspase molecule can then cleave many procaspase molecules, thereby activating them, and these can then activate even more procaspase molecules. In this way, an initial activation of a small number of protease molecules can lead, via an amplifying chain reaction (a cascade), to the explosive activation of a large number of protease molecules. Some of the activated caspases then break down a number of key proteins in the cell, such as nuclear lamins, leading to the controlled death of the cell.

The machinery that is responsible for this kind of controlled cell suicide seems to be similar in all animal cells. Apoptosis is carried out by a family of proteases—enzymes that cut up other proteins—called **caspases**. The caspases are made as inactive precursors called *procaspases*, which are themselves activated by proteolytic cleavage in response to signals that induce apoptosis. The activated caspases cleave, and thereby activate, other members of the family, resulting in an amplifying proteolytic cascade (Figure 18–21). They also cleave other key proteins in the cell. One of the caspases, for example, cleaves the lamin proteins, which form the nuclear lamina underlying the nuclear envelope; this causes the irreversible breakdown of the nuclear lamina (see Figure 18–21). In this way, the cell dismantles itself quickly and cleanly, and its corpse is rapidly taken up and digested by another cell.

Activation of the apoptotic program, like entry into a new stage of the cell cycle, is usually triggered in an all-or-none fashion. The proteolytic cascade is not only destructive and self-amplifying, but also irreversible; once a cell reaches a critical point along the path to destruction, it cannot turn back. Thus, it is important that the decision to die is tightly controlled.

The Death Program Is Regulated by the Bcl-2 Family of Intracellular Proteins

All nucleated animal cells contain the seeds of their own destruction: in these cells, inactive procaspases lie waiting for a signal to destroy the cell. It is therefore not surprising that caspase activity is tightly regulated inside the cell to ensure that the death program is held in check until it is needed.

Question 18–6

Why do you think apoptosis occurs by a different mechanism from the cell death that occurs in cell necrosis? What might be the consequences if apoptosis were not achieved in so neat and orderly a fashion, whereby the cell destroys itself from within and avoids leakage of its contents into the extracellular space?

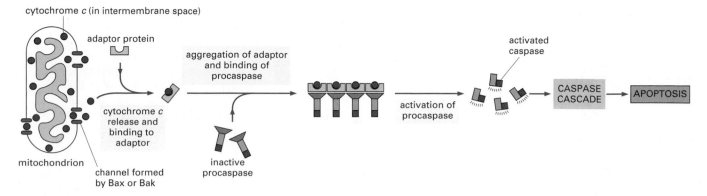

Figure 18–22 Apoptosis is regulated by members of the Bcl-2 family of intracellular proteins.
The death-promoting Bcl-2 family proteins Bak and Bax are thought to help form channels in the outer mitochondrial membrane, allowing cytochrome *c* to be released into the cytosol. Cytochrome *c* then binds to an adaptor protein that promotes the aggregation and activation of a particular procaspase molecule. Once activated, this caspase triggers a caspase cascade, leading to apoptosis.

The main proteins that regulate the activation of procaspases are members of the **Bcl-2 family** of intracellular proteins. Some members of this protein family promote procaspase activation and cell death, whereas others inhibit these processes. Two of the most important death-promoting family members are proteins called *Bax* and *Bak*. These proteins activate procaspases indirectly, by inducing the release of cytochrome *c* from mitochondria into the cytosol. Cytochrome *c* binds to an adaptor protein, which then activates a specific procaspase. This activated procaspase initiates the caspase cascade that leads to apoptosis (Figure 18–22). Bax and Bak proteins are themselves activated by other death-promoting members of the Bcl-2 family, which are produced or activated by various insults to the cell, such as DNA damage.

Other members of the Bcl-2 family, including Bcl-2 itself, act to inhibit procaspase activation and apoptosis, rather than promote them. One way they do so is by blocking the ability of Bax and Bak to release cytochrome *c* from mitochondria. Some of the Bcl-2 family members that promote apoptosis do so by binding to and blocking the activity of Bcl-2 and other death-suppressing members of the Bcl-2 family.

The intracellular death program is also regulated by signals from other cells, which can either activate or suppress the program. Indeed, cell survival, cell division, and cell growth are all regulated by extracellular signals, which, together, help multicellular organisms control cell number and cell size, as we now discuss.

Extracellular Control of Cell Numbers and Cell Size

A fertilized mouse egg and a fertilized human egg are similar in size, and yet an adult mouse is much smaller than an adult human. What are the differences in the control of cell behavior in humans and mice that generate such differences in size? The same fundamental question can be asked about each organ and tissue in an individual's body. What adjustment of cell behavior explains the length of an elephant's trunk or the size of its brain or its liver? These questions are largely unanswered, but it is at least possible to say what the ingredients of an answer must be. Organ and body size are determined by three fundamental processes: cell growth, cell division, and cell death. Each of these processes, in turn, is regulated by signals from other cells in the body, combined with programs intrinsic to the individual cell.

In this section, we discuss how these extracellular signals stimulate cell division, cell growth, and cell survival and thereby help control the size of an animal and its organs. We conclude the section with a brief discussion of inhibitory extracellular signals that also help regulate these processes.

Animal Cells Require Extracellular Signals To Divide, Grow, and Survive

Unicellular organisms such as bacteria and yeasts tend to grow and divide as fast as they can, and their rate of proliferation depends largely on the availability of nutrients in the environment. The cells in a multicellular organism, by contrast, must be controlled so that an individual cell divides only when another cell is required by the organism—either to allow tissue growth or to replace cell loss. Thus, for an animal cell to divide or grow, or even to survive, nutrients are not enough. It must also receive chemical signals from other cells, usually its neighbors.

Most of the extracellular signal molecules that influence cell division, cell growth, and cell survival are soluble proteins secreted by other cells or proteins bound to the surface of other cells or the extracellular matrix. Although most act positively to stimulate one or more of these cell processes, some act negatively to inhibit a particular process. The positively acting signal proteins can be divided, based on their function, into three major classes:

1. **Mitogens** stimulate cell division, primarily by overcoming the intracellular braking mechanisms that tend to block progression through the cell cycle.

2. **Growth factors** stimulate cell growth (an increase in cell mass) by promoting the synthesis and inhibiting the degradation of proteins and other macromolecules.

3. **Survival factors** promote cell survival by suppressing apoptosis.

These categories are not mutually exclusive, as many signal molecules have more than one of these functions. Unfortunately, the term "growth factor" is often used as a catch-all phrase to describe a protein with any of these roles. Indeed, the phrase "cell growth" is often used incorrectly to mean an increase in cell number, which is more correctly termed "cell proliferation."

In the following sections, we examine each of these types of signal molecules in turn.

Mitogens Stimulate Cell Division

Most mitogens are secreted signal proteins that bind to cell-surface receptors. When activated by mitogen binding, these receptors activate various intracellular signaling pathways (discussed in Chapter 16) that stimulate cell division. These signaling pathways act mainly by releasing the intracellular molecular brakes that block the transition from the G$_1$ phase of the cell cycle into S phase.

An important example of such a molecular brake is the *Retinoblastoma (Rb) protein,* first identified through studies of a rare childhood eye tumor called retinoblastoma, in which the Rb protein is missing or defective. The Rb protein is abundant in the nucleus of all vertebrate cells. It binds to particular gene regulatory proteins, preventing them from stimulating the transcription of genes required for cell proliferation. Mitogens release the Rb brake in the following way. They activate intracellular signaling pathways that lead to the activation of

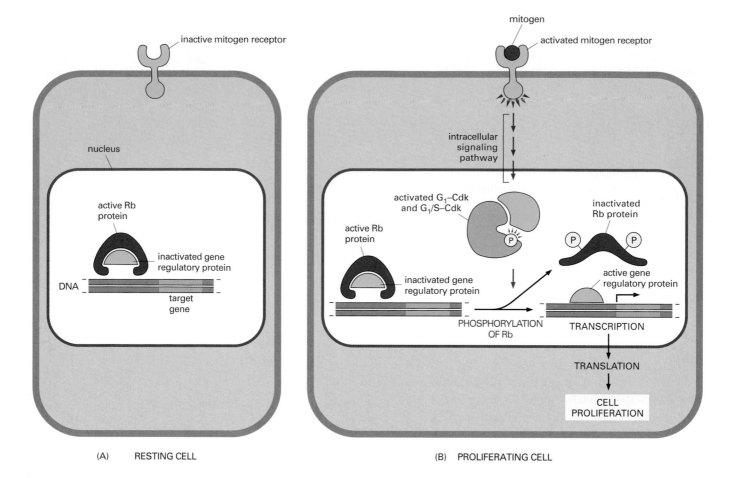

Figure 18–23 One way that mitogens stimulate cell proliferation is by inhibiting the Rb protein.
(A) In the absence of mitogens, dephosphorylated Rb protein holds specific gene regulatory proteins in an inactive state; these gene regulatory proteins are required to stimulate the transcription of target genes that encode proteins needed for cell proliferation. (B) Mitogens bind to cell-surface receptors and activate intracellular signaling pathways that lead to the formation and activation of the G_1-Cdk and G_1/S-Cdk complexes mentioned earlier. These complexes phosphorylate, and thereby inactivate, the Rb protein. The gene regulatory proteins are now free to activate the transcription of their target genes, leading to cell proliferation.

the G_1-Cdk and G_1/S-Cdk complexes mentioned earlier. These kinases phosphorylate the Rb protein, altering its conformation so that it releases its bound gene regulatory proteins, which are then free to activate the genes required for cell proliferation to proceed (Figure 18–23).

Most mitogens have been identified and characterized by their effects on cells in culture (Figure 18–24). One of the first mitogens identified in this way was *platelet-derived growth factor,* or *PDGF,* whose effects are typical of many others discovered since. When blood clots form (in a wound, for example), blood platelets incorporated in the clots are triggered to release PDGF. PDGF then binds to receptor tyrosine kinases (discussed in Chapter 16) in surviving cells at the wound site, thereby stimulating them to proliferate and help heal the wound. Similarly, if part of the liver is lost through surgery or acute injury, cells

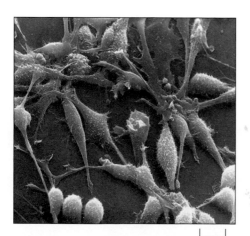

Figure 18–24 This scanning electron micrograph shows rat fibroblasts proliferating in culture. The cells are cultured in the presence of calf serum, which contains growth factors and mitogens that stimulate the cells to grow and multiply. The spherical cells at the bottom of the micrograph have rounded up in preparation for cell division. (Courtesy of Guenter Albrecht-Buehler.)

10 µm

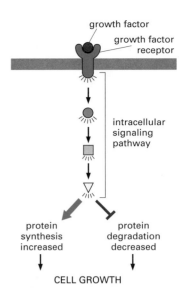

growth factor

growth factor receptor

intracellular signaling pathway

protein synthesis increased

protein degradation decreased

CELL GROWTH

Figure 18–25 **Extracellular growth factors increase the synthesis and decrease the degradation of macromolecules.** This leads to a net increase in macromolecules and thereby all growth.

in the liver and elsewhere produce a protein called *hepatocyte growth factor,* which helps stimulate the surviving liver cells to proliferate.

Extracellular Growth Factors Stimulate Cells to Grow

The growth of an organism or organ depends on cell growth as much as on cell division. If cells divided without growing, they would get progressively smaller, and there would be no increase in total cell mass. In single-celled organisms such as yeasts, cell growth (like cell division) requires only nutrients. In animals, by contrast, both cell growth and cell division depend on signals from other cells. Cell growth, however, unlike cell division, does not depend on the cell-cycle control system, in either yeasts or animal cells. Indeed, many animal cells, including nerve cells and most muscle cells, do most of their growing after they have permanently stopped dividing.

Like most mitogens, most extracellular growth factors bind to cell-surface receptors, which activate various intracellular signaling pathways. These pathways lead to the accumulation of proteins and other macromolecules, and they do so by both increasing the rate of synthesis of these molecules, and decreasing their rate of degradation (Figure 18–25). Some extracellular signal proteins, including PDGF, can act as both growth factors and mitogens, stimulating both cell growth and cell-cycle progression. Such proteins help ensure that cells maintain their appropriate size as they proliferate.

Compared to cell division, there has been surprisingly little study of how cell size is controlled in animals. As a result, it remains a mystery how different cell types in the same animal grow to be so different in size (Figure 18–26).

Animal Cells Require Survival Factors to Avoid Apoptosis

Animal cells need signals from other cells not only to grow and proliferate, but even to survive. If deprived of such survival factors, cells activate their intracellular suicide program and die by apoptosis. This requirement for signals from other cells for survival helps to ensure that cells survive only when and where they are needed. Nerve cells, for example, are produced in excess in the developing nervous system and then compete for limited amounts of survival factors that are secreted by the target cells they contact. Nerve cells that receive enough survival factor live, while the others die by apoptosis (Figure 18–27). A similar dependence on survival signals from neighboring cells is thought to control cell numbers in other tissues, both during development and in adulthood.

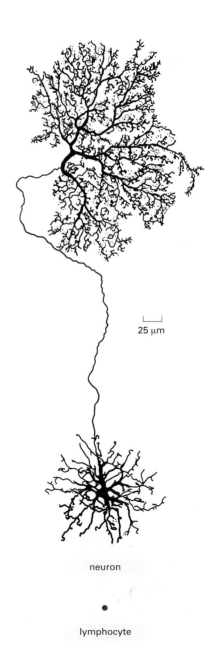

25 μm

neuron

lymphocyte

Figure 18–26 **A nerve cell and a lymphocyte are very different in size.** These two cell types, which are drawn at the same scale, both come from the same species of monkey and contain the same amount of DNA. A neuron grows progressively larger after it has permanently withdrawn from the cell cycle. During this time, the ratio of cytoplasm to DNA increases enormously (by a factor of more than 10^5 for some neurons). (Neuron from B.B. Boycott in Essays on the Nervous System [R. Bellairs and E.G. Gray, eds.]. Oxford, U.K.: Clarendon Press, 1974. © Oxford University Press.)

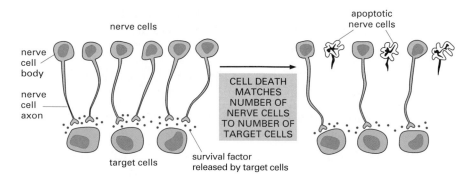

Figure 18–27 Cell death helps match the number of developing nerve cells to the number of target cells they contact. More nerve cells are produced than can be supported by the limited amount of survival factor released by the target cells. Therefore, some cells receive insufficient amounts of survival factor to keep their suicide program suppressed and, as a consequence, undergo apoptosis. This strategy of overproduction followed by culling ensures that all target cells are contacted by nerve cells and that the "extra" nerve cells are automatically eliminated.

Survival factors, like mitogens and growth factors, usually bind to cell-surface receptors. Binding of survival factors activates intracellular signaling pathways that keep the death program suppressed, usually by regulating members of the Bcl-2 family of proteins. Some survival factors, for example, increase the production of apoptosis-suppressing members of this family (Figure 18–28).

Some Extracellular Signal Proteins Inhibit Cell Growth, Division, or Survival

The extracellular signal proteins that we have discussed so far—mitogens, growth factors, and survival factors—act positively to increase the size of organs and organisms. Some extracellular signal proteins, however, act to oppose these positive regulators and thereby inhibit tissue growth. *Myostatin,* for example, is a secreted signal protein that normally inhibits the growth and proliferation of the myoblasts that fuse to form skeletal muscle cells. When the gene that encodes myostatin is deleted in mice, their muscles grow to be several times larger than normal, because both the number and the size of muscle cells is increased. Remarkably, two breeds of cattle that were bred for large muscles turned out to have mutations in the gene encoding myostatin (Figure 18–29).

Figure 18–28 Survival factors often suppress apoptosis by regulating Bcl-2 family members. In this case, the activated receptor activates a gene regulatory protein at the cell surface. The protein then moves to the nucleus where it activates the gene encoding the Bcl-2 protein, which inhibits apoptosis.

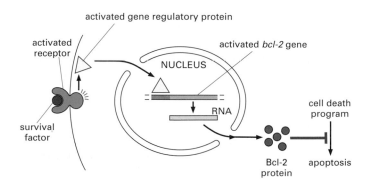

Figure 18–29 **Mutation of the *myostatin* gene leads to a dramatic increase in muscle mass.** This Belgium Blue was produced by cattle breeders and was only recently found to have a mutation in the *myostatin* gene. Mice purposely made deficient in the same gene also have remarkably big muscles. (From A.C. McPherron and S.-J. Lee, *Proc. Natl. Acad. Sci. U.S.A.* 94:12457–12461, 1997. © National Academy of Sciences.)

As we discuss in the final chapter, cancers are similarly the products of mutations that set cells free from the normal "social" controls that operate on cell growth, proliferation, and survival. Because cancer cells are generally less dependent than normal cells on signals from other cells they can outgrow, out-divide, and out-survive their normal neighbors, producing tumors that can kill an animal.

Essential Concepts

- The eucaryotic cell cycle consists of several distinct phases. These include S phase, during which the nuclear DNA is replicated, and M phase, during which the nucleus divides (mitosis) and then the cytoplasm divides (cytokinesis).
- In most cells there is one gap phase (G_1) between M phase and S phase, and another (G_2) between S phase and M phase. These gaps give the cell more time to grow.
- The cell-cycle control system coordinates the events of the cell cycle by cyclically switching on the appropriate parts of the cell-cycle machinery and then switching them off.
- The control system depends on a set of protein kinases, each composed of a regulatory subunit called a cyclin and a catalytic subunit called a cyclin-dependent protein kinase (Cdk).
- The control system also depends on protein complexes such as APC that trigger the proteolysis of specific cell-cycle regulators at particular stages of the cycle by ubiquitinating the regulators.
- The Cdks are cyclically activated by both cyclin binding and the phosphorylation of some amino acids and the dephosphorylation of others; when activated, Cdks phosphorylate key proteins in the cell.
- Cyclin concentrations rise and fall at specific times in the cell cycle, helping to time events of the cycle; the rise results from steady synthesis, while the sudden fall results from rapid proteolysis triggered by either APC or other protein complexes that ubiquitinate the cyclin.
- Different cyclin–Cdk complexes trigger different steps of the cell cycle: M-Cdk drives the cell into mitosis; G_1-Cdk drives it through G_1; G_1/S-Cdk and S-Cdk drive it into S phase.
- The cell-cycle control system can halt the cycle at specific checkpoints to ensure that the next step in the cycle does not begin before the previous one has finished, and intracellular and extracellular conditions are favorable.
- The cell cycle can be halted by at least two mechanisms: (1) Cdk inhibitor proteins can block the assembly or activity of one or more cyclin–Cdk complexes, or (2) components of the control system can stop being made, for example when cells enter G_0.

- Animal cell numbers are regulated by a combination of intracellular programs and extracellular signals that control cell proliferation, cell survival, and cell death.
- Animal cells proliferate only if stimulated by mitogens, which activate intracellular signaling pathways to override the normal brakes that otherwise block cell-cycle progression; this mechanism ensures that a cell divides only when another cell is needed.
- For an organism or an organ to grow, cells must grow as well as divide. Animal cell growth depends on extracellular growth factors, which stimulate protein synthesis and inhibit protein degradation.
- Many normal cells die during the lifetime of an animal by activating an internal suicide program and killing themselves—a process called apoptosis.
- Apoptosis depends on a family of proteolytic enzymes called caspases, which are made as inactive precursors (procaspases), which are themselves activated by proteolytic cleavage.
- Most animal cells require continuous signaling from other cells to avoid apoptosis; this may be a mechanism to ensure that cells survive only when and where they are needed.
- Cell and tissue size can also be influenced by inhibitory extracellular signal proteins that oppose the positive regulators of cell growth, cell division, and cell survival.
- Cancer cells fail to obey these normal "social" controls on cell behavior and therefore outgrow, out-divide and out-survive their normal neighbors.

Key Terms

anaphase promoting complex (APC)	G_1 phase
apoptosis	G_2 phase
Bcl-2 family	growth factor
caspase	interphase
Cdk inhibitor protein	M-Cdk
cell cycle	M-cyclin
cell-cycle control system	M phase
checkpoint	mitogen
cyclin	origin recognition complex (ORC)
cyclin-dependent protein kinase (Cdk)	programmed cell death
	S phase
	survival factor

Questions

Question 18–7

Which of the following statements are correct? Explain your answers.

A. Cells do not pass from G_1 into M phase of the cell cycle unless there are sufficient nutrients to complete an entire cell cycle.

B. Apoptosis is mediated by special intracellular proteases, one of which cleaves nuclear lamins.

C. Developing neurons compete for limited amounts of survival factors.

D. Some vertebrate cell-cycle control proteins function when expressed in yeast cells.

E. It is possible to study yeast mutants that are defective in cell-cycle control proteins, despite the fact that these proteins are essential for the cells to live.

F. Both the presence of a bound cyclin and its phosphorylation state determine whether a Cdk protein is enzymatically active.

Question 18–8

One of the functions of M-Cdk is to cause a precipitous drop in cyclin concentration halfway through M phase. Describe the consequences of this sudden decrease and suggest possible mechanisms by which it might occur.

Question 18–9

Compare the rules of cell proliferation in an animal with the rules that govern human behavior in society. What would happen to an animal if its cells behaved like people normally behave in our society? Could the rules that govern cell proliferation be applied to human societies?

Question 18–10

Figure 18–6 shows the rise of cyclin concentration and the rise of M-Cdk activity in cells as they progress through the cell cycle. It is remarkable that the cyclin concentration rises slowly and steadily, whereas M-Cdk activity increases suddenly. How do you think this difference arises?

Question 18–11

In his highly classified research laboratory Dr. Lawrence M. is charged with the task of developing a strain of dog-sized rats to be deployed behind enemy lines. In your opinion, which of the following strategies should Dr. M. pursue to increase the size of rats?

A. Block apoptosis.
B. Block p53 function.
C. Overproduce growth factors, mitogens, or survival factors.
D. Obtain a taxi driver's license and switch careers.

Explain the likely consequences of each option.

Question 18–12

PDGF is encoded by a gene that can cause cancer when expressed inappropriately. Why do cancers not arise at wounds in which PDGF is released from platelets?

Question 18–13

One important biological effect of a large dose of ionizing radiation is to halt cell division.

A. How does this occur?
B. What happens if a cell has a mutation that prevents it from halting cell division after being irradiated?
C. What might be the effects of such a mutation if the cell is not irradiated?
D. An adult human who has reached maturity will die within a few days of receiving a radiation dose large enough to stop cell division. What does that tell you (other than that one should avoid large doses of radiation)?

Question 18–14

What do you suppose happens in mutant cells that

A. cannot degrade cyclins?
B. always express high levels of p21?
C. cannot phosphorylate Rb?

Question 18–15

Many mutant yeasts have been isolated that are defective in the control of their cell cycle. They proliferate normally at low temperatures (30°C) but show abnormal patterns of cell growth and division when grown at a higher temperature (37°C). Two mutant strains (called "gee" and "wee") with defects at different sites in the same gene have very different responses to elevated temperatures. Gee strain cells grow until they become enormous but no longer divide. Wee strain cells have very short cell cycles and divide when they are very much smaller than usual. Suggest a possible model to explain these observations, and suggest what the normal protein encoded by this gene might do.

Question 18–16

If cells are grown in a culture medium containing radioactive thymidine, the thymidine will be covalently incorporated into the cell's DNA during S phase. The radioactive DNA can be detected in the nuclei of individual cells by autoradiography (i.e., by placing a photographic film over the cells, radioactive cells will activate the film and show up as black dots when looked at under a microscope). Consider a simple experiment in which cells are radioactively labeled by this method for only a short period of time (about 30 minutes). The radioactive thymidine medium is then replaced with one containing unlabeled thymidine, and the cells are grown for some additional time. At different time points after replacement of the medium, cells are examined in

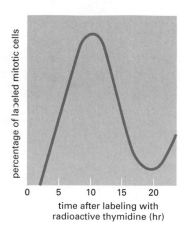

percentage of labeled mitotic cells

time after labeling with
radioactive thymidine (hr)

Figure Q18–16

Question 18–17

Liver cells proliferate both in patients with alcoholism and in patients with liver tumors. What are the differences in the mechanisms by which cell proliferation is induced in these diseases?

Question 18–18

Look carefully at the electron micrographs in Figure 18–20. Describe the differences between the cell that died by necrosis and those that died by apoptosis. How do the pictures confirm the differences between the two processes? Explain your answer.

a microscope. The fraction of cells in mitosis (that can be easily recognized because their chromosomes are condensed) that have radioactive DNA in their nuclei is then determined and plotted as a function of time after the labeling with radioactive thymidine (Figure Q18–16).

A. Would all cells (including cells at all phases of the cell cycle) contain radioactive DNA after the labeling procedure?

B. Note that initially there are no mitotic cells that contain radioactive DNA. Why is this?

C. Explain the rise and fall and then rise again of the curve.

D. Estimate the length of the G_2 phase from this graph.

Highlights from *Essential Cell Biology 2 Interactive* CD-ROM

18.2   p53–DNA Complex

18.3   Apoptosis

Cell Division

Cells reproduce by duplicating their contents and dividing in two. In unicellular organisms, such as bacteria or yeasts, each cell division produces a complete new organism, whereas many rounds of cell division are required to make a new multicellular organism from a fertilized egg. This cycle of duplication and division, central to the reproduction of all living things, is known as the *cell cycle*. In Chapter 18, we discuss how eucaryotic cells control the various phases of the cycle so that they occur at the right time and in the right sequence. In this chapter, we focus on the final phase of the cell cycle, when the cell divides its nucleus *(mitosis)* and then its cytoplasm *(cytokinesis)*. Together, mitosis and cytokincsis constitute **M phase** of the cell cycle (Figure 19–1).

Although M phase occurs over a relatively short amount of time—about one hour in a mammalian cell that divides once a day, or even once a year—it is by far the most dramatic phase of the cell cycle. During this brief period, the cell reorganizes virtually all of its components and distributes them equally into the two daughter cells. The rest of the cell cycle, in effect, serves to set the stage for the drama of M phase. In most rapidly proliferating cells, this preparative period, called *interphase*, is divided into three phases: *S phase*, when DNA is replicated, and two gap phases, G_1 and G_2, which provide additional time for the cell to grow (Figure 19–2). The events during interphase that prepare the cell for M phase, like all other events in the cell cycle, are coordinated by the *cell-cycle control system*. As we discuss in Chapter 18, the central core of the control system is a collection of proteins that are activated in sequence to trigger the various steps of the cycle. Among these regulatory proteins are the **cyclin-dependent kinases** (**Cdks**) that control entry into S phase and M phase.

For this chapter, the most important of these kinases is **M-phase Cdk (M-Cdk)**, which initiates M phase (discussed in Chapter 18). Activation of M-Cdk drives many of the morphological changes that occur during mitosis in animal cells: the chromosomes condense, the nuclear envelope breaks down, the endoplasmic reticulum and Golgi apparatus reorganize, the cell loosens its attachments to other cells and to the extracellular matrix, and the cytoskeleton undergoes a radical reorganization to form the specialized structures that will segregate the replicated chromosomes and divide the cell in two. M phase terminates when M-Cdk is inactivated.

We begin this chapter with an overview of M phase. We then discuss, in turn, the events that occur during mitosis and cytokinesis, focusing mainly on animal cells.

An Overview of M Phase

In Preparation for M Phase, DNA-binding Proteins Configure Replicated Chromosomes for Segregation

The Cytoskeleton Carries Out Both Mitosis and Cytokinesis

Centrosomes Duplicate To Help Form the Two Poles of the Mitotic Spindle

M Phase Is Conventionally Divided into Six Stages

Mitosis

Microtubule Instability Facilitates the Formation of the Mitotic Spindle

The Mitotic Spindle Starts to Assemble in Prophase

Chromosomes Attach to the Mitotic Spindle at Prometaphase

Chromosomes Line Up at the Spindle Equator at Metaphase

Daughter Chromosomes Segregate at Anaphase

The Nuclear Envelope Re-forms at Telophase

Some Organelles Fragment at Mitosis

Cytokinesis

The Mitotic Spindle Determines the Plane of Cytoplasmic Cleavage

The Contractile Ring of Animal Cells Is Made of Actin and Myosin

Cytokinesis in Plant Cells Involves New Cell-Wall Formation

Gametes Are Formed by a Specialized Kind of Cell Division

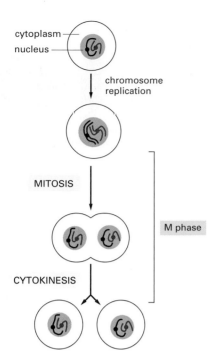

cytoplasm
nucleus

chromosome
replication

MITOSIS

M phase

CYTOKINESIS

Figure 19–1 The M phase of the cell cycle consists of nuclear division (mitosis) followed by cytoplasmic division (cytokinesis).

An Overview of M Phase

The central problem for a cell in M phase is to accurately separate and distribute (segregate) its chromosomes, which were replicated in the preceding S phase, so that each new daughter cell receives an identical copy of the genome. With minor variations, all eucaryotes solve this problem in a similar way: they assemble specialized cytoskeletal machines that pull the duplicated chromosome sets apart and split the cytoplasm into two halves. Before the duplicated chromosomes can be separated and distributed equally to the two daughter cells in M phase, however, they must be appropriately configured, and this process begins in S phase.

In Preparation for M Phase, DNA-binding Proteins Configure Replicated Chromosomes for Segregation

When the chromosomes are duplicated in S phase, the two copies of each replicated chromosome remain tightly bound together as identical **sister chromatids.** The sister chromatids are held together by protein complexes called **cohesins,** which assemble along the length of each sister chromatid as the DNA is replicated. This cohesion between sister chromatids is crucial for proper chromosome segregation, and it is only broken late in mitosis to allow the sisters chromatids to be pulled apart.

When the cell is about to enter M phase, the replicated chromosomes condense, becoming visible as threadlike structures. A set of protein complexes, called **condensins**, help carry out this **chromosome condensation**. The M-Cdk that initiates entry into M phase triggers the assembly of condensin complexes onto DNA by phosphorylating some of the condensin subunits. The accumulation of condensins on the DNA facilitates the progressive condensation of the chromosomes. Condensation makes the mitotic chromosomes more compact, reducing them to small physical packets that can be more easily segregated within the crowded confines of the dividing cell.

Figure 19–2 The eucaryotic cell cycle is comprised of four successive phases. Nuclear division and then cytoplasmic division occurs in M phase. Interphase is divided into three phases. DNA replication occurs in S phase. G_1 phase is the gap between M phase and S phase; G_2 is the gap between S phase and M phase. The cell grows continuously during interphase but stops growing during M phase.

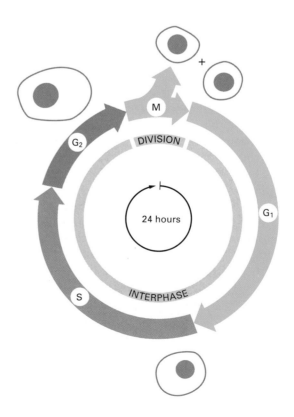

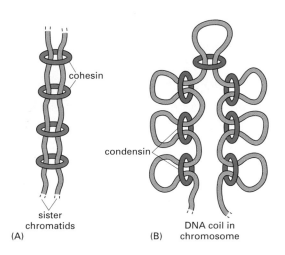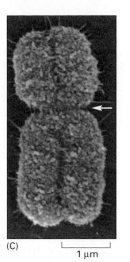

sister
chromatids
(A)

cohesin

condensin

DNA coil in
chromosome
(B)

(C)

1 µm

Figure 19–3 Cohesins and condensins help prepare the replicated chromosomes for mitosis. (A) Cohesins tie two adjacent sister chromatids together. (B) Condensins coil up single DNA molecules in the process of chromosome condensation. (C) Together, the cohesins and condensins help reduce the mitotic chromosomes to small, condensed structures that can be easily segregated during mitosis. Shown here is a scanning electron micrograph of a replicated human mitotic chromosome, consisting of two sister chromatids joined along their length. The constricted region *(arrow)* is the centromere region, where the chromosome will attach to the cytoskeletal machine that will pull the sister chromatids apart. (C, courtesy of Terry D. Allen.)

The cohesins and condensins are structurally related, and they work together to help configure the replicated chromosomes for mitosis. As illustrated in Figure 19–3, the cohesins tie two parallel DNA molecules—the identical sister chromatids—together, while the condensins tie up a single DNA molecule to help condense it.

The Cytoskeleton Carries Out Both Mitosis and Cytokinesis

After the replicated chromosomes have condensed, two separate cytoskeletal structures are assembled in sequence to carry out the two mechanical processes that occur in M phase—nuclear division (mitosis) and cytoplasmic division (cytokinesis). Both structures are rapidly disassembled after they have performed their tasks.

To produce two genetically identical daughter cells, the eucaryotic cell has to perform the delicate job of separating the replicated chromosomes and allocating one copy of each chromosome to each daughter cell. In all eucaryotic cells, this task is carried out during mitosis by a complex cytoskeletal machine called the *mitotic spindle*. The spindle is composed of microtubules and the various proteins that interact with them, including microtubule-dependent motor proteins (discussed in Chapter 17).

In animal cells and many unicellular eucaryotes, a different cytoskeletal structure is responsible for cytokinesis. It is called the *contractile ring* because it consists mainly of actin filaments and myosin filaments arranged in a ring around the equator of the cell (discussed in Chapter 17). The contractile ring starts to assemble toward the end of mitosis, just beneath the plasma membrane. As the ring contracts, it pulls the membrane inward, thereby dividing the cell in two (Figure 19–4). We discuss later how plant cells, which have a cell wall to contend with, divide their cytoplasm by a very different mechanism.

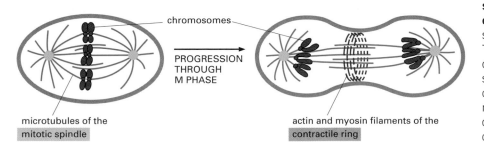

chromosomes

PROGRESSION
THROUGH
M PHASE

microtubules of the
mitotic spindle

actin and myosin filaments of the
contractile ring

Figure 19–4 Two transient cytoskeletal structures mediate M phase in animal cells. The mitotic spindle assembles first to separate the replicated chromosomes. Then, the contractile ring assembles to divide the cell in two. Whereas the mitotic spindle is based on microtubules, the contractile ring is based on actin and myosin filaments. Plant cells use a very different mechanism to divide the cytoplasm, as we discuss later.

The assembly of the mitotic spindle in animal cells depends on centrosomes, which duplicate before M phase begins and then help form the poles of the spindle, as we now discuss.

Centrosomes Duplicate To Help Form the Two Poles of the Mitotic Spindle

Before M phase begins, two critical events must be completed: DNA must be fully replicated, and, in animal cells, the centrosome must be duplicated. The **centrosome** is the principal *microtubule-organizing center* in animal cells. It must be duplicated so it can help form the two poles of the mitotic spindle, which will separate the duplicated chromosomes and distribute them to the two daughter cells. Furthermore, the centrosome must be duplicated if each daughter cell is to receive its own centrosome.

As discussed in Chapter 17, each centrosome consists of an amorphous matrix of proteins, containing hundreds of **γ-tubulin rings.** These ring complexes serve as nucleation sites for the growth of microtubules that radiate out from the centrosome. In animal cells, the centrosome also contains a pair of centrioles, each made of a cylindrical array of short microtubules (see Figure 17–11).

During interphase of each animal cell cycle, the centrosome is duplicated, and both copies remain together as a single complex on one side of the nucleus. As mitosis begins, the two centrosomes separate, and each nucleates a radial array of microtubules called an **aster**. The two asters move to opposite sides of the nucleus to form the two poles of the mitotic spindle (Figure 19–5). When the nuclear envelope breaks down, the spindle captures the chromosomes, eventually separating them later in mitosis. Centrosome duplication is triggered by the same Cdks that trigger DNA replication (discussed in Chapter 18), which explains why it starts at the beginning of S phase. As mitosis ends and the nuclear envelope re-forms around the separated chromosomes, each daughter cell receives a centrosome in association with its chromosomes. The process of centrosome duplication and separation is known as the **centrosome cycle.**

M Phase Is Conventionally Divided into Six Stages

Although M phase proceeds as a continuous sequence of events, it is traditionally divided into six stages. The first five stages of M phase—prophase, prometaphase, metaphase, anaphase, and telophase—constitute **mitosis,** which was originally defined as the period in which the chromosomes are visibly condensed. *Cytokinesis* occurs in the sixth stage, which overlaps with the end of mitosis. The six stages of M phase are summarized in Panel 19–1. Together, they form a dynamic sequence in which many independent cycles—involving the chromosomes, cytoskeleton, and centrosomes—are coordinated to produce two genetically identical daughter cells.

The five stages of mitosis occur in strict sequential order, while cytokinesis begins in anaphase and continues through telophase. During *prophase*, the replicated chromosomes condense and the mitotic spindle begins to assemble outside the nucleus. During *prometaphase*, the nuclear envelope breaks down, allowing the spindle microtubules to contact the chromosomes and bind to them. During *metaphase*, the mitotic spindle gathers all of the chromosomes to the center (equator) of the spindle. During *anaphase*, the two sister chromatids in each replicated chromosome synchronously split apart, and the spindle draws them to opposite poles of the cell. During *telophase*,

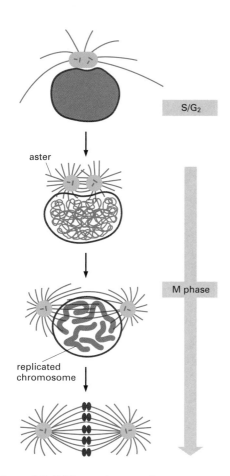

Figure 19–5 The centrosome in an interphase cell duplicates to form the two poles of a mitotic spindle. In most animal cells, a centriole pair (shown here as a pair of *dark green bars)* is associated with the centrosome matrix *(light green)* that nucleates microtubule outgrowth. (The volume of centrosome matrix is exaggerated in this diagram for clarity.) Centrosome duplication begins at the start of S phase and is completed by G_2. Initially, the two centrosomes remain together, but, in early M phase, they separate into two, each of which nucleates its own aster. The two asters then move apart, and the microtubules that interact between the two asters preferentially elongate to form a bipolar mitotic spindle, with an aster at each pole. When the nuclear envelope breaks down, the spindle microtubules are able to interact with the chromosomes.

a nuclear envelope reassembles around each of the two sets of separated chromosomes to form two nuclei. Cytokinesis is complete by the end of telophase, when the nucleus and cytoplasm of each of the daughter cells returns to interphase, signaling the end of M phase.

Mitosis

Before nuclear division, or mitosis, begins, each chromosome has been replicated and consists of two identical chromatids. These sister chromatids are held together along their length by cohesin proteins (see Figure 19–3A). During mitosis, these proteins are cleaved, and the sister chromatids split apart to become independent *daughter chromosomes*, which are pulled to opposite poles of the cell by the mitotic spindle (Figure 19–6). In this section, we examine how the mitotic spindle assembles and how it functions. We see how the dynamic instability of microtubules and the activity of microtubule-associated motor proteins facilitate the assembly of the spindle and the separation of the daughter chromosomes.

Microtubule Instability Facilitates the Formation of the Mitotic Spindle

Most animal cells in interphase contain a cytoplasmic array of microtubules radiating out from the single centrosome (see Figure 17–9A). The fast-growing ends of the microtubules, called plus ends, project outward toward the cell perimeter, while their minus ends are associated with the centrosome. As discussed in Chapter 17, these microtubules continuously polymerize and depolymerize by the addition and loss of the tubulin subunits that make up a microtubule. The individual microtubules alternate between growing and shrinking—a process called *dynamic instability* (see Figure 17–12). At the start of mitosis, the microtubules forming the cytoplasmic array disassemble and start to reassemble into a mitotic spindle. This radical change occurs abruptly and is associated with a marked increase in the dynamic instability of the microtubules, which is crucial for both the assembly and function of the spindle.

The original microtubules that radiate from each of the duplicated daughter centrosomes at the start of mitosis switch from polymerization to depolymerization at a rate that is 20 times faster than that of interphase microtubules. Moreover, many more microtubules radiate from each centrosome, and they are, on average, much shorter. Thus, at the start of mitosis, the relatively sparse, long microtubules of the interphase array are converted rapidly to a larger number of shorter, more dynamic microtubules that will form the mitotic spindle.

These differences in microtubule behavior between interphase and mitotic cells are driven by changes in the activities of various *micro-*

Question 19–1

The endoplasmic reticulum is thought to be partitioned into the daughter cells at cell division by a random distribution of fragments that are created at mitosis. Explain why random partitioning of chromosomes would not work.

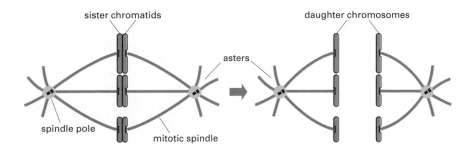

Figure 19–6 Each pair of sister chromatids separates to become two daughter chromosomes. The daughter chromosomes are then pulled to opposite poles of the cell by the mitotic spindle.

Panel 19-1 The principal stages of M phase in an animal cell

CELL DIVISION AND THE CELL CYCLE

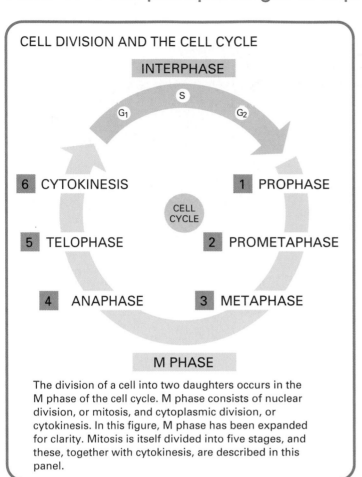

INTERPHASE

G₁   S   G₂

6 CYTOKINESIS   CELL CYCLE   1 PROPHASE

5 TELOPHASE   2 PROMETAPHASE

4 ANAPHASE   3 METAPHASE

M PHASE

The division of a cell into two daughters occurs in the M phase of the cell cycle. M phase consists of nuclear division, or mitosis, and cytoplasmic division, or cytokinesis. In this figure, M phase has been expanded for clarity. Mitosis is itself divided into five stages, and these, together with cytokinesis, are described in this panel.

INTERPHASE

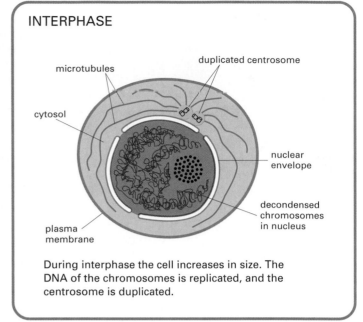

microtubules, duplicated centrosome, cytosol, nuclear envelope, decondensed chromosomes in nucleus, plasma membrane

During interphase the cell increases in size. The DNA of the chromosomes is replicated, and the centrosome is duplicated.

The light micrographs shown in this panel are of a living cell from the lung epithelium of a newt. The same cell has been photographed at different times during its division into two daughter cells. (Courtesy of Conly L. Rieder.)

1 PROPHASE

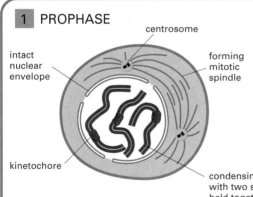

centrosome, intact nuclear envelope, forming mitotic spindle, kinetochore, condensing chromosome with two sister chromatids held together along their length

At prophase, the replicated chromosomes, each consisting of two closely associated sister chromatids, condense. Outside the nucleus, the mitotic spindle assembles between the two centrosomes, which have begun to move apart.

time = 0 min

2 PROMETAPHASE

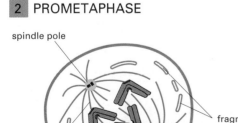

spindle pole, fragments of nuclear envelpoe, kinetochore microtubule, chromosome in active motion

Prometaphase starts abruptly with the breakdown of the nuclear envelope. Chromosomes can now attach to spindle microtubules via their kinetochores and undergo active movement.

time = 79 min

3 METAPHASE

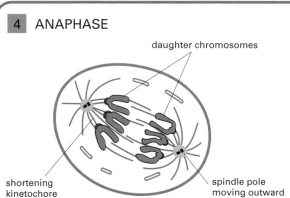

At metaphase, the chromosomes are aligned at the equator of the spindle, midway between the spindle poles. The paired kinetochore microtubules on each chromosome attach to opposite poles of the spindle.

spindle pole

astral microtubule

kinetochore microtubule

spindle pole

kinetochores of all chromosomes aligned in a plane midway between two spindle poles

time = 250 min

4 ANAPHASE

At anaphase, the paired chromatids synchronously separate to form two daughter chromosomes, and each is pulled slowly toward the spindle pole it is attached to. The kinetochore microtubules get shorter, and the spindle poles also move apart, both contributing to chromosome separation.

daughter chromosomes

shortening kinetochore microtubule

spindle pole moving outward

time = 279 min

5 TELOPHASE

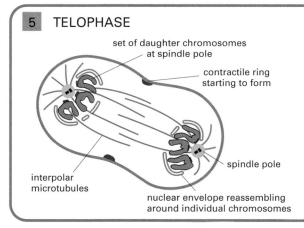

During telophase, the two sets of daughter chromo-somes arrive at the poles of the spindle. A new nuclear envelope reassembles around each set, completing the formation of two nuclei and marking the end of mitosis. The division of the cytoplasm begins with the assembly of the contractile ring.

set of daughter chromosomes at spindle pole

contractile ring starting to form

interpolar microtubules

spindle pole

nuclear envelope reassembling around individual chromosomes

time = 315 min

6 CYTOKINESIS

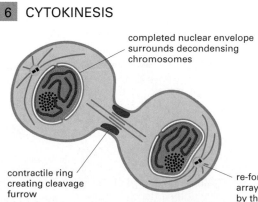

During cytokinesis of an animal cell, the cytoplasm is divided in two by a contractile ring of actin and myosin filaments, which pinches in the cell to create two daughters, each with one nucleus.

completed nuclear envelope surrounds decondensing chromosomes

contractile ring creating cleavage furrow

re-formation of interphase array of microtubules nucleated by the centrosome

time = 362 min

tubule-associated proteins (MAPs). During interphase, many different MAPs bind to and stabilize microtubules. At the onset of mitosis, the M-Cdk that triggers entry into M phase phosphorylates some of these MAPs, reducing their ability to stabilize microtubules. The presence of additional proteins called *catastrophins* further destabilizes the microtubules by promoting their sudden depolymerization. Together, these changes help to drive the massive reorganization of the cell's microtubules that takes place at the start of M phase.

The Mitotic Spindle Starts to Assemble in Prophase

As discussed earlier, at the start of S phase, the cell begins to duplicate its centrosome to produce two daughter centrosomes, which initially remain together at one side of the nucleus. At the beginning of **prophase**, the two daughter centrosomes separate. They now organize their own array of microtubules and begin to move to opposite poles of the cell (see Figure 19–5), driven, in part, by centrosome-associated motor proteins that use the energy of ATP hydrolysis to move along microtubules. (The general mechanism of action of motor proteins is discussed in Chapter 17.)

The rapidly growing and shrinking microtubules extend in all directions from the two centrosomes, exploring the interior of the cell. During prophase, some of the microtubules growing from one centrosome interact with the microtubules from the other centrosome. This interaction stabilizes the microtubules, preventing them from depolymerizing, and it joins the two sets of microtubules together to form the basic framework of the **mitotic spindle**, with its characteristic bipolar shape. The two centrosomes that give rise to these microtubules are now called **spindle poles**, and the interacting microtubules are called *interpolar microtubules* (Figure 19–7). The assembly of the spindle is driven, in part, by motor proteins associated with the interpolar microtubules that help to cross-link the two sets of microtubules that form the mitotic spindle (see How We Know, pp. 646–647).

Plant cells lack centrosomes, yet they construct fully functional, bipolar, mitotic spindles. That they do so indicates the importance of motor proteins, and of the chromosomes themselves, in spindle assembly. In cells without centrosomes, the chromosomes nucleate microtubule assembly, and motor proteins move and organize the microtubules and chromosomes into a functional bipolar spindle. This is how spindles form in the cells of plants and in animal cells that have been induced to divide without centrosomes (Figure 19–8).

In the next stage of mitosis, the replicated chromosomes attach to the spindle in such a way that, when the sister chromatids separate, they will be drawn to opposite poles.

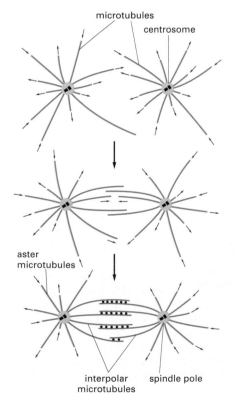

Figure 19–7 A bipolar mitotic spindle is formed by the selective stabilization of interacting microtubules. New microtubules grow out in random directions from the two centrosomes. The two ends of a microtubule, called the plus and the minus ends, have different properties, and it is the minus end that is anchored in the centrosome (discussed in Chapter 17). The free plus ends are "dynamically unstable" and switch suddenly from uniform growth (*outward-pointing red arrows*) to rapid shrinkage (*inward-pointing red arrows*). When two microtubules from opposite centrosomes interact in an overlap zone, motor proteins and other microtubule-associated proteins cross-link the microtubules together (*black dots*) in a way that stabilizes their plus ends by decreasing the probability of their depolymerization.

microtubules

centrosome

aster
microtubules

interpolar   spindle pole
microtubules

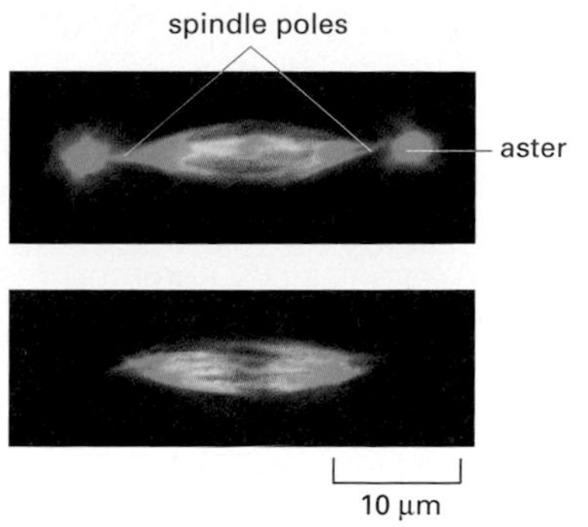

Figure 19–8 Motor proteins and chromosomes can direct the assembly of a functional bipolar spindle in the absence of centrosomes. In these fluorescence micrographs of embryos of the insect *Sciara*; the microtubules are stained *green* and the chromosomes *red*. The top micrograph shows a normal spindle formed with centrosomes in a normally fertilized embryo. The bottom micrograph shows a spindle formed without centrosomes in an embryo that initiated development without fertilization and thus lacks the centrosome normally provided by the sperm when it fertilizes the egg. Note that the spindle with centrosomes has an aster at each pole, whereas the spindle formed without centrosomes does not. Both types of spindles are able to segregate the replicated chromosomes. (From B. de Saint Phalle and W. Sullivan, *J. Cell Biol.* 141:1383–1391, 1998. © The Rockefeller University Press.)

aster

10 μm

Chromosomes Attach to the Mitotic Spindle at Prometaphase

Prometaphase starts abruptly with the disassembly of the nuclear envelope, which breaks up into small membrane vesicles. As we discuss later, this process is triggered by the phosphorylation and consequent disassembly of the intermediate filament proteins of the nuclear lamina, the network of fibrous proteins that underlies and stabilizes the nuclear envelope (see Figure 17–8). The spindle microtubules, which have been lying in wait outside the nucleus, now gain access to the replicated chromosomes and bind to them (see Panel 19–1, pp. 642–643).

The spindle microtubules end up attached to the chromosomes through specialized protein complexes called **kinetochores**, which assemble on the condensed chromosomes during late prophase. As discussed earlier, each replicated chromosome consists of two sister chromatids joined along their length, and each chromatid is constricted at a region of specialized DNA sequence called the *centromere* (see Figure 5–18). Just before prometaphase, kinetochore proteins assemble into a large complex on each centromere. Each duplicated chromosome therefore has two kinetochores (one on each sister chromatid), which face in opposite directions (Figure 19–9). Kinetochore assembly depends on the presence of the centromere DNA sequence: in the absence of this sequence, kinetochores fail to assemble and, consequently, the chromosomes fail to segregate properly during mitosis.

Once the nuclear envelope has broken down, a randomly probing microtubule encountering a chromosome will bind to it, thereby

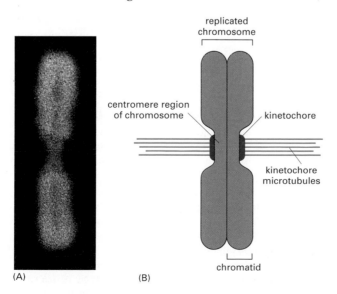

Figure 19–9 Kinetochores attach chromosomes to the mitotic spindle. (A) A fluorescence micrograph of a replicated mitotic chromosome. The DNA is stained with a fluorescent dye, and the kinetochores are stained *red* with fluorescent antibodies that recognize kinetochore proteins. These antibodies come from patients suffering from scleroderma (a disease that causes progressive overproduction of connective tissue in skin and other organs), who, for unknown reasons, produce antibodies against their own kinetochore proteins. (B) Schematic drawing of a mitotic chromosome showing its two sister chromatids attached to kinetochore microtubules, which bind by their plus ends. Each kinetochore forms a plaque on the surface of the centromere. (A, courtesy of B.R. Brinkley.)

replicated chromosome

centromere region of chromosome

kinetochore

kinetochore microtubules

chromatid

(A)   (B)

How We Know: Building the Mitotic Spindle

Scientists have been watching the dramatic dance of cell division—replicated chromosomes coming together and moving apart—for more than a century. By 1880, researchers armed with dyes that bind to DNA had described threadlike chromosomes in a wide variety of animal and plant cells. But it was not until 1882 that Walther Flemming, a German physician and microscopist, first described the movement of mitotic chromosomes. Arranging a series of stained cells like a sequential set of still photographs, Flemming could see the replicated chromosomes condense, align in the center of the cell, and then segregate into the two daughter cells.

As we now know, this chromosomal ballet is choreographed by the mitotic spindle, which captures and distributes the chromosomes, before disappearing into the cytoplasm again. Microscopy is still used to study the assembly and disassembly of the spindle, as cells move through the cell cycle from interphase to M phase and back again. Using video microscopy (see How We Know, pp. 586–588), investigators can even monitor, in real time, the growth and shrinkage of the individual microtubules that make up the spindle in a living cell.

Of course, working with whole cells presents problems of its own. Many microscopic techniques are highly sensitive to the thickness of the sample; structures in the rounded, central region of a dividing cell can be almost impossible to observe. Although one can add (or delete) a specific protein to a living cell using recombinant DNA technology or microinjection, it is much easier to manipulate proteins and to study microtubule dynamics in cytoplasmic extracts of cells, especially those prepared from *Xenopus* eggs. These cells are poised to undertake the rapid cleavage divisions that follow fertilization and are stockpiled with proteins involved in cell division, including those involved in reorganizing the microtubule cytoskeleton into a mitotic spindle. The extracts have the ability to produce many of the changes that occur in intact cells during M phase, and they have allowed researchers to identify and characterize a variety of proteins involved in spindle assembly.

Watching microtubules in cell extracts

Cytoplasmic extracts can be prepared from *Xenopus* eggs that are either in interphase or mitosis. Both of these extracts contain all of the factors needed for microtubule assembly, so that when centrosomes are added, they nucleate the assembly of tubulin into microtubules. If fluorescent tubulin is also added, one can observe fluorescent microtubules growing from the centrosomes and monitor their behavior using time-lapse, fluorescence video microscopy. This *in vitro* system confirms the observations that have been made in living cells: the microtubules in mitotic extracts, for example, are shorter than those in interphase extracts (Figure 19–10), and they are more dynamic.

But what accounts for this difference? Do microtubules in mitotic cells, for instance, grow more slowly than those in interphase cells? Do they shrink more rapidly? The answer is neither. The polymerization and depolymerization rates of individual microtubules in interphase and M phase extracts are similar. Instead, the individual microtubules in mitotic extracts show a dramatic increase in the frequency of *catastrophes*—the sudden transition from growth to shrinkage, leading to the rapid disassembly of the microtubule. This observation led researchers to search for the factors that destabilize mitotic microtubules.

Witnessing catastrophes

The first protein that promotes microtubule catastrophes was isolated from *Xenopus* egg extracts by a group of researchers who were searching for proteins related to kinesin, a motor protein thought to be involved in spindle assembly. Using an antibody that recognizes kinesins, the researchers isolated a protein that is required for normal spindle assembly in *Xenopus* egg extracts.

The protein, which they dubbed XKCM1 (for *Xenopus* kinesin central motor 1), affected microtubule dynamics quite dramatically. When the protein was removed from a mitotic extract using anti-kinesin antibodies, the resulting spindles contained abnormally long microtubules, similar in length to the microtubules formed in interphase extracts (Figure 19–11). This change in spindle formation was caused specifically by depletion of XKCM1, as when the protein was added back to the depleted extracts, the microtubule asters returned to their normal size. Furthermore, the

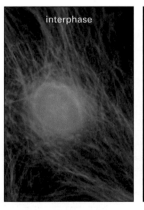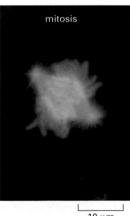

interphase | mitosis

10 μm

Figure 19–10 Microtubules are shorter in mitotic cells than in interphase cells. In the fluorescence micrographs of cultured newt lung cells in interphase (A) and metaphase (B), the chromosomes are stained *blue* and the microtubules *green*. (Courtesy of C.L. Rieder, J.C. Waters, and R.W. Cole.)

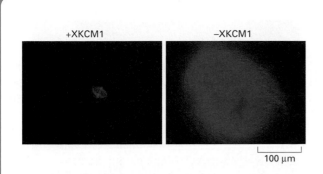

+XKCM1   −XKCM1

100 μm

Figure 19–11 A kinesin-related protein that promotes microtubule depolymerization is required for mitotic spindle assembly. When centrosomes and sperm nuclei are added to mitotic *Xenopus* egg extracts, mitotic spindles assemble (A). If, however, the kinesin-related protein XKCM1 is first removed from the extract with antibodies, huge asters form, with abnormally long microtubules, and spindles fail to assemble. This finding suggests that XKCM1 normally acts to depolymerize microtubules during mitotic spindle formation and that this function is required for spindles to assemble. In these fluorescence micrographs, microtubules are shown in *red* and DNA in *blue.* (Courtesy of Ryoma Ohi and Tim Mitchison.)

researchers demonstrated that XKCM1 promotes microtubule catastrophes in *Xenopus* extracts: depletion of the protein reduced the frequency of catastrophes, allowing the mitotic microtubules to grow to extreme lengths. Such proteins are now called *catastrophins*.

Looking at MAPs

Around the same time that researchers were isolating catastrophins, other investigators discovered that *Xenopus* egg extracts also contain proteins that have the opposite effect: they stimulate microtubule assembly. These *Xenopus* microtubule-associated proteins, or XMAPs, promote microtubule growth through a variety of mechanisms: some enhance polymerization, others inhibit depolymerization.

With all of these proteins in hand—some, like XKCM1, that promote microtubule catastrophes, others, like XMAPs, that promote microtubule assembly—researchers were left with the question of how these factors act together to control microtubule dynamics in the cell. To find out, they removed a stabilizing XMAP, a destabilizing catastrophin, or both, from *Xenopus* extracts and monitored the results.

As expected, removal of the stabilizing XMAP greatly increased the catastrophe rate and resulted in shorter, more dynamic microtubules in both interphase and mitotic preparations. Thus, this protein normally stabilizes microtubules and inhibits catastrophes in interphase and mitotic extracts (Figure 19–12). Inactivating the catastrophin XKCM1 in these extracts had the opposite effect, causing the catastrophe frequency to drop dramatically. Thus, in these extracts, it appears that microtubule length and catastrophe rates are regulated by a balance between the XMAP, which stabilizes the microtubules, and XKCM1, which promotes catastrophes. This balance is crucial for the assembly of the mitotic spindle, as microtubules that are either too long or too short are incapable of assembling into a functional spindle (see Figure 19–12B).

While microtubule behavior in extracts is not exactly the same as their behavior in living cells, these studies in egg extracts provided valuable insights into how the transition between interphase and mitotic microtubules may be mediated.

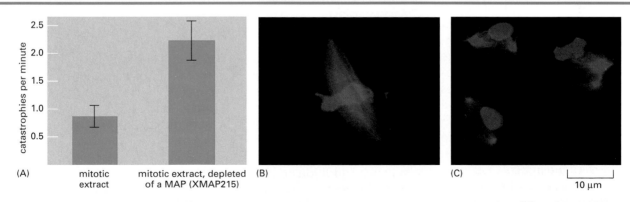

Figure 19–12 The balance between catastrophins and MAPs influences the frequency of microtubule catastrophes and microtubule length and thereby mitotic spindle formation. Mitotic *Xenopus* egg extracts were incubated with centrosomes, and the behavior of individual microtubules nucleated from the centrosomes was followed by fluorescence video microscopy. (A) The depletion of a specific MAP (XMAP215) from the mitotic extracts greatly increases the catastrophe rate, indicating that the MAP normally inhibits catastrophes in mitotic extracts. (B and C) Fluorescence micrographs of the mitotic spindles formed in an extract from normal cells (B) and in an extract in which XMAP215 has been depleted with an antibody (C). Spindle formation was initiated by the addition of centrosomes and sperm nuclei. Microtubules are shown in *red* and chromosomes in *blue.* Note that normal spindles form in normal mitotic extracts, but very abnormal spindles form when XMAP215 is depleted from the extracts, as the microtubules nucleated by the centrosomes are too short. (From R. Tournebize et al., *Nature Cell Biol.* 2:13–19, 2000. © Macmillan Magazines Ltd.)

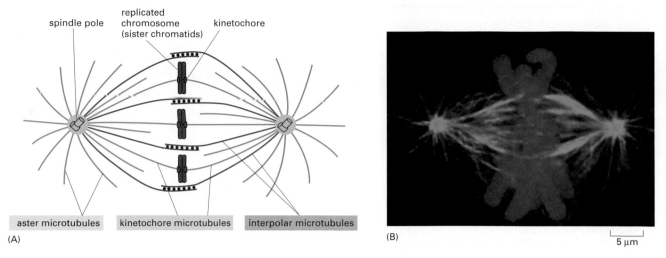

Figure 19–13 Three classes of microtubules make up the mitotic spindle. (A) Schematic drawing of a spindle with chromosomes attached, showing the three types of spindle microtubules—astral microtubules, kinetochore microtubules, and interpolar microtubules. In reality, the chromosomes are much larger than shown, and usually multiple microtubules are attached to each kinetochore. (B) Fluorescence micrograph of chromosomes at the metaphase plate of a real mitotic spindle. In this image, kinetochores are labeled in *red*, microtubules in *green*, and chromosomes in *blue*. (B, from A. Desai, *Curr. Biol.* 10:R508, 2000. © Elsevier Science.)

Question 19–2

If fine glass needles are used to manipulate a chromosome inside a living cell during early M phase, it is possible to trick the kinetochores on the two sister chromatids into attaching to the same spindle pole. This arrangement is normally unstable, but the attachments can be stabilized if the needle is used to gently pull the chromosome so that the microtubules attached to both kinetochores (and the same spindle pole) are under tension. What does this suggest to you about the mechanism by which kinetochores normally become attached and stay attached to microtubules from opposite spindle poles? Is the finding consistent with the possibility that a kinetochore is programmed to attach to microtubules from a particular spindle pole? Explain your answers.

capturing the chromosome. The microtubule eventually attaches to the kinetochore and is now called a *kinetochore microtubule*, linking the chromosome to a spindle pole (see Figure 19–9 and Panel 19–1, pp. 642–643). As kinetochores on sister chromatids face in opposite directions, they tend to attach to microtubules from opposite poles of the spindle, so that each replicated chromosome becomes linked to both spindle poles. The number of microtubules attached to each kinetochore varies among species: each human kinetochore binds 20–40 microtubules, for example, whereas a yeast kinetochore binds just one. The three classes of microtubules that form the mitotic spindle are shown in Figure 19–13.

Chromosomes Line Up at the Spindle Equator at Metaphase

During prometaphase, the chromosomes, now attached to the mitotic spindle, begin to move around, as if jerked first this way and then that. Eventually, they align at the equator of the spindle, halfway between the two spindle poles, thereby forming the *metaphase plate*. This defines the beginning of **metaphase** (Figure 19–14). Although the forces that act to bring the chromosomes to the equator are not well understood, both the continual growth and shrinkage of microtubules and the action of microtubule motor proteins are thought to be involved. A continuous balanced addition and loss of tubulin subunits is also required to maintain the metaphase spindle: when tubulin addition to the ends of microtubules is blocked by the drug colchicine, tubulin loss continues until the spindle disappears.

The chromosomes gathered at the equator of the metaphase spindle oscillate back and forth, continually adjusting their positions, indicating that the tug-of-war between the microtubules attached to opposite poles of the spindle continues to operate after the chromosomes are all aligned. If one of the pair of kinetochore attachments is artificially severed with a laser beam during metaphase, the entire chromosome immediately moves toward the pole to which it remains attached. Similarly, if the attachment between sister chromatids is cut, the two

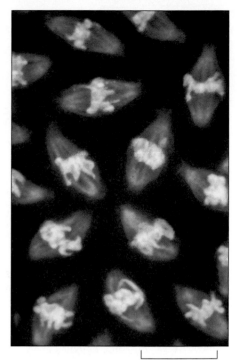

Figure 19–14 During metaphase, chromosomes gather halfway between the two spindle poles. This fluorescence micrograph shows multiple mitotic spindles at metaphase in a fruit fly *(Drosophila)* embryo. The microtubules are stained *red,* and the chromosomes are stained *green.* At this stage of *Drosophila* development, there are multiple nuclei in one large cytoplasmic compartment, and all of the nuclei divide synchronously, which is why all of the nuclei shown here are in metaphase. Although metaphase spindles are usually pictured in two dimensions, as they are here, when viewed in three dimensions the chromosomes are seen to be gathered at a platelike region at the equator of the spindle—the so-called metaphase plate. (Courtesy of William Sullivan.)

4 μm

chromatids separate and move toward opposite poles. These experiments show that the chromosomes at the metaphase plate are held there under considerable tension. Evidently, the forces that will ultimately pull the sister chromatids apart begin operating as soon as microtubules attach to the kinetochores.

Daughter Chromosomes Segregate at Anaphase

Anaphase begins abruptly with the release of the cohesin linkage that holds the sister chromatids together (see Figure 19–3A). This allows each chromatid (now called a daughter chromosome) to be gradually pulled to the spindle pole to which it is attached (Figure 19–15). This movement distributes, or *segregates,* the two identical sets of chromosomes to opposite ends of the spindle (see Panel 19–1, pp. 642–643).

The abrupt disruption of the cohesin linkage between sister chromatids is triggered by the activation of the **anaphase promoting complex** (**APC**) discussed in Chapter 18 (see p. 617). Once this proteolytic complex is activated, it cleaves an inhibitory protein, thereby releasing a proteolytic enzyme that breaks the cohesin linkage (Figure 19–16).

Once released, all of the newly separated daughter chromosomes move to the spindle poles at the same speed, typically about 1 μm per minute. The movement is the consequence of two independent processes mediated by different parts of the mitotic spindle. These processes are called *anaphase A* and *anaphase B,* and they occur more or less simultaneously. In anaphase A, the kinetochore microtubules

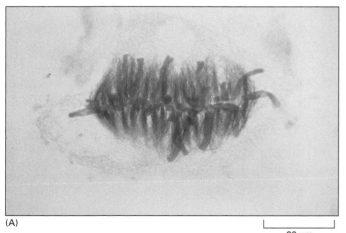

(A)

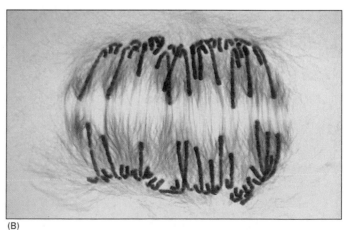

(B)

20 μm

Figure 19–15 Sister chromatids separate at anaphase. In the transition from metaphase (A) to anaphase (B), sister chromatids (stained *blue)* suddenly separate and move toward opposite poles, as seen in these plant cells stained with gold-labeled antibodies to label the microtubules. Plant cells generally do not have centrosomes and therefore have less sharply defined spindle poles than animal cells (see Figure 19–22E), but spindle poles are present here at the top and bottom of each micrograph, although they cannot be seen. (Courtesy of Andrew Bajer.)

Figure 19–16 APC triggers the separation of sister chromatids by promoting the destruction of cohesins. Activated APC indirectly triggers the cleavage of the cohesins that hold sister chromatids together. It cataylzes the ubiquitination and destruction of a protein that inhibits the activity of a proteolytic enzyme. Freed from this inhibitory protein, the proteolytic enzyme then cleaves the cohesin complexes. When the cohesins fall away, the mitotic spindle is able to pull the sister chromatids apart.

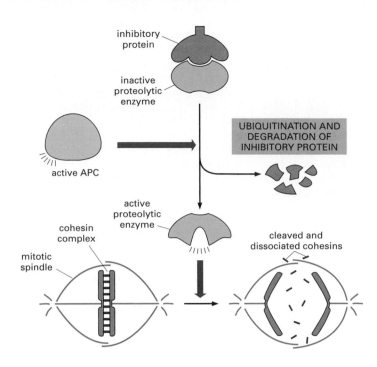

Figure 19–17 Two processes separate sister chromatids at anaphase. In *anaphase A*, the daughter chromosomes are *pulled* toward opposite poles as the kinetochore microtubules depolymerize at the kinetochore. The force driving this movement is generated mainly at the kinetochore. In *anaphase B*, the two spindle poles move apart as the result of two separate forces: (1) the elongation and sliding of the interpolar microtubules past one another pushes the two poles apart, and (2) outward forces exerted by outward-pointing astral microtubules at each spindle pole *pull* the poles away from each other, toward the cell cortex. All of these forces are thought to depend on the action of motor proteins associated with the microtubules.

shorten by depolymerization, and the attached chromosomes move poleward. In anaphase B, the spindle poles themselves move apart, further contributing to the segregation of the two sets of daughter chromosomes (Figure 19–17).

The driving force for the movements of anaphase A is thought to be provided mainly by the action of microtubule motor proteins operating at the kinetochore, aided by the loss of tubulin subunits that occurs primarily where the kinetochore microtubules attach to the chromosomes. The loss of tubulin subunits at the kinetochore depends on a catastrophphin that is bound to both the microtubule and the kinetochore and uses the energy of ATP hydrolysis to remove tubulin subunits from the microtubule.

In anaphase B, the overlapping interpolar microtubules elongate and slide past each other, pushing the spindle poles and the two sets of chromosomes farther apart. The driving forces for this process are

ANAPHASE A   CHROMOSOMES ARE PULLED POLEWARD

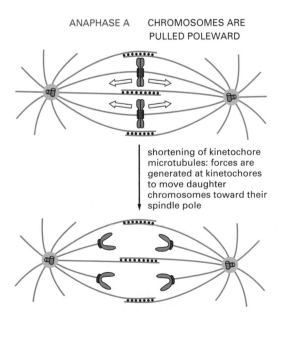

shortening of kinetochore microtubules: forces are generated at kinetochores to move daughter chromosomes toward their spindle pole

ANAPHASE B   POLES ARE PUSHED AND PULLED APART

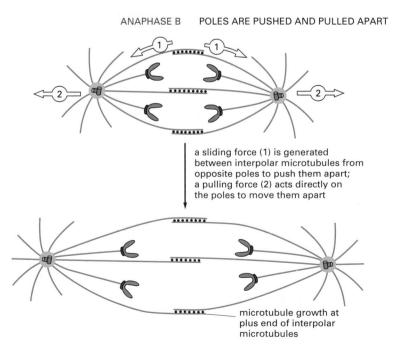

a sliding force (1) is generated between interpolar microtubules from opposite poles to push them apart; a pulling force (2) acts directly on the poles to move them apart

microtubule growth at plus end of interpolar microtubules

thought to be provided by two sets of motor proteins—members of the kinesin and dynein families (see Figure 17–17)—operating on spindle microtubules. One set acts on the long, overlapping interpolar microtubules that form the spindle itself; these motor proteins slide the interpolar microtubules from opposite poles past one another at the equator of the spindle, pushing the spindle poles apart. The other set operates on the astral microtubules that extend from the spindle poles and point away from the spindle equator and toward the cell periphery. These motor proteins are thought to be associated with the cell cortex, which underlies the plasma membrane, and they pull each pole toward the adjacent cortex and away from the other pole (see Figure 19–17).

The Nuclear Envelope Re-forms at Telophase

By the end of anaphase, the daughter chromosomes have separated into two equal groups, one at each pole of the spindle. During **telophase**, the final stage of mitosis, a nuclear envelope reassembles around each group of chromosomes to form the two daughter nuclei. Vesicles of nuclear membrane first cluster around individual chromosomes and then fuse to re-form the nuclear envelope (see Panel 19–1, pp. 642–643). During this process, the nuclear pores (discussed in Chapter 15) reassemble in the envelope, and the nuclear lamins, the intermediate filament protein subunits that were phosphorylated during prophase, are now dephosphorylated and reassociate to re-form the nuclear lamina (Figure 19–18). Once the nuclear envelope has reformed, the pores pump in nuclear proteins, the nucleus expands, the condensed mitotic chromosomes decondense into their interphase state. As a consequence of decondensation, gene transcription is able to resume. A new nucleus has been created, and mitosis is complete. All that remains is for the cell to complete its division into two.

Some Organelles Fragment at Mitosis

The process of mitosis ensures that each daughter cell receives a full complement of chromosomes. But when a eucaryotic cell divides, each daughter cell must also inherit all of the other essential cell components, including the membrane-enclosed organelles. Organelles such

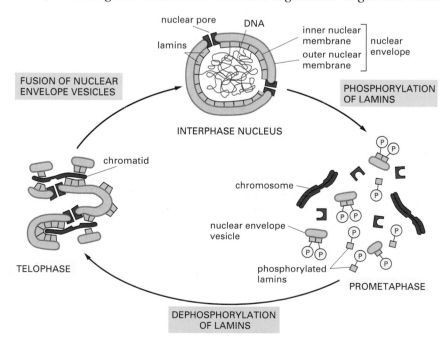

Figure 19–18 The nuclear envelope breaks down and re-forms during mitosis. The phosphorylation of the lamins helps trigger the disassembly of the nuclear lamina at prometaphase, which in turn causes the nuclear envelope to break up into vesicles. Dephosphorylation of the lamins at telophase helps reverse the process.

as mitochondria and chloroplasts cannot assemble spontaneously from their individual components; they arise only from the growth and division of the preexisting organelles. Likewise, cells cannot make a new endoplasmic reticulum or Golgi apparatus unless some part of it is already present, which can then be enlarged. How then are these various membrane-enclosed organelles segregated when the cell divides?

Organelles such as mitochondria and chloroplasts are usually present in large numbers and will be safely inherited if, on average, their numbers simply double once each cell cycle. The ER in interphase cells is continuous with the nuclear membrane and is organized by the microtubule cytoskeleton (see Figure 17–23A). Upon entry into M phase, the reorganization of the microtubules releases the ER, which fragments as the nuclear envelope breaks down. The Golgi apparatus probably fragments as well, although in some cells it seems to redistribute transiently into the ER, only to re-emerge at telophase. Some of the organelle fragments associate with the spindle microtubules via motor proteins, thereby hitching a ride into the daughter cells as the spindle elongates in anaphase. Other components of the cell, including all of the soluble proteins, are inherited randomly when the cell divides its cytoplasm in the final stage of M phase, which we now discuss.

Cytokinesis

Cytokinesis, the process by which the cytoplasm is cleaved in two. Cytokinesis usually begins in anaphase but is not completed until after the two daughter nuclei have formed. Whereas mitosis involves a transient microtubule-based structure, the mitotic spindle, cytokinesis in animal cells involves a transient structure based on actin and myosin filaments, the *contractile ring* (see Figure 19–4). Both the plane of cleavage and the timing of cytokinesis, however, are determined by the mitotic spindle.

The Mitotic Spindle Determines the Plane of Cytoplasmic Cleavage

The first visible sign of cytokinesis in animal cells is a puckering and furrowing of the plasma membrane that occurs during anaphase (Figure 19–19). The furrowing invariably occurs in a plane that runs perpendicular to the long axis of the mitotic spindle. This positioning ensures that the *cleavage furrow* cuts between the two groups of segregated daughter chromosomes so that each daughter cell receives an identical

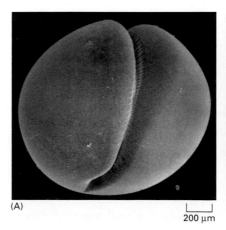

Figure 19–19 **The cleavage furrow of the plasma membrane is formed by the action of the contractile ring underneath it.** In these scanning electron micrographs of a dividing fertilized frog egg, the cleavage furrow is unusually well defined. (A) Low-magnification view of the egg surface. (B) A higher magnification view of the cleavage furrow. (From H.W. Beams and R.G. Kessel, *Am. Sci.* 36:279–290, 1976. © Reprinted by permission of *American Scientist,* journal of Sigma Xi.)

(A)   200 µm   (B)   25 µm

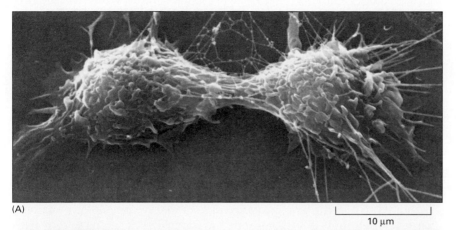

(A)

|_____|
10 µm

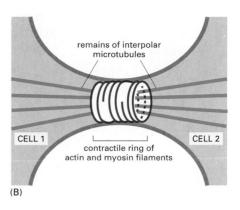

(B)

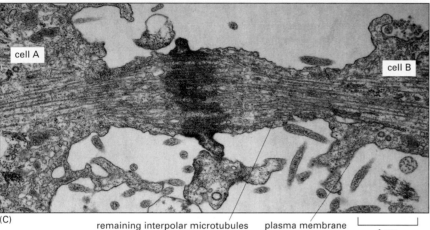

cell A

cell B

(C)

remaining interpolar microtubules   plasma membrane

|_____|
1 µm

Figure 19–20 The contractile ring divides the cell in two. (A) Scanning electron micrograph of an animal cell in culture in the last stages of dividing. (B) Schematic diagram of the midregion of a similar cell showing the contractile ring beneath the plasma membrane and the remains of the two sets of interpolar microtubules. (C) A conventional electron micrograph of a dividing animal cell. Cleavage is almost complete, but the daughter cells remain attached by a thin strand of cytoplasm containing the remains of the overlapping interpolar microtubules of the central spindle. (A, courtesy of Guenter Albrecht-Buehler; C, courtesy of J.M. Mullins.)

and complete set of chromosomes. If, as soon as the furrow appears, the mitotic spindle is deliberately displaced (using a fine glass needle inserted into the cell), the starting furrow disappears and a new one develops at a site corresponding to the new spindle location and orientation. Once the furrowing process is well under way, however, cleavage proceeds even if the mitotic spindle is artificially sucked out of the cell or depolymerized using the drug colchicine. How the mitotic spindle dictates the position of the cleavage furrow remains a mystery.

When the mitotic spindle is located centrally in the cell—the usual situation in most dividing cells—the two daughter cells produced will be of equal size. During embryonic development, however, there are some instances in which the mitotic spindle is positioned asymmetrically, and consequently, the furrow creates two cells that differ in size. In most cases, the resulting daughters also differ in the molecules they inherit, and they usually develop into different cell types. Special mechanisms are required to position the mitotic spindle eccentrically in such asymmetrical divisions.

The Contractile Ring of Animal Cells Is Made of Actin and Myosin

The contractile ring is composed mainly of an overlapping array of actin filaments and myosin filaments (Figure 19–20). It assembles at anaphase and is attached to membrane-associated proteins on the cytoplasmic face of the plasma membrane. The mechanisms responsible for timing the onset of ring assembly are still unclear. Once

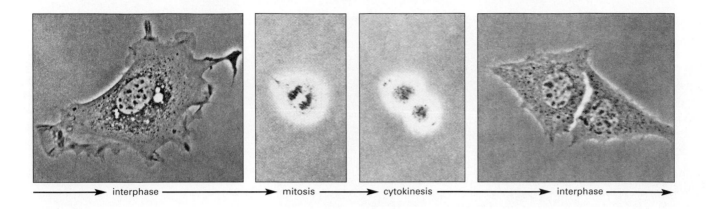

interphase ⟶ mitosis ⟶ cytokinesis ⟶ interphase ⟶

Figure 19–21 Animal cells change shape during M phase. In these micrographs of a mouse fibroblast dividing in culture, the same cell was photographed at successive times. Note how the cell rounds up as it enters mitosis; the two daughter cells then flatten out again after cytokinesis is completed. (Courtesy of Guenter Albrecht-Buehler.)

assembled, the contractile ring is capable of exerting a force strong enough to bend a fine glass needle inserted into the cell prior to cytokinesis. The force is generated by the sliding of the actin filaments against the myosin filaments (see Figure 17–41), much as occurs during muscle contraction. Unlike the contractile apparatus in muscle, however, the contractile ring is a transient structure: it assembles to carry out cytokinesis, gradually becomes smaller as cytokinesis progresses, and disassembles completely once the cell is cleaved in two.

Cell division in many animal cells is accompanied by large changes in cell shape and a decrease in the adherence of the cell to the extracellular matrix. These changes result from the reorganization of actin and myosin filaments in the cell cortex, only one aspect of which is the assembly of the contractile ring. Mammalian fibroblasts in culture, for example, spread out flat during interphase, as a result of the strong adhesive contacts they make with the surface they are growing on—called the substratum. As they enter M phase, however, the cells round up, at least partly because some of the plasma membrane proteins responsible for attaching the cells to the substratum—the *integrins* (discussed in Chapter 21)—become phosphorylated and thus weaken their grip. Once cytokinesis is complete, the daughter cells reestablish their contacts with the substratum and flatten out again (Figure 19–21). When cells divide in an animal tissue, this cycle of attachment and detachment presumably allows them to rearrange their contacts with neighboring cells and with the extracellular matrix, so that the new cells produced by cell division can be accommodated within the tissue.

Cytokinesis in Plant Cells Involves New Cell-Wall Formation

The mechanism of cytokinesis in higher plants is entirely different from that in animal cells, presumably because the plant cells are surrounded by a tough cell wall (discussed in Chapter 21). The two daughter cells are separated not by the action of a contractile ring at the cell surface but instead by the construction of a new wall inside the cell. The growing cell wall is surrounded by a membrane, and it progressively enlarges until it divides the cytoplasm in two. The positioning of this new wall precisely determines the position of the two daughter cells relative to neighboring cells. Thus, the planes of cell division, together with cell enlargement, determine the final form of the plant.

The new cell wall starts to assemble in the cytoplasm between the two sets of segregated chromosomes at the start of telophase. The assembly process is guided by a structure called the **phragmoplast**, which is formed by the remains of the interpolar microtubules at the equator of the old mitotic spindle. Small membrane-enclosed vesicles, largely derived from the Golgi aparatus and filled with polysaccharides and glycoproteins required for the cell-wall matrix, are transported along the microtubules to the equator of the phragmoplast. Here they fuse to form a disclike, membrane-enclosed structure, which expands outward by further vesicle fusion until it reaches the plasma membrane and original cell wall and divides the cell in two (Figure 19–22). Later, cellulose microfibrils are laid down within the matrix to complete the construction of the new cell wall.

Gametes Are Formed by a Specialized Kind of Cell Division

Mitosis produces a pair of daughter cells, each of which contains a full, identical complement of the parent cell's genetic material. For the cells we have been discussing in this chapter—cells that form the bodies of plants or animals—that genetic material consists of two full sets of chromosomes, one set derived from the organism's mother, the other from its father. They are therefore said to be **diploid**. A different kind of cell division, however, is required to produce gametes—the specialized cells that carry out reproduction. Unlike the diploid cells that form the body, gametes (eggs in female animals, sperm in males) contain only one set of chromosomes. They are therefore said to be **haploid**. These haploid cells are generated when a diploid cell undergoes the highly specialized process of cell division called **meiosis.** Without meiosis, the chromosome number would double with each generation produced by the fusion of an egg and sperm, and sexual reproduction would be impossible.

Although most of the mechanical features of meiosis are similar to those of mitosis, the behavior of the chromosomes is a little different. In the next chapter, we review meiosis and discuss how this specialized process of cell division underlies the genetic principles that define the laws of inheritance.

Question 19–4

Draw a detailed view of the formation of the new cell wall that separates the two daughter cells when a plant cell divides (see Figure 19–22). In particular, show where the membrane proteins of the Golgi-derived vesicles end up, indicating what happens to the part of a protein in the Golgi vesicle membrane that is exposed to the interior of the Golgi vesicle. (Refer to Chapter 11 if you need a reminder of membrane structure.)

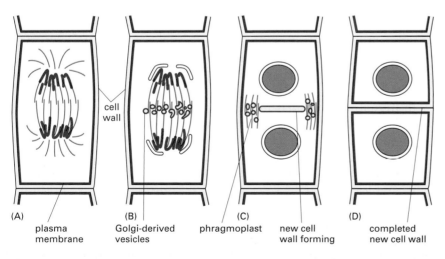

(A) plasma membrane
(B) Golgi-derived vesicles
cell wall
(C) phragmoplast
new cell wall forming
(D) completed new cell wall

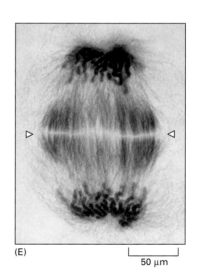

(E)
50 μm

Figure 19–22 Cytokinesis in a plant cell is guided by a specialized microtubule-based structure called the phragmoplast. At the beginning of telophase, after the chromosomes have segregated (A), a new cell wall starts to assemble inside the cell at the equator of the old spindle (B). The interpolar microtubules of the mitotic spindle remaining at telophase form the *phragmoplast* and guide the vesicles toward the equator of the spindle. Here, membrane-enclosed vesicles, derived from the Golgi apparatus and filled with cell-wall material, fuse to form the growing new cell wall (C), which grows outward to reach the plasma membrane and original cell wall. The plasma membrane and the membrane surrounding the new cell wall (both shown in *red*) fuse, completely separating the two daughter cells (D). A light micrograph of a plant cell in telophase is shown in (E) at a stage corresponding to (B). The cell has been stained to show both the microtubules and the two sets of daughter chromosomes segregated at the two poles of the spindle. The location of the growing new cell wall is indicated by the arrowheads. (E, courtesy of Andrew Bajer.)

Essential Concepts

- Cell division occurs during M phase, at which time the nucleus divides (mitosis) and then the cytoplasm divides (cytokinesis).
- DNA is replicated in S phase, before M phase begins; the two copies of each duplicated chromosome (called sister chromatids) remain tightly bound together by cohesins.
- M phase is initiated by the phosphorylations triggered by activated M-Cdk.
- M phase begins with the formation of a mitotic spindle made of microtubules, which segregates daughter chromosomes to opposite poles of the cell.
- Although the centrosome duplicates during interphase, it is only at the start of M phase that the two daughter centrosomes separate and move to opposite sides of the nucleus to form the two poles of the spindle.
- Both the separation of the centrosomes and the assembly of the spindle depend on microtubule-associated motor proteins.
- Microtubules grow out from the centrosomes, and some interact with microtubules growing from the opposite pole, thereby becoming the interpolar microtubules of the spindle.
- When the nuclear envelope breaks down, the spindle microtubules invade the nuclear area and capture the replicated chromosomes. The microtubules bind to protein complexes called kinetochores, associated with the centromere of each sister chromatid.
- Microtubules from opposite poles pull in opposite directions on each replicated chromosome, bringing the chromosomes to the equator of the mitotic spindle.
- The sudden separation of sister chromatids allows the daughter chromosomes to be pulled to opposite poles by the spindle. The two poles also move apart, further separating the two sets of chromosomes.
- The movement of chromosomes by the spindle is driven both by microtubule motor proteins and by microtubule polymerization and depolymerization.
- A nuclear envelope re-forms around the two sets of segregated chromosomes to form two new nuclei, thereby completing mitosis.
- Large membrane-enclosed organelles such as the endoplasmic reticulum and Golgi apparatus break into many smaller fragments during M phase, ensuring an even distribution between the daughter cells.
- In animal cells, cytoplasmic division is mediated by a contractile ring of actin filaments and myosin filaments, which assembles midway between the spindle poles and contracts to divide the cytoplasm in two; in plant cells, by contrast, cell division occurs by the formation of a new cell wall inside the cell, which divides the cytoplasm in two.

Key Terms

anaphase	kinetochore
anaphase promoting complex (APC)	M phase
aster	M-phase Cdk (M-Cdk)
catastrophin	metaphase
centrosome	meiosis
centrosome cycle	mitosis
chromosome condensation	mitotic spindle
chromatid	phragmoplast
cohesin	prometaphase
condensin	prophase
cyclin-dependent kinase (Cdk)	sister chromatid
cytokinesis	spindle pole
diploid	telophase
haploid	γ-tubulin ring

Questions

Question 19–5

Which of the following statements are correct? Explain your answers.

A. Centrosomes are replicated independently of chromosomes.

B. The nuclear envelope becomes fragmented at mitosis. It is thus distributed between daughter cells like other membrane-enclosed organelles such as the endoplasmic reticulum and the Golgi apparatus.

C. Two sister chromatids arise by replication of the DNA of the same chromosome and remain paired as they line up on the metaphase plate.

D. Interpolar microtubules attach end-to-end and are therefore continuous from one spindle pole to the other.

E. Microtubule polymerization and depolymerization and microtubule motor proteins are all required for DNA replication.

F. Microtubules nucleate at the centromeres and then connect to the kinetochores, which are structures at the centrosome regions of chromosomes.

Question 19–6

Roughly, how long would it take a single fertilized egg to make a cluster of cells weighing 70 kg by repeated divisions, if each cell weighs 1 nanogram just after cell division and each cell cycle takes 24 hours? Why does it take very much longer than this to make a 70-kg adult human?

Question 19–7

What is the order in which the following events occur during cell division:
- A. anaphase
- B. metaphase
- C. prometaphase
- D. telophase
- E. lunar phase
- F. mitosis
- G. prophase

Where does cytokinesis fit in?

Question 19–8

The shortest eucaryotic cell cycles of all—shorter even than those of many bacteria—occur in many early animal embryos. These divisions take place without any significant increase in the weight of the embryo. How can this be? Which phase of the cell cycle would you expect to be most reduced?

Question 19–9

The lifetime of a microtubule in a mammalian cell, between its formation by polymerization and its spontaneous disappearance by depolymerization, varies with the stage of the cell cycle. For an actively proliferating cell, the average lifetime is 5 minutes in interphase and 15 seconds in mitosis. If the average length of a microtubule in interphase is 20 μm, what will it be during mitosis, assuming that the rates of microtubule elongation due to the addition of tubulin subunits in the two phases are the same? If a typical centrosome in an interphase cell has 100 nucleation sites for microtubules, how many sites would you expect to find in centrosomes in a mitotic cell, assuming that the total number of tubulin molecules that are polymerized into microtubules is the same in both phases?

Question 19–10

An antibody that binds to myosin prevents the movement of myosin molecules along actin filaments (the interaction of actin and myosin is described in Chapter 17). How do you suppose the antibody exerts this effect? What might be the result of injecting this antibody into cells (A) on the movement of chromosomes at anaphase or (B) on cytokinesis. Explain your answers.

Question 19–11

Sketch the principal stages of mitosis, using Panel 19–1 (pp. 642–643) as a guide. Color one sister chromatid and follow it through mitosis and cytokinesis. What event commits this chromatid to a particular daughter cell? Once initially committed, can its fate be reversed? What may influence this commitment?

Question 19–12

The polar movement of chromosomes during anaphase A is associated with microtubule shortening. In particular, microtubules depolymerize at the ends at which they are attached to the kinetochores. Sketch a model that explains how a microtubule can shorten and generate force, yet remain firmly attached to the chromosome.

Question 19–13

The balance between plus-end-directed and minus-end-directed motor proteins that bind to interpolar microtubules in the overlap region of the mitotic spindle is thought to help determine the length of the spindle. How might each type of motor protein contribute to the determination of spindle length?

Question 19–14

Rarely, both sister chromatids of a replicated chromosome end up in one daughter cell. How might this happen? What could be the consequences of such a mitotic error?

Highlights from *Essential Cell Biology 2 Interactive* CD-ROM

19.1   Plant Cell Division
19.2   Animal Cell Division
19.3   Mitotic Spindles in a Fly Embryo

Genetics, Meiosis, and the Molecular Basis of Heredity

It is estimated that more than 10 million species are living on Earth today. Each species is different, and each reproduces itself faithfully, yielding progeny that belong to the same species. Dogs have puppies, cats have kittens, people have babies, and heritable characteristics are passed from parents to their offspring with some degree of predictability. The phenomenon of heredity—whereby the peculiarities of parents dictate the peculiarities of progeny—is a central feature of life.

The mechanism by which characteristics are transmitted from one generation to the next remained a mystery until the late 1800s. At that time, biologists deduced that chromosomes carry the units of inheritance, recognizing that chromosomes are doled out when the sex cells are formed and are then reunited at fertilization. Before that time, theories about inheritance entertained a variety of possibilities. For example, Aristotle, who wrote the first reputable book dealing with reproduction, *Generation of Animals*, believed that common creatures could arise from unusual unions: the giraffe, for example, was thought to be a hybrid animal resulting from a cross between a camel and a leopard. (This quaint notion is reflected in the giraffe's scientific name, *Giraffa camelopardalis*.)

Another mistaken belief was that children were the sole descendants of one parent or the other, a theory dubbed uniparental inheritance. When the Dutch physician Reinier de Graaf first encountered the female ovarian follicle in the late 1600s, he proposed that babies were preformed in the mother, and the father merely provided the "vital spark" needed to jump start the embryo's development. Around the same time, Antoni van Leeuwenhoek, looking through his simple microscope lens, first observed living sperm. His discovery spawned the opposing theory that sperm carried perfectly formed little humans, called homunculi, who simply needed to be implanted in a female incubator to grow (Figure 20–1).

By the mid-nineteenth century, people had come to recognize that traits observed in children were a mixture of traits derived from both parents. They mistakenly believed, however, that substances within the sperm and egg would mix, like paint, to produce offspring that were intermediate in character. Even, the great scientist Charles Darwin supported a form of this erroneous theory, known as blended inheritance. He supposed that every cell in an organism produces a substance corresponding to that particular cell and that these substances circulate through the body and filter into the sex cells. When a sperm and egg unite, the two sets of substances they contain would blend, and later, by an unknown mechanism, would eventually regenerate all of the cell types from which they were formed.

Today the physical basis of inheritance seems obvious. We now know that genes are carried on chromosomes. The chromosomes are portioned out into specialized sex cells, called *gametes*, the eggs and the sperm. These come together during fertilization to produce offspring with an equal number of chromosomes from each parent, explaining

The Benefits of Sex
Sexual Reproduction Involves Both Diploid and Haploid Cells
Sexual Reproduction Gives Organisms a Competitive Advantage

Meiosis
Haploid Germ Cells Are Produced From Diploid Cells Through Meiosis
Meiosis Involves a Special Process of Chromosome Pairing
Extensive Recombination Occurs Between Maternal and Paternal Chromosomes
Chromosome Pairing and Recombination Ensure the Proper Segregation of Homologs
The Second Meiotic Division Produces Haploid Daughter Cells
The Haploid Cells Contain Extensively Reassorted Genetic Information
Meiosis Is Not Flawless
Fertilization Reconstitutes a Complete Genome

Mendel and the Laws of Inheritance
Mendel Chose to Study Traits That Are Inherited in a Discrete Fashion
Mendel Could Disprove the Alternative Theories of Inheritance
Mendel's Experiments Were the First to Reveal the Discrete Nature of Heredity
Each Gamete Carries a Single Allele for Each Character
Mendel's Law of Segregation Applies to All Sexually Reproducing Organisms
Alleles for Different Traits Segregate Independently
The Behavior of Chromosomes During Meiosis Underlies Mendel's Laws of Inheritance
The Frequency of Recombination Can Be Used to Order Genes on Chromosomes
The Phenotype of a Heterozygote Reveals Whether an Allele is Dominant or Recessive
Mutant Alleles Sometimes Confer a Selective Advantage

Genetics As an Experimental Tool
The Classical Approach Begins with Random Mutagenesis
Genetic Screens Identify Mutants Deficient in Cellular Processes
A Complementation Test Reveals Whether Two Mutations Are in the Same Gene
Human Genes Are Inherited in Haplotype Blocks, Which Can Aid in the Search for Mutations That Cause Disease
Complex Traits Are Influenced by Multiple Genes

Is Our Fate Encoded in Our DNA?

Figure 20–1 **One incorrect theory of inheritance suggested that genetic traits are passed down solely from the father.** In support of this particular theory of uniparental inheritance, some microscopists fancied that they could detect a small, perfectly formed human, crouched inside the head of each sperm.

why each child possesses a unique combination of traits derived from both parents.

In this chapter, we discuss the molecular mechanisms that underlie **genetics**, the science of heredity. We have already discussed how genes are made of DNA, how they are arranged on chromosomes, and how they are copied (Chapters 5 and 6). To pass these genes on to any progeny, however, sexually reproducing organisms must first sort and distribute their chromosomes into the gametes that will carry their genes on to the next generation. Parceling out the chromosomes into sperm and egg is the job of a specialized form of cell division called meiosis.

We begin our discussion of genetics with a discussion of sexual reproduction. We then discuss how gametes are formed through meiosis, and how the random assortment of chromosomes during this process generates genetic diversity. In the third section of the chapter, we review how Gregor Mendel deduced the laws of heredity and predicted the existence of the physical units of inheritance we now call genes. We then show how the behavior of chromosomes during meiosis underlies Mendel's laws of inheritance. Finally, we discuss how genetics can be used as an experimental tool for determining how genes function together to make a whole organism, with an emphasis on the study of human genetics.

The Benefits of Sex

Most of the creatures we see around us reproduce sexually. However, many organisms, especially those invisible to the naked eye, can reproduce without sex. Bacteria and other single-celled organisms can reproduce by simple cell division (Figure 20–2). Many plants also reproduce asexually, forming multicellular offshoots that later detach from the parent to make new independent plants. Even in the animal kingdom, some worms can be split into two halves, each of which regenerates its

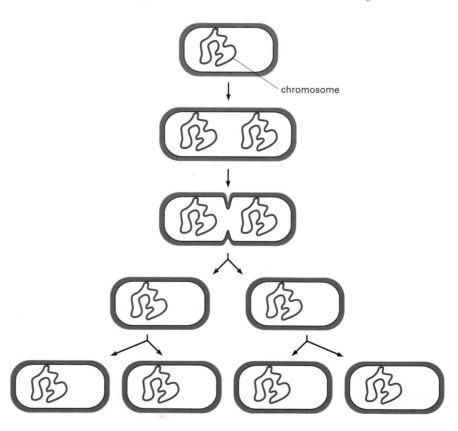

Figure 20–2 **Bacteria reproduce by simple cell division.** The division of one bacterium into two takes 20–25 minutes under ideal growth conditions.

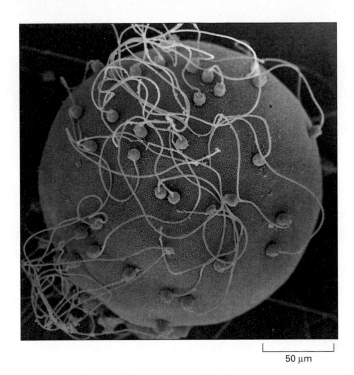

50 µm

Figure 20–3 **Despite their tremendous difference in size, sperm and egg contribute equally to the genetic character of the zygote.** This difference in size (eggs contain a large quantity of cytoplasm, whereas sperm contain almost none) is consistent with our knowledge that the cytoplasm is not the basis of inheritance. If it were, the female's contribution to the make-up of the offspring would be much greater than the male's. Shown here is a scanning electron micrograph of a section of a clam egg with sperm bound to its surface. Although many sperm are bound to the egg, only one will fertilize it. (Courtesy of David Epel.)

missing half. But while such **asexual reproduction** is simple and direct, it usually gives rise to offspring that are genetically identical to the parent organism. **Sexual reproduction**, on the other hand, involves the mixing of genomes from two individuals to produce offspring that are genetically distinct from one another and from both their parents. This mode of reproduction apparently has great advantages, as the vast majority of plants and animals have adopted it.

Sexual Reproduction Involves Both Diploid and Haploid Cells

Sexual reproduction occurs in diploid organisms, in which each cell contains two sets of chromosomes, one inherited from each parent. Each diploid cell, therefore, carries two copies of each gene (with the exception of the genes carried on the sex chromosome of males, which are often present in only one copy). Moreover, for each gene, there are numerous varieties present in the gene pool of a species, and sexual reproduction ensures that new combinations of genes are continually tried out.

Unlike the other cells in a diploid organism, the specialized cells that actually carry out sexual reproduction—the **germ cells**, or **gametes**—are haploid; that is, they each contain only one set of chromosomes. Typically, two types of gametes are produced. In animals one is large and nonmotile, and is referred to as the egg; the other is small and motile and is referred to as the sperm (Figure 20–3). These haploid germ cells are generated when a diploid cell undergoes the highly specialized process of cell division called *meiosis* (Figure 20–4). During meiosis the chromosomes of the double chromosome set are partitioned out, in fresh combinations, into single chromosome sets. The two different haploid gametes then fuse to make a diploid cell (the fertilized egg, or **zygote**) with a new combination of chromosomes. The zygote thus produced develops into a new individual with a diploid set of chromosomes that is distinct from that of either parent.

For almost all multicellular animals, including vertebrates, practically the whole life cycle is spent in the diploid state: the haploid cells exist only briefly, do not divide at all, and are highly specialized

Figure 20–4 Sexual reproduction involves both haploid and diploid cells. (A) Cells in higher eucaryotic organisms proliferate in the diploid phase to form a multicellular organism; only the gametes (the egg and the sperm) are haploid. (B) These gametes reunite at fertilization to generate a diploid zygote, which will develop into a diploid organism. For simplicity, only one chromosome is shown for each gamete, and the sperm cell has been greatly enlarged (see Figure 20–3 for actual sizes).

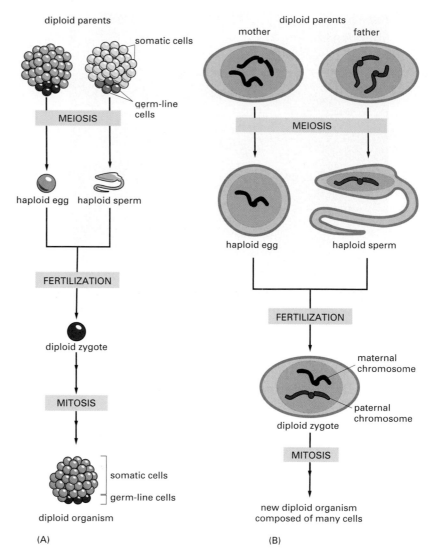

for sexual function. In most animals, therefore, a useful distinction can be drawn between the cells of the **germ line**, from which the next generation of gametes will be derived, and the **somatic cells**, which form the rest of the body and ultimately leave no progeny (Figure 20–5). In a sense, the somatic cells exist only to help the cells of the germ line (the germ cells) survive and propagate.

The sexual reproduction cycle thus involves an alternation of haploid cells, each carrying a single set of chromosomes, with generations of diploid cells, each carrying two sets of chromosomes. The mixing of genomes that characterizes sexual reproduction is achieved by fusion of two haploid cells to form a diploid cell. In this way, through cycles of diploidy, meiosis, haploidy, and cell fusion, old combinations of genes are broken up and new combinations are created.

Sexual Reproduction Gives Organisms a Competitive Advantage

The machinery of sexual reproduction is elaborate, and the resources spent on it are large. What benefits does it bring, and why did it evolve? Through the mixing of genes, sexually reproducing individuals beget unpredictably dissimilar offspring, whose patchwork genomes are at least as likely to represent a change for the worse as a change for the better. Why, then, should individuals that reproduce sexually have a competitive advantage over individuals that breed true, by an asexual

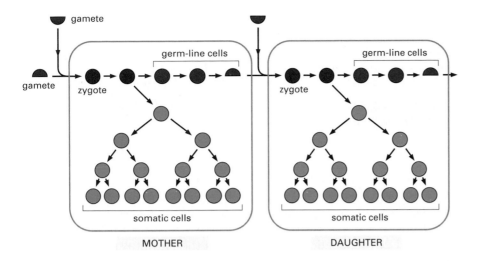

Figure 20–5 Germ-line cells and somatic cells carry out fundamentally different functions. In sexually reproducing organisms, the germ-line cells *(red)* propagate genetic information into the next generation. Somatic cells *(blue),* which form the body of the organism and are therefore necessary for sexual reproduction, leave no progeny.

process? This problem continues to perplex population geneticists, but one advantage seems to be that the reshuffling of the genes through sexual reproduction can help a species survive in an unpredictably variable environment. If two parents produce many offspring with a wide variety of gene combinations, the chance that at least one of their progeny will have the combination of features necessary for survival is increased. Sexual reproduction might also speed the elimination of deleterious genes from a population: by mating with only the fittest males, females select for "good" genes and allow "bad" genes to be lost from the population more efficiently than they would otherwise be.

Whatever its advantages, sexual reproduction has profoundly affected the diversity of life on this planet. In the following section, we review its central features, focusing on meiosis, the process by which the sex cells are formed.

Meiosis

That animals need a special process to produce their sex cells was first discovered in 1883, when it was observed that the fertilized egg of a parasitic roundworm contains four chromosomes, whereas the worm's gametes (sperm in males and eggs in females) contain only two. Thus gametes, the cells specialized for sexual reproduction, are **haploid**—they carry only a single set of chromosomes containing a single copy of the organism's genetic information (see Figure 20–4). All of the other cells of the body, including the cells that give rise to the gametes, are **diploid**—they carry two sets of chromosomes, one derived from the mother and the other from the father. Therefore, sperm and eggs must be formed by a special kind of cell division in which the number of chromosomes is precisely halved. This form of cell division is called **meiosis**, from a Greek word, meaning "diminution," or "lessening."

The recognition that chromosomes carry the units of inheritance and that each gamete possesses a haploid number of them explains how both parents can contribute equally to the character of the progeny, despite the enormous difference in size between egg and sperm (see Figure 20–3).

In this section, we follow the elaborate dance of chromosomes that occurs when a cell divides its genetic material precisely in two. We shall begin with an overview of how meiosis distributes chromosomes to the gametes. We then take a closer look at how chromosomes pair, recombine, and are segregated during the process. It is the random assortment of maternal and paternal chromosomes during meiosis that generates

Question 20–1

It is easy to see how deleterious mutations in bacteria, which have a single copy of each gene, are eliminated by natural selection; the affected bacteria die and the mutation is thereby lost from the population. Eucaryotes, however, have two copies of most genes because they are diploid. It is often the case that an individual with two normal copies of the gene (homozygous, normal) is indistinguishable in phenotype from an individual with one normal copy and one defective copy of the gene (heterozygous). In such cases, natural selection can operate only on an individual with two copies of the defective gene (homozygous, defective). Imagine the situation in which a defective form of the gene is lethal when homozygous, but without effect when heterozygous. Can such a mutation ever be eliminated from the population by natural selection? Why or why not?

gametes with novel combinations of genes. We also discuss what happens when meiosis goes awry. Finally, we consider briefly the process of fertilization whereby gametes come together to form a new and genetically distinct individual.

Haploid Germ Cells Are Produced From Diploid Cells Through Meiosis

Before diploid cells divide by mitosis, they duplicate their two sets of chromosomes precisely, which allows identical sets of chromosomes to be transmitted to each daughter cell (discussed in Chapter 19). Meiosis—the process through which germ cells are formed—is different, because only a single set of chromosomes is parceled out to each germ cell from the diploid starting cell. This process involves a single round of DNA replication that duplicates the chromosomes, followed by two successive cell divisions.

One might imagine that meiosis could occur by a simple modification of a normal mitotic cell division, in which DNA replication (S phase) was omitted. In principle, a single round of cell division would then produce two haploid cells directly. For unknown reasons, the actual meiotic process is more complicated and involves two cell divisions instead of one. Moreover, in some cells it can take very much longer than any mitosis: meiosis in a human male, for example, lasts for 24 days; in a human female, it can last for decades.

Meiosis begins in specialized diploid germ-line cells in the ovaries or testes. Each of these cells contains two copies of each chromosome, one inherited from the organism's father (the *paternal homolog*) and one from its mother (the *maternal homolog*). When meiosis begins, the chromosomes of this diploid cell are duplicated: as in any cell preparing to divide, these duplicated chromosomes remain attached to one another, like Siamese twins. The next phase of the process is unique to meiosis. Each duplicated paternal chromosome finds and pairs with the duplicated maternal homolog. This specialized pairing ensures that the homologs will segregate properly during the subsequent cell divisions, so that each of the final gametes receives a complete haploid set of chromosomes.

Two successive cell divisions, called meiotic division I and meiotic division II, now parcel out a complete set of chromosomes to each of the four haploid cells produced. Because the assignment of the homologs to each cell is random, the original maternal and paternal chromosomes are reshuffled into different combinations in the gametes that will eventually form from these haploid cells. During fertilization, two gametes will unite, generating a diploid zygote that is now genetically distinct from each of its parents (see Figure 20–4). The zygote then develops into a multicellular organism through repeated rounds of cell multiplication and cell division followed by cell specialization.

Thus, meiosis produces four cells that are genetically dissimilar and that contain exactly half as many chromosomes as the original parent cell. In contrast, mitosis produces two genetically identical daughter cells. We now discuss the molecular events in the meiotic cycle in more detail, beginning with the pairing of maternal and paternal chromosomes, a process that is central to this specialized form of cell division.

Meiosis Involves a Special Process of Chromosome Pairing

With the exception of the chromosomes that determine sex (the *sex chromosomes*), a diploid nucleus contains two very similar versions of each chromosome, one from each parent (see Figure 5–12). A diploid cell therefore contains a great deal of duplicate genetic information. The two versions of each chromosome, however, are not genetically identical, as they possess different variants of many of the genes they carry. These alternative forms of a gene, called **alleles**, differ somewhat in their nucleotide sequences; the most common difference is a substitution of a single base pair, but different alleles can also harbor deletions, insertions, and duplications. Because the parental versions of each chromosome are similar but not identical, they are called **homologous chromosomes**, or **homologs**.

Before a cell divides—by either meiosis or mitosis—it first duplicates all of its chromosomes. The twin copies of each fully replicated chromosome at first remain tightly linked along their length and are called **sister chromatids**. The way these *replicated chromosomes* are handled, however, differs in meiosis and mitosis. In mitosis, as we saw in Chapter 19, the replicated chromosomes line up in random order at the metaphase plate; as mitosis continues, the two previously joined sister chromatids then separate from each other to become individual chromosomes, and the two daughter cells produced by cytokinesis inherit one copy of each paternal chromosome and one copy of each maternal chromosome. Thus, both sets of genetic information are transmitted intact to both daughter cells, which are, therefore, diploid and genetically identical.

The events that occur in the first meiotic cell division mirror the sequence of stages that a cell goes through in mitosis: in prophase, the replicated chromosomes condense; in metaphase, they align at the equator of the meiotic spindle; and in anaphase, they are segregated to the poles. For a review of these stages, see Panel 19–1 (pp. 642–643). The need to halve the number of chromosomes during meiosis, however, makes an extra demand on the cell-division machinery and leads to the first main difference between meiosis and mitosis. In division I of meiosis, the replicated homologous paternal and maternal chromosomes (including the two replicated sex chromosomes) pair up alongside each other before they line up on the spindle (Figure 20–6). As we shall see, this physical pairing of homologs is crucial because it enables the paternal and the maternal homolog to be segregated to different daughter cells at this division. As a result, each gamete formed at the end of the process will acquire either the maternal copy or the paternal copy of a chromosome but not both. Because the assignment of maternal and paternal homologs to the gametes during meiosis is random, the original maternal and paternal chromosomes—with their different sets of alleles—are reshuffled into different combinations in each gamete.

How the homologs (and the two sex chromosomes) recognize each other is still uncertain. In many organisms, the initial association, a process called **pairing**, seems to be mediated by interaction between complementary DNA base pairs at numerous sites that are widely dispersed along the chromosomes. The structure formed when the duplicated chromosomes pair is called a **bivalent**, and it contains four chromatids (Figure 20–7A). The bivalent forms and is maintained during the long meiotic prophase, a stage that can even last for years.

Extensive Recombination Occurs Between Maternal and Paternal Chromosomes

A series of complex events occurs during the long prophase of the first meiotic division. After the duplicated homologs pair, genetic recombination is initiated. During this process of **recombination**, or **crossing-over**, parts of homologous chromosomes are exchanged. As the result of

Figure 20–6 During meiosis, homologous chromosomes pair before lining up on the spindle. In mitosis (A) the individual maternal (M) and paternal (P) chromosomes line up randomly at the metaphase plate, whereas in meiosis (B) the homologous maternal and paternal chromosomes have paired before lining up at the metaphase plate. In both cases the chromosomes have been replicated before they become aligned. The spindle is shown in *green*.

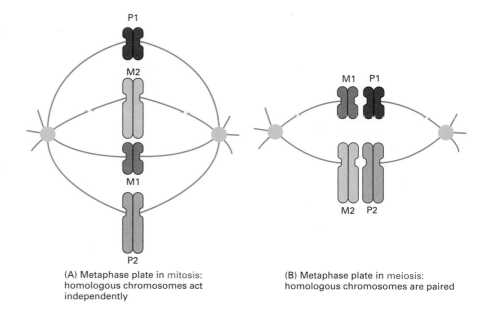

(A) Metaphase plate in mitosis: homologous chromosomes act independently

(B) Metaphase plate in meiosis: homologous chromosomes are paired

recombination, the DNA double helix is broken and then rejoined in both a maternal chromatid and a homologous paternal chromatid. Fragments of the two non-sister chromatids are thereby exchanged in a reciprocal fashion (see Figure 6–28).

The proteins responsible for carrying out the recombination process take advantage of the fact that the homologous chromosomes in the four-stranded bivalent are held together during meiotic prophase I and are closely aligned by a *synaptonemal complex*. This elaborate structure aligns the bivalents so that genetic recombination can readily occur between the non-sister chromatids; it also serves to space out the crossover events along each chromosome.

Each of the two chromatids of a duplicated chromosome can cross over with either of the two chromatids of the other chromosome in the bivalent. By the time that prophase ends, each pair of duplicated homologs is held together by at least one **chiasma**, the connection that corresponds to a crossover between two non-sister chromatids (Figure 20–7B). Many bivalents contain more than one chiasma, indicating that multiple crossovers can occur between homologous chromosomes (Figure 20–7C and D). On average, between two and three crossover events occur on each pair of human chromosomes during meiosis I.

Figure 20–7 Recombination between a maternal and a paternal chromatid in paired chromosomes forms a chiasma. (A) The structure formed when homologous chromosomes pair is called a bivalent and consists of four chromatids. (B) Crossover events create chiasmata between non-sister chromatids. In this set of paired homologs, a single crossover event has occurred during prophase, creating a single chiasma. Note that the four chromatids in the bivalent are arranged as two distinct pairs of sister chromatids. The combination of the chiasmata and the tight attachment of the sister chromatids to each other, mediated by cohesin proteins, holds the two duplicated homologs together. (C) Multiple crossovers can occur between homologous chromosomes. Shown is a light micrograph of a grasshopper bivalent with three chiasmata. (D) Diagram showing position of chiasmata in (C). (C, courtesy of Bernard John.)

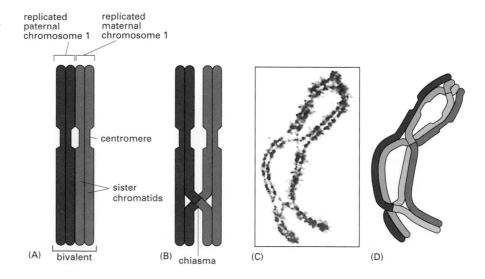

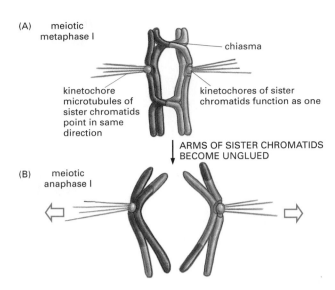

(A) meiotic metaphase I

chiasma

kinetochore microtubules of sister chromatids point in same direction

kinetochores of sister chromatids function as one

ARMS OF SISTER CHROMATIDS BECOME UNGLUED

(B) meiotic anaphase I

Figure 20–8 Chiasmata ensure proper segregation of chromosomes in meiosis. (A) In metaphase I, chiasmata created by crossing-over lock together the maternal and paternal homologs. At this stage, cohesin proteins (not shown) keep the sister chromatids glued together along their entire length. The kinetochores of sister chromatids function as a single unit, and microtubules that attach to them point in a single direction. (B) At anaphase I, the cohesins holding the arms of the sister chromatids together are degraded; sister chromatids are still held together by cohesins in the centromere. This arrangement allows the sister chromatids to remain together when the duplicated homologs separate and are pulled to opposite poles of the spindle. In contrast, at anaphase in mitosis, both the arms and the centromeres come apart at the same time.

Recombination during meiosis is a major source of genetic variation in sexually reproducing species. By scrambling the genetic constitution of each of the chromosomes in gametes, crossing-over helps to produce individuals with novel assortments of genes.

Recombination generates new combinations of maternal and paternal genes on individual chromosomes, but it also plays a second important role in meiosis. By holding homologous chromosomes together during prophase I, recombination ensures that the maternal and paternal homologs will segregate from one another correctly at the first meiotic division, as we discuss next.

Chromosome Pairing and Recombination Ensure the Proper Segregation of Homologs

In most organisms, recombination during meiosis is required for the correct segregation of the two duplicated homologs into separate daughter nuclei. The chiasmata created by crossover events have a crucial role in locking together the maternal and paternal homologs until the spindle separates them at anaphase I. Before anaphase I, the two poles of the spindle pull on the duplicated homologs in opposite directions, and the chiasmata resist this pulling (Figure 20–8A). In so doing, the chiasmata help to position and stabilize bivalents on the metaphase plate. In addition to the duplicated homologs being held together at chiasmata, the arms of sister chromatids are "glued" together along their length by *cohesin* proteins (see Figure 19–3). These cohesin proteins become unglued at the start of anaphase I, allowing the duplicated homologs to be pulled to opposite poles of the spindle (see Figure 20–8B).

The Second Meiotic Division Produces Haploid Daughter Cells

The first meiotic cell division does not produce cells with a haploid amount of DNA. To achieve this goal, each cell now proceeds through a second round of division, meiosis II, which occurs without further DNA replication and without any significant interphase period. A spindle forms, the chromosomes align at its equator, and the sister chromatids now separate to produce daughter cells with a haploid DNA content. In

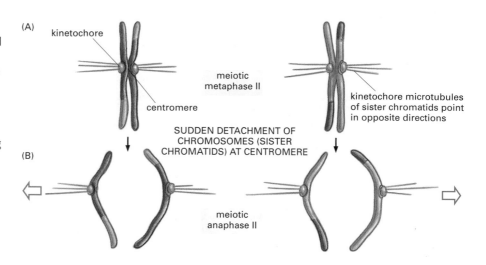

Figure 20–9 In meiosis II, as in mitosis, the kinetochores on each sister chromatid function independently, allowing the two sister chromatids to be pulled to opposite poles. (A) In metaphase II, the kinetochores of the sister chromatids point in opposite directions. (B) The cohesins holding the sister chromatids together at the centromere are now degraded, allowing kinetochore microtubules to pull the individual chromatids to opposite poles. (This figure is drawn as a continuation of Figure 20–8.)

Labels in figure: kinetochore; meiotic metaphase II; centromere; kinetochore microtubules of sister chromatids point in opposite directions; SUDDEN DETACHMENT OF CHROMOSOMES (SISTER CHROMATIDS) AT CENTROMERE; meiotic anaphase II

division II of meiosis, as in a mitotic division, the kinetochores on each sister chromatid now have attached kinetochore microtubules pointing in opposite directions, so that the chromatids can be drawn into different daughter cells at anaphase (Figure 20–9). At anaphase II, the meiotic-specific cohesins that hold the sister chromatids together at the centromere are degraded, allowing the chromatids to separate.

To summarize, meiosis consists of a single round of DNA replication followed by two cell divisions, so that four, nonidentical haploid cells are produced from each diploid cell that enters meiosis; in contrast, mitosis produces two identical diploid cells (Figure 20–10).

The Haploid Cells Contain Extensively Reassorted Genetic Information

Identical twins, which develop from a single zygote, are genetically identical; otherwise no two siblings are genetically the same. This is true because, even before fertilization takes place, meiosis has produced two kinds of randomizing genetic reassortments.

First, as we have seen, the maternal and paternal chromosomes are shuffled and dealt out among the gametes during meiosis. Although the chromosomes are carefully parceled out so that each gamete receives one and only one copy of each chromosome, each gamete receives a random mixture of paternal and maternal chromosomes (Figure 20–11A). This type of assortment is driven entirely by the way each bivalent is positioned when it lines up on the spindle during metaphase I. Whether the maternal or paternal homolog is captured by the spindles from one pole or the other depends on which way the bivalent is facing when the microtubules connect with its kinetochore. Because the orientation of each bivalent at the moment it is captured is completely random, the assortment of maternal and paternal chromosomes is random as well.

Thanks to this type of reassortment alone, an individual could in principle produce 2^n genetically different gametes, where n is the haploid number of chromosomes. Each human, for example, can in theory produce $2^{23} = 8.4 \times 10^6$ different gametes simply from the random assortment of paternal and maternal homologs that takes place in meiosis. The actual number of different gametes each person could produce, however, is much greater than 2^{23}. This is because the recombination that takes place during meiosis provides a second source of randomized genetic reassortment. As mentioned earlier, between two and

Question 20–2

Why would it not be desirable for an organism to use the first steps of meiosis (up to and including meiotic cell division I) for the ordinary mitotic division of somatic cells?

three crossovers occur on average on each pair of human chromosomes per meiosis. This process puts maternal and paternal genes that are initially on separate chromosomes onto the same chromosome, as illustrated in Figure 20–11B. Because recombination occurs at more or less

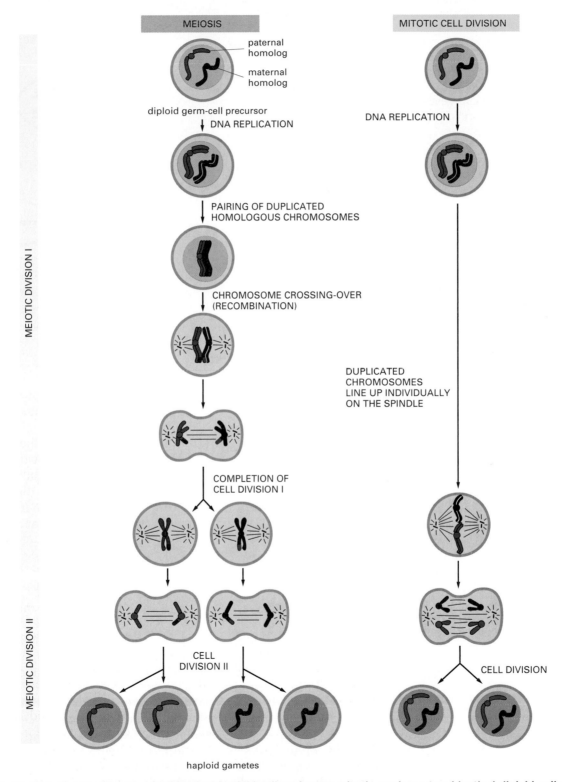

Figure 20–10 Meiosis generates four nonidentical haploid cells, whereas mitosis produces two identical diploid cells. As in Figure 20–4B, only one pair of homologous chromosomes is shown. In meiosis, two cell divisions are required after DNA replication to produce the haploid gametes. Each diploid cell that enters meiosis therefore produces four haploid cells, whereas each diploid cell that divides by mitosis produces two diploid cells. Whereas mitosis and division II of meiosis usually occur within hours, division I of meiosis can last days, months, or even years, due to the long time spent in prophase.

Figure 20–11 Two kinds of reassortment generate new chromosome combinations during meiosis. (A) The independent assortment of the maternal and paternal homologs during meiosis produces 2^n different haploid gametes for an organism with n chromosomes. Here $n = 3$, and there are 2^3, or 8, different possible gametes. For simplicity, chromosome crossing-over is not shown here. (B) Crossing-over during meiotic prophase I exchanges segments of homologous chromosomes and thereby reassorts genes on individual chromosomes. For simplicity, only a single pair of homologous chromosomes is shown. Both independent assortment and crossing over occur during every meiosis. Because of the many small differences in DNA sequences that exist between any two homologous chromosomes, both mechanisms increase the genetic variability of organisms that reproduce sexually.

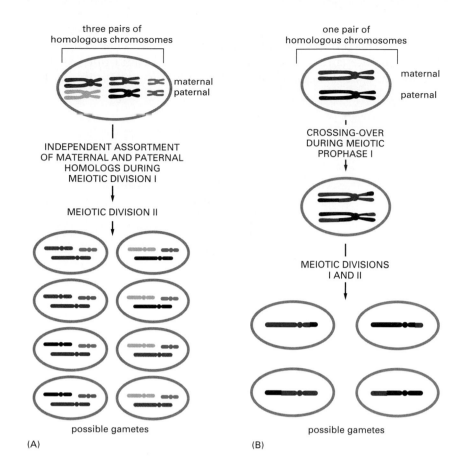

Question 20–3

Ignoring the effects of chromosome crossovers, an individual human can in principle produce 2^{23} = 8.4×10^6 genetically different gametes. How many of these possibilities can be realized in the average life of (A) a female? (B) a male?

random sites along the length of a chromosome, each meiosis will produce four sets of entirely novel chromosomes.

The reassortment of chromosomes in meiosis, together with the recombination of genes that arises from crossing-over, provides a nearly limitless source of genetic variation in the gametes produced by a single individual. Considering that each human is formed from the fusion of two such gametes, one from the father and one from the mother, the richness of human variation that we see around us is not at all surprising.

Meiosis Is Not Flawless

The sorting of chromosomes that takes place during meiosis is a remarkable feat of cellular bookkeeping: in humans, each meiosis requires that the starting cell keep track of 92 chromosomes (23 pairs, each of which has duplicated), handing out one complete set to each gamete. Not surprisingly, mistakes do occur in the distribution of chromosomes during this elaborate process.

Occasionally, homologs fail to separate properly—a phenomenon known as *nondisjunction*. As a result, some of the haploid cells that are produced lack a particular chromosome, while others have more than one copy of it. Upon fertilization, such gametes form abnormal embryos, most of which die. Some survive, however. *Down syndrome*, for example, a human disease characterized by severe mental retardation, is caused by an extra copy of Chromosome 21. This error results from nondisjunction of a Chromosome 21 pair during meiosis, giving rise to a gamete that contains two copies of Chromosome 21 instead of one copy (Figure 20–12). When this abnormal gamete fuses with a

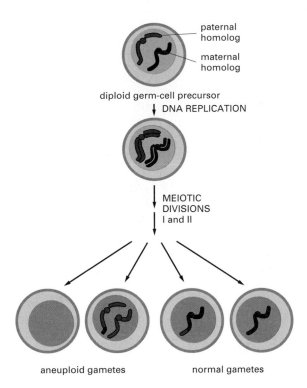

paternal homolog
maternal homolog

diploid germ-cell precursor

↓ DNA REPLICATION

↓ MEIOTIC DIVISIONS I and II

aneuploid gametes   normal gametes

Figure 20–12 Errors in chromosome segregation during meiosis can result in gametes with incorrect numbers of chromosomes. In this example, the maternal and paternal homologs fail to segregate normally during the second meiotic division. As a result, one gamete receives no copy of the chromosome, another receives an extra copy, and two receive the proper single copy. The gametes that receive an incorrect number of chromosomes are called *aneuploid* gametes. If one of them participates in the fertilization process, the resulting zygote will also have an abnormal number of chromosomes. If a gamete bearing two copies of Chromosome 21 fuses with a normal gamete, the resulting child will have Down syndrome.

normal gamete at fertilization, the resulting embryo contains three copies of Chromosome 21 instead of two. This chromosome imbalance produces an extra dose of the proteins encoded on Chromosome 21 and thereby interferes with the proper development of the embryo.

The frequency of missegregation in human gametes is remarkably high, particularly in the female: nondisjunction occurs in about 10% of the meioses in human oocytes. Regardless of whether the segregation error occurs in the sperm or the egg, nondisjunction is thought to be one reason for the high rate of miscarriages (spontaneous abortions) in early pregnancy in humans.

Whatever harm these errors might do to the fitness of the species is presumably far outweighed by the benefits that sexual reproduction brings. What's more, it must confer some benefit to the organisms that rely on it for reproduction. After all, the vast majority of plants and animals alive today reproduce sexually, using meiosis to shuffle their genetic information as they pass their characteristics on to future generations.

Fertilization Reconstitutes a Complete Genome

Having seen how chromosomes are parceled out during meiosis, we now discuss briefly how they come back together during the process of **fertilization**, when a new zygote with a complete set of chromosomes is formed. Of the 300 million human sperm ejaculated during coitus, only about 200 reach the site of fertilization in the oviduct. There is evidence that chemical signals released by cells that surround the ovulated egg attract the sperm to the egg, but the nature of the chemoattractant molecules is unknown. Once it finds an egg, the sperm must migrate through a protective layer of cells and then bind to, and tunnel into, the egg coat, called the *zona pellucida*. Finally, the sperm must bind to and fuse with the underlying egg plasma membrane (Figure 20–13). Although fertilization normally occurs by this process of sperm–egg fusion, it can also be achieved artificially by injecting the sperm directly into the egg cytoplasm; this is sometimes done in infertility clinics when there is a problem with natural sperm–egg fusion.

5 μm

Figure 20–13 A sperm binds to the plasma membrane of an egg. Shown here is a scanning electron micrograph of a human sperm contacting a hamster egg. The egg has been stripped of its zona pellucida, exposing the plasma membrane, which is covered in finger-like microvilli. Such uncoated hamster eggs are sometimes used in infertility clinics to assess whether a male's sperm is capable of penetrating an egg. The zygotes resulting from this test do not develop. (Courtesy of David M. Phillips.)

Although many sperm can bind to an egg, only one normally fuses with the egg plasma membrane and introduces its DNA into the egg cytoplasm. This control step is especially important as it ensures that a fertilized egg will contain two, and only two, sets of chromosomes. There are several mechanisms that ensure that only one sperm fertilizes an egg. In one, the first successful sperm triggers the release of a wave of Ca^{2+} ions in the egg cytoplasm, which causes a "hardening" of the zona pellucida. This prevents the "runner up" sperm from penetrating the zona pellucida and ensures that in the race to fertilize the egg there is a single winner.

Once fertilized, the egg is called a **zygote**. Fertilization is not complete, however, until the two haploid nuclei (called *pronuclei*) have come together and combined their chromosomes into a single diploid nucleus. Fertilization marks the beginning of one of the most remarkable phenomena in all of biology—the process of embryogenesis, in which the zygote divides to produce large numbers of diploid cells that develop into a new individual.

Mendel and the Laws of Inheritance

Humans, like all multicellular organisms, are composed of a large number of cells that are derived from a single fertilized egg through multiple rounds of cell division. Each of these cells contains a combination of genes, which we inherit from our mother and father, producing variations that contribute to the characteristics that make each of us unique. Everything from hair and eye colors to thrill-seeking behavior and a predisposition for developing gallstones is influenced by our genes.

Attached earlobes, insensitivity to skunk odor, an inability to taste the chemical PTC, and a condition called "uncombable hair" are each caused by alterations in a single gene (Figure 20–14). But most of the traits we use to describe ourselves are influenced by multiple genes. Physical characteristics such as eye color and height, and personal qualities such as creativity and sociability all emerge from interactions among our genes, and from interactions between our genetic makeup and the environment.

That traits like these run in families indicates that they have a genetic component, But studying inheritance in humans is difficult. Human geneticists can't dictate who mates with whom, family sizes are small, and human development is so slow that it would take 40 years just to get a couple of generations of progeny to analyze. So the laws of heredity were unraveled, instead, in species that are easy to breed, and that produce large numbers of offspring. Gregor Mendel, the father of genetics, chose to study peas, but similar experiments can be performed in fruit flies, worms, dogs, cats, or any other plant or animal that possesses characteristics of interest. We now know that the same basic laws of inheritance apply to all sexually reproducing organisms, from microscopic yeasts to peas to people.

In this section, we describe the molecular basis of genetic inheritance. We shall see how the behavior of chromosomes during meiosis—

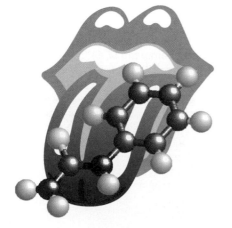

Figure 20–14 Some people taste it, some people don't. The ability to taste the chemical phenylthiocarbamide (PTC, originally called phenylthiourea) is governed by a single gene. Although geneticists have known since the 1930s that the inability to taste PTC is inherited in a Mendelian fashion, researchers did not determine the gene responsible—one that encodes a bitter taste receptor—until 2003. Nontasters produce a PTC receptor protein that contains amino acid substitutions that are thought to reduce the activity of the protein.

their segregation into gametes that then unite at random to form genetically unique offspring—explains precisely the laws of inheritance. To begin we discuss how Mendel, breeding peas in his monastery garden, first worked it all out in the nineteenth century.

Mendel Chose to Study Traits That Are Inherited in a Discrete Fashion

When designing an experiment to address a scientific question, selecting the right organism can be critical. Mendel chose to carry out his studies in pea plants. Peas, it turns out, are easy to cultivate quickly and large numbers can be raised in a small space—such as an abbey garden. Furthermore, Mendel could control which plants mated with which. Each flower on a pea plant contains both male and female structures, and left to their own devices the plants normally self-fertilize. Mendel found he could cross-pollinate the peas by removing the immature male parts from one flower and then fertilizing that emasculated plant with pollen (sperm) from another plant. Thus, Mendel could be certain of the parentage of every pea plant he examined.

Perhaps more importantly for Mendel's purposes, pea plants were available in many varieties. For example, one variety of pea plant has purple flowers, another has white flowers. One variety produces peas whose skins are smooth, another produces wrinkled peas. Mendel chose to follow sets of traits—such as flower color and seed shape—that were distinct, easily observable, and, most importantly, were inherited in a discrete fashion (Figure 20–15). In other words, the plants have either purple flowers or white flowers—nothing in between. This choice of characteristics was important for the results. If Mendel had selected traits that varied in a continuum, such as seed weight, his observations would have supported the concept of blended inheritance—whereby genetic material was thought to blend to produce offspring with intermediate traits.

Figure 20–15 Mendel studied seven different traits that are inherited in a discrete fashion. For each trait, plants display either one variation or the other. In other words, an individual pea plant produces either yellow peas or green peas—nothing in between.

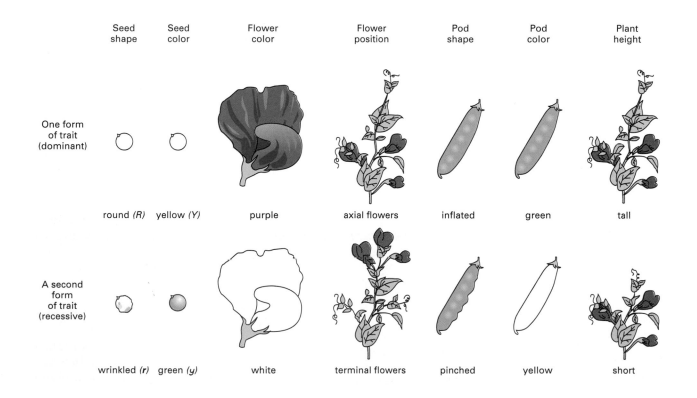

	Seed shape	Seed color	Flower color	Flower position	Pod shape	Pod color	Plant height
One form of trait (dominant)	round *(R)*	yellow *(Y)*	purple	axial flowers	inflated	green	tall
A second form of trait (recessive)	wrinkled *(r)*	green *(y)*	white	terminal flowers	pinched	yellow	short

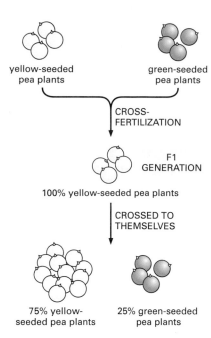

Figure 20–16 A simple experiment revealed the discrete nature of heredity. True-breeding green-pea plants, crossed with true-breeding yellow-pea plants, always produce offspring with yellow peas. But when these offspring (the F_1 plants) are bred with each other, 25% of the progeny produce green peas.

Mendel Could Disprove the Alternative Theories of Inheritance

The breeding experiments that Mendel performed were quite straight-forward. He started with stocks of genetically pure, true-breeding plants. When true-breeding plants self-pollinate, all of their offspring are of the same variety. If he were following seed color, for example, he would use plants with yellow peas that always produced offspring with yellow peas, and plants with green peas that always produced offspring with green peas.

In a typical experiment, Mendel would take two of these true-breeding varieties and cross-pollinate them. Then he would record the inheritance of the selected trait in the next generation. For example, Mendel crossed plants producing yellow peas with plants producing green peas. In this case, he discovered that the resulting hybrid off-spring, called the first filial or F_1 generation, all had yellow peas (Figure 20–16). A similar finding held true for every trait he followed: the F_1 hybrids all resembled only one of their two parents. Mendel's unique approach was to study each trait one at a time. His predecessors had focused on whole organisms that varied in many traits, and they often wound up trying to characterize offspring whose appearance varied in such a complex way that the progeny could not easily be compared to their parents.

From his experiment with green- and yellow-seeded pea plants, Mendel could conclude that seed color is not passed down by blending. If it were, the peas in the F_1 generation would have all been yellowish-green. Furthermore, Mendel determined that it did not matter whether he used the pollen from the plant with green peas to fertilize the flowers of a yellow-pea plant, or vice versa. The result—offspring with yellow peas—was always the same. This finding disproved the theory of uniparental inheritance, which required that one or other of the parents would always determine the character of the offspring. If, for example, the appearance of the offspring was dictated solely by the appearance of the father, then the peas would always be the color of the plant that was used as the source of pollen. This was clearly not the case.

Had Mendel stopped there—observing only the F_1 generation—he would never have unraveled the basic patterns of inheritance. Fortunately, Mendel took his breeding experiments to the next step: mating the F_1 plants to one another and examining the results.

Mendel's Experiments Were the First to Reveal the Discrete Nature of Heredity

Mendel's subsequent experiments with the hybrid F_1 plants were designed to address an obvious question: what happened to the traits that disappeared in the F_1 generation—such as the green peas in Figure 20–16? Did the parent plants bearing green peas somehow fail to make a genetic contribution to their offspring?

To find out, Mendel allowed the F_1 plants to self-fertilize. If the trait for green peas, for example, had been lost, then the F_1 plants would produce only plants with yellow peas in the next, F_2, generation. He used a large sample size and kept an accurate count of the results. And he found that the "disappearing trait" returned: three-quarters of the offspring in the F_2 generation had yellow peas; one-quarter had green peas (see Figure 20–16).

The result definitively laid to rest the theory of blended inheritance. There is simply no way that blending could explain how a cross between one yellow-pea plant and another yellow-pea plant could yield plants with green peas. But the data also gave Mendel a clue as to what was going on. Although the green pea trait disappeared temporarily in the F_1 generation, it reappeared in F_2. This means that at least some of the F_1 plants must have still harbored a factor that codes for green peas: it was just hidden somehow. Mendel saw the same type of behavior for each of the other six traits he examined.

To account for these observations, Mendel proposed that the inheritance of traits is governed by hereditary factors (now called genes), and that these factors act like discrete particles that remain separate instead of blending. Furthermore, he suggested that genes come in alternative versions that account for the variations seen in inherited characteristics. The gene dictating seed color, for example, exists in two "flavors": one that directs the production of yellow peas and one that produces green. Such alternative versions of a gene are today called alleles.

Mendel reasoned that for each characteristic, a plant must inherit two copies, or alleles, of each gene—one from its mother, one from its father. The appearance, or **phenotype**, of the plant depends on which versions of each allele it gets; the actual genetic makeup itself is called the **genotype**. The true-breeding parental strains, he theorized, each possessed a pair of identical alleles—the yellow-pea plants possessed two alleles for yellow peas, the green-pea plant two alleles for green peas. An individual that possesses two identical alleles is said to be **homozygous** for that trait. The F_1 hybrid plants, on the other hand, had received two dissimilar alleles—one specifying yellow peas and the other green. These plants were **heterozygous** for the trait of interest.

To explain the disappearance of one of the traits in the F_1 generation, Mendel supposed that for any pair of alleles, one allele is *dominant* and the other *recessive* or hidden. The dominant allele, whenever it is present, would dictate the plant's phenotype. In the case of seed color, the allele that specifies yellow peas is dominant; the green-pea allele is recessive.

One important consequence of heterozygosity, and of dominance and recessiveness, is that not all of the alleles an individual carries can be detected in its phenotype. Humans have about 30,000 genes, and each of us is heterozygous for a very large number of these. Thus, we all carry a great deal of genetic information that remains hidden in our personal phenotype, but that can turn up in future generations.

Each Gamete Carries a Single Allele for Each Character

Mendel's theory—that for every gene, an individual inherits one copy from its mother and one copy from its father—raised some logistical issues. If an organism has two copies of every gene, how does it pass only one copy of each to its progeny? And how do these gene sets come together again in the resulting offspring?

Mendel postulated that when sperm and eggs are formed, the two copies of each gene present in the parent separate, or segregate, so that each gamete receives only one allele for each trait. Thus, each egg and each sperm (pollen) receives only one allele for seed color (either yellow or green), one allele for seed shape (smooth or wrinkled), one allele for flower color (purple or white), and so on. During fertilization, sperm with one or the other allele unites with an egg carrying one or other allele, restoring the two copies of the gene for each trait in the fertilized egg or zygote. Which type of sperm unites with which type of egg at fertilization is entirely a matter of chance.

This principle of heredity is laid out in Mendel's first law, the **law of segregation**, which states that the two alleles for each trait separate (or segregate) during gamete formation, and that they then unite at

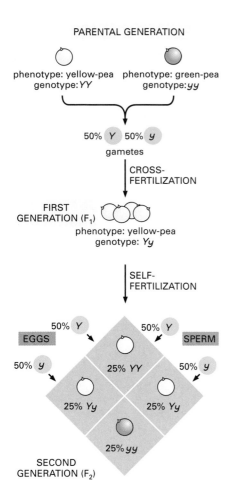

PARENTAL GENERATION

phenotype: yellow-pea   phenotype: green-pea
genotype: *YY*   genotype: *yy*

50% *Y* 50% *y*
gametes

CROSS-
FERTILIZATION

FIRST
GENERATION (F$_1$)
phenotype: yellow-pea
genotype: *Yy*

SELF-
FERTILIZATION

50% *Y*   50% *Y*
EGGS   SPERM

50% *y*   25% *YY*   50% *y*

25% *Yy*   25% *Yy*

25% *yy*

SECOND
GENERATION (F$_2$)

Figure 20–17 Parent plants produce gametes that each contain one allele for each trait; the phenotype of the offspring depends on which combination of alleles it receives. Here we see both the genotype and phenotype of the pea plants that were bred in Figure 20–16. The true-breeding yellow-pea plants produce only *Y*-bearing gametes; the true-breeding green plants produce only *y*. Their offspring, which produce yellow peas, have the genotype *Yy*. When these hybrid plants are bred with each other, 75% of the offspring have yellow peas, 25% have green. The gray box at the bottom, called a Punnett square after a British mathematician who was a follower of Mendel, allows one to track the separation of alleles during gamete formation and to predict the combinations that occur upon fertilization. According to the system invented by Mendel and still in use today, capital letters symbolize a dominant allele and lower-case letters the recessive allele.

random—one from each parent—at fertilization. This law permits us to predict the phenotypes of the plants that will result from a particular cross-breeding experiment.

According to the law of segregation, the F$_1$ hybrid plants with yellow peas will produce two classes of gametes: half the gametes will get a yellow-pea allele and half will get a green-pea allele. When the hybrid plants self-pollinate, these two classes of gametes will unite at random. Thus, an egg with a green-pea allele has an equal chance of being fertilized by a pollen grain carrying a green-pea allele or a yellow-pea allele. The same is true for an egg carrying a yellow-pea allele. There are thus four different combinations of alleles that can come together in the F$_2$ offspring (Figure 20–17). One-quarter of the plants will receive two alleles specifying green peas; these obviously will have green peas. One-quarter of the plants will receive two yellow-pea alleles and will produce yellow peas. But one-half of the plants will inherit one allele for yellow peas and one allele for green. Because the yellow allele is dominant, these plants—like their hybrid F$_1$ parents—will all produce yellow peas. All told, three-quarters of the offspring will have yellow peas and one-quarter will have green peas. Thus Mendel's law of segregation explains the 3:1 ratio that he observed in the F$_2$ generation.

Mendel's Law of Segregation Applies to All Sexually Reproducing Organisms

Mendel's law of segregation explained the data for all of the traits he examined in peas. He was also able to replicate his basic findings with corn and beans. But his rules governing inheritance are not limited to reproduction in plants. Mendel's concept of the gene and his law of segregation can be generalized to all sexually reproducing organisms, including humans.

Let's consider a human phenotype that is influenced by a single gene. Albinism is a rare condition that is inherited in a recessive manner in many animals, including humans. In other words, like the pea plants that produce green seeds, albinos are homozygous for the recessive allele (*a*) of a particular gene. Their genotype is *aa*. The dominant allele (*A*) of the gene encodes an enzyme involved in making melanin, the pigment that is responsible for most of the brown and black color present in hair, skin, and the retina of the eye. The recessive allele encodes a version of this enzyme that is less active or completely inactive. In the absence of this enzyme, albinos have white hair, white skin, and pupils that look pink because the lack of coloration reveals the red color of the hemoglobin present in blood vessels in the retina.

If an albino man (genotype *aa*) has children with an albino woman (whose genotype is also *aa*), all of their children will be albino (*aa*). However, imagine now that a nonalbino male marries and has children with an albino female. (We will assume that the male is homozygous for the dominant *A* allele; albinism is quite rare, and if no one in his family has ever had the condition, our assumption about his genotype is probably valid.) Their children should all appear "normal"—that is, none would be albino. This result mirrors Mendel's crosses between true-breeding pea plants. The father would contribute a dominant *A* allele to each gamete, and the mother would contribute a recessive *a*. The offspring, with genotype *Aa*, would all have the dominant phenotype (Figure 20–18). If one of these children someday met and started a family with an individual of a similar genotype (a man or woman who also has an albino parent), we would expect that their children would follow the pattern seen in Mendel's F$_2$ generation: for every three normally pigmented children, there would be one albino child. Or, from the point of

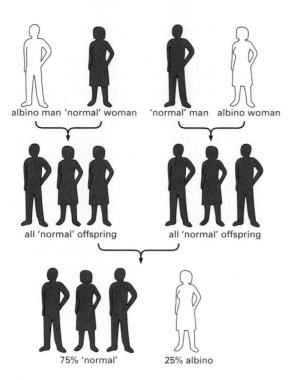

Figure 20–18 Mendel's law of segregation applies to all sexually reproducing organisms. Here we trace the inheritance of albinism, a recessive trait that is associated with a single gene in humans.

view of an individual child, each would have a 25% chance of receiving two recessive alleles.

Of course, humans generally don't have families large enough to guarantee accurate Mendelian ratios. Remember, Mendel bred and counted thousands of peas for most of his crosses. Geneticists interested in following the inheritance of specific traits in humans get around this problem by working with large numbers of families, or with several generations of a few large families. To keep track of this type of information and to help draw out the pattern of inheritance, geneticists prepare a chart that shows the phenotype of each family member for the relevant trait (Figure 20–19). This diagram, known as a **pedigree**, includes as many generations as possible.

Alleles for Different Traits Segregate Independently

Mendel deliberately simplified the problem of heredity by performing monohybrid crosses—breeding experiments that focused on the inheritance of one trait at a time. But, fortunately, he did not stop there. He continued his studies in peas, next examining the simultaneous inheritance of two or more apparently unrelated traits.

In the simplest situation, a dihybrid cross, Mendel followed the inheritance of two traits at once—seed color and seed shape, for example. For seed color, we have already seen that yellow peas are dominant over green peas. In the case of seed shape, round peas are dominant

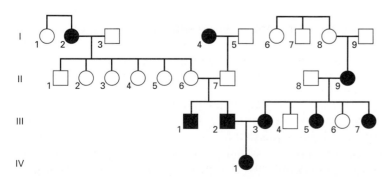

Figure 20–19 The phenotypes of entire families are often presented in standard diagrams called pedigrees. Shown here is a pedigree for a family that harbors the trait for albinism. According to convention, *squares* represent males, *circles* are females. Family members that are affected by the trait in question are indicated by a filled symbol *(black)*; those that are unaffected are represented by a hollow symbol *(white)*. A single horizontal line connecting a male and female represents a mating, and the horizontal lines above a series of symbols connect the offspring of that mating (in order of their birth from left to right). Individuals within each generation are labeled sequentially from left to right for purposes of identification. In the first generation in this pedigree, for example, individual 5, an unaffected male, marries an albino female. Their only son (individual 7 in the second generation) is unaffected. His children, however, are both albino.

over wrinkled peas. What happens, Mendel wondered, when plants that differ in both of these characters are bred? Again, he started with true-breeding parental strains: the dominant strain produced yellow round peas (its genotype is *YYRR*), the recessive strain produced green wrinkled peas (*yyrr*). One possibility is that the two characters, seed color and seed shape, would be transmitted from parents to offspring as a linked package. In other words, the dominant *Y* and *R* alleles would always appear together, producing true-breeding plants with yellow, round peas, generation after generation. (And the same would be true for the recessive *y* and *r* alleles.) The other possibility is that seed color and seed shape would be inherited independently of one another, which means that at some point plants that produce a novel mix of traits—yellow wrinkled peas or green round peas—would arise.

Mendel bred the plants and kept a careful record of his results. As expected, the F_1 generation showed a single phenotype: all plants produced peas that were yellow and round. But this would occur regardless of whether the parental alleles were linked. When these plants were cross-fertilized, the results clearly showed that each character is independently inherited; that is, the two alleles for seed color segregate independently of the two alleles for seed shape, producing four different pea phenotypes: yellow-round, yellow-wrinkled, green-round, and green-wrinkled (Figure 20–20). Mendel tried his seven pea characters in various pairwise combinations and always observed a characteristic 9:3:3:1 phenotypic ratio in the F_2 generation. The independent segregation of each pair of alleles during gamete formation is known as Mendel's second law—the **law of independent assortment**.

The Behavior of Chromosomes During Meiosis Underlies Mendel's Laws of Inheritance

Thus far, we have talked about alleles and genes as disembodied entities. As biologists, however, we are interested in heredity as more than a collection of mathematical ratios and probabilities—the likelihood that a pea plant will have purple flowers or that a child will be born an albino. We wish to understand how heredity works inside the sperm, the egg, and the resulting zygote. Mendel had assumed that genes are located in cells, but he didn't know what they were made of or where they could be found. We now know that the Mendelian factors that we call genes are carried on chromosomes, which are parceled out during the formation of gametes and then brought together in novel combinations in the zygote at fertilization. Chromosomes therefore provide the physical basis for Mendel's laws. As we shall now see, their behavior during meiosis and fertilization explains Mendel's laws perfectly.

During meiosis, as we discussed earlier, the maternal and paternal homologs—and the genes that lie on them—pair and then separate from one another on their way to being parceled out into gametes. These homologous chromosomes will possess different variants—or alleles—of many of the genes they carry. Take, for example, a pea plant that is heterozygous for yellow peas (*Yy*). During meiosis, the chromosomes bearing the *Y* and *y* alleles will be separated, producing two types of haploid gametes, ones that contain a *Y* allele; others that contain a *y*. Upon self-fertilization, these haploid gametes recombine at random to produce the diploid individuals of the next generation—which may be *YY*, *Yy*, or *yy*. The meiotic mechanisms that drive the separation of alleles into gametes and the random recombination of gametes at fertilization exactly underlie Mendel's genetic laws.

During meiosis, each set of paired homologs attaches to the spindle independently. This random arrangement of chromosomes on the metaphase spindle is reflected in Mendel's law of independent assortment, since genes on different chromosomes will be inherited independently. Although each gamete receives one, and only one, copy of each chromosome, it winds up with a random mixture of paternal and maternal homologs (see Figure 20–11A).

Figure 20–21 diagrams this process for a pea plant that is heterozygous for both seed color (*Yy*) and seed shape (*Rr*). The chromosome pair carrying the color alleles will attach to the meiotic spindle with a certain orientation. Whether the *Y*-bearing homolog or the *y*-bearing homolog is captured by the microtubules from one pole or the other depends on which way the bivalent happens to be facing at the moment of attachment (see Figure 20–21). The same is true for the chromosome pair carrying the alleles for seed shape. Whether the final gamete receives the

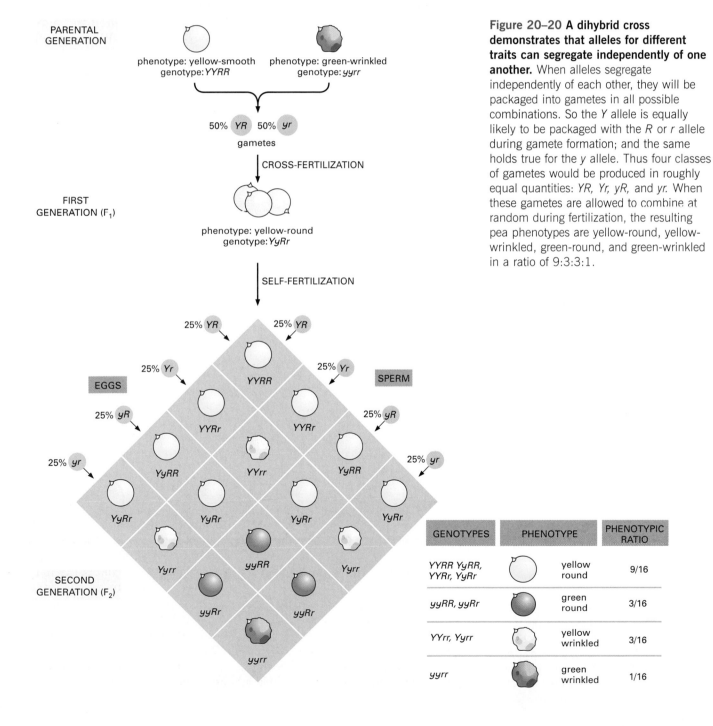

Figure 20–20 A dihybrid cross demonstrates that alleles for different traits can segregate independently of one another. When alleles segregate independently of each other, they will be packaged into gametes in all possible combinations. So the *Y* allele is equally likely to be packaged with the *R* or *r* allele during gamete formation; and the same holds true for the *y* allele. Thus four classes of gametes would be produced in roughly equal quantities: *YR*, *Yr*, *yR*, and *yr*. When these gametes are allowed to combine at random during fertilization, the resulting pea phenotypes are yellow-round, yellow-wrinkled, green-round, and green-wrinkled in a ratio of 9:3:3:1.

PARENTAL GENERATION

phenotype: yellow-smooth genotype: *YYRR*

phenotype: green-wrinkled genotype: *yyrr*

50% **YR** 50% **yr**

gametes

CROSS-FERTILIZATION

FIRST GENERATION (F₁)

phenotype: yellow-round genotype: *YyRr*

SELF-FERTILIZATION

25% **YR** 25% **YR**

25% **Yr** 25% **Yr**

EGGS | SPERM

25% **yR** 25% **yR**

25% **yr** 25% **yr**

YYRR
YYRr YYRr
YyRR YYrr YyRR
YyRr YyRr YyRr YyRr
Yyrr yyRR Yyrr
yyRr yyRr
yyrr

SECOND GENERATION (F₂)

GENOTYPES	PHENOTYPE	PHENOTYPIC RATIO
YYRR YyRR, YYRr, YyRr	yellow round	9/16
yyRR, yyRr	green round	3/16
YYrr, Yyrr	yellow wrinkled	3/16
yyrr	green wrinkled	1/16

YR, *Yr*, *yR*, or *yr* combination of alleles depends entirely on which way the two chromosome pairs were facing when they were captured by the meiotic spindle, which has the same degree of randomness as the tossing of a coin.

The Frequency of Recombination Can Be Used to Order Genes on Chromosomes

Mendel studied seven traits that were carried by seven genes (see Figure 20–15), and each trait segregated independently of the others. We now know that most of these genes were located on different chromosomes, and this situation readily explains the random assortment he observed. But, Mendel's observation that different genes assort independently does not necessarily require that the genes lie on different chromosomes. Genes that are far enough away from one another on the same chromosome will also sort independently due to the recombination that occurs during meiosis. As we discussed earlier, when the duplicated homologs form bivalents and line up on the metaphase spindle,

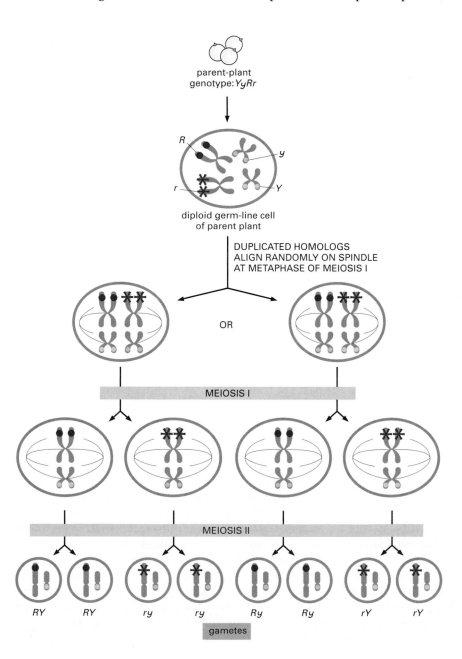

Figure 20–21 The separation of chromosomes during meiosis explains Mendel's laws of segregation and independent assortment. Here we show independent assortment of the alleles for seed color, yellow *(Y)* and green *(y)*; and seed shape, round *(R)* and wrinkled *(r)*.

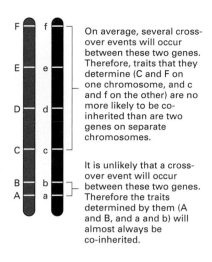

On average, several cross-over events will occur between these two genes. Therefore, traits that they determine (C and F on one chromosome, and c and f on the other) are no more likely to be co-inherited than are two genes on separate chromosomes.

It is unlikely that a cross-over event will occur between these two genes. Therefore the traits determined by them (A and B, and a and b) will almost always be co-inherited.

Figure 20–22 Because several cross-over events occur randomly along each chromosome during prophase of meiosis I, two genes on the same chromosome will obey Mendel's law of independent assortment if they are far enough apart. The cross-over events occur in the bivalents illustrated previously in Figure 20–7. Because these cross-overs are needed to form the chiasmata needed for proper chromosome segregation, several generally occur on each chromosome. As shown, genes that lie close together on a chromosome will tend to be inherited as a unit.

non-sister chromatids typically undergo several recombination events and thereby exchange genetic material. Such crossing-over events can separate alleles that were formerly together on the same chromosome, causing them to segregate into different gametes (Figure 20–22). We now know, for example, that Mendel's genes for pea shape and pod color are located on the same chromosome, but because they are far apart, they segregate independently.

On average, a human chromosome participates in two or three recombination events during meiosis, and every chromosome participates in at least one. Thus, whereas two genes that lie very close to one another on a chromosome almost always end up together in the same gametes after meiosis, two genes located at the opposite ends of a chromosome are no more likely to end up together than are genes located on different chromosomes.

As might be imagined, not all genes are inherited independently as per Mendel's second law. If genes lie close together on a chromosome, they are indeed inherited as a unit. For example genes associated with red–green colorblindness and hemophilia in humans are typically inherited as a unit. By measuring how frequently genes are co-inherited, researchers can determine whether genes reside on the same chromosome and, if so, how far apart they lie. This type of information has been extensively used to map the relative positions of the genes on each chromosome of many organisms (see How We Know, pp. 682–683). Such *genetic maps* have been crucial for the cloning of human disease genes such as the gene for cystic fibrosis.

The Phenotype of a Heterozygote Reveals Whether an Allele is Dominant or Recessive

We have seen how Mendel's observations can be explained by the movement of chromosomes during meiosis. Alleles segregate into gametes, and gametes unite to generate progeny with novel combinations of alleles. Which alleles an organism receives at fertilization then dictates its phenotype. But how do we explain the phenomenon of dominance? Why should the allele for round peas mask the expression of the allele for wrinkled peas?

In the case of Mendel's wrinkled peas, the gene that dictates seed shape encodes an enzyme that helps to convert sugars into branched starch molecules. The recessive allele, *r*, is a mutant gene that does not encode an active enzyme. Because they lack this enzyme, plants that are homozygous for the *rr* allele contain more sugar, and produce less starch, than plants that possess the dominant *R* allele. The abundance of sugar and lack of branched starch leaves the *rr* peas with a wrinkled appearance. (The frozen sweet peas available in the supermarket are wrinkled mutants, although the alleles they carry may not be the same one that Mendel used.)

Question 20–4

Imagine that each chromosome undergoes one and only one crossover event during each meiosis. How would the co-inheritance of traits that are determined by genes at the opposite end of the same chromosome compare to the co-inheritance observed for genes on two different chromosomes? How does this compare with the actual situation?

How We Know: Reading Genetic Linkage Maps

In 1913, Alfred Sturtevant, an undergraduate working with Thomas Hunt Morgan at Columbia University, published the first genetic linkage map showing the order of five genes on the *Drosophila* X chromosome. From mapping the genes for eye color and body color in fruit flies, we have moved on to the large-scale mapping of entire genomes, including those of *Drosophila,* mice, and humans. These genetic maps can be used to find new genes, trace the course of human evolution, and perhaps some day to untangle the causes of complex diseases such as cancer, diabetes, or mental illness.

Making a Map

A genetic linkage map reveals the relative locations of all of the genes on a particular chromosome. Such genetic linkage maps are based on the frequency with which two genes are co-inherited. Genes that lie close to one another on the same chromosome will be inherited together much more frequently than those that lie farther apart. If two genes are on different chromosomes—or are far apart on the same chromosome—they will have a 50–50 chance of being inherited together. By determining how often recombination separates two genes, the relative distance between them can be calculated (see Panel 20–1, p. 685).

Such recombinational analysis is most useful for constructing genetic linkage maps for experimental organisms, such as fruit flies, in which many distinctive mutants have been collected over the years. The approach is less useful for mapping genes in humans: only genes that are present in the population in at least two variant forms, that produce distinct phenotypes, can be mapped in this way. Furthermore, the true frequency of recombination between pairs of genes is difficult to determine accurately because of the small number of offspring in each generation.

This type of genetic mapping has been made much easier in humans by the collection of large numbers of DNA markers—short nucleotide sequences that exist in at least two different forms—whose location has been determined by genome sequencing. One type of marker that is proving particularly valuable for constructing detailed genetic linkage maps is the single nucleotide polymorphism (SNP). As we discuss in Chapter 9, SNPs are short sequences that differ by a single nucleotide among individuals in the population (Figure 20–23). Millions of SNPs have been found in the human genome; these polymorphisms are distributed along the length of chromosomes at a density of roughly 1 per 1000 or so nucleotide bases. The SNPs that are particularly useful for building and analyzing genetic linkage maps are those that are present at high frequency in the population—where 10% or more of the individuals in a population vary at these positions.

SNP-based genetic linkage maps can be used to identify the genes that carry mutations associated with inherited human disorders. For this purpose, DNA is collected from large family groups affected by a particular disease. These samples are then screened to determine which SNPs tend to be inherited by the individuals with the disease, but not by their unaffected relatives (Figure 20–24). Such SNPs are likely to surround the gene that carries the mutation responsible for the disease.

Finding functional SNPs

SNPs that occur within the coding or regulatory regions of a gene might alter the activity of that gene or its encoded protein. Such SNPs might account for much of the phenotypic variation seen in the human population. Researchers recently performed a detailed analysis of SNPs to explore the differences in the way that individuals respond to drugs. They focused on a set of membrane transport proteins that play a critical role in drug response; these membrane-spanning proteins control how effectively cells take in and eliminate various compounds, including drugs used to treat cancers and mental illness, for example. To determine why

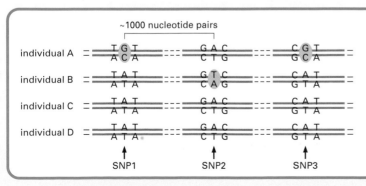

Figure 20–23 Single-nucleotide polymorphisms (SNPs) are sequences in the genome that differ by a single nucleotide between one portion of the population and another. Most, but not all, such changes in the human genome occur in regions where they do not affect function.

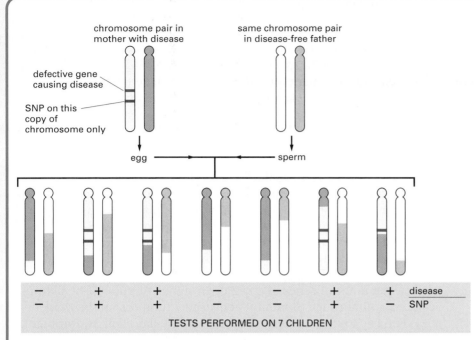

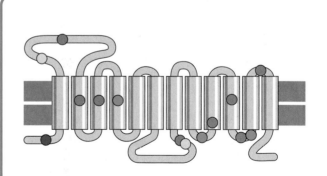

Figure 20–24 Genetic analysis can pin down the location of an unknown gene that is linked to an inherited human disease or disease susceptibility. In this procedure one studies the co-inheritance of a specific human phenotype (here a genetic disease) with a particular SNP. If individuals who inherit the disease nearly always inherit the SNP, then the gene causing the disease and the SNP are likely to be close together on the same chromosome, as shown here. To prove that an observed linkage is statistically significant, hundreds of individuals may need to be examined. Note that the linkage will not be absolute unless the SNP is located in the disease gene itself. Thus, occasionally a nearby SNP will be separated from the gene by meiotic crossing-over during the formation of the egg or sperm: this has happened in the case of the chromosome pair on the far right.

some individuals might respond poorly to such drugs, while others are exquisitely sensitive, the researchers searched hundreds of DNA samples, collected from an ethnically diverse population, for variations in a set of genes encoding 24 different membrane transport proteins. In this way, they identified 680 SNPs plus a handful of small insertions and deletions.

The researchers next set out to determine which of these SNPs affect the function of the transporters—in other words, which variations might correlate with changes in the activities of the transporter proteins, and hence with differences in human drug response. To do this, they mapped out the locations of the SNPs and found, as expected, that these polymorphisms occur less frequently at sites that are highly conserved among different animal species. The paucity of SNPs in these regions suggests that changes in the DNA sequence there are likely to affect the function of the gene, and its encoded protein. Indeed, when the researchers analyzed the function of 14 variants of a single transporter protein, the five variants that had reduced or no activity all contained SNPs that changed the identity of amino acids located in evolutionary conserved positions (Figure 20–25).

Ultimately, by correlating particular SNPs or SNP patterns with impaired transporter function, the researchers hope to be able to predict how an individual will respond to a particular drug concentration by simply looking at his or her

SNP signatures. Such SNP analyses could thereby help doctors give patients personalized treatments that are both safe and effective.

Figure 20–25 Certain SNPs alter the activity of a human membrane transport protein. Changes of this type are likely to underlie differences in the way people respond to drugs. In this diagram, the locations of amino acids affected by SNP variations are shown. Amino acid changes that reduce transporter activity appear in *light blue;* changes that eliminate activity appear in *dark blue;* the change that enhances activity appears in *red;* and changes that do not affect activity appear in *dark gray.* (Adapted from Y. Shu et al., *Proc. Natl. Acad. Sci. U.S.A.* 100:5902–5907, 2003.)

In genetics, different types of mutations are classified into the categories described in Figure 20–26. Mutations that reduce or eliminate the activity of a gene are called **loss-of-function mutations**. An organism that receives two copies of a loss-of-function allele will have a mutant phenotype—one that differs from the "normal," most commonly occurring phenotype. For pea shape, the wrinkled appearance is considered a mutant phenotype, and the round appearance is considered the normal, or wild-type, phenotype. The heterozygote, which possesses one mutant allele and one wild-type allele, makes enough active enzyme to function normally and retain the wild-type phenotype. Thus, loss-of-function mutations are usually recessive.

Another class of mutant alleles produces enzymes that are overactive, or are active in inappropriate circumstances. Such **gain-of-function mutations** are usually dominant. For example, as we saw in Chapter 16, certain mutations in the gene encoding Ras, a protein involved in cell growth and proliferation, generate a form of the protein that is always active, and therefore overshadows the wild-type allele. Thus, the mutant Ras protein can stimulate cells to divide inappropriately, even in the absence of any growth signal. About 30% of all human cancers contain such dominant activating mutations in the *ras* gene.

The key to determining whether a particular allele is dominant or recessive lies in the phenotype of the heterozygote. In such individuals, the two alleles—mutant and wild-type—are pitted against one another in a sort of functional competition. If the heterozygote resembles the wild-type organism, then the wild-type allele is dominant and the mutant allele is recessive; if the heterozygote has a mutant phenotype, then the mutant allele is dominant and the wild-type allele is recessive (Panel 20–1, p. 685).

Mutant Alleles Sometimes Confer a Selective Advantage

As we saw in Chapter 9, mutations provide the fodder for evolution. Mutations can alter the fitness of an organism, making it more (or less) likely to survive and leave progeny, and natural selection determines whether these mutations are preserved. Those changes that confer a selective advantage on an organism tend to be perpetuated, while those that seriously compromise an organism's fitness tend to be lost. Why, then, are recessive alleles that cause defects when they are present in two copies not eliminated from the gene pool?

The answer is that even deleterious mutations can benefit an organism, as individual genes can have multiple effects on phenotype. Take sickle-cell anemia in humans, for example. This disorder is caused by a mutation in the gene encoding β-globin, one of the polypeptides that make up hemoglobin (see Figure 4–23). Hemoglobin is packaged in red blood cells, where it functions to shuttle oxygen from the lungs to cells and tissues throughout the body. The sickle-cell mutation directs the formation of an abnormal polypeptide that causes red blood cells to adopt a sickled shape (see Figure 6–19). These misshapen cells clog small blood vessels, reducing the amount of oxygen that can

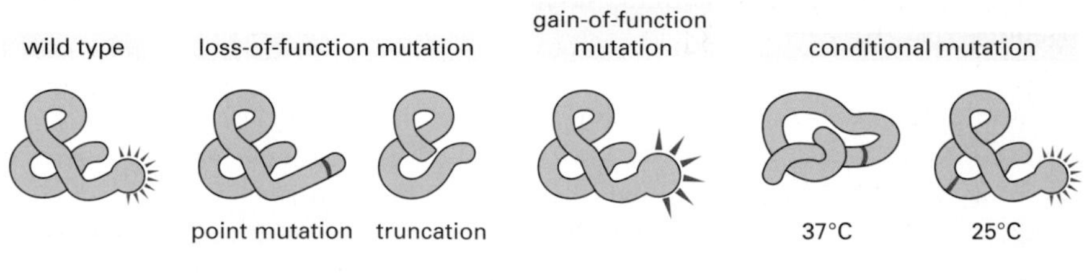

Figure 20–26 Gene mutations can affect the protein product in a variety of ways. In this example, the wild-type protein has a specific cellular function denoted by the *red rays*. Mutations that eliminate this function, make it hyperactive, or render it sensitive to higher temperatures are shown. A *temperature-sensitive mutation* is a type of *conditional mutation*: the allele produces active protein under one condition (25°C, here) but inactive protein under another. Here, the protein carries an amino acid substitution *(red)* that prevents its proper folding at 37°C.

Conditional mutations are especially useful for studying essential genes; the organism can be grown under the permissive condition and then moved to the nonpermissive condition to study the function of the gene. Loss-of-function mutations and conditional mutations are typically recessive, whereas gain-of-function mutations are most often dominant.

Panel 20–1 Some Essentials of Classical Genetics

GENES AND PHENOTYPES

Gene: a functional unit of inheritance, usually corresponding to the segment of DNA coding for a single protein.

Genome: an organism's set of genes.

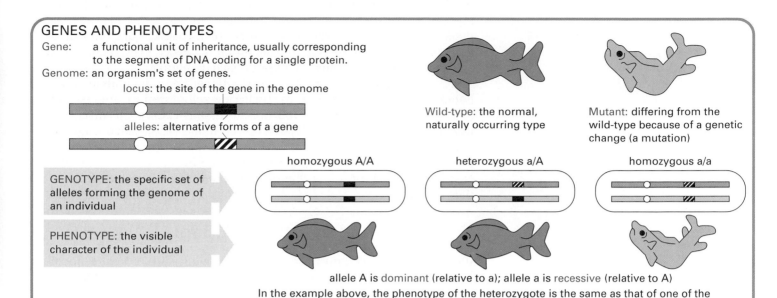

locus: the site of the gene in the genome

alleles: alternative forms of a gene

Wild-type: the normal, naturally occurring type

Mutant: differing from the wild-type because of a genetic change (a mutation)

GENOTYPE: the specific set of alleles forming the genome of an individual

PHENOTYPE: the visible character of the individual

homozygous A/A

heterozygous a/A

homozygous a/a

allele A is dominant (relative to a); allele a is recessive (relative to A)

In the example above, the phenotype of the heterozygote is the same as that of one of the homozygotes; in cases where it is different from both, the two alleles are said to be co-dominant.

TWO GENES OR ONE?

Given two mutations that produce the same phenotype, how can we tell whether they are mutations in the same gene? If the mutations are recessive (as they most often are), the answer can be found by a complementation test.

In the simplest type of complementation test, an individual who is homozygous for one mutation is mated with an individual who is homozygous for the other. The phenotype of the offspring gives the answer to the question.

COMPLEMENTATION: MUTATIONS IN TWO DIFFERENT GENES

homozygous mutant mother

homozygous mutant father

hybrid offspring shows normal phenotype: one normal copy of each gene is present

NONCOMPLEMENTATION: TWO INDEPENDENT MUTATIONS IN THE SAME GENE

homozygous mutant mother

homozygous mutant father

hybrid offspring shows mutant phenotype: no normal copies of the mutated gene are present

GENETIC MAPS

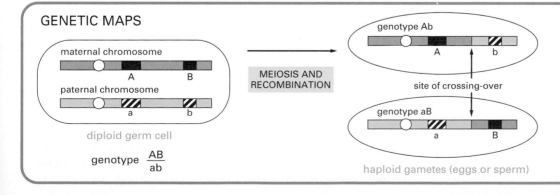

maternal chromosome

A   B

paternal chromosome

a   b

diploid germ cell

genotype $\dfrac{AB}{ab}$

MEIOSIS AND RECOMBINATION

genotype Ab

A   b

site of crossing-over

genotype aB

a   B

haploid gametes (eggs or sperm)

The greater the distance between two loci on a single chromosome, the greater is the chance that they will be separated by crossing-over occurring at a site between them. If two genes are thus reassorted in x% of gametes, they are said to be separated on a chromosome by a genetic map distance of x map units (or x centimorgans).

reach different tissues, causing a variety of symptoms—including pain, fatigue, muscle cramps, shortness of breath, and even heart failure. The defective cells are also very fragile and break easily. Phagocytic white blood cells consume these damaged and fragmented cells, leading to a shortage of red blood cells, a condition called anemia.

But the sickle-cell trait also has its benefits. Individuals who are heterozygous or homozygous for the sickle-cell mutation, are resistant to malaria. This is because the organism that causes the disease is unable to reproduce in sickle-shaped red blood cells, which fragment before the parasite has a chance to multiply. Resistance to malaria is important particularly for populations living in Africa, where the disease is still rampant—and where, presumably, the recessive allele initially arose.

Genetics as an Experimental Tool

The realization that chromosomes are the cellular structures responsible for shuttling our genes from one generation to the next did more than demystify the basis of inheritance. It united the science of genetics with other disciplines: cell biology, biochemistry, physiology, even medicine. Knowing that genes are the units of inheritance, and that genes are made of DNA, also affords us an opportunity to use genetics as a tool for new discoveries. By examining and manipulating our DNA, we can now begin to learn how our genes function together to create our phenotype, and how differences in those genes underlie the differences between individuals. Ultimately, we hope to use our knowledge of genetics to more accurately diagnose and treat human diseases, and to help us decipher who we are as individuals and as a species.

In this section, we outline the classical genetic approach to identifying genes and determining how they influence the phenotype of an organism. The process involves intentionally generating large numbers of mutant laboratory organisms, using a *genetic screen* to isolate those rare mutants among them with a phenotype of interest, and then identifying the gene or genes responsible for the observed characteristic in those selected mutant organisms. We shall also discuss how we can examine human DNA collected from populations all over the world for clues about the genetics of complex traits and diseases, as well as for hints about the evolution of our species. Finally, we ponder how much of our fate as individuals is sealed within our DNA.

The Classical Approach Begins with Random Mutagenesis

Before the advent of recombinant DNA technology (discussed in Chapter 10), most genes were identified by observing the processes disrupted when the gene was mutated. The process begins with the isolation of mutants that have an interesting or unusual appearance: fruit flies with white eyes or curly wings, for example. Working backward from this phenotype, one then determines the organism's genotype, the form of the gene responsible for that characteristic. This classical genetic approach—isolating the genes responsible for mutant phenotypes—is most easily performed in organisms that reproduce rapidly and are amenable to genetic manipulation, such as bacteria, yeasts, worms, and fruit flies.

Although spontaneous mutants can sometimes be found by examining extremely large populations—thousands or tens of thousands of individual organisms—the process of identifying interesting mutants can be made much more efficient by generating mutations with agents that

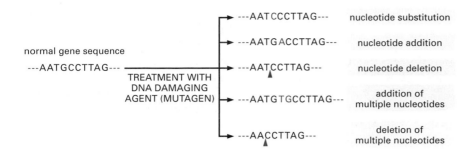

normal gene sequence

---AATGCCTTAG---

TREATMENT WITH DNA DAMAGING AGENT (MUTAGEN)

---AATCCCTTAG--- nucleotide substitution

---AATGACCTTAG--- nucleotide addition

---AATCCTTAG--- nucleotide deletion

---AATGTGCCTTAG--- addition of multiple nucleotides

---AACCTTAG--- deletion of multiple nucleotides

Figure 20–27 Mutations can be caused by a variety of alterations in DNA. Some common categories are shown here. Different mutagens tend to produce different types of changes.

damage DNA, called mutagens. Different mutagens can generate different types of DNA alterations (Figure 20–27). By treating organisms with mutagens, very large numbers of mutants can be created quickly and then screened for a particular defect of interest, as we will see shortly.

Mutagenesis is well suited for dissecting biological processes in worms and flies, but that approach obviously cannot be used for studying humans. Unlike the organisms we have been discussing, humans do not reproduce rapidly, and they are not intentionally treated with mutagens. Moreover, any human with a serious defect in an essential process, such as DNA replication, would die long before birth.

There are two answers to the question of how we study human genes. First, because genes and gene functions have been so highly conserved throughout evolution, the study of less complex model organisms has turned out to reveal critical information about similar genes and processes in humans. The corresponding human genes can then be studied further in cultured human cells. Second, many mutations that are not necessarily lethal, such the sickle-cell trait discussed above, have arisen spontaneously in the human population. Analyses of the phenotypes of the affected individuals, together with studies of their cultured cells, have provided many unique insights into important human gene functions. Although such mutations are rare, they are very efficiently discovered because of a unique human property: the mutant individuals call attention to themselves by seeking special medical care.

Genetic Screens Identify Mutants Deficient in Cellular Processes

Once a collection of mutants in a model organism such as a bacterium, a yeast, or a fruit fly has been produced, one generally must examine many thousands of individuals to find the altered phenotype of interest. Such a search is called a **genetic screen**. Because obtaining a mutation in a gene of interest depends on the likelihood that the gene will be inactivated or otherwise mutated during random mutagenesis, the larger the genome, the less likely it is that any particular gene will be mutated. Therefore, the more complex the organism, the more mutants must be examined to avoid missing genes.

The phenotype being screened for can be simple or complex. Simple phenotypes are easiest to detect: a metabolic deficiency, in which an organism is no longer able to grow in the absence of a particular amino acid or nutrient, is one such example.

Because defects in genes that are required for fundamental cell processes—RNA synthesis and processing or cell cycle control, for example—are usually lethal, the functions of these critical genes are often studied in temperature-sensitive mutants. In these mutants the protein product of the gene is only conditionally defective: it functions normally at a medium temperature, but can be inactivated by a small increase or decrease in temperature (see Figure 20–26). Thus the

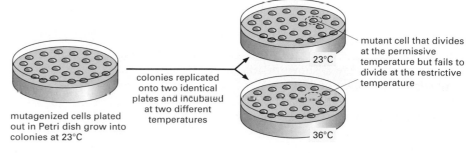

Figure 20–28 Temperature-sensitive mutants are valuable for identifying the genes and proteins involved in essential cell processes. Here yeast cells were incubated with a chemical that generates mutations in their DNA. These cells are spread onto a plate and allowed to grow at a permissive temperature, that is, one at which the cells divide normally. The colonies are then transferred to two identical Petri dishes using a technique called replica plating. One of these plates is incubated at the cooler, "permissive" temperature, the other at a higher temperature. Those cells that contain a temperature-sensitive mutation in a gene essential for proliferation can be readily identified, because they divide at the permissive temperature but fail to divide at the warmer, nonpermissive temperature.

mutagenized cells plated out in Petri dish grow into colonies at 23°C

colonies replicated onto two identical plates and incubated at two different temperatures

mutant cell that divides at the permissive temperature but fails to divide at the restrictive temperature

23°C

36°C

abnormality can be switched on and off experimentally simply by changing the temperature. A cell containing a temperature-sensitive mutation in a gene essential for survival at a nonpermissive temperature is kept alive by growing it at the normal or permissive temperature (Figure 20–28).

Many temperature-sensitive mutants were isolated to identify the genes that encode the bacterial proteins required for DNA replication. Here large populations of mutagen-treated bacteria were screened for cells that stop making DNA when they are warmed from 30°C to 42°C. Temperature-sensitive mutants have also been used to identify many of the proteins involved in regulating the cell cycle (discussed in Chapter 18) or in moving proteins through the secretory pathway in yeast (discussed in Chapter 15).

Complex phenotypes, such as changes in learning or behavior, can also be studied by carrying out genetic screens in model organisms. As an example, researchers were able to isolate a gene that affects social behavior in worms by screening for animals that feed alone (Figure 20–29).

A Complementation Test Reveals Whether Two Mutations Are in the Same Gene

A large-scale genetic screen can turn up many different mutants that have the same phenotype. These defects might lie in different genes that function in the same process, or they might represent different mutations in the same gene. How can we distinguish between the two possibilities? If the mutations are recessive—if, for example, they represent a loss of function of a particular gene—a complementation test can be used to ascertain whether the mutations fall in the same or in different genes.

Figure 20–29 Genetic screens can be used to identify mutations that affect an animal's behavior. (A) Wild-type *C. elegans* engage in social feeding. The worms swim around until they encounter their neighbors and then commence feeding. (B) Mutant worms feed alone. (Courtesy of Cornelia Bargmann, *Cell* 94:cover, 1998. © Elsevier Science.)

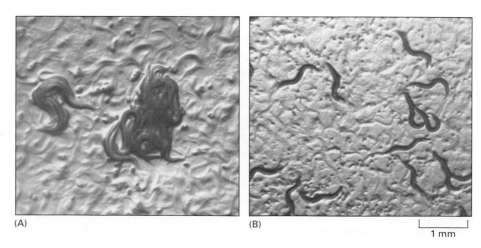

(A)

(B)

1 mm

In the simplest type of complementation test, an individual that is homozygous for one mutation—that is, it possesses two identical alleles of the mutant gene in question—is mated with an individual that is homozygous for the other mutation. If the two mutations are in the same gene, the offspring will show the mutant phenotype, because they carry only defective copies of the gene in question (see Panel 20–1, p. 685). If, in contrast, the mutations fall in different genes, the resulting offspring will show the normal, wild-type phenotype because they will have one normal copy (and one mutant copy) of each gene. The mutations thereby complement one another and restore a normal phenotype. Complementation testing of mutants identified during genetic screens has revealed, for example, that 5 genes are required for yeast to digest the sugar galactose; that 20 genes are needed for *E. coli* to build a functional flagellum; and that hundreds of genes are involved in the development of an adult nematode worm from a fertilized egg.

Human Genes Are Inherited in Haplotype Blocks, Which Can Aid in the Search for Mutations That Cause Disease

Ultimately we wish to directly examine how our own genes influence our phenotype, in the hopes that we may be better able to understand human biology. With the recent determination of the complete human genome sequence, we can now study human genetics in a way that was impossible only a few years ago. Using the human genome sequence as a starting point, we can begin to identify those DNA differences that distinguish one individual from another. No two humans (with the exception of identical twins) have the same genome. Each of us carries in our genome a set of polymorphisms—changes in nucleotide sequence—that make us unique. These polymorphisms can be used as markers for building genetic maps or for performing the genetic analyses that allow us to link particular polymorphisms with specific diseases or predispositions to disease (see How We Know, pp. 682–683).

The problem is that any two humans typically differ by about 0.1% in their nucleotide sequences (approximately one nucleotide difference every 1000 nucleotides). This translates to about 3 million differences between one person and another. Theoretically, one would need to search through all 3 million of those polymorphisms to identify the one or two changes that are responsible for a particular heritable disease or disease susceptibility. To reduce the number of polymorphisms one needs to examine, researchers are taking advantage of the recent discovery that human genes tend to be inherited in blocks.

Our species is relatively young, and it is thought that modern humans expanded from a relatively small population that existed in Africa about 10,000 years ago. Because only a few hundred generations separate us from this ancestral population, large segments of human chromosomes have passed from parent to child unaltered by the recombination events that occur in meiosis. (Recall from earlier in this chapter that only a few crossovers occur in each human chromosome per meiosis.) In fact, we observe that certain sets of alleles—and DNA markers such as the single nucleotide polymorphisms or SNPs discussed in Chapter 9—are inherited in large, linked blocks. These ancestral chromosome segments—sets of alleles and markers that have been inherited in clusters with little genetic rearrangement across the generations—are called **haplotype blocks**. Like genes and genetic markers—which exist in different alleleic forms—haplotype blocks also come in a limited number of "flavors" that are common in the human population, each of which represents an allele combination passed down from a shared ancestor long ago.

Having recently discovered the existence of haplotype blocks, researchers are now constructing a different kind of human genome map, one based on the haplotype blocks found in different human populations (initially including individuals from Africa, Asia, and Europe). Geneticists hope that the human haplotype map will make finding the genes responsible for, say, an individual's susceptibility to disease a much more manageable task. Instead of searching through each of the many millions of SNPs in the human population, the researchers need only search through a considerably smaller set of selected SNPs to identify the haplotype block that appears to be inherited by individuals with the disease. High throughput, robotic technologies for identifying SNPs can then be used to screen the appropriate populations.

If a specific haplotype block is more common among people with a certain disease than in unaffected individuals, the mutation linked to that disease will likely be located on that same segment of DNA (Figure 20–30). Researchers can then zero in on the specific region of the genome to search for one or more specific genes or genetic variants associated with the disease. This approach should, in principle, allow even disease susceptibilities that are caused by a combination of genes to be traced to the responsible set of DNA sequences.

A detailed examination of haplotype blocks can even tell us whether a particular allele has been favored by natural selection. As a rule, when a selectively neutral variation arises in a gene, it takes a long time for that allele to become common in the population. The more common—and therefore older—such an allele is, the smaller should be the haplotype block that surrounds it, because it will have had many chances of being separated from its neighboring variations by the recombination that occurs in meiosis generation after generation.

A new allele may become common quite quickly, however, if it confers some dramatic advantage on the organism. For example, mutations or variations that make an organism more resistant to an infection might be selected for because organisms with this variation would be more likely to survive and pass the mutation on to their offspring. Working with haplotype maps of individual genes, researchers have detected such positive selection for two human genes that confer

Figure 20–30 Tracing the inheritance of SNPs within haplotype blocks can reveal the location of a disease-causing gene. An ancestor who acquires a disease-causing mutation in gene 1 will pass that mutation along to his or her descendants. Part of this gene is embedded within a haplotype block *(red)*—a cluster of variations that have been passed along from the ancestor in a continuous chunk. In the 400 generations that separate the ancestor from modern descendants with the disease, SNPs located over most of the ancestral 200,000-base-pair region have been shuffled by recombination in the descendant genome *(blue)*. The 30 SNPs within the haplotype block *(red),* however, have been inherited as a group, as no crossover events have yet separated them. To locate a gene that causes the inherited disease, researchers will examine the SNP patterns in a number of people who have the disease. An individual with the disease will retain the ancestral pattern of SNPs located within the haplotype block shown, revealing that the disease-causing mutation is likely to lie within that haplotype block—thus in gene 1. The beauty of this method is that only a fraction of the total SNPs need to be examined: one should be able to locate genes after searching through only about 10% of the 3 million useful SNPs present in the human genome.

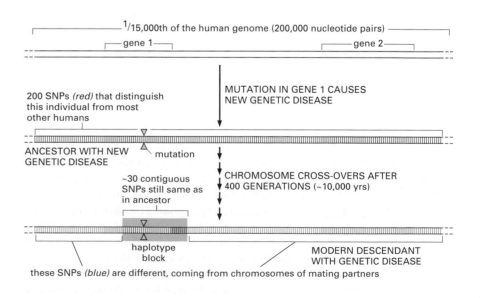

(A) size of haplotype block surrounding a typical allele

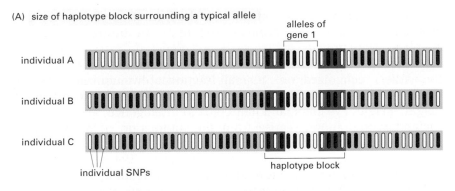

alleles of
gene 1

individual A

individual B

individual C

individual SNPs

haplotype block

(B) unusually large haplotype block surrounds a particular allele of gene 2

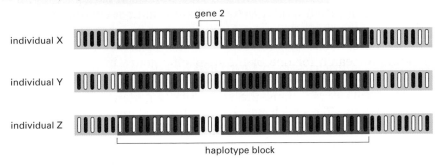

gene 2

individual X

individual Y

individual Z

haplotype block

Figure 20–31 Mutations that have been selected for in fairly recent human history can be identified by the unusually large haplotype blocks in which they are embedded. The SNPs are indicated in this diagram by vertical bars, which are shaded *white* or *blue* according to their DNA sequence. From this data one would conclude that this particular allele of gene 2 arose relatively recently in human history.

resistance to malaria. (These genes are different from the β-globin gene discussed earlier.) The alleles that confer resistance are widespread in the population, but they are embedded in unusually large haplotype blocks, suggesting that they rose to prominence recently in the human gene pool (Figure 20–31).

In revealing the paths along which humans evolved, the human haplotype map provides a new window into our past; in helping us discover the genes that make us susceptible or resistant to disease, the map may also provide a rough guide to our futures.

Complex Traits Are Influenced by Multiple Genes

A concert pianist might have an aunt who plays the violin. In another family, the parents and the children might all be fat. In a third family, the grandmother might be an alcoholic while her grandson abuses drugs. To what extent are such characteristics—musical ability, obesity, and addiction—inherited genetically? This is a very difficult question to answer. Some traits or diseases "run in families" but they appear in only a few relatives, or with no easily discernable pattern.

Characteristics that do not follow Mendel's laws but have a genetically inherited component are termed **complex traits**. These traits are often **polygenic**, that is, they are controlled by multiple genes, each of which makes a small contribution to the phenotype in question. The effects of these genes are additive, which means that together they produce a continuum of varying features within the population. Individually, the genes that correspond to a polygenic trait follow Mendel's laws, but because they all influence the phenotype, the pattern of traits inherited by offspring is often highly complex.

A simple example of a polygenic trait is that of eye color. As we mentioned in our discussion of albinism, eye color is determined by enzymes that control the distribution and production of the pigment melanin. The more melanin produced, the darker is the eye color.

Question 20–5

In a recent automated analysis, SNPs located at thousands of sites across the genome were anlayzed in pooled DNA samples from humans who had been sorted into groups according to their age. For the vast majority of these sites, there was no change in relative frequency as these humans aged. Rarely, however, a particular variant at one position was found to drop in frequency progressively for people over 50-years-old. Which of the possible explanations below seems most likely?

A. The nucleotide in that SNP at that position is unstable, and mutates with age.

B. Those people born more than 50 years ago came from a population that tended to lack the disappearing SNP.

C. The nucleotide that creates the SNP alters an important gene product in a way that shortens the human life span.

Because numerous genes contribute to the formation of melanin, eye color in humans shows enormous variation, from the palest gray to a deep chocolate brown.

Although diseases based on mutations in single genes were some of the earliest recognized, only a small fraction of human traits are dictated by single genes. The most obvious human phenotypes—from height, weight, eye color, and hair color to intelligence, temperament, sociability, and humor—arise from the interaction of many genes. Multiple genes also almost certainly underlie a propensity for the most common human diseases: diabetes, heart disease, high blood pressure, allergies, asthma, and various mental illnesses, including major depression and schizophrenia. Researchers are exploring new strategies—including the use of the haplotype maps discussed earlier—to understand the complex interplay between genes that act together to determine many of our most "human" traits.

Is Our Fate Encoded in Our DNA?

Identical twins, who share the same set of genes, do not look and behave exactly alike, which suggests that there is something more to our phenotype than our genes. One important factor is the environment. Most genes are affected to some degree by the environment. Genes that influence human height, for example, respond to nutrition: malnourished individuals will not grow to be six feet tall, no matter what genes they have. Similarly, exercise alters build, tanning darkens skin, and practice enhances athletic performance. The environment also influences which diseases we may fall prey to. Diet and exercise strongly affect whether someone with a predisposition will get heart disease, for example. The same is true for smoking and lung cancer.

One way that geneticists attempt to tease apart the relative contributions of genes and the environment to a particular trait is by comparing its inheritance in identical versus fraternal twins. Identical twins develop from the same fertilized egg; thus they share 100% of their genes. Fraternal twins, on the other hand, are no more genetically related than any other pair of siblings. A trait that occurs more frequently in identical twins than in fraternal twins must be controlled, at least in part, by genes. Such twin studies have provided information about the contributions of genes and the environment for many traits, including diseases such as schizophrenia, diabetes, and cancer (Figure 20–32).

Studies of adopted individuals and their families can also help to tease apart genetic and environmental effects. People adopted by non-relatives, for example, share an environment with their adoptive family, but are genetically distinct. At the same time, they share genes with their biological parents, but are raised in a different environment. Thus, similarities between adopted individuals and their adoptive parents likely reflect environmental influences, whereas similarities between adoptive individuals and their biological parents mostly reflect genetic

Figure 20–32 Some human traits are strongly influenced by the environment, others less so. Studies of identical and fraternal twins have suggested the relative genetic and environmental contributions to different human traits.

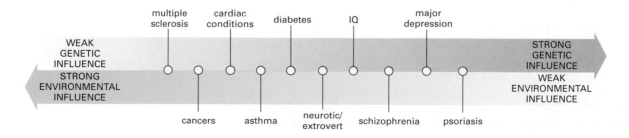

influences. One study of adopted people in Denmark demonstrated that schizophrenia was more prevalent among the biological relatives than the adoptive families of individuals diagnosed with the disease, confirming that schizophrenia has a strong genetic component.

The genes that we receive from our parents at fertilization shape our phenotype, but do not strictly dictate it. Each of us is the product of a unique genome and a unique set of experiences, and we need to appreciate the contributions of both if we are to understand the meaning of our genetic inheritance.

Essential Concepts

- Sexual reproduction involves the cyclic alternation of diploid and haploid states: diploid cells divide by meiosis to form haploid gametes, and the haploid gametes from two individuals fuse at fertilization to form a new diploid cell.
- During meiosis, the maternal and paternal chromosomes of a diploid cell are parceled out to gametes so that each gamete receives one copy of each chromosome. Because the assortment of the two members of each chromosome pair occurs at random, many genetically different gametes can be produced from a single individual.
- Crossing-over ensures the proper segregation of homologous chromosomes and enhances the genetic reassortment that occurs during meiosis by exchanging genes between them.
- Although most of the mechanical features of meiosis are similar to those of mitosis, the behavior of the chromosomes is different: meiosis produces four genetically dissimilar haploid cells by two consecutive cell divisions, whereas mitosis produces two genetically identical diploid cells by a single cell division.
- Mendel unraveled the laws of heredity by studying the inheritance of a handful of discrete traits in garden peas.
- Mendel's law of segregation states that the maternal and paternal alleles for each trait separate from one another during gamete formation, then reunite at random during fertilization.
- Mendel's law of independent assortment states that during gamete formation, different alleles segregate independently of each other.
- The behavior of chromosomes during meiosis explains Mendel's laws.
- If two genes are close to each other on a chromosome, they tend to be inherited as a unit. The frequency of recombination between them can be used to construct a genetic map that shows the order of genes on a chromosome.
- Mutant alleles can be either dominant or recessive. If the heterozygous organism has a mutant phenotype, the mutant allele is dominant; if it has a normal phenotype, the mutant allele is recessive.
- Complementation tests reveal whether two mutations that produce the same phenotype lie in the same gene or in different genes.
- Mutant organisms can be generated by treating animals with chemicals that damage DNA. Such mutants can then be screened to identify phenotypes of interest and, ultimately, to isolate the responsible genes.
- With the exception of identical twins, no two human genomes are alike. Each of us carries a unique set of polymorphisms—changes in nucleotide sequence—that shapes our individual phenotypes.
- Single-nucleotide polymorphisms (SNPs) are DNA sequences that differ by a single nucleotide base between one portion of the population and another. They provide useful markers for performing genetic analyses that link a specific trait with a particular region of DNA.

- Human genes tend to be inherited in large haplotype blocks, which contain sets of alleles and SNPs that have been inherited as a group with little geneic rearrangement. The presence of such genetic clusters make it easier to identify genes and mutations that are associated with human disease.
- Many human traits run in families but do not adhere strictly to Mendel's laws. These complex traits arise from the interactions of multiple genes, each of which contributes to the phenotype.
- Although many human traits have a strong genetic basis, some are determined primarily by the environment. It is the interactions of our genetic makeup with our environment that make each of us unique.

Key Terms

allele	heterozygous
asexual reproduction	homolog
bivalent	homologous chromosome
chiasma	homozygous
complex trait	law of independent assortment
crossing-over	loss-of-function mutation
diploid	meiosis
fertilization	pairing
gain-of-function mutation	pedigree
gamete	phenotype
genetic map	polygenic
genetic screen	recombination
genetics	segregation
genotype	sexual reproduction
haploid	sister chromatid
haplotype block	zygote

Questions

Question 20–6

Which of the following statements are correct? Explain your answers.

  A. The egg and sperm cells of animals contain haploid genomes.

  B. During meiosis, chromosomes are allocated so that each germ cell obtains one and only one copy of each of the different chromosomes.

  C. Mutations that arise during meiosis are not transmitted to the next generation.

Question 20–7

Why is it advantageous for an organism to be diploid? Why is it a disadvantage to a geneticist who might wish to study the organism? Why is it advantageous for a diploid organism to produce haploid gametes in order to reproduce itself?

Question 20–8

What might cause chromosome disjunction, where two copies of the same chromosome end up in the same daughter cell? What could be the consequences of this event occurring (a) in mitosis and (b) in meiosis?

Question 20–9

Why do sister chromatids have to remain paired in division I of meiosis? Does the answer suggest a strategy for washing your socks?

Question 20–10

Distinguish between the following genetic terms:
A. Gene and allele.
B. Homozygous and heterozygous.
C. Genotype and phenotype.
D. Dominant and recessive.

Question 20–11

You have been given three wrinkled peas, which we shall denote as A, B, and C, each of which you plant to produce a mature pea plant. Each of these three plants, once self-pollinated, produces only wrinkled peas.
A. Given that you know that the wrinkled pea phenotype is recessive, as a result of a loss-of-function mutation, what can you say about the genotype of each plant?
B. Can you safely conclude that each of the three plants carries a mutation in the same gene?
C. If not, how could you rule out the possibility that each plant carries a mutation in a different gene, each of which gives the wrinkled pea phenotype?

Question 20–12

A mutation that happened to occur during the formation of one of Susan's grandfather's sperm gave rise to a hereditary form of deafness that was passed down within Susan's family as shown in Figure Q20–12.
A. Is this mutation most likely to be dominant or recessive?
B. Is it carried on an autosome or a sex chromosome? Why?
C. A complete SNP analysis has been done for all of the eleven grandchildren (four affected, and seven disease-free). In comparing all these SNP results, how long a haplotype block would you expect to find? How would you detect it?

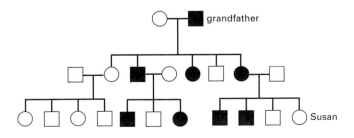

Figure Q20–12

Question 20–13

In the pedigree shown in Figure Q20–13, the first born in each of three generations is the only person effected by a dominant genetically inherited disease, X. Your friend concludes that the first child born has a greater chance of inheriting the mutant X allele than does later children.
A. According to Mendel's laws, should this be possible?
B. What is the probability of obtaining this result by chance?
C. What kind of additional data would be needed to test your friend's idea?

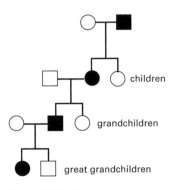

Figure Q20–13

Question 20–14

Certain mutations are called *dominant-negative mutations*. What do you think this means and how do you suppose these mutations act? Explain the difference between a dominant-negative mutation and a gain-of-function mutation.

Question 20–15

Discuss the following statement: "We would have no idea today of the importance of insulin as a regulatory hormone if its absence were not associated with the devastating human disease diabetes. It is the dramatic consequences of its absence that focused early efforts on the identification of insulin and the study of its normal role in physiology."

Question 20–16

Early genetic studies in *Drosophila* laid the foundation for our current understanding of genes. *Drosophila* geneticists were able to generate mutant flies with a variety of easily observable phenotypic changes. Alterations from the fly's normal brick-red eye color have a venerable history because the very first mutant found by Thomas Hunt Morgan was a white-eyed fly (Figure Q20–16). Since that time, a large number of mutant flies with intermediate eye colors have been isolated and given names that challenge your color sense: garnet, ruby, vermilion, cherry, coral, apricot, buff, and carnation. The mutations responsible for

Table Q20–16 Complementation Analysis of *Drosophila* Eye-color Mutations								
MUTATION	white	garnet	ruby	vermilion	cherry	coral	apricot	buff	carnation
white	–	+	+	+	–	–	–	–	+
garnet		–	+	+	+	+	+	+	+
ruby			–	+	+	+	+	+	+
vermilion				–	+	+	+	+	+
cherry					–	–	–	–	+
coral						–	–	–	+
apricot							–	–	+
buff								–	+
carnation									–

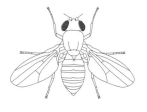
brick-red

flies with other eye colors

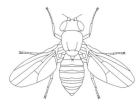
white

Figure Q20–16

these eye-color phenotypes are all recessive. To determine whether the mutations affected the same or different genes, homozygous flies for each mutation were bred to one another in pairs and the eye colors of their progeny were noted. In Table Q20–16, brick-red wild type eyes are shown as (+) and other colors are indicated as (–).

A. How is it that flies with two different eye colors—ruby and white, for example—can give rise to progeny that all have brick-red eyes?

B. Which mutations are alleles of the same gene and which affect different genes.

C. How can different alleles of the same gene give different eye colors?

Question 20–17

What are single-nucleotide polymorphisms (SNPs), and how can they be used to locate a mutant gene by linkage analysis?

Question 20–18

Tim and John have had their entire genomes sequenced. This allows a list to be made of 3 million sites where their DNA sequences differ by a single nucleotide. Why will many of these SNPs not prove to be useful DNA markers for genetic mapping experiments? What experiment could you do to distinguish between useful and nonuseful SNPs?

Highlight from *Essential Cell Biology 2 Interactive* CD-ROM

20.1   Meiosis

Tissues and Cancer

Cells are the building blocks of multicellular organisms. This seems a simple statement, but it raises deep problems. Cells are not like bricks: they are small and squishy. How can they be used to construct a giraffe or a giant redwood tree? Each cell is enclosed in a flimsy membrane less than a hundred-thousandth of a millimeter thick, and it depends on the integrity of this membrane for its survival. How then can cells be joined together robustly, with their membranes intact, to form a muscle that will lift an elephant's weight? Most mysterious of all, if cells are the building blocks, where is the builder and where are the architect's plans? How are all the different cell types in a plant or an animal produced, with each in its proper place in an elaborate pattern (Figure 21–1)?

Most of the cells in multicellular organisms are organized into cooperative assemblies called *tissues*, such as the nervous, muscle, epithelial, and connective tissues found in vertebrates (Figure 21–2). In this chapter we begin by discussing the architecture of tissues from a mechanical point of view. We shall see that tissues are composed not only of cells, with their internal framework of cytoskeletal filaments (discussed in Chapter 17), but also of *extracellular matrix*, which cells secrete around themselves; it is this matrix that gives supportive tissues such as bone or wood their strength. Cells can be bound together via the extracellular matrix, or directly by attachment to one another. We shall discuss the *cell junctions* that link cells together in the flexible, mobile tissues of animals, transmitting forces from the cytoskeleton of one cell to that of the next, or from the cytoskeleton of a cell to the extracellular matrix.

But there is more to the organization of tissues than mechanics. Just as a building needs plumbing, telephone lines, and other fittings, so an animal tissue requires blood vessels, nerves, and other components formed from a variety of specialized cell types. All the tissue components have to be coordinated correctly, and many of them require continual maintenance and renewal. Cells die and have to be replaced with new cells of the right type, in the right places, and in the right numbers. We shall consider how these processes are organized, and we shall discuss the crucial role that **stem cells** play in this tissue renewal and repair.

Disorders of tissue renewal are a major medical concern, and those due to the misbehavior of mutant cells underlie the development of **cancer**. This disease will be the topic of our final section. The study of cancer requires a synthesis of knowledge of cells and tissues at every level, from the molecular biology of DNA repair to the principles of natural selection and the social organization of cells in tissues. Many fundamental advances in cell biology have been driven by the cancer research effort, and we shall see that the basic science in return has borne fruit in a deepened understanding of the disease and a new optimism about its treatment.

Extracellular Matrix and Connective Tissues

Plant Cells Have Tough External Walls

Cellulose Fibers Give the Plant Cell Wall Its Tensile Strength

Animal Connective Tissues Consist Largely of Extracellular Matrix

Collagen Provides Tensile Strength in Animal Connective Tissues

Cells Organize the Collagen That They Secrete

Integrins Couple the Matrix Outside a Cell to the Cytoskeleton Inside It

Gels of Polysaccharide and Protein Fill Spaces and Resist Compression

Epithelial Sheets and Cell-Cell Junctions

Epithelial Sheets Are Polarized and Rest on a Basal Lamina

Tight Junctions Make an Epithelium Leak-proof and Separate Its Apical and Basal Surfaces

Cytoskeleton-linked Junctions Bind Epithelial Cells Robustly to One Another and to the Basal Lamina

Gap Junctions Allow Ions and Small Molecules to Pass from Cell to Cell

Tissue Maintenance and Renewal

Tissues Are Organized Mixtures of Many Cell Types

Different Tissues Are Renewed at Different Rates

Stem Cells Generate a Continuous Supply of Terminally Differentiated Cells

Stem Cells Can Be Used to Repair Damaged Tissues

Nuclear Transplantation Provides a Way to Generate Personalized ES Cells: the Strategy of Therapeutic Cloning

Cancer

Cancer Cells Proliferate, Invade, and Metastasize

Epidemiology Identifies Preventable Causes of Cancer

Cancers Develop by an Accumulation of Mutations

Cancers Evolve Properties That Give Them a Competitive Advantage

Many Diverse Types of Genes Are Critical for Cancer

Colorectal Cancer Illustrates How Loss of a Gene Can Lead to Growth of a Tumor

An Understanding of Cancer Cell Biology Opens the Way to New Treatments

Extracellular Matrix and Connective Tissues

Plants and animals have evolved their multicellular organization independently, and their tissues are constructed on different principles. Animals prey on other living things, and for this they must be strong and agile: they must possess tissues capable of rapid movement, and the cells that form those tissues must be able to generate and transmit forces and to change shape quickly. Plants, by contrast, are sedentary, their tissues are more or less rigid, and their cells are weak and fragile if isolated from their supporting tissue framework.

The strength of a plant tissue comes from the *cell walls*, formed like boxes, that enclose, protect, and constrain the shape of each of its cells (Figure 21–3). The cell wall is a type of extracellular matrix that the plant cell secretes around itself. The cell controls the composition of this material: it can be thick and hard, as in wood, or thin and flexible, as in a leaf. But the principle of tissue construction is the same in either case: many tiny boxes are cemented together, with a delicate cell living inside each one. Indeed, as we noted in Chapter 1, it was this close-packed mass of microscopic chambers, seen in a slice of cork by Robert Hooke three centuries ago, that originally gave rise to the term "cell."

Animal tissues are more diverse. Like plant tissues, they consist of extracellular matrix as well as cells, but these components are organized in many different ways. In some tissues, such as bone or tendon, extracellular matrix is plentiful and mechanically all-important; in others, such as muscle or epidermis, extracellular matrix is scanty, and the cytoskeleton of the cells themselves carries the mechanical load. We begin with a brief discussion of plant cells and tissues, before moving on to those of animals.

Plant Cells Have Tough External Walls

A naked plant cell, artificially stripped of its wall, is a delicate and vulnerable thing. With care, it can be kept alive in culture; but it is easily ruptured, and even a small maladjustment of the osmotic strength of the culture medium can cause it to swell and burst. Its cytoskeleton lacks the tension-bearing intermediate filaments found in animal cells, and it has virtually no tensile strength. An external wall, therefore, is essential.

The plant cell wall has to be tough, but it does not necessarily have to be rigid. Osmotic swelling of the cell, limited by the resistance of the cell wall, can keep the chamber distended, and a mass of such swollen chambers cemented together forms a semirigid tissue (Figure 21–4).

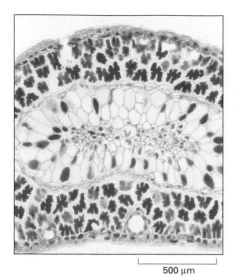

Figure 21–1 Multicellular organisms are built from organized collections of cells. Stained cross section through a leaf of a pine tree (a pine needle), showing the precisely organized pattern of different cell types.

500 μm

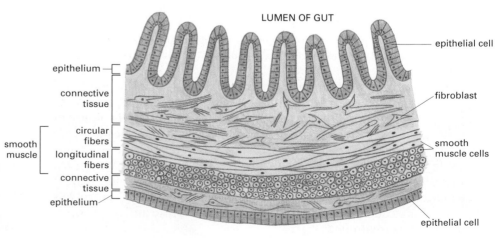

Figure 21–2 Cells are organized into tissues. Simplified drawing of a cross section through part of the wall of the intestine of a mammal. This long, tubelike organ is constructed from epithelial tissues (*red*), connective tissues (*green*), and muscle tissues (*yellow*). Each tissue is an organized assembly of cells held together by cell-cell adhesions, extracellular matrix, or both.

LUMEN OF GUT
epithelial cell
epithelium
connective tissue
fibroblast
circular fibers
smooth muscle
longitudinal fibers
smooth muscle cells
connective tissue
epithelium
epithelial cell

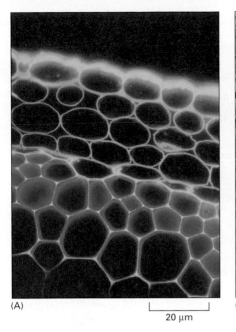
(A)

⊢————⊣
20 μm

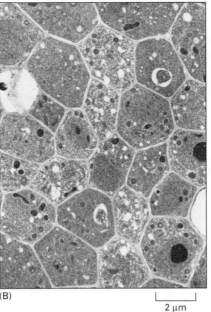
(B)

⊢————⊣
2 μm

Figure 21–3 Plant tissues are strengthened by the plant cell wall. (A) A cross section of part of the stem of the flowering plant *Arabidopsis* is shown, stained with fluorescent dyes that label two different cell wall components—cellulose in *blue,* and another polysaccharide (pectin) in *green.* The cells themselves are unstained and invisible in this preparation. Regions rich in both cellulose and pectin appear white. Pectin predominates in the outer layers of cells, which have only primary cell walls (deposited while the cell is still growing). Cellulose is more plentiful in the inner layers, which have thicker, more rigid secondary cell walls (deposited after cell growth has ceased). (B) The cells and their walls are clearly seen in this electron micrograph of young cells in the root of the same plant. (Courtesy of Paul Linstead.)

Such is the state of a crisp lettuce leaf. If water is lacking so that the cells shrink, the leaf wilts.

Most newly formed cells in a multicellular plant initially make relatively thin *primary cell walls* that are capable of slowly expanding to accommodate subsequent cell growth. The driving force for growth is the same that keeps the lettuce leaf crisp—a swelling pressure, called the *turgor pressure,* that develops due to an osmotic imbalance between the interior of the cell and its surroundings (discussed in Chapter 12). Once growth stops and the wall no longer needs to expand, a more rigid *secondary cell wall* is often produced, either by thickening of the primary wall or by deposition of new layers with a different composition underneath the old ones. When plant cells become specialized, they generally produce specially adapted types of walls: waxy, waterproof walls for the surface epidermal cells of a leaf; hard, thick, woody walls for the xylem cells of the stem; and so on, as shown in Panel 21–1 (pp. 700–701).

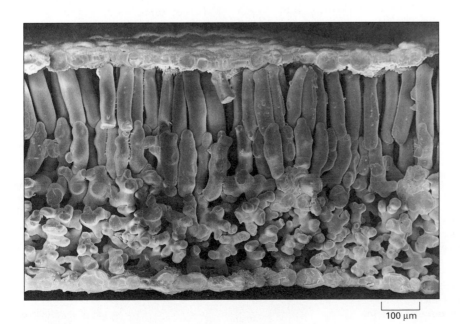

⊢————⊣
100 μm

Figure 21–4 A scanning electron micrograph shows the cells in a crisp lettuce leaf. The cells, swollen by osmotic forces, are stuck together via their walls. (Courtesy of Kim Findlay.)

Panel 21–1 The cell types and tissues from which higher plants are constructed

THE THREE TISSUE SYSTEMS

Cell division, growth, and differentiation give rise to tissue systems with specialized functions.

DERMAL TISSUE (▬): This is the plant's protective outer covering in contact with the environment. It facilitates water and ion uptake in roots and regulates gas exchange in leaves and stems.

VASCULAR TISSUE: Together, the phloem (▭) and the xylem (▬) form a continuous vascular system throughout the plant. This tissue conducts water and solutes between organs and also provides mechanical support.

GROUND TISSUE (▭): This packing and supportive tissue accounts for much of the bulk of the young plant. It also functions in food manufacture and storage.

The young flowering plant shown here is constructed from three main types of organs: leaves, stems, and roots. Each plant organ in turn is made from three tissue systems: ground (▭), dermal (▬), and vascular (▬).

All three tissue systems derive ultimately from proliferating cells of the shoot or root apical meristems, and each contains a relatively small number of specialized cell types. These three common tissue systems, and the cells that comprise them, are described in this panel.

THE PLANT

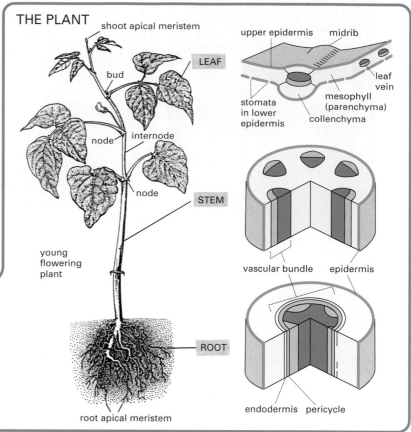

shoot apical meristem

LEAF

bud

node   internode

node

young flowering plant

ROOT

root apical meristem

upper epidermis   midrib

leaf vein

stomata in lower epidermis

mesophyll (parenchyma)

collenchyma

STEM

vascular bundle   epidermis

endodermis   pericycle

GROUND TISSUE

The ground tissue system contains three main cell types called parenchyma, collenchyma, and sclerenchyma.

Parenchyma cells are found in all tissue systems. They are generally capable of division and have a thin primary cell wall. These cells have a variety of functions. The apical and lateral meristematic cells of shoots and roots provide the new cells required for growth. Food production and storage occur in the photosynthetic cells of the leaf and stem (called mesophyll cells); storage parenchyma cells form the bulk of most fruits and vegetables. Because of their proliferative capacity, parenchyma cells also serve as stem cells for wound healing and regeneration.

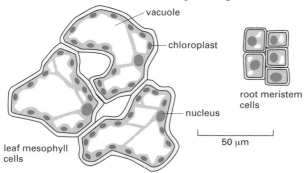

vacuole

chloroplast

root meristem cells

nucleus

leaf mesophyll cells

50 μm

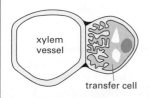

xylem vessel

transfer cell

A transfer cell, a specialized form of the parenchyma cell, is readily identified by the elaborate ingrowths of the primary cell wall. The increase in the area of the plasma membrane beneath these walls facilitates the rapid transport of solutes to and from cells of the vascular system.

Collenchyma cells are similar to parenchyma cells except that they have much thicker cell walls and are usually elongated and packed into long ropelike fibers. They are capable of stretching and provide mechanical support in the ground tissue system of the elongating regions of the plant. Collenchyma cells are especially common in subepidermal regions of stems.

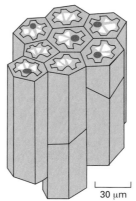

30 μm

typical locations of supporting groups of cells in a stem

sclerenchyma fibers

vascular bundle

collenchyma

Sclerenchyma cells, like collenchyma cells, have strengthening and supporting functions. However, they are usually dead and have thick, lignified secondary cell walls that prevent them from stretching as the plant grows. Two common types are fibers, which often form long bundles, and sclereids, which are shorter branched cells found in seed coats and fruit.

fiber bundle

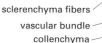

10 μm

sclereid

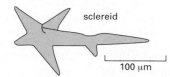

100 μm

DERMAL TISSUE

The epidermis is the primary outer protective covering of the plant body. Some cells of the epidermis are modified to form stomata and hairs of various kinds.

Epidermis

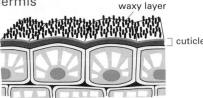

waxy layer

cuticle

The epidermis (usually one layer of cells deep) covers the entire stem, leaf, and root of the young plant. The cells have thick primary walls and are covered on their outer surface by a special cuticle with an outer waxy layer. The cells are tightly interlocked in different patterns.

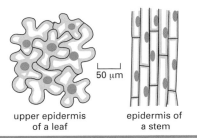

50 µm

upper epidermis of a leaf

epidermis of a stem

Stomata

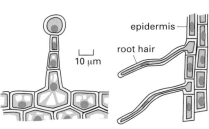

guard cells

air space

5 µm

Stomata are openings in the epidermis, mainly on the lower surface of the leaf, that regulate gas exchange in the plant. They are formed by two specialized epidermal cells called *guard cells,* which regulate the diameter of the pore. Stomata are distributed in a distinct species-specific pattern within each epidermis.

Hairs (or trichomes) are appendages derived from epidermal cells. They exist in a variety of forms and are commonly found in all plant parts. Hairs function in protection, absorption, and secretion. The developing

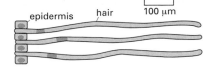

epidermis   hair   100 µm

cotton seed, for example, has many young, single-celled hairs on its epidermis. As these hairs grow, their walls become secondarily thickened with cellulose to form cotton fibers.

epidermis

root hair

10 µm

a multicellular secretory hair from a geranium leaf

Single-celled root hairs have an important function in water and ion uptake.

Vascular bundles

Roots usually have a single vascular bundle, but stems have several bundles. These are arranged with strict radial symmetry in dicots, but they are more irregularly dispersed in monocots.

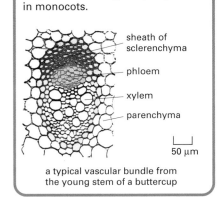

sheath of sclerenchyma

phloem

xylem

parenchyma

50 µm

a typical vascular bundle from the young stem of a buttercup

VASCULAR TISSUE

The phloem and the xylem together form a continuous vascular system throughout the plant. In young plants they are usually associated with a variety of other cell types in *vascular bundles*. Both phloem and xylem are complex tissues. Their conducting elements are associated with parenchyma cells that maintain the elements and exchange materials with them. In addition, groups of collenchyma and sclerenchyma cells provide mechanical support.

Phloem

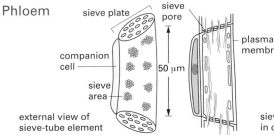

sieve plate

sieve pore

companion cell

sieve area

50 µm

plasma membrane

external view of sieve-tube element

sieve-tube element in cross-section

Phloem is involved in the transport of organic solutes in the plant. The main conducting cells (elements) are aligned to form tubes called *sieve tubes*. The sieve-tube elements at maturity are living cells that are interconnected by perforations in their end walls formed from enlarged and modified plasmodesmata (sieve plates). These cells retain their plasma membrane, but they have lost their nuclei and much of their cytoplasm; they therefore rely on associated *companion cells* for their maintenance. These companion cells have the additional function of actively transporting soluble food molecules into and out of sieve-tube elements through porous sieve areas in the wall.

Xylem

Xylem carries water and dissolved ions in the plant. In mature tissue the main conducting vessel elements (shown here) are formed from dead cells that lack a plasma membrane. The cell wall has been secondarily thickened and heavily lignified (reinforced with cross-linked lignin molecules). As shown below, its end wall is largely removed, enabling very long, continuous tubes to be formed.

small vessel element in root tip

large, mature vessel element

The vessel elements are closely associated with xylem parenchyma cells, which actively transport selected solutes into and out of the elements across the parenchyma cell plasma membrane.

xylem parenchyma cells

vessel element

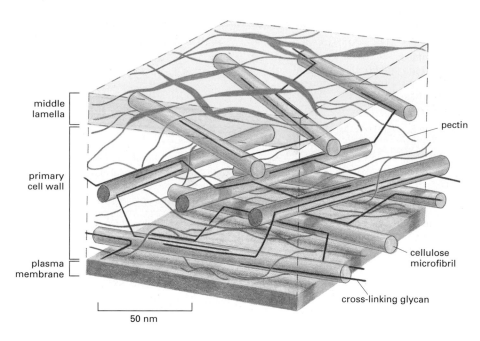

Figure 21–5 Scale model shows a portion of a primary plant cell wall. The *green bars* represent cellulose fibrils, providing tensile strength; other polysaccharides cross-link the cellulose fibrils (*red strands*) and fill the spaces between them (*blue strands*), providing resistance to compression. The middle lamella (*yellow*) is the layer that cements one cell wall to another.

Cellulose Fibers Give the Plant Cell Wall Its Tensile Strength

Like all extracellular matrices, plant cell walls derive their tensile strength from long fibers oriented along the lines of stress. In higher plants, the long fibers are generally made from the polysaccharide *cellulose,* the most abundant organic macromolecule on earth. The cellulose fibers are interwoven with other polysaccharides and some structural proteins, all bonded together to form a complex structure that resists compression as well as tension (Figure 21–5). In woody tissue, a highly cross-linked *lignin* network, composed of yet another class of molecules, is deposited within this matrix to make it more rigid and waterproof.

For a plant cell to grow or change its shape, the cell wall has to stretch or deform. Because the cellulose fibers resist stretching, their orientation governs the direction in which the growing cell enlarges: if, for example, they are arranged circumferentially as a corset, the cell will grow more readily in length than in girth (Figure 21–6). By controlling the way that it lays down its wall, the plant cell consequently controls its own shape and thus the direction of growth of the tissue to which it belongs.

Cellulose is produced in a radically different way from most other extracellular macromolecules. Instead of being made inside the cell and then exported by exocytosis (discussed in Chapter 15), it is synthesized on the outer surface of the cell by enzyme complexes embedded in the plasma membrane. These transport sugar monomers across the membrane and incorporate them into a set of growing polymer chains at their points of membrane attachment. Each set of chains forms a cellulose microfibril. The enzyme complexes move in the membrane, spinning out new polymers and laying down a trail of oriented cellulose fibers behind them.

The paths followed by the enzyme complexes dictate the orientation in which cellulose is deposited in the cell wall; but what guides the enzyme complexes? Just underneath the plasma membrane, microtubules are aligned exactly with the cellulose fibers outside the cell (Figure 21–7A and B). These microtubules are thought to serve as tracks to guide the movement of the enzyme complexes (Figure 21–7C). In this

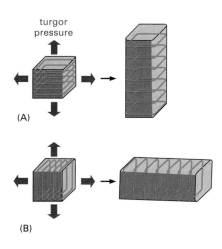

Figure 21–6 The orientation of cellulose microfibrils within the plant cell wall influences the direction in which the cell elongates. The cells in (A) and (B) start off with identical shapes (shown here as *cubes*) but with different orientations of cellulose microfibrils in their walls. Although turgor pressure is uniform in all directions, each cell tends to elongate in a direction perpendicular to the orientation of the microfibrils, which have great tensile strength. The final shape of an organ, such as a shoot, is determined by the direction in which its cells expand.

curiously indirect way, the cytoskeleton controls the shaping of the plant cell and the modeling of the plant tissues. We shall see that animal cells use their cytoskeleton to control tissue architecture in a much more direct way.

Animal Connective Tissues Consist Largely of Extracellular Matrix

It is traditional to distinguish four major types of tissues in animals—connective, epithelial, nervous, and muscular. But the basic architectural distinction is between connective tissues and the rest. In connective tissues, extracellular matrix is plentiful and carries the mechanical load. In other tissues, such as epithelia, extracellular matrix is scanty, and the cells are directly joined to one another and carry the mechanical load themselves. We discuss connective tissues first.

Animal connective tissues are enormously varied. They can be tough and flexible, like tendons or the dermis of the skin; hard and dense, like bone; resilient and shock-absorbing, like cartilage; or soft and transparent, like the jelly that fills the interior of the eye. In all these

Question 21–1

Cells in the stem of a seedling that is grown in the dark orient their microtubules horizontally. How would you expect this to affect the growth of the plant?

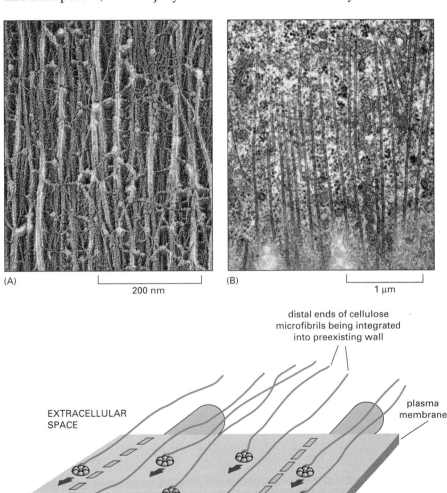

(A) |————| 200 nm

(B) |————| 1 μm

distal ends of cellulose microfibrils being integrated into preexisting wall

EXTRACELLULAR SPACE

plasma membrane

CYTOSOL

(C) |————| 0.1 μm

cellulose synthase complex

microtubule attached to plasma membrane

Figure 21–7 Microtubules direct the deposition of cellulose in the plant cell wall. (A) Oriented cellulose fibrils in a plant cell wall, shown by electron microscopy. (B) Oriented microtubules just beneath a plant cell's plasma membrane. (C) One model of how the orientation of the newly deposited extracellular cellulose microfibrils might be determined by the orientation of the intracellular microtubules. The large *cellulose synthase* enzyme complexes are integral membrane proteins that continuously synthesize cellulose microfibrils on the outer face of the plasma membrane. The outer ends of the stiff microfibrils become integrated into the texture of the wall, and their elongation at the other end pushes the synthase complex along in the plane of the membrane. Because the cortical array of microtubules is attached to the plasma membrane in a way that confines the enzyme complex to defined membrane tracks, the microtubule orientation determines the direction in which the microfibrils are laid down. (A, courtesy of Brian Wells and Keith Roberts; B, courtesy of Brian Gunning.)

Figure 21–8 Extracellular matrix is plentiful in connective tissue such as bone. Cells in this cross section of bone appear as small, dark, ant-like objects embedded in the bone matrix, which occupies most of the volume of the tissue and provides all its mechanical strength. The alternating light and dark bands are layers of matrix containing oriented collagen (made visible with the help of polarized light). Calcium phosphate crystals filling the interstices between the collagen fibrils make bone matrix resistant to compression as well as tension, like reinforced concrete.

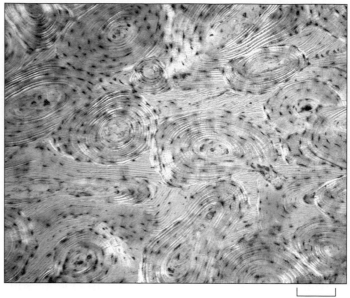

100 µm

examples, the bulk of the tissue is occupied by extracellular matrix, and the cells that produce the matrix are scattered within it like raisins in a pudding (Figure 21–8). In all of these tissues, the tensile strength—whether great or small—is provided not by a polysaccharide, as in plants, but by a fibrous protein—collagen. The various types of connective tissue owe their specific characters to the type of collagen that they contain, to its quantity, and, most importantly, to the other molecules that are interwoven with it in varying proportions.

Collagen Provides Tensile Strength in Animal Connective Tissues

Collagen is found in all multicellular animals, and it comes in many varieties. Mammals have about 20 different collagen genes, coding for the variant forms of collagen required in different tissues. Collagens are the chief proteins in bone, tendon, and skin (leather is pickled collagen), and they constitute 25% of the total protein mass in a mammal—more than any other type of protein.

The characteristic feature of a typical collagen molecule is its long, stiff, triple-stranded helical structure, in which three collagen polypeptide chains are wound around one another in a ropelike superhelix (Figure 21–9). These molecules in turn assemble into ordered polymers called *collagen fibrils,* which are thin cables 10–300 nm in diameter and many micrometers long; these can pack together into still thicker *collagen fibers* (see Figure 21–9). Other collagen molecules decorate the surface of collagen fibrils and link the fibrils to one another and to other components in the extracellular matrix.

The connective-tissue cells that manufacture and inhabit the matrix go by various names according to the tissue: in skin, tendon, and many other connective tissues they are called *fibroblasts* (Figure 21–10); in bone they are called *osteoblasts.* They make both the collagen and the other organic components of the matrix. Almost all of these molecules are synthesized intracellularly and then secreted in the standard way, by exocytosis. Outside the cell, they assemble into huge, cohesive aggregates. If assembly were to occur prematurely, before secretion, the cell would become choked with its own products. In the case of collagen, the cells avoid this risk by secreting collagen molecules in a precursor

Question 21–2

Mutations in the genes encoding collagens often have detrimental consequences, resulting in severely crippling diseases. Particularly devastating are mutations that change glycines, which are required at every third position in the collagen polypeptide chain so that it can assemble into the characteristic triple-helical rod (see Figure 21–9).

A. Would you expect that collagen mutations are detrimental if only one of the two copies of a collagen gene is defective?

B. A puzzling observation is that the change of a glycine residue into another amino acid is most detrimental if it occurs toward the amino terminus of the rod-forming domain. Suggest an explanation for this.

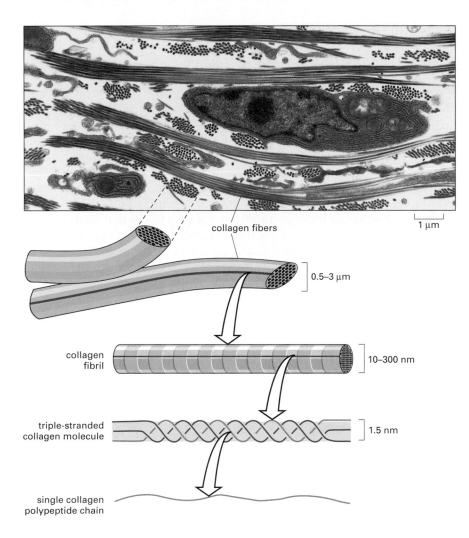

Figure 21–9 Collagen fibrils are organized into bundles. The electron micrograph shows collagen fibrils in the connective tissue of embryonic chick skin. The fibrils are organized into bundles (fibers), some running in the plane of the section, others approximately at right angles to it. The cell in the photograph is a fibroblast, which secretes the collagen as well as other extracellular matrix components. The drawings show the molecular structure of the collagen fibers. (Photograph from C. Ploetz, E.I. Zycband, and D.E. Birk, *J. Struct. Biol.* 106:73–81, 1991.)

collagen fibers

1 μm

0.5–3 μm

collagen fibril

10–300 nm

triple-stranded collagen molecule

1.5 nm

single collagen polypeptide chain

form, called *procollagen,* with additional peptides at each end that obstruct assembly into collagen fibrils. Extracellular enzymes—procollagen proteinases—cut off these terminal domains to allow assembly only after the molecules have emerged into the extracellular space.

Some people have a genetic defect in one of these proteinases, or in procollagen itself, so that their collagen fibrils do not assemble correctly. As a result, the skin and various other connective tissues have reduced tensile strength and are extraordinarily stretchable (Figure 21–11).

Cells in tissues have to be able to degrade matrix as well as make it. This is essential for tissue growth, repair, and renewal; it is also important where migratory cells, such as macrophages, need to burrow through the thicket of collagen and other extracellular matrix polymers. *Matrix proteases* that cleave extracellular proteins play a part in many disease processes, ranging from arthritis, where they contribute to the breakdown of cartilage in affected joints, to cancer, where they help the cancer cells to invade normal tissue.

Cells Organize the Collagen That They Secrete

To do their job, collagen fibrils must be correctly aligned. In skin, for example, they are woven in a wickerwork pattern, or in alternating layers with different orientations so as to resist tensile stress in multiple directions (Figure 21–12). In tendons, which attach muscles to bone, they are aligned in parallel bundles along the major axis of tension.

The connective-tissue cells control this orientation, partly by depositing the collagen in an oriented fashion and, partly by rearranging

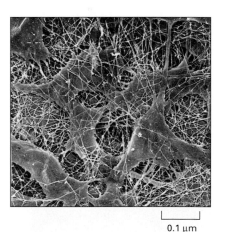

0.1 μm

Figure 21–10 Fibroblasts produce the extracellular matrix of connective tissue. Scanning electron micrograph shows fibroblasts and collagen in connective tissue from the cornea of a rat. Other components that normally form a hydrated gel filling the spaces between the collagen fibrils have been removed by enzyme and acid treatment. (From T. Nishida et al. *Invest. Ophthalmol. Vis. Sci.* 29:1887–1890, 1988.)

Figure 21–11 **Improper collagen assembly can cause hyperextensible skin.** James Morris, "the elastic skin man," from a photograph taken about 1890. Abnormally stretchable skin is part of a genetic syndrome that results from a defect in the assembly or cross-linking of collagen. In some individuals, this arises from a lack of a proteinase that converts procollagen to collagen.

it subsequently. During development of the tissue, fibroblasts work on the collagen they have secreted, crawling over it and pulling on it—helping to compact it into sheets and draw it out into cables. This mechanical role of fibroblasts in shaping collagen matrices has been demonstrated dramatically in cell culture. When fibroblasts are mixed with a meshwork of randomly oriented collagen fibrils that form a gel in a culture dish, the fibroblasts tug on the meshwork, drawing in collagen from their surroundings and compacting it. If two small pieces of embryonic tissue containing fibroblasts are placed far apart on a collagen gel, the intervening collagen becomes organized into a dense band of aligned fibers that connect the two explants (Figure 21–13). The fibroblasts migrate out from the explants along the aligned collagen fibers. Thus the fibroblasts influence the alignment of the collagen fibers, and the collagen fibers in turn affect the distribution of the fibroblasts. Fibroblasts presumably play a similar role in generating long-range order in the extracellular matrix inside the body—in helping to create tendons, for example, and the tough, dense layers of connective tissue that ensheathe and bind together most organs.

Integrins Couple the Matrix Outside a Cell to the Cytoskeleton Inside It

If cells are to pull on the matrix and crawl over it, they must be able to attach to it. Cells do not attach well to bare collagen. Another extracellular matrix protein, *fibronectin,* provides a linkage: one part of the fibronectin molecule binds to collagen, while another part forms an attachment site for a cell (Figure 21–14A).

The cell binds to the specific site in the fibronectin by means of a receptor protein, called an *integrin,* that spans the cell's plasma membrane. While the extracellular domain of the integrin binds to fibronectin, the intracellular domain binds (through a set of adaptor molecules) to actin filaments (Figure 21–14C). Thus, instead of being ripped out of the membrane when there is tension between the cell and the matrix, the integrin molecule transmits the stress from the matrix to the cytoskeleton. Muscle cells couple their contractile apparatus in a similar way to the extracellular matrix at the junction between muscle and tendon, enabling them to exert large forces while remaining enveloped in a flimsy lipid bilayer membrane.

Integrins also play a role in cell signaling: engagement with extracellular matrix molecules activates intracellular signaling cascades through protein kinases that associate with the intracellular end of the integrin molecule. In this way, the external attachments that a cell makes can help to regulate whether it grows, divides, survives, differentiates, or dies.

Gels of Polysaccharide and Protein Fill Spaces and Resist Compression

While collagen provides tensile strength to resist stretching, a completely different group of macromolecules in the extracellular matrix of animal tissues provides the complementary function, resisting compression

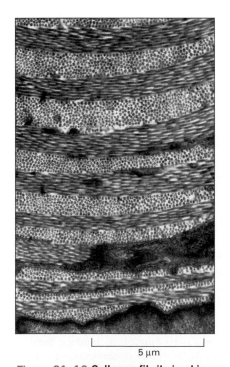

5 μm

Figure 21–12 **Collagen fibrils in skin are arranged in a plywoodlike pattern.** Successive layers of fibrils are laid down nearly at right angles to each other (see also Figure 21–9). Electron micrograph of a cross section of tadpole skin. This arrangement is also found in mature bone and in the cornea. (Courtesy of Jerome Gross.)

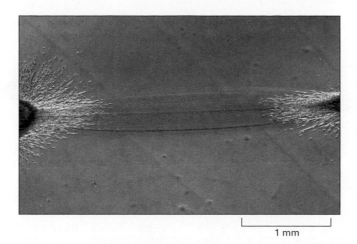

Figure 21–13 Fibroblasts influence the alignment of collagen fibers. This micrograph shows a region between two pieces of embryonic chick heart (rich in fibroblasts as well as heart muscle cells) that have grown in culture on a collagen gel for four days. A dense tract of aligned collagen fibers has formed between the explants, presumably as a result of the fibroblasts in the explants tugging on the collagen. Elsewhere in the culture dish the collagen remains disorganized and unaligned, so it appears uniformly gray. (From D. Stopak and A.K. Harris, *Dev. Biol.* 90:383–398, 1982.)

and serving as space-fillers. These are the *proteoglycans*, extracellular proteins linked to a special class of complex negatively charged polysaccharides, the glycosaminoglycans (GAGs) (Figure 21–15). Proteoglycans are extremely diverse in size, shape, and chemistry. Typically, many GAG chains are attached to a single core protein, which may in turn be linked at one end to another GAG, creating an enormous macromolecule resembling a bottlebrush, with a molecular weight in the millions (Figure 21–16).

In dense, compact connective tissues such as tendon and bone, the proportion of GAGs is small, and the matrix consists almost entirely of collagen (or, in the case of bone, of collagen plus calcium phosphate crystals). At the other extreme, the jellylike substance in the interior of the eye consists almost entirely of one particular type of GAG, plus water, with only a small amount of collagen. In general, GAGs are strongly hydrophilic and tend to adopt highly extended conformations, which occupy a huge volume relative to their mass (see Figure 21–16).

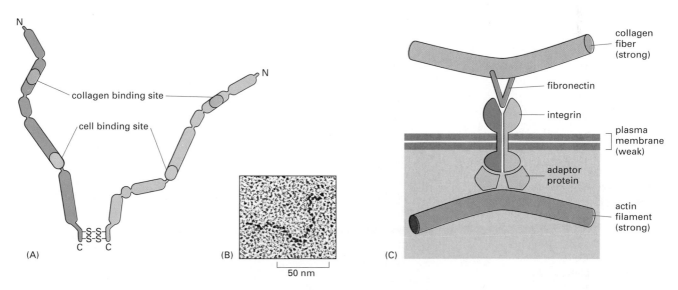

Figure 21–14 Integrins link the extracellular matrix to the cytoskeleton in an animal cell. (A) Diagram and (B) electron micrograph of a fibronectin molecule. (C) The transmembrane linkage mediated by an integrin molecule. The integrin molecule transmits tension across the plasma membrane: it is anchored inside the cell to the cytoskeleton and externally via fibronectin to the extracellular matrix. The plasma membrane itself does not have to be strong. The integrin shown links fibronectin to an actin filament inside the cell, but other integrins connect different extracellular proteins to the cytoskeleton (usually to actin filaments, but sometimes to intermediate filaments). (B, from J. Engel et al., *J. Mol. Biol.* 150:97–120, 1981. © Academic Press.)

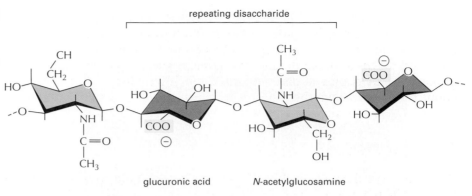

Figure 21–15 Glycosaminoglycans (GAGs) help to fill space in the extracellular matrix of connective tissues. Hyaluronan, a relatively simple GAG, consists of a single long chain of up to 25,000 repeated disaccharide units, each carrying a negative charge. As in other GAGs, one of the sugar monomers in each disaccharide unit is an amino-sugar. Many GAGs have additional negatively charged side groups, especially sulfate groups.

They form gels even at very low concentrations, their multiple negative charges attracting a cloud of cations, such as Na^+, that are osmotically active, causing large amounts of water to be sucked into the matrix. This creates a swelling pressure that is balanced by tension in the collagen fibers that are interwoven with the proteoglycans. When the matrix is rich in collagen and large quantities of GAGs are trapped in its meshes, both the swelling pressure and the counterbalancing tension are enormous. Such a matrix is tough, resilient, and resistant to compression. The cartilage matrix that lines the knee joint, for example, has this character: it can support pressures of hundreds of kilograms per square centimeter.

Proteoglycans perform many sophisticated functions in addition to providing hydrated space around cells. They can forms gels of varying pore size and charge density that act as filters to regulate the passage of molecules through the extracellular medium. They can bind

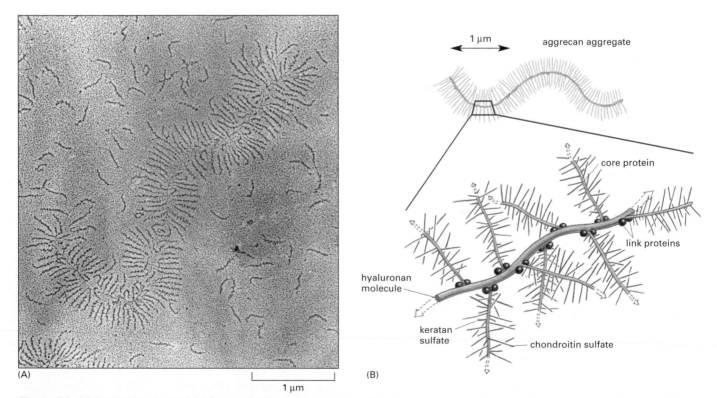

Figure 21–16 Proteoglycans and GAGs can form large aggregates. (A) Electron micrograph of an aggregate from cartilage spread out on a flat surface. Many free subunits—themselves large proteoglycan molecules—are also seen. (B) Schematic drawing of the giant aggregate illustrated in (A), showing how it is built up from GAGs and proteins. The molecular weight of such a complex can be 10^8 daltons or more, and it occupies a volume equivalent to that of a bacterium, which is about 2×10^{-12} cm^3. (A, courtesy of Lawrence Rosenberg.)

growth factors and other proteins that serve as signals for cells. They can block, encourage, or guide cell migration through the matrix. In all these ways, the matrix components influence the behavior of cells, often the same cells that make the matrix—a reciprocal interaction that has important effects on cell differentiation. Much remains to be learned about how cells weave the tapestry of matrix molecules and how the chemical messages they leave in its fabric are organized and act.

Epithelial Sheets and Cell-Cell Junctions

There are more than 200 visibly different cell types in the body of a vertebrate. The majority of these are organized into epithelia—that is, they are joined together, side to side, to form multicellular sheets. In some cases, the sheet is many cells thick, or *stratified,* as in the epidermal covering of the skin; in other cases, it is only one cell thick, or *simple,* as in the lining of the gut. The cells may be tall and *columnar,* or *cuboidal,* or squat and *squamous* (Figure 21–17). They may be all alike, or a mixture of different types. They may simply act as a protective barrier, or they may have complex biochemical functions: they may secrete specialized products such as hormones, milk, or tears; they may absorb nutrients, as in the gut lining; or they may detect signals, like the photoreceptors of the eye or the auditory hair cells of the ear. Through these and many other variations, one can recognize a standard set of structural features that virtually all animal epithelia share. The epithelial arrangement of cells is so commonplace that one easily takes it for granted; yet it requires a collection of specialized devices, as we shall see, and these are common to a wide variety of different cell types.

Epithelia cover the external surface of the body and line all its internal cavities, and they must have been an early feature in the evolution of multicellular animals. Their importance is obvious. Cells joined together in an epithelial sheet create a barrier that has the same significance for the multicellular organism that the plasma membrane has for a single cell. It keeps some molecules in, and others out; it takes up nutrients and exports wastes; it contains receptors for environmental signals; and it protects the interior of the organism from invading microorganisms and fluid loss.

Epithelial Sheets Are Polarized and Rest on a Basal Lamina

An epithelial sheet has two faces: the apical surface is free and exposed to the air or to a watery fluid; the basal surface rests on some

Question 21–3

Proteoglycans are characterized by the abundance of negative charges on their sugar chains. How would the properties of these molecules differ if the negative charges were not as abundant?

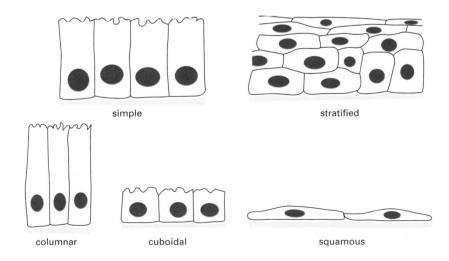

Figure 21–17 **Cells can be packed together in different ways to form an epithelial sheet.** Five basic types of epithelia are shown.

simple

stratified

columnar

cuboidal

squamous

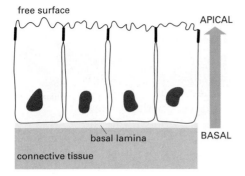

free surface

APICAL

basal lamina

BASAL

connective tissue

Figure 21–18 A sheet of epithelial cells has a polarized organization. The basal surface sits on a specialized sheet of extracellular matrix called the basal lamina, while the apical surface is free.

other tissue—usually a connective tissue—to which it is attached (Figure 21–18). Supporting the basal surface of the epithelium there lies a thin tough sheet of extracellular matrix, called the **basal lamina** (Figure 21–19), composed of a specialized type of collagen (Type IV collagen) and various other molecules. These include a protein called *laminin,* which provides adhesive sites for integrin molecules in the plasma membrane of the epithelial cells, serving a linking role that resembles that of fibronectin in connective tissues.

The apical and basal faces of an epithelium are chemically different, reflecting a polarized internal organization of the individual epithelial cells: each one has a top and a bottom, with different properties. This polarized organization is crucial for epithelial function. Consider, for example, the simple columnar epithelium that lines the small intestine of a mammal. It mainly consists of two intermingled cell types: *absorptive cells* to take up nutrients and *goblet cells* (so called from their shape) to secrete mucus that protects and lubricates the gut lining (Figure 21–20). Both cell types are polarized. The absorptive cells import food molecules through their apical surface from the gut lumen and export these molecules through their basal surface into the underlying tissues. To do this, they require different sets of membrane transport proteins in their apical and basal plasma membranes (see Figure 12–15). The goblet cells also have to be polarized, but in a different way: they have to synthesize mucus and then discharge it from their apical ends only (see Figure 21–20): the Golgi apparatus, secretory vesicles, and cytoskeleton are all asymmetrically organized so as to bring this about. While many questions remain as to how this organization is maintained, it is clear that it depends on the junctions that the epithelial cells form with one another and with the basal lamina, which in turn control the intracellular location of proteins that govern the polarized organization of the cytoplasm.

Figure 21–19 The basal lamina supports a sheet of epithelial cells. Scanning electron micrograph of a basal lamina in the cornea of a chick embryo. Some of the epithelial cells have been removed to expose the upper surface of the matlike basal lamina woven from Type IV collagen and laminin proteins. A network of other collagen fibrils in the underlying connective tissue interacts with the lower face of the lamina. (Courtesy of Robert Trelstad.)

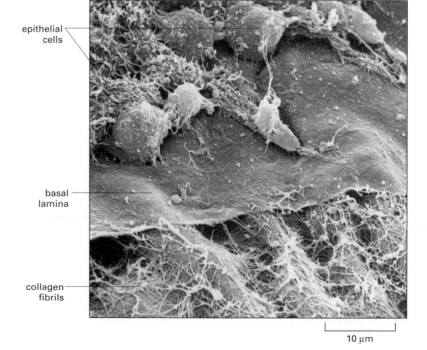

epithelial cells

basal lamina

collagen fibrils

10 μm

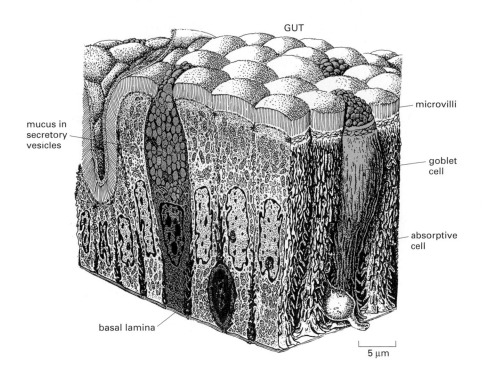

mucus in
secretory
vesicles

microvilli

goblet
cell

absorptive
cell

basal lamina

5 μm

Figure 21–20 Functionally polarized cell types line the gut. Absorptive cells, which take up nutrients from the gut, are mingled in the gut lining with goblet cells *(brown)*, which secrete mucus into the gut. The absorptive cells are often called *brush-border cells,* because of the brushlike mass of microvilli on their apical surface, serving to increase the area of membrane for transport of small molecules into the cell. The goblet cells owe their gobletlike shape to the mass of secretory vesicles that distends their apical region. (Adapted from R. Krstić, Human Microscopic Anatomy. Berlin: Springer, 1991.)

Tight Junctions Make an Epithelium Leak-proof and Separate Its Apical and Basal Surfaces

Epithelial cell junctions can be classified according to their function. Some provide a tight seal to prevent leakage of molecules across the epithelium through the gaps between its cells; some provide strong mechanical attachments; and some provide for a special type of intimate chemical communication. In most epithelia, all these types of junctions are present together (Figure 21–21). Each type of junction is characterized by its own class of membrane proteins that hold the cells together.

The sealing function is served (in vertebrates) by *tight junctions*. These seal neighboring cells together so that water-soluble molecules

Figure 21–21 Several types of cell-cell junctions are found in epithelia in animals. Tight junctions are peculiar to epithelia; the other types also occur, in modified forms, in various nonepithelial tissues.

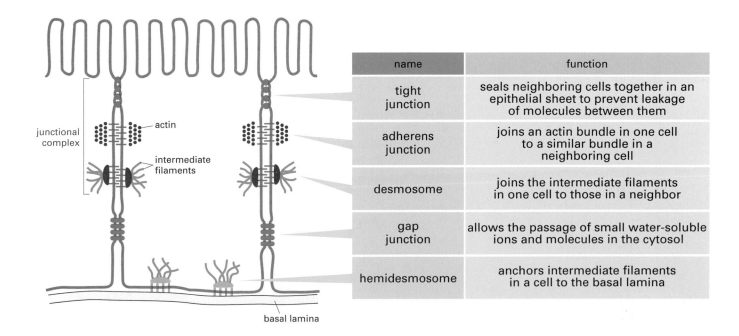

junctional
complex

actin

intermediate
filaments

basal lamina

name	function
tight junction	seals neighboring cells together in an epithelial sheet to prevent leakage of molecules between them
adherens junction	joins an actin bundle in one cell to a similar bundle in a neighboring cell
desmosome	joins the intermediate filaments in one cell to those in a neighbor
gap junction	allows the passage of small water-soluble ions and molecules in the cytosol
hemidesmosome	anchors intermediate filaments in a cell to the basal lamina

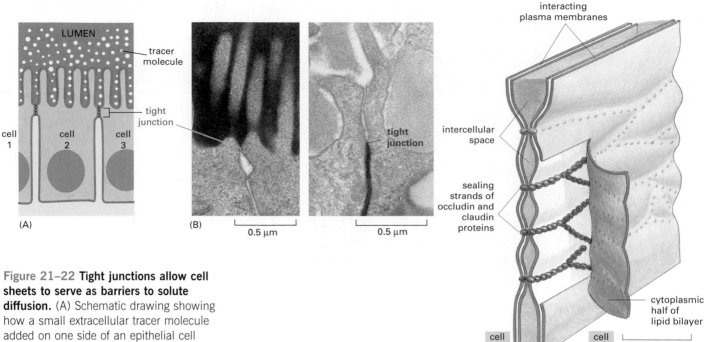

Figure 21–22 Tight junctions allow cell sheets to serve as barriers to solute diffusion. (A) Schematic drawing showing how a small extracellular tracer molecule added on one side of an epithelial cell sheet cannot traverse the tight junctions that seal adjacent cells together. (B) Electron micrographs of cells in an epithelium where a small, extracellular tracer molecule *(dark stain)* has been added to either the apical side (on the *left*) or the basolateral side (on the *right)*; in both cases the tracer is stopped by the tight junction. (C) A model of the structure of a tight junction, showing how the cells are sealed together by proteins, called claudins and occludins *(green)*, in the outer leaflet of the plasma membrane bilayer. (B, courtesy of Daniel Friend.)

cannot easily leak between the cells. If a tracer molecule is added to one side of an epithelial cell sheet, it will usually not pass beyond the tight junction (Figure 21–22). The tight junction is formed from proteins called *claudins* and *occludins*, which are arranged in strands along the lines of junction to create the seal. Without tight junctions to prevent leakage, the pumping activities of absorptive cells such as those in the gut would be futile, and the composition of the medium on the two sides of the epithelium would become uniform. As we saw in Chapter 11, tight junctions also play a key part in maintaining the polarity of the individual epithelial cells: the tight-junction complex around the apical rim of each cell prevents diffusion of membrane proteins so as to keep the apical domain of the plasma membrane different from the basal (or baso-lateral) domain (see Figure 11–39).

Cytoskeleton-linked Junctions Bind Epithelial Cells Robustly to One Another and to the Basal Lamina

The junctions that hold an epithelium together by forming mechanical attachments are of three main types. *Adherens junctions* and *desmosome junctions* bind one epithelial cell to another, while *hemidesmosomes* bind epithelial cells to the basal lamina. All of these junctions provide mechanical strength by the same strategy that we have already encountered in connective tissue (see Figure 21–14C): the molecule that forms the external adhesion spans the membrane and is linked inside the cell to strong cytoskeletal filaments. In this way, the cytoskeletal filaments are tied into a network that extends from cell to cell across the whole expanse of epithelial tissue.

Adherens junctions and desmosome junctions are both built around transmembrane proteins belonging to the same family, called *cadherins*: a cadherin molecule in the plasma membrane of one cell binds directly to an identical cadherin molecule in the plasma membrane of its neighbor

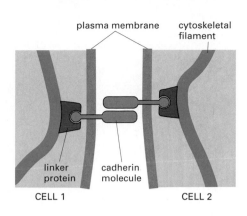

plasma membrane   cytoskeletal filament

linker protein   cadherin molecule

CELL 1   CELL 2

Figure 21–23 Cadherin molecules mediate mechanical attachment of one cell to another. Two similar cadherin molecules in the plasma membranes of adjacent cells bind to one another extracellularly; intracellularly they are attached, via linker proteins, to cytoskeletal filaments—either actin or keratin.

(Figure 21–23). Such binding of like to like is called *homophilic*. In the case of cadherins, binding requires also that Ca^{2+} be present in the extracellular medium—hence the name.

At an adherens junction, each cadherin molecule is tethered inside its cell, via several linker proteins, to actin filaments. Often, the adherens junctions form a continuous *adhesion belt* around each of the interacting epithelial cells; this belt is located near the apical end of the cell, just below the tight junctions (Figure 21–24). Actin bundles are thus connected from cell to cell across the epithelium. This actin network is potentially contractile, and it gives the epithelial sheet the capacity to develop tension and to change its shape in remarkable ways. By shrinking its apical surface along one axis, the sheet can bend so as to roll itself up into a tube (Figure 21–25A). Alternatively, by shrinking its apical surface locally along both axes at once, the sheet can develop a cup-shaped concavity and eventually create a vesicle that may pinch off from the rest of the epithelium. Epithelial movements such as these are important in embryonic development, where they create structures such as the neural tube, which gives rise to the central nervous system (Figure 21–25B) and the lens vesicle, which gives rise to the lens of the eye (Figure 21–25C).

At a desmosome junction, by contrast, different members of the family of cadherin molecules are anchored inside each cell to intermediate filaments—specifically, to keratins, which are the type of intermediate filament found in epithelia (Figure 21–26). Thick bundles of

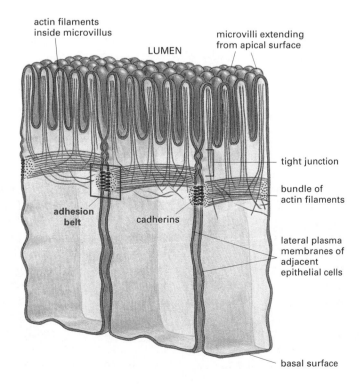

actin filaments inside microvillus

LUMEN

microvilli extending from apical surface

tight junction

bundle of actin filaments

adhesion belt   cadherins

lateral plasma membranes of adjacent epithelial cells

basal surface

Figure 21–24 Adherens junctions form adhesion belts around epithelial cells in the small intestine. A contractile bundle of actin filaments runs along the cytoplasmic surface of the plasma membrane near the apex of each cell, and these bundles of actin filaments in adjacent cells are linked to one another via cadherin molecules that span the cell membranes (see Figure 21–23).

ropelike keratin filaments, criss-crossing the cytoplasm and spot-welded via desmosome junctions to the bundles of keratin filaments in adjacent cells, confer great tensile strength and are particularly abundant in tough, exposed epithelia such as the epidermis of the skin.

Blisters are a painful reminder that it is not enough for epithelial cells to be firmly attached to one another: they must also be anchored to the underlying tissue. As we noted earlier, the anchorage is mediated by integrin proteins in the basal plasma membrane of the epithelial cells. Externally, these integrins bind to laminin in the basal lamina; inside the cell, they are linked to keratin filaments, creating a structure that looks superficially like half a desmosome. These attachments of epithelial cells to the extracellular matrix beneath them are therefore called *hemidesmosomes* (Figure 21–27).

Figure 21–25 Epithelial sheets can bend to form a tube or a vesicle. (A) Diagram showing how contraction of apical bundles of actin filaments linked from cell to cell via adherens junctions causes the epithelial cells to narrow at their apex. According to whether the contraction is oriented along one axis or is equal in all directions, the epithelium may be caused to roll up into a tube or to invaginate to form a vesicle. (B) Formation of the neural tube; the scanning electron micrograph shows a cross section through the trunk of a two-day chick embryo. Part of the epithelial sheet that covers the surface of the embryo has thickened, has rolled up into a tube by apical contraction, and is about to pinch off to become a separate internal structure. (C) Formation of the lens; a patch of surface epithelium overlying the embryonic rudiment of the retina of the eye has become concave and finally pinched off as separate vesicle—the lens vesicle—within the eye cup. (B, courtesy of Jean-Paul Revel; C, courtesy of K.W. Tosney.)

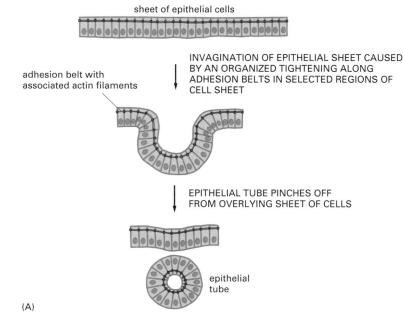

sheet of epithelial cells

adhesion belt with associated actin filaments

INVAGINATION OF EPITHELIAL SHEET CAUSED BY AN ORGANIZED TIGHTENING ALONG ADHESION BELTS IN SELECTED REGIONS OF CELL SHEET

EPITHELIAL TUBE PINCHES OFF FROM OVERLYING SHEET OF CELLS

epithelial tube

(A)

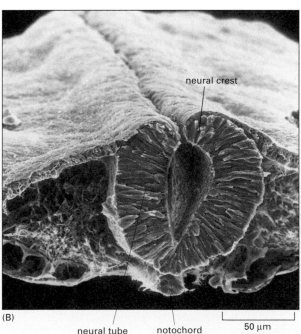

neural crest

(B)

neural tube   notochord

50 μm

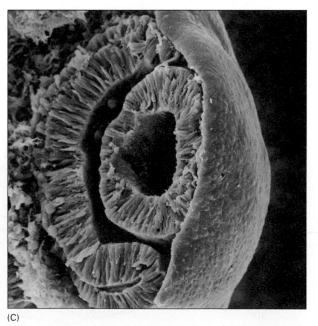

(C)

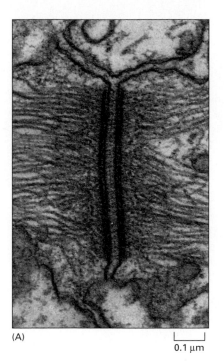

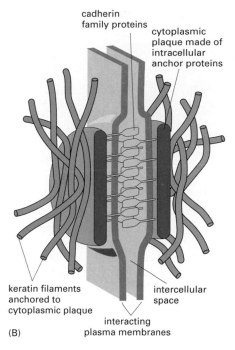

cadherin family proteins

cytoplasmic plaque made of intracellular anchor proteins

keratin filaments anchored to cytoplasmic plaque

intercellular space

interacting plasma membranes

(A)

0.1 µm

(B)

Figure 21–26 Desmosomes link the keratin filaments of one cell to another. (A) An electron micrograph of a desmosome joining two cells in the epidermis of a newt, showing the attachment of keratin filaments. (B) Schematic drawing of a desmosome. On the cytoplasmic surface of each interacting plasma membrane is a dense plaque composed of a mixture of intracellular anchor proteins. A bundle of keratin filaments is attached to the surface of each plaque. Transmembrane adhesion proteins of the cadherin family bind to the outer face of each plaque and interact through their extracellular domains to hold the adjacent cells together. (A, from D.E. Kelly, *J. Cell Biol.* 28:51–59, 1966. © Rockefeller University Press.)

Gap Junctions Allow Ions and Small Molecules to Pass from Cell to Cell

The final type of epithelial cell-cell junction, found in virtually all epithelia and in many other types of tissue, serves a totally different purpose. It is called a *gap junction*. In the electron microscope it appears as a region where the membranes of two cells lie close together and exactly parallel, with a very narrow gap of 2–4 nm between them (Figure 21–28A). The gap is not empty, but is spanned by the protruding ends of many identical protein complexes that lie in the plasma membranes of the two apposed cells. These complexes, called *connexons,* form channels across the two plasma membranes and are aligned end-to-end so as to create narrow passageways that allow inorganic ions and small water-soluble molecules (up to a molecular mass of about 1000 daltons) to move directly from the cytosol of one cell to the cytosol of another (Figure 21–28B). This creates an electrical and a metabolic coupling between the cells. Gap junctions between heart muscle cells, for example, provide the electrical coupling that allows electrical waves of excitation to spread through the tissue, triggering coordinated contraction of the cells.

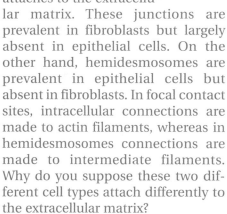

Question 21–4

Analogues of hemidesmosomes are the focal contact sites described in Chapter 17, which are also sites where the cell attaches to the extracellular matrix. These junctions are prevalent in fibroblasts but largely absent in epithelial cells. On the other hand, hemidesmosomes are prevalent in epithelial cells but absent in fibroblasts. In focal contact sites, intracellular connections are made to actin filaments, whereas in hemidesmosomes connections are made to intermediate filaments. Why do you suppose these two different cell types attach differently to the extracellular matrix?

keratin filaments

basal plasma membrane of epithelial cell

integrin molecules

basal lamina

Figure 21–27 Hemidesmosome junctions anchor the keratin filaments in an epithelial cell to the basal lamina. The linkage is mediated by integrin proteins.

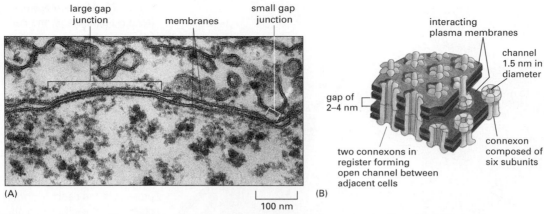

large gap junction   membranes   small gap junction

interacting plasma membranes

channel 1.5 nm in diameter

gap of 2–4 nm

two connexons in register forming open channel between adjacent cells

connexon composed of six subunits

(A)   (B)

100 nm

Figure 21–28 Gap junctions provide neighboring cells with a direct line of communication. (A) Thin-section electron micrograph of a gap junction between two cells in culture. (B) A model of a gap junction. The drawing shows the interacting plasma membranes of two adjacent cells. The apposed lipid bilayers *(red)* are penetrated by protein assemblies called *connexons (green),* each of which is thought to be formed by six identical protein subunits. Two connexons join across the intercellular gap to form an aqueous channel connecting the two cells. (A, from N.B. Gilula, in Cell Communication [R.P. Cox, ed.], pp. 1–29. New York: Wiley, 1974. © John Wiley & Sons.)

Question 21–5

Gap junctions are dynamic structures that, like conventional ion channels, are gated: they can close by a reversible conformational change in response to changes in the cell. The permeability of gap junctions decreases within seconds, for example, when the intracellular Ca^{2+} is raised. Speculate why this form of regulation might be important for the health of a tissue.

Gap junctions in many tissues can be opened or closed as needed in response to extracellular signals. The neurotransmitter dopamine, for example, reduces gap-junction communication within a class of neurons in the retina in response to an increase in light intensity (Figure 21–29). This reduction in gap-junction permeability changes the pattern of electrical signaling and helps the retina switch from using rod photoreceptors, which are good detectors of low light, to cone photoreceptors, which detect color and fine detail in bright light.

Curiously, plant tissues, though they lack all the other types of cell-cell junctions we have described earlier, have a functional counterpart of the gap junction. The cytoplasms of adjacent plant cells are connected via minute communicating channels called *plasmodesmata,* which span the intervening cell walls (Figure 21–30). In contrast with gap-junctional channels, plasmodesmata are lined with plasma membrane, which is thus continuous from one plant cell to the next. In spite of their structural differences, plasmodesmata and gap junctions allow a similarly restricted range of ions and small molecules to pass. This suggests that neighboring cells in both plants and animals have a basic need to share these components while keeping their macromolecules segregated. It is still not clear why this should be.

Figure 21–29 Extracellular signals can regulate the coupling of gap junctions. (A) A neuron in a rabbit retina was injected with the dye Lucifer yellow, which passes readily through gap junctions. The dye thus labels neurons of the same type that are connected by gap junctions. (B) The retina was first treated with the neurotransmitter dopamine, then injected with Lucifer yellow. As can be seen, the dopamine treatment greatly decreased the permeability of the gap junctions. (Courtesy of David Vaney.)

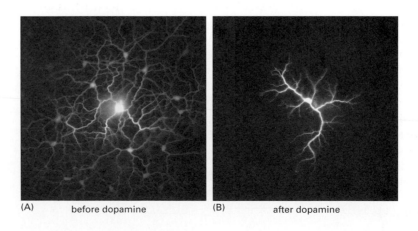

(A) before dopamine   (B) after dopamine

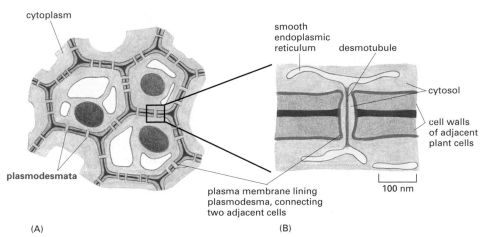

cytoplasm

plasmodesmata

(A)

smooth endoplasmic reticulum

desmotubule

cytosol

cell walls of adjacent plant cells

plasma membrane lining plasmodesma, connecting two adjacent cells

100 nm

(B)

Figure 21–30 Plant cells are connected via plasmodesmata. (A) The cytoplasmic channels of plasmodesmata pierce the plant cell wall and connect all cells in a plant together. (B) Each plasmodesma is lined with plasma membrane common to the two connected cells. It usually also contains a fine tubular structure, the desmotubule, derived from smooth endoplasmic reticulum. The plasmodesma usually allows only small molecules and ions to pass from cell to cell.

Tissue Maintenance and Renewal

One cannot contemplate the organization of tissues without wondering how these astonishingly patterned structures come into being. This raises one of the most ancient and fundamental questions in all of biology: how is a whole multicellular organism generated from a single fertilized egg?

In the process of development, the egg cell divides repeatedly to give a clone of cells—about 10,000,000,000,000 of them for a human being—all containing the same genome but specialized in different ways. This clone has a structure. It may take the form of a daisy or an oak tree, a sea urchin, a whale, or a mouse (Figure 21–31). The structure is determined by the genome that the egg contains. The linear sequence of A, G, C, and T nucleotides in the DNA directs the production of a host of distinct cell types, each expressing different sets of genes and arranged in a precise, intricate, three-dimensional pattern.

Although the final structure of an animal's body may be enormously complex, it is generated by a limited repertoire of cell activities. Examples of all these activities have been discussed in earlier pages of this book. Cells grow, divide, and die. They form mechanical attachments and generate forces for movement. They differentiate by switching on or off the production of specific sets of proteins. They produce molecular signals to influence neighboring cells, and they respond to signals that neighboring cells deliver to them. They remember the effects of previous signals they have received, and so progressively become more and more specialized in the characteristics they adopt. The genome, identical in every cell, defines the rules according to which these various possible cell activities are called into play. Through its operation in each cell individually, the genome guides the whole intricate process by which a multicellular organism is generated from a fertilized egg (Figure 21–32).

For developmental biologists, the challenge is to explain in these terms the entire sequence of interlocking events that lead from the egg to the adult organism. We shall not attempt to set out an answer to this problem here: we do not have space to do it justice, even though a great deal of the process is now understood. But the same basic activities that combine to create the organism during development continue even in the adult body, where fresh cells are continually generated in precisely controlled patterns. It is this more limited topic that we discuss in this section, focusing on the organization and maintenance of the tissues of vertebrates.

Tissues Are Organized Mixtures of Many Cell Types

Although the specialized tissues in our body differ in many ways, they all have certain basic requirements, usually provided for by a mixture of cell types, as illustrated for the skin in Figure 21–33. All tissues need mechanical strength, which is often supplied by a supporting bed or framework of connective tissue inhabited by fibroblasts. In this connective tissue, blood vessels lined with *endothelial cells* satisfy the need for oxygen, nutrients, and waste disposal. Likewise, most tissues are innervated by *nerve cell* axons, which are ensheathed by *Schwann cells.* *Macrophages* dispose of dying cells and other unwanted debris, and *lymphocytes* and other white blood cells combat infection. Most of these cell types originate outside the tissue and invade it, either early in the course of its development (endothelial cells, nerve cell axons, and Schwann cells) or continually during life (macrophages and other white blood cells). This complex supporting apparatus is required to maintain the principal specialized cells of the tissue: the contractile cells of the muscle, the secretory cells of the gland, or the blood-forming cells of the bone marrow, for example.

Almost every tissue is therefore an intricate mixture of many cell types that must remain different from one another while coexisting in the same environment. Moreover, in almost all adult tissues, cells are continually dying and being replaced; throughout this hurly-burly of cell replacement and tissue renewal, the organization of the tissue must be preserved. Three main factors contribute to make this structural stability possible (Figure 21–34).

1. *Cell communication:* each type of specialized cell continually monitors its environment for signals from other cells and adjusts its behavior accordingly; in fact, the very survival of most cells depends on such social signals (discussed in Chapter 16). These

Figure 21–31 The genome of the egg determines the structure of the clone of cells that will develop from it.
(A and B) A sea-urchin egg gives rise to a sea urchin; (C and D) a mouse egg gives rise to a mouse. (A, courtesy of David McClay; B, courtesy of M. Gibbs, Oxford Scientific Films; C, courtesy of Patricia Calarco, from G. Martin, *Science* 209:768–776, 1980; D, courtesy of O. Newman, Oxford Scientific Films.)

(A) |_____| 100 μm

(C) |_____| 50 μm

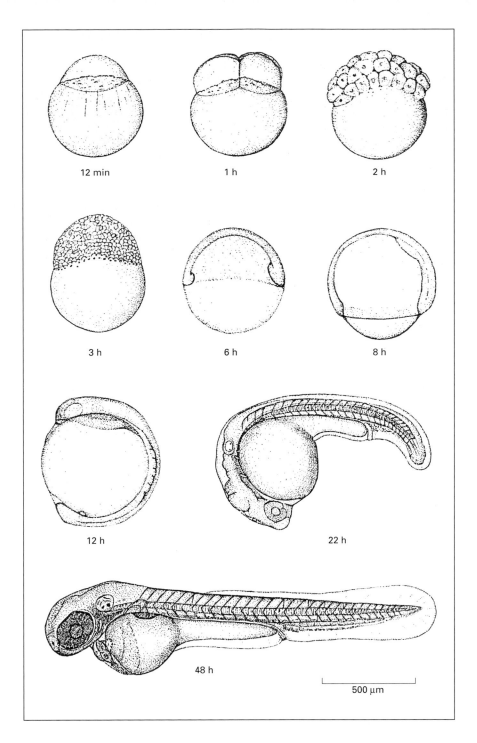

Figure 21–32 Through cell division, cell growth, cell movement, and cell specialization, a fertilized egg cell gives rise to a multicellular animal. The drawings show stages in the development of a zebrafish embryo. The fertilized egg converts itself into a recognizable baby fish within 48 hours. (Adapted from C.B. Kimmel, et al., *Developmental Dynamics* 203:256–310, 1995.)

12 min

1 h

2 h

3 h

6 h

8 h

12 h

22 h

48 h

500 μm

communications ensure that new cells are produced and survive only when and where they are required.

2. *Selective cell-cell adhesion:* because different cell types have different cadherins and other adhesion molecules in their plasma membranes, they tend to stick selectively, by homophilic binding, to other cells of the same type. They may also form selective attachments to certain other cell types or to specific extracellular matrix components. The selectivity of adhesion prevents the different cell types in a tissue from becoming chaotically mixed.

3. *Cell memory:* as we saw in Chapter 8, specialized patterns of gene expression, evoked by signals that acted during embryonic development, are afterward stably maintained, so that cells

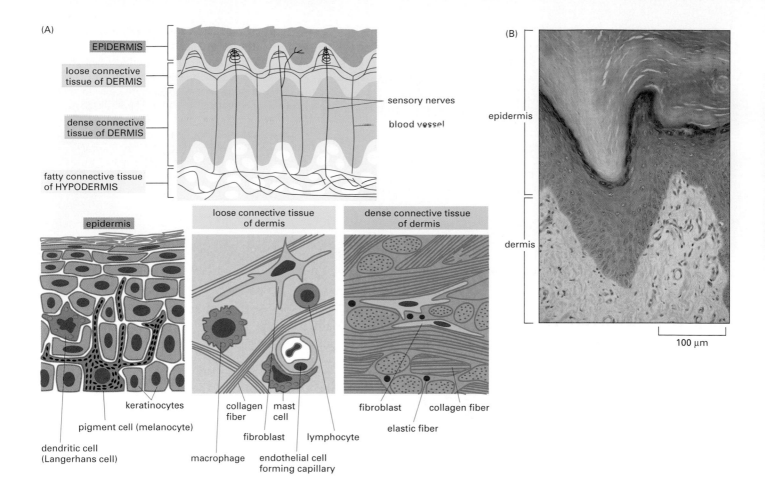

(A)

EPIDERMIS

loose connective tissue of DERMIS

dense connective tissue of DERMIS

fatty connective tissue of HYPODERMIS

sensory nerves

blood vessel

epidermis

loose connective tissue of dermis

dense connective tissue of dermis

(B)

epidermis

dermis

100 μm

keratinocytes

pigment cell (melanocyte)

dendritic cell (Langerhans cell)

macrophage

collagen fiber

fibroblast

mast cell

endothelial cell forming capillary

lymphocyte

fibroblast

elastic fiber

collagen fiber

Figure 21–33 Mammalian skin is made of a mixture of cell types. (A) Schematic diagrams showing the cellular architecture of thick skin. (B) Photograph of a cross section through the sole of a human foot, stained with hematoxylin and eosin. The skin can be viewed as a large organ composed of two main tissues: epithelial tissue (the *epidermis*), which lies outermost, and connective tissue, which consists of the tough dermis (from which leather is made) and the underlying fatty *hypodermis*. Each tissue is composed of a variety of cell types. The dermis and hypodermis are richly supplied with blood vessels and nerves. Some nerve fibers extend also into the epidermis.

autonomously preserve their distinctive character and pass it on to their progeny. The fibroblast divides to produce more fibroblasts, the endothelial cell divides to produce more endothelial cells, and so on. This principle, with elaborations that we explain later preserves the diversity of cell types in the tissue.

Different Tissues Are Renewed at Different Rates

Cells in tissues vary enormously in their rate and pattern of turnover. At one extreme are nerve cells, most of which last a lifetime without replacement. At the other extreme are the cells that line the intestine, which are replaced every few days. Between these extremes there is a

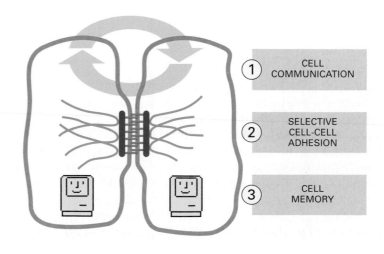

1 CELL COMMUNICATION

2 SELECTIVE CELL-CELL ADHESION

3 CELL MEMORY

Figure 21–34 Three key factors maintain the cellular organization of tissues.

spectrum of different rates and styles of cell replacement and tissue renewal. Bone, for example, has a turnover time of about ten years in humans, involving renewal of the matrix as well as of cells: old bone matrix is slowly eaten away by a set of cells called *osteoclasts,* akin to macrophages, while new matrix is deposited by another set of cells, *osteoblasts,* akin to fibroblasts. New red blood cells in humans are continually generated in the bone marrow (from yet another class of cells) and released into the circulation, from which they are removed and destroyed after 120 days. In the skin, the outer layers of the epidermis are continually flaking off and being replaced from below, so that the epidermis is renewed with a turnover time of about two months. And so on.

Our life depends on these renewal processes. A large dose of ionizing radiation, by blocking cell division, halts renewal: within a few days, the lining of the intestine, for example, becomes deprived of cells, leading to the devastating diarrhea and water loss characteristic of acute radiation sickness.

Question 21–6

Why does ionizing radiation stop cell division?

Stem Cells Generate a Continuous Supply of Terminally Differentiated Cells

Many of the differentiated cells that need continual replacement are themselves unable to divide. Red blood cells, surface epidermal cells, and the absorptive and goblet cells of the gut lining are all of this type. Such cells are referred to as *terminally differentiated:* they lie at the dead end of their developmental pathway.

Replacements for terminally differentiated cells are generated from a stock of proliferating **precursor cells**, which themselves usually derive from small numbers of more slowly dividing **stem cells**. The stem cells and proliferating precursor cells are retained in the corresponding tissues along with the differentiated cells. Stem cells are not terminally differentiated and can divide without limit (or at least for the lifetime of the animal). When a stem cell divides, though, each daughter has a choice: either it can remain a stem cell, or it can embark on a course leading irreversibly to terminal differentiation, usually via a series of precursor cell divisions (Figure 21–35). The job of the stem cell and precursor cells, therefore, is not to carry out the specialized function of the differentiated cell, but rather to produce cells that will. Stem cells are usually present in small numbers and often have a nondescript appearance, making them difficult to identify. Although they are not terminally differentiated, stem cells of adult tissues are nevertheless specialized: under normal conditions, they stably express sets of gene regulatory proteins that ensure that their differentiated progeny will be of the appropriate types.

The pattern of cell replacement varies from one stem-cell-based tissue to another. In the lining of the small intestine, for example, the absorptive and goblet cells together are arranged as a single-layered epithelium that covers the surfaces of the fingerlike *villi* that project into the lumen of the gut. This epithelium is continuous with the epithelium that lines the *crypts* that descend into the underlying connective tissue, and the stem cells lie near the bottom of the crypts. Newborn absorptive and goblet cells generated from the stem cells are carried upward by a sliding movement in the plane of the epithelial sheet until they reach the exposed surfaces of the villi; at the tips of the villi the cells die and are shed into the lumen of the gut (Figure 21–36).

A contrasting example is found in the epidermis. The epidermis is a many-layered epithelium, with stem cells and precursor cells in the basal layer, adherent to the basal lamina; the differentiating cells travel

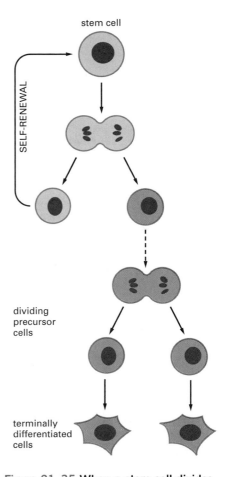

Figure 21–35 When a stem cell divides, each daughter can either remain a stem cell or go on to become terminally differentiated. The terminally differentiated cells usually develop from precursor cells that divide a limited number of times before they differentiate.

Often, a single type of stem cell gives rise to several types of differentiated progeny: the stem cells of the intestine, for example, produce absorptive cells, goblet cells, and certain other cell types. The process of blood-cell formation, or *hemopoiesis,* provides an extreme example of this phenomenon. All of the different cell types in the blood—both the red blood cells that carry oxygen and the many types of white blood cells that fight infection (Figure 21–38)—ultimately derive from a shared *hemopoietic stem cell* that normally inhabits the bone marrow (Figure 21–39).

Stem Cells Can Be Used to Repair Damaged Tissues

Because stem cells can proliferate indefinitely and produce differentiated progeny, they allow for continual renewal of normal tissue, as well as repair of tissue lost through injury. For example, by transfusing a few hemopoietic stem cells into a mouse whose own blood stem cells have been destroyed by irradiation, it is possible to fully repopulate the animal with new blood cells and rescue it from death by anemia, infection, or both. A similar approach is used in the treatment of human leukemia with irradiation (or cytotoxic drugs) followed by bone marrow transplantation.

Stem cells taken directly from adult tissues hold promise for use in tissue repair, but another type of stem cell may have even greater potential. It is possible, through cell culture, to derive from early mouse embryos an extraordinary class of stem cells called **embryonic stem cells**, or **ES cells**. Under appropriate conditions, these cells can be kept

Figure 21–36 Renewal occurs continuously in the lining of the adult intestine. (A) Cartoon showing the pattern of cell turnover and the proliferation of stem cells in the epithelium that forms the lining of the small intestine. The nondividing differentiated cells at the base of the crypts also have a finite lifetime, terminated by programmed cell death, and are continually replaced by progeny of the stem cells. (B) Photograph of a section of part of the lining of the small intestine, showing the villi and crypts. Note how mucus-secreting goblet cells (stained *red)* are interspersed among the absorptive brush-border cells in the epithelium of the villi. Smaller numbers of two other cell types—enteroendocrine cells (not shown here), which secrete gut hormones, and Paneth cells, which secrete bactericidal proteins—are also present and derive from the same stem cells. The crypt-villus organization is thought to be maintained by signals from the crypt environment that keep the crypt cells in a proliferative state, as we explain later (see How We Know, pp. 734–735).

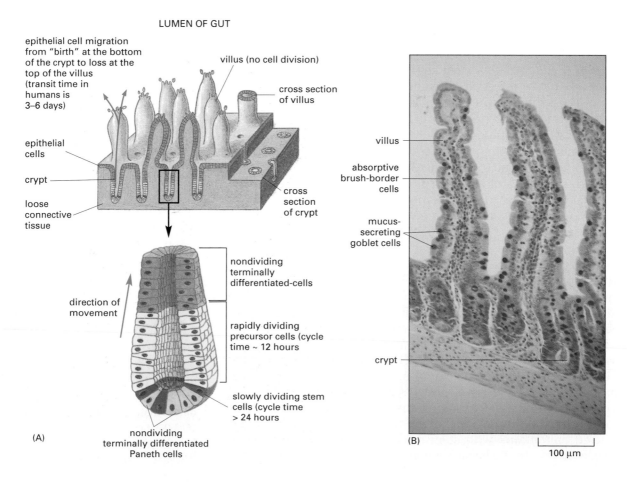

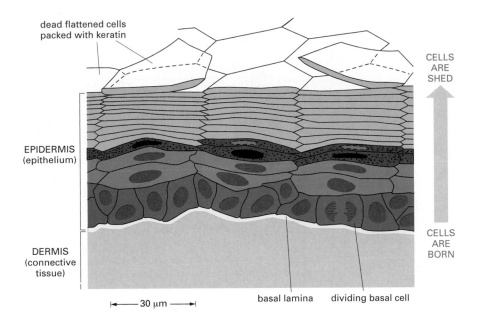

dead flattened cells packed with keratin

EPIDERMIS (epithelium)

DERMIS (connective tissue)

|← 30 μm →|

basal lamina   dividing basal cell

CELLS ARE SHED

CELLS ARE BORN

Figure 21–37 Cell replacement occurs in the epidermis. Precursor cells are produced from stem cells in the basal layer. On emerging from the basal layer, they stop dividing and move outward, differentiating as they go. Eventually, the cells undergo a special form of cell death: the nucleus disintegrates and the cell shrinks to the form of a flattened scale packed with keratin. The scale is ultimately shed from the surface of the body.

proliferating indefinitely in culture and yet retain an unrestricted developmental potential. If the cells from the culture dish are put back into an early embryonic environment, they can give rise to all the tissues and cell types in the body, including germ cells (Figure 21–40). Their descendants in the embryo are able to integrate perfectly into whatever site they come to occupy, adopting the character and behavior that normal cells would show at that site.

Cells with properties similar to those of mouse ES cells can now be derived from early human embryos, creating a potentially inexhaustible supply of cells that might be used for the replacement and repair of mature human tissues that are damaged. Experiments in mice suggest that it will be possible, in the near future, to use ES cells to replace the skeletal muscle fibers that degenerate in victims of muscular dystrophy, the nerve cells that die in patients with Parkinson's disease, the insulin-secreting cells that are destroyed in type I diabetics, and the cardiac

Question 21–7

Why do you suppose epithelial cells lining the gut are renewed frequently, whereas most neurons last for the lifetime of the organism?

Figure 21–38 The blood contains many circulating cell types, all derived from a single type of stem cell. In this scanning electron micrograph, the larger, more spherical cells with a rough surface are white blood cells; the smaller, smoother, flattened cells are red blood cells. (From R.G. Kessel and R.H. Kardon, Tissues and Organs: A Text-Atlas of Scanning Electron Microscopy. San Francisco: Freeman, 1979. © W.H. Freeman and Company.)

5 μm

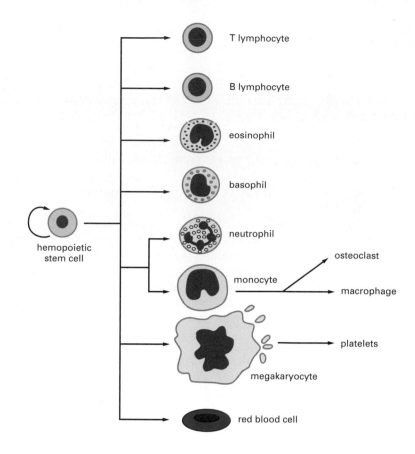

Figure 21–39 A hemopoietic stem cell divides to generate more stem cells as well as precursor cells (not shown) that proliferate and differentiate into the mature blood cell types found in the circulation. The macrophages found in many tissues of the body and the osteoclasts that eat away bone matrix originate from the same source, as do a few other types of tissue cells not shown in this scheme. A large number of different signal molecules are known to act at various points in this cell lineage to control the production of each cell type and to maintain appropriate numbers of stem cells.

muscle cells that die during a heart attack. Perhaps one day it may even become possible to grow entire organs from ES cells by a recapitulation of embryonic development.

There is, however, one major problem associated with the use of ES cells for tissue repair. If the transplanted cells are genetically different from the cells of the patient into whom they are grafted, they are likely

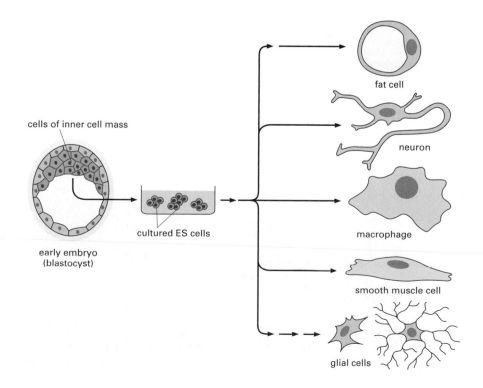

Figure 21–40 ES cells derived from an embryo can give rise to all of the tissues and cell types of the body. ES cells are harvested from the inner cell mass of an early embryo and can be maintained indefinitely as stem cells in culture. If they are put back into an embryo, they will integrate perfectly and differentiate to suit whatever environment they are placed in. The cells can also be kept in culture and supplied with different hormones or growth factors to encourage them to differentiate into specific cell types. (Based on E. Fuchs and J. A. Segré, *Cell* 100:143155, 2000.)

to be rejected and destroyed by the immune system. One possible way around this problem is to use a strategy known colloquially as "therapeutic cloning," as we now explain.

Nuclear Transplantation Provides a Way to Generate Personalized ES Cells: the Strategy of Therapeutic Cloning

The term "cloning" has been used in confusing ways as a shorthand term for several quite distinct types of procedure, particularly in public debates about the ethics of stem cell research. It is important to understand the distinctions.

As biologists define the term, a *clone* is simply a set of individuals that are genetically identical by virtue of their descent from a single ancestor. The simplest type of cloning is the cloning of cells. Thus, one can take a single epidermal stem cell from the skin and let it grow and divide in culture to obtain a large clone of genetically identical epidermal cells, which can, for example, be used to help reconstruct the skin of a badly burned patient. This kind of cloning is no more than an extension by artificial means of the processes of cell proliferation and repair that occur in a normal human body.

The cloning of entire multicellular animals, called *reproductive cloning*, is a very different enterprise, involving a far more radical departure from the ordinary course of nature. Normally, each individual animal has both a mother and a father, and is not genetically identical to either of them. In reproductive cloning, the need for two parents and sexual union is bypassed. For mammals, this difficult feat has been achieved in sheep and some other domestic animals by *nuclear transplantation*. The procedure begins with an unfertilized egg cell. The nucleus of this haploid gamete cell is sucked out, and in its place a nucleus from a regular diploid cell is introduced. The diploid donor cell can, for example, be taken from a tissue of an adult individual. The hybrid cell, consisting of a diploid donor nucleus in a host egg cytoplasm, is allowed to develop for a short while in culture. In a small proportion of cases, this can give rise to an early embryo, which is then put into the uterus of a foster mother (Figure 21–41). If the experimenter is lucky, development continues like that of a normal embryo, giving rise,

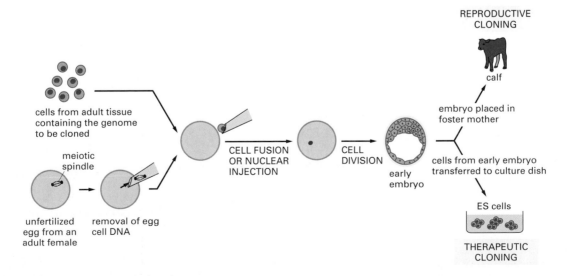

Figure 21–41 Cells from adult tissue can be used for "cloning" in two quite different senses of the word. In reproductive cloning, a whole new multicellular individual is generated; in therapeutic cloning, only cells are produced.

eventually, to a whole new animal. An individual produced in this way, by reproductive cloning, should be genetically identical to the adult individual who donated the diploid cell (except for the small amount of genetic information in mitochondria, which is inherited from the egg cytoplasm).

Another procedure, very different again from the ones just outlined, employs the technique of nuclear transplantation to produce ES cells (see Figure 21–41). In this case, the very early embryo, consisting of about 200 cells, is not transferred to the uterus of a foster mother. Instead, it is used as a source from which ES cells are derived in culture, with the aim of generating various cell types that can be used for tissue repair. This so-called *therapeutic cloning* is an elaborate technique for generating personalized ES cells, rather than whole cloned animals. Because the cells obtained by this route are genetically identical to the original donor cell, they can be grafted back into the adult from whom the donor tissue was taken, without fear of immunological rejection.

This strategy is still in its infancy and is outlawed in some countries. It remains to be seen whether it will fulfill the great hopes that medical scientists have for it.

Cancer

We pay a price for having bodies that can renew and repair themselves. The delicately adjusted mechanisms that control these processes can go wrong, leading to catastrophic disruption of the body's structure. Foremost among the diseases of tissue renewal is **cancer**, which stands alongside infectious illness, malnutrition, war, and heart disease as a major cause of death of among humans. In Europe and North America, for example, one in four of us will die of cancer.

Cancers arise from violations of the basic rules of social cell behavior. To make sense of the origins and progress of the disease, and to devise treatments, we have to draw upon almost every part of our knowledge of how cells work and interact in tissues. Conversely, much of what we know about cell and tissue biology has been discovered as a byproduct of cancer research. In this section, we examine the causes and mechanisms of cancer, the types of cell misbehavior that contribute to its progress, and the ways in which we can hope to use our understanding to defeat the cells that misbehave and stop them from killing us.

Cancer Cells Proliferate, Invade, and Metastasize

If order is to be maintained as the tissues of the body grow and renew themselves, the individual cell must adjust its behavior according to the needs of the organism as a whole. The cell must divide when new cells of its specific type are needed, and refrain from dividing when they are not; it must live as long as it is required to live, and kill itself when it is required to die; it must maintain the appropriate specialized character; and it must occupy its proper place, and not stray into inappropriate territories.

Of course, in a large organism, no significant harm is done if an occasional single cell misbehaves. But an insidious and potentially devastating breakdown of control occurs when a single cell suffers a *genetic* alteration that allows it to survive and divide when it should not, producing daughter cells that behave in the same asocial way. The organization of the tissue, and eventually that of the body as a whole, may then become disrupted by a relentlessly expanding clone of abnormal cells. It is this catastrophe that happens in cancer.

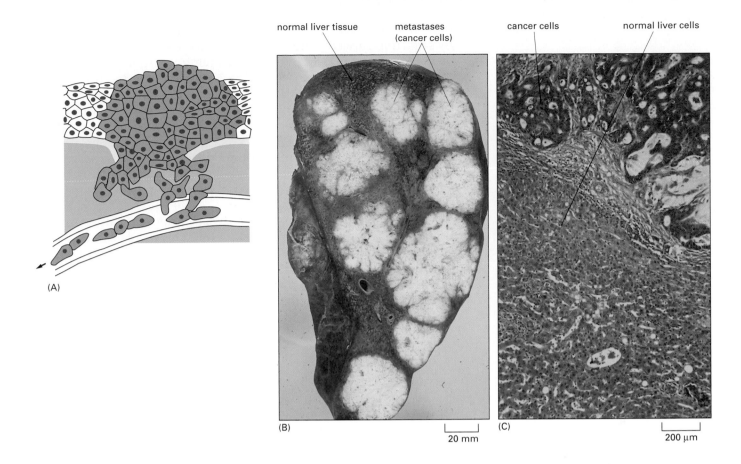

normal liver tissue   metastases (cancer cells)   cancer cells   normal liver cells

(A)

(B) |⎯⎯⎯⎯⎯⎯⎯⎯|
        20 mm

(C) |⎯⎯⎯⎯⎯⎯⎯⎯|
        200 μm

Cancer cells are defined by two heritable properties: they and their progeny (1) proliferate in defiance of the normal constraints and (2) invade and colonize territories normally reserved for other cells. It is the combination of these features that creates the lethal danger. Cells that have the first property but not the second proliferate excessively but remain clustered together in a single mass, forming a *tumor*, but the tumor in this case is said to be *benign*, and it can usually be removed cleanly and completely by surgery. A tumor is cancerous only if its cells have the ability to invade surrounding tissue, in which case it is said to be *malignant*. Malignant tumor cells with this invasive property can break loose from the primary tumor, enter the bloodstream or lymphatic vessels, and form secondary tumors, or *metastases*, at other sites in the body (Figure 21–42). The more widely the cancer spreads, the harder it becomes to eradicate.

Figure 21–42 Cancers spread by metastasis. (A) To give rise to a colony, or metastasis, in a new site, the cells of a primary tumor in an epithelium must typically cross the basal lamina, migrate through connective tissue, and get into the blood or lymphatic vessels. They then have to exit from the bloodstream or lymph and settle and survive in a new site.
(B) Metastases in a human liver, originating from a primary tumor in the colon. (C) Higher magnification view of one of the metastases, stained differently to show the contrast between the normal liver cells and the tumor cells. (B and C, courtesy of Peter Isaacson.)

Epidemiology Identifies Preventable Causes of Cancer

Prevention is better than cure, and to prevent cancer, we need to know what causes it. Do factors in our environment or features of our way of life trigger the disease and help it to progress? If so, what are they? Answers to these questions come mainly from epidemiology—the statistical analysis of human populations that is used to look for factors that correlate with disease incidence. This has provided strong evidence that the environment plays a part in the causation of most cases of cancer. The types of cancers that are common in a population, for example, vary from country to country, and studies of migrants show that it is where people live, rather than where they were born, that governs their cancer risk. Although it is still hard to discover which specific factors in the environment or life-style are critical, and many remain unknown, some of them have been identified quite precisely. Thus, it was noted long ago that cervical cancer, arising in the epithelium lining the cervix (neck) of the uterus, was much commoner in married women than in single women. This pointed to a cause related to sexual activity. We now know, through modern epidemiological studies, that most cases of cervical cancer involve infection of the cervical epithelium with certain subtypes of a common virus, called human papilloma virus. This is transmitted through sexual intercourse and can sometimes, if one is unlucky, provoke uncontrolled proliferation of the infected cells. Knowing this, we can attempt to prevent the cancer by preventing the infection—for example, by vaccination against papilloma virus. This is an active area of research.

In the great majority of human cancers, however, viruses do not appear to play a part: cancer is not an infectious disease. But epidemiology reveals other factors. Obesity, for example, is correlated with an increased cancer risk, and the relationship is suspected to be causal. By far the most important environmental cause of cancer, however, is tobacco-smoking, which is not only responsible for almost all cases of lung cancer, but also raises the incidence of several other cancers, such as those of the bladder. If we could halt the use of tobacco, it is estimated that we could prevent about 30% of all cancer deaths. No other single policy or treatment is known that would have such an impact on the cancer death rate.

As we explain below, no matter how hard we try to prevent cancer, we will never be able to eradicate it entirely; we will always be confronted with cases that demand treatment. To devise treatments that will succeed, we need to understand the biology of cancer cells and the mechanisms that underlie the growth and spread of tumors.

Cancers Develop by an Accumulation of Mutations

Cancer is fundamentally a genetic disease: it arises as a consequence of pathological changes in the information carried by DNA. It differs from other genetic diseases in that the mutations underlying cancer are mainly *somatic mutations*—those that occur in individual cells of the mature body—as opposed to *germ-line mutations,* which are handed down via the germ cells from which the entire multicellular organism develops.

Most of the identified agents known to contribute to the causation of cancer, including ionizing radiation and most chemical carcinogens, are mutagens: they cause changes in the nucleotide sequence of DNA. But even in an environment that is free of tobacco smoke, radioactivity, and all the other external mutagens that worry us, mutations will occur spontaneously as a result of fundamental limitations on the accuracy of DNA replication and DNA repair (discussed in Chapter 6). In fact, environmental carcinogens other than tobacco probably account for only a small fraction of the mutations responsible for cancer, and elimination of all these external risk factors would still leave us prone to the disease.

Spontaneous mutations occur at an estimated rate of about 10^{-6} or 10^{-7} mutations per gene per cell division, even without encouragement by external mutagens. About 10^{16} cell divisions take place in a human body in the course of a lifetime; thus every single gene is likely to have undergone mutation on more than 10^9 separate occasions in any individual. From this point of view, the problem of cancer seems to be not why it occurs, but why it occurs so infrequently.

The explanation is that it takes more than a single mutation to turn a normal cell into a cancer cell. Precisely how many mutations are required is still a matter of debate, but it is certainly more than two or

(A)

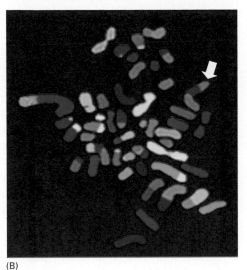

(B)

Figure 21–43 **Cancer cells often have highly abnormal chromosomes, reflecting genetic instability.** In the example shown here, chromosomes were prepared from a breast cancer cell in metaphase, spread on a glass slide, and stained with (A) a general DNA stain or (B) a combination of fluorescent stains that give a different color for each normal human chromosome. The staining (displayed in false color) shows multiple translocations, including a doubly translocated chromosome *(white arrow)* made up of two pieces of chromosome 8 *(green)* and a piece of chromosome 17 *(purple)*. The karyotype also contains 48 chromosomes, instead of the normal 46. (Courtesy of Joanne Davidson and Paul Edwards.)

three. These mutations do not all occur at once, but sequentially, usually over a period of many years.

Cancer, therefore, is typically a disease of old age, because it takes a long time for an individual line of cells to accumulate a large number of mutations (see Figure 6–20). In fact, most human cancer cells not only contain many mutations, but also are **genetically unstable**. The genetic instability results from mutations that (1) interfere with the accurate replication of the genome and thereby increase the mutation rate itself, (2) decrease the efficiency of DNA repair, or (3) increase the occurrence of chromosome breaks and rearrangements, resulting in a grossly abnormal and unstable karyotype (Figure 21–43). It is thought that the enhanced mutation rate plays an important part in facilitating the development of cancer.

Cancers Evolve Properties that Give Them a Competitive Advantage

The mutations that lead to cancer do not cripple the mutant cells. On the contrary, they give these cells a competitive advantage over their neighbors. It is this advantage enjoyed by the mutant cells that leads to disaster for the multicellular organism as a whole. Natural selection favors cells carrying mutations that enhance cell proliferation and survival, regardless of effects on neighbors, and this process culminates in the genesis of cancer cells that run riot within the population of cells that form the body, upsetting its regular structure. As an initial population of mutant cells grows, it slowly evolves: new chance mutations occur in the member cells, and some are favored by natural selection (Figure 21–44). Non-mutagenic environmental or life-style factors such as obesity may favor the development of cancer by altering the selection pressures that operate in the tissues of the body, helping mutant cells to

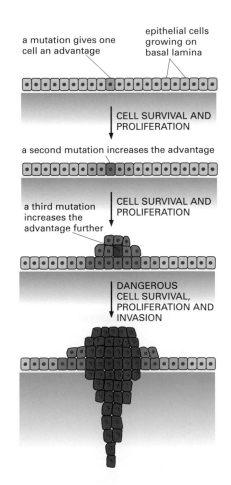

Figure 21–44 **Tumors evolve by repeated rounds of mutation and proliferation.** The final outcome is a clone of fully malignant cancer cells. At each step, a single cell undergoes a mutation that enhances its ability to proliferate, or survive, or both, so that its progeny become the dominant clone in the tumor. Proliferation of this clone then hastens occurrence of the next step of tumor progression by increasing the size of the cell population at risk of undergoing an additional mutation.

survive and proliferate. Eventually cells emerge that have all the abnormalities required for full-blown cancer.

To be successful, a cancer cell must acquire a whole range of abnormal properties—a collection of subversive new skills—as it evolves. An epithelial stem cell in the lining of the gut, for example, must undergo changes that not only permit it to carry on dividing when it should stop, but also let its progeny escape being sloughed off from the exposed surface of the epithelium. The cell and its progeny must be enabled to displace their normal neighbours and to attract a blood supply sufficient to nourish continued tumor growth. For the cells to become invasive, they must acquire the ability to digest their way through the basal lamina of the epithelium into the underlying tissue. To metastasize, they must be able to get in and out of the blood or lymph circulation and settle and survive in new sites (see Figure 21–42).

Different cancers require different combinations of properties. Nevertheless, we can draw up a general list of key behaviors of cancer cells that distinguish them from normal cells.

1. They have a reduced dependence on signals from other cells for their growth, survival, and division. Often, this is because they contain mutations in components of the cell signaling pathways through which cells respond to such social cues. A mutation in a *ras* gene (discussed in Chapter 16) can, for example, cause an intracellular signal for proliferation to be produced even in the absence of the extracellular signal that would normally be needed to trigger it, like a faulty doorbell that rings even when nobody is pressing the button.

2. Cancer cells are less prone than normal cells to kill themselves by apoptosis. This aversion to suicide is often caused by mutations in genes that regulate the intracellular death program (discussed in Chapter 18). For example, about 50% of all human cancers have lost or suffered a mutation in the *p53* gene. The p53 protein normally acts as part of a checkpoint mechanism that causes cells either to cease dividing (see Figure 18–16) or to die by apoptosis when their DNA is damaged. Chromosome breakage, for example, if not repaired, will generally cause a cell to commit suicide; but if the cell is defective in p53, it may survive and divide, creating highly abnormal daughter cells that can become more malignant.

3. Unlike most normal human cells, cancer cells can often proliferate indefinitely. Most normal human somatic cells will only divide a limited number of times in culture, after which they permanently stop, apparently because the telomeres on the ends of their chromosomes become too short (see page 326). Cancer cells typically break through this barrier by reactivating production of the telomerase enzyme that maintains telomere length.

4. Most cancer cells, as noted above, are genetically unstable, with a greatly increased mutation rate.

5. Cancer cells are abnormally invasive, and this is often in part because they lack specific cell-adhesion molecules, such as cadherins, that hold normal cells in their proper place.

6. Cancer cells can often survive and proliferate in foreign tissues to form metastases, whereas most normal cells die when misplaced. We still do not understand precisely what mutations are needed to confer this ability.

To understand the molecular biology of cancer, we have to be able to identify the mutations that give rise to these abnormal forms of behavior.

Question 21–8

About 10^{16} cell divisions take place in a human body during a lifetime, yet an adult human body consists of only about 10^{13} cells. Why are these two numbers so different?

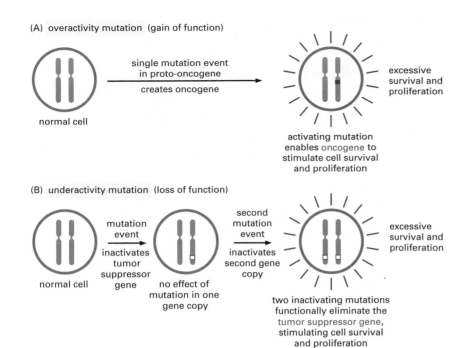

(A) overactivity mutation (gain of function)

normal cell

single mutation event in proto-oncogene creates oncogene

excessive survival and proliferation

activating mutation enables oncogene to stimulate cell survival and proliferation

(B) underactivity mutation (loss of function)

normal cell

mutation event inactivates tumor suppressor gene

no effect of mutation in one gene copy

second mutation event inactivates second gene copy

excessive survival and proliferation

two inactivating mutations functionally eliminate the tumor suppressor gene, stimulating cell survival and proliferation

Figure 21–45 Genes that are critical for cancer are classified as proto-oncogenes or tumor suppressor genes, according to whether the dangerous mutations are dominant or recessive. Oncogenes act in a dominant manner: a gain-of-function mutation in a single copy of the proto-oncogene can drive a cell toward cancer. Mutations in tumor suppressor genes, on the other hand, generally act in a recessive manner: the function of both alleles of the gene must be lost to drive a cell toward cancer. In this diagram, activating mutations are represented by *solid red boxes,* inactivating mutations by *hollow red boxes.*

Many Diverse Types of Genes Are Critical for Cancer

A great variety of approaches have been used to track down the genes and mutations that are critical for cancer. Though many of the most important of these genes have been identified, for others the hunt continues.

In some cases, the dangerous mutations are ones that make the affected gene product hyperactive. These mutations have a dominant effect—only one gene copy needs to be mutated to cause trouble—and the mutant gene is called an **oncogene** (Figure 21–45); the corresponding normal form of the gene is then called a **proto-oncogene**. Figure 21–46 shows a variety of ways in which the conversion from proto-oncogene to oncogene can occur.

For other genes, the danger lies in mutations that destroy gene function. These mutations are generally recessive—both gene copies must be lost or inactivated before an effect is seen—and the affected gene is called a **tumor suppressor gene** (see Figure 21–45). Tumor suppressor genes were first identified by a route involving human genetics. Occasionally, individuals are encountered who have inherited a mutation in a tumor-suppressor gene; although one gene copy is enough for

Figure 21–46 Several kinds of genetic change can convert a proto-oncogene into an oncogene. In each case, the change leads to an increase in the gene's function.

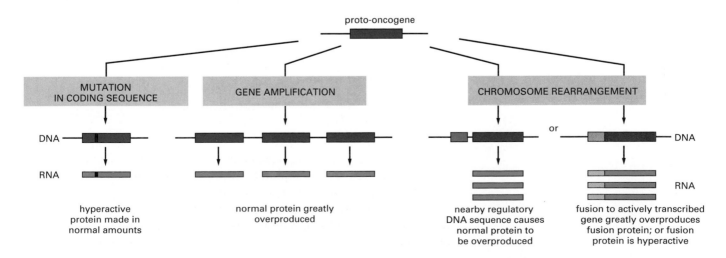

proto-oncogene

MUTATION IN CODING SEQUENCE

GENE AMPLIFICATION

CHROMOSOME REARRANGEMENT

or

DNA

RNA

DNA

RNA

hyperactive protein made in normal amounts

normal protein greatly overproduced

nearby regulatory DNA sequence causes normal protein to be overproduced

fusion to actively transcribed gene greatly overproduces fusion protein; or fusion protein is hyperactive

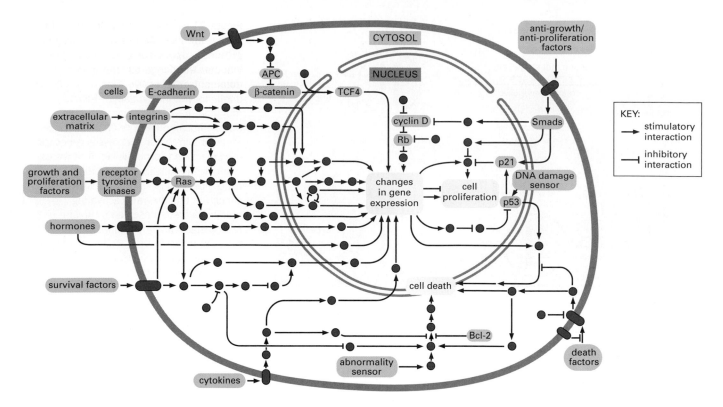

Figure 21–47 Many different types of genes are critical for cancer. The cartoon shows the major signaling pathways relevant to cancer in human cells, indicating the cellular locations of some of the proteins modified by mutation in cancers. Products of both oncogenes and tumor suppressor genes often occur within the same pathways. Individual signaling proteins are indicated by *solid red circles,* with the cancer-critical components and control mechanisms discussed in this book in *green.* Stimulatory and inhibitory interactions between components are indicated by arrows and bars, respectively, as shown in the key. (Adapted from D. Hanahan and R.A.Weinberg, *Cell* 100:57–70, 2000.)

normal cell behavior, the cells of these individuals are only one mutational step away from total loss of the gene's function (as against two steps for a normal person). Because the number of additional mutations required for cancer is less, the disease occurs with higher frequency and on average at an earlier age, sometimes in childhood. The families that carry such mutations are therefore unusually prone to cancer.

Proto-oncogenes and tumor suppressor genes are of many sorts, corresponding to the many different kinds of misbehavior that cancer cells display. Some of these genes code for growth factors, receptors, or—like *ras*—for components of the intracellular signaling pathways that growth factors activate. Others code for DNA repair proteins, for mediators of the DNA damage response such as *p53,* or for regulators of the cell cycle or of the cell death program. Still others, as we have mentioned, code for cell adhesion molecules such as cadherins. Figure 21–47 conveys some idea of this diversity.

Colorectal Cancer Illustrates How Loss of a Gene Can Lead to Growth of a Tumor

Colorectal cancer provides a well studied example to show, first, how a tumor suppressor gene can be identified and, second, how its identification leads on to an understanding of the basic molecular mechanism underlying the growth of a common type of tumor. Colorectal cancer arises from the epithelium lining the colon and rectum; most cases are seen in old people and do not have any discernible hereditary cause. A

(A)

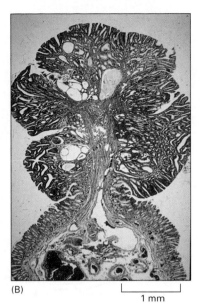
(B) ⊢——————⊣
1 mm

Figure 21–48 Colorectal cancer often begins with loss of the tumor suppressor gene *APC*, leading to growth of a polyp. (A) Thousands of small polyps, and a few much larger ones, are seen in the lining of the colon of a patient with an inherited *APC* mutation (one or two polyps might be seen in a genetically normal person). Through further mutations, some of these polyps will progress to become malignant cancers, if the tissue is not removed surgically. (B) Cross section of one such polyp; note the excessive quantities of deeply infolded epithelium, corresponding to crypts full of abnormal, proliferating cells. (A, courtesy of John Northover and Cancer Research UK; B, courtesy of Anne Campbell.)

small proportion of cases, however, occur in families that are exceptionally prone to the disease and show an unusually early onset. In one set of families, the predisposition to cancer has been traced to an inherited mutation in a DNA repair enzyme, as already discussed in Chapter 6. In another class of hereditary colorectal cancer patients, a different mutation is present, leading to a highly distinctive phenotype. The affected individuals develop colorectal cancer in early adult life, and the onset of their disease is foreshadowed by the development of hundreds or thousands of little tumorous growths, called *polyps,* in the lining of the colon and rectum. Through family studies, the abnormality can be traced to deletion or inactivation of a gene called the *adenomatous polyposis coli (APC)* gene. Affected individuals inherit one mutant copy of the gene and one normal copy; their cancers arise from cells that can be shown to have undergone a somatic mutation that inactivates the remaining good copy. But what about the great majority of colorectal cancer patients, who have inherited two good copies of *APC* and do not have the hereditary condition or any significant family history of cancer? When their tumors are analyzed, it turns out in more than 60% of cases that, while both copies of *APC* are present in the adjacent normal tissue, the tumor cells have lost both copies of this gene, presumably through two independent somatic mutations.

All this clearly identifies *APC* as a tumor suppressor gene, and knowing its sequence and mutant phenotype one can begin to decipher how its loss helps to initiate the development of cancer. As explained in the How We Know box, *APC* turns out to code for an inhibitory protein that normally restricts the activation of a cell-cell signaling pathway, called the *Wnt pathway,* which is involved in stimulating cell proliferation in the crypts of the gut lining. When *APC* is lost, the pathway is hyperactive and the cells proliferate to excess, generating a polyp (Figure 21–48). Within this growing mass of tissue, further mutations may occur, resulting in invasive cancer (Figure 21–49).

Figure 21–49 A polyp in the gut lining, caused by loss of the *APC* gene, can progress to cancer by accumulation of further mutations. The diagram shows a sequence of mutations that might underlie a typical case of colorectal cancer. A sequence of events such as that shown here would usually be spread over 10 to 20 years or more. Though most colorectal cancers are thought to begin with loss of the *APC* tumor-suppressor gene, the subsequent sequence of mutations is quite variable; indeed, many polyps never progress to cancer.

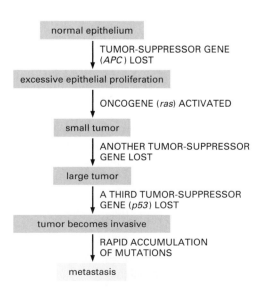

normal epithelium
↓ TUMOR-SUPPRESSOR GENE (*APC*) LOST
excessive epithelial proliferation
↓ ONCOGENE (*ras*) ACTIVATED
small tumor
↓ ANOTHER TUMOR-SUPPRESSOR GENE LOST
large tumor
↓ A THIRD TUMOR-SUPPRESSOR GENE (*p53*) LOST
tumor becomes invasive
↓ RAPID ACCUMULATION OF MUTATIONS
metastasis

How We Know: Making Sense of the Genes that Are Critical for Cancer

The search for genes that are critical for the development of cancer sometimes begins with a family that shows a hereditary predisposition to a particular form of the disease. As we have seen, *APC*—a tumor suppressor gene that is frequently deleted or inactivated in people with colorectal cancer—was tracked down by searching for genetic defects in families prone to the disease. But identifying the gene is only half the battle. The next step is determining what the gene does in a normal cell—and why alterations in that gene precipitate cancer.

Guilt by association

Determining what a gene—or its encoded protein product—does inside a cell is not a simple task. Imagine isolating an uncharacterized protein and being told that it acts as a protein kinase. That information does not reveal how the protein functions in the context of a living cell. What are its protein targets? In which tissues is it active? What role does it have in the growth or development of the organism? Additional information is required to understand the context in which the biochemical activity is used.

Most proteins do not function in isolation: they interact with other proteins inside the cell. Thus one way to begin to decipher a protein's biological role is to identify its binding partners. If an uncharacterized protein interacts with a protein whose role in the cell is understood, its function is likely to be in some way related. Perhaps the simplest method for identifying proteins that bind to one another tightly is co-immunoprecipitation (see Panel 4–6, pp. 164–165). In this technique, an antibody is used to capture and precipitate a specific target protein from an extract prepared by breaking open cells; if this target protein is associated tightly with another protein, the partner protein will precipitate as well. This is the approach that was taken to characterize APC.

Two groups of researchers used antibodies against APC to isolate the protein from extracts prepared from cultured human cells. The antibodies captured APC along with a second protein. When the researchers examined the amino acid sequence of this partner, they recognized the protein as β-catenin.

The discovery that APC interacts with β-catenin initially led to wrong guesses about the role of APC in colorectal cancer. In mammals, β-catenin was known primarily for its role at adherens junctions between cells, where it serves as a linker to connect the membrane-spanning cadherin molecules to the intracellular actin cytoskeleton (see, for example, Figure 21–23). Thus for some time scientists thought that APC might be involved in cell adhesion. But within a few years it emerged that β-catenin also has another completely differ-

ent function, and that APC's interaction with it is important in cancer for a quite different reason.

Wingless flies

Not long before the discovery that APC binds to β-catenin, developmental biologists working on *Drosophila* had noticed that the human β-catenin protein is highly similar in sequence to a *Drosophila* protein called Armadillo. Armadillo was known to be a key protein in a signaling pathway that plays an important role in normal development in fruit flies. The pathway is activated by a family of signal molecules called Wnt proteins; the founding member of the *Wnt* gene family was called *wingless,* after its mutant phenotype. Wnt proteins bind to receptors on the surface of a cell, switching on an intracellular signaling cascade that ultimately leads to the activation of a set of genes that influence cell growth, division, and differentiation. Mutations in any of the proteins in this pathway lead to developmental errors that disrupt the basic body plan of the fly. The least devastating mutations cause flies to develop without wings; most mutations, however, result in the death of the embryo. But in any case, the damage is done, it seems, through effects on gene expression. This strongly suggested that Armadillo, and hence its vertebrate homolog β-catenin, were not just nuts and bolts in the apparatus of cell adhesion, but somehow mediated the control of gene expression.

The Wnt pathway was discovered and studied intensively in fruit flies, but it turns out that a similar set of proteins controls many aspects of development in vertebrates, including mice and humans. Indeed, some of the proteins in the Wnt pathway function almost interchangeably in *Drosophila* and vertebrates. The direct link between β-catenin and gene expression became clear from work in mammalian cells. Just as APC could be used as a "bait" to catch its partner β-catenin by immunoprecipitiation, so β-catenin could be used as a bait to catch the next protein in the chain of cause and effect. This was found to be a DNA-binding gene regulatory protein called LEF-1/TCF, or TCF for short. It too was found to have a *Drosophila* counterpart in the Wnt pathway, and a combination of *Drosophila* genetics and mammalian cell biology revealed how the gene control mechanism works.

Wnt transmits its signal by promoting the accumulation of "free" β-catenin (or in flies, Armadillo)—that is, of β-catenin that is not locked up in adherens junctions. This free protein migrates from the cytoplasm into the nucleus. There it binds to the TCF gene regulatory protein, creating a complex that activates transcription of various Wnt-responsive genes, including genes whose products stimulate cell proliferation (Figure 21–50).

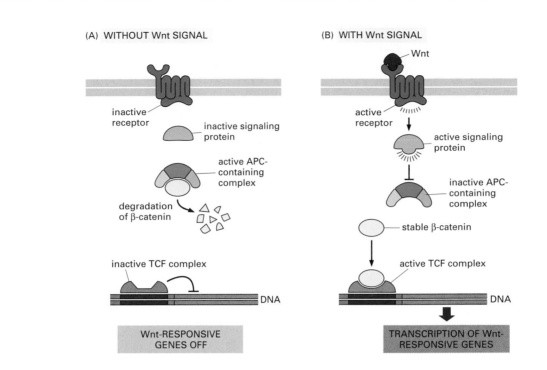

Figure 21–50 The APC protein keeps the Wnt signaling pathway inactive when the cell is not exposed to Wnt protein. It does this by promoting degradation of the signaling molecule β-catenin. In the presence of Wnt, or in the absence of active APC, free β-catenin becomes plentiful and combines with the gene regulatory protein TCF to drive transcription of Wnt target genes.

APC regulates the activity of this pathway by facilitating degradation of β-catenin and thereby preventing it from activating TCF in cells where no Wnt signal has been received (see Figure 21–50). Loss of APC allows levels of β-catenin to rise, so that TCF is activated and Wnt-responsive genes are turned on even in the absence of Wnt. But how does this cause colorectal cancer? To find out, researchers turned to mice that lack *TCF4*, a member of the *TCF* gene family that is specifically expressed in the gut lining.

Tales from the crypt

Although it may sound counterintuitive, one of the most direct ways of finding out what a gene does is to see what happens to the organism when that gene is missing. If one can pinpoint the cellular processes that are disrupted or compromised, one can begin to decipher the gene's function.

With this in mind, researchers generated "knockout" mice in which the gene encoding TCF4 was disrupted. The mutation is lethal: mice lacking TCF4 die shortly after birth. But the animals showed an interesting abnormality in their intestines. The intestinal crypts, with their populations of stem cells for renewal of the gut lining (see Figure 21–36), completely failed to develop. The researchers concluded that TCF4 is normally responsible for maintaining the pool of proliferating gut stem cells.

When APC is missing, we see the other side of the coin: without APC to promote its degradation, β-catenin accumulates in excessive quantities, binds to the TCF4 gene regulatory protein, and thereby overactivates the TCF4-responsive genes. This drives the formation of polyps by promoting the inappropriate proliferation of gut stem cells. Differentiated progeny cells continue to be produced and discarded into the gut lumen, but the crypt cell population grows too fast for this disposal mechanism to keep pace. The result is crypt enlargement and a steady increase in the number of crypts. The growing mass of tissue bulges out into the gut lumen as a polyp (see Figure 21–48). Further mutations are needed to convert this primary tumor into an invasive cancer.

More than 60% of human colorectal tumors harbor mutations in the *APC* gene. Interestingly, among the minority class of tumors that retain functional APC, about a quarter have activating mutations in β-catenin instead. These mutations tend to make the β-catenin protein more resistant to degradation and thus produce the same effect as loss of APC. In fact, mutations that enhance the activity of β-catenin have been found in a wide variety of other tumor types, including melanomas, stomach cancers, and liver cancers. Thus the Wnt signaling pathway provides multiple targets for mutations that can spur the development of cancer.

An Understanding of Cancer Cell Biology Opens the Way to New Treatments

The better we understand the tricks that cancer cells use to survive, proliferate, and spread, the better our chances of finding ways to defeat them. The task is hard, because cancer cells are mutable and, like weeds or parasites, rapidly evolve resistance to treatments used to exterminate them. Moreover, because mutations arise randomly, each case of each variety of cancer is liable to have its own unique combination of genes mutated. Thus, no single treatment is likely to work in every patient. Moreover, cancers generally are not detected until the primary tumor has reached a diameter of 1 cm or more, by which time it consists of hundreds of millions of cells that are already genetically diverse and often have already begun to metastasize (Figure 21–51).

In spite of the difficulties, many cancers can be treated effectively, and the future prospects for more and better treatments are bright. Surgery remains the most effective tactic in many cases, and surgical techniques are continually improving: if a cancer has not spread far, it can often be cured by simply cutting it out. Where surgery fails, therapies based on the intrinsic peculiarities of cancer cells can be used. As we have seen, there are some genes, such as *p53* or *ras,* whose mutation contributes to the disease in a high proportion of cases: this makes many cancers vulnerable to attack based on the absence or presence of these specific targets. Lack of p53, for example, helps to make cancer cells particularly vulnerable to DNA damage: whereas a normal cell will halt its proliferation until damage is repaired, a cell deficient in p53 will charge ahead regardless, producing daughter cells that may die because they inherit a broken, incomplete set of chromosomes. Presumably for this reason, cancer cells can often be killed by doses of radiotherapy or DNA-damaging chemotherapy that leave adjacent normal cells relatively unharmed.

These treatments are long-established, but many novel approaches are also being energetically developed. One promising strategy is to block formation of the new blood vessels that normally invade a growing tumor, and so to choke tumor growth by depriving the cells of their blood supply. Another strategy aims to use antibodies that bind to tumor-specific cell surface proteins; the antibodies can be coupled to toxins or toxin-generating enzymes that will kill the targeted cancer cells.

In some cancers, it is becoming possible to target the products of specific oncogenes directly so as to block their harmful action. Thus in *chronic myeloid leukemia* (CML), the misbehavior of the cancerous cells is known to depend on a mutant signaling protein (a tyrosine protein kinase) that causes the cells to proliferate when they should not. A small drug molecule, called Gleevec, has been designed to block the activity of this kinase (Figure 21–52). The results have been dramatically successful: in many of the patients, the abnormal proliferation and survival of the leukemic cells is strongly inhibited, and a prolonged remission of symptoms is achieved. The same drug is effective in some other cancers containing similar oncogenes.

With this example before us, we can hope that soon, equipped with our modern understanding of the molecular biology of cancer, it will be possible to devise effective rational methods of treatment for a still wider range of forms of the disease.

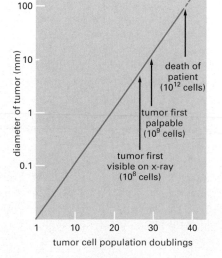

Figure 21–51 A tumor is generally not diagnosed until it has grown to contain hundreds of millions of cells. Here, the growth of a typical tumor is plotted on a logarithmic scale. Years may elapse before the tumor becomes noticeable. The doubling time of a typical breast tumor, for example, is about 100 days.

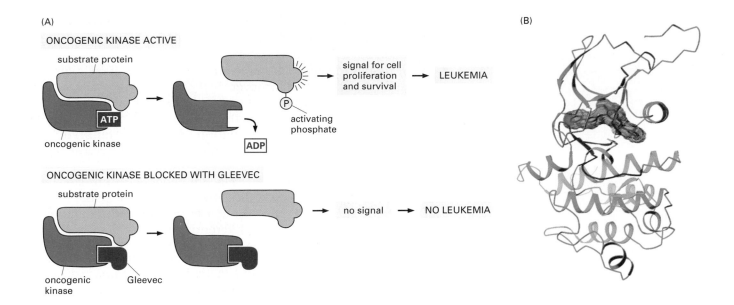

(A)

ONCOGENIC KINASE ACTIVE

substrate protein

oncogenic kinase

ATP

ADP

activating phosphate

signal for cell proliferation and survival → LEUKEMIA

ONCOGENIC KINASE BLOCKED WITH GLEEVEC

substrate protein

oncogenic kinase   Gleevec

no signal → NO LEUKEMIA

(B)

Essential Concepts

- Tissues are composed of cells plus extracellular matrix.
- In plants, each cell surrounds itself with extracellular matrix in the form of a cell wall made of cellulose and other polysaccharides.
- Naked plant cells are fragile but can exert an osmotic swelling pressure on the enveloping wall to keep the tissue to which they belong turgid.
- Cellulose fibers in the plant cell wall confer tensile strength; other cell wall components give resistance to compression.
- The orientation in which cellulose is deposited controls the orientation of plant growth.
- Animal connective tissues provide mechanical support and consist of extracellular matrix with sparsely scattered cells.
- The protein and polysaccharide components of the matrix are made by the connective-tissue cells embedded in it; in most connective tissues, these cells are called fibroblasts.
- In the extracellular matrix of animals, tensile strength is provided by the fibrous protein collagen.
- Transmembrane integrin proteins link extracellular matrix proteins such as fibronectin to the cytoskeleton.
- Glycosaminoglycans (GAGs), covalently linked to proteins to form proteoglycans, act as space-fillers and provide resistance to compression.
- Cells joined together in epithelial sheets line all external and internal surfaces of the animal body.
- In epithelial sheets, in contrast to connective tissues, tension is transmitted directly from cell to cell via cell-cell junctions.
- Proteins of the cadherin family span the epithelial cell membrane and bind to similar cadherins in the adjacent epithelial cell.
- At an adherens junction, the cadherins are linked intracellularly to actin filaments; at a desmosome junction, they are linked to keratin filaments.
- Actin bundles connected from cell to cell across an epithelium can contract, causing the epithelium to bend.
- Hemidesmosomes attach the basal face of an epithelial cell to the basal lamina.
- Tight junctions seal one epithelial cell to the next, barring diffusion of water-soluble molecules across the epithelium.

Figure 21–52 The drug Gleevec blocks the activity of an oncogenic protein and halts certain cancers. (A) In chronic myeloid leukemia, a cancer in which white blood cells are overproduced, the pathological cell behavior is almost always a result of a specific mutation (a chromosome translocation) affecting a gene called *Abl*, coding for a protein tyrosine kinase. The mutation creates an oncogenic form of *Abl*, whose protein product is hyperactive, phosphorylating other proteins when it should not and thereby generating an intracellular signal that provokes excessive production of white blood cells. Gleevec sits in the ATP-binding pocket of the hyperactive kinase and thereby prevents it from transferring a phosphate group from ATP onto a tyrosine in its target proteins. This inhibition blocks onward transmission of a signal for cell proliferation and survival. Gleevec is also effective against some other cancers involving oncogenes coding for protein tyrosine kinases similar to Abl. (B) The structure of a complex of Gleevec (solid *blue* object) with the tyrosine kinase domain of the Abl protein (ribbon diagram), as determined by X-ray crystallography. (B, from T. Schindler et al., *Science* 289:1938–1942, 2000. © AAAS.)

- Gap junctions form channels that allow passage of small molecules and ions from cell to cell; plasmodesmata in plants have the same function but a different structure.
- Most tissues in vertebrates are complex mixtures of cell types that are subject to continual turnover.
- The adult structure is maintained and renewed by the same basic processes that generate it in the embryo: cell proliferation, cell movement, and cell differentiation. As in the embryo, these processes are controlled by cell communication, selective cell-cell adhesion, and cell memory.
- New terminally differentiated cells are generated from stem cells, usually via the production of proliferating precursor cells.
- Embryonic stem cells (ES cells) can be maintained indefinitely in culture and remain capable of differentiating into any cell type in the body.
- By nuclear transplantation, personalized ES cells can in principle be produced for any adult, a technique called "therapeutic cloning."
- Cancer cells fail to obey the social constraints that normally maintain tissue organization: they proliferate when they should not, survive where they should not, and invade regions that they should keep out of.
- Tobacco smoke causes more cancers than any other environmental mutagen.
- Cancers arise from the accumulation of many mutations in a single somatic cell lineage.
- Cancer cells are genetically unstable, having increased rates of mutation; many show gross chromosomal abnormalities.
- Cancer cells typically express telomerase, enabling them to continue dividing when normal human cells would stop.
- Most human cancer cells harbor mutations in the *p53* gene, allowing them to survive and divide even when their DNA is damaged.
- The mutations that promote cancer can do so by converting proto-oncogenes into oncogenes, which are hyperactive, or by inactivating tumor suppressor genes.
- Tumor suppressor genes can sometimes be identified through studies of rare cancer-prone families in which a mutation of one gene copy is inherited.

Key Terms

adherens junction	gap junction
apical	genetic instability
basal	glycosaminoglycan
basal lamina	hemidesmosome
cadherin	metastasis
cancer	oncogene
cell junction	reproductive cloning
cell memory	stem cell
cell wall	therapeutic cloning
collagen	tight junction
desmosome junction	tissue
extracellular matrix	tumor-suppressor gene
embryonic stem (ES) cell	

Questions

Question 21–9

Which of the following statements are correct? Explain your answers.

A. Gap junctions connect the cytoskeleton of one cell to that of a neighboring cell or to the extracellular matrix.

B. A wilted plant leaf can be likened to a deflated bicycle tire.

C. Because of their rigid structure, proteoglycans can withstand a large amount of compressive force.

D. The basal lamina is a specialized layer of extracellular matrix to which sheets of epithelial cells are attached.

E. Skin cells are continually shed and are renewed every few weeks; for a permanent tattoo, it is therefore necessary to deposit pigment below the epidermis.

F. Although stem cells are not differentiated, they are specialized and therefore give rise only to specific cell types.

Question 21–10

Which of the following substances would you expect to spread from one cell to the next through (a) gap junctions and (b) plasmodesmata: glutamic acid, mRNA, cyclic AMP, Ca^{2+}, G proteins, and plasma membrane phospholipids?

Question 21–11

Discuss the following statement: "If plant cells contained intermediate filaments to provide the cells with tensile strength, their cell walls would be dispensable."

Question 21–12

Through the exchange of small metabolites and ions, gap junctions provide metabolic and electrical coupling between cells. Why, then, do you suppose that neurons communicate primarily through synapses rather than through gap junctions?

Question 21–13

Gelatin is primarily composed of collagen, which is responsible for the remarkable tensile strength of connective tissue. It is the basic ingredient of jello; yet, as you probably experienced many times yourself while consuming the strawberry-flavored variety, jello has virtually no tensile strength. Why?

Question 21–14

"The structure of an organism is determined by the genome that the egg contains." What is the evidence on which this statement is based? Indeed, a friend challenges you and suggests that you replace the DNA of a stork's egg with human DNA to see if a human baby results. How would you answer him?

Question 21–15

Leukemias—that is, cancers arising through mutations that cause excessive production of white blood cells—have an earlier average age of onset than other cancers. Propose an explanation for why this might be the case.

Question 21–16

Carefully consider the graph in Figure Q21–16, showing the number of cases of colon cancer diagnosed per 100,000 women per year as a function of age. Why is this graph so steep and curved, if mutations occur with a similar frequency throughout a person's life span?

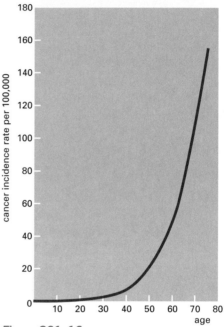

Figure Q21–16

Question 21–17

Heavy smokers or industrial workers exposed for a limited time to a chemical carcinogen that induces mutations in DNA do not usually begin to develop cancers characteristic of their habit or occupation until 10, 20, or even more years after the exposure. Suggest an explanation for this long delay.

Question 21–18

High levels of the female sex hormone estrogen increase the incidence of some forms of cancer. Thus some early types of contraceptive pills containing high concentrations of estrogen were eventually withdrawn from use because this was found to increase the risk of cancer of the lining of the uterus. Male transsexuals who use estrogen preparations to give themselves a female appearance have an increased risk of breast cancer. High levels of androgens (male sex hormones) increase the risk of some other forms of cancer, such as cancer of the prostate. Can one infer that estrogens and androgens are mutagenic?

Question 21–19

Is cancer hereditary?

Highlights from *Essential Cell Biology 2 Interactive* CD-ROM

21.2   Adhesion Junctions Between Cells
21.5   Breast Cancer Cells

Answers

Chapter 1

Answer 1–1

Trying to define life in terms of properties is an elusive business, as suggested by this scoring exercise (Table A1–1). Vacuum cleaners are highly organized objects, take matter and energy from the environment and transform the energy into motion, responding to stimuli from the operator as they do so. On the other hand, they cannot reproduce themselves, or grow and develop—but then neither can old animals. Potatoes are not particularly responsive to stimuli, and so on. It is curious that standard definitions of life usually do not mention that living organisms on Earth are largely made of organic molecules, that life is carbon based. As we now know, the key types of "informational macromolecules"—DNA, RNA, and protein—are the same in every living species.

Table A1–1 Plausible "life" scores for a vacuum cleaner, a potato, and a human			
CHARACTERISTIC	VACUUM CLEANER	POTATO	HUMAN
1. Organization	Yes	Yes	Yes
2. Homeostasis	Yes	Yes	Yes
3. Reproduction	No	Yes	Yes
4. Development	No	Yes	Yes
5. Energy	Yes	Yes	Yes
6. Responsiveness	Yes	No	Yes
7. Adaptation	No	Yes	Yes

Answer 1–2

Most random changes to the shoe design would result in objectionable defects: shoes with multiple heels, with no soles, or with awkward sizes would obviously not sell and would therefore be selected against by market forces. Other changes would be neutral, such as minor variations in color or in size. A minority of changes, however, might result in more desirable shoes: deep scratches in a previously flat sole, for example, might create shoes that would perform better in wet conditions; the loss of high heels might produce shoes that are more comfortable. The example illustrates that random changes can lead to significant improvements if the number of trials is large enough and selective pressures are imposed.

Answer 1–3

It is extremely unlikely that you created a new organism in this experiment. Far more probably, a spore from the air landed in your broth, germinated, and gave rise to the cells you observed. In the middle of the nineteenth century, Louis Pasteur invented a clever apparatus to disprove the then widely accepted belief that life could arise spontaneously. He showed that sealed flasks never grew anything if properly heat-sterilized first. He overcame the objections of those who pointed out the lack of oxygen or who suggested that his heat sterilization killed the life-generating principle, by using a special flask with a slender "swan's neck," which was designed to prevent spores carried in the air from contaminating the culture (Figure A1–3). The cultures in these flasks never showed any signs of life; however, they were capable of supporting life, as could be demonstrated by washing some of the "dust" from the neck into the culture.

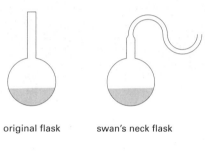

original flask   swan's neck flask

Figure A1–3

Answer 1–4

6×10^{39} ($= 6 \times 10^{27}$ g$/10^{-12}$ g) bacteria would have the same mass as the Earth. And $6 \times 10^{39} = 2^{t/20}$, according to the equation describing exponential growth. Solving this equation for t results in $t = 2642$ minutes (or 44 hours). This represents only 132 generation times(!), whereas 5×10^{14} bacterial generation times have passed during the last 3.5 billion years. Obviously, the total mass of bacteria on this planet is nowhere close to the mass of the Earth. This illustrates that exponential growth can occur only for very few generations, i.e., for minuscule periods of time compared with evolution. In any realistic scenario, food supplies become very quickly limiting.

This simple calculation shows us that the ability to grow and divide quickly when food is ample is only one factor in the survival of a species. Food is generally scarce, and individuals of the same species have to compete with one one another for the limited resources. Natural selection favors mutants that win the competition, or that find ways to exploit food sources that their neighbors are unable to use.

Answer 1–5

See Figure A1–5.

Answer 1–6

By engulfing substances, such as food particles, eucaryotic cells can sequester them and feed on them efficiently. Bacteria, in contrast, have no way of capturing lumps of food; they can export substances that help break down food substances in the environment, but the products of this labor must then be shared with other cells in the same neighborhood.

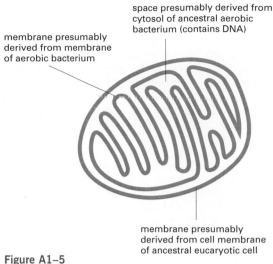

Figure A1–5

space presumably derived from cytosol of ancestral aerobic bacterium (contains DNA)

membrane presumably derived from membrane of aerobic bacterium

membrane presumably derived from cell membrane of ancestral eucaryotic cell

Answer 1–7

Light microscopy is much easier to use and requires much simpler instruments. Objects that are 1 μm in size can easily be resolved; the lower limit of resolution is 0.2 μm, which is a theoretical limit imposed by the wavelength of visible light. Visible light is nondestructive and passes readily through water, making it possible to observe living cells. Electron microscopy, on the other hand, is much more complicated, both in the preparation of the sample (which needs to be extremely thinly sliced, stained with electron-dense heavy metal, and completely dehydrated) and in the nature of the instrument. Living cells cannot be observed in an electron microscope. The resolution of electron microscopy is much higher, however, and objects as small as 10 nm can easily be resolved. To see any structural detail, microtubules, mitochondria, and bacteria would need to be analyzed by electron microscopy. It is possible, however, to stain them with specific dyes and then determine their location by light microscopy.

Answer 1–8

Because the basic workings of cells are so similar, a great deal has been learned from studying model systems. Brewer's yeast is a good model system because yeast cells are much simpler than human cancer cells. We can grow yeast inexpensively and in vast quantities, and we can manipulate yeast cells genetically and biochemically much more easily than human cells. This allows us to use yeast to decipher the ground rules governing how cells divide and grow. Cancer cells divide when they should not (and therefore give rise to tumors), and a basic understanding of how cell division is controlled is therefore directly relevant to the cancer problem. Indeed, the National Cancer Institute, the American Cancer Society, and many other institutions that are devoted to finding a cure for cancer strongly support basic research on various aspects of cell division in different model systems, such as yeast.

Answer 1–9

Check your answers using the Glossary and Panel 1–2 (p. 25).

Answer 1–10

A. False. The hereditary information is encoded in the cell's DNA, which in turn encodes its proteins.

B. True. Bacteria do not have a nucleus.

C. False. Plants are composed of eucaryotic cells that contain chloroplasts as cytoplasmic organelles. The chloroplasts are thought to be evolutionarily derived from procaryotic cells.

D. True. The number of chromosomes varies from one organism to another, but is constant in all cells of the same organism.

E. False. The cytosol is the cytoplasm excluding all organelles.

F. True. The nuclear envelope is a double membrane, and mitochondria are surrounded by both an inner and an outer membrane.

G. False. Protozoans are single-cell organisms and therefore do not have different tissues. They have a complex structure, however, that has highly specialized parts.

H. Somewhat true. Peroxisomes and lysosomes contain enzymes that catalyze the breakdown of substances produced in the cytosol or taken up by the cell. One can argue, however, that many of these substances are degraded to generate food molecules, and as such are certainly not "unwanted."

Answer 1–11

One average brain cell weighs 10^{-9} g (= 1000 g/10^{12}). Because 1 g of water occupies 1 ml = 1 cm^3 (= 10^{-6} m^3), the volume of one cell is 10^{-15} m^3 (= 10^{-9} g × 10^{-6} m^3/g). Taking the cube root yields a side length of 10^{-5} m, or 10 μm (10^6 μm = 1 m) for each cell. The page of the book has a surface of 0.057 m^2 (= 21 cm × 27.5 cm), and each cell has a footprint of 10^{-10} m^2 (10^{-5} m × 10^{-5} m). Therefore, $57 × 10^7$ (= 0.057 m^2/10^{-10} m^2) cells fit on this page when spread out as a single layer. Thus, 10^{12} cells would occupy 1750 pages (= 10^{12}/[$57 × 10^7$]).

Answer 1–12

In this plant cell, A is the nucleus, B is a vacuole, C is the cell wall, and D is a chloroplast. The scale bar is about 10 μm, the width of the nucleus.

Answer 1–13

The three major filaments are actin filaments, intermediate filaments, and microtubules. Actin filaments are involved in rapid cell movement, such as contraction of a muscle cell; intermediate filaments provide mechanical stability such as in epidermal cells of the skin; and microtubules function as "railroad tracks" for intracellular movements, and are responsible for the separation of chromosomes during cell division. Other functions of all these filaments are discussed in Chapter 17.

Answer 1–14

It takes only 20 hours, i.e., less than a day, before mutant cells become more abundant in the culture. Using the equation provided in the question, we see that the number of the original ("wild-type") bacterial cells at time t minutes after the mutation occurred is $10^6 × 2^{t/20}$. The number of mutant cells at time t is 1 × $2^{t/15}$. To find out when the mutant cells "overtake" the wild-type cells, we simply have to make these two numbers equal to each other (i.e., $10^6 × 2^{t/20} = 2^{t/15}$). Taking the logarithm to base 10 of both sides of this equation and solving it for t results in t = 1200 minutes (or 20 hours). At this time, the culture contains 2 × 10^{24} cells ($10^6 × 2^{60} + 1 × 2^{80}$). Incidentally, 2 × 10^{24} bacterial cells, each weighing 10^{-12} g, would weigh 2 × 10^{12} g (= 2 × 10^9 kg, or 2 million tons!). This can only have been a thought experiment.

Answer 1–15

Bacteria continually acquire mutations in their DNA. In the population of cells exposed to the poison, one or a few cells may acquire a mutation that makes them resistant to the action of the drug. Antibiotics that are poisonous to bacteria because they bind to certain bacterial proteins, for example, would not work if the proteins have a slightly changed surface so that binding occurs more weakly or not at all. These mutant bacteria would continue dividing rapidly while their cousins are slowed down. The antibiotic-resistant bacteria would soon become the predominant species in the culture.

Answer 1–16

$10^{13} = 2^{(t/1)}$. Therefore, it would take only 43 days [$t = 13/\log(2)$]. This explains why some cancers can progress extremely rapidly. Many cancer cells divide much more slowly, however, or die because of their internal abnormalities or because they do not have sufficient blood supply, and the actual progression of cancer is therefore usually slower.

Answer 1–17

Living cells evolved from nonliving matter, but grow and replicate. Like the material they originated from, they are governed by the laws of physics, thermodynamics, and chemistry. Thus, for example, they cannot create energy *de novo* or build ordered structures without the expenditure of free energy. We can understand virtually all cellular events, such as metabolism, catalysis, membrane assembly, and DNA replication, as complicated chemical reactions that can be experimentally reproduced, manipulated, and studied in test tubes.

Despite this fundamental reducibility, a living cell is more than the sum of its parts. We cannot randomly mix proteins, nucleic acids, and other chemicals together in a test tube, for example, and make a cell. The cell functions by virtue of its organized structure, and this is a product of its evolutionary history. Cells always come from preexisting cells, and the division of a mother cell passes both chemical constituents and structures to its daughters. The plasma membrane, for example, never has to form *de novo*, but grows by expansion of a preexisting membrane; there will always be a ribosome, in part made up of proteins whose function it is to make more proteins including those that build more ribosomes.

Answer 1–18

In a multicellular organism, different cells take on specialized functions and cooperate with one another. In this way, multicellular organisms are able to exploit food sources that are inaccessible to single-cell organisms. A plant, for example, can reach the soil with its roots to take up water and nutrients and at the same time harvest light energy and CO_2 from the air through its leaves. By protecting its reproductive cells with other specialized cells, the multicellular organism can develop new ways to survive in harsh environments or to fight off predators. When food runs out, it may be able to preserve its reproductive cells by allowing them to draw upon resources stored by their companions—or even to cannibalize relatives (a common process, in fact).

Answer 1–19

The volume and the surface area are 5.24×10^{-19} m^3 and 3.14×10^{-12} m^2 for the bacterial cell, and 1.77×10^{-15} m^3 and 7.07×10^{-10} m^2 for the animal cell, respectively. From these numbers, the surface-to-volume ratios are 6×10^6 m^{-1} and 4×10^5 m^{-1}, respectively. In other words, although the animal cell has a 3375-fold larger volume, its membrane surface is increased only 225-fold. If internal membranes are included in the calculation, however, the surface-to-volume ratios of both cells are about equal. Thus, because of their internal membranes, eucaryotic cells can grow bigger and still maintain a sufficiently large membrane area, which—as we shall discuss in more detail in later chapters—is required for many essential functions. We have already encountered the mitochondrial inner membrane as an example: it is highly invaginated (i.e., folded internally), to increase its membrane area (see Figure 1–18). This membrane is very important in the production of ATP, the energy carrier that drives most cellular reactions.

Answer 1–20

There are many lines of evidence for a common ancestor. Analyses of modern-day living cells show an amazing degree of similarity in the basic components that make up the inner workings of otherwise vastly different cells. Many metabolic pathways, for example, are conserved from one cell to another, and the compounds that make up nucleic acids and proteins are the same in all living cells, even though it is easy to imagine that a different choice of compounds (e.g., amino acids with different side chains) would have worked just as well. Similarly, it is not uncommon to find that important proteins have a closely similar detailed structure in procaryotic and eucaryotic cells. Theoretically, there would be many different ways to build proteins that could perform the same functions. The evidence overwhelmingly shows that most important processes were "invented" only once and then became fine-tuned during evolution to suit the particular needs of specialized cells.

It seems highly unlikely, however, that the first cell survived to become the primordial founder cell of today's living world. As evolution is not a directed process with a purposeful progression, it is more likely that there were a vast number of unsuccessful trial cells that replicated for a while and then became extinct because they could not adapt to changes in the environment or could not survive in competition with other types of cells. We can therefore speculate that the primordial ancestor cell was a "lucky" cell that ended up in a relatively stable environment in which it had a chance to replicate and evolve.

Answer 1–21

See Figure A1–21.

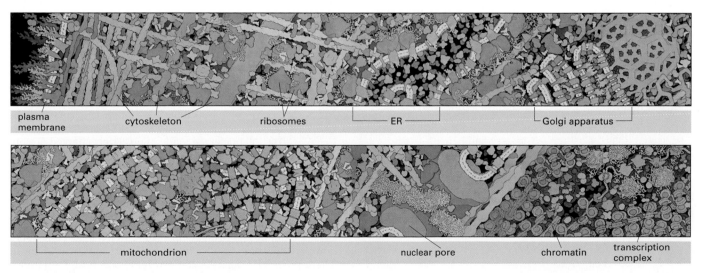

plasma membrane   cytoskeleton   ribosomes   ER   Golgi apparatus

mitochondrion   nuclear pore   chromatin   transcription complex

Figure A1–21

Chapter 2

Answer 2–1

The chances are excellent because of the enormous size of Avogadro's number. The original cup contained one mole of water, or 6×10^{23} molecules, and the volume of the world's oceans, converted to cubic centimeters, is 1.5×10^{24} cm³. After mixing, there should be on average 0.4 of a "Greek" water molecule per cm³ $(6 \times 10^{23}/1.5 \times 10^{24})$, or 7.2 molecules in 18 g of Pacific Ocean.

Answer 2–2

A. The atomic number is 6; the atomic weight is 12 (= 6 protons + 6 neutrons).

B. The number of electrons is six (= the number of protons).

C. The first shell can accommodate two and the second shell eight electrons. Carbon therefore needs four additional electrons (or would have to give up four electrons) to obtain a full outermost shell. Carbon is most stable when it shares four additional electrons with other atoms (including other carbon atoms) by forming four covalent bonds.

D. Carbon 14 has two additional neutrons in its nucleus. Because the chemical properties of an atom are determined by its electrons, carbon 14 is chemically identical to carbon 12.

Answer 2–3

The statement is correct. Both ionic and covalent bonds are based on the same principles: electrons can be shared equally between two interacting atoms, forming a nonpolar covalent bond; electrons can be shared unequally between two interacting atoms, forming a polar covalent bond; or electrons can be completely lost from one atom and gained by the other, forming an ionic bond. There are bonds of every conceivable intermediate state, and for borderline cases it becomes arbitrary whether a bond is described as a very polar covalent bond or an ionic bond.

Answer 2–4

The statement is correct. The hydrogen–oxygen bond in water molecules is polar, so that the oxygen atom carries a more negative charge than the hydrogen atoms. These partial negative charges are attracted to the positively charged sodium ions, but are repelled from the negatively charged chloride ions.

Answer 2–5

A. Hydronium (H_3O^+) ions result from water dissociating into protons and hydroxyl ions, each proton binding to a water molecule to form a hydronium ion ($2H_2O \rightarrow H_2O + H^+ + OH^- \rightarrow H_3O^+ + OH^-$). At neutral pH, i.e., in the absence of an acid providing more H_3O^+ ions or a base providing more OH^- ions, the concentrations of H_3O^+ ions and OH^- ions are equal. We know that at neutrality the pH = 7.0, and therefore, the H^+ concentration is 10^{-7} M. The H^+ concentration equals the H_3O^+ concentration.

B. To calculate the ratio of H_3O^+ ions to H_2O molecules, we need to know the concentration of water molecules. The molecular weight of water is 18 (i.e., 18 g/mole), and 1 liter of water weighs 1 kg. Therefore, the concentration of water is 55.6 M (= 1000 [g/l]/[18 g/mole]), and the ratio of H_3O^+ ions to H_2O molecules is 1.8×10^{-9} (= $10^{-7}/55.6$); i.e., only two water molecules in a billion are dissociated at neutral pH.

Answer 2–6

No mistake. Note that the small hydrogen atoms are those that are linked to an oxygen atom, whereas the ones linked to a carbon atom are larger. This reflects the polarity of the respective bonds: the H–C bond is nonpolar, whereas the H–O bond is polar. Oxygen more strongly draws the shared electrons away from the hydrogen, resulting in a smaller radius of the electron cloud of the hydrogen atom.

Answer 2–7

The synthesis of a macromolecule with a unique structure requires that in each position only one stereoisomer is used. Changing one amino acid from its L- to its D-form would result in a different protein. Thus, if for each amino acid a random mixture of the D- and L-forms were used to build a protein, its amino acid sequence could not specify a single structure, but many different structures (2^N different structures would be formed, where N is the number of amino acids in the protein).

Why L-amino acids were selected in evolution as the exclusive building blocks of proteins is a mystery; we could easily imagine a cell in which certain (or even all) amino acids were used in the D-forms to build proteins, as long as these particular stereoisomers were used exclusively.

Answer 2–8

The term "polarity" can refer to two different principles. In one meaning it refers to directional asymmetry—for example, in linear polymers such as polypeptides, which have an N-terminus and a C-terminus; or nucleic acids, which have a 3′ and a 5′ end. Because bonds form only between the amino and the carboxyl groups of the amino acids in a polypeptide, and between the 3′ and the 5′ ends of nucleotides, nucleic acids and polypeptides always have two different ends, which give the chain a defined polarity.

In the other meaning, polarity refers to a separation of electric charge in a bond or molecule. This kind of polarity allows hydrogen-bonding, and because the water solubility, or hydrophilicity, of a molecule depends upon its being polar in this sense, the term "polar" is sometimes used to indicate water solubility.

Answer 2–9

A major advantage of condensation reactions is that they are readily reversible by hydrolysis (and water is readily available in the cell). This allows cells to break down their macromolecules (or macromolecules of other organisms that were ingested as food) and to recover the subunits intact so that they can be "recycled," i.e., used to build new macromolecules.

Answer 2–10

Many of the functions that macromolecules perform rely on their ability to associate and dissociate readily. This allows cells, for example, to remodel their interior when they move or divide, and to transport components from one organelle to another. Covalent bonds would be too strong and too permanent for such a purpose.

Answer 2–11

A. True. All nuclei are made of positively charged protons and uncharged neutrons; the only exception is the hydrogen nucleus, which consists of only one proton.

B. False. Atoms are electrically neutral. The number of positively charged protons is always balanced by an equal number of negatively charged electrons.

C. True—but only for the cell nucleus (see Chapter 1), and not for the atomic nucleus discussed in this chapter.

D. False. Elements can have different isotopes, which differ only in their number of neutrons.

E. True. In certain isotopes the large number of neutrons destabilizes the nucleus, which decomposes in a process called radioactive decay.

F. True. Examples include granules of glycogen, a polymer of glucose, found in liver cells; and fat droplets, made of aggregated triacylglycerols, found in fat cells.

G. True. Individually, these bonds are weak and readily broken by thermal motion, but because interactions between two macromolecules involve a large number of such bonds, the overall binding can be quite strong, and because hydrogen bonds form only between correctly positioned groups on the interacting macromolecules, they are very specific.

Answer 2–12

A. One cellulose molecule has a molecular weight of $n \times (12[C] + 2 \times 1[H] + 16[O])$. We do not know n, but we can determine the ratio with which the individual elements contribute to the weight of cellulose. The contribution of carbon atoms is 40% [= $12/(12 + 2 + 16) \times 100\%$]. Therefore, 2 g (40% of 5 g) of carbon atoms are contained in the cellulose that makes up this page. The atomic weight of carbon is 12 g/mole, and there are 6×10^{23} atoms or molecules in a mole. Therefore, 10^{23} carbon atoms [= $(2 \text{ g}/12 \text{ [g/mole]}) \times 6 \times 10^{23}$ (molecules/mole)] make up this page.

B. The volume of the page is $4 \times 10^{-6} \text{ m}^3$ (= 21 cm × 27.5 cm × 0.07 mm), which equals a cube with a side length of 1.6 cm (= $\sqrt[3]{4 \times 10^{-6} \text{ m}^3}$). Because we know from part A that the page contains 10^{23} carbon atoms, geometry tells us that there are about 4.6×10^7 carbon atoms (= $\sqrt[3]{10^{23}}$) lined up along each side of this cube. Therefore, about 200,000 carbon atoms (= $4.6 \times 10^7 \times 0.07 \times 10^{-3}$ m/1.6×10^{-2} m) span the thickness of the page.

C. If tightly stacked, 350,000 carbon atoms with a 0.2-nm diameter would span the 0.07-mm thickness of the page.

D. The 1.7-fold difference in the two calculations reflects (1) that carbon is not the only atom in cellulose and (2) that paper is not an atomic lattice of precisely arranged molecules (as a diamond would be for precisely arranged carbon atoms), but a random meshwork of fibers containing many voids.

Answer 2–13

A. The occupancies of the three innermost electron levels are 2, 8, 8.

B.
hydrogen	gain 1 or lose 1
helium	already has full level
oxygen	gain 2
carbon	gain 4 or lose 4
sodium	lose 1
chlorine	gain 1

C. Helium with its fully occupied electron level is chemically unreactive. Sodium and chlorine, on the other hand, are extremely reactive and readily form stable Na^+ and Cl^- ions that form ionic bonds. Oxygen has to gain two electrons, whereas hydrogen can go either way, i.e., gain or lose an electron. In a covalent bond between oxygen and hydrogen, such as in water or in hydroxyl groups, electrons are pulled toward the oxygen atom. The bond is therefore polar with a relative negative charge at the oxygen atom and a relative positive charge at the hydrogen atom. An oxygen–carbon bond is similarly polar. In contrast, in a hydrogen–carbon bond

both interacting atoms share the electrons more equally and the bond is nonpolar.

Answer 2–14

A sulfur atom is much larger than an oxygen atom, and because of its larger size, the outermost electrons are not as strongly attracted to the nucleus of the sulfur atom as they are in an oxygen atom. Consequently, the hydrogen–sulfur bond is much less polar than the hydrogen–oxygen bond. Because of the reduced polarity, the sulfur in a H_2S molecule is not strongly attracted to the hydrogen atoms in an adjacent H_2S molecule, and hydrogen bonds do not form.

Answer 2–15

The reactions are diagrammed in Figure A2–15, where R_1 and R_2 are amino acid side chains.

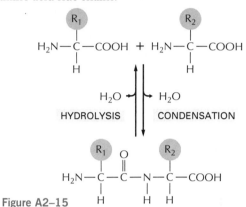

Figure A2–15

Answer 2–16

A. False. The properties of a protein depend on both the amino acids it contains and the order in which they are linked together. The diversity of proteins is due to the almost unlimited number of ways in which 20 different amino acids can be combined in a linear sequence.

B. False. Lipids assemble into bilayers by noncovalent forces. A membrane is therefore not a macromolecule.

C. True. The backbone of nucleic acids is made up of alternating ribose (or deoxyribose in DNA) and phosphate groups. Ribose and deoxyribose are sugars.

D. True. About half of the 20 naturally occurring amino acids have hydrophobic side chains. In folded proteins, many of these side chains face toward the inside of the folded-up proteins, because they are repelled from water.

E. True. Hydrophobic hydrocarbon tails contain only nonpolar bonds. Thus, they cannot participate in hydrogen-bonding and are repelled from water. We consider the underlying principles in more detail in Chapter 11.

F. False. RNA contains the four listed bases, but DNA contains T instead of U. T and U are very much alike, however, and differ only by a single methyl group.

Answer 2–17

A. (a) 400 (= 20^2); (b) 8000 (= 20^3); (c) 160,000 (= 20^4).

B. A protein with a molecular weight of 4800 is made of about 40 amino acids; thus there are 1.1×10^{52} (= 20^{40}) different ways to make such a protein. Each individual protein molecule weighs 8×10^{-21} g (= $4800/6 \times 10^{23}$); thus a mixture of one molecule each weighs 9×10^{31} g (= 8×10^{-21} g × 1.1×10^{52}), which is 15,000 times the total weight of the planet Earth, weighing 6×10^{24} kg. You would need a quite large container, indeed.

C. Given that most cellular proteins are even larger than the one used in this example, it is clear that only a minuscule fraction of the total possible amino acid sequences are used in living cells.

Answer 2–18

Because all living cells are made up of chemicals and because all chemical reactions (whether in living cells or in test tubes) follow the same rules, an understanding of the basic chemical principles is fundamentally important to the understanding of biology. In the course of this book, we will frequently refer back to these principles, on which all of the more complicated pathways and reactions that occur in cells are based.

Answer 2–19

A. Hydrogen bonds require specific groups to interact; one is always a hydrogen atom linked in a polar bond to an oxygen or a nitrogen, and the other is usually a nitrogen or an oxygen atom. Van der Waals attractions are weaker and occur between any two atoms that are in close enough proximity. Both hydrogen bonds and van der Waals attractions are short-range interactions that come into play only when two molecules are already close. Both types of bonds can therefore be thought of as means of "fine-tuning" an interaction, i.e., helping position two molecules correctly with respect to each other once they have been brought together by diffusion.

B. Van der Waals attractions would form in all three examples. Hydrogen bonds would form in (c) only.

Answer 2–20

Noncovalent interactions form between the subunits of macromolecules—e.g., the side chains of amino acids in a polypeptide chain—and cause the polypeptide chain to assume a unique shape. These interactions include hydrogen bonds, ionic interactions, van der Waals interactions, and hydrophobic interactions. Because these interactions are weak, they can be broken with relative ease; thus, most macromolecules can be unfolded by heating, which increases thermal motion.

Answer 2–21

Amphipathic molecules have both a hydrophilic and a hydrophobic end. Their hydrophilic ends can hydrogen-bond to water, but their hydrophobic ends are repelled from water because they interfere with the water structure. Consequently, the hydrophobic ends of amphipathic molecules tend to be exposed to air at air-water interfaces, or will cluster together. (See Figure A2–21.)

Answer 2–22

A,B. (A) and (B) are both correct formulas of the amino acid phenylalanine. In formula (B) phenylalanine is shown in the ionized form that exists in water solution, where the basic amino group is protonated and the acidic carboxylic group is deprotonated.

C. Incorrect. This structure of a peptide bond is missing a hydrogen atom bound to the nitrogen.

D. Incorrect. This formula of an adenine base features one double bond too many, creating a five-valent carbon atom and a four-valent nitrogen atom.

E. Incorrect. In this formula of a nucleoside triphosphate there should be two additional oxygen atoms, one between each of the phosphorus atoms.

F. This is the correct formula of ethanol.

G. Incorrect. Water does not hydrogen-bond to hydrogens bonded to carbon. The lack of the capacity to hydro-

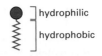

hydrophilic
hydrophobic

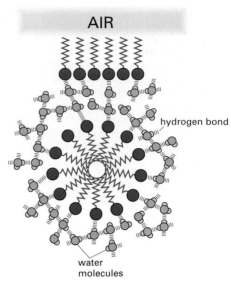

AIR

hydrogen bond

water molecules

Figure A2–21

gen-bond makes hydrocarbon chains hydrophobic, i.e., water-hating.

H. Incorrect. Na and Cl form an ionic bond, Na^+Cl^-, but a covalent bond is drawn.

I. Incorrect. The oxygen atom attracts electrons more than the carbon atom; the polarity of the two bonds should therefore be reversed.

J. This structure of glucose is correct.

K. Almost correct. It is more accurate to show that only one hydrogen is lost from the $-NH_2$ group and the $-OH$ group is lost from the $-COOH$ group.

Chapter 3

Answer 3–1

The equation represents the "bottom line" of a large set of individual reactions that are catalyzed by many individual enzymes. Because sugars are more complicated molecules than CO_2 and H_2O, the reaction generates a more ordered state inside the cell. As demanded by the second law of thermodynamics, heat is generated at many steps along the pathway of the reactions that are summarized in this equation.

Answer 3–2

Oxidation is defined as removal of electrons. Therefore, (A) is an oxidation, and (B) is a reduction. The magenta carbon atom in (C) remains largely unchanged; the neighboring black carbon atom, however, loses a hydrogen atom (i.e., an electron and a proton) and hence becomes oxidized. The magenta carbon atom in (D) becomes oxidized because it loses a hydrogen atom, whereas the carbon atom in (E) becomes reduced because it gains a hydrogen atom.

Answer 3–3

A. Both states of the coin, H and T, have an equal probability. There is therefore no driving force, i.e., no energy difference, that would favor H turning to T or vice versa. Therefore, $\Delta G° = 0$ for this reaction. However, a reaction proceeds if H and T coins are not present in the box in equal numbers. Now the concentration difference between H and T creates a driving force, a $\Delta G \neq 0$, for the reaction until it reaches equilibrium, i.e., until there are equal numbers of H and T.

B.   The amount of shaking corresponds to the temperature, as it results in the "thermal" motion of the coins. The activation energy of the reaction is the energy that needs to be expended to flip the coin, i.e., to stand it on its rim, from where it can fall back facing either side up. Jigglase would speed up the flipping by lowering the energy required for this; it could, for example, be a magnet that is suspended above the box and helps lift the coins. Jigglase would not affect where the equilibrium lies (at an equal number of H and T), but would speed up the process of reaching the equilibrium, because in the presence of jigglase more coins would flip back and forth.

Answer 3–4

See Figure A3–4. Note that $\Delta G°_{X \to Y}$ is positive, whereas $\Delta G°_{Y \to Z}$ and $\Delta G°_{X \to Z}$ are negative. The graph also shows that $\Delta G°_{X \to Z} = \Delta G°_{X \to Y} + \Delta G°_{Y \to Z}$. We do not know from the information given in Figure 3–13 how high the activation energy barriers are; they are therefore drawn to an arbitrary height (solid lines). They would be lowered by enzymes that catalyze these reactions, thereby speeding up the reaction rates (dotted lines).

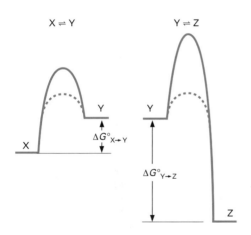

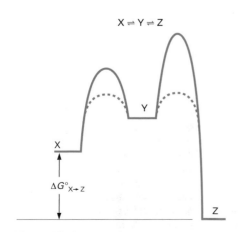

Figure A3–4

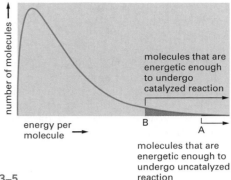

Figure A3–5

Answer 3–5

The reaction rates might be limited by (1) the concentration of the substrate, i.e., how often a molecule of CO_2 collides with the active site on the enzyme; (2) how many of these collisions are energetic enough to lead to a reaction; and (3) how fast the enzyme can release the products of the reaction and therefore be free to bind the next substrate. The diagram in Figure A3–5 shows that by lowering the activation energy barrier, more molecules have sufficient energy to undergo the reaction. The area under the curve from point A to infinite energy or from point B to infinite energy indicates the total number of molecules that will react without or with the enzyme, respectively. Although not drawn to scale, the ratio of these two areas should be 10^7.

Answer 3–6

All reactions are reversible. If the compound AB can dissociate to produce A and B, then it must also be possible for A and B to associate to form AB. Which of the two reactions predominates depends on the equilibrium constant and the concentration of A, B, and AB (as discussed in Figure 3–20). Presumably, when this enzyme was isolated its activity was detected by supplying A and B in relatively large amounts and measuring the amount of AB generated. We can suppose, however, that in the cell there is a large concentration of AB so that under normal conditions the enzyme actually catalyzes $AB \to A + B$. (This question is based on an actual example in which an enzyme was isolated and named according to the reaction in one direction, but was later shown to catalyze the reverse reaction in living cells.)

Answer 3–7

A.   The rocks in Figure 3–31B provide the energy to lift the bucket of water. ATP is driving the reaction; thus ATP corresponds to the rocks on top of the cliff. The broken debris corresponds to ADP and P_i, the products of ATP after it has released its energy and performed its work. In the reaction, ATP hydrolysis is coupled to the conversion of X to Y. X, therefore, is the starting material, the bucket on the ground, which is converted to Y, the bucket at its highest point.

B.   (i) The rock hitting the ground would be the futile hydrolysis of ATP. In the absence of an enzyme that uses the energy of ATP hydrolysis to drive an otherwise unfavorable reaction, the energy stored in the phosphoanhydride bond would be lost as heat. (ii) The energy stored in Y could be used to drive another reaction. If Y represented the activated form of an amino acid X, for example, it could undergo a condensation reaction to form a peptide bond during protein synthesis.

Answer 3–8

The free energy ΔG derived from ATP hydrolysis depends on both $\Delta G°$ and the concentrations of substrate and products. In this case,

$$\Delta G = -12 \text{ kcal/mole} = -7.3 \text{ kcal/mole} + 0.616 \ln \frac{[ADP] \times [P]}{[ATP]}$$

ΔG is smaller than $\Delta G°$, because the ATP concentration in cells is high (in the millimolar range) and the ADP concentration is low (in the 10 μM range). The concentration term of this equation is therefore smaller than 1 and its logarithm is a negative number.

$\Delta G°$ is a constant for the reaction that will not vary with reaction conditions. ΔG, in contrast, depends on the concentrations of ATP, ADP, and phosphate, which may be somewhat different between cells.

Answer 3–9

Reactions B, C, D, and E all require coupling to other, energetically favorable reactions. In each case, higher-order structures are formed that are more complicated and have higher-energy bonds than the starting materials. In contrast, reaction A is a catabolic reaction that leads to compounds in a lower energy state and will occur spontaneously.

Answer 3–10

A. Nearly true, but strictly speaking, false. Because enzymes enhance the rate but do not change the equilibrium point of a reaction, a reaction will always occur in the absence of the enzyme, though often at a minuscule rate. Moreover, competing reactions may use up the substrate more quickly, thus further impeding the desired reaction. Thus, in practical terms, without an enzyme, some reactions may never occur to an appreciable extent.

B. False. High-energy electrons are more easily transferred, i.e., more loosely bound to the donor molecule. This does not mean that they move any faster.

C. True. Hydrolysis of an ATP molecule to form AMP also produces a pyrophosphate (PP_i) molecule, which in turn is hydrolyzed into two phosphate molecules. This second reaction releases almost the same amount of energy as the initial hydrolysis of ATP, thereby approximately doubling the energy yield.

D. True. Oxidation is the removal of electrons, which reduces the diameter of the carbon atom.

E. True. ATP, for example, can donate both energy and/or a phosphate group.

F. False. Living cells have selected a particular kind of chemistry in which most oxidations are energy-releasing events; under different conditions, however, such as in a hydrogen-containing atmosphere, reductions would be energy-releasing events.

G. False. All cells, including those of cold- and warm-blooded animals, radiate comparable amounts of heat as a consequence of their metabolic reactions. For bacterial cells, for example, this becomes apparent when a compost pile heats up.

H. False. The equilibrium constant of the reaction $X \rightleftharpoons Y$ remains unchanged. If Y is removed by a second reaction, more X is converted to Y so that the ratio of X to Y remains constant.

Answer 3–11

The free-energy difference ($\Delta G°$) between Y and X due to three hydrogen bonds is –3 kcal/mole. (Note that the free energy of Y is lower than that of X, because energy would need to be expended to break the bonds to convert Y to X. The value for $\Delta G°$ for the transition X → Y is therefore negative.) The equilibrium constant for the reaction is therefore about 100 (from Table 3–1, p. 98); i.e., there are 100 times more molecules of Y than of X at equilibrium. An additional three hydrogen bonds would

increase $\Delta G°$ to –6 kcal/mole and increase the equilibrium constant another 100-fold to 10^4. Thus, relatively small differences in energy can have a major effect on equilibria.

Answer 3–12

A. The equilibrium constant is defined as $K = [AB]/([A] \times [B])$. The square brackets indicate the concentration. Thus, if A, B, and AB are each 1 μM (10^{-6} M), K will be 10^6 M^{-1} [= $10^{-6}/(10^{-6} \times 10^{-6})$].

B. Similarly, if A, B, and AB are each 1 μM (10^{-9} M), then K will be 10^9 M^{-1}.

C. This example illustrates that interacting proteins that are present in cells in lower concentrations need to bind to each other with high affinities so that a significant fraction of the molecules are bound at equilibrium. In this particular case, lowering the concentration by three orders of magnitude (from μM to nM) requires a change in the equilibrium constant by three orders of magnitude to maintain the AB protein complex (corresponding to –4.3 kcal of free energy; Figure 3–20). This corresponds to about 4–5 extra hydrogen bonds.

Answer 3–13

The statement is correct. The criterion for whether a reaction proceeds spontaneously is ΔG, not $\Delta G°$, and takes the concentrations of the reacting components into account. A reaction with a negative $\Delta G°$, for example, would not proceed spontaneously under conditions where there is already an excess of products, i.e., more than at equilibrium. Conversely, a reaction with a positive $\Delta G°$ might spontaneously go forward under conditions where there is a huge excess of substrate.

Answer 3–14

A. The energy available in 57 ATP molecules (= 686/12) corresponds to the energy released by the complete oxidation of glucose to CO_2 and H_2O.

B. The overall efficiency of ATP production would be about 53%, calculated as the ratio of actually produced ATP molecules (= 30) divided by the number of ATP molecules that could be obtained if all the energy stored in a glucose molecule could be harvested as chemical energy in ATP (= 57).

C. During the oxidation of 1 mole of glucose, 322 kcal (the remaining 47% of the available 686 kcal in one mole of glucose that is not stored as chemical energy in ATP) would be released as heat. This amount of energy would heat your body by 4.3°C (= 322 kcal/75 kg). This is a significant amount of heat, considering that 4°C of elevated temperature would be a quite incapacitating fever and that 1 mole (180 g) of glucose is no more than two cups of sugar.

D. If the energy yield were only 20%, then instead of 47% in the example above, 80% of the available energy would be released as heat and would need to be dissipated by your body. The heat production would be more than 1.7-fold higher than normal, and your body would certainly over heat.

E. The chemical formula of ATP is $C_{10}H_{12}O_{13}N_5P_3$, and its molecular weight is therefore 503 g/mole. Your resting body therefore hydrolyzes about 80 moles (= 40 kg/0.503 kg/mole) of ATP in 24 hours (this corresponds to about 1000 kcal of liberated chemical energy). Because every mole of glucose yields 30 moles of ATP, this amount of energy could be produced by oxidation of 480 g glucose (= 180 g/mole × 80 moles/30).

Answer 3–15

This scientist is definitely a fake. The 57 ATP molecules would store 684 kcal (= 57 × 12 kcal) of chemical energy, which implies that the efficiency of ATP production from glucose would have been an impossible 99.7%. This would leave virtually no energy to be released as heat, and this release is required according to the laws of thermodynamics.

Answer 3–16

A. From Table 3–1 (p. 98) we know that a free-energy difference of 4.3 kcal/mole corresponds to an equilibrium constant of 10^{-3}, i.e., $[A^*]/[A] = 10^{-3}$. The concentration of A^* is therefore 1000-fold lower than that of A.

B. The ratio of A to A^* would be unchanged. Lowering the activation energy barrier would accelerate the rate of the reaction, i.e., it would allow more molecules in a given time period to convert from $A \to A^*$ and from $A^* \to A$, but would not affect the equilibrium point.

Answer 3–17

A. The mutant organism would probably be safe to eat. ATP hydrolysis can provide approximately –12 kcal/mole of energy. This amount of energy shifts the equilibrium point of a reaction by an enormous factor: about 10^8-fold (from Table 3–1, p. 98, we see that –5.7 kcal/mole corresponds to an equilibrium constant of 10^4; thus, –12 kcal/mole corresponds to about 10^8. Note that for coupled reactions energies are additive, whereas equilibrium constants are multiplied). Therefore, if the energy of ATP hydrolysis cannot be utilized by the enzyme, 10^8-fold less poison is made. This example illustrates that coupling a reaction to the hydrolysis of an activated carrier molecule can shift the equilibrium point drastically.

B. It would be risky to consume this mutant organism. Slowing down the reaction rate would not affect its equilibrium point, and if the reaction were allowed to proceed for a long enough time, the mutant organism would likely be loaded with poison. Perhaps the equilibrium of the reaction would not be reached, but it would not be advisable to take any chances.

Answer 3–18

Enzyme A is beneficial. It allows the interconversion of two energy-carrier molecules, both of which are required as the triphosphate form for many metabolic reactions. Any ADP that is formed is quickly converted to ATP, and thus the cell maintains a high ATP/ADP ratio. Because of enzyme A, called nucleotide phosphokinase, some of the ATP is used to keep the GTP/GDP ratio similarly high.

Enzyme B would be highly detrimental to the cell. Cells use NAD^+ as an electron acceptor in catabolic reactions and must maintain high concentrations of this form of the carrier so as to break down glucose to make ATP. In contrast, NADPH is used as an electron donor in biosynthetic reactions and is kept at high concentration in the cells so as to allow the synthesis of nucleotides, fatty acids, and other essential molecules. Since enzyme B would deplete the cell's reserves of both NAD^+ and NADPH, it would reduce the rates of both catabolic and biosynthetic reactions.

Answer 3–19

Because enzymes are catalysts, enzyme reactions have to be thermodynamically feasible; the enzyme only lowers the activation energy barrier that otherwise slows the rate with which the reaction occurs. Heat, in contrast, confers more kinetic energy to the substrates so that a higher fraction of them can surmount the activation energy barrier. Many substrates, however, have many different ways in which they could react, and all of these potential pathways will be enhanced by heat, whereas an enzyme confers selectivity and facilitates only one particular pathway that, in evolution, was selected to be useful for the cell. Heat, therefore, cannot substitute for enzyme function, and chicken soup must exert its beneficial effects by other mechanisms that remain to be discovered.

Answer 3–20

A. When $[S] \ll K_M$, the term $([S] + K_M)$ approaches K_M. Therefore, the equation is simplified to rate = V_{max}/K_M [S], that is, the rate is proportional to [S].

B. When $[S] = K_M$, the term $[S]/([S] + K_M)$ equals ½. Therefore, the reaction rate is half of the maximal rate V_{max}.

C. If $[S] \gg K_M$, the term $([S] + K_M)$ approaches $[S]$. Therefore, $[S]/([S] + K_M)$ equals 1 and the reaction occurs at its maximal rate V_{max}.

Answer 3–21

The substrate concentration is 1 mM. This value can be obtained by substituting values in the equation, but it is simpler to note that the desired rate (50 μmole/sec) is exactly half of the maximum rate, V_{max}, where the substrate is typically equal to the K_M. The two plots requested are shown in Figure A3–21. A plot of 1/rate versus 1/[S] is a straight line because rearranging the standard equation yields the equation listed in Question 3–23B.

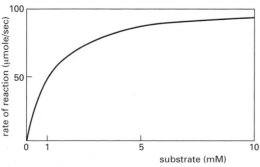

Figure A3–21

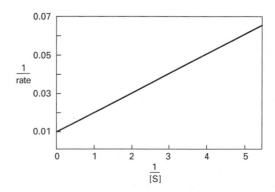

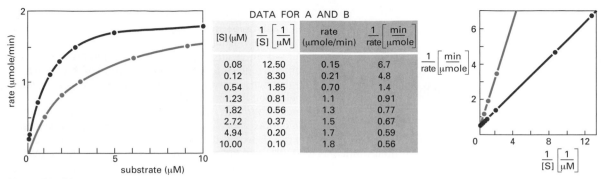

Figure A3–23

DATA FOR A AND B

[S] (μM)	$\frac{1}{[S]}\left[\frac{1}{\mu M}\right]$	rate (μmole/min)	$\frac{1}{\text{rate}}\left[\frac{\text{min}}{\mu\text{mole}}\right]$
0.08	12.50	0.15	6.7
0.12	8.30	0.21	4.8
0.54	1.85	0.70	1.4
1.23	0.81	1.1	0.91
1.82	0.56	1.3	0.77
2.72	0.37	1.5	0.67
4.94	0.20	1.7	0.59
10.00	0.10	1.8	0.56

Answer 3–22

If [S] is very much smaller than K_M, the active site of the enzyme is mostly unoccupied. If [S] is very much greater than K_M, the reaction rate is limited by the enzyme concentration (because most of the catalytic sites are fully occupied).

Answer 3–23

A,B. The data in the boxes have been used to plot the red curve and red line in Figure A3–23. From the plotted data, the K_M is 1 μM and the V_{max} is 2 μmole/min. Note that the data are much easier to interpret in the linear plot, because the curve in (A) approaches but never reaches V_{max}.

C. It is important that only a small quantity of product is made, because otherwise the rate of reaction would decrease as the substrate was depleted and product accumulated. Thus the measured rates would be lower than they should be.

D. If the K_M increases, then the concentration of substrate needed to give a half-maximal rate is increased. As more substrate is needed to produce the same rate, the enzyme-catalyzed reaction has been inhibited by the phosphorylation. The expected data plots for the phosphorylated enzyme are the green curve and the green line in Figure A3–23.

Chapter 4

Answer 4–1

Urea is a very small molecule that functions both as an efficient hydrogen-bond donor (through its –NH₂ groups) and as an efficient hydrogen-bond acceptor (through its –C=O group). As such, it can squeeze between hydrogen bonds that stabilize protein molecules and thus destabilize protein structures. In addition, nonpolar side chains are held together in the interior of folded proteins because they disrupt the structure of water if they are exposed. At high concentrations of urea, the hydrogen-bonded network of water molecules becomes disrupted so that these hydrophobic forces are significantly diminished. Proteins unfold in urea as a consequence of its effect on these two forces.

Answer 4–2

There are two α helices, and both are right-handed. The three chains that form the largest region of β sheet (green) are antiparallel. There are no knots in the polypeptide chain, presumably because a knot would interfere with the folding of the protein into its three-dimensional conformation after protein synthesis.

Answer 4–3

The amino acid sequence consists of alternating nonpolar and charged or polar amino acids. The resulting strand in a β sheet would therefore be polar on one side and hydrophobic on the other. Such a strand would probably be surrounded on either side by similar strands that together form a β sheet with a hydrophobic and a polar face. In a protein, such a β sheet (called "amphipathic," from the Greek *amphi*, "of both kinds," and *pathos*, "passion," because of its two surfaces with such different properties) would be positioned so that the hydrophobic side would face the protein's interior and the polar side would be on its surface, exposed to the water outside.

Answer 4–4

Mutations that are beneficial to an organism are selected in evolution because they confer a growth or survival advantage to the organism. Examples might be the better utilization of a food source; enhanced resistance to environmental insults, such as heat or concentrated salt; or an enhanced ability to attract a mate for sexual reproduction. In contrast, the accumulation of useless proteins is detrimental to organisms. Useless mutant proteins waste the metabolic energy required to make them. If such mutant proteins were made in excess, the synthesis of normal proteins would suffer because the capacity of the cell is limited. In more severe cases, mutant proteins interfere with the normal workings of the cell; a mutant enzyme that still binds an activated carrier molecule but does not catalyze a reaction, for example, may compete for a limited amount of this carrier and therefore inhibit normal processes. Natural selection therefore provides a strong driving force that leads to the loss of both useless and harmful proteins.

Answer 4–5

Strong reducing agents that break all of the S–S bonds would cause all of the keratin filaments to separate. Hair would therefore disintegrate. Indeed, strong reducing agents are used commercially in hair removal creams sold by your local pharmacist. However, mild reducing agents are used in treatments that either straighten or curl hair, the latter requiring hair curlers. (See Figure A4–5.)

Answer 4–6

See Figure A4–6.

Answer 4–7

A. Feedback inhibition from Z that affects the reaction B → C would increase the flux through the B → X → Y → Z pathway, because the conversion of B to C is inhibited. Thus, the more Z there is, the more production of Z would be stimulated. This is likely to result in an uncontrolled "runaway" amplification of this pathway.

B. Feedback inhibition from Z affecting Y → Z controls the production of Z. In this scheme, however, X and Y are still made at normal rates, even though both of these

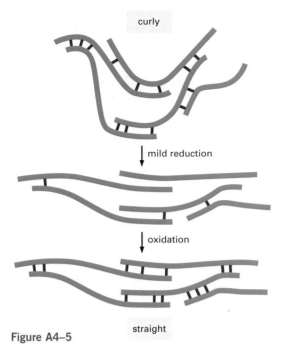

Figure A4–5

curly

↓ mild reduction

↓ oxidation

straight

intermediates are no longer needed at this level. This pathway is therefore less efficient than the one shown in Figure 4–37.

C. If Z is a positive regulator of the step B → X, then the more Z there is, the more B will be converted to X and therefore shunted into the pathway producing more Z. This would result in a runaway amplification similar to that described for (A).

D. If Z is a positive regulator of the step B → C, then accumulation of Z leads to a redirection of the pathway to make more C. This is a second possible way, in addition to that shown in the figure, to balance the distribution of compounds into the two branches of the pathway.

Answer 4–8

Both nucleotide binding and phosphorylation can induce allosteric changes in proteins. These can have a multitude of consequences, such as altered enzyme activity, drastic shape changes, and changes in affinity for other proteins or small molecules. Both mechanisms are quite versatile. An advantage of nucleotide binding is the fast rate with which a small nucleotide

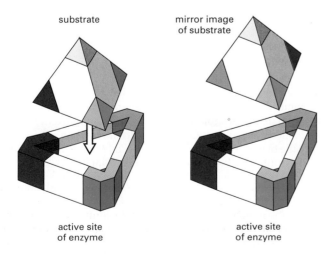

substrate

mirror image of substrate

active site of enzyme

active site of enzyme

Figure A4–6

can diffuse to the protein; the shape changes that accompany the function of motor proteins, for example, require quick nucleotide replenishment. If the different conformational states of a motor protein were controlled by phosphorylation, for example, a protein kinase would either need to diffuse into position at each step, a much slower process, or be associated permanently with each motor. One advantage of phosphorylation is that it requires only a single amino acid residue on the protein's surface, rather than a specific binding site. Phosphates can therefore be added to many different side chains on the same protein (as long as protein kinases with the proper specificities exist), thereby vastly increasing the complexity of regulation that can be achieved for a single protein.

Answer 4–9

In working as a complex, all three proteins contribute to the specificity (by binding the safe and key directly), help position one another correctly, and provide the mechanical bracing that allows them to perform a task that they could not perform individually (the key is grasped by two subunits, for example). Moreover, their functions are generally coordinated in time (for example, the binding of ATP to one subunit is likely to require that ATP has already been hydrolyzed to ADP by another).

Answer 4–10

A. True. Only a few amino acid side chains contribute to the active site. The rest of the protein is required to maintain the polypeptide chain in the correct position, provide additional binding sites for regulatory purposes, and localize the protein in the cell.

B. True. Some enzymes form covalent intermediates with their substrates (see Figure 13–5); however, in all cases the enzyme is restored to its original structure after the reaction.

C. False. β sheets can, in principle, contain any number of strands because the two strands that form the rims of the sheet are available for hydrogen-bonding to other strands. (β sheets in known proteins contain from 2 to 16 strands.)

D. False. It is true that the specificity of an antibody molecule is exclusively contained in loops on its surface; however, these loops are contributed by both the folded light- and heavy-chain domains (see Figure 4–32).

E. False. The possible linear arrangements of amino acids that lead to a stably folded protein domain are so few that most new proteins evolve by alteration of old ones.

F. True. Allosteric enzymes generally bind one or more molecules that function as regulators at sites that are distinct from the active site.

G. False. Noncovalent bonds are a major contributor to the three-dimensional structure of macromolecules.

H. False. Affinity chromatography separates specific macromolecules because of their interactions with specific ligands, not because of their charge.

I. False. The larger an organelle is, the more centrifugal force it experiences and the faster it sediments, despite an increased frictional resistance from the fluid through which it moves.

Answer 4–11

In an α helix and in the central strands of a β sheet, all of the N––H and C=O groups in the polypeptide backbone are engaged in hydrogen bonds. This gives considerable stability to these secondary structure elements, and it allows them to form from many different amino acid sequences.

Answer 4–12

No. It would not have the same or even a similar structure, because the peptide bond has a polarity. Looking at two sequential amino acids in a polypeptide chain, the amino acid that is closer to the amino-terminal end contributes the carboxyl group and the other amino acid contributes the amino group to the peptide bond that links the two. Changing their order would put the side chains into a different position with respect to the peptide backbone and therefore change their chemical environment.

Answer 4–13

As it takes 3.6 amino acid residues to complete a turn of an α helix, this sequence of 14 amino acids would make close to 4 full turns. It is remarkable because its polar and hydrophobic amino acids are spaced so that all polar residues are on one side of an α helix and all the hydrophobic residues are on the other. It is therefore likely that such an amphipathic α helix is exposed on the protein surface with its hydrophobic side facing the protein's interior. In addition, two such helices might wrap around each other as shown in Figure 4–16.

Answer 4–14

    A.   ES represents the enzyme–substrate complex.

    B.   Enzyme and substrate are in equilibrium between their free and bound states; once bound to the enzyme, a substrate molecule may either dissociate again (hence the bidirectional arrows) or be converted to product. As substrate is converted to product (with the concomitant release of free energy), however, a reaction often proceeds strongly in the forward direction, as indicated by the unidirectional arrow.

    C.   The enzyme is a catalyst and is therefore liberated in an unchanged form after the reaction; thus, E appears at both ends of the equation.

    D.   Often the products of a reaction resemble the substrates sufficiently that they can also bind to the enzyme. Any enzyme molecules that are bound to product (i.e., are part of the EP complex) are unavailable for catalysis; excess P therefore inhibits the reaction by lowering the concentration of free E.

    E.   Compound X is an inhibitor of the reaction and works similarly by forming an EX complex. However, since P has to be made before it can inhibit the reaction, it takes longer to act than X, which is present from the beginning of the reaction.

Answer 4–15

The polar amino acids Ser, Ser-P, Lys, Gln, His, and Glu are more likely to be found on a protein's surface, and the hydrophobic amino acids Leu, Phe, Val, Ile, and Met are more likely to be found in its interior. The oxidation of two cysteine residues to form a disulfide bond eliminates their potential to form hydrogen bonds and therefore makes them even more hydrophobic. Disulfide bonds are usually found in the interior of proteins. Irrespective of the nature of their side chains, the most N-terminal amino acid and the most C-terminal amino acid each contain a charged group, the amino and carboxyl groups that mark the ends of the polypeptide chain, and hence are usually found on the protein's surface.

Answer 4–16

Many secondary structure elements are not stable in isolation but require the presence of additional parts of the polypeptide chain. Hydrophobic regions that would normally be hidden in the inside of a folded domain would be exposed on the outside, and because such regions are energetically disfavored in water solution, the fragments tend to aggregate nonspecifically. Such fragments therefore would not have a defined structure, and they would be inactive for ligand binding even if they contained all of the amino acids that would normally contribute to the ligand-binding site. A protein domain, in contrast, is considered a folding unit, and fragments of a polypeptide chain that correspond to intact domains are often able to fold correctly. Thus, separated protein domains often retain their activities, such as ligand binding, if the binding site is contained entirely within this domain. Thus the most likely place in which the polypeptide chain of the protein in Figure 4–19 could be severed to give rise to stable fragments is at the boundary between the two domains (i.e., at the loop between the two α helices at the bottom right of the structure shown).

Answer 4–17

The heat inactivation of the enzyme suggests that the mutation causes the enzyme to have a less stable structure. For example, a hydrogen bond that is normally formed between two amino acid side chains might no longer be formed because the mutation replaces one of these amino acids with a different one that cannot participate in the bond. Lacking such a bond that normally helps to keep the polypeptide chain folded properly, the protein unfolds at a temperature at which it normally would be stable. Polypeptide chains that are denatured when the temperature is raised often aggregate, and they rarely refold into active proteins when the temperature is decreased.

Answer 4–18

The motor protein in the illustration can move just as easily to the left as to the right and so will not move steadily in one direction. However, if just one of the steps is coupled to ATP hydrolysis (for example, by making detachment of one foot dependent on binding of ATP and coupling the reattachment to hydrolysis of the bound ATP), then the protein will show unidirectional movement that requires the continued consumption of ATP. Note that, in principle, it does not matter which step is coupled to ATP hydrolysis (Figure A4–18).

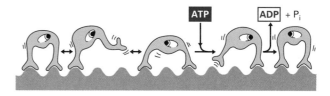

Figure A4–18

Answer 4–19

The slower migration of small molecules through a gel-filtration column results because smaller molecules have access to many more spaces in the porous beads that are packed into the column than do larger molecules. However, it is important to give the smaller molecules sufficient time to diffuse into the spaces inside the beads. At very rapid flow rates, all molecules will move rapidly around the beads, so that large and small molecules will now tend to exit together from the column.

Answer 4–20

The α helix is right-handed and the coiled-coil is left-handed. The reversal occurs because of the staggered positions of hydrophobic side chains in the α helix.

Chapter 5

Answer 5–1

A. False. The polarity of a DNA strand commonly refers to the orientation of its sugar–phosphate backbone.
B. True. G-C base pairs are held together by three hydrogen bonds, whereas A-T base pairs are held together by only two.

Answer 5–2

The scale bar in Figure 5–13 is in millions of nucleotide pairs. Using this to estimate the amount of DNA packaged into Chromosome 1 we obtain approximately 256 million nucleotide pairs. This would give a total length for the DNA of 8.7 cm ($256 \times 10^6 \times 0.34$ nm; 1 nm = $1/10^9$ m) and a compaction of 8.7 cm/10 μm = 8700-fold.

Answer 5–3

Men have only one copy of the X chromosome; a defective gene carried on it therefore has no backup copy. Women, on the other hand, have two copies of the X chromosome, one inherited from each parent, so a defective copy of the gene on one X chromosome can generally be compensated for by a normal copy on the other chromosome. This is the case with regard to the gene that causes color blindness. However, during female development, transcription from one X chromosome is shut down because it is compacted into heterochromatin (Figure 5–28). This occurs at random in each cell to one or the other of the two X chromosomes, and therefore some cells of the woman will express the defective mutant copy of the gene, whereas others will express the normal copy. This results in a retina in which on average only every other cone cell has functional color photoreceptors, and women carrying the mutant gene on one X chromosome therefore see color images with reduced resolution.

A woman who is color-blind must have two defective copies of this gene, one inherited from each parent. Her father must therefore carry the mutation on his X chromosome; because this is his only copy of the gene, he would be color-blind. Her mother could carry the defective gene on either or both of her X chromosomes. Her mother could therefore either be color-blind (defective genes on both X chromosomes) or have color vision but reduced resolution as described above. Several different types of inherited color blindness are found in the human population; this question applies to only one type.

Answer 5–4

A. The complementary strand reads 5′-TGATTGTGGA-CAAAAATCC-3′. Paired DNA strands have opposite polarity, and the convention is to write a single-stranded DNA sequence in the 5′-to-3′ direction.
B. The DNA is made of four nucleotides (100% = 13% A + x% T + y% G + z% C). Because A pairs with T, the two nucleotides are represented in equimolar proportions in DNA. Therefore, the bacterial DNA in question contains 13% thymidine. This leaves 74% [= 100% – (13% + 13%)] for G and C, which also form base pairs and hence are equimolar. Thus $y = z = 74/2 = 37$.
C. A single-stranded DNA molecule that is N nucleotides long can have any one of 4^N possible sequences, but the number of possible double-stranded DNA molecules is more difficult to calculate. Many of the 4^N single-stranded sequences will be the complement of another possible sequence in the list; for example, 5′-AGTCC-3′ and 5′-GGACT-3′ form the same double-stranded DNA molecule and therefore count as a single double-stranded possibility. If N is an odd number,

then every single-stranded sequence will complement another sequence in the list so that the number of double-stranded sequences will be 0.5×4^N. If N is an even number, then there will be slightly more than this, since some sequences will be self-complementary (such as 5′-ACTAGT-3′) and the actual value can be calculated to be $0.5 \times 4^N + 0.5 \times 4^{N/2}$.

D. To specify a unique sequence which is N nucleotides long, 4^N has to be larger than 3×10^6. Thus, 4^N /greater than 3×10^6, solved for N, gives N /greater than $\ln(3 \times 10^6)/\ln(4) = 10.7$. Thus, on average a sequence of only 11 nucleotides in length is unique in the genome. Performing the same calculation for the genome size of an animal cell yields a minimal stretch of 16 nucleotides. This shows that a relatively short sequence can mark a unique position in the genome and is sufficient, for example, to serve as an identity tag for one specific gene.

Answer 5–5

If the wrong bases were frequently incorporated during DNA replication, genetic information could not be inherited accurately. Life, as we know it, could not exist. Although the bases can form hydrogen-bonded pairs as indicated, these do not fit into the structure of the double helix. Thus, the angle with which the A residue is attached to the sugar–phosphate backbone is vastly different in the A-C pair, and the spacing between the two sugar–phosphate strands is considerably increased in the A-G pair, where two large purine rings interact. Consequently, it is energetically unfavorable to incorporate the wrong bases into the DNA chain, and such errors occur only very rarely.

Answer 5–6

A. The bases V, W, X, and Y can form a DNA-like double-helical molecule with virtually identical properties to those of bona fide DNA. V would always pair with X, and W with Y. Therefore, the macromolecules could be derived from a living organism using the same principles of replication of its genome. In principle, different bases, such as V, W, X, and Y, could have been selected during evolution as building blocks of DNA on Earth. (Similarly, there are many more conceivable amino acid side chains than the set of 20 selected in evolution that make up all proteins.)

B. None of the bases V, W, X, or Y can replace A, T, G, or C. To preserve the distance between the two sugar–phosphate strands in the double helix, a pyrimidine always has to pair with a purine (see, for example, Figure 5–6). Thus, the eight possible combinations are V-A, V-G, W-A, W-G, X-C, X-T, Y-C, and Y-T. Because of the positions of hydrogen-bond acceptors and hydrogen-bond donor groups, however, no stable base pairs would form in any of these combinations, as shown for the pairing of V-A in Figure A5–6, where only a single hydrogen bond could form.

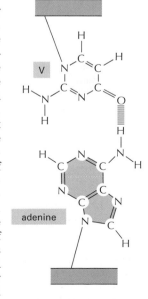

Figure A5–6

Answer 5–7

As the strands are held together by hydrogen bonds between the bases, the stability of the helix is largely dependent on the number of hydrogen bonds that can be formed. Thus two parameters determine the stability: the number of nucleotide pairs and the number of hydrogen bonds that each nucleotide pair contributes. As shown in Figure 5–6, an A-T pair contributes two hydrogen bonds, whereas a G-C pair contributes three hydrogen bonds. Therefore, helix C (containing a total of 34 hydrogen bonds) melts at the lowest temperature, helix B (containing a total of 65 hydrogen bonds) melts next, and helix A (containing a total of 78 hydrogen bonds) is the most stable, largely owing to its high GC content. Indeed, the DNA of organisms that grow in extreme temperature environments, such as certain bacteria that grow in geothermal vents, has an unusually high GC content.

Answer 5–8

The DNA would be enlarged by a factor of 2.5×10^6 (= $5 \times 10^{-3}/2 \times 10^{-9}$ m). Thus the extension cord would be 2500 km long. This is approximately the distance from London to Istanbul, San Francisco to Kansas City, Tokyo to the southern tip of Taiwan, and Melbourne to Cairns. Adjacent nucleotides would be about 0.85 mm apart (which is only about the thickness of a stack of 12 pages of this book). A gene that is 1000 nucleotide pairs long would be about 85 cm in length.

Answer 5–9

A. It takes two bits to specify each nucleotide pair (for example, 00, 01, 10, and 11 would be the binary codes for the four different nucleotides, each paired with its appropriate partner).

B. The entire human genome (3×10^9 nucleotide pairs) could be stored on two CDs ($3 \times 10^9 \times 2$ bits/4.8×10^9 bits).

Answer 5–10

A. True.

B. False. Nucleosome core proteins are approximately 11 nm in diameter. A model for the way they are packed to form a 30-nm-diameter filament is shown in Figure 5–24.

Answer 5–11

The definitions of the terms can be found in the Glossary. DNA assembles with specialized proteins to form *chromatin*. At a first level of packing, *histones* form the core of *nucleosomes*. In a nucleosome, the DNA is wrapped twice around this core. Between nuclear divisions, that is, in interphase, the *chromatin* of the *interphase chromosomes* is in a relatively extended form and is dispersed in the nucleus, although some regions of it, the *heterochromatin,* remain densely packed and are transcriptionally inactive. During nuclear division, that is, in mitosis, replicated chromosomes become condensed into *mitotic chromosomes*, which are transcriptionally inactive, and which are designed to be distributed between the daughter cells.

Answer 5–12

Colonies are clumps of cells that originate from a single founder cell and grow outward as the cells divide again and again. In the lower colony of Figure 5–27A, the *ADE2* gene is inactivated when placed near a telomere, but apparently it can become spontaneously activated in a few cells, which then turn white. Once spontaneously activated in a cell, the *ADE2* gene continues to be active in the descendants of that cell, resulting in clumps of white cells (the white sectors) in the colony. This result shows both that the inactivation of a gene positioned close to a telomere can be reversed and that this change is passed on to further generations (see Figure 8–24). This change in *ADE2* expression probably results from a spontaneous decondensation of the chromatin structure around the gene.

Answer 5–13

In the electron micrographs, one can detect chromatin regions of two different densities; the densely stained regions correspond to heterochromatin, while less condensed chromatin is more lightly stained. The chromatin in nucleus A is mostly in the form of condensed, transcriptionally inactive heterochromatin, whereas most of the chromatin in nucleus B is decondensed and therefore potentially active for transcription. Nucleus A is from a reticulocyte, a red blood cell precursor, which is largely devoted to making a single protein, hemoglobin. Nucleus B is from a lymphocyte, which is active in transcribing many different genes.

Answer 5–14

Helix A is right handed. Helix C is left handed. Helix B has one right-handed strand and one left-handed strand. There are several ways to tell the handedness of a helix. For a vertically oriented helix, like the ones in Figure Q5–14, if the strands in front point up to the right, the helix is right handed; if they point up to the left, the helix is left handed. Once you are comfortable identifying the handedness of a helix, you will be amused to note that nearly 50% of the 'DNA' helices in advertisements are left handed, as are a surprisingly high number of the ones in books. Amazingly, a version of Helix B was used in advertisements for a prominent international conference, celebrating the 30-year anniversary of the discovery of the DNA helix.

Answer 5–15

The packing ratio within a nucleosome core is 4.5 [(146 bp × 0.34 nm/bp)/(11 nm) = 4.5]. If there is an additional 54 bp of linker DNA, then the packing ratio for 'beads-on-a-string' DNA is 2.3 [(200 bp × 0.34 nm/bp)/(11 nm + {54 bp × 0.34 nm/bp}) = 2.3]. This first level of packing represents only 0.023% (2.3/10,000) of the total condensation that occurs at mitosis.

Answer 5–16

Histone octamers occupy about 9% of the volume of the nucleus. The volume of the nucleus is

$$V = 4/3 \times 3.14 \times (3 \times 10^3 \text{ nm})^3$$
$$V = 1.13 \times 10^{11} \text{ nm}^3$$

The volume of the histone octamers is

$$V = 3.14 \times (4.5 \text{ nm})^2 \times (5 \text{ nm}) \times (32 \times 10^6)$$
$$V = 1.02 \times 10^{10} \text{ nm}^3$$

The ratio of the volume of histone octamers to the nuclear volume is 0.09; thus, histone octamers occupy about 9% of the nuclear volume. Because the DNA also occupies about 9% of the nuclear volume, together they occupy about 18% of the volume of the nucleus.

Answer 5–17

In contrast to most proteins, which accumulate amino acid changes over evolutionary time, the functions of histone proteins must involve nearly all of their amino acids, so that a change in any position is deleterious to the cell. Histone proteins are exquisitely refined for their function.

Chapter 6

Answer 6–1

A. The distance between replication forks 4 and 5 is about 280 nm, corresponding to 824 nucleotides (= 280/0.34). These two replication forks would collide in about 8 seconds. Forks 7 and 8 move away from each other and would therefore never collide.

B. The total length of DNA shown in the electron micrograph is about 1.5 μm, corresponding to 4400 nucleotides. This is only about 0.002% [= (4400/1.8 × 10^8) × 100%] of the total DNA in a fly cell.

Answer 6–2

While the process may seem wasteful, it is not possible to proofread during primer formation. To start a new primer on a piece of single-stranded DNA, one nucleotide needs to be put in place and then linked to a second and then to a third, and so on. Even if these first nucleotides were perfectly matched to the template strand, such short oligonucleotides bind only with very low affinity, and it would consequently be difficult to distinguish the correct from incorrect bases by a hypothetical proofreading activity. The task of the primase is therefore to "just get anything down that binds reasonably well and don't worry about accuracy." Later these sequences are removed and replaced by DNA polymerase, which uses the accurate, correctly proofread newly synthesized DNA as its primer. The latter enzyme has the advantage—which primase does not have—of putting the new nucleotide onto an already existing strand. The newly added nucleotide is thus firmly held in place, and the accuracy of its base-pairing to the next nucleotide on the template strand can be checked. Therefore, as DNA polymerase fills the gap, it can proofread the new DNA strand that it makes.

Answer 6–3

A. Without DNA polymerase, no replication can take place at all. RNA primers will be laid down at the origin of replication.

B. DNA ligase links the DNA fragments that are produced on the lagging strand. In the absence of ligase, the newly replicated DNA strands will remain as fragments, but no nucleotides will be missing.

C. Without the sliding clamp, the DNA polymerase will frequently fall off the DNA template. In principle, it can rebind and continue, but the continual falling off and rebinding will be time-consuming and greatly slow down the rate of DNA replication.

D. In the absence of RNA excision enzymes, the RNA fragments will remain covalently attached to the newly replicated DNA fragments. No ligation will take place, because the ligase will not link DNA to RNA. The lagging strand will therefore consist of fragments composed of both RNA and DNA.

E. Without DNA helicase, the DNA polymerase will stall because it cannot separate the strands of the template DNA ahead of it. Little or no new DNA will be synthesized.

F. In the absence of primase, RNA primers cannot begin on either the leading or the lagging strand. DNA replication therefore cannot begin.

Answer 6–4

DNA defects introduced by deamination and depurination reactions occur spontaneously. They are not the result of replication and are therefore equally likely to occur on either strand. If DNA repair enzymes recognized such defects only on newly synthe-sized DNA strands, half of the defects would go uncorrected. The statement is therefore incorrect.

Answer 6–5

The AIDS virus (the human immunodeficiency virus, HIV) is a retrovirus, and thus synthesizes DNA from an RNA template using reverse transcriptase. This leads to frequent mutation of the viral genome. In fact, AIDS patients often harbor many different variants of HIV that are genetically distinct from the original virus that infected them. This poses great problems in treating the infection: drugs that block essential viral enzymes work only temporarily, because new virus strains that are resistant to these drugs arise rapidly by mutation.

RNA replicases (enzymes that synthesize RNA using RNA as a template) do not proofread either. Thus RNA viruses that replicate their RNA genomes directly (that is, without using DNA as an intermediate) also mutate frequently. In such a virus, this tends to produce changes in the coat proteins that cause the mutated virus to appear "new" to the immune system; the virus is therefore not suppressed by immunity that has arisen to the previous version. This is part of the explanation for the new strains of the influenza (flu) virus and the common cold virus that regularly appear.

Answer 6–6

If the old strand were "repaired" using the new strand that contains a replication error as the template, then the error would become a permanent mutation in the genome. The old information would be erased in the process. Therefore, if repair enzymes did not distinguish between the two strands, there would be only a 50% chance that any given replication error would be corrected.

Answer 6–7

The argument is severely flawed. You cannot transform one species into another simply by introducing 1% random changes into the DNA. It is exceedingly unlikely that the 5000 mutations that would accumulate every day in the absence of DNA repair would be in the very positions where human and ape DNA sequences are different. It is also very likely that at such a high mutation frequency many essential genes would be inactivated, leading to cell death. Furthermore, your body is made up of about 10^{13} cells. For you to turn into an ape, not just one but many of these cells would need to be changed. And even then, many of these changes would have to occur during development to effect changes in your body plan (making your arm longer than your legs, for example).

Answer 6–8

A. False. Identical DNA polymerase molecules catalyze DNA synthesis on the leading and lagging strands. The replication fork is asymmetrical because the lagging strand is synthesized in pieces that are then "stitched" together.

B. False. The RNA primers are removed by RNA nuclease; the Okazaki fragments are the pieces of newly synthesized DNA that are eventually joined to form the new lagging strand.

C. True. DNA polymerase has an error rate of one in 10^7, which is amazingly low, because of its proofreading activity; and 99% of its errors are corrected by DNA repair enzymes, bringing the final error rate to one in 10^9.

D. True. Mutations would accumulate rapidly, destroying the genes.

E. True. If an aberrant base occurred naturally in DNA, it might be recognized as a mismatch by DNA repair enzymes, but the enzymes could not tell on which strand the error was introduced. They would therefore have only a 50% chance of fixing the right strand.

F. True. Usually, multiple mutations of specific types need to accumulate before a cell turns into a cancer cell. A mutation in a gene that codes for a DNA repair enzyme can make a cell more liable to accumulate further mutations, thereby accelerating the onset of cancer.

Answer 6–9

DNA isolated from your starting cells grown under normal conditions has a light density, as you would expect. After one generation of growth in a medium containing heavy isotopes, the DNA has uniformly shifted to medium density: synthesis of the new DNA from heavy nucleotides (e.g., the *red* strand shown in Figure 6–4 that is made during the first round of replication) results in hybrid DNA molecules that contain one light, original strand and one heavy, newly synthesized strand. After another round of replication in a medium containing heavy isotopes, two forms of DNA appear in about equal proportions: one form is again a hybrid of a light and a heavy strand (the *orange/red* DNA in Figure 6–4) and has medium density, while the other form is composed of two heavy strands (the *red/green* DNA in Figure 6–4) and has a high density. During the subsequent rounds of replication, more heavy DNA is formed, and the proportion of medium-density DNA diminishes. Your results are therefore in complete agreement with the hypothesis that you set out to test.

Answer 6–10

With a single origin of replication that launches two DNA polymerases in opposite directions on the DNA each moving at 100 nucleotides per second, the number of nucleotides replicated in 24 hours will be 1.73×10^7 ($= 2 \times 100 \times 24 \times 60 \times 60$). To replicate all the 6×10^9 nucleotides of DNA in the cell in this time, therefore, will require at least 348 ($= 6 \times 10^9 / 1.73 \times 10^7$) origins of replication. The estimated 10,000 origins of replication in the human genome are therefore more than enough to satisfy this minimum requirement.

Answer 6–11

A. Compound A is dideoxycytosine triphosphate (ddCTP), identical to dCTP except that it lacks the 3′-hydroxyl group on the sugar ring. ddCTP is recognized by DNA polymerase as dCTP and becomes incorporated into DNA; because it lacks the crucial 3′-hydroxyl group, however, its addition to a growing DNA strand creates a dead end to which no further nucleotides can be added. Thus, if ddCTP is added in large excess, strands will be synthesized until the first G (the nucleotide complementary to C) is encountered on the template strand. ddCTP will then be incorporated instead of C, and the extension of this strand will be terminated.

B. If ddCTP is added at about 10% of the concentration of the available dCTP, there is a 1 in 10 chance of its being incorporated whenever a G is encountered on the template strand. Thus a population of DNA fragments will be synthesized, and from their lengths one can deduce where the G residues are located on the template strand. This experiment forms the basis of methods used to determine the sequence of nucleotides in a stretch of DNA (discussed in Chapter 10).

The same chemical phenomenon is exploited by a drug, 3′-azido-3′-deoxythymidine (AZT), that is now commonly used in HIV-infected patients to treat AIDS. AZT is converted in cells to the triphosphate form and is incorporated into the growing viral DNA. Because the drug lacks a 3′-OH group, it blocks DNA synthesis and replication of the virus. AZT inhibits viral replication preferentially because reverse transcriptase has a higher affinity for the drug than for thymidine triphosphate; human cellular DNA polymerases do not show this preference.

C. Compound B is dideoxycytosine monophosphate (ddCMP), which lacks the 5′-triphosphate group as well as the 3′-hydroxyl group of the sugar ring. It therefore cannot provide the energy that drives the polymerization reaction of nucleotides into DNA and therefore will not be incorporated into the replicating DNA. The compound is therefore expected not to affect DNA replication.

Answer 6–12

To use the energy of hydrolysis of the 3′-triphosphate group for polymerization, strand growth would need to occur in the opposite, that is, the 3′-to-5′, direction. Proofreading could then occur by a 5′-to-3′ nuclease activity. This scenario describing the hypothetical organism would be the same as that shown in Figure 6–15, except that all phosphate and triphosphate groups would be on the right sides of the DNA and nucleotide structures as drawn.

Answer 6–13

See Figure A6–13.

Answer 6–14

Both strands of the bacterial chromosome contain 6×10^6 nucleotides. During the polymerization of nucleotide triphosphates into DNA, two phosphoanhydride bonds are broken for each nucleotide added: the nucleotide triphosphate is hydrolyzed to produce the nucleotide monophosphate added to the growing DNA strand, and the released pyrophosphate is hydrolyzed to phosphate. Therefore, 1.2×10^7 high-energy bonds are hydrolyzed during each round of bacterial DNA

1. beginning of synthesis of Okazaki fragment

2. midpoint of synthesis of Okazaki fragment

Figure A6–13

replication. This requires 4×10^5 ($= 1.2 \times 10^7/30$) glucose molecules, which weigh 1.2×10^{-16} g ($= 4 \times 10^5$ molecules $\times$ 180 g/mole/6×10^{23} molecules/ mole), which is 0.01% of the total weight of the cell.

Answer 6–15

The statement is correct. If the DNA in somatic cells is not sufficiently stable (that is, if it accumulates mutations too rapidly), the organism dies (of cancer, for example), and because this may often happen before the organism can reproduce, the species will die out. If the DNA in reproductive cells is not sufficiently stable, many mutations will accumulate and be passed on to future generations, so that the species will not be maintained.

Answer 6–16

As shown in Figure A6–16, thymine and uracil contain no amino groups and therefore cannot be deaminated. Deamination of adenine and guanine produces purine rings that are not found in nucleic acids. In contrast, deamination of cytosine produces

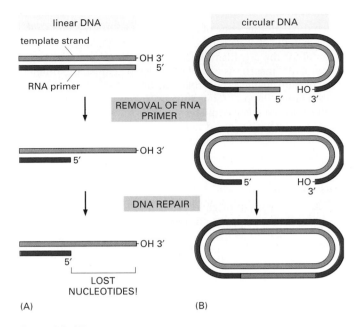

Figure A6–17

uracil. Therefore, if uracil were a naturally occurring base in DNA, repair enzymes could not distinguish whether a uracil is the appropriate base at a given position or an error that is the result of spontaneous deamination. This dilemma is not encountered, however, because thymine is used in DNA. Therefore, if a uracil base is found in DNA, it can be automatically recognized as an erroneous base and then excised and replaced by cytosine.

Answer 6–17

A. Without telomeres and telomerase, chromosome ends would shrink during each round of replication, because there would be no 3′-OH group to prime the DNA synthesis that fills in after removal of the primer of the last DNA fragment laid down on the lagging strand. Because bacterial chromosomes have no ends, the problem does not arise; there will always be a 3′-OH group available to prime the DNA polymerase that replaces the RNA primer with DNA (Figure A6–17). Telomeres and telomerase prevent the shrinking of chromosomes because the telomeres extend the 3′ end of a DNA strand with DNA repeats that are added directly by telomerase without the need for a template (see Figure 6–18).

B. As shown in Figure A6–17, telomeres and telomerase are still needed even if the last fragment of the lagging strand were initiated by primase at the very 3′ end of chromosomal DNA, inasmuch as an RNA primer must still be used.

Answer 6–18

Viruses cannot exist as free-living organisms: they have no metabolism, do not communicate with other viruses, and cannot reproduce themselves. They thus have none of the attributes that one normally associates with life. In fact, they can even be crystallized. Inside cells, they redirect normal cellular biosynthetic activities to the task of making more copies of themselves, but a virus cannot reproduce without exploiting a host cell. Thus the only aspect of "living" that viruses display is their capacity to direct their own reproduction once inside a cell.

Answer 6–19

Each time another copy of a transposon is inserted into a chromosome, the change can be either neutral, beneficial, or detrimental for the organism. Because individuals that accumulate detrimental insertions would be selected against, the proliferation of transposons is controlled by natural selection. If a transposon arose that proliferated uncontrollably, it is unlikely that a viable host organism could be maintained. For this reason, most transposons have evolved to transpose only rarely. Many transposons, for example, synthesize only infrequent bursts of very small amounts of the transposase that is required for their movement.

Answer 6–20

A. If the single origin of replication were located exactly in the center of the chromosome, it would take more than 8 days to replicate the DNA [= 75×10^6 nucleotides/(100 nucleo-tides/sec)]. The rate of replication would therefore severely limit the rate of cell division. (If the origin were located off-center, the time required for replication would be even longer.)

B. A chromosome end that is not "capped" with a telomere would lose nucleotides during each round of DNA replication and would gradually shrink. Eventually, essential genes would be lost, leading to cell death.

C. Without a centromere that attaches them to the mitotic spindle, the two new chromosomes that result from replication cannot be partitioned accurately between the two daughter cells. Therefore many daughter cells would die, because they failed to receive a full set of chromosomes.

Chapter 7

Answer 7–1

The answer is best given in a reflection of Francis Crick himself, who coined the term in the mid-1950s: "I called this idea the central dogma for two reasons, I suspect. I had already used the obvious word hypothesis in the sequence hypothesis, which proposes that genetic information is encoded in the sequence of the DNA bases, and in addition I wanted to suggest that this new assumption was more central and more powerful.... As it turned out, the use of the word dogma caused more trouble than it was worth. Many years later Jacques Monod pointed out to me that I did not appear to understand the correct use of the word dogma, which is a belief that cannot be doubted. I did appreciate this in a vague sort of way but since I thought that all religious beliefs were without serious foundation, I used the word in the way I myself thought about it, not as the world does, and simply applied it to a grand hypothesis that, however plausible, had little direct experimental support at the time." (Francis Crick, *What Mad Pursuit,* p. 109.)

Answer 7–2

Actually, the RNA polymerases are not moving at all, because they have been fixed and coated with metal to prepare the sample for viewing in the electron microscope. However, before they were fixed, they were moving from left to right, as indicated by the gradual lengthening of the RNA transcripts.

The RNA transcripts are shorter because they begin to fold up (i.e., to acquire a three-dimensional structure) as they are synthesized (see, for example, Figure 7–5), whereas the DNA is an extended double helix.

Answer 7–3

At first glance, the catalytic activities of an RNA polymerase used for transcription could replace the primase adequately. Upon further reflection, however, there are some serious problems. (1) The RNA polymerase used to make primer would need to initiate every few hundred bases, which is much more frequent than promoters are spaced on the DNA. Initiation would therefore need to occur in a promoter-independent fashion or many more promoters would have to be present in the DNA, both of which would be problematic for the control of transcription. (2) Similarly, the RNA primers used in replication are much shorter than mRNAs. The RNA polymerase would therefore need to terminate much more frequently than during transcription. Termination would need to occur spontaneously, i.e., without requiring a terminator sequence in the DNA, or many more terminators would need to be present. Again, both of these scenarios would be problematic for the control of transcription.

Although it might be possible to overcome this problem if special control proteins became attached to RNA polymerase during replication, the problem has been solved during evolution by using separate enzymes with specialized properties. Some small DNA viruses, however, do utilize the host RNA polymerase to make primers for their replication.

Answer 7–4

This experiment beautifully demonstrates that the ribosome does not check the amino acid that is attached to a tRNA. Once an amino acid has been coupled to a tRNA, the ribosome will "blindly" incorporate that amino acid into the position according to the match between the codon and anticodon. We can therefore conclude that a significant part of the correct reading of the genetic code, i.e., the matching of a codon with the correct amino acid, is performed by the synthetase enzymes that correctly match tRNAs and amino acids.

Answer 7–5

The mRNA will have a 5′-to-3′ polarity opposite to that of the DNA strand that serves as a template. Thus, the mRNA sequence will read 5′-GAAAAAAGCCGUUAA-3′. UAA specifies a stop codon. The C-terminal amino acid preceding the stop codon is therefore coded for by CGU and is an arginine residue. The N-terminal amino acid coded for by GAA is a glutamic acid residue. Note that the convention in describing the sequence of a gene is to give the sequence of the DNA strand that is *not* used as a template for RNA synthesis; this sequence is the same as that of the RNA transcript, with T written in place of U.

Answer 7–6

The first statement is factually correct: RNA is thought to have been the first self-replicating catalyst and in modern cells is no longer self-replicating. We can debate, however, whether this represents a "demotion." The role of RNA now is more than that of messenger alone: it serves as primers for DNA replication (discussed in Chapter 6) and catalyzes some of the most fundamental processes in cells.

Answer 7–7

A. False. All ribosomes are equivalent and can make any protein that is specified by the particular mRNA that they are translating. After translation, ribosomes are released from the mRNA and can then start translating a new mRNA.

B. False. mRNAs are translated as linear polymers; there is no requirement that they have any particular folded structure. In fact, such structures that are formed by mRNA can inhibit translation because the ribosome has to unfold the mRNA in order to read the message it contains.

- C. False. Ribosomal subunits exchange partners after each round of translation. After a ribosome is released from an mRNA, its two subunits dissociate and enter a pool of free small and large subunits from which new ribosomes are formed upon translation of a new mRNA.
- D. False. Ribosomes are cytoplasmic organelles, but they are not individually enclosed in a membrane.
- E. False. The position of the promoter determines the direction in which transcription proceeds and which DNA strand is used as the template. Transcription in the opposite direction would produce an mRNA with a completely different (and probably meaningless) sequence.
- F. False. RNA contains uracil but not thymine.
- G. False. The level of a protein depends on its rate of synthesis and degradation but not on its catalytic activity.

Answer 7–8

Because the deletion in the Lacheinmal mRNA is internal, it is likely that the deletion arises from a splicing defect. The simplest interpretation is that the *Lacheinmal* gene contains a 173-nucleotide-long exon (labeled "E2" in Figure A7–8), and that this exon is lost during the processing of the mutant precursor mRNA. This could occur, for example, if the mutation changed the 3′ splice site in the preceding intron ("I1") so that it was no longer recognized by the splicing machinery (a change in the CAG sequence shown in Figure 7–15 could do this). The snRNP would search for the next available 3′ splice site, which is found at the 3′ end of the next intron ("I2"), and the splicing reaction would therefore remove E2 together with I1 and I2, resulting in a shortened mRNA. The mRNA is then translated into a defective protein, resulting in the Lacheinmal deficiency.

Because 173 nucleotides do not amount to an integral number of codons, the lack of this exon in the mRNA will shift the reading frame at the splice junction. Therefore, the Lacheinmal protein would be made correctly only through exon E1. As the ribosome begins translating sequences in exon E3, it will be in a different reading frame and therefore will produce a protein sequence that is unrelated to the Lacheinmal sequence normally encoded by exon E3. Most likely, the ribosome will soon encounter a stop codon, which in RNA sequences that do not code for protein would be expected to occur on average about once in every 21 codons (there are 3 stop codons in the 64 codons of the genetic code).

Answer 7–9

Sequence 1 and sequence 4 both code for the peptide Arg-Gly-Asp. Because the genetic code is redundant, different nucleotide sequences can encode the same amino acid sequence.

Answer 7–10

- A. Incorrect. The bonds are not covalent, and their formation does not require input of energy.
- B. Correct. The aminoacyl-tRNA enters the ribosome at the A-site.
- C. Correct. The ribosome moves along the mRNA, and the tRNAs that have donated their amino acid to the growing polypeptide chain are released from the ribosome and the mRNA.

Answer 7–11

Replication. Dictionary definition: the creation of an exact copy; molecular biology definition: the act of duplicating DNA.

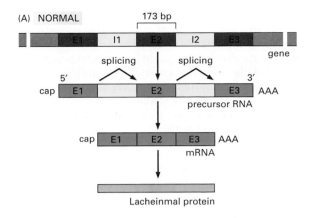

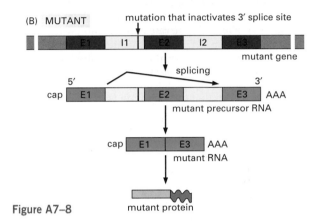

Figure A7–8

Transcription. Dictionary definition: the act of writing out a copy, especially from one physical form to another; molecular biology definition: the act of copying the information stored in DNA into RNA. *Translation.* Dictionary definition: the act of putting words into a different language; molecular biology definition: the act of polymerizing amino acids into a defined linear sequence from the information provided by the linear sequence of nucleotides in mRNA. (Note that "translation" is also used in a quite different sense, both in ordinary language and in scientific contexts, to mean a movement from one place to another.)

Answer 7–12

A code of two nucleotides could specify 16 different amino acids ($= 4^2$), and a triplet code in which the position of the nucleotides is not important could specify 20 different amino acids ($= 4$ possibilities of 3 of the same bases + 12 possibilities of 2 bases the same and one different + 4 possibilities of 3 different bases). In both cases, these maximal amino acid numbers would need to be reduced by at least 1, because of the need to specify translation stop codons. It is relatively easy to envision how a doublet code could be translated by a mechanism similar to that used in our world by providing tRNAs with only two relevant bases in the anticodon loop. It is more difficult to envision how the nucleotide composition of a stretch of three nucleotides could be translated without regard to their order, because base-pairing can then no longer be used: an AUG, for example, will not base-pair with the same anticodon as a UGA.

Answer 7–13

In present-day cells, there is some wobble in the matching of codons to anticodons: in a number of cases, the same tRNA can pair with several codons that differ slightly in their nucleotide sequence. It seems likely that in the early world, without such highly evolved ribosomes as we have now to help in the pairing process, the converse may also have been true: several different tRNAs, with slightly different anticodons, may have been able to bind to the same codon. This would have played havoc with the translation of the genetic message into protein, unless the amino acids carried by all of the tRNAs capable of binding to the same codon were chemically similar. Natural selection would thus have ensured that tRNAs with similar anticodons carried chemically similar amino acids.

Perhaps in the early world, before modern amino-acyl-tRNA synthetases had evolved, there was also some "wobble" in the matching of tRNAs with appropriate amino acids: the same tRNA might have been liable to become coupled to any of a number of amino acids that were chemically similar. One can imagine the evolution of the genetic code by refinement of a matching process that was originally imprecise and gave only a blurred relationship between sets of roughly similar codons and sets of roughly similar amino acids.

Answer 7–14

The codon for Trp is 5′-UGG-3′. Thus, a normal Trp-tRNA contains the sequence 5′-CCA-3′ in its anticodon loop. If this tRNA contains a mutation so that its anticodon is changed to U̲CA, it will recognize a UGA̲ codon and lead to the incorporation of a tryptophan residue instead of causing translation to stop. Many other protein-encoding sequences, however, contain UGA codons as their natural stop sites, and these stops would also be affected by the mutant tRNA. All of these proteins would therefore be made with additional amino acids at the C-terminal end. The additional lengths would depend on the number of codons before the ribosomes encounter a non-UGA stop codon in the mRNA in the reading frame in which the protein is translated.

Answer 7–15

One effective way of driving a reaction to completion is to remove one of the products, so that the reverse reaction cannot occur. ATP contains *two* high-energy bonds that link the three phosphate groups. In the reaction shown, PP_i is released, consisting of two phosphate groups linked by one of these high-energy bonds. Thus, PP_i can be hydrolyzed with a considerable gain of free energy, and thereby be efficiently removed. This happens rapidly in cells, and reactions that produce and further hydrolyze PP_i are therefore virtually irreversible (discussed in Chapter 3).

Answer 7–16

A. A titin molecule is made of 25,000 amino acids. It therefore takes about 3.5 hours to synthesize a single molecule of titin in muscle cells.

B. Because of its large size, the probability of making a titin molecule without any mistakes is only 0.08 [= $(1 - 10^{-4})^{25,000}$]; i.e., only 8 in 100 titin molecules synthesized are free of mistakes. In contrast, over 97% of newly synthesized proteins of average size are made correctly.

C. The error rate limits the sizes of proteins that can be synthesized accurately. Similarly, if a eucaryotic ribosomal protein were synthesized as a single molecule, a large portion (87%) of this hypothetical giant ribosomal protein would be expected to contain at least one mistake. It is more advantageous to make ribosomal proteins individually, because in this way only a small proportion of each type of protein will be defective, and these few bad molecules can be individually eliminated to ensure that there are no defects in the ribosome as a whole.

D. To calculate the time it takes to transcribe a titin mRNA, you would need to know the size of its gene, which is likely to contain many introns. Transcription of the exons alone requires about 42 minutes. Because introns can be quite large, the time required to transcribe the entire gene is likely to be considerably longer.

Answer 7–17

Mutations of the type described in (B) and (D) are often the most harmful. In both cases, the reading frame would be changed, and because this frameshift occurs near the beginning or in the middle of the coding sequence, much of the protein will contain a nonsensical and/or truncated sequence of amino acids. In contrast, a reading-frame shift that occurs toward the end of the coding sequence, as described in scenario (A), will result in a largely correct protein that may be functional. Deletion of three consecutive nucleotides, scenario (C), leads to the deletion of an amino acid but does not alter the reading frame. The deleted amino acid may or may not be important for the folding or activity of the protein; in many cases such mutations are silent, i.e., have no or only minor consequences for the organism. Substitution of one nucleotide for another, as in (E), is often completely harmless, not even causing a change in the amino acid sequence; in other cases it may change one amino acid in the protein sequence; at worst, it may create a new stop codon, giving rise to a truncated protein.

Answer 7–18

As shown in Figure 7–17, your illustration should include the following steps: (1) snRNP binding to the precursor RNA, (2) lariat formation and 5′ splice site cleavage, and (3) 3′ splice site cleavage, joining of the two exon sequences, and release of the spliced mRNA. For the reaction it is important that the snRNP complex hold on to the 5′ exon after the first cleavage reaction, so that it can be joined to the 3′ exon in the next step. If the reactions were performed by separate enzymes, exons from one primary transcript could become mixed and join with exons from another.

Chapter 8

Answer 8–1

Contacts can form between the protein and the edges of the base pairs that are exposed in the major groove of the DNA. The contacts that can form are shown in Figure A8–1. The bonds responsible for sequence-specific contacts are hydrogen bonds and a hydrophobic interaction that can occur with the methyl group on the pyrimidine ring of T. Note that the arrangement of hydrogen-bond donors and hydrogen-bond acceptors of a T-A pair is different from that of a C-G pair. Similarly, the arrangement of hydrogen-bond donors and hydrogen-bond acceptors of A-T and G-C pairs would be different from one another and from the two pairs shown in the figure. In addition to the contacts shown in the figure, ionic interactions between the positively charged amino acid side chains of the protein and the negatively charged phosphate groups in the DNA backbone usually stabilize DNA–protein interactions.

Answer 8–2

A. UV light throws the switch from the prophage to the lytic state: when cI protein is destroyed, cro is made and turns off the production of new cI. The virus starts to produce coat proteins, and new virus particles are made.

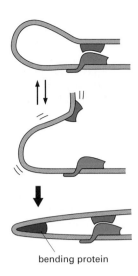

Figure A8–3

B. When the UV light is switched off, the virus remains in the lytic state. Thus, cI and cro form a gene regulatory switch that "memorizes" its previous setting.

C. This switch makes sense in the viral life cycle: UV light tends to damage the bacterial DNA (see Figure 6–24), thereby rendering the bacterium an unreliable host for the virus. A prophage virus will therefore switch to the lytic state and leave an irradiated cell in search for new host cells to infect.

Answer 8–3

Bending proteins can help to bring distant DNA regions together that normally would contact each other only inefficiently (Figure A8–3). Such proteins are found in both procaryotes and eucaryotes and are involved in many examples of transcriptional regulation.

Answer 8–4

False. Carrots can be grown from single carrot cells and tadpoles can be gotten by injecting differentiated frog nuclei into frog eggs. But carrots cannot be gotten from frog eggs no matter what.

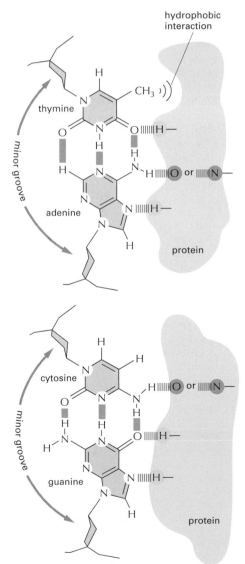

Figure A8–1

Answer 8–5

A. False. Bacteria have only one RNA polymerase that transcribes all genes; in contrast, eucaryotic cells have three different polymerases, each dedicated to one of the three classes of genes.

B. True. Procaryotic mRNAs are often transcripts of entire operons. Ribosomes can initiate translation at internal AUG start sites of these "polycistronic" mRNAs (see Figure 7–33).

C. True. The major groove of double-stranded DNA is sufficiently wide to allow a protein surface, such as one face of an α helix, access to the base pairs.

D. True. It is advantageous to exert control at the earliest possible point in a pathway. This conserves metabolic energy because unnecessary products are not made in the first place.

E. False. The zinc atoms in zinc finger domains are required for the correct folding of the protein domain; they are internal to these domains and do not contact the DNA.

Answer 8–6

From our knowledge of enhancers, one would expect their function to be relatively independent of their distance from the RNA-polymerase–binding site—possibly weakening as this distance increases. The surprising feature of the data (which have been adapted from an actual experiment) is the periodicity: the enhancer is maximally active at certain distances from the RNA-binding site (50, 60, or 70 nucleotides), but almost inactive at intermediate distances (55 or 65 nucleotides). The periodicity of 10 suggests that the mystery can be explained by considering the structure of double-helical DNA, which has very close to 10 base pairs per turn. Thus placing an enhancer on the side of the DNA opposite to that of the promoter (Figure A8–6) would make it more difficult for the activator that binds to it to interact with the proteins bound at the promoter. At longer distances, there is more DNA to absorb the twist, and the effect diminishes.

Answer 8–7

Two advantages of dimeric DNA-binding proteins are (1) that the binding affinity can be very high because the number of potential contacts with DNA is double that possible with a monomer and (2) that several different subunits can be combined in many different combinations to increase the number of

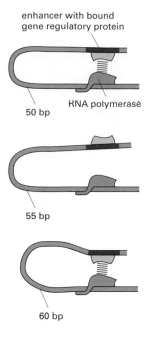

enhancer with bound
gene regulatory protein

RNA polymerase

50 bp

55 bp

60 bp

Figure A8–6

DNA-binding specificities available to cells. Three of the most frequently used protein domains involved in DNA binding are leucine zippers, homeodomains, and zinc fingers. Each provides a particularly stable fold in the polypeptide chain that positions an α helix appropriately on the protein's surface so that it can insert into the major groove of the DNA helix and contact the sides of the base pairs (see Figure 8–5).

Answer 8–8

A. If sufficient tryptophan is present in the cells, Trp repressor will block the synthesis of enzymes that would make more tryptophan. Likewise, if cells are starved for tryptophan, the unoccupied repressor would not bind to the DNA, and the enzymes that synthesize tryptophan would be induced. This simple and elegant form of feedback inhibition (see Chapter 4) allows cells to adjust the rate of tryptophan synthesis to their needs.

B. Transcription of the genes encoding the tryptophan biosynthetic enzymes would no longer be regulated by the absence or presence of tryptophan; the enzymes would be permanently on in scenario 1 and permanently off in scenario 2.

C. In scenario 1, the normal tryptophan repressor molecules would completely restore the regulation of the tryptophan biosynthesis enzymes. In contrast, expression of the normal protein would have no effect in scenario 2, because the Trp-repressor–binding sites on the DNA would remain permanently occupied by the mutant protein.

Answer 8–9

The affinity of the dimeric λ repressor for its binding site is the sum of all the interactions made by each DNA-binding domain. An individual DNA-binding domain will make just half the contacts and provide just half the binding energy as the dimer. Thus, although the concentration of binding domains is unchanged, their binding as monomers is sufficiently weak that they do not compete with the binding of RNA polymerase. As a result, the genes for lytic growth are turned on.

Answer 8–10

The function of these *arg* genes is to synthesize arginine. When arginine is abundant, expression of the biosynthetic genes should be turned off. If ArgR acts as a gene repressor (which it does in reality), then binding of arginine should increase its affinity for its regulatory sites, allowing it to bind and shut off gene expression. If ArgR acted as a gene activator instead, then the binding of arginine would be expected to reduce its affinity for its regulatory sites, preventing its binding and shutting off gene expression.

Answer 8–11

The results of this experiment favor the DNA looping model, which would not be affected by the protein bridge (so long as it allowed the DNA to bend, which it does). The scanning or entry site model, however, is likely to be affected by the nature of the linkage between the enhancer and the promoter. If the proteins enter at the enhancer and scan to the promoter, they would have to traverse the protein bridge. If such proteins are geared to scan on DNA, they are likely to have difficulty scanning across a protein.

Answer 8–12

The experiment is one that shows that a single differentiated cell taken from a specialized tissue can re-create a whole organism. This proves that the cell must contain all the information required to produce a whole organism, including all of its specialized cell types. See Figure 8–2.

Answer 8–13

You could create 16 different cell types with 4 different gene regulatory proteins (all the 8 cell types shown in Figure 8–22, plus another 8 created by adding an additional gene regulatory protein). MyoD by itself is sufficient only to induce muscle-specific gene expression in certain cell types, such as some kinds of fibroblasts. The action of MyoD is therefore consistent with the model shown in Figure 8–22: if muscle cells were specified, for example, by the combination of gene regulatory proteins 1, 3, and MyoD, then the addition of MyoD would convert only two of the cell types of Figure 8–22 (cells F and H) to muscle.

Answer 8–14

The induction of a gene activator that stimulates its own synthesis can create a positive feedback loop that can lead to cell memory. The continued self-stimulated synthesis of activator A can in principle last for many cell generations, serving as a constant reminder of an event in the distant past. By contrast, the induction of a gene repressor that inhibits its own synthesis creates a negative feedback loop that guarantees a transient response to the transient stimulus. Because repressor R shuts off its own synthesis, the cell will quickly return to the state that existed before the transient signal.

Answer 8–15

Many gene regulatory proteins are always being made in the cell; that is, their expression is constitutive and the activity of the protein is controlled by signals from inside or outside the cell (e.g., the availability of nutrients, as for the Trp repressor, or by hormones, as for the glucocorticoid receptor), thereby adjusting the transcriptional program to the physiological needs of the cell. Moreover, a given gene regulatory protein usually regulates the expression of many different genes. Gene regulatory proteins are often used in various combinations and can affect each other's activity, thereby further increasing the possible regulatory repertoire of gene expression with a limited set of proteins. Nevertheless, the cell devotes a large fraction of its genome to the control of transcription: an estimated 10% of all genes in eukaryotic cells code for gene regulatory proteins.

Chapter 9

Answer 9–1

The answer lies in the need for the cell to maintain a balance between stability and change. If the mutation rate were too high, a species would eventually die out because all its individuals would accumulate too many mutations in genes essential for survival. For a species to be successful—in evolutionary terms—it is important for individual members to have good genetic memory, that is, fidelity in DNA replication, but also to introduce occasional variations. If the change leads to an improvement, it will persist by selection; if it proves disastrous, the individual organism that was the unfortunate subject of nature's experiment will die—not the entire population.

Answer 9–2

In single-celled organisms the genome is the germline and any modification is passed on to the next generation. By contrast, in multicellular organisms most of the cells are somatic cells and make no contribution to the next generation; thus, modification of those cells by horizontal gene transfer would have no consequence for the next generation. The germline cells are usually sequestered into the interior of multicellular organisms, minimizing their contact with foreign cells, viruses, and DNA, thus, insulating the species from the effects of horizontal gene transfer.

Answer 9–3

It is unlikely that any gene came into existence perfectly optimized for its function. It is thought that highly conserved genes such as ribosomal RNA genes were optimized by more rapid evolutionary change during the evolution of the common ancestor to the archaea, eubacteria, and eucaryotes. Because ribosomal RNAs (and the products of most highly conserved genes) participate in fundamental processes that were optimized early, there has been no evolutionary pressure (and little leeway) for change. By contrast, less conserved—more rapidly evolving—genes have been continually presented with opportunities to fill new functional niches. Consider, for example, the evolution of distinct globin genes that are optimized for oxygen delivery to embryos, fetuses, and adult tissues in placental mammals.

Answer 9–4

Transposable elements could provide opportunities for homologous recombination events, thereby causing genomic rearrangements. They could insert into genes, possibly obliterating splice sites, thereby changing the gene structure. They could also insert into the regulatory region of a gene, where insertion between an enhancer and a transcription start site could block the function of the enhancer and therefore reduce the level of expression of a gene. In addition, the transposable element could itself contain an enhancer and thereby change the time and place in the organism where the gene is expressed.

Answer 9–5

It is not a simple matter to determine the function of a gene from scratch, nor is there a universal recipe for how to do it. Nevertheless, there are a variety of standard questions that help narrow down the possibilities. Below we list some of these questions.

In what tissues is the gene expressed? If the gene is expressed in all tissues, it is likely to have a general function. If it is expressed in one or a few tissues, its function is likely to be more specialized, perhaps related to the specialized functions of the tissues. If the gene is expressed in the embryo, but not the adult, it may function in development.

In what compartment of the cell is the gene expressed? Knowing the subcellular localization of the protein—nucleus, plasma membrane, mitochondria, etc.—can also help to suggest categories of potential function. For example, a protein that is localized to the plasma membrane is likely to be a transporter, a receptor or other component of a signaling pathway, a cell-adhesion molecule, etc.

What are the effects of mutations in the gene? Mutations that eliminate or modify the function of the gene product can also provide clues to function. For example, if the gene product is critical at a certain time during development, the embryo will often die at that stage or develop obvious abnormalities. Unless the abnormality is very specific, it is usually difficult to deduce the function or category of function. And often the links are very indirect, becoming apparent only after the gene's function is known.

With what other proteins does the encoded protein interact? In carrying out their function, proteins often interact with other proteins involved in the same or closely related processes. If an interacting protein can be identified, and if its function is already known (through previous research or the searching of databases), the range of possible functions can be narrowed dramatically.

Mutations in what other genes can suppress effects of mutation in the unknown gene? Looking for suppressor genes can be a very powerful approach to investigating gene function in organisms such as bacteria and yeast, which have well-developed genetic systems, but this approach is not readily applicable to mouse or most higher eucaryotes at present. The rationale for this approach is analogous to that of looking for interacting proteins: genes that interact genetically are often involved in the same or a closely related process. Identification of such an interacting gene (and knowledge of its function) would provide an important clue to the function of the unknown gene.

Addressing each of these questions requires specialized experimental expertise and a substantial time commitment from the investigator. It is no wonder that progress is made so much more rapidly when a clue to a gene's function can be found simply by identifying a similar gene of known function in the database.

Answer 9–6

With their ability to facilitate genetic recombination, transposable elements have almost certainly played an important part in the evolution of modern-day organisms. They can speed up evolution by promoting gene duplication or the creation of new genes via exon shuffling, and they can change the way that existing genes are expressed. Although the movement of a transposable element can be harmful for an individual organism—if, for example, it disrupts the activity of a critical gene—these mobile genetic elements are probably beneficial to the species as a whole.

Answer 9–7

About 7.6% of each gene is converted to mRNA [(5.4 exons/gene × 266 nucleotide pairs/exon)/(19,000 nucleotide pairs/gene) = 7.6%]. Genes occupy about 28% of chromosome 22 [(700 genes × 19,000 nucleotide pairs/gene)/(48 × 10^6 nucleotide pairs) = 27.7%].

Answer 9–8

True. Overall only a couple of percent of the human genome is present in mRNA and only about one third of the genome is transcribed into RNA. Even allowing for regulatory regions and other critical sequences, it still appears that more than half the genome has no function: it is unimportant junk.

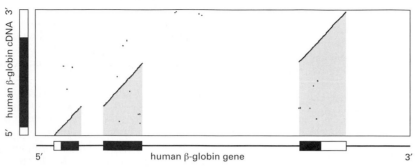

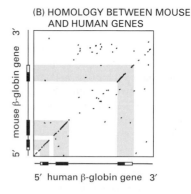

(A) EXONS IN HUMAN β-GLOBIN cDNA

(B) HOMOLOGY BETWEEN MOUSE AND HUMAN GENES

Figure A9–10

Answer 9–9

The Hox gene clusters are packed with complex and extensive regulatory sequences that ensure proper expression of individual Hox genes at the correct time and place during development. Insertion of transposable elements into the Hox clusters is thought to be selected against because it would disrupt proper regulation of the Hox genes. Comparison of the Hox cluster sequences in mouse, rat, and baboon reveals a high density of conserved noncoding segments, supporting the idea of a high density of regulatory elements.

Answer 9–10

A. The exons in the human β-globin gene correspond to the positions of homology with the cDNA, which is a direct copy of the mRNA and thus contains no introns. The introns correspond to the regions between the exons. The positions of the introns and exons in the human β-globin gene are indicated in Figure A9–10A.

B. From the positions of the exons, as defined in Figure A9–10A, it is clear that the first two exons of the human β-globin gene have homologous counterparts in the mouse β-globin gene (Figure A9–10B). However, only the first half of the third exon of the human β-globin gene is homologous to the mouse β-globin gene. The homologous portion of the third exon contains sequences that encode protein, whereas the nonhomologous portion represents the 3′ untranslated region of the gene. Because this portion of the gene does not encode protein (nor does it contain extensive regulatory sequences), it evolves at a rate similar to introns.

C. The human and mouse β-globin genes are also homologous at their 5′ ends, as indicated by the cluster of points along the same diagonal as the first exon (Figure A9–10B). These sequences correspond to the regulatory regions in front of the start sites for transcription. The regulatory function of this region has limited its evolutionary divergence, much as the coding function of exons has limited their divergence. Functional sequences, which are under selective pressure, diverge much more slowly than sequences without function.

D. The diagon plot shows that the first intron is nearly the same length in the human and mouse genes, but the length of the second intron is noticeably different (Figure A9–10B). If the introns were the same length, the line segments that represent homology would fall on the same diagonal. The easiest way to test for the colinearity of the line segments is to tilt the page and sight along the diagonal. It is impossible to tell from this comparison if the change in length is due to a shortening of the mouse intron or to a lengthening of the human intron, or some combination of those possibilities.

Answer 9–11

Computer algorithms that search for exons are complex affairs, as you might imagine. They combine statistical information derived from known genes in searching for unidentified genes. The list of features includes:

1. An exon that encodes protein will have an open reading frame, and the reading frames in adjacent exons will match up.

2   Internal exons (excluding the first and the last) will have splicing signals at each end; most of the time (98.1%) these will be AG at the 5′ ends of the exons and GT at the 3′ ends.

3. The multiple codons for most individual amino acids are not used with equal frequency. This so-called coding bias can be factored in to aid in the recognition of true exons.

4. Exons and introns have characteristic length distributions. The median length of exons in human genes is about 120 nucleotide pairs. Introns tend to be much larger: a median length of about 2 kb in genomic regions of 30–40% GC content, and a median length of about 500 nucleotide pairs in regions above 50% GC.

5. The initiation codon for protein synthesis (nearly always ATG) has a statistical association with adjacent nucleotides that seem to enhance its recognition by translation factors.

6. The terminal exon will have a signal (most commonly AATAAA) for cleavage and polyadenylation close to its 3′ end.

The statistical nature of these features coupled with the low frequency of coding information in the genome (2–3%) and the frequency of alternative splicing (an estimated 60% of human genes) makes it especially impressive that current algorithms can identify about 70% of individual exons and about 20% of complete genes in the human genome.

Answer 9–12

In a random sequence of DNA, each of the 64 different codons will be generated with equal frequency. Because 3 of the 64 are stop codons, they will be expected to occur every 21 codons (64/3 = 21.3) on average.

Answer 9–13

A gene is any DNA sequence that is transcribed as a single unit and produces a functional RNA or encodes one or a set of closely related polypeptide chains (protein isoforms). Note that the explicit requirement for transcription makes the control region a part of the gene as well.

Answer 9–14

On the surface, the extraordinary mutation resistance of the genetic code argues that it was subjected to the forces of natural selection. An underlying assumption, which seems reasonable, is that resistance to mutation is a valuable feature of a genetic code, one that would allow organisms to maintain sufficient information to specify complex phenotypes. This reasoning suggests that it would have been a lucky accident indeed—roughly a one-in-a-million chance—to stumble on a code as error proof as our own.

But all is not so simple. If resistance to mutation is an essential feature of any code that can support the complexity of organisms such as humans, then the only codes we *could* observe are ones that are error resistant. A less favorable frozen accident, giving rise to a more error-prone code, might limit the complexity of life to organisms that would never be able to contemplate their genetic code. This is akin to the anthropic principle of cosmology: many universes may be possible, but few are compatible with life that can ponder the nature of the universe.

Beyond these considerations, there is ample evidence that the code is not static, and thus could respond to the forces of natural selection. Deviant versions of the standard genetic code have been identified in the mitochondrial and nuclear genomes of several organisms. In each case one or a few codons have taken on a new meaning.

Answer 9–15

B. It is not thought that formation of genes *de novo* from the vast amount of unused, noncoding DNA typical of eucaryotic genomes is a significant process in evolution. Mutation to generate a coding sequence complete with regulatory elements is too slow a process to account for the observed rates of evolutionary change.

Answer 9–16

A. Because synonymous changes do not alter the amino acid sequence of the protein, they are not subject to selection pressures, which operate at the level of the function of the protein (and how it affects the overall fitness of the organism). By contrast, nonsynonymous changes, which substitute a new amino acid in place of the original one, have the potential to alter the function of the encoded protein (and change the fitness of the organism). Since most amino acid substitutions are deleterious to the function of the protein, they are selected against.

B. The histone H3 gene must be so exquisitely tuned to its function that virtually all amino acid substitutions are deleterious and, therefore, are selected against. The extreme conservation of histone H3 argues that its function is very tightly constrained, probably because of extensive interactions with other proteins and with its unchanging substrate, DNA.

C. Histone H3 is clearly not in a "privileged" site in the genome because it undergoes synonymous nucleotide changes at about the same rate as other genes.

Answer 9–17

A. The data in the phylogenetic tree (see Figure Q9–17) refutes the hypothesis that plant hemoglobin genes arose by horizontal transfer. Looking at the more familiar parts of the tree, we see that the vertebrates (fish to human) cluster together as a closely related set of species. Moreover, the relationships in the unrooted tree shown in Figure Q9–17 are compatible with the order of branching we know from the evolutionary relationships among these species: fish split off before amphibians, reptiles before birds, and mammals last of all in a tightly knit group. Plants also form a distinct group that displays accepted evolutionary relationships, with barley, a monocot, diverging before bean, alfalfa, and lotus, which are all dicots (and legumes). The sequences of the plant hemoglobins appear to have diverged long ago in evolution, at or before the time that mollusks, insects, and nematodes arose. The relationships in the tree indicate that the hemoglobin genes arose by descent from some common ancestor.

B. Had the plant hemoglobin genes arisen by horizontal transfer from a parasitic nematode, then the plant sequences would have clustered with the nematode sequences in the phylogenetic tree in Figure Q9–17.

Answer 9–18

In each human lineage, new mutations will be introduced at a rate of 10^{-10} alterations per nucleotide per cell generation, and the difference between two human lineages will increase at twice this rate. To accumulate 10^{-3} differences per nucleotide will thus take $10^{-3}/(2 \times 10^{-10})$ cell generations, corresponding to $(1/200) \times 10^{-3}/(2 \times 10^{-10}) = 25,000$ human generations, or 750,000 years. In reality, we are certainly not all descended from one pair of genetically identical ancestral humans; rather, it is likely that we are all descended from a relatively small founder population of humans who were already genetically diverse. More sophisticated analysis suggests that this founder population existed about 150,000 years ago, although there is also evidence for more recent "population bottlenecks" corresponding to founder subpopulations at the origin of specific subgroups of the total present-day human population.

Chapter 10

Answer 10–1

A. Digestion with Eco RI produces two products:

```
5'-AAGAATTGCGG   AATTCGAGCTTAAGGGCCGCGCCGAAGCTTTAAA-3'
3'-TTCTTAACGCCTTAA   GCTCGAATTCCCGGCGCGGCTTCGAAATTT-5'
```

B. Digestion with Alu I produces three products:

```
5'-AAGAATTGCGGAATTCGAG   CTTAAGGGCCGCGCCGAAG   CTTTAAA-3'
3'-TTCTTAACGCCTTAAGCTC   GAATTCCCGGCGCGGCTTC   GAAATTT-5'
```

C. The sequence contains no Not I cleavage site.

D. Digestion with all three enzymes therefore produces:

```
5'-AAGAATTGCGG   AATTCGAG   CTTAAGGGCCGCGCCGAAG   CTTTAAA-3'
3'-TTCTTAACGCCTTAA   GCTC   GAATTCCCGGCGCGGCTTC   GAAATTT-5'
```

Answer 10–2

If the ratio of dideoxyribonucleoside triphosphates to deoxyribonucleoside triphosphates is increased, DNA polymerization is terminated more frequently and thus shorter DNA strands are produced. Such conditions are favorable for determining short nucleotide sequences, that is, the sequences that are close to the DNA primer used in the reaction. In contrast, decreasing the ratio of dideoxyribonucleoside triphosphates to deoxyribonucleoside triphosphates allows one to determine nucleotide sequences more distant from the primer.

Answer 10–3

The presence of a mutation in a gene does not necessarily mean that the protein expressed from it is defective. For example, the mutation could change one codon into another that still specifies the same amino acid, and so does not change the amino acid sequence of the protein. Or, the mutation may cause a change from one amino acid to another in the protein, but in a position that is not important for the folding or function of the protein. In assessing the likelihood that such a mutation might cause a defective protein, information on the known β-globin mutations that are found in humans is essential. You would therefore want to know the precise nucleotide change in your mutant gene, and whether this change has any known or predictable consequences for the function of the encoded protein. If your mate has two normal copies of the globin gene, none of your children would manifest disease arising from defective hemoglobin (thalassemia); however, on average, 50% of your children would be carriers of your one defective gene.

Answer 10–4

Although several explanations are possible, the simplest is that the DNA probe has hybridized predominantly with its corresponding mRNA, which when expressed is typically present in many more copies than is the gene. The strongly hybridizing cells probably express the gene at high levels and therefore have high levels of the mRNA.

Answer 10–5

A. After an additional round of amplification there will be 2 gray, 4 green, 4 red, and 22 yellow-outlined fragments; after a second additional round there will be 2 gray, 5 green, 5 red, and 52 yellow-outlined fragments. Thus the DNA fragments outlined in yellow increase exponentially and will eventually overrun the other reaction products. Their length is precisely determined by the DNA sequence that spans the distance between the two primers used in the amplification.

B. The mass of one DNA molecule 500 nucleotide pairs long is 5.5×10^{-19} g $[= 2 \times 500 \times 330$ (g/mole)/6×10^{23} (molecules/mole)]. Ignoring the complexities of the first few steps of the amplification reaction (which produce longer products that eventually make an insignificant contribution to the total DNA amplified), this amount of product approximately doubles for every amplification step. Therefore, 100×10^{-9} g $= 2^N \times 5.5 \times 10^{-19}$ g, where N is the number of amplification steps of the reaction. Solving this equation for $N = \log(1.81 \times 10^{11})/\log(2)$ gives $N = 37.4$. Thus, only about 40 cycles of PCR amplification are sufficient to amplify DNA from a single molecule to a quantity that can be readily handled and analyzed biochemically. This whole procedure takes only about 4 hours in the laboratory.

Answer 10–6

Like the vast majority of mammalian genes, it is likely that the attractase gene contains introns. Bacteria do not have the splicing machinery required to remove introns, and therefore the correct protein cannot be expressed from the gene. For expression of most mammalian genes in bacterial cells, a cDNA version of the gene must be used.

Answer 10–7

It is the general goal of cell biology research to discover how individual cells work, but it is very difficult to study most processes in single cells. If one has a population of identical cells, however, then its analysis can yield valid conclusions about the workings of the individual cells. By contrast, if the population is a mixture of different cell types, its analysis will give properties of the mixture, which may or may not accurately describe the individual cells. Consider an analogy. We know from looking at individual human eyes that they are various shades of brown or blue or green. Yet if we could tell eye color only by looking at 1000 at a time, and if we started with a random population, we might conclude that eyes were a bluish brown—a color that doesn't apply to any single individual.

Answer 10–8

A. False. Restriction sites are found throughout the DNA, that is, within, as well as between, genes.

B. True. DNA bears a negative charge at each phosphate, giving DNA an overall negative charge.

C. False. Clones isolated from cDNA libraries never contain promoter sequences. These sequences are not transcribed and are therefore not part of the mRNAs that are used as the templates to make cDNAs.

D. True. Each polymerization reaction produces double-stranded DNA that must, at each cycle, be denatured to allow new primers to hybridize so that the DNA strand can be copied again.

E. False. Digestion of genomic DNA with restriction nucleases that recognize four-nucleotide sequences produces fragments that are *on average* 256 nucleotides long. However, the actual lengths of the fragments produced will vary considerably on both sides of the average.

F. True. Reverse transcriptase is first needed to copy the mRNA into single-stranded DNA, and DNA polymerase is then required to make the second DNA strand.

G. True. Using a sufficient number of VNTRs, individuals can be uniquely "fingerprinted" (see Figure 10–30).

H. True. If cells of the tissue do not transcribe the gene of interest, it will not be represented in a cDNA library prepared from this tissue. However, it will be represented in a genomic library prepared from the same tissue.

Answer 10–9

Human embryonic stem cell lines, which are derived from the inner cell mass of the early embryo, can proliferate indefinitely and retain the ability to give rise to any part of the body. If their differentiation can be guided appropriately in culture—an area of very intense research—they could provide a source of cells capable of replacing or repairing tissues that have been damaged by disease or injury.

Answer 10–10

A. The DNA sequence, from its 5′ end to its 3′ end, is read starting from the bottom of the gel, where the smallest DNA fragments migrate. Each band results from the incorporation of the appropriate dideoxyribonucleoside triphosphate, and as expected, there are no two bands that have the same mobility. This allows one to determine the DNA sequence by reading off the bands in strict order, proceeding upward from the bottom of the gel, and assigning the correct nucleotide according to which lane the band is in.

The nucleotide sequence of the top strand (Figure A10–10A) was obtained from Figure Q10–10, and the bottom strand was deduced from the complementary base-pairing rules.

B. The DNA sequence can then be translated into an amino acid sequence using the genetic code. However, there are two strands of DNA that could be transcribed into RNA and three possible reading frames for each strand. Thus there are six amino acid sequences that can in principle be encoded by this stretch of DNA. Of the three reading frames possible from the top strand, only one is not interrupted by a stop codon (Figure A10–10B).

From the bottom strand, two of the three reading frames also have stop codons (not shown). The third frame gives the following sequence:

SerAlaLeuGlySerSerGluAsnArgProArgThrProAlaArgThrGlyCysProValIle

(A) 5'-TATAAACTGGACAACCAGTTCGAGCTGGTGTTCGTGGTCGGTTTTCAGAAGATCCTAACGCTGACG-3'
3'-ATATTTGACCTGTTGGTCAAGCTCGACCACAAGCACCAGCCAAAAGTCTTCTAGGATTGCGACTGC-5'

(B) 5' top strand of DNA 3'

TATAAACTGGACAACCAGTTCGAGCTGGTG TTCGTGGTCGGTTTTCAGAAGATCC TAACGC TGACG
1 LeuLysLeuGluAsnGlnPheGlnLeuVal PheValValGlyPheGlnLysIleLeuThr LeuThr
2 IleAsnTrpThrThrSerSerSerTrpCys SerTrpSerValPheArgArgSer Arg
Figure A10–10 3 ThrGlyGlnProValArgAlaGlyVal ArgGlyArgPheSerGluAspProAsnAlaAsp

It is not possible from the information given to tell which of the two "open reading frames" corresponds to the actual protein encoded by this stretch of DNA.

Answer 10–11

A. Cleavage of human genomic DNA with Hae III would generate about 11×10^6 different fragments [$= 3 \times 10^9/4^4$], with Eco RI about 730,000 different fragments [$= 3 \times 10^9/4^6$] and with Not I about 46,000 different fragments [$= 3 \times 10^9/4^8$]. There will also be some additional fragments generated because the maternal and paternal chromosomes are very similar but not identical in DNA sequence.

B. A set of overlapping DNA fragments will be generated. Libraries constructed from sets of overlapping fragments of a genome are valuable because they can be used to order cloned sequences in relation to their original order on the chromosomes, and thus obtain the DNA sequence of a long stretch of DNA (see Figure 10–11). Sequences from the end of one cloned DNA are used as DNA hybridization probes to find other clones in the library that contain those sequences and that therefore might overlap. By repeating this procedure, a long stretch of continuous DNA sequence can be gradually built up (see Question 10–17). This laborious technique of "chromosome walking" has been used to build up the sequence of areas of chromosomes known to contain important genes, but where no prior information on the nucleotide sequence exists. By careful inspection of the nucleotide sequences of the areas obtained by chromosome walking, the positions and sequences of candidate genes can be identified.

Answer 10–12

By comparison with the positions of the size markers, we find that Eco RI treatment gives two fragments of 4 kb and 6 kb; Not I treatment gives one fragment of 10 kb; and treatment with Eco RI + Not I gives three fragments of 6 kb, 3 kb, and 1 kb. This gives a total length of 10 kb calculated as the sum of the fragments in each lane. Thus the original DNA molecule must be 10 kb (10,000 nucleotide pairs) long. Because treatment with Not I gives a fragment 10 kb long it could be that the original DNA is a linear molecule with no cutting site for Not I. But we can rule that out by the results of the Eco RI + Not I digestion. We know that Eco RI cleavage alone produces two fragments of 6 kb and 4 kb, and in the double digest this 4-kb fragment is further cleaved by Not I into a 3-kb and a 1-kb fragment. The DNA therefore contains a Not I cleavage site, and thus it must be circular, as only a single fragment of 10 kb is produced when it is cut with Not I alone. Arranging the cutting sites on a circular DNA to give the appropriate sizes of fragments produces the map illustrated in Figure A10–12.

Answer 10–13

If the repair enzymes act on the plasmid before it is replicated, the plasmid will indeed be repaired in cells. However, the repair enzymes cannot distinguish which strand of the DNA contains the mutation and which one contains the normal nucleotide. Therefore, in half of the cells that have been transformed with the mismatched plasmid, a normal gene is restored, whereas in the other half of the cells the normal strand is converted to match the mutated strand and the mutation is thus propagated. Cells containing a plasmid with the desired mutation can be identified, for example, by hybridization with a single-stranded DNA probe that allows one to distinguish between the normal and mutant genes.

Answer 10–14

A. The genetic code is degenerate, and there is more than one possible codon for each amino acid, with the exception of tryptophan and methionine. Therefore, to detect a nucleotide sequence known only from the amino acid sequence of the protein it encodes, many oligonucleotides must be made and pooled in order to ensure that the mixture will contain one oligonucleotide that exactly matches the DNA sequence of the gene. For the three peptide sequences given in this question, the following oligonucleotide probes need to be made (alternative bases at the same position are given in parentheses):

Peptide 1:
5'-TGGATGCA(C,T)CA(C,T)AA(A,G)-3'

Because of the three twofold degeneracies, you would need eight ($= 2^3$) different oligonucleotide sequences in the mixture.

Peptide 2:
5'(T,C)T(G,A,T,C)(A,T)(G,C)(G,A,T,C)(A,C)G-
(G,A,T,C)(T,C)T(G,A,T,C)(A,C)G(G,A,T,C)-3'

The oligonucleotide mixture representing peptide sequence #2 is much more complicated. Leu, Ser, and Arg are each encoded by six different codons; you would therefore need to synthesize a mixture of 7776 ($= 6^5$) different oligonucleotides. This could not be done, however, simply by using more than one different nucleotide in any one position because the different bases in each codon are not independent. (Ser, for example, has A or T as the first base of the codon, G or C as the second base, and G, A, T, or C as the third base; when the first base is A, however, the second base is always G and the third base can be only T or C.)

Peptide 3:
5'-TA(C,T)TT(C,T)GG(G,A,T,C)ATGCA(A,G)3'

Because of three twofold and one fourfold degeneracies, you would need 32 ($= 2^3 \times 4$) different sequences in the mixture.

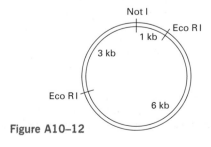

Figure A10–12

You would presumably first use probe #1 to screen your library by hybridization. Because there are only eight DNA sequences, the ratio of the one correct sequence to the incorrect ones is highest, giving you a strong signal when screening your library and thus a high likelihood that the desired sequence is among ones that you isolate from the library (see Question 10–5, p. 348). Probe #2 is practically useless, because only 1/7776 of the DNA in the mixture would perfectly hybridize to your gene of interest. You could use probe #3 to analyze your isolated sequences again by hybridization. Those sequences that hybridize to probes #1 and #3 are very likely to contain the gene of interest.

B. Knowing that peptide sequence #3 contains the last amino acid of the protein is valuable information because it tells you that the other two peptide sequences must precede it, that is, they must be located farther toward the N-terminal end of the protein. Knowing this order is important, because DNA primers can be extended by DNA polymerases only from their 3′ ends; thus, the 3′ ends of two primers need to "face" each other during a PCR amplification reaction (see Figure 10–27). A PCR primer based on peptide sequence #3 must therefore be the complementary sequence of probe #3 (so that its 3′ end corresponds to the first nucleotide of the sequence complementary to the Trp codon):

5′- (TC) TGCAT (G,A,T,C) CC (G,A) AA (G,A) TA-3′

Probe #1 could be your choice for the second primer. Probe #2, again because of its high degeneracy, would be a much less suitable choice.

C. The ends of the final amplification product are derived from the primers, which are each 15 nucleotides long. Therefore, a 270-nucleotide segment of the cDNA of the gene has been amplified. This will encode 90 amino acids; adding the amino acids encoded by the primers gives you a protein-coding sequence of 100 amino acids. This is unlikely to represent the whole protein because we do not know the location of peptide #1 in the sequence of the complete protein. To your satisfaction, however, you note that CTATCACGCTTTAGG encodes peptide sequence #2. This information therefore confirms that your PCR product indeed encodes a fragment of the protein you originally isolated.

Answer 10–15

Protein biochemistry is still very important, mostly because it provides the link between the amino acid sequence (which can be deduced from DNA sequences) and the functional properties of the protein. We are, for example, still not able to predict the folding of a polypeptide chain from its amino acid sequence, so that in many cases information regarding the function of the protein, such as its catalytic activity, cannot be deduced but must instead be obtained experimentally by analyzing the properties of proteins biochemically. Furthermore, the structural information that can be deduced from DNA sequences is necessarily incomplete. We cannot obtain information, for example, about modifications of protein side chains (such as phosphorylation), proteolytic processing, the presence of tightly bound coenzymes, or the association of the protein with other polypeptide chains.

Answer 10–16

The products will comprise a large number of different single-stranded DNA molecules, one for each nucleotide in the sequence. However, these products will be of four colors, depending on which of the four dideoxyribonucleotides terminated the polymerization reaction of the particular DNA chain. Separation by gel electrophoresis will generate a ladder of bands, each one nucleotide apart, where the color of each band indicates the nucleotide of the template at the corresponding position in the sequence. The method described here forms the basis for the DNA sequencing strategy used in most automated DNA sequencing machines (Figure A10–16).

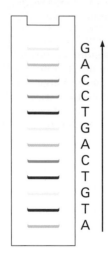

Figure A10–16

Answer 10–17

A. cDNA clones could not be used because there is no overlap between two cDNA clones even if they are derived from genes adjacent to each other in the chromosome.

B. Such repetitive DNA sequences can confuse chromosome walks, because the walk would appear to branch off in many different directions at once. The general strategy for avoiding these problems is to use genomic clones that are sufficiently long to overspan the repetitive DNA sequence.

Answer 10–18

A. Infants 2 and 8 have identical patterns and therefore must be brothers. Infants 3 and 6 also have identical patterns and must be brothers. These two sets of brothers are identical twins. The other two sets of twins must be fraternal twins because no other pairs of patterns are identical. Fraternal twins, like any pair of siblings born to the same parents, will have roughly half their genome in common. Thus, roughly half the VNTR polymorphisms in fraternal twins will be identical. Using this criterion, you can identify infants 1 and 7 as brothers and infants 4 and 5 as brothers.

B. You can match infants to their parents using the same sort of analysis of VNTR polymorphisms. Every band present in the analysis of an infant should have a matching band in one or the other of the parents, and, on average, each infant will share half of its VNTR polymorphisms with each parent. Thus the degree of match between each child and each parent will be the same as that between fraternal twins.

Answer 10–19

Mutant bacteria that do not produce ice-protein have probably arisen many times in nature. However, bacteria that produce ice-protein have a slight growth advantage over bacteria that do not, so it would be difficult to find such mutants in the wild (and it would probably be difficult for the ice-protein–free mutant bacteria to survive in the long run facing the competition of their natural counterparts). Genetic engineering, using genes deliberately mutated *in vitro*, simply makes these mutants much easier to obtain. The consequences, both advantageous and disadvantageous, of using a genetically engineered organism for a practical application, therefore, are nearly indistinguishable from those that would follow the use of a natural mutant. Indeed, bacterial and yeast strains have been selected for centuries for desirable genetic traits that make them suitable for industrial-scale applications such as cheese and wine production. The possibilities of genetic engineering are endless, however, and as with any technology, there is a finite risk that unforeseen consequences will arise. Recombinant DNA experimentation, therefore, is regulated, and the risks of individual projects are carefully assessed by review panels before permissions are granted. The state of our knowledge is sufficiently advanced that the consequences of some changes, such as the disruption of a bacterial gene in the example above, can be predicted with reasonable certainty. Other applications, such as germ-line gene therapy to correct human disease, may have far more complex outcomes, and it will take many more years of research and ethical debate to determine whether such treatments will eventually be used.

Chapter 11

Answer 11–1

Water is a liquid, and thus hydrogen bonds between water molecules are not static; they are continually formed and broken again by thermal motion. When a water molecule happens to be next to a hydrophobic molecule, it is more restricted in motion and has fewer neighbors with which it can interact, because it cannot form any hydrogen bonds in the direction of the hydrophobic molecule. It will therefore form hydrogen bonds to the more limited number of water molecules in its proximity. Bonding to fewer partners results in a more ordered water structure, which represents the cagelike structure in Figure 11–9. This structure has been likened to ice, although it is a more transient, less organized, and less extensive network than even a tiny ice crystal. The formation of any ordered structure decreases the entropy of the system (see Chapter 3) and is thus energetically unfavorable.

Answer 11–2

(B) is the correct analogy for lipid bilayer assembly because exclusion from water rather than attractive forces between the lipid molecules is involved. If the lipid molecules formed bonds with one another, the bilayer would be less fluid, and might even become rigid, depending on the strength of the interaction.

Answer 11–3

The fluidity of the bilayer is strictly confined to one plane: lipid molecules can diffuse laterally but do not readily flip from one monolayer to the other. Specific types of lipid molecules inserted into one monolayer therefore remain in it unless they are actively transferred by an enzyme—a flippase.

Answer 11–4

In both an α helix and a β barrel the polar peptide bonds of the polypeptide backbone can be completely shielded from the hydrophobic environment of the lipid bilayer by the hydrophobic amino acid side chains. Internal hydrogen bonds between the peptide bonds stabilize the α helix and β barrel.

Answer 11–5

The sulfate group in SDS is charged and therefore hydrophilic. The OH group and the C–O–C groups in Triton X-100 are polar; they can form hydrogen bonds with water and are therefore hydrophilic. In contrast, the blue portions of the molecules are either hydrocarbon chains or aromatic rings, neither of which has polar groups that could hydrogen-bond with water molecules; they are therefore hydrophobic. (See Figure A11–5.)

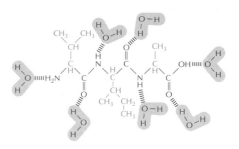

valine isoleucine alanine

Figure A11–5

Answer 11–6

Alpha helices in proteins are often used to span lipid bilayers. These structures are well suited for this purpose because they expose hydrophobic amino acid side chains to the hydrophobic interior of the lipid bilayer but sequester the polar peptide bonds of the polypeptide backbone away from the hydrophobic phase (see Figures 11–22 through 11–25). There are, however, other, less regular ways to fold up a polypeptide chain to achieve the same result, as seen in the small loop in the photosynthetic reaction center. This illustrates the importance of determining three-dimensional structures, which to date are known for only a few membrane proteins.

Answer 11–7

Some of the molecules of the two different transmembrane proteins are anchored to the spectrin filaments of the cell cortex. These molecules are not free to rotate or diffuse within the plane of the membrane. There is an excess of transmembrane proteins over the available attachment sites in the cortex, however, so that some of the transmembrane protein molecules are not anchored but rotate and diffuse freely within the plane of the membrane. Indeed, measurements of protein mobility show that there are two different populations of each transmembrane protein, corresponding to those proteins that are anchored and those that are not.

Answer 11–8

The different ways in which membrane proteins can be restricted to different regions of the membrane are summarized in Figure 11–35. The mobility of the membrane proteins is drastically reduced if they are bound to other proteins such as those of the cytoskeleton or the extracellular matrix. The mobility of proteins is not affected if the proteins are not bound to other proteins but are confined to membrane domains by barriers, such as tight junctions. The fluidity of the lipid bilayer is not significantly affected by the anchoring of membrane proteins; the sea of lipid molecules flows around anchored membrane proteins like water around the posts of a pier.

Answer 11–9

All of the statements are correct.

A, B, C, D. The lipid bilayer is fluid because the lipid molecules in the bilayer can undergo these motions.

E. Glycolipids are mostly restricted to the monolayer of membranes that faces away from the cytosol. Some special glycolipids, such as phosphatidylinositol (discussed in Chapter 16), are found specifically in the cytosolic monolayer.

F. The reduction of double bonds (by hydrogenation) allows lipid molecules to pack more tightly against one another and therefore increases the viscosity—that is, it turns oil into margarine.

G. Examples include enzymes involved in signaling (discussed in Chapter 16).

H. Polysaccharides are the main constituents of mucus and slime; the glycocalyx, which is made up of polysaccharides and oligosaccharides, is a very important lubricant—for example, for cells that line blood vessels or circulate in the bloodstream.

Answer 11–10

In a two-dimensional fluid the molecules are free to move only in one plane; the molecules in a normal fluid, in contrast, can move in three dimensions.

Answer 11–11

A. You would have a detergent. The diameter of the lipid head would be much larger than that of the hydrocarbon tail, so that the shape of the molecule would be a cone rather than a cylinder and the molecules would aggregate to form micelles rather than bilayers.

B. Lipid bilayers formed would be much more fluid. The bilayers would also be less stable, as the shorter hydrocarbon tails would be less hydrophobic, so the forces that drive the formation of the bilayer would be reduced.

C. The lipid bilayers formed would be much less fluid. Whereas a normal lipid bilayer has the viscosity of olive oil, a bilayer made of the same lipids but with saturated hydrocarbon tails would have the consistency of bacon fat.

D. The lipid bilayers formed would be much more fluid. Also, because the lipids would pack together less well, there would be more gaps and the bilayer would be more permeable to small water-soluble molecules.

E. If we assume that the lipid molecules are completely intermixed, the fluidity of the membrane would be unchanged. In such bilayers, however, the saturated lipid molecules would tend to aggregate with one another because they can pack so much more tightly and would therefore form patches of much-reduced fluidity. The bilayer would not, therefore, have uniform properties over its surface. Because normally one saturated and one unsaturated hydrocarbon tail are linked to the same hydrophilic head in lipid molecules, such segregation does not occur in cell membranes.

F. The lipid bilayers formed would have virtually unchanged properties. Each lipid molecule would now span the entire membrane, with one of its two head groups exposed at each surface. Such lipid molecules are found in the membranes of thermophilic bacteria, which can live at temperatures approaching boiling water. Their bilayers do not come apart at elevated temperatures, as usual bilayers do, because the original two monolayers are now covalently linked into a single membrane.

Answer 11–12

Lipid molecules are approximately cylindrical in shape. Detergent molecules, by contrast, are conical or wedge-shaped. A lipid molecule with only one hydrocarbon tail, for example, would be a detergent. To make a lipid molecule into a detergent, you would have to make its hydrophilic head larger or remove one of its tails so that it could form a micelle. Detergent molecules also usually have shorter hydrocarbon tails than lipid molecules. This makes them slightly water-soluble, so that detergent molecules leave and reenter micelles frequently in aqueous solution. Because of this, some monomeric detergent molecules are always present in aqueous solution and therefore can enter lipid bilayers to solubilize membrane proteins (see Figure 11–27).

Answer 11–13

When lined up, there are about 4000 lipid molecules (each 0.5 nm wide) between a lipid molecule at one end of the bacterial cell and one at the other end. Thus, if one of these molecules started to move toward the other, exchanging places with a neighboring molecule every 10^{-7} sec, it would take only 4×10^{-4} sec (= 4000×10^{-7} sec) to reach the other end. In reality, however, the lipid molecule would move in a random path rather than in a defined direction, so it would take considerably longer (1 sec) to reach the other end. If a 4-cm Ping-Pong ball exchanged places with a neighbor every 10^{-7} sec, it would travel at a speed of 1,440,000 km/hr (= 4 cm/10^{-7} sec). If its movement were only in one direction, it would reach the other wall in 1.5×10^{-5} sec; in a random walk it would take considerably longer (~2 msec).

Answer 11–14

Membrane proteins anchor the lipid bilayer to the cytoskeleton, strengthening the plasma membrane so that it can withstand the forces on it when the red blood cell is pumped through small blood vessels. Membrane proteins also transport nutrients and ions across the plasma membrane.

Answer 11–15

The hydrophilic faces of the five membrane-spanning α helices, each contributed by a different subunit, are thought to come together to form a pore across the lipid bilayer that is lined with the hydrophilic amino acid side chains (Figure A11–15). Ions can pass through this hydrophilic pore. The hydrophobic side chains interact with the hydrophobic lipid tails in the bilayer.

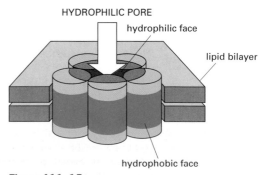

HYDROPHILIC PORE

hydrophilic face

lipid bilayer

hydrophobic face

Figure A11–15

Answer 11–16

There are about 100 lipid molecules (i.e., phospholipid + cholesterol) for every protein molecule in the membrane [= (2/50,000)/(1/800 + 1/256)]. A similar protein/lipid ratio is seen in many cell membranes.

Answer 11–17

Membrane fusion does not alter the orientation of the membrane proteins with their attached color tags: the portion of each transmembrane protein that is exposed to the cytosol always remains exposed to cytosol, and the portion exposed to the outside always remains exposed to the outside (Figure A11–17). At 0°C the fluidity of the membrane is reduced and the mixing of the membrane proteins is significantly slowed.

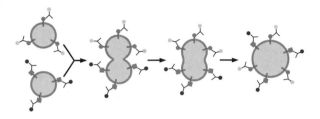

Figure A11–17

Answer 11–18

The exposure of hydrophobic amino acid side chains to water is energetically unfavorable. There are two ways that such side chains can be sequestered from water to achieve an energetically more favorable state. First, they can form transmembrane segments that span a lipid bilayer. This requires about 20 of them to be located sequentially in a polypeptide chain. Second, the hydrophobic amino acid side chains can be sequestered in the interior of the folded polypeptide chain. This is one of the major forces that lock the polypeptide chain into a unique three-dimensional structure. In either case the hydrophobic forces in the lipid bilayer or in the interior of a protein are based on the same principles.

Answer 11–19

(A) Antarctic fish live at sub-zero temperatures and are cold-blooded. To keep the membranes fluid at these temperatures, they have a high percentage of unsaturated phospholipids.

Answer 11–20

Sequence B is most likely to form a transmembrane helix. It is composed primarily of hydrophobic amino acids, and therefore can be stably integrated into a lipid bilayer. In contrast, sequence A contains many polar amino acids (S, T, N, Q), and sequence C contains many charged amino acids (K, R, H, E, D) that would be energetically disfavored in the hydrophobic interior of the lipid bilayer.

Chapter 12

Answer 12–1

A. Transport mediated by a carrier protein can be described by a strictly analogous equation:
$$(1)\ CP + S \rightleftharpoons CPS \rightarrow CP + S^*$$
where S is the solute, S* is the solute on the other side of the membrane (i.e., although it is still the same molecule, it is now located in a different environment), and CP is the carrier protein.

B. This equation is useful because it describes a binding step, followed by a delivery step. The mathematical treatment of this equation would be very similar to that described for enzymes (see Figure 3–25); thus, carrier proteins are characterized by a K_M value that describes their affinity for a solute and a V_{max} value that describes their maximal rate of transfer. To be more accurate, one could include the conformational change of the carrier protein in the reaction scheme
$$(2a)\ CP + S \rightleftharpoons CPS \rightleftharpoons CP^*S^* \rightarrow CP^* + S^*$$
$$(2b)\ CP \rightleftharpoons CP^*$$
where CP* is the carrier protein after the conformational change that exposes its solute-binding site on the other side of the membrane. This account requires a second equation (2b) that allows the carrier protein to return to its starting conformation.

C. The equations do not describe the behavior of channels because solutes passing through channels do not bind to them in the way that a substrate binds to an enzyme.

Answer 12–2

A. The properties define a symport.

B. No additional properties need to be specified. The important feature that provides the coupling of the two solutes is that the protein cannot switch its conformation if only one of the two solutes is bound. Solute B, which is driving the transport of solute A, is in excess on the side of the membrane from which transport initiates and occupies its binding site most of the time. In this state, the carrier protein, prevented from switching its conformation, waits until a solute A molecule on occasion binds. With both binding sites occupied, the carrier protein switches conformation. Now exposed to the other side of the membrane, the binding site for solute B is mostly empty because there is little of it in the solution on this side of the membrane. Although the binding site for A is now more frequently occupied, the carrier can switch back only after solute A is unloaded as well.

C. An antiport could be similarly constructed as a transmembrane protein with the following properties. It has two binding sites, one for solute A and one for solute B. The protein can undergo a conformational change to switch between two states: either both binding sites are exposed exclusively on one side of the membrane or both binding sites are exposed exclusively on the other side of the membrane. The protein can switch between the two conformational states only if one binding site is occupied, but cannot switch if both binding sites are occupied or if both binding sites are empty.

   Note that these rules provide an alternative model to that shown in Figure 12–14. Thus there are two possible ways to couple the transport of two solutes: (1) provide cooperative solute-binding sites and allow the pump to switch between the two states randomly as shown in Figure 12–14, or (2) allow independent binding of both solutes and make the switch between the two states conditional on the occupancy of the binding sites. As the structure of a coupled transporter has not yet been determined, we do not know which of the two mechanisms such pumps use.

Answer 12–3

If the Na⁺-K⁺ pump is not working at full capacity because it is partially inhibited by ouabain or digitalis, it generates an electrochemical gradient of Na⁺ that is less steep than that in untreated cells. Consequently, the Ca²⁺-Na⁺ antiporter works less efficiently, and Ca²⁺ is removed from the cell more slowly. When the next cycle of muscle contraction begins, there is still

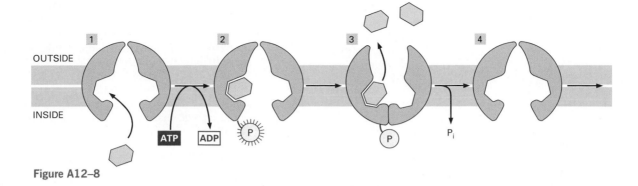

Figure A12–8

an elevated level of Ca^{2+} left in the cytosol. The entry of the same number of Ca^{2+} ions into the cell leads therefore to a higher Ca^{2+} concentration than in untreated cells, which in turn leads to a stronger and longer-lasting contraction. Because the Na^+-K^+ pump fulfills essential functions in all animal cells, both to maintain osmotic balance and to generate the Na^+ gradient used to power many transporters, the drugs are deadly poisons at higher concentrations.

Answer 12–4

Each of the rectangular peaks corresponds to the opening of a single channel that allows a small current to pass. You note from the recording that the channels present in the patch of membrane open and close frequently. Each channel remains open for a very short, somewhat variable time, averaging about 10 milliseconds. When open, the channels allow a small current with a unique amplitude (4 pA; one picoampere = 10^{-12} A) to pass. In one instance, the current doubles, indicating that two channels in the same membrane patch opened simultaneously.

If acetylcholine is omitted or added to the solution outside the pipette, you would measure only the baseline current. Acetylcholine must bind to the extracellular portion of the acetylcholine receptor molecules to allow the channel to open, and in the membrane patch shown in Figure 12–22, the cytoplasmic side of the membrane is exposed to the solution outside the microelectrode.

Answer 12–5

The equilibrium potential of K^+ is –90 mV [= 62 mV $\log_{10}$ (5 mM/140 mM)], and that of Na^+ is +72 mV [= 62 mV $\log_{10}$ (145 mM/10 mM)]. The K^+ leak channels in the plasma membrane of a resting cell allow K^+ to come to equilibrium; the membrane potential of the cell is therefore close to –90 mV. When Na^+ channels open, Na^+ rushes in, and, as a result, the membrane potential reverses its polarity to a value nearer to +72 mV, the equilibrium value for Na^+. Upon closure of the Na^+ channels, the K^+ leak channels allow K^+, now no longer at equilibrium, to exit the cell until the membrane potential is restored to the equilibrium value for K^+, about –90 mV.

Answer 12–6

When the resting membrane potential of an axon drops below a threshold value, voltage-gated Na^+ channels in the immediate neighborhood open and allow an influx of Na^+. This depolarizes the membrane further, causing more distant voltage-gated Na^+ channels to open as well. This creates a wave of depolarization that spreads rapidly along the axon, called the action potential. Because Na^+ channels become inactivated soon after they open, the flow of K^+ through voltage-gated K^+ channels and K^+ leak channels is able to restore the original resting membrane potential rapidly after the action potential has passed. (96 words)

Answer 12–7

If the number of functional acetylcholine receptors is reduced by the antibodies, the neurotransmitter (acetylcholine) that is released from the nerve terminals cannot (or can only weakly) stimulate the muscle to contract.

Answer 12–8

By analogy to the Na^+-K^+ pump shown in Figure 12–12, ATP might be hydrolyzed and donate a phosphate group to the carrier protein when—and only when—it has the solute bound on the "inside" face of the membrane (step 1 → 2). The attachment of the phosphate would trigger an immediate conformational change (step 2 → 3), thereby capturing the solute and exposing it to the "outside." The phosphate would be removed from the protein when—and only when—the solute has dissociated, and the now empty, nonphosphorylated carrier protein would switch back to the starting position (step 3 → 4) (Figure A12–8).

Answer 12–9

A. False. The plasma membrane contains proteins that confer selective permeability to many charged molecules. In contrast, a pure lipid bilayer lacking proteins is highly impermeable to all charged molecules.

B. False. Channel proteins do not bind the solute that passes through them. Selectivity of a channel protein is achieved by the size of the internal pore and by charged regions at the entrance of the pore that attract or repel ions of the appropriate charge.

C. True for animal cells. Cells contain a concentrated solution of many molecules that will cause the osmotic influx of water. Unless ions are constantly pumped out to maintain an osmotic balance, cells will eventually burst. False for plant, yeast, and bacterial cells. Although water will tend to enter them by osmosis, these cells are surrounded by a tough cell wall that prevents the plasma membrane from rupturing.

D. False. Carrier proteins are slower. They have enzymelike properties, i.e., they bind solutes and need to undergo conformational changes during their functional cycle. This limits the maximal rate of transport to about 1000 solute molecules per second, whereas channel proteins can pass up to 1,000,000 solute molecules per second.

E. True. The bacteriorhodopsin of some photosynthetic bacteria moves H^+, using energy captured from visible light.

F. True. Most animal cells contain K^+ leak channels in their plasma membrane that are predominantly open. The K^+ concentration inside the cell still remains higher than outside, because the membrane potential is negative and therefore inhibits the positively charged

K+ from leaking out. K+ is also continually pumped into the cell by the Na+-K+ pump.

G. False. A symport binds two different solutes on the same side of the membrane. Turning it around would not change it into an antiport, which must also bind to different solutes, but on opposing sides of the membrane.

H. False. The peak of an action potential corresponds to a transient shift of the membrane potential from a negative to a positive value. The influx of Na+ causes the membrane potential first to move toward zero and then to reverse, rendering the cell positively charged on its inside. Eventually, the resting potential is restored by an efflux of K+ through voltage-gated K+ channels and K+ leak channels.

Answer 12–10

The permeabilities are N_2 (small and nonpolar) > ethanol (small and slightly polar) > water (small and polar) > glucose (large and polar) > Ca^{2+} (small and charged) > RNA (very large and charged).

Answer 12–11

A. Both couple the movement of two different solutes across the membrane. Symports transport both solutes in the same direction, whereas antiports transport the solutes in opposite directions.

B. Both are mediated by membrane transport proteins. Passive transport of a solute occurs downhill, in the direction of its concentration or electrochemical gradient, whereas active transport occurs uphill and therefore needs an energy source. Active transport can be mediated by carrier proteins but not by channel proteins, whereas passive transport can be mediated by either.

C. Both terms describe energy changes involved in moving an ion from one side of a membrane to the other. The membrane potential refers to the electrical energy change; the electrochemical gradient is a composite of this electrical energy change and the chemical energy change associated with moving between a region of high concentration and a region of low concentration. The membrane potential is defined independently of the choice of ion, whereas an electrochemical gradient depends on the concentration gradient of the particular ionic solute and is therefore a solute-specific parameter.

D. A pump is a specialized carrier protein that uses energy to transport a solute uphill against an electrochemical gradient.

E. Both transmit signals by electrical activity. Wires are made of copper, axons are not. The signal passing down an axon does not diminish in strength, because it is self-amplifying, whereas the signal in a wire decreases over distance (by leakage of current across the insulating sheath).

F. Both affect the osmotic pressure of the cell. An ion is a solute that bears a charge.

Answer 12–12

A bridge allows vehicles to pass over a river in a steady stream; the entrance can be designed to exclude, for example, oversized trucks, and it can be intermittently closed to traffic by a gate. By analogy, channels allow ions to flow in a gated stream across the membrane, imposing size and charge restrictions.

A ferry, in contrast, loads vehicles on one riverbank and then, after movement of the ferry itself, unloads on the other side of the river. This process is slower. During loading, particu-

lar vehicles could be selected from the waiting line because they fit particularly well on the car deck. By analogy, carrier proteins bind solutes on one side of the membrane and then, after a conformational movement, release them on the other side. Specific binding leads to the selection of the molecules to be transported. As in the case of coupled transport, sometimes you have to wait until the ferry is full before you can go.

Answer 12–13

Acetylcholine is being transported into the vesicles by an H^+–acetylcholine antiporter in the vesicle membrane. The H^+ gradient that drives the uptake is generated by an ATP-driven H^+ pump in the vesicle membrane, which pumps H^+ into the vesicle (hence the dependence of the reaction on ATP). Raising the pH of the solution surrounding the vesicles increases the H^+ gradient: at an elevated pH there are fewer H^+ ions in the solution outside the vesicles while the number inside remains the same. This explains the observed enhanced rate of uptake.

Answer 12–14

The voltage gradient across the membrane is about 150,000 V/cm. This extremely powerful electric field is close to the limit at which insulating materials—such as the lipid bilayer—break down and cease to act as insulators. The large field corresponds to the large amount of energy that can be stored in electrical gradients across the membrane, as well as to the extreme electrical forces that proteins can experience in a membrane. A voltage of 150,000 V would instantly discharge in an arc across a 1-cm-wide gap (that is, air would be an insufficient insulator for this strength of field).

Answer 12–15

A. Nothing. You require ATP to drive the Na+-K+ pump.

B. The ATP becomes hydrolyzed, and Na+ is pumped into the vesicles, generating a concentration gradient of Na+ across the membrane. At the same time, K+ is pumped out of the vesicles, generating a concentration gradient of K+ of opposite polarity. When all the K+ had been pumped out of the vesicle or the ATP ran out, the pump would stop.

C. The Na+-K+ pump would go through states 1, 2, and 3 in Figure 12–12. Because all reaction steps must occur strictly sequentially, however, dephosphorylation and the conformation switch cannot occur in the absence of K+. The Na+-K+ pump will therefore become stuck in the phosphorylated state, waiting indefinitely for a potassium ion. The number of sodium ions transported would be minuscule, because each pump molecule would have functioned only a single time.

Similar experiments, leaving out individual ions and analyzing the consequences, were used to determine the sequence of steps by which the Na+-K+ pump works.

D. ATP would become hydrolyzed and Na+ and K+ would be pumped across the membrane as described in scenario (B). However, the pump molecules that sit in the membrane in the reverse orientation would be completely inactive (i.e., they would not—as one might have erroneously assumed—pump ions in the opposite direction), because ATP would not have access to the site on these molecules where phosphorylation occurs. This site is normally exposed to the cytosol. ATP is highly charged and cannot cross membranes without the help of specific carrier proteins.

E. ATP becomes hydrolyzed and Na+ and K+ are pumped across the membrane, as described in scenario (B). K+, however, immediately flows back into the vesicles

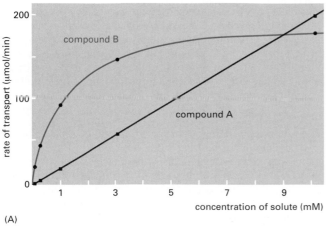

(A)

(B)

Figure A12–20

through the K⁺ leak channels. K⁺ moves down the K⁺ concentration gradient formed by the action of the Na⁺-K⁺ pump. With each K⁺ that moves into the vesicle through the leak channel, a positive charge is moved across the membrane, building a membrane potential that is positive on the inside of the vesicles. Eventually, K⁺ will stop flowing through the leak channels when the membrane potential balances the concentration gradient. The scenario described here is a slight oversimplification: the Na⁺-K⁺ pump in mammalian cells actually moves three sodium ions out of cells for each two potassium ions that it pumps into the cell, thereby driving an electric current across the membrane and making a small additional contribution to the resting membrane potential (which therefore corresponds only approximately to a state of equilibrium for K⁺ moving via K⁺ leak channels).

Answer 12–16

Ion channels can be ligand-gated, voltage-gated, or mechanically gated.

Answer 12–17

The cell has a volume of 10^{-12} liters (= 10^{-15} m³) and thus contains 6×10^4 calcium ions (= 6×10^{23} molecules/mole $\times 100 \times 10^{-9}$ moles/liter $\times 10^{-12}$ liters). Therefore, to raise the intracellular Ca²⁺ concentration fiftyfold, another 2,940,000 calcium ions have to enter the cell (note that at 5 μM concentration there are 3×10^6 ions in the cell, of which 60,000 are already present before the channels are opened). Because each of the 1000 channels allows 10^6 ions to pass per second, each channel has to stay open for only 3 milliseconds.

Answer 12–18

Animal cells drive most transport processes across the plasma membrane with the electrochemical gradient of Na⁺. ATP is needed to fuel the Na⁺-K⁺ pump to maintain the Na⁺ gradient.

Answer 12–19

A. If H⁺ is pumped across the membrane into the endosomes, an electrochemical gradient of H⁺ results—composed of both an electrical potential and a concentration gradient, with the interior of the vesicle positive. Both of these components add to the energy that is stored in the gradient and that must be supplied to generate it. The electrochemical gradient will therefore

limit the transfer of more H⁺. If, however, the membrane also contains Cl⁻ channels, the negatively charged Cl⁻ will flow into the endosomes and diminish the electrical potential. It therefore becomes energetically less expensive to pump more H⁺ across the membrane, and the interior of the endosomes can become more acidic.

B. Yes. As explained in (A), some acidification would still occur in their absence.

Answer 12–20

A. See Figure A12–20A.

B. The transport rates of compound A are proportional to its concentration, indicating that compound A can diffuse through membranes on its own. Compound A is likely to be ethanol, because it is a small and relatively nonpolar molecule that can diffuse readily through the lipid bilayer.

In contrast, the transport rates of compound B saturate at high concentrations, indicating that compound B is transported across the membrane by a transport protein. Transport rates cannot increase beyond a maximal rate at which this protein can function. Compound B is likely to be acetate, because it is a charged molecule that could not cross the membrane without the help of a membrane transport protein.

C. For ethanol, we measured a linear relationship between concentration and transport rate. Thus, at 0.5 mM the transport rate would be 10 μmol/min, and at 100 mM the transport rate would be 2 μmol/min.

For the carrier-protein–mediated transport of acetate, the relationship between concentration, S, and transport rate can be described by the Michaelis–Menten equation, which describes simple enzyme reactions:

(1) transport rate = $V_{max} \times S / [K_M + S]$

Recall from Chapter 3 (see Question 3–20, p. 117) that to determine the V_{max} and K_M, a trick is used in which the Michaelis–Menten equation is transformed so that it is possible to plot the data as a straight line. A simple transformation yields

(2) **1/rate** = (K_M/V_{max}) **(1/S)** + $1/V_{max}$

(i.e., an equation of the form $y = ax + b$)

Calculation of 1/rate and 1/S for the given data and plotting them in a new graph as in Figure A12–19B gives a straight line. The K_M (= 1.0 mM) and V_{max}

(= 200 μmol/min) are determined from the intercept of the line with the y axis ($1/V_{max}$) and from its slope (K_M/V_{max}). Knowing the values for K_M and V_{max} allows you to calculate the transport rates for 0.5 mM and 100 mM acetate using equation (1). The results are 67 μmol/min and 198 μmol/min, respectively.

Answer 12–21

The membrane potential and the high extracellular Na^+ concentration provide a large electrochemical driving force and a large reservoir of Na^+ ions, so that mostly Na^+ ions enter the cell as acetylcholine receptors open. Ca^{2+} ions will also enter the cell, but their influx is much more limited because of their lower extracellular concentration. (Most of the Ca^{2+} that enters the cytosol upon muscle activation is released from intracellular stores, as we discuss in Chapter 17). Because of their high intracellular concentration and the opposing direction of membrane potential, K^+ ions are already close to equilibrium across the membrane. For this reason, there will be little if any movement of K^+ ions upon opening of a cation channel.

Answer 12–22

The diversity of neurotransmitter-gated ion channels is a good thing for the industry as it raises the possibility of developing new drugs specific for each channel type. Each of the diverse subtypes of these channels is expressed in a narrow set of neurons. A narrow range of expression of individual subtypes so that, in principle, drugs could be discovered or designed to affect specific subtypes in a selected set of neurons, thus influencing particular brain functions specifically.

Chapter 13

Answer 13–1

In order to keep glycolysis going, cells need to regenerate NAD^+ from NADH. There is no efficient way to do this without fermentation. In the absence of regenerated NAD^+, step 6 of glycolysis (the oxidation of glyceraldehyde 3-phosphate to 1,3-bisphosphoglycerate (Panel 13–1, pp. 432–433) could not occur and the product glyceraldehyde 3-phosphate would accumulate. The same thing would happen in cells unable to make either pyruvate or ethanol: neither would be able to regenerate NAD^+, and so glycolysis would be blocked at the same step.

Answer 13–2

Arsenate instead of phosphate becomes attached in step 6 of glycolysis to form 1-arseno-3-phosphoglycerate (Figure A13–2). Because of its sensitivity to hydrolysis in water, the high-energy bond is destroyed before the molecule that contains it can diffuse to reach the next enzyme. The product of the hydrolysis, 3-

Figure A13–2

phosphoglycerate, is the same product normally formed in step 7 by the action of phosphoglycerate kinase. But because hydrolysis occurs nonenzymatically, the energy liberated by breaking the high-energy bond cannot be captured to generate ATP. In Figure 13–6, therefore, the reaction corresponding to the downward-pointing arrow would still occur, but the wheel that provides the coupling to ATP synthesis is missing. Arsenate wastes metabolic energy by uncoupling many phosphotransfer reactions by the same mechanism, which is why it is so poisonous.

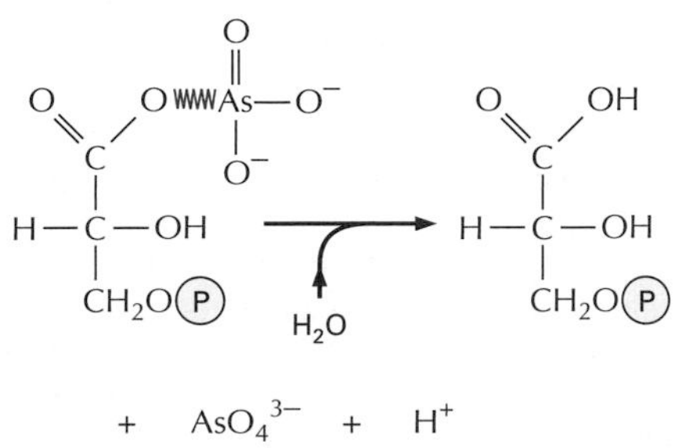

Answer 13–3

The oxidation of fatty acids breaks the carbon chain into two-carbon units (acetyl groups) that become attached to CoA. Conversely, during biogenesis fatty acids are constructed by linking together acetyl groups. Most fatty acids therefore have an even number of carbon atoms.

Answer 13–4

Because the function of the citric acid cycle is to harvest the energy released during the oxidation, it is advantageous to break the overall reaction into as many steps as possible (see Figure 13–1). Using a two-carbon compound, the available chemistry would be much more limited, and it would be impossible to generate as many intermediates.

Answer 13–5

It is true that oxygen atoms are returned as part of CO_2 to the atmosphere. The CO_2 released from the cells, however, does not contain those specific oxygen atoms that were consumed as part of the oxidative phosphorylation reaction and converted into water. One can show this directly by incubating living cells in an atmosphere that contains molecular oxygen that contains a different isotope, ^{18}O, instead of the naturally abundant isotope, ^{16}O. In such an experiment one finds that all the CO_2 released from cells contains only ^{16}O. Therefore, the oxygen atoms in the released CO_2 molecules do not come directly from the atmosphere but from organic molecules that the cell has first made and then oxidized as fuel (see Panel 13–2, pp. 450–451).

Answer 13–6

The carbon atoms in sugar molecules are already partially oxidized, in contrast to all but the very first carbon atoms in the acyl chains of fatty acids. Thus, two carbon atoms from glucose are lost as CO_2 during the conversion of pyruvate to acetyl CoA, and only four of the six carbon atoms of the sugar molecule are recovered and can enter the citric acid cycle, where most of the energy is captured. In contrast, all carbon atoms of a fatty acid are converted into acetyl CoA.

Answer 13–7

The cycle continues because intermediates are replenished as necessary by reactions leading to the citric acid cycle (instead of away from it). One of the most important reactions of this kind is the conversion of pyruvate to oxaloacetate by the enzyme pyruvate carboxylase:

$$\text{pyruvate} + CO_2 + ATP + H_2O \rightarrow \text{oxaloacetate} + ADP + P_i + 2H^+$$

This is one of the many examples of how metabolic pathways are elegantly balanced and work together to maintain appropriate concentrations of all metabolites required by the cell (see Figure A13–7).

Answer 13–8

A. False. If this were the case, then the reaction would be useless for the cell. No chemical energy would be harvested in a useful form (e.g., ATP) to be used for metabolic processes. (Cells would be nice and warm, though!)

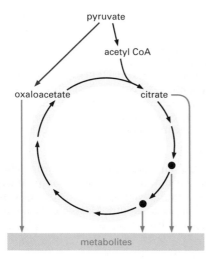

Figure A13–7

B. False. No energy conversion process can be 100% efficient. Recall that entropy in the universe always has to increase, and for most reactions this is accomplished by releasing heat.

C. True. The carbon atoms in glucose are in a reduced state compared with those in CO_2, in which they are fully oxidized.

D. False. The reaction does indeed produce some water, but water is so abundant in the biosphere that this is no more than "a drop in the bucket."

E. True. If it had occurred in only one step, then all the energy would be released at once and it would be impossible to harness it efficiently to drive other reactions, such as the synthesis of ATP.

F. False. Molecular oxygen (O_2) is used only in the very last step of the reaction.

G. True. Plants convert CO_2 into sugars by harvesting the energy of light in photosynthesis. O_2 is produced in the process and released by plant cells.

H. True. Anaerobically growing cells use glycolysis to oxidize sugars to pyruvate. Animal cells convert pyruvate to lactate, and no CO_2 is produced; yeast cells, however, convert pyruvate to ethanol and CO_2. It is this CO_2 gas, released from yeast cells during fermentation, that makes bread dough rise and that carbonates beer and champagne.

Answer 13–9

Darwin exhaled the carbon atom, which therefore must be the carbon atom of a CO_2 molecule. After spending some time in the atmosphere, the CO_2 molecule must have entered a plant cell, where it became "fixed" by photosynthesis and converted into part of a sugar molecule. While it is certain that these early steps must have happened this way, there are many different paths from there that the carbon atom could have taken. The sugar could have been broken down by the plant cell into pyruvate or acetyl CoA, for example, which then could have entered biosynthetic reactions to build an amino acid. The amino acid might have been incorporated into a plant protein, maybe an enzyme or a protein that builds the cell wall. You might have eaten the delicious leaves of the plant in your salad, and digested the protein in your gut to produce amino acids again. After circulating in your bloodstream, the amino acid might have been taken up by a developing red blood cell to make its own protein, such as the hemoglobin in question. If we wish, of course, we can make

our food chain scenario more complicated. The plant, for example, might have been eaten by an animal that in turn was consumed by you during lunch break. Moreover, because Darwin died more than 100 years ago, the carbon atom could have traveled such a route many times. In each round, however, it would have started again as fully oxidized CO_2 gas and entered the living world following its reduction during photosynthesis.

Answer 13–10

Yeast cells grow much better aerobically. Under anaerobic conditions they cannot perform oxidative phosphorylation and therefore have to produce all their ATP by glycolysis, which is less efficient. Whereas one glucose molecule yields a net gain of two ATP molecules by glycolysis, the additional use of the citric acid cycle and oxidative phosphorylation boosts the energy yield up to about 30 ATP molecules.

Answer 13–11

The amount of free energy stored in the phosphate bond in creatine phosphate is larger than that of of the anhydride bonds in ATP. Hydrolysis of creatine phosphate can therefore be directly coupled to the production of ATP.

$$\text{creatine phosphate} + \text{ADP} \rightarrow \text{creatine} + \text{ATP}$$

The ΔG^o for this reaction is –3 kcal/mole, indicating that it proceeds rapidly to the right, as written.

Answer 13–12

The extreme conservation of glycolysis is evidence that all present cells are derived from a single founder cell as discussed in Chapter 1. The elegant reactions of glycolysis would therefore have evolved only once, and then they would have been inherited as cells evolved. The later invention of oxidative phosphorylation allowed follow-up reactions to capture 15 times more energy than is possible by glycolysis alone. This remarkable efficiency is close to the theoretical limit and hence virtually eliminates the opportunity for further improvements. Thus the generation of alternative pathways would result in no obvious growth advantage that would have been selected in evolution.

Answer 13–13

As discussed in the text, 30 ATP molecules are produced from each glucose molecule that is oxidized according to the reaction $C_6H_{12}O_6$ (glucose) + $6O_2 \rightarrow 6CO_2 + 6H_2O$ + energy. Thus, one O_2 molecule is consumed for every five ATP molecules produced. The cell therefore consumes 2×10^8 O_2 molecules/min, which corresponds to the consumption of 3.3×10^{-16} moles (= 2×10^8 / 6×10^{23}) or 7.4×10^{-15} liter (= $3.3 \times 10^{-16} \times 22.4$) each minute. The volume of the cell is 10^{-15} m^3 [= $(10^{-5})^3$], which is 10^{-12} liter. The cell therefore consumes about 0.7% (= $100 \times 7 \times 10^{-15}/10^{-12}$) of its volume of O_2 gas every minute, or its own volume of O_2 gas in 2 hours and 15 minutes.

Answer 13–14

The reactions each have negative ΔG values and are therefore energetically favorable (see Figure A13–14 for energy diagrams).

Answer 13–15

A. Pyruvate is converted to acetyl CoA, and the labeled ^{14}C atom is released as $^{14}CO_2$ gas (see Figure 13–8B).

B. By following the ^{14}C-labeled atom through every reaction in the cycle, shown in Panel 13–2 (pp. 450–451), you find that the added ^{14}C label would be quantitatively recovered in oxaloacetate. The analysis also reveals, however, that it is no longer in the keto group but in the methylene group of oxaloacetate (Figure A13–15).

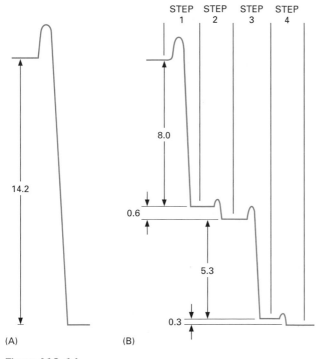

(A) (B)

Figure A13–14

Answer 13–16

In the presence of molecular oxygen, oxidative phosphorylation converts most of the cellular NADH to NAD⁺. Since fermentation requires NADH, it is severely inhibited by the availability of oxygen gas.

Chapter 14

Answer 14–1

By making membranes permeable to protons, DNP collapses— or at very small concentrations diminishes—the proton gradient across the inner mitochondrial membrane. Cells continue to oxidize food molecules to feed electrons into the electron-transport chain, but H⁺ ions pumped across the membrane flow back into the mitochondria in a futile cycle. Their energy cannot be tapped to drive ATP synthesis, and hence is released as heat. Patients who have been given small doses of DNP lose weight because their fat reserves are used more rapidly to feed the electron-transport chain, and the whole process simply "wastes" energy.

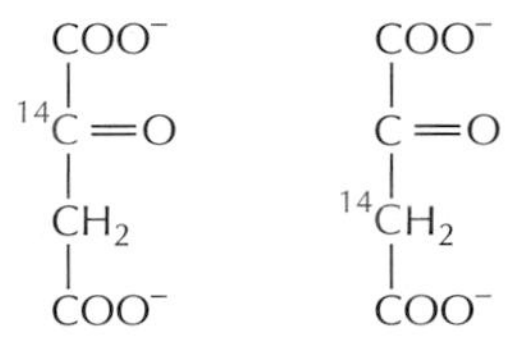

Figure A13–15

radioactive oxaloacetate added to the extract

radioactive oxaloacetate isolated after one turn of citric acid cycle

A similar mechanism of heat production is used by a specialized tissue composed of brown fat cells, which is abundant in newborn humans and in hibernating animals. These cells are packed with mitochondria that leak part of their H⁺ gradient futilely back across the membrane for the sole purpose of warming up the organism. These cells are brown because they are packed with mitochondria, which contain high concentrations of pigmented proteins, such as cytochromes.

Answer 14–2

The inner mitochondrial membrane is the site of oxidative phosphorylation, and it produces most of the cell's ATP. Cristae are portions of the mitochondrial inner membrane that are folded inward. Mitochondria that have a higher density of cristae have a larger area of inner membrane and therefore a greater capacity to carry out oxidative phosphorylation. Heart muscle expends a lot of energy during its continuous contractions, whereas skin cells have a lesser energy demand. An increased density of cristae therefore increases the ATP-production capacity of the heart muscle cell. This is a remarkable example of how cells adjust the abundance of their individual components according to need.

Answer 14–3

A. The DNP collapses the electrochemical proton gradient completely. H⁺ ions that are pumped to one side of the membrane flow back freely, and therefore no energy can be stored across the membrane.

B. An electrochemical gradient is made up of two components: a concentration gradient and an electrical potential. If the membrane is made permeable to K⁺ with nigericin, K⁺ will be driven into the matrix by the electrical potential of the inner membrane (negative inside, positive outside). The influx of positively charged K⁺ will abolish the membrane's electrical potential. In contrast, the concentration component of the H⁺ gradient (the pH difference) is unaffected by nigericin. Therefore, only part of the driving force that makes it energetically favorable for H⁺ ions to flow back into the matrix is lost.

Answer 14–4

A. Such a turbine running in reverse is an electrically driven water pump, which is analogous to what the ATP synthase becomes when it uses the energy of ATP hydrolysis to pump protons against their electrochemical gradient across the inner mitochondrial membrane.

B. The ATP synthase should stall when the energy that it can draw from the proton gradient is just equal to the free-energy change required to make ATP; at this equilibrium point there will be neither net ATP synthesis nor net ATP consumption.

C. As the cell uses up ATP, the ATP/ADP ratio in the matrix falls below the equilibrium point just described, and ATP synthase uses the energy stored in the proton gradient to synthesize ATP in order to restore the original ATP/ADP ratio. Conversely, when the electrochemical proton gradient drops below that at the equilibrium point, ATP synthase uses ATP in the matrix to restore this gradient.

Answer 14–5

A. An electron pair causes 10 H⁺ to be pumped across the membrane when passing from NADH to O₂ through the three respiratory complexes. Four H⁺ are needed to make each ATP: three for synthesis from ADP and one for ATP export to the cytosol. Therefore, 2.5 ATP molecules are synthesized from each NADH molecule.

B. Twenty ATP molecules are produced per molecule of glucose as the result of the citric acid cycle, as compared with 30 ATP molecules that are produced in total from the complete oxidation of one molecule of glucose:

Table A14–5		
PROCESS	DIRECT PRODUCT	FINAL ATP
Glycolysis	2 NADH (cytosolic)	3
	2 ATP	2
Pyruvate oxidation	2 NADH (mitochondrial matrix)	5
Acetyl-CoA oxidation	6 NADH (mitochondrial matrix)	15
(two per glucose)	2 $FADH_2$	3
	2 GTP	2
	Total yield per molecule of glucose	30

Answer 14–6

One can describe four essential roles for the proteins in the process. First, the chemical environment provided by a protein's amino acid side chains sets the redox potential of each Fe ion such that electrons can be passed in a defined order from one component to the next, giving up their energy in small steps and becoming more firmly bound as they proceed. Second, the proteins position the Fe ions so that the electrons can move efficiently between them. Third, the proteins prevent electrons from skipping an intermediate step; thus, as we have learned for other enzymes (discussed in Chapter 4), they channel the electron flow along a defined path. Fourth, the proteins couple the movement of the electrons down their energy ladder to the pumping of protons across the membrane, thereby harnessing the energy that is released and storing it in a proton gradient that is then used for ATP production.

Answer 14–7

It would not be productive to use the same carrier in two steps. If ubiquinone, for example, could transfer electrons directly to the cytochrome oxidase, the cytochrome b-c_1 complex would often be skipped when electrons are collected from NADH dehydrogenase. Given the large difference in redox potential between ubiquinone and cytochrome oxidase, a large amount of energy would be released as heat and thus be wasted. Electron transfer directly between NADH dehydrogenase and cytochrome c would similarly allow the cytochrome b-c_1 complex to be bypassed.

Answer 14–8

Protons pumped across the inner mitochondrial membrane into the intermembrane space equilibrate with the cytosol, which functions as a huge H^+ sink. Both the mitochondrial matrix and the cytosol house many metabolic reactions that require a pH around neutrality. The H^+ concentration difference, ΔpH, that can be achieved between mitochondrial matrix and cytosol is therefore relatively small (one pH unit). Much of the energy stored in the mitochondrial electrochemical proton gradient is instead due to the electrical potential of the membrane (see Figure 14–12).

In contrast, chloroplasts have a smaller, dedicated compartment into which H^+ ions are pumped. Much higher concentration differences can be achieved (up to a thousandfold, or 3 pH units), and much of the energy stored in the thylakoid H^+ gradi-

ent is due to the H^+ concentration difference between the thylakoid space and the stroma.

Answer 14–9

All statements are correct.

A. This is a necessary condition. If it were not true, electrons could not be removed from water and the reaction that splits water molecules ($H_2O \rightarrow 2H^+ + \frac{1}{2}O_2 + 2e^-$) would not occur.

B. This transfer allows the energy of the photon to be harnessed as energy that can be utilized in chemical conversions.

C. It can be argued that this is one of the most important obstacles that had to be overcome during the evolution of photosynthesis: partially reduced oxygen molecules, such as the superoxide radical O_2^-, are dangerously reactive and will attack and destroy almost any biologically active molecule. These intermediates, therefore, have to remain tightly bound to the metals in the active site of the enzyme until all four electrons have been removed from two water molecules. This requires the sequential capture of four photons by the same reaction center.

Answer 14–10

A. Photosynthesis produces sugars, most importantly sucrose, that are transported from the photosynthetic cells through the sap to root cells. There, the sugars are oxidized by glycolysis in the root cell cytoplasm and by oxidative phosphorylation in the root cell mitochondria to produce ATP, as well as being used as the building blocks for many other metabolites.

B. Mitochondria are required even during daylight hours in chloroplast-containing cells to supply the cell with ATP derived by oxidative phosphorylation. Glyceraldehyde 3-phosphate made by photosynthesis in chloroplasts moves to the cytosol and is eventually used as a source of energy to drive ATP production in mitochondria.

Answer 14–11

A. True. NAD^+ and quinones are examples of compounds that do not have metal ions but can participate in electron transfer.

B. False. The potential is due to protons (H^+) that are pumped across the membrane from the matrix to the intermembrane space. Electrons remain bound to electron carriers in the inner mitochondrial membrane.

C. True. Both components add to the driving force that makes it energetically favorable for H^+ to flow back into the matrix.

D. True. Both move rapidly in the plane of the membrane.

E. False. Not only do plants need mitochondria to make ATP in cells that do not have chloroplasts, such as root cells, but mitochondria make most of the cytosolic ATP in all plant cells.

F. True. Chlorophyll's physiological function requires it to absorb light; heme just happens to be a colored compound from which blood derives its red color.

G. False. Chlorophyll absorbs light and transfers energy in the form of an energized electron, whereas the iron in heme is a simple electron carrier.

H. False. Most of the dry weight of a tree comes from carbon derived from the CO_2 that has been fixed during photosynthesis.

Answer 14–12

It takes three protons. The precise value of the ΔG for ATP synthesis depends on the concentrations of ATP, ADP, and P_i (as described in Chapter 3). The higher the ratio of the concentration of ATP to ADP, the more energy it takes to make additional ATP. The lower value of 11 kcal/mole therefore applies to conditions where cells have expended a lot of energy and have therefore decreased the normal ATP/ADP ratio.

Answer 14–13

If no O_2 is available, all components of the mitochondrial electron-transport chain will accumulate in their *reduced* form. This is the case because electrons derived from NADH enter the chain but cannot be transferred to O_2. The electron-transport chain therefore stalls with all of its components in the reduced form. If O_2 is suddenly added again, the electron carriers in cytochrome oxidase will become *oxidized before* those in NADH dehydrogenase. This is true because, after O_2 addition, cytochrome oxidase will donate its electrons directly to O_2, thereby becoming oxidized. A wave of increasing oxidation then passes backward with time from cytochrome oxidase through the components of the electron-transport chain, as each component regains the opportunity to pass on its electrons to downstream components.

Answer 14–14

As oxidized ubiquinone becomes reduced, it picks up two electrons but also two protons from water (Figure 14–20). Upon oxidation, these protons are released. If reduction occurs on one side of the membrane and oxidation at the other side, a proton is pumped across the membrane for each electron transported. Electron transport by ubiquinone thereby contributes directly to the generation of the H^+ gradient.

Answer 14–15

Photosynthetic bacteria and plant cells use the electrons derived in the reaction $2H_2O \rightarrow 4e^- + 4H^+ + O_2$ to reduce $NADP^+$ to NADPH, which is then used to produce useful metabolites. If the electrons were used instead to produce H_2 in addition to O_2, the cells would lose any benefit they derive from carrying out the reaction, because the electrons could not take part in metabolically useful reactions.

Answer 14–16

A. The switch in solutions creates a pH gradient across the thylakoid membrane. The flow of H^+ ions down its electrochemical potential drives ATP synthase, which converts ADP to ATP.

B. No light is needed, because the H^+ gradient is established artificially without a need for the light-driven electron-transport chain.

C. Nothing. The H^+ gradient would be in the wrong direction; ATP synthase would not work.

D. The experiment provided early supporting evidence for the chemiosmotic model by showing that an H^+ gradient alone is sufficient to drive ATP synthesis.

Answer 14–17

A. When the vesicles are exposed to light, H^+ ions (derived from H_2O) pumped into the vesicles by the bacteriorhodopsin flow back out through the ATP synthase, causing ATP to be made in the solution surrounding the vesicles in response to light.

B. If the vesicles are leaky, no H^+ gradient can form and thus ATP synthase cannot work.

C. Using components from widely divergent organisms can be a very powerful experimental tool. Because two proteins come from such different sources, very unlikely that they form a direct functional interaction. The experiment therefore strongly suggests that electron transport and ATP synthesis are separate events. This approach is therefore a valid one.

Answer 14–18

The redox potential of $FADH_2$ is too low to transfer electrons to the NADH dehydrogenase complex, but high enough to transfer electrons to ubiquinone (Figure 14–21). Therefore, electrons from $FADH_2$ can enter the electron-transport chain only at this step (Figure A14–18). Because the NADH dehydrogenase complex is bypassed, fewer H^+ ions are pumped across the membrane and less ATP is made. This example shows the versatility of the electron-transport chain. The ability to use vastly different sources of electrons from the environment to feed electron transport is thought to have been an essential feature in the early evolution of life.

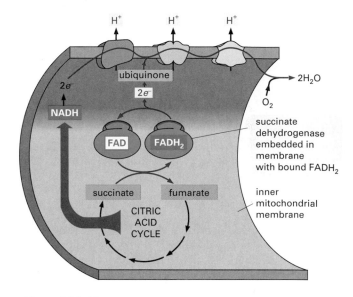

Figure A14–18

Answer 14–19

If these bacteria used a proton gradient to make their ATP in a fashion analogous to other bacteria (that is, fewer protons inside than outside), they would need to raise their cytoplasmic pH even higher than that of their environment (pH 10). Cells with a cytoplasmic pH greater than 10 would not be viable. These bacteria, therefore, must use gradients of ions other than H^+, such as Na^+ gradients, in the chemiosmotic coupling between electron transport and an ATP synthase.

Answer 14–20

Statements A and B are accurate. Statement C is incorrect, because the chemical reactions that are carried out in each cycle are completely different, even though the net effect is the same as that expected for simple reversal.

Answer 14–21

This experiment would suggest a two-step model for ATP synthase function. According to this model, the flow of protons through the base of the synthase drives rotation of the head, which in turn causes ATP synthesis. In their experiment, the authors have succeeded in uncoupling these two steps. If

rotating the head mechanically is sufficient to produce ATP in the absence of any applied proton gradient, the ATP synthase is a protein machine that indeed functions like a "molecular turbine." This would be a very exciting experiment, indeed, because it would directly demonstrate the relationship between mechanical movement and enzymatic activity. There is no doubt that it should be published and that it would become a "classic."

Answer 14–22

Only under condition (E) is electron transfer observed, with cytochrome c becoming reduced. A portion of the electron-transport chain has been reconstituted in this mixture, so that electrons can flow in the energetically favored direction from reduced ubiquinone to the cytochrome b-c_1 complex to cytochrome c. Although energetically favorable, the transfer in (A) cannot occur spontaneously in the absence of the cytochrome b-c_1 complex to catalyze this reaction. No electron flow occurs in the other experiments, whether the cytochrome b-c_1 complex is present or not: in experiments (B) and (F) both ubiquinone and cytochrome c are oxidized; in experiments (C) and (G) both are reduced; and in experiments (D) and (H) electron flow is energetically disfavored because reduced cytochrome c has a lower free energy than oxidized ubiquinone.

Chapter 15

Answer 15–1

Although the nuclear envelope forms one continuous membrane, it has specialized regions, which contain special proteins and have a characteristic appearance. One such specialized region is the inner nuclear membrane. Membrane proteins can indeed diffuse between the inner and outer nuclear membranes, at the connections formed around the nuclear pores. Those proteins with particular functions in the inner membrane, however, are usually anchored there by their interaction with other components such as chromosomes and the nuclear lamina, a protein meshwork underlying the inner nuclear membrane that helps give structural integrity to the nuclear envelope.

Answer 15–2

Eucaryotic gene expression is more complicated than procaryotic gene expression. In particular, procaryotic cells do not have introns that interrupt the coding sequences of their genes, so that an mRNA can be translated immediately after it is transcribed, without a need for further processing (discussed in Chapter 7). In fact, in procaryotic cells ribosomes start translating most mRNAs before transcription is finished. This would have disastrous consequences in eucaryotic cells, because most RNA transcripts have to be spliced before they can be translated. The nuclear envelope separates the transcription and translation processes in space and time. A primary RNA transcript is held in the nucleus until it is properly processed to form an mRNA, and only then is it allowed to leave the nucleus so that ribosomes can translate it.

Answer 15–3

An mRNA molecule is attached to the ER membrane by the ribosomes translating it. This ribosome population, however, is not static; the mRNA is continuously moved through the ribosome. Those ribosomes that have finished translation dissociate from the 3′ end of the mRNA and from the ER membrane, but the mRNA itself remains bound by other ribosomes, newly recruited from the cytosolic pool, that have attached to the 5′ end of the

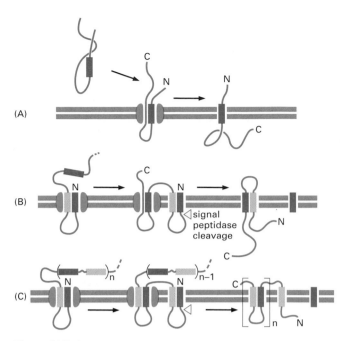

Figure A15–4

mRNA and are still translating the mRNA. Depending on its length, there are about 10–20 ribosomes attached to each membrane-bound mRNA molecule.

Answer 15–4

A. The internal signal sequence functions as a membrane anchor, as shown in Figure 15–16. Because there is no stop-transfer sequence, however, the C-terminal end of the protein continues to be translocated into the ER lumen. The resulting protein therefore has its N-terminal domain in the cytosol, followed by a single transmembrane segment, and a C-terminal domain in the ER lumen (Figure A15–4A).

B. The N-terminal signal sequence initiates translocation of the N-terminal domain of the protein until translocation is stopped by the stop-transfer sequence. A cytosolic domain is synthesized until the start-transfer sequence initiates translocation again. The situation now resembles that described in (A), and the C-terminal domain of the protein is translocated into the lumen of the ER. The resulting protein therefore spans the membrane twice. Both its N-terminal and C-terminal domains are in the ER lumen, and a loop domain between the two transmembrane regions is exposed in the cytosol (Figure A15–4B).

C. It would need a cleaved signal sequence, followed by an internal stop-transfer sequence, followed by pairs of start- and stop-transfer sequences (Figure A15–4C).

These examples demonstrate that complex protein topologies can be achieved by simple variations and combinations of the two basic mechanisms shown in Figures 15–15 and 15–16.

Answer 15–5

A. Clathrin coats cannot assemble in the absence of adaptins that link the clathrin to the membrane. At high clathrin concentrations and under the appropriate ionic conditions, clathrin cages assemble in solution, but they are empty shells, lacking other proteins, and they contain no membrane. This shows that the information to form clathrin baskets is contained in the clathrin molecules themselves, which are therefore able to self-assemble.

B. Without clathrin, adaptins still bind to receptors in the membrane, but no clathrin coat can form and thus no clathrin-coated pits or vesicles are produced.

C. Deeply invaginated clathrin-coated pits form on the membrane, but they do not pinch off to form closed vesicles.

D. Procaryotic cells do not perform endocytosis. A procaryotic cell therefore does not contain any receptors with appropriate cytosolic tails that could mediate adaptin binding. Therefore, no clathrin can bind and no clathrin coats can assemble.

Answer 15–6

The preassembled sugar chain allows for better quality control. The assembled oligosaccharide chains can be checked for accuracy before they are added to the protein; if a mistake were made in adding sugars individually to the protein, the whole protein would have to be discarded. Because far more energy is used in building a protein than in building a short oligosaccharide chain, this is a much more economical strategy. Also, once a sugar tree is added to a protein, it is more difficult for enzymes to modify its branches, compared with modifying them on the free sugar tree. This difficulty becomes apparent as the protein moves to the cell surface: although sugar chains are continually modified by enzymes in various compartments of the secretory pathway, these modifications are often incomplete and result in considerable heterogeneity of the glycoproteins that leave the cell. This heterogeneity is largely due to the restricted access that the enzymes have to the sugar trees attached to the surface of proteins. The heterogeneity also explains why glycoproteins are more difficult to study and purify than nonglycosylated proteins.

Answer 15–7

Aggregates of the secretory proteins would form in the ER, just as they do in the *trans* Golgi network. As the aggregation is specific for secretory proteins, ER proteins would be excluded from the aggregates. The aggregates eventually would be degraded.

Answer 15–8

Transferrin without Fe bound does not interact with its receptor and circulates in the bloodstream until it catches an Fe ion. Once iron is bound, the iron–transferrin complex can bind to the transferrin receptor on the surface of a cell and be endocytosed. Under the acidic conditions of the endosome, the transferrin releases its iron, but the transferrin remains bound to the transferrin receptor, which is recycled back to the cell surface, where it encounters the neutral pH environment of the blood. The neutral pH causes the receptor to release the transferrin into the circulation, where it can pick up another Fe ion to repeat the cycle. The iron released in the endosome, like the LDL in Figure 15–32, moves on to lysosomes, from where it is transported into the cytosol.

The system allows cells to take up iron efficiently even though the concentration of iron in the blood is extremely low. The iron bound to transferrin is concentrated at the cell surface by binding to transferrin receptors; it becomes further concentrated in clathrin-coated pits, which collect the transferrin receptors. In this way, transferrin cycles between the blood and endosomes, delivering the iron that cells need to grow.

Answer 15–9

A. True.

B. False. The signal sequences that direct proteins to the ER contain a core of eight or more hydrophobic amino acids. The sequence shown here contains many hydrophilic amino acid side chains, including the charged amino acids His, Arg, Asp, and Lys, and the uncharged hydrophilic amino acids Gln and Ser.

C. True. Otherwise they could not dock at the correct target membrane or recruit a fusion complex to a docking site.

D. True.

E. True. Lysosomal proteins are selected in the *trans* Golgi network and packaged into transport vesicles that deliver them to the late endosome. If not selected, they would enter by default into transport vesicles that move constitutively to the cell surface.

F. False. Lysosomes also digest internal organelles by autophagocytosis.

G. False. Mitochondria do not participate in vesicular transport, and therefore *N*-linked glycoproteins, which are exclusively assembled in the ER, cannot be transported to mitochondria.

Answer 15–10

They must contain a nuclear localization signal as well. Proteins with nuclear export signals shuttle between the nucleus and the cytosol. An example is the A1 protein, which binds to mRNAs in the nucleus and guides them through the nuclear pores. Once in the cytosol, a nuclear localization signal ensures that the A1 protein is reimported so that it can participate in the export of further mRNAs.

Answer 15–11

Influenza virus enters cells by endocytosis and is delivered to endosomes, where it encounters an acidic pH that activates its fusion protein. The viral membrane then fuses with the membrane of the endosome, releasing the viral genome into the cytosol (Figure A15–11). NH_3 is a small molecule that readily penetrates membranes. Thus it can enter all intracellular compartments, including endosomes, by diffusion. Once in a compartment that has an acidic pH, NH_3 binds H^+ to form NH_4^+, which is a charged ion and therefore cannot cross the membrane by diffusion. NH_4^+ ions therefore accumulate in acidic compartments, raising their pH. When the pH of the endosome is raised, viruses are still endocytosed, but because the viral fusion protein cannot be activated, the virus cannot enter the cytosol. Remember this the next time you have the flu and have access to a stable.

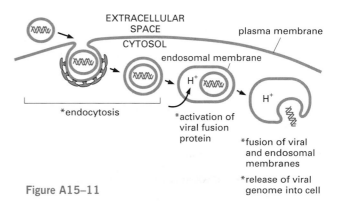

Figure A15–11

Answer 15–12

A. The problem is that vesicles having two different kinds of v-SNAREs in their membrane could dock on either of two different membranes.

B. The answer to this puzzle is presently not known, but we can predict that cells must have ways of turning the docking ability of SNAREs on and off. This may be achieved through other proteins that are, for example, copackaged in the ER with SNAREs into transport vesicles and facilitate the interactions of the correct v-SNARE with the t-SNARE in the *cis* Golgi network.

Answer 15–13

Synaptic transmission involves the release of neurotransmitters by exocytosis. During this event, the membrane of the synaptic vesicle fuses with the plasma membrane of the nerve terminals. To make new synaptic vesicles, membrane must be retrieved from the plasma membrane by endocytosis. This endocytosis step is blocked if dynamin is defective, as the protein seems to be required to pinch off the clathrin-coated endocytic vesicles. The clue to deciphering the role of dynamin came from electron micrographs of synapses of the mutant flies (Figure A15–13). Note that there are many flasklike invaginations of the plasma membrane, representing deeply invaginated clathrin-coated pits that cannot pinch off. The collars visible around the necks of these invaginations are made of mutant dynamin.

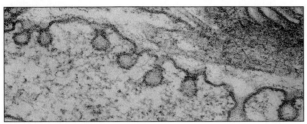

Figure A15–13

Courtesty of Kazuo Ikeda

Answer 15–14

The first two sentences are correct. The third is not. It should read: "Because the contents of the lumen of the ER or any other compartment in the secretory or endocytic pathways never mix with the cytosol, proteins that enter these pathways will never need to be imported again." When the nuclear envelope and the ER break down in mitosis, they form vesicles whose contents remain separated from the cytosol by the vesicle membrane.

Answer 15–15

The protein is translocated into the ER. Its ER signal sequence is recognized as soon as it emerges from the ribosome. The ribosome then becomes bound to the ER membrane, and the growing polypeptide chain is transferred through the ER translocation channel. The nuclear localization sequence is therefore never exposed to the cytosol. It will never encounter nuclear import receptors, and the protein will not enter the nucleus.

Answer 15–16

(1) Proteins are imported into the nucleus after they have been synthesized, folded, and, if appropriate, assembled into complexes. In contrast, unfolded polypeptide chains are translocated into the ER as they are being made by the ribosomes. Ribosomes are assembled in the nucleus yet function in the cytosol, and the enzyme complexes that catalyze RNA transcription and splicing are assembled in the cytosol yet function in the nucleus. Thus both ribosomes and these enzyme complexes need to be transported through the nuclear pores intact. (2) Nuclear pores are gates, which are always open to small molecules; in contrast, translocation channels in the ER membrane are normally closed (as indicated by the "plug" in Figure 15–14), and open only after the ribosome attaches to the membrane and the translocating polypeptide chain seals the channel from the cytosol. It is important that the ER membrane remain impermeable to small molecules during the translocation process, as the ER is a major store for Ca^{2+} in the cell, and Ca^{2+} release into the

cytosol must be tightly controlled (discussed in Chapter 16). (3) Nuclear localization signals are not cleaved off after protein import into the nucleus; in contrast, ER signal peptides are usually cleaved off. Nuclear localization signals are needed to repeatedly reimport nuclear proteins after they have been released into the cytosol during mitosis, when the nuclear envelope breaks down.

Answer 15–17

The transient intermixing of nuclear and cytosolic contents during mitosis supports the idea that the nuclear interior and the cytosol are indeed evolutionarily related. In fact, one can consider the nucleus as a subcompartment of the cytosol that has become surrounded by the nuclear envelope, with access only through the nuclear pores.

Answer 15–18

The actual explanation is that the single amino acid change causes the protein to misfold slightly so that, although it is still active as a protease inhibitor, it is prevented by chaperone proteins in the ER from exiting this organelle. It therefore accumulates in the ER lumen and is eventually degraded. Alternative interpretations might have been that (1) the mutation affects the stability of the protein in the bloodstream so that it is degraded much faster in the blood than the normal protein, or (2) the mutation inactivates the ER signal sequence and prevents the protein from entering the ER. (3) Another explanation could have been that the mutation altered the sequence to create an ER retention signal, which would have retained the mutant protein in the ER. One could distinguish between these possibilities by using fluorescent-tagged antibodies against the protein or express the protein as a fusion with GFP to follow its transport in the cells (see Panel 4–6, pp. 164–165 and How We Know, pp. 520–521).

Answer 15–19

Critique: "Dr. Outonalimb proposes to study the biosynthesis of forgettin, a protein of significant interest. The main hypothesis on which this proposal is based, however, requires further support. In particular it is questionable whether forgettin is indeed a secreted protein, as proposed. ER signal sequences are normally found at the N-terminus. C-terminal hydrophobic sequences will be exposed outside the ribosome only after protein synthesis has already terminated and can therefore not be recognized by an SRP during translation. It is therefore unlikely that forgettin will be translocated by an SRP-dependent mechanism, and it may therefore remain in the cytosol. Dr. Outonalimb should take these considerations into account when submitting a revised application."

Answer 15–20

The Golgi apparatus may have evolved from specialized patches of ER membrane. These regions of the ER might have pinched off, forming a new compartment (Figure A15–20), which still communicates with the ER by vesicular transport. For the newly evolved Golgi compartment to be useful, transport vesicles would also have to have evolved.

Answer 15–21

This is a chicken-and-egg question. In fact, the situation never arises in present-day cells, although it must have posed a considerable problem for the first cells that evolved. New cell membranes are made by expansion of existing membranes, and the ER is never made *de novo*. There will always be an existing piece of ER with translocation channels to integrate new translocation channels. Inheritance is therefore not limited to the propagation of the genome; a cell's organelles must also be passed

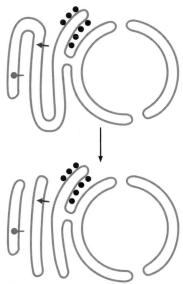

Figure A15–20

from generation to generation. In fact, the ER translocation channels can be traced back to structurally related translocation channels in the procaryotic plasma membrane.

Answer 15–22

    A.   Extracellular space
    B.   Cytosol
    C.   Plasma membrane
    D.   Clathrin coat
    E.   Membrane of deeply invaginated clathrin-coated pit
    F.   Captured cargo particles
    G.   Lumen of deeply invaginated clathrin-coated pit

Chapter 16

Answer 16–1

Most paracrine signaling molecules are very short-lived after they are released from a signaling cell: they are either degraded by extracellular enzymes or are rapidly taken up by neighboring target cells. In addition, some become attached to the extracellular matrix and are thus prevented from diffusing too far.

Answer 16–2

Each photon causes the hydrolysis of 200,000 cyclic GMP molecules; i.e., the signal is amplified 200,000-fold (= 500 × 4000 × 0.1).

Answer 16–3

Polar groups are hydrophilic, and cholesterol, having only one –OH group, would be too hydrophobic to be an effective hormone. Because it is virtually insoluble in water, it could not move readily as a messenger from one cell to another via the extracellular fluid.

Answer 16–4

In the case of the steroid hormone receptor, a one-to-one complex of steroid and receptor binds to DNA to activate or inactivate gene transcription; there is thus no amplification between ligand binding and transcriptional regulation. Amplification occurs later, because transcription of a gene gives rise to many mRNAs that are each translated to give many protein molecules (discussed in Chapter 7). For the ion-channel–linked receptors,

a single ion channel will let through thousands of ions in the time it remains open; this serves as the amplification step in this type of signaling system.

Answer 16–5

The mutant G protein would be almost continuously activated, as GDP would spontaneously dissociate, allowing GTP to bind even in the absence of an activated G-protein–linked receptor. The consequences for the cell would therefore be similar to those caused by cholera toxin, which modifies the α subunit of G_s so that it cannot hydrolyze GTP to shut itself off. In contrast to the cholera toxin case, however, the mutant G protein would not stay permanently activated: it would switch itself off normally but then instantly become activated again as the GDP dissociated and GTP re-bound.

Answer 16–6

Rapid breakdown keeps the intracellular cyclic AMP concentrations low. The lower the cAMP levels are, the larger and faster the increase achieved upon activation of adenylyl cyclase, which makes new cyclic AMP. If you have $100 in the bank and you deposit another $100, you have doubled your wealth; if you have only $10 to start with and you deposit $100, you have increased your wealth tenfold, a much larger proportional increase resulting from the same deposit.

Answer 16–7

Recall that the plasma membrane constitutes a rather small area compared with the total membrane surfaces in a cell (discussed in Chapter 15). The endoplasmic reticulum is especially abundant and spans the entire volume of the cell as a vast network of membrane tubes and sheets. The Ca^{2+} stored in the endoplasmic reticulum can therefore be released throughout the cytosol. This is important because the fast clearing of Ca^{2+} ions from the cytosol by Ca^{2+} pumps prevents Ca^{2+} from diffusing any significant distance in the cytosol.

Answer 16–8

Each reaction involved in the amplification scheme must be turned off to reset the signaling pathway to a resting level. Each of these off switches is equally important.

Answer 16–9

Because each antibody has two antigen-binding sites, binding to the receptors can induce receptor clustering on the cell surface. This is likely to activate receptor tyrosine kinases, which are usually activated by self-phosphorylation after individual kinase domains of the receptors have been brought into proximity. The activation of G-protein–linked receptors is more complicated, because the ligand has to induce a particular conformational change. Only very special antibodies mimic receptor ligands sufficiently well to induce the conformational change that activates a G-protein–linked receptor.

Answer 16–10

The more steps there are in an intracellular signaling pathway, the more places the cell has to regulate the pathway, amplify the signal, integrate signals from different pathways, and spread the signal along divergent paths (see Figure 16–8).

Answer 16–11

    A.   True. Acetylcholine, for example, decreases the beating of heart muscle cells by binding to a G-protein–linked receptor and stimulates the contraction of skeletal muscle cells by binding to a different acetylcholine receptor, which is a ligand-gated ion channel.

B. False. Acetylcholine is short-lived and exerts its effects locally. Indeed, the consequences of prolonging its lifetime can be disastrous. Compounds that inhibit the enzyme acetylcholinesterase, which normally breaks down acetylcholine at a nerve–muscle synapse, are extremely toxic: an example is the nerve gas sarin, used in chemical warfare.

C. True. Nucleotide-free βγ complexes can activate ion channels, and GTP-bound α subunits can activate enzymes. The GDP-bound form of trimeric G proteins is the inactive state.

D. True. The inositol phospholipid that is cleaved to produce IP_3 contains three phosphate groups, one of which links the sugar to the diacylglycerol lipid. IP_3 is generated by a simple hydrolysis reaction (see Figure 16–25).

E. False. Calmodulin senses but does not regulate intracellular Ca^{2+} levels.

F. True. See Figure 16–38.

G. False. *ras* is a proto-oncogene. It becomes an oncogene, i.e., promotes the development of cancer, if it harbors mutations that keep it in an active state all the time.

H. True. See Figure 16–30.

Answer 16–12

1. You would expect a high background level of Ras activity because Ras cannot be turned off efficiently.

2. As some Ras molecules are already GTP-bound, Ras activity in response to an extracellular signal would be greater than normal, but would be liable to saturate when all Ras molecules are converted to the GTP-bound form.

3. The response to a signal would be much less rapid, because the signal-dependent increase in GTP-bound Ras would occur over an elevated background of preexisting GTP-bound Ras.

4. The increase in Ras activity in response to a signal would also be prolonged compared to normal cells.

Answer 16–13

A. Both types of signaling can occur over long-range: neurons can send action potentials along very long axons (think of the axons in the neck of a giraffe, for example), and hormones are passed through the bloodstream throughout the organism. Because neurons secrete large amounts of neurotransmitters at a synapse, a small, well-defined space between two cells, the concentrations of signal molecules are high; neurotransmitter receptors, therefore, need to bind to neurotransmitters with only low affinity. Hormones, in contrast, are vastly diluted in the bloodstream, where they circulate at often minuscule concentrations; hormone receptors, therefore, generally bind their hormone with extremely high affinities.

B. Whereas neuronal signaling is a private affair, one neuron talking to a select group of target cells through specific synaptic connections, hormonal signaling is a public announcement, with target cells sensing the hormone levels in the blood. Neuronal signaling is very fast, limited only by the speed of propagation of the action potential and the workings of the synapse, whereas hormonal signaling is slower, limited by blood flow and diffusion over larger distances.

Answer 16–14

A. There are 100,000 molecules of X and 10,000 molecules of Y in the cell (= rate of synthesis × average lifetime).

B. After one second, the concentration of X will have increased by 10,000 molecules. The concentration of X, therefore, one second after its synthesis is increased, is about 110,000 molecules per cell—which is a 10% increase over the concentration of X before the boost of its synthesis. The concentration of Y will also increase by 10,000 molecules, which, in contrast to X, represents a full twofold increase in its concentration (for simplicity, we can neglect the breakdown in this estimation because X and Y are relatively stable during the one-second stimulation).

C. Because of its larger proportional increase, Y is the preferred signal molecule. This calculation illustrates the surprising but important principle that the time it takes to switch a signal on is determined by the lifetime of the signal molecule.

Answer 16–15

The information transmitted by a cell-signaling pathway is contained in the *concentration* of the messenger, be it a small molecule or a phosphorylated protein. Therefore, to allow detection of a change in concentration, the original messenger has to be rapidly destroyed. The shorter the lifetime of the messenger, the faster the system can respond to changes. Human communication relies on messages that are delivered only once and that are generally not interpreted by their abundance, but by their *content*. So it is a mistake to kill the messengers; they can be used more than once.

Answer 16–16

A. The mutant receptor tyrosine kinase lacking its extracellular ligand-binding domain is inactive. It cannot bind extracellular signals, and its presence has no consequences for the function of the normal receptor kinase (Figure A16–16A).

B. The mutant receptor lacking its intracellular domain is also inactive, but its presence will block signaling by the normal receptors. When a signal molecule binds to either receptor, it will induce their dimerization. Two normal receptors have to come together to activate each other by phosphorylation. In the presence of an excess of mutant receptor, however, normal receptors will usually form mixed dimers, in which their intracellular domain cannot be activated because their partner is a mutant and lacks a kinase domain (Figure A16–16B).

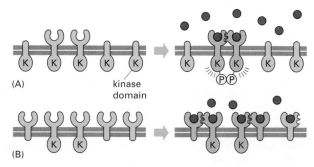

Figure A16–16

Answer 16–17

The statement is correct. Upon ligand binding, transmembrane helices of multispanning receptors, like the G-protein–linked

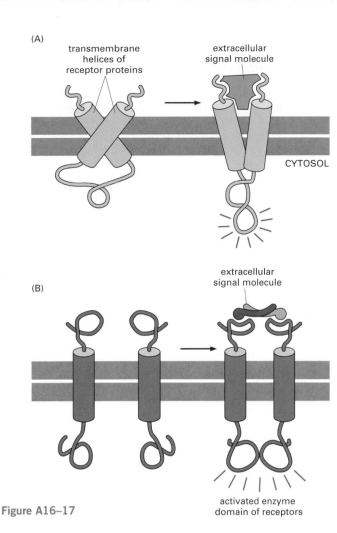

(A) transmembrane helices of receptor proteins — extracellular signal molecule — CYTOSOL

(B) extracellular signal molecule — activated enzyme domain of receptors

Figure A16–17

receptors, shift and rearrange with respect to one another (Figure A16–17A). This conformational change is sensed on the cytosolic side of the membrane because of a change in the arrangement of the cytoplasmic loops. A single transmembrane segment is not sufficient to transmit a signal across the membrane directly; no rearrangements in the membrane are possible upon ligand binding. Upon ligand binding, single-span receptors such as receptor tyrosine kinases tend to dimerize, thereby bringing their intracellular kinase domains into proximity so that they can cross-phosphorylate and activate each other (Figure A16–17B).

Answer 16–18

Activation in both cases depends on proteins that catalyze GDP–GTP exchange on the G protein or Ras protein. Whereas activated G-protein–linked receptors perform this function directly for G proteins, enzyme-linked receptors assemble multiple adaptor proteins into a signaling complex when the receptors are activated by phosphorylation, and one of these adaptors recruits a Ras-activating protein that fulfills this function for Ras.

Answer 16–19

Because the intracellular concentration of Ca^{2+} is so low, an influx of relatively few Ca^{2+} ions leads to large changes in its cytosolic concentration. Thus a tenfold increase in Ca^{2+} can be achieved by raising the concentration of Ca^{2+} into the micromolar range, which would require far fewer ions than would be required to change significantly the concentration of a more abundant ion such as Na^+. In muscle, a greater than tenfold change in cytosolic Ca^{2+} concentration can be achieved in

microseconds by releasing Ca^{2+} from the intracellular stores of the sarcoplasmic reticulum, a task that would be difficult to accomplish if changes in the millimolar range were required.

Answer 16–20

In a multicellular organism such as an animal, it is important that cells survive only when and where they are needed. Having cells depend on signals from other cells may be a simple way of ensuring this. A misplaced cell, for example, would probably fail to get the survival signals it needs (as its neighbors would be inappropriate) and would therefore kill itself. This strategy can also help regulate cell numbers: if cell type A depends on a survival signal from cell type B, the number of B cells could control the number of A cells by making a limited amount of the survival signal, so that only a certain number of A cells could survive. There is indeed evidence that such a mechanism does operate to help regulate cell numbers—both in developing and adult tissues (see Figure 18–27).

Answer 16–21

Ca^{2+}-activated Ca^{2+} channels create a positive feedback loop: the more Ca^{2+} that is released, the more Ca^{2+} channels open. The Ca^{2+} signal in the cytosol is therefore propagated explosively throughout the entire muscle cell, thereby ensuring that all myosin–actin filaments contract almost synchronously.

Answer 16–22

K2 activates K1. If K1 is permanently activated, a response is observed independent of the status of K2. If the order were reversed, K1 would need to activate K2, which cannot occur because in our example K2 contains an inactivating mutation.

Answer 16–23

A. Extracellular signal → receptor tyrosine kinase → adaptor protein → Ras-activating protein → MAP-kinase-kinase-kinase → MAP-kinase-kinase → MAP-kinase → gene regulatory protein,
or
extracellular signal → G-protein–linked receptor → G protein → phospholipase C → IP_3 → Ca^{2+} → calmodulin → CaM-kinase → gene regulatory protein,
or
extracellular signal → G-protein–linked receptor → G protein → adenylyl cyclase → cyclic AMP → PKA → gene regulatory protein.

B. TGF-β → TGF-β receptor → SMAD gene regulatory protein,
or
cytokine → cytokine receptor → JAK kinase → STAT gene regulatory protein.

Answer 16–24

Animals and plants are thought to have evolved multicellularity independently and therefore will be expected to have evolved some distinct signaling mechanisms for their cells to communicate with one another. On the other hand, animal and plant cells are thought to have evolved from a common eucaryotic ancestor cell, and so plants and animals would be expected to share some intracellular signaling mechanisms that the common ancestor cell used to respond to its environment.

Chapter 17

Answer 17–1

Cells that migrate rapidly from one place to another, such as amoebae (A) and sperm cells (F), do not in general need intermediate filaments in their cytoplasm, since they do not develop or sustain large tensile forces. Plant cells (G) are pushed and pulled by the forces of wind and water, but they resist these forces by means of their rigid cell walls, rather than by their cytoskeleton. Epithelial cells (B), smooth muscle cells (C), and the long axons of nerve cells (E) are all rich in cytoplasmic intermediate filaments, which prevent them from rupturing as they are stretched and compressed by the movements of their surrounding tissues.

All of the above eucaryotic cells possess at least intermediate filaments in their nuclear lamina. Bacteria, such as *Escherichia coli* (D), have none whatsoever.

Answer 17–2

Two tubulin dimers have a lower affinity for each other (because of a more limited number of interaction sites) than a tubulin dimer has for the end of a microtubule (where there are multiple possible interaction sites, both end-to-end of tubulin dimers adding to a protofilament and side-to-side of the tubulin dimers interacting with tubulin subunits in adjacent protofilaments forming the ringlike cross section). Thus, to initiate a microtubule from scratch, enough tubulin dimers have to come together and remain bound to one another for long enough for other tubulin molecules to add to them. Only when a number of tubulin dimers have already assembled will the binding of the next subunit be favored. The formation of these initial "nucleating sites" is therefore rare and will not occur spontaneously at cellular tubulin concentrations.

Centrosomes contain preassembled rings of γ-tubulin (in which the γ-tubulin subunits are held together in much tighter side-to-side interactions than αβ-tubulin can form) to which αβ-tubulin dimers can bind. The binding conditions of αβ-tubulin dimers resemble those of adding to the end of an assembled microtubule. The γ-tubulin rings in the centrosome can therefore be thought of as permanently preassembled nucleation sites.

Answer 17–3

A. The microtubule is shrinking because it has lost its GTP cap, i.e., the tubulin subunits at its end are all in their GDP-bound form. GTP-loaded tubulin subunits from solution will still add to this end, but they will be short-lived—either because they hydrolyze their GTP or because they fall off as the microtubule rim around them disassembles. If, however, enough GTP-loaded subunits are added quickly enough to cover up the GDP-containing tubulin subunits at the microtubule end, then a new GTP cap can form and regrowth is favored.

B. The rate of addition of GTP-tubulin will be greater at higher tubulin concentrations. The frequency with which shrinking microtubules switch to the growing mode will therefore increase with increasing tubulin concentration. The consequence of this regulation is that the system is self-balancing: the more microtubules shrink (resulting in a higher concentration of free tubulin), the more frequently microtubules will start to grow again. Conversely, the more microtubules grow, the lower the concentration of free tubulin will become and the rate of GTP-tubulin addition will slow down; at some point GTP hydrolysis will catch up with new GTP-tubulin addition, the GTP cap will be destroyed, and the microtubule will switch to the shrinking mode.

C. If only GDP were present, microtubules would continue to shrink and eventually disappear, because tubulin dimers with GDP have very low affinity for each other and will not add stably to microtubules.

D. If GTP is present but cannot be hydrolyzed, microtubules will continue to grow until all free tubulin subunits have been used up.

Answer 17–4

If all the dynein arms were equally active, there could be no significant relative motion of one microtubule to the other as required for bending (think of a circle of nine weight lifters, each trying to lift his neighbor off the ground: if they all succeeded, the group would levitate!). Thus, a few ciliary dynein molecules must be activated selectively on one side of the cilium. As they move their neighboring microtubules toward the tip of the cilium, the cilium bends away from the side containing the activated dyneins.

Answer 17–5

Any actin-binding protein that stabilizes complexes of two or more actin monomers without blocking the ends required for filament growth will facilitate the initiation of a new filament (nucleation).

Answer 17–6

Only fluorescent actin molecules assembled into filaments are visible, since unpolymerized actin molecules diffuse so rapidly they produce a dim uniform background. Since in your experiment, so few actin molecules are labeled (1:10,000), there should be at most one labeled actin monomer per filament (see Figure 17–30). The lamellipodium as a whole has many actin filaments some of which overlap and therefore show a random speckled pattern of actin molecules, each marking a different filament.

This technique (called "speckle fluorescence") can be used to follow the movement of polymerized actin in a migrating cell. If you watch this pattern with time, you will see that individual fluorescent spots move steadily back from the leading edge toward the interior of the cell, a movement that occurs whether or not the cell is actually migrating. Rearward movement takes place because actin monomers are added to filaments at the plus end and are lost from the minus end (where they are depolymerized) (see Figure 17–36). In effect actin monomers "move through" the actin filaments, a phenomenon termed "treadmilling." Treadmilling has been demonstrated to occur in isolated actin filaments in solution and also in dynamic microtubules, such as those within a mitotic spindle.

Answer 17–7

Cells contain actin-binding proteins that bundle and cross-link actin filaments (see Figure 17–32). The filaments extending from lamellipodia and filopodia become firmly connected to the filamentous meshwork of the cell cortex, thus providing the mechanical anchorage required for the growing rodlike filaments to deform the cell membrane.

Answer 17–8

Although the subunits are indeed held together by noncovalent bonds that are individually weak, there are a very large number of them, distributed among a very large number of filaments. As a result, the stress a human being exerts by lifting a heavy object is dispersed over so many subunits that their interaction strength is not exceeded. By analogy, a single thread of silk is not nearly strong enough to hold a human, but a rope woven of such fibers is.

Answer 17–9

Both filaments are composed of subunits in the form of protein dimers that are held together by coiled-coil interactions. Moreover, in both cases, the dimers polymerize through their coiled-coil domains into filaments. Whereas intermediate filament dimers assemble head-to-head, however, and thereby create a filament that has no polarity, all myosin molecules in the same half of the myosin filament are oriented with their heads pointing in the same direction. This polarity is necessary for them to be able to develop a contractile force in muscle.

Answer 17–10

A. Successive actin molecules in an actin filament are identical in position and conformation. After a first protein (such as troponin) had bound to the actin filament, there would be no way a second protein could recognize every seventh monomer in a naked actin filament. Tropomyosin, however, binds along the length of an actin filament, spanning precisely seven monomers, and thus provides a molecular "ruler" that measures the length of seven actin monomers. Troponin becomes localized by binding to the end of a tropomyosin molecule.

B. Calcium ions influence force generation in the actin–myosin system only if both troponin (to bind the calcium ions) and tropomyosin (to transmit the information that troponin has bound calcium to the actin filament) are present. (i) Troponin cannot bind to actin without tropomyosin. The actin filament would be permanently exposed to the myosin, and the system would be continuously active, independent of whether calcium ions were present or not (a muscle cell would therefore be continuously contracted with no possibility of regulation). (ii) Tropomyosin would bind to actin and block binding of myosin completely; the system would be permanently inactive, no matter whether calcium ions were present, as tropomyosin is not affected by calcium. (iii) The system will contract in response to calcium ions.

Answer 17–11

A. True. A continual outward movement of ER is required; in the absence of microtubules, the ER collapses toward the center of the cell.

B. True. Actin is needed to make the contractile ring that causes the physical cleavage between the two daughter cells, whereas the mitotic spindle that partitions the chromosomes is composed of microtubules.

C. True. Both extensions are associated with transmembrane proteins that protrude from the plasma membrane and enable the cell to form new anchor points on the substratum.

D. False. To cause bending, ATP is hydrolyzed by the dynein motor proteins that are attached to the outer microtubules in the flagellum.

E. False. Cells could not divide without rearranging their intermediate filaments, but many terminally differentiated and long-lived cells, such as nerve cells, have stable intermediate filaments that are not known to depolymerize.

F. False. The rate of growth is independent of the size of the GTP cap. The plus and minus ends have different growth rates because they have physically distinct binding sites for the incoming tubulin subunits; the rate of addition of tubulin subunits differs at the two ends.

G. True. Both are nice examples of how the same membrane can have regions that are highly specialized for a particular function.

H. False. Myosin movement is activated by phosphorylation of myosin, or by calcium binding to troponin.

Answer 17–12

The average time taken for a small molecule (such as ATP) to diffuse a distance of 10 μm is given by the calculation
$$(10^{-3})^2 / (2 \times 5 \times 10^{-6}) = 0.1 \text{ sec}$$
Similarly, a protein takes 1 second and a vesicle 10 seconds on average to travel 10 μm. A vesicle would require on average 10^9 seconds, or more than 30 years, to diffuse to the end of a 10-cm axon. This calculation makes it clear why kinesin and other motor proteins evolved to carry molecules and organelles along microtubules.

Answer 17–13

(1) Animal cells are much larger and more diversely shaped, and do not have a cell wall. Cytoskeletal elements are required to provide mechanical strength and shape in the absence of a cell wall. (2) Animal cells, and all other eucaryotic cells, have a nucleus that is shaped and held in place in the cell by intermediate filaments; the nuclear lamins attached to the inner nuclear membrane support and shape the nuclear membrane, and a meshwork of intermediate filaments surrounds the nucleus and spans the cytosol. (3) Animal cells can move by a process that requires a change in cell shape. Actin filaments and myosin motor proteins are required for these activities. (4) Animal cells have a much larger genome than bacteria; this genome is fragmented into many chromosomes. For cell division, chromosomes need to be accurately distributed to the daughter cells, requiring the function of the microtubules that form the mitotic spindle. (5) Animal cells have internal organelles. Their localization in the cell is dependent on motor proteins that move them along microtubules. A remarkable example is the long-distance travel of membrane-enclosed vesicles (organelles) along microtubules in an axon that can be up to 1 m (≈3 ft) long in the case of the nerve cells that extend from your spinal cord to your feet.

Answer 17–14

The ends of an intermediate filament are indistinguishable from each other, because the filaments are built by assembly of symmetrical tetramers made from two coiled-coil dimers. In contrast to microtubules and actin filaments, intermediate filaments therefore have no polarity.

Answer 17–15

Intermediate filaments have no polarity; their ends are chemically indistinguishable. It would therefore be difficult to envision how a hypothetical motor protein that bound to the middle of the filament could sense a defined direction. Such a motor protein would be equally likely to attach to the filament facing one end or the other.

Answer 17–16

Katanin breaks microtubules along their length, and at positions remote from their GTP caps. The fragments that form therefore contain GDP-tubulin at their exposed ends and rapidly depolymerize. Katanin thus provides a very quick means of destroying existing microtubules.

Answer 17–17

Cell division depends on the ability of microtubules both to polymerize and to depolymerize. This is most obvious when one considers that the formation of the mitotic spindle requires the prior depolymerization of other cellular microtubules to free up the tubulin required to build the spindle. This rearrangement is not possible in taxol-treated cells, whereas in colchicine-treated cells, division is blocked because a spindle cannot be assembled. On a more subtle but no less important level, both drugs block the dynamic instability of microtubules and would there-

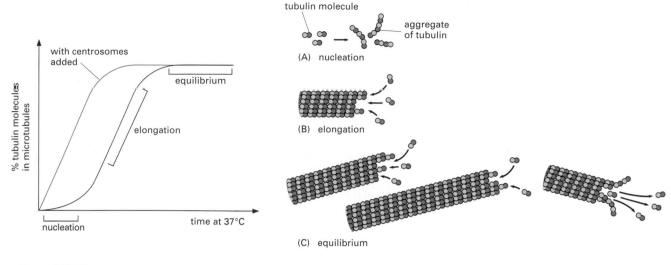

(A) nucleation

(B) elongation

(C) equilibrium

Figure A17–19

fore interfere with the workings of the mitotic spindle, even if one could be properly assembled.

Answer 17–18

Motor proteins are unidirectional in their action; kinesin always moves toward the plus end of a microtubule and dynein toward the minus end. Thus if kinesin molecules are attached to glass, only those individual motors that have the correct orientation in relation to the microtubule that settles on them can attach to the microtubule and exert force on it to propel it forward. Since kinesin moves toward the plus end of the microtubule, the microtubule will always crawl minus end first over the coverslip.

Answer 17–19

A. Phase A corresponds to a lag phase, during which tubulin molecules assemble to form nucleation centers (Figure A17–19A). Nucleation is followed by a rapid rise (phase B) to a plateau value as tubulin dimers add to the ends of the elongating microtubules (Figure A17–19B). At phase C, equilibrium is reached with some microtubules in the population growing while others are rapidly shrinking (Figure A17–19C). The concentration of free tubulin is constant at this point, because polymerization and depolymerization are balanced (see also Question 17–3, p. 583).

B. The addition of centrosomes introduces nucleation sites that eliminate the lag phase A as shown by the red curve in Figure A17–19. The rate of microtubule growth (i.e., the slope of the curve in the elongation phase B) and equilibrium level of free tubulin remain unchanged, because the presence of centrosomes does not affect the rates of polymerization and depolymerization.

Answer 17–20

The ends of the shrinking microtubule are visibly frayed, and the individual protofilaments appear to come apart and curl as the end depolymerizes. This micrograph therefore suggests that the GTP cap (which is lost from shrinking microtubules) holds the protofilaments properly aligned with each other, perhaps by strengthening the side-to-side interactions between αβ-tubulin subunits when they are in their GTP-bound form.

Answer 17–21

Cytochalasin interferes with actin filament formation, and its

effect on the cell demonstrates the importance of actin to cell locomotion. The experiment with colchicine shows that microtubules are required to give a cell a polarity that then determines which end becomes the leading edge (see Figure 17–14). In the absence of microtubules, cells still go through the motions normally associated with cell movement, such as the extension of lamellipodia, but in the absence of cell polarity these are futile exercises because they happen indiscriminately in all directions.

Antibodies bind tightly to the antigen (in this case vimentin) to which they were raised (see Panel 4–6, pp. 164–165). When bound, an antibody can interfere with the function of the antigen by preventing it from interacting properly with other cell components. The antibody injection experiment therefore suggests that intermediate filaments are not required for the maintenance of cell polarity or for the motile machinery.

Answer 17–22

Either (B) or (C) would complete the sentence correctly. The direct result of the action potential in the plasma membrane is the release of Ca^{2+} into the cytosol from the sarcoplasmic reticulum; muscle cells are triggered to contract by this rapid rise in cytosolic Ca^{2+}. Calcium ions at high concentrations bind to troponin, which in turn causes tropomyosin to move to expose myosin-binding sites on the actin filaments. (A) and (D) would be wrong because Ca^{2+} has no effect on the detachment of the myosin head from actin, which is the result of ATP hydrolysis. Nor does it have any role in maintaining the structure of the myosin filament.

Answer 17–23

Only (D) is correct. Upon contraction, the Z discs move closer together, and neither actin nor myosin filaments contract. The answer to this question will become clear if you reexamine Figures 17–43 and 17–44.

Chapter 18

Answer 18–1

Because all cells arise by division of another cell, this statement is correct, assuming that "first cell division" refers to the division of the successful founder cell from which all life as we know it has derived. There were probably many other unsuccessful attempts to start the chain of life.

Answer 18–2

Cells in peak B contain twice as much DNA as those in peak A, indicating that they contain replicated DNA whereas the cells in peak A contain unreplicated DNA. Peak A therefore contains cells that are in G_1, and peak B contains cells that are in G_2 and mitosis. Cells in S phase have begun but not finished DNA synthesis; they therefore have various intermediate amounts of DNA and are found in the region between the two peaks. Most cells are in G_1, indicating that it is the longest phase of the cell cycle (see Figure 18–2).

Answer 18–3

The cell would replicate its damaged DNA and therefore would introduce mutations to the two daughter cells when the cell divides. Such mutations could increase the chances that the progeny of the affected daughter cells would eventually become cancer cells.

Answer 18–4

The frog oocytes must contain an inactive M-Cdk. Upon injection of the M-phase cytoplasm into an oocyte, the small amount of the active M-Cdk in the injected cytoplasm activates some of the inactive M-Cdk by switching on the enzymes that cause phosphorylation and dephosphorylation of inactive M-Cdk at the appropriate sites (see Figure 18–11). An extract of the second oocyte, now in M phase itself, will therefore contain as much active M-Cdk as the original cytoplasmic extract, and so on.

Answer 18–5

For multicellular organisms, the control of cell division is extremely important. Individual cells must not proliferate unless it is to the benefit of the whole organism. The G_0 state offers protection from aberrant activation of cell division, because the cell-cycle control system is largely dismantled. If, on the other hand, a cell just paused in G_1, it would still contain all of the cell-cycle control system and might be induced to divide. The cell would also have to remake the "decision" not to divide almost continuously. To reenter the cell cycle from G_0, a cell has to resynthesize all of the components that have disappeared.

Answer 18–6

As apoptosis occurs on a large scale in both developing and adult tissues, it must not trigger alarm reactions that are normally associated with cell injury. Tissue injury, for example, leads to the release of signal molecules that stimulate the proliferation of surrounding cells so that the wound heals. It also causes the release of signals that can cause a destructive inflammatory reaction. Moreover, the release of intracellular contents could elicit an immune response against molecules that are normally not encountered by the immune system. Such reactions would be self-defeating if they occurred in response to the massive cell death that occurs in normal development.

Answer 18–7

A. False. There is no G_1 to M phase transition. The statement is correct, however, for the G_1 to S phase transition, where cells commit themselves to a division cycle.

B. True. Apoptosis is an active process carried out by special proteases (caspases).

C. True. This mechanism is thought to adjust the number of neurons to the number of specific target cells to which the neurons connect.

D. True. An amazing evolutionary conservation!

E. True. Such studies employ so-called conditional mutations, which lead to the production of proteins that usually are stable and functional at one temperature, but unstable or inactive at another temperature. Cells can be grown at the temperature at which the mutant protein functions normally, and then can be shifted to a temperature at which the protein's function is lost.

F. True. Association of a Cdk protein with a cyclin is required for its activity (hence its name cyclin-dependent kinase). Furthermore, phosphorylation at specific sites and dephosphorylation at other sites on the Cdk protein are required for the cyclin–Cdk complex to be active.

Answer 18–8

Loss of M-cyclin leads to inactivation of the M-Cdk. As a result, its target proteins become dephosphorylated by phosphatases, and the cells exit mitosis: they disassemble the mitotic spindle, reassemble the nuclear envelope, decondense their chromosomes, and so on. Cyclin is degraded by ubiquitin-dependent destruction in proteosomes, and the activation of M-Cdk leads to the activation of APC, which ubiquitinates the cyclin, but with a substantial delay. As discussed in Chapter 7, ubiquitination tags proteins for degradation in proteasomes.

Answer 18–9

Cells in an animal must behave for the good of the organism as a whole—to a much greater extent than people generally act for the good of society as a whole. In the context of an organism, unsocial behavior would lead to a loss of organization and to cancer. Many of the rules that cells have to obey would be unacceptable in a human society. Most people, for example, would be reluctant to kill themselves for the good of society, yet cells do it all the time.

Answer 18–10

M-cyclin accumulates gradually as it is steadily synthesized. As it accumulates, it will tend to form complexes with the mitotic Cdk molecules that are present. After a certain threshold level has been reached, a sufficient amount of M-Cdk has been formed so that it is activated by the appropriate kinases and phosphatases that phosphorylate and dephosphorylate it. Once activated, M-Cdk acts to enhance the activity of the activating phosphatase; this positive feedback leads to the explosive activation of M-Cdk. Thus M-cyclin accumulation acts like a slow-burning fuse, which eventually helps trigger the explosive self-activation of M-Cdk. The precipitous destruction of M-cyclin terminates M-Cdk activity, and a new round of M-cyclin accumulation begins.

Answer 18–11

The most likely approach to success (if that is what the goal should be called) is plan C, which should result in an increase in cell numbers. The problem is, of course, that cell numbers of each tissue must be increased similarly to maintain balanced proportions in the organism, yet different cells respond to different growth factors. As shown in Figure A18–11, however, the approach has indeed met with limited success. A mouse producing very large quantities of growth hormone (left)—which acts to stimulate the production of a secreted protein that acts as a survival factor, growth factor, or mitogen, depending on the cell type—grows to almost twice the size of a normal mouse (right). To achieve this twofold change in size, however, growth hormone was massively overproduced (about fiftyfold).

The other approaches have conceptual problems:

A. Blocking apoptosis might lead to defects in development, as development requires the selective death of many cells. It is unlikely that a viable animal would be obtained.

B. Blocking p53 function would eliminate an important

Courtesy of Ralph Brinster

Figure A18–11

checkpoint of the cell cycle that detects DNA damage and stops the cycle so that the cell can repair the damage; removing p53 would increase mutation rates and lead to cancer. Indeed, mice without p53 usually develop normally but die of cancer at a young age.

D.   Given the circumstances, switching careers might not be a bad option.

Answer 18–12

The on-demand, limited release of PDGF at a wound site triggers cell division of neighboring cells for a limited amount of time, until the PDGF is degraded. This is different from the continuous release of PDGF from mutant cells, where PDGF is made in an uncontrolled way at high levels. Moreover, the mutant cells that make PDGF often inappropriately express their own PDGF receptor, so that they can stimulate their own proliferation, thereby promoting the development of cancer.

Answer 18–13

A.   Radiation leads to DNA damage, which activates a checkpoint mechanism (mediated by p53 and p21; see Figure 18–15) that arrests the cell cycle until the DNA has been repaired.

B.   The cell will replicate damaged DNA and thereby introduce mutations to the daughter cells when the cell divides.

C.   The cell will be able to divide normally, but it will be prone to mutations, because some DNA damage always occurs as the result of natural irradiation caused, for example, by cosmic rays. The checkpoint mechanism mediated by p53 is mainly required as a safeguard against the devastating effects of accumulating DNA damage, but not for the natural progression of the cell cycle in undamaged cells.

D.   Cell division is an ongoing process that does not cease upon reaching maturity. Blood cells, epithelial cells in the skin or lining the gut, and the cells of the immune system, for example, are being constantly produced by cell division to meet the body's needs; each day, your body produces about 10^{11} new red blood cells alone.

Answer 18–14

All three types of mutant cells would be unable to divide. The cells

A.   would enter mitosis but would not be able to exit mitosis

B.   would arrest permanently in G_1 because the cyclin–Cdk complexes that act in G_1 would be inactivated

C.   would not be able to activate the transcription of genes required for cell division because the required gene regulatory proteins would be constantly inhibited by unphosphorylated Rb.

Answer 18–15

Normally, yeast cells divide only when they have grown to a certain size. This size control is clearly defective in the two mutant strains. In the case of gee cells, cell size increases without ever triggering cell division, suggesting that the mutant cell-cycle control protein has lost its ability to monitor cell size. It might, for example, now permanently inhibit M-Cdk, so that the cells cannot enter mitosis. In wee cells, on the other hand, the mutant control protein triggers cell division prematurely, before cells have grown to the appropriate size. This could be a control protein, for example, that no longer inhibits M-Cdk, so that M-Cdk becomes active prematurely. In fact, there is a yeast cell-cycle control protein called Wee1, which is a kinase that phosphorylates M-Cdk on a site that causes its inactivation; yeast cells with a mutation in the *wee-1* gene that inactivates Wee1 have a short cell cycle and are small.

Answer 18–16

A.   Only the cells that were in the S phase of their cell cycle (i.e., those cells making DNA) during the 30-minute labeling period contain any radioactive DNA.

B.   Initially, mitotic cells contain no radioactive DNA because these cells were not engaged in DNA synthesis during the labeling period. Indeed it takes about two hours before the first labeled mitotic cells appear.

C.   The initial rise of the curve corresponds to cells that were just finishing DNA replication when the radioactive thymidine was added. The curve rises as more labeled cells enter mitosis; the peak corresponds to those cells that had just started S phase when the radioactive thymidine was added. The labeled cells then exit mitosis, and are replaced by unlabeled mitotic cells, which were not yet in S phase during the labeling period. After 20 hours the curve starts rising again, because the labeled cells enter their second round of mitosis.

D.   The intial two-hour lag before any labeled mitotic cells appear corresponds to the G_2 phase, which is the time between the end of S phase and the beginning of mitosis. The first labeled cells seen in mitosis were those that were just finishing S phase (DNA synthesis) when the radioactive thymidine was added.

Answer 18–17

In alcoholism, liver cells proliferate because the organ is overburdened and becomes damaged by the large amounts of alcohol that have to be metabolized. This need for more liver cells activates the control mechanisms that normally regulate cell proliferation. Unless badly damaged, the liver will usually shrink back to a normal size after the patient stops drinking excessively. In a liver tumor, in contrast, mutations abolish normal cell proliferation control and, as a result, cells divide and keep on dividing in an uncontrolled manner, which is usually fatal.

Answer 18–18

The plasma membrane of the cell that died by necrosis in Figure 18–20A is ruptured; a clear break is visible, for example, at a position corresponding to the 11 o'clock mark on a watch. The cell's contents, mostly membranous and cytoskeletal debris, are seen spilling into the surroundings through these breaks. The cytosol stains lightly, as most soluble cell components were lost before the cell was fixed. In contrast, the cell that underwent apoptosis in Figure 18–20B is surrounded by an intact membrane, and its cytosol is densely stained, indicating a normal concentration of cell components. The cell's interior is remarkably different from a normal cell, however. Particularly characteristic are the large "blobs" that extrude from the nucleus, probably as the result of the breakdown of the nuclear lamina. The cytosol also contains many large, round, membrane-enclosed vesicles of unknown origin, which are not normally seen in healthy cells. The pictures visually confirm the notion that necrosis involves cell lysis, whereas cells undergoing apoptosis remain relatively intact until they are digested inside a normal cell.

Chapter 19

Answer 19–1

In a eucaryotic organism, the genetic information that the organism needs to survive and reproduce is distributed between multiple chromosomes. It is crucial therefore that each daughter cell receives a copy of each chromosome when a cell divides; if a daughter cell receives too few or too many chromosomes, the effects are usually deleterious or even lethal. Only two copies of each chromosome are produced by chromosome replication in mitosis. If the cell were to randomly distribute the chromosomes when it divided, it would be very unlikely that each daughter cell would receive precisely one copy of each chromosome. In contrast, the endoplasmic reticulum fragments into tiny vesicles that are all alike, and by random distribution it is very likely that each daughter cell will receive an approximately equal number of them.

Answer 19–2

The experiment shows that kinetochores are not preassigned to one or the other spindle pole; microtubules attach to the kinetochores that they are able to reach. For the chromosomes to remain attached, tension has to be exerted. Tension is normally achieved by the opposing pulling forces from opposite spindle poles. The requirement for such tension ensures that if two sister kinetochores ever become attached to the same spindle pole, so that tension is not generated, one or both of the connections would break, and microtubules from the opposing spindle pole would have another chance to attach properly.

Answer 19–3

Recall from Figure 19–18 that the new nuclear envelope reassembles on the surface of the chromosomes. The close apposition of the envelope to the chromosomes prevents cytosolic proteins from being trapped between the chromosomes and the envelope. Nuclear proteins are then selectively imported through the nuclear pores, causing the nucleus to expand while maintaining its characteristic protein composition.

Answer 19–4

The membranes of the Golgi vesicles fuse to form the new plasma membranes of the two daughter cells. The interiors of the vesicles, which are filled with cell-wall material, become the new cell-wall matrix separating the two daughter cells. Proteins in the membranes of the Golgi vesicles thus become plasma membrane proteins. Those parts of the proteins that were exposed to the lumen of the Golgi vesicle will end up exposed to the new cell wall (Figure A19–4).

Answer 19–5

A. True. Centrosomes replicate during interphase, before M phase begins. The mechanism by which one centro-

some gives rise to two, and only two, centrosomes is unknown.

B. True. In fact, the nuclear envelope is continuous with the ER membrane (discussed in Chapter 15), and some of the fragments therefore contain membrane derived from both origins.

C. True. Sister chromatids separate only at the start of anaphase.

D. False. The ends of interpolar microtubules overlap and attach to one another via proteins (including motor proteins) that bridge between the microtubules.

E. False. Microtubules and their motor proteins play no role in DNA replication.

F. False. To be a correct statement, the terms "centromere" and "centrosome" must be switched.

Answer 19–6

Because the cell population is growing exponentially, doubling its weight at every cell division, the weight of the cell cluster after N number of cell divisions is $2^N \times 10^{-9}$ g. Therefore, 70 kg $(70 \times 10^3$ g$) = 2^N \times 10^{-9}$ g, or $2^N = 7 \times 10^{13}$. Taking the logarithm on both sides allows you to solve the equation for N. Therefore, $N = \ln(7 \times 10^{13}) / \ln 2 = 46$—i.e., it would take only 46 days if cells proliferated exponentially. Cell division in animals is tightly controlled, however, and most cells in the human body stop dividing when they become highly specialized. The example demonstrates that exponential cell proliferation occurs only for very brief periods, even during embryonic development.

Answer 19–7

The order is G, C, B, A, D. Together, these five steps are referred to as mitosis (F). No step in mitosis is influenced by the phases of the moon (E). Cytokinesis is the last step in M phase, which overlaps with anaphase and telophase. Mitosis and cytokinesis are both part of M phase.

Answer 19–8

Many egg cells are big and contain stores of enough cell components to last for many cell divisions. The daughter cells that form during the first cell divisions after the egg is fertilized are progressively smaller in size and thus can be formed without a need for new protein or RNA synthesis. Whereas normally dividing cells would grow continuously in G_1, G_2, and S phases, until their size doubled, there is no cell growth in these early divisions, and both G_1 and G_2 are virtually absent in the first cell divisions of egg cells. As G_1 is usually longer than G_2, G_1 is the most drastically reduced in these divisions.

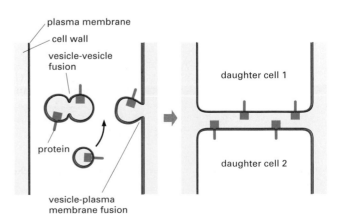

Figure A19–4

Answer 19–9

If the growth rate of microtubules is the same in mitotic and in interphase cells, their length is proportional to their lifetime. Thus, the average length of microtubules in mitosis is 1 mm (= 20 μm × 15 s/300 s). If microtubules are on average 20 times shorter, but in total contain the same number of tubulin monomers, then there must be 20 times as many microtubules, or 2000 nucleation sites per centrosome in mitosis.

Answer 19–10

Antibodies bind tightly to the antigen (in this case myosin) to which they were raised. When bound, an antibody can interfere with the function of the antigen by preventing it from interacting properly with other cell components. (A) The movement of chromosomes at anaphase depends on microtubules and their motor proteins and does not depend on actin or myosin. Injection of an anti-myosin antibody into a cell will therefore have no effect on chromosome movement during anaphase. (B) Cytokinesis, on the other hand, depends on the assembly and contraction of a ring of actin and myosin filaments, which forms the cleavage furrow that splits the cell in two. Injection of anti-myosin antibody will therefore block cytokinesis.

Answer 19–11

The sister chromatid becomes committed when a microtubule from one of the spindle poles attaches to the kinetochore of the chromatid. Microtubule attachment is still reversible until a second microtubule from the other spindle pole attaches to the kinetochore of its partner sister chromatid so that the duplicated chromosome is now put under mechanical tension by pulling forces from both poles. The tension ensures that both microtubules remain attached to the chromosome. The position of the chromosome in the cell at the time the nuclear envelope breaks down will influence which spindle pole the chromatid will be pulled to, as a kinetochore is most likely to become attached to the spindle pole toward which it is facing.

Answer 19–12

The fact is that it is not certain how this works. Two possible models of how the kinetochore may generate a poleward force on its chromosome during anaphase A are shown in Figure A19–12. In (A), microtubule motor proteins are part of the kinetochore and use the energy of ATP hydrolysis to pull the chromosome along its bound microtubules. The depolymerization of the microtubule at its kinetochore end would occur as a consequence of this movement. In (B), chromosome movement is driven by microtubule disassembly catalyzed by an enzyme that uses the energy of ATP hydrolysis to remove tubulin subunits from the attached end of the microtubule. As tubulin subunits dissociate, the kinetochore is obliged to slide poleward in order to maintain its binding to the walls of the microtubule. It is possible that both mechanisms are used.

Answer 19–13

As shown in Figure A19–13, the overlapping interpolar microtubules from opposite poles of the spindle have their plus ends pointing in opposite directions. Plus-end-directed motor proteins cross-link adjacent antiparallel microtubules together and tend to move the microtubules in the direction that will push the two poles of the spindle apart, as shown in the Figure. Minus-end-directed motor proteins also cross-link adjacent antiparallel microtubules together but move in the opposite direction, tending to pull the spindle poles together (not shown).

Answer 19–14

Both sister chromatids could end up in the same daughter cell for any of a number of reasons. (1) If the microtubules or their connections with a kinetochore were to break during anaphase, both sister chromatids could be drawn to the same pole, and hence into the the same daughter cell. (2) If microtubules from the same spindle pole attached to both kinetochores, the chromosome would be pulled to the same pole. (3) If the cohesins that link sister chromatids were not degraded, the pair of chromatids might be pulled to the same pole. (4) If a chromosome never engaged microtubules and was left out of the spindle, it would also end up in one daughter cell.

Some of these errors in the mitotic process would be expected to activate a checkpoint mechanism that delays the onset of anaphase until all chromosomes are attached properly to both poles of the spindle (discussed in Chapter 18). This checkpoint mechanism should allow most chromosome attachment errors to be corrected, which is one reason why such errors are rare.

The consequences of both sister chromatids ending up in one daughter cell are usually dire. One daughter cell would contain only one copy of all the genes carried on that chromosome and the other daughter cell would contain three copies. The altered

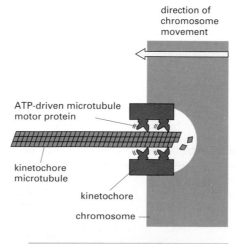

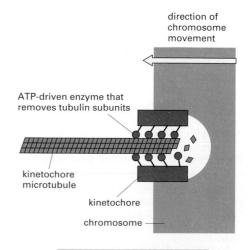

(A) ATP-driven chromosome movement drives microtubule disassembly

(B) microtubule disassembly drives chromosome movement

Figure A19–12

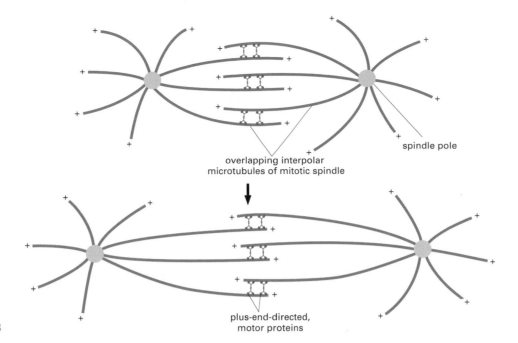

Figure A19–13

overlapping interpolar
microtubules of mitotic spindle

spindle pole

plus-end-directed,
motor proteins

gene dosage, leading to correspondingly changed amounts of the mRNAs and proteins produced, is often detrimental to the cell. In addition, there is the possibility that the cell with a single copy of the chromosome may be defective for a critical gene, a defect that would normally be hidden by the presence of a second, good copy of the gene on the other chromosome that is now missing.

Chapter 20

Answer 20–1

Natural selection alone is not sufficient to eliminate recessive lethal genes from the population. Consider the following line of reasoning. Homozygous defective individuals can arise only as the offspring of a mating between two heterozygous individuals. By the rules of Mendelian genetics, offspring of such a mating will be in the ratio of 1 homozygous, normal: 2 heterozygous: 1 homozygous, defective. Thus, statistically, heterozygous individuals should always be more numerous than the homozygous, defective individuals. And although natural selection effectively eliminates the defective genes in homozygous individuals through death, it can't touch the defective genes in heterozygous individuals because they do not affect the phenotype. Natural selection will keep the frequency of the defective gene low in the population, but in the absence of any other effect there will always be a reservoir of defective genes in the heterozygous individuals.

At low frequencies of the defective gene another important factor—chance—comes into play. Chance variation can increase or decrease the frequency of heterozygous individuals (and thereby the frequency of the defective gene). By chance, the offspring of a mating between heterozygotes could all be normal, which would eliminate the defective gene from that lineage. Increases in the frequency of a deleterious gene are opposed by natural selection; however, decreases are unopposed and can, by chance, lead to elimination of the defective gene from the population.

Answer 20–2

Although each daughter cell ends up with a diploid amount of DNA afer the first meiotic division, each cell has effectively only

a haploid set of chromosomes (albeit in two copies), representing only one or other homolog of each type of chromosome (although some mixing will have occurred during crossing-over). Because the maternal and paternal chromosomes of a pair will carry different versions of many of the genes, these daughter cells will not be genetically identical. In contrast, somatic cells dividing by mitosis inherit a diploid set of chromosomes, and all daughter cells are genetically identical. The role of gametes produced by meiosis is to mix and reassort gene pools during sexual reproduction, and thus it is a definite advantage for each of them to have a slightly different genetic constitution. The role of somatic cells on the other hand is to build an organism that contains the same genes in all its cells.

Answer 20–3

A typical human female produces fewer than 1000 mature eggs in her lifetime (12 per year over about 40 years); this is less than one-tenth of a percent of the possible gametes excluding the effects of meiotic crossing-over. A typical human male produces billions of sperm during a lifetime, so in principle, all the possible chromosome combinations are each sampled many times.

Answer 20–4

For simplicity, consider the situation where a father carries genes for two dominant traits, M and N, on one of his two copies of human chromosome 1. If these two genes were located at opposite ends of this chromosome, and there was one and only one crossover event per chromosome as postulated in the question, half of his children would express trait M and the other half would express trait N—with no child resembling the father in carrying both traits. This is very different from the actual situation, where there are multiple crossover events per chromosome, causing the traits M and N to be inherited as if they were on separate chromosomes. Thus, by constructing a Punnett square like that in Figure 20–20, one can see that we would actually expect one-fourth of the children of this father to inherit both traits, one-fourth to inherit trait M only, one-fourth to inherit trait N only, and one-fourth to inherit neither trait.

Answer 20–5

Although any one of the three explanations could in principle account for the observed result, A and B can be ruled out as being completely unreasonable.

A. There is no precedent for any instability in DNA so great as to be detectable in such a SNP analysis; in any case, the hypothesis would predict a steady decrease in the frequency of the SNP with age, not a drop in frequency that begins only at age 50.

B. Human populations change only very slowly over time; people born 50 years ago must be, on average, no different than those being born today.

C. This hypothesis is correct. A SNP with these properties has recently been used to discover a gene that, when altered by the SNP, appears to cause a substantial increase in the probability of death from cardiac problems.

Answer 20–6

A. True.
B. True.
C. False. Mutations that occur during meiosis can be propagated, unless they give rise to nonviable gametes.

Answer 20–7

Diploid organisms contain two copies of most of their genes in each cell, providing "backup copies" for genes that may become damaged by mutation. This complicates genetic studies. Many mutations in diploid cells do not cause any detectable defect in cellular processes because a normal copy of the gene is also expressed in the same cell and can often compensate for the mutated gene. If a diploid organism produced diploid gametes, its number of chromosomes would double in each generation. In addition, it has been suggested that haploid gametes can allow a diploid organism to eliminate many detrimental mutations that might not be apparent in diploid cells, because impaired gametes are likely to be less successful in fertilization. Gametes are usually made in numbers greatly exceeding what might seem to be required, and must pass competitive "fitness tests." Millions of mammalian sperm cells, for example, compete in a fierce race to reach and penetrate a single egg cell.

Answer 20–8

Two copies of the same chromosome can end up in the same daughter cell if one of the microtubule connections breaks before sister chromatids are separated. Alternatively, microtubules from the same spindle pole could attach to both kinetochores of the chromosome. As the consequence of this severe and rare error, one daughter cell would contain only one copy of all the genes carried on that chromosome and the other daughter cell would contain three copies. The changed gene dosage, leading to correspondingly changed amounts of the mRNAs and proteins produced, is in many cases detrimental to the cell. If the mistake happens during meiosis, in the process of gamete formation, it will be propagated in all cells of the organism. A severe form of mental retardation called Down syndrome, for example, is due to the presence of three copies of Chromosome 21 in all of the nucleated cells in the body.

Answer 20–9

Meiosis begins with DNA replication, producing a tetraploid cell containing four copies of each chromosome. These four copies have to be distributed equally during the two sequential meiotic divisions into four haploid cells. Sister chromatids remain paired so that (1) the cells resulting from the first division receive two complete sets of chromosomes and (2) the chromosomes can be evenly distributed again in the second meiotic division. If the sister chromatids did not remain paired, it would not be possible in the second division to distinguish which chromatids belong together, and it would therefore be difficult to ensure that precisely one copy of each chromatid is pulled into each daughter cell. Keeping two sister chromatids paired in the first meiotic division is therefore an easy way to keep track of which chromatids belong together.

This biological principle suggests that you might consider clamping your socks together in matching pairs before putting them into the laundry. In this way, the cumbersome process of sorting them out afterward—and the seemingly inevitable mistakes that occur during that process—could be avoided.

Answer 20–10

A. A gene is a stretch of DNA that codes for a protein or functional RNA. An allele is an alternative form of a gene. Within the population, there are often several "normal" alleles, whose functions are indistinguishable. In addition, there may be many rare alleles that are defective to varying degrees. An individual, however, normally has a maximum of two alleles of a gene.

B. An individual is said to be homozygous if the two alleles of a gene are the same. An individual is said to be heterozygous if the two alleles of a gene are different.

C. Genotype is the specific set of alleles forming the genome of an individual; it is an enumeration of all the particular forms of each gene in the genome. In practice, for organisms studied in a laboratory, the genotype is usually specified as a list of the known differences between the individual and the wild type, which is the standard, naturally occurring type. Phenotype is a description of the visible characteristics of the individual. In practice, phenotype is usually a list of the differences in visible characteristics between the individual and the wild type.

D. An allele is dominant (relative to a second allele) if the phenotype is the same when the allele is homozygous and when it is heterozygous. In that case the second allele, whose presence makes no difference to the phenotype, is said to be recessive (to the first allele). If the phenotype of the heterozygous individual differs from the phenotypes of individuals that are homozygous for either allele, the alleles are said to be co-dominant.

Answer 20–11

A. Since the pea plant is diploid, any true-breeding plant must carry two mutant copies of the same gene—both of which have lost their function.

B. No, the same phenotype will often be produced by several different genotypes.

C. If each plant carries a mutation in a different gene, this will be revealed by complementation tests (see Panel 20–1, p. 685). When plant A is crossed with plant B, all of the F_1 plants will produce only round peas. And the same result will be obtained when plant B is crossed with plant C, or when plant A is crossed with plant C. In contrast, a cross between any two true-breeding plants that carry loss-of-function mutations in the same gene (even if these mutations are different) should produce only plants with wrinkled peas.

Answer 20–12

A. The mutation is likely to be dominant, because half of the progeny born to an affected parent are deaf.

B. The mutation is present on an autosome. If it were instead carried on a sex chromosome, either only the female progeny should be affected (expected if the mutation arose in a gene on the grandfather's X chromosome), or only the male progeny should be affected (expected if the mutation arose in a gene on the grand-

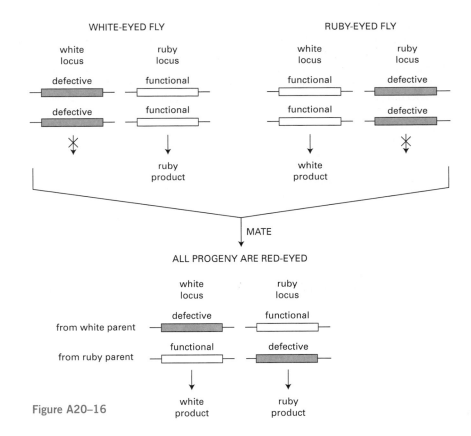

WHITE-EYED FLY

white locus — defective / defective ✗
ruby locus — functional / functional → ruby product

RUBY-EYED FLY

white locus — functional / functional → white product
ruby locus — defective / defective ✗

MATE

ALL PROGENY ARE RED-EYED

from white parent — white locus: defective → / ruby locus: functional →
from ruby parent — white locus: functional → white product / ruby locus: defective → ruby product

Figure A20–16

father's Y chromosome). In fact, the pedigree reveals that both some males and some females have inherited the mutant form of the gene.

C. Suppose that the mutation occurred on one of the two copies of the grandfather's chromosome 12. Each of these copies of chromosome 12 would be expected to carry a different pattern of SNPs, since one of them was inherited from his father and the other was inherited from his mother. Each of the copies of chromosome 12 that was passed to his grandchildren will have gone through two meioses—one meiosis per generation.

Because 2–3 crossover events occur per chromosome during a meiosis, each chromosome inherited by a grandchild will have been subjected to about five crossovers since it left the grandfather. An identical pattern of SNPs should surround whatever gene causes the deafness in each of the four affected grandchildren; moreover, this SNP pattern should be clearly different from that surrounding the same gene in each of the seven grandchildren who are normal. These SNPs would form an unusually long haplotype block—one that extends for about one-fifth of the length of chromosome 12. (One-fourth of the DNA of each grandchild will have been inherited from the grandfather, in roughly 50 to 60 segments of this length scattered among the grandchild's 46 chromosomes.)

Answer 20–13

Your friend is wrong. (A) Mendel's laws, and the clear understanding that we now have concerning the mechanisms that produce them, rule out many false ideas concerning human heredity. One of them is that a first-born child has a different chance of inheriting particular traits from its parents than its siblings. (B) The probability of this type of pedigree arising by chance is one-fourth for each generation, or one time in 64 for the three generations shown. (C) Data from an enlarged sampling of family members, or from more generations, would quickly reveal that the regular pattern observed in this particular pedigree arose by chance.

Answer 20–14

A dominant–negative mutation gives rise to a mutant gene product that interferes with the function of the normal gene product, causing a loss-of-function phenotype even in the presence of a normal copy of the gene. This ability of a single defective allele to determine the phenotype is the reason why such an allele is dominant. A gain-of-function mutation increases the activity of the gene or makes it active in inappropriate circumstances. The change in activity often has a phenotypic consequence, which is why such mutations are usually dominant.

Answer 20–15

This statement is largely true. Diabetes is one of the oldest diseases described by humans, dating at least back to the time of the ancient Greeks. Diabetes itself comes from the Greek word for siphon, which was used to describe the main symptoms "The disease was called diabetes, as though it were a siphon, because it converts the human body into a pipe for the transflux of liquid humors." If there were no human disease, the role of insulin would not have come to our attention in so demanding a way. We would have ultimately understood its role—and by now may have. Yet it is difficult to overstate the case for the role of disease in focusing our efforts toward a molecular understanding. Even today, the quest to understand and alleviate human disease is a principal driving force in biomedical research.

Answer 20–16

A. As outlined in Figure A20–16, if flies are defective in different genes their progeny will have one normal gene at each locus. In the case of a mating between a ruby-eyed fly and a white-eyed fly, every progeny fly will inherit one functional copy of the white gene from one parent and one functional copy of the ruby gene from the other parent. Because each of the mutant alleles is recessive to the corresponding wild-type allele, the progeny will have the wild-type phenotype—brick-red eyes.

B. Garnet, ruby, vermilion, and carnation complement one another and the various alleles of the *white* gene (that is, when these mutant flies are mated with each other, they produce flies with a normal eye color); thus each of these mutants defines a separate gene. In contrast, white, cherry, coral, apricot, and buff do not complement each other; thus, they must be alleles of the same gene, which has been named the *white* gene. Thus, these nine different eye-color mutants define five different genes.

C. Different alleles of the same gene, like the five alleles of the *white* gene, often have different phenotypes. Different mutations compromise the function of the gene product to different extents, depending on the location of the mutation. Different alleles of the same gene, which do not produce any functional product (null alleles) do have the same phenotype.

Answer 20–17

SNPs are single-nucleotide differences between individuals, which occur roughly once per 1000 nucleotides of sequence. Many have been collected and mapped in various organisms, including several million in the human genome. SNPs, which can be detected by oligonucleotide hybridization, serve as physical markers whose genomic locations are known. By tracking a mutant gene through different matings, and correlating the presence of the gene with the co-inheritance of particular SNPs, one can narrow down the potential location of a gene to a chromosomal region that may contain only a few genes. These candidate genes can then be tested for the presence of a mutation that could serve as the basis for the original mutant phenotype (see Figure 20–30).

Answer 20–18

What you immediately know is all of the nucleotide sequence differences between Tim and John. But SNPs that are rare in the human population are not useful for most genetic mapping analyses. Testing each SNP for its frequency in a large population of humans will reveal which of them are found in at least 10 percent of the people in that population. These are the selected SNPs that will provide useful markers for future mapping analyses.

Chapter 21

Answer 21–1

The horizontal orientation of the microtubules will be associated with a horizontal orientation of cellulose fibers deposited in the cell walls. The growth of the cells will therefore be in a vertical direction, expanding the distance between the cellulose fibers without stretching these fibers. In this way, the stem will rapidly elongate; in a typical natural environment, this will hasten emergence from darkness into light.

Answer 21–2

A. As three collagen chains have to come together to form the triple helix, a defective molecule will impair assembly, even if normal collagen chains are present at the same time. Collagen mutations are therefore dominant, i.e., they have a deleterious effect even in the presence of a normal copy of the gene.

B. The different severity of the mutations results from a polarity in the assembly process. Collagen monomers assemble into the triple-helical rod starting from their amino-terminal ends. A mutation in an "early" glycine therefore allows only short rods to form, whereas a mutation farther downstream allows for longer, more normal rods.

Answer 21–3

The remarkable ability to swell and thus occupy a large volume of space depends on the negative charges. These attract a cloud of positive ions, chiefly Na^+, which by osmosis draw in large amounts of water, thus giving proteoglycans their unique properties. Uncharged polysaccharides such as cellulose, starch, and glycogen, by contrast, are easily compacted into fibers or granules.

Answer 21–4

Focal contact sites are common in connective tissue, where fibroblasts exert traction forces on the extracellular matrix, and in cell culture, where cell crawling is observed. The forces for pulling on matrix or for driving crawling movement are generated by the actin cytoskeleton. In mature epithelium, focal contact sites are presumably less prominent because the cells are largely fixed in place and have no need to crawl over the basal lamina or actively pull on it.

Answer 21–5

Suppose a cell is damaged so that its plasma membrane becomes leaky. Ions present in high concentration in the extracellular fluid, such as Na^+ and Ca^{2+}, then rush into the cell, and valuable metabolites leak out. If the cell were to remain connected to its healthy neighbors, these too would suffer from the damage. But the influx of Ca^{2+} into the sick cell causes its gap junctions to close immediately, effectively isolating the cell and preventing damage from spreading in this way.

Answer 21–6

Ionizing (high-energy) radiation tears through matter, knocking electrons out of their orbits and breaking chemical bonds. In particular, it creates breaks and other damage in DNA, and thus causes cells to arrest in the cell cycle (see Chapter 18). If the damage is so severe that it cannot be repaired, cells become permanently arrested and undergo apoptosis, i.e., they activate a suicide program.

Answer 21–7

Cells in the gut epithelium are exposed to a quite hostile environment, containing digestive enzymes and many other substances that vary drastically from day to day depending on the food intake of the organism. The epithelial cells also form a first line of defense against potentially hazardous compounds and mutagens that are ubiquitous in our environment. The rapid turnover protects the organism from harmful consequences, as wounded and sick cells are discarded. If an epithelial cell started to divide inappropriately as the result of a mutation, for example, it and its unwanted progeny would most often simply be discarded by natural disposal from the tip of a villus: even though such mutations must occur often, they rarely give rise to a cancer.

A neuron, on the other hand, lives in a very protected environment, insulated from the outside world. Its function depends on a complex system of connections with other neurons—a system that is created during development and is not easy to reconstruct if the neuron subsequently dies.

Answer 21–8

Every cell division generates one additional cell; so if the cells were never lost or discarded from the body, the number of cells in the body should equal the number of divisions plus one. The number of divisions is 1000-fold greater than the number of cells because, in the course of a lifetime, 1000 cells are discarded and replaced for every cell that is retained in the body.

Answer 21–9

A. False. Gap junctions are not connected to the cytoskeleton; their role is to provide cell-to-cell communication by allowing small molecules to pass from one cell to another.

B. True. Upon wilting, the turgor pressure in the plant cell is reduced, and consequently, the cell walls, having tensile but little compressive strength, like a rubber tire, no longer provide rigidity.

C. False. Proteoglycans can withstand a large amount of compressive force but do not have a rigid structure. Their space-filling properties result from their tendency to absorb large amounts of water.

D. True.

E. True.

F. True. Stem cells stably express control genes that ensure that their daughter cells will be of the appropriate differentiated cell types.

Answer 21–10

Small cytosolic molecules, such as glutamic acid, cyclic AMP, and Ca^{2+} ions, pass readily through both gap junctions and plasmodesmata, whereas large cytosolic macromolecules, such as mRNA and G proteins, are excluded. Plasma membrane phospholipids diffuse in the plane of the membrane through plasmodesmata because the plasma membranes of adjacent cells are continuous through these junctions. This traffic is not possible through gap junctions, because the membranes of the connected cells remain separate.

Answer 21–11

Plants are exposed to extreme changes in the environment, which often are accompanied by huge fluctuations in the osmotic properties of their surroundings. An intermediate filament network as we know it from animal cells would not be able to provide full osmotic support for cells: the sparse rivetlike attachment points would not be able to prevent the membrane from bursting in response to a huge osmotic pressure applied from the inside of the cell.

Answer 21–12

Action potentials can, in fact, be passed from cell to cell through gap junctions. Indeed, heart muscle cells are connected this way, which ensures that they contract synchronously when stimulated. This mechanism of passing the signal from cell to cell is rather limited, however. As we discuss in Chapter 12, synapses are far more sophisticated and allow signals to be modulated and to be integrated with other signals received by the cell. Thus gap junctions are like simple soldered joints between electrical components, while synapses are like complex relay devices, enabling systems of neurons to perform computations.

Answer 21–13

To make jello, gelatin is boiled in water, which denatures the collagen fibers. Upon cooling, the disordered fibers form a tangled mess that solidifies into a gel. This gel actually resembles the collagen as it is initially secreted by fibroblasts, i.e., before the fibers become aligned and cross-bridged.

Answer 21–14

The evidence that DNA is the blueprint that specifies all the structural characteristics of an organism is based on observations that small changes in the DNA by mutation result in changes in the organism. While DNA provides the plans that specify structure, these plans need to be executed during development. This requires a suitable environment (a human baby would not fit into a stork's egg shell), suitable nourishment, suitable tools (such as the appropriate gene regulatory proteins required for early development), suitable spatial organization (such as the asymmetries in the egg cell required to allow for appropriate cell differentiation during the early cell divisions), and so on. Thus inheritance is not restricted to the passing on of the organism's DNA, because development requires appropriate conditions to be set up by the parent. Nevertheless, when all these conditions are met, the plans that are archived in the genome will determine the structure of the organism to be built.

Answer 21–15

White blood cells circulate in the bloodstream and migrate into and out of tissues in performance of their normal function of defending the body against infection: they are naturally invasive. Once mutations have occurred to upset the normal controls on production of these cells, there is no need for additional mutations to enable the cells to spread through the body. Thus the number of mutations that have to be accumulated in order to give rise to leukemia is fewer than for other types of cancer.

Answer 21–16

The shape of the curve reflects the need for multiple mutations to accumulate in a cell before a cancer results. If a single mutation were sufficient, the graph would be a straight horizontal line: the likelihood of occurrence of a particular mutation, and therefore of cancer, would be the same at any age. If two specific mutations were required, the graph would be a straight line sloping upward from the origin: the second mutation has an equal chance of occurring at any time, but will tip the cell into cancerous behavior only if the first mutation has already occurred in the same lineage; and the likelihood that the first mutation has already occurred will be proportional to the age of the individual. The steeply curved graph shown in the figure goes up approximately as the fifth power of the age, and this indicates that far more than two mutations have to be accumulated before cancer sets in. It is not easy to say precisely how many, because of the complex ways in which cancers develop. Successive mutations can alter cell numbers and cell behavior, and thereby change both the probability of subsequent mutations and the selection pressures that drive the evolution of cancer.

Answer 21–17

During exposure to the carcinogen, mutations are induced, but the number of relevant mutations in any one cell is usually not enough to convert it directly into a cancer cell. Over the years, the cells that have become predisposed to cancer through the induced mutations accumulate progressively more mutations. Eventually, one of them will turn into a cancer cell. The long delay between exposure and cancer has made it extremely difficult to hold cigarette manufacturers or producers of industrial carcinogens legally responsible for the damage that is caused by their products.

Answer 21–18

By definition, a carcinogen is any substance that promotes the occurrence of one or more types of cancer. The sex hormones can therefore be classified as naturally occurring carcinogens. Although most carcinogens act by directly causing mutations, carcinogenic effects are also often exerted in other ways. The sex hormones increase both the rate of cell division and the numbers of cells in hormone-sensitive organs such as breast, uterus, and prostate. The first effect increases the mutation rate per cell, because mutations, regardless of environmental factors, are spontaneously generated in the course of DNA replication and chromosome segregation; the second effect increases the number of cells at risk. In these and possibly other ways, the hormones can favor development of cancer, even though they do not directly cause mutations.

Answer 21–19

The short answer is no—cancer in general is not a hereditary disease. It arises from new mutations occurring in our own somatic cells, rather than mutations we inherit from our parents. In some very rare types of cancer, however, there is a strong heritable risk factor, so that parents and their children both show the same predisposition to a specific form of the disease. This occurs, for example, in families carrying a mutation that knocks out one of the two copies of the tumor suppressor gene *APC;* the children then inherit a propensity to colorectal cancer. Much weaker heritable tendencies are also seen in several other cancers, including breast cancer, but the genes responsible for these effects are still mostly unknown.

Glossary

acetyl CoA (acetyl coenzyme A)
Small water-soluble molecule that carries acetyl groups in cells. Contains an acetyl group linked to coenzyme A (CoA) by an easily hydrolyzable thioester bond.

acetyl group
Chemical group derived from acetic acid.

acid
An organic molecule that dissociates in water to generate hydronium (H_3O^+) ions (thereby producing a low pH).

actin filament
Protein filament, about 7-nm wide, formed from a chain of globular actin molecules. A major constituent of the cytoskeleton of all eucaryotic cells and especially abundant in muscle cells.

action potential
Rapid, transient, self-propagating electrical signal in the plasma membrane of a cell such as a neuron or muscle cell. A nerve impulse.

activated carrier
A small molecule used to carry energy or chemical groups in many different metabolic reactions. Examples include ATP, acetyl CoA and NADH.

activation energy
Extra energy that must be acquired by a molecule in order to undergo a particular chemical reaction.

activator
In bacteria, a protein that binds to a specific region of DNA to permit transcription of an adjacent gene.

active site
Region of an enzyme surface to which a substrate molecule binds before it undergoes a catalyzed reaction.

active transport
Movement of a molecule across a membrane driven by ATP hydrolysis or other form of metabolic energy.

acyl group
Functional group derived from a carboxylic acid. (R represents an alkyl group, such as methyl.)

adaptation
Adjustment of sensitivity of a cell or organism following repeated stimulation. Allows a response even when there is a high background level of stimulation.

adenylyl cyclase
Membrane-bound enzyme that catalyzes the formation of cyclic AMP from ATP. An important component of some intracellular signaling pathways.

adherens junction
Cell junction in which the cytoplasmic face is attached to actin filaments.

ADP (adenosine 5′-diphosphate))
Nucleotide that is produced by hydrolysis of the terminal phosphate of ATP. (*See* Figure 3–32.)

alcohol
Organic compound containing a hydroxyl group (–OH) bound to a saturated carbon atom—for example, ethyl alcohol.

aldehyde
Reactive organic compound that contains the –CH=O group, for example glyceraldehyde.

alkaline—*see* **basic**

alkane
Compound made of carbon and hydrogen atoms that has only single covalent bonds. An example is ethane.

alkene
Hydrocarbon with one or more carbon-carbon double bonds. An example is ethylene (ethene).

alkyl group
General term for a group of covalently linked carbon and hydrogen atoms such as methyl (–CH_3) or ethyl (–CH_2CH_3) groups.

allele
One of a set of alternative forms of a gene. In a diploid cell each gene will have two alleles, each occupying the same position (locus) on homologous chromosomes.

allosteric
Describes a protein that exists in two or more conformations depending on the binding of a molecule (a ligand) at a site other than the catalytic site. Allosteric proteins composed of multiple subunits often display a cooperative response to ligand binding.

alpha helix (α helix)
Common structural motif of proteins in which a linear sequence of amino acids folds into a right-handed helix stabilized by internal hydrogen bonding between backbone atoms.

amide
Molecule containing a carbonyl group linked to an amine.

aminoacyl-tRNA synthetase
Enzyme that attaches the correct amino acid to a tRNA molecule to form an aminoacyl-tRNA.

amino acid
Organic molecule containing both an amino group and a carboxyl group. α amino acids (those in which the amino and carboxyl groups are linked to the same carbon atom serve as the building blocks of proteins. (*See* Panel 2–5, pp. 74–75.)

amino group
Weakly basic functional group (—NH_2) derived from ammonia. In aqueous solution an amino group can accept a proton and carry a positive charge.

amino terminus—*see* **N-terminus**

amoeba (plural **amoebae**)

General description given to free-living single-celled carnivorous organisms that crawl. More precisely, a subdivision of protozoa. *Amoeba proteus* is a species of giant freshwater amoeba widely used in studies of cell crawling.

AMP (adenosine 5′ monophosphate)

One of the four nucleotides in an RNA molecule. AMP is produced by the energetically favorable hydrolysis of ATP. (*See* Figure 3–41.)

amphipathic

Having both hydrophobic and hydrophilic regions, as in a phospholipid or a detergent molecule.

anabolism

Reaction pathways by which large molecules are made from smaller ones. Biosynthesis.

anaerobic

Describes a cell, organism, or metabolic process that functions in the absence of air or, more precisely, in the absence of molecular oxygen.

anaphase

Stage of mitosis during which the two sets of chromosomes separate and move away from each other. Composed of anaphase A (chromosomes move toward the two spindle poles) and anaphase B (spindle poles move apart).

anaphase-promoting complex (APC)

A protein complex that promotes the destruction of specific proteins, by catalyzing their ubiquitination. It is a crucial component of the cell-cycle control system.

anion

Negatively charged ion, such as Cl⁻ or CH_3COO^-.

antibody (immunoglobulin)

Protein produced by B lymphocytes in response to a foreign molecule or invading organism. Binds to the foreign molecule or cell extremely tightly, thereby inactivating it or marking it for destruction.

anticodon

Sequence of three nucleotides in a transfer RNA molecule that is complementary to the three-nucleotide codon on a messenger RNA molecule; the anticodon is matched to a specific amino acid covalently attached to the transfer RNA molecule.

antigen

Molecule that provokes the cellular production of specific neutralizing antibodies in an immune response.

antiparallel

Describes two similar structures arranged in opposite orientations, such as the two strands of a DNA double helix.

antiport

Membrane carrier protein that transports two different ions or small molecules across a membrane in opposite directions, either simultaneously or in sequence.

APC—*see* **anaphase-promoting complex**

apical

Describes the tip of a cell, structure, or organ. The apical surface of an epithelial cell is the exposed free surface (opposite to the basal surface).

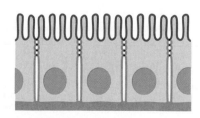

apoptosis—*see* **programmed cell death**

archaea

One of the two divisions of procaryotes, typically found in hostile environments such as hot springs or concentrated brine. (*See also* **eubacteria**.)

asexual reproduction

Any type of reproduction (such as budding in *Hydra*, binary fission in bacteria, or mitotic division in eucaryotic microorganisms) that does not involve gamete formation and fusion. It produces an individual genetically identical to the parent.

aster

Star-shaped system of microtubules emanating from a centrosome or from a pole of a mitotic spindle.

atom

The smallest particle of an element that still retains its distinctive chemical properties.

atomic number

The number of protons in the nucleus of an atom of an element.

atomic weight

Mass of an atom of an isotope expressed in daltons.

ATP (adenosine 5′-triphosphate)

Nucleoside triphosphate composed of adenine, ribose, and three phosphate groups that is the principal carrier of chemical energy in cells. The terminal phosphate groups are highly reactive in the sense that their hydrolysis, or transfer to another molecule, takes place with release of a large amount of free energy. (*See* Figure 2–23.)

ATP synthase

Membrane-associated enzyme complex that catalyzes the formation of ATP during oxidative phosphorylation and photosynthesis. Found in mitochondria, chloroplasts and bacteria.

Avogadro's number

The number of molecules in a quantity of substance equal to its molecular weight in grams. Approximately 6×10^{23}.

axon

Long thin nerve cell process capable of rapidly conducting nerve impulses over long distances so as to deliver signals to other cells.

bacteriorhodopsin

Pigmented protein found in the plasma membrane of a salt-loving bacterium, *Halobacterium halobium;* it pumps protons out of the cell in response to light.

bacteria (singular **bacterium**)

Common name for procaryotic organisms, but more precisely refers to members of the domain Bacteria (*see* **eubacteria**). Most are single-celled organisms but multicellular forms exist (such as *Streptomyces*). (*See also* **archaea**.)

basal

Situated near the base. The basal surface of a cell is opposite the apical surface.

basal body—*see* **centriole**

basal lamina

Thin mat of extracellular matrix that separates epithelial sheets, and many types of cells such as muscle cells or fat cells, from connective tissue. Sometimes called a basement membrane.

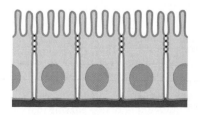

base

Molecule that accepts a proton in solution. Also used to refer to the purine or pyrimidines in DNA and RNA.

base pair

Two nucleotides in an RNA or a DNA molecule that are paired by hydrogen bonds—for example, G with C, and A with T or U.

basic

Having the properties of a base. Alkaline.

Bcl-2 family

Family of intracellular proteins that either promote or inhibit apoptosis by regulating the activation of caspases.

beta sheet (β sheet)

Folding pattern found in many proteins in which neighboring regions of the polypeptide chain associate with each other through hydrogen bonds to give a rigid, flattened structure.

binding site

Region on the surface of a protein, typically a cavity or groove, that is complementary in shape to, and forms multiple noncovalent bonds with, a second molecule (the ligand).

biochemistry

The study of the chemical compounds and reactions that occur in living organisms.

biosynthesis

The formation of complex molecules from simple substances by living cells.

bivalent

A duplicated chromosome paired with its homologous duplicated chromosome at the beginning of meiosis.

bond—*see* **chemical bond**

bond length

The distance between two atoms in a molecule, usually those linked by a covalent bond.

bond energy

The strength of the chemical linkage between two atoms, measured by the energy in kilocalories needed to break it.

C-terminus (carboxyl terminus)

That end of a polypeptide chain that carries an unattached carboxylic acid group.

cadherin

A member of a family of proteins that mediates Ca^{2+}-dependent cell–cell adhesion in animal tissues.

calmodulin (CaM)

Small Ca^{2+}-binding protein that modifies the activity of many target enzymes and membrane transport proteins in response to changes in Ca^{2+} concentration.

Ca^{2+}/calmodulin-dependent protein kinase (CaM kinase)

Enzyme that phosphorylates target proteins in response to a rise in Ca^{2+} ions, through its interaction with the Ca^{2+}-binding protein calmodulin.

calorie

Unit of heat. One calorie (small "c") is the amount of heat needed to raise the temperature of 1 gram of water by $1°C$.

cancer

Disease caused by abnormal and uncontrolled cell division resulting in localized growths, or tumors, which may spread throughout the body.

carbohydrate

General term for sugars and related compounds with the general formula $(CH_2O)_n$.

carbohydrate layer

A layer of sugar residues, including the polysaccharide portions of proteoglycans and oligosaccharides attached to protein or lipid molecules, on the outer surface of a cell.

carbon fixation

Process by which green plants incorporate carbon atoms from atmospheric carbon dioxide into sugars. The second stage of photosynthesis.

carbonyl group

Pair of atoms consisting of a carbon atom linked to an oxygen atom by a double bond.

carboxyl group

Carbon atom linked both to an oxygen atom by a double bond and to a hydroxyl group. Molecules containing a carboxyl group are weak (carboxylic) acids.

carboxyl terminus—*see* **C-terminus**

carrier protein

Membrane transport protein that binds to a solute and transports it across the membrane by undergoing a series of conformational changes.

cascade—*see* **signaling cascade**

catabolism

General term for the enzyme-catalyzed reactions in a cell by which complex molecules are degraded to simpler ones with release of energy. Intermediates in these catabolic reactions are sometimes called catabolites.

catalyst

Substance that accelerates a chemical reaction without itself undergoing a change. Enzymes are protein catalysts.

catastrophin

Protein that destabilizes microtubules by promoting their depolymerization.

cation

Positively charged ion, such as Na^+ or $CH_3NH_3^+$. (Pronounced "cat–ion")

Cdk—*see* **cyclin-dependent kinase**

Cdk inhibitor protein

Protein that inhibits cyclin-Cdk complexes, primarily to inhibit progress through the G_1 and S phases of the cell cycle.

cDNA—*see* **complementary DNA**

cDNA clone

One of a large number of identical copies of a cDNA molecule.

cDNA library

A collection of cDNA clones, usually representing most of the genes expressed in a particular cell type or tissue.

cell

The basic unit from which living organisms are made, consisting of an aqueous solution of organic molecules enclosed by a membrane. All cells arise from existing cells, usually by a process of division.

cell body

Main part of a nerve cell that contains the nucleus. The other parts are axons and dendrites.

cell cortex

Specialized layer of cytoplasm on the inner face of the plasma membrane. In animal cells it is an actin-rich layer responsible for cell-surface movements.

cell cycle

Reproductive cycle of the cell: the orderly sequence of events by which a cell duplicates its contents and divides into two.

cell-cycle control system

Network of regulatory proteins that governs progression of a eucaryotic cell through the cell cycle.

cell division
Separation of a cell into two daughter cells. In eucaryotic cells it entails division of the nucleus (mitosis) closely followed by division of the cytoplasm (cytokinesis).

cell junction
Specialized region of connection between two cells or between a cell and the extracellular matrix.

cell line
Population of cells of plant or animal origin capable of dividing indefinitely in culture.

cell locomotion (cell migration)
Active movement of a cell from one location to another. Particularly the migration of a cell over a surface.

cell memory
The ability of cells and their descendants, without undergoing any change of DNA sequence, to retain a trace of the effects of past influences, displaying the consequences in persistently altered patterns of gene expression.

cell senescence
The normal ageing of cells in a higher animal whereby, after an allotted number of divisions, they cease to divide and eventually die.

cell signaling
The molecular mechanisms by which cells detect and respond to external stimuli and send messages to other cells.

cellulose
Structural polysaccharide consisting of long chains of covalently linked glucose units. It provides tensile strength in plant cell walls.

cell wall
Mechanically strong fibrous layer deposited by a cell outside its plasma membrane. Prominent in most plants, bacteria, algae, and fungi but not present in most animal cells.

central dogma
The principle that genetic information flows from DNA to RNA to protein.

centriole
Short cylindrical array of microtubules, usually found in pairs at the center of a centrosome in animal cells. Also found at the base of cilia and flagella (and called basal bodies).

centromere
Constricted region of a mitotic chromosome that holds sister chromatids together; also the site on the DNA where the kinetochore forms and then captures microtubules from the mitotic spindle.

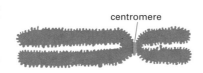
centromere

centrosome (cell center)
Centrally located organelle of animal cells that is the primary microtubule-organizing center and acts as the spindle pole during mitosis. In most animal cells it contains a pair of centrioles.

centrosome cycle
Duplication of the centrosome (during interphase) and separation of the two new centrosomes (at the beginning of mitosis), to form the poles of the mitotic spindle.

channel
An aqueous pore in a lipid membrane, with walls made of protein, through which selected ions or molecules can pass.

channel protein
Protein that forms a narrow hydrophilic pore across a membrane that allows ions and small molecules to move passively from one side to the other.

checkpoint
Point in the eucaryotic cell division cycle where progress through the cycle can be halted until conditions are suitable for the cell to proceed to the next stage.

chemical bond
Chemical affinity between two atoms that holds them together. Types found in living cells include ionic bonds, covalent bonds, polar bonds and hydrogen bonds.

chemical group
Set of covalently linked atoms, such as a hydroxyl group (–OH) or an amino group (–NH$_2$) that occurs in many different molecules and the chemical behavior of which is well characterized.

chemiosmotic coupling
Mechanism in which a gradient of hydrogen ions (a pH gradient) across a membrane is used to drive an energy-requiring process, such as ATP production or the transport of a molecule across a membrane.

chiasma (plural **chiasmata**)
X-shaped connection visible between paired homologous chromosomes in division I of meiosis, and which represents a site of crossing-over.

Chlamydomonas
Unicellular green algae with two flagella.

chlorophyll
Light-absorbing pigment that plays a central part in photosynthesis.

chloroplast
Specialized organelle in algae and plants that contains chlorophyll and in which photosynthesis takes place.

cholesterol
Lipid molecule with a characteristic four-ringed steroid structure that is an important component of the plasma membranes of animal cells. (See Figure 11–7.)

chromatid
One copy of a chromosome formed by DNA replication that is still joined at the centromere to the other copy (the sister chromatid).

chromatin
Complex of DNA, histones, and nonhistone proteins found in the nucleus of a eucaryotic cell. The material of which chromosomes are made.

chromatin remodeling complex
Enzyme (typically multisubunit) that uses ATP hydrolysis to alter histone–DNA interactions in eucaryotic chromosomes; the resulting alteration changes the accessibility of the underlying DNA to other proteins, including those involved in transcription.

chromosome
Long threadlike structure composed of DNA and associated proteins that carries part or all of the genetic information of an organism. Especially evident in plant and animal cells undergoing mitosis or meiosis.

chromosome condensation
Process by which a chromosome becomes packed up into a more compact structure prior to M phase of the cell cycle.

citric acid cycle (TCA, or tricarboxylic acid cycle; Krebs cycle)
Central metabolic pathway in all aerobic organisms that oxidizes acetyl groups derived from food molecules to CO_2. In eucaryotic cells these reactions are located in the mitochondrial matrix.

ciliate
Type of single-celled eucaryotic organism (protozoan) characterized by numerous cilia on its surface. The cilia are used for swimming, feeding, or the capture of prey.

cilium (plural **cilia**)
Hairlike extension on the surface of a cell with a core bundle of microtubules and capable of performing repeated beating movements. Cilia, in large numbers, drive the movement of fluid over epithelial sheets, as in the lungs.

cis
On the same side, for example the *cis* Golgi network is that part closest to the endoplasmic reticulum.

cloning
Making many identical copies of a cell or a DNA molecule or an organism.

cloning vector—*see* **vector**

coated vesicle
Small membrane-bounded organelle with a cage of proteins (the coat) on its cytosolic surface. It is formed by the pinching off of a coated region of membrane.

codon
Sequence of three nucleotides in a DNA or messenger RNA molecule that represents the instruction for incorporation of a specific amino acid into a growing polypeptide chain.

coenzyme A (CoA)
Small molecule used in the enzymatic transfer of acyl groups in the cell. (*See also* **acetyl CoA** and Figure 3–37.)

cohesin
Protein complex that holds sister chromatids together after DNA has been replicated in the cell cycle.

coiled-coil
Especially stable rod-like protein structure formed by two α helices coiled around each other.

collagen
Fibrous protein rich in glycine and proline that is a major component of the extracellular matrix and connective tissues. Exists in many forms: type I, the most common, is found in skin, tendon, and bone; type II is found in cartilage; type IV is present in basal laminae; and so on.

combinatorial
Describes any process that is governed by a specific combination of factors (rather than by any one single factor), with different combinations having different effects.

combinatorial control
Describes the way in which groups of proteins work together to control the expression of a single gene.

complementary
Describes two molecular surfaces that fit together closely and form noncovalent bonds with each other. Examples include complementary base pairs, such A and T, and the two complementary strands of a DNA molecule.

complementary DNA (cDNA)
DNA molecule made as a copy of mRNA and therefore lacking the introns that are present in genomic DNA. Used to determine the amino acid sequence of a protein by DNA sequencing or to make the protein in large quantities by cloning followed by expression.

complex—*see* **molecular complex**

complex trait
A heritable characteristic whose transmission to progeny does not obey Mendel's laws. Such traits often arise from the interaction of multiple genes.

condensation
Type of chemical reaction in which two organic molecules become linked to each other by a covalent bond with concomitant removal of a molecule of water. Also called a dehydration reaction.

conformation
Spatial location of the atoms of a molecule. The precise shape of a protein or other macromolecule in three dimensions.

coupled reaction
One of a linked pair of chemical reactions in which free energy released by the first serves to drive the second.

coupled transport
Membrane transport process in which the transfer of one molecule depends on the simultaneous or sequential transfer of a second molecule.

covalent bond
Stable chemical link between two atoms produced by sharing one or more pairs of electrons.

crossing-over
Process whereby two homologous chromosomes break at corresponding sites and rejoin to produce two recombined chromosomes.

cyclic AMP (cAMP)
Nucleotide generated from ATP in response to hormonal stimulation of cell-surface receptors. cAMP acts as a signaling molecule by activating protein kinase A; it is hydrolyzed to AMP by a phosphodiesterase.

cyclic AMP-dependent protein kinase (protein kinase A, PKA)
Enzyme that phosphorylates target proteins in response to a rise in intracellular cyclic AMP.

cyclin
Protein that periodically rises and falls in concentration in step with the eucaryotic cell cycle. Cyclins activate specific protein kinases (*see* **cyclin-dependent protein kinases**) and thereby help control progression from one stage of the cell cycle to the next.

cyclin-dependent protein kinase (Cdk)
Protein kinase that has to be complexed with a cyclin protein in order to act. Different Cdk-cyclin complexes trigger different steps in the cell-division cycle by phosphorylating specific target proteins.

cytochrome
Colored, heme-containing protein that transfers electrons during cellular respiration and photosynthesis.

cytokinesis
Division of the cytoplasm of a plant or animal cell into two, as distinct from the division of its nucleus (which is mitosis).

cytoplasm
Contents of a cell that are contained within its plasma membrane but, in the case of eucaryotic cells, outside the nucleus.

cytoskeleton
System of protein filaments in the cytoplasm of a eucaryotic cell that gives the cell shape and the capacity for directed movement. Its most abundant components are actin filaments, microtubules, and intermediate filaments.

cytosol
Contents of the main compartment of the cytoplasm, excluding membrane-bounded organelles such as endoplasmic reticulum and mitochondria. The cell fraction remaining after membranes, cytoskeletal components, and other organelles have been removed.

DAG—*see* **diacylglycerol**

dalton

Unit of molecular mass. Defined as one-twelfth the mass of an atom of carbon-12 (1.66×10^{-24} g); approximately equal to the mass of a hydrogen atom.

dehydration reaction—*see* **condensation reaction**

denature

To cause a dramatic change in conformation of a protein or nucleic acid by heating it or by exposing it to chemicals. Usually results in the loss of biological function.

dendrite

Extension of a nerve cell, typically branched and relatively short, that receives stimuli from other nerve cells.

deoxyribonucleic acid—*see* **DNA**

desmosome

Specialized cell–cell junction, usually formed between two epithelial cells, characterized by dense plaques of protein into which intermediate filaments in the two adjoining cells insert.

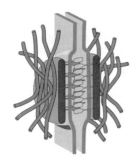

detergent

Soapy substance used by biochemists to solubilize membrane proteins.

development

Succession of changes that take place in an organism as a fertilized egg gives rise to an adult plant or animal.

diacylglycerol (DAG)

Lipid produced by the cleavage of inositol phospholipids in response to extracellular signals. Composed of two fatty acid chains linked to glycerol, it serves as a signaling molecule to help activate protein kinase C.

dideoxy method

The standard method of DNA sequencing. It utilizes DNA polymerases and chain-terminating nucleotides.

differentiation

Process by which a cell undergoes a progressive change to a more specialized and usually easily recognized cell type.

diffusion

The spread of molecules and small particles from one location to another by random, thermally driven movements.

digestion

The enzymatic breakdown of proteins, lipids and polysaccharides ingested as food. The small molecules produced then enter the cytosol of cells and are metabolized.

dimer

A structure composed of two equivalent halves. The term "heterodimer" is sometimes used when the two halves are not perfectly identical.

diploid

A cell or organism containing two sets of homologous chromosomes and hence two copies of each gene or genetic locus.

disaccharide

Carbohydrate molecule, such as sucrose, consisting of two covalently joined monosaccharide units.

disulfide bond (S–S bond)

Covalent linkage formed between two sulfhydryl groups on cysteines. Common way to join two proteins or to link together different parts of the same protein in the extracellular space.

DNA (deoxyribonucleic acid)

Double-stranded polynucleotide formed from two separate chains of covalently linked deoxyribonucleotide units; serves as the carrier of genetic information.

DNA cloning—*see* **cloning**

DNA library

Collection of cloned DNA molecules, representing either an entire genome (genomic library) or copies of the mRNA produced by a cell (cDNA library).

DNA ligase—*see* **ligase**

DNA microarray

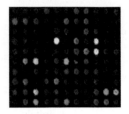

A glass slide upon which a large number of short DNA molecules (typically in the tens of thousands) have been immobilized in an orderly pattern. Each of these DNA fragments acts as a probe for a specific gene, allowing the RNA products of thousands of genes to be monitored at the same time.

DNA polymerase—*see* **polymerase**

DNA repair

Collective term for the enzymatic processes by that correct deleterious changes affecting the continuity or sequence of a DNA molecule.

DNA replication

The process by which a copy of a DNA molecule is made.

DNA transcription—*see* **transcription**

domain

Small discrete region of a structure. A protein domain is a compact and stable folded region of polypeptide. A membrane domain is a region of bilayer with a characteritistic lipid and protein composition.

double bond

A type of chemical linkage between two atoms formed by sharing four electrons.

double helix

The typical conformation of a DNA molecule in which two strands are wound around each other with base pairing between the strands.

Drosophila melanogaster

Species of small fly, commonly called a fruit fly, much used in genetic studies of development.

dynamic instability

The property shown by microtubules of growing and shrinking repeatedly through the addition and loss of tubulin subunits from their exposed ends.

dynein

Member of a family of large motor proteins that undergo ATP-dependent movement along microtubules. Dynein is responsible for the bending of cilia.

egg

The female germ cell, usually large, nonmotile and having abundant cytoplasm.

electrochemical gradient

Driving force that causes an ion to move across a membrane. Caused by differences in ion concentration and in electrical charge on either side of the membrane.

electron

Fundamental subatomic particle with a unit negative charge (e^-).

electron acceptor
Atom or molecule that takes up electrons readily, thereby gaining an electron and becoming reduced.

electron carrier
Molecule such as cytochrome *c* that transfers an electron from a donor molecule to an acceptor molecule.

electron donor
Molecule that easily gives up an electron, becoming oxidized in the process.

electron transport
Movement of electrons from a higher to a lower energy level along a series of electron carrier molecules (termed an *electron transport chain*) as in oxidative phosphorylation and photosynthesis.

element
Substance that cannot be broken down to any other chemical form; composed of a single type of atom.

embryonic stem cell (ES cell)
An undifferentiated cell type derived from the inner cell mass of an early mammalian embryo. Embryonic stem cells can be maintained indefinitely as a proliferating cell population (cell line) in culture, but remain capable of differentiating , when placed in an appropriate environment, to give any of the specialized cell types in the adult body.

endocytosis
Uptake of material into a cell by an invagination of the plasma membrane and its internalization in a membrane-bounded vesicle. (*See also* **pinocytosis** and **phagocytosis**.)

endoplasmic reticulum (ER)
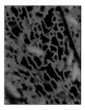
Labyrinthine, membrane-bounded compartment in the cytoplasm of eucaryotic cells, where lipids and secreted and membrane-bound proteins are made.

endosome
Membrane-bounded compartment of a eucaryotic cell through which endocytosed material passes on its way to lysosomes.

enhancer
Regulatory DNA sequence to which gene regulatory proteins bind, influencing the rate of transcription of a structural gene that can be many thousands of base pairs away.

entropy
Thermodynamic quantity that measures the degree of disorder in a system; the higher the entropy, the more the disorder.

enzyme
A protein that catalyzes a specific chemical reaction.

epithelium
Sheet of cells covering or lining an external surface or cavity.

equilibrium
In a chemical context, a state in which two or more reactions are proceeding at such a rate that they exactly balance each other and no net chemical change is occurring.

equilibrium constant (K)
A number that characterizes the steady state reached by a reversible chemical reaction. Given by the ratio of forward and reverse rate constants of a reaction. (*See* Table 3–1, p. 98.)

***Escherichia coli* (*E. coli*)**
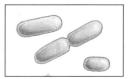
Rodlike bacterium normally found in the colon of humans and other mammals and widely used in biomedical research.

eubacteria
The proper term for the bacteria of common occurrence, used to distinguish them from archaea.

eucaryote (eukaryote)
Living organism composed of one or more cells with a distinct nucleus and cytoplasm. Includes all forms of life except viruses and bacteria (procaryotes).

evolution
The gradual change in living organisms taking place over generations that results in new species being formed.

exocytosis
Process by which most molecules are secreted from a eucaryotic cell. These molecules are packaged in membrane-bounded vesicles that fuse with the plasma membrane, releasing their contents to the outside.

exon
Segment of a eucaryotic gene that is transcribed into RNA and expressed; dictates the amino acid sequence of part of a protein.

exon shuffling
Evolutionary process by which new genes form by linking together combinations of initially separate exons encoding different protein domains.

expressed sequence tag (EST)
A nucleotide sequence (typically 300–500 nucleotides in length) derived from the mRNA of an organism. Large collections of ESTs are useful for because they highlight the protein-coding regions of genomes.

extracellular matrix
Complex network of polysaccharides (such as glycosaminoglycans or cellulose) and proteins (such as collagen) secreted by cells. A structural component of tissues that also influences their development and physiology.

FAD—*see* **FADH₂**

FADH₂ (reduced flavin adenine dinucleotide)
Major electron carrier in metabolism produced by oxidation of FAD during the oxidation of catabolites such as succinate.

fat
Lipids used by living cells to store metabolic energy. Mainly composed of triacylglycerols. (*See* Panel 2–4, pp. 72–73.)

fatty acid

Compound such as palmitic acid that has a carboxylic acid attached to a long hydrocarbon chain. Used as a major source of energy during metabolism and as a starting point for the synthesis of phospholipids. (*See* Panel 2–4, pp. 72–73.)

feedback inhibition
A form of metabolic control in which the end-product of a chain of enzymatic reactions reduces the activity of an enzyme early in the pathway.

fermentation
The breakdown of organic molecules without the involvement of molecular oxygen. Oxidation is less complete than in aerobic processes and yields less energy.

fertilization
Sequence of events that starts when a sperm cell makes contact with an egg and leads to their fusion and further development.

fibroblast
Common cell type found in connective tissue that secretes an extracellular matrix rich in collagen and other extracellular matrix macromolecules. Migrates and proliferates readily in wounded tissue and in tissue culture.

fibrous protein
A protein with an elongated shape. Typically one such as collagen or intermediate filament protein that is able to associate into long filamentous structures.

filopodium
Long thin actin-containing extension on the surface of an animal cell. Sometimes has an exploratory function, as in a growth cone.

flagellum (plural **flagella**)
Long, whiplike protrusion that drives a cell through a fluid medium by its beating. Eucaryotic flagella are longer versions of cilia; bacterial flagella are completely different, being smaller and simpler in construction.

free energy (G)
Energy that can be extracted from a system to do useful work, such as driving a chemical reaction. The standard free energy of a substance, G^0, is its free energy measured at a defined concentration, temperature and pressure.

free energy change (ΔG)
"Delta G": the difference in free energy between reactant and product molecules in a chemical reaction. A large negative value of ΔG indicates that the reaction has a strong tendency to occur.

G-protein-linked receptor
Cell-surface receptor that associates with an intracellular trimeric GTP-binding protein (G protein) after receptor activation by an extracellular ligand. These receptors are sevenpass transmembrane proteins.

gain-of-function mutation
A mutation that increases the activity of a gene, or makes it active in inappropriate circumstances; such mutations are usually dominant.

gamete—*see* **germ cell**

gap junction
Communicating cell-cell junction that allows ions and small molecules to pass from the cytoplasm of one cell to the cytoplasm of the next.

GDP (guanosine 5′-diphosphate)
Nucleotide that is produced by the hydrolysis of the terminal phosphate of GTP, a reaction that also produces inorganic phosphate. When free in solution, GDP is rapidly rephosphorylated to GTP, usually by the transfer of the terminal phosphate from ATP in the reaction, ATP + GDP → ADP + GTP.

gene
Region of DNA that controls a discrete hereditary characteristic of an organism, usually corresponding to a single protein or RNA.

gene expression
The process by which a gene makes its effect on an cell or organism, usually by directing the synthesis of a protein with a characteristic activity.

gene regulatory protein
General name for any protein that binds to a specific DNA sequence to alter the expression of a gene.

general transcription factor
Any of the proteins whose assembly around the TATA box is required for the initiation of transcription of most eucaryotic genes.

genetic code
Set of rules specifying the correspondence between nucleotide triplets (codons) in DNA or RNA and amino acids in proteins.

genetic engineering—*see* **recombinant DNA technology**

genetic map
A graphic representation of the order of genes in chromosomes, spaced according to the amount of recombination that occurs between them.

genetic screen
A search through a collection of mutants for a particular phenotype.

genetically unstable
Term used to describe cells with an enhanced mutation rate, such as cancer cells.

genetics
The study of the genes of an organism based on heredity and variation.

genome
The total genetic information carried by a cell or an organism (or the DNA molecules that carry this information).

genome annotation
Process by which scientists attempt to bridge the gap between the nucleotide sequence of an organism's genome and the organism's physiology. Genome annotation includes delineating genes and their regulatory sequences, assigning functions to gene products, specifying when in development and in which cell types each gene is expressed, and, ultimately, understanding how all the gene products work together in the intact organism.

genotype
Set of genes carried by an individual cell or organism.

germ cell (gamete)
Cell type in a diploid organism that carries only one set of chromosomes and is specialized for sexual reproduction. A sperm or an egg.

germ line
The lineage of reproductive cells which contributes to the formation of a new generation of organisms, as distinct from somatic cells, which form the body and leave no descendants.

globular protein
Any protein with an approximately rounded shape. Most enzymes are globular.

glucose
Six-carbon sugar that plays a major role in the metabolism of living cells. Stored in polymeric form as glycogen in animal cells and as starch in plant cells. (*See* Panel 2–3, pp. 70–71.)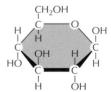

glycogen
Polysaccharide composed exclusively of glucose units used to store energy in animal cells. Large granules of glycogen are especially abundant in liver and muscle cells.

glycolipid
Membrane lipid molecule with a short carbohydrate chain attached to a hydrophobic tail.

glycolysis
Ubiquitous metabolic pathway in the cytosol in which sugars are incompletely degraded with production of ATP. (Literally, "sugar splitting.")

glycoprotein

Any protein with one or more covalently linked oligosaccharide chains. Includes most secreted proteins and most proteins exposed on the outer surface of the plasma membrane.

glycosaminoglycan (GAG)

Family of high molecular weight polysaccharides containing amino sugars found as protective coats around animal cells.

Golgi apparatus

 Membrane-bounded organelle in eucaryotic cells where the proteins and lipids made in the endoplasmic reticulum are modified and sorted. (Named after its discoverer, Camillo Golgi.)

G₁ phase

Gap 1 phase of the eucaryotic cell cycle, between the end of cytokinesis and the start of DNA synthesis.

G₂ phase

Gap 2 phase of the eucaryotic cell cycle, between the end of DNA synthesis and the beginning of mitosis.

G protein

One of a large family of GTP-binding proteins composed of three different subunits (heterotrimeric GTP-binding proteins) that are important intermediaries in intracellular signaling pathways. Usually activated by the binding of a hormone or other ligand to a transmembrane receptor.

green fluorescent protein (GFP)

Fluorescent protein (from a jellyfish) that is widely used as a marker for monitoring the movement of proteins in living cells.

group—*see* **chemical group**

growth factor

Extracellular polypeptide signaling molecule that stimulates a cell to grow or proliferate. Examples are epidermal growth factor (EGF) and platelet-derived growth factor (PDGF).

GTP (guanosine 5′-triphosphate)

Major nucleoside triphosphate used in the synthesis of RNA and in some energy-transfer reactions. Has a special role in microtubule assembly, protein synthesis, and cell signaling.

GTP-binding protein

An allosteric protein whose conformation is determined by its association with either GTP or GDP. Includes many proteins involved in cell signaling, such as Ras and G proteins.

haploid

A cell or organism with only one set of chromosomes, as in a sperm cell or a bacterium. (*See also* **diploid**.)

haplotype block

A combination of alleles and other DNA markers that has been inherited in a large, linked block—undisturbed by genetic recombination—across many generations.

helix

 An elongated structure in which a filament or thread twists in regular fashion around a central axis.

α-helix—*see* **alpha helix**

hemidesmosome

Specialized anchoring cell junction between an epithelial cell and the underlying basal lamina.

heredity

The transmission from one generation to another of genetic factors that determine individual characteristics. Responsible for the similarity between parents and children.

heterochromatin

Region of a chromosome that remains unusually condensed and transcriptionally inactive during interphase.

heterozygous

Of an organism with dissimilar alleles for any one gene.

high-energy bond

Covalent bond whose hydrolysis releases an unusually large amount of free energy under the conditions existing in a cell. Examples include the phosphodiester bonds in ATP and the thioester linkage in acetyl CoA.

histone

One of a group of basic proteins, rich in arginine and lysine, that are associated with DNA in chromosomes.

histone deacetylase

Enzyme that removes acetyl groups from lysines present in histones; the acetylation state of histones acts as a signal that attracts other proteins that activate or repress transcription.

homolog

A homologous chromosome or, more generally, a macromolecule that has a close evolutionary relationship to another.

homologous

Describes organs or molecules that are similar because of their common evolutionary origin. Specifically it describes similarities in protein or nucleic acid sequence.

homologous chromosome

One of the two copies of a particular chromosome in a diploid cell, one from the father and the other from the mother.

homologous recombination (general recombination)

Genetic exchange between a pair of homologous DNA sequences, typically those located on two copies of the same chromosome.

homophilic

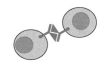 Adjective used to describe a molecule that binds to others of the same kind, especially those involved in cell–cell adhesion.

homozygous

Describes an organism having identical alleles for a given gene.

horizontal gene transfer

Process through which DNA is passed from one organism to another, permanently changing the DNA composition of the recipient. This contrasts with "vertical" gene transfer, which refers to the inheritance of genes from parent to progeny.

hormone

A chemical substance produced by one set of cells in a multicellular organism and transported via body fluids to target tissues on which it exerts a specific effect.

hybridization

Experimental process in which two complementary nucleic acid strands form a double helix; a powerful technique for detecting specific nucleotide sequences.

hydrogen bond

 A weak chemical bond between an electronegative atom such as nitrogen or oxygen and a hydrogen atom bound to another electronegative atom.

hydrogen ion

Commonly used term for a proton (H^+) in aqueous solution, the basis of acidity. Since the proton readily combines with a water molecule to form H_3O^+ it is more accurate to call it a **hydronium ion**.

hydrolysis (adjective hydrolytic)
Cleavage of a covalent bond with accompanying addition of water, –H being added to one product of the cleavage and –OH to the other.

hydronium ion (H_3O^+)
The form taken by a proton (H^+) in aqueous solution.

hydrophilic
Polar molecule or part of a molecule that forms enough hydrogen bonds to water to dissolve readily in water. (Literally, "water loving.")

hydrophobic (lipophilic)
Nonpolar molecule or part of a molecule that cannot form favorable bonding interactions with water molecules and therefore does not dissolve in water. (Literally, "water hating.")

hydroxyl (–OH)
Chemical group consisting of a hydrogen atom linked to an oxygen, as in an alcohol.

hypertonic
Describes any solution with a sufficiently high concentration of solutes to cause water to move out of a cell due to osmosis. (From Greek, *hyper,* over.)

hypotonic
Describes any solution with a sufficiently low concentration of solutes to cause water to move into a cell due to osmosis. (From Greek, *hypo,* under.)

immortalization
Spontaneous or purposeful genetic alteration of a cell so that it can undergo an unlimited number of cell divisions.

initiation factor
Protein that promotes the proper association of ribosomes with mRNA and is required for the initiation of protein synthesis.

initiator tRNA
Special tRNA that initiates translation. It always carries the amino acid methionine.

inositol
Sugar molecule with six hydroxyl groups that forms the framework for inositol phospholipids.

inositol 1, 4, 5-trisphosphate (IP$_3$)
Small intracellular signaling molecule produced during activation of the inositol phospholipid signaling pathway to release Ca^{2+} from the endoplasmic reticulum.

inositol phospholipids (phosphoinositides)
Minor lipid components of plasma membranes containing phosphorylated inositol derivatives that are important in signal transduction in eucaryotic cells.

***in situ* hybridization**
Technique in which a single-stranded RNA or DNA probe is used to locate a gene or an mRNA molecule in a cell or tissue.

integration
Process by which one DNA molecule recombines with and becomes physically part of another DNA molecule. This is the way, for example, that a plasmid or viral genome enters a host genome.

intermediate filament
Fibrous protein filament (about 10 nm in diameter) that forms ropelike networks in animal cells. Often used as a structural element that resists tension applied to the cell from outside.

internal membrane
Eucaryotic cell membrane other than the plasma membrane. The membranes of the endoplasmic reticulum and the Golgi apparatus are examples.

interphase
Long period of the cell cycle between one mitosis and the next. Includes G_1 phase, S phase, and G_2 phase.

interphase chromosome
State of a eucaryotic chromosome, as long extended active threads, seen when the cell is between divisions.

Intron
Noncoding region of a eucaryotic gene that is transcribed into an RNA molecule but is then excised by RNA splicing to produce mRNA.

in vitro
Term used by biochemists to describe a process taking place in an isolated cell-free extract. Also used by cell biologists to refer to cells growing in culture (*in vitro*), as opposed to in an organism (*in vivo*). (Latin for "in glass.")

in vivo
In an intact cell or organism. (Latin for "in life.")

ion
An atom carrying an electrical charge, either positive or negative.

ion channel
Transmembrane protein complex that forms a water-filled channel across the lipid bilayer through which specific inorganic ions can diffuse down their electrochemical gradients.

ionic bond
Attractive force that holds together two ions, one positive the other negative.

IP$_3$—*see* **inositol 1,4, 5-trisphosphate**

iron-sulfur center
One of a family of electron transporters containing iron atoms linked to sulfur atoms and cysteine side chains.

isoforms
Multiple forms of the same protein that differ somewhat in their amino acid sequence. They can be produced by different genes or by alternative splicing of RNA transcripts from the same gene.

isomer (stereoisomer)
One of two or more substances that contain the same atoms and have the same molecular formula (such as $C_6H_{12}O_6$) but differ in the spatial arrangement of these atoms. Optical isomers differ only by being mirror images of each other.

isotopes
Two or more forms of an atom that have the same chemistry but differ in atomic weight. May be either stable or radioactive.

K—*see* **equilibrium constant**

K^+
Potassium ion—a major ionic constituent of living cells.

K_M
The concentration of substrate at which an enzyme works at half its maximum rate. Large values of K_m usually indicate that enzyme binds to its substrate with relatively low affinity.

karyotype
A display of the full set of chromosomes of a cell arranged with respect to size, shape, and number.

kilocalorie (kcal)
Unit of heat equal to 1000 calories. Often used to express the energy content of food or molecules: bond strengths, for example, are measured in kcal/mole. An alternative unit in wide use is the kilojoule.

kilojoule (kJ)
Standard unit of energy equal to 0.239 kilocalories.

kinase—*see* **protein kinase**

kinesin

One member of a large family of motor protein that uses the energy of ATP hydrolysis to move along a microtubule.

kinetochore

Complex protein-containing structure on a mitotic chromosome to which microtubules attach. The kinetochore forms on the part of the chromosome known as the centromere.

knockout mouse

A genetically engineered mouse in which a specific gene has been inactivateed, for example, by introducing a deletion in its DNA.

lagging strand

One of the two newly made strands of DNA found at a replication fork. The lagging strand is made in discontinuous lengths that are later joined covalently.

lamellipodium

Dynamic sheetlike extension on the surface of an animal cell, especially one migrating over a surface.

law of independent assortment

The second law of heredity, derived by Mendel, which states that during gamete formation, the alleles for different traits segregate independently of one another.

law of segregation

The first law of heredity, derived by Mendel, which states that the maternal and paternal alleles for a trait separate from one another during gamete formation and then reunite during fertilization.

leading strand

One of the two newly made strands of DNA found at a replication fork. The leading strand is made by continuous synthesis in the 5′-to-3′ direction.

ligand

Molecule such as a hormone or a neurotransmitter that binds to a specific site on a protein.

ligand-gated channel

An ion channel that opens when it binds a small molecule such as a neurotransmitter.

ligase

Enzyme that joins two adjacent DNA strands together.

lipid

Organic molecule that is insoluble in water but dissolves readily in nonpolar organic solvents. One class, the phospholipids, forms the structural basis of biological membranes.

lipid bilayer

Thin bimolecular sheet of mainly phospholipid molecules that forms the structural basis for all cell membranes. The two layers of lipid molecules are packed with their hydrophobic tails pointing inward and their hydrophilic heads outward, exposed to water.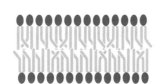

lipophilic—*see* **hydrophobic**

local mediator

Secreted signal molecule that acts at a short range on adjacent cells.

loss-of-function mutation

A mutation that reduces or eliminates the activity of a gene. Such mutations are usually recessive: the organism can function normally as long as it retains at least one normal copy of the affected gene.

lumen

Cavity enclosed by an epithelial sheet (in a tissue) or by a membrane (in a cell), as in the lumen of the endoplasmic reticulum. (From Latin, *lumen*, light or opening.)

lymphocyte

White blood cell that mediates the immune response to a foreign molecule (an antigen). Lymphocytes are either of the antibody-secreting B–cell type or the T–cell type that underwrites the cell-mediated immune response system.

lysosome

Intracellular membrane-bounded organelle containing digestive enzymes, typically those most active at the acid pH found in these organelles.

M-Cdk

Active complex formed at the start of M phase of the cell cycle by an M-cyclin and the mitotic cyclin-dependent protein kinase (Cdk).

M-cyclin

Cyclin protein that binds to mitotic Cdk to form M-Cdk at the start of M phase of the cell cycle.

M phase

Period of the eucaryotic cell cycle during which the nucleus and cytoplasm divide.

M-phase-promoting factor—*see* **MPF**

macromolecule

Molecule such as a protein, nucleic acid, or polysaccharide with a molecular mass greater than a few thousand daltons. (From Greek, *makros*, large.)

macrophage

Cell found in animal tissues that is specialized for the uptake of particulate material by phagocytosis; derived from a type of white blood cell.

MAP-kinase

Protein kinase that performs a crucial step in relaying signals from cell-surface receptors to the nucleus. It is the final kinase in a 3-kinase sequence called the MAP-kinase cascade.

MAP-kinase cascade—*see* **MAP-kinase**

matrix

Most generally, a space within which something is formed. In cell biology, this word usually refers to the large internal compartment of the mitochondrion. The mitochondrial matrix contains a concentrated mixture of special enzymes that catalyze oxidation reactions, as well as the mitochondrial genome and the proteins needed to express mitochondrial genes. (*See* Figure 14–3.)

meiosis

Special type of cell division by which eggs and sperm cells are made. Two successive nuclear divisions with only one round of DNA replication generates four haploid daughter cells from an initial diploid cell. (From Greek, *meiosis*, diminution.)

membrane

Thin sheet of lipid molecules and associated proteins that encloses all cells and forms the boundaries of many eucaryotic organelles.

membrane potential

Voltage difference across a membrane due to a slight excess of positive ions on one side and of negative ions on the other. A typical membrane potential for an animal cell plasma membrane is –60 mV (inside negative), measured relative to the surrounding fluid.

membrane protein

A protein associated with a lipid bilayer; can be transmembrane, integral or peripheral.

membrane transport protein

Any protein embedded in a membrane the serves as a carrier of ions or small molecules from one side to the other.

messenger RNA (mRNA)

RNA molecule that specifies the amino acid sequence of a protein. Produced by RNA splicing (in eucaryotes) from a larger RNA molecule made by RNA polymerase as a complementary copy of DNA. It is translated into protein in a process catalyzed by ribosomes.

metabolic pathway

Sequence of enzymatic reactions in which the product of one reaction is the substrate of the next.

metabolism

The sum total of the chemical reactions that take place in the cells of a living organism resulting in growth, division, energy production, excretion of waste and so on.

metaphase

Stage of mitosis at which chromosomes are firmly attached to the mitotic spindle at its equator but have not yet segregated toward opposite poles.

methyl (–CH$_3$) group

Hydrophobic chemical group derived from methane (CH$_4$).

micro

Prefix denoting 10^{-6}.

micrograph

Picture taken through a microscope. Either a light micrograph or an electron micrograph depending upon the type of microscope used.

micron (micrometer, or μm)

Unit of measurement often applied to cells and organelles. Equal to 10^{-6} meter or 10^{-4} centimeter.

microscope

Instrument for viewing extremely small objects. A light microscope utilizes a focused beam of visible light and is used to examine cells and organelles. An electron microscope utilizes a beam of electrons and can be used to examine objects as small as individual molecules.

microtubule

Long, stiff, cylindrical structure composed of the protein tubulin. Used by eucaryotic cells to regulate their shape and control their movements.

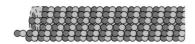

milli

Prefix denoting 10^{-3}.

mismatch repair

Important error correction mechanism in DNA replication that is triggered by the misfit ("mismatch") of noncomplementary base pairs.

mitochondrion (plural mitochondria)

Membrane-bounded organelle, about the size of a bacterium, that carries out oxidative phosphorylation and produces most of the ATP in eucaryotic cells.

mitogen

An extracellular signal molecule that stimulates cell proliferation.

mitosis

Division of the nucleus of a eucaryotic cell, involving condensation of the DNA into visible chromosomes. (From Greek, *mitos*, a thread, referring to the threadlike appearance of the condensed chromosomes.)

mitotic Cdk

Protein kinase that forms a complex with M-cyclin to form M-Cdk at the start of M phase of the cell cycle.

mitotic chromosome

Highly condensed duplicated chromosome with the two new chromosomes still held together at the centromere. A chromosome during the first stages of mitosis.

mitotic spindle

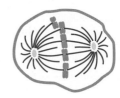

Array of microtubules and associated molecules that forms between the opposite poles of a eucaryotic cell during mitosis and serves to move the duplicated chromosomes apart.

mobile genetic element

Short segment of DNA that can move, sometimes through an RNA intermediate, from one location in a genome to another.

model organism

Organisms selected for intensive study as a representative of a large group of species.

mole

M grams of a substance, where M is its relative molecular mass (molecular weight); this will be 6×10^{23} molecules of the substance.

molecular complex

An assembly of molecules, usually macromolecules, held together by noncovalent bonds and performing a specific function, such as DNA replication or the synthesis of phospholipids.

molecular recognition

Selective binding of two molecular surfaces resulting in a specific response in a target molecule or cell. For example, an enzyme recognizes its substrate and a sperm recognizes the egg.

molecular specificity

Selective affinity of one molecule for another that permits the two to bind or react even in the presence of many unrelated molecular species.

molecular switch

Protein or protein complex that operates in an intracellular signaling pathway and can reversibly switch between an active and inactive state.

molecular weight

Mass of a molecule expressed in daltons.

molecule

Group of atoms joined together by covalent bonds.

monomer

Small molecule that can be linked to others of the same type to form a larger molecule (polymer).

motor protein

Protein such as myosin or kinesin that uses energy derived from ATP hydrolysis to propel itself along a protein filament or polymeric molecule.

MPF (M-phase-promoting factor)

Protein complex containing cyclin and a protein kinase that triggers a cell to enter M phase (originally called maturation-promoting factor).

mRNA—*see* **messenger RNA**

mutation

A randomly produced, heritable change in the nucleotide sequence of a chromosome.

myofibril

Long, highly organized bundle of actin, myosin, and other proteins in the cytoplasm of muscle cells that contracts by a sliding filament mechanism.

myosin
Type of motor protein that uses ATP to drive movements along actin filaments. Myosin II is a large protein that forms the thick filaments of skeletal muscle. Smaller myosins, such as myosin I, are widely distributed, and responsible for many actin-based movements.

N-terminus (amino terminus)
The end of a polypeptide chain that carries a free α-amino group.

Na$^+$
Sodium ion—a major ionic constituent of living cells.

NAD$^+$ (nicotine adenine dinucleotide)
Activated carrier molecule that participates in an oxidation reaction by accepting a hydride ion (H$^-$) from a donor molecule thereby producing **NADH**. Widely used in the energy-producing breakdown of sugar molecules. (*See* Figure 3–35.)

NADPH (nicotine adenine dinucleotide phosphate)
A carrier molecule closely related to NADH used as an electron donor in biosynthetic pathways. In the process it is oxidized to **NADP**.

Na$^+$-K$^+$ pump (Na$^+$-K$^+$ ATPase, sodium pump)
Transmembrane carrier protein found in the plasma membrane of most animal cells that pumps Na$^+$ out of and K$^+$ into the cell, using the energy derived from ATP hydrolysis.

natural selection
Process through which individuals with certain characteristics tend to be eliminated from a population, while individuals with other characteristics survive and reproduce. The individuals with a greater chance of survival and successful reproduction are said to have higher *fitness*. As a result of natural selection, hereditary factors conferring higher fitness are propagated, while those conferring lower fitness are eliminated. This is the central principle of evolution.

nanometer (nm)
Unit of length commonly used to measure molecules and cell organelles. 1 nm = 10^{-3} μm = 10^{-9} m.

Nernst equation
Quantitative expression that relates the equilibrium ratio of concentrations of an ion on either side of a permeable membrane to the voltage difference across the membrane.

nerve cell—*see* **neuron**

nerve terminal
The ending of an axon from which signals are sent to adjoining cells, usually at a synapse.

neuron (nerve cell)
Cell with long processes specialized to receive, conduct, and transmit signals in the nervous system.

neurotransmitter
Small signaling molecule secreted by a nerve cell at a chemical synapse to signal to the postsynaptic cell. Examples include acetylcholine, glutamate, GABA, and glycine.

neutron
Fundamental subatomic particle uncharged and found in the atomic nucleus.

nitric oxide (NO)
Small highly diffusible molecule widely used as an intracellular signal.

nitrogen fixation
Conversion of nitrogen from the atmosphere into nitrogen-containing organic molecules by soil bacteria and cyanobacteria.

NO—*see* **nitric oxide**

noncovalent bond
Chemical bond in which, in contrast with a covalent bond, no electrons are shared. Noncovalent bonds are relatively weak, but they can sum together to produce strong, highly specific interactions between molecules.

nondisjunction
An event that occurs occasionally during meiosis in which a pair of homologous chromosomes fails to separate so that the resulting germ cell has either too many or too few chromosomes.

nonpolar
Said of a molecule that lacks any local accumulation of positive or negative charge. Nonpolar molecules are generally insoluble in water.

nuclear envelope
Double membrane surrounding the nucleus. Consists of outer and inner membranes perforated by nuclear pores.

nuclear lamina
Fibrous layer on the inner surface of the inner nuclear membrane made up of a network of intermediate filaments made from nuclear lamins.

nuclear pore
Channel through the nuclear envelope that allows selected molecules to move between the nucleus and cytoplasm.

nucleic acid
RNA or DNA; consists of a chain of nucleotides joined together by phosphodiester bonds.

nucleolus
Structure in the nucleus where ribosomal RNA is transcribed and ribosomal subunits are assembled.

nucleoside
Compound composed of a purine or pyrimidine base linked to either a ribose or a deoxyribose sugar. (*See* Panel 2–6, pp. 76–77.)

nucleosome
Structural, beadlike unit of a eucaryotic chromosome composed of a short length of DNA wrapped around a core of histone proteins; the fundamental subunit of chromatin.

nucleotide
Nucleoside with one or more phosphate groups joined in ester linkages to the sugar moiety. DNA and RNA are polymers of nucleotides.

nucleus
The major organelle of a eucaryotic cell, which contains DNA organized into chromosomes. Also, when referring to an atom, the central mass built from neutrons and protons.

Okazaki fragment
Short length of DNA produced on the lagging strand during DNA replication. These fragments are rapidly joined together by DNA ligase to form a continuous DNA strand.

oligo–
Prefix that denotes a short polymer (oligomer). May be made of amino acids (oligopeptide), sugars (oligosaccharide), or nucleotides (oligonucleotide). (From Greek, *oligos,* few or little.)

oncogene
Any gene that makes a cell cancerous. Typically a mutant form of a normal gene (proto-oncogene) involved in the control of cell growth or division.

organic chemistry
The branch of chemistry concerned with compounds made of carbon. Includes essentially all of the molecules from which living cells are made, apart from water.

organelle
A discrete structure or subcompartment of a eucaryotic cell (especially one that is visible in the light microscope) that is specialized to carry out a particular function. Examples include mitochondria and the Golgi apparatus.

origin recognition complex (ORC)
Large protein complex that is bound to the DNA at origins of replication in eucaryotic chromosomes throughout the cell cycle.

osmosis
Net movement of water molecules across a semipermeable membrane driven by a difference in concentration of solute on either side. The membrane must be permeable to water but not to the solute molecules.

osmotic pressure
Pressure that must be exerted on the low-solute concentration side of a semipermeable membrane to prevent the flow of water across the membrane as a result of osmosis.

oxidation
Loss of electron density from an atom, as occurs during the addition of oxygen to a molecule or when a hydrogen is removed. The opposite of **reduction**. (*See* Figure 3–12.)

oxidative phosphorylation
Process in bacteria and mitochondria in which ATP formation is driven by the transfer of electrons from food molecules to molecular oxygen. Involves the intermediate generation of a pH gradient across a membrane and chemiosmotic coupling.

pairing
In a genetic sense, the event in meiosis at which two homologous chromosomes line up together to form a duplicated structure.

palindromic sequence
Nucleotide sequence that is identical to its complementary strand when each is read in the same chemical direction.

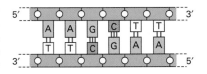

passive transport
The movement of a small molecule or ion across a membrane due to a difference in concentration or electrical charge.

patch-clamp recording
Technique in which the tip of a small glass electrode is sealed onto a patch of cell membrane, thereby making it possible to record the flow of current through individual ion channels in the patch.

PCR—*see* **polymerase chain reaction**

pedigree
The line of descent, or ancestry, of an individual animal.

peptide bond
Chemical bond between the carbonyl group of one amino acid and the amino group of a second amino acid—a special form of amide linkage. (*See* Panel 2–5, pp. 74–75.)

peroxisome
Small membrane-bounded organelle that uses molecular oxygen to oxidize organic molecules. Contains some enzymes that produce and others that degrade hydrogen peroxide (H_2O_2).

phagocytic cell
A cell such as a macrophage or neutrophil that is specialized to take up particles and microorganisms by phagocytosis.

phagocytosis
The process by which particulate material is engulfed ("eaten") by a cell. Prominent in predatory cells, such as *Amoeba proteus* and in cells of the vertebrate immune system such as macrophages.

phenotype
The observable character of a cell or organism.

phosphatidylcholine
Common phospholipid present in abundance in most biological membranes. (*See* Figure 11–6.)

phosphodiester bond
A covalent chemical bond formed when two hydroxyl groups are linked in ester linkage to the same phosphate group, such as between adjacent nucleotides in RNA or DNA. (*See* Figure 2–25.)

phospholipase C
Enzyme associated with the plasma membrane that performs a crucial step in inositol phospholipid signaling pathways.

phospholipid
Type of lipid molecule used to make biological membranes. Generally composed of two fatty acids linked through glycerol phosphate to one of a variety of polar groups.

phosphorylation—*see* **protein phosphorylation**

photosynthesis
The process by which plants and some bacteria use the energy of sunlight to drive the synthesis of organic molecules from carbon dioxide and water.

photosystem
Large multiprotein complex containing chlorophyll that captures light energy.

phragmoplast
Structure made of microtubules and membrane vesicles that forms in the equatorial region of a dividing plant cell.

pH scale
Scale used to measure the acidity of a solution: "p" refers to power of 10, "H" to hydrogen. Defined as the negative logarithm of the hydrogen ion concentration in moles per liter (M). Thus an acidic solution with pH 3 will contain 10^{-3} M hydrogen ions.

phylogenetic tree
Chart or "family tree" showing the evolutionary history of a group of organisms.

pinocytosis
Type of endocytosis in which soluble materials are taken up from the environment and incorporated into vesicles for digestion. (Literally, "cell drinking.")

plasma membrane
Membrane that surrounds a living cell.

plasmid
Small circular DNA molecule that replicates independently of the genome. Used extensively as a vector for DNA cloning.

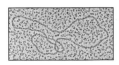

plasmodesma (plural **plasmodesmata**)
Cell–cell junction in plants in which a channel of cytoplasm lined by membrane connects two adjacent cells through a small pore in their cell walls.

polar
Describes a molecule, or a covalent bond in a molecule, in which bonding electrons are attracted more strongly to specific atoms, thereby creating an uneven (or polarized) distribution of electric charge.

polarity
Refers to a structure such as an actin filament or a fertilized egg that has an inherent direction—so that one can distinguish one end from the other.

polygenic trait
A characteristic controlled by multiple genes, each of which makes a small contribution to the phenotype.

polymer
Large and usually linear molecule made by the repetitive assembly, using covalent bonds, of multiple identical or similar units (monomers).

polymerase
General term for an enzyme that catalyzes addition of subunits to a polymer. DNA polymerase, for example, makes DNA and RNA polymerase make RNA.

polymerase chain reaction (PCR)
Technique for amplifying specific regions of DNA by multiple cycles of DNA polymerization, each followed by a brief heat treatment to separate complementary strands.

polynucleotide
A molecular chain of nucleotides chemically bonded by a series of phosphoester linkages. RNA or DNA.

polypeptide
Linear polymer composed of multiple amino acids. Proteins are large polypeptides, and the two names can be used interchangeably.

polypeptide chain
The backbone of atoms containing repeating peptide bonds that runs through a protein molecule and to which amino acid side chains are attached.

polysaccharide
Linear or branched polymer composed of sugars. Examples are glycogen, hyaluronic acid, and cellulose.

polysome (polyribosome)
Messenger RNA molecule with attached ribosomes and engaged in protein synthesis.

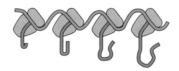

porin
One of a family of proteins that make large water-filled channels ("pores") in the outer membrane of bacteria or mitochondria.

position effect
Refers to differences in the expression of a gene dependent on its location in the genome.

positive feedback loop
Situation in which the end product of a reaction stimulates its own production.

primary transcript—*see* **transcription**

procaryote (prokaryote)
Major category of living cells distinguished by the absence of a nucleus. Bacteria.

processive
Describes a protein that performs multiple rounds of catalysis or conformational changes while still attached to a polymer. A characteristic of motor proteins involved in transport, such as kinesin.

programmed cell death (apoptosis)
Normal benign sequence of cell death in which a cell shrinks the DNA fragments and changes in the cell surface activate its phagocytosis by macrophages.

prometaphase
Stage of mitosis that precedes metaphase.

promoter
Nucleotide sequence in DNA to which RNA polymerase binds to begin transcription.

proofreading
The process by which DNA polymerase corrects its own errors as it moves along DNA.

prophase
First stage of mitosis during which the chromosomes are condensed but not yet attached to a mitotic spindle. Also a superficially similar stage in meiosis.

protease (proteinase, proteolytic enzyme)
Enzyme such as trypsin that degrades proteins by hydrolyzing some of their peptide bonds.

proteasome
Large protein complex in the cytosol that is responsible for degrading cytosolic proteins that have been marked for destruction by ubiquitination or by some other means.

proteolysis
Degradation of a protein by means of a protease.

protein
The major macromolecular constituent of cells. Linear polymer of amino acids linked together by peptide bonds in a specific sequence.

protein domain—*see* **domain**

protein family
A group of proteins in an organism with a similar amino acid sequence. The similarity is thought to reflect the evolution of the genes that encode the proteins from a common ancestor gene through a process of gene duplication followed by gene divergence. Usually, the different members of a protein family will have a related but distinct function. For example, each member of the protein kinase family carries out a similar phosphorylation reaction, but the substrates and regulation will differ for each enzyme.

protein kinase
One of a very large number of enzymes that transfers the terminal phosphate group of ATP to a specific amino acid of a target protein.

protein kinase A—*see* **cyclic AMP-dependent protein kinase**

protein kinase C (PKC)
Enzyme that phosphorylates target proteins in response to a rise in diacylglycerol and Ca^{2+} ions.

protein machine
A set of protein molecules that bind to each other in specific ways, so that concerted movements within the protein complex can carry out a sequence of reactions with unusual speed and effectiveness. A large number of the central reactions of the cell are catalyzed by such protein machines, with protein synthesis and DNA replication being particularly well understood examples.

protein phosphatase (phosphoprotein phosphatase)
Enzyme that removes, by hydrolysis, a phosphate group from a protein, usually with high specificity.

protein phosphorylation
The covalent addition of a phosphate group to a side chain of a protein catalyzed by a protein kinase.

proteoglycan
Molecule consisting of one or more glycosaminoglycan (GAG) chains attached to a core protein.

proteolysis
Degradation of a protein by means of a protease.

proton
Subatomic particle found in the atomic nucleus. Also exists as an independent chemical species as a positive hydrogen ion (H^+).

proto-oncogene—*see* **oncogene**

protozoa
Free-living, nonphotosynthetic, single-celled, motile eucaryotic organisms. Many protozoa, such as *Paramecium* or *Amoeba*, live by feeding on other organisms.

pseudopodium (plural **pseudopodia**)
Large cell-surface protrusion formed by amoeboid cells as they crawl. More generally, any dynamic actin-rich extension of the surface of an animal cell. (Latin for "false foot.")

pump
Transmembrane protein that drives the active transport of ions and small molecules across the lipid bilayer.

purine
One of the two categories of nitrogen-containing ring compounds found in DNA and RNA. Examples are adenine and guanine. (*See* Panel 2–6, pp. 76–77.)

pyrimidine
One of the two categories of nitrogen-containing ring compounds found in DNA and RNA. An example is cytosine. (*See* Panel 2–6, pp. 76–77.)

pyruvate
Metabolite formed from the breakdown of glucose that provides a crucial link to the citric acid cycle and many biosynthetic pathways.

quinone
Small, lipid soluble, mobile electron carrier molecule found in the respiratory and photosynthetic electron-transport chains. (See Figure 14–20.)

Ras
One of a large family of GTP-binding proteins (also called the small GTPases) which help relay signals from cell-surface receptors to the nucleus. Named for the *ras* gene, first identified in viruses that cause rat sarcomas.

reaction center
In photosynthetic membranes, a specialized pair of chlorophyll molecules that perform photochemical reactions.

reading frame
The set of successive triplets in which a string of nucleotides is translated into protein. An mRNA molecule is read in one of three possible reading frames, depending on the starting point.

receptor
Cell protein that binds a specific extracellular signal molecule and initiates a response in the cell. Cell-surface receptors, such as the acetylcholine receptor and the insulin receptor, are located in the plasma membrane, with their ligand-binding site exposed to the external medium. Intracellular receptors, such as steroid hormone receptors, bind ligands that diffuse into the cell across the plasma membrane.

receptor-mediated endocytosis
Mechanism of selective uptake by animal cells in which a macromolecule binds to a receptor in the plasma membrane and enters the cell in clathrin-coated vesicles.

receptor protein
Protein that detects a stimulus, usually a change in concentration of a specific molecule, and then initiates a response in the cell. Many receptor proteins are located in the plasma membrane, and interact with hormones, neurotransmitters and other molecules in the external medium.

receptor serine/threonine kinase
Cell-surface receptor with an extracellular signal-binding domain and an intracellular kinase domain that phosphorylates signaling proteins on serine or threonine.

receptor tyrosine kinase
Type of enzyme-linked receptor in which the cytoplasmic tail has intrinsic tyrosine kinase activity, which is activated on ligand binding.

recognition—*see* **molecular recognition**

recombinant DNA technology (genetic engineering)
The collection of techniques by which DNA segments from different sources are combined to make new DNA. Recombinant DNAs are widely used in the cloning of genes, in the genetic modification of organisms, and in molecular biology generally.

recombination
Process in which chromosomes or DNA molecules are broken and the fragments are rejoined in new combinations. Recombination can occur in living cells—for example, through crossing-over during meiosis—or in a test tube using purified DNA and enzymes that break and ligate DNA strands.

redox pair
Pair of molecules in which one acts as an electron donor and one as an electron acceptor in an oxidation-reduction reaction; for example, NADH (electron donor) and NAD^+ (electron acceptor).

redox potential
A measure of the tendency of a given system to donate electrons (act as a reducing agent) or to accept electrons (act as an oxidizing agent).

redox reaction
A reaction in which electrons are transferred from one chemical species to another. An oxidation-reduction reaction.

reduction
Addition of electron density to an atom, as occurs during the addition of hydrogen to a molecule or the removal of oxygen from it. The opposite of oxidation. (*See* Figure 3–12.)

regulatory DNA sequence
DNA sequence to which a gene regulatory protein binds to determine when, where, and in what quantities a gene is to be transcribed into RNA.

replication fork
Y-shaped region of a replicating DNA molecule at which the two daughter strands are formed and separate.

replication origin
Special DNA sequence on a bacterial or viral chromosome at which DNA replication begins.

repressor
In bacteria, a protein that binds to a specific region of DNA to prevent transcription of an adjacent gene.

respiration
General term for any process in a cell in which the uptake of molecular oxygen (O_2) is coupled to the production of CO_2.

restriction nuclease (restriction enzyme)
Nuclease that can cleave a DNA molecule at any site where a specific short sequence of nucleotides occurs. Different restriction nucleases cut at different sequences. Extensively used in recombinant DNA technology.

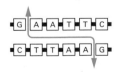

restriction map
Table or graphical representation of a DNA molecule showing its sites of cleavage by various restriction nucleases.

retrotransposan
Type of transposable element that moves by being first transcribed into an RNA copy that is then reconverted to DNA by reverse transcriptase and inserted elsewhere in the chromosomes.

retrovirus
RNA-containing virus that replicates in a cell by first making a double-stranded DNA intermediate.

reverse transcriptase
Enzyme, present in retroviruses, that makes a double-stranded DNA copy from a single-stranded RNA template molecule.

Rho protein family
Family of monomeric GTPases involved in signaling the rearrangement of the actin cytoskeleton.

ribonucleic acid—*see* **RNA**

ribosomal RNA (rRNA)
Any one of a number of specific RNA molecules that form part of the structure of a ribosome and participate in the synthesis of proteins. Often distinguished by their sedimentation coefficient, such as 28S rRNA or 5S rRNA.

ribosome
Particle composed of ribosomal RNAs and ribosomal proteins that associates with messenger RNA and catalyzes the synthesis of protein.

ribozyme
An RNA molecule possessing catalytic properties.

RNA polymerase
Enzyme that catalyzes the synthesis of an RNA molecule on a DNA template from nucleoside triphosphate precursors.

RNA primer
Short length of RNA made on the lagging strand during DNA replication and subsequently removed.

RNA processing
Broad term for the modifications an RNA undergoes as it reaches its mature form. For a eucaryotic mRNA, processing typically includes capping, splicing, and polyadenylation.

RNA splicing
Process in which intron sequences are excised from RNA molecules in the nucleus during formation of messenger RNA.

rough endoplasmic reticulum (RER)
Region of the endoplasmic reticulum associated with ribosomes and involved in the synthesis of secreted and membrane-bound proteins.

rRNA—*see* **ribosomal RNA**

S phase
Period during a eucaryotic cell cycle in which DNA is synthesized.

saccharide
Used as a suffix to denote a sugar, as in disaccharide (made of two sugars) or polysaccharide.

sarcomere
Repeating unit of a myofibril in a muscle cell, about 2.5 μm long, composed of an array of overlapping thick (myosin) and thin (actin) filaments.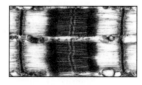

saturated
Describes an organic molecule that contains no C=C or C≡C bonds. Not unsaturated.

second messenger
Small molecule formed in or released into the cytosol in response to an extracellular signal that helps to relay the signal to the interior of the cell. Examples include cAMP, IP$_3$, and Ca^{2+}.

secondary structure
Regular local folding pattern of a polymeric molecule. In proteins, α helices and β sheets.

secretion
Production and release of a substance from a cell.

secretory vesicle
Membrane-bounded organelle in which molecules destined for secretion are stored prior to release. Sometimes called secretory granule because darkly staining contents make the organelle visible as a small solid object.

sequence
The linear order of subunits in a large molecules, for example amino acids in a protein or nucleotides in DNA. In general the sequence of a macromolecule specifies its precise biological function.

serine/threonine kinase
Enzyme that phosphorylates specific proteins on serines or threonines.

sex chromosome
Chromosome that may be present or absent, or present in a variable number of copies, according to the sex of the individual. In mammals, the X and Y chromosomes.

sexual reproduction
Type of reproduction in which the genomes of two individuals are mixed in the formation of a new organism. Individuals produced by sexual reproduction differ from either of their parents and from each other.

β sheet—*see* **beta sheet**

side chain
Portion of an amino acid not involved in making peptide bonds and which gives each amino acid its unique properties.

signaling cascade
Sequence of linked protein reactions, often including phosphorylation and dephosphorylation, which carries information within a cell.

signal sequence
Amino acid sequence that directs a protein to a specific location in the cell, such as the nucleus or mitochondria.

signal transduction
Conversion of an impulse or stimulus from one physical or chemical form to another. In cell biology, the process by which a cell responds to an extracellular signal.

single-nucleotide polymorphism (SNP)
Sequences in the genome that differ by a single nucleotide between one portion of the population and another.

sister chromatid—*see* **chromatid**

site-directed mutagenesis
Technique by which a mutation can be made at a particular site in DNA.

site-specific recombination
Type of recombination that does not require extensive similarity in the two DNA sequences. Can occur between two different DNA molecules or within a single DNA molecule.

small nuclear ribonucleoprotein particle (snRNP)
Structural unit of a spliceosome built of RNA and protein.

small nuclear RNA (snRNA)
RNA molecules of around 200 nucleotides involved in RNA splicing.

smooth endoplasmic reticulum (SER)
Region of the endoplasmic reticulum not associated with ribosomes; involved in the synthesis of lipids.

SNARE
One of a family of membrane proteins responsible for the selective fusion of vesicles with a target membrane.

sodium pump—*see* **Na⁺-K⁺ pump**

solute
Any molecule that is dissolved in a liquid. The liquid is called a *solvent*.

somatic cell
Any cell of a plant or animal other than a germ cell or germ-line precursor. (From Greek *soma*, body.)

Southern blotting
Technique in which DNA fragments separated by electrophoresis are immobilized on a paper sheet and then detected with a labeled nucleic acid probe. (Named after E.M. Southern, inventor of the technique.)

specificity—*see* **molecular specificity**

sperm (spermatozoon)
The male gamete, usually small, highly motile and produced in large numbers.

spindle pole
One of two centrosomes in a cell undergoing mitosis. Microtubules radiating from these centrosomes form the mitotic spindle.

spliceosome
Large assembly of RNA and protein molecules that performs pre-mRNA splicing in eucaryotic cells.

starch
Polysaccharide composed exclusively of glucose units, used as an energy store in plant cells.

stem cell
Relatively undifferentiated cell that can continue dividing indefinitely, throwing off daughter cells that can undergo terminal differentiation into particular cell types.

steroid hormone
Lipophilic molecule related to cholesterol that acts as a hormone. Examples include estrogen and testosterone.

stress-activated channel
Membrane protein that allows the selective entry of specific ions into a cell and is opened by mechanical force.

stroma
(1) The connective tissue in which a glandular or other epithelium is embedded. (2) The large interior space of a chloroplast, containing enzymes that incorporate CO_2 into sugars for photosynthesis.

substrate
The substance on which an enzyme acts.

substratum
Solid surface to which a cell adheres.

subunit
A chemical group or molecule that forms part of a larger molecule; a monomer. Many proteins, for example, are built from multiple polypeptide chains, each of which is a protein subunit.

sugar
A substance made of carbon, hydrogen and oxygen with the general formula $(CH_2O)_n$. A carbohydrate or saccharide. The "sugar" of everyday usage is a specific sweet-tasting disaccharide produced by beet or sugar cane.

sulfhydryl group (–SH, thiol)
Chemical group containing sulfur and hydrogen found in the amino acid cysteine and other molecules. Two sulfhydryls can join to produce a disulfide bond.

survival factor
Extracellular signaling molecule that must be present in order to prevent programmed cell death.

symbiosis
Intimate association between two organisms of different species from which both derive a long-term selective advantage.

symport
Intimate association between two organisms of different species from which both derive a long-term selective advantage.

synapse
Specialized junction between a nerve cell and another cell (nerve cell, muscle cell, gland cell) across which the nerve impulse is transferred. In most synapses the signal is carried by a neurotransmitter, which is secreted by the nerve cell and diffuses to the target cell.

synaptic vesicle
Small membrane sacs filled with neurotransmitter that releases their contents by exocytosis at a synapse.

TATA box
Consensus sequence in the promoter region rich in T's and A's of many eucaryotic genes that specifies the position where transcription is initiated.

telomerase
Enzyme that elongates telomeres, the repetitive nucleotide sequences found at the ends of eucaryotic chromosomes.

telomere
End of a chromosome, associated with a characteristic DNA sequence that is replicated in a special way. Counteracts the tendency of the chromosome otherwise to shorten with each round of replication. (From Greek, *telos*, "end".)

telophase
Final stage of mitosis in which the two sets of separated chromosomes decondense and become enclosed by nuclear envelopes.

template
A molecular structure that serves as a pattern for the production of other molecules. Thus a specific sequence of nucleotides in DNA can act as a template to direct the synthesis of a new strands of complementary DNA.

tight junction
Cell–cell junction that seals adjacent epithelial cells together, preventing the passage of most dissolved molecules from one side of the epithelial sheet to the other.

tissue
Organized mass of cells with a specific function that forms a distinctive part of a plant or animal.

thioester bond
High-energy bond formed by a condensation reaction between an acid (acyl) group and a thiol group (–SH); seen, for example, in acetyl CoA and in many enzyme-substrate complexes.

trans
Beyond, or on the other side.

transcription
Copying of one strand of DNA into a complementary RNA sequence, sometimes termed the primary transcript, by the enzyme RNA polymerase.

transcription factor
Term loosely applied to any protein required to initiate or regulate transcription in eucaryotes. Includes gene regulatory proteins as well as the general transcription factors.

transduction
Virus-mediated transfer of host DNA from one cell to another.

transfer RNA (tRNA)
Set of small RNA molecules used in protein synthesis as an interface (adaptor) between mRNA and amino acids. Each type of tRNA molecule is covalently linked to a particular amino acid.

transformation
Process by which cells take up DNA molecules from their surroundings and then express genes on that DNA.

transgenic organism
A plant or animal that has stably incorporated one or more genes from another cell or organism and can pass them on to successive generations.

trans Golgi network (TGN)
That part of the Golgi apparatus that is furthest from the endoplasmic reticulum and from which proteins and lipids leave for lysosomes, secretory vesicles or the cell surface.

transition state
Chemical structure that forms transiently in the course of a reaction and has the highest free energy of any reaction intermediate.

translation
Process by which the sequence of nucleotides in a messenger RNA molecule directs the incorporation of amino acids into protein; occurs on a ribosome.

translation initiation factor
Protein that promotes the proper association of ribosomes with mRNA and is required for the initiation of protein synthesis.

transport vesicle
Membrane vesicles that carry proteins from one intracellular compartment to another, for example from the ER to the Golgi apparatus.

transposon (transposable element)
Lengths of DNA that can move from one location to another in a chromosome or from one chromosome to another in the same cell. An important source of genetic variation in most organisms.

triacylglcerol
Glycerol ester of fatty acids. The main constituent of fat droplets in animal tissues (where the fatty acids are saturated) and of vegetable oil (where the fatty acids are mainly unsaturated).

tryptophan repressor
A bacterial protein that, in the presence of tryptophan, binds to a specific region of DNA and shuts off production of the tryptophan biosynthetic enzymes.

tubulin
Protein from which microtubules are made.

γ-tubulin ring
Protein complex in centrosomes that nucleates microtubule assembly.

tumor suppressor gene
A gene that in a normal tissue cell inhibits progress through the cell cycle. Loss or inactivation of both copies of such a gene from a diploid cell causes it to divide as a cancer cell.

tumor virus
Virus that makes the cell it infects cancerous.

turnover number
In enzyme catalysis, the number of substrate molecules processed to product per second per enzyme molecule. Although different types of enzymes can have very different turnover numbers, turnover numbers of 1000 or more are quite common—a reflection of the impressive catalytic power of enzymes.

tyrosine kinase
Enzyme that phosphorylates specific proteins on tyrosines.

unsaturated
Describes a molecule that contains one or more double or triple carbon–carbon bonds.

V_{max}
The maximum rate of an enzymatic reaction, attained immediately after addition of the substrate at saturating levels.

valence
For an atom, the number of electrons that it must either gain or lose (whether by electron sharing or by electron transfer) to achieve a filled outer shell most readily. Thus, for example, the valence of Na is one (it must lose one electron), and the valence of Cl is one (it must gain one electron). The valence of an atom is equal to the number of single bonds that the atom can form.

van der Waals force
Attractive force due to fluctuating electrical charges that comes into play between two atoms that are 0.3 to 0.4 nm apart. At a shorter distance, repulsive forces begin to operate.

vector
Genetic element, usually a bacteriophage or plasmid, that is used to carry a fragment of DNA into a recipient cell for the purpose of gene cloning.

vesicle
Small, membrane-bounded, spherical organelle in the cytoplasm of a eucaryotic cell.

virus
Particle consisting of nucleic acid (RNA or DNA) enclosed in a protein coat and capable of replicating within a host cell and spreading from cell to cell. Often the cause of disease.

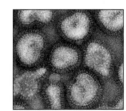

voltage-gated channel
Membrane protein that selectively allows ions such as Na$^+$ (voltage-gated Na$^+$ channel) to cross a membrane and is opened by changes in membrane potential. Found mainly in electrically excitable cells such as nerve and muscle.

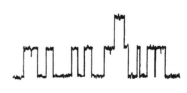

wild type
Normal, nonmutant form of a species resulting from breeding under natural conditions.

X chromosome
One of the two sex chromosomes in mammals. The cells of men possess one X and one Y chromosome.

Y chromosome
One of the two sex chromosomes of mammals. The cells of women contain two X chromosomes.

yeast

Common term for several families of unicellular fungi. Includes species used for brewing beer and making bread, as well as species that cause disease.

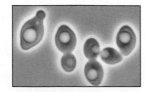

zygote

Diploid cell produced by fusion of a male and female gamete. A fertilized egg.

Index

Page numbers in **boldface** refer to a major text discussion of the entry; page numbers with an F refer to a figure, with an FF to figures that follow consecutively; page numbers with a T refer to a table.

A

Acetyl CoA, 112F, 457
 energy carrier, 111–112
 historical research, 443
 oxidation see Citric acid cycle
 production, 429, 429F, 439F, 449
  see also Pyruvate
  fatty acid oxidation, 438F, 439
  pyruvate dehydrogenase, **435,** 438F, **439**
Acetylcholine
 excitatory neurotransmission, 419
 receptors, 543–544
  conformational changes, 419F
  differential responses, 536, 538F
  G protein-linked, 544, 549, 549F
  ion channels, 543
  neuromuscular junction, 418
Acid(s), **49–50,** 49F, 69F, 469
Actin-binding proteins, 592, **594,** 595F, 607
 see also specific proteins
 cell movement role, 596–597
 extracellular regulation, 599–600
Actin filaments, 22, 23F, 25F, 140, 141F,
  573, **592–600,** 607
 accessory proteins, 592, **594,** 595F,
  596–597, 599, 603
  see also Actin-binding proteins
 actin polymerization, **593–594**
  ATP hydrolysis, 593, 594F
 extracellular signaling, **599–600,** 600F
  Cdc42 GTPase, 600
  Rho GTPase, 599–600
 functional roles, 592
  see also Contractile structures
  adherens junctions, 713, 713F
  cell cortex, **594–595,** 607, 654
  cell crawling, **595–598,** 596F, 597F,
   598F
  cell division, 592, 601, 639, 639F,
   653–654
  connective tissues, 706, 707F
  muscle contraction see Muscle
   contraction
 motor protein association, **598–599**
  see also Myosin(s)
 structure, **593,** 593F
Actin-related proteins (ARPs), 596–597, 600
Action potential, **411–421,** 412F, 422–423
 ionic mechanisms, **412–413,** 413F, **416,**
  422
  experimental analysis, **414–415,** 414F,
   415F
  voltage-gated K$^+$ channels, 413, 415,
   416
  voltage-gated Na$^+$ channels, 412–413,
   415

  membrane depolarization, 412
   muscle, 603
  propagation, 416, 416F
  shape, 415, 415F
  squid giant axons, 412, **414–415,** 414F
Activated carrier molecules, **106–114,** 116
 see also Reaction coupling; individual
   carriers
 formation by reaction coupling, 106–107,
  106F
 GTP, 440, 441F
 types, 111–112, 111T
Activation energy, enzyme lowering, 92, 92F,
  93F, 102, 115, 147, 148
Active transport, **392,** 393F, 422
 ATP-driven pumps, 395, **395–397,** 396F
 carrier proteins, **395–396**
 coupled transport, 395, 396F, **397–399**
  antiport, 397, 399F
  mitochondrial, **466,** 466F
  symport, 397, 398, 399F
 light-driven pumps, 395, 396F, 402
Acyl phosphate bond, 67F
Adaptation, signaling pathways, 556
Adaptins, 514
ADE2 gene, position effects, 188F
Adenine, 56, 171
 structure, 76F
 thymine base pairs, 175F
Adenomatous polyposis coli (APC) gene, 733,
  733F, 734–735, 735F
Adenosine diphosphate (ADP) see ADP
  (adenosine diphosphate)
Adenosine monophosphate (AMP), 114
Adenosine triphosphate (ATP) see ATP
  (adenosine triphosphate)
Adenovirus, viral coat, 222F
Adenylyl cyclase
 cAMP synthesis, 550, 550F
 G protein-mediated activation, 549
Adherens junctions, 711F, 712, 713, 713F,
  737
Adipose tissue, 445, 445F
Adoption studies, gene–environment
  interactions, 692–693
ADP (adenosine diphosphate), 107, 108F
 cellular ATP/ADP ratio, **468**
Adrenaline, cAMP pathway, 551, 551T
Aerobic organisms, 14
 energy metabolism see Respiration (cellular)
 evolution, **488,** 488F
Affinity chromatography, 162F, 165F
Age, mutations and, 210, 210F
Agrobacterium, plant transformation, 359F
AIDS, 222
Alanine, structure, 55F, 75F

Albinism, inheritance, 676–677, 677F
Alcohol, 67F
 fermentation, 434F
Alcohol dehydrogenase, 127F
Aldehydes, structure, 67F
Aldolase, 432F
Alleles, 665, 675, **681, 684,** 685F, 693
Allosteric regulation, **151–153,** 159
 motor proteins, 157F
 proton pumping, 476, 477F
 tryptophan repressor, 273
α-helix, 126, 128F
 coiled-coils, 134–135, 135F
 handedness, 134, 134F
 transmembrane proteins, 135F, **376–377,**
  376F, 377F
Alternative splicing, 241, 241F, 318, 318F
Alu sequences, 219, 220F, 301, 306
Amides, 67F, 72F
Amines, 67F
Amino acids, 4, 51, 51F, **55–56,** 74FF, 123F
 see also Protein synthesis; individual amino
  acids
 catabolism, 441
 families, 74F
  acidic, 75F
  basic, 74F
  nonpolar, 75F, 123F
  polar (uncharged), 75F, 123F
 number, 56
 peptide bonds, 56, 74F, 121, 121F, 249
 protein primary structure, 56, **121–123,**
  125
 structure, 55F, 74F
  optical isomers, 56, 74F
  side chains, 56, 121
 tRNA binding, 245, 245F, **248,** 249F
Amino groups, 50, 55F
Aminoacyl-tRNA synthetases, **248,** 249F,
  252F
Amoeba, 27F
AMP (adenosine monophosphate), 114
Amphipathic molecules, 367
 fatty acids, 53F, 54, 367
 membrane lipids, 367, 368F, 369, 369F,
  370F, 386
Anabaena cylindrica, photosynthesis, 15F
Anabolism see Biosynthesis
Anaerobic organisms, 14
 energy metabolism, **431, 434,** 434F
 evolution, 488
 formic acid oxidation, 489F
Anaphase
 meiosis
  anaphase I, 665, 667
  anaphase II, 668

mitosis, 640, 643F, **649–651**
  anaphase A, 649–650, 650F
  anaphase B, 650–651, 650F
  APC role, 649, 651F
Anaphase promoting complex
            (APC)chromosome segregation,
            649, 651F
  M-cyclin regulation, 617
Aneuploidy, 670–671, 671F
Animal(s)
  cells see Animal cell(s)
  model organisms, **29, 32–33**
  multicellularity evolution, 698
  tissues, **703**
    see also specific tissues
    plant tissues vs., 698
Animal cell(s)
  osmotic balance, 400, 401F
  structure, 10–11, 10F, 25F
Anion, definition, 44, 408
Antibiotics
  protein synthesis inhibition, **255–256,** 256T
  resistance, gene transfer, 302–303, 343
Antibodies, 164F
  antigen binding, **144–145,** 164F
  cell isolation, 325
  immunoprecipitation, 165F, 561
  monoclonal, 165F
  protein analysis, 164FF
  structure, 145, 146F, 164F
    variable domain, 137F, 146F
Anticodon, 245, 245F, 248
Antigens, 145
  antibody binding, **144–145,** 164F
Antimitotic drugs, 582, 589
Antiporters, 397, 399F
APC, see Adenomatous polyposis coli;
            Anaphase promoting complex
Apoptosis, **625–628,** 634
  cell division balance, 625–626
  developmental, 625, 625F
  necrosis vs., 626, 626F
  prevention
    cancer cells, 730
    by survival factors, 629, 631–632, 632F
  proteolytic cascade, **626–627**
    caspases, 627, 627F
  regulation by Bcl-2 family, **627–628,** 628F,
            632, 632F
Arabidopsis thaliana, 29F
  cell signaling, 567
  complete genome sequence, 331
  as model organism, **28–29**
Archaea (archaebacteria), 15
  evolution, 310, 310F
    genome analysis, 491–492
    horizontal gene transfer, 303, 303F
  Methanococcus, **490–492**
  nitrogen fixation, 491
Arginine, structure, 74F
Aristotle, Generation of Animals, 659
Asparagine, 75F
Aspartate, 75F
Aspartate transcarbamoylase, 127F
Aster, 640
Aster microtubules, 644F
Atom(s), 39, 64
  see also Chemical bonds
  isotopes, 40
  size, 41
  structure, 40–41, 40F
    electrons see Electron(s)
  valence, 43
  van der Waals radius, 78F

Atomic number, 40, 40F
Atomic weight, 40
ATP (adenosine triphosphate)
  biosynthetic functions, 108–109, 109F
  cellular ATP/ADP ratio, **468**
  as energy carrier, 56–57, 57F, 77F,
            107–109
  evolutionary significance, 453, 454
  hydrolysis see ATP hydrolysis
  phosphorylation reactions, 107–108, 108F
    protein regulation, 154
  structure, 57, 57F
  synthesis see ATP synthesis
ATP-driven pumps, 395, 396F, 401–402,
            402F
  see also specific pumps
ATP hydrolysis, 107, 108F
  see also Reaction coupling
  actin polymerization, 593, 594F
  aminoacylation reaction, 248
  ATP synthase, 465–466, 465F
  biosynthetic reaction coupling, 108–109,
            109F
  chromatin remodeling, 189
  ΔG, 114
  motor proteins, 156, 584, 585, 598
  muscle contraction, 602
  nucleic acid synthesis, 115F
  pyrophosphate intermediate, 114F
  sodium pump see Na$^+$-K$^+$ pump
ATP synthase, 379, 454, 464–465, 465F,
            492
  experimental analysis, 461, 461F
  as molecular motor, 465
  as reversible coupling device, 465–466,
            465F
ATP synthesis, 107, 108F, 428, 429F, 467T
  see also Catabolism; Chloroplasts;
            Mitochondria; Respiration
            (cellular); specific metabolic
            pathways
  anaerobic organisms (fermentation), **431,
            434,** 434F
  ATP synthase see ATP synthase
  experimental analysis, 461, 461F
  glycolytic, 428, 431, 431F, 433F, 435,
            436F, 437F, 460
  membrane-based mechanisms, **453–455,**
            492
    see also Chemiosmotic coupling; Electron-
            transport chain(s); Oxidative
            phosphorylation
  mitochondrial production, 17, **455–468**
  photosynthesis, **482–485**
  proton gradient and, 441, 454, 459F,
            464–467, 464F
  yield, 458F, 466–467, 466F, 467T
ATPase(s), 397
  calcium pumps, 401–402
  sodium pump see Na$^+$-K$^+$ pump
Autocatalysis, origin of life, **259**
Autophagosome, 528
Autophosphorylation, receptor tyrosine kinases,
            558
Avery, Oswald, 172, 174F
Avogadro's number, 41
Axonal transport, 583–584, 584F
Axons, 411
  action potentials see Action potential
  squid giant axons, 412, **414–415,** 414F,
            415F
    motor protein studies, 586, 586F
  transport along, 583–584, 584F

B

B cells, antibody production, 164F
Bacillus subtilis, genome evolution, 297F
Backbone protein models, 132F
Bacteria (eubacteria), 11, 15
  see also individual species
  cell division, 660, 660F
  cell membrane fluidity, 372
  chemical composition, 51T
  chloroplast origins, 20F, 487, 501, 501F
  diversity, 14–15, 14F
  energy sources, 14–15
    anaerobic, 431, 434
    photosynthetic, 15, 15F, 380F, **489–490**
    sulfur, 15, 15F, 489, 489F
  evolution, 310, 310F
    aerobic metabolism, 388, 389F
    anaerobic metabolism, 388
    horizontal gene transfer, 302–303, 343
    photosynthesis, **489–490**
    speed of, 14
  genomes
    "chromosome," 177
    DNA-only transposons, 218, 218F
    gene organization, 238, 238F
    operons, 273, 275, 285
  H$^+$ pumps/gradients, 402
  mitochondria origins, 17, 18F, 487, 501
  restriction–modification systems, 328
  size, 14, 24
  structure, 2, 2F, 14, 14F, 25F
    cell walls, 14, 25F
    cytoskeleton, 573
  transcription see Transcription
  transformation, **341–342,** 343F
  translation, 253–254, 254F
    antibiotic inhibition, **255–256,** 256T
    ribosomes, 249
Bacterial artificial chromosomes (BACs), 335,
            335F
Bacteriophage T2, identification of DNA as
            genetic material, 172–174, 174F
Bacteriophage T4, viral coat, 222F
Bacteriorhodopsin, 378–379, 379F
  proton pump, 379, 403T
Bak protein, apoptosis, 628, 628F
Basal body, cilia, 590
Basal lamina, 709–710, 710F
Base(s) (chemical), **49–50,** 49F, 69F, 469
Base(s) (nucleic acid), 56, 76F, 131, 131F,
            171
  see also Purine bases; Pyrimidine bases
  modification in tRNA, 245F
Base-pairing
  DNA, 58, 175, 175F
    hydrogen bonds, 78F, 171, 171F, 175,
            175F
    replication role, 196–197
  RNA, 231, 231F, 232F
    hydrogen bonds, 78F
    wobble, 248
Bax protein, apoptosis, 628, 628F
Bcl-2 family
  apoptosis regulation, **627–628,** 628F
  survival factors and, 632, 632F
Bcl-2 protein, apoptosis inhibition, 628, 628F,
            632, 632F
Bdellovibrio bacteriovorus, structure, 2F, 3
Beggiatoa, sulfur metabolism, 15F
Benzene molecule, 47, 66F
β-catenin, 734
β-sheets, 126, 128F, **135**
  antiparallel, 135, 136F

β barrels, 376–377, 377F
    parallel, 135, 136F
Bicoid, *Drosophila* anterior–posterior polarity,
    283, 283F
Biological names, 14
Biological order, 84, **85–88**, 85F, 86F
    see also Thermodynamics
Biosphere, 89
Biosynthesis, **83–118**
    see also Enzyme(s); Thermodynamics;
        specific reactions
    ATP role, 108–109, 109F, 114, 114F
    energetics, **112–114**
    macromolecules, 58, 59, 59F, 62
    metabolic pathway interaction, 85F, **447,**
        447F, 449
Biotechnology, enzyme catalysis, 105
Biotin, 150
1,3-Bisphosphoglycerate, 433F, 436F
Bivalents, 664
Blended inheritance, 659
    disproof by Mendel, 674
Blood cells, development, 722, 724F
    see also specific cell types
Body size, 628
    see also Cell death; Cell growth
Bone, 704
    turnover, 721
Brewer's yeast *see Saccharomyces cerevisiae*
Brightfield microscopy, 8F
Brown fat cells, chemiosmotic uncoupling,
    460–461

C

Cadherins, cell junctions, 712–713, 713F
Caenorhabditis elegans
    complete genome sequence, 331
    genetic screens, 688F
    as model organism, 32, 32F
    RNAi-mediated gene inactivation, 358
Calcium-binding proteins, 555
Calcium/calmodulin-dependent kinase
    (CaMKs), 555
Calcium channels, 555
    sarcoplasmic reticulum, 603, 605F
    voltage-gated, neurotransmitter release,
        417, 418F, 422
Calcium ions (Ca^{2+}), 401
    channels *see* Calcium channels
    excitatory neurotransmission, 419
    fertilization, 554, 554F, 672
    intracellular concentration, 554–555
        extracellular *vs.,* 390T
    neurotransmitter release, 417, 418F
    pumping *see* Calcium pump(s)
    signaling pathways *see* Calcium signaling
Calcium pump(s), **401–402**, 403T
    intracellular [Ca^{2+}] control, 555
    sarcoplasmic reticulum, 393, 394F, 604
Calcium signaling, 401, **554–555**
    calmodulin, 555, 555F
    fertilization role, 554, 554F, 672
    IP$_3$ pathway, **552–554**, 568
    muscle contraction, **603–604**, 605F
    synapses, 417, 418F, 419
Calmodulin, 127F, 555, 555F
Cancer, 195, 560, 697, **726–736**, 738
    cell properties, **727–728**, 730
        apoptosis prevention, 730
        colonization, 727, 727F
        competitive advantage, **729–730**, 729F
        genetic instability, 728, 728F, 730

immortality, 326, 730
        proliferation, 626–627, 634, 727, 730
    cell signaling, 730, 732F
    diagnosis, 736, 736F
    epidemiology, **727–728**
    genetic alterations, 210, 210F, 726,
        731–732, 732F, 738
        see also Mutation(s); Oncogene(s); Tumor
            suppressor gene(s)
        accumulation, **728–729**
        natural selection, 729–730, 729F
    inherited predisposition
        colorectal cancer, **732–733**, 733F
        mismatch repair defects, 211–212
    metastases, 727, 727F, 730
    prevention, 727
    treatment development, **736**
        Gleevec, 736, 737F
Cap-binding complex, 242
5′ capping, 236–237, 237F
Carbohydrates, 52
    see also Sugars
    cell surface (glycocalyx), **381–382**, 382F
Carbon
    atomic structure, 40F
    cellular component, 50–51
    compounds, 66FF
    covalent bonds, 46F, 66FF
    photosynthetic fixation *see* Carbon fixation
    skeletons, 66F
    utilization, 89, 90F
Carbon cycle, 89, 90F
Carbon fixation, 88, 446, 481, **485–486,**
    485F, 486F
    Methanococcus, 491, 491F
    ribulose bisphosphate carboxylase,
        485–486
    sucrose/starch production, **486**
Carbonyl groups, 50, 67F
Carboxyl groups, 50, 55F, 67F, 72F, 111,
    112F
Carboxylic acid, structure, 67F
Carboxypeptidase, zinc cofactor, 150
Cardiac muscle, 600, 605–606
    acetylcholine effects, 536, 544, 549, 549F
    gap junctions, 715
Cargo receptors, 514
Carrier proteins, 120F, 374F, 392, 392F,
    393–402, 403T, 422
    see also Electrochemical gradients;
        Membrane potential; *individual
        proteins*
    active transport, **395–396**
        ATP-driven, 395, **395–397**, 396F, 398F
        coupled transport, 395, 396F, **397–399,**
            399F, **466,** 466F
        light-driven, 395, 396F
    animals *vs.* plants, 400–401, 402, 402F
    antiporters, 397, 399F
    calcium homeostasis, **401–402**
    cell specific, 393F
    conformational changes, 394, 394F
    export, 393
    import, 393
    mutations, 683, 683F
    osmotic balance role, **399–401**
    passive transport, **393–395**, 393F
        concentration gradients, 394
        electrochemical gradients, 395, 395F
    specificity, 393, 394
    symporters, 397, 398, 399F
    uniporters, 397, 399F
Cartilage, 708F

Caspases, 627–628, 627F, 628F
Catabolism, 17, 83, 115, 427, **428–444**
    see also Respiration (cellular); *specific
        pathways*
    anaerobic organisms (fermentation), **431,
        434,** 434F
    biosynthesis role, 85F, **447,** 447F, 449
    efficiency, 430
    stages, **428–430**, 428F, 429F
Catabolite activator protein (CAP), 274–275,
    274F
    dimers, 139, 139F
    protein domains, 136
Catalase, 127F
Catalysis, 83, 92
    see also Enzyme(s); Thermodynamics
    autocatalysis and origin of life, **259**
Catastrophins, 644, 647, 647F
Cation, definition, 44, 408
cdc2, evolutionary conservation, 31, 31F
Cdc6, 621–622
cdc28, evolutionary conservation, 31, 31F
Cdc42, actin regulation, 600
cDNA libraries, **346–348**, 361
    genomic *versus,* 346–347, 347F
    PCR production, 349F
    reverse transcriptase, 346, 347F
Cell(s), **1–38**
    *see also individual cell types; specific
        components*
    chemical components, **39–83**
        bacterial cell, 51T
        carbon compounds, 50–51, 66F
        macromolecules *see* Macromolecules
        similarities, 3–4
        water, 48
    communication *see* Cell signaling
    definition, 1
    diversity, **1–5,** 2F
    energetics *see* Energy; Thermodynamics
    eucaryotic *see* Eucaryotic cell(s)
    evolution *see* Evolution; Origin of life
    genome *see* Genome(s)
    hybrid, 165F, 383F
    manipulation, **323–364**
        see also Cell culture; Recombinant DNA
            techniques
    osmotic pressure, **399–401**
    procaryotic *see* Procaryotic cells
    scale, 13F
    similarities, **1–5**
    specialization, 3, 267
        see also Cell differentiation; Multicellular
            organisms; Tissue(s)
    structure, 10–11, 573
    study of *see* Cell biology
Cell adhesion molecules
    see also Extracellular matrix; Cell junctions;
        individual molecules
    cancer cells, 730
    cell crawling, 597–598
    cell division, 654
    tissue maintenance/renewal, 719
Cell biology
    future, 323
    history, 24T
        macromolecules, 60–61, 60F
        microscopy, 6–7
        proteins, 158T
    model organisms, **27–35**
        see also specific organisms
Cell cortex
    actin filaments, **594–595**, 607, 654

Page numbers in **boldface** refer to a major text discussion of the entry; page numbers with an F refer to a figure, with an FF to figures that follow consecutively; page numbers with a T refer to a table.

cell movement and, 596, 596F
plasma membrane reinforcement, **380–381,**
    381F, 387
Cell crawling, **595–598,** 607
    see also Actin filaments
    actin-related proteins, 596–597
    cell adhesion, 597–598
    cortex role, 596
    filopodia/lamellipodia, 596, 597F
    forces generated, 596F
Cell culture, **324–327,** 360
    cell senescence, 326–327, 360
    differentiated properties, 326, 327F
    embryonic stem cells, 327, 358F, 723
    immortalized cell lines, 326–327, 360
    mitogens, 630, 630F
    plant cells, 359, 359F
    protein expression in, **352–353,** 353F
    RNAi-mediated gene inactivation, 358
    tissue dissociation, **325,** 360
        cell sorting, 325, 326F
    in vitro growth, **325–326**
Cell cycle, **611–636,** 612F, 637, 638F, 642F
    see also Cell growth; DNA replication
    arrest, **624–625**
    control *see* Cell cycle control
    eucaryotic cycling times, 612, 613T
    experimental analysis, **30–31,** 30F, 31F,
        618–619
    G$_0$ (resting phase), 615, 624
    G$_1$ phase, 200, 613, 613F, 622, 633
    G$_2$ phase, 613, 613F, 633
    interphase, 181–182, 613, 642F
        chromosome changes, 181–182, 182F,
            183, 183F
        chromosome organization, **183,** 183F
        chromosome structure, **187–189**
    M phase, 613, 613F, 618, 633, 637,
        638–641, 638F, 656
        see also Cell division; Cytoskeleton
        cell shape changes, 654, 654F
        centrosome changes, **640,** 640F
        chromosome changes *see* Mitotic
            chromosomes
        cytokinesis *see* Cytokinesis
        cytoskeleton role, 22, 24F, 579, 582,
            592, 601, **639–640,** 639F
        meiosis *see* Meiosis
        mitosis *see* Mitosis
        stages, **640–641,** 642FF
    mutation effects, 30
    overview, **612–615,** 613F, 642F
    protein conservation, 31, 31F, 619
    S. cerevisiae as model organism, 28
    S phase, 198, 613, 613F, 633
        see also DNA replication
    universal features, 30–31, 611
Cell cycle control, 611–612, **615–625,** 633,
        637
    Cdcs, 30–31, 621–622
    Cdks *see* Cyclin-dependent protein kinases
        (Cdks)
    checkpoints, 615, **622–624,** 633
        cell cycle arrest, **624–625,** 625F
        DNA damage, 623–624, 623F
        G$_1$, 615, 623–624, 623F, 624F
        G$_2$, 615
        mitosis, 624
    cyclins *see* Cyclin(s)
    experimental analysis, **618–619**
    feedback, 614, 615F, 620, 620F
    mutations, experimental uses, 619
    p21 role, 623–624

p53 role, 623–624, 623F
phosphorylation role, **616, 617, 620,** 620F
principles, 614–615, 614F, 615F
Cell death
    apoptotic (programmed) *see* Apoptosis
    cell division balance, 625–626
    necrosis, 626, 626F
    prevention, **629, 631–632,** 632F
        see also Cancer; Survival factors
Cell differentiation, 3, 267
    see also Stem cells; *specific cell types*
    cell polarization, 583, 583F
    in culture, 326, 327F
    embryonic stem cells, 724F
    gene expression, 5, **268–269, 280–289,**
        286, 287F, 289
        see also Cell memory; Developmental
            gene regulation; Gene expression
            regulation
    genome constancy, **268–269**
    plant cells, 699
Cell division, 182, **637–658**
    see also Cell cycle; Cell growth; DNA
        replication
    cell death balance, 625–626
    cell shape changes, 654, 654F
    chromosome condensation *see* Mitotic
        chromosomes
    cleavage divisions, 614, 618
    cytokinesis *see* Cytokinesis
    cytoskeleton role, 22, 24F, 579, 582, 592,
        601, **639–640,** 639F
        see also Actin filaments; Microtubule(s)
    experimental analysis, **30–31**
        mitotic spindles, **646–647**
    extracellular control, **628–633**
        inhibitory, **632–633**
        mitogens, **629–631**
    meiosis *see* Meiosis
    mitosis *see* Mitosis
    organelles and, 502, **651–652**
    overview, **638–641,** 638F
    rate variation, 624
    stable gene expression, 280, 286–288
Cell doctrine, 611
Cell fractionation, 160FF
Cell-free protein synthesis, 246
Cell growth, 612F, 634
    see also Cell cycle; Cell division
    cell death balance, 625–626
    extracellular control, **628–633**
        growth factors, 629, **631,** 631F
        inhibitory factors, **632–633**
Cell junction(s), 697, 711F, **712–714,**
        737–738
    see also specific types
    actin filaments, 713
    cadherins, 712–713, 713F
    intermediate filaments, 575, 575F, 577F
Cell lines, immortal, 326–327, 360
Cell lysis, 160F
Cell memory, 280, **286–288**
    see also Cell differentiation
    chromatin structure, 287–288
    positive feedback loop, 287, 288F
    tissue maintenance, 719–720
Cell movement, 23, 573, 595–598
    see also Cilia; Cytoskeleton; Flagella
    crawling *see* Cell crawling
Cell number, extracellular control, **628–633**
    see also Cell death; Cell division; Cell
        growth
Cell senescence, 326–327, 360

Cell shape
    cell division changes, 654, 654F
    cytoskeleton role, 573, 574F
    diversity, 2–3, 2F
Cell signaling, **533–572**
    see also Gene expression regulation; Signal
            transduction; *specific signaling
            molecules/pathways*
    actin filament regulation, **599–600**
    calcium role *see* Calcium signaling
    cancer cells, 730, 732F, 734–735
    complexity, 565–566, 566F
    control of cell numbers, **628–633**
        see also Cell death; Cell division; Cell
            growth
    evolution, **566–567**
    experimental analysis, **561–563**
    functional roles, 533, 538F
    general principles, **533–546**
    intracellular pathways *see* Intracellular
            signaling pathways
    plants, 567
    receptors *see* Receptor(s)
    signaling molecules *see* Signaling molecules
    time-scale, 552, 552F
    tissue maintenance/renewal, 718–719
    types of signaling, **534–536,** 534F, 535F
        contact-dependent, 535, 535F, 536F
        endocrine, 534, 535F
        neurotransmission, 535, 535F
        paracrine, 534, 535F
Cell sorting, 325, 326F
Cell survival, **628–633**
    see also Cell death
    inhibitory factors, **632–633**
    survival factors, 629, **631–632,** 632F
Cell theory, 7
Cell wall(s), 3
    bacterial, 14, 25F
    plant cells, 10F, 25F, **698–699,** 699F
        cellulose, 699F, 702F
        cytokinesis and cell division, **654–655,**
            655F
        pectin, 699F, 702F
        primary, 699, 702F
        secondary, 699
        specialization, 699
        turgor pressure, 401, 699
Cellular respiration *see* Respiration (cellular)
Cellulose, 53, 699, **702–703,** 737
    microfibril orientation, 702F
    microtubule-directed synthesis, 702–703,
        703F
Central dogma, 4, 4F, 229, 262, 330F
Centrifugation
    organelle studies, 499–500
    protein analysis, 160FF
    ultracentrifugation, 60–61, 61F
Centriole(s), 25F, 580
Centromere(s), 182, 182F, 645, 645F
Centrosome, 25F
    see also Mitosis; Mitotic spindle
    cycle (duplication and separation), **640,**
        641F
    γ-tubulin rings, 580, 640
    microtubules, 579, **580–581,** 640, 641
        aster formation, 640, 640F
    motor proteins, 644, 645F
Cervical cancer, 728
Chaperones
    see also Protein folding
    misfolded proteins in ER, 518, 518F
    protein folding, 125

protein translocation, 507
Chase, Martha, 172–174, 174F
Chelating agents, enzyme inhibition, 105
Chemical bonds, **39–50,** 64–65, 66FF
    see also Atom(s); Electron(s); *specific bonds*
    bond energy, 45
    bond length, 45, 46T
    bond strength, 46T
    electron interactions, **41–43,** 43F
    energy storage, 88, 97F
Chemical reactions
    see also specific types of reaction
    activation energy, 92, 92F, 93F, 102, 115,
        147, 148
    equilibrium, 95, 98F
        constant *see* Equilibrium constant *(K)*
    free energy changes *see* Free energy (G)
    thermodynamics *see* Thermodynamics
    transition state, 147
Chemiosmotic coupling, 454, 454F, **458–459,**
        492
    artificial ATP generation, 461, 461F
    evolution, **490–492**
    historical research, **460–461**
    photosynthesis, 459
    uncoupling agents, 460, 460F
Chemiosmotic hypothesis *see* Chemiosmotic
        coupling
Chiasma(ta), meiosis, 666, 666F, 667, 667F
Chimpanzee, genome evolution, 305, **306,**
        307F
Chitin, 53
Chloride ions (Cl⁻), 390T
    inhibitory neurotransmission, 419
Chlorophyll, 18, **481–482,** 481F
    see also Photosynthetic reaction center(s);
        Photosystem(s)
    light reactions (electron transport),
        480–481, 483F
Chloroplasts, **18–19,** 25F, 478, 499
    cell division and, 652
    chlorophyll *see* Chlorophyll
    evolution, 20F, **487–492,** 501, 501F
    functions
        energy metabolism, **446–447,** 447F
        photosynthesis *see* Photosynthesis
        storage, 446, 446F
    genome, 19, 479, 487
    growth/proliferation, 502
    protein import, **506–507,** 528–529
    structure, 18, 19F, 446F, **478–480,** 479F,
        480F, 506
        see also Photosystem(s)
        grana, 480F
        inner membrane, 478
        intermembrane space, 479
        mitochondria cf., 480F
        outer membrane, 478
        stroma, 479
        thylakoid, 479–480
Cholera toxin, 547
Cholesterol, 73F, 368F, 542F
    biosynthesis, NADPH/NADP⁺, 111F
    LDL transport, 525
    membrane stiffening, 372, 372F
    receptor-mediated endocytosis, 525–526,
        526F
Choline, 72F
Chondroitin sulfate, 708F
Chromatin, 25F, 178, 184, 191
    30-nm fiber, 184, 186–187, 186F, 187F
    euchromatin, 188
    gene expression and, 187, 276, **279–280,**
        280F

see also Gene expression regulation
    cell memory, 287–288, 288F
    position effects, 188, 188F
    X-inactivation, 188–189, 189F, 280
    heterochromatin, 188–189, 189F, 280
    interphase chromosomes, **187–189**
    levels of organization, 186–187, 186F
    mitotic chromosomes, 187F
    nucleosomes *see* Nucleosome(s)
    remodeling, **189–190,** 191, 279–280
        ATP hydrolysis, 189
        chromatin remodeling complexes, 189,
            190F
        histone modification, 189–190, 191F,
            279–280
        regulation, 190
    structure inheritance, 288F
Chromatin remodeling complexes, 189, 190F
Chromatography, 162F, 165F
Chromosome(s), 17, 170, 191, 659
    see also DNA (deoxyribonucleic acid);
        Gene(s); Genome(s)
    abnormal *see* Chromosome abnormalities
    banding patterns, 178, 179F
    condensation *see* Mitotic chromosomes
    end replication problem, 208
    haplotype blocks, **689–691,** 691F
    homologous, 178
    human *see* Human genome
    interphase *see* Interphase chromosome(s)
    karyotypes, 178, 178F
    ploidy, 655, 663
    recombination *see* Recombination
    replication, 182, 182F
        see also Cell cycle; DNA replication;
            Mitosis
    sex chromosomes, 178
    species differences, 181, 181F
    structure *see* Chromosome structure
Chromosome abnormalities, 178, 179F
    aneuploidy, 670–671, 671F
    cancer cells, 728, 728F
Chromosome paints, 178, 178F
Chromosome structure, **177–190**
    cell cycle and, **181–182**
        see also Interphase chromosome(s);
            Mitotic chromosomes
    centromeres, 182, 182F, 645, 645F
    higher order, **183–184**
        chromatin *see* Chromatin
        levels, **186–187,** 186F
    replication origins, 182, 182F
    telomeres, 182, 182F, **207–208,** 208F
Chronic myeloid leukemia (CML), treatment,
        736, 737F
Chymotrypsin, structure, 127F, 138F
Cilia, 591F
    "9+2" array, 591, 591F
    basal body, 590
    microtubules, 579, **590–591**
    motility, 590, 590F, 591
    Paramecium, 2–3, 2F
Ciliates, 27F
Citrate, 440, 440F
Citric acid cycle, 429F, 430, **439–441,** 440F,
        442F, 449
    biosynthetic role, **477,** 477F
    electron generation, **457–458**
        FADH₂, 440, 441F
        NADH, 429F, 430, 439–441
    GTP production, 440, 441F
    malonate inhibition, 442–443, 442F, 443F
    rate enhancement, 443, 443F
Clamp loader, 207F

Classical genetics *see* Genetics (classical)
Clathrin-coated vesicles, 513–514, 514F,
        514T
    membrane budding, 513F
    pinocytosis, 525
    receptor-mediated endocytosis, 525–526
Claudins, 712
Clone-by-clone sequencing, 335, 335F
Cloning, 725
    recombinant DNA technology *see* DNA
        cloning
    reproductive, 725, 725F
    therapeutic, **725–726,** 725F
    vectors, **342–343,** 344
Codons, **244,** 244F, 263
    codon–anticodon interaction, 245, 245F,
        248, 249F
    decoding *see* Genetic code
    start (AUG), 253
    stop, 254
    wobble, 248
Coenzyme A (CoA), 77F
    see also Acetyl CoA; Citric acid cycle
Coenzymes, nucleotides, 77F
Cofactors, **149–150**
Cohesins, 638, 639, 641, 667
Coiled-coils, 134–135, 135F
Colchicine, microtubule effects, 582, 589
Collagen(s), **704–705**
    organization and remodeling, **705–706,**
        706F, 707F
    procollagen, 705
    structure, 127F, 141–142, 142F
        fibers/fibrils, 704, 705F
Collagenase, 705
    hyperextensible skin, 705, 706F
Collenchyma, 700F
Colorectal cancer, 732–733, 733F, 735
Column chromatography, protein analysis,
        162F
Columnar epithelium, 709, 709F
Combinatorial gene expression control, **281,**
        281F
    cell differentiation, **285–286**
    coordination, **281, 285,** 286F
    Drosophila eve gene, 283–284, 283F, 284F
Companion cells, 701F
Comparative genome analysis, 34–35,
        214–215, 319
    see also Phylogenetics
    conservation, **306–308**
        see also Evolutionary conservation
    gene identification, 315, 333
    human genome, **316**
        human–chimpanzee, 305, **306,** 307F
        human–mouse, 308, 308F, 317F
    "junk" DNA, **308–309**
    sequence divergence rates, **305–306**
Complementary DNA libraries *see* cDNA
        libraries
Complementation tests, 685F, **688–689**
Complex traits, **691–692**
    gene–environment interactions, **692–693,**
        692F
Condensation reactions, 59F, 65, 113, 113F
    ATP coupling, 109, 109F
    sugars, 52–53, 53F
Condensins, 638, 639
Conditional mutants, 30, 30F, 31F, 684F,
        687–688, 688F
Confocal microscopy, 9F
Conjugation, 302–303, 303F
Connective tissue(s), 698F, **703–709,** 737
    see also Collagen(s); Extracellular matrix

Page numbers in **boldface** refer to a major text discussion of the entry; page numbers with an F refer to a figure, with an FF to figures that follow consecutively; page numbers with a T refer to a table.

I:5

glycosaminoglycans, 707–708, 708F
integrins, **706,** 707F
organization, 718
other tissues *vs.,* 703
proteoglycans, **706–709,** 708F
strength, collagen role, **704–705**
fibril alignment, **705–706,** 706F, 707F
Connexons, 715, 716F
Conservation, evolutionary *see* Evolutionary
conservation
Contact-dependent signaling, 535, 535F
Delta–Notch, 536F, 537T
Contractile bundles, 592, 601
Contractile ring
assembly, 653
cytokinesis role, 639, 639F, 652, 652F,
653F
structure, 592, 601, **653–654**
see also Actin filaments; Myosin(s)
Contractile structures, 592
actin–myosin interactions, 598–599, 601
contractile bundles, 592, 601
cytokinesis *see* Contractile ring
muscle *see* Muscle
COP-coated vesicles, 514, 514T
Cortisol, 541F
intracellular receptors, 541–542
Coupled transport, 395, 396F, **397–399,**
399F
Covalent bonds, **45–48,** 64–65, 66FF
bond energy, 45
bond length, 45, 46T
bond strength, 46–47, 46T
electron sharing, 43, 43F, 45, 45F
geometry, 45, 46F
hydrogen molecules, 45F
peptide bonds *see* Peptide bonds
phosphodiester bonds, 58F, 77F
types, 47–48, 66FF
double bonds, 47, 47F, 66F
double *vs.* single, 47F
intermediate bonds, 47, 66F
polar bonds, 47–48, 47F, 68F, 90, 91F
single bonds, 47, 47F, 66F
Crick, Francis, 169, 170
Crossing-over *see* Recombination
Cuboidal epithelium, 709, 709F
Cut and paste transposition, 218F
Cyanobacteria, evolution, 489
Cyclic AMP (cAMP), 77F
CAP activation, 275
degradation by cAMP phosphodiesterase,
550
G protein-mediated production, 549, 568
signaling pathway, **550–552,** 551F
cell responses, 551, 551T
gene activation, 552, 553F
PKA activation, 551–552
time-scale, 552, 552F
synthesis by adenylyl cyclase, 550, 550F
Cyclic AMP phosphodiesterase, 550
Cyclic GMP (cGMP), NO signaling, 540
Cyclin(s), **616,** 621T, 633
Cdk association *see* Cyclin–Cdk complexes
experimental analysis, **618–619**
G_1-cyclins, 620
G_1/S-cyclin, 620
M-cyclin, 617
cell cycle concentrations, 617F
regulation by APC, 617, 624
S-cyclin, 620
ubiquitin-mediated degradation, 617, 617F

Cyclin-dependent protein kinases (Cdks), **616,**
616F, 621T, 633
cyclin association *see* Cyclin–Cdk complexes
experimental analysis, **618–619**
G_1-Cdk, 620, 630
G_1 inactivity, **622**
G_1/S-Cdk, 620, 630
inhibitory proteins, 622
p21, 623–624
M-Cdk, **617,** 637
activity during cycle, 617F, 619F
cyclin-mediated regulation, **617,** 617F
discovery, 618
MAP phosphorylation, 644
positive feedback, 620, 620F
regulation by phosphorylation, **617, 620,**
620F
Rb inhibition, 630, 630F
S-Cdk, 620, **621–622,** 622F
Cyclin–Cdk complexes, 616, **620,** 621F,
621T, 633
Cysteine, structure, 75F
Cystic fibrosis, ER protein export, 518
Cytochalasin, actin effects, 593
Cytochrome b-c_1 complex, 462, 462F, 473,
476F
proton pump, 475–476
Cytochrome b_{562}, structure, 137F
Cytochrome c, 127F, 473, 473F
apoptosis role, 628
Cytochrome oxidase complex, 462, 462F,
474–475, 475F
active site, 474, 474F
metal ions, 473–474
as proton pump, 475, 476
spatial orientation, 474F
Cytochromes, 473
see also individual molecules
Cytogenetics, chromosome abnormalities, 178,
179F
Cytokines, signaling pathway, 564, 564F
Cytokinesis, 613, 637, 640, 641, 643F,
652–655
see also Meiosis; Mitosis
cleavage furrow, 652–653, 652F
contractile ring *see* Contractile ring
plane of cleavage, **652–653**
plant cells, **645–655,** 655F
symmetric *vs.* asymmetric, 653
Cytoplasm, 10, 22, 22F, 36
see also Organelle(s)
diffusion through, 100–101
dynamics, 23
Cytosine, 56, 67F, 76F, 171
deamination, 212F, 213F
guanine base pairs, 175F
Cytoskeleton, **22–23,** 25F, 36, **573–610**
bacterial, 573
filaments, 22, 23F, 573–574, 574F
see also individual types
functions, 499, 573–574, 574F
cell division, **639–640,** 639F
plant, 698
Cytosol, **22,** 36, 498

D

DAG, *see* Diacylgycerol
Darwin, Charles, 7, 659
de Graaf, Reiner, 659
Deamination, 212, 212F, 213F
Dehydrogenation reactions, 90

Delta–Notch signaling, 536F, 537T
Density gradient centrifugation, 161F
Deoxyribonuclease (DNase), structure, 127F
Deoxyribose, 76F, 231F
ribose *vs.,* 261
Depurination, 212, 212F, 213F
Dermis
mammalian skin, 720F
plant tissues, 700F, 701F
Desmosomes, 711F, 712, 713–714, 715F,
737
cadherins, 712–713
intermediate filaments, 575, 575F, 577,
713
Detergents, 378F
membrane protein isolation, **377,** 378F,
385, 385F
Development, 697, 717, 718F, 719F
see also Cell differentiation
apoptosis role, 625, 625F, 626F
gene regulation in *see* Developmental gene
regulation
neural tube formation, 714F
survival factors, 631, 632F
Developmental gene regulation, **280–289,**
286, 287F, 290, 318–319, 318F
see also Cell differentiation
armadillo and Wnt signaling, 734
cancer genetics and, 734–735
cell memory, **286–288,** 288F
chromatin role, 287–288, 288F
Delta–Notch signaling, 536F
egg polarity genes, **282–284,** 282F, 283F,
284F
eye development, 288–289, 289F
globin gene expression, 299
organogenesis, **288–289,** 289F
Diacylglycerol (DAG), 549, 553
PKC activation, 554, 554F, 568
Dideoxy DNA sequencing method, 331, 332F
Didinium, 26–27, 26F
Differential centrifugation, 161F, 499–500
Diffusion
enzyme–substrate binding, 100–101
membrane permeability, 391, 391F, 399,
400F
random walks, 100F
Digestion, 428
Dihybrid crosses, 677–678, 679F
Dihydroxyacetone phosphate, 432F
Dihydroxyacetone, structure, 70F
Dimerization
catabolite activator protein (CAP), 139,
139F
receptor tyrosine kinases, 558
thymine, 212, 213F
Dinoflagellates, 27F
Diploid cells, 655, 663
Disaccharides, 52, 53F, 71F
Disulfide bonds, 75F
ER role, 516
protein crosslinks, 142–143, 143F
DNA (deoxyribonucleic acid), 3–4, 25F,
169–194
analysis *see* DNA analysis
chloroplast, 19, 479, 487
cloning *see* DNA cloning
damage *see* DNA damage
denaturation, 336, 336F
electron microscopy, 12F
engineering *see* Recombinant DNA
techniques

functions, **170–177**
historical research
  identification as genetic material, 170,
    172–174, 173F, 174F
  Streptococcus pneumoniae experiments,
    172, 173F
  structure elucidation, 169, 170
  "transforming principle," 172, 174F
hybridization *see* DNA hybridization
information coding, 169, **176–177,** 176F,
  191
  see also Gene(s); Genetic code
  DNA to RNA to protein (central dogma),
    4, 4F, 229, 230F, 262
mitochondrial, 17, 455, 487
packaging, 177, **183–184**
  see also Chromosome(s)
  levels, **186–187,** 186F
  transcriptional control, **279–280**
"parasitic" *see* Mobile genetic elements
PCR amplification *see* Polymerase chain
  reaction (PCR)
recombination *see* Recombination
repair *see* DNA repair
repetitive *see* Repetitive DNA
replication *see* DNA replication
RNA *versus,* 231, 231F
  in origin of life, **261–262,** 262F
sequence *see* DNA sequence
stability, 197
structure *see* DNA structure
viral genomes, 221F
visualization, 329–330, 330F
DNA analysis, **327–336,** 360
electrophoretic separation, **329–330,** 330F
fragmentation, 327–328, 344
hybridization *see* DNA hybridization
model organisms, 327
restriction enzymes, **328,** 329F
restriction maps, 336
sequencing *see* DNA sequencing
DNA-binding proteins, 271, 289
motifs, 272–273, 272F
protein–DNA interactions, 271–272, 272
structure, 127F
DNA cloning, **341–351,** 360–361
see also DNA libraries
bacterial transformation/replication,
  341–342, 343, 344F
definitions, 341
DNA fragmentation for, 344
genome sequencing, 335, 335F
human genes, **343–345**
ligation, **341,** 343, 344F, 361
  blunt *vs.* staggered ends, 342F
polymerase chain reaction (PCR), 348, 349F
protein expression, 352–353, 353F
serial, **352,** 352F
vectors, 244, **342–343,** 343F
DNA damage, 195, 209, **212,** 224, 687F
see also DNA repair; Mutation(s)
cancer and, 728
cell cycle checkpoint, 623–624, 623F
deamination/depurination, 212, 212F, 213F
replication errors, 203, 210
thymine dimer formation, 212, 213F
DNA fingerprinting, polymerase chain reaction,
  349, 350F
DNA helicase(s), 206, 207F
transgenic mice, 359F
DNA hybridization, **336–341,** 360
diagnostic, **336–338,** 337F
microarrays *see* DNA microarrays

Northern blotting, 338F
principles, 336, 336F
probes, 336
screening DNA libraries, 345, 346F
in situ hybridization, **340–341**
  cells, 340–341, 340F
  chromosomes, 178, 178F, 340, 340F
Southern blotting, 338, 338F
specificity, 337
DNA libraries, 344–345, 346F, 361
see also cDNA libraries; Genomic libraries
DNA ligase
cloning use, **341,** 342F, 343, 344F, 361
repair, 214
replication, 206
DNA loops
chromatin 30-nm fiber, 187
transcription regulation, 278, 279F
DNA microarrays, **338–340,** 339F, 360
replication origin analysis, 199–200, 199F
DNA-only transposons, 218, 218F
DNA polymerase(s), 201, 207F, 223
3′ to 5′ exonuclease, 203
5′ to 3′ synthesis, 202–203, 202F, 203F
error rate, 203, 210, 210T
PCR reaction, 348
proofreading, **203–204,** 203F, 205F, 223
repair polymerase, 205, 214
RNA polymerase *vs.,* 233
structure, 204F
DNA primase, 204, 207F
DNA repair, **209–215,** 224
see also DNA damage
basic pathway, 213–214, 214F
defects, 211–212, 214
evolutionary significance, 214–215
mismatch repair, **210–212,** 211F
strand recognition, 211, 213
DNA repair polymerase, 205, 214
DNA replication, 4–5, 195, **196–208,**
  223–224
see also Cell cycle; DNA synthesis
cell cycle regulation, **621–622,** 622F
chromatin structure inheritance, 288F
current understanding, 208
DNA unwinding, 197, 197F, 206
end replication problem, 208
errors, 203, 210T, 211F
  mismatch repair, **210–212**
  polymerase proofreading, **203–204,**
    203F, 205F
machinery, 197, **206–207,** 207F, 223–224
  clamp loader, 207F
  helicase, 206, 207F
  ligase, 206
  nuclease, 205
  polymerase *see* DNA polymerase(s)
  primase, 204, 206, 207F
  primosome, 207F
  repair polymerase, 205
  single-strand binding protein, 206, 207F
  sliding clamp, 206, 207F
origins of replication *see* Replication origin(s)
replication fork *see* Replication fork
RNA primers, **204–206,** 206F, 207F
semiconservative nature, 197, 197F
speed, 196–197
telomerase, **207–208**
template, 196–197, 196F, 202F
DNA sequence, **176–177,** 176F, 229
see also Genetic code; *specific elements*
analysis, 177F, **314–315, 331–333**
  annotation, 333

comparative *see* Comparative genome
  analysis
expressed sequence tags (ESTs), 315
gene (ORF) identification, 314–315,
  314F, **333, 336**
human genome, 311, 314–315
sequencing *see* DNA sequencing
evolutionary aspects, 34, 214–215
  see also Genome evolution; Phylogenetics
  divergence rates, 305–306
variations in *see* Genetic variation
DNA sequencing, **331–333,** 360
automation, 331, 333F
complete sequences, 331
dideoxy method, 331, 332F
human genome, 314–315, **334–335**
  chromosome 22 (complete), 312F
  clone-by-clone approach, 335, 335F
  combined approach, 335
  first draft, 311, 335
  shotgun sequencing, 334–335, 334F
DNA structure, **170–177,** 191
backbone, 171, 171F
base-pairing *see* Base-pairing
double helix, 171F
  antiparallel strands, 176
  complementary chains (strands), **171,**
    175–176, 196
  major and minor grooves, 176F, 272F
  polarity, 171, 175, 201, 202, 202F
  replication template, 196, 196F, 202F
  stability, 197
  transcription template, 231–232, 232F
  Watson and Crick model, 169, 170
  X-ray diffraction, 170
negative charge, 329
nucleotides, 3, 57–58, 58F, 171, 171F
  see also Nucleotide(s)
phosphodiester bonds, 58F, 77F
DNA synthesis, 223
see also DNA polymerase(s); DNA
  replication; Replication fork
5′ to 3′ direction, 201, 202–203, 202F
  3′ to 5′ *vs.,* 205F
chemical synthesis, 336–337
initiation, 197
RNA primers, **204–206**
DNA–RNA hybrids, 206F, 233
Dolichol, glycosylation role, 517
Dolichol phosphate, 73F
Dominant alleles, 675, 685F
Double strand breaks, homologous
  recombination, 216, 216F
Down syndrome (trisomy 21), 670–671, 671F
Drosophila melanogaster, 29F
alternative splicing, DSCAM proteins, 318F
developmental gene regulation
  Armadillo and Wnt signaling, 734
  egg polarity genes, **282–284,** 282F,
    283F, 284F
  eye development, 288–289, 289F
genome
  chromosome 2, 180F
  complete sequence, 331
  insertional mutagenesis, 302F
  position effects, 188F
as model organism, 29, 32
Dynamic instability, microtubules, **581–582,**
  581F
Dynamin, 513, 514F
Dynein(s), 585, 606–607
cilia/flagella motility, 591, 592F
mechanism, 585, 585F

Page numbers in **boldface** refer to a major text discussion of the entry; page numbers with an F refer to a figure, with
an FF to figures that follow consecutively; page numbers with a T refer to a table.

E. coli, see Escherichia coli
EF-Tu elongation factor, as GTP-binding
protein, 154–155, 155F
Egg cell(s), 659, 661F, 717
calcium wave, 554, 554F, 672
fertilization *see* Fertilization
production *see* Meiosis
Xenopus laevis oocytes
cell cycle experiments, 618, 618F, 619F
mitosis research, 646–647, 646F, 647F
zona pellucida, 670
Elastase, structure, 138F
"Elastic skin man," 706F
Elastin, structure, 142, 142F
Electrical signaling, 405
see also Action potential; Membrane
potential
Electrochemical gradients, 422
see also Membrane potential
active transport, 398, 399F
chemiosmotic coupling *see* Chemiosmotic
coupling
components, 395F
passive transport, 395
protons (H$^+$), 379, 402, **462–464**, 492
see also Proton pump(s)
ATP synthesis, 379, 441, 454,
464–466, 464F, 466–467
consequences, 463–464, 464F
coupled transport, **466**, 466F
electron transfer generated, **462–464**,
468–469, 470F, 475–476
photosynthesis, 480–481
Electron(s), 64
atomic interactions, **41–43**
see also Chemical bonds
electron shells, 42, 42F
atomic structure, 40, 40F
atomic valence, 43
transfer *see* Electron transfer
Electron carriers
see also Electron transfer; Electron-transport
chain(s)
FAD/FADH$_2$, 440, 458
metal ions, **472–474**
NADH/NADPH, **109–111**, 110F, 458,
458F
Electron microscopy, 6, 9F, 11
resolution, 10, 10F
Electron transfer, 453
see also Electron carriers; Electron-transport
chain(s)
electrical batteries, 463F
oxidation reactions, **90**, 91F
proton gradient production, **462–464**,
468–469, 470F, 475–476
redox reactions, **109–111**, 462
reduction reactions, **90**, 91F
thermodynamics, **470**
Electron-transport chain(s), 453, 454,
468–477, 492
see also Electrochemical gradients; Electron
carriers; Electron transfer; Proton
pump(s)
chemiosmotic coupling *see* Chemiosmotic
coupling
energetics, **470**
oxidative phosphorylation, 430, **441**, **444**,
449, **459**, **462–464**
see also FAD/FADH$_2$; NADH/NAD$^+$;
individual components
electron donors, 457–458, 459, 462

enzyme complexes, 462, 462F
metal ions, **472–474**
proton pumps, 463, **475–476**, 477F
Q cycle, 476
redox potential, 470, 472F
ubiquinone role, 472, 472F, 476
photosynthesis, 88, 480–481, 483F, 484F
see also Chlorophyll; Photosynthetic
reaction center(s)
ATP and NADPH synthesis, **482–485**
redox potentials, 484F
redox potentials, 470, 472F, 484F
redox reactions, 462, 469
uncoupling, 460–461, 461
Electrophoresis
DNA separation, **329–330**, 330F
protein separation, 129, 163F
Elements, 39, 41, 64
chemical reactivity, 42, 42F
distribution/abundance, 41F
periodic table, 43, 44F
EM, *see* Electron microscopy
Embryonic stem cells, 722–725, 738
cell culture, 327, 723
cell differentiation (totipotent), 724F
nuclear transplantation, **725–726**
rejection, 724–725
transgenic mice, 358F
End replication problem, 208
Endocrine cell signaling, 534, 535F
Endocytosis, 21, 22F, 497, 512–513,
523–528, 529
see also Clathrin-coated vesicles;
Endosome(s)
phagocytosis, **523–524**, 524F
phagosomes, 523, 524
pseudopodia, 524, 524F
pinocytosis, 523, **525**
protein sorting, **526–527**, 527F
receptor-mediated, **525–526**, 526F
transcytosis, 527
Endomembrane system *see* Endoplasmic
reticulum (ER); Endosome(s);
Golgi apparatus; Lysosome(s)
Endoplasmic reticulum (ER), 19, 529
cell division and, 502, 652
evolution, 500–501, 501F
functions, 498–499, 499T
membrane biosynthesis, 373–374
protein modification, **516–517**, 517F
microtubule-associated movement, 589,
589F
protein export (ER to Golgi), 512, 512F,
529
see also Golgi apparatus; Vesicular
transport
quality control, **517–518**
protein import, **507–509**, 529
cotranslational, 508
ER retention signal, 517
signal-recognition particle, 509
signal sequence, 508, 509, 510F
soluble proteins (for export), 507–508,
509, 510F
translocation channels, 509, 510F
transmembrane proteins, 507, 508,
510–512
rough ER, 20F, 249F, 498, 498F, 508
see also Ribosome(s)
sarcoplasmic reticulum *see* Sarcoplasmic
reticulum
smooth ER, 498–499
structure, 20F, 498–499, 498F

electron microscopy, 12F, 20F, 21F, 508F
membrane system, 507, 508F
nuclear envelope and, 505F
Endosome(s), 499
see also Endocytosis
acidity, 526, 527
early/late, 526
evolution, 500–501, 501F
protein sorting, **526–527**
Endothelial cells, 718
nitric oxide effects, 540–541, 541F
Energy, 115–116
see also Metabolism; Thermodynamics
chemical bonds, 45, 88, 97F, 437F
forms, 87F
heat energy, **85–88**
sunlight, 88, 88F, 115, 482
Gibbs free energy *see* Free energy (G)
metabolic production *see* Energy production
metabolic utilization, **112–114**
see also Biosynthesis
storage *see* Energy storage
Energy production, 17, **453–496**
see also ATP (adenosine triphosphate);
Chloroplasts; Mitochondria
efficiency, **476–477**
food molecules, 14–15, 17, 83, 116,
427–452
see also Catabolism; *specific molecules*
oxidation reactions, **89**, 89F, 427
photosynthesis *see* Photosynthesis
release of heat energy, **85–88**
respiration *see* Respiration (cellular)
Energy storage, **444–449**
activated carriers *see* Activated carrier
molecules
animals, 438F, 444–445, 445F
chemical bonds, 88, 97F
organic molecules (small cellular), 88
plants, 446, 446F
Energy utilization, **112–114**
see also Biosynthesis; Reaction coupling
Enhancers, **278–279**, 279F
Enol phosphate bond, energy, 437F
Enolase, 433F
Entropy, 85, 86F, 96F, 115
Environment, gene interactions, **692–693**,
692F
Enzyme(s), 59, 83, 92, 115, 120F, 145, 159
active site, 93, 101, 147, 149F
allosteric, **151–153**
catalysis *see* Enzyme catalysis
cofactors, 150
functional classes, 146, 147T
inhibitors, 104–105, 105F
membrane proteins, 374F
receptor-linked *see* Enzyme-linked
receptor(s)
specificity, 92
substrates, 83, 92, 146
Enzyme catalysis, 83, 84F, 94F, 149F
biotechnology and, 105
equilibrium points, 102, 102F
kinetics *see* Enzyme kinetics
lysozyme, **146–148**, 149F
regulation, **104–105**, 151
allosteric, **151–153**, 153F
feedback inhibition, 151, 151F, 152F,
153F
substrate concentration, 101
substrate interactions *see* Enzyme–substrate
interactions
thermodynamics, **91–93**

see also Free energy (G)
activation energy reduction, 92, 92F, 93F, 102, 115, 147, 148
transition states, 147
Enzyme kinetics, **101–102**, 102F
experimental analysis, **103–105**, 103F, 104F, 105F
K_M (Michaelis' constant), 102, 102F
reaction rate, 92, **103–104**, 103F, 115
turnover number, 102
V_{max}, 102, 102F
determination, 103–104, 103F
Enzyme-linked receptor(s), 542, 543, 543F, **557–568**, 569
cytokine receptors, 564
gene regulatory proteins, **560, 564**, 565F, 569
intracellular signaling, 545, 545F, 569
see also Protein phosphorylation
protein kinase networks, **565–566**, 566F
signaling complexes, **557–559**, 558F
plants, 567
receptor serine/threonine kinases, 564, 565F
receptor tyrosine kinases see Receptor tyrosine kinases
TGF-β superfamily, 564, 565F
Enzyme–substrate interactions, **145–146**
competitive inhibition, 104–105, 105F
noncovalent bonds, 63, 98, 101, 147–148
substrate diffusion, 100–101
Epidemiology, cancer, **727–728**
Epidermis
maintenance/renewal, 721–722, 723F
mammalian skin, 720F
plant tissues, 700F, 701F
Epidermolysis bullosa simplex, 577
Epithelium, 698F, **709–716**, 737
see also specific tissues
blister formation, 714
cell–cell junctions see Cell junction(s)
contractility, 713, 714F
importance, 709
polarized organization, **709–710**, 710F
types, 709, 709F
Equilibrium, chemical, 95, 98F
Equilibrium constant (K), 95, 97F, 116
ΔG° relationship, 95, 98T, 99F
strength of reaction, **95, 98**, 99F
Equilibrium (density gradient) sedimentation, 161F
ER, see Endoplasmic reticulum
Erythrocytes see Red blood cells
ES cells, see Embryonic stem cells
Escherichia coli
DNA analysis, 327
genome, 273
horizontal gene transfer, 302
mutation rate, 296
size, 35
as model organism, **28**
structure, 14F
transcriptional control, **273–275**
lac operon, **275**, 275F
repressors, 273–274
tryptophan operon, 273–274, 273F
Esters, 67F, 72F
Estradiol, 541F
intracellular receptors, 541–542
Ethylene glycol, poisoning, 104–105
Ethylene, plant hormone, 567
Eubacteria see Bacteria (eubacteria)
Eucaryotic cell(s), 11, **16–27**, 25F, 36

see also Animal cell(s); Plant cell(s); Yeasts; individual cell types; specific species
animal cf. plant, 10F
cell cycle times, 612, 613T
see also Cell cycle
cell senescence, 326–327
chromosomes see Chromosome(s)
cytoplasm, 22F, 23
cytoskeleton see Cytoskeleton
cytosol, 22
evolution see Evolution
evolutionary conservation, 31, 31F
genomes see Genome(s)
intracellular vs. extracellular environment, **390–391**, 390T, 400
organelles see Organelle(s)
procaryotes vs., 11, 24, 276
replication origins, 197, 198
see also DNA replication
retroviruses, **221–222**
ribosomes, 249
single-celled organisms, 16, 16F, 26–27, 26F, 27F, 36
Euchromatin, 188
Euglenoids, 27F
even-skipped (eve), **282–284**, 283F, 284F
Evolution, 319
cancer cells, 729–730, 729F
chemiosmotic coupling, **490–492**
common evolutionary ancestor, 4–5, 33, 310F, 490
conservation in see Evolutionary conservation
Darwin's theory of, 7
eucaryotes, **500–501**
animal vs. plant, 698
cell signaling, **566–567**
common ancestor, 33
phylogenetic tree, 26F, 310, 310F
plant–animal divergence, 566–567
predator origins, **24, 26–27**
genes see Gene(s)
genomes see Genome evolution
horizontal gene transfer, 303, 303F
human, 689
membrane-bound organelles, **500–501**, 501F
chloroplasts, 20F, **487–492**
mitochondria, 17, 18F, **487–492**
mobile genetic elements, 219, **301–302**, 302F
mutations and, 5
see also Mutation(s); Natural selection
protein domains, 241
protein families, 138
protein machines, 157
purifying selection, 308
RNA world theory, **258–262**
Evolutionary conservation, 304–305, **306–308**
cell cycle proteins, 31, 31F, 619
distant evolutionary relationships, **309–310**, 310F
DNA repair role, 214–215
Fugu genome, **308–309**, 309F
genome organization (synteny), 307, 307F, 316
histones, 186
homologous genes (homologs), 34, 304, 315
protein structure, 138
purifying selection, 308

ribosomal RNA, 310, 310F
sex determining genes, 215F
Excitatory neurotransmission, 419, 420F, 422
Exocytosis, 21, 22F, 497, 516, **519, 522–523**, 529
constitutive pathway, 519, 522, 522F
regulated pathway, 522, 522F
Exons, 238, 238F, 262
duplication, **300–301**, 300F
shuffling, 296, **301**, 301F
Expressed sequence tags (ESTs), DNA sequence analysis, 315
Expression vectors, 353
Extracellular fluid, ion concentrations, **390–391**, 390F, 390T, 400
Extracellular matrix, 7, 697, **703–704**, 737
see also Connective tissue(s); Tissue(s)
degradation by matrix proteases, 705
fibrous proteins, 141–142
see also Collagen(s)
glycosaminoglycans, 707–708, 708F, 737
intergrins and, **706**, 707f
membrane protein anchoring, 383
production, 704, 705F
proteoglycans, **706–709**, 737
Eye color, inheritance, 691–692
Eye development, gene regulation, 288–289

F

Factor VIII
gene cloning, **343–345**
recombinant protein expression, 353
FAD/FADH₂, 449
citric acid cycle, 440, 441F
electron carrier, 440, 458
Fat(s)
see also Fatty acid(s); Lipid(s)
breakdown, **428–430**, 429F, 438F, 439
efficiency, 430
energy storage
animals, 438F, 444–445, 445F
plants, 446, 446F
Fat cells, 445, 445F
chemiosmotic uncoupling, 460–461
Fat droplets, 438F, 445F, 446F
Fatty acid(s), 51, 51F, **53–55**, 65, 72FF
see also specific molecules
cell membranes, 54–55, 73F, 368F
oxidation cycle, 438F, 439, 457, 458F
saturated, 54, 54F, 72F, 372
storage (triacylglycerols), 54, 54F, 72F, 369F, 438F
structure, 53F, 54F, 72F
amphipathic nature, 53F, 54, 55, 55F, 367
hydrocarbon tail, 66F
unsaturated, 54, 54F, 72F, 372
Feedback regulation, 151, 151F, 152F
see also Allosteric regulation
conformational change, 152, 153F
photosynthesis, 481
Fermentation, **431, 434**, 434F
Fertilization, 662F, **671–672**
calcium wave, 554, 554F, 672
egg–sperm fusion, 671, 671F
zygote formation, 661, 672
Fibroblasts
cell culture, 327F
collagen organization/remodeling, 706, 707F
conversion to muscle cells, 286F
extracellular matrix production, 704, 705F

Page numbers in **boldface** refer to a major text discussion of the entry; page numbers with an F refer to a figure, with an FF to figures that follow consecutively; page numbers with a T refer to a table.

Fibroin, β–sheets, 126
Fibronectin, 706, 707F
Fibrous proteins, 127F, **141–142**, 142F
Filopodia, 596, 597F
Fischer, Emil, 60
Flagella, 25F, 591F
  "9+2" array, 591, 591F
  Bdellovibrio bacteriovorus, 2F, 3
  microtubules, 579, **590–591**
  mitochondria, 456F, 466, 467F
  motility, 591, 591F, 592F
Flavins, respiratory chain, 473
Flemming, Walther, 646
Flippases, 373, 373F
Fluorescence-activated cell sorter, 325, 326F
Fluorescence microscopy, 8F, 355, 520–521, 521F
Fluorescence recovery after photobleaching (FRAP), 384–385, 384F
Fluorescent probes, 8F
Food molecules, 17, 83, 116, **427–452**
  see also Energy; Fatty acid(s); Sugars
  oxidation *see* Catabolism
  storage, **444–449**
Forensic medicine, PCR role, 349, 350F
Formic acid oxidation, anaerobic bacteria, 489F
Free energy (G), 86, 91, 94, 96FF
  change (ΔG), **93–94**, 96F, 115–116
    ATP hydrolysis, 114
    equilibrium constant (K), 95, 97F, 98T, 99T
    favorable *vs.* unfavorable reactions, 94, 94F
    reactant concentration and, **94–95**
    redox reactions, 469–470
    sequential reactions, **98–100**
    standard (ΔG°), 95
  predicting reactions, 97F
  reaction rates, 96F
  spontaneous reactions, 96F
Frog, metamorphosis and apoptosis, 625, 626F
Fructose 1,6-bisphosphate, 432F
Fructose 6-phosphate, 432F
Fructose, structure, 70F
Fugu rubripes, genome, **308–309**, 309F
Fungi, 28
  H^+ pumps/gradients, 402
  yeasts *see* Yeasts

G

G protein(s) (trimeric GTP-binding proteins), 543, **546–548**, 568
  α subunit, 546, 547, 547F, 548F
  βγ complex, 546, 547F
  activation, 546, 547F
  G_i, 547–548
  G_s, 547
  inactivation, 547, 548F
  membrane targets, 546–547
    ion channels, **548–549**, 549F
    membrane-bound enzymes, **549–550**
  toxin effects, 547–548
  transducin, 556
G-protein-linked receptor(s), 542, 543, 543F, **546–557, 568**
  activation, **546–548**, 547F
  adaptation, 556
  inactivation, 547, 548F
  intracellular signaling, 545–546, 549–550
    see also specific pathways
    calcium role, **552–554**

cyclic AMP production, 549, **550–552**, 550F, 568
diacyl glycerol production, 549, 554, 554F
IP_3 production, 549, 553–554, 554F, 568
speed/sensitivity, **555–557**
visual transduction, 556, 557F
regulation, 547
rhodopsin, 556
structure, 546, 546F
toxin effects, 547–548
GABA, 419
  receptors, drug targets, 420
Galactocerebroside, structure, 368F
Galactose, glucose isomer, 52, 70F
Gametes, 293, 655, 659, **661–662**
  see also Egg cell(s); Spermatozoa
  cell division *see* Meiosis
  fertilization *see* Fertilization
  genetic variation, **668–670**
  germ line *vs.* somatic cells, 294, 662, 663F
  Mendel's law of segregation, **675–676**, 676F
  mutation, 210, 294, 294F
γ-aminobutyric acid (GABA), 419, 420
Gap junctions, 711F, **715–716**, 716F, 738
Gel-filtration chromatography, protein analysis, 162F
Gels, glycosaminoglycans, 708
Gelsolin, 594
Gene(s), 3, 5, **179–181**, 659, 685F
  see also Chromosome(s); DNA (deoxyribonucleic acid); Genetics (classical)
  alleles, 665, 675, **681, 684**, 685F
  bacterial, 238, 238F
  central dogma, 4, 4F, 229, 262, 330F
  chemical nature, 169
  definition, 179, 685F
  environmental effects, **692–693**, 692F
  eucaryotic, **237–238**, 238F, 262
    see also Exons; Introns
    advantages, 240
    evolution, 242–243
    human, 311, **689–691**
    organization, 237–238, 238F, 240–241, 262
  evolution, **293–322**
    see also Genome evolution
    duplication/divergence *see* Gene duplication
    eucaryotic genes, 242–243
    introns early *vs.* introns late, 243
    pseudogenes, 299
  expression *see* Gene expression
  functional analysis, **355–356**, 356F, 734
  genome number, 179
  homologous, 34, 304, 315
  housekeeping, 268
  identification, 314–315, 314F, **333, 336**, **686–693**
  information coding, **176–177**, 176F, 177F
    see also Genetic code
  linkage, 681F
  mapping, **680–681, 682–683**, 685F
    haplotype, 690, 690F, 692
Gene duplication, 295, 295F, **297–298**, 319
  divergence following, 297
  exon expansion, **300–301**, 300F
  globin family evolution, **298–299**, 299F
  importance of in vertebrate evolution, **299–300**
  introns role, 301

pseudogenes, 299
unequal crossing-over, 297, 298F, 300F
whole-genome, 300
Gene expression, 177, 267
  see also RNA (ribonucleic acid); Transcription; Translation
  analysis, **353–355**, 354F
    see also Reporter genes; *specific techniques*
  cancer cells, **734–735**
  nuclear transplantation, 268, 269F
  change in response to external signals, 270
  chromatin effects *see under* Chromatin
  constitutive, 274
  control *see* Gene expression regulation
  developmental *see* Cell differentiation; Developmental gene regulation
  differing levels, 230, 230F
Gene expression regulation, **267–292**
  see also Cell signaling
  β-catenin role, 734–735
  cAMP-mediated signaling, 552, 553F
  combinatorial control, **281**, 281F
    cell differentiation, **285–286**
    coordination, **281, 285**, 286F
    eve, **282–284**, 283F, 284F
    developmental *see* Developmental gene regulation
    experimental analysis, **282–284**, 282F, 283F, 284F
    human genes, 317–319
    levels, **270**, 270F
    regulatory proteins *see* Gene regulatory proteins
    transcriptional *see* Transcriptional gene regulation
Gene families, 34, 35
  evolution, **297–298**
  globin genes, **298–299**, 299F
Gene inactivation, RNA interference (RNAi), 357–358
Gene regulatory proteins, 120F, 271, 289
  see also Gene regulatory sequences; Transcription factors (general); *specific proteins*
  activators *see* Transcriptional activators
  DNA-binding motifs, 272–273, 272F, 289
  eucaryotic, **278–279**, 290
    combinatorial control, **281**, 281F, 283–284, **285–286**
    coordination, **281, 285**, 286F
    developmental expression *see* Developmental gene regulation
    enzyme-linked receptors, **560, 564**, 565F, 569
    experimental analysis, **282–284**, 282F, 283F, 284F, 286, 287F
    PKA-mediated phosphorylation, 552
    regulatory modules, 283–284, 284F
    steroid hormone receptors, 541–542, 542F
  positive feedback loops, 287, 288F, 289
  protein–DNA interactions, 271–272
  repressors *see* Transcriptional repressors
Gene regulatory sequences, 271, **271–273**
  see also Gene regulatory proteins; Promoter elements
  bacterial, 289
    operators, 273
  development role, 318–319, 318F
  eucaryotic, 276, 290
    enhancers, 278, 279F
    genome location, 281
    human, 311, 313, 317–318

TATA box, 277, 278F
Gene replacement (transgenes), 357, 357F
Gene splicing see Recombinant DNA
    techniques
Gene therapy, 357

Gene–environment interactions, **692–693,**
    692F
Genetic code, 176–177, **244,** 263
    codons see Codons
    cracking the code, **246–247,** 247F
    reading frame, 244, 245F
    redundancy, 248
Genetic disease(s), 195, 209–210, 209F
    see also Mutation(s); specific disorders
    diagnosis, **336–338,** 337F
    genetic mapping, 682–683, 683F, 690,
        690F
    pedigrees, 677, 677F
    recessive mutations, 336
    selective advantage, 684, 686
Genetic drift, 305
Genetic engineering see Recombinant DNA
    techniques
Genetic instability, cancer cells, 728, 728F,
    730
Genetic linkage maps, **682–683,** 685F
    haplotype, 690, 690F, 692
    recombination frequency, **680–681,** 681F,
        682
    SNP analysis, 682–683, 682F, 683F, 689
Genetic recombination see Recombination
Genetic screens, 686, **687–688,** 688F
    signaling pathway analysis, 562, 563F
Genetic traits, 672
    see also Genetic disease(s); Inheritance
    complex, **691–692**
    discrete, **673,** 673F
    gene–environment interactions, **692–693,**
        692F
Genetic variation, **293–303,** 319, 672
    see also Genome evolution; Mutation(s)
    germ-line vs. somatic, 293–294
    human, **313, 316,** 319
    mutation rate, 296
    neutral (silent) mutation(s), 5, 297, 304,
        305, 690
    selection and, **304–305**
    sexual reproduction, 294
        meiosis, 667, **668–670,** 670F
    single-nucleotide polymorphism, 316
    SNPs see Single-nucleotide polymorphisms
        (SNPs)
Genetically modified crops, 360
Genetics (classical), **659–696,** 685F
    see also Chromosome(s); Gene(s); Heredity;
        Mutation(s); Reproduction
    early misconceptions, 659
    environment and, **692–693,** 692F
    essential concepts, **693–694**
    gene identification, 355, **686–693**
        complementation tests, 685F, **688–689**
        genetic screens, 562, 563F, 686,
            687–688, 688F
        random mutagenesis, **686–687**
    gene mapping, **680–681,** 681F, **682–683,**
        683F, 685F
    genotype, 355, 675, 685F
    haplotypes, **689–691,** 691F
    Mendelian inheritance see Inheritance
    phenotype, 355, 675, **681, 684,** 685F
Genome(s), 5, 36, 177
    see also Chromosome(s); DNA

(deoxyribonucleic acid); individual
    genomes
coding regions see Gene(s)
complexity, 179, 181
constancy during differentiation, **268–269**
definition, 685F
eucaryotic
    multicellular organisms, 268, 717, 718F
    organellar, 17, 19, 455
    sizes, 34F, 35
    species differences, 181, 181F
    yeast, 28
evolution see Genome evolution
haplotype blocks, **689–691,** 691F
noncoding DNA see Noncoding DNA
procaryotic cells, 34, 34F
sequence comparisons see Comparative
    genome analysis
sizes, 34–35, 34F, 179, 181
viral, 220–221, 221F
Genome evolution, **293–322**
    see also Gene(s); Genetic variation;
        Mutation(s); Phylogenetics;
        specific genomes
    conserved sequences/organization see
        Evolutionary conservation
    exon expansion, **300–301,** 300F
    exon shuffling, 295F, 296, **301**
    gene deletion, 295, 295F
    gene duplication/divergence see Gene
        duplication
    gene organization, significance, 240–241
    genetic drift, 305
    haplotype blocks, 689–691, 691F
    homologous genes, 34, 304
    horizontal transfer, 295F, 296, **302–303**
    "junk" DNA, **308–309**
    mechanisms, **296–297**
    point mutations within gene, 196, 295F
    pseudogenes, 299
    transposable elements role, 219, **301–302,**
        302F, 307, 308F
    types of change, **295–296,** 295F
    whole genome duplication, 300
Genome mapping, restriction maps, 336
Genomic libraries, 344–345, 361
    advantages, 348
    cDNA versus, 346–347, 347F
    construction, 345F, 349F
    disadvantages, 345
Genotype, 355, 675, 685F
Genus names, 14
Germ cells see Gametes
Giant, Drosophila anterior–posterior polarity,
    283, 283F
Gibbs free energy see Free energy (G)
Gilbert, Walter, gene number estimates, 314
Gleevec, 737F
α-globin genes, evolution, **298–299,** 298F,
    299F
β-globin genes, 238F
    evolution, **298–299,** 298F, 299F
        Alu-mediated recombination, 301
        human–mouse divergence, 308F
    mutation
        selective advantage, 684, 686
        sickle-cell anemia, 209–210, 209F,
            337–338, 337F, 684, 686
γ-globin genes, evolution, 299, 299F
Glucocorticoids, gene expression regulation,
    270, 285, 286F
Glucosamine, 71F
Glucose, 52F, 70F

activated carriers, 111
as energy source, 53, 427, 428–429, 429F
    see also Glycolysis
isomers, 52, 70F
membrane transport see Glucose
    transporter(s)
polymers, 53
Glucose 6-phosphate, 432F
Glucose transporter(s), 400F
    coupled (active) transport, 398, 399F
    passive carrier, 394, 394F, 403T
Glucuronic acid, 71F
Glutamate, 75F
    excitatory neurotransmission, 419
Glutamine, 75F
    biosynthesis, 109F
Glyceraldehyde, 70F
Glyceraldehyde 3-phosphate
    carbon fixation, 485, 485F
    glycolysis, 432F, 433F, 435, 436F
Glyceraldehyde 3-phosphate dehydrogenase,
    433F, 435, 436F
Glycerol, 54, 72F
Glycine, 75F
    inhibitory neurotransmission, 419
Glycocalyx, **381–382,** 382F
Glycogen, 53, 71F, 444, 445F
Glycolipid(s), 53, 55
    amphipathic nature, 367
    cell membranes, 367, 373, 374
    glycocalyx (cell surface), 381–382, 382F
    structure, 73F, 368F
Glycolysis, 428–429, 429F, **430–431,** 449
    ATP production, 428, 431F, 433F, 436F,
        437F, 460
    biosynthetic role, **477,** 477F
    NADH production, 428, 430, 433F, 436F,
        437F
    pathway and enzymes, 430, 430F, 432FF
    pyruvate production, 428–429, 430, 433F
    reaction coupling, **434–435,** 436F, 437F
Glycoproteins, 53
    glycocalyx (cell surface), 381–382, 382F
    glycosylation reaction, 516–517, 517F
Glycosaminoglycans (GAGs), 707–708, 708F,
    737
N-Glycosidic bond, 76F
Glycosylation, 516–517, 517F
Goblet cells, 710
Golgi apparatus, 19–20, 25F, 499T, **518–519,**
    529
    see also Vesicular transport
    cell division and, 502, 652
    cis Golgi network, 518
    cisternae, 518
    ER–Golgi transport, 512, 512F, 529
        see also Endoplasmic reticulum (ER)
    evolution, 500–501, 501F
    faces, 518
    Golgi–lysosome transport, 527–528
        see also Lysosome(s)
    membrane synthesis, 374
    microtubule-associated movement, 589,
        589F
    protein modification, 519
    structure, 21F, 518, 519F
    trans Golgi network, 518–519, 529
        lysosomal proteins, 527–528
        secretory vesicles, 522, 522F
Green fluorescent protein (GFP), 355, 355F,
    520–521, 521F
Green sulfur bacteria, 489, 489F
Griffith, Fred, 172, 173F

Page numbers in **boldface** refer to a major text discussion of the entry; page numbers with an F refer to a figure, with
an FF to figures that follow consecutively; page numbers with a T refer to a table.

Growth factors, 629, **631**, 631F
GTP (guanosine triphosphate), 441F
  citric acid cycle production, 440, 441F
  cyclic GMP production, 540
  G protein activation, 546, 547F
  GTPase regulation, 154, 155F
  hydrolysis see GTP hydrolysis
GTP-binding proteins (GTPases), **154 155**
  conformational changes, 154, 155F
  dynamin, 513
  functional roles, 154
  molecular switches, 154, 155F, 545–546, 545F
  monomeric, 559
    actin regulation, 599–600, 600F
    Ras see Ras GTPase
    Rho, 599–600
  trimeric see G protein(s) (trimeric GTP-binding proteins)
  vesicle assembly, 513, 514F
GTP hydrolysis
  motor protein movement, 156
  nuclear protein import, 506
  tubulin and dynamic instability, 581–582, 582F
Guanine, 56, 76F
  cytosine base pairs, 175F
  depurination, 212F
  DNA, 171
Guanosine triphosphate see GTP (guanosine triphosphate)
Guanylyl cyclase, 540
Guard cells, 701F
Gut epithelium, 698F
  cells, 710
  functional polarization, 710, 711F
  glucose transport, 398, 400F
  maintenance/renewal, 721, 722F
  membrane proteins, 386, 386F
  Na^+-H^+ exchanger, 399

H

Hair cells, auditory, 407, 408F
Hairs, plant (trichomes), 701F
Halobacterium halobium, bacteriorhodopsin, 378–379
Haploid cells, 655, 663
Haplotypes, **689–691,** 690F
Heat energy, **85–88**
Helices
  α-helix see α-helix
  as biological structure, 134–135, 134F
  DNA see DNA structure
  protein complexes (assemblies), 140, 140F, 141F
Heliozoan, 27F
Helix-turn-helix motif, 272F
Heme, 150, 150F
  cytochrome proteins, 473
Hemidesmosomes, 711F, 712, 714, 715F, 737
Hemoglobin, 60
  evolution, **298–299**, 299F
  sickle-cell anemia, 209–210, 209F, 337–338
    selective advantage, 684, 686
  structure, 127F, 140, 140F, 298F
    see also Globin genes
    heme group, 150, 150F
Hemophilia
  factor VIII defects, 343–344
  HIV infection in, 344

insertional mutagenesis, 302
Hemopoietic stem cells, 722, 724F
Hepatocyte growth factor, 631
Heredity, 195, 196F, 229, 659
  see also DNA (deoxyribonucleic acid); Genetics (classical)
  identification of DNA as genetic material, 170, **172–174,** 173F, 174F
  Mendel's laws of inheritance see Inheritance
Hershey, Alfred, 172–174, 174F
Heterochromatin, 188–189, 189F, 280
Heterozygotes, 675
Hexokinase, 432F
Histidine, structure, 74F
Histone(s), 184, 186
  core (octamer), 185, 185F
  modification, 186, 189–190, 191F
    acetylation, 279–280
    regulation, 190
Histone deacetylases, 279
HIV (human immunodeficiency virus), 222, 344
Holliday junction, 216, 217F
Homeodomain proteins, 272F
Homologous chromosomes, 665
Homologous genes (homologs), 34, 304, 315
Homologous recombination, **215–216,** 215F, 224
  crossing-over, 215, 215F
  double strand breaks, 216, 216F
  Holliday junction, 216, 217F
  joint molecule, 216
  meiotic, 216, **665–667,** 693
    chiasma(ta), 666, 666F
    frequency and genetic maps, **680–681**
    genetic variation, 667–670, 670F
  site-directed mutagenesis, 356, 356F
Homozygotes, 675
Hooke, Robert, 6
Horizontal gene transfer
  antibiotic resistance, 302–303, 343
  bacterial transformation, **341–342,** 343F
  conjugation, 303F
  genome evolution, 295F, 296, **302–303,** 303F
Hormones, 537T, 541F
  see also specific hormones
  endocrine signaling, 534, 535F, 568
  intracellular receptors, **541–542,** 542F
  plant, 567
Housekeeping genes/proteins, 268
Hpr phosphocarrier protein, structure, 126, 127F, 132FF
HPV, see Human papilloma virus
Human evolution, 689
Human genome, **311–319,** 312T
  chromosomes, 178, 181
    aneuploidy, 670–671, 671F
    banding patterns, 179F
    cancer cells, 728, 728F
    chromosome 2 evolution, 306
    chromosome 22, 183, 311, 312F, 312T
    DNA packaging, 183
    karyotype, 178F
  DNA content, 177, 178, 305
  evolution
    see also Comparative genome analysis
    ancestral gene sequences, 307F
    conserved synteny, 307F, 316, 317F
    exon shuffling, 301
    human–chimpanzee divergence, 305, **306,** 307F
    human–mouse divergence, 308, 308F

gene cloning, **343–345**
gene functions, 354
gene number, 313
  estimation, **314–315**
gene regulation, 311, 313, 317–319
gene size, 311
haplotype blocks, **689–691,** 691F
mobile genetic elements, **218–219,** 311, 313F
  Alu sequences, 219, 220F, 306, 307, 308F
  L1 elements, 219, 220F, 308F
mutation rate, 296
organization, 238F, 306, **311–313**
  conserved synteny, 307F, 316, 317F
  disarray, 313
  Fugu comparison, 309, 309F
repetitive DNA, 218–219, 311, 313F, 316
sequence, 311, **314–315,** 316, 333, 689
  see also DNA sequencing
  chromosome 22 (complete), 312F
  first draft, 311, 335
  genetic variation, **313, 316,** 319
  methodology, **334–335,** 334F, 335F
  size, 35, 311, 311F
Human Genome project, 311, **317–319**
Human immunodeficiency virus (HIV), 222, 344
Human papilloma virus (HPV), cervical cancer, 728
Humans
  development, 318–319
  genome see Human genome
  as model organisms, 33
Hunchback, *Drosophila* anterior–posterior polarity, 283, 283F
Hybrid cells, 165F, 383F
Hydrocarbons, 66F
Hydrogen
  abundance, 469
  atomic structure, 40F, 43
  hydrocarbons, 66F
  ion see Hydrogen ion (H^+, proton)
  molecule, 45F
Hydrogen bonds, 45, 78F
  base-pairing, 78F
  bond length, 46T, 68F
  bond strength, 46T
  macromolecular interactions, 145F
  macromolecular structure role, 48, 62
    DNA base-pairing, 78F, 171, 171F, 175F
    polypeptide chains, 78F, 122, 123F, 124F
  water, **48–49,** 68F, 78F
Hydrogen ion (H^+, proton), 40
  acids, 49, 49F
  exchange, 69F
  gradients see Electrochemical gradients
  intracellular vs. extracellular concentration, 390T
  mobility in water, 469, 469F
  pumps see Proton pump(s)
Hydrogenation reactions, 90
Hydrolysis reactions, 113F
  energetics, 97F, 113
  lysozyme, 146–147
  sugars, 53, 53F
Hydronium ion, 49, 49F
Hydrophilic molecule(s), 48, 68F, 367, 369F
Hydrophobic forces, 79F
  lipid bilayers, **367–370**
  macromolecular structure role, 63, 123, 124F

Hydrophobic molecule(s), 48, 68F, 367, 369F
Hydrothermal vents, evolutionary role, 490–491
Hydroxyl groups, 50, 67F
Hydroxyl ion, 49F, 50
Hypodermis, mammalian skin, 720F

I

Immortal cell lines, 326–327, 360
Immune response, phagocytosis, 524, 524F
Immunoglobulins see Antibodies
Immunoprecipitation, 165F, 561
In situ hybridization, 178, 178F, **340–341,** 340F
In vitro, definition, 325
In vitro motility assays, 587, 587F
In vitro mutagenesis, 355–356, 356F
In vivo, definition, 325
Inflammatory response, glycocalyx role, 382, 382F
Influenza virus, viral coat, 222F
Inheritance, **672–686,** 693
    see also Gene(s); Genetic disease(s); Genetic traits
    discrete (particulate) nature, **674–675,** 674F
    dominant alleles, 675, **681, 684**
    early misconceptions, 659
        disproving, 674
    F_1 and F_2 generations, 674, 674F
    heterozygotes, 675
    homozygotes, 675
    meiosis role, **678–680,** 679F, 680F
    Mendel's law of independent assortment, **677–678,** 680–681, 681F
    Mendel's law of segregation, **675–677,** 676F, 677F
    pedigrees, 677, 677F
    recessive alleles, 675, **681, 684**
Inhibitory neurotransmission, 419, 420F, 422
Initiator tRNA, 253, 253F
Inositol phospholipids, 374, 552
    signaling pathways, **552–554,** 553T, 554F
Inositol triphosphate (IP_3), 553
    calcium signaling, 553–554, 554F, 568
    G protein-mediated production, 549, 568
Insulin, 125, 127F
Integrins
    cell crawling, 597–598
    cell division, 654
    cell signaling, 706
    connective tissue role, **706,** 707F
    epithelial tissue role, 714
Interference-contrast microscopy, 8F
Interferon, signaling pathway, 564
Intermediate filaments, 22, 23F, 25F, 573–574, **574–578,** 574F, 606
    accessory proteins, 577–578, 578F
    cell strengthening, **576–578,** 577F
    cellular location, 575, 575F, 713
    classes, 576, 577F
    nuclear lamina, 575, **578,** 578F
    structure, 141, **575,** 576F, 713
    tensile strength, 574, 575
Interphase chromosome(s), 181–182, 182F
    chromatin structure, **187–189**
    nuclear organization, **183,** 183F
Intestine
    epithelium see Gut epithelium
    maintenance/renewal, 721, 722F
Intracellular signaling pathways, 537, **538–540**
    see also GTP-binding proteins (GTPases);

Protein phosphorylation; specific pathways
    adaptation, 556
    evolution, **566–567**
    experimental analysis, **561–563,** 562F, 563F
    inositol phospholipids, **552–554,** 553T, 554F
    modulation/control, 540
    molecular switches, **545–546,** 545F
    phosphorylation cascades, 545
        complexity, **565–566,** 566F
        integration, 566, 567F
        MAP kinase cascade, 559–560, 560F
    signal amplification, 540, 556, 557F
    signal distribution, 540
    signal relay, 539
    signal transduction, 539
    signaling cascades, 539–540
    time-scale, 552, 552F
Intracellular transport, **497–532**
    see also Cytoskeleton; Motor proteins; Organelle(s); Protein sorting; individual compartments
    across membranes see Membrane transport
    experimental analysis, **520–521**
        motor proteins, **586–588**
        video microscopy, 520–521, 521F, 586, 586F, 587–588
        in vitro assays, 520, 520F, 587, 587F
        yeast models, 520, 521F
    microtubules see Microtubule-associated transport
    organellar protein import, **502–512**
    of organelles, 499, 583–584, 584F, **585, 589,** 589F
    vesicular see Vesicular transport
Introns, 238, 238F, 262
    evolution, 242–243
    gene duplication role, 301, 319
    removal from transcript see Splicing
    splice sites, 239, 239F
Ion(s)
    see also Electron(s); individual ions
    acids/bases, **49–50,** 49F
    definitions, 44, 408
    extracellular vs. intracellular concentration, **390–391,** 390F, 390T, 400
    inorganic, 390
Ion channel(s), 389, 392F, **403–410,** 420T, 422, 542, 543, 543F
    see also specific channels
    gating, **407,** 407F, 408F
        G protein activated, **548–549,** 549F
        ligand-gated see Ligand-gated ion channels
        mechanically-gated, 407, 407F, 408F
        voltage-gated see Voltage-gated ion channels
    membrane potential role, 405, **408–410**
        see also Action potential; Membrane potential
    open vs. closed conformation, 404F, **405–407**
    patch-clamp recordings, 405–406, 406F
    selectivity, 403–404, 404F, 407
    transport rate, 404
Ion-exchange chromatography, protein analysis, 162F
Ionic bond(s), **43–45,** 79F
    aqueous solutions, 79F
    bond length, 46T
    bond strength, 46T
    electron donation, 42–43, 43F

    importance, 79F
    macromolecular interactions, 145F
    macromolecular structure role, 62, 122, 123F
    sodium chloride, 44, 44F, 68F
Iron, receptor-mediated endocytosis, 526
Iron–sulfur centers, 473, 473F
Isoelectric focusing, protein separation, 163F
Isoleucine, structure, 75F
Isomers
    amino acids, 56, 74F
    sugars, 52, 70F
Isoprene, 73F
Isotopes, 40

J

Jasplakinolides, actin effects, 594
"Junk" DNA, 179, **308–309**

K

Keratan sulfate, 708F
Keratin(s), 577
    α-helix, 126
    desmosomes, 714, 715F
    intermediate filaments, 141, 576, 577, 713
α-Ketoglutarate, 441
Ketones, structure, 67F
Khorana, Gobind, 246
Kinesin(s), 585, 585F, 606–607
    directional movement, 156, 588F
    experimental analysis, 587–588, 588F
Kinetochores, 645, 645F, 650, 650F
    meiotic chromosomes, 668, 668F
    microtubules, 648, 648F
Knockout animals, 357–358, 357F, 735
Krebs cycle see Citric acid cycle
Krebs, Hans, **442–443**
Krüppel, Drosophila anterior–posterior polarity, 283, 283F

L

L1 (LINE) elements, 219, 220F, 313F
lac operon, **275,** 275F
Lactate dehydrogenase, structure, 137F
Lactate, fermentation, 434F
Lamellipodia, 596, 597F
Leaf, 700F, 701F
Lectins, 381–382
Leder, Philip, 247
Leucine, structure, 75F
Leucine zipper motif, 272F
Leucocyte(s)
    hemopoietic stem cells, 722, 724F
    phagocytosis, 524, 524F
Life, 1, 3
    chemistry, 39
    diversity, 1
    domains, 310, 310F
    origins see Origin of life
    viruses, living or non-living?, 4
Ligand, definition, 143
Ligand-gated ion channels, 407, 407F, 542, 543, 543F, **544–545**
    drug targets, **419–420**
    neurotransmitter receptors, **417–418,** 418, 418F, 544
Light-driven pumps, active transport, 395, 396F, 402
Light microscopy, 6, 8FF, 11F
    see also individual techniques
    resolution, 10, 10F

Page numbers in **boldface** refer to a major text discussion of the entry; page numbers with an F refer to a figure, with an FF to figures that follow consecutively; page numbers with a T refer to a table.

sample preparation, 8F, 10
Lignin, 702
Linkage, **680–681**, 681F
Lipid(s), 54, 73F
  see also specific types
  aggregates, 73F
  cell membranes *see* Lipid bilayer; Membrane
      lipids
  polyisoprenoids, 73F
  steroids, 73F
  triacylglycerols, 54, 54F, 72F, 369F
  water interactions, **367–370**, 369F
Lipid bilayer, 55, 55F, **366–373**, 367F, 386
  see also Fatty acid(s); Glycolipid(s);
          Membrane lipids; Phospholipid(s)
  asymmetry, **373–374**, 373F, 374F, 386
  electron microscopy, 367F
  flexibility, 370
  fluid-mosaic model, 384
  fluidity, **370–371**
    composition effects, **371–373**
    flip-flop, 371, 372F
    importance, 372–373
    lateral diffusion, 371, 372F
    measurement, **384–385**
    rotation, 371, 372F
  formation, **367–370**, 369F
    amphipathic molecules, 367, 367F,
        368F, 369, 369F, 370F, 386
    energetics, 369, 370F
    flippases, 373, 373F
  membrane protein associations, **375–376**,
      375F
  membrane protein removal, **377**
  permeability, 391, 391F, 422
  self-sealing, 370, 370F
  synthetic, 370–371, 371F
Liposome(s), 370–371, 371F
Liver, metastases, 727F
Local mediators, 534, 535F, 537T
Locus, definition, 685F
Loligo (squid), 412F
  giant axons *see under* Axons
Low-density lipoprotein (LDL), receptor-
          mediated endocytosis, 525–526,
          526F
Lymphocytes, 718, 723F
  hemopoietic stem cells, 722, 724F
Lysine, structure, 74F
Lysosome(s), 20, 25F, 256
  electron microscopy, 12F
  evolution, 500–501, 501F
  hydrolytic enzymes, 527
    see also Lysozyme
  Golgi export, 527–528
  pathways to, 528, 528F
  proton pumps, 402, 527, 527F
Lysozyme
  catalytic reaction, **146–148**
    mechanism, 148, 149F
    transition state, 147
  structure, 127F, 146, 148F
    active site, 147, 148, 149F
    disulfide bonds, 142–143
    substrate binding, 147–148, 148F

M

M-phase *see under* Cell cycle
MacLeod, Colin, 172, 174F
Macromolecular complexes, 63–64, 64F
Macromolecular interactions, 64F
  enzyme interactions *see* Enzyme–substrate
      interactions

  noncovalent interactions, 63–64, 63F, 98,
      101
  nucleic acid hybridization, **336–341**
  protein interactions *see* Protein interactions
Macromolecular structure
  experimental analysis, **60–61**
    ultracentrifugation, 60–61, 61F
    X ray diffraction *see* X-ray diffraction
  noncovalent interactions, 62–63, 65
    hydrogen bonds, 48, 62, 78F
    hydrophobic interactions, 63
    ionic bonds, 62
    van der Waals attractions, 62–63
  stable conformations, 62, 62F
Macromolecules, **58–64**, 65
  see also individual macromolecules
  abundance in cells, 58F
  biosynthesis, 58, 59, 59F, 62
    energetics, **112–114**
  breakdown *see* Catabolism
  diffusion through cytoplasm, 101
  experimental analysis, **60–61**, 61F
  functions, 59
  interactions *see* Macromolecular interactions
  macromolecular complexes, 63–64, 64F
  specific sequences, 59, 62
  structure *see* Macromolecular structure
Macrophage(s), 718
  phagocytosis, 524, 524F
Magnesium ions, intracellular *vs.* extracellular
          concentration, 390T
Malaria, resistance, 686, 691
Malonate, 442–443, 442F, 443F
Mannose, glucose isomer, 52, 70F
MAP kinase (MAPK), 559
MAP kinase cascade, 559–560
  plant signaling, 567
  Ras activation, 559, 560F
MAP kinase-kinase (MAPKK), 559
MAP kinase-kinase-kinase (MAPKKK), 559
Mass spectrometry, protein identification, 129,
          130F
Matrix proteases, 705
Matthaei, Heinrich, 246
Maturation promoting factor (MPF, M-Cdk),
          618
McCarty, Maclyn, 172, 174F
Mechanically-gated ion channels, 407, 407F,
          408F
Meiosis, **655**, 662F, **663–672**, 693
  see also Gametes; Sexual reproduction
  errors, **670–671**, 671F
  generation of genetic diversity, 667,
      668–670, 670F
  homologous chromosomes, 665
    see also Homologous recombination
    nondisjunction, 670–671, 671F
    pairing, **664–665**, 666F
    recombination, 216, **665–667**, 666F
    segregation, **667**, 667F
  meiosis I, **664–667**, 669F
  meiosis II, 664, **667–668**, 668F, 669F
  mitosis *vs.*, 665, 669F
  role in Mendelian inheritance, **678–680**,
      679F, 680F
Membrane(s), 10
  artificial, 370, 371F, 385, 385F
  biosynthesis, **373–374**, 374F
  dynamics, 365, **370–371**, 372F
    composition effects, **371–373**
    importance, 372–373
    measurement, **384–385**
    protein movements, **383**, 383F, **386**
  electrical potential *see* Membrane potential

  functions, 365, 366F, 386
    see also Action potential; Cell signaling;
        Electron-transport chain(s);
        Organelle(s)
  energy metabolism, **453–455**
  intracellular compartmentalization, 11,
      19, 365, 366F
  selective barriers, 365, 366F, 386
  permeability, 391, 391F, 422
  plasma/cell *see* Plasma membrane
  structure *see* Membrane structure
  transport across *see* Membrane transport
Membrane domains, 383, 386, 386F, 398
Membrane lipids, 54–55, 55F, 73F, 386
  see also Fatty acid(s); Glycolipid(s); Lipid
      bilayer; Phospholipid(s)
  asymmetry, **373–374**, 373F, 386
  biosynthesis, 373–374
  fluidity, **371–373**, 372F
    cholesterol, 372, 372F
    movements, 371, 372F
    saturated cf unsaturated, 372
  water interactions, **367–370**, 367F, 368F,
      369F
Membrane potential, 395, 405, **408–410**,
          409F, 422
  see also Action potential; Electrochemical
      gradients; Ion channel(s)
  electrochemical gradients, 395, 395F
    protons, 463–464, 464F
  ion channels and, 405, **407–408**, 409,
      410, 410F
  Nernst equation, 410, 410F
  passive spread, 411–412
  patch-clamp recording, 405–406, 406F
  resting potential, 409–410
Membrane proteins, **374–386**
  see also specific proteins
  functions, 374, 374F, 375T, 386
    see also Cell signaling; Membrane
        transport; Receptor(s)
  cell cortex attachment, 380–381, 381F,
      387
  glycoproteins, 381, 382F
  integral *vs.* peripheral, 375
  lipid bilayer associations, **375–376**, 375F,
      386–387
  movement, **383**, 385F, **386**
    restriction, 383, 383F, 386, 386F, 398
  orientation, 376
  proteoglycans, 381, 382F
  solubilization in detergent, **377**, 378F, 385,
      385F
  structure determination, 377, **378–379**
  transmembrane *see* Transmembrane proteins
Membrane structure, **365–388**
  asymmetry, **373–374**, 373F
  commonalities, 366
  flexibility, 370
  lipid bilayer *see* Lipid bilayer
  lipids *see* Membrane lipids
  membrane domains, 383, 383F, 386, 386F
  proteins *see* Membrane proteins
  surface carbohydrates (glycocalyx),
      381–382, 382F
    see also Glycolipid(s); Glycoproteins;
        Proteoglycans
Membrane transport, **389–426**
  see also Electrochemical gradients;
          Membrane potential;
          Transport/translocator proteins
  active *vs.* passive transport, **392**, 393F
    see also Active transport; Passive
        transport

carrier proteins *see* Carrier proteins
channels, 389, 392, 392F, 422
  ion channels *see* Ion channel(s)
  large pores, 127F, 377, 377F
organellar protein import, 503, 503F
principles, **389–393**
  ion balance, 390–391, 390F, 390T
  membrane permeability, 391, 391F
specificity, 391
Mendel, Gregor, 672
  dihybrid crosses, 677–678, 679F
  discrete nature of heredity, **674–675**, 674F
  discrete traits (peas), **673**, 673F
  disproving of previous theories, 674
  laws of inheritance, 693
    see also Inheritance
    first (law of segregation), **675–677**, 676F, 677F
    meiosis and, **678–680**, 679F, 680F
    second (law of independent assortment), **677–678**
Meristems, 700F
Messenger RNA (mRNA), 233
  cDNA synthesis, 346, 347F
  degradation, **242**
  eucaryotic, 262
    processing *see* Messenger RNA processing
    stability, 237
  procaryotic, polycistronic transcripts, 254, 254F
  in situ hybridization, 340–341, 340F
  synthetic, cracking the genetic code, 246–247, 247F
  translation *see* Translation
Messenger RNA processing, **236–237**, 262
  3' poly A tail, 237, 237F
  5' cap, 236–237, 237F
  capping, 236–237, 237F
  nuclear export, 236, 236F, **241–242**, 242F
  splicing *see* Splicing
Metabolism, 3, 83, 85F
  see also Energy; Enzyme(s); Thermodynamics; *specific pathways*
  activated carriers *see* Activated carrier molecules
  anabolism *see* Biosynthesis
  catabolism *see* Catabolism
  complex networks, 84F, 151, 448F
  energy production, **427–444**
  energy storage, **444–449**
  enzyme-catalysis, 84F
  feedback inhibition, 151–153, 151F, 152F
  procaryotic cells, 14–15
  regulation, **448–449**, 448F
  universal, 3–4
Metamorphosis, apoptosis, 625, 626F
Metaphase
  meiosis, 665
  mitosis, 640, 643F, **648–649**, 649F
Metaphase plate, 648, 649F
Metastases, 727, 727F, 730
Methane, 66F
Methanococcus, chemiosmotic coupling, **490–492**, 491F
Methionine, 75F
Methyl groups, 50, 66F, 111
Mice
  human genome comparison, 308, 308F, 317F
  as model organisms, 33
  transgenic, 357, 358F, 359F
  see also Knockout animals

Micelles, 73F, 377, 385, 385F
Microelectrodes, 405–406, 406F
Microsatellite DNA, DNA fingerprinting, 351F
Microscopy, **5–11**, 8FF
  see also individual techniques
  historical aspects, 6–7, 7F
  resolution, 7, 10–11, 10F
  sample preparation, 8F, 10
  single-particle tracking (SPT), 385, 385F
Microtubule(s), 22, 23F, 25F, 573, **579–591**, 606–607
  accessory proteins, 584, 644
  aster, 644
  cell division, 22, 24F, 579, 582, 639, 639F
    see also Mitotic spindle
    antimitotic drugs, 582
    instability role, **641, 644**, 644F
    phragmoplast, 654, 655F
  centrioles, 580
  centrosome, 579, **580–581**
  cilia/flagella, 579, **590–591**, 591F
  dynamic instability, **581–582**, 581F, 641, 644
    catastrophins and, 644, 646
  experimental analysis, **586–588**
    cell division role, 646–647, 646F, 647F
    squid giant axon, 586, 586F
    video microscopy, 586
  growth, 579–580
    GTP role, 581–582, 582F
    nucleation sites, 580, 581, 581F
  intracellular transport role *see* Microtubule-associated transport
  plant cell wall synthesis, 702–703, 703F
  remodeling, **582–583**
  selective stabilization, 583, 583F
  structure, **579–580**, 580F
    GTP caps, 582
    polarity, 579
    tubulins, 579, 580, 581
Microtubule-associated proteins (MAPs), 584, 644, 647, 647F
Microtubule-associated transport, **583–584**
  axonal, 584F
  motor proteins, **584–585**, 585F
    see also Dynein(s); Kinesin(s)
    experimental analysis, **586–588**, 586F, 587F, 588F
  organelles, 583, 584F, **585, 589**
Microtubule organizing centers (MTOCs), 579, 579F, 580, **580–581**, 640
  see also Centrosome
Microvilli, 592
Mimosa, leaf closing response, 408, 409F
Mismatch repair, **210–212**
Mitchell, Peter, chemiosmotic hypothesis, 460
Mitochondria, **17**, 17F, 455, 492, 499
  cell division and growth, 487, 487F, 502, 652
  cellular localization, 455, 456F
  coupled active transport, **466**, 466F
  disorders, 455
  electron microscopy, 12F, 18F
  evolution, 17, 18F, **487–492**, 501, 501F
  genome, 17, 455, 487
  metabolism, 17, 429–430, 458F
    see also Citric acid cycle; Oxidative phosphorylation
    acetyl CoA production, **435**, 438F, **439**, 439F
    ATP/ADP ratio, **468**
    ATP production, 17, **441, 444**

protein import, **506–507**, 507F, 528–529
structure, 17, 18F, **455–456**, 506
  chloroplasts cf., 480F
  cristae, 456
  inner membrane, 17, 18F, 441, 456, 457F
  intermembrane space, 456, 457F
  matrix, 456, 457F
  outer membrane, 456, 457F
Mitogen-activated protein kinase(s) *see* MAP kinase cascade
Mitogens, **629–631**, 630F
Mitosis, 613, 637, **641–652**, 642FF, 656
  see also Cytokinesis
  anaphase, 640, 643F, **649–651**
    anaphase A, 649–650, 650F
    anaphase B, 650–651, 650F
    APC role, 649, 651F
  cell cycle checkpoint, 624
  centrosome cycle, **640**, 640F
  chromosomes *see* Mitotic chromosomes
  experimental analysis, **646–647**
  meiosis *vs.*, 665, 669F
  metaphase, 640, 643F, **648–649**, 649F
  microtubules role, 22, 24F, 579, 639, 639F, 656
    antimitotic drugs, 582
    instability role, **641, 644**, 644F
    MAPs, 644
    spindle formation *see* Mitotic spindle
  nuclear envelope changes, 645, 648, **651**, 651F
  organelle distribution, 502, **651–652**
  prometaphase, 640, 642F, **645**, 645F, **648**
  prophase, 640, 642F, **644**
  telophase, 640–641, 643F, **651**, 651F
Mitotic chromosomes
  chromatin structure, 187F
  condensation, 17F, 170F, 182, 182F, 614, **638–639**, 639F
  cohesins, 638, 639, 641, 649
  condensins, 638, 639
  daughter chromosomes, 641, 641F
  metaphase plate formation, 648, 649F
  segregation, **649–651**, 649F
  sister chromatids, 638, 639F, 641, 641F
    separation, **641**, 641F, **644**
  spindle attachment via kinetochores, **645**, 645F, **648**
Mitotic spindle
  antimitotic drugs and, 582
  chromosome attachment, **645**, 645F, **648**
  cleavage plane determination, **652–653**, 652F
  formation, 641, **644**
    catastrophins, 644, 647, 647F
    centrosome duplication, **640**, 640F
    MAPs, 644, 647, 647F
    motor proteins, 644, 645F
  historical research, **646–647**
  microtubules, 22, 24F, 579, 582, 639, 639F, 641
    see also Microtubule(s)
    aster, 644F, 648F
    instability, **641, 644**, 644F
    interpolar, 644F, 648F
    kinetochore, 648, 648F
  spindle poles, 644, 644F
Mobile genetic elements, 217, 224
  see also Repetitive DNA; *specific elements*
  bacterial, 218, 218F
  evolution role, 219, **301–302**, 302F
  human *see under* Human genome

Page numbers in **boldface** refer to a major text discussion of the entry; page numbers with an F refer to a figure, with an FF to figures that follow consecutively; page numbers with a T refer to a table.

insertional mutagenesis, 302, 302F
retrotransposons, 219, 219F
site-specific recombination, 217
transposition mechanisms, **217–218,** 218F,
  219F
viruses see Viruses
Model organisms, **27–35,** 36
  see also individual organisms
  classical genetics, 687
  DNA analysis, 327
Molarity, definition, 41, 41F
Molecular chaperones see Chaperones
Molecular motor, ATP synthase as, 465
Molecular switches, 154, 155F, **545–546,**
  545F
Molecular tags, antibodies, 165F
Molecular weight, 40
Molecule(s), 39
  amphipathic see Amphipathic molecules
  cellular, **50–58**
  chemical bonds see Chemical bonds
  hydrophilic, 48, 68F, 367, 369F
  hydrophobic, 48, 68F, 367, 369F
  large see Macromolecules
  organic vs. inorganic, 50
  polar, 49–50, 49F
  small organic see Organic molecules (small
    cellular)
Monoclonal antibodies, 165F
Monosaccharides, 52, 70F, 71F
Morgan, Thomas Hunt, 682
Motor neurons, synaptic connections, 421,
  421F
Motor proteins, 120F, 155
  see also individual proteins
  allostery, 157F
  centrosome, 644, 645F
  conformational changes, 156, 156F, 157F,
    604F
  experimental analysis, **586–588,** 586F,
    587F
  functions, 155–156
  mechanisms, 585, 585F
  microtubule-associated transport, **584–585,**
    606–607
  nucleotide hydrolysis, **155–156,** 157F, 584,
    585
mRNA, see Messenger RNA
Multicellular organisms, 3, 397, 698F
  see also Animal(s); Eucaryotic cell(s);
    Plant(s)
  cellular specialization see Cell differentiation
  development see Development
  evolution, animal vs. plant, 698
  gene expression control, 268
  generating genetic diversity, 293–294
  genomes, 268, 717, 718F
  size control, **628–633**
    see also Cell death; Cell division; Cell
      growth
  tissues see Tissue(s)
Multisubunit proteins, **139–140**
Muscle, 600, 698F
  see also specific types
  cells see Muscle cell(s) (myocytes)
  contraction see Muscle contraction
  myoblast differentiation, 285–286
  structure, 601–602, 607
    myofibrils, 601, 602F
    sarcomeres, 601–602, 602F
    T tubules, 603, 605F
    Z discs, 602, 602F
Muscle cell(s) (myocytes)
  anaerobic metabolism, 434

differentiation, 285–286, 286F
  mitochondria, 456, 456F
  specialization, **605–606**
Muscle contraction, **600–606**
  actin–myosin interactions, **600–601,** 607
    myosin structural changes, 604F
  calcium role, **603–604,** 605F
  sliding-filament mechanism, **601–603,**
    601F, 603F
  ATP hydrolysis, 602
Mutagenesis
  gene identification, **686–687**
  signaling pathway analysis, 561–562, 562F,
    563F
  site-directed, 355–356, 356F
Mutagens, 687, 728
Mutant, definition, 685F
Mutation(s), 5, 211F, 213F, 295, 687F
  see also DNA damage; Genetic variation;
    Genetics (classical)
  beneficial, 5, 304
  conditional, temperature-sensitive, 30, 30F,
    31F, 684F, 687–688, 688F
  consequences, 5, 195, **209–210,** 319
    see also Cancer; Evolution; Genetic
      disease(s); Natural selection
  cancer, 210, 210F, 684, **728–729,** 738
  cell cycle effects, 30
  genetic disease, 209–210, 209F
  germ-line vs. somatic, 294, 294F
  gain-of-function, 684
  gene function analysis, 355–356, 356F
  germ line, 210, 294, 294F
  harmful (deleterious), 5, 196, 304
    positive selection, 684, 686
  loss-of-function, 684
  mobile genetic elements and, 302
  neutral (silent), 5, 297, 304, 305, 690
  point mutations, 196, 295, 295F,
    296–297, 304, 687F
  rate, 296–297, 728
  recessive, 337
  selection, 5, 214, 304–305, **684, 686**
  somatic, 210, 294, 294F, 728
Mycobacterium tuberculosis, immune evasion,
  524
Mycoplasma genitalium, genome size, 34
Myoblasts, 285–286
  cell culture, 327F
Myoclonic epilepsy and ragged red fibers
  (MERRF), 455
Myocytes see Muscle cell(s) (myocytes)
MyoD gene, 286
MyoD gene regulatory protein, 286, 287
Myofibrils, 601, 602F
Myoglobin, structure, 127F
Myosin(s), 598, 607
  actin association, **598–599,** 599F,
    653–654
    see also Actin filaments
  ATP hydrolysis, 598
  directional movement, 156
  muscle contraction see Muscle contraction
  myosin-I subfamily, 599
  myosin-II subfamily, 599, 600–601, 601F
Myosin filaments, 601, 601F, 602
Myostatin, 632
  mutations, 632–633, 633 F

N-acetylglucosamine, 71F
N-linked oligosaccharide(s), protein
  glycosylation, 517

Na^+-H^+ exchanger, 399, 403T
Na^+-K^+ pump, 403T
  dam analogy, 397, 397F
  functions, 396, 396F
    coupled transport, 397
    osmotic balance, **399–401**
  inhibition by oubain, 397, 400
  mechanism, 398F
    see also Active transport
  ATP hydrolysis, 396–397
  phosphorylation role, **397**
NADH dehydrogenase, 462, 462F
  proton pump, 476
NADH/NAD$^+$
  see also Electron-transport chain(s); specific
    pathways
  citric acid cycle, 429F, 430, 439–441
  electron carriers, **109–111,** 110F, 458,
    458F
  glycolysis, 428, 430, 433F, 436F, 437F
  oxidative phosphorylation, 444F, 449
NADPH/NADP$^+$
  cholesterol biosynthesis, 111F
  electron carriers, **109–111,** 110F
  phosphate role, 110–111
  photosynthesis (light reactions), **482–485**
Natural selection
  see also Evolution
  cancer cells, 729–730, 729F
  genetic variation, **304–305**
    haplotype analysis, 690–691, 690F,
      691F
    mutations and, 5, 214, **684, 686**
  positive selection, 684, 686, 690–691
  purifying selection, 308
Nernst equation, 410, 410F
Nerve cell(s) see Neuron(s)
Nerve impulse see Action potential
Nerve terminal, 411
Neural networks, 420
Neural tube formation, 714F
Neuraminidase, multiple subunits, 139F, 140
Neurofilaments, 576
Neuromuscular junction, 418, 419F, 603
Neuron(s)
  see also Axons; Neurotransmitters; Synapses
  action potentials see Action potential
  cell signaling, 535, 535F
  connectivity, 420, 420F
  development
    cell death, 625
    Delta–Notch signaling, 536F
  DNA content, 268
  excitation/inhibition, **419,** 420F
  intracellular transport, 583–584, 584F
  size, 631, 631F
  structure, 2, 2F, 411, 411F
  survival factors, 631, 632F
Neurotransmission, 535, 535F
Neurotransmitters, 417, 535, 537T
  see also Synapses; individual
    neurotransmitters
  calcium-mediated release, **417,** 418F, 422
  excitatory vs. inhibitory, 419, 420F, 422
  receptors, **417–418,** 418F, 544
  drug targets, **419–420**
Neutral (silent) mutations, 5, 297, 304, 305,
  690
Neutrons, 40
Neutrophils, cell surface carbohydrate
  recognition, 382, 382F
Neutrophils(s), structure, 2F
Nicotinamide adenine dinucleotide (and
  reduced form) see NADH/NAD$^+$

Nicotinamide adenine dinucleotide phosphate (and reduced form) *see* NADPH/NADP⁺

Nirenberg, Marshall, 246, 247

Nitric oxide (NO), **540–541,** 540F, 541F

Nitrogen
carbon compounds, 67F
covalent bond geometry, 46F
fixation by *Methanococcus,* 491
NMR spectroscopy, 130, 131F

Noncoding DNA, 35, 179
see also Mobile genetic elements; Repetitive DNA
introns *see* Introns
"junk DNA," 179, **308–309**
mutation accumulation, 304

Noncovalent bonds, 45, 46T, 78FF
see also specific types
hydrophobic interactions, 63, 79F
macromolecular interactions, 63–64, 63F, 64F, 101
energetics, 98, 99F
enzyme–substrate, 63, 98, 101, 147–148
protein–ligand, 144, 145F
protein–protein, 48F
macromolecular structure, 48, 62–63, 65
protein folding, 121–123, 123F
weak interactions, 78F

Northern blotting, 338F

Notch, 536F

Nuclear envelope, 16–17, 16F, 20F, 21F, 25F, 498
evolution, 501F
formation, 502
interphase chromosomes, 183, 183F
lamina *see* Nuclear lamina
membranes, 504–505, 505F
mitotic changes, 645, 648, **651,** 651F
pores *see* Nuclear pore(s)

Nuclear lamina, 504–505
intermediate filaments, 575, **578,** 578F
interphase chromosomes, 183, 183F

Nuclear lamins, 576, 578

Nuclear localization signals, 504T, 506

Nuclear magnetic resonance (NMR) spectroscopy, 130, 131F

Nuclear pore(s), 25F, 241, 498, 503, 503F, 505
see also Nuclear–cytoplasmic transport
interphase chromosomes, 183, 183F
protein import, 505–506, 506F

Nuclear pore complex (NPC), 241, 505–506, 505F

Nuclear transplantation
embryonic stem cells, **725–726,** 738
gene expression studies, 268, 269F

Nuclear transport receptors, 242F, 506

Nuclear–cytoplasmic transport, 503, 503F, 528
see also Nuclear pore(s)
protein import, **504–506,** 505F, 506F
localization signals, 506
RNA export, 236, 236F, **241–242,** 242F

Nucleic acid(s), 59
see also Nucleotide(s)
base-pairing *see* Base-pairing
biosynthesis, 57–58, 77F, 113–114, 113F, 115F
DNA *see* DNA (deoxyribonucleic acid)
genetic information, 58
hybridization, **336–341**
phosphodiester bonds, 58F, 77F

RNA *see* RNA (ribonucleic acid)

Nucleolus, 25F, 183, 183F

Nucleoside(s), definition, 56, 77F

Nucleosome(s), **184–186,** 191
30-nm fiber, 184, 186–187, 186F, 187F
"beads on a string," 184F, 185, 186
core particle, 185, 185F
electron microscopy, 184F
gene expression and, 276, 279
histones *see* Histone(s)
linker DNA, 185
remodeling, **189–190**

Nucleotide(s), 51, 51F, **56–58,** 65, 76FF
see also individual molecules
base-pairing, 58, 171, 171F, 175, 175F
coenzymes, 77F
definition, 56, 77F
as energy carriers, 56–57, 57F, 77F
nomenclature, 56, 77F
nucleic acid formation, 3, 57–58, 171, 171F
sequences, 176, 177F
signaling molecules, 77F
structure, 76F, 171
bases *see* Base(s) (nucleic acid)
phosphates, 76F, 171
sugar, 56, 76F, 171

Nucleus, 10, **16–17,** 25F, 36
see also Chromosome(s); Genome(s); Nuclear–cytoplasmic transport
electron microscopy, 12F, 16F, 20F, 498F
evolution, 501F
interphase chromosomes, **183,** 183F
RNA export, 236, 236F, **241–242,** 242F
RNA processing, 236–237
structure, 498
see also Nuclear envelope; Nuclear pore(s)

Nucleus (atomic), 40, 40F

O

Obesity, cancer link, 728

Occludins, 712

Okazaki fragments, 203, 206F, 207F

Oleic acid, 72F

Oligodendrocyte precursors, cell culture, 327F

Oligonucleotides, 336–337, 348, 356F

Oligosaccharides, 52, 71F
cell surface, 381
ER processing, 517

Oncogene(s), 560, 731, 731F, 732, 732F, 738
see also specific genes

Oocytes *see* Egg cell(s)

Open reading frames (ORFs), gene identification, 314–315, 314F, 333

Operators, 273

Operons, 273, 285
lactose *(lac),* **275**
tryptophan, 273–274, 273F

Optical isomers, amino acids, 56, 74F

Organelle(s), 10–11, **19–21,** 36, **498–501,** 528
see also specific organelles
cell division, 502, **651–652**
cellular distribution, 21F, 499F
cytoskeleton interaction, 499
microtuble-associated transport, 583–584, **585, 589,** 589F
differential centrifugation, 499–500
electron microscopy, 11, 12F, 498F

evolution, 17, 18F, 20F, **500–501,** 501F
functions, 498–499, 499T
genomes, 17, 19, 455
membranes, 11, 19, 365, 366F, 498, 498F
protein import, 502–503, 503F, 528
see also Protein sorting
endoplasmic reticulum, **507–509,** 510F
mitochondria/chloroplasts, **506–507**
nuclear, **504–506,** 505F, 506F
signal sequences *see* Signal sequences
relative volume, 499, 500T
vesicular transport between *see* Vesicular transport

Organic chemistry, 39

Organic molecules (small cellular), **51–52,** 51F
see also specific molecules
diffusion through cytoplasm, 101
energy release (oxidation), 89
energy storage, 88
polymerization, 58, 59, 59F
see also Macromolecules
synthesis, 88

Organs, size control, 628

Origin of life, 490F
aerobic respiration, **488,** 488F
chemiosmotic coupling, **490–492**
common evolutionary ancestor, 4–5, 490
photosynthesis role, 478, 489–490
RNA and *see* RNA world theory
tree of life, 310F

Origin recognition complex (ORC), 621, 622F

Osmosis, 399, 400F

Osmotic balance, **399–401,** 699, 737

Osmotic pressure, 399

Osteoblasts, 704, 721

Osteoclasts, 721

Oubain, Na⁺-K⁺ pump inhibition, 397, 400

Ovaries, 664

Oxaloacetate, 440, 440F, 441

Oxidation reactions, 90
see also Activated carrier molecules; Catabolism; Electron-transport chain(s); Respiration (cellular)
cellular energy production, **89,** 106, 427
electron transfer, **90,** 91F

Oxidation-reduction (redox) reactions, **109–111,** 462, 469

Oxidation–reduction (redox) reactions, *seereactions*

Oxidative phosphorylation, 430, 440, 449, **455–468,** 459F
electron sources, 444F, 449, 457–458, 459, 462
evolution, **488,** 488F
proton-driven ATP synthesis, 379, 441, 454, 459, **464–466,** 464F
see also ATP synthase; Chemiosmotic coupling; Electrochemical gradients
respiratory chain *see* Electron-transport chain(s)
yield, 466–467, 467T

Oxygen
abundance, 89
cellular respiration, 17, 89, 430, 435, 441
covalent bond geometry, 46F
production by photosynthesis, 483, 484F, 485
see also Photosynthesis

Oxygen-binding proteins, evolution, 298

Page numbers in **boldface** refer to a major text discussion of the entry; page numbers with an F refer to a figure, with an FF to figures that follow consecutively; page numbers with a T refer to a table.

p21, Cdk inhibition, 623–624
p53
    DNA damage checkpoint, 623–624, 623F
    mutation and cancer, 624, 730, 736, 738
PAGE, see Polyacrylamide gel electrophoresis
Palmitic acid, 53F, 72F
Paracrine cell signaling, 534, 535F
Paramecium, structure, 2–3, 2F
Parasitic DNA see Mobile genetic elements
Parenchyma, 700F
Passive transport, **392**, 393F, 422
    see also Electrochemical gradients;
            Membrane potential
    carrier proteins, **393–395**
    concentration gradients, 394
    electrochemical gradients, 395, 395F
Patch-clamp recordings, 405–406, 406F
Pax-6 and eye development, 288–289
PCR, see Polymerase chain reaction
Pectin, plant cell wall, 699
Pedigrees, 677, 677F
Peptide bonds, 56, 56F, 74F, 121, 121F
    formation, large ribosomal subunit, 249
    polarity, 376, 376F
Peptidyl transferase, 250
Periodic table, 43, 44F
Peroxisomes, 12F, 20, 25F, 499, 502
Pertussis toxin, 547–548
pH, 50, 69F
    gradient due to proton pumping, 463–464,
            464F
Phagocytosis, **523–524**
Phagosomes, 523, 524
Phase-contrast microscopy, 8F
Phenotype, 355, 675, 685F
    dominant vs. recessive alleles, **681, 684**
Phenylalanine, 75F
    as first codon to be decoded, 246, 247F
Phenylthiocarbamide (PTC, phenylthiourea),
            ability to taste, 672, 672F
Phloem, 701F
Phosphate bond(s), 67F, 437F
    see also individual types
Phosphates, 67F, 76F, 171
Phosphatidyl-inositol 3-kinase, 558
Phosphatidylcholine, 72F, 367, 368F
Phosphatidylserine, 368F
Phosphoanhydride bonds, 67F
    adenosine triphosphate (ATP), 57
    energy, 437F
Phosphodiester bonds, 58F, 77F
Phosphoenolpyruvate, 433F
Phosphoester bonds, energy, 437F
Phosphofructokinase, 432F
Phosphoglucose isomerase, 432F
2-Phosphoglycerate, 433F
3-Phosphoglycerate, 433F, 435, 436F
Phosphoglycerate kinase, 433F, 435, 436F
Phosphoglycerate mutase, 433F
Phospholipase C, 549, 552, 553T
Phospholipid(s), 55
    see also specific lipids
    cell membranes, 55F, 73F, 367, 373, 374
    structure, 55, 55F, 72F, 368F
        see also Fatty acid(s)
        amphipathic nature, 367, 368F, 369,
            369F
Phosphoryl groups, 67F
Phosphorylation reactions, 107–108, 108F,
            616
Photobleaching techniques, membrane flow,
            384, 384F

Photosynthesis, 15, **18–19, 88**, 88F, 115,
            446, **478–486**, 481F, 492
    carbon fixation (dark reactions) see Carbon
            fixation
    chemiosmotic coupling, 459
    electron transfer (light reactions), 88,
            480–481, 483F, 484F
        see also Chlorophyll; Photosynthetic
            reaction center(s)
    ATP/NADPH synthesis, **482–485**
    evolutionary role, 478
    feedback regulation, 481
    procaryotic cells, 15, 15F
    evolution, **489–490**
    respiration and, 89, 89F
Photosynthetic reaction center(s), 379, 380F,
            482, 482F
    electron transfer, 383F, 483, 483F, 484F
    evolution in bacteria, 489
Photosystem(s), 481–482, 482F, 484F
    see also Chlorophyll; Chloroplasts;
            Photosynthetic reaction center(s)
    antenna complex, 481
    bacterial evolution, 489–490
    photosystems I and II, 483
    redox potentials, 484F
    water-splitting enzyme, 483, 484F, 485
Phragmoplast, 654, 655F
Phylogenetics, **304–310**
    see also Comparative genome analysis;
            Genome evolution
    distant evolutionary relationships, **309–310**
    fish–mammal divergence, 308–309
    human–chimpanzee divergence, 305, **306**,
            307F
    human–mouse divergence, 308, 308F
    sequence divergence rates, **305–306**
    tree construction, 306, 306F
    tree of life, 310F
PI 3-kinase, see Phosphatidyl-inositol 3-kinase
Pinocytosis, 523, **525**
PKA, see Protein kinase A
PKC, see Protein kinase C
Plant(s)
    Arabidopsis thaliana as model, **28–29**
    cell signaling, 567
    cells see Plant cell(s)
    evolutionary relationships, 29F
    food storage, 446, 446F
    multicellularity evolution, 698
    starch production, **486**
    structure, 700F
    sucrose production, **486**
    tissues see Plant tissue(s)
    transgenic, **359–360**
        production, 359F
Plant cell(s), 10F
    cell division
        cytokinesis, **654–655**, 655F
        mitotic spindle formation, 644
        phragmoplast, 654, 655F
    cytoskeleton, 698
        microtubules and cellulose synthesis,
            702–703, 703F
    energy metabolism, **446–447**, 447F
    growth, 699, 702
    H+ pumps/gradients, 402, 402F
    microscopy, 7F
    osmotic balance, 400–401, 401, 401F,
            699, 737
    photosynthesis see Photosynthesis
    starch granules, 446F
    structure, 2F, 3, 25F, 402F, 479F
        cell wall see Cell wall(s)

chloroplasts see Chloroplasts
    plasmodesmata, 716, 717F
Plant tissue(s), **698–703**, 700FF, 737
    see also Plant cell(s)
    animal tissues vs., 698
    dermal, 700F, 701F
    ground tissue, 700F
    leaf, 700F, 701F
    plasmodesmata, 716, 717F
    roots, 700F
    strength, **698–699**, 699F
        see also Cell wall(s); Cellulose
    cellulose role, **702–703**, 702F, 737
    lignin, 702
    vascular, 700F, 701F
Plasma membrane, 3, 11, 25F, 365
    cell cortex reinforcement, **380–381**, 381F,
            387
    electron microscopy, 12F
    vesicular transport, 20–21, 22F
Plasmids, 198, 343F
    DNA cloning vectors, **342–343**
    replication origin expression, 198–199,
            198F
Plasmodesmata, 716, 717F
Platelet-derived growth factor, 630
Plectin, 577–578
Point mutations, 687F
    evolution role, 295, 295F
    within genes, 196, 295, 295F
    noncoding regions, 304
    rates, 296–297
Poly-A-binding proteins, 242
Polyacrylamide gel electrophoresis (PAGE),
            protein separation, 163F
Polyadenylation, 237, 237F
Polycistronic mRNA, 254, 254F
Polygenic traits, 691–692
Polyisoprenoids, 73F
Polymerase chain reaction (PCR), **348–351**,
            349F, 361
Polymers, 39, 65
Polymorphism see Genetic variation
Polypeptide chains see Protein(s)
Polyribosomes (polysomes), **254–255**, 255F,
            508–509
Polysaccharides, 52, 71F
    biosynthesis, 113–114, 113F
    hydrolysis, 148, 149F
    plant cell walls, 699F, **702–703**, 702F
Porins, structure, 127F
    β barrels, 377, 377F
Position effects, 188, 188F
Posttranslational modification, **516–517**, 517F
    see also specific modifications
Potassium channels
    G protein activation, 549, 549F
    leak channels, 409, 410F
    membrane potential role, 409, 410, 410F
    selectivity, 404F
    voltage-gated, action potentials, 413, 416
Potassium ions (K+)
    electrochemical gradient, 395
    intracellular vs. extracellular concentration,
            390, 390T, 410
    ion channels see Potassium channels
    Na+-K+ pump see Na+-K+ pump
Potato virus X, viral coat, 222F
Precursor cells, 721
Premature aging, 359F
Primate genome(s), evolution, 305, **306**, 307F
Primosome, 207F
Prion proteins, "infectious misfolding," 125,
            126F

Procaryotic cells, **11–16**, 35–36
  diversity, 14–15
  domains, 15–16
    archaea see Archaea (archaebacteria)
    bacteria see Bacteria (eubacteria)
  eucaryotes *vs.*, 11, 24
  evolution, 14, 310, 310F
  food sources, 14–15
  genome sizes, 34, 34F
  habitats, 14, 15–16
  photosynthesis, 15, 15F
  ribosomes, 249
Procaspases, 627, 627F
Procollagen, 705
Profilin, 594
Programmed cell death see Apoptosis
Proline, 75F
Prometaphase, mitosis, 640, 642F, **645**, 645F, **648**
Promoter elements, 234, 271
  see also Gene regulatory sequences; RNA polymerase(s)
  asymmetry, 235–236
  bacterial, 235F, 273
  eucaryotic, 277
  RNA polymerase binding, 234–235, 277, 289
Propane, covalent bond geometry, 46F
Prophase
  meiosis, 665, 666
  mitosis, 640, 642F, **644**
Proteases, 256
Proteasome(s), 256–257, 257F
Protein(s), 59, **119–168**
  see also specific types
  aggregation see Protein aggregates
  assemblies see Protein complexes (assemblies)
  cofactors, **149–150**
  denaturation, 124
  evolution, 138, 157
    see also Genome evolution
    domains, 241, 300–301
    protein families, **297–298**
  experimental analysis see Protein analysis
  functions, 119, 120F
    analysis, 354–355
  historical landmarks, 158T
  housekeeping, 268
  interactions see Protein interactions
  isoelectric point, 163F
  membrane see Membrane proteins
  motor see Motor proteins
  multisubunit, **139–140**
  posttranslational modification, **516–517**, 517F
    see also specific modifications
  regulation, **150–158**, 159
    allosteric, **151–153**
    conformational changes, 152, 153F
    degradation, **256–257**, 263
    feedback inhibition, 151, 151F, 152F, 153F
    phosphorylation see Protein phosphorylation
  soluble, 507–508, **509**
  specialized, 268
  structure see Protein structure
  structure–function relationship, 143
  synthesis see Protein synthesis
  turnover, 256–257
Protein aggregates, 125, 256
  connective tissue, 708F

secretory vesicles, 522, 523F
Protein analysis, 125–126, **157–158**, 352
  antibodies, 164FF
  centrifugation techniques, 160FF
  chromatography, 162F
  electrophoresis, 129, 163F
  extraction, 129, 160FF
  functional, 354–355
  identification, 129, 130F
  mass spectrometry, 129, 130F
  recombinant protein expression, 129, 352–353, 353F
  sequencing, 125
    evolutionary conservation, 31, 31F
  structure determination, 126, **129–131**
    detergents, **377**, 378F, 385, 385F
    large scale, 158
    membrane proteins, 377, **378–379**, 378F
    NMR spectroscopy, 130, 131F
    X-ray diffraction, 129–130, 131F
Protein anchors, 374F, 383
Protein complexes (assemblies), **140–141**, 140F, 141F, 142F
  protein machines, **156–157**, 157F
Protein degradation, **256–257**, 263, **527–528**, 528F
  see also Lysosome(s); Proteolysis
Protein extraction, 129
Protein families, **138–139**
Protein folding, **124–125**
  see also Chaperones; Protein structure
  denaturation/renaturation, 124
    protein import into organelles, 506–507
  misfolded proteins, 125
    degradation, 256
    prion proteins, 125, 126F
    retention in ER, 518
  molecular chaperones, 125
  noncovalent interactions, 121–123, 123F, 124F
  posttranslational, 254
  stable conformation, 122, 124
Protein identification, 129
Protein interactions, **143–150**
  antibody–antigen, **144–145**
  binding sites, 144, 144F, 145F
  enzymes see Enzyme–substrate interactions
  noncovalent bonds, 48F, 144, 145F
  protein–DNA, 271–273, 272F, 289
  protein–ligand, **143–144**, 144F, 145F
  protein–protein, 48F
Protein kinase(s), 154, 154F, 545
  see also Protein phosphorylation; *individual kinases*
  cancer role, 559, 736
  cell networks, **565–566**, 566F
    integration, 566, 567F
Protein kinase A (PKA), cAMP-dependent activation, 551–552, 568
Protein kinase C (PKC), DAG activation, 554, 554F, 568
Protein motifs, DNA-binding, 272–273, 272F
Protein phosphatase(s), 154, 154F, 545
Protein phosphorylation, **153–154**, 154F
  see also Protein kinase(s); Protein phosphatase(s)
  ATP donor, 154
  cell cycle control, **616**
  conformational change, 153, 154, 155F
  experimental analysis, 561
  GTP donor, **154–155**, 155F
  intracellular signaling, 545, 545F
    complexity, **565–566**, 566F

integration, 566, 567F
  Na$^+$-K$^+$ pump, **397**
  RNA polymerase II, 277, 278F
Protein purification, 129
Protein sequencing, 125
  cell cycle proteins, 31, 31F
Protein sorting, 497, **502–512**
  chloroplast protein import, **506–507**
  endosomes, 526–527
  energy requirement, 506
  ER protein import, **507–509**
  mitochondrial protein import, **506–507**, 507F
  nuclear protein import, **504–506**, 505F, 506F
  secretory pathway, 522–523
  sorting signals, 502
    see also Signal sequences
    ER retention signal, 517
    vesicular transport, 514
Protein structure, **119–143**, **139–140**, 158–159
  conformations, **124–125**
    see also Protein folding
    representation, 132FF
  cross-linkages, **142–143**, 143F
  determination see Protein analysis
  domains, 136, 137F
    evolution, 241, 300–301
    evolutionary pressure, 138
  fibrous, 127F, **141–142**, 142F
  globular, 127F, 141
  noncovalent interactions, 121–123
    hydrogen bonds, 78F, 122, 123F, 124F
    hydrophobic interactions, 123, 124F
    ionic bonds, 122, 123F
    protein–protein interactions, 48F
    van der Waals forces, 122, 123F
  peptide bonds see Peptide bonds
  prediction, 126
  primary structure, **121–123**, 122F, 136
    see also Amino acids
    backbone, 121, 122F, 376
    side chains, 121, 122F, 376
    signal sequences, 503–504, 504T
  quaternary structure (subunits), 136, **139–140**, 139F, 140F
    see also Protein complexes (assemblies)
  secondary structure, **126**, 136
    see also α-helix; β-sheets
    transmembrane proteins, 135F, **376–377**, 376F, 377F
  structure–function relationship, 143
  tertiary structure, 136
  variety, **125–126**
Protein synthesis, 4, 4F, 56, 113–114, 113F, 229, **243–257**, 263
  see also Amino acids; Ribosome(s); Transfer RNA (tRNA)
  bacterial, 249, 253–254, 254F
    inhibition (antibiotics), **255–256**, 256T
    speed, 255
  cell-free systems, 246
  cellular localization, 502
  coding for see Genetic code
  elongation, 250
  folding see Protein folding
  initiation, 249, **253–254**, 253F
  pathway, 257, 258F
  peptide bond formation, 249, 252
  peptidyl transferase, 250
  sorting signals, 502
  speed, 254, 255

Page numbers in **boldface** refer to a major text discussion of the entry; page numbers with an F refer to a figure, with an FF to figures that follow consecutively; page numbers with a T refer to a table.

termination, 254, 254F
translation *see* Translation
Protein translocation
  see also Intracellular transport
  chaperones, 507
  cotranslational into ER, **507–509**
  mitochondria/chloroplasts, **506–507**, 507F
  nuclear import, **504–506**, 505F, 506F
Protein tyrosine phosphatase, 558
Protein–DNA interactions, gene regulatory
        proteins, 271–272
Protein–ligand interactions, **143–144**, 144F,
        145F
Protein–protein interactions, noncovalent
        bonds, 48F
Proteoglycans, 737
  connective tissue(s), **706–709**, 708F
  glycocalyx (cell surface), 381–382, 382F
Proteolysis
  apoptosis, **626–627**, 627F
  regulatory mechanism, **256–257**
  tissue dissociation, 325
Proteomics, 157–158
Proto-oncogenes, 731, 732
  see also Oncogene(s)
  conversion into oncogenes, 731F
Proton pump(s), 402, 402F, 403T, 492
  see also Chemiosmotic coupling;
        Electrochemical gradients;
        Electron-transport chain(s)
  ATP synthesis and, 379, 441, 454, 459F
  bacteriorhodopsin, 379, 379F
  electron-transport chain, **462–464,
        468–469**, 470F, 492
    mechanism, **475–476**, 477F
  endosome acidification, 526
  evolution, 488
  lysosome acidification, 402, 527, 527F
  membrane potential and, 463–464, 464F
  respiratory chain enzymes, 463, **475–476,
        477F**
Protons *see* Hydrogen ion (H$^+$, proton)
Protozoans, 26–27
  diversity, 27, 27F
  osmotic pressure, 401, 401F
  phagocytosis, 523
  as predators, 26F
Pseudogenes, 299
Pseudopodia, 524, 524F
Psychoactive drugs, **419–420**
PTC, *see* Phenythiocarbamide
Purifying selection, evolutionary conservation,
        308
Purine bases, 56, 76F
  see also Adenine; Guanine
  depurination, 212, 212F, 231F
Pyrimidine bases, 56, 76F
  see also Cytosine; Thymine; Uracil
  deamination, 212, 212F, 213F
Pyruvate
  fermentation reactions, 434, 434F
  oxidation to acetyl CoA, 435, 438F, 439,
        449
    see also Acetyl CoA; Citric acid cycle
  production, 428–429, 430, 433F, 457,
        458F
    see also Glycolysis
Pyruvate dehydrogenase complex, 435, 438F,
        439
Pyruvate kinase, 433F

Q

Q cycle, 476
Quinones, 472, 472F

R

Racker, Efraim, 461, 461F
Ras GTPase, **559–560**, 559F
  cancer role, 560, 561, 562F, 730
    gain-of-function mutations, 684
  MAP kinase cascade, 559–560, 560F
Reaction coupling, 94, 97F, 99, 100F, 107F,
        116
  see also Activated carrier molecules; Free
        energy (G); Thermodynamics
  activated carrier formation, 106–107, 106F
  glycolysis, **434–435**, 436F
Reading frame, 244, 245F
Receptor(s), 120F, 374F, 536, 538, 568
  see also Signaling molecules; *specific
        receptors*
  adaptation, 556
  cell-surface, 536, 540, 540F
    classes, **542–544**, 543F
    signal binding, 543F
    variety, 536–537
  differential responses, 536, 538F
  drug/poison targets, **419–420**, 544, 544T
  enzyme-linked *see* Enzyme-linked
        receptor(s)
  G-protein-linked *see* G-protein-linked
        receptor(s)
  intracellular, 540F, 541–542, 542F
  intracellular signals *see* Intracellular
        signaling pathways
  ion channels *see* Ion channel(s)
  neurotransmitters, **417–418**, 418F
Receptor-mediated endocytosis, **525–526,**
        526F
Receptor serine/threonine kinases, 564, 565F
Receptor tyrosine kinases, **557–559**, 569
  activation/inactivation, 558, 558F
  cancer role, 559, 560
  intracellular signaling, 558–559
    MAP kinase cascade, 559–560, 560F
    PI 3-kinase, 558
    Ras activation, **559–560**, 559F
Recessive alleles, 336, 337, 675, 685F
Recombinant DNA techniques, **352–360,**
        353F
  see also Cell culture; DNA analysis; DNA
        hybridization; *specific techniques*
  applications
    enzyme design, 105
    gene expression analysis, **353–355**
    gene function analysis, **355–356**
    gene therapy, 357
    novel DNA molecules, **352**, 352F
    protein expression, 129, **352–353**, 353F
    replication origin analysis, 199
    signaling pathway analysis, 561–562,
        562F, 563F
    transgenic organisms, **356–360**, 357F
  cloning *see* DNA cloning
  consequences, 324, 361
  expression vectors, 353
  history, 323–324
  PCR *see* Polymerase chain reaction (PCR)
  reporter genes *see* Reporter genes
  restriction nucleases, **328**, 329F
  RNA interference (RNAi), 357–358
  transgenic animals, 357

in vitro (site-directed) mutagenesis,
        355–356, 356F
Recombination, **215–222**, 224
  functions, 215
  homologous *see* Homologous recombination
  site-specific, **216–217**
  unequal crossing-over, 297, 298F, 300F
Red blood cells, 380F, 723F
  see also Hemoglobin
  cell cortex, 380–381, 381F
  genetic defects, 381
  hemopoietic stem cells, 722, 724F
  turnover, 721
Redox pairs, 470
Redox potential, **469–470**, 471F
  photosystems, 484F
  respiratory chain, 472F
Redox reactions, **109–111**, 462, 469
Reduction reactions, 90
  electron transfer, **90**, 91F
Release factors, 254, 254F
Repetitive DNA, 224
  see also Mobile genetic elements
  DNA fingerprinting, 351F
  human genome, 218–219, 311, 313F, 316
  shotgun DNA sequencing and, 334, 334F
Replication fork, **201**, 223
  see also DNA polymerase(s)
  asymmetry, **202–203**, 202F, 203F
  bidirectional movement, 201, 201F
  lagging strand, 203, 203F, 223
    Okazaki fragments, 203, 206F, 207F
    RNA primers, 205, 206F
  leading strand, 203, 203F
    RNA primers, 204–205
Replication origin(s), 182, 182F, **197**
  bacterial, 197
  DNA unwinding, 197, 197F
  eucaryotic, 197, 198
  experimental analysis, **198–200**, 198F,
        199F, 200F
  origin recognition complex, 621, 622F
  pre-replication complex, 621–622
  sequence, 197
Replicative transposition, 218F
Reporter genes
  gene expression analysis, 354–355, 354F
    Drosophila egg polarity genes, 283, 284F
  intracellular transport studies, 520–521,
        521F
Repressors *see* Transcriptional repressors
Reproduction
  asexual, 660F, 661
  cell division role, 637
  sexual *see* Sexual reproduction
Reproductive cloning, 725, 725F
RER, *see* Endoplasmic reticulum, rough
Respiration (cellular), 17, 89, 89F, **455–477**
  see also ATP synthesis; Citric acid cycle;
        Glycolysis; Oxidative
        phosphorylation
  aerobic, 431, **476–477**
  anaerobic (fermentation), **431, 434**, 434F,
        488, 489F
  evolution, 453, **488**, 488F, 490F
Resting membrane potential, 409–410
Restriction maps, 336
Restriction nucleases, **328**, 360
  blunt *vs.* staggered ends, 342F
  DNA cloning, 344
  restriction maps, 336
  target sequences, 328, 329F
Restriction–modification systems, 328

Retinal, 150, 150F, 378–379
Retinoblastoma (Rb) protein, mitogen effects, 629–630, 630F
Retrotransposons, 219, 219F
Retroviruses, **221–222,** 223F
    see also Reverse transcriptase
Reverse transcriptase, 221
    cDNA library production, 346, 347F
    retrotransposition, 219, 219F
Rho GTPase, actin regulation, 599–600
Rhodopseudomonas viridis, photosynthetic reaction center, 380F
Rhodopsin, 149–150, 556
Ribbon protein models, 132F, 137F
Ribose, 70F, 76F, 231F
    deoxyribose *vs.,* 261
Ribosomal DNA (rDNA), nucleolus, 183, 183F
Ribosomal protein(s), 248, 251
Ribosomal RNA (rRNA), 233, 248, 251, 252, 252F
    evolutionary conservation, 310, 310F
    genes, nucleolar location, 183, 183F
Ribosome(s), 22, 25F, **248–251,** 263, 529
    catalytic activity, **251–253**
        peptidyl transferase, 250, 252
    electron microscopy, 12F
    eucaryotic, 249, 249F, 502, 508, 509F
        see also Endoplasmic reticulum (ER)
        translation initiation, 253, 253F
    peptide bond formation, 249, 252
    peptidyl transfer, 250
    polyribosomes (polysomes), **254–255,** 255F, 508–509
    procaryotic, 249
        translation initiation, 253–254
    structure, 250F, 251, 251F
        see also Ribosomal RNA (rRNA)
        A-, E- and P-sites, 250, 251F
        proteins, 248, 251
        RNA role, 251, 252F
        subunits, 249, 250F
        X-ray diffraction, 130
Ribozyme(s), 252, 260, 260F
    origin of life and, 260
    reactions catalyzed, 261T
    ribosomes as, **251–253**
Ribulose 1,5-bisphosphate, 485, 485F
Ribulose bisphosphate carboxylase (rubisco), **485–486**
Ribulose, structure, 70F
RNA (ribonucleic acid), 4, 204, 229
    base-pairing see Base-pairing
    catalytic see Ribozyme(s)
    degradation, **242**
    DNA *versus,* 231, 231F, 261
    information coding, 4, 4F, 259–260
    modification, tRNAs, 245F
    origin of life (RNA world theory) see RNA world theory
    primer in DNA synthesis, **204–206,** 206F
    structure, 230–231, 231F
        base-pairing, 58, 232F
        evolutionary significance, 259–260
        nucleotides, 57, 230–231, 231F
        secondary, 231, 232F
    synthesis, **230–243**
        processing, **236–242**
        RNA polymerase see RNA polymerase(s)
        transcription see Transcription
    types, **233–234,** 234T, 262
        see also specific RNAs
        small RNAs, 234T, 238–239
    viral genomes, 221F

RNA interference (RNAi), gene inactivation, 357–358
RNA polymerase(s), 232–233, 233F, 262, 289
    bacterial, 235F
        sigma (σ) factor, 234–235
    DNA polymerase *vs.,* 233
    eucaryotic, 276, 276T, 290
        phosphorylation, 277, 278F
        RNA polymerase I, 276, 276T
        RNA polymerase II, 276, 276T, 277–278, 277F, 278F
        RNA polymerase III, 276, 276T
RNA splicing see Splicing
RNA world theory, **258–262,** 259F, 263
    autocatalysis, **259**
    RNA *vs.* DNA, **261–262,** 262F
    self-replication and catalysis, **259–260,** 259F, 261F
RNase, 242
Rod photoreceptors, 556, 556F, 557F
Root hairs, 701F
Roots, 700F
Rough endoplasmic reticulum, see Endoplasmic reticulum, rough
rRNA, see Ribosomal RNA

S

Saccharomyces cerevisiae, 28F
    genome, 28
        complete sequence, 331, 331F
    as model organism, **28**
        cell cycle analysis, 30–31, 30F, 31F
Salmonella, structure, 14F
Sarcomere(s), 601–602, 602F
Sarcoplasmic reticulum, 603
    calcium channels, 603, 605F
    calcium pumps, 393, 394F
    muscle contraction, 603–604, 605F
Scanning electron microscopy (SEM), 9F, 11
Schizosaccharomyces pombe, 30–31
    evolutionary conservation, 31, 31F
Schleiden, Matthias, 6
Schwann cells, 718
Schwann, Theodore, 6
Sclerenchyma, 700F
Second messengers, 549–550, 550F
    see also Cyclic AMP; Inositol phospholipids
Secretion, 522
Secretory cells, regulated exocytosis, 522, 522F, 529
Secretory pathway, 512, **516–523,** 529
    see also Endoplasmic reticulum (ER); Exocytosis; Golgi apparatus
    protein modification
        ER, **516–517,** 517F
        Golgi apparatus, **518–519,** 519F
    protein quality control, **517–518,** 518F
    protein release by exocytosis, **519, 522–523,** 522F
    protein sorting, 522–523
Secretory vesicles, 522–523, 523F
Seed(s), food storage, 446, 446F
Selective breeding, 323, 324F
Selfish DNA, 243
SEM, *see* Scanning electron microscopy
Serine, 75F
Serine proteases, 138–139, 138F
Serine/threonine kinases, 564, 565F
Sex chromosomes, 178
Sex determination genes, evolutionary conservation, 215F

Sexual development, testosterone role, 542
Sexual reproduction, **660–663,** 693
    see also Meiosis; Natural selection
    advantages, **662–663**
        genetic variation, 294
    cycle, **661–662,** 662F
    gametes see Gametes
Shotgun DNA sequencing, 334–335, 334F
Sickle-cell anemia, 209–210, 209F
    genetic diagnosis, 337–338, 337F
    selective advantage, 684, 686
Sieve tubes, 701F
Sigma (σ) factor, 234–235
Signal proteins, 120F
Signal-recognition particle (SRP), 509, 510F
Signal-recognition particle (SRP) receptor, 509
Signal sequences, 504F, 504T
    see also Intracellular transport; Protein sorting
    ER signal sequence, 508, 509, 510F
    nuclear localization signal, 506
    recognition, 509, 510F
    transmembrane proteins, 507
Signal transduction, 533, 534, 534F
    see also Signaling molecules
    complexity, 536–537
    intracellular pathways see Intracellular signaling pathways
    neurotransmission, 535, 535F
    receptors see Receptor(s)
    visual, 556, 557F
Signaling cascades, 539–540
Signaling molecules, 533, **536–538,** 538F, 568
    see also Receptor(s); specific signals
    contact-dependent, 535, 536F, 537T
    direct intracellular action, 540F, 568
        nitric oxide, **540–541,** 541F
        steroid hormones, **541–542,** 542F
    hormones, 534, 535F, 537T, 568
    integrins, 706
    intracellular see Intracellular signaling pathways
    local mediators, 534, 535F
    neurotransmitters see Neurotransmitters
    nucleotides, 77F
    receptor-mediated transduction, 536–538, 540, 540F, 543F
Silk fibers, 135
Simple epithelium, 709, 709F
SINE elements, 313F
Single-nucleotide polymorphisms (SNPs), 682, 682F
    human genome, 313, 316
    linkage mapping, 682–683, 683F, 689
Single-particle tracking (SPT) microscopy, membrane flow, 385, 385F
Single-stranded binding protein, DNA replication, 206
Sister chromatids
    meiotic chromosomes, 665, 668F
    mitotic chromosomes, 638, 639F, **641,** 641F, **644**
Site-directed mutagenesis, 355–356, 356F
Site-specific recombination, **216–217**
Skeletal muscle, 600
    structure, 601–602
Skin, 720F
    collagen alignment, 705F, 706F
    hyperextensible, 706F
    maintenance/renewal, 721–722, 723F
Sliding clamp, 206, 207F
SMADs, 564

Page numbers in **boldface** refer to a major text discussion of the entry; page numbers with an F refer to a figure, with an FF to figures that follow consecutively; page numbers with a T refer to a table.

I:21

Small nuclear riboproteins (snRNPs), splicing role, 238–239, 239F
Small nuclear RNAs (snRNAs), splicing role, 238–239
Smoking, cancer link, 728
Smooth muscle, 600, 605
  nitric oxide-mediated relaxation, 540–541, 541F
SNARE proteins, **515–516**, 516F, 529
SNPs, see Single-nucleotide polymorphisms
snRNAs, see Small nuclear RNAs
snRNPs, see Small nuclear riboproteins
Sodium channels, voltage-gated, 412–413, 413F
  see also Action potential
Sodium dodecyl sulfate (SDS), 378F
Sodium ions (Na+)
  active transport, 396
    Na+-H+ exchanger, 399
    Na+-K+ ATPase see Na+-K+ pump
    transport coupling, **397–399**
  electrochemical gradient, 395
  excitatory neurotransmission, 419
  intracellular vs. extracellular concentration, 390, 390T
Solubility, 69F
Solvents, water, 69F
Somatic cells, 293–294, 294F
  germ line vs., 294, 662, 663F
  mutation, 210, 294, 294F, 728
    see also Cancer
Sorting signals, 502, 514, 517
  see also Protein sorting
Southern blotting, 338, 338F
Space-filling protein models, 127F, 133F
Speciation, human–chimpanzee divergence, 305
Species names, 14
Species numbers, 659
Spectrin, 381, 381F
Spectrophotometry, enzyme reaction rates, 103, 103F
Spermatozoa, 659, 661F
  see also Fertilization; Meiosis
  flagella, 591F
  mitochondria, 456F
  uniparental inheritance theory, 659, 660F
Spliceosome, 238–239, 239F
Splicing, 230, **238–241**
  alternative, 241, 241F, 318, 318F
  lariat structure, 238, 238F
  signals, 238, 238F
  spliceosome, 238–239, 239F
Squamous epithelium, 709, 709F
Squid (Loligo), 412F
  giant axons see under Axons
Starch, 53, 446, **486**
  granules, 446F
Start-transfer sequence(s), 511, 511F
Stearic acid, 72F
Stem cells, 697, **721–722**, 721F, 738
  see also Cell differentiation; Tissue(s)
  embryonic see Embryonic stem cells
  hemopoietic, 722, 724F
  self-renewal, 721, 721F
  tissue repair, **722–725**
Steroids, 73F, 541F
  intracellular signaling, **541–542**, 542F
Sterols, 367, 368F
Stockenius, Walther, 461, 461F
Stomata, 700F, 701F
Stop-transfer sequence(s), 510, 511F
Storage proteins, 120F

Stratified epithelium, 709, 709F
Streptococcus pneumoniae, identification of DNA as genetic material, 172, 173F
Streptococcus, spherical structure, 14F
Stromatolites, 478F
Structural proteins, 120F
Sturtevant, Alfred, 682
Succinate dehydrogenase, malonate inhibition, 442–443, 443F
Succinate, malonate similarity, 442F
Sucrose, production in plants, **486**
Sugars, 51, 51F, **52–53**, 65, 70FF
  see also Carbohydrates; Glycolipid(s); Glycoproteins; Proteoglycans; specific sugars
  breakdown, 427, 428–429, 428F, 449
    see also Glycolysis
  cell surface (glycocalyx), **381–382**, 382F
  condensation reactions, 52–53, 53F
  conversion to fat, 445
  derivatives, 71F
  disaccharides, 52, 53F, 71F
  functions, 53, 427
  hydrolysis reactions, 53, 53F
  monosaccharides, 52, 70F, 71F
  nucleotide structure, 56, 76F, 171
  oligosaccharides, 52, 71F, 381
  polysaccharides see Polysaccharides
  storage as glycogen, 444
  structure, 52
    α and β linkages, 71F
    isomers, 52, 70F
    ring formation, 70F
Sulfur bacteria, 15F
Survival factors, 629, **631–632**
  neuronal development, 631, 632F
Svedberg, Theodore, 60–61
Symport, 397, 398, 399F
Synapses, 417, 417F, 418F
  see also Neurotransmitters
  cell signaling, 535, 535F, 544
  inhibitory/excitatory, 419, 420F
  neuromuscular junction, 418, 419F
  neuronal connectivity, **420–421**, 421F
  psychoactive drug targets, **419–420**
Synaptic cleft, 417
Synaptic vesicles, 417, 417F
Synaptonemal complex, 666
Synteny, conservation, 307
Szent-Györgyi, Albert, 442

T

T tubules, 603, 605F
Target SNARE proteins (t-SNAREs), 515, 515F
TATA-binding protein, 278F
TATA box, 277, 278F
Taxol, microtubule effects, 582
TCA cycle see Citric acid cycle
TCF4 knockout mice, 735
Telomerase, **207–208**, 208F, 224
  cell senescence, 326–327
Telomere(s), 182, 182F, 208
  replication, **207–208**, 208F
Telophase, mitosis, 640–641, 643F, **651**, 651F
TEM, see Transmission electron microscopy
Temperature-sensitive mutants, 684F, 688F
  cell cycle analysis, 30, 30F, 31F, 687–688
Testes, 664
Testosterone, 73F, 541F
  intracellular receptors, 541–542

sexual development, 542
TFIID transcription factor, 277
TFIIH transcription factor, 277
TGF-β superfamily, SMAD-dependent signaling, 564, 565F
Thermodynamics, **84–105**
  activation energy, 92, 92F, 102, 115
  biological order and, **85–88**, 86F
  electron transfer, **470**
  energetically favorable reactions, 94, 94F
  energetically unfavorable reactions, 94, 94F
    reaction coupling see Reaction coupling
  entropy, 85, 86F, 96F, 115
  enzyme catalysis, **91–93**
  first law of, 87, 87F
  free energy see Free energy (G)
  second law of, 85–86
Threonine, structure, 75F
Thylakoid membrane, 479–480, 479F, 506
Thymine, 56, 76F
  adenine base pairs, 175F
  dimer formation, 212, 213F
  DNA, 171
Thymosin, 594
Thyroid hormones, structure, 541F
Thyroxine, 541F
  intracellular receptors, 541–542
Tight junctions, 386, 711F, 737
  claudins/occludins, 712
  epithelial, **711–712**, 712F
Tissue(s), 697–726
  see also Cell junction(s); Extracellular matrix; specific tissues
  connective see Connective tissue(s)
  development, 697, 717, 718F, 719F
  dissociation for cell culture, **325**, 360
    cell sorting, 325, 326F
  epithelial see Epithelium
  maintenance/renewal, 697, **717–726**, 720F, 738
    see also Cancer; Cell differentiation
    cell communication, 718–719
    cell memory, 280, **286–288**, 288F, 719–720
    cell–cell adhesion, 719
    disorders, 697
    precursor cells, 721
    rates, **720–721**
    replacement patterns, variety, 721–722
    stem cells see Stem cells
  mechanical strength, 718
  muscle see Muscle
  organization, 697, 698F, **718–720**, 720F
    see also individual cell types
    cell types, 718, 720F
  plant see Plant tissue(s)
Tobacco cells, cell culture, 327F
Tobacco-smoke, cancer link, 728
Tonoplast, 25F
Toxins
  cytoskeletal, 582, 589, 593
  G protein, 547–548
  Na+-K+ pump inhibition, 397, 400
Transcription, 4, 4F, 229, **230–236**, 262
  bacterial, 235F, 273F, 289
    eucaryotes vs., 243F, 276
    initiation, 234–235, 274–275, 274F, 275F
    operators, 273
    polycistronic mRNA, 254, 254F
    repression, 273–274, 274F, 275, 275F
    sigma (σ) factor, 234–235
  basic principles, **231–233**

base pairing, 230–231
DNA template, 231–232, 232F, 236F
direction, 236, 236F
electron microscopy, 233F
elongation, 234
eucaryotic, **236–237**, 243F, **275–280**, 289
see also Transcription factors (general)
gene expression control *see* Transcriptional gene regulation
initiation
see also Promoter elements; Transcriptional activators
bacterial, 234–235, 271, 274–275, 274F, 275F
eucaryotic, **275–278**
regulation, **234–236**
RNA polymerase *see* RNA polymerase(s)
termination, 234
universal nature, 242
Transcription factors (general), **276–278**, 290
RNA polymerase II, 277–278, 277F
TATA binding, 277, 278F
Transcription initiation complex, 277
Transcriptional activators, 274F
bacterial, 274–275, 275F
eucaryotic, 278, 279, 280F
Transcriptional gene regulation, **271–280,** 289–290
see also Gene regulatory proteins; Gene regulatory sequences; Transcriptional activators; Transcriptional repressors
bacterial, **273–275,** 289
eucaryotic *vs.*, 276
eucaryotic, **275–280,** 290
chromatin structure, 187, 188–189, 276, **279–280,** 280F
DNA loops, 278, 279F
enhancers, **278–279,** 279F, 289
mediator complex, 278
RNA polymerases, 276, 276T, 289
transcription factors *see* Transcription factors (general); Gene regulatory proteins
protein–DNA interactions, **271–273,** 272F
Transcriptional repressors
bacterial
lac repressor, 275, 275F
tryptophan repressor, 273–274, 274F
chromatin structure and, 279
eucaryotic, 279
mechanism, 274F
Transcytosis, 527, 527F
Transducin, 556
Transfer RNA (tRNA), 233, **245, 248,** 263
aminoacylation, **248,** 249F, 252F
codon–anticodon interaction, 245, 245F, 248, 249F, 252F
initiator (methionine), 253, 253F
modification, 245F
ribosomal binding sites, 250, 251–252
structure, 245, 245F
wobble base-pairing, 248
Transforming growth factor-β superfamily, SMAD-dependent signaling, 564, 565F
Transgenes, 357, 357F
Transgenic organisms, **356–358,** 357F
mice, 357, 358F, 359F
plants, **359–360**
genetically modified crops, 360
production, 359, 359F
Transition state, catalytic reactions, 147

Translation, 4, 4F, 230, **243–257,** 249F, 252F, 262–263
see also Ribosome(s); Transfer RNA (tRNA)
antibiotic inhibition, **255–256,** 256T
codons and decoding, **244,** 244F
see also Codons; Genetic code
reading frame, 244, 245F
elongation, 250, 252F
initiation, 249, 253–254, 253F
bacteria, 253–254, 254F
eucaryotes, 253
initiation factors, 253
initiator tRNA, 253, 253F
polyribosomes (polysomes), **254–255,** 255F
ribosomal location, **248–251**
termination, 254, 254F
in vitro system, 246
Translation initiation factors, 253
Transmembrane proteins, 375, 375F
see also Carrier proteins; Ion channel(s)
α-helix, 135F, **376–377,** 376F, 377F
aqueous pores, 376, 377F
β barrels, 376–377, 377F
cell cortex attachment, 380–381, 381F
insertion into lipid bilayer, **510–512,** 511F
Transmission electron microscopy (TEM), 9F, 11, 12F
Transmitter-gated ion channels *see* Ligand-gated ion channels
Transport/translocator proteins, 120F, 503, 503F
endoplasmic reticulum, 509, 510F
mutations, 683, 683F
transporters (carriers) *see* Carrier proteins
Transport vesicles *see* Vesicles
Transposable elements *see* Mobile genetic elements
Transposase, 218
Transposition, **217–218,** 219, 219F
Transposons, 218, 218F, 307
Tree of life, 310F
Treponema pallidum, structure, 14F
Triacylglycerols, 54, 54F, 72F, 369F
Tricarboxylic acid cycle *see* Citric acid cycle
Trichomes, 701F
Triglycerides, 73F
Triose phosphate isomerase, 432F
Trisomy 21 (Down syndrome), 670–671, 671F
Triton X-100, 378F
Trophic factors *see* Survival factors
Tropomyosin, 603
alternative splicing, 241F
Troponin(s), 603
tRNA, *see* Transfer RNA
Tryptophan, 75F
bacterial synthesis, gene control, 273–274, 273F
Tubulin, 606
see also Microtubule(s)
α-tubulin, 579
β-tubulin, 579
γ-tubulin, 580, 640
polymerization, 579–580
cf actin, 593–594
GTP hydrolysis and, 581–582, 582F
nucleation sites, 580, 581, 581F
Tumor(s)
see also Cancer
benign *vs.* malignant, 727
genetics, **731–732,** 731F, 732F
see also Oncogene(s); Tumor suppressor gene(s)

colorectal cancer, **732–733**
growth, 736, 736F
mutation and proliferation, 729F
treatment, **736**
Tumor suppressor gene(s), 731–732, 732F, 738
APC gene, 733, 733F, 734
identification, 732, **734–735**
Turgor pressure, 401, 699
Twin studies, gene–environment interactions, 692
Two-dimensional electrophoresis, protein separation, 163F
Tyrosine, 75F
Tyrosine kinases
cancer role, 559, 736
Gleevec action, 736, 737F
receptor kinases *see* Receptor tyrosine kinases

U

Ubiquinone, 472, 472F, 476
Ubiquitin
cyclin degradation, 617, 617F
proteasome targeting, 257
Ultracentrifugation, 60–61, 61F
Ultraviolet radiation, thymine dimer formation, 212, 213F
Unequal crossing-over
exon duplication, 300F
gene duplication, 297, 298F
Uniparental inheritance, 659
disproving by Mendel, 674
Uniporters, 397, 399F
Uracil, 56, 76F, 231F
base-pairing, 231F
RNA *vs.* DNA, 231

V

Vacuoles, 25F
Valine, structure, 75F
van der Waals attraction, 78F
bond length, 46T
bond strength, 46T
macromolecular structure role, 62–63
polypeptide chains, 122, 123F
van Leeuwenhoek, Antoni, 659
Variable number tandem repeats (VNTRs), DNA fingerprinting, 351F
Vectors
cloning, **342–343,** 343F, 344
expression, 353
Velocity sedimentation, 161F
Venter, Craig, 315
Vertebrate genomes, duplications, **299–300**
Vesicles, 25F, 529
clathrin-coated, 513–514, 513F, 514F, 514T
COP-coated, 514, 514T
endocytic, 523
membrane budding, 374, 374F, 513F
clathrin-coated pits, 513
dynamin role, 513, 514F
membrane docking, 515, 515F
membrane fusion, 374, 374F
secretory vesicles, 523
SNAREs, 515–516, 515F, 516F
protein coats, **513–514,** 514T
adaptins, 514
clathrin, 513, 513F
secretory, 522–523, 523F
synaptic, 417, 417F

Page numbers in **boldface** refer to a major text discussion of the entry; page numbers with an F refer to a figure, with an FF to figures that follow consecutively; page numbers with a T refer to a table.

Introduction to Cells : Chemical Components of Cells :
Energy, Catalysis, and Biosynthesis :
Protein Structure and Function : DNA and Chromosomes :
DNA Replication, Repair, and Recombination :
From DNA to Protein : Control of Gene Expression :
How Genes and Genomes Evolve :
Manipulating Genes and Cells : Membrane Structure :
Membrane Transport : How Cells Obtain Energy from Food :
Energy Generation in Mitochondria and Chloroplasts :
Intracellular Compartments and Transport :
Cell Communication : Cytoskeleton :
Cell-Cycle Control and Cell Death : Cell Division :
Genetics, Meiosis, and the Molecular Basis of Heredity :
Tissues and Cancer